Alfalfa and Alfalfa Improvement

AGRONOMY
A Series of Monographs

The American Society of Agronomy and Academic Press published the first six books in this series. The General Editor of Monographs 1 to 6 was A. G. Norman. They are available through Academic Press, Inc., 111 Fifth Avenue, New York NY 10003.

1. C. EDMUND MARSHALL: The Colloid Chemical of the Silicate Minerals, 1949
2. BYRON T. SHAW, *Editor*: Soil Physical Conditions and Plant Growth, 1952
3. K. D. JACOB, *Editor*: Fertilizer Technology and Resources in the United States, 1953
4. W. H. PIERRE and A. G. NORMAN, *Editors*: Soil and Fertilizer Phosphate in Crop Nutrition, 1953
5. GEORGE F. SPRAGUE, *Editor*: Corn and Corn Improvement, 1955
6. J. LEVITT: The Hardiness of Plants, 1956

The Monographs published since 1957 are available from the American Society of Agronomy, 677 S. Segoe Road, Madison, WI 53711.

7. JAMES N. LUTHIN, *Editor*: Drainage of Agricultural Lands, 1957   General Editor, D. E. Gregg
8. FRANKLIN A. COFFMAN, *Editor*: Oats and Oat Improvement, 1961
    Managing Editor, H. L. Hamilton
9. A. KLUTE, *Editor*:Methods of Soil Analysis, 1986
    Part 1—Physical and Mineralogical Methods, Second Edition   Managing Editor, R. C. Dinauer
    A. L. PAGE, *Editor*: Methods of Soil Analysis, 1982
    Part 2—Chemical and Microbiological Properties, Second Edition Managing Editor, R. C. Dinauer
10. W. V. BARTHOLOMEW and F. E. CLARK, *Editors*: Soil Nitrogen, 1965
    (Out of print; replaced by no. 22)   Managing Editor, H. L. Hamilton
11. R. M. HAGAN, H. R. HAISE, and T. W. EDMINSTER, *Editors*: Irrigation of Agricultural Lands, 1967   Managing Editor, R. C. Dinauer
12. FRED ADAMS, *Editor*: Soil Acidity and Liming, Second Edition, 1984
    Managing Editor, R. C. Dinauer
13. E. G. HEYNE, *Editor*: Wheat and Wheat Improvement, Second Edition 1987
    Managing Editor, S. H. Mickelson
14. A. A. HANSON and F. V. JUSKA, *Editors*: Turfgrass Science, 1969
    Managing Editor, H. L. Hamilton
15. CLARENCE H. HANSON, *Editor*: Alfalfa Science and Technology, 1972
    Managing Editor, H. L. Hamilton
16. J. R. WILCOX, *Editor*: Soybeans: Improvement, Production, and Uses, Second Edition, 1987
    Managing Editor, D. A. Fuccillo
17. JAN VAN SCHILFGAARDE, *Editor*: Drainage for Agriculture, 1974
    Managing Editor, R. C. Dinauer
18. GEORGE F. SPRAGUE, *Editor*: Corn and Corn Improvement, 1977
    Managing Editor, D. A. Fuccillo
19. JACK F. CARTER, *Editor*: Sunflower Science and Technology, 1978
    Managing Editor, D. A. Fuccillo
20. ROBERT C. BUCKNER and L. P. BUSH, *Editors*: Tall Fescue, 1979
    Managing Editor, D. A. Fuccillo
21. M. T. BEATTY, G. W. PETERSEN, and L. D. SWINDALE, *Editors*: Planning the Uses and Management of Land, 1979   Managing Editor, R. C. Dinauer
22. F. J. STEVENSON, *Editor*: Nitrogen in Agricultural Soils, 1982
    Managing Editor, R. C. Dinauer
23. H. E. DREGNE and W. O. WILLIS, *Editors*: Dryland Agriculture, 1983
    Managing Editor, D. A. Fuccillo
24. R. J. KOHEL and C. F. LEWIS, *Editors*: Cotton, 1984
    Managing Editor, D. A. Fuccillo
25. N. L. TAYLOR, *Editor*: Clover Science and Technology, 1985
    Managing Editor, D. A. Fuccillo
26. D. C. RASMUSSON, *Editor*: Barley, 1985
    Managing Editor, D. A. Fuccillo
27. M. A. TABATABAI, *Editor*: Sulfur in Agriculture, 1986
    Managing Editor, R. C. Dinauer
28. R. A. OLSON and K. J. FREY, *Editors*: Nutritional Quality of Cereal Grains: Genetic and Agronomic Improvement. 1987
    Managing Editor, S. H. Mickelson
29. A. A. HANSON, D. K. BARNES, and R. R. HILL, JR. *Editors*: Alfalfa and Alfalfa Improvement. 1988
    Managing Editor, S. H. Mickelson

Alfalfa and Alfalfa Improvement

A. A. Hanson, *editor*
D. K. Barnes, *co-editor*
R. R. Hill, Jr. *co-editor*

Associate Editors
G. H. Heichel   K. T. Leath
O. J. Hunt   G. C. Marten
M. B. Tesar

Managing Editor: S. H. Mickelson
Assistant Editor: K. A. Holtgraver
Editor-in-Chief ASA Publications: G. H. Heichel
Editor-in-Chief CSSA Publications: C. W. Stuber
Editor-in-Chief SSSA Publications: D. E. Kissel

Number 29 in the series
AGRONOMY

American Society of Agronomy, Inc.
Crop Science Society of America, Inc.
Soil Science Society of America, Inc.
Publishers
Madison, Wisconsin, USA

1988

Copyright © 1988 by the American Society of Agronomy, Inc.
Crop Science Society of America, Inc.
Soil Science Society of America, Inc.

ALL RIGHTS RESERVED UNDER THE U.S. COPYRIGHT
LAW of 1978 (P. L. 94-553)

Any and all uses beyond the limitations of the "fair use" provision of the law require written permission from the publishers and/or author(s); not applicable to contributions prepared by officers or employees of the U.S. Government as part of their official duties.

American Society of Agronomy, Inc.
Crop Science Society of America, Inc.
Soil Science Society of America, Inc.
677 South Segoe Road, Madison, WI 53711 USA

Library of Congress Cataloging-in-Publication Data

Alfalfa and alfalfa improvement / A.A. Hanson, editor ; D.K. Barnes, co-editor ; R. R. Hill, co-editor.
    p.   cm.—(Agronomy ; no. 29)
    Includes bibliographies and index.
    ISBN 0-89118-094-X
    1. Alfalfa. 2. Alfalfa—Breeding.   I. Hanson, A. A. (Angus Alexander), 1922- . II. Barnes, D. K., 1935- . III. Hill, R. R. (Richard Ray), 1936- . IV. Series.
SB205.A4A54 1988
633.3'1—dc19   88-3350
                                                                  CIP

Printed in the United States of America

CONTENTS

FOREWORD	xv
PREFACE	xvii
CONTRIBUTORS	xix
CONVERSION FACTORS FOR SI AND NON-SI UNITS	xxiv

1 Highlights in the USA and Canada — 1
D. K. Barnes, B. P. Goplen, and J. E. Baylor

1-1	Alfalfa Distribution	2
1-2	Origins of North American Germplasm	2
1-3	Germplasm Use in Cultivar Development	3
1-4	Germplasm Collection, Preservation, and Evaluation Improved	3
1-5	Public and Industry Breeding Programs Change Emphasis	5
1-6	Cultivar Descriptions are Standardized	6
1-7	Seed Production Problems Change	6
1-8	Canadian Forage Seed Project and Seeds Canada	7
1-9	Canadian Seed Certification	8
1-10	New Disease Resistance Developed	9
1-11	New Insect Resistance Developed	10
1-12	New Seeding Techniques	11
1-13	Weed Control Increases	11
1-14	Hay Preservatives Gain in Usage	12
1-15	Feeding Value Measurements Improved	12
1-16	Hay Quality Standards Developed	13
1-17	Progress in Bloat Research	13
1-18	Dinitrogen Fixation Improved	14
1-19	Crop Growth Models Available	15
1-20	Tissue Culture Success Encouraging	16
1-21	Cytogenetic Studies Aid Breeding	16
1-22	North American Alfalfa Improvement Conference Promotes Progress	17
1-23	Keeping the Crop Competitive	18
1-24	Significant Events	18
	References	22

2 World Distribution and Historical Development — 25
Réal Michaud, W. F. Lehman, and M.D. Rumbaugh

2-1	Scientific and Vernacular Names	26
2-2	Geographical Movement	27
2-3	Distribution, Production, and Research Around the World	31
	References	82

3 The Genus *Medicago* and the Origin of the *Medicago sativa* Complex — 93
Carlos F. Quiros and Gary R. Bauchan

3–1	Centers of Diversity	93
3–2	Ploidy	93
3–3	Breeding Systems	94
3–4	Evolution	94
3–5	Taxonomy	95
3–6	The *Medicago sativa* Complex	107
3–7	Taxogenetics of *Medicago*	116
3–8	Summary	120
	Acknowledgment	121
	References	121

4 Morphology and Anatomy 125

L. R. Teuber and M. A. Brick

4–1	Seed Morphology, Anatomy, and Development	125
4–2	Roots	132
4–3	The Crown	138
4–4	The Stem	139
4–5	The Leaf	148
4–6	The Flower	150
4–7	Summary	158
	References	158

5 Environmental Physiology and Crop Growth 163

G. W. Fick, D. A. Holt, and D. G. Lugg

5–1	Seedling Growth and Development	164
5–2	Vegetative Growth	168
5–3	Reproductive Development	179
5–4	Computer Modeling	184
5–5	Summary	187
	References	188

6 Carbon Assimilation, Partitioning, and Utilization 195

G. H. Heichel, R. H. Delaney, and H. T. Cralle

6–1	Leaf Carbon Dioxide Exchange	196
6–2	Root System Carbon Dioxide Exchange	205
6–3	Crop Community Carbon Dioxide Exchange	206
6–4	Carbon Dioxide Exchange and Yield	211
6–5	Photosynthate Partitioning and Utilization	213
6–6	Prospects for Improvement of Carbon Economy and Yield	220
	References	222

7 Nodulation and Symbiotic Dinitrogen Fixation 229

C. P. Vance, G. H. Heichel, and D. A. Phillips

7–1	Establishing the Symbiosis	231
7–2	Nodule Development and Structure	234
7–3	Biochemistry of Dinitrogen Fixation	240

	7-4 Factors Affecting Dinitrogen Fixation	242
	7-5 Dinitrogen Fixation by Alfalfa Communities	245
	References	251

8 Cold and Heat Tolerance — 259
J. S. McKenzie, Roger Paquin, and S. H. Duke

	8-1 Cold Tolerance	260
	8-2 Heat Tolerance	285
	8-3 Prospects for Improvement	291
	References	292

9 Alfalfa Establishment — 303
M. B. Tesar and V. L. Marble

	9-1 Stand Density	303
	9-2 Mixtures of Alfalfa and Grasses	304
	9-3 Soil and Seedbed Preparation	306
	9-4 Soil Preparation and Influence of Prior Crops	308
	9-5 Seeding	310
	9-6 Special Seeding Situations	325
	9-7 Management During Seeding Year	327
	References	329

10 Nutrition and Fertilizer Use — 333
L. E. Lanyon and W. K. Griffith

	10-1 Factors Affecting Nutrient Need	334
	10-2 Soils and their Nutrient-Supplying Power	339
	10-3 Soil Acidity and Liming	341
	10-4 Nitrogen	345
	10-5 Phosphorus	349
	10-6 Potassium	352
	10-7 Calcium and Magnesium	354
	10-8 Sulfur	355
	10-9 Boron	359
	10-10 Other Micronutrients	361
	10-11 Animal Manures and Municipal Wastes	362
	Acknowledgment	364
	References	364

11 Alfalfa Water Relations and Irrigation — 373
C. C. Sheaffer, C. B. Tanner, and M. B. Kirkham

	11-1 Physiological Effects of Water Deficits	373
	11-2 Pathway of Water Movement	377
	11-3 Crop Water Requirements	385
	11-4 Crop Productivity and Water Use	388
	11-5 Irrigation Scheduling	392
	11-6 Irrigation Application Status	401

	11–7	Summary	401
		References	402
12	**Cutting Schedules and Stands**	411
	C. C. Sheaffer, G. D. Lacefield, and V. L. Marble	
	12–1	The Importance of Cultivar Development	412
	12–2	Crop Development and Cutting Systems	413
	12–3	Harvesting on a Fixed Interval	415
	12–4	Harvesting by Stage of Growth	419
	12–5	Harvesting by Crown Shoot Development	420
	12–6	Height of Cutting	422
	12–7	Fall Harvest Management	423
	12–8	Spring Management	427
	12–9	Insect Relations	428
	12–10	Stand Decline	429
	12–11	Summary	430
		References	430
13	**Relationships with Other Species in a Mixture**	439
	Douglas S. Chamblee and Michael Collins	
	13–1	Principles of Competition	439
	13–2	Specific Competitive Effects	441
	13–3	Mutual Effects of Above- and Below-Ground Associations of Alfalfa-Grass Mixtures	446
	13–4	Management Effects on Alfalfa-Grass Mixtures	450
	13–5	Summary	456
		References	457
14	**Feeding Value (Forage Quality)**	463
	G. C. Marten, D. R. Buxton, and R. F Barnes	
	14–1	Characteristics of Alfalfa-Feeding Value	465
	14–2	Preharvest Factors that Influence Feeding Value	472
	14–3	Summary	483
		References	484
15	**Antiquality Factors and Nonnutritive Chemical Components**	493
	R. E. Howarth	
	15–1	Herbage Proteins	494
	15–2	Ruminant Bloat	495
	15–3	Phenolic Constituents	498
	15–4	Tannins	500
	15–5	Lignin	502
	15–6	Saponins	503
	15–7	Estrogenic Activity	505
	15–8	Volatile Components	507
	15–9	Minerals	507

CONTENTS

15-10	Alkaloids	508
15-11	Purines and Pyrimidines	508
15-12	Photosensitization	508
15-13	Protease Inhibitors	508
15-14	Allergenic Compounds	509
15-15	Summary	509
	References	510

16 Pasture Production and Utilization — 515
R. W. Van Keuren and A. G. Matches

16-1	Adaptation and Distribution of Alfalfa for Pasture	515
16-2	Alfalfa Pasture Management	516
16-3	Alfalfa as Pasture for Dairy Cows	521
16-4	Alfalfa as Pasture for Beef Cattle	523
16-5	Alfalfa as Pasture for Sheep	526
16-6	Effect of Alfalfa Estrogens	529
16-7	Alfalfa as Pasture for Swine	529
16-8	Alfalfa as Pasture for Poultry	530
16-9	Alfalfa as Pasture for Horses	531
16-10	Bloat in Animals on Alfalfa Pasture	531
16-11	Summary	532
	References	532

17 Role in Livestock Feeding—Greenchop, Silage, Hay, and Dehy — 539
H. R. Conrad and T. J. Klopfenstein

17-1	Postharvest Nutritional Changes	540
17-2	Nutritional Quality of Alfalfa Hay	542
17-3	Nutritional Value of Alfalfa Silage	544
17-4	Nutritive Value of Greenchopped Alfalfa	546
17-5	Dehy for Protein in Ruminants	546
17-6	Summary	549
	References	550

18 Wet Fractionation Processes and Products — 553
Neal A. Jorgensen and Richard G. Koegel

18-1	Reasons for Using the System	553
18-2	Production	555
18-3	Cell Rupture	555
18-4	Expression of Juice	556
18-5	Protein Separation	558
18-6	Protein Preservation	558
18-7	Deproteinized Juice	559
18-8	Utilization	559
18-9	Summary	564
	References	564

19 Equipment for Harvesting, Transporting, Storing, and Feeding 567
A. W. Pauli, V. L. Lechtenberg, and W. F. Wedin

- 19-1 Harvesting 568
- 19-2 Transporting 580
- 19-3 Storage and Feeding 584
- 19-4 Systems Analysis 593
- References 593

20 Geographic Adaptation and Cultivar Selection 595
Bill Melton, Jim B. Moutray, and Joe H. Bouton

- 20-1 Adaptation and Cultivar Development 596
- 20-2 Cultivar Release Procedures 597
- 20-3 Adaptation in Nonhumid Regions 599
- 20-4 Adaptation in Humid Regions 608
- 20-5 Adaptation in Canada 614
- References 618

21 Diseases and Nematodes 621
Kenneth T. Leath, Donald C. Erwin, and Gerald D. Griffin

- 21-1 Diseases of Leaves and Stems 622
- 21-2 Diseases of Roots 640
- 21-3 Diseases of the Vascular System 656
- 21-4 The Alfalfa Disease Situation—An Overview 661
- References 662

22 Insects and Mites 671
George R. Manglitz and Roger H. Ratcliffe

- 22-1 Insects that Consume Foliage 671
- 22-2 Insects that Suck Sap 678
- 22-3 Insects that Feed on Roots 684
- 22-4 Insects that Affect Seed Production 686
- 22-5 Control 690
- 22-6 Control Systems 694
- 22-7 Summary 695
- References 695

23 Weeds and Weed Control 705
Elroy J. Peters and Dean L. Linscott

- 23-1 Losses Caused by Weeds 705
- 23-2 Weed Problems 707
- 23-3 Cultural Weed Control Methods: Alfalfa Seedings 710
- 23-4 Chemical Weed Control Methods: Alfalfa Seedings 715
- 23-5 No-Till and Other Methods of Establishment of Alfalfa 720
- 23-6 Weed Control Methods: Established Stands 723
- 23-7 Summary 729

CONTENTS

References 729

24 Cytology and Cytogenetics of Alfalfa   737
T. J. McCoy and E. T. Bingham

24-1 Microsporogenesis, Megasporogenesis, and Gametophyte Development 737
24-2 Fertilization and Embryo Development 739
24-3 Cytology of Meiotic and Gametophytic Abnormalities 741
24-4 Origin and Cytology of Euploids of Alfalfa 746
24-5 Origin, Cytology, and Uses of Aneuploids 754
24-6 Uses of Haploids 756
24-7 Cytogenetic and Genetic Confirmation of Autopolyploidy   756
24-8 Uses of the Ploidy Series 758
24-9 Cytogenetics of the *Medicago* Genus 760
24-10 Cytogenetics of Interspecific Hybrids Between *M. sativa* and Other *Medicago* Species 762
References 771

25 Breeding and Quantitative Genetics   777
M. D. Rumbaugh, J. L. Caddel, and D. E. Rowe

25-1 Breeding 777
25-2 Qualitative Genetics 794
25-3 Quantitative Genetics 794
References 805

26 Breeding for Yield and Quality   809
R. R. Hill, Jr., J. S. Shenk, and R. F Barnes

26-1 Yield 809
26-2 Quality 816
26-3 Summary 823
References 823

27 Breeding for Disease and Nematode Resistance   827
J. H. Elgin, Jr., R.E. Welty, and D.B. Gilchrist

27-1 Principles of Disease and Nematode Resistance Breeding ..   828
27-2 Breeding for Resistance to Specific Diseases   830
27-3 Summary 848
References 848

28 Breeding for Insect Resistance   859
E. L. Sorensen, R. A. Byers, and E. K. Horber

28-1 Categories and Causes of Resistance 860
28-2 Durability of Resistance 862
28-3 Role of Resistance in Integrated Control 863
28-4 Development of Plant Resistance 864

	28–5 Summary	888
	References	889
29	**Alfalfa Tissue Culture**	**903**
	E. T. Bingham, T. J. McCoy, and K. A. Walker	
	29–1 Role of the Genotype in Regeneration from Callus, Suspension, and Protoplasts	905
	29–2 Regeneration Physiology	909
	29–3 Variability Among Regenerated Plants	913
	29–4 Somatic Cell Selection	919
	29–5 Somatic Cell Fusion	924
	29–6 Recent Developments	924
	29–7 Conclusion	925
	References	925
30	**Pollination Control: Mechanical and Sterility**	**931**
	D. R. Viands, P. Sun, and D. K. Barnes	
	30–1 Floral Morphology, Tripping, Pollination, and Fertilization	931
	30–2 Mechanical Pollination Control	938
	30–3 Self-Incompatibility and Self-Sterility	939
	30–4 In Vitro Pollen Germination and Pollen-Tube Growth	945
	30–5 Male Sterility	947
	30–6 Pollination Control in Hybrid Production in the Field	950
	30–7 Female Sterility	954
	30–8 Summary	955
	References	956
31	**Seed Physiology, Seedling Performance, and Seed Sprouting**	**961**
	L. N. Bass, C. R. Gunn, O. B. Hesterman, and E. E. Roos	
	31–1 Seed Physiology	962
	31–2 Seed Storage	967
	31–3 Deterioration	971
	31–4 Seed and Seedling Performance	974
	31–5 Seed Sprouting	975
	References	979
32	**Seed Production Practices**	**985**
	Clarence M. Rincker, V. L. Marble, D. E. Brown, and Carl A. Johansen	
	32–1 Areas of Seed Production	985
	32–2 Stand Establishment	986
	32–3 Managing Established Stands	990
	32–4 Insect Control	994
	32–5 Pollination	1001
	32–6 Harvesting	1009
	32–7 Combine Harvesting	1011
	32–8 Postharvest Cultural Operations	1014
	32–9 Seed Yields and Factors Affecting Yield	1014
	References	1017

33 The Seed Industry — 1023
Donald L. Smith

33-1	Development of the Industry	1023
33-2	Unique Characteristics of the Industry	1026
33-3	Research	1026
33-4	Production	1029
33-5	Conditioning	1030
33-6	Marketing	1032
33-7	Product Information	1034
33-8	Trade Organizations	1035
33-9	Summary	1035
	References	1036

34 Future Trends in North America — 1037
G. E. Carlson and A. A. Hanson

34-1	Changing Demands for Alfalfa	1037
34-2	Changing Science and Technology	1038
34-3	Changing Institutional Relationships	1043
34-4	Summary	1045
	References	1045
APPENDIX		1046
SUBJECT INDEX		1055

FOREWORD

This is a remarkable book about a remarkable crop. The book began as a revision of American Society of Agronomy Monograph 15, "Alfalfa Science and Technology", but expanded into the most comprehensive treatment of the accumulated knowledge on alfalfa (*Medicago sativa* L.) currently available.

The subject matter of the book ranges from the most fundamental aspects of alfalfa genetics to practical information on alfalfa management. The authors have extremely diverse backgrounds, perspectives, and areas of expertise. This book will be a valuable research and teaching reference, and it should form the basis upon which to launch new research efforts on alfalfa production and use. Hopefully, it will stimulate expanded effort to exploit the still untapped potential of this important species.

We particularly appreciate the efforts of Dr. A. A. Hanson in managing the final review of the book and the important contributions of others in making this a truly excellent product, one in which the American Society of Agronomy, the Crop Science Society of America, and the Soil Science Society of America can be justly proud. As officers of the tri-societies, we wish to thank W-L Research, Inc., a firm with major involvement in alfalfa research and seed marketing, for permitting A. A. Hanson to edit the monograph as part of his regular employment, and for covering expenses involved in his communication with others.

D. A. Holt, *president*, 1988
American Society of Agronomy

C. J. Nelson, *president*, 1988
Crop Science Society of America

D. R. Keeney, *president*, 1988
Soil Science Society of America

PREFACE

This monograph is a testament to the oldest cultivated forage crop, alfalfa; a crop that has increased in importance over time as a mainstay in the production of livestock products. In many countries, including those in North America, alfalfa is the basic component in feeding programs for dairy cattle, as well as an important feed for beef cattle, horses, sheep, and other classes of livestock. Furthermore, alfalfa enhances the stability of agricultural production systems by improving soil productivity and reducing losses of soil and water.

The importance of alfalfa in world agriculture can be attributed to a number of morphological and physiological characteristics that contribute to its high yield of nutritious herbage, rapid recovery after cutting, longevity, and tolerance to environmental stress. Also, symbiotic N_2 fixation in alfalfa eliminates the need for chemical N, and adds a beneficial carryover effect in crop rotations. In spite of the recognized merits of alfalfa, the crop would not enjoy a competitive advantage in many regions and agricultural zones in the absence of research. Thus, the success of alfalfa production in many countries documents the research efforts of scientists who have adapted germplasm, improved understanding of crop requirements, developed superior cultivars resistant to major pest insects and diseases, advanced seed production technology, and increased efficiency in animal feeding. Recent history suggests that research efforts to maintain and enhance the contribution of alfalfa must be strengthened, in response to new problems and opportunities that will arise with the anticipated trend to more intensive agricultural production systems.

Many major developments have occurred since the publication in 1972 of Monograph 15, *Alfalfa Science and Technology* edited by Clarence H. Hanson. These numerous changes provide the basis for this revision and a new title, *Alfalfa and Alfalfa Improvement*.

Authors exercised judgment in preparing material for inclusion in the revised edition. In some chapters they have included salient information from the previous publication, while in others the authors present a solid inventory of recent achievements with limited reference to older literature. This approach has been effective in providing a comprehensive overview of new and emerging areas of science, together with a balanced assessment of other topics in which new developments have had a modest impact. A significant feature and strength of the new monograph rests on the inclusion of complete citations to the most recent literature. In addition, much new information is presented from such diverse fields as interspecific hybridization, tissue culture and genetic engineering, N_2 fixation, carbon assimilation and partitioning, pest control, crop management, crop utilization, and breeding.

The revision of *Alfalfa Science and Technology* was initiated under the leadership of co-editors D. K. Barnes and R. R. Hill, Jr., who, with an editorial committee consisting of G. H. Heichel, O. J. Hunt, K. T. Leath, G. C. Marten, and M. B. Tesar, identified topics for inclusion and contacted specialists to prepare authoritative chapters for publication. Co-editors and members of the editorial committee were assigned responsibility for chapters in their respective areas of expertise. In this capacity they received manuscripts, arranged for peer reviews, and corresponded with authors in the revision of manuscripts. Subsequently, I was asked to finalize the monograph in an effort to expedite publication. Efforts of the editorial committee are recognized by identifying them as associate editors.

As senior editor, I assumed sole responsibility for final editorial clearance and submitting all chapters to the Society for publication. Full credit must be given to authors and coauthors who prepared the 34 informative chapters; to my co-editors and associate editors for their efforts in arranging for reviews and revisions; to numerous scientists who reviewed individual chapters; and to the managing editor and Society staff for bringing the monograph to press. Although credit must be assigned as stated, I remain accountable for any oversights and deficiencies that should have been corrected during the final review.

Grateful acknowledgment is extended to my employer, W-L Research, Inc., who provided me with time to edit the monograph as a part of my regular duties, and for covering expenses involved in communicating with authors. My contribution would have been impossible in the absence of support from W-L Research, Inc.

A. A. Hanson, *editor*

CONTRIBUTORS

D. K. Barnes	Research Geneticist, USDA-ARS, Minnesota Agricultural Experiment Station, St. Paul, Minnesota
R. F. Barnes	Executive Vice President, American Society of Agronomy, Crop Science Society of America, and Soil Science of America, Madison, Wisconsin
L. N. Bass	Director (deceased), National Seed Storage Laboratory, USDA-ARS, Fort Collins, Colorado
Gary R. Bauchan	Research Geneticist, Germplasm Quality and Enhancement Laboratory, USDA-ARS, Beltsville, Maryland
J. E. Baylor	Director of Market Development, Beachley-Hardy Seed Company, Camp Hill, Pennsylvania
R. A. Byers	Research Entomologist, U.S. Regional Pasture Research Laboratory, USDA-ARS, University Park, Pennsylvania
E. T. Bingham	Professor, Department of Agronomy, University of Wisconsin, Madison, Wisconsin
Joe H. Bouton	Associate Professor, Department of Agronomy, University of Georgia, Athens, Georgia
M. A. Brick	Associate Professor, Department of Agronomy, Colorado State University, Fort Collins, Colorado
D. E. Brown	Senior Agronomist, Land O'Lakes, Caldwell, Idaho
D. R. Buxton	Research Plant Physiologist, USDA-ARS, Iowa Agricultural Experiment Station, Ames, Iowa
J. L. Caddel	Professor, Department of Agronomy, Oklahoma State University, Stillwater, Oklahoma
G. E. Carlson	Area Director, Midwest Area, USDA-ARS, Peoria, Illinois
Douglas S. Chamblee	Professor, Department of Crop Science, North Carolina State University, Raleigh, North Carolina
Michael Collins	Associate Professor, Department of Agronomy, University of Kentucky, Lexington, Kentucky
H. R. Conrad	Professor and Associate Director, Ohio State University, Ohio Agricultural Research and Development Center, Wooster, Ohio
H. T. Cralle	Assistant Professor, Department of Soil and Crop Science, Texas A&M University, College Station, Texas
R. H. Delaney	Professor, Plant Science Division, University of Wyoming, Laramie, Wyoming
Stanley H. Duke	Associate Professor, Department of Agronomy, University of Wisconsin, Madison, Wisconsin
J. H. Elgin, Jr.	Research Geneticist, National Program Staff, USDA-ARS, Beltsville, Maryland
Donald E. Erwin	Professor, Department of Plant Pathology, University of California, Riverside, California
G. W. Fick	Professor, Department of Agronomy, Cornell University, Ithaca, New York

CONTRIBUTORS

D. G. Gilchrist	Professor, Department of Plant Pathology, University of California, Davis, California
B. P. Goplen	Head-Forage Section, Agriculture Canada Research Station, Saskatoon, Saskatchewan
Gerald D. Griffin	Research Plant Pathologist, USDA-ARS, Utah Agricultural Experiment Station, Logan, Utah
W. K. Griffith	Eastern Director, Potash and Phosphate Institute, Great Falls, Virginia
C. R. Gunn	Botanist, Systematic Botany, Mycology and Nematology Laboratory, USDA-ARS, Beltsville, Maryland
A. A. Hanson	Vice-President-Research, W-L Research, Highland, Maryland
G. H. Heichel	Plant Physiologist, USDA-ARS, Minnesota Agricultural Experiment Station, St. Paul, Minnesota
O. B. Hesterman	Assistant Professor, Department of Crop and Soil Sciences, Michigan State University, East Lansing, Michigan
R. R. Hill, Jr.	Research Leader, U.S. Regional Pasture Research Laboratory, USDA-ARS, University Park, Pennsylvania
D. A. Holt	Director, Illinois Agricultural Experiment Station, and Associate Dean, College of Agriculture, University of Illinois, Urbana, Illinois
E. K. Horber	Entomologist, Department of Entomology, Kansas State University, Manhattan, Kansas
R. E. Howarth	Senior Research Scientist, Agriculture Canada Research Station, Saskatoon, Saskatchewan
Carl A. Johansen	Entomologist (retired), Department of Entomology, Washington State University, Pullman, Washington
Neal A. Jorgensen	Associate Dean and Director, Wisconsin Agricultural Experiment Station, University of Wisconsin, Madison
M. B. Kirkham	Professor, Evapotranspiration Laboratory, Department of Agronomy, Kansas State University, Manhattan, Kansas
T. Klopfenstein	Professor, Department of Animal Science, University of Nebraska, Lincoln, Nebraska
Richard G. Koegel	Research Agricultural Engineer, Dairy Forage Research Center, USDA-ARS, Madison, Wisconsin
G. D. Lacefield	Professor, University of Kentucky, West Kentucky Research and Education Center, Princeton, Kentucky
Les Lanyon	Associate Professor, Department of Agronomy, Pennsylvania State University, University Park, Pennsylvania
Kenneth T. Leath	Research Plant Pathologist, U.S. Regional Pasture Research Laboratory, USDA-ARS, University Park, Pennsylvania
V. L. Lechtenberg	Associate Director, Indiana Agricultural Experiment Station, Purdue University, West Lafayette, Indiana
W. F. Lehman	Agronomist, Department of Agronomy and Range Science, University of California, El Centro, California
Dean L. Linscott	Lead Scientist, USDA-ARS, Ithaca, New York
D. G. Lugg	Agricultural Consultant, Compton Marsh Farm, Wiltshire, England (formerly New Mexico State University, Las Cruces, New Mexico

CONTRIBUTORS

George R. Manglitz	Research Entomologist, USDA-ARS, Nebraska Agricultural Experiment Station, Lincoln, Nebraska
V. L. Marble	Extension Agronomist, Department of Agronomy and Range Science, University of California, Davis, California
G. C. Marten	Lead Scientist, USDA-ARS, Minnesota Agricultural Experiment Station, St. Paul, Minnesota
N. P. Martin	Professor of Extension, Department of Agronomy and Genetics, University of Minnesota, St. Paul, Minnesota
A. G. Matches	Thornton Professor of Plant and Soil Science, Department of Plant and Soil Sciences, Texas Tech. University, Lubbock, Texas
T. J. McCoy	Assistant Professor, Department of Plant Sciences, University of Arizona, Tucson, Arizona
James S. McKenzie	Research Scientist, Agriculture Canada Research Station, Beaverlodge, Alberta
Bill Melton	Professor, Department of Crop and Soil Sciences, New Mexico State University, Las Cruces, New Mexico
Réal Michaud	Research Scientist, Agriculture Canada Research Station, Ste-Foy, Quebec
Jim B. Moutray	Forage Research Director, AgriPro, Ames, Iowa
Roger Paquin	Research Scientist, Agriculture Canada Research Station, Ste-Foy, Quebec
A. W. Pauli	Agronomist (retired), Deere and Company, East Moline, Illinois
Elroy J. Peters	Professor, Department of Agronomy, University of Missouri, Columbia, Missouri
D. A. Phillips	Professor, Department of Agronomy and Range Science, University of California, Davis, California
Carlos F. Quiros	Assistant Professor, Department of Vegetable Crops, University of California, Davis, California
Roger H. Ratcliffe	Research Entomologist, Germplasm Quality and Enhancement Laboratory, USDA-ARS, Beltsville, Maryland
Clarence M. Rincker	Research Agronomist (retired), USDA-ARS, Irrigated Agriculture Research and Extension Center, Prosser, Washington
E. E. Roos	Plant Physiologist, National Seed Storage Laboratory, USDA-ARS, Ft. Collins, Colorado
D. E. Rowe	Research Leader, USDA-ARS, Mississippi Agricultural Experiment Station, Starkville, Mississippi
M. D. Rumbaugh	Research Geneticist, USDA-ARS, Utah Agricultural Experiment Station, Logan, Utah
C. C. Shaeffer	Professor, Department of Agronomy and Plant Genetics, University of Minnesota, St. Paul, Minnesota
J. S. Shenk	Professor, Department of Agronomy, Pennsylvania State University, University Park, Pennsylvania
Donald L. Smith	Plant Breeder and Consultant, 2 Loma Vista Place, Woodland, California
E. L. Sorensen	Research Agronomist, USDA-ARS, Kansas Agricultural Experiment Station, Manhattan, Kansas

CONTRIBUTORS

Paul Sun	Research Director, Dairyland Research International, Clinton, Wisconsin
C. B. Tanner	Professor, Department of Soil Science, University of Wisconsin, Madison, Wisconsin
M. B. Tesar	Professor, Department of Crop and Soil Sciences, Michigan State University, East Lansing, Michigan
L. R. Teuber	Associate Professor, Department of Agronomy and Range Science, University of California, Davis, California
C. P. Vance	Research Plant Physiologist, USDA-ARS, Minnesota Agricultural Experiment Station, St. Paul, Minnesota
R. W. Van Keuren	Professor, Department of Agronomy, Ohio State University, Ohio Agricultural Research and Development Center, Wooster, Ohio
D. R. Viands	Associate Professor, Department of Plant Breeding and Biometry, Cornell University, Ithaca, New York
K. Walker	Vice-President, Plant Genetics, Davis, California
W. F. Wedin	Professor, Department of Agronomy, Iowa State University, Ames, Iowa
R. E. Welty	Research Plant Pathologist, USDA-ARS, Oregon Agricultural Experiment Station, Corvallis, Oregon

Conversion Factors for SI and non-SI Units

To convert Column 1 into Column 2, multiply by	Column 1 SI Unit	Column 2 non-SI Unit	To convert Column 2 into Column 1 multiply by
Length			
0.621	kilometer, km (10^3 m)	mile, mi	1.609
1.094	meter, m	yard, yd	0.914
3.28	meter, m	foot, ft	0.304
1.0	micrometer, μm (10^{-6} m)	micron, μ	1.0
3.94×10^{-2}	millimeter, mm (10^{-3} m)	inch, in	25.4
10	nanometer, nm (10^{-9} m)	Angstrom, Å	0.1
Area			
2.47	hectare, ha	acre	0.405
247	square kilometer, km^2 (10^3 m)^2	acre	4.05×10^{-3}
0.386	square kilometer, km^2 (10^3 m)^2	square mile, mi^2	2.590
2.47×10^{-4}	square meter, m^2	acre	4.05×10^3
10.76	square meter, m^2	square foot, ft^2	9.29×10^{-2}
1.55×10^{-3}	square millimeter, mm^2 (10^{-6} m)^2	square inch, in^2	645
Volume			
9.73×10^{-3}	cubic meter, m^3	acre-inch	102.8
35.3	cubic meter, m^3	cubic foot, ft^3	2.83×10^{-2}
6.10×10^4	cubic meter, m^3	cubic inch, in^3	1.64×10^{-5}
2.84×10^{-2}	liter, L (10^{-3} m^3)	bushel, bu	35.24
1.057	liter, L (10^{-3} m^3)	quart (liquid), qt	0.946
3.53×10^{-2}	liter, L (10^{-3} m^3)	cubic foot, ft^3	28.3
0.265	liter, L (10^{-3} m^3)	gallon	3.78
33.78	liter, L (10^{-3} m^3)	ounce (fluid), oz	2.96×10^{-2}
2.11	liter, L (10^{-3} m^3)	pint (fluid), pt	0.473

continued on next page

Conversion Factors for SI and non-SI Units

To convert Column 1 into Column 2, multiply by	Column 1 SI Unit	Column 2 non-SI Unit	To convert Column 2 into Column 1 multiply by
Mass			
2.20×10^{-3}	gram, g (10^{-3} kg)	pound, lb	454
3.52×10^{-2}	gram, g (10^{-3} kg)	ounce (avdp), oz	28.4
2.205	kilogram, kg	pound, lb	0.454
10^{-2}	kilogram, kg	quintal (metric), q	10^2
1.10×10^{-3}	kilogram, kg	ton (2000 lb), ton	907
1.102	megagram, Mg (tonne)	ton (U.S.), ton	0.907
1.102	tonne, t	ton (U.S.), ton	0.907
Yield and Rate			
0.893	kilogram per hectare, kg ha^{-1}	pound per acre, lb acre^{-1}	1.12
7.77×10^{-2}	kilogram per cubic meter, kg m^{-3}	pound per bushel, lb bu^{-1}	12.87
1.49×10^{-2}	kilogram per hectare, kg ha^{-1}	bushel per acre, 60 lb	67.19
1.59×10^{-2}	kilogram per hectare, kg ha^{-1}	bushel per acre, 56 lb	62.71
1.86×10^{-2}	kilogram per hectare, kg ha^{-1}	bushel per acre, 48 lb	53.75
0.107	liter per hectare, L ha^{-1}	gallon per acre	9.35
893	tonnes per hectare, t ha^{-1}	pound per acre, lb acre^{-1}	1.12×10^{-3}
893	megagram per hectare, Mg ha^{-1}	pound per acre, lb acre^{-1}	1.12×10^{-3}
0.446	megagram per hectare, Mg ha^{-1}	ton (2000 lb) per acre, ton acre^{-1}	2.24
2.24	meter per second, m s^{-1}	mile per hour	0.447
Specific Surface			
10	square meter per kilogram, m^2 kg^{-1}	square centimeter per gram, cm^2 g^{-1}	0.1
10^3	square meter per kilogram, m^2 kg^{-1}	square millimeter per gram, mm^2 g^{-1}	10^{-3}

CONVERSION FACTORS FOR SI UNIT

Pressure

9.90	megapascal, MPa (10^6 Pa)	atmosphere	0.101
10	megapascal, MPa (10^6 Pa)	bar	0.1
1.00	megagram per cubic meter, Mg m^{-3}	gram per cubic centimeter, g cm^{-3}	1.00
2.09×10^{-2}	pascal, Pa	pound per square foot, lb ft^{-2}	47.9
1.45×10^{-4}	pascal, Pa	pound per square inch, lb in^{-2}	6.90×10^3

Temperature

1.00 (K − 273)	Kelvin, K	Celsius, °C	1.00 (°C + 273)
(9/5 °C) + 32	Celsius, °C	Fahrenheit, °F	5/9 (°F − 32)

Energy, Work, Quantity of Heat

9.52×10^{-4}	joule, J	British thermal unit, Btu	1.05×10^3
0.239	joule, J	calorie, cal	4.19
10^7	joule, J	erg	10^{-7}
0.735	joule, J	foot-pound	1.36
2.387×10^{-5}	joule per square meter, J m^{-2}	calorie per square centimeter (langley)	4.19×10^4
10^5	newton, N	dyne	10^{-5}
1.43×10^{-3}	watt per square meter, W m^{-2}	calorie per square centimeter minute (irradiance), cal cm^{-2} min^{-1}	698

Transpiration and Photosynthesis

3.60×10^{-2}	milligram per square meter second, mg m^{-2} s^{-1}	gram per square decimeter hour, g dm^{-2} h^{-1}	27.8
5.56×10^{-3}	milligram (H$_2$O) per square meter second, mg m^{-2} s^{-1}	micromole (H$_2$O) per square centimeter second, μmol cm^{-2} s^{-1}	180
10^{-4}	milligram per square meter second, mg m^{-2} s^{-1}	milligram per square centimeter second, mg cm^{-2} s^{-1}	10^4
35.97	milligram per square meter second, mg m^{-2} s^{-1}	milligram per square decimeter hour, mg dm^{-2} h^{-1}	2.78×10^{-2}

Plane Angle

57.3	radian, rad	degrees (angle), °	1.75×10^{-2}

continued on next page

Conversion Factors for SI and non-SI Units

To convert Column 1 into Column 2, multiply by	Column 1 SI Unit	Column 2 non-SI Unit	To convert Column 2 into Column 1 multiply by
Electrical Conductivity			
10	siemen per meter, S m^{-1}	millimho per centimeter, mmho cm^{-1}	0.1
Water Measurement			
9.73×10^{-3}	cubic meter, m^3	acre-inches, acre-in	102.8
9.81×10^{-3}	cubic meter per hour, m^3 h^{-1}	cubic feet per second, ft^3 s^{-1}	101.9
4.40	cubic meter per hour, m^3 h^{-1}	U.S. gallons per minute, gal min^{-1}	0.227
8.11	hectare-meters, ha-m	acre-feet, acre-ft	0.123
97.28	hectare-meters, ha-m	acre-inches, acre-in	1.03×10^{-2}
8.1×10^{-2}	hectare-centimeters, ha-cm	acre-feet, acre-ft	12.33
Concentrations			
1	centimole per kilogram, cmol kg^{-1} (ion exchange capacity)	milliequivalents per 100 grams, meq 100 g^{-1}	1
0.1	gram per kilogram, g kg^{-1}	percent, %	10
1	milligram per kilogram, mg kg^{-1}	parts per million, ppm	1
Radioactivity			
2.7×10^{-11}	bequerel, Bq	curie, Ci	3.7×10^{10}
2.7×10^{-2}	bequerel per kilogram, Bq kg^{-1}	picocurie per gram, pCi g^{-1}	37
100	gray, Gy (absorbed dose)	rad, rd	0.01
100	sievert, Sv (equivalent dose)	rem (roentgen equivalent man)	0.01
Plant Nutrient Conversion			
	Elemental	*Oxide*	
2.29	P	P$_2$O$_5$	0.437
1.20	K	K$_2$O	0.830
1.39	Ca	CaO	0.715
1.66	Mg	MgO	0.602

15 June 1987

1 Highlights in the USA and Canada

D. K. BARNES
*USDA-ARS and University of Minnesota
St. Paul, Minnesota*

B. P. GOPLEN
*Agriculture Canada Research Station
Saskatoon, Saskatchewan*

J. E. BAYLOR
*Beachley-Hardy Seed Company
Camp Hill, Pennsylvania*

Alfalfa (*Medicago sativa* L.), often called "Queen of the Forages", is the most important forage crop species in the USA and Canada. It is a remarkable crop in comparison with others. Alfalfa is recognized as the most

—widely adapted agronomic crop;
—effective source of biological nitrogen (N_2) fixation
—energy-efficient crop to grow;
—important source of protein yield/ha; and
—attractive source of nectar for honey bees.

In addition to being an excellent source of vitamins and minerals, alfalfa is important for improving soil tilth. Because of these impressive credits, alfalfa has achieved the status of being a primary agricultural crop in the USA and Canada. It also provides a model system for many types of research on other forage crops and for species characterized by autotetraploid inheritance.

In 1972, publication of the ASA monograph 15, *Alfalfa Science and Technology*, was a significant highlight in the history of alfalfa in North America. It assembled all available information about alfalfa and provided an impetus for scientists to improve the breeding, production, and utilization of alfalfa, and for industry to increase the potential usefulness of the crop. Hanson and Davis (13) authored chapter 2 "Highlights in the United States of America" in the monograph. This chapter is a sequel to theirs; it summarizes past highlights, and describes major advances in technology achieved in the USA and Canada since the early 1970s. We have concentrated on the development of improved breeding and crop production practices, and on factors that affect germplasm improvement, seed availability, crop utilization, and research productivity.

Copyright 1988 © ASA-CSSA-SSSA, 677 South Segoe Road, Madison, WI 53711, USA. *Alfalfa and Alfalfa Improvement*—Agronomy Monograph no. 29.

1-1 ALFALFA DISTRIBUTION

As early as 1736, colonists brought alfalfa to the eastern USA. These and subsequent introductions were not successful, except on a few well-drained, limestone soils (36). Alfalfa was well suited to the dry climates and irrigated soils of western USA, where it was introduced in the mid-1850s. Alfalfa eventually spread to the Southern Great Plains, and with the introduction of winter-hardy types to the North Central States. The area of alfalfa cut for hay in the USA increased from a few hectares in 1854 to 0.8 million ha (2 million acres) in 1900, 4.1 million ha (10 million acres) in 1942, 8.1 million ha (20 million acres) in 1950, and to a peak of 12.2 million ha (30 million acres) in 1958. Since 1980, the area has varied between 10.5 and 10.9 million ha (26 and 27 million acres). Wisconsin, South Dakota, and Minnesota each harvested more than 0.8 million ha (2 million acres) in 1980. California, Idaho, Iowa, Michigan, Montana, Nebraska, and North Dakota had 0.4 to 0.8 million ha (1 to 2 million acres).

Alfalfa was first introduced into eastern Canada in 1871, and its use spread gradually throughout Ontario, Quebec, and the Atlantic provinces. In western Canada, with its more severe winters, alfalfa was scarcely grown until the introduction and further selection of extremely winter-hardy types. Dr. L.E. Kirk of the Univ. of Saskatchewan accomplished this by developing strain 666 of 'Grimm' alfalfa which was first distributed in 1926. Alfalfa cut for hay in Canada in 1941 and 1951 was <0.4 million ha (1 million acres). This was followed by a rapid expansion to 1.8 million ha (4.5 million acres) in 1961, 2.2 million ha (5.4 million acres) in 1971 and 2.6 million ha (6.3 million acres) in 1981. Distribution of alfalfa grown for hay across Canada shows most of the alfalfa grown in the prairie provinces of Alberta, Saskatchewan, and Manitoba in the west, and in Ontario and Quebec in the east (about 0.7, 0.6, 0.4, 0.6, and 0.2 million ha [1.8, 1.4, 1.0, 1.4, and 0.4 million acres] per province, respectively). Acid soils and problems with winter icing and heaving restrict alfalfa production in Quebec and the Atlantic provinces. Canadian hectarages cited do not include alfalfa grown for seed, pasture, the dehydration industry, silage, and green manure. In Canada, alfalfa is grown on an estimated 4.9 to 5.7 million ha (12–14 million acres).

1-2 ORIGINS OF NORTH AMERICAN GERMPLASM

Alfalfa is an immigrant to North America. It originated near Iran, but related forms and species are found scattered over central Asia as far north as Siberia. Alfalfa was brought to Europe and South America by invading armies, explorers, and missionaries as a valuable source of feed for horses (*Equus caballus*) and other animals.

In about 1850, Spanish types of alfalfa germplasm were introduced into southwestern USA from South America. Between 1858 and 1910, three winter-hardy germplasm sources from Europe and Russia were brought into the upper midwestern USA and eastern Canada. There were two intermediate winter-hardy introductions. One was from a broad area in the near east (southern Russia, Iran, Afghanistan, and Turkey) between 1898 and 1925 and the other from France in 1947. Nonwinter-hardy types were introduced from Peru (1899), India (1913 and 1956), and Africa (1924). Essentially all basic germplasm used in the development of present North American cultivars traces to these nine sources of alfalfa (2).

1-3 GERMPLASM USE IN CULTIVAR DEVELOPMENT

Before 1925, most alfalfa breeding efforts in North America were directed toward selecting strains that were more winterhardy. In 1925, bacterial wilt was described as a new alfalfa disease (16). During the next 30 yr, emphasis was placed on developing cultivars that combined winterhardiness and bacterial-wilt resistance. During the late 1950s, it became apparent that there was an urgent need to develop cultivars that were resistant to other diseases and several insect pests. The need for improved pest resistance increased the diversity of germplasm required in cultivars.

Before 1955, about 33 recognized cultivars were grown in the USA and Canada. About half of these were introduced from Europe, Asia, and the Near East. Between 1958 and 1985, the number of recognized cultivars increased to more than 250. None of the cultivars released between 1965 and 1985 were direct plant introductions. Barnes et al. (2) made a genetic analysis of the chronological change in germplasm diversity of North American alfalfa cultivars. They found that before 1930 cultivars were developed primarily from a single germplasm source. Between 1941 and 1960, cultivars were developed by combining two or three of the original nine germplasm sources. Between 1961 and 1970, new cultivars often included three or four germplasm sources. Many recent cultivars have genes contributed by all nine recognized germplasm sources. The trend to increase the genetic diversity in alfalfa cultivars should make alfalfa less vulnerable to a genetic disaster than it was 50 yr ago. The effect of this broad germplasm base on future breeding strategies remains to be determined.

1-4 GERMPLASM COLLECTION, PRESERVATION, AND EVALUATION IMPROVED

Following the 1970 corn (*Zea mays* L.) leaf blight epidemic, much concern was expressed about the genetic vulnerability of all major crops grown in the USA. In 1972, the North American Alfalfa Improvement Conference (NAAIC) formed a Committee on Preservation of Germplasm (18). The committee addressed the many issues involved in seed increase of plant introductions, the incorporation of Canadian germplasm into the plant introduction (PI) system in the USA, germplasm exchange with foreign countries, development of germplasm pools, plant exploration, simultaneous collection of both seed and nodule-forming rhizobia, and domestic seed collection from old stands, particularly those found in western rangelands and adjacent parts of Canada.

Alfalfa was one of eight crop species that was included originally in the *Germplasm Resources Information Project* sponsored by the USDA and initiated in 1978 to develop an information system to service the National Plant Germplasm System (NPGS). The system became functional for alfalfa in 1986. It stores all available data on individual PI's and permits queries for information and seed from interested scientists. The USDA-ARS Regional Plant Introduction Station, Washington State University, Pullman, is the alfalfa center. An Alfalfa Crop Advisory Committee (ACAC) was appointed, with representatives from federal and state governments, universities, private industry, and Canada. Membership of the NAAIC Committee on preservation of Germplasm is identical to that of the Alfalfa Crop Advisory Committee. The NAAIC/ACAC developed descriptor lists, defined the descriptors, and set priorities for germplasm collection and research on PI's.

Plant explorations identified and initiated by the committee were: Andes mountains of northern Chile (1980); domestic collection in old stands of western USA and adjacent areas of Canada (1980); Peru, Bolivia, and Ecuador (1981); Turkey (1981); USSR (1982); Romania (1984); Morocco (1984); Pakistan (1985); and Spain (1985). The goal has been to collect native germplasm in all of the world's important alfalfa-growing areas before the local ecotypes and/or species are lost or contaminated with introduced germplasm. Seed collected in each country is shared with the germplasm maintenance agency of that country. The collected seed is then entered as a PI in the USDA-ARS collection at Pullman. The recent germplasm collections have been the most intensive since N.E. Hanson conducted eight collection trips in Europe and Asia from 1894 to 1934 (34).

Before 1979, seed increases of PI's were done under open-pollinated conditions. In 1979, a new system was organized for increasing original seed of about 150 to 200 perennial alfalfa PI's/yr. The increases are made under cages using leafcutting bees as pollinators. Beginning in 1983, a program was established to evaluate newly increased PI's seed for resistance to nine insects, 10 diseases, three nematodes, and several agronomic traits. These evaluations were conducted by scientists at eight locations. Additional traits will be included in the evaluation program as new evaluation procedures become available. This USDA-ARS supported program should ensure the long-term maintenance, evaluation, description, availability, and distribution of basic germplasm sources for use by all alfalfa scientists.

1-5 PUBLIC AND INDUSTRY BREEDING PROGRAMS CHANGE EMPHASIS

During the late 1950s, private alfalfa breeding programs were established in the USA. These programs were responsible for the rapid increase in the numbers of recognized alfalfa cultivars. The rate of cultivar release increased from about 0.3/yr between 1901 and 1940 to about 1.0/yr between 1941 and 1960 to about 17.0/yr between 1981 and 1985. The percentage of cultivars developed by private industry in the USA increased steadily during 1955 to 1985. The proportion of new cultivars released that were privately developed cultivars for 1955 to 1960, 1961 to 1965, 1966 to 1970, 1971 to 1975, 1976 to 1980, and 1981 to 1985 were 20, 59, 66, 72, 90, and 93%, respectively. The increased effort by private breeders was accompanied by a reduction in the number of public breeding programs, and by a change in the direction of public research from breeding new cultivars to developing new breeding procedures and improved germplasm.

The transition from applied alfalfa breeding to more fundamental research within public agencies has contributed to the development of comprehensive multidisciplinary research efforts in many problem areas of: plant morphology and anatomy (see Chapter 4 in this book), breeding for improved N_2 fixation (see Chapter 7 in this book), breeding for resistance to bloat (see Chapter 15 in this book), developing cytological methods for the transfer of genes between ploidy levels and *Medicago* spp. (see Chapter 24 in this book), developing models for heterosis to maximize its effects when breeding for yield (see Chapter 26 in this book), breeding for increased glandular hairs as a type of host plant insect resistance (see Chapter 28 in this book), and using tissue culture in selection programs (see Chapter 29 in this book). In addition, public research has provided new insights into the general areas of plant physiology, plant growth, temperature stress, and pest resistance. An impressive amount of basic information essential to the future of alfalfa improvement was produced from 1975 to 1985.

In the USA, industry breeding programs increased in number, size, and sophistication from 1965 to 1985. Most programs screen for multiple pest resistance, often in multiple-step procedures in which they can begin with several hundred thousand plants per population. Industry research programs have been quick to adopt new screening procedures developed in public research programs. Some industry research programs also have the expertise to modify and improve procedures and to develop new technologies. The high quality of the best proprietary cultivars demonstrates the efficacy of industry breeding programs, and suggests that an appropriate balance has been reached between fundamental public research and more applied industry research.

1-6 CULTIVAR DESCRIPTIONS ARE STANDARDIZED

The National Alfalfa Variety Review Board (NAVRB) first met in 1962. It was organized to provide a forum where public and private agencies could standardize the review and description of new cultivars. The composition of the NAVRB is one voting member representing each: Association of Official Seed Certifying Agencies (AOSCA); American Seed Trade Association (ASTA)—Research; ASTA—Management; NAAIC (formerly the National Alfalfa Improvement Conference); USDA-ARS; and Ex Officio—USDA Plant Variety Protection Office. The NAVRB does not judge cultivars on merit. Favorable action by the NAVRB indicates that the cultivar is described accurately according to the available data and is eligible for the production of certified seed. The cultivar description then becomes an open record, and is available to AOSCA for use by seed-certifying agencies.

The development of pest-resistant alfalfa cultivars has made a significant contribution to agriculture. The NAVRB, growers, extension personnel, marketing specialists, and research scientists needed uniform procedures for describing levels of pest resistance of cultivars. The Plant Variety Protection Act of 1970 also increased the need for standard descriptions of pest resistance. An NAAIC committee on "Standard Tests for Characterizing Disease and Insect Resistance of Alfalfa Cultivars" prepared a publication in 1974 which identified and described all pest evaluation procedures that could be accepted as standard tests. This publication was revised in 1984 (8).

A 1983 survey of alfalfa breeders, extension specialists, and production personnel showed near unanimous support for the NAVRB system and the nature of data required (4). They also recommended development of a national publication describing cultivar dormancy and pest resistance. In 1986, the Certified Alfalfa Seed Council published a listing of currently available certified alfalfa cultivars together with their respective dormancy and pest-resistance ratings. Plans have been made to update this publication annually. The Certified Alfalfa Seed Council represents growers of certified alfalfa seed who are interested in promoting the use of certified alfalfa seed.

1-7 SEED PRODUCTION PROBLEMS CHANGE

Development of an alfalfa seed industry in the western USA has been an important accomplishment. Specialized production of alfalfa seed was practically nonexistent until the mid-1940s. The industry expanded to produce 73 million kg (161 million lb) of seed in 1963. Currently, the USA uses about 36 million kg (80 million lb) of alfalfa seed annually compared to about 58 million kg (128 million lb) in the mid-1960s. The 1985 area planted to alfalfa for forage was only about 7% smaller than that planted in the 1960s. New cultivars as well as improved management practices have contributed to the more efficient use of seed. New pest-resistant cultivars produce more forage and reduce stand establishment problems compared with older cultivars. Future trends in seed usage will be associated closely with livestock production, the use of alfalfa in crop rotations, and the total area devoted to alfalfa.

The increased numbers of new cultivars have created problems in finding adequate isolation during seed production. The large numbers of proprietary cultivars have reduced opportunities for speculative seed production of public cultivars. Seed production is shifting among production areas in response to changes in production costs and the relative value of competing crops. In brief, the USA seed industry with its high-production capacity is in transition. This is illustrated by the termination of the National Foundation Seed Project in 1985. In 1949, the USDA and other government agencies organized this project to improve certified seed supplies of public cultivars of alfalfa and other forages by ensuring a reliable supply of foundation seed (13). Current low numbers of public cultivar releases and exclusive release procedures adopted by some public agencies eliminated the need for the National Foundation Seed Project.

In Canada, alfalfa seed production is concentrated in the western provinces. Production of seed has changed dramatically through the years. A high of 5.4 million kg (12 million lb) was produced annually in the 5-yr period from 1945 to 1949 when wild bees (*Bombus* and *Megachile* spp.) were plentiful and prices good. This dropped to 1.5 million kg (3.4 million lb) during 1964 to 1968, then increased to 2.1 million kg (4.7 million lb) during 1978 to 1982. Production in 1983 reached 3.8 million kg (8.4 million lb). This recovery in seed production coincided with the introduction and commercial use of the alfalfa leafcutting bee [*Megachile rotundata* (Fabricius)] (30).

1-8 CANADIAN FORAGE SEED PROJECT AND SEEDS CANADA

The Canadian Forage Seed Project was formed in 1951 "to encourage production, distribution, continuity of supply, and limited stockpiling of breeder and foundation stocks of recommended forage varieties" (12). Before this time, there was minimal seed production of pedigreed forage cultivars. New cultivars that plant breeders developed were virtually lost because a system to multiply seed of public cultivars for farm use was nonexistent. The project stimulated production and multiplication of public forage crop cultivars, organized distribution channels through the provincial governments and the seed trade, and established reliable seed supplies with controlled production and maintenance of breeder and foundation seed stocks.

The project acted as the sole agent for the distribution of breeder seed and contracts with growers for the production of foundation seed. Foundation seed produced under contract was purchased by the project and resold on an allocation basis to certified seed growers through provincial departments of agriculture and, beginning in 1963, through the Canadian Seed Trade Association. Agriculture Canada underwrote production and administrative costs and the chair of the project was the director of the Food Production and Inspection Branch. Despite worthy objectives, the Canadian Forage Seed Project met with limited success. In 1976 in an attempt to overcome the inherent problems of the project, Seeds Canada (SeCan) was organized.

The SeCan Association was developed to correct an obvious deficiency, namely, adequate promotion of cultivars developed by public plant-breeding institutions, the major source of new cultivars. In contrast to the Canadian Forage Seed Project, SeCan Association is a private, nonprofit corporation funded entirely by its members. SeCan provides a mechanism for promoting and sharing promotion costs by all merchandisers of a cultivar in proportion to their sale of certified seed. On request, the association will collect a royalty on behalf of the breeder. Almost all new public cultivars of the major agricultural crops licensed in Canada are now assigned to SeCan.

1–9 CANADIAN SEED CERTIFICATION

Seed certification in Canada has some differences from that of the USA. The Canadian Seed Growers' Association is designated by the Canada Seeds Act and Regulations as the official pedigreeing agency responsible for prescribing standards and issuing crop certificates for Canadian-produced pedigreed seed of all agricultural crops, with the exception of potatoes (*Solanum tuberosum* L.). The association is national in scope, but with provincial branches concerned with various aspects of local production. The association originated in 1904, and since 1926 has been staffed entirely by nongovernment employees. Although changes in methods and operations have occurred throughout time, the objectives of the association have remained the same: to improve pedigreed seed production and usage. The association cooperates closely with the Food Production and Inspection Branch and the Research Branch of Agriculture Canada, the agricultural faculties of the Universities, the provincial Departments and Ministries of Agriculture, and the seed trade. Canadian forage cultivars have only three classes of pedigreed seed; breeder, foundation, and certified. Genetic purity in perennial crops is maintained by age of stand limitations (number of seed harvests), isolation distance and land requirements, and other regulations established by the association. All cultivars in Canada must be submitted to the Food Production and Inspection Branch for approval and licensing. Various expert committees furnish recommendations for licensing to the Food Production and Inspection Branch. Unless alfalfa has pedigreed status, it cannot be advertised and sold by cultivar name in Canada.

1-10 NEW DISEASE RESISTANCE DEVELOPED

Alfalfa breeders' primary emphasis continues to be to develop resistance to diseases, insects, and nematodes. During the mid-1970s, resistance to Phytophthora root rot [caused by *Phytophthora megasperma* (Drechs.) f. sp. *medicaginis*] and anthracnose (caused by *Colletotrichum trifolii* Bain.) began to be routinely incorporated into new cultivars, in addition to resistance to bacterial wilt [caused by *Corynebacterium insidiosum* (McCull) H.L. Jens]. In the early 1980s, resistance to Fusarium wilt [caused by *Fusarium oxysporum* Schl. f. sp. *medicaginis* Weimer) Sny. & Hans.] and Verticillium wilt (caused by *Verticillium albo-atrum* Reinke & Berth.) became new objectives in most breeding programs.

The Verticillium wilt story was unusual because the pathogen was identified first in the seed-producing area of Washington state, rather than in the major forage-producing areas of the Midwest and East. Found in 1976, Verticillium wilt is thought to have been introduced into North America from Europe in some basic seed stocks planted for seed production in Washington. The disease was identified only in Washington, Oregon, Idaho, and British Columbia, Canada until it was reported in Wisconsin in 1980. Because the disease was found first in a seed production area, there were concerns expressed over the possible spread of the disease with seed. Before 1981, essentially all Verticillium research and breeding had to be done in the Pacific Northwest. There was no opportunity to determine how "resistant" germplasm performed when exposed to the pathogen in humid alfalfa-growing areas. For all previous "new" pests, the first resistant cultivars were developed by public agencies. Those programs developed life history and control data for the pest in conjunction with the breeding programs. The first public cultivar often demonstrated the merits of resistance, and met the initial demand for resistance for several years before the release of industry-developed cultivars. The private cultivars often represented further improvements in yield and multiple pest resistance.

In response to the discovery of Verticillium wilt in the major forage-growing area of Wisconsin, the first resistant cultivars were simultaneously released by both a public and several industry-breeding programs. Some of the cultivars were released for use on the basis of limited field testing and with essentially no data illustrating economic losses from the disease nor potential gain from growing resistant cultivars. Most projections of economic loss were based on experience in the Pacific Northwest and in Europe rather than on data generated in the production areas of the Midwest and East.

Verticillium wilt is now established in many northern and northeastern production areas. It has not spread as rapidly and is not as devastating as first predicted. Nevertheless, the disease has become serious in local areas confirming the need to incorporate resistance into cultivars for use in infested areas. The Verticillium wilt experience illustrates the importance of maintaining a strong balance between fundamental and applied research in all aspects of alfalfa research, especially that devoted to new pests. The credibility of both public and private organizations suffers when inadequate data are available to support claims and recommendations.

1-11 NEW INSECT RESISTANCE DEVELOPED

During 1975 to 1985, breeding for insect resistance concentrated on the spotted alfalfa aphid [*Therioaphis maculata* (Buckton)] and pea aphid [*Acyrthosiphon pisum* (Harris)]. Biotypes were observed for both aphid species so it became necessary to continually monitor biotypes used in screening and evaluation tests. In 1974, the blue alfalfa aphid (*A. kondoi* Shinji) was first identified as a new alfalfa pest in California. It quickly spread throughout the southwestern USA. Also, the blotch leafminer [*Agromyza frontella* (Rondani)] was identified as a new pest in the northeastern USA. Further research will be required to define the magnitude of losses and to identify sources of resistance to the blotch leafminer. Host plant resistance to the blue alfalfa aphid was found and resistant cultivars are available.

Research on alfalfa weevil resistance diminished with the success of biological control from the introduction of parasites and predators and the demonstrated effectiveness of timely harvesting combined with chemical treatments. The potato leafhopper [*Empoasca fabae* (Harris)] continues to be the most serious insect pest to forage production in the north-central USA, while *Lygus* spp. and the alfalfa seed chalcid [*Bruchophagus roddi* (Gussakovsky)] remain the most serious insect pests in seed production.

Research at Kansas State University and Purdue University on the glandular hairs found in several *Medicago* spp. showed that they provided a high level of resistance to several insect species (see Chapter 8 in this book). Glandular hairs produce an adhesive that can immobilize and prevent feeding by small insects, including alfalfa weevil larvae, aphids, and potato leafhopper nymphs. This type of mechanical resistance should not be influenced by insect biotypes, thus providing a permanent type of resistance to several pest insects. Glandular hairs are not found in most cultivated alfalfas, but it has been possible to breed for increased hair numbers in intercrosses among *Medicago* spp.

In many areas of the USA, integrated pest management (IPM) programs have been successful in reducing losses from alfalfa pests. Crop growth modeling programs have been combined with host plant resistance and insect life history information to minimize the use of chemical control measures while maximizing alfalfa yields and forage quality (41).

1-12 NEW SEEDING TECHNIQUES

High alfalfa yields begin at seeding. Proven techniques for seedling establishment on firm, well-prepared seedbeds are still popular. These include band seeding using special drills with press or packer wheels or broadcast seedings using a cultipacker or a similar seeder. Fluid or suspension seeding is a new effective custom method of broadcast seeding large acreages. However, it requires cultipacking before and after seeding.

No-till seedings of alfalfa in either small grain or corn stubble or in sod has increased in many areas. Success of no-till seedings depends on: eliminating or minimizing competition from other plants, maintaining medium-to-high soil fertility with optimum pH, protecting seedlings from pests, and using an appropriate no-till seeder at the appropriate time of year (31, 40).

Seed treatments with systemic fungicides have provided some protection during seeding establishment against *Pythium* and *Phytophthora* root-rot organisms. Lime coating along with inoculating seed has been accepted in some areas as an aid to better stands.

1-13 WEED CONTROL INCREASES

Weed control at planting time can help ensure better alfalfa stands and reduce weed problems in subsequent years. Growers have increased use of herbicides to establish clear seedings of alfalfa, in preference to seeding with a small grain companion crop. Several new preplant-incorporated herbicides can control most annual grasses and many broadleaf weeds (see Chapter 23 in this book). Postemergence herbicides can control broadleaf weeds in new stands of alfalfa-grass mixtures. Herbicides are also available to remove many weeds from established stands; however, the benefits of removing weeds must be judged on the basis of economic return to the grower. Many common weeds in alfalfa can result in reduced protein content, reduced intake by livestock, lower digestibility, and reduced palatability (7). However, other weeds, especially when immature, do not decrease the nutritive value of hay and pasture (22).

1-13.1 Yield Barriers Broken

The development of cultivars with multiple pest resistance and high yield potential and the use of improved production practices, has resulted in yields at one time thought impossible. Seasonal yields of 22 Mg/ha (10 tons/acre) or better of hay equivalent have been produced without irrigation in research trials in several states including Michigan and Kentucky. The Michigan 2-yr average yield of 24 Mg/ha was the highest documented 2-yr average research yield for nonirrigated alfalfa in the USA (37). The high yields were the result of combining a series of essential inputs, including use of a multiple pest-resistant cultivar with the genetic potential for high forage yield, appropriate soil pH, high soil fertility, early harvests and good insect control along with adequate and well-distributed precipitation. Similar management practices and yields were observed on the top farms in the Wisconsin and Pennsylvania state alfalfa grower programs (32).

A yield of 54 Mg/ha (24 tons/acre) was reported from an irrigated yield study in Arizona (1).

1-14 HAY PRESERVATIVES GAIN IN USAGE

Dry matter and quality losses in alfalfa during hay making and storage operations can be high (26, 39). Mechanical methods of conditioning by crushing, breaking, and abrading forage are used by alfalfa growers to reduce field drying time. Chemical treatments can reduce heat damage and further increase drying rate of hay. Organic acid compounds containing high percentages of propionic acid, applied at baling time, are the most reliable preservatives. Heat damage can be minimized by mixing propionic acid with acetic acid, inorganic acids, formaldehyde, water, flavoring ingredients, or antioxidants. To be effective, propionic acid must be applied at the proper rate, depending on hay moisture, and uniformly distributed throughout the hay mass. Anhydrous ammonia, while still somewhat experimental, shows promise for use as a preservative on low-quality hay.

Potassium carbonate (K_2CO_3) and sodium carbonate (Na_2CO_3) carbonate have received attention as chemical drying agents. When properly applied to the stems and used in conjunction with the newer conditioners, these agents have reduced field drying time on legumes, such as alfalfa (33, 39). The drying agents affect the waxy cutin on the plant's surface, thereby increasing the rate of moisture loss from stems. They are most effective on second and third cuttings when drying conditions are more favorable. Materials are applied at cutting with a special bar used to lay the crop over to improve spray contact with stems.

1-15 FEEDING VALUE MEASUREMENTS IMPROVED

The feeding value of alfalfa exceeds that of perennial grasses at similar stages of maturity primarily because alfalfa has greater intake potential associated with faster digestibility. Alfalfa can provide the sole plant component in many livestock feeding programs when supplemented with proper minerals (see Chapter 14 in this book). However, many factors reduce feeding value, such as: mature growth stage, pests that affect leaf/stem ratio, and adverse climatic and edaphic factors. Before 1976, measurements of forage quality could only be made from wet chemistry analyses or from animal-feeding trials. Both types of measurements were slow and expensive.

Norris et al. (28) published the first research report on predicting forage quality by near-infrared reflectance spectroscopy (NIRS). The NIRS method of analysis is a nondestructive instrumental method for rapidly and reproducibly measuring the chemical composition of forage with minimal sample preparation. It is based on the observation that each major chemical component in a sample has near-infrared absorption properties that can differentiate one component from others (27). Although NIRS technology is still at the developmental stage, it has been applied successfully to the analysis of forages. The method allows technicians to sample a truckload of hay and provide a reliable estimate of crude protein, acid detergent fiber, neutral detergent fiber, dry matter, lignin, and in-vitro digestibility in < 3 min.

The NIRS equipment has been installed in mobile vans for on-farm testing of forages and for pricing hay at auctions according to relative feed value (25). Furthermore, NIRS has become an educational tool to increase grower interest in improving forage quality for specific levels of livestock performance. The rapid, low cost, nondestructive nature of NIRS makes it an invaluable tool to conduct plant selection programs for improved forage quality, and to measure the effectiveness of various forage management systems.

1-16 HAY QUALITY STANDARDS DEVELOPED

In 1976, a task force on the American Forage and Grasslands Council (AFGC), consisting of University and USDA-ARS research scientists and extension specialists, developed a forage evaluation system based on chemical analyses that approached results obtained from feeding trials. The task force used those data to develop new hay standards based on the feeding value of hay. Included in the new standards was a relative feed value rating that related feed value to each of five hay grades and one sample grade. The goal was to provide a uniform marketing system for hay in which price would be a function of feeding value. In 1983, a common standard was accepted by research, extension, and commercial interests (14). The standards include acid detergent fiber (ADF) used for estimating digestible dry matter (EDDM), crude protein (CP), and dry matter (DM). The NIRS technology has provided a rapid method for analyzing and for computing relative feed value. The hay-quality standards have been used successfully in many hay auctions.

1-17 PROGRESS IN BLOAT RESEARCH

The value of alfalfa as a pasture species has been restricted by the prospect of bloat in grazing animals. The promising role of tannins in preventing bloat was reported by Goplen et al. (11). The reputed role of saponins as a causal factor in bloat was proven invalid in pasture trials by Majak et al. (21). Agriculture Canada has sponsored a major multidisciplinary research effort to develop a bloat-safe alfalfa cultivar. Present strategy for breeding a bloat-safe alfalfa cultivar is to slow the rapid breakdown of alfalfa leaf cells in the animal rumen, a characteristic of bloat-causing herbages. Screening procedures involve partial digestion in nylon bags in fistulated cattle (*Bos* spp.). Breeding has been carried to the third cycle of recurrent selection at Saskatoon, Saskatchewan (10). Genetic progress has been demonstrated and an alfalfa cultivar with reduced bloat potential is expected to be released within 8 yr (see Chapter 15 in this book).

1-18 DINITROGEN FIXATION IMPROVED

On a seasonal basis, alfalfa symbiotically fixes greater amounts of atmospheric N_2 and has a larger percentage of its total N content derived from symbiosis than most other legume species. The symbiosis of alfalfa with *Rhizobium meliloti* Dang. has been described in detail (see Chapter 7 in this book). Dinitrogen fixation research with alfalfa was prompted by the energy crises in the 1970s and the prospect of introducing larger quantities of fixed N into agricultural cropping systems as a replacement for fertilizer N. A major research effort was initiated by a team of USDA-ARS scientists in cooperation with the Univ. of Minnesota. Other research programs on N_2 fixation in alfalfa were conducted at the Univ. of California and New Mexico State University.

Since about 1975, a comprehensive framework of basic knowledge on N_2 fixation in alfalfa was developed where none existed before. It was demonstrated that no massive loss of nodules occurs after harvest (38). Nodules from harvested plants showed degeneration only at the proximal end. The meristem and vascular bundles remained intact during the period after harvest when the nodule was not fixing N_2. The partly degenerated nodules began to regrow and fix N_2 after shoot regrowth began. These results demonstrated the importance of harvest management in managing N_2 fixation in alfalfa.

Studies were conducted on the interrelationships among plant morphological, plant physiological, and *Rhizobium* effectiveness traits. It was concluded that N_2 fixation in alfalfa is affected by coordinated responses among many physiological and biochemical traits. A multiple-step breeding program was developed that permitted simultaneous selection for improved seedling vigor, *Rhizobium* preference, shoot growth, nodule mass, root growth, nitrogenase enzyme activity (as measured by acetylene reduction), and other nodule enzyme activities (3). Also, methods were developed to measure N_2 fixation in field plantings with ^{15}N isotope-dilution techniques. Significant progress was made in improving N_2 fixation in alfalfa. In addition to improving N_2 fixation, the program was effective in the bidirectional selection for activities of the nodule enzymes, phosphoenolpyruvate carboxylase and glutamate synthetase, that are associated with C and NH_3 assimilation. Thus, the possibility of selecting simultaneously for more than one enzyme in a plant organ was demonstrated.

The N_2-fixation research produced many useful techniques and germplasms. This includes the discovery and selection of alfalfa populations with plant conditioned nonnodulation and ineffective nodulation characteristics (29). These germplasm sources provide plant breeders and other scientists with a method for determining N_2 fixation of alfalfa in the field by the relatively inexpensive difference method (see Chapter 7 in this book). A nonwinter-dormant cv. Nitro was selected in the Minnesota program for enhanced N storage capacity and N accumulation in roots. Nondormant alfalfa was chosen as a parent because it provided a longer duration of growth and N_2 fixation during the autumn than dormant germplasm. In Minnesota trials, Nitro was planted in late April and harvested three times for hay or green chop by early September. The average forage yield was about 7.5 Mg/ha hay (3.4 tons/acre) for both Nitro and the dormant check cultivars. When all cultivars were incorporated by tillage about 20 October, Nitro provided 139 kg/ha N (124 lb/acre), 105 kg (94 lb) were from symbiotic N_2 fixation; compared to 95 kg/ha N (85 lb/acre) with 66 kg (59 lb) from symbiotic N_2 fixation for the best dormant cultivar. The 1986 Nitro release represented the first commercial application of alfalfa with specialized N accumulation attributes for use in cropping systems. The use of 1-yr alfalfa stands could create new opportunities for modification of cropping systems.

1-19 CROP GROWTH MODELS AVAILABLE

Advances in computer technology have provided the tools for the development of rational simulation models for predicting crop responses to environmental conditions. In 1975, scientists at Purdue University developed the SIMED model, the first basic crop model for alfalfa (15). It uses weather data to predict the accumulation of dry matter in leaves, stems, and roots. The model was verified and validated against field measurements of alfalfa grown during a 2-yr period (35). It has been used as the crop model in alfalfa pest management systems.

Scientists at Cornell University developed the ALSIM 1 (LEVEL 0) model (9). This model was devised as a simple alfalfa production simulation model suitable for exploring alternative management programs for alfalfa. The model can account for a large proportion of the variation in alfalfa yield in humid environments from such variables as genetic potential, soil moisture, temperature, and cutting management. It has been more difficult to develop growth simulation models for drought conditions. However, simplified water-crop yield models developed in the Netherlands have been satisfactory for estimating the effects of alternative water management strategies on alfalfa yields. A similar model was developed in Utah for predicting the effect of different irrigation schedules on seasonal yield of alfalfa. French researchers have developed a model for the dynamics of N uptake during regrowth (19).

The value of computer simulation models is that they provide an opportunity to predict and study the influence of management strategies and environmental variables on crop growth without actually conducting the more costly field experiments. Simulation models also can be used to identify voids in knowledge that need to be researched. Only the most critical factors need be validated by field experimentation. Advances in crop modeling should have many applied and basic applications to alfalfa science and production.

1-20 TISSUE CULTURE SUCCESS ENCOURAGING

In 1972, alfalfa was one of the first major crop plants regenerated from tissue culture. Tobacco (*Nicotiana tabacum* L.) and carrot (*Daucus carota* L.) had been regenerated about 10 yr earlier. Essentially all somatic cells of alfalfa can produce callus and regenerate tissue (see Chapter 29 in this book). A population of alfalfa, Regen-S, was selected for increased regeneration (5). The capability for regeneration was inherited as a dominant trait. The ability of plants to form adventitious shoots appeared to be associated with regeneration in vitro. Alfalfa cell lines have been selected for resistance to diseases and salt tolerance. Somaclonal variants of alfalfa are produced in tissue culture. Advancements in alfalfa tissue-culture technology have made it a model system. Tissue culture represents a promising tool for eventual use in alfalfa improvement.

1-21 CYTOGENETIC STUDIES AID BREEDING

Numerous traits have been identified in *Medicago* spp. that would be useful in *M. sativa* breeding programs. Examples of these traits include glandular hairs, pest resistances, increased seedling vigor, and salt tolerance. In 1979, Lesins and Lesins (20) published a book on the *Medicago* genus that summarized all available taxogenetic information. The Lesins' seed collection described in the book and maintained at the Univ. of Alberta, Edmonton, Canada, provided much of the basic germplasm for cytogenetic research in the early 1980s (see Chapter 24 in this book). This was supplemented by additional germplasm collections during the 1980s.

New interspecific crosses have been made between *M. sativa* and perennial *Medicago* spp. Many of these crosses were successful because of new research tools. Diploid clones of *M. sativa* from the improved CADL population (6) served as parents. This germplasm has extremely low frequencies of deleterious genes, thereby increasing the rate of success and plant vigor of interspecific crosses. Two new reproductive mutants have also been useful. The *rp* gene conditions $2n$ pollen formation in alfalfa (23) and the *jp* gene lacks a postmeiotic cytokinesis which results in 4n pollen (24). Plants homozygous for *jp* are effectively male-sterile and represent useful female parents. These new germplasms led to the recovery of several new interspecific crosses. The success rate also increased dramatically when interspecific hybrid embryos were excised at 7 to 20 d postpollination and cultured. Many new *Medicago* spp. can now be considered potential germplasm donors for alfalfa improvement.

1-22 NORTH AMERICAN ALFALFA IMPROVEMENT CONFERENCE PROMOTES PROGRESS

The NAAIC began in 1934 at Lincoln, NE when 27 scientists met to discuss alfalfa research. During the early years participants were primarily plant breeders and plant pathologists. The breadth of interest gradually expanded so that all disciplines are now represented. The NAAIC has provided a forum for the exchange of information. When issues and significant questions have arisen, committees have been formed to study specific problems, make recommendations for solutions, and in many situations initiate research required to provide background information to formulate solutions. The NAAIC has prepared numerous reports and special publications on a wide range of topics. Much of the progress in alfalfa improvement can be credited to the activities of the NAAIC.

The NAAIC is organized as three regional conferences: the Eastern Forage Improvement Conference, the Central Alfalfa Improvement Conference, and the Western Alfalfa Improvement Conference. The regional conferences conduct meetings in odd-numbered years. Total membership of the NAAIC meets in even-numbered years. The NAAIC has served as a catalyst for the alfalfa research community, providing the impetus for change and progress. The NAAIC has been a recognized model for crop commodity groups. It deserves credit for organizing the first variety review board, providing leadership for germplasm research, encouraging a high level of cooperation among private and public scientists, and greatly contributing to the standardization of cultivar and pest-resistance descriptions.

For 50 yr the affairs of the NAAIC were the responsibility of only two officers, a chairperson that was elected every 2 yr and a permanent secretary. The chairperson was a public scientist, usually a plant breeder, and the permanent secretary was the USDA-ARS alfalfa investigations leader. The growth in numbers of alfalfa scientists, changes in the affiliations and expertise of the membership, and a reorganization of USDA-ARS created a need for change. In 1984, the first bylaws for the NAAIC were drafted and approved (17), the number of officers were increased, and the first industry scientist served as chairperson. These changes plus the continued spirit of cooperation and willingness to solve problems will ensure a bright future for alfalfa. We consider the effectiveness of the

NAAIC as one of the most important highlights in the development and use of alfalfa in North America.

1-23 KEEPING THE CROP COMPETITIVE

Alfalfa is a perennial, autotetraploid, partially self-sterile species that requires an insect pollinator for seed production. A planting of alfalfa is exposed to 3 or more years of summer and winter climatic stresses, at least 10 harvests, and numerous pest insects, diseases, and nematodes. The crop produces several regrowths, goes through a dormancy period, and fixes N_2 each year. Seed production is generally in a different geographic area than where the crop is grown for forage production. Alfalfa is grown over a diverse range of soil types and must tolerate excesses and deficiencies of moisture. Thus, the task of improving alfalfa is formidable. Several years of multiple annual harvests are needed to collect yield and adaptation data, and the value of the resulting forage must be determined by animal performance.

Throughout the years, a series of production crises caused by new pests, such as the spotted alfalfa aphid, alfalfa weevil, and Verticillium wilt, have necessitated the diversion of scarce resources from fundamental research. The change in emphasis by public researchers from cultivar development to more fundamental or basic research is beginning to pay off. Since about 1970, alfalfa has become more competitive with other crops as a source of energy and nutrients for livestock, and as a source of N and improved soil tilth when used in crop rotations. The high amount of multiple pest resistance achieved in new cultivars has contributed to improved winter survival, yield, and tolerance to frequent cutting. Tissue culture research has been successful, and offers much promise in the application of biotechnology to alfalfa improvement.

Hanson and Davis (13) viewed the future based on past progress, potential for further improvement, and the assumption of reasonable support for future research. They summarized the situation with the statement that "alfalfa is coming of age and that the future is bright". Alfalfa has become of age, and the progress has been remarkable despite vast changes within and among public and private research organizations. The breadth of alfalfa research has increased markedly from 1972 to 1986. New ideas and disciplines have been merged with those of experienced alfalfa scientists. This has contributed to a virtual renaissance for alfalfa.

We endorse Hanson and Davis's concluding statement that, "the future is challenging". Based on past performance, the authors predict that alfalfa will experience further gains in productivity, profitability, competitiveness, and importance in North American agriculture.

1-24 SIGNIFICANT EVENTS

The following dates pinpoint some of the major events associated with the culture and use of alfalfa in the USA and Canada:

About 1850—Introduction of "Chilean alfalfa" to California from South America and subsequent spread into other western states to give rise to the "commons".

1897—Plant explorers begin collecting alfalfa germplasm from other parts of the world.

1901—Minnesota Agricultural Experiment Station conducts first field trials of Grimm alfalfa. Seed brought to Minnesota from Germany in 1858, but did not come to the attention of the Minnesota station until 1900. Grimm marked the beginning of winter-hardy alfalfa in the USA.

1904—Detection of alfalfa weevil near Salt Lake City, UT.

1910—Introduction of 'Ladak' alfalfa from Kashmir.

Mid-1910s—Cutting management studies initiated that contributed to present-day concepts.

1924—Introduction of 'African' alfalfa from Egypt.

1925—Bacterial wilt of alfalfa described.

1930s—Beginning of the alfalfa dehydration industry.

1934—First meeting of Alfalfa Improvement Conference held at Lincoln, NE.

1934–1936—Drought years in central USA stimulated interest in alfalfa.

1940—Need for tripping of flower in seed production firmly established.

1940s—Shortage of winter-hardy seed affected hectarage (acreage) trends.

1942—Automatic, self-tying pickup balers introduced.

—Release of the first two bacterial wilt-resistant cultivars: 'Ranger' for the Northern Great Plains, and 'Buffalo' for the Central Plains.

1943—Importance of lygus bugs in seed production established.

Mid-1940s—Seed production begins shift to specialized seed-growing areas of the western states.

—Introduction of mechanized hay conditioners.

1948—Major expansion of alfalfa in the North Central States begins, reaching 7.9 million ha (19.6 million acres) in 1956.

1949—Establishment of the National Foundation Seed Project.

Late 1940s—Forced-air hay dryers introduced.

—Wilted silage generally recommended.

1950—Experimental data published on genetic shifts during seed multiplication.

Early 1950s—Moisture concentration of silage (haylage) as low as 5.0 g/kg (50%) or lower recommended in conjunction with use of "gas-tight" silos.

1951—Detection of the "eastern strain" of the alfalfa weevil near Baltimore. First report of the alfalfa weevil east of Mississippi River. Heavy crop losses follow.

—Autotetraploid inheritance demonstrated to explain genetic behavior of alfalfa.

—Canadian Forage Seed Project formed to promote and stabilize supplies of pedigreed seed of recommended cultivars in Canada.

1952—Genetic male sterility described.

—Research on band seeding promising.

1953—Release of 'Vernal' alfalfa with high winterhardiness and resistance to bacterial wilt. It gained wide acceptance and was the leading cultivar for nearly 25 yr.

—Certified Alfalfa Seed Council initiates educational program to promote use of alfalfa.

1954—Discovery of the spotted alfalfa aphid in New Mexico.

—Release of 'Lahontan' alfalfa, resistant to stem nematode and bacterial wilt. Later found resistant to spotted alfalfa aphid.

Mid-1950s—Direct-cut silage increases in popularity associated with development of new equipment.

—Research on use of selective herbicides for spring establishment appears promising.

1957—Release of 'Moapa' and 'Zia' resistant to spotted alfalfa aphid for southwestern USA.

1958—Rules and procedures established by Organization for Economic Cooperation and Development (OECD) for movement of certified seed into international trade.

—Seed industry initiates research on alfalfa breeding.

1959—Leafcutting bee becomes major pollinator in the Pacific Northwest, USA.

Early 1960s—Interest increases in additives for improving direct-cut silage.

1962—Introduction of field-cuber machines for forage "packaging".

—National Alfalfa Variety Review Board established.

1963—Release of 'Saranac', a Flamande-type (Flemish) cultivar resistant to bacterial wilt.

1964—Foliar diseases linked with buildup of estrogens in alfalfa.

1967—Cytoplasmic male sterility reported.

1968—Release of 'Team' alfalfa for Maryland, Virginia, and North Carolina—moderately resistant to alfalfa weevil.

Late 1960s—New era of multiple disease and insect resistance begins. Increasing numbers of private cultivars released.

—Contribution of pea aphid resistance determined on statewide basis in Kansas. Resistant cv. Kanza produced two to three times more forage than susceptible cultivars during natural infestations, eliminating need for insecticides.

—First cultivar utilizing cytoplasmic male sterility marketed.

—Concern developed regarding heat damage in haylage—new recommended moisture concentration increased to 5.5 to 6.0 g/kg (55 or 60%).

—New biotypes of spotted alfalfa aphid capable of attacking previously resistant cultivars become a problem in southwestern USA.

—Pilot plant in operation in California for wet processing of alfalfa to produce leaf protein concentrates for monogastric and ruminant animals.

1970—Beginning use of hay and silage preservatives.

—Releases of 'Hayden' alfalfa in Arizona, resistant to four biotypes of the spotted alfalfa aphid.

1972—Publication of monograph *Alfalfa Science and Technology*.

—NAAIC forms committee to study germplasm situation.

—Alfalfa regenerated from tissue culture.

—Release of 'Agate' alfalfa, resistant to Phytophthora root rot.

—Beginning of multidisciplinary team to breed nonbloating alfalfa.

1973—Release of 'Arc' alfalfa, resistant to anthracnose.

1974—Publication of *Standard Tests to Characterize Resistance in Alfalfa Cultivars*; updated in 1984.

—Blue alfalfa aphid identified in California, quickly spread throughout southwestern USA.

—Chalkbrood disease appears in leafcutting bees in Nevada and quickly spread through northwestern USA and western Canada.

1975—SIMED growth model published and subsequently used as basis for pest-management systems.

Mid-1970s—Public research begins to place more emphasis on basic problems in physiology, genetics, and cytogenetics with concomitant reduction in applied plant breeding, particularly in the USA.

—Leafcutting bee becomes major pollinator in western Canada.

1976—Near-infrared reflectance spectroscopy described as method for predicting forage quality.

—Verticillium wilt first identified in Washington and later in Idaho, Oregon, and British Columbia, Canada.

—Release of 'CUF-101' alfalfa, resistant to blue alfalfa aphid.

—Beginning of multidisciplinary research efforts to breed for increased N_2 fixation.

—SeCan Association formed to increase and promote sale of certified seed in Canada.

1978—Major cause of pasture bloat identified by Canadian researchers.

1979—Program established to increase plant introductions under cage isolation to preserve genetic integrity.

—Glandular trichomes reported among *Medicago* spp. Later shown to be an effective defense against insects.

1980—Verticillium wilt identified in Wisconsin followed by reports in other midwest and northeastern states.

1980–1985—Nine plant explorations conducted to collect native germplasm from important alfalfa-growing areas.

—Cytogenetic studies increase with both annual and perennial *Medicago* spp.

1981—Release of 'Apollo II', 'DK-135', 'Trumpetor', 'WL316' and 'Vernema' having resistance to Verticillium wilt.

1982—Ten-ton acre (22 Mg/ha) hay yield barrier broken for 2 yr without irrigation in experimental plots at East Lansing, MI.

—First private industry scientist served as NAAIC president (chairperson), previous chairpersons were all public scientists.

1983—Program established to systematically evaluate plant introductions for pest resistances and agronomic traits.

—USA hay quality standards developed for ADF, DDM, CP, and DM.

1984—NAAIC established first bylaws.

1985—National Foundation Seed Stocks Project terminated because of lack of new public cultivars.

1986—Release of 'Nitro' alfalfa, nondormant alfalfa with increased N_2 fixation and storage of N in roots for use in short-term crop rotations.

—Certified Alfalfa Seed Council published list of dormancy and pest resistance traits for cultivars eligible for certification.

REFERENCES

1. Anonymous. 1984. Alfalfa yield of 24 tons per acre. Winter 1983–84 Report. Univ. of Arizona Yuma Mesa Exp. Stn.
2. Barnes, D.K., E.T. Bingham, R.P. Murphy, O.J. Hunt, D.F. Beard, W.H. Skrdla, and L.R. Teuber. 1977. Alfalfa germplasm in the United States: genetic vulnerability, use, improvement, and maintenance. USDA Tech. Bull. 1571. U.S. Government Printing Office, Washington, DC.
3. ----, G.H. Heichel, C.P. Vance, and W.R. Ellis. 1984. A multiple-trait breeding program for improving the symbiosis for N_2 fixation between *Medicago sativa* L. and *Rhizobium meliloti*. Plant Soil 82:303–314.
4. ----, and D.L. Smith. 1984. Review and description of alfalfa varieties. p. 115–120. *In* Rep. 29th Alfalfa Improve. Conf., Lethbridge, AB, Canada. 15–20 July. North American Alfalfa Improvement Conference, Alberta.
5. Bingham, E.T., L.V. Hurley, D.M. Kaatz, and J.W. Saunders. 1975. Breeding alfalfa which regenerates from callus tissue in culture. Crop Sci. 15:719–721.
6. ----, and T.J. McCoy. 1979. Cultivated alfalfa at the diploid level: origin, reproductive stability, and yield of seed and forage. Crop Sci. 19:97–100.
7. Doll, J.D. 1986. Do weeds affect forage quality? p. 167–170. *In* Proc. 16th Natl. Alfalfa Symp., Fort Wayne, IN. 4–5 March. Certified Alfalfa Seed Council, Davis, CA.
8. Elgin, J.H. Jr., et al. 1984. Standard tests to characterize pest resistance in alfalfa cultivars. USDA-ARS Misc. Pub. 1434. U.S. Government Printing Office, Washington, DC.
9. Fick, G.W. 1984. Simple simulation models for yield prediction applied to alfalfa in the northeast. Agron. J. 76:235–239.
10. Goplen, B.P., R.E. Howarth, G.L. Lees, W. Majak, J.P. Fay, and K.-J. Cheng. 1983. Evolution of selection techniques in breeding bloat-safe alfalfa. p. 221–223. *In* J.A.

Smith and V.W. Hays (ed.) Proc. 14th Int. Grassl. Congr., Lexington, KY. 15-24 June 1981. Westview Press, Boulder, CO.
11. ----, ----, S.K. Sarkar, and K. Lesins. 1980. A search for condensed tannins in annual and perennial species of *Medicago, Trigonella,* and *Onobrychis.* Crop Sci. 20:801-804.
12. Hamilton, D.G. 1963. The Canadian Forage Sced Project. Forage Notes 9:8-11.
13. Hanson, C.H., and R.L. Davis. 1972. Highlights in the United States. p. 35-51. *In* C.H. Hanson (ed.) Alfalfa science and technology. ASA, Madison, WI.
14. Hannaway, D.B. 1984. Hay testing laboratory certification manual. p. 27. *In* Spec. Pub. 2. The National Hay Assoc. and Am. Forage and Grassl. Counc., Oregon State University, Corvalis.
15. Holt, D.A., R.J. Bula, G.W. Miles, M.M. Schreiber, and R.M. Peart. 1975. Environmental physiology, modeling, and simulation of alfalfa growth. I. Conceptual development of SIMED. Purdue Agric. Exp. Stn. Res. Bull. 907.
16. Jones, F.R. 1925. A new bacterial disease of alfalfa. Phytopathology 15:243-244.
17. Kalton, R.R. 1984. By-laws of the North American Alfalfa Improvement Conference. p. 128-130. *In* Rep. 29th Alfalfa Improve. Conf., Lethbridge, AB, Canada. 15-20 July. North American Alfalfa Improvement Conference, Alberta
18. Kehr, W.R. 1983. History of germplasm involvement by the National Alfalfa Improvement Conference. p. 54-57. *In* Rep. 28th Alfalfa Improve. Conf., Davis, CA. 13-16 July 1982. North American Alfalfa Improvement Conference, Alberta
19. Lemaire, G., P. Cruz, G. Gosse, and M. Chartier. 1985. Relationship between dynamics of nitrogen uptake and dry matter growth for lucerne (*Medicago sativa*). Agronomie 5:685-692.
20. Lesins, K.A., and I. Lesins. 1979. Genus *Medicago* (*Leguminosae*): A taxogenetic study. Junk, The Hague, Netherlands.
21. Majak, W., R.E. Howarth, A.C. Fesser, B.P. Goplen, and M.W. Pedersen. 1980. Relationship between ruminant bloat and the composition of alfalfa herbage. II. Saponins. Can. J. Anim. Sci. 60:699-708.
22. Marten, G.C., and R.N. Andersen. 1975. Forage nutritive value and palatability of 12 common annual weeds. Crop Sci. 15:821-827.
23. McCoy, T.J. 1982. The inheritance of 2n pollen formation in diploid alfalfa, *Medicago sativa.* Can. J. Genet. Cytol. 24:315-323.
24. ----, and L.Y. Smith. 1984. Uneven ploidy levels and a reproductive mutant required for interspecific hybridization of *Medicago sativa* L × *M. dzhawakhetica* Bordz. Can. J. Genet. Cytol. 26:511-518.
25. Martin, N.P., and J.G. Linn. 1985. Extension application in NIRS technology transfer. p. 48-53. *In* G.C. Marten et al. (ed). Near infrared reflectance spectroscopy (NIRS): Analysis of forage quality. USDA-ARS Agric. Handb. 643. U.S. Government Printing Office, Washington, DC.
26. Moser, L.E. 1980. Quality of forages affected by post-harvest storage and processing. p. 227-260. *In* C.S. Hoveland (ed.) Crop quality, storage and utilization. American Society of Agronomy, Madison, WI.
27. Norris, K.H. 1985. Definition of NIRS analysis. p. 6. *In* G.C. Marten et al. (ed.) Near infrared reflectance spectroscopy (NIRS): Analysis of forage quality. USDA-ARS, Agric. Handb. 643. U.S. Government Printing Office, Washington, DC.
28. ----, R.F. Barnes, J.E. Moore, and J.S. Shenk. 1976. Predicting forage quality by infrared reflectance spectroscopy. J. Anim. Sci. 43:889-897.
29. Peterson, M.A., and D.K. Barnes. 1981. Inheritance of ineffective nodulation and non-nodulation traits in alfalfa. Crop Sci. 21:611-616.
30. Richards, K.W. 1984. Alfalfa leafcutter bee management in Western Canada. Agric. Canada Pub. 1495/E.
31. Rogers, D.D., D.S. Chamblee, J.P. Mueller, and W.V. Campbell. 1985. Fall no-till seeding of alfalfa into tall fescue as influenced by time of seeding and grass and insect suppression. Agron. J. 77:150-157.
32. Rohweder, D.A. 1986. WFC green gold project produced highest yields in humid U.S. p. 65-79. *In* Proc. 10th Forage Production and Use Symp. Wisconsin Dells, WI. 28-29 January. Wisconsin Forage Council, Madison, WI.
33. Rotz, C.A., and R.J. Davis. 1986. Drying and field losses of alfalfa as influenced by mechanical and chemical conditioning. p. 157-160. *In* Proc. 1986 Forage Grassl. Conf. 15-17 April. American Forage and Grassland Council, Athens, GA.
34. Rumbaugh, M.D. 1979. N.E. Hanson's contributions to alfalfa breeding in North America. South Dakota State Univ. Pub. 3538A.
35. Schreiber, M.M., G.E. Miles, D.A. Holt, and R.J. Bula. 1978. Sensitivity analysis of SIMED. Agron. J. 70:105-108.
36. Stewart, H. 1926. Alfalfa-growing in the United States and Canada. Macmillan Publishing Co., New York.
37. Tesar, M.B. 1985. Fertilization and management for a yield of ten tons of alfalfa without irrigation. p. 327-333. *In* Proc. 1985 Forage and Grassland Conf. American Forage and Grassland Council, Hershey, PA.

38. Vance, C.P., G.H. Heichel, D.K. Barnes, J.W. Bryan, and L.E. Johnson. 1979. Nitrogen fixation, nodule development, and vegetative regrowth of alfalfa (*Medicago sativa* L.) following harvest. Plant Physiol. 64:18.
39. Vough, L.R., and T.H. Miller. 1985. Advances in preserving hay quality. p. 45–56. *In* Proc. 1985 Forage and Grassl. Conf. American Forage and Grassland Council, Hershey, PA.
40. White, H.E., D.D. Wolf, and E.S. Hagood. 1985. Forage establishment innovations. p. 19–25. *In* Proc. 1985 Forage and Grassland Conference, American Forage and Grassland Council, Hershey, PA.
41. Wilson, M.C. 1986. Manage the alfalfa weevil. Spec. Pub. Certified Alfalfa Seed Council, Davis, CA.

2 World Distribution and Historical Development

RÉAL MICHAUD

Agriculture Canada Research Station
Sainte-Foy, Quebec

W. F. LEHMAN

University of California
El Centro, California

M. D. RUMBAUGH

USDA-ARS
Logan, Utah

Alfalfa (*Medicago sativa* L.) has had a long and rich history. It is recognized as the oldest plant grown solely for forage. Bolton et al. (29) prepared information on the history and development of alfalfa for *Alfalfa Science and Technology* (Agronomy monograph 15). Highlights from their discussion are summarized in this volume to provide essential background for the reader.

Alfalfa was cultivated before recorded history and is now found growing wild from China to Spain and from Sweden to North Africa. In addition, it has become acclimatized in South Africa, Australia, New Zealand, and North and South America.

It is generally agreed that common alfalfa originated in Vavilov's "Near Eastern Center"—Asia Minor, Transcaucasia, Iran, and the highlands of Turkmenistan (28, 178, 272, 273). These areas have cold winters and hot, dry summers. Soils are typically well drained, near neutral in pH, with subsoils having a high lime content (149, 235).

Sinskaya (235) concluded from studies of alfalfa in the USSR that it had two distinct centers of origin. The first center was the mountainous region of Transcaucasia and Asia Minor and adjoining areas of northwestern Iran. Alfalfa from this area is a good source of winterhardiness. The second center was Central Asia. Her reasons were that alfalfa under these conditions was susceptible to fungal diseases. In Turkestan and some other parts of Central Asia, irrigation had been practiced before recorded history, and the climate characterized by low humidity, hot, dry summers and moderately cold winters (216). Hence, evolution of Central Asiatic alfalfas in an area of low humidity would account for their lack

---

Copyright 1988 © ASA-CSSA-SSSA, 677 South Segoe Road, Madison, WI 53711, USA. *Alfalfa and Alfalfa Improvement*—Agronomy Monograph no. 29.

of resistance to leaf diseases and comparatively low drought resistance. However, a valuable series of characteristics evolved under these conditions; namely, resistance to diseases and insects, such as bacterial wilt [*Corynebacterium insidiosum* (McCull.) H.L. Jens.], stem nematode [*Ditylenchus dipsaci* (Kühn) Filipjev], other root pathogens, and aphids.

Medicago falcata L. has played an important role in the evolution of common alfalfa. This yellow-flowered species has a wider natural distribution than common alfalfa. It is a cold-tolerant alfalfa with a range of adaptation extending far north into Siberia and to Eurasian regions of comparable climate. According to Sinskaya (235), it is predominantly a plant of the steppe and forest steppe zones, that may be found under semidesert conditions. Sinskaya states that when alfalfa was introduced to Germany and northern France in the 16th century, hybridization with the local wild yellow-flowered medics began to play a significant role in its evolution and distribution. The most important characteristics contributed by *M. falcata* germplasm were winterhardiness, drought resistance, disease resistance, and creeping roots or rhizomes. The resultant hybrid forms (*M. media*) gave rise to many hardy and useful variegated alfalfa crops (i.e., German 'Franconian' and French 'Provence' types) and were responsible for a great expansion of the crop into northern areas of Europe and North America. Introgression of many local endemic ecotypes of *M. falcata* (yellow medic) undoubtedly played a major role in enriching the gene pool for natural selection under a great variety of edaphic and climatic conditions.

The history of alfalfa is a story of the world's most important forage crop. Alfalfa was recognized and domesticated by early man as a valuable crop plant. In an evolutionary sense, alfalfa has been extremely successful. Probably the main explanation can be found in the plant's root system, and in the evolution of a symbiotic relationship with *Rhizobium*, that reduces dependence on soil N. The plant's deep taproot has made it possible to use soil moisture from depths of 6 m or more; an obvious advantage to escape prolonged drought. In addition, the plant has the ability to become dormant in periods of drought and cold, and to resume growth when conditions become favorable. Creeping roots, rhizomes, and deep-set crowns provide additional protection against frost heaving and cold winters. The existence and maintenance of a large store of genetic variability by cross-pollination and interfertility among species within the Falcago complex contributed to successful evolution.

2–1 SCIENTIFIC AND VERNACULAR NAMES

The most complete account of the derivation of the scientific and common names of alfalfa has been given by Scofield (225). Tournefort (260) described alfalfa and its related forms under the group name *Medica*. In the same publication, he described a plant native to southern Europe, very similar to alfalfa, which he named *Medicago* (later to be known as *Trigonella radiata*). Linnaeus (162), in his first edition of *Systema Naturae* published in 1735, followed the example of Tournefort and earlier botanists and applied the name *Medica* to alfalfa and related species. He apparently regarded the *Medicago* of Tournefort as congeneric with alfalfa because he used the name *Medicago* as a synonym of *Medica*. However, in his *Species Plantarum* (163), published in 1753, the standing of these names was reversed without comment or explantion; then, the name *Medicago* was applied to the genus and *Medica* relegated to a synonym. Thus, Linnaeus named alfalfa *Medicago sativa* and the species *Medicago* described by Tournefort was named *M. radiata*. In 1872, Boissier (26) described *M. radiata* under the new binomial *Trigonella radiata*, and since then it has been known as either *T. radiata* or *M. radiata*. Botanists have accepted the classification of Linnaeus, and common alfalfa is known as *M. sativa* L. The taxonomy of *Medicago* spp. is described in more detail in Chapter 3 in this book.

Common and local names given to alfalfa indicate its widespread use and complement the chronicle of ancient peoples and their activities. Some common names include purple medic, snail clover, median herb, Burgundy hay/clover, Chilean clover, and Bourgoens Hoy. In spite of the various names applied to alfalfa, the crop is most commonly called *alfalfa* or known as *lucerne* throughout Europe (except Spain and Portugal), South Africa, Australia, and New Zealand.

The closely related yellow-flowered alfalfa is known by various names. Some of the more common are yellow medic, yellow-flowered medic/alfalfa, sickle medic, Siberian alfalfa, Swedish alfalfa (235), and sholteek. Klinkowski (149) suggested that the name "sholteek", which is used exclusively in the Semipalatinsk area of the USSR, is probably a Kirghizian corruption of two Russian words denoting "yellowish". Most common names for *M. falcata* are descriptive in nature and describe the main characteristics of the species; i.e., yellow flower color, sickle (falcate) pod, and cold hardiness.

2-2 GEOGRAPHICAL MOVEMENT

2-2.1 Early Progress Through Europe, Asia, and Africa

Although there is a temptation to be definitive in recording when and how alfalfa reached various countries and areas, this is impossible on the basis of available historic records. For example, the oldest known reference to alfalfa is from Turkey (1300 B.C.) and Babylonia (700 B.C.). However, Hendry (109) indicates that maritime trade was well developed in the eastern Mediterranean as early as 4000 B.C. This trade could have contributed to the spread of alfalfa. In addition, the plain of Mesopotamia (Iraq) was the traditional meeting place for ancient races of Asia, Africa, and Europe. Hence the triumphal progress of the "Queen of the Forage

Plants", stretches far into the preChristian era, and follows in the path of civilizations from east to west.

The oldest recorded reference to date indicates that alfalfa was used as a forage more than 3300 yr ago. During archaeological excavations in the Corum/Alacahöyük regions in Turkey, Hittite (1400–1200 B.C.) brick tablets were discovered. These tablets indicated that the animals were fed alfalfa through the winter season and that alfalfa was regarded as a highly nutritious animal feed. Sinskaya (235) states that historical evidence testifies to the widespread distribution of alfalfa in Media (northwestern Persia) in the 1st millennium B.C.. In the 4th century B.C., Theophrastus described how alfalfa was brought to Greece by the invading Median armies in order to feed their chariot horses and other animals. Many writers, such as Aristophanes (440–380 B.C.), Aristotle (384–322 B.C.) and others, either mentioned alfalfa or discussed it at length (109, 149). For the next 200 or 300 yr, however, no further references on alfalfa are found until Roman times.

In the 2nd century B.C., the Romans acquired alfalfa, and this new crop thrived and spread throughout Italy. Ahlgren (5) credits the Romans with being the fathers of forage culture, because of their advanced knowledge of forage crops, including seeding, management, and haymaking practices.

Simultaneous with the arrival of alfalfa into Italy, the eastward trend in its worldwide dissemination began. In 126 B.C., the Chinese Emperor Wu dispatched an expedition under General Zhang Quian into the (present) Russian Turkestan area to secure specimens of highly prized Iranian horses (109). While securing the horses, General Zhang Quian collected seed of alfalfa that was used by the Arabs as the main fodder for horses. Soon thereafter, alfalfa appeared in the gardens of the Imperial palaces of China. In succeeding years, alfalfa became established throughout northern China as an important forage crop. Thus at the beginning of the Christian Era, alfalfa had been disseminated over a broad area from its supposed center of origin (Iran). Klinkowski (149) and Sinskaya (235) claim that by this time it was grown in the oases of North Africa.

During the period of the Roman Empire (27 B.C.–395 A.D.), Roman colonists established the crop in their newly acquired provinces. Columella planted alfalfa in Andalusia in southern Spain in the 1st century, and at the same time its culture became general in the Lucerne Lake region in central Switzerland (106, 109). Although it may have reached southern France at this time, it did not flourish there until the 13th century. Hendry (109) notes the possibility of a separate Moslem introduction of alfalfa into Spain via North Africa during the Moorish invasions. The Spanish acceptance of the Arabic word *alfalfa* in preference to the Roman words, *medica* or *lucerne*, gives credence to this alternative route.

The fall of the Roman Empire marked the virtual disappearance of alfalfa from Europe. Hendry (109) states that Italian authors of the Renaissance period affirm that alfalfa virtually disappeared from Italy following the barbaric invasions in the 5th century. Further evidence is that Crescenz, writing in 1478 and cited by Hendry, does not mention alfalfa among Italian field crops. In the 16th century, alfalfa was reintroduced to Italy from Spain and again spread throughout the country. According to Klinkowski (149), alfalfa spread from Spain to France in 1550, to Belgium and Holland in 1565, to England in 1650, to Germany and Austria around 1750, to Sweden in 1770, and to Russia during the 18th century. The history of alfalfa in USSR is given by Klinkowski (149) and Sinskaya (235). In the 18th century, alfalfa was distributed worldwide when it was taken from Europe to the New World, Australia, and New Zealand. The advent of alfalfa to the Americas hailed the beginning of an era of rapid expansion and acceptance of the crop.

2-2.2 To the Americas

The discovery of the Americas and their colonization by Portuguese and Spaniards in the 16th century led to the introduction of alfalfa into Mexico and Peru. Alfalfa thrived in its new environment and soon spread from Peru to Chile, Argentina, and finally to Uruguay by 1775 (149).

Alfalfa was probably brought from Mexico to Texas, Arizona, New Mexico, and California by the early missionaries. Stewart (248) claims that many areas were producing alfalfa in the southwestern USA by 1836. However, it was the introduction of the "Chilean clover" to California during the days of the gold rush that proved of major importance. Hendry (109) notes this event as follows, "... the first parcel of seed probably arrived unlisted in some cargo of some trading vessel from South America between 1847 and 1850.... The earliest mention of *alfalfa* in any California paper, which I have found, appears in the "California Farmer" for 15 Mar. 1855 ..., and it is unlikely that eastern seed reached the State at an earlier date, because the name *lucerne* (as European colonists called alfalfa) does not appear in the early agricultural literature of California ... alfalfa was then (1858) becoming popular among California stockmen".

Alfalfa proved particularly suited to the sunny and dry climate and irrigated soils of the southwestern USA. Hence, it soon spread eastward to Utah, where the Mormon settlers grew their new lucerne crop for seed. From there it spread to neighboring states and by 1894 was grown extensively in Kansas. Meanwhile, additional introductions of Spanish alfalfas were brought into Colorado from Mexico (272). In the late 1800s, alfalfa was being grown to some extent in Montana, Iowa, Missouri, and Ohio. While the Spaniards introduced alfalfa to Mexico and South America, the New England colonists and other West European colonists brought it under the name *lucerne* to eastern North America (225). Hence, more than 100 yr before alfalfa made its important entry into California from Chile, the crop had been recorded in Georgia (1736), North Carolina (1739), and New York (1791). However, the crop thrived in only a few places, e.g., on the calcareous soils in New York. Bolton (28) suggested that the lack of success could be attributed to typically acid soils and humid climate in the eastern USA. Thus in spite of a much earlier entry of alfalfa into eastern USA, by 1899 only 1% of the acreage was grown east of the Mississippi River (271).

The Spanish sources of alfalfa were *M. sativa* types that, as a group, were not very winterhardy. Climatic conditions (cold winters) restricted their northward penetration. Successful culture of alfalfa in northern USA and Canada was not possible until the introduction of the more hardy *M. media* (*M. sativa* × *M. falcata*) strains. The earliest and most important of these introductions was by the German immigrant Wendelin Grimm, who brought about 8 kg of seed of an old Franconian hybrid alfalfa from his native Germany when he settled in Minnesota in 1857, at about the same time as the Spanish alfalfas were brought to California (5).

In 1858, Grimm planted his *Ewiger Klee* (everlasting clover). Even though his early plantings were unsuccessful, he persisted in his efforts to grow the crop (248). Over the years and generations, natural selection played a major role in eliminating the less hardy plants until a very hardy strain resulted (270). Major attention and impetus were given to this strain when it was brought to the attention of the Minnesota Exp. Stn. and the USDA (41). Subsequently, the winter-hardy 'Grimm' alfalfa, advanced successful alfalfa culture into the northern USA and Canada.

Although there were other notable introductions of hardy alfalfas, they were not as significant as Grimm. Thus, 'Baltic' alfalfa, a hardy variegated introduction from Europe was grown at Baltic, SD (271). It traces to seed obtained in 1896 by a dealer at Hartford, SD. Additional information on the origin of alfalfa germplasm in the USA is provided in Chapters 1 and 25 in this book.

Alfalfa was first introduced into Canada in 1871 in the province of Ontario (15). A shepherd accompanying an importation of sheep from Lorraine, France, brought about 1 kg of seed to a farm in Welland, Ontario. This seed was the source of 'Ontario Variegated'. Successful alfalfa culture in western Canada, with its severe winters, began with the introduction of Grimm alfalfa in 1908. At that time, an importation of 57 kg of a Grimm strain was made by a farmer at Suffield, Alberta (28). By 1914, this Grimm strain (known as '19-A' or 'Disco 19-A') formed the basis for most Grimm grown in western Canada. However, the strain of Grimm alfalfa licensed in Canada was developed by mass selection of surviving plants from introductions of Grimm made in 1910. L.E. Kirk developed 'Grimm 666' from the best of 1300 single plant selections made in 1912 (28). Grimm 666 was distributed in 1926, and became the source of breeder seed for Grimm.

2–2.3 To South Africa and Australasia

Alfalfa was brought from France to Cape Colony in South Africa around 1850, and initially was of major importance on large ostrich farms.

WORLD DISTRIBUTION AND DEVELOPMENT

When ostrich farming declined, alfalfa remained and is widely grown on irrigated land in arid and semiarid districts. The 'Provence' alfalfa was the most popular strain grown for many years. 'Chinese' alfalfa, originating from Tibet, was grown to some extent and valued for its winter-hardiness (28).

Alfalfa is thought to have been introduced to New Zealand around 1800 from Europe, although Argentina has been suggested as a source by Palmer (193). 'Marlborough' was the principal strain that evolved in the South Island. It is thought to result from natural selection of the Provence or 'Hunter River' types. Another alfalfa of hybrid origin, such as Grimm, may also have contributed germplasm to account for variegated and yellow-flowered plants and a degree of winter dormancy in the Marlborough strain (193).

Alfalfa was brought to Australia soon after settlement and received favorable comment by Governor King as early as 1806 (216). It was first grown commercially on the alluvial flats of the Hunter and Peel Rivers. About 800 ha were grown in New South Wales in 1833 and by 1920 this had increased to 40 500 ha. Hunter River, the main cultivar grown in Australia until 1965, is believed to trace to the Provence alfalfa from France, although it has been suggested that it could represent 'Smooth Peruvian', 'Arabian', and perhaps even 'American Common' (216). Natural selection played a major part in the development of Hunter River alfalfa.

2-3 DISTRIBUTION, PRODUCTION, AND RESEARCH AROUND THE WORLD

The collection of statistics on the distribution and production of alfalfa presents many difficulties. In a number of countries, agricultural statistics are limited and alfalfa is not reported, or simply included under "hay crops". In reports from other countries, such as the USA, Argentina, Canada, and Australia, alfalfa-grass mixtures are combined with pure-stands of alfalfa. The data presented herein represents our best estimate of the world distribution of alfalfa.

In some countries, information was obtained directly through official sources, while in others, statistics were provided by private correspondents. Their assistance in assessing the current status of alfalfa production was invaluable, and we acknowledge their kind collaboration.

Examination of the current distribution of cultivated alfalfa indicates that the crop is concentrated in certain zones within the northern hemisphere, i.e., USA, Canada, Italy, France, China and southern USSR, and in selected countries in the southern hemisphere, i.e., Argentina, Chile, South Africa, Australia, and New Zealand. Alfalfa can be grown as far as the 60° N Lat but it is essentially a crop of temperate regions as indicated by its origin. The USA, USSR, and Argentina contribute about

70% to the worldwide estimate of about 32 million ha. France, Italy, Canada, and China combine to account for an additional 17%.

This chapter summarizes some of the research in progress. Only major alfalfa research centers are listed for individual countries. Also, brief references are made to climatic conditions, utilization of the crop, cultivars generally grown, and problems encountered with growing alfalfa.

2–3.1 Europe

Despite a decrease of more than 1 million ha in total area seeded to alfalfa since last reported by Bolton et al. (29); Europe remains one of the major alfalfa-producing areas of the world with an estimated total of nearly 8 million ha (Table 2–1). The largest decline occurred in France and Italy, with smaller decreases in Austria, Czechoslovakia, and West Germany. Conversely, alfalfa-producing areas were either maintained or increased in Romania, Bulgaria, Hungary, Spain, and Yugoslavia.

In the mid-1970s, the alfalfa dehydration industry suffered greatly from the energy crisis and disappeared almost entirely in some countries. Alfalfa was replaced by more important cash crops. In addition, alfalfa pests such as Verticillium wilt (caused by *Verticillium albo-atrum* Reinke & Berth.) added to the difficulties in producing the crop, and might be partly responsible for the decline in alfalfa production.

Many European countries have active research programs with the development of multiple pest resistant cultivars as a primary objective. Seed production is an important industry, concentrated primarily below 46° N Lat, in the southern and southeastern countries of Europe.

2–3.1.1 Austria

In 1968, there were 45 000 ha of cultivated alfalfa in Austria (29) and by 1979, only 15 350 ha were reported (276). Since then the area seeded to pure stands of alfalfa has continued to decline with 12 630 ha reported in 1983 (277).

Alfalfa is grown throughout Austria but the heaviest concentration (81%) is in the northeast part of the country near Wien and in lower Austria. The eastern regions have a drier, more continental type of climate than the western regions. Yearly rainfall is about 550 mm and the median temperature ranges between -2.0 in January and 20.0°C in July.

Alfalfa is used in short rotation (3–4 yr) because Verticillium wilt is becoming a problem in older stands. Alfalfa is harvested mainly for hay or green fodder. In some years, common leafspot [caused by *Pseudopeziza medicaginis* (Lib.) Sacc.] and field mice (*Mus* spp.) may be a problem.

Flemish-type cv., such as, Europe, Orchésienne, DuPuits, and Everest are popular, with little use as yet of newer Verticillium wilt resistant cultivars. Seed of local ecotypes is unavailable.

Table 2-1. World areas in cultivated alfalfa.

Continent and country	Year	Hectares
Europe		
Austria	1983	12 630
Belgium	1982	1 780
Bulgaria	1982	399 000
Czechoslovakia	1983	200 000
Denmark	1983	5 000
East Germany	1982	190 000
Finland	1983	500
France	1983	566 000
Greece	1980	198 700
Hungary	1982	337 500
Italy	1982	1 300 000
Luxembourg	1983	400
Netherlands	1983	2 000
Norway	1983	200
Poland	1981–1983	258 000
Romania	1981	400 000
Spain	1981	332 600
Sweden	1983	25 000
Switzerland	1983	6 000
USSR (European Russia)	1971	3 375 000
United Kingdom	1983	16 000
West Germany	1983	31 000
Yugoslavia	1984	337 000
Total Europe		7 994 310
North America		
Canada	1981	2 544 300
Mexico	1982	245 000
USA	1981	10 559 025
Total North America		13 348 325
South America		
Argentina	1981	7 500 000
Bolivia	1983	9 500
Brazil	1983	26 000
Chile	1983	60 000
Ecuador	1969	30 000
Peru	1981	120 000
Uruguay	1983	25 000
Total South America		7 770 500
Africa		
Algeria	1981	10 000
Egypt	1983	81 000
Kenya	1984	120
Morocco	1981	30 000
South Africa	1985	300 000
Tunisia	1981	12 300
Zimbabwe	1969	1 550
Total Africa		434 970
Asia		
India	1976	58 000
Iran	1977	270 000
Iraq	1970	4 800

(continued on next page)

Table 2-1. Continued.

Continent and country	Year	Hectares
Israel	1984	2 000
People's Republic of China	1983	960 000
Saudi Arabia	1977	8 300
Turkey	1969	73 700
USSR (Siberia)	1971	1 125 000
Total Asia		2 501 800
Oceania		
Australia	1981-1982	115 500
New Zealand	1984	101 200
Total Oceania		216 700
World Total		32 266 605

Research on alfalfa is conducted at Bundesanstalt für Pflanzenbau, Alliiertenstrasse 1, (Wien), with emphasis on cultivar evaluation and alfalfa-grass mixtures.

2–3.1.2 Belgium

Alfalfa has never been an important crop in Belgium. From the 11 000 ha recorded in 1960, official statistics show a decline to 1780 ha in 1982. It occupies only 0.13% of the total agricultural area (21).

A median temperature range from 2.7°C in January to 17°C in summer and an average annual rainfall of more than 700 mm are most conducive for grass growth. In addition, soils with a high water table during winter and high pH are not well suited for growing alfalfa. In the past, alfalfa was grown mainly for dehydration products. Now, artificial drying has become too expensive, and alfalfa production has declined (10).

Alfalfa is generally seeded in pure stands (30 kg/ha) under a cover crop or mixed with orchardgrass (*Dactylis glomerata* L.), or tall fescue (*Festuca elatior* L.) (10). Alfalfa is harvested three times/yr with average yields of 10 to 15 Mg/ha. The most commonly grown cultivars are Orca (65%), Europe (25%), and DuPuits.

The most important production problems are forage conservation, weeds in pure stands of alfalfa, especially *Poa annua* L. and *Rumex* spp., and Verticillium wilt.

Little research on alfalfa is conducted except for cultivar evaluation at The Rijksstation voor Plantenveredeling, (Merelbeke).

2-3.1.3 Bulgaria

The greater part of Bulgaria has a moderate continental climate with relatively cold winters ($-2.0\,°C$ in January) and warm summers (20°C). The area near the Black Sea coast has warmer winters but cooler summers. The annual rainfall is 450 to 600 mm, often with a summer drought.

Alfalfa is an important crop. Between 1940 and 1982, the area in alfalfa increased from 45 600 to 399 000 ha and average yields rose from 4.6 to 6.3 Mg/ha. The crop is used mainly for hay, silage, and vitamin flour (alfalfa meal) with about 25 000 ha left annually for seed production (174).

The local cv. Dunavka is by far the most commonly used at 90%. It is a genetically broad-based, widely adapted population. It has good wintersurvival, fast regrowth, and high hay yield. Seed yields are somewhat unsatisfactory (174). Other cultivars are Nadezda 2 and Nadezda 1 (10%); and Pleven 1, Pleven 6, and Palava that occupy small areas.

Common leafspot is the most important alfalfa pathogen in Bulgaria (209). Other diseases are spring black stem [caused by *Phoma medicaginis* Malbr. & Roum. var. *medicaginis* (œq *Ascochyta imperfecta* Pk.)], bacterial wilt, and *Fusarium* spp. Alfalfa mosaic virus is reported.

The objective of the breeding program is to develop cultivars resistant to diseases and adverse climatic conditions. Other problems dealing with irrigation, pollinating insects, and utilization of alfalfa are investigated at a number of institutes: The Pleven Institute for Forages, The Russe Institute of Seed and Seed Production, and The Vassil Kolarov Institute of Plovdiv.

2-3.1.4 Czechoslovakia

The climate in Czechoslovakia is variable with no recognizable climatic zones. The median temperature ranges between $-5.0\,°C$ in February and 20.0°C in July. Alfalfa is grown as fodder crop on about 200 000 ha in the warmer regions, at altitudes up to 300 m where the annual precipitation amounts to 450 to 650 mm.

About 24 000 ha of alfalfa is left for seed at elevations up to 200 m. Seed production is centered south of Bnro and near Bratislava in the south and near Kosice in the southeast. Ecological factors have a pronounced effect on seed yield (53) with yield varying from 50 to 500 kg/ha. Infestations of downy mildew (caused by *Peronospora trifoliorum* d'By), gall midges (*Contarinia medicaginis* Kieffer) and plant bugs (Lygus and *Adelphocoris* have been reported as affecting seed production (55).

Alfalfa is generally seeded with fodder crops, such as, broad horsebean (*Vicia faba* L.), oat (*Avena sativa* L.), or ryegrass (*Lolium multiflorum* Lam.). In the 2 to 4 yr following establishment, three to four cuttings are harvested annually, with more cuttings taken under irrigation. Alfalfa is used mainly for hay and greed fodder and to a lesser extent for silage and feed mixtures. When grown for seed production, the first cutting in the seeding year is left for seed if alfalfa is seeded without a cover crop. In the following 2 or 3 yr, seed is harvested from the first regrowth (54, 213).

Czechoslovak cultivars (Bobrava-1978, Nitranka-1963, Palava-1967 and Prerovska-1939) are grown, with some imports of European cultivars (Europe and Orca). When tested in Czechoslovakia, no foreign cultivars gave higher dry matter or seed yields than Bobrava and Palava (71, 103, 214). The four Czechoslovak cultivars are winterhardy, and tolerant to frequent cuttings. None of them has been selected specifically for resistance to local pathogens, such as Verticillium wilt, common leafspot, and spring black stem (209). Nevertheless, Palava contains about 40% of plants tolerant to Verticillium wilt, and exhibits some resistance to stem nematode (1). New synthetics have been developed for tolerance to bacterial wilt and Verticillium wilt (153), and stem nematode (98).

The cultivation of alfalfa for fodder does not face any major problems; however, soil acidity, low soil organic matter content, soil compaction, and high levels of N applied to cover crops may reduce stand establishment (152).

Research is conducted on crop management, seed production, diseases and pests, breeding procedures (215), root systems (52) and fodder conservation and utilization. The main institutions are: The Research and Breeding Institute for Fodder Plants, Troubsko u Brna, The Research Institute of Plant Production-Piestany (Slovakia), and The Plant Breeding Station-Zelesice u Brna.

2–3.1.5 Denmark

The cold and rainy climate of Jutland is unfavorable and alfalfa is rarely grown there. Despite evidence that the crop does well where conditions are favorable, the area under alfalfa declined to about 5000 ha in 1983. It is being replaced with various cash crops (129).

Alfalfa is used mainly for dehydration. Resis and Vela are the most commonly grown cultivars. Both are resistant to Verticillium wilt, the most severe alfalfa disease in the country (209).

Research on alfalfa is conducted at The Royal Veterinary and Agricultural University, Dep. of Crop Husbandry and Plant Breeding, Hojbakkegard (Tastrup), The Research Institute of Commercial and Industrial Plants, Holbergvej (Kolding), The Odum Forsogsstation (Hadsten), and The Landskontoret for Planteavl, Kongsgardsvej (Viby).

2–3.1.6 East Germany

Statistics show that since 1965, alfalfa has been cultivated on about 140 000 ha in East Germany (German Democratic Republic, GDR), with an additional 50 000 ha occupied by alfalfa-grass mixtures. Alfalfa growing is concentrated in the districts of Halle, Erfurt, and Magdeburg. In these areas, where rainfall is <550 mm, alfalfa is the most important multi-cut forage plant (275).

Although the principal use of alfalfa in GDR is for hay, green forage, and silage, high-quality, early cut alfalfa is used in the production of protein flour (alfalfa meal) for cattle, hog, and poultry feeds. Alfalfa is sometimes used in land reclamation on coal mine sites (170). Three to four harvests/yr yield between 7 and 9 Mg/ha of dry matter.

Verticillium wilt is the most important alfalfa disease in the country. Under severe infestations, trials have shown that the disease can reduce yield by 25% and reduce herbage quality by depressing crude protein and digestibility (244). Two Verticillium-resistant cv., Vertibenda (1973) and Verko (1978), were released cooperatively by alfalfa breeders in GDR and Hungary (245). These two cultivars and 'Vertus' are on the official GDR list. The cv. Europe is grown in short rotations or in areas where Verticillium wilt is less severe (224). Most seed is imported.

Research is conducted on methods of breeding (246), breedings for high yield and Verticillium wilt resistance (245), stand establishment, and soil compaction problems. The main research institutions are: Sektion Pflanzenproduktion der Martin-Luther-Universität (Halle-Wittenberg), The VEG Saatzucht, Gotha-Friedrichswerth (Gotha), and The VEG Saatzucht Plaussig (Leipzig).

2–3.1.7 Finland

The crop is grown on about 500 ha in Finland (179). It has been shown, however, that in areas where alfalfa thrives, its yield has surpassed that of red clover (*Trifolium pratense* L.) and fescues (*Festuca* spp.). Factors that hinder alfalfa production include low soil pH, low soil fertility, high water tables, lack of adapted cultivars, short growing seasons, low temperatures, and winterkilling (201).

Alfalfa is used in rotation with sugarbeet (*Beta vulgaris* L.). The crop is dehydrated by factories owned by sugarbeet companies (179). The winterhardy cv. Jokioinen was developed by the Institute of Plant Breeding of Jokioinen. Additional research is conducted at the Dep. of Plant Husbandry, Univ. of Helsinki.

2–3.1.8 France

Statistics in France show a significant decline in the area devoted to alfalfa; from 1 588 000 ha in 1965 to 566 000 ha in 1983. A portion of this decline has been offset by an increase in average yield from 6.4 Mg/ha (1965) to 7.6 Mg/ha (1983). The crop is grown throughout the country with major concentrations in the Departments of Midi-Pyrénées, Poitou-

Charentes, and Champagne-Ardenne. In general, climate is favorable for alfalfa production.

Alfalfa is used for hay, silage, green fodder, and dehydration (199). The production of dehydrated alfalfa increased by 20 to 30% annually between 1960 and 1978, with production stabilizing in 1980. Nearly 100 000 ha are devoted to alfalfa for dehydration. Production of dehydrated products is concentrated in Champagne-Ardenne with a production of 640 000 Mg, of the national total of 800 000 Mg.

The area in seed production decreased from more than 30 000 ha in 1970 to about 20 000 ha in 1984, concentrated primarily in southwestern France. Average seed yields are about 250 to 300 kg/ha. Since 1980, about 3000 Mg of certified seed were used in France annually while nearly 3500 Mg were exported. Several cultivars are included on the French Official List. Europe occupies between 35 to 40% of the area with Resis, Sitel, Hybride Milfeuil, Lutèce, Kara, and Magali listed as important cultivars (14).

Verticillium wilt is the most important alfalfa disease north of the Loire River Valley (90, 207). Yield reductions of 15% have been reported (90).

Sclerotinia crown and stem rot (caused by *Sclerotinia trifoliorum* Eriks.) also occurs in the northwestern regions. Other diseases which have been reported are: downy mildew, rust (caused by *Uromyces striatus* Schroet.), common leafspot, spring black stem, Leptosphaerulina leafspot [caused by *Leptosphaerulina briosiana* (Poll. Graham & Luttrell)], anthracnose (caused by *Colletotrichum trifolii* Bain. & Essary), crown wart [caused by *Urophlyctis alfalfae* (Lagh.) Magn.], violet root rot (caused by *Rhizoctonia violaceae* Tul.) and *Pythium ultimum* Trow (207, 208). Stem nematode is extremely important in the northern regions (51).

Several insects affect forage yield and seed production. The foliage pests: pea aphid (*Acyrthosiphon pisum* Harris), *Apion pisi* F., alfalfa weevil [*Hypera postica* Gyd (œq *H. variabilis* Hbst.)] and *Colaspidema atrum* Oliv. are not considered as important as foliar diseases. Insects inducing galls: *Dasyneura ignorata* Wachtl. and *Contarinia medicaginis* Kieffer; and seed pests: *Laspeyresia medicaginis* Kuzn, *Tychius aureolus* Kiesw., *Adelphocoris lineolatus* Goeze, *Lygus rugulipennis* Popp., alfalfa seed chalcid (*Bruchophagus roddi* Guss.) and *Sitona* spp. have been reported (30, 31).

Research on alfalfa is extremely comprehensive. Pollination problems, both in breeding and in practical seed production, are under investigation (23, 31, 86, 102). Schemes for breeding (81) and selection of alfalfa for resistance to various pests, including Verticillium wilt, stem nematode and spring black stem have been developed (11, 50, 101). Research is conducted on alfalfa-orchardgrass mixtures, N_2 fixation, saponin content, and tolerance to frequent cuttings.

Most research is centered at The Station d'Amélioration des Plantes Fourragères, National Institute of Agronomy Research (INRA) (Lusignan). Several private plant breeding institutions are involved with the development of new cultivars.

2-3.1.9 Greece

Alfalfa is the most important forage crop in Greece. It is grown throughout the country for hay production, with seed production concentrated mainly in Central Greece (251).

The basically Mediterranean climate of Greece is subject to a number of regional and local variations occasioned by the country's physical diversity. The annual rainfall is lower than 400 mm in some southeast regions and above 1000 mm in some western areas. The mean temperature varies between 1.0 to 12.0°C in January and 23 to 27°C in July.

The area in alfalfa has increased from 10 121 ha in 1943 to 198 700 ha in 1980. There has been a corresponding increase in average yields from 5.3 Mg/ha (1945) to 10.2 Mg/ha (1980). Although irrigation is required for maximum hay production, about 30% of the alfalfa crop is grown without irrigation. In 1979, average yields were estimated at 11.7 Mg/ha with irrigation, and 6.6 Mg/ha without irrigation. For the same year, seed yields were 450 and 350 kg/ha with and without irrigation, respectively.

Alfalfa is seeded in pure stands in early spring in narrow spacing of 20 cm or broadcast at a rate of 20 to 25 kg/ha for hay production, and in 40 cm spaced rows at a rate of 5 to 15 kg/ha for seed production. Hay is harvested five to six times per year and usually grazed by sheep (*Ovis aries*) during autumn and winter. Regrowth after the first cut may be left for seed production. Seed production is an important industry but the total amount of seed produced varies considerably from year to year. It is used domestically and for export (about 10-20%). Weed control is required for seed production (251).

The Greek cv. Hypati and Hyliki, followed by the U.S. cv. Talent and African are most commonly grown. Old Greek landraces are cultivated in small areas. European cv. Alfa, Orca, DuPuits, Europe, Verneuil, Everest, etc., are cultivated for seed production, under special agreement for export (251).

The most important alfalfa diseases in Greece are: alfalfa mosaic virus, common leafspot, anthracnose, *Fusarium* spp., *Rhizoctonia* spp., and Sclerotinia crown and stem rot. The most important insects are: *Hypera postica, Apion apricans, Contarinia medicaginis, Lygus* spp., *Aphis laburni*, and *A. fabae* Scop.

The research program conducted at The Fodder Crops and Pastures Institute (Larissa) includes breeding, management, and practices to improve hay production, with or without irrigation. Current breeding objectives include drought resistance, disease resistance, persistence, and higher dry matter yields.

2-3.1.10 Hungary

Alfalfa was cultivated on about 337 500 ha in 1982, ranking fourth after wheat (*Triticum aestivum* L.), corn (*Zea mays* L.), and barley (*Hordeum vulgare* L.). This is a loss compared with the 420 000 ha recorded in 1975. Increasing demand for cereals and a reduction in dehydration because of the rising cost of energy are responsible for much of this decrease (32, 258).

Alfalfa is distributed throughout the country. About 60 to 70% of the crop is concentrated in the Alföld (Great Hungarian Plains) where it is grown everywhere except on very salt soils. Alfalfa for seed is grown on 30 to 40 000 ha in the arid parts of the Alföld.

In the Alföld, the mean annual temperature is 10.5°C and ranges from −4.0°C in January to 21.0°C in July. The yearly mean precipitation is 550 to 600 mm.

Alfalfa in pure stands is used mainly for hay (60%), green fodder (15%), dehydration (10%), silage (10%), and seed. The crop is harvested four to five times per year, with average dry matter yields of 5.5 Mg/ha. Frequent cuttings and wilt caused by *Fusarium* spp. and *Verticillium* often lead to drastic thinning of the stands in the 3rd yr; other diseases include common leafspot, spring black stem, and viruses (209). Since the energy crisis, dehydration has decreased while silage from pre-wilted first cut plus additives has increased in popularity (258).

Hungary is generally self-sufficient in seed production. Average seed yield for 1978 to 1983 was 130 kg/ha. Seed is harvested from the first regrowth. Dodder (*Cuscuta* spp.) is one of the major problems in growing alfalfa for seed or hay. Presently, no effective farm scale control is available. The great number of parasites on the pollinating leafcutting bees as well as other harmful insects, such as *Contarinia medicaginis*, *Bruchophagus roddi*, *Tychius flavus* Becker, *Sitona*, *Lygus*, and *Adelphocoris*, are detrimental to seed production (169).

Twelve cultivars are registered in Hungary. A number of them are of sativa type: 'Nagyszenasi' (1961), 'Tapioszelei-1' (1970), 'Szarvasi-1' (1971), 'Szarvasi-2' (1978), 'Verko' (1978), 'Szarvasi-4' (1981), 'KM Hybridalfa' (1981), or of media type: 'Ovari tarka' (1944), 'Ovari Kuszo' (1964), 'Kisvardai-1' (1978), and 'Vertibenda' (1974). The Verticillium wilt resistant cv. Verko and Vertibenda were released jointly by Hungary and GDR. 'KM Hybridalfa' was the first European hybrid alfalfa produced using cytoplasmic male sterility (185).

Research on alfalfa has concentrated on breeding but other problems related to production and forage conservation receive attention as well. Breeding for resistance to Fusarium and Verticillium wilt (18, 259), tolerance to frequent cuttings (35), low saponin levels (34), creeping growth habit, high dry matter yield, and seed yield are important priorities. Other studies involve the use of male sterility to produce hybrid alfalfa (27, 185), weed control, soil fertility (195), N_2 fixation, and seed production (33, 169).

Primary research centers are; The Gate Agricultural Research Institute (Kompolt), The Research Institute for Irrigation (Szarvas), The Research Institute of the University of Agricultural Sciences (Godollo Kompolt), The Seed Growing Company (Budapest), The Kate, Mosonmagyarovar Faculty, The MTA Research Institute for Phytopathology (Budapest), and The Research Station (Szentes).

2-3.1.11 Italy

Alfalfa is an extremely important crop in Italy. It is grown on an estimated 1 300 000 ha distributed throughout the country, but mostly in the northern and central provinces that have favorable climatic and edaphic conditions (80).

Italy can be divided into three climatic zones: Continental, Peninsular, and Island. The Alpine regions and the Po valley have severe winters and hot summers. Rain falls mainly in the spring and autumn and increases with altitude. The Peninsular and Island areas have very hot dry summers. The rainfall from April to June is considered decisive for good yields.

Alfalfa, a typical forage crop on the hilly lands in central Italy, is used mostly as hay. Alfalfa, as pasture, is limited to about 14 000 ha in central and southern Italy (217). Dehydration has declined to 120 000 Mg in recent years. Seed production (20 000 ha: 18 500 ha of ecotypes and 1500 ha for cultivars) is erratic, and seed supply is uncertain. Because of variable climatic conditions, seed yields from 60 to 600 kg/ha may be obtained from the same field in successive years (80). Diseases are not considered important although *Fusarium* spp. have been reported (209).

Alfalfa is harvested, at early bloom, three to five times per year according to the area and precipitation. Seed is harvested from the second crop. Alfalfa fields are plowed in the 2nd or 3rd yr in rotation with wheat or barley.

Local ecotypes have a dominant place in cultivation (90%), and no one strain or cultivar has broad acceptance. Although these ecotypes differ according to area of origin, they are generally winter dormant and summer active in the north, and exhibit high yields and good leaf/stem ratios. In order of importance, cultivars and ecotypes grown in the continental district are: 'Bresaola', 'Canè', 'La Rocca', 'Equipe', 'Europe', 'Delta', 'Ezzelina', 'Garisenda', and 'Prospera' (217).

Research programs include simultaneous selection for seed and forage yield (165), selection for yield stability, tolerance to frequent cutting, and low saponin content (218, 267).

Research centers with interest in alfalfa are: The Istituto Sperimentale per la Colture Foraggere (Lodi), The Istituto di Allevamento Vegetale (Perugia), and The Sisforaggera Industry (Bologna).

2-3.1.12 Netherlands

The Netherlands have a temperate maritime climate, with an average rainfall of about 700 mm. Average temperatures in January and July are 1.8 and 17.0°C, respectively.

Alfalfa was grown on about 2000 ha in 1983, with a total annual yield of about 10 to 11 Mg/ha of dry product (70, 254). It is grown primarily on good clay soils, and seeded in pure stand under cover crops. Three to four harvests are taken each year, and the product used primarily for dehydration. In 1983, the cv. Resis and Europe were used on 61 and 36% of the area, respectively (65).

Lately, there has been renewed interest in alfalfa production, especially for silage. Research is conducted at the Agricultural University of Wageningen, (Centre for Agrobiological Research and The Institute for Research on Varieties of Cultivated Plants), and The Research and Advisory Station for Cattle Husbandry, Lelystad.

2-3.1.13 Norway

Alfalfa is grown on about 200 ha in Norway but interest for the crop is increasing in the southeast. Climatic conditions and acid soils have limited the use of alfalfa, with winterkilling a serious problem (172).

Alfalfa is generally seeded in mixtures with timothy (*Phleum pratense* L.) or smooth bromegrass (*Bromus inermis* Leyss.) and to a lesser extent in pure stands. The crop harvested twice a year is utilized mainly for silage.

Recent cultivar trials indicate that some Canadian cultivars may be better than the standard cv. Sverre (172). Some breeding work is conducted at the state agricultural experiment station of Loken, with cultivar trials and management studies at state agricultural experiment stations at Loken and Apelsvoll.

2-3.1.14 Poland

In Poland, alfalfa is grown on 258 000 ha (1981-1983). The crop is distributed throughout the country, but mainly in dryer parts of the western, central, and southern Poland where it occupies about 4.3% of the arable land. Climatic conditions in Poland are highly variable. The annual average precipitation is about 600 mm dropping to about 450 mm in central Poland.

Alfalfa grown in pure stands or in mixtures with grasses is used mainly for hay, green fodder, dehydration, and to a lesser extent for silage and pasture. For dehydration and swine feed, alfalfa is cut several times a year at the early bud stage while late bud or early bloom stages are selected for making hay (132). The first, third, and fourth harvests are generally used for green fodder while hay is made from the second cut. Average yields of green matter are 30.9, 36.0, and 27.0 Mg/ha for the south, west, and northeast Poland, respectively. Productivity and life of stand are related to resistance to Verticillium wilt, the most important disease, and tolerance to frequent cuttings. Other diseases that have been reported are common leafspot, spring black stem, and bacterial wilt (209).

The cv. Europe, Euver, Everest, Orchésienne, Franken Neu, Luna, Dalava, Verko, Vertibenda, Vertus, and Tuna are generally grown, with most seed imported (nearly 90%). The most productive cultivars are characterized as semierect, rapid growing, and resistant to Verticillium wilt. The old Polish cv. Kleszczewska, Miechowska, Warminska, and Kujawska are more dormant and winterhardy than imported cultivars. They are better adapted under extreme climatic conditions. Recently, three cultivars have been released: 'Kometa' for resistance to Verticillium wilt, 'Tula' for tolerance to stem nematode, and 'Boja' for high yield and persistence (132, 243).

Seed production is difficult because of poor pollination. The multiplication of effective pollinating insects, such as *Bombus* spp., *Anthophora parietina* F., and *Megachile pacifica* Panz., is receiving attention.

Research is conducted on the biology of flowering, pollen vitality under different climatic conditions, influence of growth regulators on seed development and production, superior seed set combined with high forage yield, male sterility, and inbreeding and self-incompatibility in relation to hybrid breeding (131, 239, 242). Other studies include work on resistance to Verticillium wilt, common leafspot, and diseases caused by *Fusarium* spp.; time and rates of seeding, frequency of harvest, nutritive value (130, 138), and soil fertility.

The main research centers are: The Institute of Plant Breeding and Acclimatization (Radzikov), The Dep. of Forage Crops of the Institute of Soil Science and Plant Cultivation (Pulawy), The Dep. of Bee Research of the Institute of Horticultural Science (Pulawy), The Plant Breeding Stations of Szelejevo and Nagradowice, and The Agricultural Colleges of Poznan, Olsztyn, Wroclaw, and Lublin.

2–3.1.15 Romania

Alfalfa occupied 400 000 ha or nearly 4.4% of the arable land in the country in 1981, compared with only 0.91% in 1927. In the past, alfalfa was grown mainly in Transylvania and the Western Plains. By the early 1970s, the alfalfa-producing area had moved progressively towards the Southwestern Plains (Timis, 21 285 ha) and into the chernozem and brown soils of the Plain of the Danube. In certain districts, more than 50% of the area devoted to forage crops is cultivated with alfalfa (265). In recent years, alfalfa production has been extended into the arid and nonirrigated region of Dobruja. Alfalfa grown under irrigation has increased from the 35 200 ha in 1965 to 113 900 ha in 1971.

The Romanian climate is transitional between temperate regions and the harsher extremes of the continental interior. The average annual rainfall is about 700 mm in the west, and about 450 to 500 mm in the Dobruja region.

The crop is grown in rotations for green-cut forage, silage, and alfalfa meal. New technologies and better cultivars have contributed to the increase in dry matter yield production from an average 2.3 Mg/ha in 1960 to 6.1 Mg/ha in 1978 (264). The average life of stands is about 4 yr. Under irrigation, pure stands of alfalfa are shorter lived, so it is seeded with grasses, such as *Dactylis glomerata, Lolium perenne* L., *Bromus inermis*, or *Festuca* spp.

Fusarium wilt [*Fusarium oxysporum* f. sp. *medicaginis* (Weimer) Sny. & Hans.] is by far the most important alfalfa disease (209). Other diseases include common leafspot, spring black stem, *Pythium* spp., and rust. Some insects are also found: *Subcoccinella 24-punctata* L., *Phytodecta fornicata* Brügg, *Aphis* spp., *Sitona, Phytonomus variabilis* Herbst (= *H. postica*) (265).

The Romanian-bred cultivars: H652 (1963), Luxin (1973), Lutetia (1981), and Gloria (1982) are grown most commonly. Lutetia is characterized by resistance to Fusarium wilt and high yields of forage and seed (264).

The development of high-yielding cultivars is an important breeding objective, together with breeding for resistance to Fusarium wilt, earliness, tolerance to frequent cuttings, resistance to lodging, persistence, improved quality, and nonbloat characteristics (262, 263, 265). Inbreeding followed by crossing of inbred lines is the breeding method used to obtain these objectives (265). Research is conducted at the Research Institute for Cereals and Industrial Crops (Fundulea).

2-3.1.16 Spain

Three basic climatic zones are found within the country. The mountain climate of northern Spain from Galacia to northern Catalonia is characterized by moderate summer and low winter temperatures, with annual rainfall averaging between 750 and 1725 mm. The central plateau (Ebro, Center, and Duero) has a continental climate with lower rainfall (250–500 mm) and high summer and low winter temperatures. The coastal climate of the Mediterranean and Atlantic seaboards is dry with high summer and moderate winter temperatures.

The area planted to alfalfa was 332 600 ha in 1981, with nearly 75% of the crop grown under irrigation in arid regions. Local ecotypes of alfalfa are cultivated in specific climatic zones (110). The ecotypes Aragon, Urgel, and Mediterranean are used in irrigated areas, while Ampurdan and Tierra de Campos are cultivated under dryland conditions.

Aragon is the most important ecotype in Spain. It is cultivated on about 250 000 ha on irrigated lands of the Ebro region, where it originated, and in the agrarian regions of Duero, Center, Andalusia, and Extremadura as well as in more humid regions. This ecotype is characterized by a rapid growth that can be harvested five to six times per year. It is winter dormant, and susceptible to *Colaspidema atrum* Oliv. and common leafspot. Several cultivars have been derived from Aragon for use on irrigated land: Logrono, San Isidro, Adamar, and Aima (111, 118).

Urgel is cultivated under irrigation as well as on the dryland on about 20 000 ha concentrated exclusively in its center of origin. The Mediterranean ecotype is cultivated exclusively in the Mediterranean region (Levant) where it originated and occupies nearly 18 000 ha. It is a nondormant type producing 10 to 12 cuttings per year. Ampurdan is cultivated on about 19 500 ha of dryland in the Catalonia region. It tolerates drought, low temperatures, and long periods of winter and summer dormancy (116). Tierra de Campos originated in the zone Tierra de Campos of Central and Duero regions. It is cultivated mostly in its center of origin, on about 25 000 ha of dryland. It is very summer and winter dormant and produces one to three cuts per year.

Other cultivars have been selected from dryland populations (Adalfa) or from foreign material (Adyta) (114, 115, 182), but their contribution to the alfalfa production in Spain is limited.

The crop is used mainly for hay, with about for 50% of the Tierra de Campos and Ampurdan ecotypes used as pasture or pasture-hay production, and 25% of the Mediterranean alfalfas for green forage. Dehydration is not significant. Fall and winter regrowth is grazed by sheep (119). Good alfalfa stands last about 4 to 5 yr, with dry matter yields of 12 and 15 Mg/ha harvested under irrigation.

Nearly 10 000 ha of alfalfa is grown for seed, for domestic use and export to Mexico. Seed yields as high as 1000 kg/ha has been reported.

The main disease is common leafspot that appears in the fall. Other diseases include rust, downy mildew, violet root rot, and bacterial wilt. The most widespread pest is *Colaspidema atrum*; other insects with some importance are: alfalfa weevil, *Apion pisi* F., and *Aphis fabae* Scop. (117). Dodder [*Cuscuta epitymum* (L.) Marc] is a serious problem in seed production.

The development of new cultivars for the dryland and irrigated areas, with resistance to drought and diseases, and with high yield are the main breeding objectives. Research is conducted on evaluation of Spanish and foreign ecotypes and cultivars, management practices and utilization, and N_2 fixation.

Alfalfa research is conducted at the following centers: The Estacion Experimental de Aula Dei (Zaragoza), The Centro de Edafologia y Biologia Aplicada de Salamanca (Salamanca), Numerous research centers belonging to the Instituto Nacional de Investigaciones Agrarias throughout Spain, The Asociation de Investigacion y Mejora de Alfalfa (Zaragoza), and The Actividades Agricolas Aragonesas, Cogullada (Zaragoza).

2-3.1.17 Sweden

There are no official statistics on the total area of alfalfa in Sweden. It is estimated to occupy about 25 000 ha (135).

The area where alfalfa can be grown lies between the 55 and 61° N Lat. Because Sweden benefits from the influence of the warm Gulf Stream, its maritime type of climate is much milder than expected. Yearly rainfall is between 525 and 630 mm and the mean temperature ranges between −3.0 in February and 17.5°C in July.

Since the 1960s and early 1970s, the main utilization of alfalfa in Sweden has changed from dehydration to hay and silage production. When used for hay and silage, there has been a change from complex mixtures with red clover and grasses to more simple mixture with only 20 to 40% grass, mainly timothy, meadow fescue, or orchardgrass. Alfalfa and alfalfa-grass mixtures usually produce more dry matter and protein per ha than red clover-grass mixtures. Alfalfa is cut three times per year for two or three harvest years (135).

The Swedish cv. Alfa II, Lesina, Sverre, and Vertus are grown, with the first two the most common. Breeding work with Flemish germplasm has been oriented toward increasing winterhardiness and improving resistance to both stem nematode and Verticillium wilt, the two most devastating alfalfa pests in the country. The release of Sverre and Vertus, combining resistance to both pests, has reduced these problems (136).

Two swedish companies, namely, Svalöf AB at Svalöv and Weibullsholm Plant Breeding Institute at Landskrona conduct breeding work. Investigations on management, inoculation, liming, pesticides, mixtures, forage production, and conservation are conducted at The Swedish University of Agricultural Science (Uppsala).

2-3.1.18 Switzerland

Alfalfa is cultivated on about 6000 ha in an area between Geneva and Schaffhausen on the southern side of the Jura mountains. The variation in soil quality within small areas produced by geological conditions and by the relief makes large-scale single crop farming impossible. Precipitation varies from 700 to 1000 mm, with mean temperatures of −0.9 in January and 17.7°C in July.

In general, alfalfa is grown in mixture with red clover and grasses, such as timothy, ryegrass (*L. perenne* L.), and orchardgrass. Good stands last 2 to 3 yr. The crop is used mainly for green fodder and hay (134). Cultivars include Europe, Euver, Everest, Franken Neu, Resis, Vela, and Vertus (133).

Verticillium wilt is the most important disease. Other diseases include crown and stem rot, spring black stem, *Fusarium* spp., etc. (274).

Research is conducted at Eidgenössische Forchungsanstalt für Landwirtschaftlichen Pflanzenbau (Zürich-Rechenholz) and The Station Fédérale de Recherches Agronomiques de Changins/Lyon. Emphasis is placed on cultivar testing in pure stands and mixtures, cutting management, inoculation, and N fertilization.

2-3.1.19 United Kingdom

Alfalfa is cultivated on an estimated 16 000 ha in all parts of England and Wales but more than half is concentrated in east and southeast England. The annual rainfall in this area does not exceed 750 mm.

Alfalfa is used for silage, hay, and dehydration. Three to four cuts are taken at intervals of about 6 weeks. Both wilting and additives are required for silage (63). Further expansion of alfalfa hectarage in the United Kingdom will be limited not only by climatic conditions but also by the lack of flexibility compared with herbage grasses (73).

The cv. Europe, Sabilt, and Vertus are recommended, with provisional recommendations granted to Euver, Vela, and Maris Phoenix. All recommended cultivars except Europe have resistance to Verticillium wilt and/or to stem nematode; two important pests in the United Kingdom. Maris Phoenix is the only recommended cultivar with good resistance to bacterial wilt, a disease that is becoming more frequent. Other diseases include sclerotinia crown and stem rot, common leafspot, and spring black stem (209).

Alfalfa breeding is conducted at The Welsh Plant Breeding Station (Aberystwyth) and cultivar assessment at The National Institute Agriculture Botany (Cambridge). The value of alfalfa in animal nutrition is investigated at The Grassland Research Institute (Hurley). The Dep. of Botany, Univ. of Nottingham, is conducting research on protoplast regeneration and fusion (166).

2-3.1.20 USSR

Statistics on the area and distribution of alfalfa in the USSR were not available to the authors. However, the area of alfalfa in the USSR has been estimated at 4 500 000 ha (29, 148, 265).

Alfalfa is the most important perennial legume. It is used as hay and pasture not only in the traditional production zone (Central Asia, Transcaucasus, and Ukraine) but also in other regions (Nonchernozem zone, Baltic Republics, and Belorussia) (196). The Southwestern Asian center of origin (235) has had a decisive role in the evolution and the distribution of alfalfa species. Among the 13 alfalfa species originating in that center, *M. falcata* is certainly the most widely distributed in the steppes of Russia and Siberia, where it contributes substantially to the production of native grasslands.

Alfalfa breeders in the USSR have taken advantage of the important and highly diverse gene pool situated in that country. High-yielding regionalized alfalfa cultivars, with a number of excellent economic and biological parameters, are now available for cultivation in all zones of the USSR (187).

Two aspects are emphasized in developing new cultivars, either fast growth and high yield for hay production or early growth in spring, regrowth after grazing, and persistence for pasture use. Other breeding objectives include response to fertilizers, suitability for irrigation, resistance to diseases (*F. oxysporum*) and other pests (187), and tolerance to multiple cuttings (196). Attention is given as well to the development of cultivars for extreme conditions, that may require combinations of the following: improved winterhardiness, cold resistance, tolerance to prolonged flooding, fibrous root systems, and persistence.

Interspecific and intraspecific hybridization with subsequent selection is used in the breeding work, especially in the development of cultivars for grazing (197). Deep set crowns and creeping-rooted forms of alfalfa are well suited for grazing.

Physiological studies and research on seed production, management practices, fertilizers, and breeding are conducted at All-Union Williams Forage Research Institute (Lugovaya-Moscow), All-Union Vavilov Research Institute for Plant Growing (Leningrad), All-Union Research Institutes for Agriculture (Novosibresk and Alma Ata), Krasnodar Agricultural Research Institute (Krasnodar), and numerous experimental stations and institutes of agriculture throughout the agricultural regions of the USSR.

2–3.1.21 West Germany

West Germany has a temperate climate and well-distributed precipitation (500–1000 mm). Official statistics show a decrease from 63 000 ha in 1975 to 31 000 ha in 1982. Alfalfa is concentrated in the southern part of the country, particularly on well-drained loam soils with high pH. In 1982, the leading state was Bavaria with 16 600 ha (53% of the total hectarage) followed by Baden-Württemberg with 10 200 ha (32%). The total production was 271 000 Mg or 8.7 Mg/ha (233).

Alfalfa is harvested for hay three times per year. The West German list of cultivars contains 15 cultivars, eight of foreign origin. The cv. Franken Neu, Luna, and Europe are grown most commonly.

Verticillium wilt is the most important alfalfa disease. It affects longevity because the commonly used cultivars are only moderately resistant. Common leafspot, downy mildew, and spring black stem have been reported (209).

Alfalfa research has been concentrated on breeding synthetic and hybrid cultivars (232, 234). Additional research involves disease problems, management practices, crop utilization, seed multiplication, and tolerance to stresses.

The main research institutions are: The Technische Universität

Munchen Lehrstuhl für Grünland und Futterbau (Freisen-Werhenstephan), and The Bayerische Landesanstalt für Bodenkultur und Pflanzenbau (Freisen).

2–3.1.22 Yugoslavia

Alfalfa was cultivated on about 337 000 ha in 1984. The main areas of alfalfa production are the central and southern part of Serbia (117 000 ha), Vojvodina (68 000 ha), and Croatia (60 000 ha) (240). Only 11 000 ha are grown in Slovenia because of heavy soils and the lack of adapted cultivars (151). The main alfalfa region in the northeastern plains is characterized by a continental climate with seasonal extremes ranging from −25 to 33°C. Annual rainfall varies from 600 to 1200 mm.

The principal use is for hay, and more recently as silage with grasses and corn. The primary breeding objectives involve the development of cultivars for marginal areas, with tolerance to cold and high rainfall. Breeding material includes the common 'Panonian' alfalfa and foreign cultivars (37, 39, 231). Several cultivars have been released [Osjecka 66 and Osjecka 70 (36, 38), NS Mediana, Olimpik 84, Kruseuacka 23, Krajina, Debarska, and Stela]. Breeding for a shorter flower tube is considered as one approach to improving seed set along with the use of adapted bees (240).

The main alfalfa diseases include: Stemphyllium leaf spot (*Stemphyllium botryosum* Wallr.), spring black stem, *Gleosporium morianum* Sacc., and rust.

Institutions with a major interest in alfalfa are: The Poljoprivredni Institut (Osijek), The Poljoprivredni Fak. oour Institut Za Ratarstvo (Novi Sad), The Poljoprivredni Fak. (Sarajevo), The Zavod za Krmno Bilje (Kruseuac), The Zavod za Poljoprivredni (Zazecar), The Zemjodelski Fak. (Skopje), The Biotehniska Fak. vtozd za Agronomijo (Ljubljana), The Kmetijski Institut (Ljubljana), and The Semenarski Kombinat (Ljubljana).

2–3.2 North America

The main production center of cultivated alfalfa is in North America where the USA, Canada, and Mexico grow nearly 42% of the total world hectarage. The greatest concentration occurs in the western and north-central regions with 20 and 60% of the total area in USA, respectively, and Alberta, Ontario, and Saskatchewan which combine for more than 70% of the total area in Canada. The overall area devoted to alfalfa has increased slightly during the past decade in North America; with the small decrease in USA compensated largely by increases in Canada and Mexico.

A more complete description of alfalfa-growing areas and production in the USA and Canada appears in Chapter 20 in this book. A brief summary is provided here to complete the worldwide treatment of the crop.

2-3.2.1 Canada

Alfalfa, the most important forage legume in Canada, is grown in all provinces. Recent statistics which included both pure stands and alfalfa-grass mixtures seeded for hay production showed a total area of about 2 544 300 ha in 1981, which represents a major increase from 1 809 675 ha reported in 1961. There is probably a greater area used for pasture. In addition, large areas are devoted to the crop for green manure, silage, dehydration, and seed production. Hence the area in alfalfa has been estimated at 4 to 6 million ha in Canada (93).

In 1981, alfalfa occupied nearly 50% of the total area devoted to tame hay production in Canada, as compared with 37 and 44% in 1961 and 1971, respectively.

2-3.2.2 United States of America

As in many other countries, statistics make no distinction between alfalfa grown in pure stand and that grown in mixtures with perennial grasses. The expansion and location of alfalfa production in the 20th century is shown in Fig. 2-1. In 1981, average yields were as follows: North Central, 6.8 Mg/ha; Western, 8.4 Mg/ha; Northeastern, 6.2 Mg/ha; South Central, 7.1 Mg/ha; and Southeastern, 5.7 Mg/ha. More than 60% (6 498 225 ha) of the area harvested for hay and 58% (44 447 859 Mg) of the total production is concentrated in the northcentral part of the country. Alfalfa hay constitutes about 60% of all hay harvested in the USA. An additional unreported and smaller land area of alfalfa-grass mixtures is used for grazing.

More than 174 000 ha of alfalfa was devoted to the production of seed in 1980, with an average yield of 267 kg/ha. Seed production is concentrated in the western states (see Chapters 32 and 33 in this book).

2-3.2.3 Mexico

Alfalfa was introduced into Mexico by the Spaniards in 1519 and rapidly became one of the most important cultivated forage crops. It is grown extensively at higher elevations in the central part of the country and in the dry irrigated valleys along the Pacific coast. The crop is largely produced on sandy, well-drained volcanic soils.

The area in alfalfa increased from about 47 000 ha in 1925 to 1929 to an estimated 245 000 ha in 1983; with an average yield of 10.0 Mg/

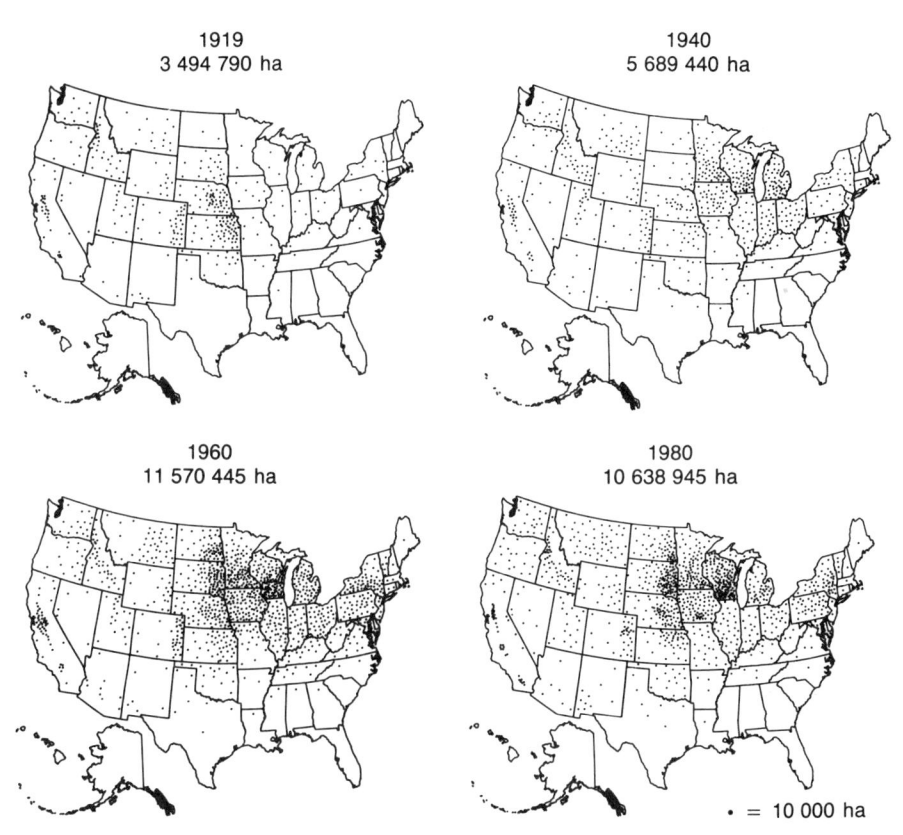

Fig. 2-1. Distribution of U.S. alfalfa production areas during the 20th century.

ha and a total production of 2.5 million Mg. In 1981, the area of alfalfa in the four leading states was: Guanajuato, 39 216 ha; Hidalgo, 28 097 ha; Chihuahua, 26 358 ha; and Coahuila, 25 619 ha. Alfalfa has tripled in sown area since 1970 in some regions, such as Comarca Lagunera (171). This region is the principal supplier of milk to Mexico City about 1000 km to the south. Nine harvests can be made annually with irrigated yields exceeding 18.0 Mg/ha.

A common practice in southern Mexico is to cut daily and feed green, especially during the rainy season. In the hot, dry environment of the north, alfalfa is usually irrigated and harvested 6 to 12 times annually for hay. Grazing is seldom practiced because of bloat. Stands may persist 20 yr but the average life is 4 to 6 yr.

Seeding rates of 20 to 30 kg/ha are commonly recommended (49, 210). Establishment is favored by low temperatures, and time of seeding is between 15 September and 15 January depending upon location. Most alfalfa is grown in pure stands but it has been tested with several adapted grass species (85). Alfalfa seeded with wheat in November or early December, irrigated after the wheat harvest, and plowed under 8 weeks prior to wheat seeding produced 3.4 Mg/ha dry matter and 100 to 120 kg of N/ha (198). Application of phosphate fertilizer has resulted in marked yield increases at some locations (212). Use of 40–180–40 gave the optimum economic response at the Pabellon Experimental Station in the North Central Region (180).

Local nondormant ecotypes produced and persisted quite well in regional yield trials (43, 44). The U.S. cv. Arizona Chilean, Caliverde, and Moapa in the same trials yielded well but were short-lived. Valencia, introduced from Spain, is adapted to the high-altitude valleys (42, 221). In the northern part of the country, Moapa, Mesa Sirsa, and Espanola (Aragon) occupy more than three-fourths of the alfalfa production area. Recommended cv. include Astro, Bajio 76, INIA 76, Matador, Mesa Sirsa, Mexteca 76, NK-819, Puebla 76, and San Miguelito (49, 83, 84, 188, 210). These yield 10 to 20% more than the more commonly used Moapa, Joaquin 11, Valencia, and the native criolla ecotypes. Mexican ecotypes are not used in the northern part of the country (46).

Ten percent of the seed required in Mexico is produced locally but much is imported each year from Spain and the USA (46). Seed production studies have included seeding methods, seeding rates, fertilizers, weed control, harvesting, and conditioning. The use of modern technology has resulted in yields of 500 to 800 kg/ha (89, 221).

Diseases occur frequently on alfalfa and many have been identified in Mexico (104). These include common leafspot, downy mildew, root and crown diseases caused by *F. oxysporum* Schl.; Rhizoctonia crown and root rot (*Rhizoctonia solani* Kuehn); bacterial wilt, Verticillium wilt; spring black stem (*Ascochyta imperfecta* Pk. = *P. medicaginis*); summer black stem (*Cercospora zebrina* Pass); and others.

Major destructive insects include the spotted alfalfa aphid (*Therioaphis maculata* Buckton); the root-feeding weevil (*Epicaerus aurifer* Boh.); *Lygus lineolaris* de Beauvois; *Colias eurytheme* Boisduval; *Spissistilus festinus* Say; *Acyrthosiphon pisum*; *Caliothrips phaseoli* Hood; and *Spodoptera exigua* Hübner (126, 192, 226).

The demand for many basic agricultural commodities is shifting to include more animal protein. Mexico has insufficient dairy production and relies heavily on imported nonfat dry milk to meet consumers needs (108). These factors will result in an increasing emphasis on alfalfa production where it is adapted. Approximately 12% of the land is cropland and 40% is pasturage. Half of the country is deficient in moisture for crop production throughout the year, with only 12% of the land receiving adequate rainfall in all seasons. The 25% of cropland which is irrigated provides an important stabilizing influence and irrigated alfalfa production is an important asset to the agricultural economy. Factors that significantly limit alfalfa yields are diseases, insects, a scarcity of seed of adapted cultivars, and lack of adequate technology for the management of improved cultivars (48).

Research on all aspects of alfalfa breeding, production, and utilization is conducted at experiment stations in Chapingo, Chihuahua, Coahuila, Guanajuato, Guerrero, Santa Elena, Sinaloa, and Sonora.

2-3.3 Central America

Alfalfa is not important in Central America, but small fields of <1 ha are grown occasionally at elevations of from 1500 to 3000 m. The crop is customarily cut daily and fed green to both small animals and cattle. Stands are used for hay, occasionally for seed production, and seldom for grazing. On fertile, well-drained volcanic soils, alfalfa stands may last as long as 10 yr, with 8 to 10 cuttings annually. The first harvest can be made within 16 to 20 weeks after sowing and every 40 d thereafter. Average yields in the temperate zone range from 12 to 14 Mg/ha dry matter. Ecotypes derived from 'Hairy Peruvian' are grown in Costa Rica. Indigenous alfalfas give higher yields than foreign cultivars because of their greater tolerance to rust.

Kudzu is the only forage legume used in Puerto Rico (4). However, there is great interest in the introduction and evaluation of new alfalfa cultivars for use on the southern coast (266).

Alfalfa is a relatively new crop in Cuba. Increasing alfalfa utilization is limited by lack of adapted cultivars and availability of seed. The cv. Gilboa and Galilée both introduced from Israël, and Oaxaca and Tanverde from Mexico are commonly used. When cut at 10% bloom, yields of about 7.0 Mg/ha dry matter may be obtained from four to six cuttings. Yield decreases during the rainy season due to slower growth and stand deterioration (289, 290). Seed production is difficult. Attempts to improve seed yields through natural selection (293) were more successful than cultural practices (291, 292). Most research on alfalfa is conducted at the Institute of Animal Science (Havana).

2-3.4 South America

The greatest concentration of alfalfa in South America is in north-central Argentina, with substantial production in several other countries. Most of the crop is used for forage but seed production is important in some areas. A large proportion of the alfalfa seed crop is exported.

2-3.4.1 Argentina

Alfalfa has been grown in Argentina since the early 17th century. An "explosion" in the use of the crop occurred late in the 19th and early in the 20th centuries (127). Alfalfa is well adapted throughout the country. Principal regions of production are in the provinces of Buenos Aires,

Cordoba, La Pampa, and Santa Fe. These regions represent 94% of the alfalfa cropland in Argentina. Other important regions for irrigated alfalfa are Mendoza, San Juan, the valleys of the Andes, Santiago del Estero, Tucuman, and from the Rio Negro southward.

The area in production reached a peak of 8.5 million ha in 1921 to 1922 and then declined to 5.3 million ha in 1934. Recently alfalfa production in pure stands has fallen to < 2.0 million ha. Alfalfa-grass mixtures became popular during this period of decline and if they are taken into account the area in alfalfa has remained relatively stable at about 7.5 million ha. Much of this is located in the Great Pampean Region. The Extension Service of the National Institute of Agricultural Technology (INTA) estimated that 4.5 million ha of alfalfa were grown there without irrigation. Approximately 70% of this area was planted with alfalfa-grass mixtures utilized primarily for grazing. Major grasses include: *Festuca arundinacea, Dactylis glomerata, Bromus unioloides, Phalaris aquatica*, and *Agropyron elongatum*. In the remainder of this region, alfalfa is established in pure stands with a companion crop of rye, oat, wheat, or barley as protection against wind erosion, and used for grazing, hay, or seed production. The crop is planted in autumn at a rate of 4 to 6 kg alfalfa seed/ha and with <15 kg/ha of the companion crop. Drought is a problem in the Pampean region and forage yields are directly related to precipitation. Rainfall varies from 550 to 750 mm in the semiarid and subhumid areas to more than 800 mm in the humid area (128).

In the irrigated northwest and south of the Rio Colorado and Rio Negro valleys, alfalfa is sown by broadcasting 20 to more than 30 kg seed/ha. Seed is usually not inoculated and most growers consider this practice unnecessary. Use of fertilizers is limited even where serious phosphate deficiencies have been detected, as in the southwest part of the Province of Buenos Aires. In experimental plots, nondormant cultivars yield 16 to 20 Mg of dry matter/ha when harvested six to eight times per year, in the northwest; and semidormant cultivars yield 13 to 20 Mg of dry matter/ha with four or five cuttings in the south. Most growers average about 6.2 Mg/ha.

Seed production has been of secondary importance to forage production in Argentina. In the Great Pampean Region, alfalfa is cultivated in dryland conditions for pasture but in favorable years growers remove cattle from fields and harvest seed. Seed yields vary from 60 to 150 kg/ha (128).

In some irrigated areas, principally in the west and north of the country, seed production is an important enterprise. High potential yields are possible because of favorable climatic and soil conditions. Where native pollinators are abundant, it is possible to obtain 500 to 600 kg/ha. Prior to 1967, alfalfa seed exports were neglibible but since then seed exports have exceeded 1000 Mg/yr.

Local ecotypes of alfalfa are important in Argentina, especially in the Pampean Region. These ecotypes developed after more than a century of natural selection under grazing. They are pasture types with wide, submerged crowns, thin stems with many basal leaves, and tolerance to low temperatures. The 'Pampera' and 'Cordobesa' ecotypes are typical of the region.

Variegated alfalfas occur in Buenos Aires, Rio Negro, and La Pampa although most Argentine alfalfas are pure *M. sativa* types. In Santiago del Estero, a glabrous Peruvian type called 'Saladina' or 'Inverniza' is grown. In Cordoba, Santa Fe, and other provinces, an alfalfa strain named 'Scantamburlo' is well adapted and high yielding. The cv. Cordobesa INTA is very similar to Scantamburlo in adaptation and agronomic performance. Other national cultivars are Bordenave INTA and Varsat INTA, both of which are adapted to the calcareous soils in the dry, cold parts of southwestern Buenos Aires and La Pampa.

In 1970, INTA organized the Programa Alfalfa to strengthen research on this important crop. Breeding was emphasized and several improved cultivars have been developed. 'Paine INTA', released in 1975, was the first local cultivar resistant to pea aphid. It and a later release, 'Anguil INTA', perform well in the semiarid and subhumid parts of the Pampean Region. For the well-watered humid Pampa, the cv. Fortin Pergamino, Fortinera INTA (resistant to pea aphid), and San Martin (resistant to stem nematode) are recommended. 'Saladina Sintetica', 'Seleccion Salta', and 'Salinera INTA' (resistant to pea aphid and saline soils) are recommended for Santiago del Estero, Tucuman, Salta, and other provinces in the northwest.

After the detection of the pea aphid in Argentina in 1969 and the subsequent development of this insect as a major alfalfa pest, the resistant U.S. cv. Dawson, Kanza, and Washoe were introduced. These performed reasonably well in the central part of the Pampean Region and very well south of the Rio Colorado Valley. More recently, 'CUF 101', 'WL 309', 'WL 318', and 'WL 514' have been successful commercially. Because of its resistance to blue aphid (*Acyrthosiphon kondoi* Shinji), CUF 101 was registered in Argentina in 1976.

The new Programa Alfalfa received economic and technical assistance from 1972 to 1979 through an FAO/INTA agreement. Research was directed toward breeding cultivars with multiple pest resistance, evaluation of national and foreign cultivars under various cutting and grazing conditions, various aspects of establishment and management, pest control, identification and evaluation of principal diseases, and seed production technology.

At present, damage by pea and blue alfalfa aphids has declined in importance because of efficient biological control. Currently, emphasis is placed on providing sufficient quantities of well-conditioned seed of the improved cultivars which have been released by INTA, and on encouraging better inoculation practices (128). Bloat of cattle remains an important problem despite the trend to grow alfalfa in mixture with grasses. Increasing use is made of poloxalene, rotational grazing, efficient weed and pest control measures, and improved cultivars. The seed certification system of Argentina is governed by the Ley de Semillas y Creaciones Fitogeneticas no. 20. 247 promulgated by the national government in 1978.

Alfalfa research is conducted by INTA; some eight experimental stations and substations located in the main production zone; various departments of the Centro Nacional de Investigaciones Agropecuarias, Castelar (Buenos Aires); the Universidad Nacional de Buenos Aires, Facultad de Agronomia; and the Universidad Nacional de Cordoba, Facultad de Agronomia, Rio Cuarto.

2-3.4.2 Bolivia

Approximately 9500 ha of alfalfa are grown in Bolivia with an average fresh forage yield of 19.3 Mg/ha. About 90% of the crop is produced in the southern provinces of Cochabamba, Santa Cruz, Chuquisaca, and Tarija. Alfalfa is cut by hand close to the ground and fed to all classes of livestock including rabbits (*Oryctolagus cuniculus*) and guinea pigs (*Cavia* spp.). In the altiplano, both irrigated and dryland alfalfa fields are attacked by green pea aphids. 'Ranger' and 'Saranac' are grown although most of the fields are of the native criolla ecotypes. The role of alfalfa in N_2 fixation is not understood by most farmers. The importance of root and crown rot diseases are not recognized and stand loss is usually attributed to poor water management or mineral deficiencies. Blue aphids, pea aphids, and leafhoppers are serious problems at low elevations. Two to three cuttings per year are harvested in the altiplano, and as many as seven to eight cuttings at lower elevations. Alfalfa cultivar adaptation trials are conducted at experiment stations near Chinoli, Cochabamba, and Patacamaya.

2-3.4.3 Brazil

The total area of alfalfa in Brazil is 26 000 ha, with an average yield of 6.5 Mg/ha. The crop is grown in the states of Parana, Santa Catarina, Sao Paulo, and Rio Grande do Sul. Climatic conditions in the last named state are favorable for alfalfa and about 70% of the total area is centered there. Topographically, some of the area resembles the Argentine Pampas, and natural grasslands prevail. Rainfall is well distributed throughout the year, and frosts are common on uplands during winter. A thriving livestock industry has developed along with diversified farming (241). In Sao Paulo, four to six harvests can be made annually with yields of 8.0 to 13.0 Mg/ha dry matter possible. Alfalfa is hand sown or drilled into a well-prepared seedbed at a rate of 30 kg/ha (22). Phosphorus and K fertilizers are required at seeding but seldom applied thereafter. Locally developed strains of *R. meliloti* are available in Rio Grande do Sul (82).

Adaptation trials with European strains and variegated types were conducted as early as 1899 (270). Research on alfalfa has been conducted at experiment stations near Piracicaba and Sao Paulo.

2-3.4.4 Chile

Alfalfa cultivation in Chile is practiced under a wide range of ecological conditions. It can be grown from the valleys of the extreme north (15° S Lat) to the cold zones of the Magellanic Steppe (54° S Lat) (123, 125). However, the principal growing areas are from Region IV (Coquimbo Province, 30° S Lat) to Region VII (Talca Province, 33.5° S Lat), in the central irrigated zone. The area seeded to alfalfa represents 7.8% of the total sown pastures of the country and ranks third after mixed species pastures (302 301 ha) and red clover. Irrigated pastures represent 34% of the total sown area with alfalfa occupying 22.4% of these irrigated sown pastures.

Recent estimates indicate an annual seed sales volume of 350 000 kg. Considering an average life span of 5 yr for alfalfa pastures and a seeding rate of 20 kg/ha, it is estimated that there were 60 000 ha of alfalfa in Chile by the end of 1983. Alfalfa production has increased in the area from Linares to Los Angeles south of Regions VII and VIII, and in nonirrigated sections with a yearly rainfall of 1000 to 1200 mm. This recent expansion is due primarily to the introduction of cultivars that are winter dormant and adapted to colder temperatures (164).

The total area of irrigated alfalfa is almost 40% lower than that reported in 1965 (124). There are three main reasons for this reduction: (i) the expansion of urban development on lands with high-quality irrigated soils, especially in the vicinity of metropolitain Santiago; (ii) an increase in orchards and vineyards because of policies favoring agricultural export products rather than alfalfa; and (iii) changes in the land tenure system after recent land reforms.

The principal use of alfalfa in Chile is for dairy and beef cattle feeding, both for grazing and hay. Part of the alfalfa production is exported as baled hay and pellets. In 1980 to 1981, 7260 Mg of bales and 12 815 Mg of pellets were exported (190). Grazing may be intensive and regulated by tethering animals in fields. Rivadeneira and Davidovich (211) reported that 'Moapa' produced 8.3 Mg dry matter/ha in 142 d, with a meat production potential of 980 kg/ha when grazed by 6.6 animal units of 400 kg each. Fear of bloat restricts the grazing of irrigated alfalfa.

Nondormant cultivars yield 20% more than semidormant cultivars where adapted. Local ecotypes were popular until the mid-1950s. The Chilean cv. Rayen, developed by the Instituto de Investigaciones Agropecuarias (INIAO) from U.S. germplasm, is well adapted in the central zone. It is high yielding with an average production of 20 to 22 Mg/ha of dry matter; resistant to stem nematode, highly tolerant to root-knot nematodes, moderately susceptible to pea aphid and blue alfalfa aphid, and nondormant (238). Some American cultivars have been found to yield well in Chile if soils are not heavily infested with nematodes (100). 'WL 518', 'WL 514', 'WL 512', 'WL 321', and 'WL 318 are used extensively (164), together with American sources identified as 'California 40', '50', and '60'.

Persistence is an important problem in Chile, with stands normally lasting from 3 to 5 yr. Detrimental factors include pests, diseases, and improper management (16, 45, 92, 212). *Meloidogyne* spp. and *Pratylenchus* spp. are the most important root-knot nematodes, while *D. dipsaci* is the principal stem nematode (100). Although all three are widely dispersed within the main cropping area, their impact on production has not been evaluated fully. Insects such as pea aphid, blue aphid, and *Epinotia* spp. are important in some areas and in some years (92, 212). Management practices such as the height and frequency of cutting, phosphate fertilization, and irrigation procedures are of special importance; together with problems associated with establishment of alfalfa on marginal soils (low pH, high Al concentration, heavy soils with poor drainage) and weed control.

Present agronomic research is directed toward testing of newly introduced cultivars in different environments, establishment and management of alfalfa-grass mixtures (with tall fescue and orchardgrass), and evaluation of mixtures under grazing by beef and dairy cattle (238). Previous research on grazing, cultural practices, and seed production has been conducted at several experiment stations (25, 61, 69, 97, 189). National Institute of Agricultural Research, through its experiment stations at Santiago (La Platina), Chillan (Guilampapu), and Temuco (Carillanca), and the Agricultural Faculties of the Catholic Universities of Santiago and Valparaiso are the principal centers that conduct research with alfalfa in Chile. Private enterprises interested in alfalfa seed production, introduction, and sales as well as in technological improvements are Semilleros A. C. Baldrich and ANASAC (Agricola Nacional S.A.C.I., Santiago, Chile).

2-3.4.5 Colombia

Alfalfa is not widely grown in Colombia. A few large plantings are found in the savannah of Bogota and in the Cauca Valley at elevations of from 1000 to 2800 m. Historically, it was a principal crop of the densely populated plateau of the Department of Boyaca (77). Most of the small plantings, 1 ha or less, are harvested daily for green feed (60). Some silage is made from alfalfa alone or in grass mixtures (59). Irrigated alfalfa grows throughout the year and the erratic appearance of flowers is not a reliable indicator of time of harvest. Harvesting when the new shoots were 5 cm high resulted in 4 to 12 weeks between cuttings (60). Clipping according to shoot height provided high-quality forage and yields equal to those from other cutting frequencies.

2-3.4.6 Ecuador

Alfalfa production occurs mainly at elevations above 1000 m near large cities where it is used as green feed for dairy herds. Historically, 'Guaranda' alfalfa was an important high elevation ecotype in the province of Bolivar (40). Three local types are 'Abatoriana', 'La Nacional', and 'Morada', which probably trace to Hairy Peruvian germplasm. They are tall, erect, purple flowered, with considerable anthocyanin development on the stems. 'African' and 'Diablo Verde' are used for grazing and appear to tolerate trampling by livestock better than native alfalfa. The crop is sometimes grown with Italian ryegrass (*L. multiflorum* Lam.) which helps to reduce invasion of fields by the weedy kikuyugrass (*Pennisetum clandestinum* Hochst. ex Chiov).

A crop is harvested every 35 d in the northern part of the country and every 45 d in the south, with up to 12 cuttings per year. Stand life averages about 3 to 4 yr. Crown and root rots, crown wart [caused by *Physoderma alfalfae* (Lagh.) Karling], common leafspot, downy mildew, pea aphids, and nematodes are major problems.

Alfalfa seed is grown on marginal lands by small farmers who may produce as little as 10 to 30 kg of seed. Seed is rarely inoculated prior to planting. Research with alfalfa is conducted at the Santa Catalina experiment station near Quito.

2-3.4.7 Peru

Alfalfa was introduced in 1535 by a Spanish soldier, Cristobal Gago (58). It has since become the most important forage crop in the country with approximately 120 000 ha grown each year. There are two distinct growth types, a "summer" type used in warmer low elevation sites, and a "winter" type grown at higher elevations under cooler climatic conditions (295). The summer ecotypes are very pubescent or hairy, much like alfalfas from Arabia. These are often collectively identified as 'Hairy Peruvian' alfalfa. The finer stemmed winter types lack this pubescence. Gradations between these two extremes exist and an estimated 150 distinct ecotypes of alfalfa are believed to exist in Peru (161). These ecotypes range from the winter type, 'Llaclla', that is grown at 3500 m, to summer types, such as 'Tambo,' grown at about 200 m in the Tambo River Valley. They include strains resistant to salinity ('Agua Salada'), winter types with resistance to leaf disease ('Moro Moro'), as well as the typical Hairy Peruvian ('San Pedro').

Alfalfas most commonly grown in the coastal plain region include the Peruvian summer ecotypes, 'San Pedro', 'Monsefu', 'Santa Lucia', and 'Mochona', and the introduced cv., Moapa. In the highlands, the winter ecotypes 'Alta Sierra' and 'Yaraqua' are widely used. Three improved populations have recently been developed at the La Molina Experiment Station (294). These are the synthetics 'Chola' and 'Andina', and a broad-based selection, 'Improved San Pedro'.

Alfalfa management is dependent on elevation. Most of the crop is either fed green or grazed closely by tethered animals. When cut, up to eight harvests per year are possible at many lower elevation sites and up to 14 harvests are taken at locations near sea level. Irrigated and highly fertilized alfalfa in the southern desert region has a maximum growth rate of 200 kg of dry forage/(ha d) during summer and 150 kg during winter. These high rates are maintained for 10 to 20 d. Yields of 5.0 Mg/ha of dry forage in summer and 3.5 Mg/ha in winter were obtained in 53-d growth periods (268). In the south, alfalfa is used intensively with stands persisting as long as 6 yr.

'Wairau' alfalfa from New Zealand was introduced successfully into improved native pastures in the Altiplano near Puno by oversowing, mechanical seeding, and hand sowing (247). Dry matter production increased from about 0.6 to 6.0 Mg/ha in 2 yr and up to 10.0 Mg/ha in older swards. Inoculation was essential to achieve these yield levels. Other research with alfalfa has been conducted at experiment stations near Arequipa, Huancayo, Lambayeque, La Molina, Mal Paso, and Puno.

Major destructive diseases on alfalfa in Peru include *Cercospora medicaginis* Ell. & Er., downy mildew, and common leafspot (294). Downy mildew occurs mostly in the highlands, but the other two are nationwide problems. Forty-six insect species are known to cause forage yield losses in Peru. The most damaging are *Acythosiphon pisum* Harris, *Anticarsia gemmatilis* Hubner, *Contarinia medicaginis* Kieffer, *Empoasca kraeneri* Ross and Moore, *Epinotia opposita* Heinr, *Feltia experta* Walker, *F. subterranea* Fabricius, *Liriomyza huidobrensis* Blanchard, *Prodenia eridania* Cramer, and *Tetranychus telarius* L.

2–3.4.8 Uruguay

Alfalfa is the most widely grown legume in Uruguay. More than 25 000 ha of alfalfa is produced, with two-thirds utilized as pasture and the balance as hay. Production is concentrated in the provinces of Cancelones and Montevideo. It represents only 0.17% of the million ha in agricultural production, of which 85% is in pasture crops.

Alfalfa has not gained the prominence in Uruguay that it has in neighboring Argentina. In many areas, compacted, acidic soils and impermeable subsoils prevent satisfactory stands and reduce yields (24). Where it is seeded, alfalfa seldom persists more than 4 yr. The crop is harvested three to four times per year and produces 10.0 to 12.0 Mg/ha. Cultivars adapted to Argentina do well in Uruguay.

2–3.5 Africa

Alfalfa is grown in the northern, eastern, and southern regions of Africa. Most is grown under irrigation, with some dryland alfalfa in southeastern and southern areas. Information on alfalfa is limited. No written information was found on alfalfa in Sudan, Ethiopia, and Somalia. However, some alfalfa might be expected because it is grown in regions that have similar climatic conditions.

Diseases, insects, and usage appear similar over all of north Africa while a different set of problems and usage patterns prevail in southeast Africa. Only limited research is conducted on alfalfa in Africa. Although Egypt has a relatively small area planted to alfalfa, it seems to conduct more research than many other countries. In the near future, the countries of Africa may have to depend on Saudi Arabia, India, Australia, New Zealand, and USA for new information on alfalfa production.

2-3.5.1 Algeria

Five main geographic areas can be identified in Algeria: Coastal and subcoastal plains with sufficient rainfall for production without irrigation, Interior plains where irrigation is practiced, Dryland areas, Mountainous areas, and the vast interior desert with the occasional oasis. Most alfalfa is grown under irrigation in the interior plains. However, some alfalfa is grown in the dryland areas mainly in rows spaced 1 to 1.5 m apart, and at scattered oases. Annual medics are grown in areas with 150 to 500 mm annual precipitation.

Algeria has an estimated 10 000 ha of alfalfa which is planted in the spring and the fall. Areas of about 5 to 10 ha are planted by most private farmers with border-type irrigation. Larger plantings using borders and sprinklers are found on government farms. At the other extreme are the small, 1- to 2-ha plantings made at interior oases. Alfalfa is fed as long, fresh hay mainly to dairy animals in the north and to sheep, goats (*Capra hircus*), and donkeys (*Equus asinus*) in the south. Yields of 80 to 100 Mg/ha green weight can be expected. Most of the summer hay is used as green chop, and some alfalfa may be pastured during the winter (96).

Most of the old, original alfalfa cultivars have been replaced by foreign cultivars except in the oases. This replacement of local ecotypes probably took place more rapidly in Algeria than in other North African countries because the area was dominated by progressive French farmers for many years. At first, cultivars were mainly from France ('Provence', which has some dormancy) but now include introductions from Spain, Italy, and southwestern USA. African, Sonora, and Moapa from southwestern USA are some of the main cultivars under production

The main alfalfa problems include downy mildew, rust, aphids, *Sitona* spp., and salinity in the oases. An experimental program with the medics which appears promising and is being tested in growers' fields in Algeria and Tunisia is patterned after the *Medicago*/cereal rotation used in Australia (19, 47). When wheat is grown with self-seeding and self-regenerating medics, forage and cereal seed yields are high.

The National Institute for Agronomic Research (INRA), Algiers, is responsible for research, and The National Institute for Agronomy (INA), Algiers, is responsible for instruction.

2–3.5.2 Egypt

Alfalfa is grown only on newly developed land and at old oases. Some larger areas of production are the New Valley (near the Kharga Oasis west of the Nile River), Nubaria, and North Tahrir (south of Alexandria), the Aswan Region, and Ismaliya (88). Alfalfa in these areas is used for land reclamation and to provide forage for livestock. Egyptian clover or berseem (*Trifolium alexandrinum* L.) is essentially the only forage crop grown in the old lands of the Nile River Valley and Delta.

The 81 000 ha planted to alfalfa are used for pasture, green feed, and hay. About eight cuttings are obtained per year with a green weight of 81 Mg/ha (88). Frequent cuttings improve quality (206) but result in lower yields and reduced persistence.

Most alfalfa seed is imported from other countries, such as USA, Italy, and Argentina. Sonora, from southwestern USA, is the dominant imported cultivar. Cultivar use is expected to change depending on performance and seed availability. The local cv. Sewa and Hegazi have performed well in experiments, and a breeding program is in progress using these and other cultivars as parental material. Emphasis in the breeding program is being placed on production and pest resistance (205).

Diseases important on alfalfa are: Rhizoctonia root rot, Fusarium root rot (*Fusarium* spp.), Sclerotinia crown rot (*Sclerotium balaticola* Taub.), Stemphyllium leafspot (*Stemphyllium* spp.), downy mildew, rust, and common leafspot (204, 227). Major insects include: pea aphid, spotted alfalfa aphid, Egyptian alfalfa weevil (*Hypera brunneipennis* Boh.), and *Sitona* spp.

Alfalfa research is conducted at the Ministry of Agriculture Agricultural Research Center, The University of Cairo, and The Ain-Shams University (Cairo) and at research stations primarily located in the newly developed lands.

2–3.5.3 Kenya

Kenya lies on the equator where temperatures are fairly constant, and legume adaptation is a function of altitude. Alfalfa grows well at elevations up to about 3000 m. Alfalfa was introduced in Kenya about 1910 and cultivated around Kabete and Naivasha. Today, it is planted on about 120 ha around Naivasha, Narok, and Kitale (269).

Dryland alfalfa is planted at the start of the wet season, but irrigated alfalfa can be planted at any time of the year. All seed is inoculated with Rhizobium prior to planting and pelleted with lime when planted in soils with a pH of <6. A 60-kg preplant application of phosphorous (P_2O_5) is used. Seed may be planted in pure stands at 6 to 9 kg/ha or at 3 kg/ha with Rhodes or setaria grass at 5 kg/ha. The cv. Hunter River and Diablo Verde are planted at the lower elevations (0–2000 m) with rainfall between 650 and 1400 mm/yr, and Hairy Peruvian is planted at the higher elevations, 2000 and 3000 m (269).

Alfalfa is fed primarily to dairy cows as hay, silage, and green chop. It is also grazed by livestock. Alfalfa can be cut every 5 to 6 weeks in the cooler parts of the country. Stands of alfalfa were reported by Strange (250) to produce well under grazing for up to 4 yr. Rust and aphids have been the only plant pests recorded on alfalfa. Seed has been produced around Kitale at yields up to 100 kg/ha (269). Research is conducted on adaptation of cultivars, weed control, and grazing at the National Agricultural Research Station (Kitale).

2–3.5.4 Libya

Libya has a desert-type climate except for relatively small areas of low rainfall (250–500 mm) near the Mediterranean Sea in the vicinity of Tripoli and Benghazi. Some alfalfa is grown under irrigation in the two coastal areas and at oases located in the interior desert area. In one of these oases, Kufra, located in the southeastern part of the country, a highly productive, water-bearing aquifer is utilized to irrigate alfalfa and other field crops in a large land development program. Seeding rates of 28 kg/ha were used at Kufra for alfalfa, and 31.4 Mg/ha of dry hay obtained in nine cuttings per year (278). The cv. Moapa and Mesa Sirsa from southwestern USA were used by Worker (278), but local cultivars are used at most oases. Libyan cv. are Azzawia, Tajuria, and Megali. Alfalfa is fed dry or green to all types of large animals, including cattle, sheep, goats, and donkeys.

2–3.5.5 Morocco

Alfalfa was brought to Morocco from the east during the expansion of Islam. It is well adapted and grown under irrigation in all parts of the country. In addition, annual *Medicago* spp. are well represented in areas receiving 150 mm or more annual precipitation. Also, perennial species are found in certain parts of the country. The geography and climate of Morocco encourage species diversification because climatic zones vary from a Mediterrean-type climate near sea level, to a relatively cold climate at 4000 m where considerable snow may accumulate in winter. Annual precipitation ranges between 50 mm in the desert areas, 500 mm in the valley zones in the north, and to more than 1500 mm in the mountain zones.

Alfalfa is seeded on an estimated 30 000 ha in pure stands, in either the spring or fall. Field size varies from 1 to 2 ha in the oases to larger fields in other areas. The most important characteristic of alfalfa cultivars is nondormancy, but one or more cold-resistant cultivars, such as Demnat, are used. Local ecotypes are used in the oases, while in the north, imported cultivars from Spain, Italy, France, and southwestern USA have replaced most of these local strains. Local strains have been classified by Rumbaugh and Graves (219) into five groups designated by the geographic areas in which they are grown: Draa, Tafilalt, Sous, Mountain, and Haouz. As yet, there is no program designed to utilize the annual medics or the other *Medicago* spp.

Alfalfa hay is fed long (unchopped) and green primarily to dairy animals in the northern part of the country. It is fed in the same form in other parts of the country, but mainly to sheep and goats. Some pasturing is practiced in winter.

Production problems may include rust, downy mildew, a black aphid, and soil salinity (219). A heavy infestation of alfalfa by *Sitona humeralis* Steph. was found by Aeschlimann (2).

Alfalfa pollination by honeybees (*Apis mellifera* L.) in the irrigated arid lands was studied by Tasei (253). He found that pollen was collected by 86% of the foragers of the local honeybee, (*A. mellifera intermissa*), but that the tripping rate was only 20%. Bees from the relatively temperate region near Rabat foraged only on other plants, such as *Zea mays* L., *Amaranthus*, and *Datura*. Most seed is imported, except for that planted at oases. At oases, where about 45% of the land is planted to alfalfa, farmers produce much of their own seed.

The National Institute for Agronomic Research (Rabat), which has an extensive field station system, conducts much of the research of alfalfa. Hassan II Agronomic Institute, an educational institution, conducts work on the development of local Moroccan cultivars.

2–3.5.6 South Africa

The climate of South Africa varies from Mediterranean along the coast to subtropical inland. Alfalfa was first cultivated in South Africa during 1858 to 1861 along the Hex River and used as a forage crop for ostriches (67). Now, approximately 300 000 ha of irrigated and dryland alfalfa are under cultivation, with the highest concentrations in the Southwestern and Eastern Cape provinces (72%), Orange Free State (17%), Transvaal (9%), and Natal (2%) (67). In dry regions, irrigated alfalfa is used for hay production. In other parts of the country it is used for hay, pasture, green feed, and meal. Some stands are reputed to be about 80-yr old. In the southwestern part of Cape Province, alfalfa is grown in rotation with wheat and used for pasture. Seed production is concentrated in the Karoo region.

Alfalfa production increased from 448 000 Mg in 1951 to 1952 to 1 254 000 Mg in 1979 to 1980. Production per cutting in experiments cut at 3-, 4-, and 5-week intervals was 13.3, 16.5, and 19.2 Mg/ha of dry hay, respectively (144). The most widely grown cultivar is South African Standard, a product of selection among introductions of 'Provence', 'Hunter River', and 'Chinese' which were brought in prior to a seed embargo imposed in 1921 to keep out bacterial wilt and virus diseases (66). However, the embargo was partially lifted after the blue alfalfa aphid was reported for the first time in April 1981 in the Barkly West and Dordrecht districts of South Africa (76). Damage from the blue alfalfa aphid was severe and cultivars resistant to this insect, such as CUF 101 from southwestern USA, were imported. Importation of commercial quantities of seed into South Africa is controlled by the Lucerne Seed Control Board. This board maintains a list of acceptable cultivars that may be imported. In mid-1986, CUF 101, 'Granada', 'Baronet', and 'Pierce' were on the approved list (237).

About 95% of the alfalfa seed is produced in the Klein Karoo region of Cape Province with most of the remaining seed coming from the Orange Free State (237). Total annual seed production varies from below 1000 to 3000 Mg (67). The annual consumption is about 1800 Mg, and a substantial carry-over is desirable because seed production is highly variable from year to year. The average yield of alfalfa seed is 320 kg/ha.

Phoma medicaginis Malbr. & Roum. var. *medicaginis* Boerema was the most common disease reported on annual *Medicago* spp. (155). Other diseases include Leptosphaerulina leafspot, anthracnose, and *Fusarium* spp. (156, 157, 256). Restrictions on imported seed has been one approach to controlling seed-borne pests. Bacterial wilt was found in about 1969, and efforts have been made to find resistant cultivars through testing of introduced cultivars. Also, cultivars are evaluated for resistance to Verticillium wilt. 'Atlantic', 'Beaver', 'ATCal5', 'Poiton', 'Kane', and 'A44920' were found to have an appreciable degree of resistance (13). The blue alfalfa aphid was reported to be especially abundant in the fall (76) and to cause complete defoliation at times. Weeds become a problem after the deterioration of alfalfa stands, but often some weeds are tolerated and the reduction in forage quality accepted. However, the highly objectionable nutgrass (*Cyperus rotundus* L.) has been reported in the Pretoria area. Strong stands of alfalfa effectively controlled this weed through shading. No other biological factors investigated seemed of value in control.

Most research reports pertain to the feeding and utilization of alfalfa by cattle, sheep, and pig (*Sus scrofa domesticus*). Research on cultivars has dealt primarily with testing cultivars for resistance to important diseases and insects. Research is conducted by the many Agricultural Research Institutes located in the alfalfa-producing areas, and at the Agricultural College, Univ. of Stellenbosch, Stellenbosch, Cape Province.

2-3.5.7 Tanzania

The climate of Tanzania is mostly tropical. However, alfalfa is becoming an important crop in some of the cooler and dryer areas where livestock are raised. In East Africa, alfalfa is usually managed independently from the other crops and crop rotations (249). It may be grown in pure stands or temporary pasture mixtures with grasses on well-drained, fertile soils. As a rule, it is cut green and fed to dairy cows or other animals to supplement forages of lower nutritive value. Some alfalfa or alfalfa-grass mixtures may be cut at an early stage and dried to provide supplemental feed of high quality.

Rhizobial inoculation is needed at seeding (56). Alfalfa is grown alone as well as with other grasses such as *Chloris gayana* Kunth. and *Ceuchrus ciliaris* L. The yields of alfalfa ranged from 2 to 12 Mg/ha of dry matter. When alfalfa was sown in mixtures, it made little contribution to the yield of the mixture (9). In the central and northern parts of Tanzania, alfalfa produced much less than *P. purpureum* Schumach and *Setaria splendida* Stapf.

Alfalfa mosaic virus was reported by Kaiser and Robinson (142) as a potentially important virus in Tanzania, but field surveys in Kenya and Tanzania indicated its distribution might be restricted. Root nematodes (*Vanderlindia venata* n. sp. and *Xiphinema heynisi* n. sp.) were found in alfalfa soils and around roots of alfalfa at south Kilimanjaro by Siddiqi (228, 229). Results of research on alfalfa have been reported by The Dar es Salaam Univ., (Morogoro) and the Northern Research Center (Tengeru).

2–3.5.8 Tunisia

The climate of Tunisia is mild and rainfall limited except in the north and northwest. Annual precipitation ranges from 900 mm in the northwest to <100 mm in the south with most of this coming in the winter and spring. Alfalfa is grown primarily in the dryer irrigated areas, whereas annual medics are found in areas where rainfall ranges from 150 to 500 mm per year. Some alfalfa is grown in unirrigated areas in northern Tunisia. Yields of 11 to 33 Mg/ha fresh wt. were obtained under dryland conditions. Alfalfa is grown at oases in the desert areas of the south.

Spring and fall seedings are made on areas varying in size from large government projects, to the 5- to 10-ha areas on private farms, and 1- to 2-ha areas of basin-irrigated alfalfa at oases. About 12 300 ha of alfalfa were grown in 1981 using cultivars imported from France, Italy, Spain, and the Southwestern USA. However, locally produced seed of native ecotypes is grown at oases. Introduced cultivars started replacing local cultivars about 20 yr ago and have replaced most of them in the northern part of the country. Cultivars include Gabes, a southern ecotype, and the introduced cv. African, Moapa, Sonora, UC Salton, and CUF 101 from southwestern USA.

Alfalfa is fed primarily to dairy cows in the green, long (unchopped) form, with some pasturing in the winter. Production of green alfalfa can be as high as 80 to 100 Mg/ha. Production problems include downy mildew, rust, spotted alfalfa aphid, black aphid (*Aphis* spp.), dodder, salinity in the oases (96), and *Sitona* spp. (3).

Research has been implemented on wheat/medic rotations similar to that tried in Algeria where wheat is grown with self-generating medics which provide fodder and supplemental N to the wheat (7). Teyssendier de la Serve and Boutin (255) selected within the cv. Gabes for germination under high levels of salinity. As germination increased, there was an increase in vigor and recovery, and a decrease in cold tolerance.

Research in Tunisia is conducted by the National Institute for Agronomic Research (INRAT), with instruction at the National Institutes for Agronomic Teaching (INAT).

2-3.5.9 Zambia

Alfalfa is recommended for planting on a wide range of soil types, primarily in the dryer areas of Zambia where it is grown under irrigation, with limited use elsewhere. Planting is done in the fall (March–June) at seeding rates of 15 to 25 kg/ha of inoculated seed. Cuttings are made 8 to 12 weeks after emergence (257). Good stands of alfalfa will produce 2.0 to 4.0 Mg/ha of dry hay per cut, and produce from 12 to 20 Mg/ha of green feed. Six to eight cuttings are obtained each year, at 4-week intervals during the dry period and at 8-week intervals or more during the wet season. In general, P and B are needed, and N may be applied if a deficiency is suspected. Most alfalfa is cut and fed green to livestock, with limited use of dry hay. Bloat can be a problem when animals are fed green hay. Diseases most likely to occur during the rainy season are common leafspot, pseudoplea leaf spot (*Pseudoplea medicaginis* Miles), anthracnose and, occasionally, rust.

2-3.5.10 Zimbabwe

Although alfalfa can be grown through much of the country, it is produced primarily under irrigation in the dryer, cooler regions. About 1550 ha of alfalfa were grown in 1969. Hay and alfalfa meal are the principal uses of alfalfa, with some green-cut forage used on farms. Alfalfa is seeded in rows spaced 20 to 50 cm apart at rates of 10 to 12 kg/ha. Yields range from 18.0 to 22.5 Mg/ha. Davis (64) has shown that 17 to 30 cm of water were required for optimum production. Efficient irrigation provided extra cuttings during the cooler months, but reduced stand longevity (158). Hairy Peruvian, Galilee, Hunter River, Provence, and South African Standard are leading cultivars. Diseases prevalent in the country are common leafspot, rust, anthracnose, spring black stem, and downy mildew.

2–3.6 Asia

Siberia probably is the largest alfalfa-producing area in Asia. Although records on forage crops are difficult to obtain, the area planted to alfalfa seems to be increasing in other parts of Asia. Only countries where information on alfalfa could be obtained are discussed. Fragmentary information from Afghanistan, Syria, Lebanon, Oman, and other countries was insufficient to develop reports. Almost all alfalfa in Asia is produced in the colder and/or dryer parts of the continent. In many of these areas it is grown under irrigation, especially at southern latitudes. No alfalfa is reported from the warm, wet southeastern areas of Asia, except for one area in northeastern India. New lands have been developed in the oil-rich countries of southwest Asia. Alfalfa is an important crop because of their dependence on animals for transportation, field work, meat, and milk.

Information on diseases and insects was difficult to obtain from southwestern Asia. Other *Medicago* spp. are regular components of the endemic vegetation over the vast regions of the Asian continent. Leaders in research in Asia are the Soviet Union, India, Japan, and Saudi Arabia.

2–3.6.1 India

Alfalfa was introduced into India in about 1900. It is now a popular and widespread forage capable of withstanding hot climates (186). Alfalfa is grown in two main areas. The largest, most productive area is in the northeastern plains of Punjab, Haranya, Western Uttar Pradesh, Rajasthan, Maharashtra, Gujarat, and parts of Madhya Pradesh where it grows from October to the end of June at the low elevations and from May to October in the high elevations. In the northeastern plains, alfalfa is grown as a winter crop because of stand deterioration in the hot, humid environment of the rainy season (143). Most alfalfa is grown under irrigation, but is usually rain-fed in the temperate areas.

About 58 000 ha are planted to alfalfa. Seedings at a rate of 7 to 9 kg/ha are made from September to November with an optimum of mid-October in the Punjab. First cuttings are made about 10 to 12 weeks after planting, with subsequent cuttings at intervals of 30 to 40 d. An average of 8 to 10 cuttings may be obtained, with dry matter yields of about 8.0 Mg/ha.

The old, traditional cv. are Kandhar or Quetta, Persian or Arabian, and Meerut. The first two Indian-bred cv., Sirsa 8 and Sirsa 9, were released by the Fodder Research Station, Sirsa, Haryana. Sirsa 9 is used widely, with several new cultivars available for planting i.e., Anand-2, IGFRI 244, NDRI Selection No. 1, LLI, C4 and LL Composite-5.

Alfalfa is used mainly as green fodder for cattle, horses, and other livestock. In the vicinity of villages, it is in great demand for horses. Because alfalfa is nutritious, with 15 to 20% protein on a dry weight basis, it is used in small quantities when fed to livestock. A daily ration of 1 to 2 kg of green alfalfa fed with low-protein roughages, such as cereal straw, is considered ample for a milk cow or working bullock (186). Farmers began to recognize alfalfa as a good feed for dairy cows when it was shown that danger of bloat is removed by limiting alfalfa to no more than one-third of the forage.

Important diseases in India are: downy mildew, bacterial wilt, rust, common leafspot, and Fusarium root rot (*Fusarium* spp.). The latter root rot was found important in stand deterioration in the temperate northeastern region by Kamalakar et al. (143). Alfalfa in this region is grown as a winter annual because of stand deterioration in the hot and humid rainy season. When alfalfa was planted in November, no disease injury was apparent during the winter period. However, lesion damage was apparent at the onset of the high summer temperatures and humidity. Stands were lost by the end of two or three summers.

Insects of economic importance are: the alfalfa weevil, spotted alfalfa aphid, leafhopper (*Empoasca* spp.), aphid (*A. craccivora* Koch), and the alfalfa treehopper (*Tricentrus bicolor* Dist.). The alfalfa treehopper is most damaging during September and October. It is controlled with insecticides (222).

In research on the mechanism of antibiosis and tolerance to the alfalfa weevil, Gerewal and Dhaliwal (87) found the life cycle from egg to pupa was longest on the cv. LLI, with no pupation on 'Arc' (which exhibited the highest level of antibiosis) and LLI. Arc was most resistant to the alfalfa weevil, and 'T9' was highly susceptible.

Research is conducted on forage quality, leaf protein, fertilizers, nodulation, alfalfa pests, seed production, and cultivar improvement and testing. In research on the yield of extractable leaf protein, Dev et al. (68) found that protein yield was affected by season and dry matter production. Factors that increased dry matter yield increased leaf protein. Joshi (137) found that alfalfa produced more leaf protein than other crops, such as, mustard, wheat, and berseem clover. Batra et al. (20) found extractable N varied with the fiber/juice ratio, time of day, temperature of storage, and length of storage. Extracted N was found to decrease 12 to 45% when fractionation was delayed 4 h at room temperature. In related work, Sundaram et al. (252) found that crude protein increased to about the fourth cutting and then decreased progressively, and that applied P and K had no effect on crude protein.

In experiments on nodulation, Lakshmi-Kumari et al. (154) found that increasing levels of salinity and alkalinity in the root medium resulted in a root system devoid of hairs, mucilaginous layers, and infection thread formation despite good growth of *R. meliloti*. Also, they demonstrated the existence of alkali-sensitive stages in the early periods of nodulation.

Seed production is a problem in India. Seed research, centered around the search for effective pollinators, is conducted at the Punjab Agricultural

University, Ludhiana, Punjab. Research at the Fodder Research Institute, Jhansi, Uttar Pradesh, involves many subject matter areas, including work on insects, diseases, and cultivars. Research on leaf protein and nodulation has been conducted at Marathivada University, Aurangabad and the Agricultural Research Institute, New Delhi, respectively.

2-3.6.2 Iran

Alfalfa is grown throughout most of Iran, with major producing areas in the central province (Teheran) and in the provinces to the west and northwest of Teheran. Little alfalfa is grown along the Caspian Sea and Persian Gulf. An estimated 270 000 ha were planted to alfalfa in 1977 with an average production of about 3.8 Mg/ha. Yield was highest in the warmer regions of the southwest and lowest in the northwest. Production on experimental plots at Karaj (near Teheran) ranged between 18.5 and 30.0 Mg/ha of green matter over fertilizer and cropping treatments during the 1st yr, and totaled 77.3 Mg/ha in 3 yr (17).

Alfalfa is fed green on small farms and oases to all types of farm animals, i.e., sheep, goats, cattle, and camels (*Camelus* spp.). Near large cities and in large land developments much of the crop is dried and used on dairies, or sold as a cash crop. Local Iranian cv., such as, Yazd, Hamadan, and Yasdi are used, together with seed imported from other countries. Major pest problems are: alfalfa stem nematode, *Sitona* spp., Egyptian alfalfa weevil, blue alfalfa aphid, pea aphid (*A. craccivora* Koch), alfalfa mosaic virus, Phytophthora root rot (*Phytophthora megasperma* Drech.), Rhizoctonia root rot, Fusarium root rot, and dodder (*C. approximata* Bab.). Research on alfalfa is conducted at the College of Agriculture, Teheran University (Karaj) and other universities and experiment stations. Much of the research involves utilization by various animals of alfalfa produced, handled, or fed in different ways. A second large area of investigation deals with the biology, ecology, morphology, and/or control of alfalfa pests. Limited research has been reported on cultivar evaluation, soil fertility, and water usage.

2-3.6.3 Iraq

The climate of Iraq ranges from the dry, desert region in the south, to a central steppe zone, and a mountainous, forested region in the north and northeast. Most alfalfa is grown in the latter region which is adjacent to the good alfalfa-producing areas in Turkey and Iran. Alfalfa, with a reported green weight production of 45 to 101 Mg/ha, is considered one of the major forage crops in Iraq along with berseem clover, maize, sorghum [*Sorghum bicolor* (L.) Moench], and barley (107, 230).

The medics with reported dry weight yields from 2.1 to 2.6 Mg/ha have shown promise as forage crops in northern Iraq, in an area with a rainfall of 300 to 500 mm/yr (202). Alfalfa is cut and fed green or grazed by cattle, camels, sheep, and goats.

Inoculation with *Rhizobium* is recommended for best production (105). Cultivars mentioned in the literature are Moapa, Sonora, India, St. Louis No. 7, and Local (6). When these cultivars were tested for their reaction to salinity, none performed well at levels of salinity above 4dS m^{-1}, but Sonora and St. Louis no. 7 gave improved production with less water uptake. Hassawy (107) found that stands irrigated weekly produced well for the first 2 yr, but less during the 3rd yr because of loss of stand and invasion of weeds. Stands irrigated bi- and tri-weekly gave slightly lower production, but good stands were maintained. In addition to soil salinity, alfalfa weevil (*Hypera* spp.) and dodder (*C. chinensis* Lam.) were considered problems in alfalfa production.

Research on alfalfa and the medics is conducted at Mosul University Hamman Al-Alill (Mosul) and The Univ. of Baghdad (Abu-Ghraib).

2–3.6.4 Israel

Alfalfa usage is limited in Israel because its water requirements are high. However, 1700 ha of alfalfa were grown in 1982 and an estimated 2000 ha were in production in 1984. About 1100 of the 1982 hectarage was on large cooperative farms, and the remainder on small farms. Alfalfa plantings are distributed through the nonmountainous areas of the country with the largest portion concentrated in the Jordan Valley (north and south) and extending southward to the "Beit-She'an" and "Bikat-Hayarden" regions. Another concentration is in the "Arava" desert, and since 1980 alfalfa production has been extended into the Negev, an arid area in southern Israel (99).

About 80% of the alfalfa plantings are made in the fall (October and November) with the remainder seeded in the spring (February to mid-April). Rates of planting are from 30 to 45 kg/ha. All alfalfa is grown under sprinkler irrigation, and experimental work has been started on subterranean drip irrigation. Phosphate is used on all alfalfa plantings, and K must be applied on most fields. The only cultivar used in Israel is Gilboa which was developed from 'Hairy Peruvian'. Cultivar trials are being conducted, but no cultivars have been found superior to Gilboa (99.)

An estimated 85% of the fields are cut for hay in summer, with the remainder green chopped. During the winter when sun curing is difficult, alfalfa is harvested either as green forage or ensiled. An annual production of 14 to 20 Mg/ha of dry matter is obtained from 8 to 11 cuttings. Almost all alfalfa is fed to dairy cows, with small amounts used by sheep, horses, and zoo animals. Dehydrating of alfalfa was discontinued as the cost of energy increased (99).

Seed production is adequate for domestic needs, and surplus seed is exported. Seed is imported only when supplies are short. Seed is produced on hay fields at a time when cotton uses all the available water on farms. Seed yields range from 300 to 800 kg/ha with a mean yield of 420 kg/ha. Forage production of fields used for seed production is 10 to 14 Mg/ha dry matter (99).

The most important pests of alfalfa are field mice (*Microtus guentheri*) and the Egyptian armyworm (*Spodoptera littoralis*). Commercial damage from other pests is marginal except in the Arava region where aphids are a problem (99). Research is concentrated on irrigation, cultivar testing, and weed control. Alfalfa research is conducted at the Hebrew University (Rehovot), and research stations where alfalfa is grown.

2-3.6.5 Japan

Alfalfa and grasslands are becoming increasingly important in Japan. Part of this is due to a reduction in the area planted to rice (*Oryza sativa* L.). Recommendations for the future are to convert rice fields to pasture, to produce feeds, and to use alfalfa in dairy rations (191). Alfalfa can be grown from north to south in Japan, with the largest area found in the cool climate of the northern island of Hokkaido. Alfalfa is grown in the lowland plains of northeastern Honshu around Sendai, the central plain surrounding Tokyo Bay area, the southern Houshu plain near Osaka-Kyoto and Nagoya, the southwestern Honshu area near Hiroshima, and southern Kyushu.

Experiments with introduced cultivars have been conducted to determine their adaptation to various areas of Japan. Natsuwakaba and Kitawakaba are cultivars that were developed in Japan for north and south Japan, respectively (121, 173). Most alfalfa is used in mixtures with grasses and fed to animals as hay or silage, primarily to dairy animals (191).

Insects reported on alfalfa are: blue alfalfa aphid, spotted alfalfa-like aphid (*Therioaphis trifolii* Monell), weevil (*Hypera* spp.), and sitona (*Sitona* spp.). The blue alfalfa aphid was found on all three main islands (91), the spotted alfalfa aphid in Yamaguchi Prefecture near Fukuoka City (287), and the weevil and Sitona in Hokkaido (220). Diseases reported on alfalfa are: Fusarium root rot, Verticillium wilt, Phytophthora root rot, southern blight (*Corticum rolfsii*), and alfalfa mosaic virus (121, 122, 147, 175, 176, 223).

A large amount of research can be identified at several research organizations and universities located near areas where alfalfa is of interest. Research is conducted at the Hokkaido National Agricultural Experiment Station on seed production and seed characteristics, quality of forage, preservatives in silage, forage production, and alfalfa breeding and diseases. The Shintoku Animal Husbandry Experimental Station and the Obihiro University of Agriculture and Veterinary Medicine are conducting research on alfalfa winter injury (150, 261). The College of Dairying, Ebetsu (Hokkaido) has a primary interest in forage quality. The

National Grassland Research Institute, Nishinasuno (Tochigi) has conducted research on alfalfa growth and physiology, cultivars, diseases, and the effect of soil conditions on plant composition. The University of Matsudo (Chiba) conducts research on the effects of cultural practices on alfalfa, physiology, and production. Research conducted on the southern island of Kyushu by different organizations involves alfalfa physiology and the use of phosphate and lime (120). In addition, research on alfalfa is found at universities and public research organizations in the cities of Sendai, Nagoya, Kyoto, and Hiroshima. Current and projected research can be expected to expand the use of alfalfa.

2-3.6.6 Pakistan

Alfalfa is grown to a limited extent in Pakistan, in spite of its potential as a forage crop (146). Major drawbacks to increasing alfalfa production have been poor seed yields and poor adaptation of imported cultivars. It is grown mainly on the irrigated plains and in some rainfed areas. Alfalfa is cut five to seven times per year with an annual production of 70 to 80 and 50 to 60 Mg/ha under irrigated and rainfed conditions, respectively. Forage is fed green and utilized primarily by draft animals (145).

Alfalfa is seeded at 12 to 15 kg/ha in October and November, and again from mid-February to March. Cultivars include Type 8, Type 9, and Synthetic 78 in the irrigated plains; El Unico, Kandhari, and Mesa Sirsa in the rainfed Pothowar; and Type 18, Punjab Selection, and Mesa Sirsa in the Mediterranean-type climate (145). The aphid (*A. craccivora*) is an important pest on alfalfa. Alfalfa research is conducted at The National Agricultural Research Center (Islamabad), and The Fodder Research Institute (Sargodha).

2-3.6.7 People's Republic of China

Alfalfa, introduced from Iran (Persia), has been grown as a cultivated crop for more than 2000 yr. It was first grown in the Yellow River Valley and is now cultivated in 14 provinces (Xinjiang, Gansu, Ningxia, Qinghai, Liaoning, Jilin, Heilongjiang, Shaanxi, Shanxi, Shandong, Hebei, Henan, Jiangsu, and Anhui), and in the Neimenggu Autonomous Region (112, 281). Alfalfa was planted on an area of 960 000 ha in 1983. It is grown in the semiarid, and alpine climates of northwestern (64%) and northern China (4%), in a broad area of the northeast (29%), and in central and southeastern China (3%) (281).

The climatic conditions prevailing in the alfalfa-growing areas vary considerably from winter to summer. In general, the temperature decreases from south to north and rainfall decreases from the southeast to the northwest. For instance, the mean annual temperature in the northeast is about 5°C with variations for −30°C in January to 20°C in July, while in the north and northwest, the mean annual temperature is around 7 to 10°C. The annual average rainfall fluctuates between 300 and 600 mm (281, 296).

About half of the grown alfalfa is used for hay and half for pasture. In some areas, alfalfa is used in rotation with grain crops to improve soil fertility and structure and as feed for livestock. Fertilizers used for alfalfa are either solid manure or calcium superphosphate or both. In experimental plantings which may indicate grower practice, Wu Renrun (279) planted 8 to 25 kg/ha from 1 April to 15 May in Hubei Province. Hay cut two to four times per year produced between 3.8 and 15.0 Mg/ha. Seed yields are 150 to 225 kg/ha with high percentages of hard seed (112).

In the Northern and Western Plateau regions, the winter-hardy cv. Gansu, Yimeng, Zhaodong, and Variegated and the introduced Canadian cv. Rambler have done well. Gongnong No. '1' and '2' are grown in the northeast. Yellow alfalfa (*M. falcata*) is adapted to most of the same areas as common alfalfa. Since it is more resistant to cold and drought, it is grown at high altitudes and areas with cold winters, such as, the Ili River Valley, West Xinjiang, and northern areas of Neimenggu, Heilongjiang, and Hebei. The cv. Xinjiang Large Leaved and Gongnong No. 1 are fast growing and high producing. Five species of *Medicago* are found in China, *M. archiducis-nicolai* G. Sirjaev (Gansu, Qinghai, and Xinjiang Provinces), *M. edgeworthii* Sirjaev (northern Yunnan), *M. lupulina* L. and *M. minima* L. (throughout China), and *M. polymorpha* (all regions except north China) (281).

From 1976 to 1983, 147 regional and local cultivars have been identified (List of National Forage Grasses and Feed Crop Varieties and Germplasm Resources, 1983, p. 233–268). In support of alfalfa breeding work, research has been conducted on male sterility in alfalfa (282). These studies led to the development of a hybrid strain of yellow alfalfa with drought resistance, winterhardiness, high production and adaptation to the Tongde Range of the Qinghai Plateau at an elevation of 3200 m. In addition, yellow-flowered alfalfa has been crossed with *M. sativa* to develop cultivars with improved winterhardiness and drought resistance (112). In other breeding research, tissue culture is used to produce plants from radiated and untreated pollen (286). To preserve germplasm, the People's Republic of China maintains a collection of *Medicago* spp., cultivars, and ecotypes. This collection was increased by 221% from 1976 to 1981 (281).

Common leafspot, rust, and downy mildew (*Peronospora aestivalis* Syd.) are found throughout China. Yellow leafspot (*Pyronopeziza medicaginis* Fckl.) is found primarily in the northwest, whereas Sclerotinia crown rot and alfalfa mosaic are found in the central-producing area in the provinces of Hubei and Henan (281). The stem nematode (*Ditylenchus triformis*) and root rots have been reported recently by Mi Shuzheng (181). Insects important in Gansu, northwest China are black bean aphid

(*A. craccivora* Koch), green leafhopper (*Cicadella viridis* L.), pasture mirid (*Lygus pratensis* L.) and alfalfa bud worm (*Heliothis dipsacea* L.) (167). In Hubei, blister beetle (*Epicauta gorhami* Marsel) and leaf beetle (Rhaphidopalpa femoralis Motschulsky) are important insect pests. The black striped leafbug (*Adelphocoris suturalis* Jakovlev) is broadly distributed. Dodder (*C. chinensis* Lam.) is found in many producing areas (280, 281, 288).

Research is conducted on improving the production of carbohydrates and protein in alfalfa at different stages of growth. Increasing plant longevity is an important objective of this work, and transpiration rates have been investigated (284, 285). Applied research is conducted on planting methods to improve survival and forage production in the severe climates of northwest China. In comparison of trench vs. flat planting, winter survival was increased by about 60% when plantings were made in trenches. Also, winter survival was better on sandy soils, when alfalfa was uncut, and where yellow alfalfa cultivars were used (280). In comparisons of winter coverings of alfalfa with snow, sheep manure, and uncovered check, it was found that winter survival was greatest under sheep manure. Winter survival was best in the yellow alfalfa (62). Interplanting alfalfa with grass species (*Dactylis*, *Festuca*, and *Elymus*) increased production from 12 to 26% (281).

Research is conducted by three institutes under the administration of the Chinese Academy of Agricultural Sciences Beijing: The Lanzhou Institute of Animal Science (Lanzhou), The Institute of Animal Science (Beijing), and The Institute of Grasslands (Huhehot). Additional research is conducted at many colleges and universities located in provinces where alfalfa is grown (281).

2-3.6.8 Saudi Arabia

Alfalfa has been cultivated in Saudi Arabia for a long time, and it is the only forage crop of consequence. It is planted on about 8300 ha of newly reclaimed land and oases. Alfalfa is cut 12 to 14 times/yr and produces about 35 Mg/ha of dry matter per year (113). The crop fits well into rotation with cereals and other field crops, and is used as green forage, silage, and dry hay.

For maximum production, plantings are made in the autumn; early October is optimum. Seeding rates are high and range between 50 and 60 kg/ha (113). A local oasis ecotype, Hasawi, has performed best in experiments with introduced cultivars, but seed is in short supply. Hasawi alfalfa appears to be better adapted to the high levels of soil salinity present in most soils, the local strains of *Rhizobium*, and perhaps the diseases and insects prevalent in the area (79, 113).

The most important problem in the production of alfalfa on new desert land is reclamation, which involves leveling, leaching, and possible deep tillage in order to produce good crops of alfalfa. Reducing levels of salinity is important in older fields. Because irrigation water is scarce, preliminary investigations are being conducted on the use of highly saline water for irrigation.

One of the most important insect pests is *Tripips tabaci* Lind which can cause most damage in the spring. Pea aphid, cowpea aphid (*Aphis craccivora* Koch), and the spotted alfalfa aphid are common. Other important pest species are *Autographa gamma* (L.), *Hypera postica*, root knot nematodes (*Meloidogyne* spp.) and stem nematode (*D. dipsaci*) (12).

Disease seems to be less important than insects in Saudi Arabia. The most commonly recorded disease is rust. Common leafspot, downy mildew, damping off incited by *Fusarium* spp., and Rhizoctonia root rot have been reported (8, 113, 184).

An excellent research program has been in progress under the direction of the Ministry of Agriculture and Water, in cooperation with the Univ. of North Wales and FAO. All aspects of production and utilization of alfalfa and other forages are investigated. Most reported research has been conducted at the Hofuf Research Center located about 75 km from the Persian Gulf.

2-3.6.9 Turkey

Scientists are in general agreement that eastern Turkey is on the fringe of the distribution range of wild alfalfa. There is an abundance of wild alfalfas in this area, but only *M. sativa* L. has been cultivated. Alfalfa, called *yonca* or horse hay, has had a long history as a cultivated crop in Turkey.

Alfalfa is grown throughout Turkey. The old area of cultivation has a continental climate, but recently the cultivation of alfalfa has extended into the subtropical coastal region. In 1969, the total area planted to alfalfa was estimated at 73 700 ha. Green forage yields from 8 to 10 cuttings per year in coastal areas may reach 26.8 to 35.8 Mg/ha, but are relatively low at higher elevations. About 95% of the hay is hand harvested. Hay is stored as loose hay or in hand-rolled bales in stacks for winter feeding.

There are two main ecotypes in Turkey; the best known is 'Kayseri' which is a mixture of types thought to have been introduced from Iran, between 1400 and 1600 B.C.. It is productive, moderately winterhardy, drought resistant, long lived, with good recovery after cutting. It has broad leaves and exhibits some disease and frost resistance within its range of adaptation. The wide use of Kayseri has led to the abandonment of land races in many villages (236). The other ecotype, 'Yerli Yonca' is grown in eastern Turkey. It is characterized by slower growth and lower yields in comparison with Kayseri. The first Turkish-developed cv., Bilensoy 80 was released in 1981. It was selected from the Kayseri-type alfalfas.

Seed production of alfalfa is most important in the central province of Kayseri where ecological factors (low relative humidity, sunny days, great differences between day and night temperatures, and a high degree of insect activity) are suitable for seed set. Some seed is produced in eastern regions.

Insects observed in the field in Turkey, and not positively identified were: alfalfa weevil, pea aphid (green form) (*Acyrthosiphon pisum* Harris), pea aphid (red form) (genus unknown), snout beetles (*Sitona* spp.), and plant bugs (*Lygus* spp.) (78). Severe infestations of stem nematode and dodder have been observed in forage and seed fields in Kayseri.

The main center for research activity in Turkey is Cayir Mera Yem Bitkileri ve Zootekni Arastirma Ensitusu, P.K. Y53, Ankara.

2-3.7 Oceania

The period since the early 1970s has been one of change for alfalfa in Australia and New Zealand. After a long period during which only local strains of alfalfa were grown, these areas were confronted with a series of crises resulting from a build-up, often gradual, of introduced and native diseases and insects. In addition, severe drought affected parts of both countries. Research initiated in response to these problems led to an improved understanding of alfalfa pests, the development of resistant cultivars, and the adoption of better management practices.

These countries have organized effective research programs that will support the expansion and use of alfalfa.

2-3.7.1 Australia

Alfalfa was introduced into Australia in about 1806, soon after the first settlement of the colony of New South Wales. These first introductions were probably of French origin since France was the most important seed-producing country at that time (177). Hunter River, which was the principal cultivar in Australia for many years, seems most closely related to the French or Mediterranean-type 'Provence'. It has wide adaptation in Australia, probably due to many years of natural selection. It performed well until the introduction of the spotted alfalfa aphid and the blue alfalfa aphid in early 1977.

Alfalfa is grown widely in the eastern, southeastern, and southwestern areas of the country from the tropics in the north to a Mediterranean-type climate in the south. In 1982, production from pure stands utilized for hay was concentrated in New South Wales (58 200 ha), Victoria (20 300 ha), Queensland (18 600 ha) and South Australia (14 400 ha). No data were available for alfalfa planted in mixtures with other species, except for a reported 15 700 ha in South Australia. As indicated by these data from South Australia, inclusion of alfalfa in mixtures would result in higher figures for hectares.

In southern Queensland where the annual rainfall exceeds 430 mm and where winter frosts are common, alfalfa is valued for its cool-season growth. Its ability to grow at low temperatures when soil moisture is adequate gives alfalfa an advantage over subtropical grasses in a mixture, but during the summer these species are very competitive. This competition combined with root, crown, and leaf diseases reduces stands of alfalfa. To avoid the dominance of associated grasses, alfalfa is frequently grown alone.

Alfalfa grows well on the tableland of New South Wales. Typical of this environment is Canberra at 35° 17′ S Lat and an annual precipitation of 590 mm. Highly satisfactory yields are obtained from rotationally grazed alfalfa-grass pastures. In semiarid areas west of Canberra (350 mm annual precipitation), alfalfa persists for many years if provided with long rest periods after grazing. In the Mediterranean climate of Victoria and South Australia, alfalfa is grown extensively on the Mallee and heath country for both hay and pasture. However, on the floodplains of rivers and streams its main use is for hay. Alfalfa is one of the forage crops frequently grown under irrigation in many areas of Australia.

Because alfalfa is planted over a wide range of conditions in Australia, seedlings may be established in almost every month of the year, with spring and fall sowing preferred. Similarly, treatment for soil conditions limiting production will vary considerably, but inoculation is usually considered essential. Seeding rates of 3 to 15 kg/ha are used, depending on rainfall. Under irrigation, higher rates (as high as 20 kg/ha) were formerly used to reduce weed competition, but now selective herbicides may be used to maximize production in the 1st yr. Lower seeding rates of about 3 to 8 kg/ha are used when alfalfa is planted for grazing. Most of these plantings are sown without herbicides, and the seedlings usually protected with one or two applications of insectides to reduce populations of mites, fleas, and blue alfalfa aphid. Established stands seldom receive chemical protection because costs are uneconomical (94, 139.).

Hunter River was essentially the only cultivar grown in Australia until 1965. About this time four other cultivars (Siro Peruvian, African, DuPuits, and Cancreep) were made available in an effort to extend the range of adaptation for alfalfa. Later 'Demnat', 'Paravivo', and 'Falkiner' were released, and an extensive cultivar-testing program was started in Victoria. After the spotted alfalfa aphid and blue alfalfa aphid were discovered, it was found that all of the available cultivars were susceptible to the two new aphids. The search for adapted aphid-resistant cultivars and the breeding of superior cultivars were expanded and intensified. Restrictions on importation of large volumes of seed of resistant cultivars were lifted, and seed of these cultivars were brought into the country, mainly from USA. In 1979, Nova, a spotted alfalfa aphid resistant cultivar, was released in New South Wales. Since then, other resistant cultivars such as Sheffield, Springfield, Wakefield, Siriver, Trifecta, and Sequel have been released.

The importance of rotational grazing to maintain alfalfa as pasture has been demonstrated. There are practical problems, however, in managing rotational grazing on extensive grazing lands where alfalfa must endure long periods of intensive grazing, especially during drought when animals consume all available feed (139).

Alfalfa seed is produced in many alfalfa-growing areas. Hay is the main crop with seed produced after one of the summer cuttings. Yields from these fields may be as low as 90 kg/ha, but generally range from about 200 to 280 kg/ha. Since 1975, row cropping has developed in irrigated fields, usually sown with 2 to 5 kg/ha of seed in rows spaced 35 cm apart. This is a compromise between the optimum density for seed production and the use of these fields for hay and grazing at other times of the year. In these fields, seed yields may range for 400 to 600 kg/ha, and yields of more than 1000 kg/ha have been achieved with flood irrigation and management similar to that in western USA. Consequently, seed production is now concentrated in regions with assured irrigation water and a low incidence of late summer rainfall. The most intensive area of seed production is around Keith, South Australia (139). The honeybee is the main pollinator of alfalfa, but competition from other pollen sources, especially the river red gum, can lure bees away from alfalfa. Only one set of seed has been obtained in Australia. The major pests of alfalfa seed are alfalfa seed chalcid (*Bruchophagus* spp.), pod-borer (*Etiella* spp.), thrips, Lygus (*Lygus* spp.), and the larva of *Heliothis* spp. Also, aphids can affect seed production (168). At harvest, plants are usually chemically dessicated and combine harvested as a standing crop.

Australia enjoyed a long period when alfalfa pests were of minor importance. This situation began to change when bacterial wilt was found in Victoria, and abruptly ended when the spotted alfalfa aphid and the blue alfalfa aphid were introduced. *Acyrthosiphon pisum* was identified in early 1980 (159). Other insect pests identified as important are: native budworm (*Heliothis* spp.), pink cutworm, red-legged earth mite (*Halotydens destructor* Tucker), lucerne flea (*Sminthuris viridus* L.), sitona weevil (*Sitona humeralis* Stephens), wingless grasshopper, Etiella moth (*Etiella* spp.), the two-spotted mite (*Tetranychus telarius* Linne), and alfalfa leaf roller (*Merophyas divulsana* Walk.) (95, 140). Diseases considered important in addition to bacterial wilt are: southern anthracnose, Rhizoctonia root canker, Phytophthora root rot, Stagonospora root rot (*Stagonospora* spp.), common leafspot, Stemphyllium leaf spot, pepper spot (*Leptosphaerulina* spp.), Phoma (black stem) crown, root, and stem rot (*Phoma* spp.), stem nematode, and root knot nematode (*Meloidogyne* spp.) (140).

Some of the numerous organizations conducting research on alfalfa in Australia are: The Commonwealth Scientific and Industrial Research Organization (CSIRO), Canberra, which has work at many stations in alfalfa-producing areas throughout the country, the Departments of Agriculture in the major alfalfa- producing states, and universities located near alfalfa-producing areas. Research on alfalfa increased dramatically after the spotted and blue alfalfa aphids were found. It appears that the devastating effect of these two aphids on alfalfa demonstrated the value of the alfalfa crop to the economy of many areas in Australia. Research was initiated to reduce losses from these aphids, including, breeding resistant cultivars, biology of the plant and aphids, as well as chemical and biological control. In addition, work is conducted on the complex of other organisms that affect this valuable crop. Australia can be ranked among the leaders in alfalfa science and technology.

2-3.7.2 New Zealand

The area seeded to alfalfa peaked at 220 000 ha in 1975 and has declined to approximately 101 200 ha in 1984, with nearly 80% on the South Island (72). Most of the alfalfa on the South Island is grown in areas where the annual rainfall averages 500 to 1180 mm. About 45% of the alfalfa in New Zealand is grown in the east/central province of Canterbury with lesser amounts in the provinces on either side of it. Almost all the alfalfa grown on the North Island is found in the South Auckland/Bay of Plenty which has a rainfall of 1575 mm, frequent summer droughts, and porous, pumice soils.

The area cut for hay and silage in New Zealand increased by about 125% from 1960 to 1976. However, the trend was reversed from 1977 to 1979 (75). Palmer (194) implies this reduction may be temporary because the reduced plantings seem to be associated with a series of solvable or changeable problems such as diseases, insects, and a period of high rainfall.

Most alfalfa is planted from September through February, with October and November being the favored months. However, some planting is done in almost every month of the year except July (283). The rate of seeding used by growers may vary from 4 to 20 kg/ha with an average rate of 12 to 20 kg/ha. Application of an inoculant is recommended. Certified alfalfa inoculants have been made available and are tested for approval by the Seed Testing Station, MAF, Palmerston North. Over half the alfalfa planted in New Zealand is sown without a cover crop, but if one is used, it is recommended that the soil be prepared for the alfalfa and the alfalfa planted last. To reduce competition from weeds the new alfalfa stands may be grazed when they reach the height of about 15 cm. Oversowing or overdrilling alfalfa into old pastures has been practiced with success when specified requirements are met, i.e., inoculation, lime, and suppression of other vegetation (183).

The cultivar situation in New Zealand was very simple prior to 1974. The main cultivar for many years was Wairau which was released in 1950 and bred from selections out of 'Marlborough', 'Grimm', 'Ontario Variegated', and an introduction from USA. Hunter River was the next most significant cultivar. Other seed sources and cultivars used on a much smaller scale were Provence, Chanticleer, American Commons, College Glutinosa, and Rhizoma. After 1974, a number of new cultivars (Saranac,

Washoe, AS13R, Pr521, WL311, WL318, and Pr524) were introduced primarily in response to disease and pest problems (74). In late 1975, the blue alfalfa aphid was found and caused extensive damage on all cultivars. In answer to this problem, an intensive breeding program was initiated at the Crop Research Division, Dep. of Scientific and Industrial Research (DSRI), Lincoln. The blue alfalfa aphid resistant cv. Rere, derived from U.S. material was released from this program and placed on the "Acceptable Cultivar" list in 1979. Work is continuing on the development of other improved cultivars. In addition to the breeding program at Christchurch, the Grasslands Division DSIR, Palmerston North, has an alfalfa-breeding program. Evaluation of introduced cultivars is continuing.

About 64% of the alfalfa planted in New Zealand is grazed for periods of the year, and is used for production of either hay or silage. The other 36% of the planted area is used solely for grazing (75). A greater percentage of the alfalfa (39%) is used for grazing only on the South Island than on the North Island (17%). There is considerable interest in developing better grazing systems and stocking rates for both cattle and sheep.

Certified alfalfa seed production increased from >200 Mg in 1963 to about 450 Mg in 1969 and then decreased rapidly to <100 mg in 1979. Dunbier et al. (75) indicated that this reduction in seed production may be due to a rapid change in cultivars and the importation of larger quantities of seed. According to Dunbier and Easton (74), new seed production techniques will have to be developed and implemented to meet the present requirements of a changing, multiple cultivar seed production system.

Bacterial wilt was identified in New Zealand in 1970. Other diseases are: Verticillium wilt, alfalfa mosaic, Australian lucerne latent virus, lucerne transient streak virus, subterranean clover red leaf virus, Phytophthora root rot, Rhizoctonia, Stagonospora root rot, Fusarium root rot, Sclerotinia root rot, *Thielaviopsis basicola* Berk. 4 Br. Ferr., *Phoma herbarium* var. *medicaginis* West., crown wart, common leafspot, pepper spot, yellow leaf bloch (*Pseudopeziza jonesii* Nannf.), and stemphyllium leaf spot (57). Two nematodes have been found on alfalfa: stem nematode and root-knot nematode (*M. hapla*).

In 1967, Pottinger and McFarlane (200) reported that 49 insect and nematode species were known on alfalfa in New Zealand, but only the common grass grub (*Costelytra zealandica* White), the white fringed weevil (*Graphognathus leucoloma* Boheman), and stem nematode were considered of major importance. Additional major insect pests of alfalfa described by Kain and Trought (141) were: the blue alfalfa aphid, pea aphid, and Sitona weevil (*Sitona* spp.). Spotted alfalfa aphid, recently discovered on the northern North Island, has caused no significant damage.

Tremendous progress has been made in alfalfa production since 1965. Part or much of this progress appears to have been stimulated by the need to combat important pests, and the continuing need for the production of high-quality forage to support livestock.

REFERENCES

1. Adamova, R. 1981. Hodnecini rezistence odrud vojtesky (*Medicago sativa* L.) vuci hadatku zhoubnemu *Ditylenchus dipsaci* (Kühn) Filipjev. Sbornik vêdeckych praci Vyzkumného a slechtitelského ustavu picninarského. Troubsko u Brna 7: 161–166.
2. Aeschlimann, J.P. 1978. Heavy infestations of *Sitona humeralis* Stephens (Col.: Curculionidae) on lucerne in southern Morocco. Ann. Zool. Ecol. Anim. 10:221–225.
3. ----. 1980. The *Sitona* (Col.: Curculionidae) species occuring on *Medicago* and their natural enemies in the Mediterranean region. Entomophaga 25: 139–153.
4. Agricultura al Dia (AAD). 1970. Agric. al Dia 17(3–4): 21–22.
5. Ahlgren, G.H. 1949. Forage crops. McGraw-Hill Book Co., New York.
6. Al-Kawaz, G.M. 1975. A report on salinity effects on yield and water consumption of local and introduced alfalfa cultivars grown under greenhouse conditions. Iraqi J. Agric. Sci. 10: 105–115.
7. Ali, H. 1982. The *Medicago*-wheat rotation system and its evaluation in Tunisia. p. 336–374. *In* Special Report. Oreg. State Univ. Agric. Exp.
8. Amad, S.M.F. 1978. Studies on damping-off disease of alfalfa in Saudi Arabia. p. 229–243. *In* M.A. Migahid et al. (ed.) Proc. of the First Conf. on the Biological Aspects of Saudi Arabia. Riyad University Press, Saudi Arabia.
9. Anderson, G.D., and Z. Navah. 1968. Promising pasture plants for Northern Tanzania. East Afr. Agric. For. J. 34: 84–105.
10. Andries, A. 1984. Personal Communication., Rijksstation voor Plantenveredeling, Merelbeke, Belgium.
11. Angevin, M. 1983. Méthode d'infection artificielle pour la sélection de la luzerne contre *Phoma medicaginis* Malbr & Roum. Agronomie 3: 911–916.
12. Anonymous. 1972. Report to the Government of Saudi Arabia on research in plant protection based on the work of H.E. Martin, FAO Entomologist. FAO, Rome.
13. ----. 1977. Lucerne. South African Department of Agricultural Technical Services. Agricultural Research, 1976.
14. ----. 1986. Semences fourragères "Special Distribution". p. 26–51. *In* Semences et Progrès. No. 47. SEDIS, Paris.
15. Armstrong, J.M., F.W. Nowosad, P.O. Ripley, and W. Kalbfleish. 1948. Alfalfa for hay silage and pasture. Can. Dep. Agric. Pub. 735.
16. Avendano, T.T., and A. Guinez S. 1966. Comportamiento de diez variedades de alfalfa bajo una poblacion de nematodos. Agric. Tec. (Chile) 26: 158–160.
17. Baghestani, M. 1976. Influence of lucerne on the soil level of fertility in Karaj. Agron. Tropic 34: 417–420.
18. Bakheit, B.R., and E. Toth. 1983. Results of resistance breeding on alfalfa. I. Resistance to Fusarium wilt. Acta Agron. Acad. Sci. Hung. 32: 424–429.
19. Bakhtri, M.N. 1978. Wheat/forage legume rotation and integration of crop and sheep husbandries in Algeria. p. 161–166. *In* J.C. Homes (ed.) Technology for increasing food production, FAO, Rome.
20. Batra, U.R., M.G. Dshmukh, and R.N. Joshi. 1976. Factors affecting extractability of protein from green plants. Indian J. Plant Pathol. 19: 211–216.
21. Behaeghe, T. 1984. Personal communication, Univ. of Ghent, Belgium.
22. Bierrenbach De Castro, J. 1961. Cultura de alfalfa. Sup. Agric. Sao Paulo, Brazil 7, 313: 8–9.
23. Blondon, F., M. Ghesquiere, and P. Guy. 1981. Variation de la fertilité pollinique en fonction de la température chez des luzernes de différentes origines (*Medicago sativa* L. et *M. media* Pers). Agronomie 1: 383–388.
24. Boerger, A. 1949. Ecologia de *Medicago sativa* L. en el continente Americano. Arch. Fitotec. (Uruguay) 4: 107–121.
25. Bofarrul, N., and J. De Pozo. 1965. Control de malezas en el establecimiento de alfalfa en la provincia de Nuble. Agric. Tec. (Chile) 25: 109–114.
26. Boissier, E. 1872. Fl. Orient., Vol. 2, p. 90. cited by Scofield (225).
27. Bojtos, Z. 1981. Ten years research work on hybrid alfalfa in Hungary. p. 81–82. *In* Rep. 27th Alfalfa Improve. Conf., Madison, WI. 8–10 July 1980. USDA-SEA, University of Wisconsin, Madison. Peoria, IL.
28. Bolton, J.L. 1962. Alfalfa botany, cultivation and utilization. World Crops Books, Leonard Hill, London.
29. ----, B.P. Goplen, and H. Baenziger. 1972. World distribution and historical developments. *In* C.H. Hanson (ed.) Alfalfa science and technology. Agronomy 15: 1–34.
30. Bournoville, R. 1983. Les principaux ravageurs des luzernières. Phytoma 348: 43–46.
31. ----, P. Guy, J.N. Tasei, and A. Delaude. 1977. Problems of phytophagous and pollinating insect control of seed lucerne in France. p. 451–454. *In* E. Wojahn and H. Thöns (ed.) Proc. 13th Int. Grassl. Congr., Leipzig, German Democratic Republic 18–27 May. Akademie-Verlag, Berlin.

32. Bocsa, I. 1984. Personal communication, Agric. Res. Inst., Kompolt, Hungary.
33. ----, and J. Buglos. 1983. Seed yield and some factors influencing seed setting at the variety level in lucerne. Z. Pflanzenzuecht. 90: 172–176.
34. ----, ----, and Z. Majko. 1975. Breeding for quantitative factors in lucerne. *In* Breeding of lucerne and development of its production. Institute for Crop Growing and Soil Conservation, Kompolt, Hungary.
35. ----, ----, and I. Sziraki. 1983. Breeding lucerne for tolerance to frequent cutting, and physiological aspects of tolerance. Z. Pflanzenzuecht. 90: 222–228.
36. Bosnjak, D. 1973. Nova selekcionirana sinteticna sorta lucerne Osjecka 66. Z. Rad. Poljopr. Inst. Osijek 3: 123–128.
37. ----. 1980. Ispitivanje prinosa i sadrzaja proteina domacih i inozemnih sorti lucerne u istocnoj Hovatskoj. Z. Rad. Poljopr. Inst. Osijek 10: 23–31.
38. ----, and I. Sikora. 1973. Nova sorta lucerne Osjecka 70. Z. Rad,. Poljopr. Inst. Osijek 3: 129–134.
39. ----, and M. Stjepanovic. 1977. Productivity of cultivated native and foreign alfalfa varieties on two soil types in Yugoslavia. p. 359–363. *In* E. Wojahn and H. Thöns (ed.) Proc. of 13th Int. Grassl. Congr., Leipzig, German Democratic Republic 18–27 May. Akademie-Verlag, Berlin.
40. Brand, C.J. 1907. Peruvian alfalfa: A new long-season variety for the southwest. USDA Bureau of Plant Industry Bull. 118. U.S. Government Printing Office, Washington, DC.
41. ----. 1911. Grimm alfalfa and its utilization in the Northwest. USDA Bureau of Plant Industry Bull 209. U.S. Government Printing Office, Washington, DC.
42. Buller, R.E., and M. Gonzalez. 1958. Performance of alfalfa varieties, red clover, and alsike clover grown under irrigation at approximately 8,800 feet above sea level in Mexico. Agron. J. 50: 19–22.
43. ----, J.B. Pitner, and M. Ramirez. 1955. Behavior of alfalfa varieties in the valley of Mexico. Agron. J. 47: 510–512.
44. ----, R. Valdivieso, and R. Garza. 1958. Comportamiento de variedades selecionadas de alfalfa y recommendaciones para su mejoramiento en Mexico. Folleto Tec. 32. Ofcna. Est. Esp. Sec. Agric. Ganaderia, Mexico.
45. Cafati, K.C., and R. Avendano T. 1968. Capacidad combinatoria general y especifica de seis clones de alfalfa (*Medicago sativa* L.) para resistencia a *Meloidogyne* species y rendimiento en forraje. Agric. Tec. (Chile) 28: 57–65.
46. Carmona, S.S. 1984. Personal communication, Instituto Nacional de Investigaciones Agricolas, Centro de Investigaciones Agricolas de el Bajio (CIAB), Celaya, Gto., Mexico.
47. Carter, E.D. 1975. The potential for increasing cereal and livestock production in Algeria. CIMMYT, Mexico DF, Mexico.
48. Castro, A.L. 1978. Alfalfa. p. 165–170. *In* Analisis de los recursos geneticos disponibles a Mexico. Soc. Mex. Fitogenetica.
49. ----. 1982. Guia para cultivar alfalfa en los estados de Mexico e Hidalgo. Folleto para Productores 15, Campo Agric. Exp. Valle de Mexico, Chapingo.
50. Caubel, G., M. Bossis, G. Génier, and P. Guy. 1977. Mise au point d'un test de sélection de luzernes résistantes à *D. dipsaci*. Sci. Agron. Rennes, p. 25–32.
51. ----, R. Champion, and R. Marre. 1983. Le nématode des tiges des légumineuses fourragères: fumigation des semences, variétés résistantes. Perspect. Agricole (Suppl.) 75: 23–27.
52. Chloupek, O. 1982. Combining ability for growth of young alfalfa plants as related to size of the root system. Z. Pflanzensuecht 88: 54–60.
53. ----. 1984a. Analyza vynosu semen u vojtesky seté. Rostl. Vyroba 30: 107–114.
54. ----. 1984b. Personal communication, Plant Breeding Stn., Zelesice, Czechoslavakia.
55. ----, and J. Babinec. 1982. Vliv chorob a skudcu na semenarskou productivnost vojtesky. Ochr. Rostl. 18: 133–139.
56. Chowdhury, M.S. 1978. Rhizobium studies in Tanzania. *In* J. Dobereiner et al. (ed.) Limitations and potentials for biological nitrogen fixation in the tropics. Basic Life Sciences, Vol. 10.
57. Close, R.C., I.C. Harvey, and F.R. Sanderson. 1982. Lucerne diseases in New Zealand and their control. p. 61–69. *In* R.B. Wynn-Williams (ed.) Lucerne for the 80s. Spec. Pub. 1. Agronomy Society of New Zealand, Palmerston North.
58. Cobo, P.B. 1891. Historia del Nuevo Mundo. Tomo II. Imprenta de E. Rasco, Bustes Tavera.
59. Crowder, L.V. 1959. Recomendaciones para el cultivo de pastos y forrajes de clima frio. Agric. Trop. 15: 35–50.
60. ----, J. Vanegas, and J. Silva. 1960. The influence of cutting interval on alfalfa production in the high Andes. Agron. J. 52: 128–130.
61. Cuarta Memoria Anual. 1967–1968 (CMA). 1968. Instituto de Investigaciones Agropecuria. Ministry of Agriculture, Santiago, Chile.

62. Daomiti. 1978. A summary of alfalfa growing tests in Yuerdus Region of Xinjiang Province. p. 321-327. *In* Symp. 3rd Grassl. Meet. in Northwest China.
63. Davies, W.E. 1984. Personal communication, Welsh Plant Breeding Stn., Aberystwyth, Wales.
64. Davis, A.G. 1962. Lucerne. Rhod. Agric. J. 59: 68-84.
65. Deinum, I.B. 1984. Personal communication, Dep. of Field Crops and Grassland Science, Wageningen, Netherlands.
66. DeKock, G.C. 1965. p. 141-143. *In* Proc. 9th Int. Grassl. Congr., Brazil.
67. ----. 1978. Historical survey of lucerne cultivation. Farming in South Africa. Lucerne A1/1978. Department of Agricultural and Technological Services, Pretoria.
68. Dev, D.V., U.R. Batra, and R.N. Joshi. 1974. The yields of extracted leaf protein from lucerne (*Medicago sativa* L.). J. Sci. Food Agric. 25: 725-733.
69. Devilat, B.J. 1967. Valoracion con novillos de los ensilages de alfalfa y trebol con y sin marchitamiento. Agric. Tec. (Chile) 27: 21-28.
70. Dijk, G.E. 1984. Personal communication, Foundation for Agricultural Plant Breeding, Wageningen, Netherlands.
71. Dobias, A. 1983. The seed mass per plant of some American lucerne varieties in Czechoslovakia. p. 64-65. *In* Rep. 28th Alfalfa Improve. Conf., Davis, CA. 13-16 July 1982. Curlie Printing Co., Minneapolis.
72. Douglas, J.A. 1986. The production and utilization of lucerne in New Zealand. Grass Forage Sci. 41: 81-128.
73. Doyle, C.J., and D.J. Thompson. 1985. The future of lucerne in British agriculture: an economic assessment. Grass Forage Sci. 40: 57-68.
74. Dunbier, M.W., and H.S. Easton. 1982. Cultivar development. p. 117-120. *In* R.B. Wynn-Williams (ed.) Lucerne for the 80's. Spec. Pub. 1. Agronomy Society of New Zealand, Palmerston North.
75. ----, R.B. Wynn-Williams, and P.A. Burnett. 1982. Lucerne in the 70s. p. 3-7. *In* R.B. Wynn-Williams (ed.) Lucerne for the 80's. Spec. Pub. 1. Agronomy Society of New Zealand, Palmerston North.
76. Durr, H.J.R. 1981. The first occurrence of *Acyrthosiphon kondoi* Shinji (Hemiptera: Aphididae), a new threat to lucerne pastures in South Africa. Phytophylactica 13: 113-115.
77. Eder, J.P. 1913. Colombia. Unwin, London.
78. Elgin, J.H., and R.H. Ratcliffe. 1981. Alfalfa exploration in Turkey. 15 July—15 August. Mimeograph Pub.
79. Farnworth, J., and I.B. Ruxton. 1983. A comparison of Hasawi alfalfa with five imported varieties. Publication, Joint Agricultural Research and Development Project, University College of North Wales, Bangor and Ministry of Agriculture and Water, Saudi Arabia, No. 13.
80. Falcinelli, M. 1984. Personal communication, Istituto de Miglioramento Genetica, Perugia, Italy.
81. Gallais, A. 1984. An analysis of heterosis *vs* inbreeding effects with an autotetraploid cross-fertilized plant: *Medicago sativa* L. Genetics 106: 123-137.
82. Gandarillas, H., T. Rios, G. Barja, and M. Duran. 1965. Estudio de adaptacion de plantas forrajeras en el altiplano Boliviano. p. 339-344. *In* Proc. 9th Int. Grassl. Congr., Brazil.
83. Garcia, C.A., and C. Sanchez. 1982. Evaluacion de variedades de alfalfa bajo condiciones de riego. p. 9. *In* Resumes de Investigacion-Forrajes 1981. Secretaria de Agricultura y Recursos Hidraulicos CIANOC, Mexico.
84. ----, ----, and G.J. Garcia. 1982. Ensayo comparativo de veinte variedades de alfalfa. p. 15. *In* Resumes de Investigacion-Forrajes 1980. Secretaria de Agricultura y Recursos Hidraulicos. CIANOC, Mexico.
85. Garza, R., and R.E. Buller. 1960. Comportamiento de leguminoses y zacates forrajeros en la region del Bajio. Folleto Tec. 41. Ofcna. Est. Sec. Agric. Ganaderia, Mexico.
86. Génier, G., and P. Guy. 1981. Use of male sterility for dry matter yield improvement in lucerne. Genetic variability for seed yield. p. 43-48. *In* Breeding high yielding forage varieties combined with high seed yield. Centre of Agricultural Research, Merelbeke-Gent, Belgium.
87. Gerewal, G.S., and J.S. Dhaliwal. 1983. Antibiosis and tolerance in lucerne to lucerne weevil, *Hypera postica* (Gyll.). Indian J. Agric. Sci. 53: 73-77.
88. Ghobrial, K.M. 1978. Vegetative growth and yield of four varieties of alfalfa in the New Valley. Agric. Res. Rev. 56: 39-42.
89. Gleason, L.S. 1957. Defoliacion de la alfalfa para la produccion de semilla. Agric. Tec. Mex. 21 (4): 37-39.
90. Gondran, J. 1976. Répartition en France de la verticilliose de la luzerne. Ann. Phytopathol. 8: 203-212.
91. Gonzalez, D., M. Miyazaki, W. White, H. Takada, R.C. Dickson, and J.C. Hall. 1979.

WORLD DISTRIBUTION AND DEVELOPMENT

Geographical distribution of *Acyrthosiphon kondoi* Shinji (Homoptera: Aphidiae) and some of its parasites and hyperparasites in Japan. Kontyu 47: 1–7.
92. Gonzalez, G. 1961. Agric. Tec. (Chile) 21(1–2): 100–107.
93. Goplen, B.P., H. Baenziger, L.D. Bailey, A.T.H. Gross, M.R. Hanna, R. Michaud, K.W. Richards, and J. Waddinton. 1982. Growing and managing alfalfa in Canada. Pub. 1705/E. Agriculture, Canada.
94. Gramshaw, D. 1981a. Growing new lucernes in Queensland. Queensl. Agric. J. 107: 249–254.
95. ———. 1981b. Factors limiting lucerne production. Queensl. Agric. J. 107: 25–28.
96. Graves, W.L. 1984. Personal communication, California Agric. Ext., San Diego, CA.
97. Grove, V.H., D.L. McCune, and J. Obrador. 1961. Agric. Ganad. 6(26–27): 48–50.
98. Gubis, V. 1982. Slachtenie lucerny na odolnosti proti hadalku zhubnemué. Uroda 8: 360–361.
99. Guggenheim, J. 1984. Personal communication, Ministry of Agriculture, Extension Service Forage Crops Division, Hakirya, Tel-Aviv, Israel.
100. Guinez, A. 1983. Nematodo en alfalfa. Invest. Prog. Agropecuario La Platina. 15: 17.
101. Guy, P. 1975. L'amélioration de la luzerne pour la résistance à ses ennemis végétaux et animaux. Fourrages 64: 19–34.
102. ———, and Y. Dattée. 1982. Reproduction patterns of *Medicago sativa*. p. 125–141. *In* Seed regeneration in cross pollinated species. Proc. C.E.C./Eucarpia, Nyberg Denmark. 15–17 July 1981.
103. Hajek, D. 1980. Vykonnost a vyznamnejsi hospodarské vlastonoti ceskoslovenskych a nekterych zahranicnich odrud vojtesky v CSSR v letech 1974–1977. Acta Univ. Agric., Fac. Agron. (Brno) 28: 199–203.
104. Halpin, J.D., and R. Osario. 1957. Agric. Tec. Mex. 5 (4–5): 42–43.
105. Hamdi, Y.A., A.N. Yousef, A. Al-Tai, S.K. Azawi, and M. Al-Baquari. 1978. Distribution of *Rhizobium meliloti* and *R. trifolii* in Iraq soils. Soil Biol. Biochem. 10: 149–150.
106. Harte, W. 1770. Essays on husbandry. W. Frederick, London. Cited by Hendry (109).
107. Hassawy, G.S. 1975. Effect of irrigation frequency on yield, root development, and stand of alfalfa and on weed growth in Central Iraq. Summary of research papers. The Second Scientific Conference of the Scientific Research Foundation, 6–11 December.
108. Heimpel, G. 1981. Mexico: Agricultural and trade policies. USDA FAS-306. U.S. Government Printing Office, Washington, DC.
109. Hendry, G.W. 1923. Alfalfa in history. J. Am. Soc. Agron. 15: 171–176.
110. Hidalgo, M.F. 1966. Classification de les alfalfas espanolas. Bull. Assoc. Invest. Mejona. Alf. Zaragoza.
111. ———. 1979. La alfalfa Aragon y su mejora de conservacion. Pastos 9: 58–71.
112. Huang, W. 1981. Adaptation and utilization of three legumes in China. p. 187–189. *In* J.A. Smith and V.W. Hays (ed.) Proc. 14th Int. Grassl. Congr., Lexington, KY. 15–24 June. Westview Press, Boulder, CO.
113. Hussain, Z. 1978. Alfalfa cultivation in Saudi Arabia. World Crops 30: 260–261.
114. Hycka, M. 1971. Alfalfa Adyta. An. Aula Dei 11: 58–68.
115. ———. 1975. Alfalfa for the dryland of the mediterranean region. p. 131–134. *In* Proc. 6th General Meet. European Grassl. Fed. Madrid.
116. ———. 1976. Alfalfas en el secano aragonés. An. Aula Dei 13: 290–300.
117. ———. 1982. Alfalfa en el regadio de la zona del Ebro. Pub. Exp. Stn. Aula Dei, Zaragoza.
118. ———, and J.M. Benitez-Sidon. 1979. Algunas caracteristicas de nuevos cultivares espanoles de alfalfa. An. Aula Dei 14: 558–574.
119. ———. 1984. Personal communication, Estacion Exp. de Aula Dei, Zaragoza, Spain.
120. Igarashi, T., S. Kitajima, J.H. Hashimito, and A. Kishita. 1973. Effects of phosphate and calcareous fertilizers on the growth of alfalfa cultivated on volcanic-ash soil in the warm region of Japan. Bull Kyushu Agric. Exp. Stn. 16: 625–648.
121. Inami, S., F. Fujimoto, K. Nakashima, and S. Suzuki. 1981. Breeding alfalfa, *Medicago sativa* L., for southern blight resistance. II. Evidence of resistance in selected strains. J. Jpn. Soc. Grassl. Sci. 26: 365–371.
122. ———, and S. Suzuki. 1981. Breeding alfalfa, *Medicago sativa* L., for southern blight resistance. I. Varietal difference of disease injury. J. Jpn. Soc. Grassl. Sci. 26: 360–364.
123. Instituto Geografico Militar (IGM). 1970. Atlas de la Republica de Chile. 2 ed.
124. Instituto Nacional de Estadisticas (INE). 1969. IV Censo Nacional del pais. INE, Santiago.
125. ———. 1976. V Censo Nacional Agroperuario 1975–1976. Total pais. INE, Santiago, Chile.
126. Instituto Nacional de Investigaciones Agricolas (INIA). 1969. Guia para la Asistencia Tecnia Agric. en CIAB, CIANE, CIANO, CIAS, CIB. Sec. Agric. Ganaderia, Mexico.

127. Itria, C.D. 1969. La alfalfa en la Republica Argentina. IDIA Suplemento 21.
128. ——. 1983. Personal communication, Centro Nacional Investigaciones Agropecurias, Buenos Aires, Argentina.
129. Jacobsen, A. 1984. Personal communication, Landskontoret for Planteavl, Viby, Denmark.
130. Jelinowska, A. 1979. Suitability of some lucerne cultivars to obtain leaf protein. Eucarpia fodder crops section meeting, Radzikow, Poland. 4–9 Sept. 1978. Institute of Soil Science and Plant Culture, Pulawy, Poland.
131. ——. 1980. Biologisches extragspotential und möglichkeiten der luzernesamener zeugung. Wissenschaftliche Beiträge, Martin-Luther-Universität Halle-Wittenberg 20: 588–603.
132. ——. Personal communication, IUNG, Pulawy, Poland.
133. Joggi, D., J. Lehmann, and H.U. Briner. 1982a. Considérations sur la culture de la luzerne (*Medicago sativa* L.) et essais variétaux 1978–1980. Rev. Suisse Agric. 14: 197–201.
134. ——, ——, and E. Meister. 1982b. Sortenversuche mit luzerne (1978 bis 1980). Mitt. Schweiz. Landw. 30: 109–117.
135. Jönsson, H.A. 1984. Personal communication, Weibullsholm Plant Breeding Inst., Landskrora, Sweden.
136. Jönsson, N. 1984. Personal communication, Univ. of Agric. Sci., Uppsala, Sweden.
137. Joshi, R.N. 1979. Status of green crop fractionation in India. p. 22–23. *In* Proc. 2nd Int. Green Crop Drying Congr.
138. Jurzysta, M. 1979. Biochemical characteristic and feeding value of alfalfa selected for saponin content. Eucarpia fodder crops section meeting, Radzikow, Poland. 4–9 Sept. 1978. Institute Uprawy, Pulawy, Poland.
139. Kaehne, I.D. 1984. Personal communication, Dep. of Agriculture and Fisheries, Adelaide, South Australia.
140. ——, and A. Lake. 1982. Lucerne varieties—1982. Department of Agriculture, South Australia, Fact Sheet 9/82. AGDEX 121/32.
141. Kain, W.M., and T.E. Trought. 1982. Insect pests of lucerne in New Zealand. p. 49–59. *In* R.B. Wynn-Williams (ed.) Lucerne for the 80's. Spec. Pub. 1. Agronomy Society of New Zealand, Palmerston North.
142. Kaiser, W.J., and D.G. Robertson. 1976. Notes on East African plant virus disease. II. Alfalfa mosaic virus. East Afr. Agric. For. J. 42: 47–54.
143. Kamalakar, B., S. Maiti, and B.N. Chetterjee. 1982. Deterioration of alfalfa swards in the hot, humid subtropics of India. Agron. J. 74: 827–830.
144. Kemm, E.H., and M.N. Ras. 1981. The effect of stage of cutting on the value of dehydrated lucerne meal in growing diets for pigs. South Afr. J. Anim. Sci. 11: 285–286.
145. Khan, M. 1984. Personal communciation, Pakistan Agric. Res. Counc., Islamabad, Pakistan.
146. ——, and M.S. Bhatti. 1962. Agric. Pak. XIII 4: 1–9.
147. Kitazawa, K., and R. Sato. 1981. Wilt caused by *Verticillium albo-atrum* Reinke and Berthold. Ann. Phytopathol. Soc. Jpn. 47: 272–274.
148. Klesnil, A., J. Velich, and V. Regal. 1965. Alfalfa. p. 40. *In* State Agric. Pub. House, Prague (Translated from Czechoslovakian.)
149. Klinkowski, M. 1933. Lucerne: Its ecological position and distribution in the world. Imperial Bureau of Plant Genetics: Herbage Plants, Bull. 12, Aberystwyth, Wales. (Translated by G.M. Roseveare.)
150. Komatsu, T., J. Maruyama, Y. Horikama, and F. Tsuchiya. 1985. Winter injury of alfalfa (*Medicago sativa* L.) in soil freezing area of Japan. *In* Proc. 15th Int. Grassl. Congr., Kyoto, Japan. 24–31 August. Science Council of Japan and the Japanese Society of Grassland Science, Tochigi-Ken, Japan.
151. Korosec, J., and T. Mekeln. 1984. Personal communication, Biotehniska Fakulteto Vtozd za Agronomijo, Ljubljana, Yugoslavia.
152. Kopriva, J., and J. Prochazka. 1980. Srovnani sememarsk produktivnosti vojtesky u ruznych method zakladani porostu. Rostl. Vyroba 26: 1147–1155.
153. Kudela, V., and E. Fikesova. 1980. Perspektivy selekce vojtesky na toleranci k bacterial niimu a verticiliovemu vadnuti. Genet. Slechteni 16: 215–224.
154. Lakshmi-Kumari, M., C.S. Singh, and N.S.S. Rao. 1974. Root hair infection and nodulation in lucerne (*Medicago sativa* L.) as influenced by salinity and alkalinity. Plant Soil 40: 261–268.
155. Lamprecht, S.C., and P.S. Knox-Davies. 1984a. Preliminary survey of foliage diseases of annual *Medicago* spp. in South Africa. Phytophylactica 16: 177–183.
156. ——, and ——. 1984b. *Colletrotrichum trifolii* on annual *Medicago* spp. in South Africa. Phytophylactica 16: 185–188.
157. ——, ——, and W.F.O. Marasas. 1984. *Fusarium* spp. associated with diseased root and crown tissue of annual *Medicago* spp. Phytophylactica 16: 195–200.

158. Landsberg, J.J. 1964. Lucerne moisture requirements. Rhod. Agric. J. 61: 48–50.
159. Lehane, L. 1982. Biological control of lucerne aphids. Rural Res. (Autumn): 4–10.
160. Lehmann, J. 1984. Personal communication, Swiss Federal Res. Stn. for Agronomy, Zurich, Switzerland.
161. Lehman, W.F., M.D. Rumbaugh, R. Zambrano, L.P. Spiaggi, and V. Otazu. 1981. Collecting alfalfa and other plants in Bolivia, Peru and Ecuador. p. 34–40. *In* Proc. Western Alfalfa Improv. Conf., Las Cruces, New Mexico.
162. Linnaeus, C.H. 1735. Systema naturae, Ludg, Bat. Stockholm. Cited by Scofield (225).
163. ----. 1753. Species plantarum, Stockholm. Cited by Scofield (225).
164. Lopez, H. 1984. Personal communication, Estacion Exp. La Platino, Santiago, Chile.
165. Lorenzetti, F. 1981. Relationship between dry matter and seed yield in leguminous forage plants. p. 57–74. *In* Breeding high yielding forage varieties combined with high seed yield. Eucarpia meeting. Centre of Agricultural Research, Merelbeke-Gent, Belgium.
166. Lu, D.Y., M.R. Davey, and E.C. Cocking. 1983. A comparison of the cultural behavior of protoplasts from leaves, cotyledons and roots of *Medicago sativa*. Plant Sci. Lett. 31: 87–99.
167. Lu, T. 1984. A study on the community of insects in cultivated forage fields. p. 137–139. *In* Proc. 1st Symp. Grassl. Ecology by the Chinese Grassl. Soc.
168. MacKay, J.H.E. 1978. Lucerne aphids and the problem of seed production. Lucerne Aphid Workshop. 27–30 November. Department of Agriculture, New South Wales.
169. Manninger, S., K. Manninger, Cs. Erdelyi, J. Buglos, L. Martinovich, and A. Dobrovolszky. 1981. Relationship between growth habit and some factors affecting alfalfa seed yield in Hungary. p. 79–81. *In* Rep. 17th Alfalfa Improve. Conf., University of Wisconsin, Madison. 8–10 July. USDA-SEA, Peoria, IL.
170. Martin, B. 1984. Personal communication, Martin-Luther-Universität, Halle-Wittinberg.
171. Martinez, R.A., M. Quiroga, N. Thomas, and K.F. Byerley. 1983. Alternative forage-production pattern for dairy cattle under conditions of limited water supply in northern Mexico. p. 804–806. *In* J.A. Smith and V.W. Hays (ed.) Proc. 14th Int. Grassl. Congr., Lexington, Ky. 15–24 June 1981 Westview Press, Boulder, CO.
172. Marum, P. 1984. Personal communication, State Agric. Exp. Stn., Löken, Norway.
173. Maruyama, J. 1986. Personal communication, Obihiro Univ. of Agric. and Veterinary Medicine, Japan.
174. Maslinkov, M. 1984. Personal communication, Vassil Kolarov Inst. Plovdiv, Bulgaria.
175. Matsumoto, N., and T. Araki. 1978. Alfalfa root rot caused by *Phytophthora megasperma* Drech. in Japan. Ann. Phytopathol. Soc. Jpn. 44: 214–217.
176. Matsumoto, T., T. Kodama, and D. Murayama. 1968. On the alfalfa mosaic virus strains occurring on forage crops in Hokkaido. J. Fac. Agric. Hokkaido Univ. 56: 1–16.
177. McMaster, G.S., and M.H. Walker. 1970. A brief history of lucerne and its present distribution in New South Wales. Bull. P407. New South Wales Dep. Agric., Division of Plant Industry.
178. McWilliam, J.R. 1968. Lucerne, the plant. J. Aust. Inst. Agric. Sci. 34: 191–193.
179. Mela, T. 1984. Personal communication, Inst. of Plant Husbandry, Helsinki, Finland.
180. Mendoza, E.J.B. 1982. Determinacion de la dosis economica de fertilizacion de alfalfa. p. 21. *In* Resumes de Investigacion-Forrajes 1980. Secretaria de Agricultura y Recursos Hidraulicos. CIANOC, Mexico.
181. Mi, Shuzheng. 1982. The epidemic disease of common alfalfa in Hulunbeier Region of Inner Mongolia. p. 483–488. *In* Symp. 4th Grassl. Meet. Northeast China.
182. Molina Frances, J.A. 1975. Alfalfa in the dryland of Aragon. p. 135–140. *In* Proc. 6th General Meet. European Grassl. Fed. Madrid.
183. Musgrave, D.J. 1982. Lucerne establishment-oversowing and overdrilling. p. 21–31. *In* R.B. Wynn-Williams (ed.) Lucerne for the 80's. Spec. Pub. 1. Agronomy Society of New Zealand, Palmerston North.
184. Nagieb, M.A., H. Elsaid, N. Jahlin, and H. Elarosi. 1980. New plant diseases in Al-Hassa Oasis. p. 279–284. *In* Proc. 4th Conf. on the Biological Aspects of Saudi Arabia. University of Riyadh, Saudi Arabia.
185. Nagy, B. 1983. Research on combining ability of back-crossed alfalfa cms-lines. p. 78–79. *In* Rep. 28th Alfalfa Improve. Conf., Davis, CA. 13–16 July. Curlie Printing Co., Minneapolis.
186. Narayanan, T.R., and P.M. Dabadghao. 1972. Forage crops of India. Indian Counc. of Agric. Res., New Delhi.
187. Novoselova, A. 1977. The gene pool and its use in breeding grassland plants. p. 51–59. *In* E. Wojahn and H. Thöns (ed.) Proc. 13th Int. Grassl. Congr. Leipzig, German Democratic Republic. 18–27 May. Akademie-Verlag, Berlin.
188. Nunez, G. 1982. Evaluacion de variadades de alfalfa a nivel semicomercial. p. 15. *In*

Resumes de investigaciones-forrajes 1981. Secretaria de Agricultura y Recursos Hidraulicos. CIANOC, Mexico.
189. Obrador, R.J. 1966. Descantes en produccion de semilla de plantas forrajeras. Agric. Tec. (Chile) 26(4): 164–168.
190. Oficina de Planificacion Nacional (ODEPA). 1981. Ministero de Agricultura. Estadisticas Agropecuarias 1980–1981, Santiago, ODEPA.
191. Ohara, H. 1982. The present state and future prospects of grassland dairying in Japan. Jpn. J. Dairy Food Sci. 31: A187–A191.
192. Padilla, R., and W.R. Young. 1958. El pulgon manchado de la alfalfa en Mexico. *Therioaphis* (Pterocallidium) *maculata* (Buckton). Folleto Tec. 25. Ofcna. Est. Esp. Sec. Agric. Ganaderia, Mexico.
193. Palmer, T.P. 1967. Lucerne breeding in New Zealand. p. 85–93. *In* R.H.M. Langer et al. (ed.) The lucerne crop. Reed, Wellington, New Zealand.
194. ----. 1982. Lucerne for the 80's, conclusions from the symposium. p. 129–131. *In* R.B. Wynn-Williams (ed.) Lucerne for the 80s. Spec. Pub. 1. Agronomy Society of New Zealand, Palmerston North.
195. Pekary, K. 1975. Supply of nutrients of lucerne. p. 167–177. *In* Breeding of lucerne and development of its production. Institute for Crop Growing and Soil Conservation, Kompolt, Hungary.
196. Piskovatski, Y.M. 1982. Breeding alfalfa for tolerance to multiple cutting. *In* Rep. 28th Alfalfa Improve. Conf., Davis, CA. 13–16 July. Curlie Printing Co., Minneapolis.
197. ----, Y. Nenarokov, and G. Stepanova. 1977. The development of grazable alfalfa varieties suitable for the conditions in the nonchernozem zone of the RSFSR. p. 353–354. *In* E. Wojahn and H. Thöns (ed.) Proc. 13th Int. Grassl. Congr., Leipzig, German Democratic Republic. 18–27 May. Akademie-Verlag, Berlin.
198. Pitner, J.B., N. Sanchez, and R.P. Peregrina. 1955. Aumento en la produccion de maiz y trigo mediante abonos verde. Folleto Tec. 17. Ofcna Est. Esp. Sec. Agric. Ganaderia, Mexico.
199. Plancquaert, Ph. 1976. La luzerne; culture-utilisation. Bull. ITCF, Paris.
200. Pottinger, R.P., and R.P. MacFarlane. 1967. Insect pests and nematodes of lucerne. p. 229–284. *In* R.H.M. Langer et al. (ed.) The lucerne crop.. Reed, Wellington, New Zealand.
201. Pulli, S. 1980. Growth factors and management technique in relation to the development rhythm and yield formation pattern of a seeding year lucerne stand. J. Sci. Agric. Soc. Finl. 52: 477–494.
202. Radwan, M.S., A.K. Al-Fakhry, and A.M. Al-Hasan. 1978. Some observations on the performance of annual medics in northern Iraq. Mesopotamia J. Agric. 13: 55–67.
203. Ram, S., and M.P. Gupta. 1976. Assessment of losses in the fodder yield of lucerne due to leaf-hopper and weevil. Indian J. Agric. Sci. 46: 278–280.
204. Rammah, A.M. 1981. Survey of alfalfa and clover diseases. Final Report to the USDA on Grant FG-EG-183, Survey and identification of alfalfa diseases and a search for resistant and tolerant cultivars.
205. ----. 1984. Personal communication, Field Crops Res. Inst., Giza, Egypt.
206. ----, and A.S. Hamza. 1980. Cutting schedules and seasonal effects on yield and quality of alfalfa. Ann. Agric. Sci. 14: 61–74.
207. Raynal, G. 1982a. Répartition géographique et importance relative des maladies de la luzerne en France: résultats d'une enquête nationale et conseils pour la lutte. Le Sélectionneur Français 30: 49–56.
208. ----. 1982b. Comportement au champ de quelques cultivars de luzerne vis-à-vis d'*Urophlyctis alfalfae* (Lagh) Magn. Agronomie 2: 197–200.
209. ----, and P. Guy. 1977. Répartition et importance des maladies de la luzerne en France et en Europe. Fourrages 71: 5–14.
210. Reyes, F.E. 1979. Guia para cultivar alfalfa en el Valle del Guadiana, Dgo. Desplegable para Productores CAEVAG 1. Campo Agric. Exp. Valle del Guadiana, Durango, Mexico.
211. Rivadeneira, H., and A. Davidovich. 1970. Incidencia de meteorismo en animales que pastorean una pradera de alfalfa. I. Efecto del grado de madurez de la pradera y de la suplementacion con heno de avena. Agric. Tec. (Chile) 30: 96–100.
212. Rockfeller Foundation Annual Report (RFAR). 1961. Program in Agric. Sci. Ser. 1960–61. p. 140–141, 210.
213. Rod, J. 1984. Personal communication, Research and Breeding Inst. for Fodder Plants, Troubsko, Czechoslovakia.
214. ----, and Pelikan. 1983. International variety trial with lucerne for seed (Eucarpia). Final report. Vyzkumny a slechtitelsky ustav picninarsky, Troubsko.
215. ----, and J. Vondracek. 1981. Contribution to component selection for synthetic varieties of lucerne. p. 179–186. *In* A. Gallais (ed.) Quantitative genetics and breeding methods. INRA, Versailles, France.

216. Rogers, U.E. 1967. Adaptability of lucerne to soil and climate. p. 36–46. *In* R.H.M. Langer, et al. (ed.) The lucerne crop. Reed, Wellington, New Zealand.
217. Rotilli, P. 1984. Personal communication, Istituto sperimentale per le Colture Foraggere, Lodi, Italy.
218. ----, and L. Zannone. 1981. Alfalfa breeding for forage yield and low saponin content. p. 229-232. *In* J.A. Smith and V.W. Hays (ed.) Proc. 14th Int. Grassl. Congr., Lexington, KY. Westview Press, Boulder, CO.
219. Rumbaugh, M.D., and W.L. Graves. 1983. The *Medicago* germplasm resources of Morocco. Abstracts Western Alfalfa Improvement Conference. p. 18-21.
220. Sakamoto, Y. 1980. Ecological studies of clover weevils. J. Coll. Dairy. 8: 113-248.
221. Sanchez, D.A., and D.M. Ramirez. 1963. Folleto de Divilgacion no. 32. Sec. Agric. Ganaderia, INIA, Mexico.
222. Sankhla, G.R., and K.S. Kushwaha. 1969. Biology and external morphology of forage pests. II. The common tree-hopper, *Tricentrus bicolor* Distant. Indian J. Agric. Sci. 39: 132-161.
223. Sawada, Y. 1982. Interactions of rhizobial nodulation of alfalfa and root rot caused by *Fusarium oxysporum*. Bull. Natl. Grassl. Res. Inst. 22: 19-26.
224. Schmidt, L., and B. Martin. 1977. Empfehlungen zum Luzerne–und Luzernegrasanbau in der DDR. Publ. Agra. Markkleeberg DDR.
225. Scofield, C.S. 1908. The botanical history and classification of alfalfa. USDA Bureau of Plant Industry. Bull. 131., p. 11–19.
226. Secretaria de Agricultura y Recursos Hidraulicos (SARH). 1981. Guia para cultivar alfalfa en Guanajuato. Folleto para Productores 4. Campo Agric. Exp. de el Bajio, Mexico.
227. Seif El-Nasar, H.I., and K.T. Leath. 1983. Crown and root fungal diseases of alfalfa in Egypt. Plant Dis. 65: 509-511.
228. Siddiqi, M.R. 1963. Descriptions of *Chitwoodis brasiliensis* n. sp. *Chitwoodis rusticulus* n. sp. and *Vanderlindia venata* n. sp. (Dorylamida Tylencholaimidae: Vanderlindiinae). J. Nematol. 15: 192-197.
229. ----. 1979. Four new species of *Xihinema* Coob, 1913 (Nematoda: Dorylaimida) from East Africa. Rev. Nematol. 2: 51-64.
230. Siebert, M. 1977. Experiences from the introduction of efficient field forage growing in Iraq. p. 364-369. *In* E. Wojahn and H. Thöns (ed.) Proc. 13th Int Grassl. Congr., Leipzig, German Democratic Republic. 18–27 May. Akademie-Verlag, Berlin.
231. Sikora, I. 1974. Procjena fenotipskih i genetskih parametara u panonskoj lucerni i njihovo koristenhe u procesu oplemenjuvanja. Z. Rad. Poljopr. Inst. Osijek 4.
232. Simon, U. 1971. Einfluss der Zusammensetzung auf den Ertrag synthetischer Luzernesurten. Z. Acker Pflanzenbau. 134: 165-174.
233. ----. 1984. Personal communication, Lehrstuhl fur Grunland und Futterbau Technische Univ., Munchen, West Germany.
234. ----, A. Spanakakis, and H. Scheller. 1975. Combining effects in a complete reciprocal diallel cross of nine non-inbred lucerne clones. Z. Planzenzuecht. 75: 322-330.
235. Sinskaya, E.N. 1950. Flora of cultivated plants of the U.S.R.R. XIII Perennial Leguminous plants. Part I. Medic, sweetclover, fenugreek. Translated by Israel Program for Scientific Translations, Jerusalem 1961.
236. Small, E. 1982. Medicago collecting in Turkey. Plant Genet. Resources, Newsl. 49: 11-12.
237. Smith, A. 1986. Alfalfa evaluation in South Africa. p. 61. *In* Rep. 30th North Am. Alfalfa Improve. Conf., Univ. of Minnesota, St. Paul. 27-31 July. Curlie Printing Co., Minneapolis.
238. Soto, K. 1979-1983. Informes Tecnicos Anuales. Area de Producion Animal. Instituto de Investigaciones Agropecuarias, Estacion Experimental La Platina, Santiago, Chile.
239. Sowa, S., I. Mostowska, and L. Grygorczyk. 1980. Wplyw niektorych czynnikow na zapylanie i plonowanie lucerny nasiennej w warunkach wojewodztwa dsztynskiego. Zesz. Nauk. Akad. Roln.-Tech. Olsztynie, Roln. 30: 247-258.
240. Spanring, J. 1986. Personal communication, INDOL Center za biotehniko, Ljubljana, Yugoslavia.
241. Stancil, M. 1981. Brazil: Agriculture and trade policies. USDA FAS M-305,. U.S. Government Printing Office, Washington, DC.
242. Staszewski, Z. 1980. Metoda hodowli heterozyjnych mieszancow lucerny. Biul. Inst. Hodowli Aklim. Rosl. 142: 6.
243. ----. 1984. Personal communication, Inst. of Plant Breeding, Radzikow, Poland.
244. Steuckardt, R., K. Hempel, H. Knüpfer, and A. Meinel. 1980. Der Einfluß desβefalls mit *Verticillium albo- atrum* Rke et Berth auf den Ertrag und verschiedene Qualitatsmerkmale von Luzernevarietäten und-stammen mit unterschiedlichem Resistenzgrad. Arch. Zuechtungsforsch. 10: 383-393.
245. ----, and D. Kohls. 1981. The breeding of the lucerne variety 'Verco' and the value of resistant varieties for fodder production. p. 247-252. *In* Proc. Eucarpia Conf., Kompolt, Hungary.

246. ——, and C. Schiefer. 1981. Comparison of results from test-crosses to diallel-crossing scheme and of experimental plan I of Comstock and Robinson to test the gca-and sca-effects in lucerne; and conclusions for an effective breeding method. p. 36–48. *In* Proc. Eucarpia Conf., Kompolt, Hungary.
247. Stevens, E.J., and P. Villalta. 1983. Grassland development in the southern highlands of Peru. p. 370–371. *In* J.A. Smith and V.W. Hays (ed.) Proc. 14th Int. Grassl. Congr., Westview Press, Boulder, CO.
248. Stewart, G. 1926. Alfalfa growing in the United States and Canada. Macmillan Publishing Co., New York.
249. Strange, R. 1958. Preliminary trials of grasses and legumes under grazing. East Afr. Agric. J. 24: 92–102.
250. Strange, L.R.N. 1980. An introduction to African pastureland production. FAO, Rome.
251. Stylopoulos, E., and T. Vaitsis. 1984. Personal communication, Inst. for Fodder Crops, Larissa, Greece.
252. Sundaram, P., P. Kandasamy, G.S. Thangamuthu, S.S. Rao, and A. Gopalan. 1973. Uptake of nitrogen and phosphorus by lucerne (*Medicago sativa* L.). Madras Agric. J. 60: 973–976.
253. Tasei, J.N. 1972. Lucerne pollination by honey bees in irrigated arid lands of Morocco. Apidologie 3: 105–124.
254. Te Velde, H.A. 1967. Production of lucerne in the Netherlands. Acta Univ. Agric. Fac. Agron. (Brno): 125–129.
255. Teyssendier de la Serve, B., and J.P. Boutin. 1972. Polymorphism of lucerne: the effects of a selection for germination with sodium chloride on Gabes cultivar. Ann. Amélior. Plant. 22: 225–232.
256. Thompson, A.H. 1985. A preliminary survey of fungal diseases of lucerne in the Republic of South Africa. Tech. Commun. 197. Dep. of Agric. and Water Supply.
257. Thorp, T.K. 1979. A Zambian handbook of pasture and fodder crops. FAO, Rome.
258. Toth, E. 1984. Personal communication, Res. Inst. for Irrigation, Szarvas, Hungary.
259. ——, and B.R. Bakheit. 1983. Results of resistance breeding in alfalfa. II. Resistance to Verticillium wilt. Acta Agron. Acad. Sci. Hung. 32: 78–85.
260. Tournefort, J.P. De. 1700. Medicago, id est planta ad Medicam accedens. Inst. Herb. p. 412. Cited by Scofield (225).
261. Tsuchiya, F., J. Maruyama, and T. Komatsu. 1985. Seasonal ground freezing in agricultural land and root breakage of alfalfa. p. 77–81. *In* Fourth Int. Symp. on Ground Freezing, Sapporo, Japan.
262. Varga, P., and S. Culica. 1980. La digestibilité "in vitro" un objectif de perspective dans l'amélioration de la luzerne. Bull. Acad. Sci. Agric. For. 10: 43–47.
263. ——, L. Gumaniuc, and M. Ittu. 1981a. Identificarea unor noi surse de precocitate, resistenta la vestejirea fusariana si la cosiri precvente, la lucerna. An. Inst. Cercet. Cereale Plante Teh. 48: 99–105.
264. ——, ——, and ——. 1981b. Soiul de lucerna Lutetia. An. Inst. Cercet. Cereale Plante Teh. 48: 107–112.
265. ——, I. Moga, E. Kellner, C. Balan, and M. Ionescu. 1973. Lucerna. Ed. Ceres. Bucarest.
266. Velez-Santiago, J., J.A. Arroyo-Aguilar, S. Torres-Rivera, and N. Corchado Juarde. 1983. Performance and chemical composition of 18 nondormant alfalfa cultivars at the Lajas Valley. J. Agric. Univ. P. R. 67: 204–212.
267. Verezoni, F., S. Arcioni, A. Mariani, and M. Falcinelli. 1982. Risposta alla selezione per la produzione foraggera in erba medica. Riv. Agron. 16: 366–372.
268. Versteeg, M.N., I. Zipori, J. Medina, and H. Valdivia. 1982. Potential growth of alfalfa (*Medicago sativa* L.) in the desert of Southern Peru and its response to high NPK fertilization. Plant Soil 67: 157–165.
269. Wanjala, B.W.K. 1986. Personal communication, National Agric. Res. Stn., Kitale, Kenya.
270. Westgate, J.M. 1910. Variegated alfalfa. USDA Bureau of Plant Industry Bull. 169. U.S. Government Printing Office, Washington, DC.
271. Wheeler, W.A. 1951. Beginnings of hardy alfalfa in Northern America. Seed World 69(5): 8, 10; (6): 16, 18, 47; (7): 20, 22, 23; (8): 20, 22, 45; (9): 24, 49–50; (10): 10–12, 46; (11): 24, 44–45.
272. Whyte, R.O., G. Nilsson-Leissner, and H.C. Trumble. 1953. Legumes in agriculture. FAO Agric. Studies, Series 21. Rome, Italy.
273. Wilsie, C.P. 1962. Crop adaptation and distribution. Freeman, San Francisco.
274. Winter, W. 1982. Untersuchungen über Luzernekrankhuten in der deutschprachigen Schweiz in den Jahren 1978 bis 1980. Mitt. Schweiz. Landwirtsch 30: 118–124.
275. Wojahn, E. 1977. Review of 50 years grassland farming and trends of development in the further intensification of forage production in the German Democratic Republic. p. 3–29. *In* E. Wojahn and H. Thöns (ed.) Proc. 13th Int. Grassl. Congr. Leipzig, German Democratic Republic. 18–27 May. Akademie-Verlag, Berlin.

276. Wolffhardt, V.D. 1981. Zum anbau von Feldfutter in Osterreich-ein Uberblick über die Entwicklung. p. 113-132. *In* Festschrift 100 Jahre Bundesanstalt für Pflanzenbau und Samenprüfung in Wien, 1881-1981.
277. ----. 1984. Personal communication, Federal Institute for Plant Production and Seed Testing, Vienna.
278. Worker, G.F. Jr. 1975. Field crops and variety comparison between Imperial Valley, California, and Kufra Oasis, Libya. J. Agric. Sci. 84: 215-219.
279. Wu, Renrun. 1956. A report of preliminary tests of alfalfa cultivation in Wuhan Region. Agron. J. Central China 4: 203-217.
280. ----. 1964. Research of alfalfa winter survival. p. 1-19. *In* The 1964 Annual Test Rep. of Northwestern Animal Husbandry and Veterinary Institute.
281. ----. 1986. Personal communication, Chinese Academy of Agricultural Sciences, Lanzhou Institute of Animal Science, Lanzhou, Gansu, People's Republic of China.
282. Wu, Yongfu. 1982. A preliminary report of a male sterile line. p. 183-184. *In* Symp. of the 2nd Meet. Chinese Grassl. Soc.
283. Wynn-Williams, R.B. 1982. Lucerne establishment—conventional. p. 11-19. *In* R.B. Wynn-Williams (ed.) Lucerne for the 80's. Spec. Pub. 1. Agronomy Society of New Zealand, Palmerston North.
284. Xu, Lingren. 1982a. The transpiration of alfalfa and its effect on growth of alfalfa. p. 150-151. *In* Symp. 2nd Meet. Chinese Grassl. Soc.
285. ----. 1982b. Research on variation of nutrition in some common alfalfas from seedling to harvest. p. 147-148. *In* Symp. 2nd Meet. Chinese Grassl. Soc.
286. Xu, Su. 1979. The induced success of alfalfa plants from pollen. p. 84-88. *In* Symp. 1st Meet. Chinese Grassl. Soc.
287. Yano, K., T. Miyake, and S. Hamasaki. 1982. Discovery of a spotted alfalfa-like aphid *Therioaphis trifolii* Monell s. lat. in Japan. Jpn. J. Appl. Entomol. Zool. 26: 35-40.
288. Yu, Lile, Yue Wu, and Renrun Wu. 1982. A test report of the introduction of new varieties of herbage. The Soil and Fertilizer Institute of the Academy of Agricultural Science of Hubei Province. p. 36-37.
289. Zambrana, T. 1970. Effect of sowing distance on alfalfa yield. Rev. Cubana Cienc. Agric. 4: 131-135.
290. ----. 1971. Influence of the stage of maturity at harvest on the productivity and population density of alfalfa. Rev. Cubana Cienc. Agric. 5: 119-123.
291. ----. 1972a. Optimal sowing date for alfalfa seed production in Cuba. Rev. Cubana Cienc. Agric. 6: 139-142.
292. ----. 1972b. Effect of the population density on the components of yield in alfalfa. Rev. Cubana Cienc. Agric. 6: 143-151.
293. ----, and M. Oduardo. 1972. Evaluation of the agronomic performance of advanced generations of alfalfa seed. Rev. Cubana Cienc. Agric. 6: 129-137.
294. Zambrano, R. 1977. Herencia del caracter pubescencia de hojas y tallos en ecotipos peruanos de alfalfa. Invest. Agropecu. 7(1-2): 3-9.
295. ----. 1984. Personal communication, Pueblo Libre, Lima, Peru.
296. Zhang, Xueshang, and Yuan Ruiming. 1957. The survey and research on alfalfa in Northwest China. The Shaanx: Peoples Press, p. 79.

3 The Genus *Medicago* and the Origin of the *Medicago sativa* Complex[1]

CARLOS F. QUIROS

University of California
Davis, California

GARY R. BAUCHAN

USDA-ARS
Beltsville, Maryland

The most recent analysis of the genus *Medicago*, is the comprehensive taxogenetic study of Lesins and Lesins (40), which revised and expanded the chapter written by Lesins and Gillies (35) for *Alfalfa Science and Technology* (Agronomy monograph 15). In this chapter, we attempt to update and complement these two treatises focusing on recent research developments dealing with the origin and evolution of *Medicago*.

3-1 CENTERS OF DIVERSITY

The genus *Medicago* is very extensive comprising more than 60 different species, two-thirds of which are annuals and one-third perennials (40). *Medicago* is endemic to the Mediterranean region, spreading to Spain and to the Canary Islands toward the West, to People's Republic of China toward the East, to Siberia toward the North and to Yemen toward the South. The primary center for the genus is found in the Caucasus, northwestern Iran and northeastern Turkey (20).

3-2 PLOIDY

The basic genomic number of *Medicago* is $x=8$, except for the annual species *M. constricta* Dur., *M. praecox* DC., *M. polymorpha* L., *M. rigidula* (L.) All., and *M. murex* Willd. which have a genomic number of $x=7$. Cytological studies (41) indicate that the last four species might have arisen independently by chromosomal translocations involving two chromosomes and the subsequent loss of the resulting centric fragment.

---

[1] Dedicated to the memory of the late Professor Karl Lesins.

Copyright 1988 © ASA-CSSA-SSSA, 677 South Segoe Road, Madison, WI 53711, USA. *Alfalfa and Alfalfa Improvement*—Agronomy Monograph no. 29.

On the other hand, *M. constricta* seems to have evolved from *M. murex* by additional chromosomal rearrangements.

Three ploidy levels are found among the *Medicago* spp., diploids $2n=2x=14$ and $2n=2x=16$, tetraploids ($2n=4x=32$), and hexaploids $2n=6x=48$. Recently, Bauchan and Elgin (4) reported that the species *M. scutellata* (L.) Mill. and *M. rugosa* Desr. have $2n=30$. These might be allotetraploids resulting from the hybridization of $2n=14$ and $2n=16$ species, followed by polyploidization. Since the majority of the *Medicago* spp. are diploids, it is likely that the basic evolution of the genus has taken place at this level. The tetraploid species probably arose by unreduced gametes (46,51,78), giving rise to vigorous and highly heterozygous individuals, aggressive enough to colonize new habitats and surpass the range of distribution of diploids (55). Two alloautoploid hexaploids species have been reported, *M. cancellata* M.B. and *M. saxatilis* M.B., both perennials. Hexaploid accessions of the otherwise tetraploid species *M. arborea* L. have been reported (40). Presumably these are autohexaploids.

3-3 BREEDING SYSTEMS

Annual species of *Medicago* are autogamous, as a result of the presence of self-tripping in their flowers. In conformity with their breeding system, progenies of annual plants are remarkably uniform (57). Conversely, most perennial species are allogamous, with different degrees of self-incompatibility. Although some self-tripping occurs in these species, as a rule they rely on insect pollinators to activate their flower tripping and pollination (40). Most of these species are quite polymorphic, because of their high degree of outcrossing (54). Their pollinating agents include numerous species of bees.

3-4 EVOLUTION

According to Lesins and Lesins (40), the evolution of the genus took place during the Tertiary Period. Important geological events occurred at that time, including the formation of prominent mountain ranges, such as the Alps, Pyrenees, Apennines, Himalayas, and the Tien Shan. The northern coast of the Mediterranean appears to be the area of origin of perennial species based on their present range of distribution. During part of that time, the Mediterranean basin was a hot desert, resulting from the intermittent closure of Gibraltar. This provided new habitats for the evolution of annuals from the perennial species. The annual species with a short life cycle and seed dormancy may have colonized newly opened areas subject to intermittent flooding. Also, the change in longevity from perennial to annual, was accompanied by the self-pollinating trait, an essential reproductive strategy because of their isolation and lack of pollinators in newly colonized habitats. The final opening of the Gibraltar resulted in the extinction of numerous species, leaving those that remained on mountain peaks, which now form several Mediterranean islands. The annual species, as a younger group, may not contribute much to understanding the area of origin of the genus. The shrub *M. arborea*, the oldest member of the genus, is genetically too distant in its present tetraploid and hexaploid forms to be considered an immediate progenitor of the diploid perennial species (40). *Medicago carstiensis* Wulf. may be a possible relic, because of its unique physiological and morphological traits (19). It thrives in shade and requires exposure to a cold treatment in order to flower.

3-5 TAXONOMY

3-5.1 Identification of the *Medicago* Species

3-5.1.1 Useful Traits to Distinguish Different Taxa

Legume and seed characteristics are considered of great taxonomic value in distinguishing *Medicago* spp. (40). Other important traits for taxonomic purposes are growth habit and longevity, organ hairness, inflorescenses: bracts, stipules, and florets; leaves, cotyledons, and chromosome numbers. Lesins and Gillies (34) and Lesins and Lesins (40) provide excellent photographic records of most important morphological traits used in the identification of *Medicago* spp. On the basis of these traits, useful keys have been constructed as guides for species identification. However, these classification keys using macroscopic characteristics fail in several instances to group related taxa. Thus, *M. sativa* spp. *falcata* Arcangeli., although a subspecies of *M. sativa* L., is grouped with other less related perennials, because of its yellow flowers and uncoiled pods.

There seems to be enough distinction in the characteristics of pollen grains to use them as a criteria in differentiating many species (37) and related genera (68). Under the light microscope it is possible to distinguish two broad groups of pollen grains on the basis of shape: cylindrical, including spindle and block shapes, and pyramidal, including triangular and tetragonal bisphenoids. Electron microscopy reveals more subtle differences for each of these two categories. For example, *M. arborea* has cylindrical pollen grains (Fig. 3–1A), while *M. sativa* have elliptical grains (Fig. 3–1C). The grains of *M. rhodopea* Velen., although cylindrical, are wider at one end (Fig. 3–1E). Some accessions of *M. tornata* Mill. show wing-like protuberances in the colpi (Fig. 3–2A and 3–2B). Within the pyramidal shapes, the pollen of *M. soleirolii* Duby and *M. hybrida* Trautv. are good representatives of triangular pyramids (Fig. 3–1D and 3–1F), while that of *M. leiocarpa* Benth., is representative of tetragonal bisphenoids (Fig. 3–1B). Although differences in exine sculpting are evident

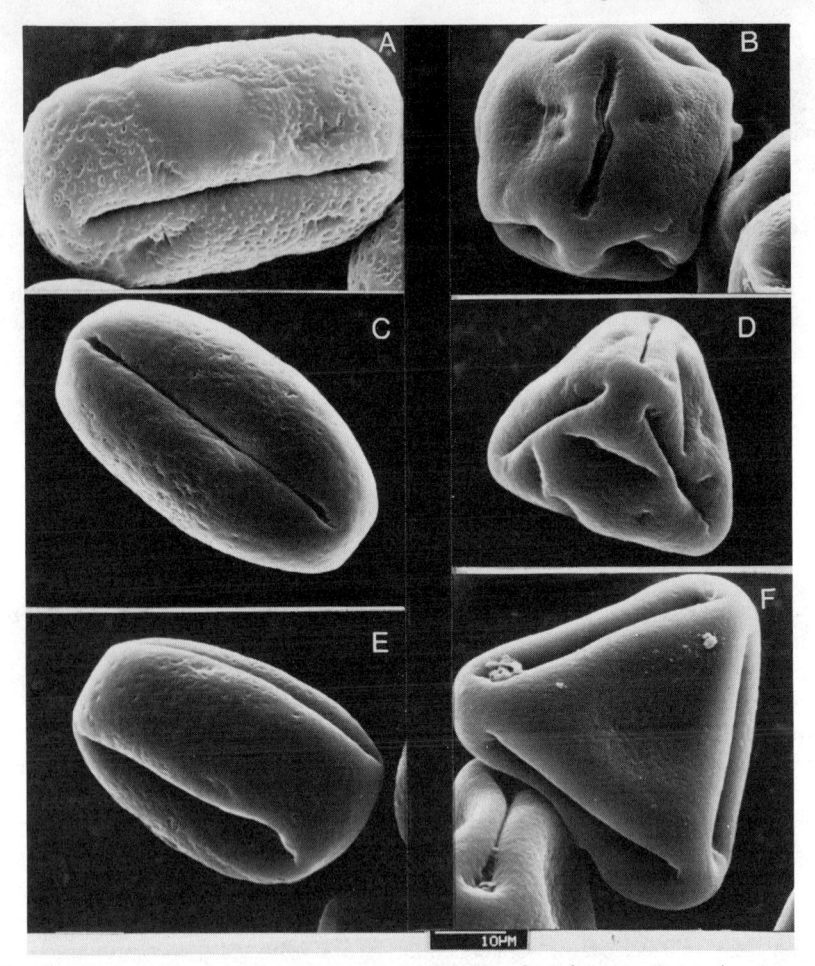

Fig. 1A–1F. Scanning electron microscope photographs of *Medicago* pollen grains, 1 cm = 10 μm. *Fig. 1A. M. arborea*, UAG 31, cylindrical. *Fig. 1B. M. leiocarpa*, UAG 555, tetrahedral bisphenoid. *Fig. 1C. M. sativa* ssp. *sativa* elliptical. *Fig. 1D. M. hybrida*, UAG 1637, pyramidal. *Fig. 1E. M. rhodopea*, UAG 493, cylindrical, with wider end. *Fig. 1F. M. soleirolii*, UAG 2399, pyramidal.

among various species and some accessions of the same species (Fig. 3-2C to 3-2F), they are not elaborate.

Numerical taxonomy has been applied to study the *Medicago* spp. Small (67) recognized 12 species groupings in the genus, by numerical taxonomic analysis based on 75 characters, mostly vegetative and pod traits. The classification of the annual *Medicago* spp. was in good agreement with taxonomic conclusions reached by other researchers, although several discrepancies were found. This might be due to the fact that Small conducted his study on herbarium specimens of plants grown in their natural habitats. Many traits are known to be influenced by environment, thus lacking constant expression, i.e., leaflet length and narrowness, height,

ORIGIN AND EVOLUTION OF *MEDICAGO* 97

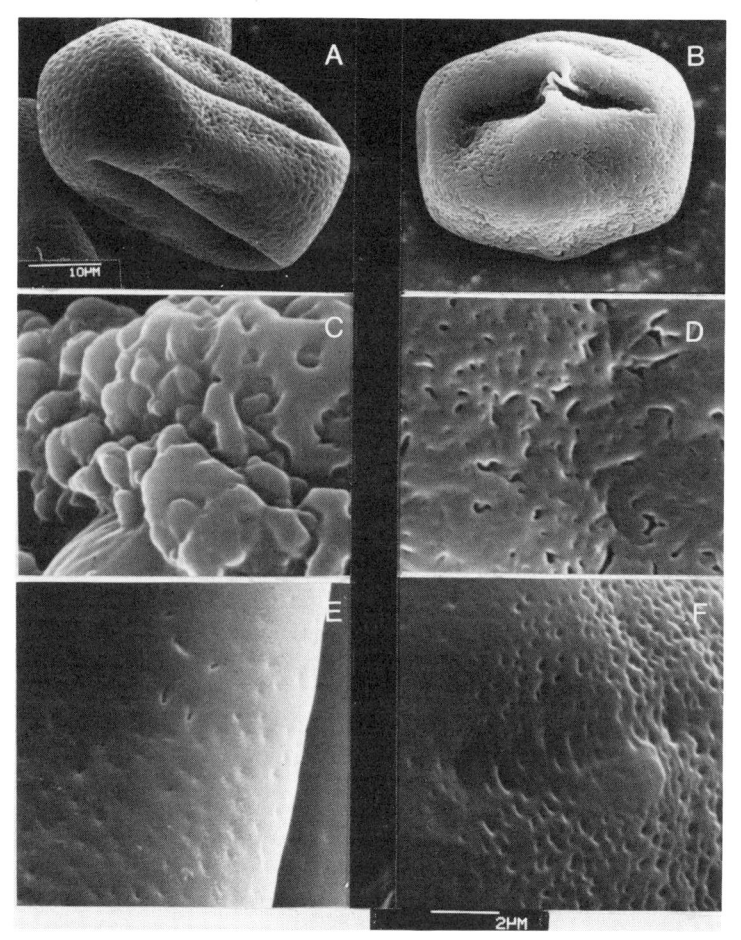

Fig. 2A–2F. Scanning electron microscope photographs of *Medicago* pollen grains. *Fig. 2A–2B*: 1 cm = 10 μm, *Fig. 2C–2F*: 1 cm = 2 μm. *Fig. 2A. M. tornata*, UAG 1590, cylindrical. *Fig. 2B. M. tornata*, UAG 1324. *Fig. 2D. M. sativa* ssp. *sativa* exine. *Fig. 2E. M. soleirolii* exine, UAG 2399. *Fig. 2F. M. leiocarpa* exine, UAG 555).

and other quantitative characteristics. Therefore, it is likely that part of the observed variation was environmental.

Small and Lefkovitch (73) attempted to discriminate the subspecies of *M. sativa* by agrochemotaxometry. That is, by numerical analysis involving morphological and chemical variables. They analyzed plants of each subspecies for various chemical elements and for amounts of structural components. They found significant differences in S and Ca content among subspecies, but this measurement was not as reliable in making separations on the basis of pod and flower characteristics. More recently, Small and Brookes (70) used canonical analysis to define intermediate types in the *M. sativa* complex. Numerical taxonomy has been used to reclassify the annual *Trigonella asclersoniana* Urb. as *M. hypogaea* Small under the section Spirocarpus (71). In another study, Small and Brookes

(72) reclassified the $x=8$ accessions of *M. murex* as *M. lesinsii* Small by discriminant analysis.

The ideal approach to elucidate taxonomic relationships in *Medicago* would involve the use of monogenic traits of constant expression in most environments. The technique of enzyme electrophoresis discloses many genetic markers that comply with these requirements. Furthermore, isozymes allow the detection of homozygous and heterozygous individuals, facilitating the determination of genetic variability and amount of heterozygosity in natural populations. These markers have proved useful in the identification of the *Medicago* spp. and their hybrids. For example, the species *M. turbinata* (L.) All. and *M. truncatula* Gaertn. have different alleles at three loci, while *M. soleirolii* and *M. tornata* can be distinguished from each other on the basis of at least one locus (55,57). Furthermore, *M. sativa* ssp. *falcata* has a few specific allozymes not found in the closely related *M. sativa* ssp. *sativa* (L.) L.&L. and *M. sativa* ssp. *coerulea* Schmalh. (57). It is likely that with the development of additional enzyme systems in *Medicago*, the number of criteria to separate species will increase substantially. Another powerful tool that is becoming available to taxonomist involves the analysis of DNA restriction fragment length polymorphisms (RFLP analysis). This technique may prove to be more sensitive than the isozymes due to its ability to detect variation directly at the DNA level (5).

3–5.1.2 Key to *Medicago* Species

The general key constructed by Lesins and Lesins (40) is the best available guide for the identification of the *Medicago* spp. Their key is presented for reference purposes, together with a few modifications, mostly additions.

1	Perennials	2
—	Annuals; yellow flowers	25
2	Shrubs	*M. arborea* L.
—	Herbs	3
3	Corolla yellow	4
—	Corolla anthocyanin-colored; Violet, or variegated (i.e., all shades between violet and yellow)	23
4	Pods small (<3.5 mm long), kidney-shaped nutlets, containing one seed. Florets small (1–4 mm long)	*M. lupulina* L.
—	Pods and florets larger	5

5	Pods straight or sickle-shaped, coiled in not more than one half-circle	6
–	Pods coiled in more than one half-circle	10
6	Pods narrow, 1 to 3 mm in width	*M. sativa* ssp. *falcata* Arcengeli
–	Pods more than 3 mm in width	7
7	Radicles and cotyledons of seeds with their long axes almost at right angles to pod's ventral suture	8
–	Radicles and cotyledons of seeds with their long axes almost parallel to the ventral suture	*M. hybrida* Trautv.
8	Pods 14 to 22 mm long, elliptical leaflets	*M. platycarpa* Trautv.
–	Pods 8 to 12 mm long	9
9	Pods wide, length <1.5 times width, florets with wings larger than the keel.	*M. cretacea* M.B.
–	Pods elongated, length two or more times width, florets with wings almost as large as the standard	*M. ruthenica* Ledeb.
10	Pods with spines or tubercles	11
–	Pods without spines or tubercles	15
11	Plants with rhizomes. Spines uniformly thick, 3 to 7 mm long. Seed with radicle at 45° angle to pod's ventral suture	*M. carstiensis* Wulf.
–	Plants without rhizomes. Spines thick at the base, short and rigid, or tubercles only. Seed with radicle almost parallel to the ventral suture	12
12	Pods and whole plants covered thickly with felted hairs; seashore plants	*M. marina* L.
–	Pods and plants not covered with felted hairs; not seashore plants	13
13	Pods 5 to 7 mm in diameter	14

—	Pods <5 mm in diameter. Usually entire, awl-shaped stipules.	*M. rhodopea* Velen.
14	Pods with glandular, articulated hairs. Standard oblong with sides parallel in the middle part $2n=16$.	*M. pironeae* Vis.
—	Pods without glandular, articulated hairs. Standard oval $2n=48$	*M. saxatilis* M.B.
15	Pod surface corrugated by a net of elevated veins	16
—	Pod surface smooth (after removal of hairs), or with only slightly elevated veins	17
16	Pods with one to three coils, 4 to 6 mm in diameter. $2n=48$, often narrow leaflets	*M. cancellata* M.B.
—	Pods with 1 to 1.5 coils, 3 to 4 mm in diameter. $2n=16$.	*M. rupestris* M.B.
17	Pods covered with articulate, semitransparent, as if membranous, glandular hairs. $2n=16, 32$.	*M. papillosa* Boiss.
—	Pod surface without glandular hairs. $2n=16, 2n=32$.	*M. dzhawakhetica* Bordz.
18	Leaflets nearly round, broadly ovate to obcordate even at upper nodes, (length/width, 3:2), upper side of leaflets glabrous. Pollen grains triangular-pyramid-shaped.	*M. suffruticosa* Raymond
—	Leaflets oblong, narrowly ovate, at least at upper nodes, (length/width, more than 2:1); upper side of leaflets always somewhat hairy. Pollen grains spindle-cylindrical	19
19	Pods small, 2 to 4 mm in diameter, with or without glandular hairs	*M. prostrata* Jacq.
—	Pods larger, 5 to 9 mm in diameter, covered with glandular hairs	20

20	Pods large, 8 to 9 mm in diameter, with one to two loose coils, open in the center. In the keel and in the middle of the standard the yellow color is often of different intensity from the rest of the corolla.	*M. sativa* ssp. *glutinosa* M.B.
—	Pods smaller, 5 to 8 mm in diameter, coiled in 1.5 to 4 tight coils with only a small opening in the center. Corolla color uniform.	*M. glomerata* Balb.
21	Pods with spines	*M. daghestanica* Rupr.
—	Pods without spines	22
22	Corolla violet; standard and petals with parallel sides. Pods coiled in 1.5 to 4.0 tight coils, 2 to 5 mm diam, $2n=16$.	*M. sativa* ssp. *coerulea* Schmalh.
—	Flowers variegated, standard petals oval, pods coiled in 0.5 to 2.5 loose coils $2n=16, 32$	*M. sativa* ssp. × *varia* Martin, *M. sativa* ssp. × *hemicycla* Grossh.
23	Corolla violet, standard petals with parallel sides, pods coiled in two to five tight coils, 5 to 9 mm diam, $2n=32$	*M. sativa* ssp. *sativa* (L.) L.&L.
—	Flowers variegated, standard petals oval.	24
24	Pods with two to five coils, with a large opening in the center, covered with glandular hairs	*M. sativa* ssp. × *tunetana* Murbeck
—	Pods twisted with one to two coils glandular hairs, with a small opening in the center, covered with glandular hairs	*M. polychroa* Grossh.
25	Veins on face of pod running obliquely from pod's ventral suture, do not change direction before joining the dorsal suture. Pods small, one-seeded nutlets. Pollen grains spindle-cylindrical	*M. lupulina* L.
—	Veins on face of pod change direction before joining the dorsal suture. Pollen grains triangular pyramid-shaped	*M. secundiflora* Durieu

26	Seeds with long axes almost at right angle to pod's ventral suture; seedcoats ridged or verrucose, pods larger than 3.5 mm long with at least one-fourth coils	27
–	Seeds with long axes almost parallel to the ventral suture; seedcoats smooth	29
27	Pods spiny. Seedcoats ridged	28
–	Pods flat and spineless. Seedcoats verrucose	*M. orbicularis* Bart.
28	Coils with one row of spines, coil edge paper-thin	*M. radiata* L.
–	Coils with two rows of spines, coil edge wide.	*M. heyniana* Greuter
29	Seeds black or red brown	30
–	Seeds yellow or yellow brown	33
30	Pods large, with 8 to 10 coils, florets large, 8 to 10 mm long; seeds large, 13 to 17 g/1000	31
–	Pods smaller, with five to seven coils; florets smaller, 5 to 7 mm long, seeds smaller, 7.5 to 10 g/1000	32
31	Pods and spines with glandular, articulated hairs	*M. ciliaris* All.
–	Pods glabrous or with a few simple hairs, leaflets often with anthocyanin spot	*M. intertexta* Mill.
32	Pods cylindrical or disk-shaped with truncate apex and base	*M. muricoleptis* Tineo
–	Pods more round, barrel-shaped	*M. granadensis* Willd.
33	Pods densely covered with long hairs, resembling small cotton balls	*M. lanigera* Winkl. & Fedtsch.
–	Pods without long hairs	34
34	Coils imbricate like a set of bowls, with their convex parts towards apex and base, or towards base only	35
–	Coils not markedly imbricate	36

35	Coils with their convex parts towards pod base only. $2n=30$	*M. scutellata* Mill.
—	Coils with their convex parts towards both base and apex $2n=16$	*M. blancheana* Boiss.
36	Coil edges with ridges, or with wing-like elevations running obliquely or at right angle to the dorsal suture	37
—	Coil edges smooth, or spined, or with tubercles	39
37	Wing-like elevations on coil edges running at right angle to pod's dorsal suture. Pods small, 3 to 5 mm in diameter, usually with 1.5 coils	*M. shepardii* Post
—	Ridges on coil edges running obliquely towards the dorsal suture. Pods more than 5 mm in diameter with 2.5 to 5 coils	38
38	Dorsal suture usually in a groove in the middle of the coil edge. Calyx appressed to the base of the pod as a regular star. $2n=16$	*M. noeana* Boiss.
—	Dorsal suture elevated in the middle of the coil edge. Calyx appressed sideways to the base of the pod. $2n=30$	*M. rugosa* Desr.
39	Pollen grains shaped like irregular blocks, pods cylindrical, spines short (0.5–2 mm), inserted in the margins of the coil edges almost at right angle to the face of the coil; apical coil concave. Leaflets often notched and with anthocyanin basal fleck.	*M. rotata* Boiss.
—	Pollen grains spindle-cylinder, or triangular pyramid-shaped.	40
40	Pods soft-walled. Central part of each coil consisting mainly of veins with thin membranous tissue between them (coils may be pulled apart releasing the	

	seed). Spines, if present, slender, their base with two prongs (roots) connected by a membrane, one prong inserted in the dorsal suture, the other in the lateral vein or in a veinless zone	41
—	Pods hard-walled (for release of seed, crushing of pod may be necessary). Spines, if present, stocky, their base conical, often embedded in spongy tissue. Venation on the face of the coil usually not clearly discernible.	49
41	Face of coil with radial veins running into a veinless zone	42
—	Face of coil with radial veins running into a lateral vein	43
42	Pod edge grooved; spines slanted away from the apical coil; apical coil spineless	*M. disciformis* DC.
—	Pod edge level; spines pointing to both apical and basal end	*M. tenoreana* Ser.
43	Pod edge level or slightly concave, completely or almost completely covering grooves between the edge and lateral veins	44
—	Pod edge grooved, grooves between dorsal suture and lateral veins observable in edge on view	45
44	Peduncle usually many-flowered (up to 17 florets). $2n=16$	*M. coronata* Bart.
—	Peduncle few-flowered (1–2 florets). $2n=14$	*M. praecox* DC.
45	Dorsal suture of pod lying in a groove; on pod edge alternate four ridges with three grooves. Leaflets usually with dark spot	*M. arabica* Huds.
—	Dorsal suture of pod elevated above lateral veins.	46

46	Lateral veins on face of coil at one-third to two-fifths of the radius below the dorsal suture. Plants densely hairy. Stipules entire or slightly toothed.	*M. minima* Bart.
–	Lateral veins on face of coil at one-third or less of the radius below the dorsal suture. Stipules deeply incised	47
47	Florets with wings longer than the keel. $2n=14$	*M. polymorpha* L.
–	Florets with wings shorter than the keel. $2n=16$	48
48	Apical coil spiny; lateral veins on coil face joining as shoulders at right angle to the elevated dorsal suture	*M. laciniata* Mill.
–	Apical coil spineless; lateral veins only slightly protruding from the coil face	*M. sauvagei* Nègre
49	Coils spineless; coil face without lateral vein or veinless zone; coils tightly appressed	*M. soleirolii* Duby
–	Coil face with lateral veins or veinless zone	50
50	Radial veins ending in a veinless zone on coil face	51
–	Radial veins ending in a lateral vein on coil face	53
51	Veinless zone wide, about one-third of coil radius; upper side of leaves completely glabrous	52
–	Veinless zone narrower, one-fourth to one-fifth of coil radius, upper side of leaves at least sparsely hairy	*M. turbinata* All.
52	Leaflets with small whitish patches, longer and thinner stems, edge of pod coils with none to three ridges, $2n=16$	*M. lesinsii* Small
–	Leaflets without patches, shorter and thicker stems, edge of pod coil with three ridges, $2n=14$.	*M. murex* Willd.

53	Pods convex at both ends, subspherical or oval in shape. Young pod contracted and concealed within calyx.	*M. doliata* Carmign.
—	Pod ends truncate, pods cylindrical or subcylindrical	54
54	Coils of pod tightly appressed, with no slits between them in dry mature pods; juncture between individual coil edges not markedly depressed	55
—	A continuous or interrupted slit between coil in dry mature pods; coil edges usually sloped towards their juncture	57
55	No groove on pod edge between dorsal suture and lateral veins. Radial veins on pod face strongly curved, running almost concentric before joining the lateral vein. $2n=14$	*M. constricta* Dur.
—	On edge of immature pods a shallow groove between dorsal suture and lateral veins, disappearing at pod maturity. Radial veins only slightly curved. $2n=16$	56
56	Pods glabrous; dorsal suture usually not higher than margins of the edge; spines, if present, inserted at 180° to the plant of coil face or obliquely to it.	*M. littoralis* Rohde
—	Pods with sparse hairs; dorsal suture usually strongly protruding in the middle of coil edge; spines inserted at 90° or obliquely to the coil face	*M. truncatula* Gaertn.
57	Dorsal suture in the middle of an evenly convex pod edge. Radial veins somewhat curved. $2n=16$	*M. tornata* Mill.
—	Shallow groove between dorsal suture and lateral veins that disappears at pod maturity. Radial veins strongly curved. $2n=14$.	*M. rigidula* (L.) All.

3-6 THE *MEDICAGO SATIVA* COMPLEX

3-6.1 Species Forming the Complex

The taxa constituting the *M. sativa* complex belong to the section *Falcago*, subsection *Falcatae* within the genus *Medicago*, which includes diploid and/or tetraploid forms of the species, *M. sativa* ssp. *sativa*, *M. sativa* ssp. *falcata*, and *M. sativa* ssp. *glutinosa*. They intercross with each other and share the same karyotypes (14). These taxa have been given the hierarchy of species by some authors (40) and of subspecies by others (15). Considering that the main barrier for gene exchange among them is ploidy, which might be often broken by diploids generating unreduced gametes (46,68), it is justifiable to classify them as subspecies, especially when no hybridization barriers are present and where genetic evidence supports common ancestry. Other species closely related to the *M. sativa* complex forming part of the same gene pool are *M. glomerata* and *M. prostrata*.

On the basis of legume and flower characters, Gunn et al. (15) considered all taxa within the complex as subspecies of a single species, *M. sativa*. They recognized a total of nine subspecies. Some of these are questionable, however, because the range of variation for some of the traits used to separate the taxa, either overlap or did not agree with previous descriptions. For example, their descriptions for *M. sativa* ssp. × *varia* Arcangeli and *M. sativa* ssp. *ambigua* Tutin overlap for corolla color, being differentiated mainly on the basis of pilosity, pod shape, flower size; C-shaped to spiraled pilose for the former and, glabrous, falcate pod shape for the latter. Furthermore, Gunn et al. (15) recognized *M. sativa* ssp. *ambigua* as a diploid form. They apparently chose the name *ambigua* over *trautvetteri*, an epithet commonly used to describe this particular taxa, reported to occur in diploid and tetraploids forms (40). Gunn et al. (15) described *M. sativa* ssp. *glomerata* (Balbis) Tutin as lacking glandular hairs in the legume. However, Lesins and Lesins (40), Ivanov (20), Lubenets (44), and Sinskaya (66) among others, agreed in their description of *M. glomerata* as a species characterized by legumes covered with glandular hairs. Furthermore, changing the status of *M. glomerata* to *M. sativa* ssp. *glomerata* is unjustifiable because of the incipient genetic barrier, reflected in abnormal meiocytes and low fertility in F_1 hybrids between *M. sativa* and *M. glomerata* (26). Gunn et al. (15) apparently renamed *M. glutinosa*, a tetraploid with yellow flowers and coiled pods as *M. sativa* ssp. *praefalcata* Sinsk., and included under this taxa diploid and tetraploid accessions. The subspecies name *praefalcata* is misleading because it implies that it is the ancestor of *M. sativa* ssp. *falcata*.

The morphological variability in species of the *M. sativa* complex accounts for the proliferation of epithets to name them. In many situations, only subtle morphological differences which might have been the product of genetic recombination, have been accepted as sufficient evidence to create a new species or subspecies. With respect to this problem, Lesins and Gillies (35) wrote: "A most confusing situation arises when taxa interbreed freely and the offspring have unimpaired viability. On recombination of parental characters, hybrid swarms are produced covering the range of variability between the parental species. This is the situation with the *M. sativa*, *M. falcata*, and *M. glutinosa*." Evidence for this is the large number of separate species names assigned to hybrids which include more than 50 described species and close to 100 names below the species rank. Table 3-1, adapted from Lesins and Lesins (40) lists most of the epithets for each of the species in the complex and their hybrids. On the basis of the present taxogenetic evidence, eight subspecies

Table 3-1. Species epithets used for subspecies of the *Medicago sativa* complex. Adapted from Lesins and Lesins (39).

ssp. *sativa*, ssp. *coerulea*	ssp. *falcata*	ssp. *glutinosa*	*M. glomerata*
afghanica Vass.	*altissima* Grossh.	*gunibica* Vass.	*sativa* ssp. *glomerata* Rouy
agropyretorum Vass.	*borealis* Grossh.		
asiatica Sinsk.	*difalcata* Sinsk.		*sativa* ssp. *faurei* Maire
asiatica Sinsk.	*difalcata* Sinsk.		
coerulea Gunn et al.	*erecta* Kotov		
coerulea Less.	*glandulosa* David.		
jemenemsis Sinsk.	*quasifalcata* Sinsk.		
kopetdaghi Vass.	*romanica* Prod.		
lavrenkoi Vass.	*tenderensis* Opperm.		
mesopotamica Vass.			
orientalis Vass.			
polia Vass.			
sogdiana Vass.			
subdicycla Vass.			
tadzhicorum Vass.			
transoxana Vass.			

ssp. × *hemicycla*, ssp. × *varia*: Hybrids between ssp. *sativa* or ssp. *coerulea* and ssp. *falcata*		ssp. × *polychroa*: Hybrids between ssp. *sativa* and ssp. *glutinosa*	ssp. × *tunetana*: Hybrids between ssp. *sativa* or ssp. *coerulea* and *M. glomerata*
alaschanica Vass.	*lavrenkoi* Vass.	*glutinosa* M.B.	*tunetana* Vass.
alatavica Vass.	*media* Pers.	*grossheimii* Vass.	
beipinensis Vass.	*ochroleuca* M. Kult.	*polychroa* Grossh.	
caucasica Vass.	*roborovskii* Vass.	*virescens* Grossh.	
falcata L.	*rivularis* Vass.		
gaetula Trbut.	*sativa* L.		
grandiflora Vass.	*schischkinii* Sumn.		
hemicoerulea Sinsk.	*tianschanica* Vass.		
hemicycla Grossh.	*tibetana* Vass.		
komarovii Vass.	*trautvetteri* Sumn.		
kultiassovii Vass.	*vardanis* Vass.		
ladak Vass.	*varia* Mart.		

of *M. sativa* can be considered the members of the *M. sativa* complex. Cytological (13,14,18) and genetic evidence (54,55,75) based on a number of accessions for both diploid and tetraploid forms of *M. sativa* and *M. falcata* indicate that they behave like varieties of a common species. This evidence justifies the treatment by Gunn et al. (15) of *M. falcata* as *M. sativa* ssp. *falcata*. Similarly, hybridization and cytological studies involving *M. glutinosa* and *M. sativa* (1,14,40,45) support the treatment of the former as *M. sativa* ssp. *glutinosa*.

Medicago sativa ssp. *falcata* is characterized by yellow flowers with straight to sickle-shaped pods; occasionally the pods are twisted like a corkscrew. Mainly because of these traits, which are not present in *M. sativa* ssp. *sativa*, *M. sativa* ssp. *falcata* was classified as a separate species. There are diploid and tetraploid forms of this species. It is highly variable in morphological and biochemical traits. Because of this variability, the diploid forms may have received species or subspecies status, such as *borealis*, *romanica*, *altissima*, *glandulosa*, *difalcata*, *tenderensis*, and *erecta* (Table 3-1) (3,8,40).

The diploids of *M. sativa* ssp. *falcata* are distributed over a wide geographical range: from south Germany in the west, to Siberia in the east, and from the Black Sea coast of Bulgaria in the south, to Leningrad in the north. It is a species well adapted to cold regions. The tetraploid form of this subspecies is found in the same areas of distribution as the diploids. Contrary to Gunn et al. (15), tetraploid *M. sativa* ssp. *falcata* occurs at higher frequencies than the diploids, which are considered evolutionary relics (38,40).

Diploid and tetraploid forms of purple flowered *M. sativa* are named *M. sativa* ssp. *coerulea* and *M. sativa* ssp. *sativa*, respectively. They are characterized by violet or lavender flowers, and coiled pods. These species are adapted to temperate regions below 92° N L. The range of distribution of both diploid and tetraploid forms covers an extensive territory, including the Mediterranean, the Near and Middle East, the Caucasus, Middle, Central and South Asia. The highest viability of the species is concentrated in the foothills and mountain valleys of Armenia, Eastern Anatolia, Iran, Afghanistan, Central Asia, Jamm and Kashmir (20). *M. sativa* ssp. *glutinosa* is a tetraploid taxon, characterized by bright yellow or cream corolla color at the bud stage or in recently opened flowers, changing to full yellow several hours after opening. The pods are coiled and covered with glandular hairs. It is adapted to the moist, subalpine regions of Caucasia, along river valleys. *Medicago glomerata* and *M. sativa* ssp. *falcata* might be the progenitors of this subspecies (40).

The putative hybrid subspecies of *M. sativa*: *hemicycla*, *varia*, *tunetana*, and *polychroa* are found within the range of distribution of *M. sativa* ssp. *sativa*, *M. sativa* ssp. *coerulea*, *M. sativa* ssp. *falcata*, *M. sativa* ssp. *glutinosa*, and *M. glomerata*. They are considered hybrids because of their wide polymorphism for mixed flower color and degree of legume coiling (26). The tetraploid *M. sativa* ssp. × *tunetana* might have originated from tetraploidized hybrids of the diploids *M. sativa* ssp. *coerulea* and *M. glomerata* (40). *Medicago sativa* ssp. × *polychroa* Grossh. has been described as a tetraploid, which seems to have originated from the hybridization of the tetraploid *M. sativa* ssp. *sativa* and *M. sativa* ssp. *glutinosa* (40). These two hybrid subspecies have been studied very little, and their classification is tentative. The other two subspecies, × *varia* and × *hemicycla* may have resulted from the hybridization of *M. sativa* ssp. *falcata* with *M. sativa* ssp. *sativa* and *M. sativa* ssp. *coerulea*, respectively. The former is tetraploid and the latter diploid.

Medicago glomerata is a species related to the *M. sativa* complex, characterized by bright yellow flowers and coiled pods covered with glandular hairs. It is found in the diploid form in southern Europe, Mountain Alps, and North Africa. In North Africa, however, a tetraploid species described as *M. sativa* ssp. *faurei* by Moire (40) has been considered on morphological grounds, as the tetraploid form of *M. glomerata*. *Medicago glomerata* is more closely related to the species *M. prostrata* than to taxa of the *M. sativa* complex (26).

Medicago prostrata, a species found in diploid and tetraploid forms, is characterized also by yellow flowers and coiled pods. Its pods are similar to those of *M. sativa* ssp. *coerulea*, and has yellow flowers in common with *M. sativa* ssp. *falcata*. Inflorescences of *M. prostrata* are less dense and have fewer florets, however, than those of *M. sativa* (35). *Medicago prostrata* crosses with the species of the *M. sativa* complex when it is used as the pollen parent. In reciprocal crosses, only poorly developed seeds are produced, thus a postzygotic barrier seems to exist between these species (25).

3-6.2 Evolutionary Trends of the *M. sativa* Complex

Much speculation is involved in the events that lead to the morphological differentiation of the diploid *M. sativa* ssp. *coerulea* and *M. sativa* ssp. *falcata*. Lesins and Lesins (38) proposed that diploid *M. glomerata*, a species characterized by yellow flowers and coiled pods, was once distributed farther to the east colonizing a large territory spreading to the Caucasus where it served as the probable ancestor of the diploids in the complex. It is projected as giving rise to both subspecies, *coerulea* and *falcata*, by spatial isolation of the ancestral population which was divided by Parathethys, (they connect Black and Caspian Seas), during the Tertiary Period. The southern population resulting from this division, ssp. *coerulea*, suffered a loss of carotenoids and again in anthocyanin in their flowers, due to selection pressures imposed by competition for pollinators in the newly isolated area. The northern population, ssp. *falcata*, involved selective pressure for straight pods, thus shatter for easy seed dispersal. This characteristic might have been favored in the north, where thick grasslands would prevent the dispersal of coiled pods, adapted to rolling on open ground. The change must have been gradual, because pod shape bution of diploid *M. sativa* ssp. *falcata* over a wide geographical range, the great distances between existing populations, and the extensive variability observed in this subspecies, indicate that it was widespread at one time over a large territory. The distribution pattern might result from glaciation during the ice age in Pleistocene. During this geological event, only those populations occupying unglaciated sites were able to survive (38).

Recent evidence demonstrating the occurrence of unreduced gametes (46,53,78) in diploid subspecies of *M. sativa*, indicates that this might have been a significant event in the origin of the tetraploid subspecies. Previous reports had indicated the presence of a ploidy barrier between diploid and tetraploids, preventing intercrossing in nature. These studies were based on crosses among a limited number of diploid and tetraploid accessions. In any event, $2x$-$4x$ or $4x$-$2x$ crosses are possible when the diploid parent has the ability to generate unreduced ($2n$) gametes, resulting in tetraploid progeny. Spontaneous chromosome doubling cannot be discounted in the generational tetraploids, although it seems to be more the exception than the rule in the evolution of polyploid species (16). The hypothetical evolution of species of the *M. sativa* complex is summarized in Fig. 3-3 on the basis of available data.

3-6.2.1 Diploids

Quiros and Morgan (56) reported the inheritance of several isozymes in diploid *M. sativa* ssp. *falcata*, *M. sativa* ssp. *coerulea*, and *M. sativa* ssp. × *hemicycla* Grossh. On the basis of these genetic markers, the genetic variability of natural diploid populations was summarized by Quiros (55). Results obtained for each of these three diploid subspecies and for *M. glomerata* for four polymorphic loci, *Prx-1*, *Prx-2*, *Lap-1*, and *Lap-2* coding for peroxidase (PRX) and leucine amino peptidase (LAP) isozymes are shown in Table 3-2. Thirteen accessions of *M. sativa* ssp. *falcata* and seven accessions of *M. sativa* ssp. *coerulea* were sampled from the entire range of distribution of the species. Only three accessions of *M. sativa* ssp. × *hemicycla*, two from Armenia and one from the Transcaucasus, and three accessions of *M. glomerata*, all from Italy, were included in the study. In agreement with the morphological data, *M.*

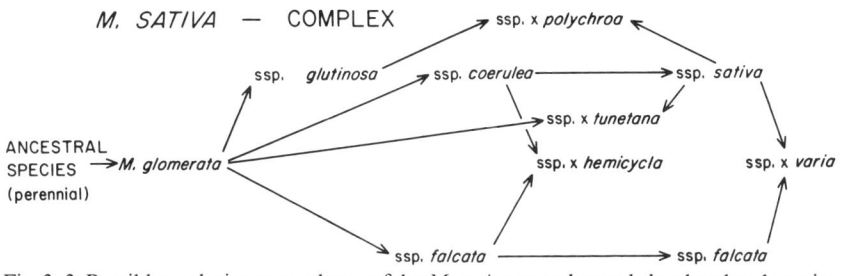

Fig. 3-3. Possible evolutionary pathway of the *M. sativa* complex and closely related species.

Table 3-2. Genetic variability of four polymorphic isozyme loci Prx-1, Prx-2, Lap-1, and Lap-2 found in diploid and tetraploid subspecies of the Medicago sativa-complex and in M. glomerata.†

Ploidy	Species	No. accessions	No. plants	No. of alleles $\bar{x}$	Range	Percentage heterozygosity
Diploid	M. glomerata	3	95	8.3	8–9	16.7
Diploid	M. sativa ssp. coerulea	7	368	8.6	5–10	20.5
Diploid	M. sativa ssp. falcata	13	447	12.0	7–17	27.7
Diploid	M. sativa ssp. × hemicycla	3	190	16.3	14–20	50.9
Tetraploid	M. sativa ssp. sativa	6	173	9.8	9–14	37.5
Tetraploid	M. sativa ssp. × varia	5	165	11.2	9–14	49.2
Tetraploid	M. sativa ssp. falcata	10	378	14.1	10–20	57.92
Tetraploid	Cultivated alfalfa	17	776	10.9	7–16	34.72

† Prx = peroxidase, Lap = leucine amino peptidase.

sativa ssp. *falcata* was found to be significantly more variable than *M. sativa* ssp. *coerulea*, measured by the number of alleles and the percentage of heterozygous individuals in the population. Although the sample of *M. sativa* ssp. × *hemicycla* was small, its variability measured by both criteria vastly surpassed that of *M. sativa* ssp. *falcata*. This was consistent for the three accessions of *M. sativa* ssp. × *hemicycla* sampled. Only one accession of *M. sativa* ssp. *falcata* (UAG 127 from Bulgaria) carrying a total of 17 alleles and 47.75% heterozygosity came close to the values obtained for *M. sativa* ssp. × *hemicycla*. On the other hand, the degree of genetic variability for *M. glomerata* was considerably lower. Although these data are not extensive enough to support solid conclusions on the evolution of the *M. sativa* complex, they provide further evidence on the origin of the species. According to Sinskaya (66), *M. sativa* ssp. × *hemicycla* is the ancestor of both *M. sativa* ssp. *falcata* and *M. sativa* ssp. *coerulea*. Lesins and Lesins (40) and Ivanov (20) argued instead that *M. sativa* ssp. *hemicycla* is of recent origin, being the product of hybridization between *M. sativa* ssp. *falcata* and *M. sativa* ssp. *coerulea*. They based their conclusion in the fact that artificial hybrids between both species fall within the spectrum of flower color and pod coiling of *M. sativa* ssp. × *hemicycla* and in the present range of distribution of the three species (40). This natural hybridization might have taken place at more than one location, where the two subspecies were in contact. The intermixing of two gene pools would increase the variability of the hybrid subspecies.

The electrophoretic data demonstrates that *M. sativa* ssp. × *hemicycla* includes accessions that are the most variable of all the taxa tested. It shares specific allozymes with *M. sativa* ssp. *falcata* which are not found in other subspecies of *M. sativa* (55). Thus, it is conceivable that *M. sativa* ssp. × *hemicycla* has resulted from the hybridization of *M. sativa* ssp. *coerulea* and *M. sativa* ssp. *falcata*. Sinskaya's hypothesis, however, cannot be discounted until additional accessions and loci are sampled.

3-6.2.2 Tetraploids

Tetraploid subspecies in the *M. sativa* complex, *M. sativa* ssp. *sativa*, *M. sativa* ssp. × *varia*, *M. sativa* ssp. *falcata*, and *M. sativa* ssp. *glutinosa* are distinguished from their diploid counterparts by the larger size of their flowers, pods and seeds. *Medicago sativa* ssp. *varia* arose by hybridization of *M. sativa* ssp. *sativa* with *M. sativa* ssp. *falcata* (40) which still occurs in the numerous locations where the two subspecies are sympatric. This takes place at diploid, tetraploid, and most likely at diploid-tetraploid levels with unreduced gametes generated by diploid accessions (46,51,53,78). The numerous epithets given to *M. sativa* ssp. × *varia* are listed in Table 3-1.

Stanford (76) provided conclusive evidence of autotetraploidy in alfalfa, after finding tetrasomic segregation for flower color in progenies of diallelic plants. Quiros (54) proved the existence of multiple alleles for several isozyme loci. On the basis of these genetic markers tetrasomic segregation in di-,tri-, and tetraallelic plants has been documented in alfalfa. In that study, it was observed that segregating progenies of *sativa-falcata* hybrids did not deviate from expected ratios, supporting their classification as subspecies of a common species.

Together with data obtained from diploid subspecies, Quiros (55) surveyed the genetic variability of both commercial cultivars of alfalfa from diverse origins, and of "natural" tetraploids obtained from the entire range of distribution of *M. sativa*. These "natural" tetraploids included feral collections and primitive cultigens of the *M. sativa* ssp. *sativa*, *M. sativa* ssp. × *varia* (forms *tianschianica* and *hemicoerulea*) and *M. sativa* ssp. *falcata*. They were collected by the later professor Karl Lesins, forming part of the *Medicago* collection housed at the Univ. of Alberta. Commercial alfalfa was represented by 16 cultivars (Table 3-4), from seven different sources, as identified by Barnes et al. (2). In general, natural tetraploids were found to be more variable than their diploid counterparts, measured by percentage heterozygosity (Table 3-2). This is expected, because the former have four copies for each chromosome. Tetraploid *M. sativa* ssp. *falcata* was found to have the highest values for

Table 3-3. Number of tetraallelic plants found in alfalfa cultivars and in tetraploid accessions of *Medicago sativa* ssp. *falcata* from various origins (55).

Cutlivar/Accession	No. of plants	Tetraallelic plants for: Prx-1†	Lap-2‡	Percentage
Hairy Peruvian	25	4	0	16.0
Sonora	106	3	0	2.8
Leningrad (UAG 110)	46	2	0	4.3
Czechoslovakia (UAG 102)	70	3	0	4.3
Latvia (UAG 105)	51	4	0	7.8
Siberia (UAG 113)	48	4	9	27.1
Turkey (UAG 2008)	18	0	6	33.3

† Prx = peroxidase. ‡ Lap = leucine amino peptidase.

Table 3-4. Characteristic isozymes present in various cultivars of different origins (55).

Origin	Cultivar	PRX† Allozymes	LAP‡	Subspecies involved according to: Barnes et al (2)	Quiros (55)
Turkistan	Deseret	1^3 2^3 2^n 3^3	2^2	Mostly sativa	sativa
	Lahontan	2^3 2^n 3^3			
	Nemastan	3^3			
Indian	Mesa Sirsa	1^8 2^3 3^3	2^2 2^3	Only sativa	sativa
Peruvian	Hairy Peruvian		2^3	Only sativa	sativa
	Monsefu		2^2 2^3		
Ladak	Ladak	1^n 2^n	2^5	Mostly sativa	sativa and falcata
Flemish	Tuna	1^3 1^6 1^8 2^3	2^2 2^3	Only sativa	sativa and falcata
	Du Puits	1^3 1^8 2^3	1^3 2^2 2^3 2^4 2^5		
	Socheville	1^3 3^3			
	Europe	1^3 3^3	1^3 2^2 2^3		
African	A-435	2^3 3^3	1^5 2^5	Unreported	sativa and falcata
	Sonora	1^3 1^6 2^3 2^5 3^3	2^2		
	Moapa	1^3 2^3	2^2 2^6		
Chilean	Buffalo	2^3	1^2	2^6 Only sativa	sativa and falcata
	Chilean	1^8 3^3		2^6	

† PRX = peroxidase.   ‡ LAP = leucine amino peptidase.

both, percentage heterozygosity and number of alleles, among all subspecies including both diploids and tetraploids. Interestingly, the percentage heterozygosity for the *M. sativa* ssp. × *hemicycla*, a diploid, was similar to that of the hybrid tetraploid *M. sativa* ssp. × *varia*. This can be explained by the fact that the former, on the average, has twice as many alleles as the latter. The level of variability of the cultivated alfalfa, which included cultivars known to carry *M. sativa* ssp. *falcata* genes, was comparable to that observed for the natural populations of *M. sativa* ssp. *sativa* but far below the values obtained for *M. sativa* ssp. *falcata* and *M. sativa* ssp. × *varia*. When the percent heterozygosity in the tetraploid populations was reduced to the level of individual plant heterozygosity, it was found that there were twice as many tri- and tetraallelic plants in the natural tetraploids as in the cultivars. In either case, the average frequency of tetraallelic plants was <1.5%, while that of triallelic plants was 9% for the cultivars and 19% for the natural tetraploids. Table 3-3 lists the accessions of natural tetraploids and cultivars where tetraallelic plants were found for the two most polymorphic loci. Tetraallelic plants in the natural tetraploids were found only in *M. sativa* ssp. *falcata*. In particular, two accessions, one from Siberia, another from Turkey and the cv. Hairy Peruvian, were the richest in tetraallelic plants, considering the small expected probability of drawing tetraallelic plants from random populations (6,10). These results support the hypothesis that maximum heterozygosity, reflected in tri- and tetraallelic plants, plays an important adaptive role in autotetraploids (6).

3–6.2.3 Origin of Cultivated Alfalfa

According to Lesins (31), the hybridization of the *M. sativa* ssp. *sativa* and *M. sativa* ssp. *falcata* might have contributed to the cultivation of alfalfa throughout much of the temperate zone.

Ivanov (20) suggests that alfalfa was cultivated 8000 to 9000 yr ago. The most likely centers for alfalfa domestication are the Armenia Highland, including the trans-Caucasus, Turkey, and Iran; and Southern Turkistan. Its domestication was perhaps concurrent with that of the horse (31). From the original centers of cultivation, alfalfa spread to Mesopotamia, the Old World, People's Republic of China, and India. In the 16th century, it was introduced to Mexico and Peru by the Spanish. It reached North America after several introductions in the 1800s from Mexico by the missionaries, and from Chile to California, as "Chilean clover" during the gold rush.

Barnes et al. (2) summarized the sources of alfalfa introductions to North America. They gave an account of the subspecies of *M. sativa* contributing to each of these germplasm sources, on the basis of historical information and morphological and physiological attributes. Specific isozymes can be used to differentiate *M. sativa* ssp. *falcata* from *M. sativa* ssp. *sativa* and to trace the ancestors of alfalfa cultivars (55). The alleles $Prx-1^7$, $Prx-1^{15}$, $Prx-1^n$, $Prx-2^5$, $Lap-1^5$, $Lap-2^4$, $Lap-2^5$, and $Lap-2^6$, have been found only in *M. sativa* ssp. *falcata* and in the hybrid ssp. × *hemicycla* and × *varia*. Many other isozymes are common to all subspecies of the *M. sativa* complex. Thus, it is possible to determine if ssp. *falcata* genes are present in a given cultivar on the basis of the alleles listed above. Sixteen cultivars of diverse origins were tested (Table 3–4). Good agreement was found between the isozyme data and the conclusions reached by Barnes et al. (2). The only exceptions were the cultivars of Flemish and Chilean origins, which are assumed to have only ssp. *sativa* genes. The isozyme data indicates that *M. sativa* ssp. *falcata* is also involved in their genetic makeup. In addition, it was found that *M. sativa* ssp. *falcata* had contributed to the development of African alfalfas. In general, cultivars of the same origin shared the same alleles. A few cultivars could be distinguished by rare alleles, such as A-435 (African) by allele $Lap-1^{15}$, and Ladak by the null allele $Prx-1^n$. Several alleles found in diploid and tetraploid *M. sativa* ssp. *falcata* accessions from northern latitudes were not encountered in any of the cultivars tested. This could suggest the availability of germplasm which has not been exploited in alfalfa breeding. Conversely, this assessment is based on only two isozyme systems and additional data are needed.

3-7 TAXOGENETICS OF *MEDICAGO*

3-7.1 Hybridization and Crossing Relationships

The ability of species to be crossed with one another is perhaps the best indicator of their relationship. The crossing relationship of the subspecies in the *M. sativa* complex are described in the previous section of this chapter and in Chapter 24 in this book. Hybridization of species in the *M. sativa* complex with other wild perennial *Medicago* spp. is also reported in Chapter 24 in this book. A summary of the successful interspecific crosses between *Medicago* spp. can be found in Table 3-6.

Little hybridization work has been done with the annual *Medicago* spp. Most of the taxonomic classification is based on morphological traits. Hybridization between *M. rotata* and *M. blancheana* occurs frequently in nature. This is supported by the fact that the hybrids between these two species have been described as several subspecies (40).

Successful hybridizations between species in the section *Pachyspirae* (Table 3-5) have been limited because of the occurrence of chlorophyll-deficient plants and early embryo abortion. Crosses between *M. truncatula* (as *M. striata*) and *M. littoralis* are possible with good chromosome pairing in the hybrid. However, when *M. littoralis* is used as the maternal parent the resulting hybrids are chlorophyll deficient, with many of the plants occurring as chimeras. The reciprocal cross results in hybrid plants with normal green foliage and morphological characteristics intermediate between the parents (33,64). *Medicago tornata* can be hybridized with *M. littoralis* but the hybrid displays the same type of chlorophyll deficiency (33,63). *Medicago tornata* is more distantly related to *M. littoralis* than *M. truncatula* judged by the presence of a chromosomal translocation (63). Likewise, hybrids between *M. soleirolii* and *M. tornata* are chlorophyll deficient (40). Lesins et al. (32) were able to hybridize *M. turbinata* with *M. truncatula* only when *M. truncatula* was used as the maternal parent. The hybrids were sterile with lighter green foliage than either of the parents. Fertility was recovered when the chromosome number of the hybrid was doubled. Lack of segregation for several traits including isozymes in the hybrid and the occurrence of preferential pairing, indicates that these two species have different genomes (32,57).

The chlorophyll-deficiency syndrome was noted, as well, in hybrids between *M. laciniata* and *M. sauvagei* which is taxonomically in a different section (Leptospirae) (Table 3-5) (65). These two species are very closely related to each other as normal bivalent pairing occurs in their hybrid.

Medicago intertexta, M. ciliaris, and *M. muricoleptis* can be intercrossed with each other. *Medicago muricoleptis* is more distantly related to *M. intertexta* than *M. ciliaris* judged by the reduced pollen fertility and chromosome pairing observed in hybrids between *M. muricoleptis* and either *M. intertexta* or *M. ciliaris*. Some taxonomists consider *M.*

Table 3-5. *Medicago* spp., grouped by subgenus and sections.

Subgenus *Lupularia*	Subgenus *Spirocarpos*
M. lupulina	Section *Rotatae*
M. secundiflora	*M. rotata*
	M. blancheana
Subgenus *Orbicularia*	*M. noeana*
Section *Carstiensae*	*M. shepardii*
M. carstiensis	*M. rugosa*
Section *Platycarpae*	*M. scutellata*
M. platycarpa	Section *Pachyspirae*
M. ruthenica	*M. soleirolii*
Section *Orbiculares*	*M. tornata*
M. orbicularis	*M. littoralis*
Section *Hymenocarpos*	*M. truncatula*
M. radiata	*M. rigidula*
Section *Heynianae*	*M. murex*
M. heyniana	*M. lesnsii*
Section *Cretaceae*	*M. constricta*
M. cretacea	*M. turbinata*
	M. doliata
Subgenus Medicago	Section *Leptospirae*
Section *Falcago*	*M. sauvagei*
Subsection *Falcatae*	*M. laciniata*
M. sativa	*M. minima*
M. glomerata	*M. praecox*
M. prostrata	*M. coronata*
Subsection *Rupestres*	*M. polymorpha*
M. rhodopea	*M. arabica*
M. saxatillis	*M. lanigera*
M. rupestris	*M. disciformis*
M. cancellata	*M. tenoreana*
Subsection *Daghnestanicae*	Section *Intertextae*
M. daghestanica	*M. intertexta*
M. pironae	*M. ciliaris*
Subsection *Papillosae*	*M. muricoleptis*
M. dzhawakhetica	*M. granadensis*
M. papillosa	
Section *Arborea*	
M. arborea	
Section *Marinae*	
M. marina	
Section *Suffruticosae*	
M. suffruticosa	
M. hybrida	

ciliaris to be a subspecies of *M. intertexta* (40). Refer to Table 3-6 for a summary of the successful hybridizations among the annual species.

The only hybrid which has been obtained between a perennial and an annual *Medicago* is that of *M. sativa* ssp. *sativa* × *M. scutellata* (59). Sangduen et al. (59) proposed that the *M. sativa* clone, a male sterile used as the female parent, generated unreduced eggs which united with normal haploid gametes from *M. scutellata* producing a mixaploid with the primary stem being hexaploid. The hybrid was both male and female sterile. The hybrid was perennial and purple flowered like the *M. sativa* parent.

Table 3-6. Successful interspecific hybridization between *Medicago* spp.†

Interspecific hybrids	References
Perennials	
cancellata × sativa & reciprocal	23, 74, 80
cancellata × saxatilis	28
daghestanica × pironae	34
F_1 (daghestanica × pironae) × sativa	29
dzhawakhetica × sativa & reciprocal	8, 23, 39, 50
glomerata × falcata & reciprocal	22, 26
glomerata × prostrata	26
glomerata × sativa	26
hybrida × suffruticosa	27
platycarpa × ruthenica	52
prostrata × falcata	25
prostrata × sativa & reciprocal	25
rhodopea × rupestris & reciprocal	14, 30
F_1 [sativa × cancellata] × saxatilis	28
sativa × hybrida	47
sativa × marina	47
sativa × papillosa	47
sativa × rhodopea	47
sativa × rupestris	47
F_1 [sativa × rhodopea] × saxatilis	28
sativa × saxatilis	28
F_1 [sativa × saxatilis] × F_1 (sativa × rhodopea)	28
Perennial × Annual	
sativa × scutellata	59
Annuals	
bonarotiana × rotata & reciprocal = (× blancheana)	42
ciliaris × intertexta	43
littoralis × tornata & reciprocal	33, 64
littoralis × truncatula & reciprocal	17, 33, 63, 64
muricoleptis × intertexta	43
muricoleptis × ciliaris	43
lesinsii × turbinata	41
laciniata × sauvagei	65
sauvagei × laciniata	65
soleirolii × tornata	32
truncatula × turbinata & reciprocal	32

† Exchange hybridizations between subspecies of the *M. sativa* complex.

The color of its vegetative parts and its self-tripping mechanism were like those found in the annual parent.

3-7.2 Potential Uses of Wild *Medicago* Species as Alfalfa Genetic Resources

Alfalfa, the world's most important forage legume, is subject to varying degrees of loss from a large number of diseases and pest insects. In addition, expanding the cultivation of alfalfa to include less suitable areas would be desirable, and further improvements in quality would be beneficial. All of these problems can be solved best by plant breeding.

The use of wild (exotic, weedy, or unimproved) germplasm in incorporating useful agronomic traits into the cultivated germplasm has been demonstrated countless times. Thus, genes for the agronomic traits, including winter hardiness, creeping root habit, and resistance to some foliar diseases have been added to cultivated alfalfa from wild *M. sativa* ssp. *falcata* (2). The use of this subspecies, adapted to colder habitats, has been invaluable in extending alfalfa production to more northern regions.

Vasilchenko (77) described in detail the potential uses of various *Medicago* spp. in alfalfa improvement. Accessions of *M. sativa* ssp. × *varia* from Tien Shan (People's Republic of China) may be resistant to environmental stress, because of their adaptation to a harsh environment and to rocky soils; while those found growing in Tibet are adapted to altitudes of 5000 m. Resistance to drought is present in accessions of *M. sativa* ssp. *falcata*, *M. sativa* ssp. *polychroa*, *M. dzhawaketica*, and *M. cancellata*. *Medicago marina* and *M. littoralis*, endemic to seashores, might be resistant to salinity. The former species is characterized by creeping, subterranean stems, suggesting its use in dune stabilization. The annual species, *M. truncatula*, *M. littoralis*, *M. rugosa*, *M. scutellata*, and *M. tornata* have good potential as annual forages. Some of these are cultivated for this purpose in Australia (79).

Most breeding efforts to date have concentrated on species within the *M. sativa* complex. The use of less related species in breeding work will require a substantial effort on the collection, preservation, description, and preliminary screening of germplasm. Some *Medicago* spp. have been screened, especially for disease and insect resistance. Renfro and Sprague (58) found that *M. dzhawakhetica* and *M. suffruticosa* have a high level of resistance to spring blackstem (*Phoma medicaginis*), and that *M. arborea* had resistance to bacterial wilt (*Corynebacterium insidiosum*). Resistance to *Stemphylium* leafspot was reported in *M. cancellata* (7). Among the annual *Medicago* spp., *M. littoralis*, *M. murex*, *M. rigidula*, *M. tenoreana*, and *M. truncatula* exhibited a very high level of anthracnose resistance to both race 1 and 2; and a high level of resistance to both races of anthracnose was found in *M. arborea* (11). See Chapters 21 and 27 in this book for additional information on sources of disease resistance.

High levels of resistance to the potato leafhopper [*Empoasca fabae* (Harris)] and alfalfa weevil [*Hypera postica* (Gyllenhal)], major insect pests of alfalfa, has been shown in a few annual *Medicago* spp.; namely *M. disciformis*, *M. minima*, *M. polymorpha*, *M. rugosa*, *M. scutellata*, and *M. truncatula* (3,61,62). The mechanism for this insect resistance is the presence of gland tipped hairs (62). See Chapters 22 and 28 in this book for additional information on sources of insect resistance.

All of these sources of agronomic traits is worthless unless these traits can be transferred to cultivated alfalfa. A major problem in the utilization of the wild *Medicago* spp. is that little is known about their crossing relationships to *M. sativa*. Several crossing barriers have been identified such as unequal ploidy levels (23,50), chromosome rearrangement (35), nonlethal chlorophyll deficiency (23,33,65), very slow pollen tube growth (60), and poorly developed seeds (25,33,59).

Methods for overcoming these barriers have involved the use of colchicine to double the chromosome number: (i) for hybridization to occur at equal ploidy levels (24,28,30); (ii) to accomplish hybridization at unequal ploidy levels (23,50); and (iii) to restore fertility in nonfertile hybrids (32). The discovery of a diploid male-sterile *M. sativa* (CADL) (49) has enhanced the interspecific hybridization efforts (46,49) (see Chapter 24 in this book). Also, phytohormones have been used to overcome crossing barriers. Sangduen et al. (59) applied gibberellic acid to the pedicels and peduncles immediately after pollination so that the hybrid pods were retained long enough for the development of mature seed. Researchers in other crops have used phytohormones to enhance pollen tube growth (9,12,21). The use of tissue culture and genetic engineering technology such as in vitro pollination, embryo rescue (81), protoplast fusion (85), and gene splicing (82, 83) will be useful in the near future for breaking crossing barriers and transferring agronomically useful traits. Recently, McCoy (48) reported the use of ovule-embryo culture techniques to rescue aborting interspecific hybrid embryos. See Chapter 24 in this book for further information on the use of these techniques.

No matter which method is used to unlock genetic barriers, the potential exists for the transfer of new and unique sources of pest resistance and for significant improvement in adaptation and yield characteristics.

3–8 SUMMARY

The genus *Medicago* comprise more than 60 different species among annuals and perennials. The primary center of diversity for the genus is found in the Caucasus, northwestern Iran and northeastern Turkey. The basic genomic number of *Medicago* is $x=8$, except for a few species which have a genomic number of $x=7$. Diploid, tetraploid, and hexaploids species occur in the genus.

The evolution of *Medicago* might have taken place in the Tertiary period. The annual species which are autogamous might have originated from allogamous perennial species. Taxonomically, the *Medicago* species are distinguished mainly by legume and seed characteristics, pubescence, pollen grain morphology, and chromosome numbers. Other biochemical traits might be used successfully for this purpose. The key to *Medicago* spp. of Lesins and Lesins (40), with a few modifications is included as a guide for species identification.

The *M. sativa* complex includes the taxa that hybridize freely with alfalfa. Therefore, they can be classified as subspecies of the species *M. sativa* on the basis of morphological and ploidy criteria. The *M. sativa* ssp. recognized in this chapter are: ssp. *sativa*, ssp. *coerulea*, ssp. *falcata*, ssp. × *varia*, ssp. × *hemicycla*, ssp. × *polychroa*, ssp. × *tunetana*, and ssp. *glutinosa*. *Medicago glomerata* and *M. prostrata* are two other species closely related to the complex. The former species has been considered ancestral to *M. sativa* giving rise to the diploids *M. sativa* ssp. *coerulea* and *M. sativa* ssp. *falcata*. The tetraploid subspecies *M. sativa* ssp. *sativa* and *M. sativa* ssp. *falcata* likely originated from the diploids via unreduced gametes. On the basis of isozyme studies, it was found that among the diploid taxa, the most variable are the subspecies *M. sativa* ssp. × *hemicycla* and *M. sativa* ssp. *falcata* as measured by numbers of alleles and percentage heterozygosity. In general, tetraploids subspecies were found to be more variable than diploids. Nevertheless, the percentage heterozygosity of the diploid *M. sativa* ssp. × *hemicycla* was similar to that of the tetraploid *M. sativa* ssp. × *varia* but higher than that found in cultivated alfalfa. Twice as many tri- and tetraallelic plants were found in natural tetraploids of *M. sativa* than in cultivated alfalfa. The Hairy Peruvian strain was the richest in tetraallelic plants among those tested. *Medicago sativa* ssp. *falcata* besides its characteristic yellow flowers and straight pods, has unique isozymes that can be used to determine if this subspecies is involved in the genetic composition of alfalfa cultivars that differ in origin.

An account of the interspecific hybridization in the genus is given in this chapter. The use of unimproved germplasm as an alfalfa genetic resource is documented. The potential of various wild species as sources of disease, insect, and stress resistance is mentioned. Possible strategies to achieve interspecific hybridization, especially for wide crosses are suggested.

ACKNOWLEDGMENT

The authors are indebted to the late Ernest H. Stanford and James H. Elgin, Jr. for reviewing the manuscript. Part of this work was supported by the Alberta Research Council grant DB102 to R.C. von Borstel and grant NSERC GR2 to the president of the Univ. of Alberta, Canada.

REFERENCES

1. Armstrong, J.M., and D.R. Gibson. 1941. Inheritance of certain characters in the hybrid of *Medicago media* and *M. glutinosa*. Sci. Agric. 22:1–10.
2. Barnes, D.K., E.T. Bingham, R.P. Murphy, O.J. Hunt, D.F. Beard, W.H. Skrdla, and L.R. Teuber. 1977. Alfalfa germplasm in the United States: Genetic vulnerability, use, improvement, and maintenance. USDA-ARS Tech. Bull. 1571. U.S. Government Printing Office, Washington, DC.
3. ----, and R.H. Ratcliffe. 1969. Evaluation of annual species of *Medicago* as sources of alfalfa weevil resistance. Crop Sci. 9:640–642.
4. Bauchan, G.R., and J.H. Elgin, Jr. 1984. A new chromosome number for the genus *Medicago*. Crop Sci. 24:193–195.
5. Beckmann, J.S., and M. Soller. 1983. Restriction fragment length polymorphisms in genetic improvement: methodologies, mapping and costs. Theor. Appl. Genet. 67:35–43.

6. Bingham, E.T. 1980. Maximizing heterozygosity in autotetraploids. *In* W.H. Lewis (ed.) Polyploidy: Biological relevance. Plenum Publishing Corp., New York.
7. Borges, O.L., E.H. Stanford, and R.K. Webster. 1976. Sources and inheritance of resistance to Stemphyllium leafspot of alfalfa. Crop Sci. 16:458–461.
8. Clement, Jr., W.M. 1963. Chromosome relationships and diploid hybrid between *Medicago sativa* L. and *M. dzhawakhetica* Bordz. Can. J. Genet. Cytol. 5:427–435.
9. Crane, M.B., and E. Marks. 1952. Pear-apple hybrid. Nature (London) 170:1017.
10. Dubier, M.W., and E.T. Bingham. 1975. Maximum heterozygosity in alfalfa: results using haploid derived autotetraploids. Crop Sci. 15:527–531.
11. Elgin, Jr., J.H., and S.A. Ostazeski. 1982. Evaluation of selected alfalfa cultivars and related *Medicago* species for resistance to race 1 and race 2 anthracnose. Crop Sci. 22:39–42.
12. Emsweller, S.L., and N.W. Stuart. 1948. Use of growth regulation substances to overcome incompatibilities in *Lilium*. Proc. Am. Soc. Hortic. Sci. 51:581–589.
13. Gillies, C.B. 1970. Alfalfa chromosomes. I. Pachytene karyotype of a diploid *Medicago falcata* L. and its relationship to *M. sativa* L. Crop Sci. 10:169–171.
14. ——. 1972. Pachytene chromosomes of perennial species. II. Species closely related to *M. sativa*. Heredity 72:277–288.
15. Gunn, C.R., W.H. Skrdla, and H.C. Spencer. 1978. Classification of *Medicago sativa* L. using legume characters and flower colors. USDA-ARS Tech. Bull. 1574. U.S. Government Printing Office, Washington, DC.
16. Harlan, J.R., and J.M.J. de Wet. 1975. On O. Winge and a prayer: The origins of polyploidy. Bot. Rev. 41:361–390.
17. Heyn, C.C. 1963. The annual species of *Medicago*. Scr. Hierosolymitana 12:1–154. Magnes Press, Jerusalem.
18. Ho, K.M., and K.J. Kasha. 1972. Chromosome homology at pachytene in diploid *Medicago sativa*, *M. falcata* and their hybrids. Can. J. Genet. Cytol. 14:829–838.
19. Ignasiak, T., and K. Lesins. 1975. Carotenoids in petals of perennial *Medicago* species. Biochem. Syst. Ecol. 2:177–180.
20. Ivanov, A.I. 1977. History, origin and evolution of the genus *Medicago*, subgenus *Falcago*. Bull. Appl. Bot. Genet. Select. 59:3–40 (Trudy po prikladnoy botanique, genetike i selektsii. Translation by the Multilingual Services Div. Dep. of the Secretary of State, Canada).
21. Larter, E., and C. Chaubey. 1965. Use of exogenous growth substances on promoting pollen tube growth and fertilization in barley-rye crosses. Can. J. Genet. Cytol. 7:511–518.
22. Lesins, K. 1952. Some data on the cytogenetics of alfalfa. J. Hered. 43:287–291.
23. —— 1961a. Interspecific crosses involving alfalfa. I. *Medicago dzhawakhetica* (Bordz.) × *M. sativa* L. and its peculiarities. Can. J. Genet. Cytol. 3:135–152.
24. ——. 1961b. Interspecific crosses involving alfalfa. II. *Medicago cancellata* M. B. × *M. sativa* L. Can. J. Genet. Cytol. 3:316–324.
25. ——. 1962. Interspecific crosses involving alfalfa. III. *Medicago sativa* L. × *M. prostrata* Jacq. Can. J. Genet. Cytol. 4:14–23.
26. ——. 1968. Interspecific crosses involving alfalfa. IV. *Medicago glomerata* × *M. sativa* with reference to *M. prostrata*. Can. J. Genet. Cytol. 10:536–544.
27. ——. 1969. Relationship to taxa in genus *Medicago* as revealed by hybridization. IV. *M. hybrida* × *M. suffruticosa*. Can. J. Genet. Cytol. 11:340–345.
28. ——. 1970. Interspecific crosses involving alfalfa. *Medicago saxatilis* × *M. sativa* with reference to *M. cancellata* and *M. rhodopea*. Can. J. Genet. Cytol. 14:221–226.
29. ——. 1971. Interspecific crosses involving alfalfa. VI. Ineffectiveness of alloploidy in induction fertility in *Medicago pironae* × *M. daghestanica* hybrids. Can. J. Genet. Cytol. 13:437–442.
30. ——. 1972. Interspecific crosses involving alfalfa. VII. *Medicago sativa* × *M. rhodopea*. Can. J. Genet. Cytol. 14:221–226.
31. ——. 1970. Alfalfa, lucerne, *Medicago sativa* (Leguminosae-Papilionatae). p. 165–168. *In* N.W. Simmonds (ed.) Evolution of crop plants. Longman Group, London.
32. ——, J. Dickson, and L. Ostafichuk. 1980. Relationship of taxa in *Medicago* as revealed by hybridization. IX. *M. turbinata* × *M. truncatula*. Can. J. Genet. Cytol. 22:137–142.
33. ——, and A. Erac. 1968. Relationship of taxa in the genus *Medicago* as revealed by hybridization. I. *M. striata* × *M. littoralis*. Can. J. Genet. Cytol. 10:263–275.
34. ——, and C.B. Gillies. 1968. Relationship of taxa in the genus *Medicago* as revealed by hybridization. II. *M. pironae* × *M. daghestanica* with reference to *M. sativa*. Can. J. Genet. Cytol. 10:454–459.
35. ——, and C.B. Gillies. 1972. Taxonomy and cytogenetics of *Medicago*. *In* C.H. Hanson (ed.) Alfalfa science and technology. Agronomy 15:53–86.
36. ——, and I. Lesins. 1960. Sibling species in *Medicago prostrata* Jacq. Can. J. Genet. Cytol. 2:416–417.

37. ——, and ——. 1963. Pollen morphology and species relationship in *Medicago* L. Can. J. Genet. Cytol. 5:270–280.
38. ——, and ——. 1964. Diploid *Medicago falcata* L. Can. J. Genet. Cytol. 6:152–163.
39. ——, and ——. 1966. Little known *Medicagos* and their chromosome complements. IV. Some mountain species. Can. J. Genet. Cytol. 8:8–13.
40. ——, and ——. 1979. Genus *Medicago* (*Leguminosae*). A taxogenetic study. Junk, The Hague, Netherlands.
41. ——, ——, and C.B. Gillies. 1970. *Medicago murex* with 2n=16 and 2n=14 chromosome complements. Chromosoma 30:109–122.
42. ——, R.S. Sadasivaiah, and S.M. Singh. 1976. Relationship of taxa in the genus *Medicago* as revealed by hybridization. VIII. Section *Rotatae*. Can. J. Genet. Cytol. 18:345–355.
43. ——, S.M. Singh, and A. Erac. 1971. Relationship of taxa in the genus *Medicago* as revealed by hybridization. V. Section *Intertextae*. Can. J. Genet. Cytol. 13:335–346.
44. Lubenets, P.A. 1972. Alfalfa-*Medicago* L. A brief survey of the genus and the classification of the subgenus *Falcago* (Rehb.) Grossh. Bull. Appl. Bot. Genet. Select: Forage Crops 47:3–68. (Trudy po prikladnoi botanike, genetike i selektsii; kormovye kul'tury. Translation by the Multilingual Service Div., Dep. of the Secretary of State, Canada).
45. Mariani, A., and F. Veronesi. 1979. Cytological and fertility relationships of different *Medicago* species and cytogenetic behavior of their hybrids. Genet. Agric. 33:245–268.
46. McCoy, T.J. 1982. The inheritance of 2n pollen formation in diploid alfalfa. *Medicago sativa*. Can. J. Genet. Cytol. 24:315–323.
47. ——. 1984. Interspecific hybrids of perennial *Medicago* species produced via ovule-embryo culture. p. 56. *In* D.R. Viands (ed.) 29th National Alfalfa Improvement Conf., Lethbridge, AB, Canada. 15–20 July. National Alfalfa Improvement Conf., Lethbridge, AB, Canada.
48. ——. 1985. Interspecific hybridization of *Medicago sativa* and *M. rupestris* MB. using ovule-embryo culture. Can. J. Genet. Cytol. 27:238–245.
49. ——, and L.Y. Smith. 1983. Genetics, cytology, and crossing behavior of an alfalfa (*Medicago sativa*) mutant resulting in failure of the postmeiotic cytokinesis. Can. J. Genet. Cytol. 25:390–397.
50. ——, and ——. 1984. Uneven ploidy levels and a reproductive mutant required for interspecific hybridization of *Medicago sativa* and *M. dzhawakhetica* Bordz. Can. J. Genet. Cytol. 26:511–518.
51. McLennan, H.A., J.M. Amstrong, and K.J. Kasha. 1966. Cytogenetic behavior of alfalfa hybrids from tetraploid by diploid crosses. Can. J. Genet. Cytol. 8:544–555.
52. Oldemeyer, R.K. 1956. Interspecific hybridization in *Medicago*. Agron. J. 48:584–585.
53. Pfeiffer, T.W., and E.T. Bingham. 1983. Abnormal meiosis in alfalfa, *Medicago sativa*: Cytology of 2n egg and 4n pollen formation. Can. J. Genet. Cytol. 25:107–112.
54. Quiros, C.F. 1982. Tetrasomic segregation for multiple alleles in alfalfa. Genetics 101:117–127.
55. ——. 1983. Alfalfa, luzerne (*Medicago sativa* L.) p. 253–294. *In* S.D. Tanksley and T.J. Orton (ed.) Isozymes in plant genetics and breeding, Part B. Elsevier Scientific Publishing Co., Amsterdam.
56. ——, and K. Morgan. 1981. Peroxidase and leucine-aminopeptidase in diploid *Medicago* species closely related to alfalfa: Multiple gene loci, multiple allelism and linkage. Theor. Appl. Genet. 50:221–228.
57. ——, and L. Ostafichuk. 1983. Allozymes and genetic variability in *Medicago turbinata*, *M. truncatula*, and their hybrid. Can. J. Genet. Cytol. 25:286–291.
58. Renfro, B.L., and E.W. Sprague. 1959. Reaction of *Medicago* species to eight alfalfa pathogens. Agron. J. 51:481–483.
59. Sangduen, N., E.L. Sorensen, and G.H. Liang. 1982. A perennial × annual *Medicago* cross. Can. J. Genet. Cytol. 24:361–365.
60. ——, ——, and ——. 1983. Pollen germination and pollen tube growth following self-pollination and intra- and interspecific pollination of *Medicago* species. Euphytica 32:527–534.
61. Shade, R.E., M.J. Doskocil, and N.O. Maxon. 1979. Potato leafhopper resistance in glandular-haired alfalfa species. Crop Sci. 19:287–288.
62. ——, T.E. Thompson, and W.R. Campbell. 1975. An alfalfa weevil larval resistance mechanism detected in *Medicago*. J. Econ. Entomol. 8:399–494.
63. Simon, J.P. 1965. Relationship in annual species of *Medicago*. II. Interspecific crosses between *M. tornata* (L.) Mill. and *M. littoralis* Rhode. Austr. J. Agric. Res. 16:51–60.
64. ——, and A.J. Millington. 1967. Relationship in annual species of *Medicago*. III. The complex *M. littoralis-M. truncatula* Gaertn. Aust. J. Bot. 15:35–73.
65. Singh, S.M., and K. Lesins. 1972. Relationship of taxa in the genus *Medicago* as revealed by hybridization. VI. *M. laciniata* × *M. sauvagei*. Can. J. Genet. Cytol. 14:823–828.
66. Sinskaya, E.N. 1961. Flora of cultivated plants of the U.S.S.R. XIII. Perennial leg-

uminous plants. Israel Program for Scientific Translations, Jerusalem, National Science Foundation Pub., Washington, DC.
67. Small, E. 1981. A numerical analysis of major groupings in *Medicago* employing traditionally used characters. Can. J. Bot. 59:1553–1577.
68. ——, I.J. Bassett, and C.W. Crompton. 1981. Pollen variation in tribe *Trigonella* (Leguminosae) with special reference to *Medicago*. Pollen Spores 23:295–320.
69. ——, and G.R. Bauchan. 1984. Chromosome numbers of the *Medicago sativa* complex in Turkey. Can. J. Bot. 62:749–752.
70. ——, and B.S. Brookes. 1984a. Taxonomic circumscription and identification in the *Medicago sativa-falcata* (alfalfa) continuum. Econ. Bot. 38:83–96.
71. ——, and ——. 1984b. Reduction of the geocarpic Factorovskya to *Medicago*. Taxon 33:622–635.
72. ——, and ——. 1985. *Medicago lesinsii*, a new Mediterranean species. Can. J. Bot. 63:728–734.
73. ——, and P. Lefkovitch. 1982. Agrochemotaxometry of alfalfa. Can. J. Plant Sci. 62:919–928.
74. Smith, S.E., R.P. Murphy, and D.R. Viands. 1984. Reproductive characteristics of hexaploid alfalfa derived from 3x-6x crosses. Crop Sci. 24:169–172.
75. Sprague, E.W. 1959. Cytology and fertility relationships of *Medicago sativa*, *M. falcata*, and *M. gaetula*. Agron. J. 51:249–252.
76. Stanford, E.H. 1951. Tetrasomic inheritance in alfalfa. Agron. J. 43:222–225.
77. Vasilchenko, I.T. 1949. Flora and systematics of higher plants: Alfalfa—the best forage plant. Trans. V. L. Komarov Bot. Inst. Acad. Sci. USSR 8:9–421. (Trudy Botanicheskogo Instituta Imeni V. C. Komarova Akademic Mauk Soyuza Sovitskikh Sotsialticheskikh Republick. Translation by the Multilingual Service Div. Dep. of the Secretary of State, Canada).
78. Vorsa, N., and E.T. Bingham. 1979. Cytology of 2n pollen formation in diploid alfalfa, *Medicago sativa*. Can. J. Genet. Cytol. 21:525–530.
79. von Borstel, R.C., and K. Lesins. 1977. On germplasm conservations with special reference to the genus *Medicago*. In A. Muhammed, et al. (ed.) Genetic diversity in plants. Basic Life Sciences, Vol. 8. Plenum Press, New York.
80. Yen, S.T., and R.P. Murphy. 1979. Cytology and breeding of hexaploid alfalfa. I. Stability of chromosome number. Crop Sci. 19:389–393.

ADDITIONAL REFERENCES

81. Bauchan, G.R. 1987. Embryo culture of *Medicago scutellata* and *M. sativa*. Plant Cell Tissue Organ Cult. 10:21–29.
82. Deak, M., G.B. Kiss, C. Koncz, and D. Dudits. 1986. Transformation of *Medicago* by *Agrobacterium* mediated gene transfer. Plant Cell Rep. 5:97–100.
83. Shahin, E.A., A. Spielmann, K. Sukhapinda, R.B. Simpson, and M. Yashar. 1986 Transformation of cultivated alfalfa using disarmed *Agrobacterium tumefaciens*. Crop Sci. 26:1235–1239.
84. Small, E. 1984. A clarification of the *Medicago blancheana-bonarotiana-rotate* complex. Can. J. Bot. 62:1693–1696.
85. Téoulé, E. 1983. Hybridation somatique entre *Medicago sativa* L. et *M. falcata* L. C.R. Acad. Sci. Ser. 3, 297:13–16.

4 Morphology and Anatomy

L. R. TEUBER

University of California
Davis, California

M. A. BRICK

Colorado State University
Ft. Collins, Colorado

The morphology and anatomy of the alfalfa (*Medicago sativa* L.) plant was described prior to 1940. Since 1940, much of the research in this area has been initiated in attempts to solve problems encountered in the production of alfalfa. These findings have refined our knowledge of the structure of alfalfa, and revealed that a considerable amount of genetic variation exists among alfalfa plants at the organ level. Additional studies describing plant form and function will ultimately lead to the exploitation of genetic variation in such structures as the root, crown, and flower. This chapter reviews and updates the literature on the morphology and anatomy of alfalfa, highlighting genetic variation. The general features of seed germination, seedling growth, the structure of the seed, root, crown, leaves, and flower are discussed.

4-1 SEED MORPHOLOGY, ANATOMY, AND DEVELOPMENT

The alfalfa seed consists of the embryo, endosperm, and testa (seed coat) (Fig. 4-1). The seeds are formed in the fruit (pod) and usually are somewhat kidney shaped. However, many seeds are angular as a result of internal and external forces in the pod that shape the seed during maturation (Fig. 4-2). Mature seeds are approximately 1 to 2 mm long, 1 to 2 mm wide, and 1 mm thick; usually, they are about twice as long as wide. The average count among 418 seed lots representing 39 cultivars was 464.5 seeds/g (12).

The color of the seed is usually either yellow or olive green to brown. However, white- and black-seeded genotypes have been reported (11). The black-seeded characteristic appears to be controlled by three genes. The white-seed trait is controlled by a single recessive gene, that also conditions white flower color (11).

The entire testa is covered by a cuticle that is composed of cutin, a complex mixture of fatty acids and waxes (73). The cuticle conforms to

---

Copyright 1988 © ASA-CSSA-SSSA, 677 South Segoe Road, Madison, WI 53711, USA.
Alfalfa and Alfalfa Improvement—Agronomy Monograph no. 29.

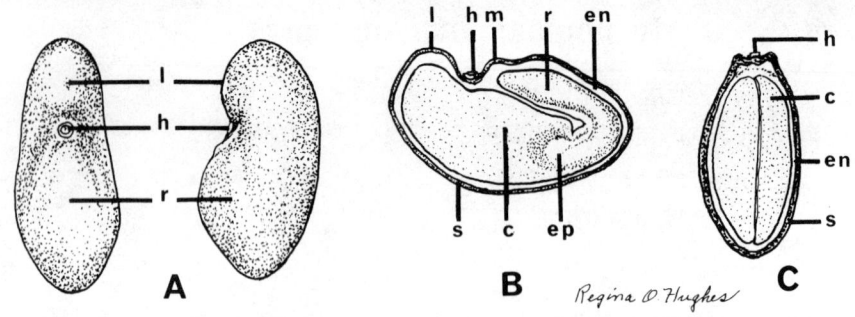

Fig. 4-1. Diagrammatic view of alfalfa seeds. (A) External, (B) Internal with one cotyledon removed, and (C) Cross-section. c = cotyledon, en = endosperm, ep = epicotyl, h = hilum, l = lens, m = micropyle, r = radicle (hypocotyl), and s = seed coat (testa).

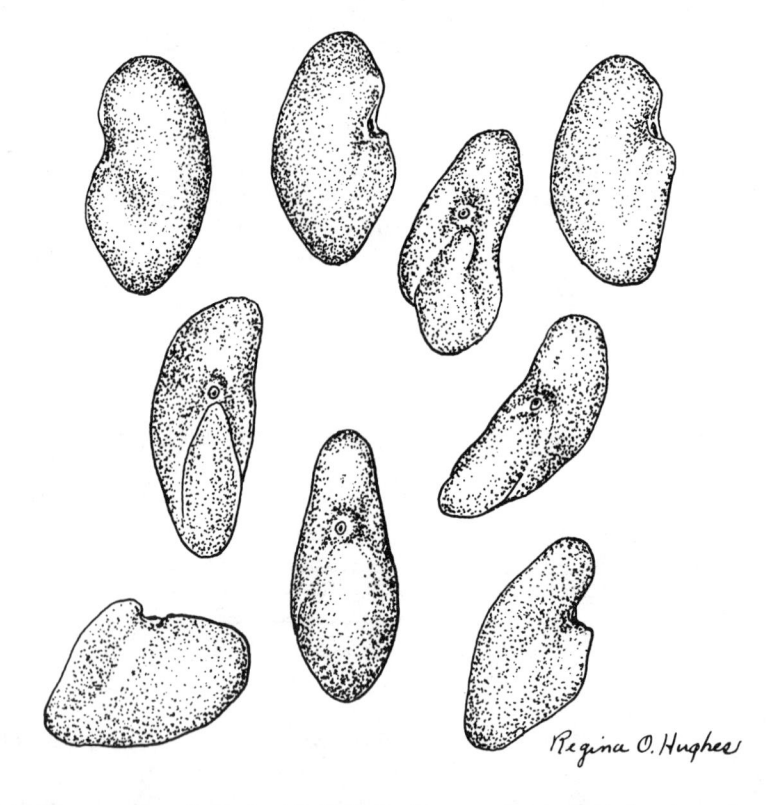

Fig. 4-2. Fully matured, well-developed alfalfa seeds are variously shaped and angular because of internal and external forces that shape the seeds during maturation.

the configuration of the rounded outer palisade cells of the testa. It varies in thickness over the surface of the seed, frequently decreasing to a very thin layer (73).

There are several distinct surface features of the testa, including a centrally located hilum, a lens located about 0.4 mm from the hilum opposite the radicle, and a micropyle located adjacent to the hilum on the radicle side (Fig. 4–3A). The hilum results from the abscission of the mature seed from the funiculus. The abscission layer is formed between the palisade cells of the outer integuments across the base of the seed and extends to the inner vascular bundles.

Beneath the hilar surface, a tracheid bar (Fig. 4–3B) extends from the micropyle at one end of the hilum to the ovular bundle at the opposite end (71). The tracheid bar is a strip of vertically oriented large tracheid-like cells that are lignified and pitted. The lens, which is the weakest point in the palisade layer of the seed coat, is located at the apex of the cotyledons. This opening provides a point of entry for water into the seed during the imbibition phase of germination. This can be illustrated by the movement of chromic-nitric stain into the seed (Fig. 4–4).

The testa is composed of three cell layers: the outer palisade cells, intermediate hour-glass cells, and inner parenchymatous cells. The palisade layer (Fig. 4–5) (Malpighian cells) consists of a single layer of prismatic, thick-walled contiguous cells, and is more or less hexagonal in cross section (37). Near the tip of the palisade cells is the light line, an optical effect that gives the false impression that the palisade cells consists of two cell layers (73). Corner (37) proposed that this optical effect may result from the papillate nature of the cuticle that extends deeply into the crevices between the rounded ends of the palisade cells. The presence of the cuticle in these crevices causes light to be refracted differently than from the rest of the cell wall. Beneath the palisade cells are the hour-glass cells (osteoscleroids) that are broad at the base and frequently separated by intercellular spaces at their outer limits. These cells may be one to several layers thick (73). A zone of parenchymatous cells occurs under the hour-glass cells; these are elongated periclinally and have thin cell walls. The outermost layers lack intercellular spaces, while the inner layers are often loosely arranged (52).

A layer of endosperm lies between the seed coat and the cotyledons (Fig. 4–1). The endosperm has a distinct outer layer of aleurone cells and an underlying zone of thin-walled parenchyma cells that become mucilaginous following imbibition of water (52). The cotyledons, which lie inside the endosperm, have small epidermal cells to the surface and mesophyll tissue to the interior. The cotyledonary mesophyll consists of three upper layers of palisade cells and several rows of subadjacent spongy cells (52).

Impermeable or "hard seeds" are common. These seeds fail to imbibe water when placed in a moist environment. Lute (73) investigated the cause of hard seed and demonstrated that the thickened outer walls of the palisade cells rather than the cuticle constitute the moisture barrier. Furthermore, she found no relationship between seed permeability and either structural differences in the seed or altitude where the seed was produced. Seed dormancy that is caused by an impermeable seed coat

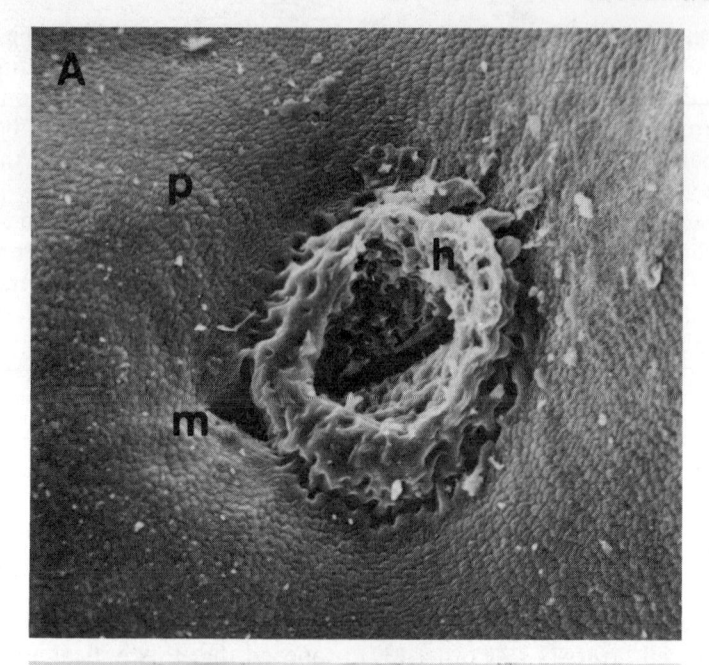

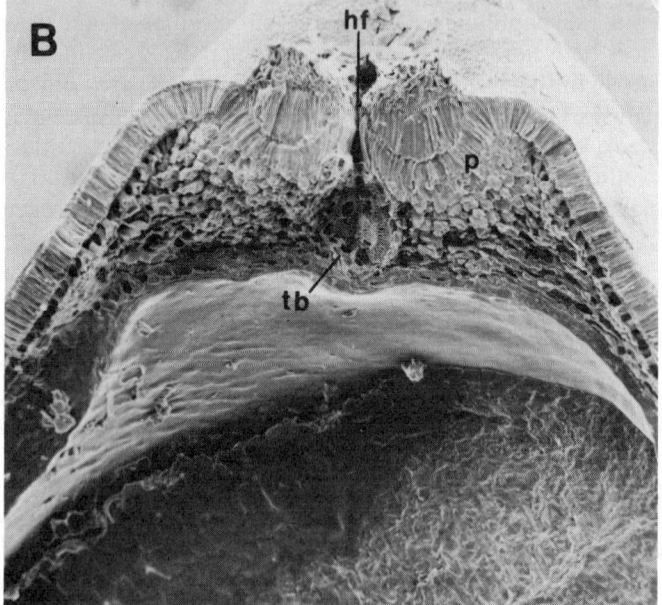

Fig. 4-3. Scanning election micrographs of the hilar region. (A) surface (× 200). h = hilum, m = micropyle, and p = papillose seed surface. Courtesy of P. Miklas, Colorado State Univ., Fort Collins. (B) Median longitudinal section of the hilum (× 253). hf = hilar fissure, p = palisade epidermis, and tb = tracheid bar. Courtesy of L. R. Teuber, Univ. of California, Davis.

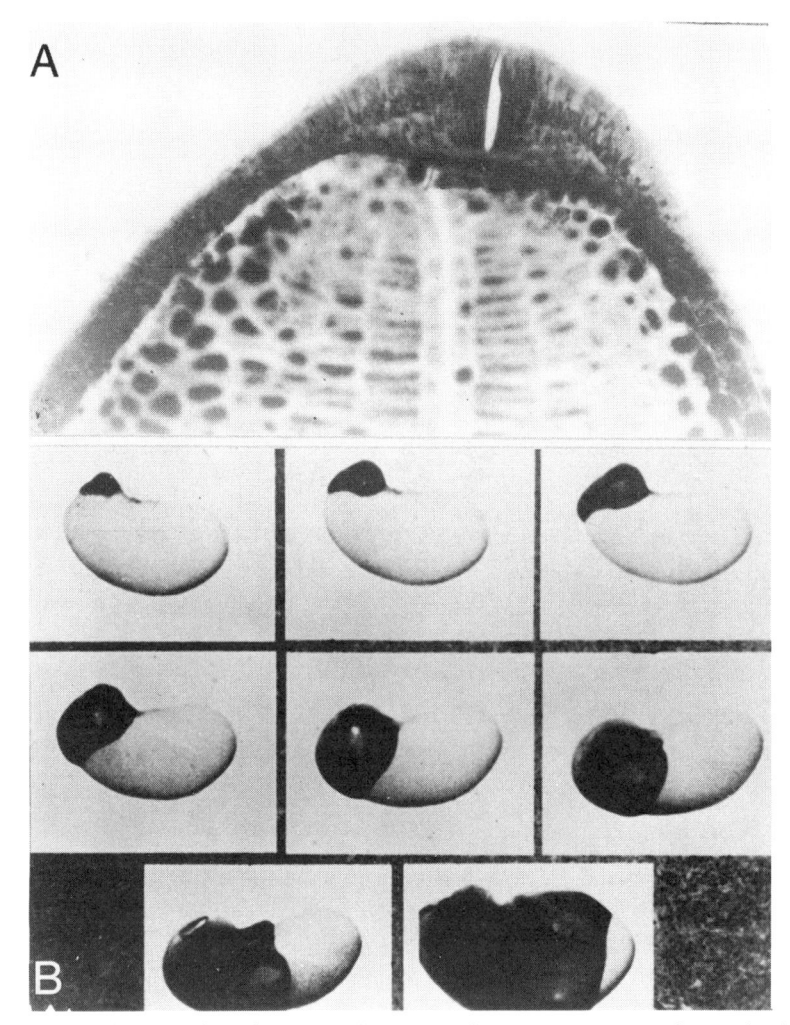

Fig. 4–4. (*A*) Cross section of lens showing an opening. (*B*) Movement of chromic-nitric acid stain through lens and under seed coat. Time-lapse photographs: 3, 6, 9, 12, 15, 18, 30, 60 min, from left to right. Courtesy of G.F. Simmons, Washington State University, Pullman.

has been reported for many members of the Leguminosae family (5, 107, 108). In general, it is agreed that the best method to overcome hardseededness is to mechanically scarify the seed by abrading the seed coat.

Alfalfa seed may remain viable for an extensive period of time. Wilton et al. (110) germinated 228 lots of alfalfa seed varying in age from 23 to 70 yr and found that one seed lot germinated 7% after 70 yr of storage, and a second seed lot germinated 30% after 62 yr. Rincker (89) stored three seed lots for 20 yr in continuous subfeezing temperatures. He found the average decline in germination ranged from 3 to 13% (see Chapter 31 in this book).

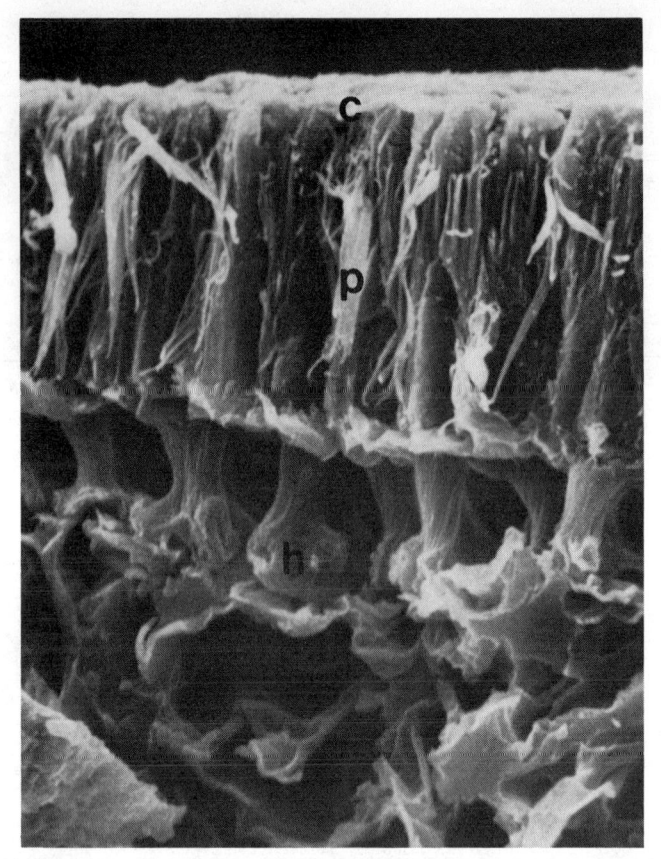

Fig. 4–5. Scanning electron micrograph of the alfalfa seed coat (× 510). p = palisade layer, h = hour-glass cells, and c = cuticle. Courtesy of P. Miklas, Colorado State University, Fort Collins.

4–1.1 Seed Germination and Seedling Development

Grove and Carlson (51) state that "the primary root emerges near the hilum and penetrates the soil as an unbranched tap root. As the hypocotyledonary area straightens and elongates the cotyledonary leaves (seed leaves) emerge aboveground. The first foliar leaf (unifoliolate leaf) is simple with a slender petiole". Subsequent leaves produced on the primary stem are trifoliolate.

The seedling or primary axis continues to elongate and develop alternately arranged leaves. The hypocotyledonary region concurrently undergoes contractile growth pulling the cotyledons closer to and sometimes beneath the soil surface. The axillary bud of the unifoliolate leaf develops into the first secondary stem. Subsequent secondary stems arise from the axillary buds of the cotyledons. Seedlings of nondormant cultivars have been shown to produce more secondary stems at the cotyledonary nodes than dormant cultivars. Stems arising from the axillary

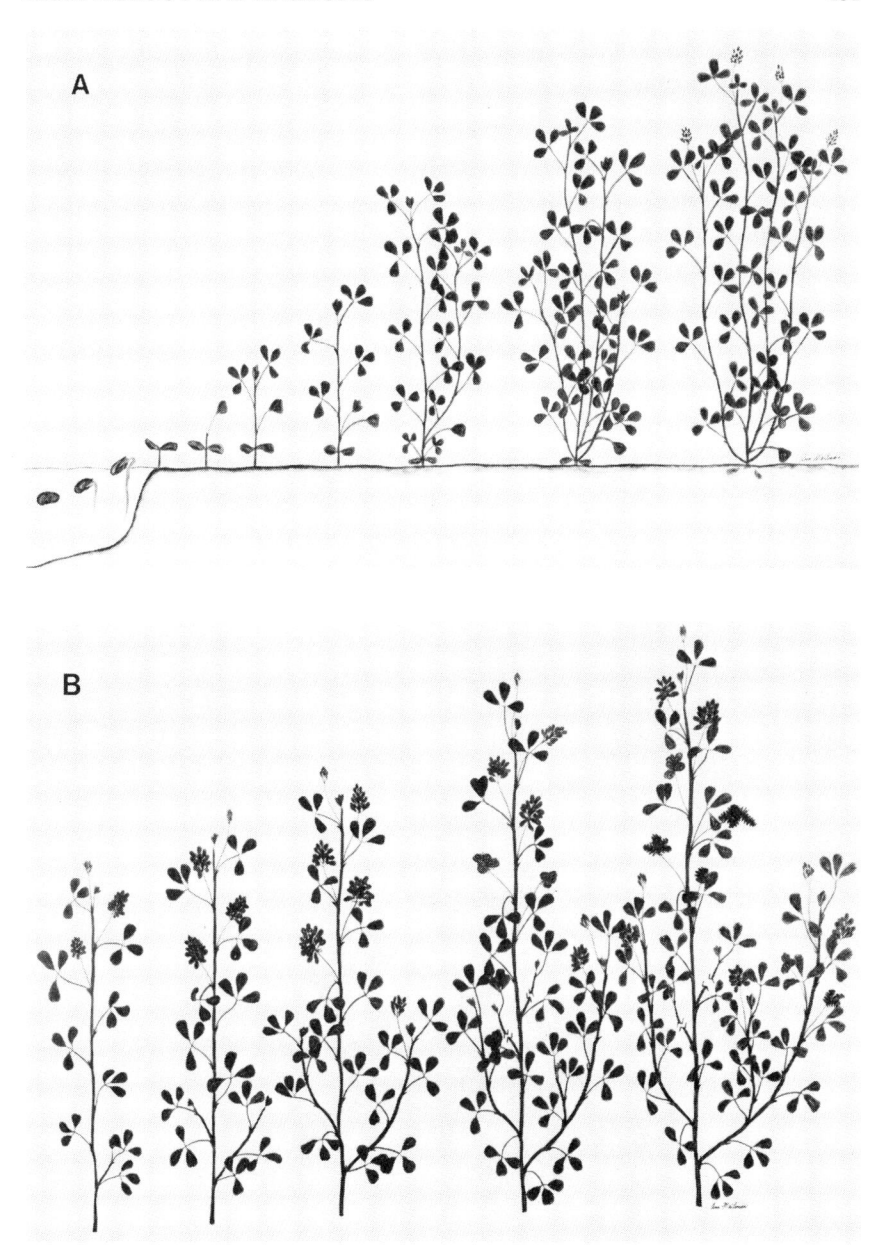

Fig. 4–6. (A) Sequence of seed germination and initial seedling development in alfalfa depicting contractile growth and development of secondary stems from axils of the unifoliolate leaf and cotyledons. (B) Stem branching sequence in seedling alfalfa and approximate relationship of secondary and tertary branching to floral development. Courtesy of A. N. Martensen, Univ. of California, Davis.

buds of the unifoliolate leaf and the cotyledons from the structure that becomes the primary crown (43) (Fig. 4–6A). Seedlings left to grow without harvest produce additional secondary stems from the axillary buds of trifoliolate leaves proceeding upward on the primary stem, and eventually tertiary stems arise from the axils of leaves on the secondary stems (27) (Fig. 4–6B) (see Chapter 5 in this book).

4-2 ROOTS

4-2.1 Primary Root Growth

The structures of the primary root are derived from the apical meristem of the radicle. They consist of a well-defined epidermis, a cortex, and the stele, which is a solid cylinder of vascular tissue in the central axis of the root. Differentiation of vascular elements (protoxylem) usually begins from three points in the stele near the pericycle (Fig. 4–7). Vascular elements are recognizable <1 mm behind the growing tip (96). Differentiation proceeds centripetally to form three rows of protoxylem cells. Roots that form three rows of protoxylem cells are termed *triarch*. However, tetrarch and diarch forms also occur in alfalfa (35; 111; 96; 80). The protoxylem vessels usually have annular or spiral secondary wall thickening, although Simonds (96) reported reticulate and pitted protoxylem vessels in alfalfa. A large reticulate mexaxylem vessel forms in the

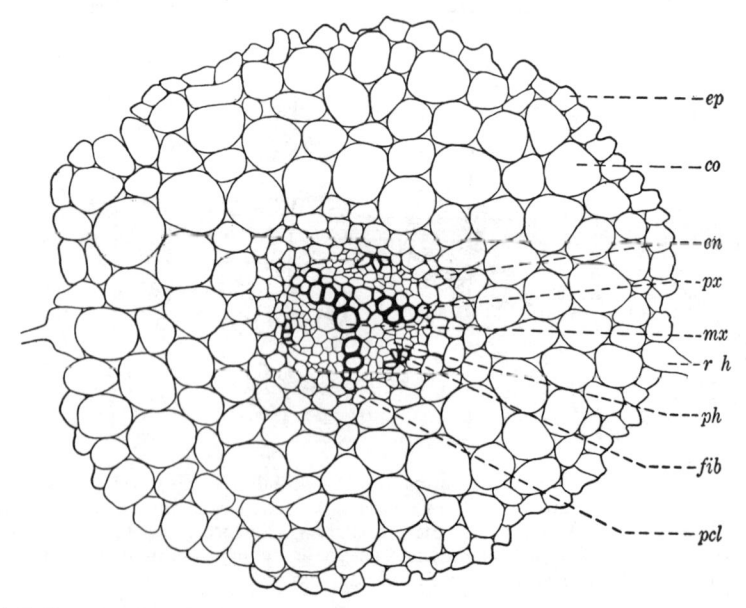

Fig. 4–7. Cross section of primary root; co = cortex, en = endodermis, ep = epidermis, fib = phloem fibers, mx = metaxylem, pcl = pericycle, ph = phloem, px = protoxylem, and rh = root hair. After Hayward (52).

center of the stele at the point where the three rows of primary xylem converge (Fig. 4–7).

Primary phloem groups develop in contact with the pericycle between the rows of protoxylem vessels (96). Peripheral cells differentiate as phloem fibers soon after protoxylem formation (Fig. 4–7) (52).

The pericycle forms the peripheral nonvascular tissue of the stele. It arises from the same initials of the apical meristem as the vascular tissues, and, therefore, is considered part of the stele. The pericycle is uniseriate at first, but becomes multiseriate outside the protoxylem points by undergoing tangential divisions (52).

The cortex lies outside the stele and occupies three-fourths of the cross-sectional area in the young root. It consists of the endodermis, which lies external to the pericycle, and several layers of large thin-walled parenchyma cells. Small casparian strips in the endodermal cells are recognizable in cross sections of well-developed tissue (96). Simonds (96) observed that in the outer region of the primary cortex, cells were relatively large and loosely joined, whereas adjacent to the endodermis the cells were much smaller and more compact. The epidermis consists of a single layer of cells that delineates the outer limits of the primary root. Some epidermal cells have root hairs. Root hairs are tubular extensions of epidermal cells that markedly extend the absorbing surface of the root (41). Root hairs are also the site of attachment for *Rhizobium* bacteria (see Chapter 7 in this book).

4–2.2 Secondary Root Growth

Procambial cells that remain undifferentiated, between the primary xylem and phloem, begin to divide upon completion of primary root growth. Hence, at the onset of secondary growth, the vascular cambium is formed in strips that lie between rows of primary xylem (Fig. 4–8). Subsequently, the pericyclic cells located directly outside the primary xylem rows become active as a cambium, and the cambium encircles the xylem core. This activity results in the formation of a continuous vascular cambium that forms the same outline as the primary xylem (i.e., triangular in triarch roots). As secondary growth continues, the vascular cambium produces the first secondary xylem between rows of primary xylem. This causes the cambium to be displaced outwardly until its circumference becomes circular in cross section. The early secondary phloem is formed by activity of cambial cells contiguous with the primary phloem (96).

The vascular cambium continues to produce secondary xylem to the inside and secondary phloem to the outside throughout secondary growth. The rows of primary xylem may be crushed during secondary growth. Secondary xylem is composed of vessel elements, parenchyma cells, and libriform fibers. The xylem vessel elements are short with scalariformly and reticulately pitted secondary walls (69). Vessel elements vary in diameter within secondary tissues. The vessels of winter-hardy cultivars

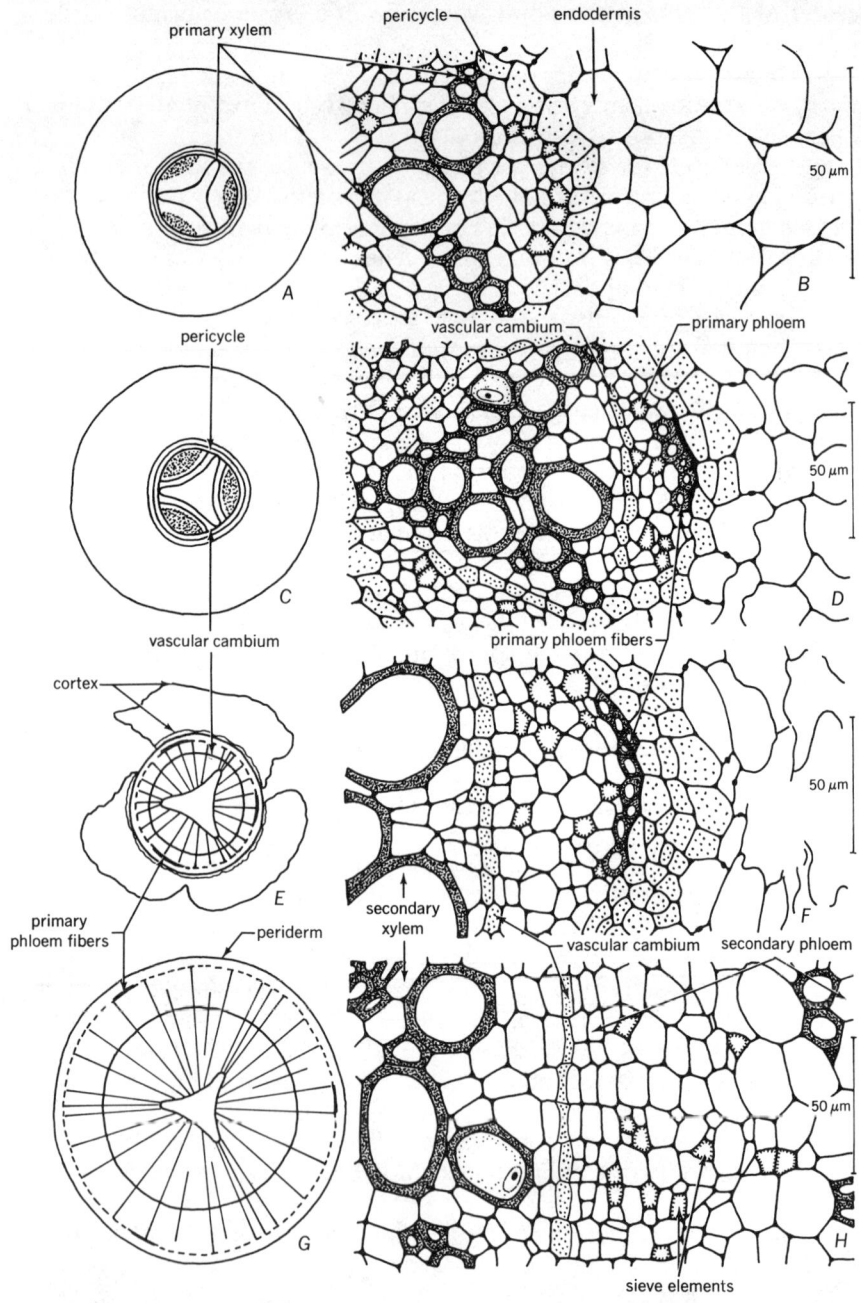

Fig. 4–8. Diagrams and detailed drawings of cross sections of root of alfalfa in different stages of development A, B, primary stage of growth. C, D, initiation of vascular cambium. E, F, secondary growth of vascular cylinder, cell division in pericycle, and rupture of cortex. G, H, secondary growth is established. After Esau (41). From *Anatomy of Seed Plants* (2nd Edition) by Katherine Esau. Reprinted by permission of John Wiley and Sons, Inc.

produced early in the season are larger in diameter than those produced later in the season (60, 93). Furthermore, early season, woody tissue has fewer parenchyma cells and more libriform fibers. These differences can be observed as annual growth rings in cross sections of older root tissue (60).

Secondary phloem is composed of sieve-tube members with companion cells, parenchyma cells, and fibers. Simonds (96) found that secondary phloem fibers were derived from cells present during primary growth that were contiguous to the primary phloem fibers. Jones (60) noted that during the 1st yr of secondary phloem development fewer fibers are produced than in succeeding years. Annual increments of secondary phloem are well defined, but are not as large as those of the xylem. The annual increments of secondary phloem can be observed because the first cells to differentiate are phloem fibers, followed by sieve tube members and phloem parenchyma (Fig. 4-9) (52). Most of the preceding years' sieve tube members are crushed during secondary growth; however, the thick-walled fibers remain intact and can be used to identify successive years of secondary phloem production.

Following the initiation of secondary root growth, the pericycle undergoes periclinal and anticlinal divisions to give rise to the periderm.

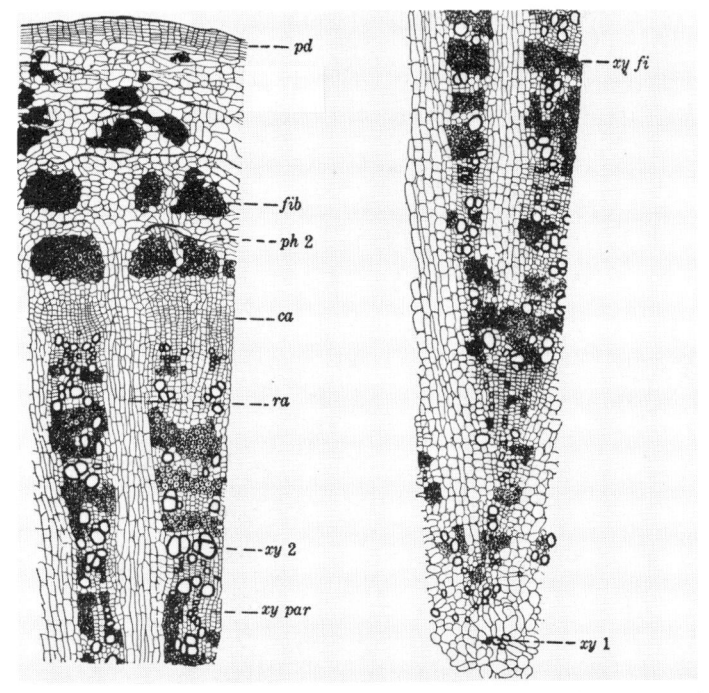

Fig. 4-9. Sectors of 4-yr-old taproot; at left, showing the outer portion; and at right, the inner portion of root; ca = cambium, fib = phloem fibers, pd = periderm, ph2 = secondary phloem, ra = xylem ray, xy 1 = primary xylem, and xy 2 = secondary xylem, xy fi = xylem fibers, xy par = xylem parenchyma. From Hayward after Jones (60).

This activity and that of the vascular cambium places outward pressure on the primary cortex. Since the cortex does not undergo secondary growth, it (including the endodermis) is ruptured and sloughed off with the epidermis (Fig. 4–10). The external cell layers of the pericycle became meristematic and function as an active *phellogen*. The phellogen produces *phellem* (cork) to the outside and *phelloderm* (secondary cortex) to the inside. The phellum, phellogen, and phelloderm form the outer tissue layer collectively known as the *periderm*. The phellem is arranged as compact radial rows usually either two or three cells thick. As new phellum cells are formed from the activity of the phellogen, the outer cells are sloughed off. Wound periderm often forms in root (and crown) tissue following either environmental or mechanical injury (68).

Lenticels, which are masses of loosely arranged cells that protrude above the periderm, may occur on the root surface. Lenticels often form at the site of future lateral root emergence (96), with the lateral root growing out through the lenticel. However, many roots emerge prior to lenticel formation, with subsequent lenticel formation at the site of emergence. Lenticels are thought to function in gas exchange through the periderm. Their role in gas exchange is substantiated by the observation that alfalfa plants grown under saturated soil conditions have more lenticels than those grown under low soil moisture.

4–2.3 Lateral Root Growth

Lateral roots originate in the region of the pericycle on the same radii as the rows of protoxylem. Hence, the lateral roots formed during early

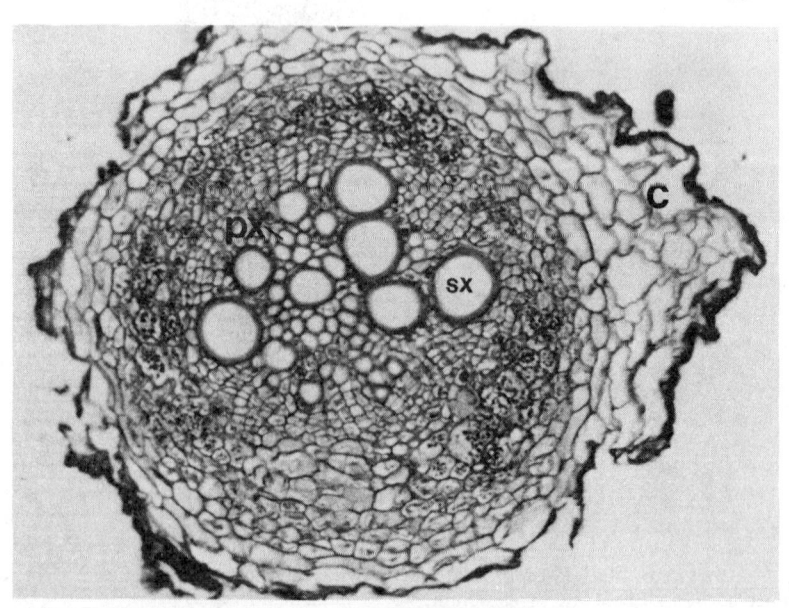

Fig. 4–10. Cross section of alfalfa root during initial stage of secondary growth. c = cortex, px = primary xylem, sx = secondary xylem. Courtesy M. A. Brick, Colorado State Univ., Ft. Collins.

secondary growth are arranged in longitudinal rows corresponding to the number of primary xylem lobes (52). The apical meristem of the lateral root initially produces undifferentiated cells on the radial axis in order to increase length. The vascular tissue of the lateral root differentiates to connect the vascular system of the lateral root to that of the main axis (Fig. 4–11). Differentiation of the nonvascular tissue proceeds in the same manner as in the primary root. There is no continuity of either cortical or epidermal cells (51).

The degree of root branching among alfalfa genotypes has been investigated by several scientists (48, 97, 102, 75, 76). Garver (48) reported that the nonwinter-hardy genotypes had fewer branch roots than the winter-hardy types. Smith (97) compared root branching in seven alfalfa cultivars and reported that the winter-hardy cultivars were more heavily branched than the moderately hardy cultivars. McIntosh (75) conducted genetic studies of the root-branching trait in three cultivars. She obtained broad-sense heritability estimates for propensity to form branch roots of 38 and 9% for greenhouse and field-grown propagules, respectively.

4–2.4 Agronomic Significance of Root Structure and Function

Many root traits have been studied to determine the inheritance and/or associations between root structure and function. Some of these characteristics include N accretion and storage (74, 46, 67, 9) (see Chapter 7 in this book), soil moisture extraction (39, 57) (see Chapter 11 in this book), disease reaction (93, 30, 47), root pigmentation (32, 21), root growth and distribution (25, 62, 105, 50, 17, 26, 87) (see Chapter 11 in this book), nutrient uptake (91, 24), cork formation (61, 68), and other anatomical characteristics (18, 19).

LeClerg and Durrell (69) investigated the relationship between root anatomy and vascular plugging of alfalfa roots due to an unidentified organism, which was most likely bacterial wilt [*Corynebacterium insidiosum* (McCull.) H.L. Jens.]. They reported that the vessels near the crown were smaller than those in the root, and, therefore, were subject

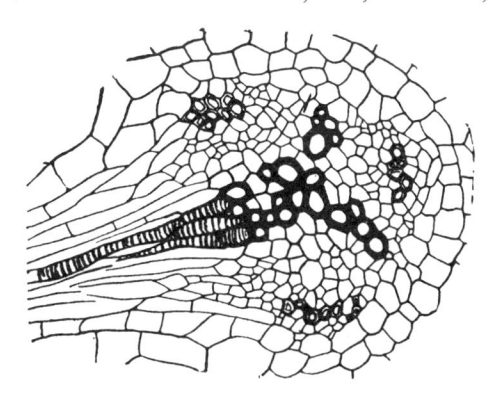

Fig. 4–11. Cross section of primary root showing vascular connection of a lateral root. From Hayward after Winter (111).

to plugging with "gum" deposits. Cho et al. (30) reported that roots and stems of plants resistant to bacterial wilt had fewer vascular bundles, shorter vessel elements, and a thicker cortex than plants of susceptible cultivars. However, they found that the number, shape, and compactness of vessel elements were not correlated with resistance to bacterial wilt.

4-3 THE CROWN

The seedling or primary crown develops from the axils of the unifoliolate leaf and the cotyledons (43) (Fig. 4–6A). Although the general pattern of development has been described previously in this chapter, considerable variation exists in the relative activity of the sites of primary crown development. Foord (43) determined that the nondormant cv. Moapa 69 and Wadi-Quaryat produced more stems from the cotyledonary nodes than more dormant cv., such as Lahontan and Norseman. Differences in rapidity of crown development among cultivars nearly disappear approximately 16.5 weeks after planting (81).

Contractile growth and unifoliolate internode length have been shown to influence the position of the unifoliolate node. Dormant cultivars have more pronounced contractile growth (86), and shorter average unifoliolate internode lengths (43, 44, 45, 92) than nondormant cultivars. Perry and Larson (86) reported that dormant cv. Travois had unifoliolate nodes as much as 1.25 cm below the soil surface. Both "shallow" and "deep" crown types have been reported. According to Shimada and Murakami (90), plants with deep crowns show greater persistence and have many more stems originating from below the soil surface than plants with shallow crowns. Such characteristics may have considerable value in developing plant types that can tolerate physical stress, such as wheel traffic, and temperature extremes, such as heat and freezing (see Chapter 8 in this book). Unifoliolate internode length and contractile growth may be useful selection criteria in breeding for deep crowns.

Further development of the crown occurs from nodes located <6 mm apart on the primary crown and on the stubble zone of primary shoots below the cutting height. However, few of the latter buds survive the winter (81) (Fig. 4–12). Grove and Carlson (51) cite Hayward (52) to the effect that no buds arise from the root, and Stewart (101) as including only perennial portions of the stem in a definition of the crown. They go on to state that "the crown ... Is neither a simple nor a single morphological structure but is an area including several separate structures."

The crown and the structures included in the crown (Fig. 4–13) can be altered substantially by a number of cultural and environmental conditions (see Chapter 5 in this book). This complex structure is extremely important to the development of new stems and is subject to damage from a number of organisms constituting the crown rot complex (see Chapter 21 in this book).

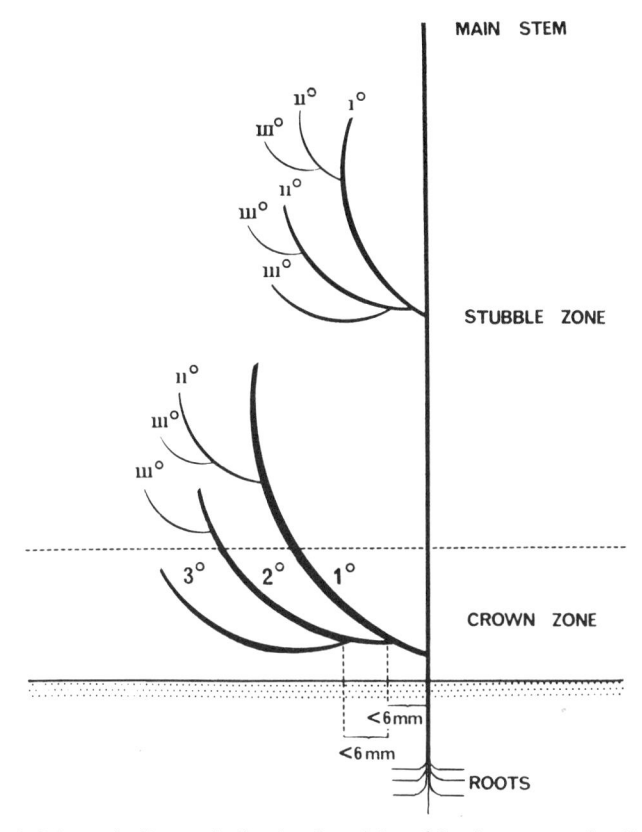

Fig. 4–12. Schematic diagram indicating the origins of the shoot categories. (Arabic numerals refer to crown shoots, Roman numerals to stubble shoots). From Musgrave and Langer (81).

4–4 THE STEM

4–4.1 Stem Anatomy

The stem arises through meristematic activity of the shoot apex. Cell division and differentiation in the region of the shoot apical meristem give rise to epidermal, cortical, and vascular tissues. The stem is defined longitudinally by nodes and internodes. Lateral appendages including leaves, axillary branches, and flowers arise along the stem at the nodes with an alternate phyllotaxy.

The young stem is nearly square in cross section. Major vascular bundles are located in the angles of the axis with smaller bundles between (Fig. 4–14). The epidermis surrounds the stem and is underlain by chlorenchyma (chloroplast-containing) and collenchyma (mechanical) tissue (Fig. 4–15). The innermost layer of the cortex is the starch sheath, and the cortex is bounded on the outside by the epidermis. The center of the

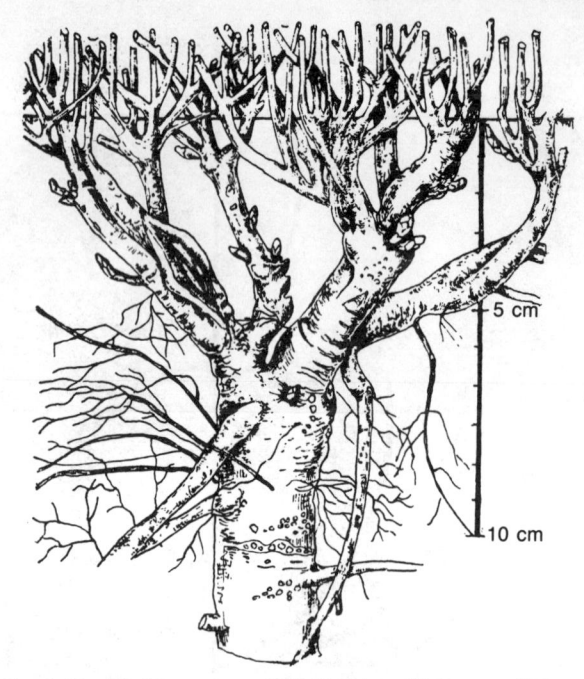

Fig. 4-13. An older crown of alfalfa. From McKeyeva (78).

stem is occupied by the pith which is composed of large, compactly arranged parenchyma cells.

The cortex usually contains one layer of collenchyma cells to the inside of the epidermis; however, at the angles of the stem, collenchyma cells are more numerous and may form a pronounced longitudinal ridge. In mature stem tissue, collenchyma may occupy the area from the epidermis to the starch sheath. Chlorenchyma forms in the area between the collenchyma and the starch sheath. The starch sheath is a continuous row of either rectangular or oval, nonchlorophyllous cells that are larger than the perivascular cells located central to them (Fig. 4-16). Starch sheath cells often contain starch and calcium oxalate crystals that appear late in the ontogeny of the stem, and usually do not have casparian strips (52).

Pith rays (interfascicular regions) that connect the pith and cortex provide lateral movement of ergastic substances. The large parenchyma cells in the central portion of the pith are unlignified, in contrast to the smaller lignified cells near the vascular bundles (109). The parenchymatous cells near the center of the pith may disintegrate during maturation, so that the stem becomes hollow.

All of the vascular bundles in alfalfa are collateral. Primary xylem elements formed in the young shoot have spiral thickenings. Primary phloem elements are long, narrow sieve tubes with simple sieve plates in their end walls (51). Companion cells and phloem parenchyma are also present in the phloem. The phloem is limited outwardly by strands

MORPHOLOGY AND ANATOMY

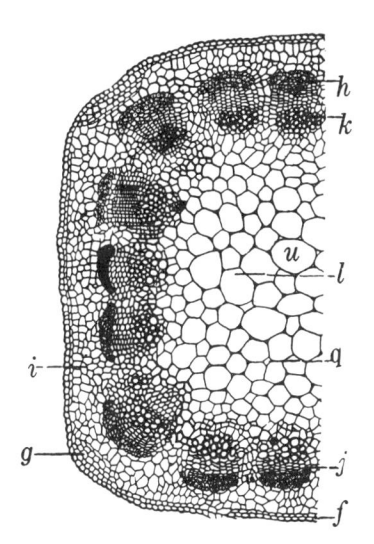

Fig. 4–14. Cross section of primary stem; l = pith, q = intercellular space, u = pith cell, k = xylem, j = phloem, h = phloem fiber, f = epidermis; g = collenchyma tissue, and i = chlorenchyma tissue. After Wilson (109).

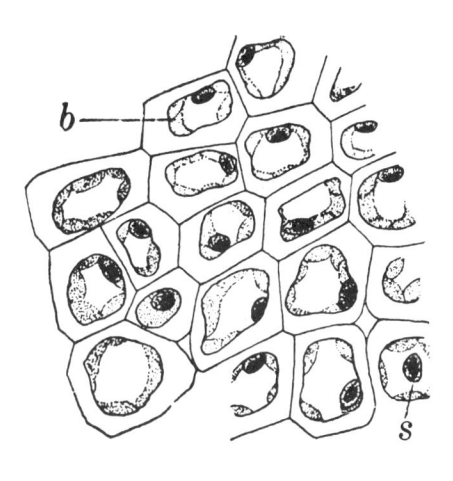

Fig. 4–15. Section of chlorenchyma tissue in stem; b = chloroplasts and s = nucleus. After Wilson (109).

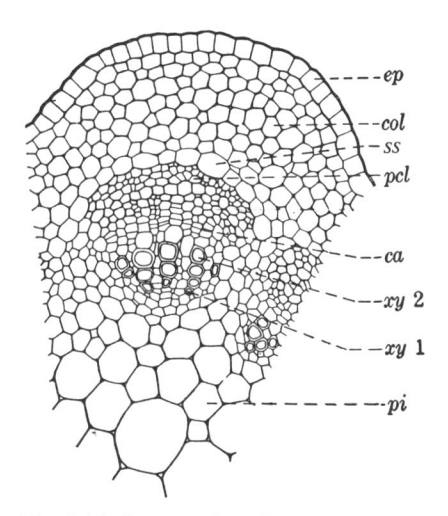

Fig. 4–16. Cross section of stem; ca = cambium, col = collenchyma, ss = starch sheath, ep = epidermis, ph = phloem, pi = pith, xy 1 = primary xylem, and xy 2 = secondary xylem. [After Hayward (1938).]

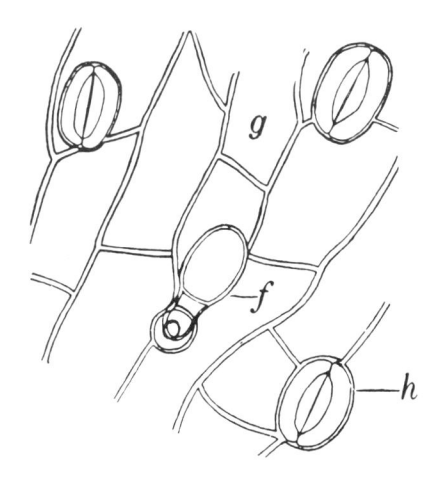

Fig. 4–17. Stem epidermis; g = epidermal cell, h = stoma, and f = glandular hair. After Wilson (109).

of elongated thick-walled perivascular fibers that supplement the xylem vessels and collenchyma tissue for mechanical support (52). The vascular cambium is recognizable at the initiation of secondary thickening and is partially derived from parenchyma cells located between the primary xylem and phloem (Fig. 4-16). During secondary thickening, an interfascicular cambium develops to complete the continuous cambial cylinder. The interfascicular cambium primarily produces parenchyma cells, some of which become lignified. The fascicular cambium produces fibers and phloem tissue to the outside vessels, fibers and parenchyma to the inside during maturation. Some of the smaller parenchyma cells central to the inner face of the bundles become lignified and form an uninterrupted zone of supporting tissue with the lignified tissue produced by the interfascicular cambium (52).

Numerous epidermal trichomes may occur along the stem surface, some of which are club-shaped multicellular structures (Fig. 4-17). More numerous are the single-celled, elongated trichomes similar to those found on leaves (109). Trichomes have been implicated as a mechanical defense, especially against phytophagous insects. Levin (72) reported that trichome density was negatively correlated with insect feeding, oviposition response, and larval nutrition in may species. Erect glandular trichomes on certain annual *Medicago* spp. provide a mechanism of resistance to the larvae of the alfalfa weevil [*Hypera postica* (Gyllenhal)] (95, 58). Exudates from the glandular trichomes act as a glue to immobilize the larvae.

Kreitner and Sorenson (64) studied the morphology of glandular trichomes among several *Medicago* spp. in order to understand the mechanisms of resistance to the alfalfa weevil. They found two types of glandular trichomes, erect and procumbent (Fig. 4-18). The erect type had a globose head and multicellular stalk; whereas, the procumbent trichome had a club-shaped head and unicellular stalk. Shade et al. (94) reported a higher rate of adult mortality in potato leafhoppers [*Empoasca fabae* (Harris)] on five glandular-haired *Medicago* spp. than on the glabrous control. With one species (*M. noëana* Boiss.), there was 100% insect mortality within 3 d of contact with the plant. Kitch et al. (63) studied the inheritance of erect glandular trichomes in alfalfa and reported a narrow-sense heritability estimate of 55%.

4-4.2 Vascular Transition from the Root to the Stem

The vascular system forms a continuous series of conducting channels from the root through the stem into the leaves and axillary branches. Grove and Carlson (51) reviewed the literature regarding the vascular system in alfalfa as follows:

Vascular transition is completed in the hypocotyl, and the orientation of the vascular tissues from the exarch radial arrangement of the root to the collateral bundles of the stem is complete in the veins of the cotyledonary leaves. Interpretation of vascular transition has varied with in-

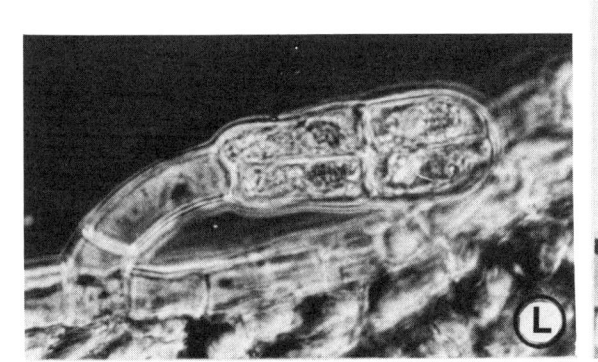

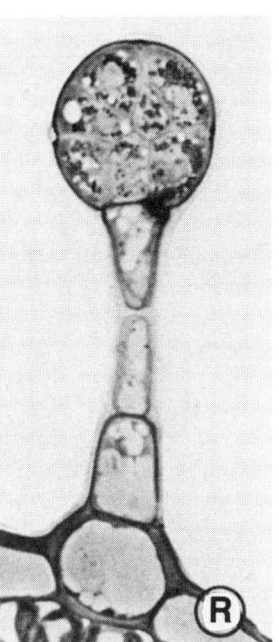

Fig. 4-18. (*Left*) Light micrograph of a fresh procumbent gland on a stipule of *Medicago sativa* ssp. *sativa* (× 850). (*Right*) Scanning election micrograph of erect glandular trichome on *M. prostrata* (× 1050). After Kreitner and Sorenson (64).

vestigators (106, 49, 35, 111, 96). Gerard (49) and Simonds (96) concluded that a direct connection of primary vascular tissue exists from the root to the first foliar leaf, but according to Winter (111) no primary vascular tissue of the root is continuous with the epicotyledonary region of the primary axis. Such differences may be ascribed to relatively early additions of secondary vascular tissue in the hypocotyl and differentiation of foliar vascular tissue in the epicotyledonary region. Continuous vascularization could result from the basipetal development of foliar traces, anastomosing with the hypocotyledonary portion of the cotyledonary traces at the time of secondary thickening.

At the base of the hypocotyl, the root exhibits a tetrarch-exarch-radial arrangement of the primary vascular tissues, with two of the larger xylem arms lying in the same vertical plane as the cotyledons and two located in the intercotyledonary plane. The two larger xylem arms are referred to as *polar xylem*. Four groups of phloem lie between the xylem arms on alternate radii (Fig. 4-19A).

At a higher level of the hypocotyl, eight groups of phloem cells are oriented in a position collateral to the four xylem arms (Fig. 4-19B and 4-19C). Immediately below the cotyledonary plate, the position occupied by differentiated polar metaxylem cells forms two V-shaped groups, which along with the polar protoxylem, make up the triad vascular structure of each cotyledon (Fig. 4-19E and 4-19F).

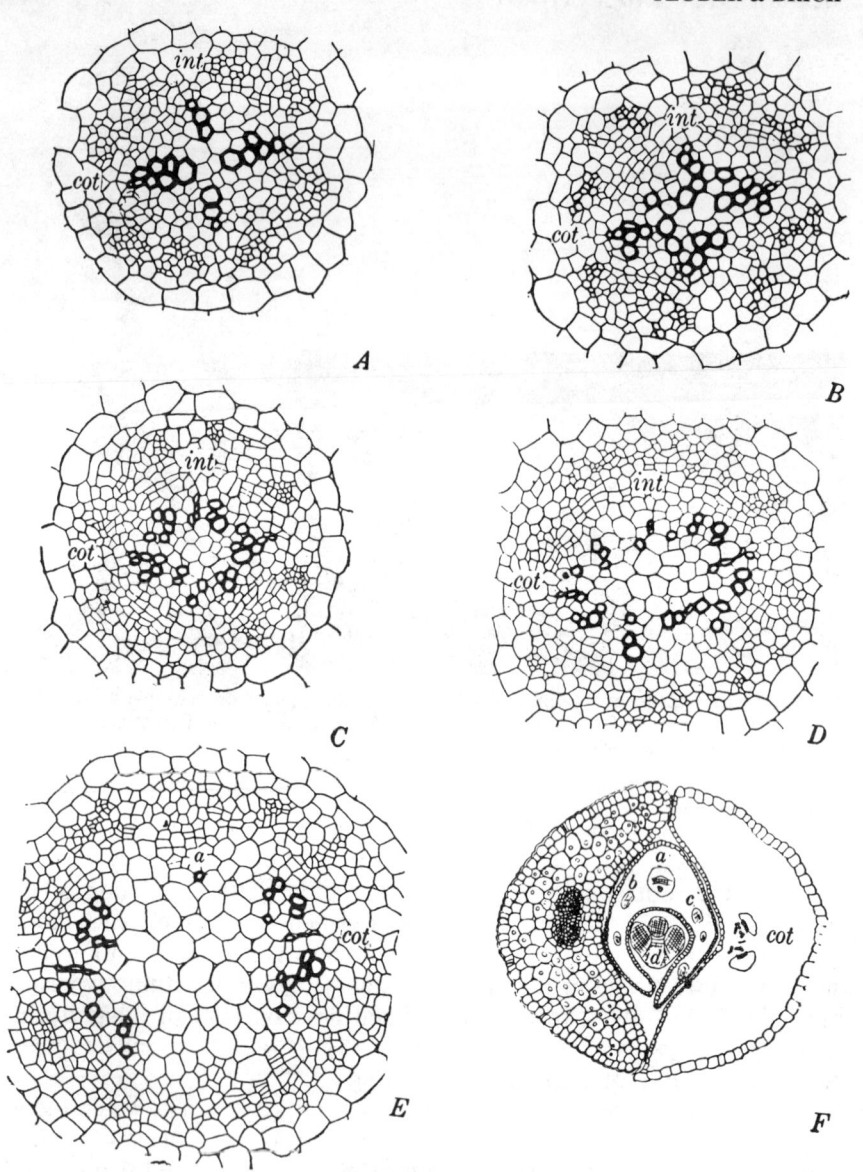

Fig. 4–19. (A) Tetrarch stage. (B), (C), (D) Bifurcation of metaxylem and formation of pith. (E) Cotyledonary node below point of divergence of cotyledons. (F) Transection of cotyledons and epicotyl; cot = cotyledon, a, b, c, = bundles of first foliage leaf, and d = median bundle of second foliage leaf. In (A) through (E) *cot* and *int* indicate cotyledonary and intercotyledonary planes, respectively. From Hayward after Winter (111).

As subsequent vascular cells differentiate in a more peripheral position in the upper hypocotyl, the relatively undifferentiated central parenchyma is recognizable as the pith. Enlargement of these cells in a radial direction accounts for an increased diameter of the hypocotyl. Jones (60) suggested that this lateral enlargement results in the shortening of the hypocotyl (contractile growth). Following radial expansion, other cells are forced into a variety of convolutions, resulting in a general shortening of this portion of the axis. Essentially all primary xylem of the root is continuous with that of the hypocotyl. Winter (111) reported that the protoxylem cells of the root, located in the intercotyledonary plane, become fragmented and are not continuous.

According to Hayward (52), the collateral, median bundle of the first foliar leaf is located in the intercotyledonary plane (Fig. 4-19F). The primary xylem of this median bundle differentiates to become continuous with the vascular tissue of the hypocotyl. The two lateral bundles of the first foliar leaf lie on either side of the stele and anastomose with the secondary tissue of the hypocotyl. Branches of these lateral traces supply the stipules of the first foliar leaf. Above the level of divergence of the stipular traces, the median and lateral traces fuse in the petiole only to separate again (Fig. 4-20). At about this level, there are vascular branches that connect the median and lateral traces.

Vascularization of the young stem has been investigated by Nageli (82), Winter (111), and Mokeyeva (78). The three petiolar bundles of each leaf are easily recognized in the stem, and their arrangement is roughly triangular. The median and lateral bundles of the next successive leaf form a similar triangular arrangement turned 180°.

Studies by Nageli (82), Mokeyeva (78), and Winter (111) indicate that on entering the stem the median bundle of the leaf extends downward through nearly two internodes without branching (Fig. 4-21G, 4-21L, and 4-21O). After entering the stem, lateral leaf bundles pass downward through a single internode (Fig. 4-21I, 4-21N, and 4-21Q) before anastomosing with branches of median bundles that are a part of the leaf vascularization, which diverge on the alternate side of the stem.

The median bundle of each leaf separates into two branches immediately above the median bundle of the next lower leaf in the same rank. The discontinuity of vascular tissue at this level, caused by the presence of parenchyma tissues, is a leaf gap.

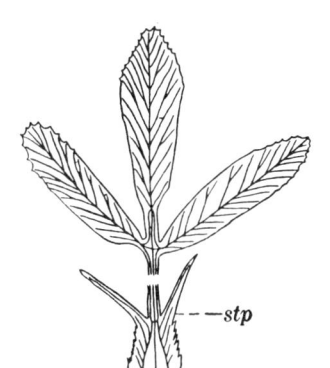

Fig. 4-20. Foliage leaf showing stipules and venation; stp = stipule. After Hayward (52).

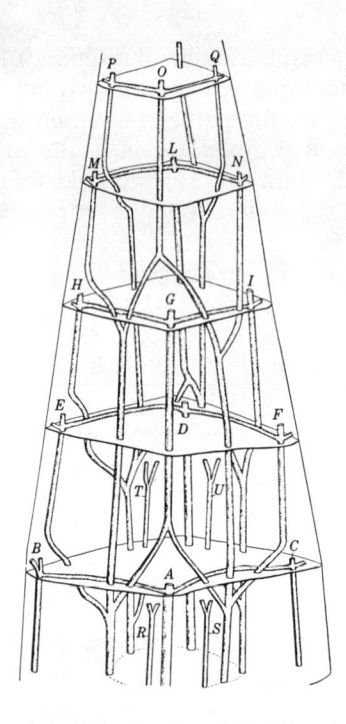

Fig. 4–21. Schematic representation of bundles near apex of stem showing vascular anatomy of five nodes and internodes in *Medicago Lathyrus* type. Each leaf is supplied by a median bundle and two laterals, and the alternate phyllotaxy of leaves is indicated by departure of traces ABC, DEF, GHI, LMN, and OPQ. The bundles R, S, T, and U supply axillary buds. From Hayward after Nageli (82).

Lateral buds that form in the axil of each leaf are vascularized by two bundles that extend through a single internode (Fig. 4–21T). Parenchyma cells in the vascular ring central to the bundles diverging to the bud (stem) are recognized as the branch gap.

4–4.3 Adventitious Stems

The ontogeny of adventitious stems on the roots of alfalfa was described by Murray (80). She distinguished between rhizomes and adventitious stems by their points of origin. She stated that rhizomes are initiated only from either the original root axis or the crown, whereas adventitious stems may occur on either the primary root or secondary roots. Plants with adventitious stems on their roots spread laterally and are "creeping-rooted."

Adventitious stems generally only form following secondary thickening of the lateral roots. The first phase of formation is the initiation of a primordial dome in the phellogen region (Fig. 4–22). The cell nuclei in this region become enlarged, and some of the cells initiate division in the tangential plane. Initial meristematic activity is not confined to a single group of phellogen cells, but involves two or three groups of cells, often in the area of a small lateral root. Meristematic activity eventually extends to the phelloderm and phloem parenchyma beneath the phellogen. Subsequently, the meristematic zone enlarges until the zone extends to the vascular cambium of the root. Procambial strands are formed

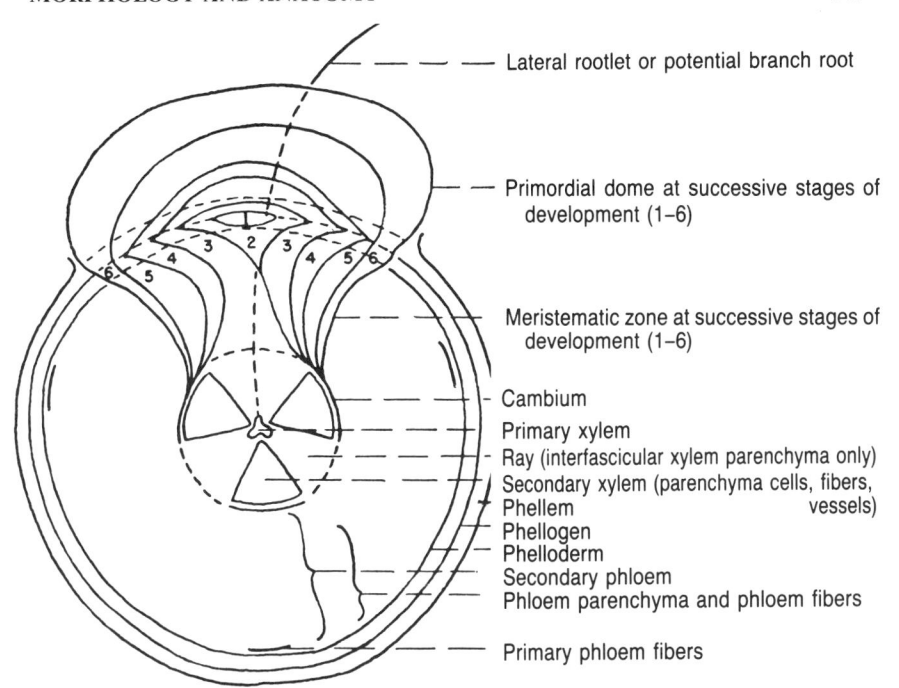

Fig. 4-22. A diagrammatic illustration of successive stages in the ontogeny of the adventitious shoot primordial dome and meristematic zone. Stage 1 = 1 to 3 d, stage 2 = 3 to 10 d, stage 3 = 10 to 15 d, stage 4 = 15 to 20 d, stage 5 = 20 to 30 d, and stage 6 = more than 30 d. From Murray (80).

either as soon as or before the primordia are visible on the root surface. A continuous series of procambial strands develops in groups or clusters at the periphery of the meristematic zone. This arrangement somewhat resembles the cross-sectional organization of a young stem. Subsequent vascular differentiation is initiated midway between the primordial dome and vascular cambium of the root, and proceeds bidirectionally.

Stem primordia differentiate in the region of the dome to produce as many as 15 stem apices. However, only one to four stem primordia continue to develop, while the others retain their rudimentary form. As maturation continues, the adventitious shoots emerge from the dome. Ultimately, the procambial strands in the shoots become continuous with the vascular system in the root.

Three to four leaves form before the adventitious stems elongate. Generally, the first leaf to emerge from adventitious stems is not the typical trifoliolate or unifoliolate leaf. The first leaf often has such anomalies as fused stipules, an extended petiole, or leaflets on an otherwise normal petiole that lacks stipules.

Selection for the creeping root trait was initiated by Heinrichs (53) in Canada. Genetic studies indicate that both additive and nonadditive genetic effects are involved in the expression of creeping roots (79, 56,

38). The time required to express the creeping root trait varies among clones and the environment in which they are grown (54, 55).

4–5 THE LEAF

Alfalfa leaves originate at the shoot apex through the activity of the apical meristem. The shoot apex produces lateral leaf buttresses to the side of the apex, which develop into the leaf primordia. The leaf primordium differentiates into three tissue systems, the dermal, vascular, and fundamental tissue systems, which give rise to the epidermis, the vascular system, and mesophyll tissue, respectively.

The first true leaf arising form the epicotyl is usually unifoliolate with an orbicular blade and stipules. The second and subsequent leaves are normally pinnately trifoliolate with slender stipules adnate to the petiole (Fig. 4–20). Leaflet form is usually either oblong or obovate with serrations toward the apex. Several anomalies in leaf form have been observed. Bingham (15) reported that some plants produce trifoliolate first leaves, and others unifoliolate or multifoliolate secondary leaves. Other leaf anomalies include: crinkled leaf (83), folded and mottled leaf (100), torn leaf (4), sticky leaf (99), yellow leaf (28), virdis leaf (29), double unifoliolate leaf (85), zebra leaf (98), narrow leaflet (20), and spotted leaf (3) traits.

The primary features of the leaflet include an upper and lower epidermis, palisade and spongy mesophyll, and vascular tissue (Fig. 4–23). The epidermal cells on the adaxial (*upper*) surface are approximately 35 μ in diameter with sinuous radial walls (112). The epidermal cells on the abaxial (*lower*) leaf surface are similar to those on the adaxial surface, except there are more trichomes present, especially along the veins (52).

Both leaf surfaces have stomata. Average stomatal density ranged from 146 to 265 mm^{-2} among five cultivars studied by Cole and Dobrenz (34). They reported greater stomatal density on the adaxial surface, and that the terminal leaves have more stomata per unit area than the basal leaves. Chloroplasts occur in the stomatal guard cells. Bingham (16) reported that the ploidy level of the plant could be distinguished by the number of chloroplasts present in the guard cells.

Epidermal trichomes similar to those found on the stem are also present on the leaf. Wilson (109) reported that as many as 700 trichomes mm^{-2} occur on leaf surfaces. Unicellular trichomes are more numerous and are usually more dense along the veins of the leaf. Also, multicellular capitate trichomes may also be found uniformly distributed over the entire leaf surface (112).

The mesophyll tissue consists of chlorenchyma tissue, separated into a distinct palisade zone in the upper portion of the leaf, and spongy mesophyll in the lower portion. The palisade zone usually consists of either two or three rows of chlorenchyma cells, approximately twice as high as they are broad. The palisade layer usually constitutes about one

MORPHOLOGY AND ANATOMY

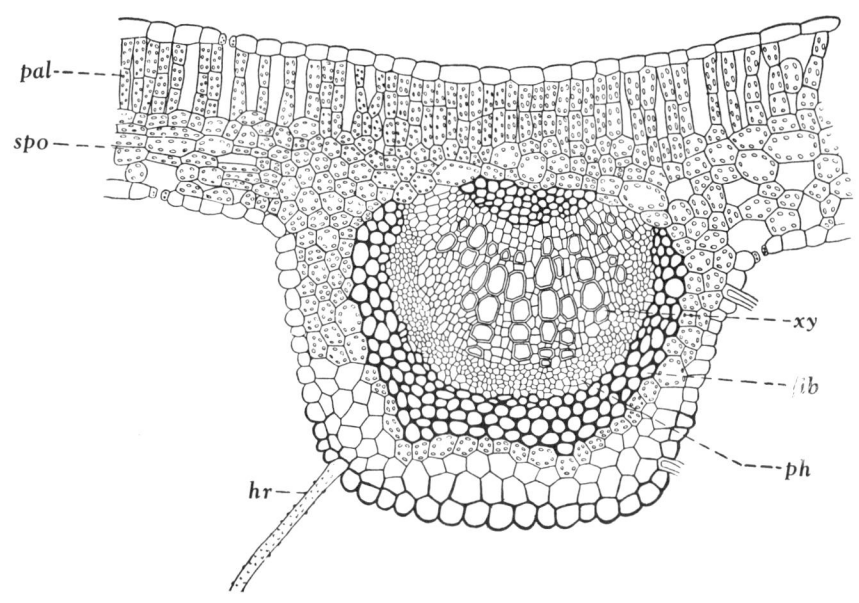

Fig. 4–23. Transection of portion of leaflet cut through midvein showing distribution of chloroplasts and intercellular spaces; hr = hair, fib = fibers, pal = palisade region, ph = phloem, spo = spongy tissue, and xy = xylem. After Hayward (52).

half the leaf thickness. The spongy mesophyll extends from below the palisade layer to the lower epidermis. It contains loosely arranged thin-walled chlorenchyma cells with numerous intercellular spaces. These intercellular spaces form a conducting network for gas exchange with stomata on the leaf surface.

The vascular network is made up of a prominent midvein and lateral veins extending to the margins of the leaflet. The lateral veins give rise to smaller veins forming a pinnately veined system (Fig. 4–20). The small distal ends of the lateral veins terminate in the leaf mesophyll, but some extend directly to the margin of the leaf. Wilson (109) observed that beads of water accumulated on the terminal margin of the leaf under humid conditions. He suggested that these terminal margins may function as hydathodes. Spirally thickened xylem elements occur at the terminal ends of the veins. The vascular elements are identical to those found in the stem, with secondary thickening occurring in the larger veins (51).

Bingham (15) investigated the vascular structure of the petiole and found that the stipules are supplied by veins from the lateral bundles. The lateral bundles anastomose with the central bundle in the region of the upper stipular attachment (Fig. 4–24A). The bundles remain separate in the basal portion of the petiole to about 1 mm below the lateral leaflets (Fig. 4–24B). At that point, the large central bundle trifurcates in the rachis. Part of the central bundle continues up the rachis to supply the terminal leaflet. The lateral branches of the central bundle fuse with the lateral bundles before entering the leaflets. Each lateral bundle supplies

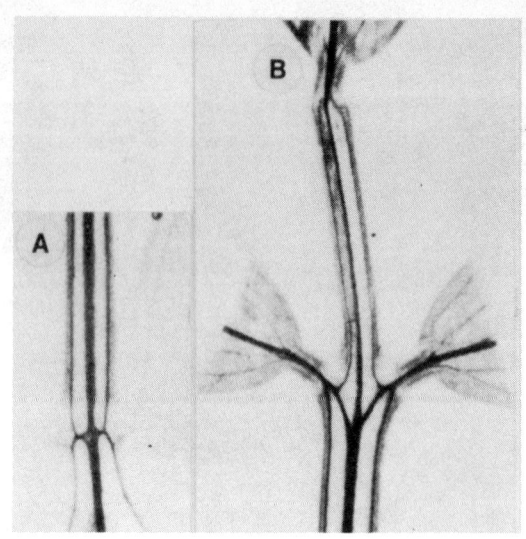

Fig. 4-24. Photomicrographs of vascular bundles in cleared petioles and leaf rachis. (A) Stipular region. (B) Stipular region in area of lateral leaflet divergence. After Bingham (15).

a branch for a lateral leaflet. A portion of the lateral bundles curves adaxially over the lateral branch derived from the central bundle and continues up the rachis to supply the terminal leaflet.

The midvein of the leaflet extends the entire length of the leaflet and forms a prominent ridge on the abaxial surface. The midvein is surrounded almost completely by a sheath of thick-walled fibers (Fig. 4-23). These cells become lignified in the abaxial sector and contain numerous calcium oxalate crystals (52). The degree of lignification in bundle sheath cells is less pronounced toward the distal portion of the midvein and in lateral midveins. The vascular tissue of the midvein undergoes some secondary growth and is similar to that in the stems (52).

Leaves with more than three leaflets (multifoliolate) are not uncommon (Fig. 4-25), and have been reported extensively in the literature (78, 13, 14, 42, 22). Multifoliolate leaves with as many as 11 leaflets have been reported (15). Bingham and Murphy (14) reported that the multifoliolate leaf trait was controlled by at least three genes in diploid alfalfa. Bauder (13) proposed a disomic mode of inheritance for two independent dominant genes in tetraploid alfalfa. A cultivar with approximately 99% of the plants exhibiting the multifoliolate character was released in 1980 by Murphy and Lowe (2).

4-6 THE FLOWER

According to Barnes et al. (8), "Development of the alfalfa flower begins at the shoot apex with the transition from vegetative to repro-

Fig. 4-25. Multifoliolate leaves exhibiting four, five, six, and seven leaflets. Courtesy, M. A. Brick, Colorado State Univ., Ft. Collins, CO.

ductive growth." This transition takes place between the 6th and 14th node (40, 77) depending on both environmental and genetic factors. The transition is recognized as a protuberance of meristematic tissue in the axil of the leaf primordium adjacent to the shoot apex (Fig. 4-26) (40). Each primordium gives rise to a simple raceme. Flower buds of these racemes go through four visible stages of development during maturation: straight bud, pointed bud, hooded bud, and erect standard (mature flower) (33). The shoot is normally indeterminate, and the shoot apex continues to produce both vegetative and floral organs at between 1.3 to 2.0 per 7 d from primary and secondary stems until the stem either senesces or is removed (Fig. 4-6B) (27).

"Each flower primordium is determinate and produces a calyx, a corolla, 10 stamens, and a pistil. The calyx tube consists of five undiverged sepals terminated by five lobes or teeth that exceed the length of the tube (Fig. 4-27). The papilionaceous corolla is highly evolved (70) and consists of five petals: a large standard or banner, two lateral wing petals, and two fused petals that form the keel" (Fig. 4-28) (8). The keel petals are held together by a series of ridges and grooves. These ridges

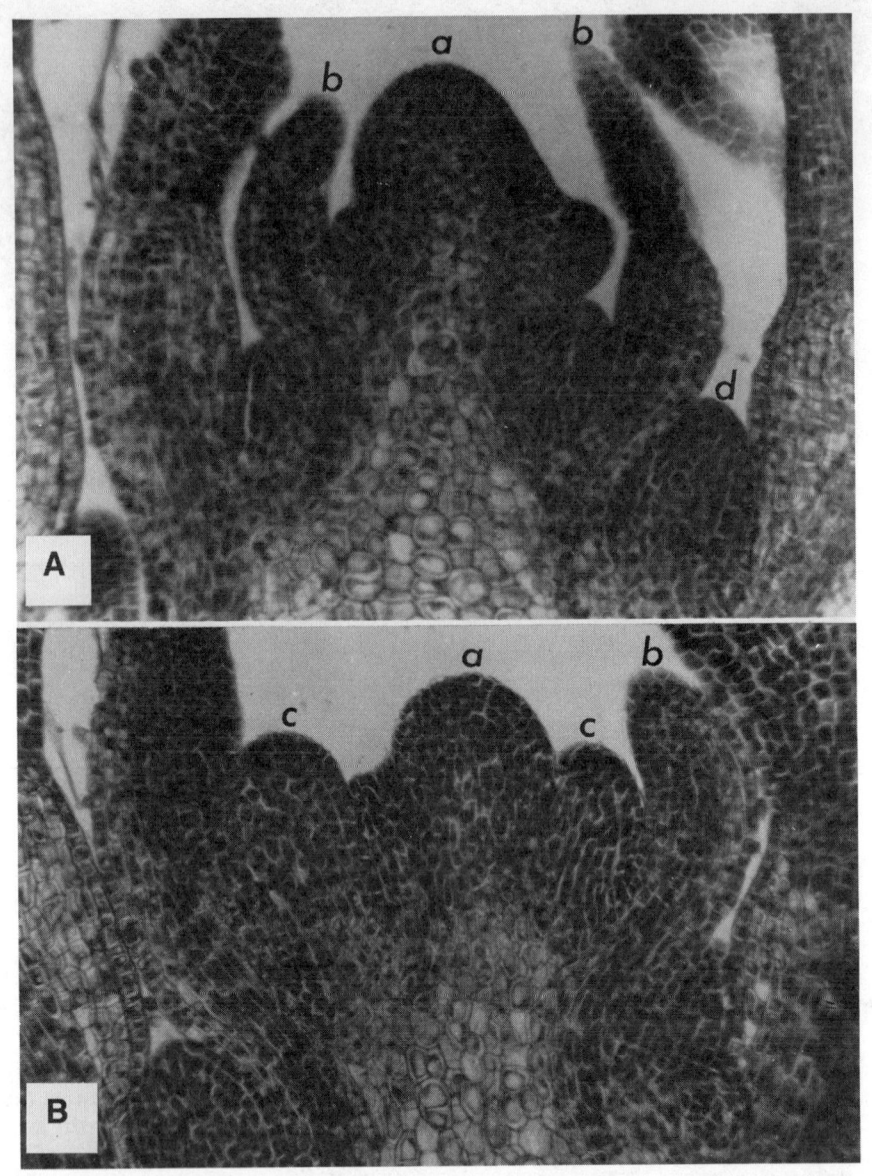

Fig. 4–26. (A) Transition from vegetative shoot to (B) floral shoot; a = apical meristem, b = leaf primordia, c = raceme primorida, and d = axillary bud meristem. From Dobrenz et al. (40).

and grooves on the two petals mesh together, locking the petals in place (66) (Fig. 4–29 and 4–30).

Flowers of cultivated alfalfa are most frequently purple, while flowers on *M. falcata* are yellow. Germplasm which traces to both *M. sativa* and *M. falcata* will express a range of flower colors including purple, variated,

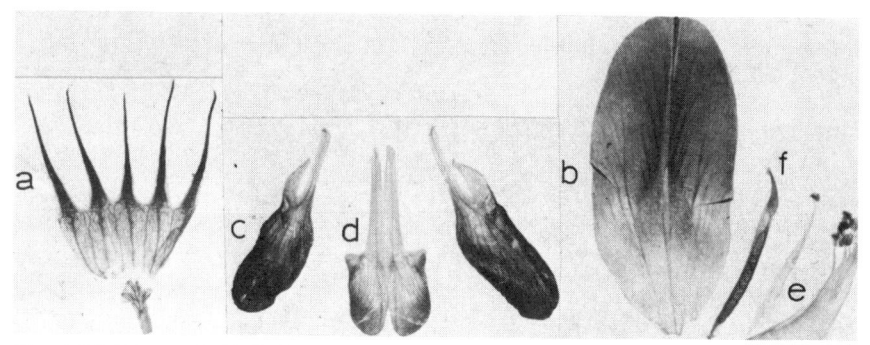

Fig. 4-27. Dissected alfalfa flower; a = sepals, b = standard or banner petal, c = wing petals, d = fused keel petals, e = staminal column, and f = carpel. From Barnes et al. (8).

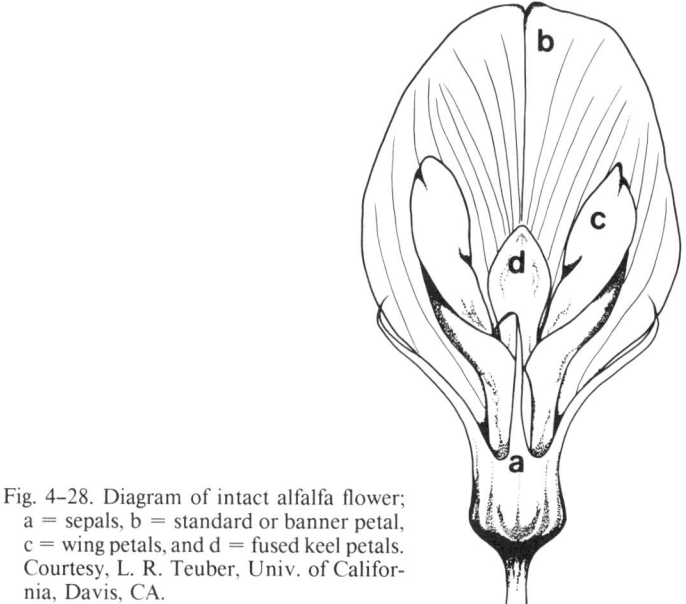

Fig. 4-28. Diagram of intact alfalfa flower; a = sepals, b = standard or banner petal, c = wing petals, and d = fused keel petals. Courtesy, L. R. Teuber, Univ. of California, Davis, CA.

cream, yellow, and white (7). Variegated flowers are produced on plants possessing genes for both purple and yellow flower color. This results in shades of blue and green (6, 7). Particularly in the variegated types, the degree of color expression may vary with flower age.

"The pistil consists of a single carpel that develops a superior ovary, a smooth awl-shaped hollow style, and a well defined stigma" (8) (Fig. 4-27). The stigma develops prior to the straight bud stage. It is composed of cylindrical cells with centrally located nuclei that are covered with a cuticle (stigmatic membrane) (23) which is about 120 μm thick (65). The junction between the stigmatic surface and the style is demarcated by a distinct fringe of unicellular hairs (Fig. 4-31) (65). Transfer cells have

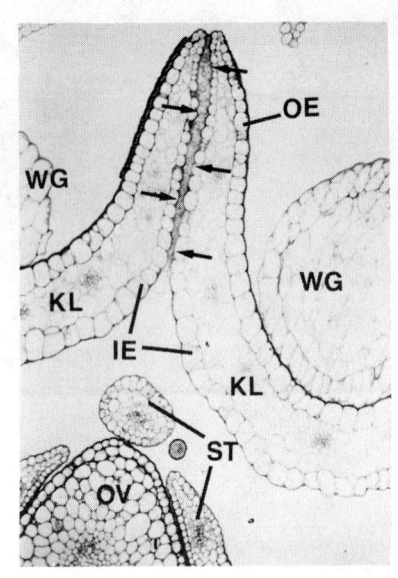

Fig. 4-29. Light microscope cross section (× 7.5) through the zone of interlocked keel petals. Wing horns (WG) rest against the outer epidermis (OE) of the keel (KL), while locking of the keel petals (*arrows*) is on the inner epidermis (IE). From Kreitner and Sorensen (66).

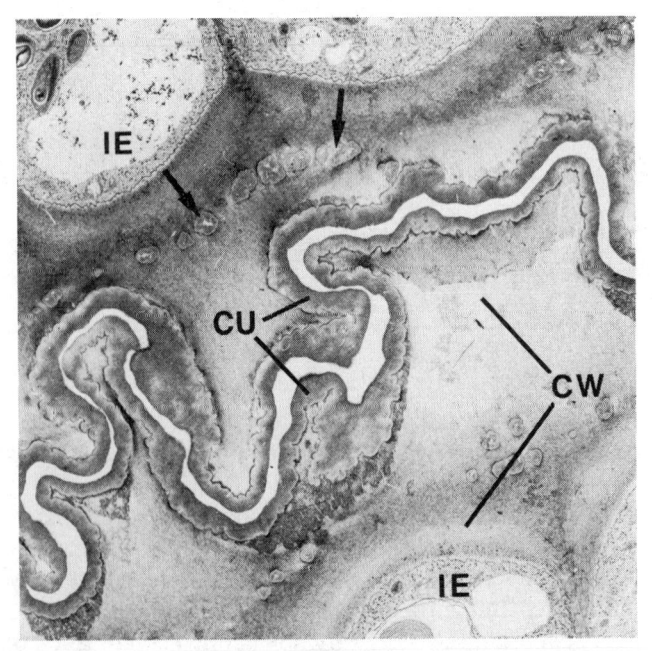

Fig. 4-30. Transmission electron micrograph (× 2800) of the interlocking zone. Cuticle (CU) covers cell wall ridges (CW) arising from the inner epidermis (IE). From Kreitner and Sorensen (66).

MORPHOLOGY AND ANATOMY

Fig. 4-31. Juncture between the (a) stigmatic surface and the (b) style showing the fringe of unicellular hairs. From Kreitner and Sorensen (66).

been observed in the style and ovary (59). "Ovules are campylotropous (Fig. 4-32) (see Chapter 30 in this book has details on gametogenesis and fertilization). Cooper et al. (36) reported that the uppermost ovule at the stylar end usually curves toward the style, whereas the remainder of the ovules usually curve toward the base of the ovary. Curvature of most ovules is such that the micropyle faces the base of the ovary.

"Ten to 12 ovules per ovary usually develop in alfalfa, but the number may range from 6 to 18 (10). Variability in ovule number at the diploid level appeared to be under the control of four genes with incomplete dominance. The 10 stamens form a diadelphous tube in which nine are fused. The 10th one, nearest the standard, is free. The fused filaments alternate long and short such that the anthers fit tightly around the stigma in a double ring. At the time the pollen mother cells are formed, the anther wall consists of an epidermal layer, two layers of parietal cells, and a layer of uninucleate tapetal cells (88). The outer parietal cells differentiate into an endothecium. While the pollen develops, the tatetum"—which is composed of transfer cells (59)—"and the inner layer of the parietal tissue degenerate, providing nourishment to the developing pollen. At maturity the anther wall consists only of an epidermis and endothecium" (8). Alfalfa pollen is binucleate and has three germination pores (colpi) (88). When observed with the light microscope, all alfalfa pollen grains appear similar (Fig. 4-33). However, scanning electron microscopy (SEM) reveals that substantial variation exists among pollen grains in the sculpturing of the exine (Fig. 4-34) (1, 31, 84). Measurements taken using the SEM show that substantial variation exists both among

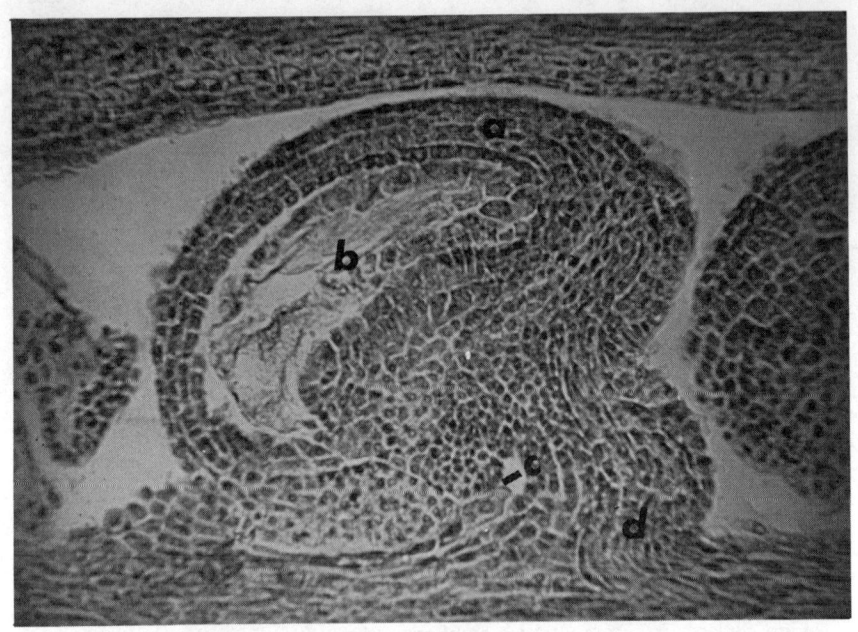

Fig. 4-32. Cross section alfalfa ovule; a = integuments, b = nucellus, c = micropyle, and d = funiculus.

Fig. 4-33. Alfalfa pollen grain, LM. From Chaney (27).

and within alfalfa cultivars for polar intercolpal distance, equatorial intercolpal distance, and equatorial diameter, in addition to surface sculpturing (84). None of this variation was associated closely with germplasm origin.

The alfalfa nectary is an annular (discoid) structure located at the base of the staminal column (fused filaments), primarily on the receptacle (Fig. 4-35), and around the ovary (103). The nectar-producing cells are a group of smaller cells adjacent to the cells of the receptacle and are subtended by branches of the vascular bundles to the staminal column. The nectariferous cells extend a short distance up the staminal column on the adaxial side of the ovary. At its maximum thickness, this tissue

MORPHOLOGY AND ANATOMY

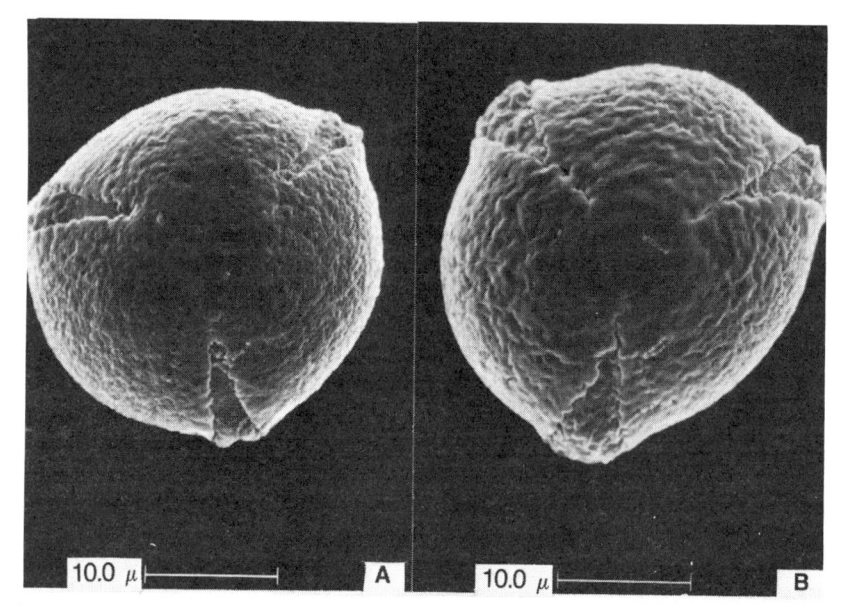

Fig. 4-34. Alfalfa pollen grains, SEM. Courtesy, Y. S. C. Peng, Univ. of California, Davis, CA.

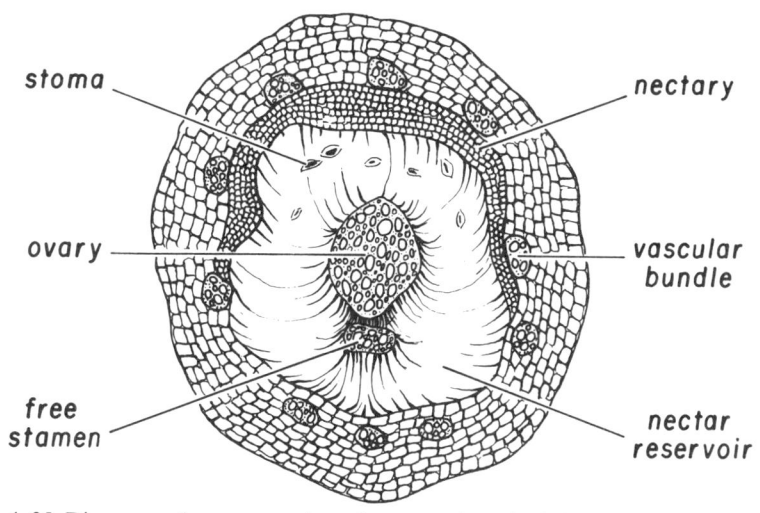

Fig. 4-35. Diagrammatic representation of cross section of alfalfa flower receptacle. Courtesy, L. R. Teuber, Univ. of California, Davis, CA.

is seven to nine cells thick. The nectary is covered by a cuticle and nectar is secreted into the nectar reservoir (nectar holder) by stomata which are permanently open and are found on the abaxial side of the ovary where the nectariferous tissue is thickest. On the abaxial side of the ovary, the nectary is thinner and passes between the ovary and the free stamen (Fig. 4-36). Both nectar reservoir diameter and number of stomata on the

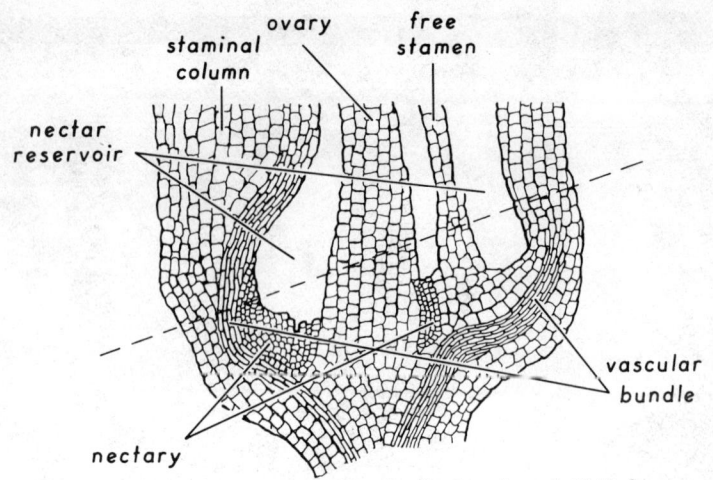

Fig. 4-36. Diagrammatic representation of longitudinal section of alfalfa flower receptacle. Dashed line represents plane of Fig. 4-35. Courtesy, L. R. Teuber, Univ. of California, Davis, CA.

surface of the nectary can be altered by selection and would appear to be controlled by gene complexes that are largely independent (104).

4-7 SUMMARY

This chapter reviews the ontogeny, morphology, and anatomy of the alfalfa plant from the seed through development to the mature plant. Descriptions of the growth and development of the root, crown, stem, and leaf are presented. Differentiation of the vascular tissue and patterns of vascularization of the entire plant are reviewed. Floral development is described from the onset of bud primordia to seed pod formation. Discussions of genetic variability, agronomic suitability, and heritability of many traits are included.

There is a need for more information regarding the associations between plant form and productivity. Much of the research to date has been directed toward the aboveground plant parts with comparatively little effort on underground portions of the plant. It is important for future progress to improve our understanding of underground growth and development of alfalfa. Many studies have demonstrated genetic control of plant form. Plant scientists must continue to search for new genetic variability in order to develop germplasm that is more productive and better adapted to specific environmental conditions.

REFERENCES

1. Adams, R.J., and M.V. Smith. 1977. Scanning electron and light microscope studies of pollens of some leguminosae. J. Agric. Res. 16:99-106.

2. Anonymous. 1980. Report of the national certified alfalfa variety review board. AOSCA, Raleigh, NC.
3. Azizi, M.R., and D.K. Barnes. 1977. Characterization and inheritance of a spotted leaf trait in alfalfa. Crop Sci. 17:126-132.
4. Baenziger, H. 1977. Inheritance of the torn-leaf mutant in alfalfa. Can. J. Plant Sci. 57:47-50.
5. Ballard, L.A.T. 1973. Physical barriers to germination. Seed Sci. Tech. 1:285-303.
6. Barnes, D.K. 1966. Flower color inheritance in diploid and tetraploid alfalfa: a reevaluation. USDA Tech. Bull. 1353. U.S. Government Printing Office, Washington, DC.
7. ----. 1972. A system for visually classifying alfalfa flower color. USDA Handb. 424. U.S. Government Printing Office, Washington, DC.
8. ----, E.T. Bingham, J.D. Axtell, and W.H. Davis. 1972. The flower, sterility mechanism, and pollination control. In C.H. Hanson (ed.) Alfalfa science and technology. Agronomy 15:123-141.
9. ----, M.A. Brick, and G.H. Heichel. 1978. Increasing the nitrogen content in alfalfa roots and crowns. Agron. Abstr. American Society of Agronomy, Madison, WI, p. 69.
10. ----, and R.W. Cleveland. 1963. Inheritance of ovule number in diploid alfalfa. Crop Sci. 3:499-504.
11. ----, and C.H. Hanson. 1967. An illustrated summary of genetic traits in tetraploid and diploid alfalfa. USDA Tech. Bull. 1370. U.S. Government Printing Office, Washington, DC.
12. Bass, L. 1972. As cited by C.R. Gunn In C.H. Hanson (ed.) Alfalfa science and technology. Agronomy 15:677-687.
13. Bauder, W.W. 1938. Inheritance of odd-leaf character in *Medicago* sp. M.S. thesis. University of Nebraska, Lincoln.
14. Bingham, E.T., and R.P. Murphy. 1965. Breeding and morphological studies on multifoliolate selections of alfalfa *Medicago sativa* L. Crop Sci. 5:233-235.
15. ----. 1966. Morphology and petiole vasculature of five heritable leaf forms in *Medicago sativa* L. Bot. Gaz. 127:221-225.
16. ----. 1968. Stomatal chloroplasts in alfalfa at four ploidy levels. Crop Sci. 8:509-510.
17. Brick, M.A. 1980. Morphology and inheritance of several root characteristics in alfalfa. Ph.D. thesis. Univ. of Minnesota, St. Paul (Diss. Abstr. 81-02069).
18. ----, and D.K. Barnes. 1981. Inheritance and ontogeny of a lobed-cambium trait in alfalfa roots. J. Hered. 72:419-422.
19. ----, and ----. 1982. Inheritance and anatomy of root bark area in alfalfa. Crop Sci. 22:747-752.
20. ----, ----, and A.K. Dobrenz. 1984. Inheritance and anatomy of a narrow leaflet trait in alfalfa. Crop Sci. 24:787-790.
21. ----, ----, and C.P. Vance. 1981. Inheritance and identification of a carotenoid pigment in orange alfalfa roots. Crop Sci. 21:748-750.
22. ----, A.K. Dobrenz, and M.H. Schonhorst. 1976. Transmittance of the multifoliolate leaf characteristic into non-dormant alfalfa. Agron. J. 68:134-136.
23. Brink, R.A., and D.C. Cooper. 1936. The mechanism of pollination in alfalfa (*Medicago sativa*). Am. J. Bot. 23:678-683.
24. Buss, G.R., J.A. Lutz, Jr., and G.W. Hawkins. 1975. Effect of soil pH and plant genotype on element concentration and uptake by alfalfa. Crop Sci. 15:614-616.
25. Carlson, F.A. 1925. The effects of soil structure on the character of alfalfa root-systems. J. Am. Soc. Agron. 17:336-345.
26. Carter, P.R., C.C. Shaeffer, and W.B. Voorhees. 1982. Root growth herbage yield and plant water status of alfalfa *Medicago sativa* cultivars. Crop Sci. 22:425-427.
27. Chaney, D.C. 1985. Bloom dynamics in alfalfa (*Medicago sativa* L.): Implications for pollination and seed production. M.S. thesis. Univ. of California, Davis.
28. Childers, W.R. 1962. The nature and inheritance of a yellow-leaf character in *Medicago sativa* L. Can. J. Bot. 40:89-93.
29. ----, and H.A. McLennan. 1961. The inheritance and histology of a chlorophyll-deficient character in *Medicago sativa* L. Can. J. Bot. 39:847-853.
30. Cho, Y.S., R.D. Wilcoxson, and F.I. Frosheiser. 1973. Difference in anatomy, plant-extracts, and movement of bacteria in plants of bacterial wilt resistant and susceptible varieties of alfalfa. Phytopathology 63:760-765.
31. Clarke, G.C.S., and F.K. Kupicha. 1976. The relationships of the genus *Cicer* L. (Leguminosae): the evidence from pollen morphology. Bot. J. Linn. Soc. 72:35-44.
32. Clement, W.M., Jr., and E.H. Stanford. 1966. Red-root in alfalfa: Inheritance and relationship with flower color. Crop Sci. 6:569-570.
33. Coffman, F.A. 1922. Pollination in alfalfa. Bot. Gaz. 74:197-203.
34. Cole, D.F., and A.K. Dobrenz. 1970. Stomate density of alfalfa (*Medicago sativa* L.) Crop Sci. 10:61-63.
35. Compton, R.H. 1912. Investigation of seedling structure of the Leguminosae. J. Linn. Soc. London Bot. 41:1-122.

36. Cooper, D.C., R.A. Brink, and H.R. Albrecht. 1937. Embryo mortality in relation to seed formation in alfalfa (*Medicago sativa*). Am. J. Bot. 24:203-213.
37. Corner, E.J.H. 1951. The leguminous seed. Phytomorphology 1:117-150.
38. Daday, H., A. Grassia, and F.E. Binet. 1977. Non-additive gene effects for creeping-root in lucerne. Theor. Appl. Genet. 50:23-27.
39. Daigger, L.A., L.S. Axthelm, and C.L. Ashburn. 1970. Consumptive use of water by alfalfa in western Nebraska. Agron. J. 62:507-508.
40. Dobrenz, A.K., M.A. Messengale, and W.S. Phillips. 1965. Floral initiation in alfalfa (*Medicago sativa* L.). Crop Sci. 15:572-575.
41. Esau, K. 1977. Anatomy of seed plants. John Wiley and Sons, New York.
42. Ferguson, J.E., and R.P. Murphy. 1973. Comparison of trifoliolate and multifoliolate phenotypes of alfalfa (*Medicago sativa* L.) Crop Sci. 13:463-465.
43. Foord, K.E. 1985. Physiological, environmental, and genetic determinants of seedling growth in *Medicago sativa* L. Ph.D. Diss. University of California, Davis (Diss. Abstr. 85:21203).
44. ----, and L.R. Teuber. 1982. Variation in alfalfa seedling development. p. 88. *In* Forage and Grassl. Conf. Proc., Rochester, MN. 21-24 February, American Forage and Grassland Council, Lexington, KY.
45. ----, and ----. 1984. Unifoliolate internode length of alfalfa. Environmental effects and inheritance. Agron. Abstr, American Society of Agronomy, Madison, WI.
46. Fribourg, H.A., and I.J. Johnson. 1955. Dry matter and nitrogen yield of legume tops and roots in the fall of the seeding year. Agron. J.: 47:73-77.
47. Frosheiser, F.I. 1977. Effect of alfalfa mosaic virus strains on root development of alfalfa stem cuttings. Plant Dis. Rep. 61:115-118.
48. Garver, S. 1922. Alfalfa root studies. USDA Bull. 1087. U.S. Government Printing Office, Washington, DC.
49. Gerard, R. 1881. Recherches sur la passage de la racine a la tige. Ann. Sci. Nat. Bot. Ser. VI. 11:279-430.
50. Grimes, D.W., W.R. Sheesley, and P.L. Wiley. 1978. Alfalfa root development and shoot regrowth in compact soil of wheel traffic patterns. Agron. J. 70:955-958.
51. Grove, A.R., Jr., and G.E. Carlson. 1972. Morphology and anatomy. *In* C.H. Hanson (ed.) Alfalfa science and technology. Agronomy 15:103-122.
52. Hayward, H.E. 1938. Leguminosae, *Medicago sativa*. p. 309-338. *In* The structure of economic plants. Macmillan Publishing Co., New York.
53. Heinrichs, D.H. 1954. Developing creeping-rooted alfalfa for pasture. Can. J. Agric. Sci. 34:269-280.
54. ----. 1973. Time factor in expression of the creeping rooted character in alfalfa. Can. J. Plant Sci. 53:511-514.
55. ----, B.P. Goplen, and M.R. Hanna. 1977. Selection for and against creeping-rootedness within creeping-rooted alfalfa cultivars. Can. J. Plant Sci. 57:221-225.
56. ----, and F.H. Morley. 1962. Quantitative inheritance of creeping root in alfalfa. Can. J. Genet. Cytol. 4:79-89.
57. Jodari-Karimi, F., V. Watson, H. Hodges, and F. Whisler. 1983. Root distribution and water use efficiency of alfalfa *Medicago sativa* cultivar Cimarron as influenced by depth of irrigation. Agron. J. 75:207-211.
58. Johnson, K.K., E.L. Sorensen, and E.K. Horber. 1980. Resistance in glandular-haired cannual *Medicago* species to feeding by adult alfalfa weevil (*Hypera pastica*). Environ. Entomol. 9:133-136.
59. Johnson, L.E.B., R.D. Wilcoxson, and F.I. Frosheiser. 1975. Transfer cells in tissues of the reproductive system of alfalfa. Can. J. Bot. 53:952-956.
60. Jones, F.R. 1928. Winter injury of alfalfa. J. Agric. Res. 30:189-211.
61. ----. 1949. Rough root: a heritable character in alfalfa. Agron. J. 41:559-561.
62. Kiesselbach, T.A., J.C. Russel, and A. Anderson. 1929. The significance of subsoil moisture in alfalfa production. J. Am. Soc. Agron. 21:241-268.
63. Kitch, L.W., R.E. Shade, W.E. Nyquist, and J.D. Axtell. 1985. Inheritance of density of erect glandular trichomes in the genus *Medicago*. Crop Sci. 25:607-611.
64. Kreitner, G.L., and E.L. Sorenson. 1979. Glandular trichomes on *Medicago* species. Crop Sci. 19:380-384.
65. ----, and ----. 1984. Stigma development and stigmatic cuticle of alfalfa, *Medicago sativa* L. Bot. Gaz. 145:436-443.
66. ----, and ----. 1985. Structure of the real-looking mechanism in insect-pollinated and self-pollinated alfalfa species. Crop Sci. 25:631-634.
67. Kroontje, W., and W.R. Kehr. 1956. Legume top and root yields in the year of seeding and subsequent barley yields. Agron. J. 48:127-131.
68. Labanauskas, C.K., and J.A. Jackobs. 1957. Cork formation in tap-roots and crowns of alfalfa. Agron. J. 49:95-97.
69. LeClerg, E.L., and L.W. Durrell. 1928. Vascular structure and plugging of alfalfa roots. Colo. Exp. Stn. no. 339.

70. Leppik, E.E. 1966. Floral evolution and pollination in the Leguminosae. Ann. Bot. Fenn. 3:299–308.
71. Lersten, N.R. 1982. Tracheid bar vestured pits in legume seeds (Leguminoae:Papilloinoideae). Am. J. Bot. 69:98–107.
72. Levin, D.A. 1973. The role of trichomes in plant defense. Q. Rev. Biol. 48:3–15.
73. Lute, A.M. 1928. Impermeable seed of alfalfa. Colo. Agric. Exp. Stn. Bull. 326.
74. Lyon, T.L., and J.A. Bizzell. 1934. A comparison of several legumes with respect to nitrogen accretion. J. Am. Soc. Agron. 26:651–656.
75. McIntosh, M.S. 1979. Genetic and related studies on the branching-root habit in alfalfa. Ph.D. thesis. Univ. of Illinois, Urbana.
76. ----, and D.A. Miller. 1980. Development of root-branching in three alfalfa cultivars. Crop Sci. 20:807–809.
77. Medler, J.T., M.A. Massengale, and Barrow. 1955. Flowering habit of alfalfa clones during the first and second growth. Agron. J. 47:216–217.
78. Mokeyeva, E.V. 1940. Biological-anatomical studies of alfalfa (*Medicago sativa* L.). English translation. State Agric. Pub. House of Literature, S.S.R., Tashkent.
79. Morley, F.H., and D.H. Heinrichs. 1960. Breeding for creeping-root in alfalfa. Can. J. Plant Sci. 40:424–433.
80. Murray, B.E. 1957. The ontogeny of adventitious stems on roots of creeping-rooted alfalfa. Can. J. Bot. 35:463–475.
81. Musgrave, D.J., and R.H.M. Langer. 1977. Crown development of two diverse genotypes of lucerne. N.Z.J. Agric. Res. 20:453–458.
82. Nageli, C. 1858. Beitrage Zur Wissenschaftlichen. Botanik, Part 1, Englemann, Leipzig.
83. Odland, T.E., and R. Lepper, Jr. 1939. A crinkled leaf mutation in alfalfa. J. Am. Soc. Agron. 31:128–130.
84. Peng, Y.S., D.M. Gordon, M.E. Nasr, and L.R. Teuber. 1987. Pollen variation in alfalfa (*Medicago sativa* L.) cultivars. Pollen et Spores 29:45–58.
85. Pergament, E. 1958. Inheritance of two unifoliolate leaves in alfalfa. Agron. Abstr. American Society of Agronomy, Madison, WI, p. 50.
86. Perry, L.T., Jr., and K.L. Larson. 1974. Influence of drought on tillering and internode number and length in alfalfa. Crop Sci. 14:693–696.
87. Reid, J.B. 1982. Observations on root hair production by lucerne *Medicago sativa* cultivar Europe, maize *Zea mays*, cultivar LG-11 and perennial rye grass *Lolium perenne*, cultivar Aberystwyth-S-24 grown in sandy soil. Plant Soil 62:319–322.
88. Reeves, R.G. 1930. Nuclear and cytoplasmic division in the microsporogenesis of alfalfa. Am. J. Bot. 17:29–40.
89. Rincker, C.M. 1983. Germination of forage crop seed after 20 years of subfreezing storage. Crop Sci. 23:229–231.
90. Schimada, T., and K. Morakami. 1976. Principle component analysis of root and crown characteristics of alfalfa in relation to their persistence. Obihiro Chikusan Daigoku Res. Bull. 10:203–210.
91. Schnappinger, M.G., Jr., V.A. Bandel, and C.B. Kresge. 1969. Effect of phosphorus and potassium on alfalfa root anatomy. Agron. J. 61:805–808.
92. Schneider, M., L.R. Teuber, and K.E. Foord. 1984. Unifoliolate internode length of alfalfa: relationship to origin and fall dormancy. Agron. Abstr., American Society of Agronomy, Madison, WI, p. 87.
93. Seth, J., and S.T. Dexter. 1958. Root anatomy and growth habit of some alfalfa varieties in relation to wilt resistance and winter-hardiness. Agron. J. 50:141–144.
94. Shade, R.E., M.J. Doskocil, and N.P. Maxon. 1979. Potato leafhopper resistance in glandular-haired alfalfa species. Crop Sci. 19:287–289.
95. ----, T.E. Thompson, and W.R. Campbell. 1975. An alfalfa weevil larval resistance mechanism detected in *Medicago*. J. Econ. Entomol. 68:399–404.
96. Simonds, A.O. 1935. Histological studies on the development of the root and crown of alfalfa. Iowa State Coll. J. Sci. 9(4):641–659.
97. Smith, O. 1951. Root branching of alfalfa varieties and strains. Agron. J. 43:573–575.
98. Stanford, E.H. 1959. The zebra leaf character in alfalfa and its dosage-dominance relationship. Agron. J. 51:274–277.
99. ----. 1965. Inheritance of sticky leaf character in alfalfa. Crop Sci. 5:281.
100. ----, and R.W. Cleveland. 1954. The inheritance of two leaf abnormalities in alfalfa. Agron. J. 48:203–206.
101. Stewart, G. 1926. Alfalfa growing in the United States and Canada. Macmillan Publishing Co., New York.
102. Sullivan, J.A., and W.E. Tossell. 1979. Genetic variation for root and crown type and root growth during the winter dormancy period in alfalfa. p. 30–39. *In* Proc. 3rd Eastern Forage Improvement Conf., Carlton University, Ottawa, Onatrio, Canada. 10–12 July. Eastern Forage Improvement Conference, Ottawa, Ontario.
103. Teuber, L.R., M.C. Albertsen, D.K. Barnes, and G.H. Heichel. 1980. Structure of floral

nectaries of alfalfa (*Medicago sativa* L.) in relation to nectar production. Am. J. Bot. 67:433-439.
104. ----. 1984. Effect of selection for nectar volume on nectary stomata in alfalfa. Agron. Abstr., American Society of Agronomy, Madison, WI, p. 92.
105. Upchurch, R.P., and R.L. Lovvorn. 1951. Grass morphological root habits of alfalfa in North Carolina. Agron. J. 43:493-498.
106. van Tieghem, P. 1971. Recherches sur la symetrie de structure des plantes vasculaires. Ann. Sci. Nat. Bot. 13:5-314.
107. Watson, D.P. 1948. Structure of the testa and its relation to germination in the Papilionacea tribes Trifoliae and Loteae. Ann. Bot. 12:385-409.
108. Werker, E. 1980. Seed dormancy as explained by anatomy of embryo envelopes. Isr. J. Bot. 29:22-44.
109. Wilson, O.T. 1913. Studies on the anatomy of alfalfa. Kansas Univ. Sci. Bull. VII(17):291-299.
110. Wilton, A.C., C.E. Townsend, R.J. Lorenz, and G.A. Rogler. 1978. Longevity of alfalfa seed. Crop Sci. 18:1091-1093.
111. Winter, C.W. 1932. Vascular system of young plants of *Medicago sativa*. Bot. Gaz. 94:152-167.
112. Winton, K.B. 1914. Comparative histology of alfalfa and clovers. Bot. Gaz. 57:53-63.

5 Environmental Physiology and Crop Growth

G. W. FICK
Cornell University
Ithaca, New York

D. A. HOLT
University of Illinois
Urbana, Illinois

D. G. LUGG
New Mexico State University
Las Cruces, New Mexico

Environmental physiology is the study of the relationship between the crop and its environment, especially the modification of physiological processes by environmental factors. In the field, the environment is constantly changing, and the alfalfa (*Medicago sativa* L.) crop shows a corresponding, genetically programmed response to environmental signals. Seed germination, elongation of basal buds, flowering, and cold hardening are specific examples of plant responses modified by the environment. At the basic level, environmental physiology provides an understanding of how crop growth and development are related to the environment. At the applied level, understanding alfalfa growth and development is used in both breeding and managing alfalfa for higher yields, better nutritive quality, and longer stand life.

The successful application of environmental physiology to breeding and managing alfalfa is clearest with the biotic environment (e.g., pests and symbionts) and with those aspects of the physical environment that can be influenced by practices such as irrigation, fertilization, time of seeding, and cutting schedules. The significance of those variables to production warrant separate treatment in other chapters. In this chapter, consideration is given to those environmental factors that cannot be managed practically under most field conditions. The effect of temperature, light, water, and salt have received the most attention. These factors are considered primarily from the standpoint of explaining crop growth and development, with some direct applications to breeding and management. Interest in yield prediction and computer modeling of alfalfa production has stimulated study of environmental physiology, and consid-

---

Copyright 1988 © ASA-CSSA-SSSA, 677 South Segoe Road, Madison, WI 53711, USA.
Alfalfa and Alfalfa Improvement—Agronomy Monograph no. 29.

erable work has been done since the earlier edition of the alfalfa monograph. In those areas in which the literature is extensive, the authors have chosen to concentrate on more recent work. For additional background, the reader is advised to consult previous reviews by Bula and Massengale (18), Christian (25), and Field et al. (53).

Physiological responses to environmental variation depend on age, growth stage, prior conditioning of the crop, and genotype, as well as the physiological processes under investigation. In this chapter, the stages of the life cycle of the crop are considered in sequence. Plant growth responses are emphasized because specific physiological processes appear in Chapters 9 through 12. A final section is devoted to alfalfa growth modeling.

5-1 SEEDLING GROWTH AND DEVELOPMENT

5-1.1 Germination and Emergence

The germination and emergence of alfalfa seed have been studied in relation to water supply, salt concentration, temperature, and light. Variation in water supply was achieved by using osmotica, such as mannitol (39, 40, 139, 181, 191), polyethylene glycol (110, 139), and sucrose (74, 146). In addition, NaCl and other salts have been used to determine the effects of salt concentration on germination (22, 86, 139, 146, 174, 181, 182). In all studies, germination was reduced to essentially zero between approximately -1.0 (110) and -1.5 MPa (139, 174, 181). The reduced germination was associated with reductions in the rate and amount of water uptake by seed.

Germination also may be reduced by toxic effects of ions. Sodium chloride (NaCl) reduced germination more than mannitol (181) or sucrose (74) at equal osmotic concentrations. Of the individual salts, magnesium chloride ($MgCl_2$) has been rated both more (86) and less (139) inhibitory of germination than NaCl. Harris (70) showed that NaCl reduced emergence more than sodium sulfate (Na_2SO_4) or sodium carbonate (Na_2CO_3). Recovery after treatment was greater with NaCl and potassium chloride (KCl) than with $MgCl_2$ and Na_2SO_4 (139), and combinations were found to be less toxic than single salts (86).

Differences in response to ionic and moisture stress among cultivars and lines have been reported (22, 39, 40, 74, 139, 146, 174, 191). In general, germination of seed from dormant cultivars is more sensitive to increasing osmotic concentration than that of seed from nondormant cultivars (39, 74, 146, 191), although there are many exceptions. Rumbaugh and Johnson (148) found that alfalfa containing *M. sativa* germplasm had greater emergence under moisture stress than plants containing *M. falcata* germplasm. Dotzenko and Haus (40) reported that the ability to germinate at -1.2 MPa in a mannitol solution appeared to be heritable and, recently, programs to develop salt tolerance of alfalfa at the germination stage have been undertaken in Arizona (3) and California (96). After five selection cycles, germination at −1.3 MPa increased from 3% in the original 'Mesa-Sirsa' population to 86% (3).

Triplett and Tesar (179) found significant correlations between emergence and soil water potential. No emergence occurred below −1.1 MPa in silt loam soil and −1.0 MPa in sandy loam soil. Robinson (144) found no significant differences in emergence with water containing 877 and 1350 mg/L total dissolved solids (TDS) applied via a sprinkler irrigation system. With water containing 850 mg/L TDS, emergence was almost four times higher with sprinkler than with flood irrigation in a clay soil. In another study, a salt concentration of 10 000 mg/L reduced both the rate and final percentage of emergence (70).

Townsend and McGinnies (178) showed that the rate of alfalfa seed germination depended on temperature, but the final germination percentage after 7 d was relatively insensitive over the 5 to 35°C range. Generally, alfalfa germinated between 2 and 40°C with an optimum about 19 to 25°C when no salt was present (107, 174, 182). McElgunn (107) found that a 13/2°C alternating temperature treatment (12 h/12 h) reduced the final germination percentage compared with constant temperatures of 7, 10, 13, and 21°C, and alternating temperatures of 15/4, 18/7, and 27/16°C. Salt reduced germination more at high (33-39°C) than at moderate (21-27°C) temperatures (Fig. 5-1), and there was some indication that salt slightly reduced the optimal temperature for germination (174, 182).

Upon germination of the alfalfa seed, the hypocotyl elongates, pushing the hypocotyledonary hook upward and pulling the cotyledons to the soil surface. When the hypocotyledonary hook penetrates the soil surface, it is exposed to light which activates the enzymatic destruction of elongation-stimulating auxin. This causes growth on the lighted side of the hypocotyl to slow or stop. Growth on the unlighted side continues,

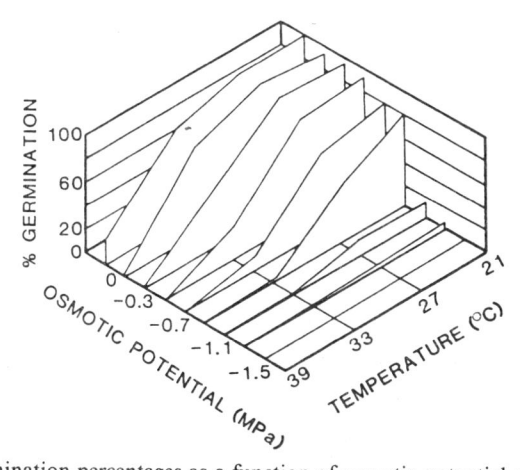

Fig. 5-1. Germination percentages as a function of osmotic potential and temperature for 'Ladak 65' alfalfa after 7 d. Redrawn from Stone et al. (174).

straightening the hook, and turning the cotyledons upward. This phenomenon is typical of a phototropic response. The lower the irradiance at the soil surface when seedlings emerge, the taller and more etiolated are the seedlings. This is in part because of the auxin effect and in part because of the generally lower temperature and higher humidity associated with lower irradiance. If seedlings emerge under a leaf canopy, changes in light quality (red enrichment) are involved in etiolation. Thus, cotyledons of seedlings emerging under a heavy companion crop canopy would be expected to be somewhat higher above the soil surface than those of seedlings emerging from bare soil.

Stem height and the rate of stem elongation of recently emerged 'Ranger' seedlings (135) were greatest at 4.5 klux (a PPFD of roughly 65 μmol m^{-2} s^{-1})[1] and decreased as light increased to 30 klux (roughly 425 μmol m^{-2} s^{-1}). There was little or no stem elongation at luminous flux densities lower than 4.5 klux. Stems that developed at lower light levels tended to be thinner than ones developing under higher light.

5-1.2 Seedling Growth

The small initial size of the alfalfa seedling reduces its competitive ability, and, at the soil surface, it can be exposed to a highly variable microenvironment. Thus, there are some unique effects of environment on primary growth. Even in controlled environments with little competition, seedling growth and development are typically slower than regrowth from established plants, probably because the seedling lacks the crown, root, and nodule system of an older plant (45, 126, 127).

Available soil moisture greatly affects growth of alfalfa seedlings. (See also Chapter 11 in this book.) Gist and Mott (60) found that both top and root growth were reduced with increasing moisture stress, and Cowett and Sprague (29) found that number of stems per plant, buds per plant, plant height, and top and root weight were reduced with increasing stress. Root to top weight ratio increased with greater moisture stress in one study (29), but not in the other (60). Over the establishment year, Janson (82) found that both root and top growth increased with greater amount and frequency of irrigation and that root to top weight ratio decreased with higher available moisture. Again, differences in seedling growth under moisture stress have been found among genotypes. Rumbaugh et al. (149) reported that average shoot weights of 8- to 12-week-old seedlings under high irrigation were 1.7 times greater than those under low irrigation, using a line-source sprinkler system. Average water application was 27.7 cm with high irrigation and 4.3 cm with low. Improved cultivars were higher yielding than wild populations when compared at all levels of irrigation.

---

[1] Depending on the light source, there are from 12 to 21 μmol m^{-2} s^{-1} of incident photosynthetic photon flux density (PPFD) per klux (106). We report the older units because, in many examples, they cannot be converted accurately into PPFD in the SI system.

Excess soil water decreases the growth of alfalfa seedlings. Wet soil conditions may lead to the development of fungal diseases, especially diseases in the *Pythium* (damping off) and *Phytophthora* (root rot) complexes. Wet soils also reduced growth directly through the development of anaerobic conditions in the root zone (7) with injury occurring more rapidly as temperatures increased (177). There was some evidence that sensitivity to soil flooding increased until the seedling was about 6-weeks old and that greater rooting depth before flooding resulted in greater sensitivity, probably because deeper roots had less O_2 than roots near the surface.

Salt has a deleterious effect on the growth of alfalfa seedlings. Using 4- to 5-week-old seedlings in the greenhouse, injury and killing from exposure to different salts, including KCl, NaCl, potassium sulfate (K_2SO_4), Na_2SO_4, and sodium hydrogen phosphate (Na_2HPO_4), were greatest with the salts containing Cl^- (166). Chloride damage increased as application rate and as temperature or soil moisture supply became higher. Plants exposed to low levels of Cl^- developed tolerance to higher rates applied later.

Growth of recently germinated seedlings is most rapid at 20 to 30°C. For example, fresh weight of alfalfa sprouts after 6 d was greater at 21 and 27°C than at 16°C (78). As seedlings grew, the optimum temperature declined to 15 to 25°C (20, 58, 126, 180). Garza et al. (58) showed that lowering of night temperature was particularly important in effecting this change in response. Growth rates were greatly reduced outside the temperature range of 10 to 37°C (5, 31, 69).

Field et al. (53) attributed the decline in optimal temperature to a change in the process that most strongly influences dry matter accumulation. Initially, rapid leaf expansion and establishment of autotrophy is the dominant influence. Later, the difference between photosynthesis and maintenance respiration becomes more important, and this second phase is favored by lower temperatures, especially at night. In addition, the capacity for N_2 fixation may also be involved in the change. Macdowall (101, 102) observed that the relative growth rate (RGR) of alfalfa seedlings increased with increasing nitrate (NO_3^-) supply but was unresponsive to CO_2 enrichment prior to 4 weeks of age. Thereafter, the response was reversed. For non-nodulated seedlings in Hoagland's solution, RGR declined after about 4 weeks of growth, but there was an increase in RGR at about the same time for nodulated alfalfa without added NO_3^-. Baysdorfer and Bassham (9) studied the effect of CO_2 enrichment during primary growth and concluded that there is a shift from a source to a sink limitation on photosynthesis in seedling alfalfa.

Partitioning to roots and root morphology of seedlings depend on temperature. With a 12-h photoperiod, Daday and Williams (31) found that a warm regime (24/19°C, light/dark) gave taproot to lateral root ratios greater than 3.5. A cool dark period (21/4, 15/4, and 9/4°C) always gave ratios <1.0, with 'Rambler' having lower ratios than 'Hairy Peruvian'. Pearson and Hunt (126) pointed out that root yield was a linear function of top yield over the temperature range of 15/10 to 35/30°C. In the first growth from seeding, root to top ratio at 50% flowering declined from 0.8 at the lowest to 0.4 at the highest temperatures. The ratio was initially dependent on plant age but reached a stable value before flowering in environments tested. Partitioning to roots of the weekly dry matter increment varied from about 50% at the low temperature to about 20% at the higher temperatures (126, 127).

Newly emerged alfalfa seedlings are relatively intolerant of low irradiances. In contrast to red clover (*Trifolium pratense* L.), which had larger leaflets at low light, alfalfa leaflets tended to be smaller at low than at moderate irradiances (105, 142). Under the low irradiances characteristic of seedling environments in companion crops, alfalfa seedlings accumulated less dry weight than when unshaded. The weight of roots was reduced even more than the weight of aboveground parts under low light (19, 60). Under low irradiance, alfalfa seedlings partitioned relatively more dry matter to leaves than to stems, as evidenced by higher leaf to stem ratios under such conditions. Hesterman et al. (78) found that the fresh weight of alfalfa sprouts grown for 6 d with no light was greater than those grown in light durations of 12 and 24 h.

There is some evidence that light quality influences stem growth of seedling alfalfa plants. Green and red light promoted greater stem elongation in nondormant alfalfa cultivars grown under 8-h photoperiods than in dormant types (118).

Phenological development of alfalfa seedlings is slower than that of regrowth. Pearson and Hunt (127) demonstrated that plants from seedlings of 'Vernal' and 'Moapa' took 47 to 61 d to reach 50% bloom, or 24 to 33 d longer than did regrowth on the same plants. A similar pattern was found in *M. falcata* (183). Plants from seedlings also produced four to eight more nodes to reach that stage than did their regrowth. Macdowall (101) observed initiation of floral primordia on primary growth about 20 d later on nodulated alfalfa without added NO_3^- than with non-nodulated plants given NO_3^-.

5-2 VEGETATIVE GROWTH

As wilt-resistant and winter-hardy cultivars of alfalfa became available in the 1950s and 1960s, emphasis was given to earlier and more frequent harvests. (See also Chapter 12 in this book.) Environmental effects on vegetative growth have received attention because rapid vegetative growth is important for high yields under such harvest management. Some investigators compared environmental effects on plants of common chronological age while others compared plants of common developmental stage. The distinction is important because growth rates (measured by yield) and development rate (measured by morphology) respond differently to the environment. Thus, Harada (68) found that the optimal temperature for yield at a common age (27 d) was 27/21°C, but at a common stage (first flower) it was 21/15°C. At a common stage, plants exposed to lower temperature developed more slowly and had at least 10 more days to accumulate yield than plants exposed to higher temperature. Careful distinction must be made between age and stage of growth in order to avoid confusion.

The balance between photosynthesis and respiration controls the rate of dry matter increase in any crop. The environmental variables of light, temperature, and moisture have important effects on these basic physiological processes, as discussed in other chapters. A summary of relevant points is included here for reference. First, the high degree of genetic variation, typical of alfalfa, was found with CO_2 exchange rate (CER) and for acclimation of CER to environmental changes (24, 54, 89, 90). Second, light interacts with other environmental factors to alter alfalfa physiology. For example, Ku and Hunt (89) found that the optimal pretreatment temperature for CER was 20/15°C at a PPFD of 760 μmol m^{-2} s^{-1}, but at 1160 μmol m^{-2} s^{-1} it was 30/25°C for at least one cultivar. Since many studies in controlled environments are done at relatively low light, extrapolation to the field must be done with caution. Third, physiological characteristics of CER in individual leaves were found to be poorly correlated with dry matter production in the crop canopy (54, 71). Morphological characteristics such as leaf area were needed to relate individual leaf physiology to field productivity.

5–2.1 Number and Size of Vegetative Shoots

Yield components are a function of vegetative growth rate and plant morphology. The yield components of a forage crop are (i) number of plants per unit area, (ii) number of shoots (or tillers) per plant, and (iii) mass per individual shoot. Leaf area and leaf mass per shoot are commonly measured because they link yield components to studies of leaf physiology and nutritional value. Sometimes, crown width is recorded as an indirect measure of shoots per plant (55, 113).

In the year of establishment, the number of plants per unit area is strongly dependent on the seeding rate. Pulli (136, 137) recorded a range of 20 plants m^{-2} with 1 kg ha^{-1} of seed to 865 plants m^{-2} with 40 kg ha^{-1} of seed. There were 565 plants m^{-2} with 20 kg ha^{-2} of seed. After the 1st yr, the plant population declines and becomes less variable. For example, Heichel et al. (73) working in Minnesota reported 100 to 150, 60 to 90, and 40 to 70 plants m^{-2} in the 1st, 2nd, and 3rd yr following seeding, respectively. In contrast, Porter and Reynolds (134) working in Tennessee found a decline from 200 to 55 plants m^{-2} during the year following seeding. Cultivars differed in the rate of attrition, probably because of adaptation to stresses from pests and winter conditions (29, 73, 116).

The number of shoots per plant typically increased with age, but in any growth cycle the maximum was usually set within 14 d of the start of regrowth (92), and then declined as the canopy matured (116, 162). In the seeding year, McLaughlin and Christie (109) found the mean number of shoots to increase from 2.4 to 3.9 to 6.2 to 6.5 stems plant^{-1} in four sequential harvests. In the year following seeding, Singh and Winch (162) reported that Vernal had more shoots per plant than 'Saranac' at the early vegetative stage, but differential shoot mortality was responsible for the absence of differences by the bud stage, when there were 5.4 to 7.4 shoots plant^{-1}. Smith (163, 164, 165) and Cowett and Sprague (29) observed no consistent temperature effect on alfalfa tillering, but in a study of 2000 genotypes, McLaughlin and Christie (109) observed that plants with higher shoot numbers matured earlier and had lower optimal growth temperatures than plants with lower shoot numbers. Pulli (137) measured increased stem numbers per plant as plant populations decreased.

The number of basal buds and the number of shoots per plant is reduced substantially by moisture stress (27, 29, 131). Brown and Tanner (15) showed that the effect was expressed in the first 14 d of regrowth and that water stress after the first 14 d had no influence on stems per plant. Cameron (20) pointed out that flooded soil conditions reduced stems per plant, with a greater reduction at 33 than at 21°C. Alfalfa flooded immediately after harvest was more severely affected than alfalfa that grew for 5 d before flooding. Leach (94) recorded an increase in stems per plant with 'Rhizoma' at 33 or 27°C compared to 21 or 15°C. Stem numbers in 'Totana' were not influenced by temperature, while at 15°C the rate of appearance of new stems was reduced in both cultivars.

Alfalfa plants growing in full sunlight in the field produced greater stem numbers than those growing under partially shaded conditions, provided the initial plant population was low (29, 30). This suggests a quantitative relationship between solar radiation and stem number, presumably reflecting high rates of canopy photosynthesis and consequent substrate production with increasing irradiance.

The origin, number, and size of individual alfalfa shoots was studied in a series of experiments by Leach (91, 92, 93, 94) and by Singh and Winch (162). In these investigations, regrowth following a harvest generally came from axillary buds. Removal of axillary buds caused crown buds to elongate, delayed regrowth, and reduced the number of stems. Leach (94) found that the size of individual shoots did not depend on the number of shoots per plant except under the most favorable growing conditions. At 33°C, the most favorable growing condition studied, four or eight shoots plant^{-1} gave smaller individual shoots than one or two shoots plant^{-1} after 10 d of regrowth. Singh and Winch (162) measured masses up to 0.74 g stem^{-1} in 40 d of regrowth and up to about 5 g stem^{-1} in unharvested canopies. They found that yield per shoot depended on cultivar and decreased in successive harvests during the growing season.

Yield per shoot has not received sufficient study for us to fully understand the components of high yield. Researchers have focused instead on yield per unit area, which may be assumed to parallel yield per shoot if there is relatively low variation in number of shoots per unit area. This is most likely true at time of harvest in high-yielding stands. As defined by Offutt (121), high-yielding stands score at least 45% alfalfa measured by point analysis at 10 to 14 d after the first harvest. In stands of Saranac giving near record yields, we measured a range of 520 to 780 shoots m^{-2} at 42 d of age in 60 observations over the entire growing season. However, Singh and Winch (162) reported data with a range of 450 to 1350 stems m^{-2} when plants were at first flower to full bloom. Thus, the assumed correlation between yield per shoot and yield per unit area may not hold, even in high-yielding stands. Analysis of yield components could help us understand how environment affects alfalfa development if the required information were available.

Turning to environmental effects on yield per unit area, there have been numerous studies of the influence of irrigation and evapotranspiration on alfalfa (8, 32, 37, 65, 87, 141, 152, 170). In general, yield increases linearly as the amount of water applied increases up to field capacity. Sammis (152) found that alfalfa yields and evapotranspiration from a number of western states were always positively correlated.

At higher levels of moisture, yield tends to level off or decline (7, 20, 132, 147, 170, 185). Reduction in growth can be attributed partly to anaerobic conditions in the root zone (177) and to diseases that infect the plant after exposure to flooded conditions (57, 138, 185). (See also Chapter 11 in this book.)

Salinity also affects growth and yield of alfalfa (10, 12, 14, 56, 79, 81, 150). Alfalfa is considered to be moderately sensitive to salinity. Yield started to decline at an electrolytic conductivity of saturation extract (EC$_e$) of 2.0 dS m^{-1}, and declined to zero at an EC$_e$ of approximately 16 dS m^{-1} (13, 100). There was some evidence that alfalfa yield increased slightly at low soil salinity (12, 79).

As with seedling growth, the optimal temperature for the growth of shoots shifts from 30 to 33°C in the 1st week of regrowth (20, 44, 94) to the 10 to 27°C range for older shoots (36, 52, 126, 153, 154, 155). When shoots were harvested at a fixed morphological stage, the greatest yield commonly occurred in the lowest temperature regime, provided the minimum temperature exceeded 10°C (45, 58, 64, 68, 95, 117, 127, 164). The longer period of development at lower temperatures probably accounts for the greater yield. When crop growth rate and RGR were measured, the optimal temperature was in the 15 to 27°C range (4, 16, 24, 68, 155). Harada (68) reported that the highest crop growth rate for Vernal and 'Florida 66' was at about 21/15°C, but for 'Cody' it was near 27/21°C. Arbi et al. (4) investigated the effects of nuclear ploidy and reported a wider range of temperature adaptation for 6× than for 4× or 2× plants. Evans and Peaden (43) speculated that the optimal temperatures under high light conditions in the field might be higher than those measured in growth chambers: 30 to 32°C (day) and 15 to 20°C (night). However,

Stock (173) using German weather and growth data estimated optimal field temperatures at 24 to 25°C (day) and 18 to 19°C (night).

Sato (153, 154) showed inconsistent photoperiod effects on the optimal temperature for herbage and whole plant yields, but Susuki et al. (175) determined that photoperiod was highly correlated to stem elongation rate, with less dormant cultivars having lower correlations than more dormant cultivars. The marginal (minimal) temperature for stem elongation in the autumn ranged from 6.9 to 13.6°C for the least (e.g., Moapa) to the most (e.g., 'Ladak') dormant cultivars. The corresponding marginal photoperiods for stem elongation were 9.2 h for the least to 10.4 h for the most dormant of cultivars. Guy et al. (67) found that yield increased as photoperiod increased from 9 to 24 h of light per day at 18 klux (roughly 250 μmol m^{-2} s^{-1} of PPFD). Chatterton and Carlson (24) selected high- and low-yielding clones in the field and then measured yield under controlled conditions. The high-yielding clone was superior at 14-h days, but the low-yielding clone was superior at 10-h days and 20/15°C. Again, genotype must account for the variability.

Although partitioning of dry matter in alfalfa is not completely understood, it is obviously influenced by environment. In the following sections, environmental effects on specific organs are examined.

5–2.2 Leaf Characteristics

Alfalfa leaves from plants grown under saline conditions generally were dark green and had a high carotene concentration (14). However, high rates of KCl resulted in yellowing and thickening of leaflets (168), and under sprinkler irrigation Maas et al. (99) observed leaf margin necrosis associated with excess salt. Moisture deficits also affect leaf characteristics. Brown and Tanner (15) concluded that there is little leaf growth below a leaf water potential of −1.0 MPa. Leaf area, leaf size, leaf dry weight, and leaf growth rate were all reduced under moisture stress (23, 87). However, in one study, the number of leaves on 4-week-old canopies stressed for only the last 2 weeks of growth was not affected (15). Soil flooding caused yellowing of leaves (177).

The proportion of herbage represented by leaves decreases with the developmental age of the canopy. Onstad and Fick (122) used growing degree days with a 5°C base temperature to scale developmental age, and confirmed earlier work (120, 190) that showed lower leaf proportions in spring growth than in alfalfa regrowth (Fig. 5–2). Leaf fraction is generally highest at the lowest environmental temperatures (if >10°C) when measured at a given age (95, 180) but at intermediate temperatures (15 to 30°C) when measured at a given stage (4, 127, 189). However, Bula (16) showed decreasing leaf yields and increasing leaf to stem ratios in both seedlings and regrowth harvested at 22 or 25 d of age for the following sequence of controlled temperatures: 25, 20, 30, and 35°C (Fig. 5–3).

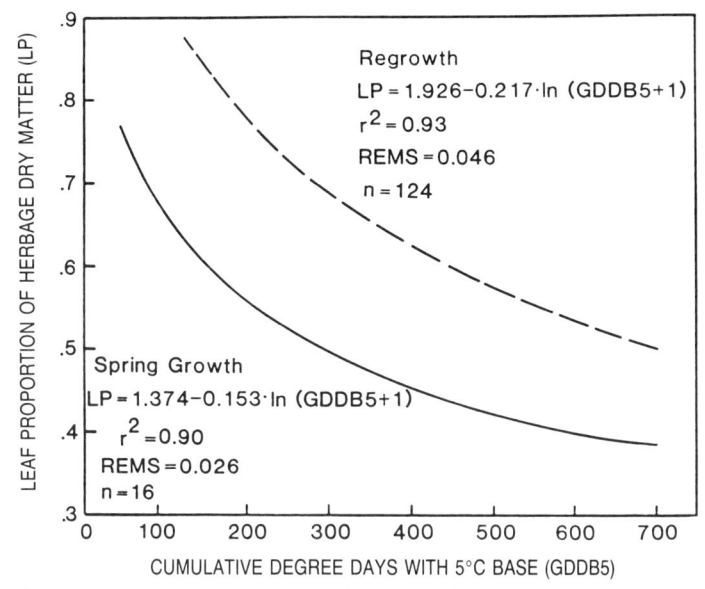

Fig. 5-2. The relationship between cumulative degree days of alfalfa growth and the leaf proportion of alfalfa herbage changes from spring growth to regrowth canopies. From Onstad and Fick (122). Root error mean square for the estimate is coded REMS.

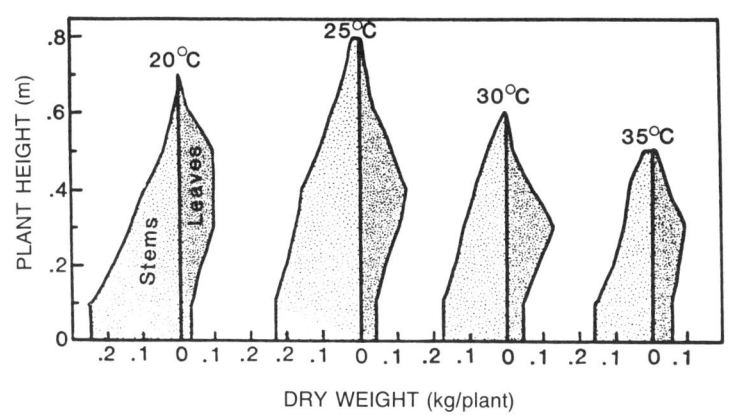

Fig. 5-3. Dry weight of leaves and stems in 0.1-m segments of plant height of alfalfa grown at four temperatures for 22 d of regrowth. Redrawn from Bula (16).

(The difference between 20 and 25°C was not statistically significant.) Wolf and Blaser (189) reported decreasing rates of leaf area accumulation with controlled temperature in the sequence of 21, 32, and 10°C.

Increasing temperature from the 15 to 20°C range to the 25 to 30°C range decreased (i) area of fully expanded leaflets (16, 89, 126, 155, 168, 189), (ii) leaf thickness (155), (iii) intercellular space in the leaves (89, 155), and (iv) leaflet density (kg m^{-3}) (89). As temperature increased, leaflet shape changed from generally obovate to generally oblanceolate resulting in a decrease in breadth to length ratio (154). In the 10 to 15°C range, less intercellular space and leaflet area occurred than at about 20°C (155, 189). Leaf enlargement did not occur below about 5°C (69).

Specific leaf weight (SLW, g m^{-2}) has been found to increase (127, 168), decrease (89), or remain unchanged (16, 17) in response to temperature changes. These differences are probably due to differences in either starch accumulation, which was as high as 40% of the leaflet dry matter with low night temperatures (167), leaf age and position in the canopy (124), or constant (16) vs. alternating temperature regimes (168).

Pearson and Hunt (127) found that the rate of leaf area accumulation for primary growth was about equal in 20/15 and 30/25°C environments, but was faster for regrowth in the 20/15°C environment, possibly because of higher total nonstructural carbohydrates (TNC) at the lower temperature. For the first 7 d of regrowth, Totana had maximal rates at about 27°C, and Rhizoma at about 33°C (74). Bula (16) showed that leaf area and leaf mass responded similarly to temperature, with the optimum at 20 to 25°C for three cultivars after 22 to 25 d of growth. The optimal temperature was higher when measured at a given age than when measured at a given stage (4).

The rate of appearance of new leaves increases with temperature to about 30°C and then begins to decrease (126, 127, 153, 154, 189). The number of leaves expanding at one time did not depend on temperature (189). The increase in the rate of leaf appearance compensated for the decrease in the area of individual leaflets over the temperature range of 10 to 20°C, but not above 20°C, except for the 1st week of regrowth (53). As a result, maximum leaf areas were reached at day temperatures of about 20 to 25°C (16, 127, 189). Compared with plants at the same stage but grown at higher temperatures, longer duration of growth to a given morphological stage under cooler temperatures gave a higher proportion of leaves in the herbage (64).

The nature of the light regime under which leaves develop has an important effect on several physiological and morphological characteristics of leaves. Cooper and Qualls (28) observed that the ratio of leaf area to leaf weight was more than 50% greater in alfalfa leaves that developed in the shade than in those that developed in the sun. Leaves exposed to the sun had about 20% higher stomatal frequency, and stomatal frequency was greater on both the top and bottom surfaces than on shaded leaves. The quantity of chlorophyll per unit leaf weight was greater in shaded leaves, but the quantity of chlorophyll per unit leaf area was greater in the sun. Thus, plants appear to partition more dry matter to the energy-capturing apparatus and spread it over greater leaf area when the irradiance is low. In these experiments, the ratio of leaf weight to plant weight did not change as a result of shading.

Lower leaves in an alfalfa canopy are shaded by both upper leaves and upper main stems, with a possible concomitant reduction in photosynthetic capacity. Wolf and Blaser (188) found that upper green leaves on main stems in an alfalfa canopy fixed CO_2 twice as rapidly as lower green leaves on main stems. Carbon exchange rates were compared in these experiments by exposing each class of leaves to the same luminous flux densities ranging from 0 to 65 klux (roughly 900 μmol m^{-2} s^{-1} of PPFD) during the measurements. (See also Chapter 6 in this book.)

The thicker leaves that develop under higher radiant energy had more and larger palisade and mesophyll cells (28). Their potential contribution to higher rates of photosynthesis led Pearce and Lee (125) to suggest the use of specific leaf weight (SLW) as a selection criterion in alfalfa breeding. Structural features of the leaflets may influence CER (34, 71), but Ku and Hunt (89) could not show an effect when differences were induced by temperature treatment. Likewise, Porter and Reynolds (134) found no correlation between SLW and yield in the field, but SLW tended to increase when air temperatures were higher. Foutz et al. (54) and Hart et al. (71) reported that yield variation among clones was associated more closely with differences in leaf area and leaf yield per plant than with CER or SLW.

Photoperiod also has an impact on leaf development. Sato (155) demonstrated that plants growing under short photoperiods (8 h) had slightly thicker leaves than those exposed to longer photoperiods (16 h). Hesterman and Teuber (77) noted that SLW increased and leaf size decreased as photoperiod increased from 8 to 16 h.

5–2.2 Stem Characteristics

Stem dry weight and internode length are reduced with increasing moisture stress (15, 37, 89, 131). Both Brown and Tanner (15) and Lucey and Tesar (98) showed that the stem extension rate was nearly constant throughout the vegetative regrowth period under well-watered conditions. Water stress reduced the growth rate of stems (Fig. 5–4); and, as was also found for leaves, little growth occurred when leaf water potential dropped below -1.0 MPa (15).

Stem growth usually is reduced more than leaf growth with increasing moisture stress so that the leaf/stem ratio increases with increasing moisture stress (23, 171, 184). Some researchers have observed, however, situations in which both leaf and stem growth were reduced equally, with leaf/stem weight ratio remaining the same under different moisture stresses (87, 186). Brown and Tanner (15) showed that the variable response in the leaf/stem ratio depended on the actual stem lengths realized. Above about 50 cm, the negative relationship between stem length and leaf fraction in the herbage disappeared.

Stem growth is also reduced under flooded conditions. Cameron (20) reported that plant height was decreased and that regrowth was severely etiolated when alfalfa was maintained under flooded conditions in greenhouse pots. However, Peterschmidt et al. (132) stated that flooding had no effect on leaf/stem ratio when field-grown alfalfa was allowed to deplete available soil moisture to one-half field capacity before being irrigated

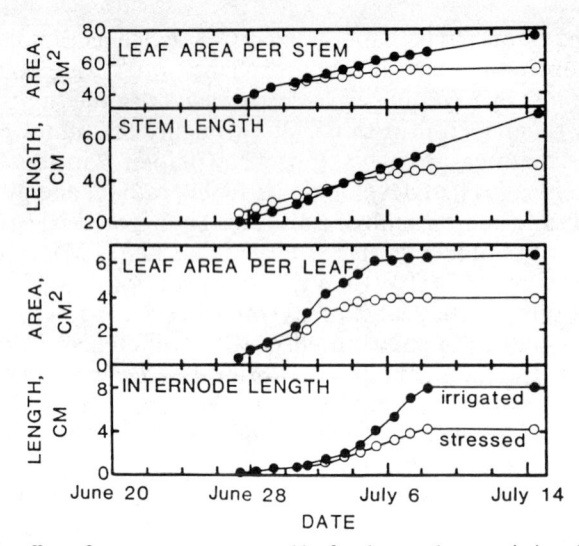

Fig. 5-4. The effect of water stress on several leaf and stem characteristics of alfalfa. Redrawn from Brown and Tanner (15).

with up to four times the amount of water required to return the soil to field capacity.

Salinity causes a general stunting of the whole plant and stem growth is usually more affected than leaf growth. Hoffman et al. (79) reported that salinity consistently increased the leaf/stem ratio at all cuttings.

Greenfield and Smith (64) observed that alfalfa stems yielded more at 27/18°C and leaves at 21/12°C, so that similar dry matter yields were obtained in both environments. As with leaves, stems generally produced highest yields at the lowest temperature (if >10°C) when measured at a given stage of growth (4, 68, 95) because of longer growth periods at lower temperatures. However, when measured at a given age, stem growth rate was maximal in the range of 20 to 27°C with highest stem yields at moderate temperatures (4, 16). It has been shown that stem growth rates decreased below 18°C (15), or above 30°C (16). Extension rates of stems were linearly related to night temperature but not to day temperature (160). Generally, extension rates were greater at night than during the day because of lower leaf water potentials during the day. Brown and Tanner (15) also found lower growth rates during the day than at night for water-stressed alfalfa, but the reverse was generally true for non-stressed (irrigated) alfalfa.

Although plant height and stem length may not be clearly distinguished in the literature, both field and growth chamber studies indicate that accelerated development associated with high summer temperatures is also associated with shorter plants (117, 126, 127, 153, 163, 164, 165, 184). Pearson and Hunt (127) observed that stems were more erect at 20/15°C than at 30/25°C. Sato and Itoh (156) concluded that the optimum temperature for internode length was about 23/17°C. Stem branching was greatest over the 20 to 28°C range (153, 156).

The structure and composition of the stem is modified by temperature. (See Chapter 14 in this book.) Faix (45) found significant positive correlations between the lignin to cellulose ratio of alfalfa stems and growth temperature over the range of 17 to 32°C, with increasing temperature increasing the lignification of cell walls. Sato (155) showed that stem diameter decreased with increasing temperature, a change which was probably associated with more lignification. The mass to length ratio of the basal internodes increased at lower temperatures (16, 165). This could be due to thicker cell walls with relatively more cellulose, or to a longer growth period because of slower leaf appearance at low temperatures.

As photoperiod increased from 8 to 16 h, shoots were longer and more upright, with longer internodes and larger diameters (153, 155). The dry weight of tops was little affected by photoperiod, while the proportion of stems in the herbage decreased under shorter photoperiods. Conversely, when clones from Moapa, 'Lahontan', and 'Norseman' were subjected to photoperiods varying from 8 to 16 h at 743 and 382 μmol m^{-2} s^{-1} in growth chambers held constant at 24 ± 3°C, stem length, number of crown buds, and internode length were influenced significantly by irradiance, but not photoperiod (77). Shoots were significantly heavier under longer photoperiods. The variation among cultivars may account for the difference in results reported by these two laboratories.

In Arizona, Massengale et al. (104) found that interrupting the dark period for 30 min once or twice, with a minimum luminous flux density of 107 lux (roughly 1.5 μmol m^{-2} s^{-1} of PPFD) during the interruption, increased height of plants growing in March and April over that of plants grown with an uninterrupted dark period. During May, June, and July, when seasonal temperatures are normally highest, stem height of regrowth was greater without dark period interruption. After July, interruption again produced taller regrowth than controls. Crown diameter and forage production were reduced by dark period interruption, suggesting that short nights decrease yield per stem and/or the number of stems.

Dormant alfalfa cultivars responded to shortening photoperiods with reduced shoot and enhanced root growth (17, 175). This response was absent in nondormant cultivars. The pattern seems to be reversed as days lengthen in the spring, but the reverse process is apparently less photoperiod sensitive. In the field, seasonal changes in soil temperatures lag behind changes in air temperature, creating a different environment for roots and shoots. Also, the influence of temperature on source-sink relationships in those organs may influence the relative amounts of stem growth and photosynthate movement to roots.

5-2.3 Root Characteristics

Drier soil under rainfed conditions leads to deeper and more branched rooting patterns in alfalfa than do more moist conditions (187). Root mass has been shown to be positively related to the amount of evapotranspiration if water was applied in light, frequent irrigations using a sprinkler system (1). Salter et al. (151) also found that root mass decreased and fibrousness increased as less water was applied. Flooding of the soil almost immediately stopped root growth, and prolonged flooding caused deterioration of the root system (177). Increasing temperature hastened and increased the amount of injury from flooding (20, 75, 177). Thompson and Fick (177) reported that 14 d of flooding at 16°C prevented regrowth, but only 6 d at 32°C. Cameron (20) demonstrated that the soil temperature was the critical factor. Anaerobic conditions caused by flooding were associated with increased ethyl alcohol production in the roots and decreased TNC accumulation (7). (See Chapter 11 in this book.)

In general, roots show the same responses to soil temperature as do tops to air temperature (68, 95, 153). Relative to tops, roots did better at cooler temperatures (as low as 9°C), but maximum root yields were realized at 23°C (soil temperature) in a study by Sato and Itoh (156). Daday and Williams (31) compared root growth at night temperatures of 19 and 4°C, and found that higher temperature favored taproot development while temperature had less influence on lateral root development. As a result, the taproot to lateral root ratio decreased to <1.0 at 4°C.

Relatively low temperatures increased the accumulation of TNC in taproots and crowns (64). Pearson and Hunt (127) measured more dry matter accretion by roots and stubble at 20/15 than at 30/25°C. During regrowth, TNC utilization was greater at higher temperatures (127). In taproots, lower TNC fractions have been associated with hot summer weather (46). However, with seedlings, temperatures of about 4°C depressed taproot development, and presumably, TNC accumulation (31). The optimal root zone temperature for N_2 fixation (acetylene reduction) was about 15 to 27°C (69, 114), and temperatures greater than 30°C had significant adverse effects on root nodule function (6, 114).

Decreasing the photoperiod from 12 to 8 h reduced root growth in the studies of Sato (154), but root to total plant weight ratios were increased (153, 155). Mannetje and Pritchard (103) found a photoperiod decrease from 14 to 11 h did not reduce root growth unless it was associated with temperature reduction. Carlson (21) and Hesterman and Teuber (77) also found no effect of photoperiod on root dry weight, though increasing irradiance increased root dry weight.

5-2.4 The Summer Decline

When alfalfa is grown in temperate climates, yields usually decline in successive harvests within a growing season. This phenomenon was first called the "summer slump" by Stanberry et al. (172). In a three-cut system, the ratio of yield in successive harvests is usually 7:5:3 (50, 158, 162); with a four-cut system, about 9:7:5:3 (23, 43). The same phenomenon has been studied in the Southwest with 7 to 10 harvests per season (145, 175). Singh and Winch (162) suggested a combination of changes in temperature, photoperiod, and moisture deficit caused the yield decline. Aging and lower TNC accumulations may contribute to the decline in yield (46, 126). When harvests are made at a fixed developmental stage, the shorter summer growth period associated with faster maturation at higher temperatures is an important consideration.

This last factor was studied in detail by the group at the Univ. of Guelph (52, 53, 126, 127, 128). By growth analysis and modeling, Field (52) showed that part of the decline was due to higher summer temperatures. McLaughlin and Christie (109) found genotypic variation for response to temperature and suggested the possibility of breeding cultivars with less temperature-induced summer decline. However, the genotypic differences selected in the growth chamber disappeared in the field, possibly because of less extreme temperatures in the field. More recent work has focused on the role of moisture deficits. Carter and Sheaffer (23) and Evans and Peaden (43) showed that irrigated alfalfa did not show a summer decline. Brown and Tanner (15) observed that evapotranspiration of alfalfa was not reduced by soil water depletion until a week after growth rates were reduced. Thus, some yield reduction could occur even when water use was not restricted. In unirrigated plots, Carter and Sheaffer (23) found yield ratios typical of the summer decline. Even under irrigated conditions, yields decreased in sequential harvests in Washington (43), but the lower yields were associated with shorter growth intervals and faster phenological development in the summer. Growth rates and energy-use efficiency did not decline (Table 5-1).

5-3. REPRODUCTIVE DEVELOPMENT

5-3.1 Alfalfa Phenology

Traditionally, the phenological development of alfalfa has been described in terms of the morphological stage of reproductive locations on

Table 5-1. Forage growth rate and forage conversion efficiency of photosynthetically active radiation (PAR) for irrigated 'Vernal' alfalfa in Washington State. Adapted from Evans and Peaden (43).

	\multicolumn{4}{c}{Growth period within the year†}
	First	Second	Third	Fourth
	\multicolumn{4}{c}{1980}
g m^{-2} d^{-1}	15.8 ± 1.1	18.6 ± 0.7	20.7 ± 0.8	10.8 ± 1.1
Percentage PAR as forage	3.0 ± 0.2	3.3 ± 0.1	3.5 ± 0.2	2.6 ± 0.3
Harvest date	21 May	9 July	19 Aug.	22 Sept.
	\multicolumn{4}{c}{1981}
g m^{-2} d^{-1}	12.3 ± 3.4	17.7 ± 1.6	18.4 ± 2.0	10.9 ± 2.7
Percentage PAR as forage	2.4 ± 1.7	3.3 ± 0.2	2.8 ± 0.3	2.1 ± 0.6
Harvest date	22 May	29 June	3 Aug.	3 Sept.

† Confidence intervals are at the 0.05 level. The PAR conversion efficiency is for the development period after the canopy reached a height of 10 to 20 cm.

dominant shoots of the canopy. Thus, the canopy is said to be at the vegetative, bud, flowering, or seed pod stage. Because the canopy always exhibits a range in size and reproductive stage of shoots, such a system is imprecise, especially if phenology is to be related to environmental history of the canopy (85). Gengenbach and Miller (59) used a four-stage system based on the frequency distribution of shoots in different stages, but the system lacked the detail needed to study possible changes in the vegetative stage related to damage from insect pests or diseases. Kalu and Fick (84) developed a 10-stage classification with the mean stage by weight (MSW) for all shoots calculated as the average stage, weighted by the dry mass of shoots within each stage (Table 5-2). The system was used to relate MSW to the nutritive value of temperate-zone cultivars used for hay (84, 85). Other cultivars, or other environments which inhibit flowering, might display unique relationships between phenology and nutritive value. However, the MSW system is sufficiently detailed to precisely define morphological stage of development for most studies.

By definition, phenology relates morphological development to seasonal changes in the environment. Temperature and photoperiod appear to be the main factors involved in phenological development in alfalfa. A few controlled environment studies have considered the combined effects of temperature and photoperiod (45, 67, 153); and in summary, development rate is accelerated by increasing temperature and increasing photoperiod (Fig. 5-5). In the field, development was faster in the summer

Table 5-2. Equation and stage definitions for measuring mean stage by weight (MSW). Adapted from Kalu and Fick (84).

Equation: $MSW = \sum_{S=0}^{9} S \cdot D/W$

where   S = morphological stage number,
D = dry wt. of shoots in stage S, and
W = total dry wt. of all shoots in forage sample of at least 50 random shoots.

Stage definitions:

Stage no.	Stage name	Stage definition
0	Early vegetative	Stems ≤15 cm long†; no buds, flowers, or seed pods
1	Mid-vegetative	Stems 16 to 30 cm long; no buds, flowers, or seed pods
2	Late vegetative	Stems >30 cm long; no buds, flowers, or seed pods
3	Early bud	One or two nodes with visible buds, no flowers or seed pods
4	Late bud	≥ three nodes with buds, no flowers or seed pods
5	Early flower	One node with one open flower (standard open); no seed pods
6	Late flower	≥ two nodes with open flowers; no seed pods
7	Early seed pod	One to three nodes with green seed pods
8	Late seed pod	≥ four nodes with green seed pods
9	Ripe seed pod	Nodes with mostly brown mature seed pods

† Above a 2.5-cm stubble.

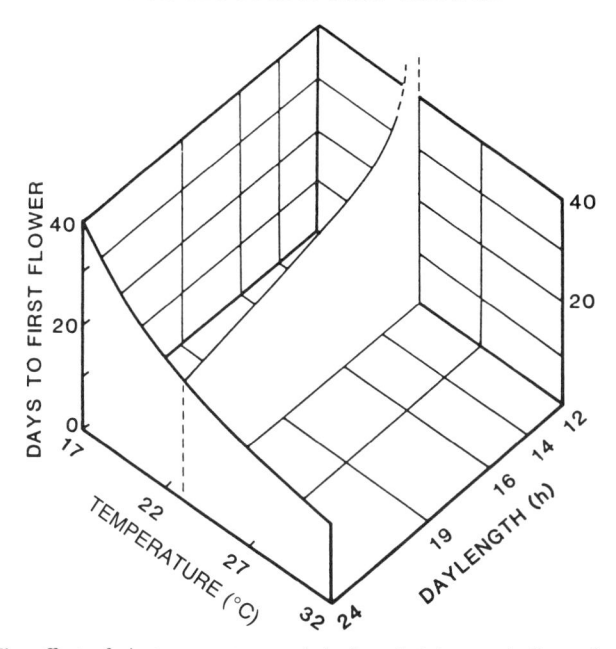

Fig. 5-5. The effect of air temperature and daylength (photoperiod) on the days to first flower with 'Iroquois' alfalfa. Smoothed data from Faix (45).

than in the spring or autumn (43, 84, 145), and light interruption of the dark period promoted flowering (104). In general, nonlethal environmental stresses appear to slow phenological development, but the relationship between alfalfa phenology and environmental stress from pests, mineral nutrients, and water supply needs more thorough study.

The general shortening of the prereproductive period as temperature increases has been noted in several reports (4, 45, 64, 67, 68, 95, 117, 126, 153, 163). A typical response (Fig. 5-5) shows flowering about 3 weeks earlier at 32°C than at 17°C. However, at about 35°C, there was a delay in maturation observed with both dormant (Vernal) and nondormant (Moapa) cultivars (126). When air temperatures were held at 25/18°C, increasing the soil temperature from 10 to 20°C caused flowering to occur about 1 week sooner (108).

Greenfield and Smith (64) switched temperature environments at first bud stage, and the resulting times to first flower suggested that heat summation could be a basis for predicting alfalfa phenology. Jeney (83) had used such an approach in Hungary, with about 700, 880, 1075, and 1900 growing degree days to bud, first flower, full bloom, and ripe seed stages. In this study, the base temperature was 0°C, while other workers have used a base temperature of 5°C (50, 80, 158). Recently, Sharratt (159) concluded that the apparent base temperature for alfalfa changed with time of year.

Photoperiod and quality of light are sensed by the phytochrome mechanism of plants. Although alfalfa is regarded as a long day plant, that is, it tends to flower when photoperiods are longest (Fig. 5–5), it is not completely photoperiod sensitive. There are considerable differences among genotypes in photoperiod sensitivity, and sensitivity is complicated by important interactions with irradiance and temperature. Furthermore, the indeterminate nature of flowering in alfalfa may present a problem in interpreting results. A 30-min interruption of the dark period at a minimum of 107 lux (about 1.5 μmol m^{-2} s^{-1} of PPFD) increased flowering at all harvests (104). The effect was greater in the naturally shorter days and cooler temperatures of April and November. The degree of response differed among genotypes.

Other research (45, 115, 119, 153) confirmed the identification of alfalfa as a long day plant, with the highest degree of flowering observed under continuous light. White light interruption of the dark period stimulated the development of floral initials (115). Returning plants to short photoperiods after the formation of floral initials tended to reverse the process, causing racemes to atrophy and turn white. Faix (45) observed similar bud abortion at 12-h days. Light interruption was not as effective in triggering flowering as continuous light. Two equally spaced 80-min, far-red light interruptions, one 4-h red light interruption, or no interruption of a 15-h dark period, produced no evidence of floral development (115).

A yellowish green band that darkened when exposed to far-red light and lightened when exposed to red light appeared on calcium phosphate (CaPO$_4$) columns after extraction of alfalfa with Tris buffers (115). This is the only direct evidence that photoperiodic responses in alfalfa are mediated by the phytochrome system.

5–3.2 Flowering, Pollination, and Seed Set

Moisture stress has a significant effect on seed production in alfalfa, although not in the same way as it affects forage yield. According to Pedersen et al. (129), the highest seed yields are produced when there is enough soil water available to prevent severe moisture stress, while moisture conditions causing excessive vegetative growth reduce seed yield. (see Chapter 32 in this book.) Under field conditions in Alberta, Krogman and Hobbs (88) reported that seed yield was not increased when alfalfa was irrigated after the bud to early bloom stage, while in Utah, Taylor et al. (176) found that seed yield was reduced with irrigation during flowering. There was a range of optimum soil moisture potentials for seed production, with maximum seed yields produced when potentials were between -0.2 and -0.8 MPa (176). High water applications reduced seed yields (2, 27, 61, 111, 176). Conversely, severe water stress reduced seed yields (Table 5–3) (2, 176).

Generally, the number of stems and the number of racemes per stem were reduced with increasing moisture stress (27, 61). The number of flowers per raceme was generally not affected over a moderate range of moisture stress. However, pod set, or the number of pods per raceme,

Table 5-3. Effect of irrigation interval and quantity applied during the growing season on seed yield and other characteristics. Adapted from Abu-Shakra et al. (2).

Irrigation interval	Irrigation quantity	Seed yield	Pods/ raceme	Seeds/ pod	Hard seeds	Plant height
weeks	m	kg/ha	— no. —		%	m
1	0.50	685	8.6	6.40	4.5	1.22
2	0.26	755	9.6	6.35	8.5	1.17
3	0.17	610	8.3	6.00	11.0	0.99
4	0.09	510	7.3	5.65	15.0	0.96

was reduced with either high or low amounts of irrigation (Table 5-3), with the highest number of pods per raceme at intermediate levels (2, 27, 61). Relative humidity above 50% also appeared to reduce pod set, especially at high temperatures (63). The number of seeds per pod and 1000-seed weight were reduced with severe moisture stress, while the hard seed percentage increased (Table 5-3) (2).

The effect of temperature on seed production has not received much attention in recent years. Guy et al. (67) concluded that the effect of temperature on flower number is complex. Over the 17 to 27°C range, increasing temperatures generally increased the number of racemes per shoot but decreased the number of flowers per raceme. Blondon et al. (11) showed increased pollen fertility from 17 to 27°C, but a genotype × environment interaction was observed. In selections from dry regions, increasing temperature and decreasing relative humidity increased the amount of automatic flower tripping for some genotypes, but this did not include the cv. Rambler (76). In addition, the amount of seed abortion nearly doubled between 27 and 32°C (33). Minimum temperatures above 20°C are known to favor seed production (42). High temperature during seed formation increased initial amount of hard seed and may decrease seedling vigor (38).

If photoperiods are long enough to promote flowering, more flowers will be produced at higher irradiance (119). Sixty-four percent of seedlings grown under 20-h photoperiods flowered after 5 weeks at 23.7 klux (roughly 350 μmol m^{-2} s^{-1} of PPFD), but only 5% flowered at 11.9 klux (roughly 175 μmol m^{-2} s^{-1}). Under the naturally high irradiance prevailing in the field, flowers were produced when photoperiods were extended at luminous flux densities as low as 342 to 854 lux (about 5-10 μmol m^{-2} s^{-1}) (143).

Genetic and cytoplasmic pollen sterility in male-sterile alfalfa is strongly influenced by environmental factors such as temperature and light. Mean pollen sterility was 7.6% higher under 19 klux than under 55 klux (roughly 250 vs. 750 μmol m^{-2} s^{-1}) (161). Temperature, had a much greater effect, with 30% more sterile pollen from plants grown at 15°C than from plants grown at 24/19°C.

Volatiles released from flowers attract pollinating insects to alfalfa.

The production of volatiles reached a maximum of 6.5 ng floret^{-1} during a 30-min period 6 to 8 h after the beginning of illumination of plants growing under 18 h of light and 6 h of darkness (97). Thereafter, production of volatiles decreased to 0.1 ng at 11 h after the beginning of the light period. Continuous illumination caused little change in volatile compound production.

5-4 COMPUTER MODELING

Computer modeling provides another tool for studying the effect of the environment on crop production and on the potential success of management plans. Although sunlight, temperature, and precipitation are not ordinarily managed, they affect alfalfa growth in a predictable manner, and when physiological responses to plants are sufficiently understood, weather data can be used to develop mechanistic crop growth models. Other environmental factors, such as soil and pest conditions, and environmental modifications caused by management, can be included in computer models. Thus, computer models have become the vehicles for applying knowledge about environmental physiology.

By slightly modifying the ideas of Pielou (133), the objectives of crop growth models can be summarized as (i) explanation, (ii) prediction, and (iii) hypothesis comparison. To be explanatory, computer models incorporate cause-and-effect mechanisms, usually by linking environmental causes to physiological effects that modify calculated growth processes. Such models simulate growth by a technique called *dynamic simulation*, usually by treating time as the independent variable. Explanatory models are also predictive, but when the primary goal of modeling is prediction, the cause-and-effect mechanism may be greatly simplified, as in the extreme case of regression analysis. Although regression models may be useful for interpolation, a mechanistic model is more reliable for extrapolation. Thus, several alfalfa growth models that were developed mainly for prediction retain simplified cause-and-effect mechanisms taken from environmental physiology. Alternative cause-and-effect mechanisms can be specified and compared by computer modeling. The comparison can identify either unacceptable hypotheses or rank alternative models according to accuracy. Again, dynamic simulation is the usual technique.

Since the first dynamic simulation model of alfalfa growth was published in 1974, new models have appeared at a rate of more than one per year (Table 5-4). All of the models considered here use information about the environmental physiology of alfalfa and predict growth by means of dynamic simulation. Each can be classified as focusing on one of the primary objectives of modeling, although all probably have multiple functions. Two of the models, SIMFOY of Selirio and Brown (158) and GROWIT of Smith and Loewer (169), were developed for perennial forages and can be applied to simulate alfalfa growth with inclusion of

Table 5-4. A chronological listing of computer models that simulate alfalfa growth, classified according to primary objective.

Model	Authors	Reference no.	Primary objective
The Field model	Field, 1974	52	Explanation
ALSIM 1 (LEVEL 1)	Fick, 1975	47	Prediction
SIMED	Holt et al., 1975	80	Explanation
The California model I	Gutierrez et al., 1976	66	Prediction
The California model II	Regev et al., 1976	140	Prediction
SIMED 2	Dougherty, 1977	41	Explanation
REGROW	Fick, 1977	48	Hypothesis comparison
SIMFOY	Selirio and Brown, 1979	158	Prediction
The Canberra model	Christian and Milthorpe, 1981	26	Explanation
ALSIM 1 (LEVEL 2)	Fick, 1981	49	Prediction
YIELD	Hayes et al., 1982	72	Prediction
ALSIM 1 (LEVEL 0)	Fick and Onstad, 1983	51	Prediction
GROWIT	Smith and Loewer, 1983	169	Prediction
ALFALFA	Denison et al., 1984	35	Explanation
ALSIM 1 (LEVEL ZERO)	Fick, 1984	50	Hypothesis comparison
The Gosse model	Gosse et al., 1984	62	Prediction
ALFMAN	Onstad and Shoemaker, 1984	123	Prediction

model data describing alfalfa. The model named YIELD (72) was set up for alfalfa as well as for several food crops.

Penning de Vries (130) has classified crop growth models as being preliminary, comprehensive, or summary in nature. The LEVEL 1 and LEVEL 2 versions of ALSIM 1 (47, 49) and SIMED (80) and SIMED 2 (41) show how models become more comprehensive. The second member of each set includes a soil water budget which was not a part of the first version. The LEVEL ZERO version of ALSIM 1 (50) includes a refined forage quality predictor not in the version called LEVEL 0 (51). The LEVEL ZERO version is a good example of a preliminary model, in which growth rate depends on environmental temperature (growing degree days), soil water holding capacity, and regrowth potential as a function of growing degree days between harvests. Although the model was developed for comparison of simple, environmentally based mechanisms for predicting alfalfa growth rate, it appears useful in relating cutting management to yield (50). The Gosse model (62) is another preliminary model in which the effect of temperature on leaf area, leaf area on light interception, and light interception on yield are reduced to a single equation. The REGROW model (48) was employed to compare possible mechanisms of alfalfa regrowth, and the findings were used to develop ALSIM 1 (LEVEL 2) (49). A comprehensive model of the alfalfa regrowth mechanism is being formulated in ALFALFA (35).

By definition, comprehensive models include much of the detail known to influence the processes being simulated, and thus, models developed for explanation are usually comprehensive. In general, such models also simulate the diurnal patterns of change in the crop by calculating conditions with simulated time steps of about 1 h. The model of Field (52) treated the effect of temperature on the growth of the individual tiller, and it helped to demonstrate the role of temperature in the observed decline in summer production. The more recent Canberra model (26) simulated the short-term dynamics of heat, momentum, CO_2, water, carbohydrate, and P in the alfalfa field. The complete documentation of SIMED makes it a good example of a comprehensive model (80, 112, 157). Thirty functions of the type shown in Fig. 5-6 related environmental and physiological conditions to the rates of photosynthesis, respiration, translocation, and the growth of leaves, stems, and roots. As a result, the mechanism of carbohydrate translocation was well simulated.

Summary models are condensations of comprehensive models that retain most of the generality useful in extrapolation and accuracy. The model of Gutierrez et al. (66) and ALSIM 1 (LEVEL 1 and LEVEL 2) are examples developed for modeling alfalfa weevil (*Hypera* spp.) management programs. Both versions of ALSIM 1 were based on SIMED. In these summary models, the simulation time step was increased to 1 d. As a result, seasonal yield patterns, but not diurnal patterns, can be simulated. Structural detail was also reduced; for example ALSIM 1 has less than half the rate equations of SIMED and does not predict either root growth or root yields. Systems analyses of the economics of alfalfa

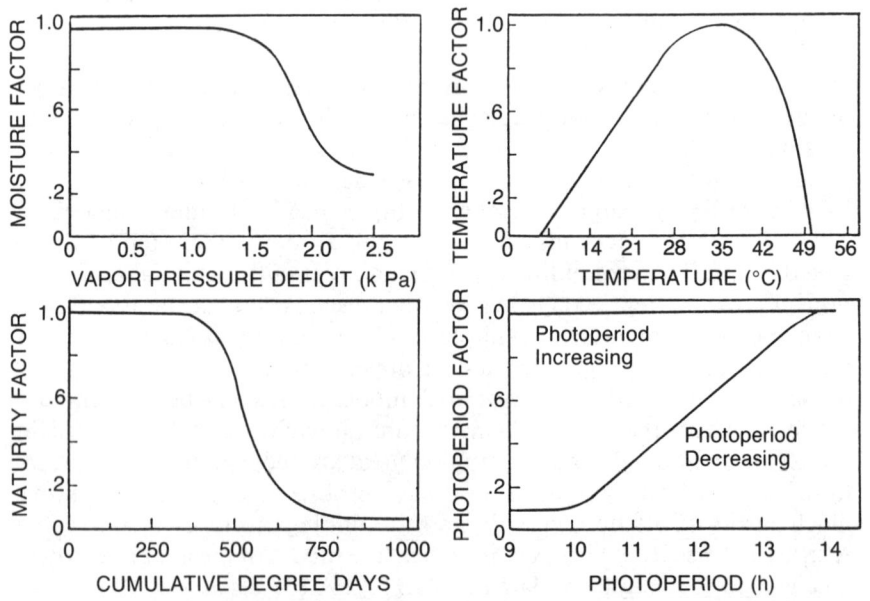

Fig. 5-6. The relative effect (0.0 to 1.0 multiplier) of the level of four environmental factors (atmospheric vapor pressure deficit, air temperature, cumulative degree days with a 5°C base, and photoperiod) on the growth rate of alfalfa leaves. From the SIMED model, Holt et al. (80).

weevil [*Hypera postica* (Gyllenhall)] control have used summary versions of California model I for irrigated alfalfa (140) and of LEVEL 1 for nonirrigated alfalfa (see 123). In theory, summary models are most reliable for management applications, but preliminary models like LEVEL ZERO appear to be about as good in some comparisons (49, 50).

Computer modeling is useful because it involves the simultaneous consideration of more factors than would be possible without the aid of a computer. However, models can be no better than the data and theory on which they are based. Therefore, conventional study of environmental physiology has been stimulated and not replaced by computer modeling. For example, the modeling of Field (52) led to the thorough analysis of the effect of temperature on alfalfa (53). More recently, the models of Fick (49) and Gosse et al. (62) showed the need for a more quantitative understanding of the effect of photoperiod on alfalfa partitioning. The absence of alfalfa models describing population dynamics of plants and stems in stands may result from lack of attention given to the environmental physiology of the yield components of alfalfa.

Computer models, of necessity, are based on simplifications of reality. As simplifications, there are always situations for which models are inappropriate, for which they will not "work." Computer models have helped the study of environmental physiology, but studies of environmental physiology are also needed to help evaluate the appropriateness of model applications. Thus, the development and the evaluation of computer models are closely linked to traditional research work in environmental physiology.

5-5 SUMMARY

The study of environmental effects on alfalfa growth and development has resulted in a number of important and advances in management practices and breeding programs. The physical environment is partially regulated by irrigation and fertilization; the biological environment by symbiosis and pest resistance. The very success of these programs has led to their separation from the domain of environmental physiology, and what remains are those environmental factors not yet readily controlled: temperature, light, and the natural water supply, including salt effects. For these factors, emphasis is placed largely on explaining and predicting plant response. Much has been learned about each of these factors and their role in every phase of alfalfa development. Dynamic simulation modeling has provided a means of integrating and evaluating the concepts of environmental physiology while providing a stimulus for research in this field. However, the problems are difficult and much remains to be accomplished.

The inherent genetic variability of alfalfa makes generalization risky without comprehensive studies of divergent genetic material. In addition, the physiological responses of alfalfa depend on plant age and acclimation to past and present environments. More comprehensive work is needed to develop generalizations that encompass this variability. Basic concepts like yield components, phenological classification, and heat summation as describers of growth and development need more work before the mechanisms of alfalfa development can be understood. Temperature, light, and water supply are known to interact in their effects on alfalfa physiology, but environmental interactions have yet to receive much attention in alfalfa research. As the future provides more complete understanding of the relationship between environment and alfalfa growth, chemical composition, and longevity, it should be possible to refine our alfalfa management programs and enhance the value of the crop.

REFERENCES

1. Abdul-Jabbar, A.S., T.W. Sammis, and D.G. Lugg. 1982. Effect of moisture level on the root pattern of alfalfa. Irrig. Sci. 3:197–207.
2. Abu-Shakra, S., M. Akhar, and D.W. Bray, 1969. Influence of irrigation interval and plant density on alfalfa seed production. Agron. J. 61:569–571.
3. Allen, S.G., A.K. Dobrenz, M.H. Schonhorst, and J.E. Stoner. 1985. Heritability of NaCl tolerance in germinating alfalfa seeds. Agron. J. 77:99–101.
4. Arbi, N., Dale Smith, and E.T. Bingham. 1979. Dry matter and morphological responses to temperature of alfalfa strains with differing ploidy levels. Agron. J. 71:573–577.
5. Baldocchi, D.D., S.B. Verma, and N.J. Rosenberg. 1981. Seasonal and diurnal variation in CO_2 flux and CO_2 water flux ratio of alfalfa. Agric. Meteorol. 23:231–244.
6. Barta, A.L. 1978. Effect of root temperature on dry matter distribution, carbohydrate accumulation, and acetylene reduction activity in alfalfa and birdsfoot trefoil. Crop Sci. 18:637–640.
7. ----. 1980. Regrowth and alcohol dehydrogenase activity in waterlogged alfalfa and birdsfoot trefoil. Agron. J. 72:1017–1020.
8. Bauder, J.W., A. Bauer, J.M. Ramirez, and D.K. Cassel. 1978. Alfalfa water use and production on dryland and irrigated sandy loam. Agron. J. 70:95–99.
9. Baysdorfer, Chris, and J.A. Bassham. 1985. Photosynthate supply and utilization in alfalfa. A developmental shift from a source to a sink limitation of photosynthesis. Plant Physiol. 77:313–317.
10. Bernstein, L., and L.E. Francois. 1973. Leaching requirement studies: sensitivity of alfalfa to salinity of irrigation and drainage waters. Soil Sci. Soc. Am. Proc. 37:931–943.
11. Blondon, F., B. Cambier, Y. Dattee, and P. Guy. 1979. Influence de la temperature sur la fertilite male et femelle de luzerne: Temoins, male-steriles et "mainteneurs". Ann. Amelior. Plant. 29:89–96.
12. Bower, C.A., G. Ogata, and J.M. Tucker. 1969. Rootzone soil profiles and alfalfa growth as influenced by irrigation water salinity and leaching fraction. Agron. J. 61:783–785.
13. Bresler, E., B.L. McNeal, and D.L. Carter. 1982. Saline and sodic soils. Principles-dynamics-modeling. Advanced series in agricultural sciences 10. Springer-Verlag New York, New York.
14. Brown, J.W., and H.E. Hayward. 1956. Salt tolerance of alfalfa varieties. Agron. J. 48:18–20.
15. ----, and C.B. Tanner. 1983. Alfalfa stem and leaf growth during water stress. Agron. J. 75:799–805.
16. Bula, R.J. 1972. Morphological characteristics of alfalfa plants grown at several temperatures. Crop Sci. 12:683–686.
17. ----, and C.S. Garrison. 1962. Fall regrowth response of Ranger and Vernal alfalfa as related to generation of increase and area of seed production. Crop Sci. 2:156–159.
18. ----, and M.A. Massengale. 1972. Environmental physiology. In C.H. Hanson (ed.) Alfalfa science and technology. Agronomy 15:167–184.
19. ----, C.L. Rhykerd, and R.C. Langston. 1959. Growth response of alfalfa seedlings under various light regimes. Agron. J. 51:84–86.

20. Cameron, D.G. 1973. Lucerne in wet soils—the effect of stage of regrowth, cultivar, air temperature, and root temperature. Aust. J. Agric. Res. 24:851–861.
21. Carlson, G.E. 1965. Photoperiodic control of adventitious stem initiation on roots. Crop Sci. 5:248–250.
22. Carlson, J.R., Jr., R.L. Ditterline, J. Martin, D.C. Sands, and R.E. Lund. 1983. Alfalfa seed germination in antibiotic agar containing NaCl. Crop Sci. 23:882–885.
23. Carter, P.R., and C.C. Sheaffer. 1983. Alfalfa response to soil water deficits. I. Growth, forage quality, yield, water use, and water use efficiency. Crop Sci. 23:669–675.
24. Chatterton, N.J., and G.E. Carlson. 1981. Growth and photosynthate partitioning in alfalfa under eight temperature-photosynthetic period combinations. Agron. J. 73:392–394.
25. Christian, K.R. 1977. Effects of the environment on the growth of alfalfa. Adv. Agron. 29:183–227.
26. ----, and F.L. Milthorpe. 1981. A systematic approach to the simulation of short-term processes in the plant-environment complex. Plant Cell Environ. 4:275–284.
27. Cohen, Y., H. Bielorai, and A. Dovrat. 1972. Effect of timing of irrigation on total nonstructural carbohydrate level in roots and on seed yield of alfalfa (*Medicago sativa* L.). Crop Sci. 12:634–636.
28. Cooper, C.S., and M. Qualls. 1967. Morphology and chlorophyll content of shade and sun leaves of two legumes. Crop Sci. 7:66–72.
29. Cowett, E.R., and M.A. Sprague. 1962. Factors affecting tillering in alfalfa. Agron. J. 54:294–297.
30. ----, and ----. 1963. Effect of stand density and light intensity on the microenvironment and stem production of alfalfa. Agron. J. 55:432–434.
31. Daday, H., and J.D. Williams. 1976. Changes in lateral root growth of lucerne cultivars with temperature. J. Aust. Inst. Agric. Sci. 42:119–121.
32. Daigger, L.A., L.S. Axthelm, and C.L. Ashburn. 1970. Consumptive use of water by alfalfa in western Nebraska. Agron. J. 62:507–508.
33. Dane, F., and B. Melton. 1973. Effects of temperature and method of pollination on the frequency of aborted seed in alfalfa. Crop Sci. 13:753–754.
34. Delaney, R.H., and A.K. Dobrenz. 1974. Morphological and anatomical features of alfalfa leaves as related to CO_2 exchange. Crop Sci. 14:444–447.
35. Denison, R.R., R.S. Loomis, and J.L. Rouanet. 1984. Simulation of regrowth and overwintering in alfalfa. p. 24. In Seventh Ann. Workshop on Crop Simul., Lincoln, NE. 19-21 March. Univ. of Nebraska, Lincoln.
36. Dermine, P., M. Hidiroglou, and K.A. Hamilton. 1967. Effects of temperature on yields and hydrolysable carbohydrate content of alfalfa and timothy seedlings. Can. J. Plant Sci. 47:523–531.
37. Donovan, T.J., and B.D. Meek. 1983. Alfalfa response to irrigation treatment and environment. Agron. J. 75:461–464.
38. Dotzenko, A.D., C.S. Cooper, A.K. Dobrenz, H.M. Laude, M.A. Massengale, and K.C. Feltner. 1967. Temperature stress on growth and seed characteristics of grasses and legumes. Colo. Agric. Exp. Stn. Tech. Bull. 97.
39. ----, and J.G. Dean. 1959. Germination of six alfalfa varieties at three levels of osmotic pressure. Agron. J. 51:308–309.
40. ----, and T.E. Haus. 1960. Selection of alfalfa lines for their ability to germinate under high osmotic pressure. Agron. J. 52:200–201.
41. Dougherty, C.T. 1977. Water in the crop model SIMED 2. p. 103–110. In DSIR Inf. Ser. 126, Wellington, New Zealand.
42. Doull, K.M. 1967. A review of the factors affecting seed production in lucerne. In R.H.M. Langer (ed.) The lucerne crop. Reed, Wellington, New Zealand.
43. Evans, D.W., and R.N. Peaden. 1984. Seasonal forage growth rate and solar energy conversion of irrigated Vernal alfalfa. Crop Sci. 24:981–984.
44. Evenson, P.D. 1979. Optimum crown temperatures for maximum alfalfa growth. Agron. J. 71:798–800.
45. Faix, J.J. 1974. The effect of temperature and daylength on the quality and morphological components of three legumes. Ph.D. thesis. Cornell Univ., Ithaca, NY (Diss. Abstr. Int. 35:2021-B).
46. Feltner, K.C., and M.A. Massengale. 1965. Influence of temperature and harvest management on growth, level of carbohydrates in the roots, and survival of alfalfa (*Medicago sativa* L.). Crop Sci. 5:585–588.
47. Fick, G.W. 1975. ALSIM 1 (LEVEL 1) user's manual. Agron. Mimeo 75-20. Cornell Univ., Ithaca, NY.
48. ----. 1977. The mechanism of alfalfa regrowth: a computer simulation approach. Search: Agriculture 7(3):1–28. Cornell Univ., Ithaca, NY.
49. ----. 1981. ALSIM 1 (LEVEL 2) user's manual. Agron. Mimeo 81-35. Dep. of Agron., Cornell Univ., Ithaca, NY.

50. ----. 1984. Simple simulation models for yield prediction applied to alfalfa in the Northeast. Agron. J. 76:235-239.
51. ----, and David Onstad. 1983. Simple computer simulation models for forage management applications. p. 483-485. In J.A. Smith and V.W. Hays (ed.) Proc. 14th Int. Grassl. Congr., Lexington, KY. 15-24 June 1981. Westview Press, Boulder, CO.
52. Field, T.R.O. 1974. Analysis and simulation of the effect of temperature on the growth and development of alfalfa (Medicago sativa L.). Ph.D. thesis. Univ. of Guelph, Ontario, Canada (Diss. Abstr. Int. 35:4749-B).
53. ----, C.J. Pearson, and L.A. Hunt. 1976. Effects of temperature on the growth and development of alfalfa. Herb. Abstr. 46:145-150.
54. Foutz, L., W.W. Wilhelm, and A.K. Dobrenz. 1976. Relationship between physiological and morphological characteristics and yield of nondormant alfalfa clones. Agron. J. 68:587-591.
55. Frakes, R.V., R.L. Davis, and F.L. Patterson. 1961. The breeding behavior of yield and related variables in alfalfa. II. Associations between characters. Crop Sci. 1:207-209.
56. Francois, L.E. 1981. Alfalfa management under saline conditions with zero leaching. Agron. J. 73:1042-1046.
57. Frosheiser, F.I., and D.K. Barnes. 1973. Field and greenhouse selection for *Phytophthora* root rot. Crop Sci. 13:735-738.
58. Garza, R.T., R.F. Barnes, G.O. Mott, and C.L. Rhykerd. 1965. Influence of light intensity, temperature and growing period on the growth and chemical composition and digestibility of Culver and Tanverde alfalfa seedlings. Agron. J. 57:417-420.
59. Gengenbach, B.G., and D.A. Miller. 1972. Variation and heritability of protein concentration in various alfalfa plant parts. Crop Sci. 12:767-769.
60. Gist, G.R., and G.O. Mott. 1957. Some effects of light intensity, temperature, and soil moisture on the growth of alfalfa, red clover, and birdsfoot trefoil seedlings. Agron. J. 49:33-36.
61. Goldman, A., and A. Dovrat. 1980. Irrigation regime and honeybee activity as related to seed yield in alfalfa. Agron. J. 72:961-965.
62. Gosse, G., M. Chartier, and G. Lemaire. 1984. Physiologie vegetale—Mise au point d'un modele de prevision de production pour une culture de Luzerne. C. R. Acad. Sci. Ser. 3 298:541-544.
63. Grandfield, C.O. 1945. Alfalfa seed production as affected by organic reserves, air temperature, humidity, and soil moisture. J. Agric. Res. 70:123-132.
64. Greenfield, P.L., and Dale Smith. 1973. Influence of temperature change at bud on composition of alfalfa at first flower. Agron. J. 65:871-874.
65. Guitjens, J.C. 1982. Models of alfalfa yield and evapotranspiration. J. Irrig. Drain. Div., Am. Soc. Civ. Eng. 108:212-222.
66. Gutierrez, A.P., J.B. Christensen, C.M. Merritt, W.B. Loew, C.G. Summers, and W.R. Cothran. 1976. Alfalfa and the Egyptian alfalfa weevil (Coleoptera: Curculionidae). Can. Entomol. 108:635-648.
67. Guy, P., F. Blondon, and J. Durand. 1971. Action de la temperature et de la duree d'eclairement sur la croissance et la floraison de deux types eloignes de luzerne cultivee, *Medicago sativa* L. Ann. Amelior. Plant. 21:409-422.
68. Harada, I. 1975. Influence of temperature on the growth of three cultivars of alfalfa (*Medicago sativa* L.). J. Jpn. Soc. Grassl. Sci. 21:169-179.
69. Harding, S.C., and J.E. Sheehy. 1980. Influence of shoot and root temperature on leaf growth, photosynthesis and nitrogen fixation of lucerne. Ann. Bot. 45:229-233.
70. Harris, F.S. 1915. Effect of alkali salts in soils on the germination and growth of crops. J. Agric. Res. 5:1-53.
71. Hart, R.H., R.B. Pearce, N.J. Chatterton, G.E. Carlson, D.K. Barnes, and C.H. Hanson. 1978. Alfalfa yield, specific leaf weight, CO_2 exchange rate, and morphology. Crop Sci. 18:649-653.
72. Hayes, J.T., P.A. O'Rourke, W.H. Terjung, and P.E. Todhunter. 1982. A feasible crop yield model for worldwide international food production. Int. J. Biometeorol. 26:239-257.
73. Heichel, G.H., D.K. Barnes, C.P. Vance, and K.I. Henjum. 1984. N_2 fixation, and N and dry matter partitioning during a 4-year alfalfa stand. Crop Sci. 24:811-815.
74. Heinrichs, D.H. 1959. Germination in alfalfa varieties in solutions of varying osmotic pressure and relationship to winter hardiness. Can. J. Plant Sci. 39:384-394.
75. ----. 1972. Root-zone temperature effects on flooding tolerance of legumes. Can. J. Plant Sci. 52:985-990.
76. Hely, F.W., and M. Zorin. 1977. Influence of temperature and humidity on tripping of lucerne flowers. Aust. J. Agric. Res. 28:1015-1027.
77. Hesterman, O.B., and L.R. Teuber. 1981. Effect of photoperiod and irradiance on fall dormancy of alfalfa. Agron. Abstr. American Society of Agronomy, Madison, WI, p. 87.

78. ----, ----, and A.L. Livingston. 1981. Effect of environment and genotype on alfalfa sprout production. Crop Sci. 21:720-726.
79. Hoffman, G.J., E.V. Maas, and S.L. Rawlins. 1975. Salinity-ozone interactive effects on alfalfa yield and water relations. J. Environ. Qual. 4:326-331.
80. Holt, D.A., R.J. Bula, G.E. Miles, M.M. Schreiber, and R.M. Peart. 1975. Environmental physiology, modeling and simulation of alfalfa growth: I. Conceptual development of SIMED. Purdue Agric. Exp. Stn. Res. Bull. 907.
81. Ingvalson, R.D., J.D. Rhoades, and A.L. Page. 1976. Correlation of alfalfa yield with various index of salinity. Soil Sci. 122:145-153.
82. Janson, C.G. 1975. Irrigation of lucerne in its establishment season. N. Z. J. Exp. Agric. 3:223-228.
83. Jeney, C. 1972. The influence of air temperature on the growth and development of lucerne. Takarmany-bazis 12(2):19-36.
84. Kalu, B.A., and G.W. Fick. 1981. Quantifying morphological development of alfalfa for studies of herbage quality. Crop Sci. 21:267-271.
85. ----, and ----. 1983. Morphological stage of development as a predictor of alfalfa herbage quality. Crop Sci. 23:1167-1172.
86. Khatib, K.H., and M.A. Massengale. 1966. Effect of certain salts on germination of alfalfa and berseem clover seed. Prog. Agric. Arizona 18:21-23.
87. Kirkham, M.B., D.E. Johnson, Jr., E.T. Kanemasu, and L.R. Stone. 1983. Canopy temperature and growth of differentially irrigated alfalfa. Agric. Meteorol. 29:235-246.
88. Krogman, K.K., and E.H. Hobbs. 1977. Irrigation management of alfalfa for seed. Can. J. Plant Sci. 57:891-896.
89. Ku, S.B., and L.A. Hunt. 1973. Effects of temperature on the morphology and photosynthetic activity of newly matured leaves of alfalfa. Can. J. Bot. 51:1907-1916.
90. ----, and ----. 1977. Effects of temperature on the photosynthesis-irradiance response curves of newly matured leaves of alfalfa. Can. J. Bot. 55:872-879.
91. Leach, G.J. 1968. The growth of the lucerne plant after cutting: the effects of cutting at different stages of maturity and at different intensities. Aust. J. Agric. Res. 19:517-530.
92. ----. 1969. Shoot numbers, shoot size, and yield of regrowth in three lucerne cultivars. Aust. J. Agric. Res. 20:425-434.
93. ----. 1970. Growth of the lucerne plant after defoliation. p. 562-566. In M.J.T. Norman (ed.) Proc. 11th Int. Grassl. Congr., Queensland, Australia. 12-23 April. Univ. of Queensland Press, St. Lucia, Queensland.
94. ----. 1971. The relation between lucerne shoot growth and temperature. Aust. J. Agric. Res. 22:49-59.
95. Lee, C.-T., and Dale Smith. 1972. Influence of soil nitrogen and potassium levels on the growth and composition of lucerne grown to first flower in four temperature regimes. J. Sci. Food Agric. 23:1169-1181.
96. Lehman, W.F., and F.E. Robinson. 1979. Progress in developing salt tolerance in alfalfa. p. 73-75. In Proc. 9th California Alfalfa Symp. Agronomy and Range Science Extension, University of California, Davis.
97. Loper, G.M., and A.M. Lopioli. 1972. Photoperiodic effects on the emanation of volatiles from alfalfa (*Medicago sativa* L.) florets. Plant Physiol. 49:729-732.
98. Lucey, R.F., and M.B. Tesar. 1965. Frequency and rate of irrigation as factors in forage growth and water absorption. Agron. J. 57:519-523.
99. Maas, E.V., S.R. Grattan, and G. Ogata. 1982. Foliar salt accumulation and injury in crops sprinkled with saline water. Irrig. Sci. 3:157-168.
100. ----, and G.J. Hoffman. 1977. Crop salt tolerance—current assessment. J. Irrig. Drain. Div., Am. Soc. Civ. Eng. 103:115-134.
101. Macdowall, F.D.H. 1981. Photosynthesis and growth limitation in alfalfa. p. 81-86. In George Akoyunoglou (ed.) Photosynthesis VI. Photosynthesis and productivity, photosynthesis and environment. Balaban Int. Sci. Services, Philadelphia.
102. ----. 1982. Effect of light intensity and CO_2 concentration on the kinetics of 1st month growth and nitrogen fixation of alfalfa. Can. J. Bot. 61:731-740.
103. Mannetje, L.'t, and A.J. Pritchard. 1974. The effect of daylength and temperature on introduced legumes and grasses for the tropics and subtropics of coastal Australia. 1. Dry matter production, tillering and leaf area. Aust. J. Exp. Agric. Anim. Husb. 14:173-181.
104. Massengale, M.A., A.K. Dobrenz, H.A. Brubaker, and A.E. Bard, Jr. 1971. Response of alfalfa (*Medicago* spp.) to light interruption of the dark period. Crop Sci. 11:9-12.
105. Matches, A.G., G.O. Mott, and R.J. Bula. 1962. Vegetative development of alfalfa seedlings under varying levels of shading and potassium fertilization. Agron. J. 54:541-543.
106. McCree, K.J. 1972. Test of current definitions of photosynthetically active radiation against leaf photosynthesis data. Agric. Meteorol. 10:443-453.

107. McElgunn, J.D. 1973. Germination response of forage legumes to constant and alternating temperatures. Can. J. Plant Sci. 53:797-800.
108. ----, and D.H. Heinrichs. 1975. Water use of alfalfa genotypes of diverse genetic origin at three soil temperatures. Can. J. Plant Sci. 55:705-708.
109. McLaughlin, R.J., and B.R. Christie. 1980. Genetic variation for temperature response in alfalfa (*Medicago sativa* L.). Can. J. Plant Sci. 60:547-554.
110. McWilliam, J.R., R.J. Clements, and P.M. Dowling. 1970. Some factors influencing the germination and early seedling development of pasture plants. Aust. J. Agric. Res. 21:19-32.
111. Melton, B.A. 1972. Alfalfa seed production studies. N. M. State Univ. Agric. Exp. Stn. Bull. 597.
112. Miles, G.E., and R.M. Peart. 1975. Environmental physiology, modeling and simulation of alfalfa growth. IV. SIMED user's guide. Purdue Univ. Agric. Exp. Stn. Bull. 78.
113. Miller, D.A., J.P. Shrivastava, and J.A. Jacobs. 1969. Alfalfa yield components in solid seedings. Crop Sci. 9:440-443.
114. Munns, D.N., V.W. Fogle, and B.G. Hallock. 1977. Alfalfa root nodule distribution and inhibition of nitrogen fixation by heat. Agron. J. 69:377-380.
115. Murray, G.A. 1967. The relationship of light quality, duration, and intensity to vegetative and reproductive growth in alfalfa (*Medicago sativa* L.). Ph.D. thesis, Univ. of Arizona (Diss. Abstr. Int. 28:136-B).
116. Nelson, C.J., and Dale Smith. 1968. Growth of birdsfoot trefoil and alfalfa. II. Morphological development and dry matter distribution. Crop Sci. 8:21-25.
117. ----, and ----. 1969. Growth of birdsfoot trefoil and alfalfa. IV: Carbohydrate reserve levels and growth analysis under two temperature regimes. Crop Sci. 9:589-591.
118. Nittler, L.W., and G.H. Gibbs. 1959. The response of alfalfa varieties to photoperiod, color of light and temperature. Agron. J. 51:727-730.
119. ----, and T.J. Kenny. 1964. Induction of flowering in alfalfa, birdsfoot trefoil, and red clover as an aid in testing for varietal purity. Crop Sci. 4:187-190.
120. Norton, J.E. 1931. Irrigated alfalfa in Montana. Mont. Agric. Exp. Stn. Bull. 245.
121. Offutt, M.S. 1979. Effect of stand density on yield of alfalfa. Arkansas Farm Res. 28(5):14.
122. Onstad, D.W., and G.W. Fick. 1983. Predicting crude protein, in vitro true digestibility, and leaf proportion in alfalfa herbage. Crop Sci. 23:961-964.
123. ----, and C.A. Shoemaker. 1984. Management of alfalfa and the alfalfa weevil (*Hypera postica*): an example of systems analysis in forage production. Agric. Systems 14:1-30.
124. Pearce, R.B., R.H. Brown, and R.E. Blaser. 1968. Photosynthesis of alfalfa leaves as influenced by age and environment. Crop Sci. 8:677-680.
125. ----, and S.R. Lee. 1969. Photosynthetic and morphological adaptation of alfalfa leaves to light intensity at different stages of maturity. Crop Sci. 9:791-794.
126. Pearson, C.J., and L.A. Hunt. 1972a. Effects of temperature on primary growth of alfalfa. Can. J. Plant Sci. 52:1007-1015.
127. ----, and ----. 1972b. Effects of temperature on primary growth and regrowth of alfalfa. Can. J. Plant Sci. 52:1017-1027.
128. ----, and ----. 1972c. Effects of pretreatment temperature on carbon dioxide exchange in alfalfa. Can. J. Bot. 50:1925-1930.
129. Pedersen, M.W., G.E. Bohart, V.L. Marble, and E.C. Klostermeyer. 1972. Seed production practices. *In* C.H. Hanson (ed.) Alfalfa science and technology. Agronomy 15:689-720.
130. Penning de Vries, F.W.T. 1982. Phases of development of models. p. 20-25. *In* F.W.T. Penning de Vries and H.H. van Laar (ed.) Simulation of plant growth and production. Pudoc, Wageningen, Netherlands.
131. Perry, L.J., Jr., and K.L. Larson. 1974. Influence of drought on tillering and internode number and length in alfalfa. Crop Sci. 14:693-696.
132. Peterschmidt, N.A., R.H. Delaney, and M.C. Greene. 1979. Effects of overirrigation on growth and quality of alfalfa. Agron. J. 71:752-754.
133. Pielou, E.C. 1981. Usefulness of ecological models: a stock-taking. Q. Rev. Biol. 56:17-31.
134. Porter, T.K., and J.H. Reynolds. 1975. Relationship of alfalfa cultivar yield to specific leaf weight, plant density, and chemical composition. Agron. J. 67:625-629.
135. Pritchett, W.L., and L.B. Nelson. 1951. The effect of light intensity on the growth characteristics of alfalfa and bromegrass. Agron. J. 43:172-177.
136. Pulli, Seppo. 1980. Seedling year alfalfa population development as influenced by weed competition and density of establishment. J. Sci. Agric. Soc. Finl. 52:403-422.
137. ----. 1980. Growth factors and management technique in relation to the developmental rhythm yield formation pattern of a seeding year lucerne stand. J. Sci. Agric. Soc. Finl. 52:177-194.

138. Pulli, S.K., and M.B. Tesar. 1975. *Phytophthora* root rot in seeding year alfalfa as affected by management practices inducing stress. Crop Sci. 15:861–864.
139. Redmann, R.E. 1974. Osmotic and specific ion effects on the germination of alfalfa. Can. J. Bot. 52:803–808.
140. Regev, Uri, A.P. Gutierrez, and G. Feder. 1976. Pests as a common property resource: a case study of alfalfa weevil control. Am. J. Agric. Econ. 58:186–197.
141. Retta, A., and R.J. Hanks. 1980. Corn and alfalfa production as influenced by limited irrigation. Irrig. Sci. 1:135–147.
142. Rhykerd, C.L., Ruble Langston, and J.B. Peterson. 1959. Influence of light on the foliar growth of alfalfa, red clover, and birdsfoot trefoil. Agron. J. 51:199–201.
143. Roberts, R.H., and B.E. Struckmeyer. 1939. Further studies of the effects of temperature and other environmental factors upon the photoperiodic responses of plants. J. Agric. Res. 59:699–709.
144. Robinson, F.E. 1977. Sprinkler irrigation of alfalfa with saline water. p. 28–30. *In* Proc. 7th California Alfalfa Symp. Agronomy and Range Science Extension, University of California, Davis.
145. Robinson, G.D., and M.A. Massengale. 1968. Effect of harvest management and temperature on forage yield, root carbohydrates, plant density and leaf area relationships in alfalfa (*Medicago sativa* L. cultivar 'Moapa'). Crop Sci. 8:147–151.
146. Rodger, J.B.A., G.G. Williams, and R.L. Davis. 1957. A rapid method for determining winterhardiness in alfalfa. Agron. J. 49:88–92.
147. Rogers, V.E. 1974. The response of lucerne cultivars to levels of waterlogging. Aust. J. Exp. Agric. Anim. Husb. 14:520–525.
148. Rumbaugh, M.D., and D.A. Johnson. 1981. Screening alfalfa germplasm for seedling drought resistance. Crop Sci. 21:709–713.
149. ----, ----, and D.N. Rinehart. 1983. Stand density, shoot weight, and acetylene reduction activity of alfalfa populations subjected to field and greenhouse moisture gradients. Crop Sci. 23:784–789.
150. Russell, J.S. 1976. Comparative salt tolerance of some tropical and temperate legumes and tropical grasses. Aust. J. Exp. Agric. Anim. Husb. 16:103–109.
151. Salter, RoseMary, Bill Melton, Marvin Wilson, and Cliff Currier. 1984. Selection in alfalfa for forage yield with three moisture levels in drought boxes. Crop Sci. 24:345–349.
152. Sammis, T.W. 1981. Yield of alfalfa and cotton as influenced by irrigation. Agron. J. 73:323–329.
153. Sato, K. 1971a. Growth and development of lucerne plants in a controlled environment. 1. The effects of daylength and temperature on growth and chemical composition. Proc. Crop Sci. Soc. Jpn. 40:120–126.
154. ----. 1971b. Growth and development of lucerne plants in a controlled environment. 2. The combined effects of daylength and temperature before and after cutting on the regrowth process. J. Jpn. Soc. Grassl. Sci. 17:127–132.
155. ----. 1974. Growth and development of lucerne plants in a controlled environment. 3. The effects of photoperiod and temperature on the growth and anatomical features of photosynthetic tissues. Proc. Crop Sci. Soc. Jpn. 43:59–67.
156. ----, and M. Itoh. 1974. The effects of air and soil temperatures upon the growth and chemical composition of lucerne and red clover. J. Jpn. Soc. Grassl. Sci. 20:211–216.
157. Schreiber, M.M., G.E. Miles, D.A. Holt, and R.J. Bula. 1978. Sensitivity analysis of SIMED. Agron. J. 70:105–108.
158. Selirio, I.S., and D.M. Brown. 1979. Soil moisture-based simulation of forage yield. Agric. Meteorol. 20:99–114.
159. Sharratt, B.S. 1984. Influence of the winter and growing season environments on the production of alfalfa. Ph.D. thesis, Univ. of Minnesota, St. Paul (Diss. Abstr. Int. 45:2376-B).
160. Sheehy, J.E., and S.C. Popple. 1981. Photosynthesis, water relations, temperature and canopy structure as factors influencing the growth of sainfoin (*Onobrychis viciifolia* Scop.) and lucerne (*Medicago sativa* L.). Ann. Bot. 48:113–128.
161. Siginobu, K., and Y. Maki. 1979. Environmental effects of the expression of male sterility in alfalfa. Agric. Rev. and Manuals, USDA-SEA, p. 23.
162. Singh, Y., and J.E. Winch. 1974. Morphological development of two alfalfa cultivars under various harvesting schedules. Can. J. Plant Sci. 54:79–87.
163. Smith, Dale. 1969. Influence of temperature on the yield and chemical composition of 'Vernal' alfalfa at first flower. Agron. J. 61:470–472.
164. ----. 1970. Influence of temperature on the yield and chemical composition of five forage legume species. Agron. J. 62:520–523.
165. ----. 1970. Yield and chemical composition of leaves and stems of alfalfa at intervals up the shoots. J. Agric. Food Chem. 18:652–656.

166. ——, A.K. Dobrenz, and M.H. Schonhorst. 1981. Response of alfalfa seedlings to high levels of chloride-salts. J. Plant Nutr. 4:143–147.
167. ——, and B.E. Struckmeyer. 1974. Gross morphology and starch accumulation in leaves of alfalfa plants grown at high and low temperatures. Crop Sci. 14:433–436.
168. ——, and ——. 1977. Effects of high levels of chlorine in alfalfa shoots. Can. J. Plant Sci. 57:293–296.
169. Smith, E.M., and O.J. Loewer, Jr. 1983. Mathematical-logic to simulate growth of two perennial grasses. Trans. ASAE 26:878–883.
170. Snaydon, R.W. 1972a. The effect of total water supply, and of frequency of application, upon lucerne. I. Dry matter production. Aust. J. Agric. Res. 23:239–251.
171. ——. 1972b. The effect of total water supply, and of frequency of application, upon lucerne. II. Chemical composition. Aust. J. Agric. Res. 23:253–256.
172. Stanberry, C.O., C.D. Converse, H.R. Hiase, and O.J. Kelley. 1955. Effect of moisture and phosphate variables on alfalfa hay production on the Yuma Mesa. Soil Sci. Soc. Am. Proc. 19:303–310.
173. Stock, H.G. 1971. Die Wirkung von Temperatur, Bodenfeuchte und Windgeschwindigkeit auf das Wachstum von Rotklee und Luzerne. Arch. Acker Pflanzenbau. Bodenkd. 15:951–962.
174. Stone, J.E., D.B. Marx, and A.K. Dobrenz. 1979. Interaction of sodium chloride and temperature on germination of two alfalfa cultivars. Agron. J. 71:425–427.
175. Susuki, S., S. Inami, and Y. Sakurai. 1975. Influence of daylength and temperature on plant growth in the classified groups of lucerne cultivars. J. Jpn. Soc. Grassl. Sci. 21:245–251.
176. Taylor, S.A., J.L. Haddock, and M.W. Pedersen. 1959. Alfalfa irrigation for maximum seed production. Agron. J. 51:357–360.
177. Thompson, T.E., and G.W. Fick. 1981. Growth response of alfalfa to duration of soil flooding and to temperature. Agron. J. 73:329–332.
178. Townsend, C.E., and W.J. McGinnies. 1972. Temperature requirements for seed germination of several forage legumes. Agron. J. 64:809–812.
179. Triplett, G.B., Jr., and M.B. Tesar. 1960. Effects of compaction, depth of planting, and soil moisture tension on seedling emergence of alfalfa. Agron. J. 52:681–684.
180. Ueno, M., and Dale Smith. 1970. Influence of temperature on seedling growth and carbohydrate composition of three alfalfa cultivars. Agron. J. 62:764–767.
181. Uhvits, R. 1946. Effect of osmotic pressure on water absorption and germination of alfalfa seeds. Am. J. Bot. 33:278–285.
182. Ungar, I.A. 1967. Influence of salinity and temperature on seed germination. Ohio J. Sci. 67:120–123.
183. Vitkus, A.A., and N.A. Lapinskiene. 1977. Biological characteristics and chemical composition of *Medicago falcata*. 7. Growth and development. Liet. TSR Mokslu Akad. Darb. Ser. C (3):39–46.
184. Vough, L.R., and G.C. Marten. 1971. Influence of soil moisture and ambient temperature on yield and quality of alfalfa forage. Agron. J. 63:40–42.
185. Wahab, H.A., and D.S. Chamblee. 1972. Influence of irrigation on the yield and persistence of forage legumes. Agron. J. 64:713–716.
186. Walgenbach, R.P., G.C. Marten, and G.R. Blake. 1981. Release of soluble protein and nitrogen in alfalfa. I. Influence of growth temperature and soil moisture. Crop Sci. 21:843–849.
187. Weaver, J.E. 1926. Root development of field crops. McGraw-Hill Book Co., New York.
188. Wolf, D.D., and R.E. Blaser. 1971a. Photosynthesis of plant parts of alfalfa canopies. Crop Sci. 11:55–58.
189. ——, and ——. 1971b. Leaf development of alfalfa at several temperatures. Crop Sci. 11:479–482.
190. Woodman, H.E., and R.E. Evans. 1935. Nutritive value of lucerne. The leaf stem ratio. J. Agric. Sci. 25:578–597.
191. Younis, M.A., F.C. Stickler, and E.L. Sorensen. 1963. Reactions of seven alfalfa varieties under simulated moisture stresses in the seedling stage. Agron. J. 55:177–182.

6 Carbon Assimilation, Partitioning, and Utilization

G. H. HEICHEL

USDA-ARS
St. Paul, Minnesota

R. H. DELANEY

University of Wyoming
Laramie, Wyoming

H. T. CRALLE

Texas A&M University
College Station, Texas

Plants obtain C for growth and for generation of metabolic energy by the photosynthetic reduction of atmospheric carbon dioxide (CO_2). On a global basis, some 80 Gt (100 billion short tons) of C are annually deposited in plant material, two-thirds of this on land. About 5% of the annual CO_2 fixation is by agricultural plants which, by processes of assimilation, partitioning, and utilization, produce the food, feed, and fiber to feed, shelter, and clothe humankind.

The process of assimilation is the incorporation of randomly and often sparsely distributed components of the environment into the structure of complex molecules within the plant body. Assimilatory processes require some external form of energy to bond relatively unstructured and widely dispersed atoms, molecules, or ions into the highly structured, initial products of metabolism. Carbon assimilation, therefore, is the photosynthetic reduction of atmospheric CO_2 and its synthesis into the initial products of photosynthesis.

Partitioning encompasses the allocation, distribution, or transport of assimilates (molecules or ions) from their sites of synthesis (*sources*) to their sites of utilization (*sinks*). Partitioning involves active or passive movement of C in molecular or ionic form from organelle to organelle within a cell, from cell to cell within a tissue, from tissue to tissue within an organ, or from one organ to another within the plant (60).

Utilization encompasses the constructive benefits realized by a plant as a result of assimilation and partitioning. Carbon is utilized for plant structure and economic yield. Also, C is utilized beneficially and returned to the atmosphere by respiration using high-energy C forms to release

---

Copyright 1988 © ASA-CSSA-SSSA, 677 South Segoe Road, Madison, WI 53711, USA.
Alfalfa and Alfalfa Improvement—Agronomy Monograph no. 29.

energy for synthetic processes. In addition, C is essential in the formation of metabolites that are stored as reserves until needed for a variety of metabolic purposes. Mobilization is a form of partitioning that refers to the movement of an assimilate from a storage site to a new site of utilization.

This chapter addresses the assimilation, partitioning, and utilization of C by alfalfa (*Medicago sativa* L.). These processes will be examined on a leaf, organismal, and community basis to underscore recognized principles and voids in our understanding of the C economy of the crop—the balance sheet of C income and loss. No attempt has been made to summarize the excellent earlier version of this chapter (17), which should be consulted for a thorough analysis of literature before 1972, as well as important unpublished data. The chapter closes with a forecast of how and where future improvements in alfalfa C assimilation might occur.

6-1 LEAF CARBON DIOXIDE EXCHANGE

Photosynthesis (Pn), photorespiration (PR), and dark respiration (DR) are the component processes by which CO_2 is exchanged between leaves, the principal assimilatory organs of alfalfa, and the ambient environment. Rates of these component processes are influenced by environmental conditions during growth, environmental conditions at the time of measurement, genotype, and the anatomical and morphological characteristics of the plant.

6-1.1 Photosynthesis of Leaves

6-1.1.1 Key Metabolic Pathways

The CO_2 exchange characteristics of alfalfa are typical of C_3 (reductive pentose phosphate cycle) species. The cycle begins with the carboxylation of ribulose 1,5-bisphosphate (9) by ribulose 1,5-bisphosphate carboxylase/oxygenase (rubisco) located in the mesophyll cell chloroplasts (3, 72). The carboxylase and oxygenase activities of crystallized alfalfa rubisco are similar to those of other C_3 species (74). The rubisco photosynthetic activity is controlled by the CO_2/O_2 enzyme specificity factor and internal concentrations of CO_2 and O_2. Jordan and Ogren (75) reported an alfalfa specificity factor of 77 which is equivalent to that of C_3 plants and higher than that of some C_4 species. Alfalfa rubisco contains one large polypeptide subunit and two to three small subunits depending on the genotype (43). The rubisco, synonymous to Fraction I protein, accounts for 32 to 39% of the total protein (17) in alfalfa leaves.

6-1.1.2 Light and Carbon Dioxide Response

The flux of photosynthetically active sunlight limits Pn of alfalfa canopies. Although light saturation of Pn of young leaves occurs at approximately one-third to one-half of full sunlight, older leaves within a developing canopy are exposed to suboptimal photosynthetic photon flux densities (PPFD) (88, 108, 110, 164, 165). Shading of leaves within a canopy also reduces their photosynthetic capability if the leaves are subsequently subjected to higher irradiance, which may occur after grazing (69, 110, 164, 165). There is conflicting evidence on the extent of photosynthetic recovery after leaf shading.

Alfalfa leaves require 1 to 4.5 h of exposure to full sunlight to achieve 90% of maximum Pn following overnight darkness (24, 111, 113). The time needed to achieve the photosynthetic maximum is affected by temperature and genotype. The light compensation point, where CO_2 intake equals CO_2 output, for alfalfa leaves occurs at a PPFD approximately 1% of full sunlight (17, 69, 88, 164, 165).

The responses of alfalfa leaves to elevated CO_2 are associated with Pn, PR, leaf age, and carboxylation efficiency. Concentrations of CO_2 above ambient (ca. 340 μL L^{-1}) increase Pn of individual leaves (69, 79, 88) and yields of forage (10, 88, 99, 100). Baysdorfer and Bassham (10) observed that elevated CO_2 increased alfalfa growth in the seedling stage, but Pn/unit leaf area decreased after 5 weeks of growth. Along with increased growth, elevated CO_2 also decreased transpiration (99) and increased the water-use efficiency (100). The increase in Pn with elevated CO_2 is usually associated with a decrease in PR, which is consistent with the displacement of O_2 by CO_2 on active sites of rubisco. Flux of CO_2 decreases in aging leaves due to greater internal resistance to CO_2 transfer (69). Sheehy et al. (139) reported internal carboxylation efficiencies from 7.5 to 19.7 μg CO_2 m^{-2} s^{-1} and external efficiencies from 3.6 to 12.2 μg CO_2 m^{-2} s^{-1} for alfalfa grown in controlled environments.

6-1.1.3 Rates; Variation with Season and Ontogeny

The highest rates of Pn are usually reported for young, fully expanded leaves of plants grown in controlled environments; Pn rates of leaves of field-grown plants are frequently 25 to 30% less (Table 6-1). Despite the

Table 6-1. Alfalfa photosynthetic rates from plants grown in the field and in controlled environments.

Leaf source	Photosynthetic rate	References
	mg CO_2 m^{-2} leaf area s^{-1}	
	Field	
Whole plant	1.41–1.72	12
Upper 0.3 m of stem	0.12–0.96	46, 49, 58, 86
Young leaves	0.42–1.45	15, 107, 165
	Controlled environment	
Whole plant	0.53–2.28	109, 115, 138
Leaves (six nodes)	1.12	80
Young leaves	0.56–2.38	24, 25, 66, 69, 79, 101, 107, 108

occasional exception (107), the relatively low rates of field-grown leaves are probably attributable to the environmental history of the leaf during ontogeny, as well as to rate of leaf development and time of senescence.

Photosynthetic rates of individual plants expressed on a leaf area basis vary throughout the growing season. Delaney et al. (49) and Leavitt et al. (86) found that Pn was 37% lower during the hot July and August growth periods of Arizona compared to June (Fig. 6-1), and concluded the response to high temperatures may contribute to the "summer yield slump" observed in alfalfa. Baldocchi et al. (4) observed similar seasonal responses to temperature through three successive harvests in Nebraska.

In general, young fully expanded leaves have about double the Pn per leaf area of older green leaves within the canopy (15, 69, 80, 110, 164, 167). Older basal leaves remaining on the plant at harvest may be photosynthetically active and benefit alfalfa regrowth (84). Reduced Pn activity of aging leaves deep in the canopy is attributable to low irradiance. Photosynthetic rates of 26-d-old leaves remained high in a thinned stand in which leaves were somewhat better illuminated during ontogeny (165).

6-1.1.4 Nutrition

There is little information on the effects of mineral nutrition on Pn of alfalfa, although the general responses are undoubtedly similar to those in other species. Whether Pn responds to supplemental mineral nutrition depends upon the existing nutritional status of the plant at the time of treatment. Photosynthetic rates per plant and per leaf area respond to K fertilization (31, 114, 167), and high tissue K concentrations prolonged Pn during leaf aging (167). However, rubisco specific activity was apparently unrelated to leaf K concentration (31, 114). Potassium nutrition may be more important to long distance transport of photoassimilates than to leaf Pn. Response of Pn to S nutrition has not been observed (31, 44). The effects of N nutrition on Pn are confounded by light and specific leaf weight (12).

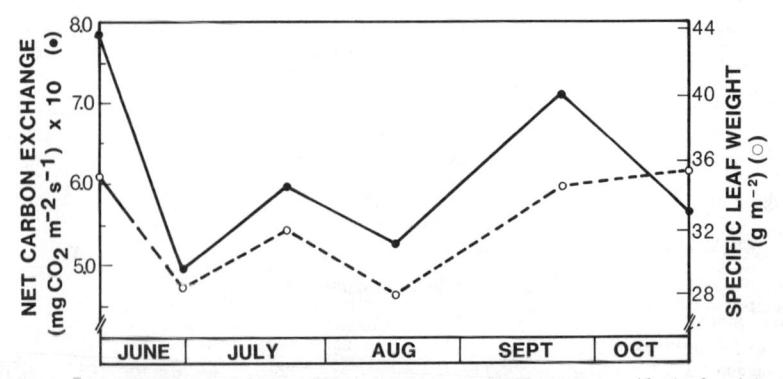

Fig. 6-1. Seasonal patterns of net carbon exchange rate (●) and specific leaf weight (○). Values are means of 12 genotypes. Adapted from Delaney et al. (49).

6–1.1.5 Genetic Variability

Considerable genotypic variation has been observed for Pn expressed per unit leaf area (46, 58, 66, 81, 108, 113, 115). However, limited attempts to improve Pn through genetic selection have not been successful. Many reports of genotypic differences in Pn show evidence of interactions with environmental conditions or with associated morphological characteristics.

The CO_2 assimilation of isolated chloroplasts increased with nuclear ploidy, and is apparently attributable to an increase of cell volume, chlorophyll content, and rubisco content (95, 97). The larger cells associated with increased nuclear ploidy and the potential for inbreeding depression may explain the inconsistent effect on Pn per area of leaf (118, 137).

6–1.2 Photorespiration

Photorespiration is thought to reduce phytomass accretion of C_3 species, although this accepted interpretation has been questioned. For this reason, considerable research has been conducted recently to characterize the environmental, physiological, morphological, and genetic characteristics related to PR of alfalfa.

6–1.2.1 Key Metabolic Pathway

The metabolic (glycolate) pathway begins with the oxygenase activity of rubisco in which phosphoglycolate is produced (13). Glycerate, a potential PR intermediate, reached maximum concentration in alfalfa after about 12 min of steady-state Pn (119).

The glycolate pathway involves the loss of CO_2 when two glycine molecules are metabolized to serine (122). These amino acids can be produced from 3-P-glycerate without PR (123). This alternate pathway in alfalfa is supported by observations (120) that serine and glycerate production were not reduced when high CO_2 was used to inhibit PR. The results indicated that the conversion of glycine to serine may not be a major source of photorespiratory CO_2 in alfalfa, but rather that CO_2 was primarily released by glycolate oxidation without serine formation.

Oxygen inhibits Pn in C_3 plants, and competes with CO_2 as a substrate for rubisco in alfalfa (79). The Pn of alfalfa at 20°C increased 27% at low O_2 (1.5%) compared to ambient O_2 concentration, while at 32°C the enhancement of Pn by low O_2 was 34%. The increase in PR with temperature, and subsequent decrease in Pn, was attributed to an increase in the O_2/CO_2 solubility ratio (79).

6–1.2.2 Rates; Variation with Season and Ontogeny

Photorespiration rates in alfalfa (Table 6–2) are generally measured by indirect methods with considerable variation reported. Most methods do not evaluate the refixation of photorespiratory CO_2 by Pn within the

Table 6-2. Rate of photorespiration of alfalfa leaves, and the percentage of net photosynthesis (Pn) consumed by photorespiratory processes.

Photorespiration		Reference
mg CO_2 m^{-2} s^{-1}	% of Pn	
0.20	20	69
0.09–0.39	13–39	86
0.23–0.46	22–31	79
0.07–0.21	40–60	116
0.19–0.56	25–28	138
0.12–0.48	10–50	81

leaf, which results in an underestimation of the PR rate. Despite the acknowledged problems of accuracy and precision of measurement, rates of PR are frequently two- to fourfold those of DR and 20 to 60% of apparent Pn.

Photorespiration responds to the same environmental and ontogenetic factors as Pn. This is illustrated by the close association of PR with Pn (69, 112, 138). For example, a highly significant correlation between PR and Pn over five harvests was observed (86). The decline of PR with leaf age was similar to that of Pn, and was temporarily reversed in aged leaves by partial defoliation (69).

The effect of temperature on PR of alfalfa is similar to that of other C_3 species. In general, PR increases with temperature and is consistently higher at 30 than 20°C (79, 81, 112). The linear increase in the PR/Pn ratio with temperature may indicate that the PR and Pn coupling is temperature dependent (112).

6–1.2.3 Genetic Variability

There is limited evidence of genetic variability for PR in alfalfa. Small-leafed 'Hayden' genotypes had higher PR per unit leaf area than large-leafed genotypes (86). Although Pn was also higher for the small-leafed plants, the PR/Pn ratio was larger for the small-leafed types. Foutz et al. (58) observed significant differences for PR expressed per leaf area and per plant in both spaced and densely planted genotypes. Likewise, genotypic differences in PR were reported by Peterschmidt and Dobrenz (117). One 'Saranac' and one 'Vernal' genotype exhibited different PR rates when subjected to various temperature conditions (81).

6–1.3 Dark Respiration

Dark respiration is essential to growth and maintenance, but inevitably results in a loss of C that might otherwise be devoted to economic yield. This has stimulated interest in the relationship of DR to growth and yield, and a search for genotypes that partition less fixed C to DR processes.

6-1.3.1 Key Metabolic Pathways

The major DR metabolic pathway for alfalfa is assumed to be catabolism of TCA cycle intermediates coupled to oxidative phosphorylation by electron transport in the mitochondria. Dark respiration by other pathways including the pentose phosphate shunt and anaerobic respiration are thought to be of minor importance to total C loss. Cyanide-insensitive respiration has been observed in alfalfa nodules (141), but its significance is not understood.

6-1.3.2 Mitochondrial Efficiency

The efficiency of DR in alfalfa can be discussed from two approaches, mitochondrial efficiency measured as ADP/O ratios, and Pn/DR ratios of whole plants. Schneiter et al. (133) observed ADP/O ratios ranging from 1.97 to 3.07 for mitochondria extracted from 5-d-old seedlings of alfalfa, and a range of 1.98 to 2.50 for etiolated tissues from seven field-grown genotypes (134). We have calculated the proportion of gross photosynthesis (Pn + DR) lost via DR from observations (49, 58, 86) on field-grown plants. Assuming active Pn for 60% of the growing season and a steady DR rate during night and day, DR ranged from 21 to 58% of (Pn + DR) with an extreme of 83% for one genotype during a high-temperature growth period. Pearson and Hunt (111) reported that total plant CO_2 loss by DR was 23 to 30% of net C intake for alfalfa in a controlled environment.

6-1.3.3 Rates; Variation with Season and Ontogeny

Individual leaf DR rates observed for field-grown alfalfa typically ranged from 0.04 to 0.19 mg CO_2 m^{-2} s^{-1} (49, 58, 86, 139). The DR rate is highly sensitive to temperature during both plant growth and experimental measurement, and affected by photosynthate supply, leaf ontogeny, and CO_2 concentration. The response of DR to temperature was sensitive to pretreatment temperature (112). The increase of DR with temperature gave a Q_{10} at 10°C of 2.7 and 4.5 for pretreatments of 20/15°C and 30/25°C, respectively. At 40°C the Q_{10} was 1.3 for both pretreatment temperatures. Others (101) found that DR was unaffected by pretreatment temperature between 0 and 20°C. Delaney et al. (49) and Leavitt et al. (86) observed large decreases in DR of field-grown plants as summer temperatures increased (Fig. 6-2). They attributed the reduction of DR to associated decreases in photosynthate production and perhaps to temperature inhibition.

Dark respiration consistently decreases as leaves age (138, 164), which would be expected as a consequence of reduced Pn and photosynthate supply. Rates of DR approached zero for yellow, senescent leaves in the lower canopy (164). This suggests that it may be possible to extend the maintenance of leaf area and the effective vegetative growth period of alfalfa without the bottom leaves becoming a sink for photosynthates.

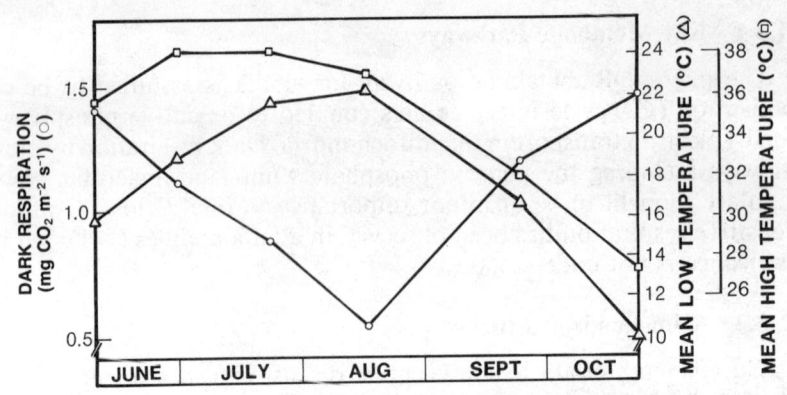

Fig. 6-2. Seasonal patterns of dark respiration rates (O) in relation to the mean daily high (□) and low (△) temperatures. Dark respiration rates were measured after 13 h of darkness and are means of 13 genotypes. Adapted from Delaney et al. (49).

Elevated CO_2 at night decreased DR with the greatest effect on maintenance respiration (127). A CO_2 concentration of 1200 μL L^{-1} increased the dry weight of seedlings by 22%. At ambient CO_2, maintenance respiration was 20% of total DR for tops and 34% for roots. Maintenance DR associated with the large alfalfa taproot may be a significant C loss component, especially in northern latitudes where the root system is maintained during long periods in the absence of herbage growth.

6-1.3.4 Genetic Variability

Various alfalfa clones and genetic lines exhibit significant differences in mitochondrial efficiency (92, 133, 134) while genotypic differences in leaf DR have been reported (45, 58, 111). However, genotypes with low DR typically have low Pn (58), so it is unclear whether DR and Pn are genetically independent.

6-1.4 Leaf Anatomical and Morphological Effects on Carbon Dioxide Exchange

6-1.4.1 Stomatal Characteristics

Alfalfa has a 25 to 40% higher stomatal frequency on the adaxial than abaxial leaf surface (23, 30, 34), with adaxial stomatal frequencies ranging from 217 to 266 mm^{-2}. Shading of leaves reduced stomatal frequency (34). A wide range of stomatal diffusive resistances to CO_2 has been reported for alfalfa (65, 69, 81, 113, 114, 140); environmental effects usually explain the great range in reported values. Increased stomatal diffusive resistance was associated with reduced Pn following exposure to low temperatures (81, 113). Irradiance and leaf age have little effect on stomatal diffusive resistance except under extreme conditions (69, 140).

6-1.4.2 Specific Leaf Weight

Specific leaf weight (SLW, leaf mass/leaf area), measures leaf thickness and density. Because of numerous reports of a positive correlation between SLW and Pn (47, 49, 66, 86, 108, 110, 165), SLW has been evaluated as one approach to improving Pn and forage yield. A diurnal effect was observed with SLW decreasing the first 1 to 4 h of the photoperiod when Pn was increasing (24, 27). The negative relationship between diurnal patterns of SLW and Pn of individual leaves has been attributed to product inhibition (24). Rate of photoassimilate translocation presumably lags behind Pn for most of the photoperiod.

A significant relationship is usually absent between SLW and forage yield per plant (58, 66, 121, 150). However, a positive relationship under controlled environmental conditions was found (25), and Chilcote et al. (28) reported a positive relationship when leaves from eight nodes were used to determine SLW. Hart et al. (66) observed a significant negative relationship between SLW and yield for widely spaced field-grown plants.

Specific leaf weight is sensitive to PPFD and decreases with depth into the canopy (6, 33, 49, 107, 108, 110, 151, 164, 165). Leaf ontogeny effects on SLW appear to be associated with PPFD and location on the plant rather than age alone. Leaves at the bottom of the canopy had a lower SLW before the leaves aged and were shaded (28).

Inconsistent effects of air temperature on SLW have been observed during leaf growth. High temperature resulted in lower SLW in Arizona (Fig. 6-1); SLW increased with temperature in Tennessee (121). The SLW was about 50% lower on plants grown at 30/25°C (D/N) compared with 20/15°C (80). In contrast, SLW was higher for plants grown under a constant regime of 35°C than at 20, 25, or 30°C (19). Thus, a diurnal temperature cycle may have important effects on leaf growth, SLW, and resultant correlations of SLW with Pn.

6-1.4.3 Leaflet Area

The area of individual leaflets is related to C uptake, forage yield, SLW and genotype (Fig. 6-3). Although leaflet size is highly influenced by environmental factors, relative genotypic differences (Fig. 6-4) are maintained throughout the growing season in both northern and southern latitudes (48, 55, 86).

The Pn per leaf area is usually lower for genotypes with large compared to those with small leaflet areas (47, 86). The negative relationship is attributable to the relatively higher SLW and smaller cell volume of small leaflets. However, leaflet size and SLW are probably independent traits (6, 150).

6-1.4.4 Genetic Variability

Genetic variability for stomatal frequency among cultivars and among individual plants has been identified. The stomatal frequency of a non-

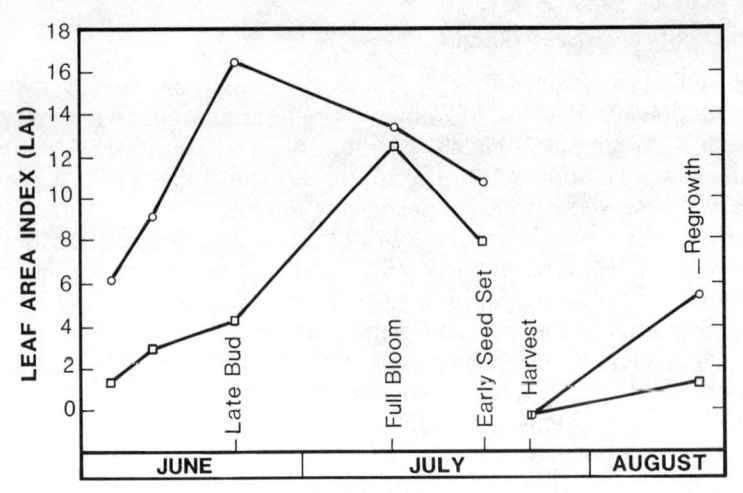

Fig. 6-3. Mean leaf area index (LAI) of six small (□) and six large (○) leaflet 'Ladak 65' genotypes at successive stages during the growing season. Data were collected from plants established in the field for 3 yr on 10-cm centers. Adapted from Delaney and Dobrenz (48).

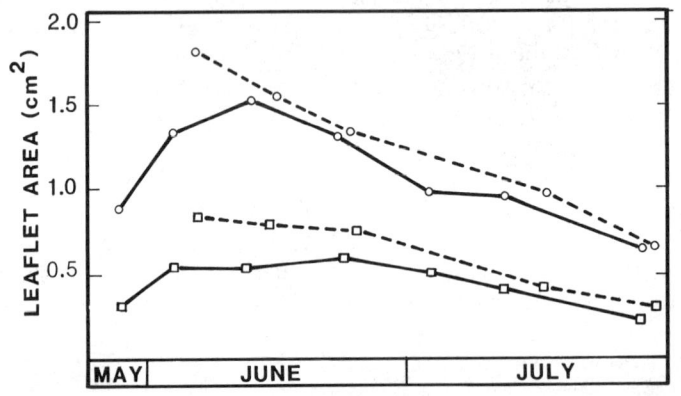

Fig. 6-4. Mean leaflet area of six small (□) and six large (○) leaflet 'Ladak 65' genotypes during 1978(---) and 1979(_____). The field-grown plants were at the bud stage of development after 12 June. Adapted from Delaney and Dobrenz (48).

dormant cultivar was 20 to 55% below that of a dormant-type cultivar (30). Carlson et al. (23) observed significant differences in stomatal frequency among five field-grown genotypes. However, they concluded that within-plant variability limited use of the trait in a breeding program. Hanscom and Ting (65) observed significant differences in the stomatal diffusive resistance among 15 alfalfa selections.

There is considerable genetic variation for SLW (6, 14, 28, 47, 58, 86, 121, 151) among and within cultivars. The inheritance of SLW is sufficient to expect progress in a breeding program (14, 150). Song and Walton (150) concluded that leaflet area would respond to selection. Leavitt et al. (86) observed an association between leaflet area and total Pn per plant. The increased total Pn by large-leaflet genotypes typically resulted in higher forage yields (48, 86). The rapid rates of C uptake of large-leaflet plants result from rapid leaf area development, which appears to be more important than SLW to total C assimilation.

6–2 ROOT SYSTEM CARBON DIOXIDE EXCHANGE

6–2.1 Carbon Metabolism and Respiration

6–2.1.1 Key Metabolic Pathways

Dark respiration (discussed above) and nonphotosynthetic CO_2 fixation by phosphoenolpyruvate carboxylase (PEPc) occur in both roots and nodules of alfalfa (2, 50, 91, 159). Dark respiration, the coupling of TCA cycle metabolites with oxidative phosphorylation, provides reduced pyridine nucleotides and adenosine triphosphate for synthetic activities such as root growth and nodule N_2 fixation. Nodule respiration may comprise 20 to 42% of total plant DR (96, 130).

Part of the CO_2 released by respiration of roots and nodules is assimilated into organic acids by PEPc. In roots and nodules, the organic acids so produced may replenish intermediates of the TCA cycle previously utilized in DR, or maintain pH gradients important for ion transport within cells and tissues (160). There is substantial evidence that the CO_2 fixed by root nodules provides C skeletons for assimilation of symbiotically fixed N_2 into amides, which are subsequently transported throughout the plant (91).

6–2.1.2 Rates of Respiration and Nonphotosynthetic Carbon Dioxide Fixation

Alfalfa nodules respire 5- to 10-fold faster and fix CO_2 about seven fold faster per unit mass at saturating CO_2 concentration than do roots (2, 50, 91, 159). On a nodulated root system basis, the relative distribution of respiratory and CO_2 fixation activities between roots and nodules depends upon the distribution of mass. In plants with 2% of their total phytomass devoted to nodules, nodule respiration may comprise 30 to 40% of the total root system respiration, and nodule CO_2 fixation a similar amount of total root system CO_2 fixation. Carbon balance calculations suggest that both organs fix about 20 to 40% of their gross (apparent respiration plus nonphotosynthetic CO_2 fixation) respiratory CO_2 efflux (2). These results have implications for the photosynthetic and energy costs of N_2 fixation, which have not yet accounted for the consequences of CO_2 recycling. Whether the CO_2 fixation phenomenon of nodules represents a C conservation process or consumes additional photosynthate remains to be investigated. Nevertheless, it is apparent that the C fixed by nodule PEPc may provide up to 25% of that needed for nodule ammonia (NH_3) assimilation of alfalfa (91, 160).

6-2.1.3 Rates; Variation with Season and Ontogeny

The temperature environment of the plant during growth, and during subsequent measurement have major influences on root respiratory activity (53, 93, 111, 152). Root respiration was steady during the light and dark period for cool (20/15°C) air temperatures, but at 30/25°C respiratory rates increased for 4 to 8 h during the light period and decreased throughout the dark period (112). Preferential use of photosynthate by tops at the high temperature may have caused this response. Rates of respiration ranged from 0.55 to 1.05 mg CO_2 kg^{-1} dry wt. s^{-1} during the light period and 0.39 to 0.72 mg CO_2 kg^{-1} dry wt. s^{-1} during the dark period. Temperatures that induce winter hardening increase root respiration (53, 152). In contrast to results from earlier experiments (93, 143), it has been shown that dormant cultivars have higher root respiration rates under hardening conditions than nondormant types (53, 142, 152). Root respiration rates are confounded by root mass. Larger roots have lower rates than smaller roots per gram of tissue (111).

6-3. CROP COMMUNITY CARBON DIOXIDE EXCHANGE

Phytomass production of alfalfa should be greatest when the daily CO_2 assimilation, the difference between shoot Pn and DR by the entire plant, is maximized. Efficient interception and utilization of photosynthetically active radiation (PAR) is crucial to high rates of CO_2 assimilation.

6-3.1 Canopy Structure Characteristics

Efficient use of sunlight on a crop community basis requires a leaf canopy with a high photosynthetic capacity that intercepts virtually all (>95%) the incident light, and a favorable sunlight distribution on leaves throughout the canopy. The leaf area index (LAI), and the leaf distribution, orientation, and light interception in the canopy may be important factors in efficient use of sunlight. We examine the role of each of these factors below.

6-3.1.1 Leaf Area Index

Leaf area index, the ratio of leaf surface to land surface, is a key determinant of alfalfa community CO_2 exchange and growth. In general, increased LAI leads to greater sunlight interception (Fig. 6-5) and canopy CO_2 assimilation (Fig. 6-6) in otherwise favorable environments.

In winterdormant and non-winterdormant cultivars, the LAI is usually greatest during the first one or two forage harvest intervals of the growing season and then decreases during successive harvests (49, 59, 73, 103, 129) with concomitant declines in herbage yield (20, 59, 129).

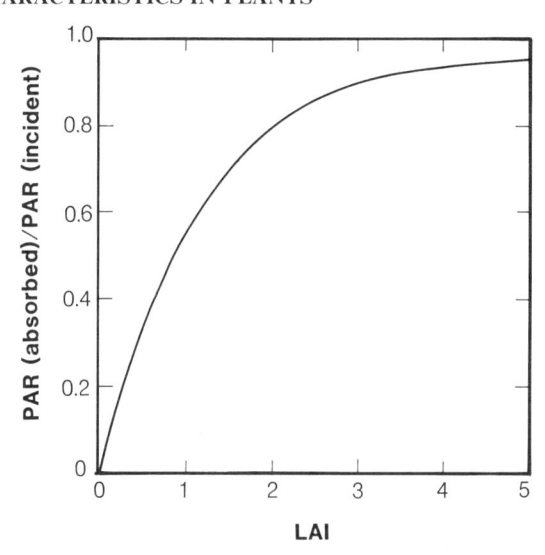

Fig. 6-5. Interception of photosynthetically active radiation (PAR) in relation to LAI for 'DuPuits' alfalfa. Measurements were taken over 325 d for spring growth following establishment, and for summer and autumn regrowths. Adapted from Gosse et al. (62).

Leaf area indices of developed alfalfa canopies are usually limited by foliar diseases and mutual leaf shading that leads to senescence and abscission of lower leaves. Early in the growing season, LAIs of 4 to 6 are common, with decreases to about 1 during the last harvest interval. In controlled environments (78) or in cool, irrigated field environments (48) LAI may reach 12 to 15.

6-3.1.2 Leaf Area Distribution

The distribution of leaf area within the crop canopy varies with growth stage of the crop. Vegetative canopies 2 weeks before herbage harvest have the largest concentration of leaf area in the upper stratum of the canopy with a corresponding decline through successively lower strata to the base of the plant (12, 168). At early to mid-flower, leaf area assumes an approximately normal distribution from the top to the base of the canopy (Fig. 6-7), with a pronounced bulge in the middle leaf strata of the crop (12, 20, 168). The change in leaf area distribution between vegetative and reproductive growth reflects the suppression of new leaf formation in the upper canopy by flowering, the increase in leaf drop in the lower canopy because of foliar diseases, and coincides with a reduction in crop growth rates (e.g., 102).

6-3.1.3 Leaf Orientation

Leaves of alfalfa track the sun by changes in leaflet angle from the vertical and by changes in azimuth or compass direction (136, 158). The angle and azimuth of leaflets in a closed canopy follow the sun throughout

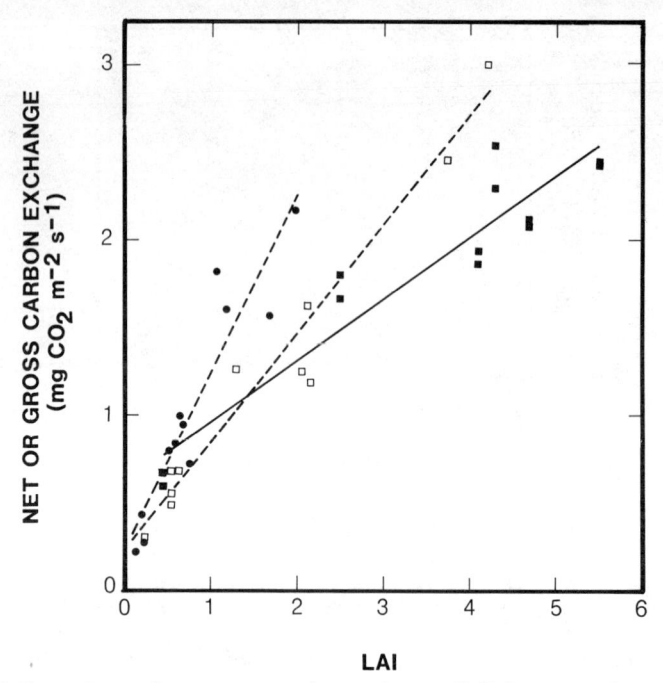

Fig. 6-6. Dependence of net or gross carbon exchange of alfalfa measured at ambient CO_2 concentration on LAI. Net carbon exchange during three harvest-growth cycles of vegetative canopies measured 2 weeks before harvest (●) or of reproductive canopies measured at 5% flower (□) with PAR = 850 μmol m^{-2} s^{-1} (unpublished data of Boller and Heichel, summarized in 12). Gross carbon exchange (■) during regrowth to flowering after herbage harvest, with PAR = 1400 μmol m^{-2} s^{-1}. Adapted from Sheehy et al. (140).

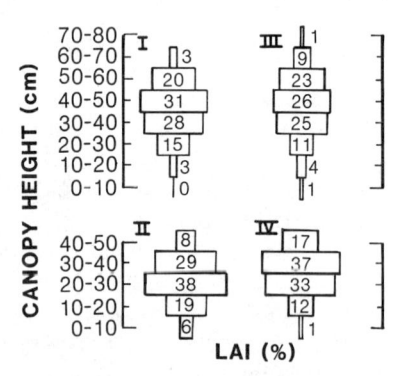

Fig. 6-7. Distribution of LAI in 10-cm strata of the alfalfa canopy on the last sampling date of the first (I) through fourth (IV) regrowth cycles after herbage harvest. Adapted from Bula et al. (20).

the day owing to light sensitization of a photoreceptor in the pulvinus of each leaflet (125). Leaflets deep within the canopy follow the sun more effectively than do upper leaves (158). Average leaflet angles of 35 to 50° from the horizontal have been measured in well-developed canopies (62, 136). However, leaflet angle may differ in upper than in lower canopy strata (153)

The solar-tracking phenomenon may contribute to canopy light interception except during water stress. Under water stress upper leaflet surfaces tend to fold together and align parallel to the incoming light, a phenomenon that may reduce the radiant energy load on the leaf and reduce photosynthesis. Although the solar-tracking phenomenon has become better understood, its importance to CO_2 assimilation and growth of alfalfa is not yet clear.

6-3.1.4 Light Interception

Closed alfalfa canopies reflect 10 to 12% of the incident PAR and up to 50% of the shortwave infrared radiation (139, 140). Light absorption varies with wavelength but on average 20% of the incident PAR and 10% of the shortwave infrared is absorbed by the leaf.

The attenuation of radiation within an alfalfa canopy owing to interception by herbage can be described by $I = Io \exp[-k \cdot LAI]$ (98), where I and Io represent the downward fluxes of radiation at the soil surface and above the canopy, respectively, and k is the extinction coefficient or slope of ln (I/Io) vs. LAI. The extinction coefficient is a function of leaf orientation (discussed above) and depth of the leaf canopy. Canopies with largely horizontal and regular mean leaf orientation have $k \geq 1$, while those with more vertical leaves have $k = 0.4$ to 0.7. See Brown and Blaser (15) for a further discussion of factors contributing to variation in k.

Changes in the extinction coefficient among successive herbage harvests measured with anthrazene benzene meters (73) or selenium photocells (162) have been interpreted as evidence that the effectiveness of light interception changed with growth stage and season. More recent evidence obtained with radiation sensors having the spectral response of photosynthesis (62) suggests that attenuation of PAR by foliage is primarily dependent upon LAI (Fig. 6-5) and only secondarily by growth stage or season. Nevertheless, there is some evidence (62, 73) that a unit of LAI is more efficient in intercepting PAR earlier than later in the growing season, an outcome which is consistent with measurements of Fuess and Tesar (59) beneath a canopy subjected to a three or four harvest cutting schedule. Changes in light interception during the season owing to variation in leaf/stem ratio among regrowths, or to variation in stem and leaflet angles among cultivars with contrasting winterdormancy characteristics, have not been investigated.

6-3.2 Canopy Photosynthesis

Relatively little information is available on the CO_2 exchange of alfalfa crop communities. The classic series of papers by Thomas and Hill (154, 155, 156) remain the benchmark investigations, and the chapter by Brown et al. (17) should be consulted for a thorough review of this early work. Recent experiments in the USA and Europe are consistent with the classical work in showing that CO_2 concentration, PPFD, and leaf area are the main factors limiting alfalfa community Pn in otherwise favorable environments.

6-3.2.1 Carbon Dioxide, Light, and Leaf Area Index

Photosynthesis of alfalfa canopies increases with CO_2 concentration between ambient and about 4 g CO_2 m^{-3} (Fig. 6-8), which provides clear evidence that CO_2 supply to the crop limits dry matter accumulation. Over the long term, some CO_2-induced increases in alfalfa photosynthesis and productivity might be anticipated from the slowly increasing concentrations of CO_2 in the atmosphere (77). There are no practical means to increase photosynthesis by artificially increasing CO_2 concentrations in the field.

The daily course of canopy Pn closely follows the diurnal pattern of solar radiation (4, 17, 132, 154, 155). Canopy Pn peaks about solar noon and is dependent upon leaf area development of the crop, absorption of PAR, and temperature. In alfalfa crops with well-developed leaf canopies, light compensation of Pn (the equilibrium between CO_2 uptake and efflux) occurred at about 5% of full sunlight (62). Rate of canopy Pn increased with PPFD to about 0.8 full sunlight (Fig. 6-9), when maximum rates of Pn were achieved at LAIs of about 5 to 6 (5, 62, 132, 162).

Rate of Pn of alfalfa canopies measured at a constant PPFD increased until a LAI of 5 to 6 was achieved (12, 140) when about 95% of incident PAR was absorbed (Fig. 6-5 and 6-6). Leaf area indices above those needed for about 95% interception of PPFD neither increased nor decreased photosynthesis because of changes in leaf illumination and respiratory activity within the crop canopy. High values of LAI result in internal shading as well as cooler temperatures in the lower canopy strata. This causes the canopy Pn to plateau with increased LAI. Since respiration of lower leaves declines in proportion to photosynthetic capacity,

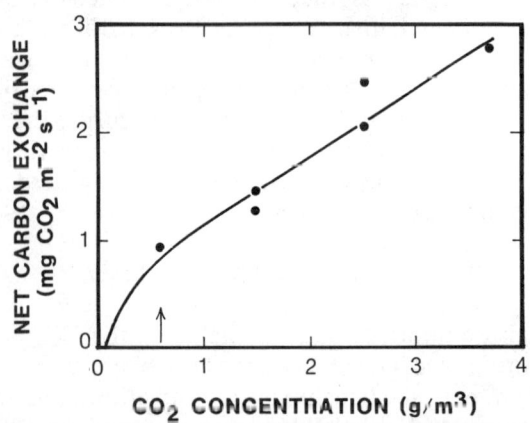

Fig. 6-8. Response of net carbon exchange of alfalfa canopies at high PAR to multifold increases in atmospheric CO_2 concentration above ambient, designated by arrow. Adapted from Thomas and Hill (156).

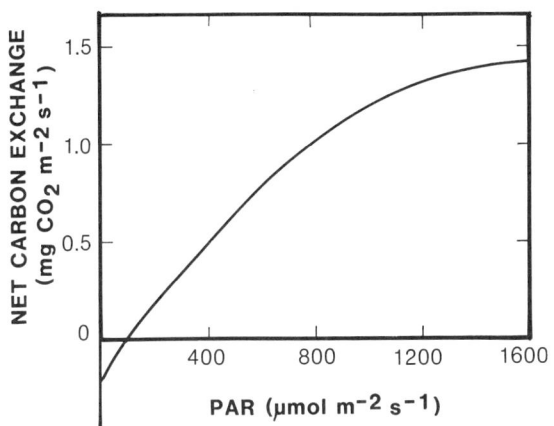

Fig. 6-9. Response of net carbon exchange of alfalfa canopies to PAR at ambient CO_2 concentration. Adapted from Wilfong et al. (162) for LAI >2.7, from Baldocchi et al. (5) for LAI >2.4, and from Gosse et al. (61) for several regrowths.

the lower shaded leaves do not become parasitic, or a source of net carbohydrate drain, upon the plant (162).

6-3.2.2 Variation with Season and Ontogeny

Canopy Pn varies with time over the growing season because of periodic leaf area removal by harvesting, seasonal changes in temperature, and photoperiod (154, 155, 156). Within a regrowth cycle following herbage removal, daily canopy photosynthesis increases with the increase in leaf area (140). Although the ratio of leaf to nonleaf tissue varies with regrowth, it is likely that petioles and stipules contribute no more than 2% of the net CO_2 assimilation of the crop canopy. Because of the unfavorable exposure to PAR, stems are probably net sinks for photosynthate (164).

6-3.2.3 Canopy Respiration

Dark respiration of alfalfa includes both that of shoots and roots. Shoot DR increases with LAI during a regrowth cycle (Table 6-3), with daily shoot respiration ranging from 31 to 67% of the daily gross photosynthesis. Respiration of alfalfa roots (uncorrected for soil respiration) varies with stage of herbage regrowth, and is strongly dependent upon the photosynthetic activity of the shoots (156). On a land area basis, root respiration of alfalfa ranged from 70 to 145% of the respiration of shoots across growth stages and stands of various ages. Shoot plus root respiration was 30 to 50% of the gross photosynthesis of the crop over the growing season.

6-4 CARBON DIOXIDE EXCHANGE AND YIELD

Extensive interest in the CO_2 exchange-yield relationship in alfalfa resulted from a search for new tools to improve forage yield. A positive

Table 6-3. Leaf area index (LAI), gross photosynthesis of shoots and shoot respiration of 'Europe' alfalfa during a regrowth cycle following harvest. Adapted from Woodward and Sheehy (168).

Characteristic	Days after herbage harvest					
	1	26	41	55†	69‡	77
LAI	0.45	2.2	4.3	5.9	4.5	4.3
Daily gross photosynthesis (g CO_2 m^{-2} d^{-1})	15	39	88	102	101	82
Daily shoot respiration (g CO_2 m^{-2} d^{-1})	9	26	21	31	36	39
Respiration/photosynthesis	0.60	0.67	0.24	0.30	0.36	0.48

† Early bud.   ‡ First flower.

relationship of CO_2 assimilation to yield might be anticipated because of the high harvest index (grams of yield/grams of phytomass) of alfalfa as a forage crop. Evidence of product inhibition of CO_2 assimilation raises the counter argument that translocation and not Pn capability is a limiting factor for phytomass accumulation in alfalfa (24). Conversely, yield responses to CO_2 enrichment support the hypothesis that Pn limits growth (88), especially in the seedling or early regrowth stages of plant development (10).

6.4.1 Mitochondrial Respiration

Mitochondrial efficiency measured as the ADP/O ratio is related to forage yield (92, 133, 134). Low mitochondrial respiration is related to a high yield but is of less predictive value for genotypic selection than are ADP/O ratios (92).

6-4.2 Leaf Carbon Dioxide Exchange

An association between Pn per unit leaf area and forage yield has not been observed (25, 46, 58, 66, 86, 138); the same is true for DR (46, 57). The lack of association is probably due to plants with a low CO_2 exchange per leaf area having a high leaf area per plant, and vice versa.

6-4.3 Plant Carbon Dioxide Exchange

A consistent association has been observed between forage yield and total Pn per plant (46, 57, 58, 86, 138); total DR per plant (46, 58, 96), and total PR (58). Total (aerial) plant C exchange accounts for the cumulative effects of rate per unit leaf area, leaf age, leaf shading, leaf area per plant, and C exchange of other aboveground plant parts. Selection of genotypes with improved total CO_2 fixation of shoots may result in accumulation of genes for yield improvement. However, this approach would not accommodate genetic variation for partitioning to roots, or CO_2 losses by root respiration.

6–5 PHOTOSYNTHATE PARTITIONING AND UTILIZATION

6–5.1 Source-Sink Concepts

Photosynthate assimilation and partitioning within plants are functions of a complex system of sources and sinks. Sources are plant parts that exhibit a net export of assimilates, while sinks are parts that exhibit a net import of these assimilates. This movement of photosynthate involves loading of sucrose at the source into the sieve tubes of the phloem, translocation in the phloem, and phloem unloading at the sink.

Source-sink relations in alfalfa are highly dynamic. During most of the growing period, the principal source of C assimilates is recent photosynthate produced by the fully expanded leaves of the shoot. The principal sinks are the meristematic regions of the plant, the N_2-fixing nodules, and the carbohydrate storage regions in the roots and crown. However, during early regrowth following harvest or winter dormancy, the major source of C assimilates is the carbohydrates stored in the crown and roots. During this time the shoots are the principal sinks. Thus, source-sink relations in alfalfa involve the partitioning of both recent photosynthate and stored carbohydrates as influenced by stage of plant growth and environmental factors.

6–5.2 Partitioning of Recent Photosynthate Among Organs

6–5.2.1 Time Course and Diurnal Variations

The partitioning of photosynthate from source to sink has temporal and spatial characteristics. The time course of photosynthate export from source leaves and import by sink organs has been measured during specific periods. Export of photosynthate in glasshouse-grown alfalfa varied with time following labeling with $^{14}CO_2$, leaf position, and stage of plant growth (36, 39, 41). In these experiments the proportion of ^{14}C exported by individual source leaves ranged from 45 to 70%. The lowermost leaf (LL) on the main stem (MS) required 12 h for completion of ^{14}C export, while the unshaded uppermost fully expanded leaf (TL) on the MS required only 3 h. However, the LL exported a greater proportion of ^{14}C assimilates at most growth stages than did the TL. These differences likely reflect a positive relationship between rates of Pn and C export, and a greater carbohydrate requirement of younger leaves for their continued activity. Reduced export of ^{14}C by the LL from 70% before herbage removal to 50% 2 d following harvest suggests a positive relationship between photosynthate export by leaves and sink demand in alfalfa. The pattern of interorgan partitioning of ^{14}C from the TL was similar when labeling occurred at the onset and end of the photoperiod (39). However, it required a period of 24 to 48 h following exposure of $^{14}CO_2$ for final structural incorporation into target organs of ^{14}C from the TL or LL. For example, the percentage of total plant radioactivity declined from 88 to

23% in the MS and increased from 3 to 47% in the root during 48 h of measurements.

Increased concentrations of reducing sugars (1), total sugars (94, 147), water-soluble carbohydrates (71, 87, 147), and total nonstructural carbohydrates (42, 87, 94, 147) in herbage have been observed during daylight. Sucrose and starch concentration of leaves and sucrose concentration of stems also increased during the light period and decreased during darkness (87). These carbohydrate analyses and the foregoing ^{14}C labeling studies indicated that C was photosynthetically assimilated more rapidly than it was translocated. Hence, the starch content of the leaves increased during the light period as a temporary storage pool for photoassimilates.

6-5.2.2 Effect of Leaf Position

Source-to-sink distance is an important factor affecting photosynthate partitioning in many species. The generalization that source-sink proximity favors partitioning and remoteness limits partitioning must be applied with caution to alfalfa. Studies of photosynthate partitioning have not demonstrated a consistent relationship between source-sink distance and the pattern of partitioning (36, 39, 40, 70).

The influence of source-sink distance on photosynthate partitioning was mediated by stage of plant growth (36, 41). At first flower and at 10 d following herbage removal, the patterns of ^{14}C-labeled photosynthate partitioning to various organs were similar from the TL and LL on the MS. However, the patterns on Day 22 following harvest were different and clearly related to sink proximity. At this stage, the LL was the primary source of photosynthate for young shoots growing from crown buds, the crown, and the roots. The TL was the main source for axillary shoots growing from buds on the MS and the unexpanded leaves on the MS.

There was similar partitioning of photosynthate from upper and lower source leaves on the MS to nodulated roots of seedlings at first flower and of more mature plants at flowering (36, 39, 40). This similarity in photosynthate distribution to below-ground organs from these leaf positions, despite differences in source-sink distance, indicated a different pattern of partitioning in alfalfa than that observed in many annual species. The lower leaves in annuals (32, 105, 106, 157) were the primary source of photoassimilates for the roots.

However, the observations on alfalfa agreed with those from studies of other perennial forage legumes (131). The substantial partitioning of photosynthate from the upper leaves of alfalfa and other perennial forage legumes to the nodulated roots may be explained by the carbohydrate requirements of continued root growth in these species, and by the reduced leaf area and Pn activity of the lower leaves in a closed canopy.

6-5.2.3 Effect of Development and Interorgan Competition

The pattern of photosynthate partitioning changes with plant development (11, 36, 41, 109, 154). Hence, the pattern during vegetative regrowth following herbage removal is different from that during flowering or fruit development (Table 6-4). Allocation of ^{14}C from the middle leaf (ML) was greater to the organs of the regrowing MS than to the MS at flowering or fruiting. Conversely, the allocation was less to the other crown shoots, the crown, and the roots during regrowth than at other stages (Table 6-4) (109, 163). These results reflect the high C requirement of the organs of the MS during vigorous vegetative regrowth, the less rapid growth of the crown and root during this period, and the independence of the other crown shoots from the leaves of the MS.

Leaves remaining on stubble following herbage harvest may be an important source of C during the early phase of vegetative regrowth. Hodgkinson et al. (70) and Cralle (36) found shoots from axillary buds remaining on the cut stem to be the main sinks for ^{14}C from labeled lower leaves following harvest.

Since there was no difference between the flowering and fruiting periods in allocation of photosynthate to any organ except to seed pods (Table 6-4), the pods apparently do not receive current photosynthate largely at the expense of any one sink or set of sinks. This pattern is different from that found in many annual legumes. The allocation of photosynthate to nodules, roots, and vegetative shoots declined during seed pod development in annual legumes (67, 82, 83). This difference may be explained by the relatively small proportion (about 5%) of the total dry mass contained by the pods of alfalfa compared with those of annual legumes.

Table 6-4. Partitioning of ^{14}C-photosynthate from the middle leaf (ML) on main (longest) stem, measured as relative specific activity (RSA), to organs of alfalfa at three stages of development. Values are means of 12 replicates. Adapted from Cralle (36).

Organ†	First flower	Fruiting	Regrowth following herbage harvest
		RSA‡	
UL on MS	4.8b*	3.9b	26.7a
L on MS	0.1b	0.2b	0.5a
ASA-ML	8.0b	6.1b	28.4a
ASMS	2.4b	2.0b	11.1a
MS	2.6b	3.2b	13.0a
Pods on MS	--	7.8	--
Pods on OS	--	0.1	
OS	0.4a	0.4a	0.1b
Nodules	1.2ab	2.5a	1.0b
Crown	1.2a	1.2a	0.4b
Roots	1.2a	1.0a	0.4b

* Values between columns followed by the same letter are not significantly different at the 5% level (HSD).
† Organ abbreviations: MS = main (longest) stem; UL = unexpanded leaves plus shoot apex on MS; L = fully expanded leaves on MS (except the labeled ML); ASA-ML = the axillary shoot growing from the bud adjacent to the ML; ASMS = other axillary shoots growing from buds on the MS; OS = other shoots growing from crown buds.
‡ RSA = [dpm/dry mass (organ)]/dpm/dry mass (whole plant)].

Nodulated roots imported more photosynthate than did those without nodules (22). The quantities of photosynthate partitioned to the nodules on a unit mass basis, based on individual leaf (36, 39, 40, 41) and whole shoot (11) labeling experiments, were similar to those partitioned at most growth stages to the crown and roots, but less than that partitioned to young, unexpanded leaves. However, alfalfa nodules retained < 5% total plant radioactivity exported by TL, LL, or ML leaves (36, 39, 40, 41) and, therefore, accumulated little total photosynthate compared to other organs. Alfalfa populations developed during two cycles of recurrent selection for low, intermediate, and high nodule mass did not differ in the pattern of photosynthate partitioning from the ML to any organ on the basis of ^{14}C per unit mass or percentage of total plant radioactivity (41). Thus, the relatively low requirement of alfalfa nodules for recent photosynthate would not detract from the development of alfalfa populations for greater N_2-fixation capability, by recurrent selection for increased nodule mass, without a corresponding reduction in herbage yield or profound alteration in C assimilation.

6–5.3 Storage, Mobilization, and Utilization

6–5.3.1 Storage Forms and Organs

Sugars in alfalfa such as glucose, fructose, and sucrose exist principally as intermediates in the synthesis and degradation of starch (161). Carbon is primarily transported in the form of sucrose in the phloem sieve tubes. The main pathway of sucrose synthesis in plants is:
 1. Uridine triphosphate (UTP) + glucose-1-phosphate = uridine diphosphoglucose (UDPG) + pyrophosphate.
 2. UDPG + fructose-6-phosphate = sucrose-6-phosphate + uridine diphosphate (UDP).
 3. Sucrose-6-phosphate + H_2O = sucrose + $H_2PO_4^-$.

These reactions are enzymatically catalyzed by a pyrophosphorylase, sucrose phosphate synthetase, and a phosphatase. An alternate pathway of sucrose synthesis is catalyzed by sucrose synthetase: UDPG + fructose = sucrose + UDP + H^+. Sucrose can be degraded by reactions catalyzed by invertases (sucrose + H_2O = glucose + fructose), or the reversal of the sucrose synthetase reaction given above.

Starch is the major storage carbohydrate in alfalfa roots. Typical analyses of total nonstructural carbohydrates (TNC) in roots revealed a composition of 90% starch and 10% sugars (149, 161). However, sucrose became the major carbohydrate fraction during periods of drought, the onset of cold hardening, and early spring regrowth following winter dormancy (103).

Starch consists of unbranched (amylose form) and branched (amylopectin form) chains of D glucose units connected in α-1,4- and α-1,6-linkages. Starch synthesis occurs in a reversible reaction catalyzed by starch phosphorylase: glucose-1-phosphate + $(starch)_n \rightleftharpoons (starch)_{n+1}$ +

CARBON CHARACTERISTICS IN PLANTS

H_2PO_4. It also occurs in an irreversible reaction catalyzed by ADPG- or UDPG-starch transglucosylase: Adenosine diphosphoglucose (ADPG) or UDPT + (starch)$_n \rightarrow$ (starch)$_{n+1}$ + ADP or UDP.

A number of enzymes participate in starch degradation. Amylases hydrolyze amylose to glucose and maltose. Maltase hydrolyzes maltose to glucose. Starch phosphorylase converts starch to glucose-1-phosphate. The actions of amylases and starch phosphorylase only partly degrade amylopectin. These enzymes convert amylopectin into dextrin. The branched 1,6-linkages in dextrin are hydrolyzed by a debranching enzyme and by dextrinases into units which can be completely degraded by amylases or starch phosphorylase.

The accumulation of TNC is localized in various tissues, and regions within a tissue. Most TNC storage occurred in the upper 10 cm segment of the root and less progressively downward (54). Ueno and Smith (161) found 55% of the TNC in the root wood, 20% in the root bark, and 25% in the crown.

6-5.3.2 Variation with Season and Environment

The seasonal pattern of carbohydrate accumulation and depletion for unharvested and thrice harvested alfalfas appears in Fig. 6-10. Warming temperatures in spring promote vegetative growth of shoots from crown buds. This growth is initially supported by TNC in the roots and crown. Once the expansion of leaf area permits photosynthetic C assimilation to exceed the needs of shoot growth and whole plant respiration, the decline in TNC ceases and the accumulation in the roots and crown commences. The accumulation of maximum TNC in the roots occurs between 10% flower and full bloom (35, 128).

The small decline in TNC concentrations following 10% flower has

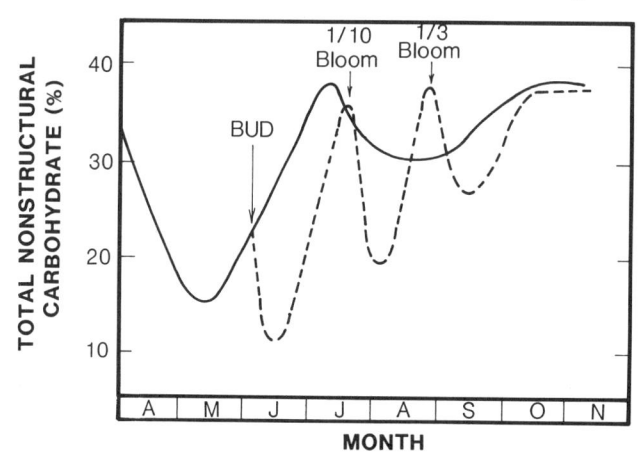

Fig. 6-10. Seasonal trends of total nonstructural carbohydrates in roots of 'Vernal' alfalfa with three-cut management imposed at bud, 1/10, and 1/3 bloom (---), and without cutting (———). Adapted by Brown et al. (17) from Smith (144).

been attributed to the activity of two sinks. Dobrenz and Massengale (51) suggested that the decline is a consequence of utilization by developing fruit. In support of this interpretation, they cite a decrease in content of root TNC coincident with fruit development, and a positive correlation between TNC and the following indices of fruit production: number of pods per stem, number of seeds per pod, and percent pod set. Other authors (29, 52) reported similar results. Conversely, Brown et al. (17) suggested that the postflowering decline in root TNC concentration was a consequence of the growth of new crown buds. This interpretation is favored on the basis of source-to-sink distance. The two interpretations are not mutually exclusive and await further investigations.

Temperature is a critical environmental factor affecting carbohydrate storage. As temperature decreases in autumn, shoot growth is slowed and carbohydrates accumulate in the root. Although one study (7) found no significant difference in root TNC concentrations between cool (16°C) and warm (30°C) environments, several investigations (56, 104, 129, 145, 146) have demonstrated that warm temperatures prevent and cool temperatures enhance storage of TNC in roots. Temperature also affects the chemical composition of TNC in roots. As temperatures decline during late fall and winter, starch is converted to sugar in the root (21, 76).

Although the effects of other environmental factors on carbohydrate accumulation in alfalfa have received little attention, some evidence suggests the importance of light intensity, water, and mineral nutrition. While reduced light intensity did not decrease root TNC concentration, it did decrease proportionately both quantity of TNC and root growth (90). The rate of TNC accumulation in roots of water-stressed alfalfa plants exceeded the rate of that in nonstressed plants (29). This result may be a consequence of reduced sink competition of the slow-growing shoots of the water-stressed plants. Conversely, the concentration of TNC in alfalfa roots declined in response to K deficiency (89, 126). However, Barta (8) found that low soil K concentrations reduced ^{14}C partitioning to nodules within 1 h of labeling, but not after 24 h. Nitrogen fertilization had no effect on the TNC concentration of roots (89).

6-5.3.3 Variation with Herbage Removal

Many studies have demonstrated that carbohydrates stored in roots and crowns of alfalfa support vegetative regrowth following herbage harvest. Evidence for this utilization includes the regrowth in darkness of alfalfa shoots (36, 63), the depletion of TNC in crowns and roots after harvest (7, 26, 35, 37, 54, 56, 64, 128, 129, 144, 161, 164), and the mobilization of stored ^{14}C from crowns and roots to new shoots during regrowth (36, 68, 109, 148, 149). When ^{14}C labeled starch was injected into roots 3 d after harvest, 67% was partitioned to leaves and stems within 48 h (36).

The characteristic pattern of carbohydrate accumulation and depletion in the roots and crown (Fig. 6-10) occurs in response to herbage removal and subsequent regrowth (166). Although the decline in stored carbohydrate of the roots and crown measured as depletion of TNC or ^{14}C-labeled carbohydrates was greatest during the early phase of vegetative regrowth (149), bidirectional transport of C compounds still occurs. Partitioning of photosynthate from leaves to the root was observed within the 1st week following harvest (36, 68, 109). Despite this partitioning, the decline in the concentration of root TNC during early vegetative regrowth was accompaned by a reduction of root cambial activity with a consequent cessation of root growth (124). Root cambial activity and growth resumed only with the recovery in TNC concentration (124).

The magnitude of the decline in carbohydrate reserves following harvest depends upon the physiological condition of the plant and the proportion of leaf area removed. Robison and Massengale (129) found a greater decline in root TNC concentrations after harvest at 50% flower than at 10% flower. When alfalfa plants were shaded to retard shoot growth after harvest, less ^{14}C-labeled compounds were partitioned from the roots to the shoots than in unshaded plants (68). This result indicated a close relationship between rate of shoot growth and rate of carbohydrate utilization. There may be intershoot competition for these reserves and for recent photosynthate after herbage harvest. Individual shoots on plants with only one or two shoots were longer than those on plants with four or eight shoots (85). The decline in TNC concentrations of roots and crowns was greater following total than partial shoot removal (37, 164). Frequent harvesting of alfalfa can prevent replenishment of sufficient TNC in storage organs and, thereby, reduce plant yields and stand survival (e.g., 18, 26, 35, 56, 128).

After reviewing data from various studies, Brown et al. (17) concluded that 25% of the reserve C compounds consumed following harvest were lost in root respiration, 39 to 45% were lost in respiration of the crown and shoots, and 30 to 36% were incorporated into shoot growth. Escalada and Smith (54) found 63% of the root TNC utilized during alfalfa regrowth came from the root wood and 37% from the root bark. Most of the utilized TNC originated in the upper 10-cm segment of the taproot with progressively less from lower segments. While starch and sugars accounted for 62 and 32%, respectively, of the TNC consumed during regrowth, the accumulation of TNC in roots during later vegetative growth consisted of 88% starch and 12% sugar. Thus, the primary source of C for shoot regrowth following herbage harvest is starch in the root wood of the upper 10-cm segment of the taproot.

Nodules are an additional sink for TNC. These carbohydrates permit maintenance of low rates of nitrogenase activity following total shoot removal (36, 37, 38) and during weeks of continuous darkness (36). Cralle (36) observed the partitioning of ^{14}C-labeled root reserves to new nodule dry matter and ^{14}C-labeled starch injected into the root to established nodules during vegetative regrowth following harvest. However, nodules were relatively weak sinks for these labeled compounds as compared with shoots.

6-5.3.4 Genetic Variability

Few investigations of genetic differences in carbohydrate storage in alfalfa have been conducted except with reference to cold hardiness. Selection for tolerance to frequent harvest in 'Moapa' resulted in a germplasm with 24% greater TNC concentration in roots during the critical summer growth period, and a forage yield superior to the parent source (135). Chatterton et al. (25) identified clones that were either tolerant or intolerant to frequent herbage harvests. The tolerant clones exhibited a more rapid rate of TNC accumulation in the roots and crowns than did the intolerant clones. They further identified clones that were tillering or nontillering. The TNC concentration of the roots and crowns of the tillering clones was greater than that of the nontillering. Hence, genetic differences in the rate of accumulation and final concentration of reserve carbohydrates in the roots and crowns appear to be important in determining survival, growth rate, and tillering following herbage removal.

6-6 PROSPECTS FOR IMPROVEMENT OF CARBON ECONOMY AND YIELD

6-6.1 Selection for Associated Traits

The impressive surge in research on the cellular and organismal (whole-plant) physiology of C assimilation, partitioning, and utilization by alfalfa provided one focus of this chapter. Numerous examples are available of the apparent genetic control of physiological and biochemical traits, which has fostered attempts to include selection for such traits in programs to develop improved cultivars. It was difficult to find clear evidence, however, that breeding for physiological and biochemical traits associated with C assimilation, partitioning, and utilization has improved the efficiency of cultivar development or yielding ability. Available physiological, biochemical, and genetic analyses suggest that traits associated with the C economy of alfalfa are: composed of complex, interdependent components; quantitatively inherited; often poorly suited to phenotypic evaluation; sensitive to the environment; and variably expressed depending upon stage of plant development. These traits can be modified by selection procedures, but without improving yield, quality or agronomic performance more efficiently than conventional breeding methods; and selection is expensive when conducted with the large plant numbers required in a breeding program.

Although breeding for physiological and biochemical traits has had limited practical or agronomic impact, there have been two distinct scientific benefits to alfalfa improvement from this research. Breeding for physiological and biochemical traits associated with C assimilation, partitioning, and utilization has increased the understanding of how these processes are interrelated and regulated. Furthermore, breeding for physiological and biochemical traits is an important experimental approach to improved understanding and analysis of developmental and organismal biology of alfalfa.

6-6.2 Application of Tissue Culture and Molecular Biology

The potential for possible future improvements in the C metabolism of alfalfa would be incomplete without brief consideration to the role of tissue culture and molecular biology methodologies. Although these methods are covered in Chapter 29 in this book, it is important to recognize the identification of alfalfa strains amenable to tissue, protoplast, and nodule culture; the use of tissue culture in selection for salinity tolerance; selection for herbicide resistance; and selection for the overproduction of amino acids. These are important initiatives whose time to application cannot be forecast with certainty. Nevertheless, like breeding for physiological and biochemical traits, the application of tissue culture and molecular biology are certain to influence approaches to improving the C economy of alfalfa.

6-6.3 Crop Communities

On a community basis the greatest benefits to C accumulation may ensue from attempts to increase the utilization of PAR, and to optimize the partitioning of photoassimilates between herbage and non-economic structures (e.g., roots and crowns). Faster regrowth of leaf area after harvest, and greater retention of photosynthetically active leaves by the entire canopy, will increase the use efficiency of incident solar radiation for CO_2 fixation. Faster leaf regrowth after harvest may portend a shortening of the regrowth periods between successive harvests, a necessary prerequisite to minimize the metabolic penalties of leaf aging and to optimize forage quality. Furthermore, greater seasonal forage yields may result from shortening the interval between harvests.

Despite the numerous investigations of nonstructural carbohydrate storage and its subsequent mobilization for new shoot growth, information is lacking on the optimum partitioning of assimilates among herbage production, growth of noneconomic parts of the plant and TNC storage in roots and crowns. What are the minimum concentrations and contents of TNC necessary for overwintering, or necessary to support leaf area renewal after harvest? Are there opportunities to reduce partitioning to root structure, storage, and maintenance for the benefit of herbage production?

Another proposal for improvement of C yield is to tailor the winter-dormancy characteristics of alfalfa to increasingly narrow increments of latitude. Little is known of how photoperiod and temperature act separately and in concert to trigger the onset of spring growth and the termination of autumn growth. Understanding of the factors that govern the dormancy reaction, hence the total seasonal period of potential photosynthetic activity, will contribute to improved forage yields.

The future improvement of yield of alfalfa will undoubtedly rely on a clearer understanding of how the expression of desirable genes is controlled by stage of growth and environment. Closer scientific collaboration between conventional genetics and breeding, and physiology, biochemistry, and molecular biology is anticipated. This should be a fruitful approach to understanding the genetic control of quantitative traits and manipulating them for improvement of C assimilation, partitioning, and utilization. We envision advancement from the current status of recognition of potential to actual proof of concept within the coming decade.

REFERENCES

1. Allen, R.S., R.E. Worthington, N.R. Gould, M.L. Jacobson, and A.E. Freeman. 1961. Diurnal variation in composition of alfalfa. J. Agric. Food Chem. 9:406-408.
2. Anderson, M.P., G.H. Heichel, and C.P. Vance. 1984. Nonphotosynthetic CO_2 assimilation by roots and nodules of alfalfa. Agron. Abstr. American Society of Agronomy, Madison, WI, p. 97.
3. Arnon, D.I., M.B. Allen, and F.R. Whatley. 1954. Photosynthesis by isolated chloroplasts. Nature (London) 174:394-396.
4. Baldocchi, D.D., S.B. Verma, and N.J. Rosenberg. 1981a. Seasonal and diurnal variation in the CO_2 flux and CO_2 water flux ratio of alfalfa. Agric. Meteorol. 23:231-244.
5. ----, ----, and ----. 1981b. Environmental effects on the CO_2 flux and CO_2 water flux ratio of alfalfa. Agric. Meteorol. 24:175-184.
6. Barnes, D.K., R.B. Pearce, G.E. Carlson, R.H. Hart, and C.H. Hanson. 1969. Specific leaf weight differences in alfalfa associated with variety and plant age. Crop Sci. 9:421-423.
7. Barta, A.L. 1978. Effect of root temperature on dry matter distribution, carbohydrate accumulation, and acetylene reduction activity in alfalfa and birdsfoot trefoil. Crop Sci. 18:637-640.
8. ----. 1982. Response of symbiotic N fixation and assimilate partitioning to K supply in alfalfa. Crop Sci. 22:89-92.
9. Bassham, J.A. 1977. Increasing crop production through more controlled photosynthesis. Science 197:630-638.
10. Baysdorfer, C., and J.A. Bassham. 1985. Photosynthate supply and utilization in alfalfa. Plant Physiol. 77:313-317.
11. Boller, B.C., and G.H. Heichel. 1983. Photosynthate partitioning in relation to N fixation capability of alfalfa. Crop Sci. 23:655-659.
12. ----, and ----. 1984. Canopy structure and photosynthesis of alfalfa genotypes differing in nodule effectiveness. Crop Sci. 24:91-96.
13. Bowes, G., W.L. Ogren, and R.H. Hageman. 1971. Phosphoglycolate production catalyzed by ribulose diphosphate carboxylase. Biochem. Biophys. Res. Commun. 45:716-722.
14. Brick, M.A., A.K. Dobrenz, and M.H. Schonhorst. 1976. Transmittance of the multifoliolate leaf characteristic into nondormant alfalfa. Agron. J. 68:134-136.
15. Brown, R.H., and R.E. Blaser. 1968. Leaf area index in pasture growth. Herb. Abstr. 38:1-9.
16. ----, R.B. Cooper, and R.E. Blaser. 1966. Effects of leaf age on efficiency. Crop Sci. 6:206-209.
17. ----, R.B. Pearce, D.D. Wolf, and R.E. Blaser. 1972. Energy accumulation and utilization. *In* C.H. Hanson (ed.) Alfalfa science and technology. Agronomy 15:143-166.
18. Bryant, H.T., and R.E. Blaser. 1964. Yield and persistence of an alfalfa-orchardgrass mixture as affected by cutting treatment. Va. Agric. Exp. Stn. Bull. 555.
19. Bula, R.J. 1972. Morphological characteristics of alfalfa plants grown at several temperatures. Crop Sci. 12:683-686.
20. ----, D.A. Holt, R.G. May, and M.M. Schreiber. 1975. Environmental physiology, modeling, and simulation of alfalfa growth. II. Biomass accumulation characteristics of an alfalfa canopy. Purdue Univ. Agric. Exp. Stn. Bull 76.
21. ----, and D. Smith. 1954. Cold resistance and chemical composition in overwintering alfalfa, red clover, and sweetclover. Agron. J. 46:397-401.

22. Caldwell, C.D., D.S. Fensom, L. Bordeleau, R.G. Thompson, R. Drouin, and R. Didsbury. 1984. Translocation of ^{13}N and ^{11}C between nodulated roots and leaves in alfalfa seedlings. J. Exp. Bot. 35:431–443.
23. Carlson, J.R., Jr., R.L. Ditterline, J.M. Martin, and R.E. Lund. 1981. Sampling stomatal density in alfalfa. Crop Sci. 21:467–469.
24. Chatterton, N.J. 1973. Product inhibition of photosynthesis in alfalfa leaves as related to specific leaf weight. Crop Sci. 13:284–285.
25. ----, and G.E. Carlson. 1981. Growth and photosynthate partitioning in alfalfa under eight temperature-photosynthetic period combinations. Agron. J. 73:392–394.
26. ----, ----, R.H. Hart, and W.E. Hungerford. 1974. Tillering, nonstructural carbohydrates, and survival relationships in alfalfa. Crop Sci. 14:783–787.
27. ----, D.R. Lee, and W.E. Hungerford. 1972. Diurnal change in specific leaf weight of *Medicago sativa* L. and *Zea mays* L. Crop Sci. 12:576–578.
28. Chilcote, D.O., R.V. Frakes, and R.C. Ackerson. 1981. Specific leaf weight profiles of selected alfalfa genotypes. p. 79–86. *In* R.H. Delaney (ed.) Physiological and morphological criteria for alfalfa plant breeding. Wyoming Agric. Exp. Stn. Res. J. 164.
29. Cohen, Y., H. Bielorai, and A. Dovrat. 1972. Effect of timing of irrigation on total nonstructural carbohydrate level in roots and on seed yield of alfalfa (*Medicago sativa* L.) Crop Sci. 12:634–636.
30. Cole, D.F., and A.K. Dobrenz. 1970. Stomate density of alfalfa (*Medicago sativa* L.). Crop Sci. 10:61–63.
31. Collins, M., and S.H. Duke. 1981. Influence of potassium-fertilization rate and form on photosynthesis and N_2 fixation of alfalfa. Crop Sci. 21:481–485.
32. Cook, M.G., and L.T. Evans. 1978. Effect of relative size and distance of competing sinks on the distribution of photosynthetic assimilates in wheat. Aust. J. Plant Physiol. 5:495–509.
33. Cooper, C.S. 1966. Response of birdsfoot trefoil and alfalfa to various levels of shade. Crop Sci. 6:63–66.
34. ----, and M. Qualls. 1967. Morphology and chlorophyll content of shade and sun leaves of two legumes. Crop Sci. 7:672–673.
35. ----, and C.A. Watson. 1968. Total available carbohydrates in roots of sainfoin (*Onobrychis viciaefolia* Scop.) and alfalfa (*Medicago sativa* L.) when grown under several management regimes. Crop Sci. 8:83–85.
36. Cralle, H.T. 1983. Photosynthate partitioning in alfalfa populations selected for high nitrogen fixation capability. Ph.D. thesis. Univ. of Minnesota, St. Paul (Diss. Abstr. DA 8329508).
37. ----, and G.H. Heichel. 1981. Nitrogen fixation and vegetative regrowth of alfalfa and birdsfoot trefoil after successive harvests or floral debudding. Plant Physiol. 67:898–905.
38. ----, and ----. 1982. Temperature and chilling sensitivity of nodule nitrogenase activity of unhardened alfalfa. Crop Sci. 2:300–304.
39. ----, and ----. 1985. Interorgan photosynthate partitioning in alfalfa. Plant Physiol. 79:381–385.
40. ----, and ----. 1986. Photosynthate and dry matter distribution in effectively and ineffectively nodulated alfalfa. Crop Sci. 26:117–121.
41. ----, ----, and D.K. Barnes. 1987. Photosynthate partitioning in plants of alfalfa populations selected for high and low nodule mass. Crop Sci. 27:96–100.
42. Curtis, O.F. 1944. The food content of forage crops as influenced by the time of day at which they are cut. J. Am. Soc. Agron. 36:401.
43. Dady, H.V., J.I. Whitecross, and D.C. Shaw. 1986. One large subunit of ribulose 1,5-bisphosphate carboxylase oxygenase in *Medicago, Spinacia* and *Nicotina*. Theor. Appl. Genet. 71:708–715.
44. DeBoer, D.L., and S.H. Duke. 1982. Effects of sulphur nutrition on nitrogen and carbon metabolism in lucerne (*Medicago sativa* L.). Physiol. Plant. 54:343–350.
45. Delaney, R.H. 1972. Morphological features of alfalfa (*Medicago sativa* L.) clones and their relation to photosynthesis and respiration. Ph.D. thesis. Univ. of Arizona, Tucson (Diss. Abstr. 72-31845).
46. ----, and A.K. Dobrenz. 1974a. Yield of alfalfa as related to carbon exchange. Agron. J. 66:498–500.
47. ----, and ----. 1974b. Morphological and anatomical features of alfalfa leaves as related to CO_2 exchange. Crop Sci. 14:444–447.
48. ----, and ----. 1981. Carbon exchange and leaf morphology and anatomical characteristics. p. 15–26. *In* R.H. Delaney (ed.) Physiological and morphological criteria for alfalfa plant breeding. Wyom. Agric. Exp. Stn. Res. J. 164.
49. ----, ----, and H.T. Poole. 1974. Seasonal variation in photosynthesis, respiration, and growth components of nondormant alfalfa (*Medicago sativa* L.). Crop Sci. 14:58–61.

50. Deroche, M.E., E. Carrayol, G. Gosse, O. Bethenod, and E. Jolivet. 1981. PEP carboxylase in a leguminous plant: *Medicago sativa*. p. 115-121. *In* George Akoyunoglou (ed.) Proc. 5th Int. Congr. Photosynthesis, Sect. 4, Halkidiki, Greece, 7-13 Sept. 1980. Balaban International Science Services, Philadelphia.
51. Dobrenz, A.K., and M.A. Massengale. 1966. Change in carbohydrates in alfalfa (*Medicago sativa* L.) roots during the period of floral initiation and seed development. Crop Sci. 6:604-607.
52. Dovrat, A., D. Levanon, and M. Waldman. 1969. Effect of plant spacing on carbohydrates in roots and on components of seed yield in alfalfa (*Medicago sativa* L.) Crop Sci. 9:33-34.
53. Duke, S.H., and D.C. Doehlert. 1981. Root respiration, nodulation, and enzyme activities in alfalfa during cold acclimation. Crop Sci. 21:489-495.
54. Escalada, J.A., and D. Smith. 1972. Changes in nonstructural carbohydrate fractions at intervals down the tap root bark and wood of alfalfa (*Medicago sativa* L.) during regrowth. Crop Sci. 12:745-749.
55. Evans, D.W., and R.N. Peaden. 1981. Alfalfa leaf size over harvests and season. p. 94-102. *In* R.H. Delaney (ed.) Physiological and morphological criteria for alfalfa plant breeding. Wyom. Agric. Exp. Stn. Res. J. 164.
56. Feltner, K.C., and M.A. Massengale. 1965. Influence of temperature and harvest management on growth, level of carbohydrates in the roots, and survival of alfalfa (*Medicago sativa* L.). Crop Sci. 5:585-588.
57. Fishbeck, K.A., and D.A. Phillips. 1980. Apparent photosynthesis, root carbohydrates, and nitrogen fixation in alfalfa following harvest. Plant Physiol. (Suppl. 6) 65:109.
58. Foutz, A.L., W.W. Wilhelm, and A.K. Dobrenz. 1976. Relationship between physiological and morphological characteristics and yield of nondormant alfalfa clones. Agron. J. 68:587-591.
59. Fuess, R.W., and M.B. Tesar. 1968. Photosynthetic efficiency, yields, and leaf loss in alfalfa. Crop Sci. 8:159-163.
60. Gifford, R.M., J.H. Thorne, W.D. Hitz, and R.T. Giaquinta. 1984. Crop productivity and photoassimilate partitioning. Science 225:801-807.
61. Gosse, G., M. Chartier, and G. Lemaire. 1984. Mise au point d'un modele de prevision de production pour une culture de lucerne. C.R. Acad. Sci. (Paris). Ser. III. 298:541-544.
62. ----, ----, C. Varlet-Grancher, and R. Bonhomme. 1982. Interception du rayonnement utile a la photosynthese chez la luzerne; variations et modelisation. Agronomie 2:583-588.
63. Graber, L.F., N.T. Nelson, W.A. Levkel, and W.B. Albert. 1927. Organic food reserves in relation to the growth of alfalfa and other perennial herbaceous plants. Wis. Agric. Exp. Stn. Bull. 80.
64. Greub, L.V., and W.F. Wedin. 1971. Leaf area, dry matter accumulation, and carbohydrate reserves of alfalfa and birdsfoot trefoil under a three cut management. Crop Sci. 11:341-344.
65. Hanscom, Z., III, and I.P. Ting. 1981. The use of tritiated water vapor as a tracer for transpiration and $^{14}CO_2$ as a tracer for photosynthesis in the study of alfalfa and other economic crop plants. p. 47-64. *In* R.H. Delaney (ed.) Physiological and morphological criteria for alfalfa plant breeding. Wyo. Agric. Exp. Stn. Res. J. 164.
66. Hart, R.H., R.B. Pearce, N.J. Chatterton, G.E. Carlson, D.K. Barnes, and C.H. Hanson. Alfalfa yield, specific leaf weight, CO_2 exchange rate, and morphology. Crop Sci. 18:649-653.
67. Herridge, D.F., and J.S. Pate. 1977. Utilization of net photosynthate for nitrogen fixation and protein production in an annual legume. Plant Physiol. 60:759-764.
68. Hodgkinson, K.C. 1969. The utilization of root organic compounds during the regeneration of lucerne. Aust. J. Biol. Sci. 22:1113-1123.
69. ----. 1974. Influence of partial defoliation on photosynthesis, photorespiration and transpiration by lucerne leaves of different age. Aust. J. Plant Physiol. 1:561-578.
70. ----, N.G. Smith, and G.E. Miles. 1972. The photosynthetic capacity of stubble leaves and their contribution to growth of the lucerne plant after high level cutting. Aust. J. Agric. Res. 23:225-238.
71. Holt, D.A., and A.R. Hilst. 1969. Daily variation in carbohydrate content of selected forage crops. Agron. J. 61:239-242.
72. Huffaker, R.C., E.L. Cox, G.E. Kleinkopf, and E.H. Stanford. 1970. Regulation of synthesis of chlorophyll, carotene, ribulose-1,5diP carboxylase and phosphoribulokinase in a temperature sensitive chlorophyll mutant of *Medicago sativa*. Physiol. Plant. 23:404-411.
73. Hunt, L.A., C.E. Moore, and J.E. Winch. 1970. Light attentuation coefficient and productivity in 'Vernal' alfalfa. Can. J. Plant Sci. 50:469-474.
74. Johal, S., D.P. Bourque, W.W. Smith, S. Won Suh, and D. Eisenberg. 1980. Crystal-

lization and characterization of ribulose 1,5bisphosphate carboxylase/oxygenase from eight plant species. J. Biol. Chem. 255:8873-8880.
75. Jordon, D.B., and W.L. Ogren. 1983. Species variation in kinetic properties of ribulose 1,5-bisphosphate carboxylase oxygenase. Arch. Biochem. Biophys. 227:425-433.
76. Jung, G.A., and D. Smith. 1961. Trends in cold resistance and chemical changes over winter in the roots and crowns of alfalfa and medium red clover. I. Changes in certain nitrogen and carbohydrate fractions. Agron. J. 53:359-366.
77. Kimball, B.A. 1983. Carbon dioxide and agricultural yield: an assemblage and analysis of 430 prior observations. Agron. J. 75:779-788.
78. King, R.W., and L.T. Evans. 1967. Photosynthesis in artificial communities of wheat, lucerne, and subterranean clover plants. Aust. J. Biol. Sci. 20:623-635.
79. Ku, S.B., and G.E. Edwards. 1977. Oxygen inhibition of photosynthesis. Plant Physiol. 59:986-990.
80. ----, and L.A. Hunt. 1973. Effects of temperature on the morphology and photosynthetic activity of newly matured leaves of alfalfa. Can. J. Bot. 51:1907-1916.
81. ----, and ----. 1977. Effects of temperature on the photosynthesis irradiance response curves of newly matured leaves of alfalfa. Can. J. Bot. 55:872-879.
82. Latimore, M.L., J. Giddens, and D.A. Ashley. 1977. Effect of ammonium and nitrate nitrogen upon photosynthate supply and nitrogen fixation by soybeans. Crop Sci. 17:399-404.
83. Lawrie, A.C., and C.T. Wheeler. 1974. The effects of flowering and fruit formation on the supply of photosynthetic assimilates to the nodules of *Pisum sativum* L. in relation to the fixation of nitrogen. New Phytol. 73:1119-1127.
84. Leach, G.J. 1970. Shoot growth of lucerne plants cut at different heights. Aust. J. Agric. Res. 21:583-591.
85. ----. 1971. The relation between lucerne shoot growth and temperature. Aust. J. Agric. Res. 22:49-59.
86. Leavitt, J.R.C., A.K. Dobrenz, and J.E. Stone. 1979. Physiological and morphological characteristics of large and small leaflet alfalfa genotypes. Agron. J. 71:529-532.
87. Lechtenberg, V.L., D.A. Holt, and H.W. Youngberg. 1971. Diurnal variation in nonstructural carbohydrates, *in vitro* digestibility, and leaf to stem ratio of alfalfa. Crop Sci. 63:719-724.
88. Macdowall, F.D. H. 1983. Effects of light intensity and CO_2 concentration on the kinetics of 1st month growth and nitrogen fixation of alfalfa. Can. J. Bot. 61:731-740.
89. MacLeod, L.B. 1965. Effect of nitrogen and potassium fertilization on the yield, regrowth, and carbohydrate content of the storage organs of alfalfa and grasses. Agron. J. 57:345-350.
90. Matches, A.G., G.O. Mott, and R.J. Bula. 1963. The development of carbohydrate reserves in alfalfa seedlings under various levels of shading and potassium fertilization. Agron. J. 55:185-188.
91. Maxwell, C.A., C.P. Vance, G.H. Heichel, and S. Stade. 1984. CO_2 fixation in alfalfa and birdsfoot trefoil root nodules and partitioning of ^{14}C to the plant. Crop Sci. 24:257-264.
92. McDaniel, R.G., A.K. Dobrenz, and M.H. Schonhorst. 1981. Mitochondrial criteria for alfalfa breeding. p. 2-14. *In* R.H. Delaney (ed.) Physiological and morphological criteria for alfalfa plant breeding. Wyom. Agric. Exp. Stn. Res. J. 164.
93. Megee, C.R. 1935. A search for factors determining winter hardiness in alfalfa. J. Am. Soc. Agron. 27:685-698.
94. Melvin, J.F. 1965. Variations in the carbohydrate content of lucerne and the effect on ensilage. Aust. J. Agric. Res. 16:951-959.
95. Meyers, S.P., S.L. Nichols, G.R. Baer, W.T. Molin, and L.E. Schrader. 1982. Ploidy effects in isogenic populations of alfalfa. I. Ribulose-1,5-bisphosphate carboxylase, soluble protein, chlorophyll, and DNA in leaves. Plant Physiol. 70:1704-1709.
96. Miller, R.W., and J.C. Sirois. 1982. Relative efficacy of different alfalfa cultivar *Rhizobium meliloti* strain combinations for symbiotic nitrogen fixation. Appl. Environ. Microbiol. 43:764-768.
97. Molin, W.T., S.P. Meyers, G.R. Baer, and L.E. Schrader. 1982. Ploidy effects in isogenic populations of alfalfa. II. Photosynthesis, chloroplast number, ribulose-1,5-bisphosphate carboxylase, chlorophyll, and DNA in protoplasts. Plant Physiol. 70:1710-1714.
98. Monsi, M., and T. Saeki. 1953. The light factor in plant communities and its significance in biomass production. Jpn. J. Bot. 14:22-52.
99. Morison, J.I.L., and R.M. Gifford. 1984. Plant growth and water use with limited water supply in high CO_2 concentrations. I. Leaf area, water use and transpiration. Aust. J. Plant Physiol. 11:361-374.
100. ----, and ----. 1984. Plant growth and water use with limited water supply in high CO_2 concentrations. II. Plant dry weight, partitioning and water use efficiency. Aust. J. Plant Physiol. 11:375-384.

101. Murata, Y., J. Iyama, and T. Honma. 1963. Studies of the photosynthesis of forage crops. IV. Influence of air temperature upon the photosynthesis and respiration of alfalfa and several southern type forage crops. Proc. Crop Sci. Soc. Jpn. 34:154–158.
102. Nelson, C.J., and D. Smith. 1968a. Growth of birdsfoot trefoil and alfalfa. II. Morphological development and dry matter distribution. Crop Sci. 8:21–25.
103. ----, and ----. 1968b. Growth of birdsfoot trefoil and alfalfa. III. Changes in carbohydrate reserves and growth analysis under field conditions. Crop Sci. 8:25–28.
104. ----, and----. 1969. Growth of birdsfoot trefoil and alfalfa. IV. Carbohydrate reserve levels and growth analysis under two temperature regimes. Crop Sci. 9:589–591.
105. Palmer, A.F.E., G.H. Heichel, and R.B. Musgrave. 1973. Patterns of translocation, respiratory loss, and redistribution of ^{14}C in maize labelled after flowering. Crop Sci. 13:371–376.
106. Pate, J.S. 1966. Photosynthesizing leaves and nodulated roots as donors of carbon to protein of the shoot of the field pea (*Pisum arvense* L.). Ann. Bot. 30:93–109.
107. Pearce, R.B., R.H. Brown, and R.E. Blaser. 1968. Photosynthesis of alfalfa leaves as influenced by age and environment. Crop Sci. 8:677–680.
108. ----, G.E. Carlson, D.K. Barnes, R.H. Hart, and C.H. Hanson. 1969. Specific leaf weight and photosynthesis in alfalfa. Crop Sci. 9:756–759.
109. ----, G. Fissel, and G.E. Carlson. 1969. Carbon uptake and distribution before and after defoliation of alfalfa. Crop Sci. 9:756–759.
110. ----, and D.R. Lee. 1969. Photosynthetic and morphological adaptation of alfalfa leaves to light intensity at different stages of maturity. Crop Sci. 9:791–794.
111. Pearson, C.J. and L.A. Hunt. 1972a. Studies on the daily course of carbon exchange in alfalfa plants. Can. J. Bot. 50:1377–1384.
112. ----, and ----. 1972b. Effects of pretreatment temperature on carbon dioxide exchange in alfalfa. Can. J. Bot. 50:1925–1930.
113. Peoples, T.R., and D.W. Koch. 1978. Physiological response of three alfalfa cultivars to one chilling night. Crop Sci. 18:255–258.
114. ----, and ----. 1979. Role of potassium in carbon dioxide assimilation in *Medicago sativa* L. Plant Physiol. 63:878–881.
115. ----, ----, and S.C. Smith. 1978. Relationship between chloroplast membrane fatty acid composition and photosynthetic response to a chilling temperature in four alfalfa cultivars. Plant Physiol. 61:472–473.
116. Peterschmidt, N.A. 1980. Photorespiration in alfalfa (*Medicago sativa* L.). Ph.D. thesis. Univ. of Arizona, Tucson. (Diss. Abstr. 81-08323).
117. ----, and A.K. Dobrenz. 1981. Genetic variability of photorespiration in an F_1 generation of alfalfa. p. 27–35. *In* R.H. Delaney (ed.) Physiological and morphological criteria for alfalfa plant breeding. Wyo. Agric. Exp. Stn. Res. J. 164.
118. Pfeiffer, T., L. E. Schrader, and E.T. Bingham. 1980. Physiological comparisons of isogenic diploid-tetraploid, tetraploid-octoploid alfalfa populations. Crop Sci. 20:299–303.
119. Platt, S.G., Z. Plaut, and J.A. Bassham. 1976. Analysis of steady state photosynthesis in alfalfa leaves. Plant Physiol. 57:69–73.
120. ----, ----, and ----. 1977. Steady-state photosynthesis in alfalfa leaflets. Plant Physiol. 60:230–234.
121. Porter, T.K., and J.H. Reynolds. 1975. Relationship of alfalfa cultivar yields to specific leaf weight, plant density, and chemical composition. Agron. J. 67:625–629.
122. Rabson, R., N.E. Tolbert, and P.C. Kearney. 1962. Formation of serine and glycerine acid by the glycolate pathway. Arch. Biochem. Biophys. 98:154–161.
123. Randall, D.D., N.E. Tolbert, and D. Gremel. 1971. 3-Phosphoglycerate phosphatase in plants. Plant Physiol. 48:480–487.
124. Rapoport, H.F., and R.L. Travis. 1984. Alfalfa root growth, cambial activity, and carbohydrate dynamics during the regrowth cycle. Crop Sci. 24:899–902.
125. Reed, R., and R.L. Travis. 1984. Phototropic leaflet reorientation in alfalfa in response to vectorial light. Crop Sci. 24:593–597.
126. Reid, D.J., D.J. Lathwell, and M.J. Wright. 1965. Yield and carbohydrate responses of alfalfa seedlings grown at several levels of potassium fertilization. Agron. J. 57:434–437.
127. Reuveni, J., and J. Gale. 1985. The effect of high levels of carbon dioxide on dark respiration and growth of plants. Plant Cell Environ. 8:623–628.
128. Reynolds, J.H., and D. Smith. 1962. Trend of carbohydrate reserves in alfalfa, smooth bromegrass, and timothy grown under various cutting schedules. Crop Sci. 2:333–336.
129. Robison, G.D., and M.A. Massengale. 1968. Effect of harvest management and temperature on forage yield, root carbohydrates, plant density and leaf area relationships in alfalfa (*Medicago sativa* L. cultivar 'Moapa'). Crop Sci. 8:147–151.
130. Ryle, G.J.A., R.A. Arnott, C.E. Powell, and A.J. Gordon. 1983. Comparison of the respiratory effluxes of nodules and roots in six temperate legumes. Ann. Bot. 52:469–477.

131. ----, C.E. Powell, and A.J. Gordon. 1981. Patterns of ^{14}C-labelled assimilate partitioning in red and white clover during vegetative growth. Ann. Bot. 47:505-514.
132. Saugier, B. 1970. Transports turbulents de CO_2 et de vapeur d'eau audessus et a l'interieur de la vegetation. Methods de mesure micrometeorologiques. Ocol. Plant. 5:171-223.
133. Schneiter, A.A., R.G. McDaniel, A.K. Dobrenz, and M.H. Schonhorst. 1974. Relationship of mitochondrial efficiency to forage yield in alfalfa. Crop Sci. 14:821-824.
134. ----, ----, ----, and ----. 1976. Mitochondrial efficiency of individual alfalfa plants as related to forage yield. Agron. J. 68:511-513.
135. Schonhorst, M.H., A.K. Dobrenz, R.K. Thompson, and M.A. Massengale. 1980. Registration of AZ-RON alfalfa germplasm. Crop Sci. 20:831.
136. Scott, D., and J.S. Wells. 1969. Leaf orientation in barley, lupin, and lucerne stands. N.Z.J. Bot. 7:372-388.
137. Setter, T.L., L.E. Schrader, and E.T. Bingham. 1978. Carbon dioxide exchange rates, transpiration, and leaf characters in genetically equivalent ploidy levels of alfalfa. Crop Sci. 18:327-332.
138. Sheehy, J.E., K.A. Fishbeck, and D.A. Phillips. 1980. Relationships between apparent nitrogen fixation and carbon exchange rate in alfalfa. Crop Sci. 20:491-495.
139. ----, and S.C. Popple. 1981. Photosynthesis, water relations, temperature and canopy structure as factors influencing the growth of sainfoin (*Onobrychis viciifolia* Scop.) and lucerne (*Medicago sativa* L.). Ann. Bot. 48:113-128.
140. ----, F.I. Woodward, M.B. Jones, and A. Windram. 1979. Microclimate, photosynthesis and growth of lucerne (*Medicago sativa* L.). I. Microclimate and photosynthesis. Ann. Bot. 44:693-707.
141. Shieh, Wen-Jang. 1985. Photosynthate partitioning and nitrogen fixation of alfalfa and birdsfoot trefoil. Ph.D. thesis. Ohio State Univ., Columbus (Diss. Abst. 8510633).
142. Shih, S.C., G.A. Jung, and D.C. Shelton. 1965. Influence of purines and pyrimidines on cold hardiness of plants. II. Respiration rate of alfalfa in relation to cold hardiness. Crop Sci. 5:307-310.
143. Silkett, V.W., C.R. Megee, and H.C. Rather. 1937. The effect of late summer and early fall cutting on crown bud formation and winterhardiness of alfalfa. J. Am. Soc. Agron. 29:53-62.
144. Smith, D. 1962. Carbohydrate root reserves in alfalfa, red clover, and birdsfoot trefoil under several management schedules. Crop Sci. 2:75-78.
145. ----. 1969. Influence of temperature on the yield and chemical composition of 'Vernal' alfalfa at first flower. Agron. J. 61:470-472.
146. ----. 1970. Influence of temperature on the yield and chemical composition of five forage legume species. Agron. J. 62:520-523.
147. ----. 1973. The nonstructural carbohydrates. p. 105-155. *In* G.W. Butler and R.W. Bailey (ed.) Chemistry and biochemistry of herbage, Vol. I. Academic Press, New York.
148. ----, and J.P. Silva. 1969. Use of carbohydrate and nitrogen root reserves in the regrowth of alfalfa from greenhouse experiments under light and dark conditions. Crop Sci. 9:464-467.
149. Smith, L.H., and G.C. Marten. 1970. Foliar regrowth of alfalfa utilizing ^{14}C-labeled carbohydrates stored in roots. Crop Sci. 10:146-150.
150. Song, S.P., and P.D. Walton. 1975. Inheritance of leaflet size and specific leaf weight in alfalfa. Crop Sci. 15:649-652.
151. Straley, C.S., and C.S. Cooper. 1972. Effect of shading mature leaves of alfalfa and sainfoin plants on specific leaf weight of leaves formed in sunlight. Crop Sci. 12:703-706.
152. Swanson, C.R., and M.W. Adams. 1959. Metabolic changes induced in alfalfa during cold hardening and freezing. Agron. J. 51:397-400.
153. Teare, I.D. 1972. Canopy structure and its effect on the attenuation of incident light. Phyton 29:37-42.
154. Thomas, M.D., and G.R. Hill. 1937. The continuous measurement of photosynthesis, respiration, and transpiration of alfalfa and wheat growing under field conditions. Plant Physiol. 12:285-307.
155. ----, and ----. 1937. Relation of sulfur dioxide in the atmosphere to photosynthesis and respiration of alfalfa. Plant Physiol. 12:309-383.
156. ----, and ----. 1949. Photosynthesis under field conditions. p. 19-52. *In* J. Franck and W.E. Loomis (ed.) Photosynthesis in plants. Iowa State College Press, Ames.
157. Thrower, S.L. 1962. Translocation of labelled assimilates in the soybean. Aust. J. Biol. Sci. 15:629-649.
158. Travis, R.L., and R. Reed. 1983. The solar tracking pattern in a closed alfalfa canopy. Crop Sci. 23:664-668.
159. Vance, C.P., K.L.M. Boylan, C.A. Maxwell, G.H. Heichel, and L.L. Hardman. 1985.

Transport and partitioning of CO_2 fixed by root nodules of ureide and amide producing legumes. Plant Physiol. 78:774–778.
160. ----, S. Stade, and C.A. Maxwell. 1983. Alfalfa root nodule carbon dioxide fixation. Plant Physiol. 72:469–473.
161. Ueno, M., and D. Smith. 1970. Growth and carbohydrate changes in the root wood and bark of different sized alfalfa plants during regrowth after cutting. Crop Sci. 10:396–399.
162. Wilfong, R.T., R.H. Brown, and R.E. Blaser. 1967. Relationships between leaf area index and apparent photosynthesis in alfalfa (*Medicago sativa* L.) and ladino clover (*Trifolium repens* L.). Crop Sci. 7:27–30.
163. Wolf, D.D. 1967. Assimilation and movement of radioactive carbon in alfalfa and reed canarygrass. Crop Sci. 7:317–320.
164. ----, and R.E. Blaser. 1971. Photosynthesis of plant parts of alfalfa canopies. Crop Sci. 11:55–58.
165. ----, and ----. 1972. Growth rate and physiology of alfalfa as influenced by canopy and light. Crop Sci. 12:23–26.
166. ----, and ----. 1981. Flexible alfalfa management: early spring utilization. Crop Sci. 21:90–93.
167. ----, E.L. Kimbrough, and R.E. Blaser. 1976. Photosynthetic efficiency of alfalfa with increasing potassium nutrition. Crop Sci. 16:292–294.
168. Woodward, F.I., and J.E. Sheehy. 1979. Microclimate, photosynthesis, and growth of lucerne (*Medicago sativa* L.). II. Canopy structure and growth. Ann. Bot. 44:709–719.

7 Nodulation and Symbiotic Dinitrogen Fixation

C. P. VANCE
USDA-ARS
St. Paul, Minnesota

G. H. HEICHEL
USDA-ARS
St. Paul, Minnesota

D. A. PHILLIPS
University of California
Davis, California

Biological nitrogen (N_2) fixation is second only to photosynthesis as the most important biochemical process on earth. Biological fixation contributes an estimated 140 Tg (140 million tonnes) of N annually to the earth, of which 80% is fixed via symbiotic associations and 20% by free-living organisms (51, 136). The biological cost of fixed N is substantial (108), amounting to about 2.5% of the primary photosynthesis on land (51). To produce an equivalent amount of N through the Haber-Bosch process would require about 2.58×10 m^3 of natural gas. The high cost of N either as photosynthate or as fossil fuel is a persuasive reason to understand and improve biological N_2 fixation.

Legumes are a major source of protein, oil, and forage for human and animal consumption and are among the world's most important crops. About 85% of N_2 fixation in agricultural soils comes from pulse and forage legumes (51). Because of great diversity in plant characteristics legumes are widely adapted. One of the primary reasons for this wide adaptation is the association of legumes with *Rhizobium* bacteria. When effective, this association enables the plant to fix gaseous N_2 and depend less on soil N.

Alfalfa has gained recognition because of its high yields and superior palatability. The capacity for N_2 fixation in association with *R. meliloti* has been a primary factor contributing to the excellence of alfalfa (*Medicago sativa* L.) (18, 136). Alfalfa consistently shows greater amounts of N_2 fixation and percentage N derived from symbiosis than most other legume species on a seasonal basis (Table 7–1). Estimates of N_2 fixation in alfalfa vary from 50 to 463 kg of N_2 fixed ha^{-1} yr^{-1} with about 200

Copyright 1988 © ASA-CSSA-SSSA, 677 South Segoe Road, Madison, WI 53711, USA.
Alfalfa and Alfalfa Improvement—Agronomy Monograph no. 29.

Table 7-1. Seasonal total of N_2 fixation by hay and pasture legumes and legume-grass swards measured by the difference method and isotope dilution technique.

Species	Rate (kg of N ha^{-1} per growing season)	Location	Method†	Reference
Alfalfa	212	Lexington, KY	Difference (*Poa pratensis* L.)	84
	0–342	Three sites in Great Britain	Difference (*Lolium perenne*)	8
	114–224	Rosemount, MN	Isotope dilution (*Phalaris arundinacea* L. and *Festuca arundinacea* Schreb.)	66
Alfalfa-Orchardgrass sward	15–136	Lucas Co., IA	Isotope dilution (*Dactylis glomerata* L.)	150
Birdsfoot trefoil	49–112	Rosemount, MN	Isotope dilution (*P. arundinacea* L. and *F. arundinacea* Schreb.)	66
Red clover	50–61	Rothamsted, UK	Isotope dilution (*L. perenne* L. and *Brassica rapa* L.)	154
	22–59	Rothamsted, UK	Difference (*L. perenne* L. and *B. rapa* L.)	154
	69–133	Rosemount, MN	Isotope dilution (*P. arundinacea* L. and *F. arundinacea* Schreb.)	66
Subterranean clover	58–183	Hopland, CA	Isotope dilution (*Bromus mollis* L.)	110
Subterranean clover-soft chess sward	21–103	Hopland, CA	Isotope dilution (*B. mollis* L.)	110
White clover	128	Lexington, KY	Difference (*P. pratensis* L.)	84

kg of N_2 fixed ha^{-1} yr^{-1} average (65, 66, 136). The variation in N_2 fixation of alfalfa is the result of a number of factors, the most important being: (i) bacterial strain by plant genotype interactions (2, 5, 14, 18, 20, 21, 42, 47, 94) and (ii) differences in management practices in various climates and locations (18, 23, 24, 59, 63, 138).

7-1 ESTABLISHING THE SYMBIOSIS

7-1.1 The Microsymbiont

Initially, a single species of bacterium was thought to infect and induce nodules on all legumes (69, 149). By 1932, however, studies had demonstrated that only certain strains of bacteria induced nodules on a specific legume host. Fred et al. (46) recognized eight major cross-inoculation groups in the family Rhizobiaceae: alfalfa (*R. meliloti*), clover (*R. trifolii*), pea (*R. leguminosarum*), bean (*R. phaseoli*), lupine (*R. lupini*), soybean (*R. japonicum*), cowpea (*Rhizobium* spp.), and Lotus (*Rhizobium* spp.).

In 1926, the binomial *R. meliloti* was applied by Danegeard (27) to those gram-negative nonspore forming bacteria capable of forming nodules on alfalfa. The bacterium is also capable of nodule formation with *Melilotus* (white and yellow sweet clover) and *Trigonella* (Greek clover) but not on other Leguminoseae. Colony growth of *R. meliloti* on yeast extract-mannitol agar is fast, of buttery consistency giving an acid reaction in litmus milk, and is accompanied by serum zone formation. Young cells are peritrichously flagellated. These rhizobia are easily cultured on defined media with mannitol as a preferred source of C (9, 41). *Rhizobium meliloti* prefers a neutral pH for maximum growth and grows poorly if at all below pH 5.0 (18, 41, 101).

The genes controlling nodulation and N_2 fixation of *R. meliloti* were originally thought to reside on the chromosome. Recent evidence, however, has shown that these genes are located extrachromosomally on plasmids (4, 33, 87). Site-directed mutations of plasmid DNA have been successful in producing strains of rhizobia that will not nodulate as well as strains that nodulate yet will not fix N (73, 85).

7-1.2 Inoculation

Inoculation of seeds with *Rhizobium* has been examined in detail by Burton (18) in a previous volume and will be considered only briefly in this chapter. In order to establish the symbiosis, plant roots must come into contact with and be infected by an effective strain of *Rhizobium*. In soils with high populations (10^4 cells g soil^{-1}) of effective bacteria, contact and infection are readily achieved (83). However in acid soils and in soils with low populations of effective bacteria, the *Rhizobium* is brought into contact with the plant through inoculation of seed and soil (8, 11, 46,

52, 80, 81). Inoculation is performed usually by either mixing seed with a moist peat-base powder containing the bacterium or coating the seed with a suspension of $CaCO_3$, acacia gum, and *Rhizobium* bacteria (18, 19). Populations of rhizobia in soil may also be increased by drenching the soil with a liquid or broth culture of the appropriate strain (18, 19, 106).

Although displacement of existing populations of *R. meliloti* in soils is difficult, inoculation of seed proved effective in increasing stand establishment and yield in acid soils (11, 95, 97, 98, 102, 106, 130, 150) and in soils infested with ineffective strains (13, 46, 81, 106, 130). The need for inoculation varies from location to location but is generally required in soils new to alfalfa (13, 18, 46, 130, 151). Burton (18) summarized his position on inoculation as follows; "Inoculation can be beneficial without being necessary. Necessary implies possible crop failure without inoculation, whereas need connotes enhancement of growth or quality". Inoculation is an inexpensive form of insurance for providing adequate nodulation.

7-1.3 Specificity and Infection

Legumes allow homologous rhizobial strains to penetrate and subsequently to develop nodules, whereas heterologous rhizobial strains and other soil bacteria are not allowed entry (6, 28, 46, 104). This specificity of infection, the basis of the cross-inoculation concept of *Rhizobium*—legume infection (20, 21, 28, 41, 46), usually is limited to a single species or closely related species of legumes (4, 28, 104). There are major exceptions: some strains of *R. leguminosarum* nodulate *Trifolium* spp., whereas some strains of *R. trifolii* nodulate *Pisum* (137).

Processes that regulate specificity are thought to occur upon initial contact of legume-host and *Rhizobium* (6, 15, 137). Bohlool and Schmidt (10) hypothesized that specificity of infection involved the binding of a *Rhizobium* spp. to its particular host plant through the attachment of root hair proteins (lectins) to specific sugars located on the bacterium. A model of this selective attachment was presented by Dazzo and Hubbell (29), who suggested that polyvalent lectins cross-bridge common antigen sites on the host root hair and bacterial surface. Whether this binding is specific or general remains in question.

Pectin degrading enzymes have been strongly implicated in host-*Rhizobium* specificity of alfalfa and clover (22, 43, 100, 101). *Rhizobium* may either produce or induce in the host plant pectolytic enzymes that are active in degrading the root hair tip, thus allowing bacteria to enter (43, 75, 100). The well-documented Ca requirement for infection and nodulation of alfalfa has been related to a role for Ca in activation of pectolytic enzyme activity (98, 99, 100, 101, 102). At acid pH levels Ca becomes unavailable and pectolytic activity at the infection site may be reduced, thus resulting in reduced nodulation (98, 99, 100, 101, 102).

Another hypothesis suggests that the root hair wall invaginates around the bacteria, enclosing them in host tissue (28, 93, 104, 131). The above mechanisms may all interact or be involved sequentially and should not be considered mutually exclusive. In any event, infection is highly regulated with only 1 to 5% of root hairs becoming infected and only a low proportion (30%) of those infected resulting in nodules (6, 103, 104, 131).

Root hairs often curl before being penetrated by *Rhizobium* (28, 43, 75, 93). However, nodulation apparently is not restricted by curling because chemicals that increase root hair curling do not increase infection (28, 103). Many infections occur through straight root hairs and epidermal cells. Plant growth hormones (e.g., indoleacetic acid and cytokinin) produced by both the bacteria and the plant, and the extracellular polysaccharide coat of the bacterium are thought to be responsible for curling (28, 39, 43, 75, 105).

The first microscopically visible sign of infection is swelling and formation of a bright spot in the root hair wall (28, 82, 103, 131). Cytoplasmic streaming increases near the infection site and the infection thread becomes visible (Fig. 7–1). The infection thread then grows down the root hair toward the root cortex at a rate of 7 to 10 μm h^{-1} (28, 104). McCoy's classical study (93) of alfalfa showed that the infection thread is similar in chemistry to the root hair wall.

Nutman (104) and Bauer (6) discussed in detail the evidence indicating that infection thread formation is self-regulated by the plant: (i) only a few root hairs become infected; (ii) root cells susceptible to infection are limited primarily to the zone just below the smallest emergent root hair; (iii) nodule formation inhibits further infection, whereas removal of nodules, nodule tips, and root tips stimulate infection; and (iv)

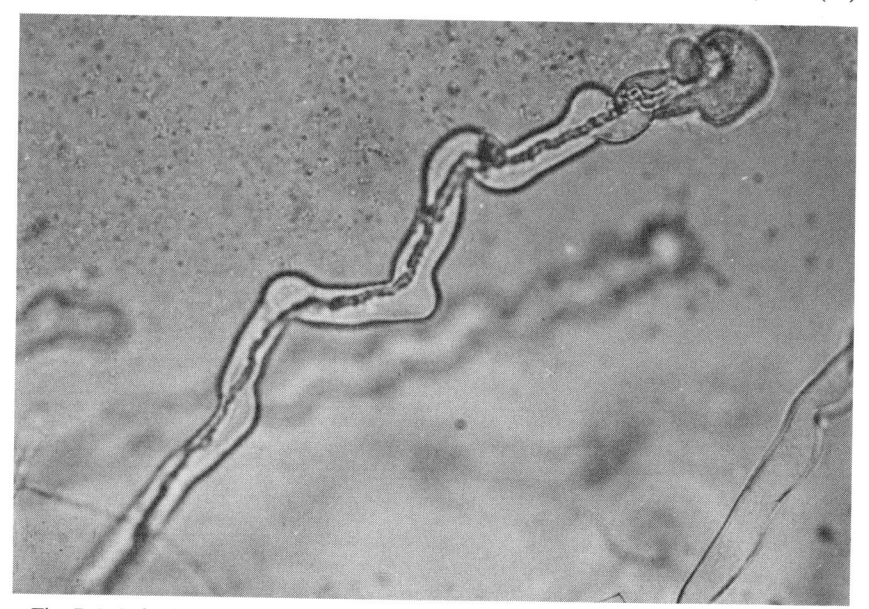

Fig. 7–1. Infection thread and curled root hair of alfalfa 5 d after inoculation. × 640.

applied N reduces infection thread formation and stimulates abortion of infection threads (99, 133, 136). The infection thread may be analogous to a disease resistance response in which the epidermal wall enlarges in reaction to attempted penetration or to mechanical damage (137).

7-2 NODULE DEVELOPMENT AND STRUCTURE

7-2.1 Initiation

Nodule initiation occurs soon after root hair infection (28, 82, 123). The infection thread can be found in the root cortex about 24 h after penetration (Fig. 7-2). As the infection thread passes through the cortex, the nuclei of the cortical cells enlarge and adjacent cells may divide. It is not known if cortical cells must by polyploid for infection to occur or if infection induces polyploidy. Meristematic activity is initiated usually in advance of the infection threads (28, 103). Infection threads may produce or stimulate auxin and cytokinin production in the infected areas. Studies of pea (*Pisum sativum* L.), lupine (*Lupinus* spp.) and alfalfa have shown that auxin and cytokinin concentrations are higher in nodule tissue than in root tissue (28, 103, 105, 136). The meristematic tissue in the root cortex, induced by the infection thread, divides to produce the nodule (Fig. 7-3).

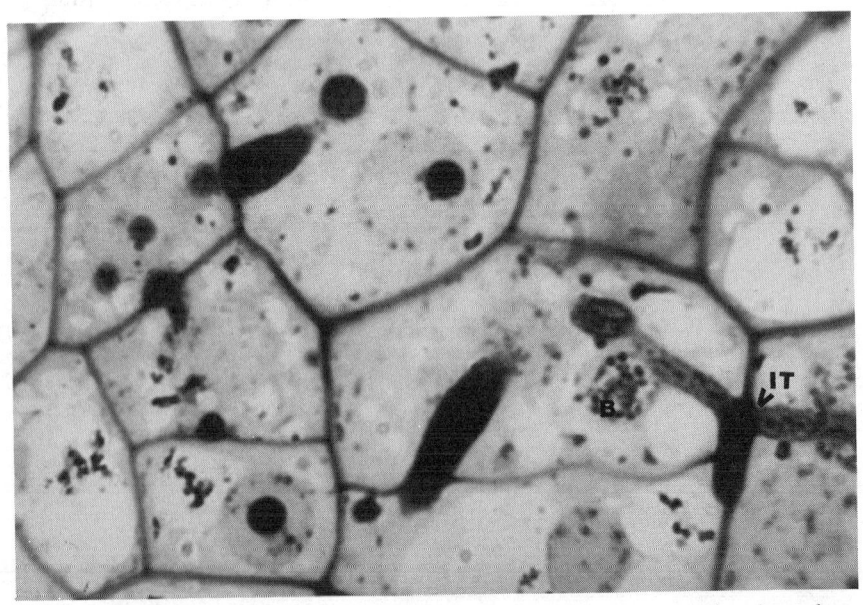

Fig. 7-2. Release of *Rhizobium* bacteria (B) from the infection thread (IT) into host plant cortical cells. × 900.

Fig. 7–3. Effective nodules of alfalfa. Photo courtesy of Dr. J. Burton.

7–2.2 Nodule Morphology

Effective alfalfa nodules are corraloid, elongate, and cylindrical and have apical meristems (135, 138, 141) at their distal end (Fig. 7–4). Nodules have two colored regions: (i) a white one that includes the nodule meristem, cortex, and zone of infection thread invasion; and (ii) a pink one, due to the presence of leghemoglobin, that is the site of active N_2 fixation and contains bacteroids in various stages of development. Older nodules have a green area indicative of senescence at their base (123, 141). Vascular bundles, surrounded by cortical cells, form a network around the nodule periphery and anastomose at the base of the nodule. The hemispherically shaped meristems are localized at the distal end of the nodule and may remain functional over a prolonged period. Nodule cells proximal to and contiguous with the meristem are invaded by in-

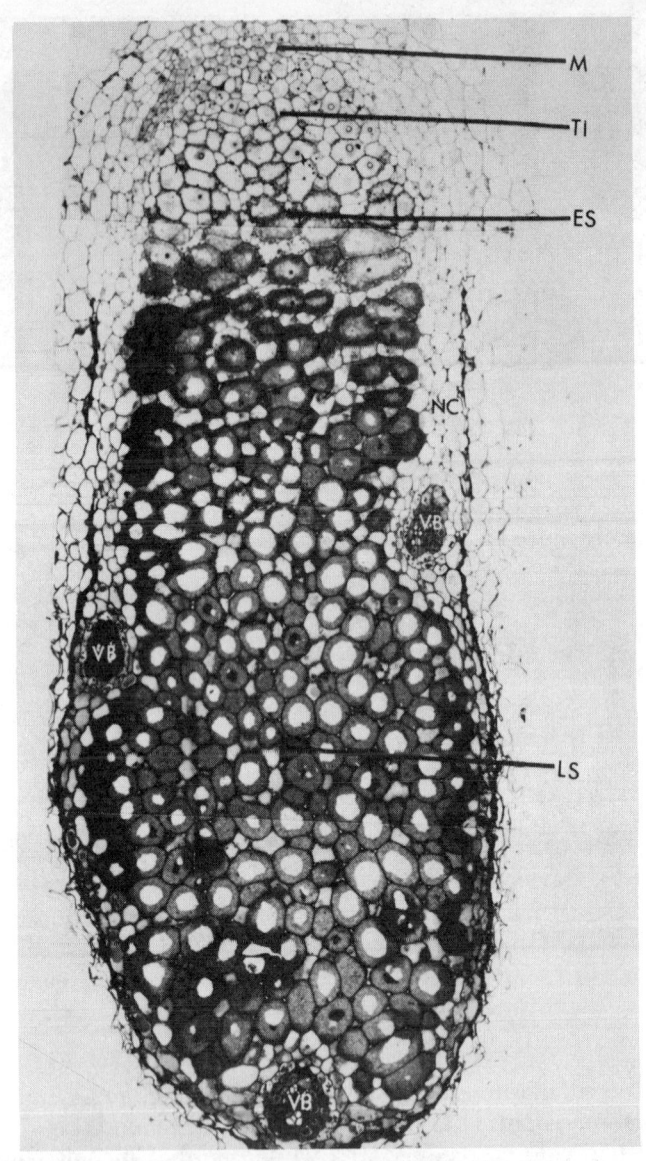

Fig. 7-4. Longitudinal section of an effective nodule. Zones illustrated are meristem (M), thread invasion (TI), early symbiotic (ES), and late symbiotic (LS). Nodule vascular bundles (VB) are outside the central dark staining mass of cells containing bacteroids and are enclosed by nodule cortex cells (NC). × 100.

fection threads (Fig. 7-2). In the thread invasion zone, rhizobia (0.5 × 1 μm) are released into the host cell cytoplasm. The bacteria, individually enclosed in host cell membranes, differentiate into bacteroids as evidenced by an increase in size (1.5 × 4.1 μm), induction of nitrogenase activity and altered cell wall components. Ultimately bacteroids fill the cytoplasm of the infected cells (Fig. 7-5).

As nodules senesce, a transition zone containing both normal and senescent bacteroid containing cells becomes apparent toward the base of the nodule (123, 141). Macroscopic evidence of senescence is a green-colored region, resulting from degradation of leghemoglobin, extending over a large portion of the nodule. Bacteroids in the senescent zone appear to aggregate and then lyse (Fig. 7-6). Eventually, senescent nodule cells contain membrane fragments and a few deteriorating bacteroids. Maintenance of nodule growth requires a fine balance between meristematic activity at the distal end of the nodule and senescence at the proximal end.

7-2.3 Ineffective Nodules

Alterations in either the host or *Rhizobium* genomes and the environment can result in the development of ineffective nodules (i.e., nodules that do not fix N_2) (74, 81, 107, 139, 140, 142). Ineffective nodules can differ from effective nodules in both physiology and structure (73, 134, 140). Ineffective nodules may also appear similar in structure to effective

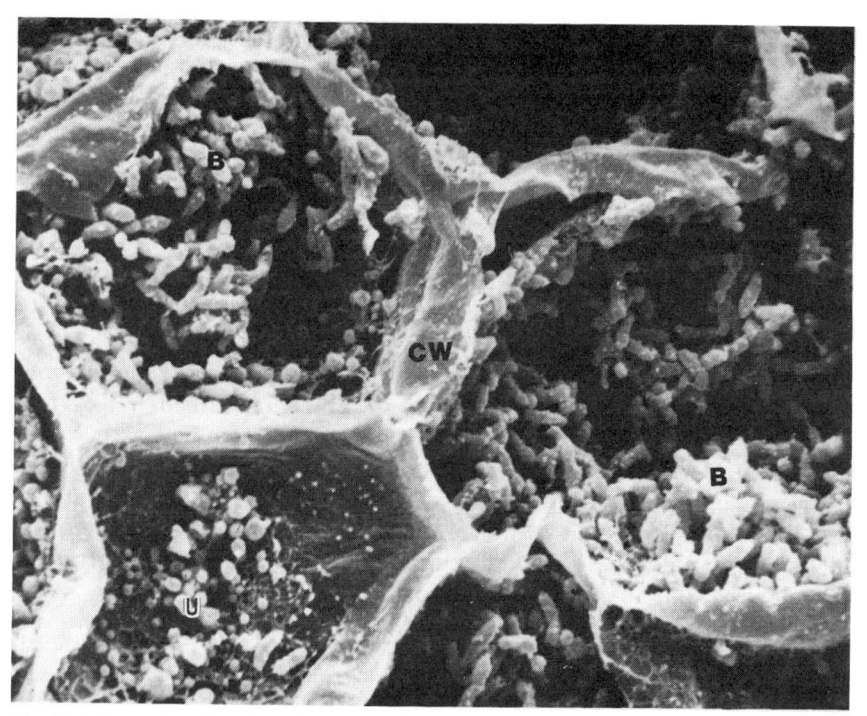

Fig. 7-5. Scanning electron micrograph of nodule cells containing bacteroids in the late symbiotic zone of an effective nodule. Bacteroids (B) fill most nodule cells, with occasional uninvaded cells (U). CW = cell wall. × 1850.

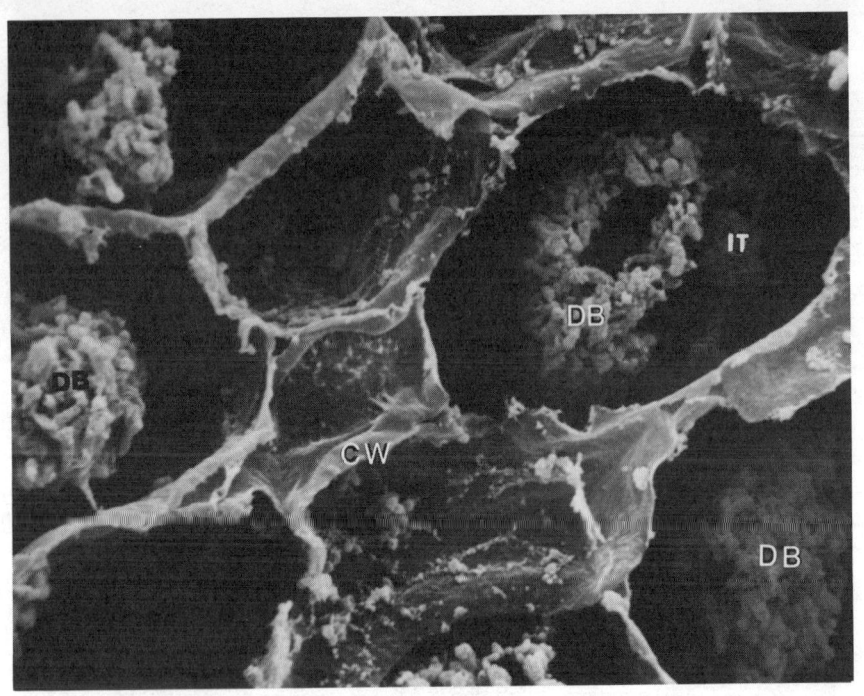

Fig. 7-6. Scanning electron micrograph of cells containing bacteroids in the senescent zone of an ineffective nodule. Deteriorating bacteroids (DB) aggregate and lyse as premature senescence occurs. IT = infection thread, CW = cell wall. × 1850.

nodules but not fix N_2 (73, 89, 139). Both *Rhizobium* and plant controlled ineffective nodules have been demonstrated in alfalfa (70, 73, 107, 134, 140).

A viomycin-resistant strain of *R. meliloti* had altered cell wall synthesis and formed ineffective nodules on alfalfa (70). The bacteria rapidly disintegrated after release into host plant nodule cells (81, 89). A leucine-requiring mutant of *R. meliloti* induced tumor-like nodules on alfalfa (134). Bacteria were not released from the infection thread, yet nodule meristems continued to divide. When plants were supplied with L-leucine, the effective condition was restored. Hirsch et al. (73) examined ineffective nodules induced by Tn5/mutants of *R. meliloti*. These mutants were altered in the *nifDK* region of the megaplasmid. Nodule bacteroids senesced prematurely. Cells containing bacteroids had large starch accumulations and meristematic activity ceased sooner than in nodules induced by the wild type of *R. meliloti*.

Ineffective nodulation regulated by the plant has been comprehensively evaluated in alfalfa (107, 139, 140, 142, 147). At least four different genetic systems in alfalfa (designated in_1-in_5) condition ineffective nodules (Table 7-2). All are simply inherited and recessive to effective symbiosis (107). These plants are ineffective with all strains of *R. meliloti* tested to date. Nodules that form on in_1, designated MnSa(In) or MnAg(In),

Table 7-2. Designation, proposed genotype, and nodule phenotype of ineffective and nonodulating alfalfa genotypes.

Designation	Genotype	Nodule phenotype
Mn Saranac (In)†	$in_1 in_1 in_1 in_1$	Numerous, large pale
Mn Agate (In)		Numerous, small white
MnNC-3226 (In)	$in_2 in_2 in_2 in_2$	Tumor-like
MnNC-3811 (In)	$in_3 in_3 in_3 in_3$	Tumor-like
MnPL-480 (In)	$in_4 in_4 in_4 in_4 in_5 in_5 in_5 in_5$	Tumor-like
MnNC-1008 (NN)†	$nn_1 nn_1 nn_1 nn_1 nn_2 nn_2 nn_2 nn_2$	Nonnodulating

† In = ineffective nodules, NN = no nodules. Nonnodulation trait (MnNC-1008) and tumor-like ineffective nodule trait (Mn PL-480) are each conditioned by two genes.

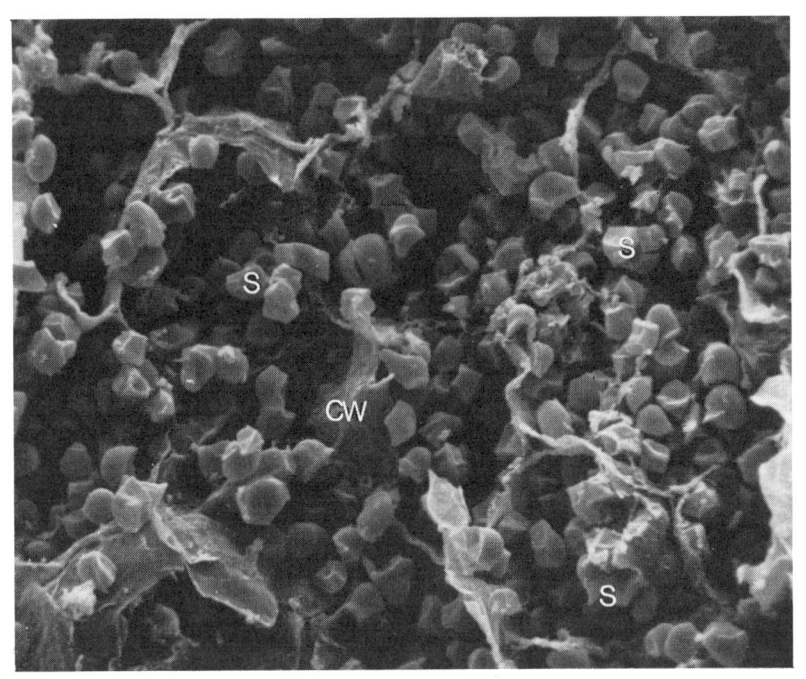

Fig. 7-7. Uninfected cells make up the major portion of tumor-like ineffective nodules. Massive starch accumulations, with no evidence of bacteria and bacteroids. CW = cell wall. × 1850.

appear normal in shape and size (137, 139). However, these nodules senesce much sooner than effective nodules. Ineffective nodules resulting from in_2 [designated McNC 3226(In)], in_3 [designated MnNC 3811(In)], and in_4, in_5 [designated MnPL 480(In)] are strikingly different from effective nodules (140). Although early development of in_2 to in_5 nodules is similar to effective nodules, mature nodules of these ineffective genotypes are tumor-like, contain few bacteria and nodule cells are filled with starch (Fig. 7-7). A nonnodulating genotype in which expression is controlled by two recessive genes has been documented (107).

The development and deterioration of both *Rhizobium* and plant controlled ineffective nodules have many features in common: (i) incomplete development and rapid senescence of bacteroids and nodules; (ii) excess starch accumulation in nodule cells; (iii) altered nodule membranes; and (iv) formation of tumor-like growths. These common features indicate that ineffective nodule formation whether plant or *Rhizobium* controlled may be regulated by similar mechanisms.

7-3 BIOCHEMISTRY OF DINITROGEN FIXATION

7-3.1 Nitrogenase and Hydrogenase

A description of biochemical properties associated with alfalfa root nodules requires discussion of the fixation, assimilation, and translocation of N as well as mention of closely related processes such as O_2 control and metabolism of C substrates and H_2. All processes have not been investigated in alfalfa, and therefore it is necessary to extrapolate, in part, from information available for other legumes (108).

The actual reduction of N_2 to NH_3 is catalyzed by the nitrogenase enzyme complex within rhizobial cells. The proteins involved in the reaction are physically and functionally quite uniform in diverse N_2-fixing organisms (17), and there is no reason to suggest that rhizobia in alfalfa nodules differ significantly in this regard. The NH_4^+ which appears as the final product from N_2 reduction apparently is secreted by the rhizobia into the surrounding alfalfa cell cytoplasm where it is assimilated by the plant enzymes glutamine synthetase and glutamate synthase (49, 125). Asparagine and aspartate (Table 7-3), the major N-containing compounds in alfalfa xylem sap (91) are produced by further metabolic reactions (122, 125). No significant concentrations of ureides, the major fixed N_2 transport compound in tropical legumes (120), have been reported in alfalfa xylem sap. Since the C/N ratio of amides is greater than

Table 7-3. Relative abundance of amino acids in alfalfa xylem sap.

Amino acid	Effectively nodulated alfalfa†	Ineffectively nodulated alfalfa‡
	% of total amino acids	
Aspartate	29.7 ± 9.4	24.5 ± 1.0
Aspargine	39.0 ± 7.8§	1.0 ± 1.0
Glutamate	1.6 ± 0.4	6.4 ± 0.1
Glutamine	13.7 ± 4.2	6.9 ± 6.9
Alanine	5.8 ± 4.7	8.2 ± 0.4
Other	10.1 ± 2.1	53.1 ± 7.2

† Each value is the mean SE of five replicates.
‡ Each value is the mean SE of two replicates.
§ The concentration of asparagine in effectively nodulated alfalfa and ineffectively nodulated alfalfa 3731 and 1, nmol/mL xylem sap^{-1}, respectively.

that of ureides, alfalfa root nodules may use more C than soybean nodules for transporting an equivalent amount of fixed N.

Although most C in the root nodule comes from photosynthetic reduction of CO_2, a recent report suggests that alfalfa provides relatively large quantities of C substrates for N assimilation by means of an extremely active phosphoenolpyruvate carboxylase. This enzyme resides in the plant cell cytoplasm surrounding the bacterial cells and fixes a portion of CO_2 that might otherwise be lost from the nodule through respiration (91, 143). The fundamental importance of this reaction for N_2 fixation is supported by a positive correlation between nodule CO_2 and N_2 fixation with various treatments which greatly affected N_2-fixation activity.

The nitrogenase enzyme complex is extremely sensitive to O_2, and yet the large ATP requirement for nitrogenase activity is believed to be derived from O_2 dependent oxidative phosphorylation within the *Rhizobium* cell (17). An important solution to this problem in leguminous root nodules is provided by leghemoglobin, a plant protein found outside the rhizobial cell. Leghemoglobin is thought to facilitate the diffusion of O_2 through the plant cell cytoplasm to the bacterial cells at concentrations that allow oxidative phosphorylation to occur without destroying nitrogenase activity (153). Recent serological data from leghemoglobin components in alfalfa root nodules suggest that these proteins are more similar to leghemoglobins found in pea and clover than to those in soybean or lupine (78).

Another important feature of N_2-fixing rhizobial cells is the metabolism of H_2. Nitrogenase allocates a significant fraction of reductant to protons to form H_2 during the process of N_2 fixation (16). This reduction of protons is an ATP-dependent process which seemingly represents an energy loss from the nodule. Some strains of *Rhizobium* have evolved a separate uptake hydrogenase system that can oxidize H_2 to water and, in some cases, couple that oxidation to ATP formation (35, 40). Although the uptake hydrogenase system confers several potential biochemical advantages to *Rhizobium*, including energy conservation and additional protection of nitrogenase from inhibition by H_2 and O_2 (36, 40), rigorous attempts to demonstrate such advantages at the agronomic level in soybean have not been reproducibly successful (40). No genotypes of *R. meliloti* with physiologically significant activities of uptake hydrogenase have been found in nature (118), and a strain of *R. meliloti* constructed by transferring plasmid-borne genes for uptake hydrogenase activity from *R. leguminosarum* showed extremely low $^3H_2$-incorporation activity in symbiosis with 12 cultivars of alfalfa (7). Additional work on this poorly understood area of metabolism in N_2-fixing rhizobia will increase our understanding of this seemingly important process. Whether such information will have any effect on alfalfa production remains to be determined.

7-4 FACTORS AFFECTING DINITROGEN FIXATION

7-4.1 *Rhizobium* × Cultivar Interactions

All strains of *R. meliloti* do not stimulate equivalent plant growth in a given alfalfa cultivar grown on an N-free rooting substrate (Fig. 7-8). A strain inducing superior performance of one cultivar may produce a suboptimal response on another cultivar. This interaction between the two organisms means that the rhizobial strain and the host cultivar must be matched carefully for optimum N_2 fixation (13, 14, 20, 21, 42, 47). These interactions can seldom be explained at the basic biological level, yet they cause obvious differences in dry matter production, reduced N content, and plant height and vigor. The most extreme case in which a particular strain of *Rhizobium* fails to nodulate a specific cultivar is seldom observed. A more common interaction, when one screens indigenous soil isolates of *R. meliloti*, is the production of small white ineffective nodules.

The problem of optimizing the *Rhizobium* × cultivar interactions in fields where alfalfa is planted for the first time has been solved by using *Rhizobium* inoculum containing a mixture of effective strains (18). In general, such inocula effectively nodulate plants, if precautions given on the package are followed.

Optimization of *Rhizobium* × cultivar interactions can be more difficult in fields that have a previous history of alfalfa production. A classic study that assessed the forage production benefits of rhizobial inoculation

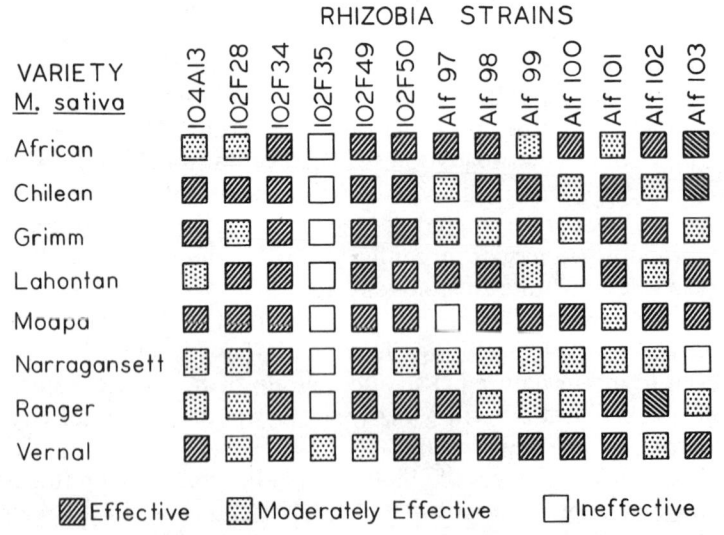

Fig. 7-8. Reactions of eight cultivars of alfalfa to inoculation with 13 strains of *R. meliloti*. Ratings are based on total N content of 6-week plants as follows: Effective 100 to 75% of N in most effective variety association; moderately effective, 74 to 50% or N in most effective variety association. All other reactions were considered ineffective. Taken with permission from Burton (18).

in alfalfa found that for many sites in the United Kingdom inoculation produced no increase in forage yield because effective rhizobia already were present in the soil (130). The experience of agronomists in the USA suggests that a similar situation exists here.

The present situation, however, does not mean that optimum symbiotic associations are established in most production fields. Research suggests that the native or indigenous populations of *R. meliloti* present in the soil, whether they form poor, moderate, or highly effective symbioses, cannot be displaced by more effective strains available in commercial inoculum preparations (53, 61). Because many of the indigenous strains found in U.S. soils form suboptimal symbioses with common alfalfa cultivars, significant improvement in alfalfa production could result if methods were developed for establishing highly effective strains in specific plant cultivars. Experience from soybean, where the same problem exists, suggests that increasing the number of rhizobia applied to the seed may result in a larger fraction of root nodules being formed by the desired strain (83). Generally, however, the indigenous strains present on most sites are so competitive that inoculated strains form an average of only 5% of the root nodules (52, 80). More recent studies designed to identify highly competitive strains of *R. meliloti* (144) may produce inocula that provide a temporary solution, but the long term need to introduce such strains over large areas remains an intractable problem. Genetic techniques that have facilitated the development of symbiotically superior rhizobial strains for other legumes (30, 129, 152) should be useful for increasing alfalfa production if the new strains of *R. meliloti* can be established under field conditions.

7-4.2 Forage Harvest

Cutting alfalfa forage has a rapid negative effect on symbiotic N_2 fixation (25, 63, 138). Within 2 d of harvest, N_2 fixation estimated by the indirect acetylene (C_2H_2) reduction assay declines 78 to 88% in dormant alfalfa. A survey of 10 cultivars from different dormancy classifications in combination with 10 strains of *R. meliloti* showed similar 70 to 96% declines in C_2H_2 reduction activity within 3 d (45). The decrease in N_2 fixation appears to be linked to a reduction in available photosynthate because partial leaf removal has less effect on N_2 fixation than total shoot harvest, and leaf area expansion after harvest parallels increases in C_2H_2 reduction (25). Direct measurements of C-exchange rates in the developing shoots showed increases that paralleled C_2H_2 reduction rates (45). Although much of the increase in shoot N content during regrowth probably can be attributed to N derived from N_2 fixation, mobilization of root and crown N reserves to developing shoots, similar to that measured in *Trifolium subterraneum* and *Bromus mollis* during regrowth (111), probably occurs in alfalfa.

7-4.3 Environmental Factors

The single most important environmental factor affecting N_2 fixation under field conditions is probably soil N. Adding fertilizer N to legumes definitely decreases the fraction of plant N derived from N_2, but adding low levels fertilizer N sometimes increases the total amount of N_2 fixed (1). This apparent discrepancy may be explained by the fact that legumes totally dependent on N_2 fixation lack the N required for optimum photosynthesis during seedling growth (31). Such observations imply that a small amount of N fertilizer might increase alfalfa forage yield during the seeding year, but interactions involving plant cultivar, *Rhizobium* strain, and available soil N (102) suggest that it will be difficult to demonstrate a reproducible practical application of this concept. Numerous studies show that inhibition of N_2 fixation by combined N results from specific effects of combined N on root nodule formation or function (62, 102). Deficiencies of other mineral elements essential for plant growth can decrease N_2 fixation, but in most cases such effects probably are produced by a general depression of plant growth. Specific reports indicate that suboptimum levels of K, Ca, and Mg can decrease N_2 fixation in alfalfa (24, 44, 95). The best recommendation for maximizing N_2 fixation under field conditions is to provide optimum plant nutrients required for growth and to allow *Rhizobium* to supply the N needs to the crop.

Soil pH is an important environmental parameter that can affect root nodule formation. With adequate Ca and nitrate (NO_3^-), alfalfa can grow normally at pH values as low as 4.0 (98). With similar concentrations of Ca in N-free nutrient solutions, however, root nodules are not formed below pH 4.5, and growth ceases (101). The same study demonstrated that root nodules failed to form with lower Ca treatments at pH values as high as 5.5. Traditional liming treatments can prevent pH problems in alfalfa production, and thus this environmental factor need not limit N_2 fixation.

Treatments unfavorable for plant growth can affect symbiotic N_2 fixation. Root temperatures of 30°C decreased apparent N_2 fixation, as measured by C_2H_2 reduction, relative to alfalfa plants maintained at 16°C (3). Lower temperatures (5–10°C), associated with hardening in alfalfa, decreased C_2H_2 reduction relative to higher temperatures (20–25°C), but very hardy cultivars were less affected than cultivars with low to moderate hardiness (38).

Experiments with irradiance and CO_2 concentration have shown that reductions in these parameters can decrease N_2 fixation in alfalfa (88). The data suggest that field-grown plants have adequate light energy but inadequate CO_2 for optimum growth and N_2 fixation. In practice, these particular variables cannot be altered economically under field conditions.

7-5 DINITROGEN FIXATION BY ALFALFA COMMUNITIES

7-5.1 Methods of Measurements

Direct estimates of N_2 fixation of crop communities have been emphasized since 1976. This research has been in response to the need for information on the variation of symbiotic activity with genotype and crop development; and the response of N_2 fixation to both environmental stresses and crop management strategies. The investigations have been facilitated by modification and application to the field of methods previously developed in the laboratory or glasshouse, viz. the C_2H_2-reduction assay (34, 55), the difference method (8), and the ^{15}N isotope dilution method (92).

7-5.1.1 Acetylene Reduction Assay

The basis for this method is that the reduction of acetylene to ethylene mimics the reduction of N_2 to NH_3 by the nitrogenase enzyme. The chromatographic assay for ethylene is highly sensitive and diagnostic for the occurrence of nitrogenase activity in nodulated roots. In addition to sensitivity, the method is attractive because of its relative simplicity, comparatively low cost, and rapid execution.

One of the principal shortfalls of the C_2H_2 reduction assay for field use is the large variability among replicate samples of the same treatment (23, 48, 110, 124). This variability can be attributed to environmental differences in assay conditions such as temperature (26), the environmental history of the plants before assay, and the difficulties in obtaining representative nodulated root samples. Additional difficulties include the difficulty in quantitatively converting C_2H_2 reduction to N_2 reduced (54), and in predicting long-term performance from short-term measurements (86). In field investigations, the C_2H_2 reduction assay has found greatest utility in differentiating the effects of management and environmental treatments upon nitrogenase activities (23, 79).

7-5.1.2 Difference Method

This method is based on determining the difference in yield of total reduced N (kg of N/ha) between a N_2-fixing crop and an appropriate non-N_2-fixing control (8). The only required measurements are Kjeldahl N analyses and dry matter yields. Uninoculated legumes grown in the absence of an indigenous rhizobial populations (117, 148) have been used as controls. Other controls include legume genotypes which are nonnodulating (71, 117) or ineffectively nodulating (71, 107) because of recessive host genes. Annual or perennial monocots with a seasonal growth pattern similar to that of the legume are also used when nonnodulating or ineffectively nodulating legume genotypes are unavailable (8). Compared with the C_2H_2-reduction assay, the difference method requires less sophisticated analytical techniques, integrates crop performance over several weeks of the growing season between samplings, and involves smaller errors of measurements. The requirements of relatively large plot areas for serial sampling (8, 71), and assumptions based on similar N uptake characteristics of the fixing and nonfixing counterparts (71, 117, 154) are among the disadvantages of this method.

7-5.1.3 Isotope Dilution Technique

This is a stable isotope procedure which relies upon labeling the organic matter of the soil in which the legume is growing with the mass 15 isotope of N. This isotope is scarce (<0.4%) in the soil and atmosphere compared with the abundant (>99.6%) mass 14 isotope. The N-15 enrichment of plant samples is measured by mass spectrometry or emission spectroscopy for the normal, effectively nodulated legume and a suitable, non-N_2-fixing control. The N-15 enrichment of the tissue of the fixing plant is less than that of the non-N_2-fixing control owing to dilution of the soil-derived tracer by N-14 fixed by nodules. The protocol and calculation methods for this technique have been published (12, 59, 92, 116, 154).

The isotope-dilution technique is the most expensive of the three field methods because of the need to purchase the stable isotope and specialized instrumentation. The operation of a mass spectrometer or emission spectrograph demands adequate physical facilities and skilled analytical personnel.

The isotope-dilution technique is perhaps more suited to serial sampling than the difference method because N_2 fixation can be estimated by the former independent of yield measurements, which reduces sample size and plot space. In comparisons with the difference method, several investigators concluded that the isotope dilution technique was the more precise (71, 119, 126, 154). Similarly, in comparisons of field measurements with the C_2H_2-reduction assay, both the difference method and isotope-dilution method have smaller measurement errors (48, 90, 110, 132, 145). Table 7-1 summarizes seasonal comparisons of alfalfa N_2 fixation obtained by isotope dilution of difference methods with those of other forage and pasture legumes, and grass-legume mixtures. More extensive comparisons of alfalfa with grain and pulse legumes are not available (57).

7-5.2 Determinants of Community Performance

Several of the legume species in Table 7-1 show a substantial range in measured values of N_2 fixation. Because there are so few field measurement of alfalfa N_2 fixation, the following examples are more applicable to the environments and localities where the experiments were conducted than to broad geographical areas.

7-5.2.1 Location Effects

There have been few attempts to compare annual or intra-annual values of alfalfa N_2 fixation across geographical locations. The large interlocation variation in N_2 fixation measured for 'Du Puits' in Great Britain (8) was attributable to effectiveness for rhizobial inoculant, fertility status of the site, and frequency of harvest. Thus, it is clear that many factors affecting dry matter productivity will affect N_2 fixation. Properly fertilized alfalfa grown year round in Great Britain (8) fixed more N_2 than that grown in the shorter Minnesota season (59, 60). For a specific growing season, total dry matter production and N_2 fixation for a particular hay or pasture legume are often highly correlated (8, 60). However, this correlation does not necessarily hold among harvests within a growing season.

7-5.2.2 Stability of Fixation

Comparisons of alfalfa N_2 fixation across locations or among years within a location are often confounded with cultivar or genotype. The changing genotypic composition of an alfalfa cultivar in successive years of the stand also contributes to the confounding of genotype and environmental effects on N_2 fixation. A recent comparison of N_2 fixation among ramets of 12 clones across 4 yr showed highly reproducible rankings over years for high and low N_2-fixing clones, with the remainder showing considerable interannual variation (61).

7-5.2.3 Stage of Growth

The nitrogenase activity of individual alfalfa plants grown in pots in controlled environments usually increases from seeding through later growth stages owing to accretion of nodule mass (25). In contrast, differences in N_2 fixation of a crop community on a land area basis among growth stages within a year follow a different pattern (Fig. 7-9).

The pattern of low symbiotic activity early in the establishment year in comparison with midseason values (Fig. 7-9) may be attributable to a delay in the establishment of nodules while seedlings utilize soil N mineralized after seedbed preparation. Because a large proportion of established nodules may overwinter in frozen soil without the necessity of reinfection and nodule development the following spring, this pattern is not evident in subsequent years. The tendency for low symbiotic activity late in the growing season (Fig. 7-9) of all years may be a response to the combined effects of cool soil temperatures, herbage damage by frost, and the onset of dormancy. Dormant alfalfa cultivars cease N_2 fixation in the fall sooner than moderately dormant cultivars (59).

Although N_2 fixation and herbage yield are closely associated on a seasonal basis, yields at individual harvests are not necessarily correlated with concurrent N_2 fixation. This may indicate the changing dependence of the legume on soil N (59), the effects of abnormal precipitation (65) as well as other environmental factors.

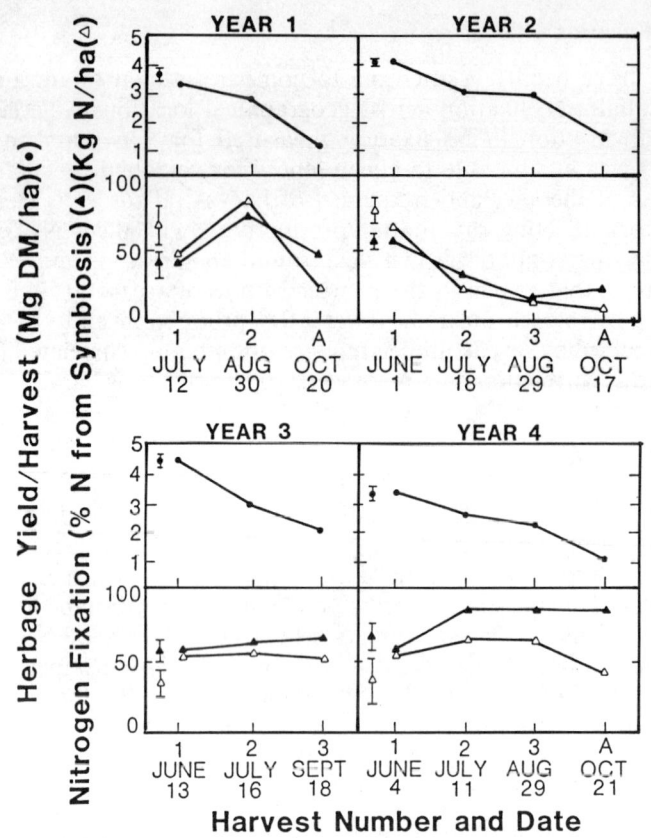

Fig. 7-9. Relation of herbage yields to N_2 fixation capability of 'Agate' alfalfa within and among years over a 4-yr life stand. Year 1 is the seeding year. Numbered harvests correspond to complete herbage removal to 7 cm height, with A designating the aftermath sampling. Adapted from Heichel et al. (60).

7-5.2.4 Age of Stand

Plant populations typically decline over the lifetime of an alfalfa stand. Despite this continual attrition of plants from the community, N_2 fixation was maintained throughout the stand life (Fig. 7-9) because of yield compensation by the surviving plants. Differences in N_2 fixation between cultivars were evident only in the 4th yr of the stand, when the dormant, disease-resistant cultivar performed 30% better than the moderately dormant, less disease-resistant cultivar (60). Thus, the principle effects of stand age on alfalfa N_2 fixation are through mechanisms affecting cultivar persistence.

7-5.2.5 Cultivar or Genotype

There have been several reports of cultivar or genotype differences in N_2 fixation for plants grown in controlled environments (47, 121, 127, 146). For two cultivars grown in crop communities with normal management, differences in N_2 fixation were evident only in the 4th yr of the stand (60). Comparisons of cultivars or genotypes grown in the field in simulated swards (59, 60, 64, 77) further confirm the existence of host-conditioned differences in alfalfa N_2 fixation that may be subject to genetic manipulation.

7-5.3 Significance in Crop Rotations

Legumes have long been known to improve the yield of subsequent nonlegume crops (115). The role of N_2 fixation by legumes as one factor in this yield improvement became known early in this century (46), although other non-N factors have been implicated (58, 72). Only the symbiotically fixed N in alfalfa dry matter can be considered as a net input to the soil-plant system in crop rotations, as the soil N contribution to alfalfa N content represents a temporary storage until it is recycled to the soil N pool.

A variable proportion of the N_2 fixed by alfalfa is typically returned to the soil for possible use by the following crop (58). This is because a portion of the symbiotically fixed N_2 is removed from the land when alfalfa is harvested, with the balance remaining in unharvested roots and crowns. The symbiotically derived N that is available for soil incorporation depends upon the time of the season when tillage occurs, and the proportion of the plowed down plant that is N-rich herbage compared with relatively N-poor crown and roots. The N budget in Table 7-4

Table 7-4. Nitrogen budget for seeding-year alfalfa illustrating the allocation of soil and symbiotic N among crop components, and the net return of N to the soil with two moldboard plowing practices. Adapted from Heichel and Barnes (58).

	Harvest		
	First (12 July)	Second (30 August)	Third (20 October)
Herbage yield (kg of dry matter ha^{-1})	3503	3054	1156
Total reduced N yield (herbage and crown and roots) (kg of N ha^{-1})	118	127	59
Total N_2 fixed (kg of N ha^{-1})	57	102	34
Herbage	52	74	22
Roots and crown	5	28	12
N from soil	61	25	25
Herbage	54	18	16
Roots and crown	7	7	9
Management options			
Plow 20 October			
N return/harvest (kg ha^{-1})	−49	+10	+34
Cumulative N return (kg ha^{-1})	−49	−39	−5
Plow 30 August			
N return/harvest (kg ha^{-1})	−49	+102	--
Cumulative N return (kg ha^{-1})	−49	+53	--

illustrates the net N return to the soil that is possible when an establishment-year stand of alfalfa is incorporated by either fall or late summer moldboard plowing (58).

If two herbage harvests are taken, followed by herbage regrowth before plowing in late October, the early season N deficit is nearly replaced by late season N_2 fixation so that only an inconsequential loss of 5 kg of N/ha occurred. In contrast, removal of one herbage harvest followed by moldboard plowing of a lush herbage regrowth in late August allowed a net input of 53 kg of N/ha. The benefit of the alfalfa N in a crop rotation depends upon its rate of mineralization and recovery by the subsequent nonlegume crop. Emerging evidence suggests that only 25% of the incorporated N is recovered in the next crop year (56). Clearly, the benefit of growing alfalfa as a N source for a subsequent nonlegume will be influenced substantially by both harvest schedule and tillage management of the legume.

7–5.4 Improvement

Crop management and plant genetics have roles in the improvement of alfalfa N_2 fixation. Dinitrogen fixation can be optimized by ensuring nodulation by a highly effective rhizobial strain, and by fertilizer applications to minimize the limitations of mineral nutrition. These topics have been discussed previously.

Plant genetics could have an important role in the future improvement of N_2 fixation because many traits associated with N_2 fixation and N accumulation are under genetic control. Host control of functionality in nodules formed by effective rhizobial strains has long been known (47) and is well-documented in alfalfa (53). Recessive genes (107) condition the formation of ineffective nodules, or entirely prevent nodulation by bacteria that effectively nodule other genotypes of alfalfa. Knowledge of the genetic control of nodule formation and development will be important to the future breeding of germplasm allowing controlled nodulation by specific rhizobial strain.

Considerable genetic variability exists in alfalfa for rate of nitrogenase activity per plant (37, 107), and this trait is highly heritable (146). The control of nitrogenase activity on a per plant basis is largely through nodule mass and number per plant (76, 146). There is no convincing evidence of intraspecies variation in nitrogenase activity per nodule mass except that conditioned by host genes for ineffective nodulation. The interdependence of nodule mass and nitrogenase activity per plant has led to relatively easy manipulation of the latter trait in alfalfa breeding programs (5, 146). Modification of root phenotype by selection for an increase of lateral or fibrous roots, and thereby the number of sites for nodulation, is important to increasing nodule mass (146).

Nodule nitrogenase activity is often associated with the activities of other nodule enzymes of N and C assimilation (49, 50). The activities of nodule phosphoenolpyruvate carboxylase (PEPC) and glutamate synthase (GOGAT) show substantial genetic variability (76, 77), but apparently do not limit growth or N_2 fixation in the field. However, the observation that alfalfa germplasm selected for a reduced PEPC activity fixed less N shows that a certain amount or activity of this nodule enzyme is essential for optimal N_2 fixation (76).

Nitrogen storage as protein in alfalfa herbage is an important determinant of forage quality. Variation among cultivars and strains for crude protein concentration (67, 68) and Fraction I protein concentration (96) is known. Selection for increased N_2 fixation has produced alfalfa populations with increased forage yield and quality (114, 128). Greenhouse selection for forage dry weight and Kjeldahl N concentration under both N_2- and $NH_4NO_3^-$ dependent growth conditions in the nondormant 'Hairy Peruvian' resulted in an increase in shoot crude protein and shoot total N (114, 128). The selected line also fixed an average of 38% more N_2 under all N regimes and appeared to have physiological traits that resulted in improved N assimilation (109, 112). The improved plants increased N_2 fixation by strains of rhizobia that were not present during the selection process (109, 112). Selection resulted in more root nodules and stems per plant and increased dry matter was not associated with decreased quality (32). Preliminary field data showed that selected material produced 10% more forage dry matter containing 22% more crude protein than the original population (109, 113).

Nitrogen stored in subterranean organs and residues of alfalfa becomes available to companion or succeeding crops after phytomass degradation. Two cycles of recurrent phenotypic selection for an increase of total reduced N concentration in roots and crowns of alfalfa increased whole-plant N concentration in two of three nondormant alfalfa germplasm (58). Selection for increased crown and root mass was also successful. Thus, potential exists within alfalfa for development of germplasm with increased N storage capability for improving forage quality and/or as an improved green manure crop for use in crop rotation.

REFERENCES

1. Allos, H.F., and W.V. Bartholomew. 1959. Replacement of symbiotic fixation by available nitrogen. Soil Sci. 87:61–66.
2. Aughtry, J.D. 1948. Effect of genetic factors in *Medicago* on symbiosis with *Rhizobium*. Cornell Univ. Agric. Exp. Stn. Memoir 280.
3. Barta, A.L. 1978. Effect of root temperature on dry matter distribution, carbohydrate accumulation and acetylene reduction in alfalfa and birdsfoot trefoil. Crop Sci. 18:637–640.
4. Banfalvi, Z., V.K. Sakanyan, C. Konz, A. Kiss, I. Dusha, and A. Kondorosi. 1981. Location of nodulation and nitrogen fixation genes on a high molecular weight plasmid of *R. meliloti*. Mol. Gen. Genet. 184:318–325.
5. Barnes, D.K., G.H. Heichel, C.P. Vance, and W.R. Ellis. 1984. A multiple trait breeding program for improving the symbiosis for N_2 fixation between *Medicago sativa* L. and *Rhizobium meliloti*. Plant Soil 82:303–314.
6. Bauer, W.D. 1981. Infection of legumes by rhizobia. Annu. Rev. Plant Physiol. 32:407–449.
7. Bedmar, E.J., N.J. Brewin, and D.A. Phillips. 1984. Effect of plasmid pIJ1008 from

Rhizobium leguminosarum on symbiotic function of *Rhizobium meliloti*. Appl. Environ. Microbiol. 47:876–878.
8. Bell, F., and P.S. Nutman. 1971. Experiments on nitrogen fixation by nodulated lucerne. Plant Soil Spec. Vol. 1971: 231–264.
9. Bergersen, F.J. 1961. The growth of *Rhizobium* in synthetic media. Aust. J. Biol. Sci. 14:349–360.
10. Bohlool, B.B., and E.L. Schmidt. 1974. Lectins: A possible basis for specificity in the *Rhizobium*–legume root nodule symbiosis. Science 185:269–271.
11. Bouton, J.H., M.E. Sumner, and J.E. Giddens. 1981. Alfalfa, *Medicago sativa* L. In highly weathered soils. I. Effect of lime and P application on yield and acetylene reduction of young plants. Plant Soil 60:205–211.
12. Broadbent, F.E., T. Nakashima, and G.Y. Chang. 1982. Estimation of nitrogen fixation by isotope dilution in field and greenhouse experiments. Agron. J. 74:625–628.
13. Brockwell, J., and R.W. Hely. 1961. Symbiotic characteristics of *Rhizobium meliloti* from the brown acid soils of MacQuarie Region of New South Wales. Aust. J. Agric. Res. 12:630–643.
14. ----. 1966. Symbiotic characteristics of *Rhizobium meliloti*: An appraisal of the systematic treatment of nodulation and nitrogen fixation interactions between hosts and rhizobia of diverse origins. Aust. J. Agric. Res. 17:885–899.
15. Broughton, W.J. 1978. Control of specificity in legume-*Rhizobium* interactions. J. Appl. Bacteriol. 45:165–194.
16. Bulen, W.A., and J.R. Le Comte. 1966. The nitrogenase system from *Azotobacter*: Two enzyme requirement for N_2 reduction, ATP-dependent H_2 evolution, and ATP hydrolysis. Proc. Natl. Acad. Sci. USA 56:979–986.
17. Burns, R.C., and R.W.F. Hardy. 1975. Nitrogen fixation in bacteria and higher plants. Springer-Verlag, New York, New York.
18. Burton, J.C. 1972. Nodulation and nitrogen fixation in alfalfa. p. 229–246. *In* C.H. Hansen (ed.) Alfalfa science and technology. Agronomy 15:229–246.
19. ----, and R.L. Curley. 1965. Comparative efficiency of liquid and peat-base inoculants on field grown soybeans (*Glycine max*). Agron. J. 57:379–381.
20. ----, and L.W. Erdman. 1940. A division of the alfalfa cross-inoculation group correlating efficiency in nitrogen fixation with source of *Rhizobium meliloti*. J. Am. Soc. Agron. 32:439–450.
21. ----, and P.W. Wilson. 1939. Host plant specificity among the *Medicago* in association with root-nodule bacteria. Soil Sci. 47:293–303.
22. Callaham, D.A., and J.G. Torrey. 1981. The structural basis for infection of root hairs of *Trifolium repens* by *Rhizobium*. Can. J. Bot. 59:1647–1664.
23. Carter, P.R., and C.C. Sheaffer. 1983. Alfalfa response to soil water deficits. III. Nodulation and N_2 fixation. Crop Sci. 23:985–990.
24. Collins, M., and S.H. Duke. 1981. Influence of potassium-fertilization rate and form of photosynthesis and N_2 fixation on alfalfa. Crop Sci. 21:481–485.
25. Cralle, H.T., and G.H. Heichel. 1981. Nitrogen fixation and vegetative regrowth of alfalfa and birdsfoot trefoil after successive harvests or floral debudding. Plant Physiol. 67:898–905.
26. ----, and ----. 1982. Temperature and chilling sensitivity of nodule nitrogenase activity of unhardened alfalfa. Crop Sci. 22:300–304.
27. Danegeard, P.A. 1926. Recherches sur des tubercules radicaux des. Legumineuses. Botaniste Ser. 16.
28. Dart, P.J. 1975. Legume root nodule initiation and development. p. 467–506. *In* J.G. Torrey and D.T. Clarkson (ed.) The development and function of roots. Academic Press, New York.
29. Dazzo, F.B., and D.H. Hubbell. 1975. Cross-reactive antigens and lectins as determinants of symbiotic specificity in *Rhizobium trifolii*–clover association. Appl. Microbiol. 30:1017–1033.
30. DeJong, T.M., N.J. Brewin, A.W.B. Johnston, and D.A. Phillips. 1982. Improvement of symbiotic properties of *Rhizobium leguminosarum* by plasmid transfer. J. Genet. Microbiol. 128:1829–1838.
31. ----, and D.A. Phillips. 1981. Nitrogen stress and apparent photosynthesis in symbiotically grown *Pisum sativum* L. Plant Physiol. 68:309–313.
32. Demment, M.W., L.R. Teuber, D.P. Borque, and D.A. Phillips. 1986. Changes in forage quality of improved alfalfa populations. Crop Sci. 26:1137–1143.
33. Denarie, J., P. Boistard, F. Casse-Delbart, A.G. Atherly, J.D. Berry, and P. Russell. 1981. Indigenous plasmids of *Rhizobium*. p. 225–246. *In* K.L. Giles and A.G. Atherly (ed.) Biology of the Rhizobiaceae. Int. Rev. Cytol. Suppl. 13. Academic Press, New York.
34. Dilworth, M.J. 1966. Acetylene reduction by nitrogen fixing preparations from *Clostridium pasteurianum*. Biochem. Biophys. Acta 127:285–294.

35. Dixon, R.O.D. 1968. Hydrogenase in root nodule bacteroids. Arch. Mikrobiol. 62:272–283.
36. ----. 1972. Hydrogenase in legume root nodule bacteroids: Occurrence and properties. Arch. Mikrobiol. 85:193–201.
37. Duhigg, P., B.A. Melton, and A.A. Baltensperger. 1978. Selection for acetylene reduction rates in 'Mesilla' alfalfa. Crop Sci. 18:813–816.
38. Duke, S.H., and D.C. Doehlert. 1981. Root respiration, nodulation, and enzyme activities in alfalfa during cold acclimation. Crop Sci. 21:489–495.
39. Dullart, J. 1967. Quantitative estimation of indole acetic acid and indole carboxylic acid in root nodules and roots of *Lupinus luteus* L. Acta Bot. Neerl. 16:222–230.
40. Eisbrenner, G., and H.J. Evans. 1983. Aspects of hydrogen metabolism in nitrogen-fixing legumes and other plant-microbe associations. Annu. Rev. Plant Physiol. 34:105–136.
41. Elkan, G.H. 1981. The taxonomy of Rhizobiaceae. p. 1–14. *In* K.L. Giles and A.G. Atherly (ed.) Biology of the Rhizobiaceae, Int. Rev. Cytol. Suppl. 13. Academic Press, New York.
42. Erdman, L.W., and U.M. Means. 1953. Strain variation of *Rhizobium meliloti* on three varieties of *Medicago sativa*. Agron. J. 45:625–629.
43. Fahraeus, G., and K. Sahlman. 1977. The infection of root hairs of leguminous plants by nodule bacteria. Annl. Acad. Reg. Sci. Upsaliensis. 20:103–131.
44. Feigenbaum, S., and K. Mengel. 1979. The effect of reduced light intensity and suboptimal potassium supply on N_2 fixation and N turnover in *Rhizobium* infected lucerne. Physiol. Plant. 45:245–249.
45. Fishbeck, K.A., and D.A. Phillips. 1982. Host plant and *Rhizobium* effects on acetylene reduction in alfalfa during regrowth. Crop Sci. 22:251–254.
46. Fred, E.B., I.L. Baldwin, and E. McCoy. 1932. Root nodule bacteria and leguminous plants. Studies in Science. No. 5. University of Wisconsin Press, Madison.
47. Gibson, A.H. 1962. Genetic variation in the effectiveness of nodulation of lucerne varieties. Aust. J. Agric. Res. 13:388–399.
48. Goh, K.M., D.C. Edmeades, and B.W. Robinson. 1978. Field measurements of symbiotic nitrogen fixation in an established pasture using acetylene reduction and a ^{15}N method. Soil Biol. Biochem. 10:13–20.
49. Groat, R.G., and C.P. Vance. 1981. Root nodule enzymes of ammonia assimilation in alfalfa (*Medicago sativa* L.): Developmental patterns and response to applied nitrogen. Plant Physiol. 67:1198–1203.
50. ----, ----, and D.K. Barnes. 1984. Host plant nodule enzymes associated with selection for increased N_2 fixation in alfalfa. Crop Sci. 24:895–898.
51. Gutschick, V.P. 1980. Energy flow in the nitrogen cycle, especially in fixation. p. 17–27. *In* W.E. Newton and W.H. Orme-Johnson (ed.) Nitrogen fixation, Vol. I. University Park Press, Baltimore.
52. Ham, G.E., V.B. Cardwell, and H.W. Johnson. 1971. Evaluation of *Rhizobium japonicum* inoculants in soils containing naturalized populations of rhizobia. Agron. J. 63:301–303.
53. Hardarson, G., G.H. Heichel, D.K. Barnes, and C.P. Vance. 1982. Rhizobial strain preference of alfalfa populations selected for characteristics associated with nitrogen fixation. Crop Sci. 22:55–58.
54. Hardy, R.W.F., R.C. Burns, and R.D. Holsten. 1973. Applications of the acetylene-ethylene assay for measurement of nitrogen-fixation. Soil Biol. Biochem. 5:47–81.
55. ----, R.O. Holsten, E.J. Jackson, and R.C. Burns. 1968. The acetylene-ethylene assay for nitrogen fixation; laboratory and field evaluation. Plant Physiol. 43:1185–1207.
56. Hiechel, G.H. 1985. Nitrogen recovery by crops that follow legumes. p. 290–295. *In* R.F Barnes et al. (ed.) Forage legumes for energy-efficient animal production. Proc. of a Trilateral Workshop, Palmerston North, New Zealand. USDA-ARS, Washington, DC.
57. ----. 1988. Legume nitrogen: Symbiotic fixation and recovery by subsequent crops. *In* Z. Helsel (ed.) Fertilizers and pesticides. World Agric. Handb. Ser. Elsevier, Amsterdam, Netherlands (In press.)
58. ----, and D.K. Barnes. 1984. Opportunities for meeting crop nitrogen needs from symbiotic nitrogen fixation. p. 49–59. *In* D. Bezdicek and J. Power (ed.) Organic farming: Current technology and its role in a sustainable agriculture. Spec. Publ. 46, American Society of Agronomy, Madison, WI.
59. ----, ----, and C.P. Vance. 1981. Nitrogen fixation of alfalfa in the seedling year. Crop Sci. 21:330–335.
60. ----, ----, ----, and K.I. Henjum. 1984. Dinitrogen fixation, and N and dry matter partitioning during a four year alfalfa stand. Crop Sci. 24:811–815.
61. ----, G. Hardarson, D.K. Barnes, and C.P. Vance. 1984. Nitrogen fixation, herbage yield, and rhizobial preference of selected alfalfa clones. Crop Sci. 24:1093–1097.

62. ----, and C.P. Vance. 1979. Nitrate-N and *Rhizobium* strain roles in alfalfa seedling nodulation and growth. Crop Sci. 19:512-518.
63. ----, and ----. 1983. Physiology and morphology of perennial legumes. p. 99-142. *In* W.J. Broughton (ed.) Nitrogen fixation, Vol. 3. Legumes. Clarendon Press, Oxford.
64. ----, ----, and D.K. Barnes. 1981. Evaluating elite alfalfa lines for N_2-fixations under field conditions. p. 217-232. *In* J.M. Lyons et al. (ed.) Genetic Engineering of Symbiotic Nitrogen Fixation and Conservation of Fixed Nitrogen. Plenum Publishing Corp., New York.
65. ----, ----, and ----. 1983. Symbiotic nitrogen fixation of alfalfa, birdsfoot trefoil, and red clover. p. 336-339. *In* J.A. Smith and V.W. Hays (ed.) Proc. 14th Int. Grassl. Congr. Westview Press, Boulder, CO.
66. ----, ----, ----, and K.I. Henjum. 1985. Dinitrogen fixation, and N and dry matter distribution during four year stands of birdsfoot trefoil and red clover. Crop Sci. 25:101-105.
67. Heinrichs, D.H., and J.E. Troelsen. 1965. Variability of chemical constituents in an alfalfa population. Can. J. Plant Sci. 45:405-412.
68. ----, ----, and F.H. Warder. 1969. Variation of chemical constituents and morphological characters within and between alfalfa populations. Can. J. Plant Sci. 49:293-305.
69. Hellriegel, H., and H. Wilfarth. 1888. Beilageheft zu der Ztschr. Ver. Rubenzucher-Industrie, Deutchen Reichs. 234.
70. Hendry, G.S., and D.C. Jordan. 1969. Ineffectiveness of viomycin-resistant mutant of *Rhizobium meliloti*. Can. J. Microbiol. 15:671-675.
71. Henson, R.A., and G.II. Heichel. 1984. Dinitrogen fixation of soybean and alfalfa: comparison of the isotope dilution and difference methods. Field Crops Res. 9:333-346.
72. Hesterman, O.B., C.C. Sheaffer, G.H. Heichel, and M.P. Russelle. 1987. Evaluation of N contribution and rotation effects in legume-corn rotations. Agron. J. 79: (In press.)
73. Hirsch, A.M., M. Bang, and F.W. Ausubel. 1983. Ultrastructural analysis of ineffective alfalfa nodules formed by *nif*::Tn5 mutants of *Rhizobium meliloti*. J. Bacteriol. 151:411-419.
74. Holl, F.B. 1975. Host plant control of the inheritance of dinitrogen fixation in the *Pisum—Rhizobium* symbiosis. Euphytica. 24:767-770.
75. Hubbell, D.H. 1981. Legume infection by *Rhizobium*: A conceptual approach. Bioscience 31:832-836.
76. Jessen, D.L. 1984. Selection for activity of nitrogen and carbon assimilating enzymes in alfalfa (*Medicago sativa* L.) root nodules. Ph.D. thesis, Univ. of Minnesota, St. Paul.
77. ----, D.K. Barnes, C.P. Vance, and G.H. Heichel. 1987. Variation for activity of nodule nitrogen and carbon assimilating enzymes in alfalfa (*Medicago sativa* L.). Crop Sci. 27:627-631.
78. Jing, Y., A.S. Paau, and W.J. Brill. 1982. Leghemoglobins from alfalfa (*Medicago sativa* L. Vernal) root nodules. I. Purification and in vitro synthesis of five leghemoglobin components. Plant Sci. Lett. 25:119-132.
79. Johnson, D.A., and M.D. Rumbaugh. 1986. Field nodulation and acetylene reduction activity of high-altitude legumes in the western United States. Arctic Alpin Res. 18:171-179.
80. Johnson, H.W., U.M. Means, and C.R. Weber. 1965. Competition for nodule sites between strains of *Rhizobium japonicum* applied as inoculum and strains in soil. Agron. J. 57:179-185.
81. Jordan, D.C. 1974. Ineffectiveness in *Rhizobium* leguminous plant association. Proc. Indian Natl. Sci. Acad. Part B. 40:713-740.
82. ----, I. Grinyer, and W.H. Coulter. 1963. Electron microscopy of infection threads and bacteria in young root nodules of *Medicago sativa*. J. Bacteriol. 86:125-137.
83. Kapusta, G., and D.L. Rowenhorst. 1973. Influence of inoculum size on *Rhizobium japonicum* serogroup distribution frequency in soybean nodules. Agron. J. 65:916-919.
84. Karraker, P.E., C.E. Bartner, and E.N. Fergus. 1950. Nitrogen balance in lysimeters as affected by growing Kentucky bluegrass and certain legumes separately and together. Ky. Agric. Exp. Stn. Bull. 557.
85. Kondrosi, A., E. Kondrosi, C.E. Pankhurst, W.J. Broughton, and Z. Banfalvi. 1982. Mobilization of a *Rhizobium meliloti* megaplasmid carrying nodulation and nitrogen fixation genes into other rhizobia and *Agrobacterium*. Mol. Gen. Genet. 188:433-439.
86. Legg, J.O., and C. Sloger. 1975. A tracer method for determining symbiotic nitrogen fixation in field studies. p. 661-666. *In* E.R. Klein and P.D. Klein (ed.) Proc. 2nd Int. Conf. Stable Isotopes, Oak Brook, IL. U.S. Energy Research and Development Administration, Washington, DC.

87. Long, S.R., W.J. Buikema, and F.M. Ausubel. 1982. Cloning of *Rhizobium meliloti* nodulation genes by direct complementation of nod mutants. Nature (London) 298:485–488.
88. MacDowell, F.D.H. 1983. Effects of light intensity and CO_2 concentration on the kinetics of first month growth and nitrogen fixation of alfalfa. Can. J. Bot. 61:731–740.
89. MacKenzie, C.R., and D.C. Jordan. 1974. Ultrastructure of root nodules formed by ineffective strains of *Rhizobium meliloti*. Can. J. Bot. 20:755–759.
90. Martensson, A.M., and H.D. Ljunggren. 1984. A comparison between the acetylene reduction method, the isotope dilution method, and the total N difference method for measuring nitrogen fixation of lucerne (*Medicago sativa* L.). Plant Soil 81:177–184.
91. Maxwell, C.A., C.P. Vance, G.H. Heichel, and S. Stade. 1984. CO_2 fixation in alfalfa and birdsfoot trefoil nodules and partitioning of ^{14}C to the plant. Crop Sci. 24:257–264.
92. McAuliffe, D., D.S. Chamblee, H. Uribe-Arango, and W.W. Woodhouse. 1958. Influence of inorganic nitrogen on nitrogen fixation of legumes as revealed by ^{15}N. Agron. J. 50:334–337.
93. McCoy, E. 1932. Infection by *Bact. radicicola* in relation to the microchemistry of the host cells walls. Proc. R. Soc. London, B. 110:514–533.
94. Miller, R.W., and J.C. Sirois. 1982. Relative efficiency of different alfalfa cultivar-*Rhizobium meliloti* strain combinations for symbiotic nitrogen fixation. Appl. Environ. Microbiol. 43:764–768.
95. ----, and ----. 1982. Calcium and magnesium effects on symbiotic nitrogen fixation in alfalfa (*M. sativa*)-*Rhizobium meliloti* system. Physiol. Plant 58:464–470.
96. Miltimore, J.E., J.M. McAuthur, B.P. Goplen, W. Majak, and R.E. Howarth. 1974. Variability of fraction I protein and heritability estimates for fraction I protein and total phenolic constituents in alfalfa. Agron. J. 66:384–386.
97. Munns, D.N. 1965. Soil acidity and growth of a legume. I. Interactions of lime with nitrogen and phosphate on growth of *Medicago sativa* L. and *Trifolium subterraneum* L. Aust. J. Agric. Res. 16:733–741.
98. ----. 1965. Soil acidity and growth of a legume II. Reactions of aluminum and phosphate in solution and effects of aluminum, phosphate, calcium, and pH on *Medicago sativa* L. and *Trifolium subterraneum* L. in solution culture. Aust. J. Agric. Res. 16:743–755.
99. ----. 1968. Nodulation of *Medicago sativa* in solution culture III. Effects of nitrate on root hairs and infection. Plant Soil 29:33–47.
100. ----. 1969. Enzymatic breakdown of pectin and acid-inhibition of the infection of *Medicago* roots by *Rhizobium*. Plant Soil 30:117–119.
101. ----. 1970. Nodulation of *Medicago sativa* in solution culture V. Calcium and pH requirements during infection. Plant Soil 32:90–102.
102. ----. 1977. Mineral nutrition and the legume symbiosis. p. 353–391. *In* R.W.F. Hardy and A.H. Gibson (ed.) A treatise on dinitrogen fixation, IV: Agronomy and ecology. John Wiley and Sons, New York.
103. Newcomb, W. 1981. Nodule morphogenesis and differentiation. p. 247–297. *In* K.L. Giles and A.G. Atherly (ed.) Biology of the Rhizobiaceae. Int. Rev. Cytol. Suppl. 13. Academic Press, New York.
104. Nutman, P.S. 1963. Factors influencing the balance of mutual advantage in legume symbiosis. p. 51–71. *In* Soc. Gen. Microbiol. 13th Symp.
105. Pate, J.S. 1958. Studies on growth substances of legume nodules using paper chromatography. Aust. J. Biol. Sci. 11:516–528.
106. Petersen, H.L., and T.E. Loynachan. 1981. The significance and application of *Rhizobium* in agriculture. p. 311–331. *In* K.L. Giles and A.G. Atherly (ed.) Biology of the Rhizobiaceae. Int. Rev. Cytol. Suppl. 13. Academic Press, New York.
107. Peterson, M.A., and D.K. Barnes. 1981. Inheritance of ineffective nodulation and non-nodulation traits in alfalfa. Crop Sci. 21:611–616.
108. Phillips, D.A. 1980. Efficiency of symbiotic nitrogen fixation in legumes. Annu. Rev. Plant Physiol. 31:29–49.
109. ----, E.J. Bedmar, C.O. Qualset, and L.R. Teuber. 1985. Host legume control of *Rhizobium* function. p. 203–212. *In* P.W. Ludden and J.E. Burris (ed.) Nitrogen fixation and CO_2 metabolism. Elsevier Science Publishing Co., New York.
110. ----, and J.P. Bennett. 1978. Measuring symbiotic nitrogen fixation in rangeland plots of *Trifolium subterraneum* L. and *Bromus mollis* L. Agron. J. 70:671–674.
111. ----, D.M. Center, and M.B. Jones. 1983. Nitrogen turnover and assimilation during regrowth in *Trifolium subterraneum* L. and *Bromus mollis* L. Plant Physiol. 71:472–476.
112. ----, S.D. Cunningham, E.J. Bedmar, T.C. Sweeney, and L.R. Teuber. 1985. Nitrogen assimilation in an improved alfalfa population. Crop Sci. 25:1011–1015.

113. ----, and L.R. Teuber. 1985. Genetic improvement of symbiotic nitrogen fixation in legumes. p. 11-17. *In* H.J. Evans et al. (ed.) Nitrogen fixation research progress. Martinus Nijhoff, Dordrecht, Netherlands.
114. ----, ----, and S.S. Jue. 1981. Variation among alfalfa genotypes for reduced nitrogen concentration. Crop Sci. 22:606-610.
115. Pieters, A.J. 1927. Green manuring, principles and practice. John Wiley and Sons, New York.
116. Rennie, R.J. 1986. Comparison of methods of enriching a soil with nitrogen-15 to estimate dinitrogen fixation by isotope dilution. Agron. J. 78:158-163.
117. ----, S. Dubetz, J.B. Bole, and H.H. Muendel. 1982. Dinitrogen fixation measured by ^{15}N isotope dilution in two Canadian soybean cultivars. Agron. J. 74:725-730.
118. Ruiz-Argueso, T., R.J. Maier, and H.J. Evans. 1979. Hydrogen evolution from alfalfa and clover nodules and hydrogen uptake by free-living *Rhizobium meliloti*. Appl. Environ. Microbiol. 37:582-587.
119. Ruschel, A.P., P.B. Vose, R.L. Victoria, and E. Salatai. 1979. Comparison of isotope techniques and non-nodulating isolines to study the effect of ammonium fertilization on dinitrogen fixation in soybean, *Glycine max*. Plant Soil 53:513-525.
120. Schubert, K.R., and M.J. Boland. 1984. The cellular and intracellular organization of the reactions of ureide biogenesis in nodules of tropical legumes. p. 445-451. *In* C. Veeger and W.E. Newton (ed.) Advances in nitrogen fixation research. Nijhoff/Jung, The Hague.
121. Seetin, M.W., and D.K. Barnes. 1977. Variation among alfalfa genotypes for rate of acetylene reduction. Crop Sci. 17:783-787.
122. Snapp, S.S., and C.P. Vance. 1986. Asparagine biosynthesis in alfalfa nodules (*Medicago sativa* L.) root nodules. Plant Physiol. 82:390-395.
123. Sutton, W.D. 1983. Nodule development and senescence. p. 144-212. *In* W.J. Broughton (ed.) Nitrogen fixation, Vol. 3: Legumes. Clarendon Press, Oxford.
124. Sinclair, A.G., R.B. Hannagan, and W.H. Risk. 1976. Evaluation of the acetylene reduction assay of nitrogen fixation in pastures using small core samples. N. Z. J. Agric. Res. 19:451-458.
125. Ta, T.C., M.A. Faris, and F.D.H. Macdowall. 1986. Pathways of nitrogen metabolism in nodules of alfalfa (*Medicago sativa* L.). Plant Physiol. 80:1002-1005.
126. Talbott, H.J., W.J. Kenworthy, and J.O. Legg. 1982. Field comparison of nitrogen-15 and difference methods of measuring nitrogen fixation. Agron. J. 74:799-804.
127. Tan, G.Y. 1981. Genetic variation for acetylene reduction rate and other characters in alfalfa. Crop Sci. 21:485-488.
128. Teuber, L.R., R.P. Levin, T.C. Sweeney, and D.A. Phillips. 1984. Selection for N concentration and forage yield in alfalfa. Crop Sci. 24:553-558.
129. Thomas, R.J., D. Jokinen, and L.E. Schrader. 1983. Effect of *Rhizobium japonicum* mutants with enhanced N_2 fixation on N transport and photosynthesis of soybeans during vegetative growth. Crop Sci. 23:453-456.
130. Thornton, H.G. 1929. The "inoculation" of lucerne (*Medicago sativa* L.) in Great Britain. J. Agric. Sci. 19:48-70.
131. ----. 1930. The early development of the root nodules of lucerne (*Medicago sativa* L.). Ann. Bot. (London) 44:385-392.
132. ----. 1930. The influence of the host plant in inducing parasitism in lucerne and clover nodules. Proc. R. Soc. London, B. 106:110-122.
133. ----. 1936. The action of sodium nitrate upon the infection of lucerne root hairs by nodule bacteria. Proc. R. Soc. London B. 119:474-492.
134. Truchet, G., M. Michel, and J. Denarie. 1980. Sequential analysis of the organogenesis of lucerne (*Medicago sativa*) root nodules using symbiotically defective mutants of *Rhizobium meliloti*. Differentiation 26:163-172.
135. Tu, J.C. 1977. Structural organization of the rhizobial root nodule of alfalfa. Can. J. Bot. 55:35-43.
136. Vance, C.P. 1978. Nitrogen fixation in alfalfa: An overview. p. 34-41. *In* Proc. 8th Annu. Alfalfa Symp.
137. ----. 1983. Rhizobium infection and nodulation: A beneficial plant disease? Annu. Rev. Microbiol. 37:399-424.
138. ----, G.H. Heichel, D.K. Barnes, J.W. Bryan, and L.E. Johnson. 1979. Nitrogen fixation, nodule development and vegetative regrowth of alfalfa (*Medicago sativa* L.) following harvest. Plant Physiol. 64:1-9.
139. ----, and L.E.B. Johnson. 1981. Nodulation: A plant disease perspective. Plant Dis. 65:118-124.
140. ----, and ----. 1983. Plant determined ineffective nodules in alfalfa (*Medicago sativa*): structural and biochemical comparisons. Can. J. Bot. 61:93-106.
141. ----, ----, A.M. Halvorsen, G.H. Heichel, and D.K. Barnes. 1980. Histological and ultrastructural observations of *Medicago sativa* root nodule senescence after foliage removal. Can. J. Bot. 58:295-309.

142. ----, ----, and G. Hardarson. 1980. Histological comparisons of plant and *Rhizobium* induced ineffective nodules in alfalfa. Physiol. Plant Pathol. 17:167–173.
143. ----, S. Stade, and C.A. Maxwell. 1983. Alfalfa root nodule carbon dioxide fixation I. Association with nitrogen fixation and incorporation into amino acids. Plant Physiol. 72:469–473.
144. Van Rensburg, H.J., and B.W. Strijdom. 1982. Competitive abilities of *Rhizobium meliloti* strains considered to have potential as inoculants. Appl. Environ. Microbiol. 44:98–106.
145. Vasilas, B.L. 1981. A tracer technique for the quantitative measurement of symbiotic dinitrogen fixation by soybeans in field studies. Ph.D. diss. Univ. of Minnesota (Diss. Abstr. DA 8211559).
146. Viands, D.R., D.K. Barnes, and G.H. Heichel. 1981. Nitrogen fixation in alfalfa: Response to bidirectional selection for associated characteristics. USDA Tech. Bull. 1643. U.S. Government Printing Office, Washington, DC.
147. ----, C.P. Vance, G.H. Heichel, and D.K. Barnes. 1979. An ineffective nitrogen fixation trait in alfalfa (*Medicago sativa* L.). Crop Sci. 19:905–908.
148. Wagner, G.H., and F. Zapata. 1982. Field evaluation of reference crops in the study of nitrogen fixation by legumes using isotope techniques. Agron. J. 74:607–612.
149. Ward, H.M. 1887. On the tubercular swellings on the roots of *Vicia faba*. Phil. Trans. R. Soc. London B. 178:539–561.
150. West, C.P., and W.F. Wedin. 1985. Dinitrogen fixation in alfalfa-orchardgrass pastures. Agron. J. 77:89–94.
151. White, J.G.H. 1967. Establishment of lucerne on acid soils. p. 105–114. *In* The lucerne crop. Reed, Wellington, New Zealand.
152. Williams, L.E., and D.A. Phillips. 1983. Increased soybean productivity with a *Rhizobium japonicum* mutant. Crop Sci. 23:246–250.
153. Wittenberg, J.B., F.J. Bergersen, C.A. Appleby, and G.L. Turner. 1974. Facilitated diffusion. The role of leghemoglobin in nitrogen fixation by bacteroids isolated from soybean root nodules. J. Biol. Chem. 249:4057–4066.
154. Witty, J.F. 1983. Estimating N_2-fixation in the field using ^{15}N-labelled fertilizer: Some problems and solutions. Soil Biol. Biochem. 15:631–639.

8 Cold and Heat Tolerance

J. S. McKENZIE

Agriculture Canada Research Station
Beaverlodge, Alberta

ROGER PAQUIN

Agriculture Canada Research Station
Ste. Foy, Quebec

STANLEY H. DUKE

University of Wisconsin
Madison, Wisconsin

Alfalfa (*Medicago sativa* L.) grows under many diverse environmental conditions. Nevertheless, temperature stress is one of the primary factors limiting its expansion into new regions. High temperatures can inhibit growth, reduce yield, and shorten stand longevity. Cold temperatures limit areas of adaptation through sublethal or lethal winter injury. Sublethal winter injury can decrease vigor during the subsequent growing season and lethal winter injury decreases the stand. Sublethal injury is difficult to detect in the field and often goes unrecognized, even though production losses from reduced growth may be significant (273).

Dormancy is a condition where physiological activities associated with growth cease, but in a reversible manner (143). Associated with dormancy is a heightened resistance to stress via factors associated with a reduction in the water content and the respiratory capacity of the cells. Dormancy is of survival value when it spans unfavorable periods for growth caused by cold, heat, or drought.

Alfalfa, unlike many tree species, has no true physiological rest period (278, 289). Alfalfa can be forced into or out of dormancy at any time by unfavorable or favorable environmental conditions. Plants brought into the greenhouse may resume growth within 3 d during every month of the year unless they have been stressed (256).

Improvements in plant productivity are possible without changing the plant's genetic potential (24). This approach to improving plant productivity, however, depends upon a better understanding of how the plant responds to its environment and on a more complete understanding of the complex stresses which occur in the field. Progress by genetic means has been limited by the polyploid nature of alfalfa, by the lack of infor-

Copyright 1988 © ASA-CSSA-SSSA, 677 South Segoe Road, Madison, WI 53711, USA. *Alfalfa and Alfalfa Improvement*—Agronomy Monograph no. 29.

mation on the genetic control of stress tolerance, and by the close association between high stress tolerance and low crop yield.

8–1 COLD TOLERANCE

Cold tolerance or cold hardiness refers to the ability of a plant to survive the effects of freezing temperature stress. Winterhardiness on the other hand involves the ability of plants to survive all factors influencing survival during the winter. This includes freezing temperatures, diseases, insects, moisture, etc. The total overwintering complex cannot be neglected, however, in discussions on cold tolerance because temperature stresses which are insufficient to kill the plant may still weaken it and make the plant more susceptible to other winter stresses.

Increases in the winterhardiness of alfalfa have been derived primarily from the hybridization of *M. sativa* with *M. falcata*. Although improved disease and pest resistance have significantly increased winter survival in many areas, resistance to cold temperature stress is by far the most important component of the winter-hardiness complex in northern latitudes.

Alfalfa cannot survive freezing temperatures below −2 to −5°C in mid-summer, but during the fall hardening period, changes occur within the plant to enable the roots and crowns to survive temperatures as low as −20°C. Alfalfa undergoes biochemical, biophysical, and morphological changes in the fall that increase tolerance to low temperature stresses. The overwintering behavior of plants is determined by factors such as time of initiation of hardening, rate of hardening, maximum midwinter-hardiness level, hardiness stability under widely fluctuating conditions in mid-winter, and time that dehardening occurs in the spring. These parameters are under complex genetic and environmental control.

8–1.1 Development of Cold Tolerance

Cold tolerance is governed by numerous genetic factors, the expression of which is affected by temperature and photoperiod, and by the environment in the soil.

8–1.1.1 Influence of Environmental Factors

8–1.1.1.1 Light.
Light affects the development of cold tolerance in two ways. First, through photosynthesis it provides the energy essential for cold hardening and for the accumulation of carbohydrates and other reserve substances in the roots and crowns. Second, the duration of the light period is important in controlling the autumn growth habit and in triggering the initiation of dormancy in the fall (8, 113, 242).

A short photoperiod (long night period) is essential to initiate the development of cold tolerance in cold tolerant cultivars (114), but once hardening has initiated, plants will continue to harden at cool temperatures regardless of the photoperiod (129, 231). However, by artificially shortening the photoperiod, Hodgson (114) found that the cold-tolerant cv. Ranger, adapted to more southern latitudes, was not able to develop additional cold tolerance in Alaska even though it was exposed to cold acclimating temperatures. On the other hand, *M. falcata*, adapted to northern latitudes, developed cold tolerance equally well under normal or shortened photoperiods. The cold sensitive, nonwinter dormant germplasm of Arizona Common, however, failed to attain an appreciable amount of cold tolerance under either photoperiod. Accordingly, the genetic potential for the development of cold resistance in a particular area is realized only after exposure to the photoperiod and temperature to which the cultivar is adapted (31, 114, 115, 140). Since many southern ecotypes are photoperiodic insensitive (113), they may be incapable of appreciable acclimation.

The response to photoperiod is often not to absolute daylength, but to an increasing or decreasing daylength above or below a critical level (219). *Medicago falcata* may be more sensitive to decreasing photoperiod under longer day conditions than cultivars adapted to more southern latitudes. In general, cold-sensitive plants appear to lack the photoperiodic sensing mechanism to induce essential internal physiological changes required for good winter survival (289). In alfalfa, this may relate to a change in endogenous growth regulators (214, 282) that trigger further processes and facilitate a response to cool temperatures.

Diurnal and seasonal changes in light quality also may influence alfalfa's cold acclimation. Woody plant species collected from different latitudes differ in their sensitivity to red (R) and far-red (FR) light (56), and under long-day (LD) conditions, end-of-day FR light triggers cold acclimation in woody species (171). This response is phytochrome mediated since the proportion of blue and far-red energy increases with decreasing solar elevation until the beginning of civil twilight (249). The long periods of twilight rich in blue and FR light in northern latitudes may influence plant development in ways not observed at lower latitudes.

8-1.1.1.2 Temperature. Kacperska-Palacz (131) described three stages of hardening in wintering herbacious plants using winter rape (*Brassica napus* var. *oleifera*) as an example. In the first stage, air temperatures from 2 to 5°C are responsible for an initial 4 to 5°C increase in cold tolerance. The second stage requires air temperatures from 0 to −2°C and is responsible for the maximum freezing tolerance in the fully hydrated cell. The third stage involves prolonged frost, cell dehydration, and an accompanying increase in cold tolerance. Similar temperature requirements for hardening alfalfa have been shown by Paquin (195) and Bula and Smith (30).

Temperature fluctuations and the length of the hardening period affect cold acclimation. Although alternating, rather than constant temperatures appear to be more conducive to hardening, such treatments have not always increased cold tolerance (51). Cold hardening in the field begins in autumn when mean air temperatures are near 10°C and it appears to accelerate when temperatures approach 5°C (31, 168, 202, 276) (Fig. 8-1). Cold hardy cultivars develop cold tolerance earlier during autumn than cold-sensitive cultivars (31, 202, 231). Although the maximum mid-winter cold-hardiness potential is normally attained by the time the soil becomes permanently frozen for winter (31), cold hardening can continue under snow in the apparent absence of photosynthesis when soil temperatures are −1 to −2°C (202). Freezing temperatures appear necessary to attain maximum cold tolerance, because alfalfa plants develop less cold tolerance in regions where snow cover prevents the soil from freezing (129) than in regions where soil begins to freeze in early December (202). Vezina and Paquin (280) observed that alfalfa roots should be maintained at 1 to 2°C, irrespective of air temperature, to acquire optimum cold tolerance.

Cold air temperatures are important in the retention of cold tolerance. Dehardening may occur if temperatures rise above 10°C for a few days during the autumn hardening period (Fig. 8-1). In the fall or winter, cold-sensitive cultivars deharden more rapidly and initiate growth faster than cold-tolerant cultivars during periods of warm weather (202, 229). Rehardening can occur under cold temperatures provided that no growth has occurred and that sufficient food reserves are available as an energy source (129).

Alfalfa dehardens normally in response to the warming of soil temperatures in the spring (130) and dehardening accelerates when the mean soil temperatures rise above 5°C (168, 202) (Fig. 8-1). Dehardening is normally complete by the time green shoots appear (126).

Seasonal changes in cold tolerance closely follow soil temperature (130) and under some conditions, cold tolerance has been predicted from soil and air temperature data (196, 298). The physiological status of the plant, however, can limit or prevent the development of cold tolerance regardless of temperature. This occurs when food reserves in the plant are reduced to critical levels (53, 129, 167, 232), when soils become saturated with water (168, 200), or when rapid and substantial growth occurs in the fall (168). Accordingly, the survival of alfalfa depends upon the physiological ability of the plant to respond to climatic factors during critical phases in its life cycle (Fig. 8-2).

Plants kept at hardening temperatures for periods longer than a normal winter are unable to maintain maximum hardiness indefinitely because of a gradual loss in reserves (129). Metabolic changes that occur before spring growth also lead to a loss of freezing tolerance even though plants are exposed to temperature and light conditions optimum for hardening (159).

Soil or plant thawing rates also affect survival. Peltier and Tysdal (206) observed that hardy plants survived −18°C for 1.5 h when thawed slowly over a 20-h period, but only 36% of the plants survived rapid

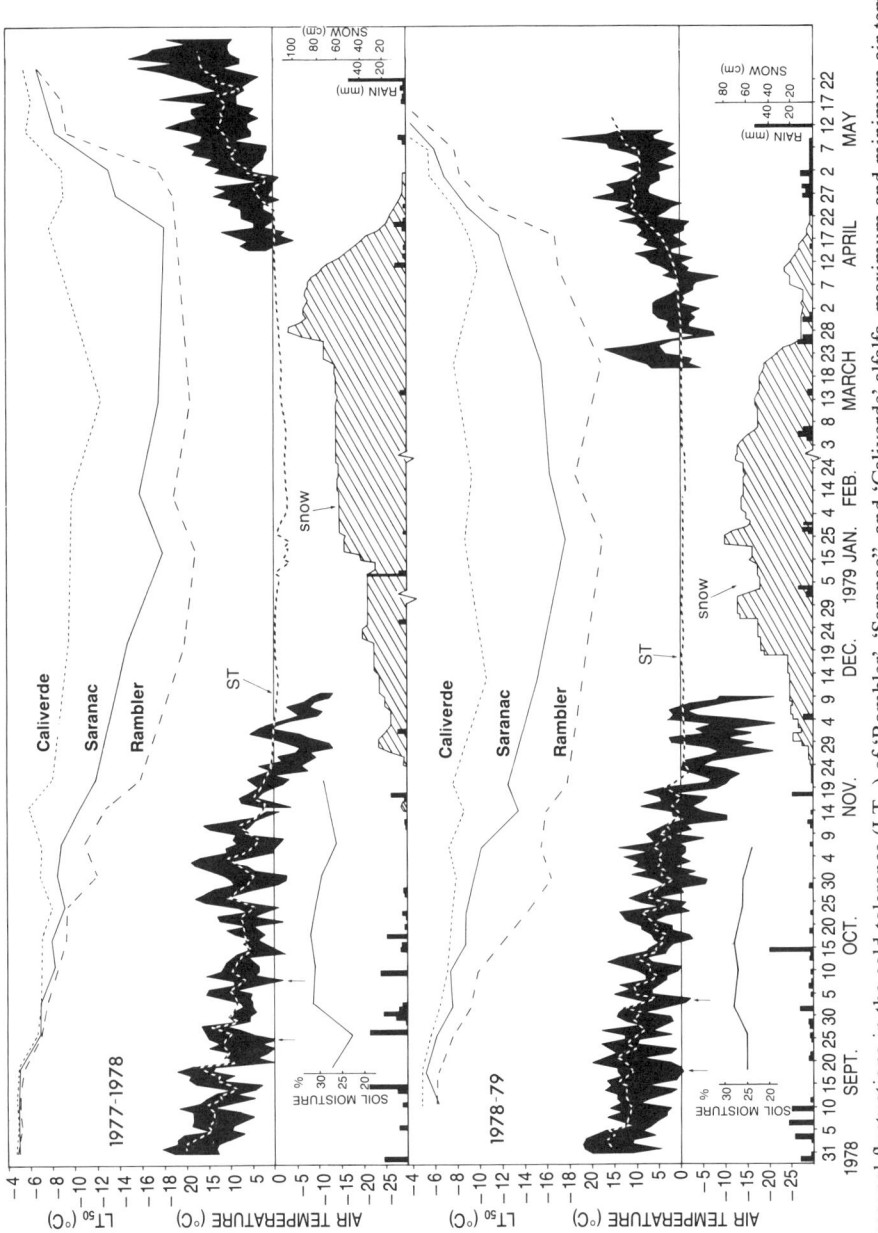

Fig. 8-1 Seasonal fluctuations in the cold tolerance (LT_{50}) of 'Rambler', 'Saranac', and 'Caliverde' alfalfa, maximum and minimum air temperatures, mean daily soil temperatures (ST), and snow depth and rainfall distribution during two winter seasons at La Pocatiere, Quebec. Arrows on air temperature profile indicate the first nights of frost. From Paquin and Pelletier (202).

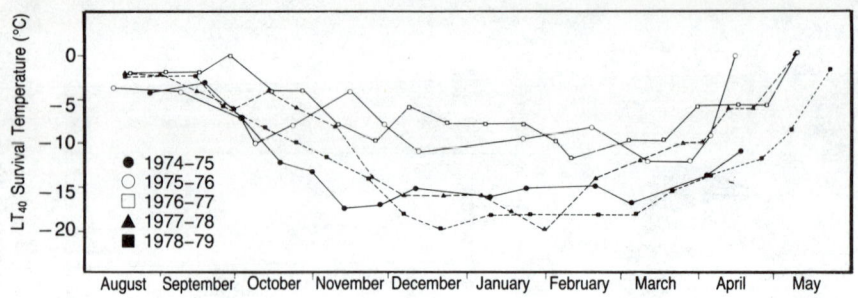

Fig. 8-2. Seasonal fluctuations in the cold tolerance (LT$_{40}$) of *M. falcata* cv. Anik during five consecutive winters in northwestern Canada. From McKenzie and McLean (168).

thawing in water at 18°C. Normally, slow thawing rates occur under field conditions (240), but rates will vary with soil moisture content, snow cover, vegetative cover, slope, surface color, air temperature, and solar radiation.

8-1.1.1.3 Soil. Soil moisture affects the survival of alfalfa by influencing the rate of freezing. The rate at which the soil freezes or thaws is governed by the amount of water in the soil and by ambient air and soil temperatures. When an unspecified soil type with 12, 17, and 27% moisture was exposed to −19°C, soil temperature decreased to −4°C within 3, 4, and 11 h, respectively (207). At 33% moisture, the temperature of the soil was only −2°C after 16 h.

Soil moisture can directly affect hardening and survival of alfalfa. Soil moisture levels below 50% of field capacity are favorable (200, 262) and high moisture levels near saturation are unfavorable for hardening (200, 267, 269). High soil moisture levels appear to have a greater influence on plants in bloom than on younger plants in the vegetative state (37, 267). Drought conditions during hardening help increase cold tolerance (276). As a result, in some regions more winter injury has been observed under irrigated conditions than under dryland conditions (108).

The influence of minerals on cold tolerance is not well understood. Variations in response may be attributed to a number of variables that are not reported or reported in full, i.e., differences in the availability of certain minerals in soils, differences in anion-cation exchange capacities and differences in mineral uptake by plants. Concentrations of cations are lower and anions higher in crowns and roots during winter than summer (130). Their role in cold tolerance, if any, remains obscure.

Soil acidity and fertility have an important effect on the survival of plants in winter. Thus, conditions favorable for fall growth, including high soil N (52, 275), are known to retard the development of cold tolerance (243), while nutrient deficiencies in soil contribute to reduced growth, poor acclimation and reduced cold tolerance.

Increased cold tolerance has been obtained with high levels of K and/or P (36, 100, 141), but reduced cold tolerance has also been reported with high levels of P (275). Others concluded that the level of K is not critical for cold acclimation (13, 46). A solution concentration with a P/K ratio of 2:5 was found to be optimum for cold tolerance in sand culture (128). An increase in total nonstructural carbohydrates (TNC) and improved recovery after cold stress has been observed following K fertilization (213, 261). Increases in winter survival with P, K, and lime application appear to be associated with an increase in starch, nonreducing sugars, and total proteins in the crowns and roots (287). This may be explained in part by K stimulated carbohydrate transport to roots and nodules of alfalfa, and subsequent increases in N_2 fixation (64). In nutrient solution studies, the K/N ratio has been shown to influence the amino acid pool and protein synthesis (162).

8–1.1.2 Influence of Anatomical and Morphological Factors

All alfalfa tissues are capable of attaining some degree of cold tolerance. Tolerance is usually greatest for crowns, intermediate for roots, and least for leaves (126). Crown buds are normally the most cold tolerant tissue until they turn green, at which point they lose their cold hardiness. Seedlings are susceptible to frost injury until four or five leaves have formed (Fig. 8–3). (5, 207, 277, 292).

8–1.1.2.1 Fall Dormancy. Fall dormancy is classified on the basis

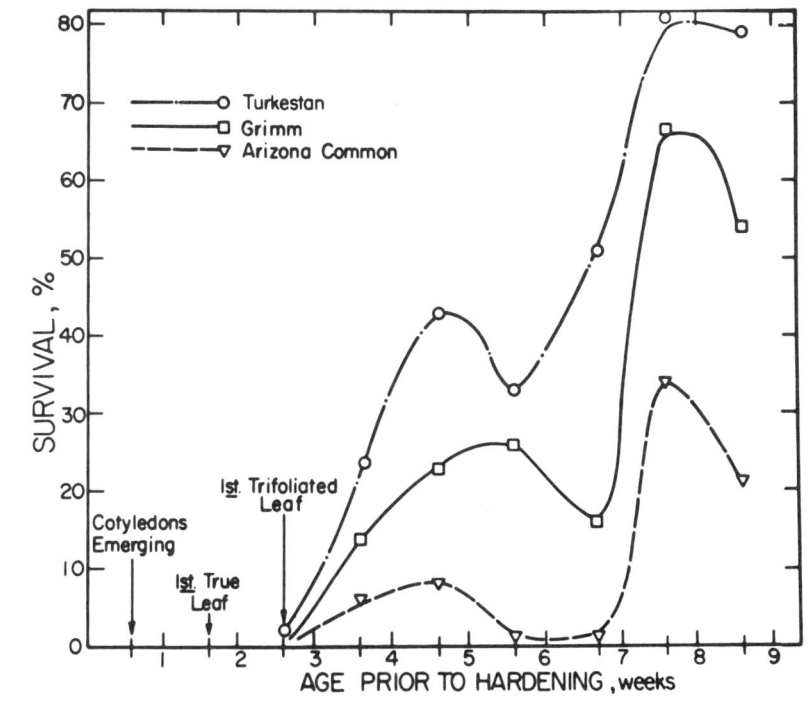

Fig. 8–3. The survival of seedlings that were hardened for 15 d at 2 to 4°C, then exposed to −16°C for 6 h. From Peltier and Tysdal (207).

of vegetative growth observed in the autumn, particularly in northern latitudes (252). The decreasing photoperiod and low temperatures in the fall contribute to the appearance of plant morphological types that are not observed either in the spring or in regrowth following cutting in early summer. Cultivars adapted to southern regions are distinguished readily from northern cultivars by their more erect, taller growth habit in the fall (184, 252). Southern cultivars respond to short photoperiods and lower temperatures by producing tall, erect stems while northern cultivars produce either long or short, prostrate stems (186, 242). For this reason, southern cultivars are called nonwinter dormant, nonfall dormant, or nondormant and northern cultivars are called winter dormant, fall dormant, or dormant. Intermediate types have also been identified (163).

Fall dormancy is closely associated with winterhardiness. Cultivars that produce little top growth during the fall because of their dormancy are generally more winterhardy than nondormant cultivars (54, 111, 152, 242).

The expression of fall dormancy results from the combined effects of short days and cool temperatures (8, 180). Under short-day conditions, differences among dormant and nondormant cultivars are more pronounced at low temperatures (44, 185, 226). At cool temperatures, hardy cultivars have the greatest dormancy response and nonhardy cultivars the least response (276). Maximum dormancy appears to be induced by a temperature of 15.5°C and a photoperiod of 12 h (8, 226). Accordingly, a decrease in photoperiod and temperature causes a greater decrease in top growth of fall dormant cultivars than in the nondormant (nonwinterhardy) cultivars (84, 113, 228). Under long-day conditions there is little difference in regrowth between dormant and nondormant cultivars.

In general, alfalfa cultivars in North America trace to nine distinct sources of germplasm introduced from different regions of the world (7). These germplasm sources: *M. falcata*, Ladak, *M. varia*, Turkistan, Flemish, Chilean, Peruvian, Indian, and African are listed in their approximate descending order of winterhardiness and fall dormancy characteristics. A 10th source of very nondormant germplasm from Saudi Arabia has generally gone unrecognized (291).

8–1.1.2.2 Leaves. Leafy, healthy plants have a greater capacity to become cold tolerant than either poorly developed or weak plants. This may relate to their capacity to accumulate carbohydrates, amino acids, and other compounds that contribute to cold acclimation (126). Consequently, factors such as insects, disease, cutting, or frosts, that reduce the photosynthetic activity of the leaves or the ability of the crowns and roots to transport photosynthates and water, can have an indirect effect on cold tolerance (246).

Time of cutting foliage in the fall may influence the survival of plants by affecting the status of food reserves. In the northern USA, Smith (245) defined the "critical harvest period" for alfalfa as being 4 to 6 weeks before the average date of the first killing frost in the fall. Cutting during this period interferes with the accumulation of food reserves because new growth is produced at the expense of winter reserves.

If food reserves are low in the fall, the physiological status of the plant may limit or prevent development of adequate cold tolerance (51, 53, 129, 167, 232). Jung and Smith (128) reported that the critical level for TNC may be about 14 to 16% of the dry weight of the root and crown. Conversely, the presence of high food reserves does not necessarily assure a corresponding high level of hardiness even under optimum hardening conditions (31, 115).

Alfalfa leaves have the capacity to develop a considerable degree of cold tolerance in the fall (Fig. 8-4) (19, 169, 180). Although plant survival does not depend on the overwintering of leaves, the cold tolerance of

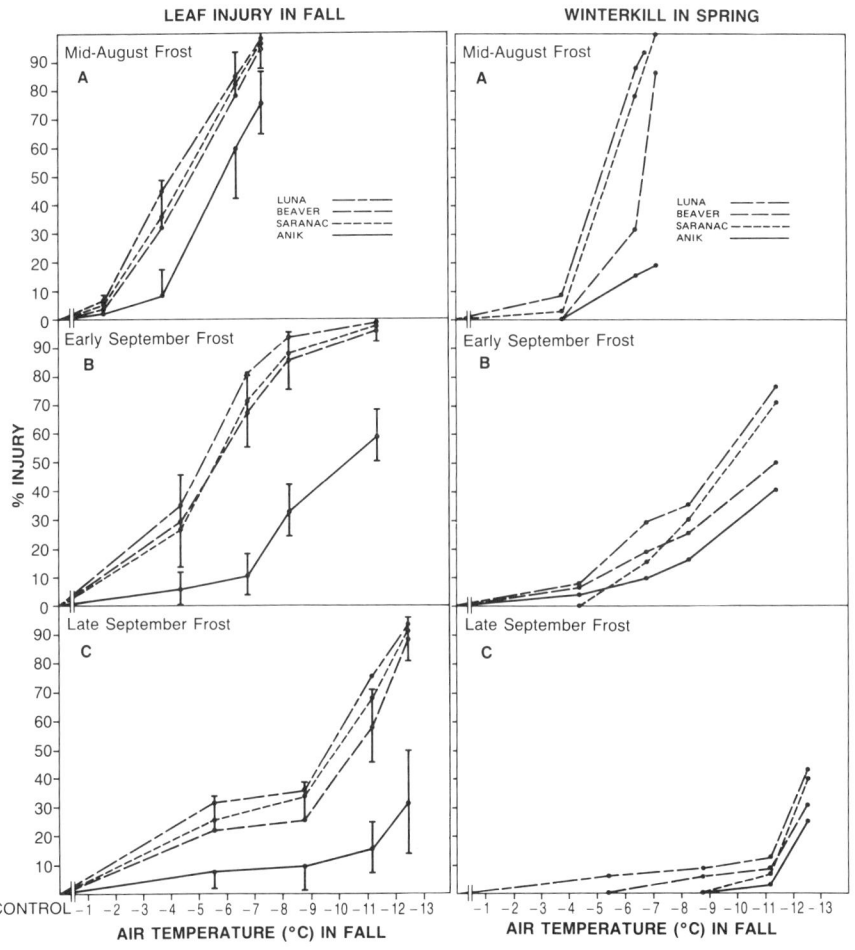

Fig. 8-4. Leaf injury in the fall in four alfalfa cultivars subjected to various freezing temperatures during the fall and evaluated for winterkill in the spring following a severe winter in northwestern Canada. Means and standard deviations are shown for *M. falcata* cv. Anik and *M. media* cv. Beaver. From McKenzie and McLean (169).

leaves determines the length of time in autumn when leaves synthesize substances that are translocated to crowns and roots to potentially increase the tolerance of plants to stress. Frost injury during the critical harvest period can be very detrimental the following spring, if food reserves are depleted as a result of regrowth and if there is sufficient stress during the winter to damage the plants (169). In northwestern Canada, it was reported that an early fall frost would seldom be of sufficient intensity to cause leaf injury in seedling stands of adapted dormant cultivars (169). Leaf frost resistance may have more importance in regions where nondormant or intermediate dormant cultivars are grown to extend production during the winter months. Morley et al. (180) have shown that the nondormant Hairy Peruvian strain of alfalfa is considerably more susceptible to frost than dormant cultivars.

8–1.1.2.3 Bud Development. The stage of development of the crown buds during autumn may influence the time of initiation of fall hardening, the rate of hardening, and the level of cold hardiness in mid-winter (168). Crown buds are normally initiated late in the fall, remain dormant during winter and begin growth when environmental conditions become favourable in the spring (93). Following a dry growing season, crown buds may initiate earlier than normal during autumn and, when environmental conditions are conducive to growth in the fall, shoots will develop from the buds. This late flush of growth has been associated with delayed hardening in the fall and a reduced mid-winter cold hardiness potential (168). Furthermore, conditions conducive to the elongation of underground crown buds during the fall have also been associated with a reduced rate of hardening (168).

Factors which increase or decrease food reserves in roots, such as the frequency and time of cutting during the growing season (47, 91, 92, 93, 232) and soil moisture (296), have a pronounced effect on crown bud initiation. Also, the survival of plants infected with anthracnose may be reduced because of the smaller number of new crown buds for recovery in the spring (122). However, vigorous plants of winter-hardy cultivars occasionally exhibit unexpected recovery from apparent "winterkilling" because, although severely damaged, these plants may retain living buds on the lower portion of the crown. New growth initiates slowly and eventually the plants recover completely (26, 164).

Although crown buds are not initiated under low soil moisture conditions, carbohydrates continue to accumulate in crowns and roots. Thus, when heavy rains break a prolonged drought, these plants produce more stems and buds per unit of crown than plants not subjected to moisture stress (47, 156, 296).

Measurable differences in the number of crown buds exist among alfalfa cultivars. Compared with *M. sativa*, *M. falcata* types produce a larger number of underground buds (240). These underground buds provide a measure of winter protection from freezing, thawing, drying, and ice sheets which form on the soil surface (239).

Although crown buds will elongate and develop in frozen soil during late winter (182), they normally initiate during fall or spring. Plants with a large number of crown buds in the fall have a greater chance of recovering from winter stresses in the spring (206). Plants that do not initiate crown buds during the autumn, or plants in which crown buds have been damaged during winter stress, are normally late in emerging the following spring. As a result, these plants are usually less competitive with grasses and weeds.

Competition among plants in the spring and recovery from winter stress depends upon morphological characteristics and the capacity of the plant to extract nutrients and moisture from the soil. Early development of numerous crown buds in the fall, associated with high levels of food reserves, is important for rapid spring growth and early attainment of a large leaf area index. Competitive ability depends on the overall plant capacity to fix and assimilate CO_2, acquire and assimilate N, and to use these metabolites to increase its size or extend its foliage (18, 38).

8–1.1.2.4 Decumbency of Shoots. It is difficult to determine whether a particular phenotype characteristic contributes to winter survival (42). Cold hardiness is associated with the *M. falcata*-type of growth (241) in that hardy cultivars are slow to regrow after cutting and fall growth is prostrate, branching, or rosette-like. *Medicago sativa*-type cultivars are characterized by having more rapid regrowth after cutting and fall growth that is more erect and nonbranching. However, in the spring, both *M. falcata* and *M. sativa* cultivars have similar erect nonbranching morphology (42). Some root and crown characteristics have been associated with winterhardiness (109, 240, 250), but discrepancies in the ranking of cultivars based on these characteristics limit their usefulness for selecting hardy cultivars (152).

8–1.1.3 Influence of Physiological Factors

8–1.1.3.1 Water. For a large part of this century, it was generally accepted that increases in bound water and decreases in total water in plant tissues were important in the cold hardiness of alfalfa (126) and other plant species (159). A number of studies with alfalfa have demonstrated a positive correlation between bound water in root and crown tissues, and the degree of cold hardiness (93, 206, 288). Such studies led to the hypothesis that bound water increases hardiness by decreasing the amount of water in tissues which may be frozen. However, in many of these early studies, bound water was not measured correctly (159). More recently, studies measuring bound water in dogwood (*Cornus stolonifera* Mich.) and wheat (*Triticum aestivum* L.) with nuclear magnetic resonance (NMR) spectroscopy indicate that there is no correlation between bound water and hardiness (32, 101). Bound water in plant tissues consists of water molecules H-bonded to sugars, proteins, nucleic acids, and other organic compounds. The term *bound water*, used to describe such water, may be inappropriate in that the bonds formed between water molecules and cellular hydrophilic compounds are quite weak, having lifespans of <1 ns (85). However, this water usually freezes at discernably lower temperatures than required for so-called *free water* in plant tissues (159). Differences in the freezing temperature of free and bound water may vary from little to great, depending upon how "tightly" water is bound.

Correlations between hardiness and bound water may reflect other factors which impart hardiness and do not necessarily implicate bound water as a major factor in hardening. Both sugar and protein content in alfalfa increase dramatically with hardening (65), thus more water will be bound in hardened alfalfa tissues. The capacity of alfalfa proteins to bind water does not increase with hardiness (27), indicating that an increased ability to bind water is a quantitative trait (i.e., more protein binds more water).

Total water content of alfalfa roots and crown buds decrease with hardening (93, 206, 288); however, at this time, it is not known whether this phenomenon imparts any degree of hardiness to alfalfa. A reduction in the total water content of plant tissues may be necessary for hardiness in some tissues of certain plant species, but not in others (159).

8-1.1.3.2 Carbon Metabolism

8-1.1.3.2.1 Sugars

Intracellular sugar concentrations increase during hardening in virtually all plant tissues that harden (159). Sucrose, glucose, total reducing sugar, and total sugar concentrations in alfalfa roots and crowns increase dramatically with hardening and are maintained at a high level throughout the winter (30, 37, 65, 120, 130, 146, 199). Recent investigations in eastern Canada (199) have shown that total sugars begin to increase in crowns of alfalfa in early September, several weeks before plants acquire some cold tolerance (Fig. 8–5). Total sugars did not increase during October and early November but cold hardiness increased rapidly. During that period, fluctuations in the level of total sugars were related to the mean air temperature. The most cold-tolerant alfalfa cultivars accumulate much higher levels of sugars during autumn than those with less cold tolerance. In a field study in Wisconsin, Duke and Doehlert (65) found that fall concentrations of sucrose and glucose in the two most hardy cultivars were six- and twofold higher, respectively, than in the least hardy cultivars (Table 8–1). In eastern Canada, a second surge in total sugars occurred in late November and continued until a maximum was obtained between January and March (Fig. 8–5). A maximum total sugar level coincides with the maximum cold hardiness level in the plant.

Sucrose accounts for about 90% of the total pool of sugars in hardened alfalfa (126). An increasing sucrose concentration in alfalfa is directly correlated with hardiness (65). Wang et al. (287) found that additions of K increased alfalfa tap root sucrose accumulation 2.5 times and winter survival by 60%. Potassium dramatically increases translocation of pho-

COLD AND HEAT TOLERANCE

Table 8-1. Sugar concentrations of the tap roots and hardiness, as measured by electrolyte leakage after freezing at −8°C for 2 h, of several alfalfa cultivars removed from the field on 18 Oct. 1979 near Madison, WI. From Duke and Doehlert (65).

Cultivar	Sucrose	Glucose	Hardiness
	— nmol g^{-1} of fresh wt —		total electrolyte leakage, %
Roamer	122a*	13.1ab	4.8b
Vernal	120a	14.5a	5.7b
Saranac	99ab	9.9b	7.9ab
Caliverde	81ab	9.6b	12.0a
Sonora	50bc	7.9bc	13.7a
Hairy Peruvian	21c	5.2c	13.2a

* Mean values in columns followed by the same letter are not significantly different at the $P = 0.05$ level according to Duncan's multiple range test.

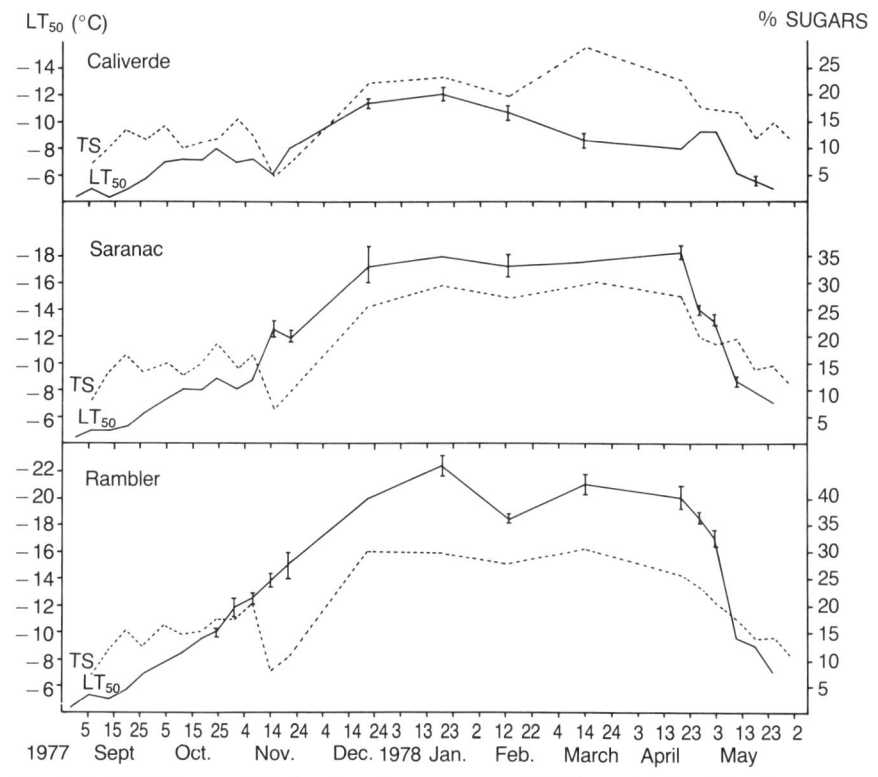

Fig. 8-5. Cold tolerance (LT$_{50}$) and total sugar content (TS, dry weight basis) in the crowns of three alfalfa cultivars, Caliverde, Saranac, and Rambler during the 1977-78 winter season at La Pocatiere, Quebec. Vertical lines indicate standard deviations > 0.2. From Paquin and Lechasseur (199).

tosynthates from shoots to roots in alfalfa (64). The initial source of alfalfa tap root and crown bud sucrose during hardening may be from the transport of current photosynthate from leaves or from the degradation of previously formed starch. In wheat, cool temperatures and decreasing photoperiod during August and September reduced growth and sugar consumption, thus favoring the accumulation of sugars and starch in the crowns and roots (135). In alfalfa, sugar accumulation stopped completely after photosynthesis ceased and leaves abscised following frost in late October and early November (199). The second surge in sugar accumulation started in late November and lasted from January until March. This is probably caused by starch degradation at temperatures near freezing (6, 41, 247).

In alfalfa, it appears that high sucrose concentrations are maintained during hardening at the expense of starch. Duke and Doehlert (65) removed four alfalfa cultivars from the field in the early autumn and placed them in growth chambers under either cold-hardening temperatures (8-h photoperiod, 10°C day, 5°C night) and low light or warm nonhardening temperatures (16-h photoperiod, 25°C day, 20°C night) and low light. Root starch levels decreased and total sugars increased under hardening temperatures, whereas starch levels were maintained at a high level and total sugar levels were decreased under nonhardening temperatures. Under hardening temperatures, sugar concentrations were highest and starch concentrations lowest in the most hardy cultivars. In this study, it appeared that there was insufficient translocation of current photosynthate to maintain high root sugar concentrations in plants grown under hardening temperatures and low light, hence starch was degraded to maintain high sugar concentrations especially in very hardy cultivars of alfalfa.

The role of sugars in hardiness is not completely understood. It is obvious that in some plant species, sugar accumulation alone will not impart hardiness. Tropical species, such as sugarcane (*Saccharum officinarum* L.), have high concentrations of accumulated sugars and no tolerance to freezing. It has been suggested that sugars may increase hardiness by being metabolized at low temperatures, causing unknown protective changes or by the osmotic effect (i.e., decreasing the amount of ice formed during extracellular ice formation) (159). In the latter situation, sugars will only impart hardiness if plant tissues have some resistance to dehydration caused by freezing. Sugars also stabilize cell structures (220, 274), and protect both chloroplast membranes (133, 221) and cell protoplasm from an increase in electrolytes (134).

An alternative hypothesis is that sugars directly alter the metabolism of plants by altering enzyme kinetics and that these alterations affect cold tolerance. Studies by Duke et al. (74) and Duke and Henson (68) have demonstrated that various sugars can have a dramatic effect on the temperature kinetics of certain plant enzymes and that at low temperatures, sugars directly affect plant metabolic pathways quite removed from sugar metabolism.

8-1.1.3.2.2 Starch

Starch accumulates in amyloplasts of root and crown bud cells of alfalfa during autumn (246). Alfalfa tap roots have very high levels of amylase activity (57, 58). Hence, it is likely that alfalfa tap roots rapidly convert starch to sugars when necessary. It appears probable that α-amylase controls starch degradation in alfalfa by acting as a catalyst to β-amylase and starch debranching enzyme activity (67). However, there is no consistent correlation between starch content and either seasonal or cultivar differences in cold tolerance (126). It is unlikely, therefore, that starch has a direct effect on the cold tolerance of alfalfa. The importance of starch in alfalfa cold tolerance rests largely on its contribution to pools of sucrose and other sugars over winter and its contribution to spring regrowth.

8–1.1.3.2.3 *Respiration*

Most studies on alfalfa root respiration and cold tolerance have indicated that cultivars with a high potential for cold tolerance have higher root respiration rates under hardening conditions than less cold-tolerant cultivars (65, 230, 271). Under hardening conditions, the cold-hardy cv. Vernal was found to have a lower energy of activation for respiration than the less hardy cv. Sonora (65). This could contribute in part to the higher respiration rate for Vernal under these conditions. A high level of energy production during hardening may be necessary for the synthesis of the high concentrations of proteins which accompany alfalfa hardiness (65). Late in the autumn and during winter, low rates of alfalfa root respiration may be associated with hardiness (30). This would decrease the prospect of plant loss due to carbohydrate depletion over winter, and would increase spring growth.

8–1.1.3.3 Nitrogen Metabolism

8–1.1.3.3.1 *Nitrogen Acquisition*

Under most conditions, alfalfa is largely dependent on symbiotic N_2 fixation for its supply of N (see Chapter 7 in this book). In two studies, temperatures below 15°C have been shown to completely inhibit or severely restrict nodulation of alfalfa (65, 124). The same relationship was observed with chilling-sensitive species, such as soybean [*Glycine max* (L.) Merr.] (73, 124). However, Duke and Doehlert (65) found that under hardening conditions (i.e., short day, 10°C day, 5°C night), alfalfa cultivars can nodulate and fix N_2. In contrast, nonhardy cultivars either failed to nodulate or nodulated very poorly and had no detectable N_2 fixation. In addition, it was found that less hardy alfalfa cultivars nodulated to a lesser extent and fixed less N_2 than hardy cultivars under nonhardening conditions. However, if *Rhizobium meliloti* nitrogenase is as cold labile as that of other N_2-fixing bacteria (61, 103, 302), it is doubtful that the alfalfa/*R. meliloti* symbiosis could fix much if any N_2 at chilling temperatures.

8–1.1.3.3.2 Amino Acids

Proline is the free amino acid that most frequently accumulates in plants exposed to various types of environmental stress (259). During acclimation, more proline accumulates in the crowns and roots in cold tolerant cultivars than in cold sensitive cultivars (Fig. 8–6). Free amino acids increased by 20% during cold acclimation in the roots of Vernal, a hardy cultivar, while no increase was observed in Caliverde, a nonhardy cultivar (295). Proline accounted for 73% of the increase in free amino acids (294) under hardening conditions.

Proline is apparently synthesized in leaves and translocated to crowns and roots. Removing leaves or cooling the lower section of the stem stops the accumulation of proline in crowns and roots (281). Light is necessary for proline synthesis in leaves (157, 255, 301), but roots can synthesize proline in darkness when supplied with sucrose and glutamic acid (147).

Hardening temperatures influence the accumulation and translocation of proline (195). At 1.5°C, proline accumulates slowly in the crowns and roots. At 5 and 10°C, proline accumulates rapidly for the first 2 to 3 weeks then decreases rapidly. This suggests the existence of a mecha-

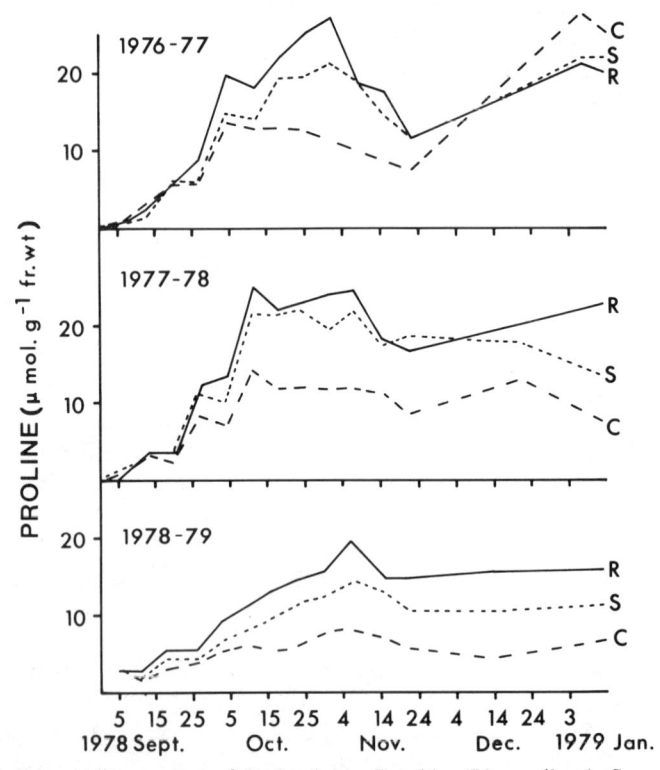

Fig. 8–6. Free proline content of the hardy cv Rambler (R), semihardy Saranac (S), and the nonhardy Caliverde (C) during fall hardening at La Pocatiere, Quebec from 1976 to 1979. From Paquin and Pelletier (203).

nism by which the plants adapt to these temperatures and a possible explanation for some of the fluctuations observed in proline accumulation in crowns of alfalfa during cold acclimation (Fig. 8–6) (203).

The role of proline in plant stress tolerance is not clear. Proline has been suggested as a reservoir of organic nitrogen for regrowth following stress (9), as a means to increase the sugar content and thus cold tolerance (157), as an osmoticum (260), as a protector of thylakoids against freezing injury (105), as a protector of membranes against salt-induced alterations (223), as an antagonist to protein denaturation (225), or as a means to regulate the cellular water structure by converting hydrophobic groups of biopolymers into hydrophilic groups (224).

Arginine has also been shown to greatly increase in concentration in alfalfa roots with hardening. Wilding et al. (295) found a ca. 300% increase in free arginine accumulation in Vernal alfalfa with hardening. As with proline, arginine accumulation appears to be associated with stress (49).

8–1.1.3.3.3 Proteins and Peptides

Protein synthesis is generally enhanced during the cold hardening of plant tissues (159, 233). There have been numerous studies demonstrating that soluble protein content increases in alfalfa during hardening (65, 81, 90, 127). Only certain isozymes of a number of alfalfa dehydrogenases and hydrolases increase during hardening (144, 145). This suggests that either certain genes are induced or that posttranslational modification of proteins occurs during hardening. Duke and Doehlert (65) found that the activities of 6 of 10 tap root enzymes assayed from hardy and nonhardy alfalfa cultivars harvested in the fall were higher in hardy cultivars (Table 8–2), suggesting the the stimulation of protein synthesis under hardening conditions is not restricted to one or a few enzymes, but is a somewhat general phenomenon. However, certain enzymes such as malate dehydrogenase and glutamate oxaloacetate transaminase may contribute to increases in soluble proteins to a much greater extent than other enzymes. It is not known whether increases in cellular soluble proteins cause or enhance cold hardiness in plants or are simply a result of the acquisition of cold tolerance (159). Changes in enzyme activities and isozymes during hardening may reflect necessary changes in plant metabolism and/or physical characteristics of enzymes if plants are to survive freezing temperatures. Such changes may directly or indirectly impart resistance to freezing. Studies by Huner and Macdowall (117, 118) have demonstrated that the number of free sulfhydryl (SH) groups and the K_m for CO_2 of ribulose bisphosphate carboxylase/oxygenase (Rubisco) are reduced during hardening of wheat (*Triticum aestivum* L.) and rye (*Secale cereale* L.). The presence of fewer exposed cysteinyl SH groups on the protein may reduce disulfide bond (cys-S-S-cys) formation during freezing, thereby lessening damage to the plant. In this way, changes in protein conformation and/or isozymes in the plant during hardening would directly impart freezing resistance. However, there is considerable uncertainty

Table 8-2. Activities of soluble enzymes and protein concentrations from roots of several field-grown alfalfa cultivars on 18 October 1979 of the seeding year near Madison, WI. Enzyme activities are expressed as μmol product produced min^{-1} g^{-1} of fresh wt. From Duke and Doehlert (65).

Cultivar	NAD-MDH†	NADP-MDH	NADP-isocitrate dehydrogenase	LDH	G6P dehydrogenase	Amylase	Invertase	NAD-GDH	Glutamate oxaloacetate transaminase	Glutamine synthetase	Percentile ranking for activity of 10 enzymes	Protein (mg g^{-1} of fresh wt)
Hairy Peruvian	26.6c*	0.265d	2.90b	0.0140c	0.0031bc	225.0a	1.13a	0.239ab	1.63b	0.150a	55.9	1.34c
Sonora	28.8bc	0.195d	3.07b	0.0207c	0.0040b	280.0a	0.98a	0.184b	2.11b	0.142a	64.0	1.79c
Caliverde	48.4b	0.362c	5.00a	0.0427ab	0.0022cd	241.0a	1.31a	0.322ab	1.64b	0.147a	73.6	2.00b
Saranac	54.0ab	0.359c	4.42a	0.0376ab	0.0010d	248.0a	1.21a	0.247ab	2.13b	0.137a	67.3	2.38b
Vernal	72.8a	0.445b	2.63b	0.0490a	0.0059a	292.0a	1.03a	0.226ab	3.80a	0.184a	85.2	3.52a
Roamer	69.6a	0.548a	4.23a	0.0341b	0.0068a	184.0a	1.53a	0.350a	3.74a	0.169a	90.4	3.55a

* Mean values within columns followed by the same letter are not significantly different at P = 0.05.
† MDH = malate dehydrogenase, LDH = lactate dehydrogenase, and GDH = glutamate dehydrogenase.

that disulfide bond formation during freezing is a major cause of freezing injury (72, 159).

A lowering of the K_m of Rubisco during hardening could result in more CO_2 fixation at low temperature and an enhanced accumulation of carbohydrates during hardening. If sugars impart hardiness, such changes in the kinetics of Rubisco would indirectly impart hardiness to the plant. It may be that changes in proteins act synergistically with changes in other factors, such as glutathione concentrations, to impart hardiness. Glutathione is a tripeptide with a free cysteinyl SH group that is capable of reducing disulfide bonds between and within proteins. Glutathione also increases in concentration during the hardening of many plants (72).

8-1.1.3.4 Lipid Metabolism. Plants that harden in response to low temperature increase in lipid content (159). Total root fatty acid concentrations double during hardening in alfalfa (89). Several studies with alfalfa have shown that fatty acid synthesis is more active in cultivars of alfalfa that are quite hardy than in those with less hardiness (96, 97, 98). The lipids which accumulate in hardening alfalfa roots are primarily the polyunsaturated linoleic and linolenic acids (89) which have lower melting points than saturated fatty acids with an equal number of carbons. Also, phospholipids, the constituents of membranes, accumulate to a greater extent in the most cold-tolerant alfalfa cultivars (98, 99). An increase in unsaturated membrane lipids decreases the temperature at which membranes undergo phase transitions from their normally physiologically active liquid-crystalline state to the crystalline (solid gel) state (212). Membrane crystallization or solidification at low temperatures may result in the loss of membrane integrity, leakage of intracellular solutes and dysfunction of membrane-bound enzymes (75, 161, 211, 212). The resulting damage to alfalfa is dependent entirely on the temperature of membrane crystallization and not on the freezing of water. Various regions of membranes and various membranes within a plant may also have lipid constituents with very different temperatures of crystallization; hence, any damage to a plant because of membrane phase transitions at low temperature might be localized.

8-1.1.3.5 Growth and Differentiation. Growth and cellular differentiation gradually slow down during the fall as the plants become dormant and increase in cold tolerance (65, 226, 242). This may contribute in part, to increased concentrations of sugars, certain amino acids, and proteins. Without significant growth, but with continual N_2 fixation and photosynthesis, various storage compounds may accumulate until feedback inhibition of N_2 fixation and photosynthesis occur. One could certainly view alfalfa tap root soluble proteins as storage proteins because they are rapidly hydrolyzed in the spring. Hence, it is possible that the accumulation of various polymers and readily metabolized monomers are a direct result of dormancy and slowed growth rather than as a direct result of low temperature.

8–1.1.3.6 Hormones.
Some growth inhibitors promote cold hardiness. This could be explained by the strong relationship between growth cessation and cold hardening (39, 119, 178). Hairy Peruvian, a nondormant, cold-sensitive cultivar of alfalfa, forms a rosette and shows an increase in freezing tolerance when grown in nutrient solutions containing abscisic acid (ABA) under short days and cool temperatures (214). When the dormant, cold-tolerant cv. Ranger was treated with gibberellic acid (GA) under short days and cool temperatures, it responded like the cold-sensitive Hairy Peruvian cultivar, developed an elongated shoot and failed to acquire significant cold tolerance (282). Since endogenous levels of ABA were not affected by the cold-hardening treatment in either cultivar, the authors concluded that the ABA/GA ratio determines the degree of cold acclimation and that this ratio is altered by changes in the GA level (282).

Abscisic acid may influence cold tolerance by increasing membrane permeability (45, 190), changing cell water and osmotic potential (139, 238), decreasing stomatal aperture (23, 76) and inducing a moisture deficit (80).

CCC (2-chloroethyl trimethyl ammonium) has been shown to increase cold tolerance in alfalfa (197, 229), with the effect nullified by GA (197). Applications of CCC and Alar-85 in the field have resulted in variable effects from year to year on growth, survival, and yield (197) probably as a result of the influence of environmental factors on endogenous levels of hormones (35).

Growth promoting hormones, such as GA and auxins (e.g., indole-3-acetic acid, napthalene acetic acid), inhibit the development of cold tolerance (119, 178), but this does not always occur (102, 132). Accordingly, the role of growth regulators in the development of cold tolerance remains to be elucidated.

8–1.2 Freezing Process and Injury

This subject has been extensively reviewed by several authors (33, 159, 160, 188, 257). Most of the fundamental work on the freezing process and injury in cells has been conducted with cereals (102, 189, 237) and with trees and shrubs (3, 88, 234). Few investigations have been published on the freezing process and injury in alfalfa.

8–1.2.1 Changes Observed in the Cells

Freezing injury in alfalfa occurs both during the freezing process and while cells are in the frozen state (94). Injury results from the pressure exerted from intercellular ice crystals on cell membranes and from cellular dehydration that disturbs the membrane structure. Further disorganization may occur in membranes and in the cytoplasm during and following the thawing process. This results in leakage from the plasmalemma and vacuole, and may result in the subsequent release of toxic compounds within the cell as well as the release of vital compounds from the cell.

Injury first occurs in older parenchyma cells of the phloem and xylem, and in the central pith-like structure of the upper part of the taproot (123). This is made evident by the presence of meristematic tissues surrounded by a cork layer that seals off the injured area (246).

Under field conditions, the crown and regenerative buds in alfalfa are below the soil surface and are protected against rapid freezing in winter by snow cover and by the low thermal conductivity of soil. In the laboratory, a cooling rate of 1°C/h is less injurious to alfalfa than one of 4 to 6°C/h (251) (Fig. 8-7). At a very rapid cooling rate, (1°C/min), ice crystal formation can result in the separation of the leaf and bud epidermis from the rest of the tissues (181). Under these conditions, histological manifestations of damage will be greatest in the less cold tolerant cultivars. Such damage to leaf tissues can occur during the fall when air

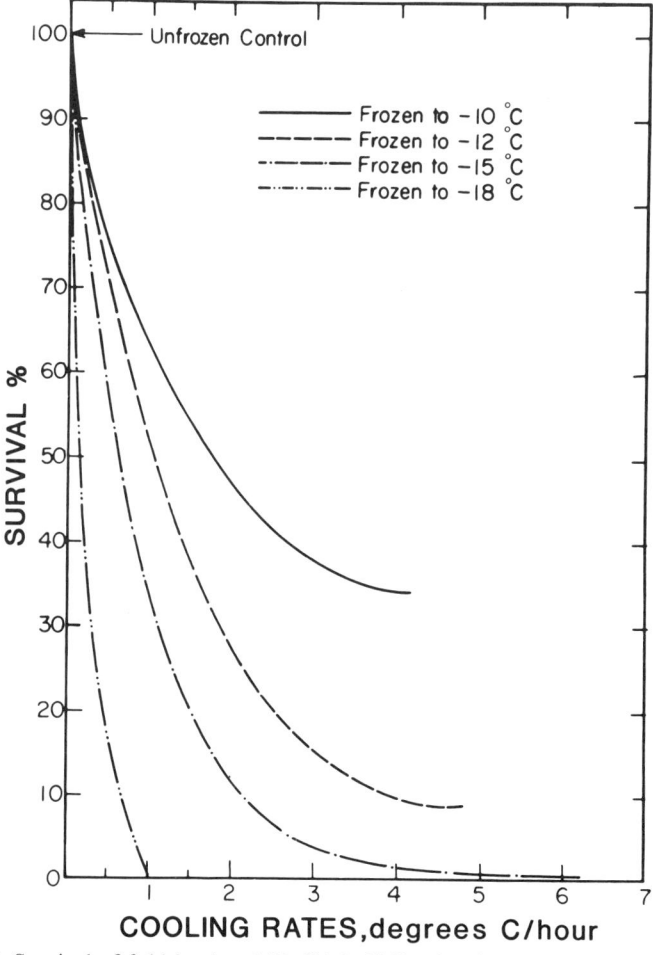

Fig. 8-7. Survival of field hardened 'Buffalo' alfalfa after freezing to different minimum temperatures and at different cooling rates. From Sprague (251).

temperatures drop rapidly. Again, in the laboratory, alfalfa tissues usually supercool to −3°C, (181, 251). Therefore, damage from ice crystal formation may not occur above −3°C.

8–1.2.2 Factors Affecting Injury in the Field

Survival of plants in the field, following winter stresses, should be envisioned as survival from multiple stresses. Although some cultivars may have an acceptable tolerance to most of the major stresses, exposure to normally sublethal winter stresses could modify tolerance to another stress factor, or to successive exposure to the same stress (173).

Not all factors contribute to injury each year. Ouellet (191, 192) correlated the field survival of alfalfa with numerous climatological parameters at several sites in Canada. He found that the climatological contribution to winter injury and the time at which specific climatological events occur differ in various parts of the country. At Swift Current, Saskatchewan, winter survival is largely dependent upon the ability of alfalfa to resist cold air temperatures to −40°C, 26-week long dormant periods and winter desiccation of the crowns by alternate freezing and thawing (108). Only cultivars with low-set crowns and cold-hardiness genes derived from *M. falcata* can resist those conditions. At Charlottetown, Prince Edward Island, mean soil temperatures below −4°C in February cause severe winterkill in all perennial legumes (265). At that location, *M. falcata* does not acquire the resistance to survive mid-winter thaws (268). In northwestern Canada, the lengthy snow cover and soil temperatures of −1 to −3°C contributed to extensive snow mold injury caused by numerous pathogens during one winter (165), while severe cold weather and −19°C soil temperatures contributed to injury during another winter (167).

Alfalfa can be killed directly by extremely low soil temperatures. This condition can occur when cold air temperatures precede sufficient snow cover for protection or when an ice sheet, which is several times more frost conducive than snow or air, is formed (167, 196, 248). In Wisconsin, the most tolerant cultivars cannot withstand long exposure to soil temperatures as low as −10 to −15°C without incurring injury or even death (248). In more northern latitudes, adapted cultivars can tolerate soil temperatures near −20°C (108, 168, 202, 239).

Water stress and low temperature effects are additive in increasing the freezing tolerance of alfalfa (200, 262). A soil moisture level of 40% of field capacity has been found to be optimum for alfalfa survival (172). Watering at 2°C before freezing can be more detrimental to alfalfa than watering at subfreezing temperatures (267). Winterkill is severe on poorly drained soils (150, 198, 218) and high soil moisture leads to ice encasement in winter, a major contributing factor in alfalfa winterkill. Ice sheets are not always injurious when formed on frozen dry soils and covered with snow (196).

The ability to develop and maintain an adequate level of freezing resistance is imperative in order to withstand stressful mid-winter temperatures (268). This is associated with two factors: resistance to alternate freezing and thawing, and waterlogging or ice encasement which contribute to an anaerobic environment and the build-up of toxic metabolites (11, 268). Ice encasement produces toxic effects which resemble those of waterlogging (86, 253, 254). The phenomenon is called *asphyxiation, smothering,* or *anoxia*. Root anoxia can favor fungal root infection by *Phytophthora megasperma* (11). Anoxia which causes an accumulation of toxic metabolites, such as acetaldehyde, ethanol, methanol and acetylene, is accompanied by the solubilization of proteins and a decrease in carbohydrates (11, 268). Furthermore, accumulated ethanol and/or acetaldehyde may remain in the tissue for a long time and affect viability and yield potential during the growing season (266). Alfalfa is unable to resist anoxia for more than 7 to 21 d at low temperatures (23, 107, 172).

Ice sheets are often lethal, but differences in survival between hardy and nonhardy cultivars are not easily evaluated under field conditions (246). Intermittent thawing of ice sheets can reduce or eliminate alfalfa injury, but stubble protruding through an ice sheet will not prevent injury (86). Contact with ice, duration under ice, porosity of ice, fluctuations in air temperatures, and the depth of crowns will complicate survival observations. Survival under these stressful conditions is dependent upon the capacity of the plant to maintain cold tolerance and to remove or metabolize products of anaerobic metabolism and reduce acetylene (268).

While there is much information on the ranking of cultivars for resistance to cold stress during controlled freezing tests, little is know of their ability to survive widely fluctuating temperatures in mid-winter. A measure of this ability could provide another approach to characterizing the capacity of the plant to harden and/or reharden during intermittent periods of unfavorable weather (149).

8-1.2.3 Means of Reducing Injury in the Field

Winter injury can be reduced by selecting suitable sites and avoiding high risk areas, by selecting hardy and adapted cultivars, and by using appropriate management techniques (246). Management techniques to improve snow cover will reduce the chance of low temperature injury. Parallel strips of tall wheatgrass [*Agropyron elongatum* (Host) Beauv.] stubble (227), swathing the crop at different heights (183), snow fences (198), or alfalfa regrowth (244) are effective in retaining snow on fields. A mixed stand of alfalfa with grasses is another approach to trap snow, and offers the additional advantage of reducing injury from soil heaving (246). Snow accumulation in spring can help prevent damage from freeze-thaw cycles. Proper cutting schedules and fertilizer application, to permit maximum accumulation of TNC and other food reserve in the overwintering crowns and roots in the fall, will increase the cold tolerance of the plant and prevent freezing injury (116, 246). Drainage of heavy soils will reduce injury from ice sheets and soil heaving.

8-1.3 Procedures for Measuring Cold Tolerance and Winter Survival

8-1.3.1 Laboratory Methods

Methods for measuring cold hardiness in the laboratory are based on an evaluation of the damage sustained by tissues during a cold treatment. Although some success has been obtained using a single freezing temperature for various time periods (151), the standard method is to subject a number of plants to a series of decreasing temperatures. Survival is evaluated following 2 to 3 weeks of regrowth and the cold tolerance of the cultivar is expressed as the temperature required to kill 50% of the plants (LT_{50}) (207, 209, 297).

Rapid methods of evaluating injury following freezing and thawing are based on measuring cellular leachates. Leached ions are measured by electrolyte conductivity (37, 55, 286) and amino acids are measured colorimetrically (235, 293). Such positive measurements of leachates have been assumed to be an indication of freeze-induced cellular rupture; however, recent evidence suggests that the leakage of ions or low molecular weight compounds may not necessarily be associated with cellular rupture (71, 193, 194). Vital staining with tetrazolium salts (264) is also a common method of measuring injury following freezing and thawing. Electrical impedance measurements have been used (104) at high and low frequency (95), but this method has not gained wide acceptance. Damage to membranes by freezing has led to the development of methods that determine free acid phosphatases liberated from membranes injured by freezing (21, 22).

Ranking the winterhardiness of alfalfa cultivars has been attempted by indirect methods such as germinating seeds in different osmotic concentrations of sodium chloride, sucrose, or mannitol (59, 60, 216). However, these methods are unreliable for detecting small differences in hardiness between cultivars (216). Other indirect measurements such as crown diameter, weight of roots (106, 152), moisture equivalent, freezing point, chemical composition, respiration, and the amount and rate of moisture loss in the roots have been found unreliable in rating the cold tolerance of alfalfa (175).

Methods based on biochemical changes have been used with crown and root tissues during cold hardening. The contents of sucrose and other sugars have shown good correlations with cold tolerance of cultivars and of selections from F_2 generations of alfalfa (65, 111, 130). The level of the amino acid proline is highly correlated with the cold hardiness of alfalfa cultivars under field conditions, but unknown factors contributing to variations in concentrations from year to year preclude its acceptance as a useful method of evaluating the cold hardiness in alfalfa (203, 295).

Laboratory freezing tests have proven useful in evaluating the cold hardiness of crop cultivars. In general, results have correlated well with field survival because freezing tolerance is a major factor in the overwintering of plants (159). However, laboratory methods have failed to distinguish between genotypes with similar degrees of hardiness (111), because differentiation is inadequate within narrow ranges and because freezing tolerance may not always be the only limiting factor in nature. Injury occurring under the standardized conditions in the laboratory may have little relationship to the types of stresses encountered in the field (159). Furthermore, seedling stands of alfalfa are less susceptible to winter stresses than established stands (108).

Before a simple and practical method of assessing winterhardiness is found, much remains to be known on what constitutes cold hardiness and how it develops in plants. Until then, the freezing test followed by regrowth using the LT_{50} remains the standard for measuring cold hardiness in alfalfa.

8-1.3.2 Field Methods

Crowns and roots have been dug from frozen soils with jackhammers (4, 167), chain saws (201), tubular drills (20, 204), and concrete cut-off saws (270). Cold hardiness can then be evaluated by freezing plants directly in, or separating them from, the soil core, and evaluating regrowth potential in a greenhouse.

Portable freezers have been used to freeze plants directly in the field (121, 169). Their use is limited by small dimensions and the requirement for a power source.

Alfalfa winterkill has been evaluated using aerial infrared photography (198, 284). The technique provides a good estimation of damage and the photographs provide a permanent record (283). However, widespread use is limited by the excessive cost, by improper timing due to poor weather conditions (198), and by difficulties in photo-interpretation when alfalfa is mixed with grasses (12).

Field survival during a "test winter" has long been used as the ultimate evaluation of a hardy cultivar. A test winter is one which is severe enough to kill most tender cultivars, with various degrees of damage to those with intermediate hardiness. However, test winters seldom occur with sufficient frequency and intensity at any one location to be used reliably for testing purposes. For this reason, the need for a quick and accurate method of evaluating cold hardiness or predicting the potential for winterkill has become imperative. Attempts to rate cultivars for field survival following cold stress through snow removal have been useful (170, 240). Also, methods which attempt to integrate many factors, such as food reserves, cold tolerance, bud development, plant vigor, soil environment, cultivar characteristics, and management have been useful in predicting winterkill over broad areas (166).

8-1.3.3 Fall Regrowth

Plant growth habit is of fundamental importance in governing the productivity and winter survival of alfalfa. For this reason, fall growth scores may be useful to predict winterhardiness and cultivar adaptation, since there is a high correlation between fall dormancy and winterhardiness (8, 242).

An effective method for measuring fall dormancy in the field has been developed in Minnesota (8). In the spring, a nursery is established with 9-week-old transplants. Regrowth following a clipping in early September is measured for plant height in mid-October at 5-cm increments on a 1 to 9 scale where 1 is scored for height greater than 40 cm. The six standard cultivars traditionally used in Minnesota and their 11-yr average fall growth scores are: African (nondormant) 3.6, Dupuits (semidormant) 5.0, Saranac (semidormant) 6.1, Vernal (dormant) 7.1, Ranger (dormant) 6.5, and Norseman (very dormant) 8.1 (8). Annual ratings will change depending upon environmental conditions: temperature, moisture, soil fertility, etc. The most productive cultivars in Minnesota have fall dormancy scores similar to Ranger.

Teuber et al. (272) in California utilized 18, rather than 9, 5-cm rankings, and classes were inverted (1 = most dormant) so that the scale can be expanded to accommodate tall plants in "extremely non-dormant environments." Schneider (222) has determined that a measure of the unifoliolate internode length could be used as a very rapid technique to index germplasm for fall dormancy.

Cultivars derived from less-dormant germplasm pools have the potential for increasing annual yield in regions where temperatures are suitable for plant growth at the time of year that photoperiod becomes limiting (113, 180). In some areas, excessive fall dormancy will reduce forage production because the entire growing season is not utilized. In other areas, insufficient dormancy will result in winterkill or reduced persistence. An advantage of using fall growth scores to predict winterhardiness over directly measuring winter survival is that fall growth can be measured every year, whereas winter survival can only be measured following a test winter. Within each dormancy group, however, cultivars have different capabilities to tolerate diseases, insects, nematodes, and short regrowth intervals. Therefore, their potential adaptation in a specific region must be evaluated in long-term variety trials under specific management practices. At present, fall dormancy is the single most useful characteristic for predicting areas of adaptation of untested cultivars and experimental germplasm (272).

Fall dormancy is positively associated with the vigor of summer regrowth (34, 252). As a result, caution must be exercised in selecting for winterhardiness bases on fall dormancy alone because this may result in selection against yield potential. There is some suggestion that this close association may be partially coincidental and that there is potential for developing plants with rapid regrowth and high cold tolerance (34, 48, 77, 142). However, more recent evidence suggests that selection for high levels of both cold hardiness and fall growth would be difficult (208).

8-2 HEAT TOLERANCE

Plants grown in the temperate regions of the world rarely tolerate continuous temperatures above 35°C (273). In many alfalfa-growing areas of the southern USA, it is not unusual for maximum air temperatures to exceed 40°C (78, 176, 215). Soil temperatures in alfalfa fields have reached 35 to 42°C in Australia (217), California (78), and South Dakota (79). These temperatures are considerably above the 27°C temperatures established for optimum herbage growth, and the 12°C soil temperature required for optimum root growth (110). Accordingly, mechanisms must have evolved to protect plants from the injurious action of high-temperature stress.

The decline in production of alfalfa during hot weather is commonly referred to as *summer slump* (29). Although the maximum rate for CO_2 fixation occurs at 30°C (see Chapter 6 in this book), respiration continues to increase with increasing temperatures and may reduce vigor and cause injury or death (174). Temperatures not sufficiently high to kill cells may still inhibit growth; impair vigor; and suppress production as a result of high respiration losses (43, 159), reduced N_2 fixation (217), reduced carbohydrate reserves in roots and crowns (215), or an increased resistance to CO_2 diffusion because of smaller cells and leaves (28). Frequent clipping (290) and excessive soil moisture (78), combined with high soil temperatures, can augment the deleterious effects of heat stress. For these reasons, the reduction in forage yield of alfalfa at high temperature cannot be ascribed to a single factor (50). Growth responses may be difficult to detect in the field and often go unrecognized, even though production losses from reduced growth may be significant (273).

8-2.1 High Temperature Injury

8-2.1.1 Direct (Acute) and Indirect (Chronic) Injury

Heat injury to plants may be categorized as either direct or indirect (159). Direct heat injury occurs within seconds to about 30 min and usually requires temperatures considerably greater than T_{max} (the highest temperature at which a plant will grow). Indirect heat injury requires extended periods, hours to many days, and occurs just below the T_{opt} (temperature at which optimal growth occurs) to near the T_{max}. Indirect heat injury appears to be the major cause of heat injury in alfalfa in its areas of cultivation.

8-2.1.2 Mechanism of Heat Injury

Metabolic dysfunctions due to imbalances in reaction rates of functional proteins (e.g., enzymes and electron carriers), physical perturbations of cellular structure, or a combination of both of these phenomena have been advanced to explain heat injury to plants. Currently, however, few heat-caused injuries to plants can be conclusively attributed to any specific mechanism.

8–2.1.2.1 Metabolic Dysfunction Due to Imbalances in Reaction Rates of Functional Proteins.

Plant metabolic dysfunctions at high temperatures may often be due to a change in the relative maximum velocity for the enzyme reaction (V_{max}'s) of enzymes or other functional proteins within a metabolic process, or within a related metabolic processes. Relative changes in V_{max}'s may be caused by limitations in in vitro substrate availability, by differences in the Arrhenius E_a's of enzymes in metabolic pathways, by proteins affecting metabolic pathways, or by the direct and selective physical denaturation of proteins by heat.

An example of a heat caused reduction of substrate availability is that of O_2 for aerobic respiration. As temperature is increased, O_2 is less soluble in aqueous solutions. This may result in a decrease in aerobic respiration and an increase in anaerobic respiration (159). Increasing the temperature in order to lower the cellular O_2 concentrations will drastically decrease the V_{max} of mitochondrial respiration as O_2 becomes limiting. In contrast, glycolysis may increase if starch breakdown and/or photosynthesis increases sufficiently to supply glucose. Anaerobic conditions will also cause the induction of pyruvate decarboxylase and alcohol dehydrogenase (187), allowing pyruvate produced via glycolysis to by shunted into acetaldehyde and ethanol (anaeobic respiration). This form of respiration is much less efficient in energy production than aerobic respiration. Also, acetaldehyde is a very reactive and toxic compound. Furthermore, many carbon skeletons produced in aerobic respiration which are necessary for the production of amino acids, amides, purines, and other essential compounds, may become scarce due to the shut-down or a slowing of aerobic respiration.

When substrates are not limiting, differences in the Arrehnius E_a's of enzymes within a metabolic process may cause metabolic dysfunctions at high temperatures. If an enzyme with a toxic substrate (e.g., alcohol dehydrogenase, glutamine synthetase and nitrite reductase) becomes limiting to a metabolic process at high temperature, due to an Arrhenius E_a lower than other enzymes in their pathway, the toxic intermediate could accumulate and cause a reduction in plant growth or death. Although there are no proven examples of this phenomenon in the literature, it is known that toxic substrates of several enzymes accumulate at high temperatures (159). This suggests that such phenomena may occur.

Over time, heat denaturation of the functional portions of metabolically active proteins will cause a complete loss of function. This results from the unfolding of polypeptide "arms" of the protein due to the breaking of hydrogen bonds. A high degree of hydration favors this phenomenon. Enzymes in plant tissues that have a low-moisture content have been found to withstand denaturation at temperatures of 100°C for 2 d, whereas fully hydrated enzymes are completely inactivated by 15 min at

100°C (70). Although it has been hypothesized that this phenomenon is only important in direct heat injury in plants (159), there is considerable evidence that both direct and "so-called" indirect heat injury may be caused by protein denaturation. Henson et al. (112) found that 40 and 35°C completely inhibited germination and caused death of 'Chippewa' and 'Wells' soybean seeds, respectively. No recoverable activity of glutamate dehydrogenase could be found in Wells seeds germinated at 35°C, whereas the activities of four other mitochondrial enzymes were either increased or were relatively unaffected. This suggests that relatively moderate heat selectively prevents synthesis, promotes degradation, and/or causes inactivation of plant enzymes.

In soybean seedlings, Duke et al. (74) have shown that glucose-6-phosphate dehydrogenase (a key enzyme in the pentose phosphate shunt) and NAD-malate dehydrogenase in crude extracts are rapidly inactivated in vitro at temperatures between 30 and 40°C. Inactivation of soybean and alfalfa nitrate reductases (in vitro assay) and nitrogenase (in vivo assay) from the bacteroids of both species also occurs rapidly at temperatures between 30 and 40°C (63). Accordingly, a number of key plant enzymes may be inactivated or denatured at temperatures that may cause chronic heat injury.

Another example of heat-induced reductions in plant metabolic reactions may be termed heat-induced plant starvation. In C_3 plants, such as alfalfa, which have photorespiration as well as dark respiration, the temperature compensation point (the temperature at which respiration and photosynthesis are equal) is relatively low. Since photosynthesis is more sensitive to high temperature than respiration (15, 16) plants may die of starvation at temperatures above the temperature compensation point. The heat sensitivity of photosynthesis has been attributed to heat inhibition of D-aminolevulinic acid synthase (an enzyme involved in chlorophyll and carotenoid synthesis) and the subsequent photo oxidation of photosynthetic pigments synthesized before high temperature treatments (82), electron transport in Photosystem II (15), and several other phenomena dependent on active proteins (159). The lowered activity of these proteins appears to be selective and suggests that protein denaturation may be involved. Starvation and the inhibition of plant growth by heat, however, may also result from a lower rate of translocation of assimilates from source tissues to sink tissues (62).

Although the mechanism(s) of heat damage to alfalfa have not been elucidated, several heat-induced phenomena could provide clues as to which of the aforementioned mechanisms are responsible for heat damage to alfalfa. Alfalfa plants that are heat tolerant had leaf respiration rates equal to heat susceptible plants at 30°C (290). However, at 40°C, heat-tolerant plants had considerably lower leaf respiration rates than those that were susceptible. If photosynthesis rates were constant for both heat resistant and heat tolerant alfalfa, the temperature compensation point would have been considerably higher for the heat-tolerant plants. This would mean, barring death due to any other heat-induced dysfunctions, the death of heat susceptible alfalfa would be from starvation at lower temperatures than heat-tolerant plants. If heat tolerance in alfalfa results from a low respiration rate at high temperature, this would suggest that the mechanism of heat damage involves metabolic dysfunctions due to imbalances between respiration and other metabolic pathways (e.g., photosynthesis).

8-2.1.2.2 Physical Pertubations of Cellular Structure.

There is considerable evidence that plant membranes (i.e., plasmalemma, tonoplast, endoplasmic reticulum, mitrochondrial, and plastid membranes) are sites for heat injury. Ben Zioni and Itai (14) found that tobacco leaves (*Nicotiana tabacum* L.), exposed to 47.5°C for 2 min, exhibited an increase in the rate of leakage of solutes. However, leaves returned to their original low rates of solute leakage when cooled. Profuse leakage of intracellular solutes and macromolecules from plant tissues is an indication of membrane damage and/or cellular rupture (71, 236). The study with tobacco leaves suggests that early heat injury to plant membranes is easily detectable and reversible. Numerous studies with light and electron microscopes indicate that loss of membrane integrity is concomitant with heat injury (159).

8-2.1.2.3 Heat Effects on the Alfalfa-Rhizobia Symbiosis: An Unexplained Phenomenon.

There are many heat-induced injuries in plants that have not been explained by any mechanism. In alfalfa, heat may cause N deficiencies when soil temperatures are moderately high for prolonged periods. In a previous edition of this monograph, it has been stated that prolonged high temperatures are especially detrimental to alfalfa. Duke and Henson (66) have observed that alfalfa fails to nodulate or nodulates poorly when grown under greenhouse conditions at air temperatures of 33°C or above. Nodules on plants previously grown under optimal conditions for nodulation will deteriorate rapidly when transferred to temperatures of 33°C or above. The same phenomenon has been observed in soybean grown under strictly controlled environmental conditions (69). In both alfalfa and soybean, this phenomenon is consistant with reduced plant growth and N deficiency (visual assessment). It may be explained by thermal inactivation of alfalfa root respiration at near 30°C (65). Although the mechanisms remain unclear, it is obvious that a chronic inability to fix N_2 under continuous high soil temperatures could severely limit alfalfa productivity, quality, and survival. Research is needed to determine if this problem occurs in the field, and if it does, whether it is of sufficient magnitude to cause plant injury.

8-2.2 Adaptive Mechanisms Against Heat Stress

Heat avoidance mechanisms are characterized by a decrease in the absorption of radiant energy through various processes. Reflectance of radiant energy depends upon leaf color and morphology, and angle of inclination (159). Plants also regulate their temperature through convection, radiation, and transpiration (87). Transpirational cooling can remove up to 23% of the incoming heat during midday and may lower leaf temperature by 5 to 10°C depending upon the species (159) and relative humidity (137).

Cellular characteristics may serve, as well, to increase heat adaptation through increased saturation of lipids and increased stability of proteins. Increased saturation of lipids is associated with a substantial increase in the thermal stability of key components of the photosynthetic apparatus (205). Photosynthesis is one of the most heat sensitive aspects of growth and at high temperatures, limitations to photosynthesis are imposed primarily by the thermal instability of the chloroplasts. This involves changes in the thylakoid membrane and perhaps soluble enzymes located outside these membranes (17). The increased thermal stability of key components of the photosynthetic apparatus results in the adjustment of the plant's photosynthetic temperature response characteristics under changing temperature conditions (179). Differences in the thermostability of various enzymes have been reported in numerous species within contrasting thermal environments (258) and in cultivars within the same environment (137).

Other factors influence plant response to high temperature stress. In bromegrasses (*Bromus* spp.), high-temperature injury in the field is less severe when temperatures rise gradually over several days than when there is an abrupt change to hot weather (155). Furthermore, while light increases the heat tolerance of older seedlings, it has no effect on emerging seedlings. Under controlled conditions, the daily maximum heat tolerance in alfalfa is attained within 4 h after exposure to light (153). Accordingly, high temperature may be extremely injurious, particularly when it occurs during critical stages in the development of the plant.

8–2.2.1 Heat Shock Proteins

Cultivated plants adapted to warm temperatures have an acquired tolerance to heat when exposed to sublethal temperatures for a short time (300). Thus, the plants can continue to function at otherwise lethal or "nonpermissive" temperatures. The response to heat stress, that protects the cell against excessive injury, appears to be universal from bacteria to man and includes higher plants (136). The first mild heat shock that induces the synthesis of heat shock proteins appears to protect the organism from injury that would otherwise be lethal. However, the response to the stress is different under a gradual increase in temperature (2). When temperatures rise gradually, normal protein synthesis occurs up to 43°C. Heat shock proteins are synthesized from 43 to 52°C. Under a rapid temperature change from 25 to 40°C, normal protein synthesis ceases. If the temperature is raised higher, the synthesis of heat shock proteins cease at 45°C. Thus under field conditions, normal protein synthesis is protected at high temperatures by a gradual temperature change.

There is some suggestion that heat treatment affects membrane intregrity, and following heat shock, the protein composition of membranes change. The main feature of the heat shock response is the activation of specific genes in which new messenger RNA's are actively transcribed and translated into heat shock proteins. The translation of messages made prior to the stress is curtailed unless the process proceeds gradually (136). Also, it has been suggested that the synthesis of heat shock proteins is induced during the process of recovery from sudden shock (10); however, the mechanism of protection remains obscure (2).

8-2.2.2 Heat Hardening and Recovery

As temperatures increase, various cellular functions are gradually disturbed in the epidermal and parenchyma cells in leaves of many plant species (1). Photosynthesis and protoplasmic streaming are inhibited and the viscosity of the protoplasm is decreased. Next, absorption of vital dyes change, the selective permeability of the protoplast is disrupted, and respiration is suppressed. The degree of damage following this process depends upon the thermo resistance of the proteins and membranes, and, on the reparative ability of the cells. During prolonged and moderate heating, injury and the reparative activity of the cell proceed simultaneously and an adaptive level of heat resistance (heat hardening) may occur.

An increase in the stability of cells to heat during heat hardening is associated with a decrease in growth and photosynthesis (1, 83). Heat hardening is disadvantageous for the maintenance of active cellular metabolism and is advantageous only during high temperature stress. Fortunately, the hardening response is lost when the cells return to optimal growth conditions.

Feldman et al. (83) concluded that the ability of plant cells to change their heat hardiness under the influence of heat stress is universal. Cellular heat resistance may be an essential factor in plant adaptation for extreme temperatures. However, an individual plant has rather limited means to avoid overheating, namely, transpiration and movement of leaves.

Moderately high temperatures may inhibit growth reversibly without any sign of injury (159). Pulgar and Laude (210) stressed alfalfa at 52°C for 2.5 h with no visible sign of tissue mortality or damage. However, the treatment depressed the number and length of shoots in the regrowth for up to 6 weeks. On the other hand, if tops were killed by exposure to 52°C for 6 h, crown buds survived to regrow. Heat-stressed plants were only delayed from 7 to 14 d in attaining the number of shoots in plants not stressed (279). Older plants produced shoots in regrowth more rapidly than younger plants. Plants in flower when stressed produced more shoots than those not yet showing flower buds. Optimal temperatures for growth favor repair from heat stress (159).

8–2.3 Selection for Heat Tolerance

Heat tolerance does not change in the acclimating temperature range of 20 to 30°C, but above 30°C, it increases dramatically and differences between tolerant and susceptible genotypes become more apparent (40, 125). Accordingly, the relative heat tolerance of different species or cultivars should be compared only when plants are in the hardened state (40, 125, 159).

Heat acclimation can occur rapidly as a result of short-term exposures from a few seconds (299), to a few hours (40), or days (138). Deacclimation, however, can be complete within 12 h at a non-acclimating temperature (40).

Heat tolerance is evaluated under 100% humidity to ensure that transpirational cooling does not lower tissue temperature (137). This can be accomplished by immersing whole plants (177), leaves (148, 299), or leaf discs (138), in a hot water bath. Heat killing time at a specified temperature appears to be the primary and most practical way of exposing plants to high temperature stress (40, 138, 263, 285). The conductivity test has been used most frequently to assess the effect of heat stress on plants (40, 263). It evaluates heat damage by measuring the amount of electrolyte leakage associated with damage.

Heat tolerance can be influenced by the stage of development, preconditioning temperatures, water status, level of irradiance, management conditions, rate of temperature change, and duration and degree of exposure (148, 154, 155, 158, 174).

8–3 PROSPECTS FOR IMPROVEMENT

Increases in cold or heat tolerance will improve alfalfa productivity significantly in the cold and hot temperate marginal environments. Considerable success in plant improvement is expected if genotypes are identified and selected under the adverse conditions likely to be encountered in the field. The integration of basic research and field studies must be expanded to improve our understanding of how plant and environmental factors interact in the field. More effort is needed to define specific factors associated with cold or heat stress resistance and sensitivity in order to understand and alter plant characteristics associated with survival. With sufficient understanding of these factors, more reliable selection procedures will become available, with the prospect of concomitant gains through the application of genetic engineering.

Significant progress has been made in alfalfa tissue culture techniques since 1979. Plant regeneration has been achieved from callus cultures, suspension cultures and protoplasts. Techniques have been applied to select amino acid analog resistance in alfalfa, and somatic hybrids of *M. sativa* and *M. falcata* have been obtained by protoplast fusion (see Chapter 29 in this book). These and future techniques to manipulate genes may provide alternatives to classical breeding techniques in selection for tolerance to temperature stress.

The importance of selection for high yield cannot be neglected in northern latitudes. It is widely believed that yield potential in these regions may be increased if winter-hardy genotypes can be developed with reduced fall dormancy. Improvements could require further advances in disease or pest resistance because a sufficient level of winterhardiness must be maintained for the region of adaptation.

Future progress to improve the performance of alfalfa in climates with extreme temperatures awaits the cooperative efforts of physiologists, pathologists, molecular biologists, geneticists and breeders in order to fully understand problems in the field, to develop suitable selection techniques and to identify germplasm pools with acceptable yield and stress tolerance. Progress awaits the development of new cultivars that will optimize yield and minimize risk of environmental stress.

REFERENCES

1. Alexandrov, V.Y. 1967. A study of the changes in resistance of plant cells to the action of various agents in the light of cytoecological considerations. p. 142–151. In A.S. Troshin (ed.) The cell and environment temperature. Pergamon Press, Elmsford, NY.
2. Altschuler, M., and J.P. Mascarenhas. 1982. The synthesis of heat shock and normal proteins at high temperatures in plants and their possible roles in survival under heat stress. p. 321–327. In M.J. Schlesinger et al (ed.) Heat shock from bacteria to man. Cold Spring Harbor Laboratory, Cold Spring Harbor, NY.
3. Anderson, J.A., L.V. Gusta, D.W. Buchanan, and M.J. Burke. 1983. Freezing of water in citrus leaves. J. Am. Soc. Hortic. Sci. 108: 397–400.
4. Andrews, C.J., M.K. Pomeroy, and I.A. de la Roche. 1974. Changes in cold hardiness of overwintering winter wheat. Can. J. Plant Sci. 54 :9–15.
5. Arakeri, H.R., and A.R. Schmid. 1949. Cold resistance of various legumes and grasses in early stages of growth. Agron. J. 41: 182–185.
6. Babenko, V.I. 1968. Some peculiarities of metabolism of soluble carbohydrates and free acids in winter wheat induced by negative temperatures. Sov. Plant Physiol. 15:710–715.
7. Barnes, D.K., E.T. Bingham, R.P. Murphy, O.J. Hunt, D.F. Beard, W.H. Skrdla, and L.R. Teuber. 1977. Alfalfa germplasm in the United States: Genetic vulnerability, use, improvement, and maintenance. USDA Tech. Bull. 1571. U.S. Government Printing Office, Washington, DC.
8. ----, D.M. Smith, R.E. Stucker, and L.J. Elling. 1979. Fall dormancy in alfalfa; A valuable predictive tool. p. 34. In D.K. Barnes (ed.) Report of the 26th Alfalfa Improvement Conf., Brookings, SD. 6–8 June 1978. Agricultural Reviews and manuals. USDA-SEA, St. Paul.
9. Barnett, N.M., and A.W. Naylor. 1966. Amino acid and protein metabolism in Bermuda grass during water stress. Plant Physiol. 41:1222–1230.
10. Barnett, T.M., M. Altschuler, C.N. McDaniel, and J.P. Mascarenhas. 1980. Heat shock proteins in plant cells. Dev. Genet. (NY) 1: 331–340.
11. Barta, A.L. 1980. Regrowth and alcohol dehydrogenase activity in waterlogged alfalfa and birdsfoot trefoil. Agron. J. 72:1017–1020.
12. Basu, P.K. 1981. Color infrared aerial photography to identify forage legumes in hay fields. Can J. Plant Sci. 61: 331–336.
13. Beattie, D.J., and H.L. Flint. 1973. Effect of K level on frost hardiness of stems of Forsythia × intermedia Zab. 'Lynwood'. J. Am. Soc. Hortic. Sci. 98: 539–541.
14. Ben Zioni, A., and C. Itai. 1973. Short- and long-term effects of high temperatures (47–49°C) on tobacco leaves. III. Efflux and ^{32}P incorporation into phospholipids. Physiol. Plant 28: 493–497.
15. Berry, J.A., D.C. Fork, and S. Garrison. 1975. Mechanistic studies of thermal damage to leaves. Year Book Carnegie Inst. Washington 74: 751–759.

16. Björkman, O. 1975. Thermal stability of the photosynthetic apparatus in intact leaves. Year Book—Carnegie Inst. Washington 74: 748-751.
17. ----, M.R. Badger, and P.A. Armond. 1980. Response and adaptation of photosynthesis to high temperatures. p. 233-249. In N.C. Turner and P.J. Kramer (ed.) Adaptation of plants to water and high temperature stress. John Wiley and Sons. New York.
18. Black, C.C., T.M. Chen, and R.H. Brown. 1969. Biochemical basis for plant competition. Weed Sci. 17: 338-344.
19. Blinn, P.K. 1911. Alfalfa: Relation of type to hardiness. Colo. Agric. Exp. Stn. Bull. 181.
20. Bolduc, R. 1976. Technique pour échantillonner les racines de plantes dans le sol gelé et enneigé. Can. J. Plant Sci. 56: 633-638.
21. ----. 1980. Une méthode enzymologique à appliquer pour la sélection de plantes résistantes au froid. Can. J. Plant Sci. 60: 1303-1308.
22. ----. L. Rancourt, P. Dolbec, and L. Chouinard-Lavoie. 1978. Mesure de l'endurcissement au froid et de la viabilité des plantes exposées au gel par le dosage des phosphatases acides libres. Can. J. Plant Sci. 58: 1007-1018.
23. Bolton, J.L., and R.E. McKenzie. 1946. The effect of early spring flooding on certain forage crops. Sci. Agric. 26: 99-105.
24. Boyer, J.S. 1982. Plant productivity and environment. Science 218: 443-448.
25. Bray, E., M.L. Brenner, and L.R. Parsons. 1979. Abscisic acid levels and stomatal resistance of red-osier dogwood during cold acclimation. Plant Physiol. (Suppl.) 63: 104.
26. Brown, C.S., and R.F. Stafford. Fall 1969. Alfalfa renewal following winter injury. Res. Life Sci. 17: 20-23.
27. Brown, J.H., R.J. Bula, and P.F. Low. 1970. Physical properties of cytoplasmic protein-water extracts from roots of hardy and nonhardy *Medicago sativa* ecotypes. Cryobiology 6: 309-314.
28. Bula, R.J. 1972. Morphological characteristics of alfalfa plants grown at several temperatures. Crop Sci. 12: 683-686.
29. ----, and M.A. Massengale. 1972. Environmental physiology. In C.H. Hanson (ed.) Alfalfa science and technology. Agronomy 15: 167-184.
30. ----, and D. Smith. 1954. Cold resistance and chemical composition in overwintering alfalfa, red clover and sweetclover. Agron. J. 46: 397-401.
31. ----, ----, and H.J. Hodgson. 1956. Cold resistance in alfalfa at two diverse latitudes. Agron. J. 48: 153-156.
32. Burke, M.J., R.G. Bryant, and C.J. Weiser. 1974. Nuclear magnetic resonance of water in cold acclimating red osier dogwood stem. Plant Physiol. 54: 392-398.
33. ----, L.V. Gusta, H.A. Quamme, C.J. Weiser, and P.H. Li. 1976. Freezing and injury in plants. Annu. Rev. Plant Physiol. 27: 507-528.
34. Busbice, T.H., and C.P. Wilsie. 1968. Fall growth, winterhardiness, recovery after cutting and wilt resistance in F_2 progenies of Vernal × DuPuits alfalfa crosses. Crop Sci. 5: 429-432.
35. Calder, F.W., W.D. Canham, and D.S. Fensom. 1973. Some effects of Alar-85 on the physiology of alfalfa and ladino clover. Can. J. Plant Sci. 53: 269-278.
36. ----, and L.B. MacLeod. 1966. Effect of cold treatment on alfalfa as influenced by harvesting system and rate of potassium application. Can. J. Plant Sci. 46: 17-26.
37. ----, ----, and L.P. Jackson. 1965. Effect of soil moisture content and stage of development on cold-hardiness of the alfalfa plant. Can. J. Plant Sci. 45: 211-218.
38. Chatterton, N.J., G.E. Carlson, R.H. Hart, and W.E. Hungerford. 1974. Tillering, nonstructural carbohydrates, and survival relationship in alfalfa. Crop Sci. 14:783-787.
39. Chen, T.H.H., and L.V. Gusta. 1983. Abscisic acid-induced freezing resistance in cultured plant cells. Plant Physiol. 73: 71-75.
40. ----, Z.Y. Shen, and P.H. Li. 1982. Adaptability of crop plants to high temperature stress. Crop Sci. 22: 719-725.
41. Chernomorets, M.V. 1969. Change in the main components of the carbohydrate complex in grape shoots in connection with their frost resistance. Sov. Plant Physiol. 16: 382-386.
42. Christian, K.R. 1977. Effects of the environment on the growth of alfalfa. Adv. Agron. 29: 183-227.
43. Christiansen, M.N. 1978. The physiology of plant tolerance to temperature extremes. p. 173-191. In G.A. Jung (ed.) Crop tolerance to suboptimal land conditions. American Society of Agronomy, Madison, WI.
44. Coffindaffer, B.L., and O.J. Burger. 1958. Response of alfalfa varieties to daylength. Agron. J. 50: 389-392.
45. Collins, J.C., and M. Morgan. 1980. The influence of temperature on the abscisic acid stimulated water flow from excised maize roots. New Phytol. 84: 19-26.

46. Cook, T.W., and D.T. Duff. 1976. Effects of K fertilization on freezing tolerance and carbohydrate content of *Festuca arundinacea* Schreb. maintained as turf. Agron. J. 68: 116–119.
47. Cowett, E.R., and M.A. Sprague. 1962. Factors affecting tillering in alfalfa. Agron. J. 54: 294–297.
48. Daday, H. 1964. Genetic relationships between cold hardiness and growth at low temperature in *Medicago sativa*. Heredity 19: 173–179.
49. DeBoer, V.L., and S.H. Duke. 1982. Effects of sulfur nutrition on nitrogen and carbon metabolism in lucerne (*Medicago sativa* L.). Physiol. Plant. 54: 343–350.
50. Delaney, R.H., A.K. Dobrenz, and H.T. Poole. 1974. Seasonal variation in photosynthesis, respiration and growth components of non-dormant alfalfa. Crop Sci. 14: 58–61.
51. Dexter, S.T. 1933. Effect of several environmental factors on the hardening of plants. Plant Physiol. 8: 123–139.
52. ----. 1935. Growth, organic nitrogen fractions and buffer capacity in relation to hardiness of plants. Plant Physiol. 10: 149–158.
53. ----. 1941. Effect of periods of warm weather upon the winter hardened condition of a plant. Plant Physiol. 16: 181–188.
54. ----, W.E. Tottingham, and L.F. Graber. 1930. Preliminary results in measuring the hardiness of plants. Plant Physiol. 5: 215–223.
55. ----, ----, and ----. 1932. Investigations of hardiness of plants by measurement of electrical conductivity. Plant Physiol. 7: 63–78.
56. Dinus, R.J. 1968. Effect of red and far red light upon growth of Douglas fir (*Pseudotsuga menzisii* (Mirb.) Franco) seedlings. Ph.D. thesis. Oregon State University, Corvallis (Diss. Abstr. AAD68-09593).
57. Doehlert, D.C., and S.H. Duke. 1983. Specific determination of α-amylase activity in crude plant extracts containing β-amylase. Plant Physiol. 71: 229–234.
58. ----, ----, and L. Anderson. 1982. Beta-amylases from alfalfa (*Medicago sativa* L.) roots. Plant Physiol. 69: 1096–1102.
59. Dotzenko, A.D., and J.G. Dean. 1959. Germination of six alfalfa varieties at three levels of osmotic pressure. Agron. J. 51: 308–309.
60. Dovrat, A., and M. Waldman. 1967. Rating of alfalfa varieties for cold resistance by germination in solutions of different osmotic values. Crop Sci. 7: 1–2.
61. Dua, R.D., and R.H. Burris. 1963. Stability of nitrogen-fixing enzymes and the reactivation of a cold labile enzyme. Proc. Natl. Acad. Sci. USA. 50: 169–175.
62. Duff, D.T., and J.B. Beard. 1974. Supraoptimal temperature effects upon *Agrostis palustris*. Part II. Influence of carbohydrate levels, photosynthetic rate, and respiration rate. Physiol. Plant. 32: 18–22.
63. Duke, S.H. 1982. Unpublished data. University of Wisconsin, Madison.
64. ----, and M. Collins. 1985. Role of potassium in legume dinitrogen fixation. p. 443–465. *In* R.D. Munson (ed.) Potassium in agriculture. American Society of Agronomy, Crop Science Society of America, and Soil Science Society of America, Madison, WI.
65. ----, and D.C. Doehlert. 1981. Root respiration, nodulation, and enzyme activities in alfalfa during cold acclimation. Crop Sci. 21: 489–495.
66. ----, and C.A. Henson. 1983. Unpublished data. University of Wisconsin, Madison.
67. ----, and ----. 1985a. Legume nodule carbon utilization in the synthesis of organic acids for the production of transport amides and amino acids. p. 293–302. *In* P.W. Ludden and J.E. Burris (ed.) Nitrogen fixation and CO_2 metabolism. Elsevier Science Publishing Co., New York.
68. ----, and ----. 1985b. Chilling stress and plant enzyme dysfunction: An overview. Plant Physiol. (Suppl.) 77: 66.
69. ----, ----, L.E. Schrader, and R.D. Vogelzang. 1978. Unpublished data. University of Wisconsin, Madison.
70. ----, and G. Kakefuda. 1981. Role of the testa in preventing cellular rupture during imbibition of legume seeds. Plant Physiol. 67: 449–456.
71. ----, ----, and T.M. Harvey. 1983. Differential leakage of intracellular substances from imbibing soybean seeds. Plant Physiol. 72: 919–924.
72. ----, and H.M. Reisenauer. 1985. Roles and requirements of sulfur in plant nutrition. *In* M.A. Tabatabai (ed.) Sulfur in agriculture. Agronomy 27: 123–168.
73. ----, L.E. Schrader, C.A. Henson, J.C. Servaites, R.D. Vogelzang, and J.W. Pendleton. 1979. Low root temperature effects on soybean nitrogen metabolism and photosynthesis. Plant Physiol. 63: 956–962.
74. ----, ----, and M.G. Miller. 1977. Low temperature effects on soybean (*Glycine max* L. Merr. cv. Wells) mitochondrial respiration and several dehydrogenases during imbibition and germination. Plant Physiol. 60: 716–722.
75. ----, ----, ----, and R.L. Niece. 1978. Low temperature effects on soybean (*Glycine max* L. Merr. cv. Wells) free amino acid pools during germination. Plant Physiol. 62: 642–647.

76. Eamus, D., and J.M. Wilson. 1983. ABA levels and effects in chilled and hardened *Phaseolus vulgaris*. J. Exp. Bot. 34: 1000–1006.
77. Elling, L.J., C.H. Hanson, and H.O. Graumann. 1960. Winterkilling in diallel crosses. p. 23–26. *In* Report of the 17th Alfalfa Improvement Conf., Saskatoon, SK. Canada. 27–29 June. USDA-ARS, Washington, DC.
78. Erwin, D.C., B.W. Kennedy, and W.F. Lehman. 1959. Xylem necrosis and root rot of alfalfa associated with excessive irrigation and high temperatures. Phytopathology 49: 572–578.
79. Evenson, P.D., and M.D. Rumbaugh. 1972. Influence of mulch on postharvest soil temperatures and subsequent regrowth of alfalfa (*Medicago sative* L.). Agron. J. 64: 154–157.
80. Eze, J.M.O., E.B. Dumbroff, and J.E. Thompson. 1983. Effects of temperature and moisture stress on the accumulation of abscisic acid in bean. Physiol. Plant. 58: 179–183.
81. Faw, W.F., S.C. Shih, and G.A. Jung. 1976. Extractant influence on the relationship between extractable proteins and cold tolerance of alfalfa. Plant Physiol. 57: 720–723.
82. Feierabend, J. 1977. Capacity for chlorophyll synthesis in heat-bleached 70-S ribosome-deficient rye leaves. Planta 135: 83–88.
83. Feldman, N.L., V.Y. Alexandrov, I.G. Zavadskaya, I.M. Kislyuk, A.G. Lomagin, M.I. Lyutova, and A. Jaskuliev. 1967. Heat hardening of plant cells under natural and experimental conditions. p. 152–160. *In* A.S. Troshin (ed.) The cell and environmental temperature. Pergamon Press, Elmsford, New York.
84. Foord, K.E. 1984. Physiological, environmental, and genetic determinants of seedling growth in *Medicago sativa*. Ph.D. thesis. Univ. of California, Davis (Diss. Abstr. DA 8521203).
85. Franks, F. 1983. Editorial. Bound water: Fact and fiction. Cryo-Lett. 4: 73–74.
86. Freyman, S., and V.C. Brink. 1967. Nature of ice-sheet injury to alfalfa. Agron. J. 59: 557–560.
87. Gates, D.M. 1965. Heat transfer in plants. Sci. Am. 213: 76–84.
88. George, M.F., and M.J. Burke. 1977. Cold hardiness and deep supercooling in xylem of shagbark hickory. Plant Physiol. 59: 319–325.
89. Gerloff, E.D., T. Richardson, and M.A. Stahmann. 1966. Changes in fatty acids of alfalfa roots during cold hardening. Plant Physiol. 41: 1280–1284.
90. ----, M.A. Stahmann, and D. Smith. 1967. Soluble proteins in alfalfa roots as related to cold hardiness. Plant Physiol. 42: 895–899.
91. Graber, L.F., N.T. Nelson, W.A. Luekel, and W.B. Albert. 1927. Organic food reserves in relation to the growth of alfalfa and other perennial berbaceous plants. Univ. of Wis. Agric. Exp. Stn. Res. Bull. 80.
92. Grandfield, C.O. 1935. The trend of organic food reserves in alfalfa roots as affected by cutting practices. J. Agric. Res. 50: 697–709.
93. ----. 1943. Food reserves and their translocation to the crown buds as related to cold and drought resistance in alfalfa. J. Agric. Res. 67: 33–47.
94. Greenham, C.G. 1966. The stages at which frost injury occurs in alfalfa. Can. J. Bot. 44: 1471–1483.
95. ----, and H. Daday. 1960. Further studies on the determination of cold hardiness in *Trifolium repense* L. and *Medicago sativa* L. Aust. J. Agric. Res. 11: 1–15.
96. Grenier, G., H.J. Hope, C. Willemot, and H.P. Therrien. 1975. Sodium-1, 2^{-14}C acetate incorporation in roots of frost-hardy and less hardy alfalfa varieties under hardening conditions. Plant Physiol. 55: 906–912.
97. ----, P. Mazliak, A. Trémolières, and C. Willemot. 1973. Cold influence on fatty acid synthesis in the roots of two alfalfa varieties, one cold-resistant and the other less resistant. Physiol. Veg. 11: 253–265.
98. ----, and C. Willemot. 1974. Lipid changes in roots of frost hardy and less hardy alfalfa varieties under hardening conditions. Cryobiology 11: 324–331.
99. ----, and ----. 1975. Lipid phosphorous content and $^{33}P_i$ incorporation in roots of alfalfa varieties during frost hardening. Can. J. Bot. 53: 1473–1477.
100. Gross, H.D., E.R. Purvis, and G.H. Ahlgren. 1953. The response of alfalfa varieties to different soil fertility levels. Agron. J. 45: 118–120.
101. Gusta, L.V., M.J. Burke, and A.C. Kapoor. 1975. Determination of unfrozen water in winter cereals at subfreezing temperature. Plant Physiol. 56: 707–709.
102. ----, D.B. Fowler, and N.J. Tyler. 1982. The effect of abscisic acid and cytokinins on the cold hardiness of winter wheat. Can. J. Bot. 60: 301–305.
103. Hardy, R.W.F., R.D. Holsten, E,K. Jackson, and R.C. Burns. 1968. The acetylene–ethylene assay for N_2 fixation: laboratory and field evaluation. Plant Physiol. 43: 1185–1207.
104. Hayden, R.E., C.A. Moyse, F.W. Calder, D.P. Crawford, and D.S. Fensom. 1969. Electrical impedance studies on potato and alfalfa tissue. J. Exp. Bot. 20: 177–200.

105. Heber, U., L. Tyankova, and K.A. Santarius. 1971. Stabilization and inactivation of biological membranes during freezing in the presence of amino acids. Biochim. Biophys. Acta 241: 578–592.
106. Heinrichs, D.H. 1959. Germination of alfalfa varieties in solutions of varying osmotic pressure and relationship to winter hardiness. Can. J. Plant Sci. 39: 384–394.
107. ----. 1970. Flooding tolerance of legumes. Can. J. Plant Sci. 50: 435–438.
108. ----. 1973. Winterhardiness of alfalfa cultivars in southern Saskatchewan. Can. J. Plant Sci. 53:773–777.
109. ----, and F.H.W. Morley. 1960. Inheritance of resistance to winter injury and its correlation with creeping rootedness in alfalfa. Can. J. Plant Sci. 40: 487–489.
110. ----, and K.F. Nielsen. 1966. Growth response of alfalfa varieties of diverse genetic origin to different root zone temp. Can. J. Plant Sci. 46: 291–298.
111. ----, J.E. Troelsen, and K.W. Clark. 1960. Winter hardiness evaluation in alfalfa. Can. J. Plant Sci. 40: 638–644.
112. Henson, C.A., L.E. Schrader, and S.H. Duke. 1980. Effects of temperature on germination and mitochondrial dehydrogenases in two soybean (*Glycine max*) cultivars. Physiol. Plant. 48: 168–174.
113. Hesterman, O.B. 1981. Effect of photoperiod and irradiance on fall dormancy response of alfalfa (*Medicago sativa* L.) M.S. thesis. Univ. of California, Davis.
114. Hodgson, H.J. 1964. Effect of photoperiod on development of cold resistance in alfalfa. Crop Sci.1 4: 302–305.
115. ----, and R.J. Bula. 1956. Hardening behavior of sweet clover (*Melilotus* spp.) varieties in a subarctic environment. Agron. J. 48: 157–160.
116. Howell, G.S., Jr., and F.G. Dennis, Jr. 1981. Cultural management of perennial plants to maximize resistance to cold stress. p. 175–204. *In* C.R. Olien and N. Smith (ed.) Analysis and improvement of plant cold hardiness. CRC Press, Boca Raton, FL.
117. Huner, N.P.A., and F.D.H. Macdowall. 1978. Evidence for an *in vivo* conformational change in ribulose bisphosphate carboxylase-oxygenase from Puma rye during cold adaptation. Can. J. Biochem. 56: 1154–1161.
118. ----, and ----. 1979. Change in the net charge and subunit properties of ribulose bisphosphate carboxylase-oxygenase during cold hardening of Puma rye. Can. J. Biochem. 57: 155–164.
119. Irving, R.M., and F.O. Lanphear. 1968. Regulation of cold hardiness in *Acer negundo*. Plant Physiol. 43: 9–13.
120. Janssen, G. 1929. The relationship of organic root reserves and other factors to the permanency of alfalfa stands. J. Am. Soc. Agron. 21: 895–911.
121. Johansson, N.O., and B. Torssell. 1956. Field trials with a portable refrigerator. Acta Agric. Scand. 6: 81–99.
122. Jones, E.R., R.B. Carroll, R.H. Swain, and K.W. Bell. 1978. Role of anthracnose in stand thinning of alfalfa in Delaware, USA. Agron. J. 70: 351–353.
123. Jones, F.R. 1928. Winter injury of alfalfa. J. Agric. Res. 37: 189–212.
124. ----, and W.B. Tisdale. 1921. Effect of soil temperature upon the development of nodules on the roots of certain legumes. J. Agric. Res. 22: 17–31.
125. Julander, O. 1945. Drought resistance in range and pasture grasses. Plant Physiol. 20: 573–599.
126. Jung, G.A., and K.L. Larson. 1972. Cold, drought and heat tolerance. *In* C.H. Hanson (ed.) Alfalfa science and technology. Agronomy 15: 185–209.
127. ----, S.C. Shih, and D.C. Shelton. 1967. Seasonal changes in soluble protein, nucleic acids, and tissue pH related to cold hardiness of alfalfa. Cryobiology 4: 11–16.
128. ----, and D. Smith. 1959. Influence of soil potassium and phosphorus content on the cold resistance of alfalfa. Agron. J. 51: 585–587.
129. ----, and ----. 1960. Influence of extended storage at constant low temperature on cold resistance and carbohydrate reserves of alfalfa and medium red clover. Plant Physiol. 35: 123–125.
130. ----, and ----. 1961. Trends of cold resistance and chemical changes over winter in the roots and crowns of alfalfa and medium red clover. I. Changes in certain nitrogen and carbohydrate fractions. Agron. J. 53: 359–364.
131. Kacperska-Palacz, A. 1978. Mechanism of cold acclimation in herbaceous plants. p. 139–152. *In* P.H. Li and A. Sakai (ed.) Plant cold hardiness and freezing stress. Mechanisms and crop implications. Academic Press, New York.
132. ----. Z. Debska, and A. Jakubowska. 1975. The phytochrome involvement in the frost hardening process of rape seedlings. Bot. Gaz. (Chicago) 136: 137–140.
133. Kappen, L. 1979. Tolerance of halophytes against freezing and salt stress and its possible biochemical causes. Ber. Dtsch. Bot. Ges. 92: 55–71.
134. ----, and W.R. Ullrich. 1970. Verteilung von Chlorid und Zuckern in Blattzellen halophiler Pflanzen bei verschieden hoher Frostresistenz. Ber. Dtsch. Bot. Ges. 83:265–

135. Kemp, D.R., and W.M. Blacklow. 1980. Diurnal extension rates of wheat leaves in relation to temperatures and carbohydrate concentrations of the extension zone. J. Exp. Bot. 31: 821–828.
136. Key, J.L., C.Y. Lin, E. Ceglarz, and F. Schoffl. 1982. Heat shock-response in plants. p. 329–336. In M.J. Schlesinger et al. (ed.) Heat shock from bacteria to man. Cold Spring Harbor Laboratory, Cold Spring Harbor, NY.
137. Kinbacher, E.J. 1963. Relative high temperature resistance of winter oats at different relative humidities. Crop Sci. 3: 466–468.
138. ----, C.Y. Sullivan, and H.R. Knull. 1967. Thermal stability of malic dehydrogenase from heat hardened *Phaseolus acutifolius* 'Tepary Buff'. Crop Sci. 7: 148–151.
139. Kirkham, M.B. 1983. Effect of ABA on the water relations of winter-wheat cultivars varying in drought resistance. Physiol. Plant. 59: 153–157.
140. Klebesadel, L.J. 1971. Selective modification of alfalfa towards acclimatization in a subarctic area of severe winter stress. Crop Sci. 11: 609–614.
141. Knapp, W.R., and J.S. Knapp. 1980. Interaction of planting date and fall fertilization on winter barley performance. Agron. J. 72: 440–445.
142. Kohel, R.J., and R.L. Davis. 1960. The inheritance of cold resistance and its relation to fall growth in F_2 and BC_1 generations of alfalfa. Agron. J. 52: 234–237.
143. Koller, D. 1969. The physiology of dormancy and survival of plants in desert environment. Proc. Soc. Exp. Biol. 23: 449–469.
144. Krasnuk, M., G.A. Jung, and F.H. Witham. 1978a. Dehydrogenase levels in cold-tolerant and cold-sensitive alfalfa. Agron. J. 70: 605–613.
145. ----, ----, and ----. 1978b. Hydrolytic enzyme differences in cold-tolerant and cold-sensitive alfalfa. Agron. J. 70: 597–605.
146. Lacefield, G.D. 1975. Effects of low temperature and drought on various morphological, biochemical, and mineral characters of alfalfa. Missouri Univ., Columbia (Diss. Abstr. 35: 4316B).
147. Laliberté, G., and R. Paquin. 1984. Effets des basses temperatures, avec ou sans apport d'acide glutamique et de saccharose, sur la teneur en proline libre et la tolérance au gel du Blé d'hiver. Physiol. Veg. 22: 305–313.
148. Lange, O.L. 1965. The heat resistance of plants, its determination and variability. p. 399–405. In F. Eckhardt (ed.) Methodology of Plant Ecophysiology. Proc. of the Montpellier Symp., Montpellier. UNESCO, Paris.
149. Larcher, W., and H. Bauer. 1981. Ecological significance of resistance to low temperature. p. 403–437. In O.L. Lange et al. (ed.) Encyclopedia of Plant Physiology, New Series, Vol. 12A, Physiol. Plant Ecology I. Springer-Verlag New York, New York.
150. Larsen, A. 1978. Freezing tolerance in grasses. Effect of different water contents in growth media. Meld. Nor. Landbrokshogsk. 57: (15) 1–19.
151. ----. 1978. Freezing tolerance in grasses. Methods for testing in controlled environments. Meld. Nor. Landbrokshogsk. 57: (23) 1–56.
152. Larson, K.L., and D. Smith. 1963. Association of various morphological characters and seed germination with the winterhardiness of alfalfa. Crop Sci. 3: 234–237.
153. Laude, H.M. 1939. Diurnal cycle of heat resistance in plants. Science 89: 556–557.
154. ----. 1964. Plant response to high temperatures. p. 15–31. In Forage plant physiology and soil-range relationships. American Society of Agronomy Madison, WI.
155. ----, and B.A. Chaugule. 1953. Effect of stage of seedling development upon heat tolerance in bromegrasses. J. Range Manage. 6: 320–324.
156. Leach, G.J. 1967. Growth and development of lucerne. p. 15–21. In R.H.M. Langer (ed.) The lucerne crop. Reed, Wellington, New Zealand.
157. Le Saint, A.M. 1966. Observations sur le gel et l'endurcissement au gel chez le chou de Milan. Thèse de Doct. University of Paris, (Série A, No. 4669).
158. Levitt, J. 1978. Crop tolerance to suboptimal land conditions—a historical overview. p. 161–171. In G.A. Jung (ed.) Crop tolerance to suboptimal land conditions. Spec. Pub. 32. American Society of Agronomy, Crop Science Society of America, and Soil Science Society of America, Madison, WI.
159. ----. 1980. Response of plants to environmental stresses. Chilling, freezing, and high temperature stresses. 2nd ed. Academic Press, New York.
160. Li, P.H., and A. Sakai. 1982. Plant cold hardiness and freezing stress. Mechanism and crop implications. Vol. 2. Academic Press, New York.
161. Londesborough, J. 1980. The causes of sharply bent or discontinuous Arrhenius plots for enzyme-catylysed reactions. Eur. J. Biochem. 105: 211–215.
162. MacLeod, L.B., and M. Suzuki. 1967. Effect of potassium on the content of amino acids in alfalfa and orchardgrass grown with NO_3 and NH_4 nitrogen in nutrient solution culture. Crop Sci. 7: 599–605.
163. Marble, V.I., and G. Peterson. 1982. Alfalfa varieties and brands for California's different soils and climates. p. 15–38. In twelfth California Alfalfa Symp., Fresno, CA. 8–9 December, University of California, Davis.

164. Mark, J.J. 1936. The relation of reserves to cold resistance in alfalfa. p. 305-335. Iowa Agric. Exp. Stn. Bull. 208.
165. McKenzie, J.S., and J.G.N. Davidson. 1975. Prevalence of alfalfa crown and root diseases in the Peace River region of Alberta and British Columbia. Can. Plant Dis. Surv. 55: 121-125.
166. ——, K. Lopetinsky, and G.E. McLean. 1984. Predicting alfalfa winterkill in northern Alberta. Agric. Can. Forage Notes 27: 5-10.
167. ——, and G.E. McLean. 1980. Some factors associated with injury to alfalfa during the 1977-78 winter at Beaverlodge, Alberta. Can. J. Plant Sci. 60: 103-112.
168. ——, and ——. 1980. Changes in the cold hardiness of alfalfa during five consecutive winters at Beaverlodge, Alberta. Can. J. Plant Sci. 60: 703-712.
169. ——, and ——. 1982. The importance of leaf frost resistance to the winter survival of seedling stands of alfalfa. Can. J. Plant Sci. 62: 399-405.
170. ——, and ——. 1984. A stress test for assessing the winter hardiness of alfalfa in northwestern Canada. Can. J. Plant Sci. 64: 917-924.
171. ——, C.J. Weiser, and M.J. Burke. 1974. Effects of red and far red light on the initiation of cold acclimation in *Cornus stolonifera* Mich. Plant Physiol. 53: 783-789.
172. McKenzie, R.E. 1951. The ability of forage plants to survive early spring flooding. Sci. Agric. 31: 358-367.
173. McKersie, B.D. 1983. Types of winter stresses—Do winter wheat cultivars respond differently? p. 26-38. *In* D.B. Fowler et al. (ed.) New frontiers in winter wheat production. Proc. West. Canada Winter Wheat Conf. Saskatoon, SK. 20-23 June. University of Saskatchewan, Saskatoon.
174. McWilliams, J.R. 1980. Adaptation of plants to water and high temperature stress. Summary and synthesis adaptation to high temperature stress. p. 444-447. *In* N.C. Turner and P.J. Kramer (ed.) Adaptation of plants to water and high temperature stress. John Wiley and Sons, New York.
175. Megee, C.R. 1935. A search for factors determining winter hardiness in alfalfa. J. Am. Soc. Agron. 27: 685-698.
176. Metochis, C., and P.I. Orphanos. 1981. Alfalfa yield and water use when forced into dormancy by withholding water during the summer. Agron. J. 73: 1048-1050.
177. Minner, D.D., P.H. Dernoedon, D.J. Wehner, and M.S. McIntosh. 1983. Heat tolerance screening of field grown cultivars of Kentucky bluegrass and perennial ryegrass. Agron. J. 75: 772-775.
178. Modlibowska, I. 1968. Effects of some growth regulators on frost damage. Cryobiology 5: 175-187.
179. Mooney, H.A., O. Björkman, and G.J. Collatz. 1978. Photosynthetic acclimation to temperature in the desert shrub, *Larrea divaricata*. I. Carbon dioxide exchange characteristics of intact leaves. Plant Physiol. 61: 406-410.
180. Morley, F.H.W., H. Daday, and J.W. Peak. 1957. Quantitative inheritance in lucerne, *Medicago sativa* L. I. Inheritance and selection for winter yield. Aust. J. Agric. Res. 8: 635-651.
181. Nath, J., and T.C. Fisher. 1971. Anatomical study of freezing injury in hardy and nonhardy alfalfa varieties treated with cytosine and quanine. Cryobiology 8: 420-430.
182. Nelson, C.J., and D. Smith. 1968. Growth of birdsfoot trefoil and alfalfa. II. Morphological development and dry matter distribution. Crop Sci. 8: 21-25.
183. Nicholaichuk, W., and D.W.L. Read. 1981. Snow management by swathing at alternate heights. Can. Agric. 26: 25-26.
184. Nittler, L.W. 1954. A possible greenhouse procedure for testing alfalfa seedlings for trueness to type. Proc. Assoc. Off. Seed Anal. 44: 114-116.
185. ——, and G.H. Gibbs. 1959. The response of alfalfa varieties to photoperiod, color of light and temperature. Agron. J. 51: 727-730.
186. Oakley, R.A., and H.L. Westover. 1921. Effect of the length of day on seedlings of alfalfa varieties and the possibility of utilizing this as a practical means of identification. J. Agric. Res. 21: 599-608.
187. Okimoto, R., M.M. Sachs, E.K. Porter, and M. Freeling. 1980. Patterns of polypeptide synthesis in various maize organs under anaerobiosis. Planta 150: 89-94.
188. Olien, C.R. 1967. Freezing stresses and survival. Ann. Rev. Plant Physiol. 18: 387-408.
189. ——, and M.N. Smith. 1981. Analysis and improvement of plant cold hardiness. CRC Press, Boca Raton, FL.
190. Osborne, D. 1966. Protecting plants from cold. New Sci. 29: 773-774.
191. Ouellet, C.E. 1976. Winter hardiness and survival of forage crops in Canada. Can. J. Plant Sci. 56: 679-689.
192. ——. 1977. Monthly climatic contribution to the winter injury of alfalfa. Can. J. Plant Sci. 57: 419-426.
193. Rahn, J.P., K.O. Jensen, and P.H. Li. 1982. Cell membrane alterations following a

slow freeze-thaw cycle: ion leakage, injury and recovery. p. 221–242. *In* P.H. Li and A. Sakai (ed.) Plant cold hardiness and freezing stress. Mechanizing and crop implications, Vol. 2. Academic Press, New York.
194. ----, and P.H. Li. 1980. Alterations in membrane transport properties by freezing injury in herbaceous plants: evidence against rupture theory. Physiol. Plant. 50: 164–175.
195. Paquin, Roger. 1977. Effet des basses températures sur la résistance au gel de la Luzerne (*Medicago media* Pers.) et son contenu en proline libre. Physiol. Veg. 15: 657–665.
196. ----, 1984. Influence of the environment on cold hardening and winter survival of forage and cereal species with consideration of proline as a metabolic marker of hardening. p. 137–154. *In* W.S. Margaris et al. (ed.) Being alive on land. Junk The Hague.
197. ----, L. Belzile, C. Willemot, and J.C. St-Pierre. 1976. Effets de quelques retardants de croissance et de l'acide gibberellique sur la résistance au froid de la luzerne (*Medicago sativa*). Can. J. Plant Sci. 56: 79–86.
198. ----, G. Ladouceur, R. Desrosiers, and A. Mack. 1977. Etude sur la survie de la luzerne au Québec au moyen de photos couleurs et infra-rouges à des des échelles de 1:6000 à 1:40000. p. 506–515. *In* Fourth Canadian Symp. Remote Sensing, Quebec City, Quebec. 16–18 May. Canadian Aeronautics and Space Institute, Ottawa.
199. ----. and P. Lechasseur. 1982. Acclimatation naturelle de la luzerne (*Medicago media* Pers.) au froid II. Variations de la teneur en sucres totaux des feuilles et des collets. Oecol. Plant. 3: 27–38.
200. ----, and G.R. Mehuys. 1980. Influence of soil moisture on cold tolerance of alfalfa. Can. J. Plant Sci. 60: 139–147.
201. ----, and H. Pelletier. 1975. Technique de prélèvement de plantes dans un sol gelé. Can. J. Plant Sci. 55: 327–330.
202. ----, and ----. 1980. Influence de l'environnement sur l'acclimatation au froid de la luzerne (*Medicago media* Pers.) et sa résistance au gel. Can. J. Plant Sci. 60: 1351–1366.
203. ----, and Guy Pelletier. 1981. Acclimatation naturelle de la Luzerne (*Medicago media* Pers.) au froid. I. Variations de la teneur en proline libre des feuilles et des collets. Physiol. Veg. 19: 103–117.
204. Peake, R.W. 1964. Evaluation of cold-hardiness by controlled freezing of field-hardened forage crops. Can. J. Plant Sci. 44: 538–543.
205. Pearcy, R.M. 1977. Effects of growth temperature on the thermal stability of the photosynthetic apparatus of *Atriplex Lentiformis* (Torr.) Wats. Plant Physiol. 59: 873–878.
206. Peltier, G.L., and H.M. Tysdal. 1931. Hardiness studies with 2-year-old alfalfa plants. J. Agric. Res. 43: 931–955.
207. ----, and ----. 1932. A method for the determination of comparative hardiness in seedling alfalfas by controlled hardening and artificial freezing. J. Agric. Res. 44: 429–444.
208. Perry, M.C., and M.S. McIntosh. 1984. Genetic analysis of cold hardiness and dormancy in alfalfa. Agron. Abstr. American Society of Agronomy, Madison, WI, p. 82–83.
209. Pomeroy, M.K., and D.B. Fowler. 1973. Use of lethal dose temperature estimates as indices of frost tolerance for wheat cold acclimated under natural and controlled environments. Can. J. Plant Sci. 53: 489–494.
210. Pulgar, C.E., and H.M. Laude. 1974. Regrowth of alfalfa after heat stress. Crop. Sci. 14: 28–30.
211. Quinn, P.J., and W.P. Williams. 1978. Plant lipids and their role in membrane function. Prog. Biophys. Molec. Biol. 34: 109–173.
212. Raison, J.K. 1980. Membrane lipids: Structure and function. p. 57–83. *In* P.K. Stumpf (ed.) The biochemistry of plants, Vol. 4. Academic Press, New York.
213. Reid, D.J., D.J. Lathwell, and M.J. Wright. 1965. Yield and carbohydrate responses of alfalfa seedlings grown at several levels of potassium fertilization. Agron. J. 57: 434–437.
214. Rikin, A., M. Waldman, A.E. Richmond, and A. Dovrat. 1975. Hormonal regulation of morphogenesis and cold-resistance. I. Modifications by abscisic acid and by gibberellic acid in alfalfa (*Medicago sativa* L.) seedlings. J. Exp. Bot. 26: 175–183.
215. Robison, G.D., and M.A. Massengale. 1968. Effect of harvest management and temperature on forage yield, root carbohydrates, plant density, and leaf area relationships on alfalfa. Crop Sci. 8: 147–151.
216. Rodger, J.B.A., G.G. Williams, and R.L. Davis. 1957. A rapid method for determining winterhardiness in alfalfa. Agron. J. 49: 88–92.
217. Rogers, V.E. 1969. Depression of nitrogen uptake and growth of lucerne at high soil temp. Field Stn. Rec. (Aust. CSIRO Div. Plant Ind.) 8: 37–44.
218. Russell, W.E., F.J. Olsen, and J.H. Jones. 1978. Frost heaving in alfalfa establishment on soils with different drainage characteristics. Agron. J. 70: 869–872.

219. Salisbury, F.B. 1981. Response to photoperiod. p. 135–167. *In* O.L. Lange et al. (ed.) Encylopedia of plant physiology, Vol. 12A. Physiological plant ecology. I. Responses to the physical environment. Springer-Verlag, Berlin.
220. Santarius, K.A. 1969. Der einfluss von elektrolyten auf chloroplasten beim gefrieren und trocknen. Planta 89: 23–46.
221. ——, and H. Milde. 1977. Sugar compartmentation in frost-hardy and partially dehardened cabbage leaf cells. Planta 136: 163–166.
222. Schneider, M. 1984. Relationship between unifoliolate internode length and fall dormancy in alfalfa. M.S. thesis. Univ. of California, Davis.
223. Schobert, B. 1977. The influence of water stress on the metabolism of diatoms. II. Proline accumulation under different conditions of stress and light. Z. Pflanzenphysiol. 85: 451–462.
224. ——. 1980. The function of proline accumulation in the diatom *Phaeodactylum tricornutum* during water stress. p. 487–488. *In* R.M. Spanswick et al. (ed.) Plant membrane transport: Current conceptual issues. Elsevier/North-Holland Biomedical Press, New York.
225. ——. 1981. Evidence for a protein stabilizing mechanism in plant cells during water stress conditions. Hoppe-Seyler's Z. Physiol. Chem. 362: (9) 1193.
226. Schonhorst, M.H., R.L. Davis, and A.S. Carter. 1957. Response of alfalfa varieties to temperature and day lengths. Agron. J. 49: 142–143.
227. Senft, D.H. 1980. Vegetative snow fences. Agric. Res. 28: (11) 15.
228. Eoth, J., and S.T. Dexter. 1958. Root anatomy and growth habit of some alfalfa varieties in relation to wilt resistance and winter-hardiness. Agron. J. 50: 141–144.
229. Shih, S.C., and G.A. Jung. 1971. Influence of purines and pyrimidines on cold hardiness of plants. IV. An analysis of the chemistry of cold hardiness in alfalfa when growth is regulated by chemicals. Cryobiology 7: 200–208.
230. ——, ——, and D.C. Shelton. 1965. Influences of purines and pyrimidines on cold hardiness of plants. II. Respiration rate of alfalfa in relation to cold hardiness. Crop Sci. 4: 307–310.
231. ——, ——, and ——. 1967. Effects of temperature and photoperiod on metabolic changes in alfalfa in relation to cold hardiness. Crop Sci. 7: 385–389.
232. Silkett, V.W., C.R. Megee, and H.C. Rather. 1937. The effect of late summer and early fall cutting on crown bud formation and winterhardiness of alfalfa. J. Am. Soc. Agron. 29: 53–62.
233. Siminovitch, D. 1963. Evidence from increase in ribonucleic acid and protein synthesis in autumn for increase in protoplasm during the frost-hardening of black locust bark cells. Can. J. Bot. 41: 1301–1308.
234. ——. 1979. Protoplasts surviving freezing to −196 C and osmotic dehydration in 5 molar salt solutions prepared from the bark of winter black locust trees. Plant Physiol. 63: 722–725.
235. ——, H. Therrien, J. Wilner, and F. Gfeller. 1962. The release of amino acids and other ninhydrin-reacting substances from plant cells after injury by freezing; a sensitive criterion for the estimation of frost injury in plant tissues. Can. J. Bot. 40: 1267–1269.
236. Simon, E.W. 1974. Phospholipids and plant membrane permeability. New Phytol. 73: 377–420.
237. Singh, J., A.I. de la Roche, and D. Siminovitch. 1977. Relative insensitivity of mitochondria in hardened and nonhardened rye coleoptile cells to freezing *in situ*. Plant Plysiol. 60: 713–715.
238. Singh, T.N., D. Aspinall, and L.G. Paleg. 1973. Stress metabolism. IV. The influence of (2-chloroethyl) trimethylammonium chloride and gibberellic acid on the growth and proline accumulation of wheat plants during water stress. Aust. J. Biol. Sci. 26: 77–86.
239. Smith, D. 1952. The survival of winter-hardened legumes encased in ice. Agron. J. 44: 469–473.
240. ——. 1955. Underground development of alfalfa crowns. Agron. J. 47: 588–589.
241. ——. 1957. Flowering response and winter survival in seedling stands of medium red clover. Agron. J. 49: 126–129.
242. ——. 1961. Association of fall growth habit and winter survival in alfalfa. Can. J. Plant Sci. 41: 244–251.
243. ——. 1964a. Winter injury and the survival of forage plants. Herb. Abstr. 34: 203–209.
244. ——. 1964b. Freezing injury of forage plants. p. 32–56. *In* Forage plant physiology and soil-range relationships. Spec. Pub. 5. American Society of Agronomy, Madison, WI.
245. ——. 1972. Cutting schedules and maintaining pure stands. *In* C.H. Hanson (ed.) Alfalfa science and technology. Agronomy 15: 481–496.
246. ——. 1981. Forage management in the north. 4th ed. Kendall/Hunt Publishing Co., Dubuque IA.

247. ——. and B.E. Struckmeyer. 1974. Gross morphology and starch accumulation in leaves of alfalfa plants grown at high and low temperatures. Crop Sci. 14: 433–436.
248. Smith, H. 1975. Phytochrome and photomorphogenesis. McGraw-Hill, Maidenhead, UK.
249. ——, and D.C. Morgan. 1981. The spectral characteristics of the visible radiation incident upon the surface of the earth. p. 3–20. In H. Smith (ed.) Plants and the daylight spectrum. Academic Press, New York.
250. Southworth, W. 1921. A study of the influence of the root system in promoting hardiness in alfalfa. Sci. Agric. 1: 5–9.
251. Sprague, M.A. 1955. The influence of rate of cooling and winter cover on the winter survival of ladino clover and alfalfa. Plant Physiol. 30: 447–451.
252. ——, and R.F. Fuelleman. 1941. Measurements of recovery after cutting and fall dormancy of varieties and strains of alfalfa (*Medicago sativa*). J. Am. Soc. Agron. 33: 437–447.
253. ——, and L.F. Graber. 1943. Ice sheet injury to alfalfa. J. Am. Soc. Agron. 35: 881–894.
254. Sprague, V.G., and L.F. Graber. 1940. Physiological factors operative in ice-sheet injury of alfalfa. Plant Physiol. 15: 661–673.
255. Stefl, M., I. Trcka, and P. Vratny. 1978. Proline biosynthesis in winter plants due to exposure to low temperatures. Biol. Plant. 20: 119–128.
256. Steinmetz, F.H. 1926. Winter hardiness in alfalfa varieties. Univ. of Minn. Agric. Exp. Stn. Tech. Bull. 38.
257. Steponkus, P.L. 1978. Cold hardiness and freezing injury of agronomic crops. Adv. Agron. 30: 51–98.
258. ——. 1981. Responses to extreme temperatures. Cellular and sub-cellular bases. p. 371–402. In O.L. Lange et al. (ed.) Physiological plant ecology. I. Response to the physical environment. Encyclopedia Plant Physiol. 12A. Springer-Verlag, Berlin.
259. Stewart, C.R., and F. Larher. 1980. Accumulation of amino acids and related compounds in relation to environmental stress. p. 609–635. In B.J. Miflin (ed.) Biochemistry of plants, Vol. 5. Academic Press, New York.
260. ——, and J.A. Lee. 1974. The role of proline accumulation in halophytes. Planta 120: 279–289.
261. Stivers, R.K., and A.J. Ohlrogge. 1952. Influence of phosphorus and potassium fertilization of two soil types on alfalfa yield, stand, and content of these elements. Agron. J. 44: 618–621.
262. Stout, D.G. 1980. Alfalfa water status and cold hardiness as influenced by cold acclimation and water stress. Plant Cell Environ. 3: 237–241.
263. Sullivan, C.Y., and W.M. Ross. 1979. Selecting for drought and heat resistance in grain sorghum. p. 263–282. In H. Mussell and R.C. Staples (ed.) Stress physiology in crop plants. Wiley-Interscience, New York.
264. Suzuki, M. 1968. Evaluation of regrowth potential of perennial forage crops using triphenyltetrazolium chloride. Can. J. Plant Sci. 48: 113.
265. ——. 1972. Winterkill patterns of forage crops and winter wheat in P.E.I. in 1972. Can. Plant Dis. Surv. 52: 156–159.
266. ——. 1973. Is winterkill predictable? Can. Agric. 18: 10–11.
267. ——. 1977. Effects of soil moisture on cold resistance of alfalfa. Can. J. Plant Sci. 57: 315.
268. ——. 1981. Responses of alfalfa to a simulated midwinter thaw. p. 390–393. In J.A. Smith and V.W. Hays (ed.) Proc. 14th Int. Grassl. Congr., Lexington, KY. 15–24 June. Westview Press, Boulder, CO.
269. ——, W.N. Black, J.A. Cutcliffe, and J.D.E. Sterling. 1975. Frequency of occurrence of winter injury to forage legume winter cereal and strawberry plants in Prince Edward Island, Canada 1901–1975. Can. J. Plant Sci. 55: 1085–1088.
270. ——, and D.N. MacKenzie. 1979. A new method for sampling overwintering plants in frozen soil. Can. J. Plant Sci. 59: 549–550.
271. Swanson, C.R., and M.W. Adams. 1959. Metabolic changes induced in alfalfa during cold hardening and freezing. Agron. 51: 397–400.
272. Teuber, L.R., V.L. Marble, W.F. Lehman, I.I. Kawaguchi, M.K. Miller, B.J. Hartman, O.J. Hunt, D.K. Barnes, B. Burrows, D.L. Lancaster, and R.H. Gripp. 1984. Climatic and dormancy data reduces need for many regional alfalfa trials. Calif. Agric. 38: 12–14.
273. Treshow, M. 1970. Environment and plant response. McGraw-Hill Book Co., New York.
274. Trunova, T.I. 1970. Sugar accumulation in chloroplasts of winter wheat during frost-hardening. Fiziol. Rast. 17: 902–906.
275. Tyler, N.J., L.V. Gusta, and D.B. Fowler. 1981. The influence of nitrogen, phosphorus and potassium on the cold acclimation of winter wheat (*Triticum aestivum* L.). Can. J. Plant Sci. 61: 879–885.

276. Tysdal, H.M. 1933. Influence of light, temperature, and soil moisture on the hardening process in alfalfa. J. Agric. Res. 46: 483–515.
277. ——, and A.J. Pieters. 1934. Cold resistance of three species of lespedeza compared to that of alfalfa, red clover, and crown vetch. J. Am. Soc. Agron. 26: 923–928.
278. Vegis, A. 1964. Dormancy in higher plants. Ann. Rev. Plant Physiol. 15: 185–224.
279. Venuto, B.C., and H.M. Laude. 1979. Heat stress as a special purpose alternative to clipping in alfalfa. Agron. J. 71: 458–460.
280. Vezina, L., and R. Paquin. 1981. Effect of low temperatures on proline translocation and frost hardening in alfalfa. Plant Physiol. (Suppl. 4) 67: 343.
281. —— and ——. 1982. Effet des basses températures sur la distribution de la proline libre dans les plantes de Luzerne (*Medicago media* Pers.). Physiol. Veg. 20: 101–109.
282. Waldman, M., A. Rikin, A. Dovrat, and A.E. Richmond. 1975. Hormonal regulation of morphogenesis and cold-resistance. II. Effect of cold-acclimation and of exogenous abscisic acid on gibberellic acid and abscisic acid activities in alfalfa (*Medicago sativa* L.) seedlings. J. Exp. Bot. 26: 853–859.
283. Wallen, V.R., and H.R. Jackson. 1978. Alfalfa winter injury, survival, and vigor determined from aerial photographs. Agron. J. 70: 922–924.
284. ——, ——, P.K. Basu, H. Baenziger, and R.G. Dixon. 1977. An electronically scanned aerial photographic technique to measure winter injury in alfalfa. Can. J. Plant Sci. 57: 647–651.
285. Wallner, S.J., M.R. Becwar, and J.D. Butler. 1982. Measuremem of turf grass heat tolerance *in vitro*. J. Am. Soc. Hortic. Sci. 107: 608–613.
286. Walton, P.D. 1975. Methods for evaluating frost hardiness in alfalfa. Can. J. Plant Sci. 55: 823–826.
287. Wang, L.C., O.J. Attoe, and E. Truog. 1953. Effect of lime and fertility levels on the chemical composition and winter survival of alfalfa. Agron. J. 45: 381–384.
288. Weimer, J.L. 1929. Some factors involved in the winterkilling of alfalfa. J. Agric. Res. 39: 263–283.
289. Weiser, C.J. 1970. Cold resistance and injury in woody plants. Science 169: 1269–1278.
290. West, S.H., and G.M. Prine. 1960. Alfalfa persistence studies. Soil Crop Sci. Soc. Fla. Proc. 20: 93–98.
291. Westover, H.L. 1931. Alfalfa varieties in the United States. USDA Farmer's Bull. 1731: 1–13.
292. White, W.J., and W.H. Horner. 1943. The winter survival of grass and legume plants in fall sown plots. Sci. Agric. 23: 399–408.
293. Wiest, S.C., and P.L. Steponkus. 1976. Acclimation of pyracantha tissues and differential thermal analysis of the freezing process. J. Am. Soc. Hortic. Sci. 101: 273–277.
294. Wilding, M.D., and M.A. Stahmann. 1962. Hydroxylysine from alfalfa roots. Phytochemistry 1: 263–265.
295. ——, ——, and D. Smith. 1960. Free amino acids in alfalfa as related to cold hardiness. Plant Physiol. 35: 726–732.
296. Willard, C.J., L.E. Thatcher, and J.S. Cutler. 1934. Ohio Agric. Exp. Stn. Bull. 540.
297. Willemot, C. 1975. Stimulation of phospholipid biosynthesis during frost hardening of winter wheat. Plant Physiol. 55: 356–359.
298. Woolley, D.G., and C.P. Wilsie. 1961. Cold unit accumulation and cold hardiness of alfalfa. Crop Sci. 1: 165–167.
299. Yarwood, C.E. 1961. Acquired tolerance of leaves to heat. Science 134: 941–942.
300. ——. 1967. Adaptation of plants and plant pathogens to heat. p. 75–89. *In* C.L. Prosser (ed.) Molecular mechanisms of temperature adaptation. American Association of Advanced Science Washington, DC.
301. Yelenosky, G. 1979. Accumulation of free proline in citrus leaves during cold hardening of young trees in controlled temperature regimes. Plant Physiol. 64: 425–427.
302. Zumft, W.G., and L.E. Mortenson. 1975. The nitrogen-fixing complex of bacteria. Biochim. Biophys. Acta 416: 1–52.

9 Alfalfa Establishment

M. B. TESAR

Michigan State, University
East Lansing, Michigan

V. L. MARBLE

University of California
Davis, California

Reliable methods for seedling establishment are of paramount importance to successful production of alfalfa (*Medicago sativa* L.). Frequency of total and partial failures in establishment can be reduced significantly if better seeding practices are followed. Good establishment (Fig. 9-1) is necessary for a productive, long-lived stand.

9-1 STAND DENSITY

The failure of a viable seed to produce a plant may result from: poor seedbed preparation; seeding too deep; inadequate moisture after germination; freezing; disease; insects; competition for light and nutrients with seedlings of other alfalfa plants, companion species, and weeds, herbicide or insecticide damage; and to "smothering" by straw from the companion crop. Seedbed preparation, seeding techniques, and postseeding management should be patterned to minimize losses from these causes. Partial and complete seeding failures involve losses associated with the cost of seed and seedbed preparation, interruption of crop rotations, soil erosion from nonsod crops, cost of producing emergency crops, low hay yields, and lower quality hay from weed encroachment.

Uniformly distributed plants are necessary for maximum yield. Kramer and Davis (37) found correlations of 0.88 and 0.80 in the 1st and 2nd yr, respectively, between percent occupied 15-cm segments in a row and yields.

Density of stand required for high yields varies depending on the area and climate. Fourteen plants per 0.1 m^2 (1.1 ft.) in the West (47), 14 in Michigan (84, 85), 23 in Ohio (101), and 26 in Illinois (28) produced maximum yields in the first harvest year after the seeding year.

As a stand gets older fewer plants are necessary for maximum yields (94). Tesar (84) in Michigan found that yields in the 1st yr after seeding

---

Copyright 1988 © ASA-CSSA-SSSA, 677 South Segoe Road, Madison, WI 53711, USA. *Alfalfa and Alfalfa Improvement*—Agronomy Monograph no. 29.

Fig. 9–1. Good stand of clear seeded alfalfa band seeded with 10 kg ha^{-1} seed.

increased as density increased to 14 plants per 0.1 m^2. Seventh-year yields were 80% higher with 3.0 plants than with 1.0 plant per 0.1 m^2. In the 10th yr, Tesar (86) reported a yield of 19.1 Mg ha^{-1} with only 2.3 plants per 0.1 m^2.

Our summary of seeding rates is that approximately 15 to 25 plants per 0.1 m^2 are necessary for maximum yields in the year after seeding. Four to six plants may be adequate for maximum yields in older stands provided the stands are uniform.

9–2 MIXTURES OF ALFALFA AND GRASSES

Other legume species and grasses are frequently sown in mixtures with alfalfa. Whenever alfalfa is grazed, a vigorous, palatable grass with a similar heading date should be included in the mixture to reduce the danger of bloat. Jacobs (26) found that alfalfa grown without grass may produce bloat in ruminants when grazed. The fibrous roots of grasses resist soil erosion better than alfalfa roots so grasses should be included with alfalfa on sites subject to erosion. In western Canada, yields of alfalfa-grass mixtures (32) are nearly constant over a wide range of proportions of grass to alfalfa. Grasses can be seeded with alfalfa to achieve certain objectives with little or no effect on yield. As the stand becomes older, the grasses generally invade and become dominant as the legume kills out (74, 100), especially on imperfectly drained soils, and help to maintain forage yield and reduce invasion of weeds, particularly dandelions *Taraxacum officinale* Weber) (100).

When the suitability of a field for alfalfa is questionable, another legume or a grass that is better suited for growing under the limiting conditions, such as red clover (*Trifolium pratense* L.), can be added to the seeding mixture without seriously jeopardizing yield (74). Tesar (88), however, found that the addition of 'Viking' trefoil (*Lotus corniculatus* L.) to bacterial wilt-resistant 'Vernal' alfalfa did not increase yields in a 5-yr period on an imperfectly drained soil. Likewise, on Illinois soils of near-optimum fertility, alfalfa seeded alone was slightly more productive than when seeded with red clover and/or ladino clover (*T. repens* L.) (27). Mixtures that were two-thirds alfalfa and one-third red clover produced as much forage as a pure seeding of alfalfa.

There is considerable latitude in compounding seeding mixtures. In general, Jackobs (27) found that the total yield of a mixture is generally determined by the legume component. When alfalfa was included in a mixture at the rate of only 3.4 kg ha^{-1}, yield was nearly the same as where three times as much alfalfa was seeded alone.

The inclusion of a grass usually reduces hay quality. This may not be true in the preflowering stage, but with onset of flowering the quality of grass hay, particularly orchardgrass (*Dactylis glomerata* L.) and reed canarygrass (*Phalaris arundinacea* L.), declines much faster than that of alfalfa (74).

When alfalfa was first introduced into western USA, it was generally grown alone for hay. Hay was frequently sold off farms and grass admixtures lowered market grade. The successful production of alfalfa in the midwestern and eastern USA often required liming and special seedbed preparations that encouraged the seeding of alfalfa in pure stands. Stands were left as long as possible because of the cost of establishment. In time the widespread use of lime for other crops made it feasible to introduce alfalfa in short rotations. In the North Central States in the 1940s, most of the alfalfa was sown with smooth bromegrass (*Bromus inermis* L.). It was first grown in mixture with alfalfa in the USA in Michigan in 1937 by Rather and Harrison (62). In Illinois in 1953, Ewing (13) reported that approximately 24% of the total alfalfa acreage was seeded to alfalfa alone and the remainder in association with other species. Seventeen percent of the area had alfalfa with other legumes but no grass, 26% had one or more grasses, and 33% had at least one grass and one other legume.

The present trend, however, is to grow alfalfa in pure stands. At least 80% of the alfalfa in the North Central States (74) and about 97% in California is seeded alone (46). Approximately two-thirds of the western alfalfa hay crop is sold off the farm (46). In New York in 1980, nearly four-fifths of the alfalfa was seeded with a grass (59). This is a much higher percentage than in the North Central states, probably because of less well-drained land. Grass mixtures lower market grades and price in both the Western and North Central regions.

9-3 SOIL AND SEEDBED PREPARATION

9-3.1 pH, Ca, Mg, and Lime

Lime increases soil pH and supplies Ca and Mg (if dolomitic) in addition to improving soil conditions for alfalfa (see Chapter 10 in this book). The importance of using lime to increase the soil pH to 6.8 has been stressed by Woodhouse (107) and others. Mixing lime into an acid soil is important, otherwise finely branched feeder roots may be confined almost completely to the limed layers. Mixing lime into the plow layer has been considered superior to spreading it on the surface (see Chapter 10 in this book). If possible, agricultural limestone should be applied at least 26 weeks before seeding because neutralization of soil acidity occurs slowly. If the interval between liming and seeding is shorter, then part of the lime application should consist of a more readily available source, such as hydrated lime. If the pH is 5.5 or less, one-half should be applied before and the remainder after plowing followed by incorporation (103).

When legumes are seeded without tillage, lime cannot be incorporated in the soil. Recent evidence indicates, however, that surface-applied lime satisfactorily increases alfalfa yields. In Michigan (23, 94), recommended applications of surface-applied lime [11 Mg ha^{-1} (5 tons/acre)] were as effective in increasing yields as when incorporated on two acid soils (pH 4.8 and 5.2). Soil pH increased to 6.0 or more in the top 2.5 cm of soil after 1 yr and to 5.8 in the 2.5- to 5.0-cm layer after 2 yr. By the end of the 4th yr, the top 5 cm ranged between pH 6.0 to 6.5 with the top 2.5-cm layer equal to the pH 6.5 of 11 Mg ha^{-1} incorporated as recommended. In Virginia, Rechcigl et al. (64) reported that surface-applied lime doubled the yield of alfalfa in the 3rd yr as compared to no lime but there was no comparison with yields following the incorporation of lime.

In arid zones of the West and Southwest, gypsum (CaSO$_4$ · 2H$_2$O), S, or sulfuric acid (H$_2$SO$_4$) may be applied to increase soil permeability (78). The choice of amendment depends on the presence of lime in the soil, cost, and the speed of action required.

9-3.2 Nutrient Requirement

9-3.2.1 Nutrients Most Likely Deficient

In addition to Ca supplied by lime, the most important nutrients for stand establishment are P and K. Phosphorus is particularly important because of its role in root development of seedlings (39, 73, 96). When a soil is deficient in K, this element must be applied in adequate quantities for high-forage yields and winter survival (90) (see Chapter 10 in this book). Many western soils respond to S, a few to Mo; B may be deficient in a few soils in both Western and Eastern states. Soil test results are the best guide in determining the quantities of P and K that should be applied for establishment. These tests are less reliable, however, in predicting the response to S and B responses(49).

Generally, N is not necessary in alfalfa establishment in the humid northern states (40, 51). In Virginia, Ward and Blaser (102) found that alfalfa seedling numbers were reduced and seedling growth was not improved as N rates were increased from 0 to 90 kg ha^{-1}. Apparently, part of the reason lies in the inhibitory effect of available soil N on nodulation and N_2 fixation. Nitrate is more inhibitory than ammonium (NH_4^+), and the degree of inhibition increases with concentration (57). This effect is temporary and new nodules form after the nitrate (NO_3^-) is depleted. Moreover, the effect appears to be local with nodulation inhibited only in that part of the root system that is exposed to NO_3^- (58). Hallock (16) found in sand box culture that NO_3^- applied at the surface above the zone of nodulation improved growth. Nitrate placed at 25 cm depressed growth because of interference with nodulation and the distance of applied N from the bulk of the absorbing roots.

In parts of the southeastern USA, the application of 11 to 22 kg ha^{-1} of N at seeding is beneficial (54). Roth et al. (67) in Arizona, however, found that N applied through a sprinkler system to the first cutting on a very sandy soil gave large yield increases in all 10 cuttings made the 1st yr after seeding. Greatest yields were obtained with the highest annual application of water (378 cm) and N (559 kg ha^{-1}). Under normal conditions with good inoculation, N is not beneficial in California (46).

9-3.2.2 Application of Nutrients

Tesar et al. (96) and Sheard et al. (73) found that placement of alfalfa seed over a band of P gives maximum seedling stimulation from fertilizer (Fig. 9-2 and 9-4). Seedlings 5 or more cm from the fertilizer receive little benefit during the first month (96). Excellent responses have been obtained in the West from disking preplant broadcast applications of P and Sulfate (SO_4^{2-}) (51).

Potassium can be applied with P as a mixed fertilizer to reduce application costs. Amounts should not be excessive since salts of K are highly soluble and may be toxic to seedlings. Jackobs and Miller (28) observed seedling injury when 836 kg ha^{-1} of K were incorporated into the soil prior to seeding. Even P may be toxic in high amounts, as shown by Tesar et al. (96). In greenhouse trials, they noted about 100% of seedling loss over a band of fertilizer containing 220 kg ha^{-1} of P (based on a normal 18-cm row width).

9-3.3 Improving Soil Moisture Conditions

Alfalfa requires a well-drained, reasonably deep (>1 m) soil, relatively salt- and alkali-free with good moisture-holding capacity (3.5 to 5.5 cm/30 cm) for maximum production (11, 44). In arid regions, alfalfa may be seeded in a moist seedbed or in dry soil and "irrigated-up" according to Dennis et al. (9). Irrigation prior to seeding saturates the soil profile and prevents loss of stand from infrequent, light rains; stores moisture and thereby reduces water demand when water is needed by annual crops; firms the seedbed; improves inoculation; and germinates weeds prior to seeding thus reducing possible weed contamination of the new stand (44). On dry seedbeds, alfalfa should be placed as shallowly as possible and irrigated. On a moist seedbed the seed should be placed deep enough to be in moist soil. If a crust forms after irrigation, additional light irrigations may be necessary to soften the crust and permit seedlings to emerge.

Using sprinkler irrigation to germinate alfalfa has proven superior to flood irrigation in the arid West, particularly when making fall seedings on saline soils (66). Irrigation with sprinklers increased emergence and establishment by nearly 400%, increased uniformity of establishment, and decreased bulk density of the soil. Sprinkler irrigation also permits establishing a stand during hot summer months, early in the fall, or late in the spring and also permits the use of water with a higher salt content (65).

Erwin (12) first associated root rot of alfalfa (*Phytophthora megasperma* sp. f. *medicaginis*) with poor drainage or excessive soil moisture in California (17) (see Chapter 11 in this book). Since then, this disease has been reported throughout the USA and in many foreign countries. It is most prevalent on poorly drained soils, during periods of heavy rainfall, and with excessive irrigation. Frosheiser and Barnes (14) have shown that resistance to Phytophthora root rot can be incorporated into improved cultivars. New cultivars have expanded alfalfa acreage to sites previously considered too poorly drained for alfalfa.

9–4 SOIL PREPARATION AND INFLUENCE OF PRIOR CROPS

9–4.1 Plowing, Disking, and Tillage

An ideal seedbed is moist and firm, especially when seedings are made in summer. It should be sufficiently fine and granular, not powdery, for good seed coverage when compacted.

Plowing may not be necessary when alfalfa follows most cereal crops because a satisfactory seedbed can be prepared rapidly and at low cost by disking and harrowing. Also, satisfactory stands can be established without tillage by using special no-till drills adapted to place seed shallowly and accurately in grass or grass-legume sods, or following harvest of cereal small grains or corn.

The principal functions of plowing are to bury existing vegetation and plant residues that may interfere with seeding, to incorporate lime into the soil, and to provide a level surface for harvesting operations. In western desert soils, particularly stratified, sandy-loam soils, deep tillage with a subsoiler is a routine part of seedbed preparation prior to seeding alfalfa (44). Deep tillage below the traditional 20-cm depth of disking increased yields from 25 to 40% (11) and from 60 to 240% (30, 31). Deep tillage increases root development and infiltration rate and decreases soil stratification resulting in decreased soil strength (11). Irrigated sandy soils with low organic matter content and bulk densities of 1.6 or above have reduced porosity and poor root development. Meyer (50) reported that the addition of manure and fertilizer, when plowed to a depth of 70 cm compared to a 20-cm disking, increased seeding-year yields by 59%. Most of this response came from the combination of manure with deep plowing. Porosity was increased significantly in the upper 10 cm, enhancing water penetration. Slip plowing (a single bottom plow) once, slip plowing in two directions, and moldboard plowing each to a depth of 90 cm increased yields by 57, 143, and 147%, respectively, over disking to a normal 20-cm depth (31). On a nonirrigated sandy loam soil in Michigan (88), 22 Mg ha^{-1} of manure incorporated into the soil to a depth of 20 cm prior to plowing increased annual alfalfa yield in a 5-yr period from 8.3 to 9.0 Mg ha^{-1}.

9–4.2 Compaction Prior to Seeding

Because shallow placement of alfalfa seed is required for good emergence, firming of the seedbed before planting usually results in better stands when soil moisture is limiting (74). In Minnesota, the use of a corrugated roller before seeding gave 48 and 66% more seedlings in the 1st and 2nd yr after seeding, respectively, as compared with harrowing only (68). When soil moisture is not limiting, as during the spring in humid areas, the use of a corrugated roller may not be necessary and could lead to excessive soil compaction, especially on clay soils. On coarse-textured loamy sandy soils (83), the use of the corrugated roller prior to seeding is recommended, and essential on dry soils.

9–4.3 Control of Quackgrass

Quackgrass (*Agropyron repens* L.) is a perennial, noxious rhizomatous grassy weed in northern USA which is best controlled with the herbicide glyphosate, N-(phosphonomethyl) glycine (trade name Roundup) applied at recommended rates (see Chapter 23 in this book). Also tillage every 7 to 10 d after small grain harvest through October will kill or suppress quackgrass for a spring seeding. For summer seeding in late July and early August in the northern states, quackgrass can be controlled by tillage every 7 to 10 d starting in late May after heavy pasturing of the first spring growth (92).

9–4.4 Autotoxicity in Alfalfa

Autotoxicity exists when alfalfa has lower germination, poorer establishment, and/or lower production when grown immediately following alfalfa than after another crop, such as maize (*Zea mays* L.) or grass.

Tesar found that alfalfa can be reestablished without autotoxicity, as successfully after alfalfa as after maize or Kentucky bluegrass (*Poa pratensis* L.) if there is a period of 2 or more weeks between plowing and seeding or 3 or more weeks after herbicide killing of alfalfa and seeding (91). (Table 9–1). Mueller-Warrant and Koch (56) in New Hampshire also report the absence of autotoxicity when there is a "2- to 3-week wait until seeding" after spraying alfalfa with glyphosate. Seedings made in two field experiments in Michigan (Table 9–1) provide the basis for successful alfalfa establishment without evidence of autotoxicity after a preceding crop of alfalfa. The following recommendations are suggested, in priority order, when alfalfa follows alfalfa: (i) use conventional seeding as early in spring as possible on the tilled seedbed after fall plowing of alfalfa; (ii) use a conventional spring or summer seeding on a tilled seedbed 2 weeks or more after plowing of alfalfa; and (iii) use sod seeding in spring or summer 3 weeks or more after killing existing alfalfa with a herbicide.

Some researchers, however, report autotoxicity if alfalfa follows alfalfa and recommend an intervening crop (21, 29, 32). Unfortunately, all except Miller (52) report only greenhouse data and seeded immediately after top growth was incorporated in the soil [which also caused autoxocity (91)] (Table 9–1). Miller (52) in Illinois reported some autotoxicity, however, when alfalfa was spring-seeded after plowed-under alfalfa as compared to alfalfa after a maize-soybean rotation. He did not specify the interval between plowing alfalfa under and reseeding; any interval <14 d results in autotoxicity as reported in Michigan (91) (Table 9–1).

9–5 SEEDING

9–5.1 Seed Quality

Seed should be as free from weed seed as possible (74) with high germination and a low percentage of hard seed. Some early workers con-

Table 9-1. Alfalfa yields in the year of spring seeding after alfalfa or maize in Michigan (91).

Treatment of previous alfalfa†	No. of days between treatment and seeding‡						Check, seeding after maize	LSD$_{0.05}$
	0	7	14	21	28	195		
	———————— Mg ha^{-1} ————————							
	Exp. I, 1984, seeded 5 June, two cuts							
Plowing	5.4	6.0	6.0	6.3	6.2	--	5.9	0.6
Spraying	5.7	5.8	5.6	6.4	5.9	--	5.9	0.6
	Exp. II, 1985, seeded 29 May, one cut							
Plowing	4.2	4.6	5.3	5.5	5.6	5.2	5.2	0.5
Spraying	4.1	4.4	4.7	5.3	4.9	5.2	5.2	0.5

† Seeding on tilled soil after plowing or no-till seeding after spraying alfalfa.
‡ 0, 7, 14, 21, or 28 d after spring spraying or plowing, or 195 d after fall plowing alfalfa.

sidered hard seed as insurance against frost, weeds, insects, or drought (1). Today, preference is given to seed with a lower percentage of hard seeds, because late-emerging seedlings are likely to perish because of competition or winter injury (10).

9-5.2 Seed Treatment

In general, the application of fungicides to control seedborne diseases is generally not recommended (see Chapter 21 in this book).

Alfalfa seed should be inoculated with the proper *Rhizobium* species to help ensure N_2 fixation (19) even though alfalfa has been grown on the area previously. The effectiveness of nodulation is generally improved with additional rhizobia (see Chapter 7 in this book). Although inoculation is recommended, alfalfa yields are not necessarily increased with inoculation on soils where alfalfa has been grown previously, particularly in northern latitudes of the USA (93). Coating alfalfa seed with lime after the seed is treated with rhizobia has helped in the establishment of airplane-seeded alfalfa on acid soils in New Zealand (104).

In the USA, lime coating of alfalfa seed after inoculation with rhizobia (often with adhesive, such as gum arabic) has not been helpful in improving stands or yields on soils that are suitable for alfalfa production (2, 93). Lime-coated seed is about one-third lime and two-thirds seed. A seeding rate of 9 kg ha^{-1} (equivalent to 13.5 kg ha^{-1} lime-coated seed) produced fewer plants and a lower 4-yr yield than 13.5 kg ha^{-1} of pure seed (84). Seed firms will lime-coat alfalfa seed at the request of a customer.

Many western farmers seed alfalfa by air. Some prefer coated seed because of its improved "drop" characteristics.

9-5.3 Seed Placement in Relation to Plant Nutrients

Haynes and Thatcher (18) first noted the beneficial effect of placing seed directly over a band of P fertilizer. Tesar et al. (96) supported these findings with radioactive P (Fig. 9-2, 9-3, and 9-4). Alfalfa seedlings had to be directly over or no more than 2.5 cm away from P banded 3.8 cm deep in order to obtain 60% or more of their P from the fertilizer in 8 weeks. In the same period, seedlings 7.6 cm away from banded P received < 3% of their P from the fertilizer. Plants directly over the band of fertilizer produced 52 and 66% more top growth than when 2.5 or 5.1 cm away from the fertilizer, respectively.

9-5.4 Depth and Compaction

The optimum seeding depth is influenced by planting time and by soil moisture, soil type, and soil compaction. Shallow seeding is the single most important factor in getting good legume stands in humid areas. Planting seed too deeply was a frequent cause of seeding failures in the early culture of alfalfa in the USA. This problem led to the development

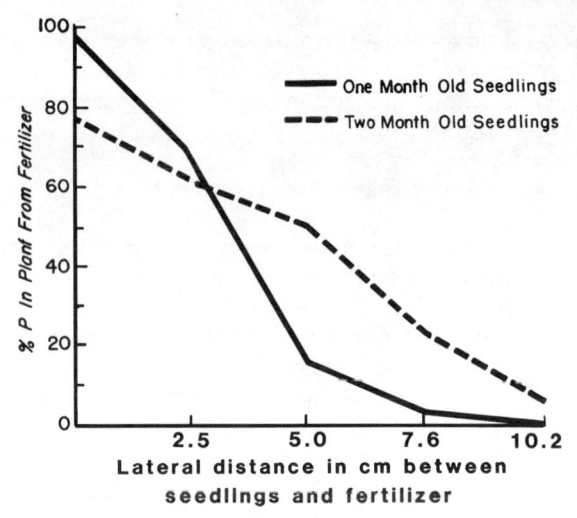

Fig. 9-2. Distance between seedlings and fertilizers.

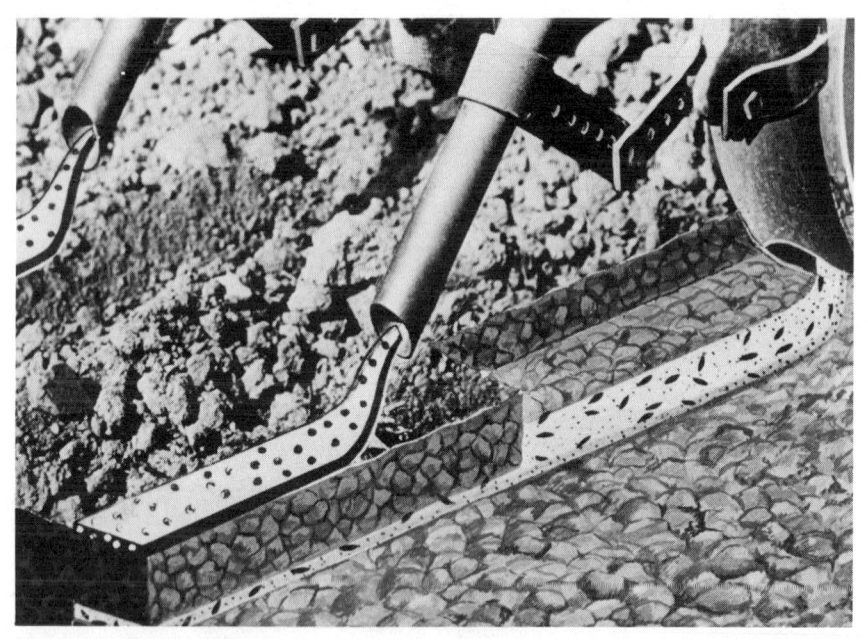

Fig. 9-3. In band seeding, legume seed is placed in a band on or near the surface (0–1.25 cm) directly over a band of fertilizer (and small grain) placed 2.5 to 5.0 cm deep. The P in the fertilizer stimulates rapid root and seedling growth.

in Wisconsin of a corrugated roller seeder primarily to effect shallow seeding (1). In Ohio, Willard and Lewis (105) emphasized shallow seedings, except on sandy soils where seed should be planted deeper. Triplett and Tesar (99) determined in three field trials on sandy loam and silt loam soils that 0.64 to 1.27 cm was the optimum planting depth for total

ALFALFA ESTABLISHMENT

Table 9-2. Emergence† of alfalfa planted at four depths and compacted at four levels in the field in Michigan (95).

Depth, cm (inch)	Compaction, 9 g dm^{-2} (psi)				Avg
	0	213 (3)	426 (6)	852 (12)	
	%				
0	10	16	24	40	22
0.64 (0.25)	60	58	58	59	59
1.27 (0.50)	60	62	64	63	62
2.54 (1.0)	50	52	52	50	51

† Average of two soil types, irrigated and nonirrigated.

Fig. 9-4. Alfalfa banded directly over P fertilizer is stimulated by the P, but alfalfa 7.5 to 10.0 cm away (as in broadcast seeding over banded fertilizer) gets very little benefit from the fertilizer in the first 8 weeks of growth.

emergence (Table 9-2). These results support current seeding recommendations of shallow depths of 0.64 to 1.91 cm on most fine-textured soils (74). Depths of 1.27 to 3.91 cm are recommended for coarse-textured soils that are subject to drought, and in arid areas (74, 98). Shallow depths in the above ranges are best when moisture conditions are favorable, especially in spring. Greater depths are preferred however, when moisture conditions are less favorable.

Press wheels or a cultipacker (Fig. 9-5) towed behind the bandseeder

Fig. 9-5. The fertilizer-grain drill with band seeder attachment places the seed shallowly over a band of fertilizer (Fig. 9-3) to optimize stimulation from P (Fig. 9-4). Press wheels (*top*) cover the seed shallowly with soil and firm the soil around the seed for good germination. If press wheels are not available, a cultipacker (*bottom*) gives almost as good a stand as press wheels (99).

ALFALFA ESTABLISHMENT

drill improve stands in the spring (68, 99) but considerably more so in the summer (Table 9-3). With no cultipacker, spring seedings were 75% as good as when a cultipacker followed the band seeder. In summer, seedings without the cultipacker were only 54% as good, indicating the great need for covering the seed and firming the soil over the seed during the drier part of the season (Table 9-3 and Fig. 9-6).

Press wheels were slightly more effective than a cultipacker in spring (6%) and more effective in summer (12%) (Table 9-3 and Fig. 9-6). Both types of equipment are excellent in achieving shallow seed coverage and firming the soil around the seed.

Table 9-3. Three-year average stands of alfalfa seeded at 10 kg ha^{-1} on two soil types and 1-yr dry weights of 9-week-old seedlings/m^2 (89).

Seeding machinery	Method of compaction	Stands seeded in† Spring	Summer	Avg	Seeded 26 Aug., weight on 1 Nov.
			%		
Band seeder drill (Fig. 9-5)	None	75	54	65	45
Band seeder drill (Fig. 9-5)	Press wheels (Fig. 9-5)	106	112	109	167
Band seeder drill (Fig. 9-5)	Cultipacker (Fig. 9-5)	100	100	100	100
Cultipacker seeder (Fig. 9-7)	Cultipacker seeder (Fig. 9-7)	84	74	89	59

† Six trials in spring, subsequent rains; five in summer, three without rain for 3 weeks.
‡ Seed banded over fertilizer except broadcast over fertilizer with cultipacker seeder.

Fig. 9-6. Summer-seeded alfalfa generally has the best stands if seeded with a band seeder drill with press wheels (*right*). A cultipacker instead of the press wheels gives almost as good a stand (*center*). Broadcast seeding with the cultipacker seeder produced a somewhat poorer but satisfactory stand (*left*).

9-5.5 Seeding Methods

The two general methods of seeding alfalfa on a prepared seedbed are broadcasting on the soil surface or band seeding.

9-5.5.1 Broadcasting

In the humid East, seed is often broadcast on the surface with a fertilizer-grain drill with a small legume and grass seed attachment or with a tailgate seeder (74). The seed tubes broadcast seed from a height of about 60 cm above the soil. Some fields are sown with broadcast seeders. In California and Arizona, most fields larger than 15 to 20 ha are seeded by air, followed by harrowing or rolling with a corrugated, flexible ring roller over broadcast fertilizer (9, 43). Smaller and irregularly shaped fields are sown with a tailgate broadcast unit, a corrugated roller seeder, or a fertilizer grain drill with legume seeder.

The corrugated roller seeder (Fig. 9-7) broadcasts the seed on the soil surface between the front and rear rollers and gives near-optimum coverage of seed except on powdery seedbeds where rolling of soil may result in excessive coverage. The front roller compacts the seedbed, a requirement that is considered essential to good seeding (68, 74). The rear roller covers the seed and further compacts the soil. This type of seeder has given excellent stands except on some heavy clay soils (74) and on some soils in summer (Table 9-3 and Fig. 9-6). Grass seeds of all sizes are sown shallowly from a separate seed box. The seeder has the disadvan-

Fig 9-7 A cultipacker seeder is an excellent seeding machine. The seed is placed between the front and rear rollers and covered shallowly by the rear roller. Fertilizer must be applied with other equipment prior to seeding if required.

tages of requiring a separate fertilizer application and the purchase of additional fertilizer equipment. The corrugated roller seeder, accepted as an improved, reliable seeder (89), is recommended by most states in humid (74) and arid areas (9).

9-5.5.2 Band Seeding

Band seeding (18) was the first significant advance in seeding establishment after the importance of shallow seeding was demonstrated in the 1930s (1). A fertilizer-grain drill with a small seed attachment is modified by extending the seed tubes to within 5 to 10 cm of the soil surface (96) (Fig. 9-3 and 9-5). The seeds fall in a band on top of the soil about 30 to 40 cm behind a disk opener and directly over a band of fertilizer placed 3 to 5 cm deep. Some of the seed is covered up to 0.64 cm (0.25 inch) deep but most is covered about 0.64 to 1.27 cm (0.25-0.50 inch) deep by press wheels or a corrugated roller. Tesar et al (96) reported 20% more alfalfa seedlings from banding seed over banded P compared with broadcasting seed over banded fertilizer. These data were obtained from over a 3-yr period in six trials on four soil types.

As a result of stimulation from P in the fertilizer, the seedlings established by banding are more vigorous and competitive to weeds, insects, and companion crops and more resistant to winter injury (5, 73, 96) than when established by broadcasting. Frequently, the better stands are reflected in greater yields in subsequent years (18). Weeds are stimulated less than if P is broadcast, and P is used more efficiently since it is less apt to become "fixed" when in a band than when broadcast (81).

Brown (4), Brown et al. (5) in the Northeast, Carmer and Jackobs (8) in Illinois, Sund et al. (79) in Wisconsin, Sheard et al. (73) in Canada, and Tesar et al. (96) in Michigan indicate that band seeding is superior to broadcast seeding under less-than-optimum conditions, i.e., on soils of low fertility or when dry periods follow seeding. Reliability in obtaining productive stands, especially under adverse conditions, explains the acceptance of band seeding as an improved seeding method in many states in the North Central and Northeastern regions. Many new drills are equipped with band seeding attachments (Fig. 9-5). Older drills have been changed to band seeders with attachments sold by farm implement companies or with modifications made by farmers.

When the fertilizer-grain drill with seeder attachment is used for broadcast or band seeding, large grass seeds, such as bromegrass, are sown at the same time by mixing with the companion crops or by mixing with the fertilizer and sowing <2.5 cm deep (89). When no companion crop is sown, the large grass seed can be sown through the grain box. Timothy (*Phleum pratense* L.) is similar in size as alfalfa and is generally mixed and seeded with alfalfa in the small-seeded attachment.

9-5.6 Companion Crops

A companion crop provides some income during the seeding year, helps control wind and water erosion, and competes less than weeds.

Table 9-4. Seeding methods as percentage of seeded acreage in the North Central and Northeastern regions of the USA from a 1984 mail survey by the authors.

Seeding method	North Central States										Northeastern States						Avg[†]
	IA	IL	IN	KS	MI	MN	MO	NE	ND	OH	WI	CT	NH	NJ	NY	PA	VA	VT	
	%																		
Spring mostly with oat[‡]	85	30	35	0	40	65	3	65	88	25	80	5	0	0	40	20	15	10	59
Clear, in spring[§]	5	23	20	75	27	25	10	15	2	30	17	30	50	50	21	40	25	50	20
Summer[¶]	5	25	20	25	24	5	75	15	8	40	1	50	40	40	15	15	25	5	15
With winter wheat[#]	4	15	15	0	3	0	3	2	0	5	1	0	0	0	0	0	0	0	2
No-till[††]	0	7	5	0	3	5	2	3	2	20	1	15	10	9	0	20	35	5	3
Other[‡‡][§§]	1	0	5	0	3	0	7	0	0	0	0	0	0	1	0	5	0	0	1

[†] Weighted average based on state acreages.
[‡] Includes barley, flax, or spring wheat.
[§] With preplant or postemergence herbicide.
[¶] With or without preplant or postemergence herbicide.
[#] Fall- or spring-planted alfalfa.
[††] Includes seeding on a herbicide-treated sod and/or no-till seeding after small grains or maize.
[‡‡] Includes tillage of a sod and seeding, seeding after tillage in spring without a herbicide, and fluid seedings.
[§§] 0 is "trace" or <1%.

Companion crops are not necessary to help the seeding as once believed, hence the obsolete term "nurse" crops. When alfalfa is seeded without a companion crop, however, weed control is essential to avoid the prospect of stand failures (89).

Companion crops compete for water, light, and plant nutrients (7, 89). Light becomes limiting with excessive lodging of small grain on fertile soils, especially under good moisture conditions. Moisture usually becomes limiting on coarse-textured soils when the companion crop becomes too competitive, especially under droughty conditions.

The best companion crop is one that is least competitive to alfalfa seedlings. Competition can be reduced by crop choice, selection of early maturing cultivars, early removal of the small grain as hay or silage (see Chapter 19 in this book), and lower sowing rates.

9-5.6.1 Choice of Companion Crops

Oat (*Avena sativa* L.) is one of the best companion crops because it is removed early and is not as leafy and competitive as other companion crops, e.g., winter wheat. Short, stiff-strawed, early maturing, lodging-resistant cultivars are recommended. Oat is the most widely used companion crop in the USA (Table 9-4), providing a valuable source of feed as well as bedding for livestock.

Flax (*Linum usitatissimum* L.) and pea (*Pisum sativum* L.) are satisfactory companion crops. They are harvested early and provide less competition for light and moisture than many other crops (74).

Spring wheat (*Triticum aestivum* L.), spring barley (*Hordeum vulgare* L.) and early maturing spring oats are fairly similar in their competitive effects (7) and are more satisfactory companion crops than either winter wheat or late-maturing cultivars of oat (89).

Winter rye (*Secale cereale* L.) and winter wheat, especially when fertilized with high rates of N, are poor companion crops because of their strong competition for light and water (7, 82). Establishing alfalfa in these fall-sown grain crops is not recommended because alfalfa must be sown the following spring when the companion crop is well established and competitive (74, 89).

9-5.6.2 Removing Small Grains Early

Competition from a small grain crop, such as oat or barley, can be reduced in spring seedings by grazing, cutting for hay, or ensiling, particularly on droughty soils. Harvested early for silage in the early boot stage, the feeding value of oat is increased 50% over the harvested grain. If oat lodges severely, early removal is vital to avoid losing the stand. Klebesadel and Smith (35, 36) found that removal of oat when 30 to 48 cm tall would reduce competition for light. Following early removal of a companion crop, it may be necessary to clip weeds during the summer.

Oat should not be used in summer seedings because of excessive competition for water, especially on droughty soils (89). Seedings made in late July or August (Table 9–5) were only 55 and 65% as productive the next year when summer-seeded with oat not removed or removed, respectively, as when seeded alone. Cash grain crops, e.g., wheat and winter barley, cannot be removed early because of reduced grain yields.

9–5.6.3 Reducing Sowing Rates of Companion Crop

Competition from spring-sown oat can be reduced by using a lower seeding rate. The recommended seeding rate for oat is about 72 to 103 kg ha^{-1} on fine-textured soil (89), similar to that of oat sown alone. In general, oat should not be sown below this rate on fine-textured soils, because weeds may increase and provide more competition than oat (75) if not controlled with a postemergence herbicide. On coarse-textured soil, however, sowing rates of oat should be reduced to minimize competition for moisture. In the irrigated West, 30 to 40 kg ha^{-1} of oat are recommended to provide protection to fall-sown alfalfa and to produce winter and early spring oat forage/silage crops. Recent experiences show that 8 Kg ha^{-1} of oat have little negative effect on establishment or the following season's yield, and provide superior first-harvest yields the following spring (44).

9–5.7 Time and Methods of Seeding

9–5.7.1 Spring Seedings with a Companion Crop

Spring seedings are made with a companion crop, generally oat, on three-fifths of the alfalfa in the North Central and Northeastern States (Table 9–4) and in the colder sections in Northwestern States (44). Flax and spring wheat are used to a limited extent in the extreme northern USA and in Canada. Companion crops may be harvested for seed, hay, or silage in summer, and alfalfa harvested the following year. In years of favorable moisture, a late-season cutting of alfalfa may be harvested in the seeding year (74, 89), particularly if oat is removed early as silage. With increasing emphasis on high N rates for high crop yields, many farmers have reported poor stands and failures of alfalfa seedings in oat on both fine-textured soils (74, 82) and coarse-textured, droughty soils

Table 9–5. Effect of summer seeding dates on 2-yr average Vernal alfalfa yields in the next year in Michigan (89).

Method of seeding in summer	Seeding dates			
	27 July	13 Aug.	27 Aug.	11 Sept.
	Mg ha^{-1}			
Alone	11.3	10.8	8.4	5.0
Plus oat, not removed	6.1	5.7	5.0	3.9
Plus oat, removed 2 Oct	6.9	7.4	5.8	4.8
Avg	8.0	7.9	6.4	1.6

(89). As a result, many of these farmers now make clear seedings of alfalfa in spring (Table 9–4).

In the northern Corn Belt and in the Northeastern States, over-seeding winter wheat in spring is too erratic for reliable stands (81, 89) (Table 9–4). This practice is satisfactory, however, in the southern part of the Corn and Winter Wheat Belt. Increased use of N on wheat may produce erratic alfalfa stands (60, 81) and consequently this seeding method has declined (Table 9–3).

9–5.7.2 Clear Seedings in Spring, No Companion Crop

Clear seeding with preplant or postemergence herbicides is the most significant advance in alfalfa establishment following the introduction of band seeding and press wheels in the mid-1950s (81,82). About one-fifth of the seeded area in the North Central and Northeastern States is clear seeded (Table 9–4). Clear seeding involves intensive management. It is used primarily on good soils with herbicides used preplant or postemergence (see Chapter 23 in this book) for broad-leaved and grassy weed control. This may provide farmers with two to three harvests in the seeding year (25, 54, 61, 70, 82, 101).

Clear seedings are an integral part of intensive management of alfalfa for high yields based on superior cultivars, good drainage, high pH, high fertility, and an intensive cutting management system (82) (see Chapter 12 in this book). Important requirements for successful clear seedings and the production of high yields of high-quality alfalfa in the range of 6 to 12 Mg ha^{-1} in the seeding year (70, 82) (Table 9–6) are: use a pre- or postemergence herbicide; seed as early as possible in spring; fertilize at seeding with P and K according to soil test; seed at high rates of about 18 kg ha^{-1} of a productive, disease-resistant cultivar; cut three times starting at early flower in the first two harvests and in mid-October for the last harvest (Table 9–6); and control leafhoppers (*Empoasca fabae* Harris) with insecticides. Pulli (61), Sheaffer (70), and Tesar (82) showed

Table 9-6. Seeding-year yields of Saranac alfalfa seeded at four rates with early and late spring seedings cut two or three times in the same year, Conover loam in East Lansing, MI (82).

	Seeded 10 April		Seeded 9 May		Avg	
Rate	Three cuts†	Two cuts‡	Three cuts§	Two cuts¶	Three cuts	Two cuts
kg ha^{-1}			Mg ha^{-1}			
4.5	9.6	7.1	6.6	5.1	8.1	6.1
9.0	9.9	8.1	7.9	5.9	8.9	7.0
18.0	10.3	6.9	7.3	6.4	8.8	6.6
36.0	8.8	6.6	7.6	6.4	8.2	6.5
Avg	9.6	7.6	7.6	5.9	8.5	6.5
Avg	8.6		6.3		7.5	

† Three cuts—9 July, 20 Aug., 27 Oct.
‡ Two cuts—20 July, 1 Oct.
§ Three cuts—21 July, 30 Aug., 27 Oct.
¶ Two cuts—3 Aug., 1 Oct.

good persistence in the next year, after following these practices in the year of seeding. In humid areas, yield goals of 6 to 13 Mg ha^{-1} in the seeding year on fine-textured soil are feasible, as reported by Sheaffer (70) in Minnesota and Tesar (82) in Michigan. In 1985, 13 Mg ha^{-1} were harvested in three cuts in Michigan (97).

Clear seeding by band seeding in early spring is the best method of establishing reliable stands of alfalfa on sandy loam or loamy sand soils (82). On these sites, spring seedings are likely to fail when seeded with oat, particularly in dry summers.

Three harvests generally yield more than two in humid areas on fine-textured soil according to the work reported above. Tesar (82) (Table 9–6) in Michigan reported that: seedings made on 10 April yielded 37% more than 9 May seedings; three cuts yielded 27% more than two cuttings; the maximum yield of 10.4 Mg ha^{-1} was obtained with 18 kg ha^{-1} seed planted 10 April; and persistence with the last cutting on 27 October was high as measured by next year's plant counts and yields. Under drier conditions in Nebraska (54) or on droughty soils (89), two cuttings is the maximum recommended. Sometimes only one cutting will be made in the seeding year on a droughty soil where the primary objective is to ensure a good stand in subsequent years (89).

The number of seeding-year harvests in the humid East is generally about one fewer than that under normal management, i.e., three where four cuttings are made or two where three cuttings represent standard practice. Under an intensive management system, the first of three cuttings is made in early July, 8 to 10 weeks after seeding, the second in mid- to late August, and the last in mid- to late October. Persistence has been high under such a three-cut system in the seeding year (61, 70, 82).

Clear seeding has made it possible to consider the use of spring-seeded alfalfa with a herbicide, and plowing down all or part of the top growth, plus the roots in the fall or spring as a source of N and organic matter preceding a cash crop (22, 33, 71, 87). Yields of maize were increased 45% in Minnesota (71) after plowing under clear-seeded alfalfa 1 yr after seeding. Heichel (19) (see Chapter 7 in this book) reports using alfalfa selections (Nitro) with high N_2-fixing ability developed by Barnes et al. (3) in an attempt to make the "1-yr" alfalfa plowdown profitable and economical. It is more likely to be economical preceding a high-income crop, such as tomato, (*Lycopersicon esculentum* Mill.), than preceding a low-income crop, such as maize. (87)

9–5.7.3 Summer Seedings

Seeding alfalfa in late July or early August after summer fallow to control weeds or after winter wheat (sometimes oat) is the third important method of alfalfa establishment in the Corn Belt and some of the Northern States (15% of the seeded acreage, Table 9–4). In the area south of the 39th parallel (southern Indiana), late summer seedings are more successful than spring seedings in oat which often crowds out the alfalfa. In states where summer seedings are popular (Table 9-4), alfalfa summer seeded alone generally yields less in the subsequent year than if summer seeded with oat (Table 9-5) (89). Since the early 1960s, summer seedings have replaced some spring seedings in oat and nearly all seedings in winter wheat. This is in response to the decline in oat acreages and to the high N rates used on winter wheat.

Clear seedings or excellent spring seedings made in oat in humid states such as Indiana, Illinois, Michigan, New York, Pennsylvania and Wisconsin, however, generally outyield excellent summer seeding in the 1st yr after seeding as shown by the 16% greater yield of the spring seeding in oat (84) and of clear seeding (61).

Seedbeds for summer seedings should be well tilled and firm, preferably compacted with a corrugated roller before seeding (68, 89). If seeded after wheat, the competition from any volunteer wheat is reduced by deep plowing rather than disking (81). Band seeding followed by compaction with a corrugated roller, or preferably by press wheels, will improve vigor of the stands and promote strong seedling development for good winter survival (Table 9-3 and Fig. 9-6).

Summer seedings are successful if seeded early enough for good seedling development prior to the onset of winter. The critical last date for successful summer seedings is mid-August in northern USA (74, 89). Seedings made in late July and early August are more vigorous and more productive the next year than those made in late August or early September (Table 9-5) because later seedings are more susceptible to winter injury and heaving. Farther south, the last safe seeding date ranges from late August near the 42nd parallel (Chicago) to mid-September near the 39th parallel (southern Indiana).

9-5.7.4 Fall Seedings

Fall seedings are successful in North Carolina (55) and in other southern states. Seedings made after mid-September are not likely to succeed in the middle part of the Corn Belt and are certain to fail in the northern part.

In California, successful seedings can be made throughout the fall months. Mid-September was the optimum planting time for the Central Valley with 1st-year yields 20 to 30% higher than in mid-fall, 60% higher than in late fall, and 62 to 80% higher than spring planting (47). This advantage persisted into the 2nd and 3rd yr for the September and mid-October planting dates. There were no differences among four cultivars, representative of the major dormancy groups, in their response to dates or rates of seeding. Traditionally, seedings in the West have been made when the soil is dry with germination beginning with fall rains in November and December. Many late fall and winter seedings are made after cotton (*Gossypium hirsutum* L.) harvest. Seeding after a cereal crop provides sufficient time for adequate soil preparation, forming of irrigation borders, and leveling of alfalfa fields in time for a late summer-fall seeding. In desert areas of the Southwest (41), alfalfa has traditionally been seeded in October. Recently, many successful seedings have been made in August and September with flood irrigation and solid-set sprinkler systems.

9–5.8 Rate of Seeding Dependent on Use

9–5.8.1 Seeding Rates

Amount and distribution of rainfall are among the most important factors in determining optimum seeding rates. Other important factors are seeding methods, seedbed preparation, soil type, soil tilth, and use of the seeding. In general, seeding rates are lower for pasture than for hay because grasses are included in seeding mixtures to reduce the incidence of bloat. Although recommended seeding rates often appear higher than necessary, they can be justified as insurance in obtaining good stand establishment.

The number of seedlings surviving the 1st yr after seeding is generally between 40 to 50% of the seed sown as a result of competition, diseases, insects, winter injury, and other causes. Seedling survival, frequently, is as low as 20% (5). A seeding rate of 18 kg ha^{-1} gives about 80 seeds per 0.1 m^2. Thirty to 40 seedlings from 80 seeds is considered good establishment. In the humid East, survival of about 20 to 30 plants in the 1st yr, and about 12 to 18 plants in the 2nd yr are considered an adequate density for maximum yield (53). In California under irrigation, 20 to 30 kg ha^{-1} are required to establish 32 to 64 plants per 0.1 m^2. Seedling survival through the first winter was dependent upon seeding rates, with 66, 43, 36, and 32% surviving the 1st yr at 10, 20, 30, and at 40 kg ha^{-1}, respectively (46).

9–5.8.2 Harvested First in Year After Establishment

Seeding rates have increased about 50% in the North Central and Northeastern regions since 1972 (95). Recommended rates in kg ha^{-1} for seedings to be harvested first in the year after establishment range from 4.5 to 9 in arid prairie areas of the western USA and Canada (20), 13 to 18 in the humid North Central and Northeastern regions of the USA and eastern Canada (53, 74, 89), and 22 to 34 in the southern and western USA (41, 46, 55, 48).

9–5.8.3 Clear Seedings Harvested in Seeding Year

Increasing seeding rates up to 18 kg ha^{-1} (now generally recommended in most states) in clear seedings made in spring without a companion crop and treated with herbicides have contributed to higher yields in the seeding year because of better use of sunlight, moisture, and plant nutrients (94). In Maine, Brown and Stafford (6) increased seeding-year yields from 4.6 to 5.8 Mg ha^{-1} by increasing seed rates from 9 to 18 kg ha^{-1}. Tesar (82), in Michigan, obtained an increase of 0.4 Mg ha^{-1} between rates of 9- and 18-kg ha^{-1} when alfalfa was seeded early (10 April) and cut three times in the seeding year (Table 9-6). No consistent increases were obtained in Illinois in a 3-yr period by increasing seed rates from 13 to 20 kg ha^{-1} (53) or in Nebraska (54) when rates were increased from 10 to 17 kg ha^{-1}.

9-6 SPECIAL SEEDING SITUATIONS

9-6.1 Pasture Renovation

Pasture renovation, introduced over a half century ago in Wisconsin (15), is still popular in incorporating alfalfa into unproductive pastures and hayfields (74, 92). It appeals to small cow-calf operators who can reduce grass competition (primarily quackgrass in northern states) with tillage without the investment in a grass-killing or suppressing herbicide (see Chapter 12 in this book).

9-6.2 Sod and No-Till Seeding

Sprague (76) in New Jersey was the first to experiment with sod seeding and described nearly all the principles used today (77). Sod seeding involves the use of a herbicide(s) to suppress or kill grasses (and broad-leaved weeds) for successful establishment of a legume by drilling in a sod without tillage (77). Seeding alfalfa in an old alfalfa (plus grass) sod is included under sod seeding.

No-till seedings include those made in a sod or in the stubble of small grains or row crops. Sod seedings are adapted primarily to sites that are too stony, erosive, or steep for conventional tillage (77, 83). Although no-till seedings made after small grain or corn can be more energy efficient than conventional tillage, their primary advantage rests with the reduction in soil erosion on hilly sites. No-till seedings are used on only about 3% of alfalfa sown in the North Central and Northeastern States (Table 9-4). They have become increasingly popular with the recent emphasis on soil and energy conservation (24, 56, 70, 80, 106).

Successful no-till establishment requires the following steps:

1. Correct soil pH. Lime must be applied on the surface prior to seeding. Sufficient lime should be applied on the surface to increase the top 2.5 to 5.0 cm of soil to pH 6.0 or above (64, 94).

2. Kill or suppress existing vegetation. Most perennial grasses, particularly quackgrass, must be killed or suppressed to permit adequate seedling establishment and development (23, 24). The systemic herbicide glyphosate, cleared for use in 1981, is now the only herbicide which kills both grassy weeds (including quackgrass) and broad-leaved weeds (24) (see Chapter 12 in this book). Paraquat, 1, 1'-dimethyl-4,4'-bipyridinium

ion, is satisfactory for suppressing all grasses (except orchardgrass and, especially, quackgrass) and annual broad-leaved weeds (24). When paraquat is used, annual or perennial broad-leaved weeds must be killed earlier with 2,4-D (2,4-dichlorophenoxy) acetic acid (42, 82, 106). Seedings can be made immediately after an application of glyphosate while at least a 12-week waiting period is necessary before seeding after 2,4-D in order to prevent seedling injury.

3. Select the proper time of seeding. Satisfactory seedings can be made on a sod in either spring or summer. In general, spring seedings on a sod are generally preferred (83, 106) because of better moisture conditions in spring than in summer particularly on coarse-textured soils. Spring seedings, however, are more likely to have severe annual weed competition unless weeds are controlled with a postemergence broad-leaved herbicide not toxic to alfalfa such as 2,4-DB [4-(2,4-dichlorophenoxyl) butyric acid] (see Chapter 12 in this book). Summer seedings on sod are satisfactory in areas where summer seedings are recommended. Although summer sod seedings may be more subject to failure than spring sod seedings, competition from annual broad-leaved weeds is reduced by killing frosts in the fall. If winter-annual broadleaved weeds are present in summer seedings, an application of 2,4-DB may be necessary to reduce competition and improve the stand (see Chapter 12 in this book).

No-till summer seedings after a small grain in the North Central States should be made as early as possible following harvest (usually early August), and after required weed control with paraquat or glyphosate (106) (see Chapter 12 in this book). No-till seedings after corn are made in early spring following control of perennial quackgrass. Spring no-till seeding may be made on either corn stubble or stover after corn harvested for grain.

4. Control soil pests. Insecticides, such as Carbofuran (Furadan), may be required to control insects in some Northeastern and Southern states where problems do not occur with conventional tillage. A beneficial response has been reported from Virginia (106) and North Carolina (55). Insecticide has not been considered necessary in other regions where sod seedings are made (Table 9–4). Recent work in Minnesota (72), however, indicates that nematode control is necessary for successful alfalfa establishment on selected sites (see Chapter 21 in this book).

5. Use appropriate seeding equipment. Special no-till drills are recommended for satisfactory alfalfa establishment (Fig. 9–8) (83, 106). The seed must be placed no deeper than 2.5 cm and soil compacted above the seed for maximum emergence (99).

6. Control leafhoppers. Leafhopper injury on spring no-till seedings may be severe unless an insecticide is applied in July and August (see Chapter 22 in this book). If leafhoppers and grass competition are controlled, two, and possibly three, cuttings in the seeding year are possible (24) from spring seedings.

ALFALFA ESTABLISHMENT

Fig. 9-8. Side view (*top*) of no-till seeder in maize stubble shows coulters to provide slit for seed and fertilizer and press wheels to provide compaction over seed. Rear view (*bottom*) of no-till seeder on a killed grass sod shows shallow seed placement (1-2 cm) and compaction by press wheels of soil over seed.

9-7 MANAGEMENT DURING SEEDING YEAR

Management of seedlings during the year of establishment will involve anticipated use of the alfalfa crop, management of the companion crop, if any, and control of weeds, insects, and disease.

9–7.1 Reducing Competition from Weeds

Weed competition in spring seedings can be reduced by early seeding because alfalfa germinates at a lower temperature than most annual weeds (74, 88).

Broad-leaved annual weeds are frequently a serious problem in spring seedings. To control such weeds in seeding with oat, the seeding may be sprayed when the oat is about 15 to 20 cm or the weeds 5 to 7 cm tall. The most commonly used herbicides are MCPA [4-chloro-2-methylphenoxy) acetic acid], dinoseb [2-(sec-butyl)-4,6-dinitrophenol], and 2,4-DB (see Chapter 23 in this book).

Weed control in spring seedings of clear-seeded alfalfa must be excellent if this practice is to serve as an acceptable alternative to seeding with a companion crop (25, 70, 82).

When seedings are made in mid- to late summer in the temperate part of the USA, a herbicide may not be needed because annual weeds are killed by frost. There are situations where competition from broad-leaved weeds, especially winter annuals, may reduce stands and reduce subsequent alfalfa yields. If winter annuals pose a threat, treatment with EPTC (S-ethyl dipropyl carbamothioate) (preplant) or 2,4- DB (postemergence) has been effective (see Chapter 23 in this book). In other parts of the USA where frosts are less likely to kill weeds in summer seedings, as in California, these herbicides are in common use for the control of broad-leaved weeds (46).

When a small grain companion crop is removed for silage or grain in July in the temperate region of the USA, weeds may grow rapidly in late summer and provide excessive competition unless clipped, generally in August (75). Thick stands of broad-leaved annual weeds can be controlled by close mowing to cut below buds on stems. This can be done without injury to the basal buds of alfalfa (88). However, mowing to control annual grass is generally ineffective. If mowed in August, alfalfa has adequate time for recovery and replenishment of carbohydrate reserves before winter. If alfalfa growth from a spring seeding is heavy in late August, mowing or late-autumn grazing is recommended to remove excessive growth and to avoid smothering (74, 89).

9–7.2 Control of Volunteer Grain

Frequently, excessive volunteer growth of small grain follows combining and will endanger a new seeding. Volunteer oat should be removed by grazing or mowing in September or killed by a suitable herbicide (see Chapter 12 in this book). Volunteer oat has been successfully controlled with an herbicide applied in August or September without appreciably reducing yield of alfalfa in subsequent years (38).

9–7.3 Straw Removal

Straw remaining after combining a companion crop may be detrimental to alfalfa seedlings. Willard and Lewis (105) determined that leaving wheat straw on the stubble after combining injured the stand and reduced hay quality the following year. Mowing the stubble, especially if removed, reduced damage. Removal of straw was not necessary if there was <2.2 Mg ha^{-1} (1 ton/acre).

Klebesadel and Smith (34, 35, 36) in Wisconsin observed similar effects where spring oat was the companion crop. Clipping in late August under droughty conditions was more harmful than clipping in July but, with abundant rainfall, clipping and removing the straw were beneficial to the alfalfa.

9–7.4 Insect Control

Insect pests may be a problem in seedling stands in some parts of the USA, particularly in sod seedings in some states (see Chapter 22 in this book). Cultivars resistant to some insects are now available (see Chapter 20 in this book). New stands should be inspected frequently to determine whether an insect problem is developing and requires treatment.

Leafhoppers must be controlled with insecticides in the seeding year, on both clear seedings and on no-till seedings made in the spring, if acceptable yields are to be expected in the seeding year in areas where this insect is a problem.

REFERENCES

1. Ahlgren, H.L., and L.F. Graber. 1940. Safeguarding new seedings. Wis. Coop. Ext. Serv. Circ. 300.
2. Barnes, D.K. 1978. Effects of treated seed on inoculation, establishment, and yield of alfalfa. p. 52–55. *In* Eighth Annu. Alfalfa Symp., St. Paul. 11–12 March. Certified Alfalfa Seed Council, Davis, CA.
3. ----, G.H. Heichel, C.C. Sheaffer, and G.C. Marten. 1984. Maximizing forage and nitrogen yield of seeding year alfalfa-genotype and fall dormancy effects. Agron. Abstr. American Society of Agronomy, Madison, WI, p. 105.
4. Brown, B.A. 1959. Band versus broadcast fertilization in alfalfa. Agron. J. 51:708–710.
5. ----, A.M. Decker, M.A. Sprague, H.A. MacDonald, and M.R. Teel. 1960. Band and broadcast seeding of alfalfa and bromegrass in the northeastern states. Md. Agric. Exp. Stn. Northeast Regional Pub. 41. Bull. A-108.
6. Brown, C.S., and R.E. Stafford. 1970. Get top yields from alfalfa seedings. Better Crops Plant Food 54(1): 16–18.
7. Bula, R.J., D. Smith, and E.E. Miller. 1954. Measurements of light beneath a small grain companion crop as related to legume establishment. Bot. Gaz. 115:271–278.
8. Carmer, S.G., and J.A. Jackobs. 1963. Establishment and yield of late summer alfalfa seedings as influenced by placement of seed and phosphate fertilizer, seeding rate and row spacing. Agron. J. 55:38–30.
9. Dennis, R.E., K.C. Hamilton, M.A. Massengale, and M.H. Schonhorst. 1966. Alfalfa for forage production in Arizona. Ariz. Coop. Ext. Serv. Bull. A-16.
10. Dexter, S.T. 1953. Environmental and cultural variables that influence the establishment and yield of meadow seedings. Mich. Agric. Exp. Stn. Q. Bull. 36:138–147.
11. Eck, H.V., T. Martinez, and G.C. Wilson. 1977. Alfalfa production on a profile-modified slowly permeable soil. Soil Sci. Soc. Am. J. 41:1181–1186.
12. Erwin, D.C. 1954. Root rot of alfalfa caused by *Phytophthora cryptogea*. Phytopathology 44:700–704.
13. Ewing, J.A. 1953. Crop acreages in Illinois. Ill. Agric. Stn. Coop. Crop Rep. Serv.
14. Frosheiser, F.A., and D.K. Barnes. 1973. Field and greenhouse selection for *Phytophthora* root rot resistance in alfalfa. Crop Sci. 13:735–738.

15. Graber, L.F. 1936. Renovating bluegrass pastures. Wisc. Agric. Exp. Stn. Circ. 227.
16. Hallock, B.G. 1976. Nitrogen fixation in alfalfa. Ph.D. thesis. Univ. of California, Davis (Diss. Abstr. 77-6337).
17. Hancock, J.G. 1983. Seedling diseases of alfalfa in California. Plant Dis. 67(11):1203–1208.
18. Haynes, J.L., and L.E. Thatcher. 1950. Band seeding method for meadow crops. Ohio Farm Home Res. 262:3–5.
19. Heichel, G.H. 1985. Dinitrogen fixation and nitrogen transfer in legume-grass swards. Agron. Abstr. American Society of Agronomy, Madison, WI, p. 81.
20. Heinrichs, D.H. 1968. Alfalfa in Canada. Canada Dep. of Agric. Pub. 1377.
21. Henderlong, P.P., and Hwei-Yiing Li. 1981. Autotoxicity in alfalfa. Agron. Abstr. American Society Agronomy, Madison, WI, p. 101.
22. Hesterman, O.B., C.C. Sheaffer, D.K. Barnes, W.E. Leuchen, and J.H. Ford. 1986. Alfalfa dry matter and N production, and fertilizer N response in legume-corn rotations. Agron. J. 78:19–23.
23. Holland, C. 1983. Evaluation of problems in the improvement of grass pastures by sod seeding. Ph.D. thesis. Michigan State University, East Lansing.
24. ———, and M.B. Tesar. 1981. Establishment of alfalfa in quackgrass sods with herbicides using conventional and sod seeding methods. Mich. Agric. Exp. Stn. Res. Rep. 420:9–18.
25. Hume, D.J., R.S. Fulkerson, and W.E. Tossell. 1969. Seedling year management of alfalfa-grass mixtures established without a companion crop. Can. J. Plant Sci. 49:477–481.
26. Jackobs, J.A. 1963. A measurement of the contribution of ten species to pasture mixtures. Agron. J. 55:127–131.
27. ———. 1967. One hundred forage seeding mixtures. Agron. J. 59:435–438.
28. ———, and D.A. Miller. 1970. Varying seeding rates of alfalfa. Agron. Abstr. American Society of Agronomy, Madison, WI, p. 80.
29. Jensen, E.H., G.J. Hartman, F. Lundin, S. Knapp, and B. Brookerd. 1981. Autotoxicity of alfalfa. Nev. Agric. Exp. Stn. R144.
30. Kaddah, M.T. 1975. Deep tillage to enhance crop growth on a stratified fine sandy soil. ASAE Paper 75-1566, American Society of Agricultural Engineering, St. Joseph, MI.
31. ———. 1977. Soil profile modification to enhance alfalfa production. p. 22–24. In Proc., 7th California Alfalfa Symp., Fresno, CA. 7–8 December. University of California Cooperative Extension, Davis.
32. Kilcher, M.R., and D.H. Heinrichs. 1960. The use of cereal grains as companion crops in dryland forage crop establishment. Can. J. Plant Sci. 40:81–93.
33. Kissiwa, A.K. 1983. One-year clear seeded alfalfa as a nitrogen source for corn. M.S. thesis. Michigan State Univ., East Lansing.
34. Klebesadel, L.J., and D. Smith. 1958. The influence of oat stubble management on the establishment of alfalfa and red clover. Agron. J. 50:680–683.
35. ———, and ———. 1959. Light and soil moisture beneath several companion crops as related to the establishment of alfalfa and red clover. Bot. Gaz. 121:39–46.
36. ———, and ———. 1960. Effects of harvesting an oat companion crop at four stages of maturity on the yield of oats, on light near the soil surface, on soil moisture, and on the establishment of alfalfa. Agron. J. 52:627–630.
37. Kramer, H.H., and R.L. Davis. 1949. The effect of stand and moisture content on computed yields of alfalfa. Agron. J. 41:470–473.
38. Kust, C.A. 1971. Control of volunteer oat in alfalfa. Agron. J. 63:394–396.
39. Lathwell, D.J. 1966. Balanced diet for alfalfa. Plant Food Rev. (Spring):4–5, 16.
40. Lee, C., and D. Smith. 1972. Influence of N fertilizer on stands, yields of herbage, and protein and nitrogenous fractions of field-grown alfalfa. Agron. J. 64:527–530.
41. Lehman, W.F. 1979. Alfalfa production in the low desert valley areas of California. Univ. of Calif. Div. of Agric. Sciences Leafl. 21097.
42. Linscott, D.L., A.A. Akhavein, and R.D. Hagin. 1969. Paraquat for weed control prior to establishing legumes. Weed Sci. 17:428–431.
43. Marble, V.L. 1972. Optimizing alfalfa production in California. p. 41–55. In Proc. 2nd California Alfalfa Symp. Fresno, CA. 5–6December. University of California Cooperative Extension, Davis.
44. ———. 1974. Optimizing alfalfa production in desert areas of the southwestern United States. p. 75–91. In Proc., California/Arizona Low Desert Alfalfa Symp. El Centro, CA. 16–17 January. University of California Cooperative Extension, Davis.
45. ———. 1983. New developments in predicting the quality of alfalfa hay. p. 1–14. In Proc., 1983 California Animal Nutrition Conf., Fresno, CA. 9–10 March. University of California Cooperative Extension, Davis.
46. ———. 1984. Unpublished data, Univ. of Calif., Davis.

47. ----, and G. Peterson. 1981. Planting dates and seeding rates for Central California. p. 22–26. In Proc., 11th California Alfalfa Symp., Fresno. 9–10 December. University of California Cooperative Extension, Davis.
48. McClellan, W.D. 1975. Alfalfa: effects of seeding rates and rhizobium inoculations. California Agric. 65(2):13.
49. Martin, W.E., and R.L. Luckhart. 1971. How to predict fertilizer needs of alfalfa. p. 16–18. In Proc., 1st California Alfalfa Symp., Fresno. 7–8 December. University of California Cooperative Extension, Davis.
50. Meyer, J.L. 1971. Soil management practices as they relate to soil types in alfalfa growth. p. 1–3. In Proc., 1st California Alfalfa Symp. Fresno, CA. 7–8 December. University of California Cooperative Extension, Davis.
51. Meyer, R.D., R. Benton, and D.B. Marcum. 1984. Alfalfa response to nitrogen, phosphorus, potassium and other treatments. p. 51–55. In Proc., 31st Annu. California Fertilizer Conf., Anaheim, CA. 24–25 February. California Fertilizer Association, Sacramento, C.
52. Miller, D.A. 1983. Allelopathic effects of alfalfa. J. Chem. Ecol. 9:1059–1072.
53. ----. 1984. Forage crops. McGraw-Hill Book Co., New York.
54. Moline, W.J., and L.R. Robison. 1971. Effects of herbicides and seed-rates on the production of alfalfa. Agron. J. 63:614–616.
55. Mueller, J.P., J.T. Green, Jr., D.S. Chamblee, and J.S. Bailey. 1984. Alfalfa production in North Carolina. N.C. Agric. Coop. Ext. Serv. AG-344.
56. Mueller-Warrant, G.W., and D.W. Koch. 1981. Renovation of old alfalfa stands without tillage. Agron. Abstr. American Society of Agronomy, Madison, WI, p. 110.
57. Munns, D.N. 1967. Nodulation of *Medicago sativa* III. Effects of nitrate on root hairs and infection. Plant Soil 29:33–47.
58. ----. 1976. Mineral nutrition and the legume symbiosis. p. 353–392. In R.W. Hardy and A.H. Gibson (ed.) A treatise on dinitrogen fixation. Vol. 3. John Wiley and Sons, New York.
59. Pardee, W.D. 1980. Personal communication, Cornell Univ., Ithaca, NY.
60. Pendleton, J.W. 1957. Effect of clover, row spacing, and rate of planting on spring oat yields. Agron. J. 49:567–568.
61. Pulli, S.K. 1973. Yields, root development carbohydrate reserves and *in vitro* dry matter disappearance of spring-seeded alfalfa. Ph.D. thesis. Mich. State Univ., East Lansing (Diss. Abstr. 34-2397-B).
62. Rather, H.C., and C.M. Harrison. 1951. Field crops. McGraw-Hill Book Co., New York.
63. ----, and C.M. Harrison. 1937. A mixture of alfalfa and smooth bromegrass for pasture. Mich. Coop. Ext. Serv. Circ. 159.
64. Rechligl, J.E., D.D. Wolf, R.B., Reneau, Jr., and W. Kroontje. 1985. Influence of surface liming on the yield and nutrient concentration of alfalfa established using no-tillage techniques. Agron. J. 77:956–959.
65. Robinson, F.E. 1977. Sprinkler germination of alfalfa with saline water. p. 28–30. In Proc., 7th California Alfalfa Symp., Fresno, CA. 7–8 December. University of California Cooperative Extension, Davis.
66. ----, O.D. McCoy, G.F. Worker, Jr., and W.F. Lehman. 1968. Sprinkler and surface irrigation of vegetables and field crops in an arid environment. Agron. J. 60:696–699.
67. Roth, R.L., B.R. Gardner, G.K. Tritz, and E.A. Lakatos. 1983. Alfalfa response to water and nitrogen variables. p. 83–88. In Proc., 13th California Alfalfa Symp., Holtvill, CA. 7–8 December, University of California Cooperative Extension, Davis.
68. Schmid, A.R., and O.E. Rud. 1952. Seeding methods and your crops. Minn. Farm Home Sci. 9:3–10.
69. Schoner, C., W. Knipe, and F. Autio. 1979. Alfalfa time of planting. p. 1–6. In Proc., 9th California Alfalfa Symp. Fresno. 12–13 December.
70. Sheaffer, C.C. 1983. Seeding-year harvest management of alfalfa. Agron. J. 75:115–119.
71. ----, G.H. Heichel, D.K. Barnes, and G.C. Marten. 1985. Maximizing forage and N yield of seeding year alfalfa: establishment and cutting management effects. Agron. Abstr. American Society of Agronomy, Madison, WI, p. 105.
72. ----, D.L. Rabas, R.I. Frosheiser, and D.L. Nelson. 1982. Nematicides and fungicides improve legume establishment. Agron. J. 74:536–538.
73. Sheard, R.W., G.J. Bradshaw, and D.L. Massey. 1971. Phosphorus placement for the establishment of alfalfa and bromegrass. Agron. J. 63:22–27.
74. Smith, D. 1981. Seeding establishment of legumes and grasses. p. 15–29. In Forage management in the North. 3rd ed. Kendall/Hunt, Dubuque, IA.
75. ----, H.J. Lowe, A.M. Strommen, and G.N. Brooks. 1954. Establishment of legumes as influenced by the rate of sowing the oat companion crop. Agron. J. 46:449–451.
76. Sprague, M.A. 1952. The substitution of chemicals for tillage in pasture renovation. Agron. J. 44:405–409.

77. ———, R.D. Ilnicki, R.J. Aldrich, A.H. Kates, T.O. Evrard, and R.W. Chase. 1962. Pasture improvement and seedbed preparation with herbicides. N.J. Agric. Exp. Stn. Bull. 803.
78. Stromberg, L.K. 1975. Managing saline and alkali soils. p. 64–68. *In* Proc. 5th California Alfalfa Symp. Fresno, CA. 10–11 December. University of California Cooperative Extension, Davis.
79. Sund, J.M., G.P. Barrington, and J.M. Scholl. 1966. Methods and depths of sowing forage grasses and legumes. p. 319–322. *In* A.G.G. Hill (ed.) Proc. 10th Int. Grassl. Congr., Helsinki, Finland. Finnish Grassland Association, Helsinki.
80. Taylor, T.H., E.M. Smith, and W.C. Templeton, Jr. 1969. Use of minimum tillage and herbicide for establishing legumes in Kentucky bluegrass (*Poa pratensis* L.) swards. Agron. J. 61:761–766.
81. Tesar, M.B. 1959. Recent advances in forage crop establishment. p. 30–40. *In* Proc. 4th Annu. Farm Seed Industry—Research Conf., Chicago. 27–28 November.
82. ———. 1976a. Clear seeding of alfalfa. Mich. Coop. Ext. Serv. Bull. 961.
83. ———. 1976b. Sod seeding of birdsfoot trefoil and alfalfa. Mich. Coop. Ext. Serv. Bull. 956.
84. ———. 1977. Productivity and longevity of alfalfa as affected by date and rate of seeding, variety, and annual topdressing. p. 12–14. *In* Proc. 15th Central Alfalfa Improve. Conf., Ames, IA. 22–23 June.
85. ———. 1978. Yield and persistence of alfalfa as affected by rate and date of seeding and annual fertilization. Mich. Agric. Exp. Stn. Res. Rep. 353:5–12.
86. ———. 1979. Alfalfa variety trials. p. 49. *In* Rep. Central Alfalfa Improve. Conf., Lincoln, NE.
87. ———. 1981. p. 32–33. *In* Proc. 17th Central Alfalfa Improve. Conf., East Lansing, MI. 30 June–1 July.
88. ———. 1982. Forage management. Michigan State University Press, East Lansing.
89. ———. 1984a. Good stands for top alfalfa production. Mich. Coop. Ext. Serv. Bull. 1017.
90. ———. 1985. Fertilization and management for a yield of ten tons of alfalfa without irrigation. *In* Proc. 1985 Forage and Grassl. Conf., Hershey, PA. 4–6 March. American Forage and Grassland Council Hershey, PA.
91. ———. 1986. Re-establishing alfalfa without autotoxicity. *In* Proc. Int. Symp. Establishment of forage crops by conservation tillage methods: pest management, University Park, PA. 15–19 June.
92. ———, and S.C. Hildebrand. 1975. Reestablishment of pastures and hay fields in one year. Mich. Coop. Ext. Serv. Ext. Bull. 527.
93. ———, and D. Huset. 1979. Lime coating and inoculation of alfalfa. p. 1. *In* 16th Central Alfalfa Improv. Conf., St. Paul. 11–12 September.
94. ———, D. Huset, and C. Holland. 1981. Surface broadcast lime for alfalfa on two acid soils. Agron. Abstr. American Society of Agronomy, Madison, WI, p. 115.
95. ———, and J.A. Jackobs. 1972. Establishing the stand. p. 415–435. *In* C.H. Hansen (ed.) Alfalfa science and technology. Agronomy 15:415–435.
96. ———, K. Lawton, and B. Kawin. 1954. Comparison of band seeding and other methods of seeding legumes. Agron. J. 46:189–194.
97. ———, R.H. Leep, and B. Graff. 1985. Alfalfa variety tests. p. 50–59. *In* Rep. Central Alfalfa Improve. Conf., Stillwater, OK.
98. ———, and G.B. Triplett. 1960. Improvements in methods of establishing alfalfa. p. 348–352. *In* Proc., 8th Int. Grassl. Congr., Reading, UK.
99. Triplett, G.B., and M.B. Tesar. 1960. Effects of compaction, depth of planting, and soil moisture tension on seeding emergence of alfalfa. Agron. J. 52:681–684.
100. ———, Jr., R.W. Van Keuren, and J.D. Walker. 1977. Influence of 2,4-D, pronamide and simazine on dry matter production and botanical composition of an alfalfa-grass sward. Crop Sci. 17:61–65.
101. Van Keuren, R.W. 1973. Alfalfa establishment and seeding rate studies. Ohio Rep. 58(2):52–54.
102. Ward, C., and R.E. Blaser. 1961. Effects of Nitrogen fertilizer on emergence and seedling growth of forage plants and subsequent production. Agron. J. 53:115–120.
103. Warncke, D.L. 1985. Fertilizer recommendations for field crops. Mich. Coop. Ext. Serv. Bull. 550.
104. White, J.G.H. 1970. Establishment of lucerne in uncultivated country by sod seeding and oversowing. p. 134–138. *In* Proc., 11th Int. Grassl. Congr., Surfer's Paradise, Australia.
105. Willard, C.J., and R.D. Lewis. 1947. Reduction of stands and yields of clover and alfalfa after combined wheat. Ohio Agric. Exp. Stn. Farm Home Res. 32:64–70.
106. Wolf, D.D., and H.E. White, 1984. No till alfalfa establishment. p. 261–264. *In* Proc. 1984 American Forage and Grassl. Counc., Lexington, KY. 23–26 January.
107. Woodhouse, W.W. 1966. Soil fertility and the fertilization of forages. p. 389–400. *In* Forages. 2nd rev. Iowa State University Press, Ames, IA.

10 Nutrition and Fertilizer Use

L. E. LANYON

The Pennsylvania State University
University Park, Pennsylvania

W. K. GRIFFITH

Potash & Phosphate Institute
Great Falls, Virginia

Plant nutrition management has a vital role in the success or failure of modern alfalfa (*Medicago sativa* L.) production. Even at average yield levels, large amounts of nutrients are removed from soils, much greater amounts than by crops harvested for grain. The alfalfa crop harvested each year in the USA is estimated to contain 1.7 Tg (1.7 million t) of K (158). This amount is equivalent to about 40% of the total fertilizer K applied annually for all purposes. It is more than twice the amount contained in the grain portion of the total U.S. maize (*Zea mays* L.) grain crop. Because maize for grain is grown on three times the production area of alfalfa, there is more than a 10-fold difference between the ratios of the estimated K applied to that removed per hectare for alfalfa and corn grain (Table 10–1).

Although estimated average alfalfa yields are relatively low, very high yields are obtained regularly by top growers. Similarly, dry matter yields of 19.5 to 21.5 Mg ha^{-1} have been reported by research workers in humid regions (16, 112, 204). Irrigated alfalfa in California yielded 47.5 Mg ha^{-1} (175) and yields of 26 to 34 Mg ha^{-1} have been achieved by producers in Arizona (191). A similar situation exists in Europe where average annual yields range from 5 to 8 Mg ha^{-1} (106). The combination of production practices into an effective management system creates the opportunity for positive interactions among various components such as, cultivars, soil fertility levels, cutting schedules, pest control, and fa-

Table 10-1. Estimated ratio of fertilizer K applied to K removed for alfalfa and maize grain in the USA (158).

	Alfalfa	Maize grain
K applied, kg ha^{-1}	43	76
K removed, kg ha^{-1}	152	27
Ratio	0.28	2.8

Copyright 1988 © ASA-CSSA-SSSA, 677 South Segoe Road, Madison, WI 53711, USA. *Alfalfa and Alfalfa Improvement*—Agronomy Monograph no. 29.

Table 10-2. Three-year average alfalfa yields from two K and two cutting management systems (199).

K rate	Two-cut system	Three-cut system
kg ha^{-1}	Mg ha^{-1}	
89.6	8.5	10.5
358.4	9.0	11.4

vorable environmental conditions or, in some regions, the use of irrigation to achieve higher yields than would be possible with the implementation of any single practice.

Research since the 1950s by Tesar (199, 202, 204) in Michigan illustrates the interaction among specific management practices in achieving high yields. In 1959 to 1961, a positive K-rate by cutting-frequency interaction produced 9% greater yields of 'Vernal' alfalfa (Table 10-2) than either main effect. With the identification of additional beneficial management factors and their integration into a production system, he harvests 18 to 20 Mg ha^{-1} (203). A four-cut system with the last cutting after mid-October and annual K applications of 560 kg ha^{-1} are part of the management associated with the top yields. Other components in this particular production system are: a well-drained soil of pH 6.9; annual application of 64 kg ha^{-1} of P and 3.4 kg ha^{-1} of B; cultivars with high genetic yield potential, good winterhardiness, fast regrowth potential, and improved disease resistance; and intensive pest control. Eleven alfalfa cultivars under this management had a 2-yr average yield of 19 Mg ha^{-1}.

This chapter considers the importance of plant nutrients as they interact with other high yield production inputs, the role of these nutrients in plants, their behavior in soils, and fertilizer application methods in a high yield system. Previous reviews related to these topics were provided by: Baylor (14), Blaser and Kimbrough (21), Duell (45), Griffith (64), and Rhykerd and Overdahl (169).

10-1 FACTORS AFFECTING NUTRIENT NEED

10-1.1 Yield Level

Two fertilizer rates were compared on alfalfa grown in a rather unproductive soil in the southeastern USA where the research objective included the control of other yield-limiting factors (141). The results (Table 10-3) were a substantial increase in yield in response to nutrient applications and a nutrient uptake which increased with yield.

Plant samples taken at each harvest from fields of producers participating in the Pennsylvania Alfalfa Growers Program provide a large nutrient uptake data base (113). Uptake of all nutrients increases as yield increases (Table 10-4). To obtain high yield levels, soil fertility status

Table 10-3. Alfalfa yields and apparent nutient removal under two fertilizer regimes (141).

Fertilizer treatment		Two-year avg yields, dry matter	Apparent nutrient removal		
P	K		N	P	K
		kg ha⁻¹			
34	179	15 735	521	50.4	266
68	358	18 398	582	55.3	371

Table 10-4. Nutrient uptake by alfalfa in the Pennsylvania Alfalfa Growers Program from 1977-1981 (113).

Yield group	Nutrient									
	N	P	K	Ca	Mg	S	B	Cu	Zn	Mn	Fe
Mg ha⁻¹	kg ha⁻¹									
Up to 9	227	25	205	99	17	18	0.22	0.06	0.18	0.40	1.09
9-11.2	253	32	270	121	21	22	0.28	0.07	0.24	0.53	1.16
11.2-13.4	351	38	315	148	27	28	0.34	0.08	0.29	0.57	1.58
13.4-15.7	418	45	379	162	29	32	0.37	0.09	0.31	0.74	1.76
15.7-17.9	480	53	451	187	34	38	0.41	0.10	0.34	0.90	1.80
>17.9	559	61	524	226	39	47	0.48	0.12	0.40	0.87	2.15

and plant nutrient concentrations must be monitored and adjustments made to assure adequate nutrient availability. Tesar (202) and Baylor (15) suggest that annual fertilizer applications of up to 59, 538, and 4.5 kg ha⁻¹ of P, K, and B, respectively, may be needed on soils where alfalfa is adapted for the production of high yields under intensive management.

10-1.2 Cutting Schedule

A close relationship exists between alfalfa maturity and nutrient concentration (186). Alfalfa is harvested at vegetative to early reproductive growth stages in high-yielding, intensively managed systems. When alfalfa is harvested at a less mature growth stage, such as full bud rather than 10% blossom, the leaf-stem ratio is greater (99) with a consistent increase in the concentration of P, K, Ca, and Mg in the herbage. The trend in concentration differences for nutrients such as K (108) is independent of the nutrient fertilization rate (Table 10-5). However, K concentration and yield did increase as K application rates increased. Potassium nutrition was suggested as a key factor in achieving rapid regrowth after clipping, which, in turn, influenced the success of more frequent cutting schedules. The relationship of K and alfalfa to the regrowth of alfalfa has been an important consideration in developing successful high yield alfalfa systems (203).

Table 10-5. Potassium effects on alfalfa regrowth and K concentration (108).

K rate	K in total herbage, days regrowth 18	32	Yields (dry matter) days regrowth 18	32
kg ha^{-1}	kg ha^{-1}		Mg ha^{-1}	
0	14.9	11.0	0.82	1.77
47	15.1	11.2	0.94	2.17
93	19.2	15.2	1.19	2.50
186	24.5	21.0	1.38	2.64
372	39.1	31.0	1.44	2.75

Table 10-6. Plant nutrient concentration in fertilized (F) and unfertilized (NF) alfalfa harvested at bud stage (173).

Nutrient	Stems NF	F	Leaves NF	F	Roots† NF	F
			g kg^{-1}			
N‡	19.7	18.6	56.0	53.0	16.3	17.3
P	2.5	2.6	3.6	3.5	2.6	2.5
K	15.1	27.9	13.7	21.2	8.2	11.1
Ca‡	7.6	6.5	24.1	20.5	7.0	6.7
Mg	3.5	2.9	4.4	4.3	2.2	2.0
S	0.6	0.9	4.9	3.5	0.9	1.1
			mg kg^{-1}			
Zn	20	21	29	29	13	16
B	17	18	67	63	18	18
Cu	14	10	26	21	8	9
Mn	13	18	78	84	13	18
Fe‡	88	102	130	140	136	184
			g plant^{-1}			
Dry matter	70	78	33	41	23	27

† Upper 15 cm of root system.
‡ Nutrient not applied in the fertilizer treatment.

10–1.3 Plant Part

Alfalfa plant parts have different nutrient concentrations (173). The concentration of most nutrients is greatest in leaves (Table 10–6),, with the greater concentration of K in stems a significant exception. Differences exist among leaves according to their position on the plant. Concentrations of N, P, and Zn are less in basal leaves than in those at the top of stems. In contrast, basal leaves have higher concentrations at Ca, Mg, B, and Mn. Significant differences among the leaves do not exist for either K or S. Stem concentrations of N, P, Ca, Mg S, Zn and B and, sometimes, Cu, decrease from the top to the bottom of the shoot. The concentration of K increases progressively to near the top of the plant then decreases slightly.

10-1.4 Climate

The nature of the growing season has an influence on alfalfa growth and total nutrient requirements (187). Three to four harvests per season are common in the northern areas of the USA. However, in parts of southwestern USA, seven to eight harvests are common. In general, total yield and nutrient removal increase with number of harvests.

Temperature, light intensity, rainfall patterns, and daylength change within and among the harvest intervals of the production year. The variation in environmental conditions will influence nutrient concentrations in forage, because of changes in rate of dry matter production, ion movement in soil, root activity, and the uptake of nutrients by the plant (29, 73, 157).

Long days can increase the stem number per plant and/or the internode length (33) and decrease the leaf/stem ratio while increasing yield (35). Greater leaf/stem ratios occur when alfalfa is grown under low or high temperature than at intermediate temperatures of 25°C (28, 127). The aboveground portion of the alfalfa plant increases more rapidly than the roots as the temperature increases from low to high (152). Differences in the nutrient concentrations of forage can result from variations in climatic factors and their effects on the relative quantities of plant parts found in alfalfa (172).

Cool temperatures slow the release of N, P, and S from soil organic matter, reduce nutrient movement in soils by diffusion, lower absorption and translocation rates in plants, and lower plant metabolic activities when compared with process rates achieved at normal temperatures. Nutrient concentrations in the herbage are lower for P, K, Fe, B, Cu, Zn, and Mn and higher for Ca and Mg at low vs. high temperatures (66, 188). For example, the concentration of K increased from 15 g kg^{-1} at 15/10°C to 26 g kg^{-1} at 27/21°C. This suggests that higher levels of exchangeable soil K are needed in cool as compared to warm environments in order to provide sufficient concentration of K in the herbage for maximum growth. Nutrient composition of the plants changes when the last harvest in the fall is delayed with Ca, and Mg increasing and N, P, K and to some extent S decreasing for all or part of the period (36). In this Wisconsin field study, K concentration decreased 45% over the harvest period, a larger change than any of the other nutrients.

Patterns and amounts of rainfall and/or irrigation are important considerations in the production of alfalfa. Nutrient uptake generally increases as soil moisture tension is decreased from the permanent wilting percentage to field capacity. Kilmer et al. (107) found that most of the increased uptake of all the major and secondary nutrients with increasing water availability could be accounted for by increased top growth (Table 10-7). However, excess soil moisture can lead to a significant reduction in the potential growth of alfalfa. Increasing the availability of soil nutrients cannot eliminate the adverse effects of excessive soil moisture (2,

Table 10-7. Relative uptake of nutrients by alfalfa at four moisture levels at third harvest (107).

Water applied	Nutrient					
	N	P	K	Ca	Mg	S
mm	relative uptake					
67	55	44	52	50	62	60
101	70	59	72	59	62	60
146	92	84	82	90	100	80
184	100	100	100	100	100	100

Table 10-8. Three-year average alfalfa yield and water-use efficiency (WUE) as affected by irrigation treatment and P fertilization (209).

Annual P rate	Irrigation treatment					
	Dry		Medium		Wet	
	Yield	WUE†	Yield	WUE	Yield	WUE
kg ha^{-1}	Mg ha^{-1}					
48.9	17.0	100	18.8	108	21.9	108
97.8	18.8	111	20.2	115	26.6	123
195.6	22.6	133	24.6	141	29.3	146
293.4	22.8	135	26.7	152	30.9	154

† WUE = A relative water-use efficiency with yield evapotranspiration and the 48.9 kg ha^{-1} rate at low irrigation rate = 100.

3). The possibility of variations in soil wetness and the importance of water for dry matter production emphasize the need to plan fertilizer use and other production inputs, with knowledge of the limitations imposed by water availability at the site where alfalfa is to be grown. Rainfall on cut forage can have a direct effect on the nutrient concentrations in the plant tissue. Collins (37) measured decreased K and increased P, Ca, and Mg when early maturity alfalfa was wetted in excess of 62 mm after drying. More mature hay was not affected and all nutrient concentrations either decreased slightly or were essentially constant.

In many areas, irrigation is essential in maintaining adequate water and soil moisture levels. Also, some soils require subsurface drainage, with tiles or other practices, to achieve satisfactory levels of production. Water management may interact positively with other production factors. Viets (209) reported a positive interaction between irrigation and P fertilization as measured by higher yields and greater water-use efficiency (Table 10-8). Potassium response on a K-deficient soil in Minnesota almost doubled when adequate irrigation was supplied as compared to no irrigation (181).

10-1.5 Management

Yield and quality of alfalfa depend on numerous management decisions and inputs. A good manager who achieves high yields increases nutrient uptake and the nutrient applications required for crop growth and the maintenance of soil nutrient reserves. Therefore, nutrient needs will depend on other management practices and fertilizer use must be consistent with the overall level of management.

Nutrient applications can influence the uptake of nutrients that were not included in the applications. Potassium applications have reduced Ca and Mg concentrations (200), while added P has been associated with a decrease in micronutrient concentrations. Although the P effect on micronutrients can result from dilution (81), other factors may be involved (29, 177, 192). Attention should be given to the effect of reduced micronutrient concentrations on ruminant animal nutrition when alfalfa is a major portion of the ration (167).

Genetic manipulation of alfalfa cultivars to achieve an improved nutrient concentration or efficiency of use may have undesirable effects on other nutrients. Hill and Jung (80) and Buss et al. (30) agree that selection for specific nutrient concentrations should proceed with caution because the concentration of one nutrient cannot be assumed to be manipulated independently of the concentrations of other nutrients.

10-2 SOILS AND THEIR NUTRIENT-SUPPLYING POWER

10-2.1 Soils Vary in Nutrient-Supplying Power

The nutrient-supplying capabilities of soils vary. The composition, potential rooting volume, successful exploitation of that volume, and previous management can influence the quantity and rate of nutrient supply to crops. The composition of the soil not only influences the nutrients present but also the fate of the applied nutrients. Soils that differ in mineralogical composition require different fertilizer application rates to maintain the available nutrient levels (Fig. 10-1) (75). Those soils that are most effective in supplying K for crop growth with no added K, can be the same soils that have the greatest potential to fix added K so that it is unavailable for crop use (123). Further, high K concentrations can be depleted more rapidly in comparison with less concentrated levels of K in the same soil (93).

10-2.2 Root Growth Patterns

Knowledge of the depth and growth pattern of alfalfa roots is an important consideration in the management of soil fertility. Most lateral roots develop from the taproot near the soil surface, but more laterals develop at greater depths as plants age. Taproot diameter and total length, and the diameter and number of lateral roots increase with time. The relative length of individual lateral roots remains constant (207). Alfalfa has less root length per unit of surface soil than many grasses, such as orchardgrass (*Dactylis glomerata* L.) or maize (51). When this low density is combined with a two to three times greater rooting depth, alfalfa can

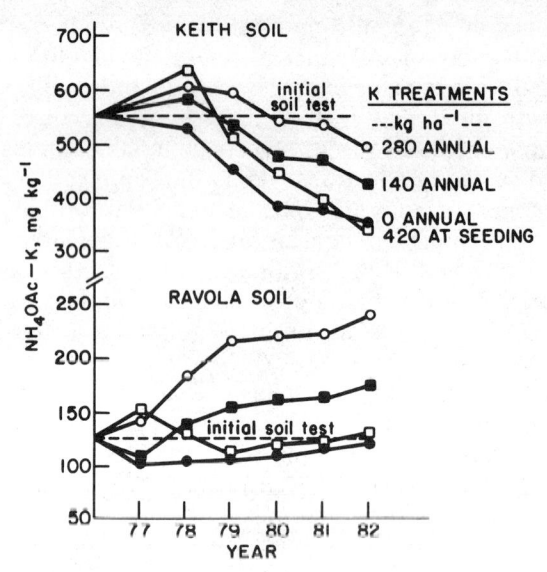

Fig. 10-1. Soil test trends during alfalfa production with several K fertilization rates and methods on two soils of contrasting composition (75).

be expected to have a lesser effect than orchardgrass on the concentration of water and nutrients, such as K, in the surface layer (93).

Phosphorus and K increase alfalfa root weight per unit volume of soil (65). Potassium applications to low K soils can increase root growth even when shoot growth is relatively unaffected (129). Nevertheless, the expected reaction is a decrease in the root/shoot ratio with higher K levels (173). The beneficial effects of P and K on root growth interact positively with other variables, thereby enabling alfalfa to obtain other nutrients and water from a greater volume of soil. Deep root systems are in contact with moist soil for a longer period during dry weather, which benefits root activity and plant growth. The beneficial effect of K on the physical and chemical properties of roots has been associated with improved winterhardiness (100, 166) and stand persistence (151).

Alfalfa may not respond as expected to surface K and S applications when adequate levels of these nutrients are available in the subsoil (74, 98, 195). Both the chemical (31) and physical (52) characteristics of the subsoil affect root growth. Alfalfa may use less water in soil surface layers than orchardgrass, but it will extract more subsurface water (18, 34, 51, 208). Subsurface water can be extracted to depths of 3.3 m, but 80% of the total absorbed may be from the first meter (109). Adequate water in the surface soil layer can reduce the uptake of nutrients from the subsoil (185).

10-2.3 Soil Testing and Plant Analysis

Soil testing and plant analysis are important management tools in determining the soil nutrient supplying potential for alfalfa production.

Sufficiency levels of nutrients in plant tissue have been used most frequently in the interpretation of the results from plant analysis. Kelling (103), Martin and Matocha (128) and Melsted et al. (136) are common sources of values for samples and will be referred to in this chapter. However, the concentrations in plant tissue are influenced by many factors and an alternate approach to plant status assessment, the Diagnostic and Recommendation Interpretation System (DRIS), has been proposed. Kelling et al. (104) developed norms for alfalfa using this approach and found the N/P, N/K, P/K, and S/K indices to be most discriminating in the diagnosis of alfalfa nutrient status. None of the indices were adequate for addressing growth-limiting factors when the yield exceeded 3.6 t ha^{-1} for an individual harvest. The DRIS approach attempts to reduce the impact of plant part, time of sampling and plant variety on the results of the interpretations. In a P×K×S factorial study, Kelling et al. (105) were able to correctly identify limiting nutrients in sequence using DRIS on each of three cuts and a sample taken 2 weeks before one of the harvests. This was possible even though the actual compositions of the plant samples were quite different. Russelle and Sheaffer (176) successfully identified K deficiencies and the changes in limiting nutrients for plant growth when K was applied. Results from soil tests also require careful interpretation. The tests should be supported by research that will ensure that test results can be correlated with the desired levels of production at a given location. These tests when combined with good management records are even more useful when used as guidelines for fertilizer applications.

10-3 SOIL ACIDITY AND LIMING

10-3.1 Desirable pH Levels

A soil pH range of 6.6 to 7.5 is a standard recommendation for alfalfa production (220). In 1984, McLean and Brown (135) summarized a group of midwestern USA studies in which 93% of maximum yield was attained at pH 6 and 100% at pH 6.6. However, the optimum pH can vary with soil texture, organic matter, and other soil chemical properties (101, 133). Some yield depressions have been recorded when heavy rates of lime have been applied (31, 60), when B is potentially limiting (40), or when P is marginal and lime is applied at normal rates (78, 94). Munns and Fox (143) suggested that reduced growth associated with liming may be a temporary problem. In the absence of added N, the response of the crop to liming will depend on both the ability of the plant and associated rhizobium to tolerate factors associated with soil acidity. If no other nutrient toxicity or deficiency exists, as in solution culture, alfalfa plants can grow at pH 4 if adequate Ca is present (142). Plant growth in soil can be expected to decrease when the pH is <5.2 in the presence of Al and/or Mn (78, 171) and possibly high levels of B (60), even when N is provided. Rhizobium survival (171) and Mo deficiency (140) at pH <6 can restrict effective symbiosis, thus yield can be increased with applications of either lime or N (Fig. 10–2). The best approach to maintaining an optimum pH range is to apply lime according to soil test recommendations developed for the region where the alfalfa is to be grown. Lime should be applied prior to stand establishment and then as indicated by soil tests during the life of the stand (see Chapter 9).

Although genetic variation in tolerance to acidity exists in alfalfa, it is generally less than that found in other crops (30). High plant-to-plant variation for Al tolerance within alfalfa populations can be exploited with recurrent selection by screening for top growth under stress conditions (41). Elliot et al. (50) suggested that the soil conditions can be inventoried for Al and Mn, and the requirement for soil amendments reduced through the use of then appropriate legume species. Similarly, tailoring the alfalfa plant to soil conditions would be possible with the development of tolerant cultivars.

10–3.2 Benefits of Liming

The benefits of liming are: (i) decreased solubility of toxic elements, (ii), increased availability of essential nutrients, and (iii) increased soil microorganism activity (153).

Aluminum, Fe, and Mn toxicities have been implicated as reducing

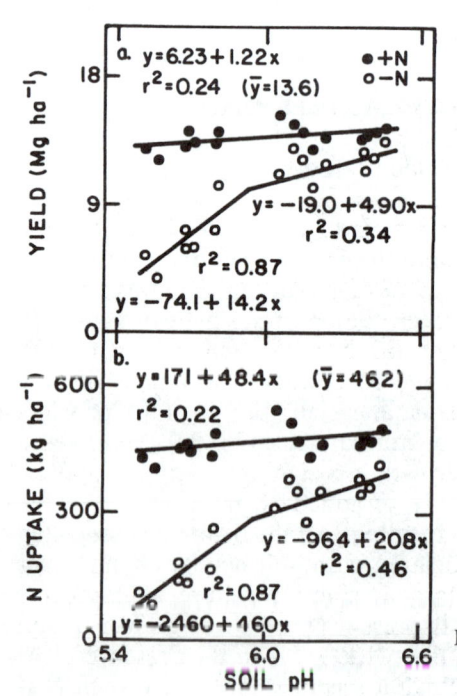

Fig. 10–2. Relationships of yield (a) and N uptake (b) for symbiotic and N fertilized alfalfa to soil pH (140).

the potential yield of alfalfa on acid soils. An increase in pH reduces the solubility of all three of these elements (215). The amount of soil Al extracted by salt solutions may be a better indicator of the potential for alfalfa production than pH (131). Root growth will be reduced if Al extracted by a $CaCl_2$ solution is greater than 0.005 to 0.008 cmol kg^{-1} and will decrease by 50% if greater than 0.05 cmol kg^{-1}. Organic matter can reduce the negative effects of Al at moderate levels of Al stress but not at Al levels of 0.57 cmol kg^{-1} or more (102). In the production of alfalfa, Helyar and Anderson (78) suggest Al levels of <20% of the cation exchange capacity (CEC). Liming to reduce Al will often lead to the reduction of Mn, but toxic levels of Mn may persist in the presence of large quantities in the soil (114, 123). Manganese leaf symptoms can be severe before yield decreases occur (78). Alfalfa yield ceased to increase at different pH levels ranging from 5.81 to 6.96 in a liming study of six soils in Oregon (94). Aluminum extractable with 1 M KCl was a better indicator of potential alfalfa growth than soil pH or Al in plant tops. Unlike Mn, Al is not readily translocated from plant roots to tops (24). Mahler (124) studied a group of high allophane Idaho soils. These soils have high Al saturations relative to most Idaho soils at similar pH levels. They required a pH of at least 6.6 (and Al saturation <12%) for best alfalfa growth; in contrast to most Idaho soils on which alfalfa does not respond to additional lime beyond pH 5.6.

Lime application increases the availability of Ca, Mg, P, and Mo. Liming materials correct acidity and supply Ca and Mg (221). Calcium deficiency may aggravate the effects of soil acidity (144). Providing adequate Ca can be more difficult for soils of the tropics which range from negative to positive change depending upon soil pH, than in soils of temperate regions (144). Although relatively large amounts of Ca are taken up by alfalfa (Table 10-4), reports of Ca deficiencies are rare. Andrews (4) measured some improvement in plant growth when Ca was augmented in solution culture of pH 5, while Doerge et al. (44) found no advantage from Ca applications in field soils. Methods for the prediction of Ca deficiency in the nutrition of alfalfa are not well developed. In general, the Ca in most liming materials will reduce potential deficiency problems when these materials are supplied to achieve a desired pH range and to inactivate toxic substances. Sumner (197) has suggested that liming-induced Mg deficiency could reduce crop yields when Mg availability is marginal and the availability of P is low. These conditions may be found in intensively weathered soils.

Molybdenum is the only micronutrient that increases in availability as soil pH increases. Molybdenum has an essential role in symbiotic N_2 fixation. Low yields of alfalfa at marginal levels of Mo may be associated with the preferential uptake of Mo by plants prior to meeting the needs of nodules (44). This results in a reduction of N_2 fixation by nodules. Although alfalfa yields may increase with an increase in pH or the addition of Mo (140), Mo concentrations in alfalfa can exceed the recommended level of 10 mg kg^{-1} when Mo is applied and the pH increased to 6 or above (140).

10–3.3 Herbicide Efficacy

The activity and persistence of the commonly used triazine herbicide are influenced by soil pH (82). These herbicides are less active at low pH because of absorption on soil colloids. Subsequent reactions on the soil colloid promote their decomposition (145). Liming a low pH soil to which triazines have been applied may cause desorption of undecomposed triazines and potentially toxic effects on alfalfa planted in the soil. Applications of EPTC (S-ethyl dipropylthiocarbamate) for weed control in clear-seeding of alfalfa can increase the toxic effect of triazine residues (48) (see Chapter 23 in this book).

10–3.4 Application of Lime

The time required to modify soil pH depends on the initial soil pH level, liming material particle size, and the amount of mixing with the soil. In most situations, liming materials should be applied 26 weeks or more prior to seeding (11). Rechcigl et al. (164) measured no difference in the effect of lime when it was applied either 32 or 90 weeks before to no-till seeding. When a large lime application is required, standard recommendations call for a split application, one-half plowdown, and one-half surface incorporated (116). No-till alfalfa establishment is well adapted for use on highly erosive soils that are unsuitable for tillage. However, lime cannot be mixed with the soil as in conventional tillage methods. Surface applications at least 26 weeks prior to seeding (219) or immediately after seeding (86) have been effective. In the latter situation, the 4-yr average yields on two soils with initial pH of 4.8 to 5.2 were comparable (16–18 Mg ha^{-1}) to yields obtained with lime incorporation.

Results from studies of deep placement of lime to promote alfalfa root growth have been variable. In general, root penetration into an unlimed acid "subsoil" is observed infrequently in short-term greenhouse studies (20, 31, 131, 184). Therefore, additions of lime to the subsoil layer would be expected to improve alfalfa root growth and performance especially under conditions of imposed water deficiency (184). Although results of deep placement under field conditions are variable, increased rooting has been observed in alfalfa (44, 131, 166). Deep placed liming materials can have a beneficial effect where subsoils are highly acid. Alfalfa yields were higher, subsoil pH levels greater, and exchangeable soil Al levels lower on plots treated with deep-placed limestone (164). The alfalfa growth response was different than that measured with soybean (*Glycine max* L.) that has a shallower root system when grown in the same soil. In a different, less-weathered soil, Rechcigl and Reneau (162) found little difference between surface applications of lime and those placed at 30 cm. The surface applications contributed to increased subsoil root growth even in the presence of high levels of exchangeable Al (263 cmol kg⁻¹). More vigorously growing plants resulting from liming of surface layers can over a period of years "bio-cycle" P to the subsoil, and thereby create more favorable microenvironments for plant root growth (161). In general, the deep placement of lime, even when yield advantages have been measured, is considered prohibitive in cost of materials and/or effort (119, 131, 184).

In regions of high rainfall and with coarse-textured soil, there is a good possibility that lime will need to be applied during the productive life of an alfalfa stand. When lime is needed on an established alfalfa stand, fall applications should be made when weather and crop conditions will minimize stand injury from traffic during application.

10-4 NITROGEN

Well-inoculated alfalfa fixes large amounts of atmospheric N by symbiotic N_2 fixation. Nitrogen concentration and removal equal or exceed that of any other nutrient, including K (Tables 10-3 and 10-4). Ball and TenEyck (8) and Roth et al. (175) grew 21.3 and 47.5 Mg ha⁻¹ yr⁻¹ of irrigated alfalfa, respectively with corresponding N removals of 784 and 1120 kg ha⁻¹. While large quantities of N are removed from the field in alfalfa harvests, nonlegume crops rotated with alfalfa benefit from the release of residual N from roots, nodules, and unharvested tops. The benefits of including alfalfa in crop rotations have been documented by Fox and Piekielek (59) and many others. Fox and Piekeilek (59) determined that the economically optimum N fertilizer rate needed for maize production was reduced more than 70% when maize followed alfalfa (Table 10-9).

The sufficiency concentration range for N is similar to K. Kelling (103) suggests that a concentration of 25 to 37 g kg⁻¹ in the upper 152

Table 10-9. Maize grain yield and N response parameters as influenced by alfalfa and field management history (59).

Management and crop history	Relative yield of check plots†	Economically optimum N rate‡	N-supplying capacity§
	%	kg ha⁻¹	
No manure, no N fertilizer, and no legume within 2 yr (11 experiments)	55	134	62
First year after an alfalfa sod with greater than 25% stand (6 experiments)	93	40	147

† [0 broadcast N yield/maximum yield] × 100.
‡ Five rates of N used with range from 0 to 179.2 kg ha⁻¹.
§ Determined by check plot uptake minus 0.75 × starter N rate of 16 kg ha⁻¹.

mm of alfalfa at the time of the first cutting is sufficient to ensure top yields.

10-4.1 Nitrogen in Plants

Heichel et al. (77) estimated that 43 to 64% of the N in alfalfa is obtained through the symbiotic process. Symbiotic N_2 fixation varies with many factors, including crop management practices. Soil pH, K, and P levels have been related to N_2-fixing capacity. For example, an increase in K availability has been found to increase nodule number, carbon exchange rate, and carbohydrate translocation from stems to nodules (13, 47, 54). These effects are important especially after clipping when adequate K stimulates regrowth and thus increases the potential for more N_2 fixation (see Chapter 7 in this book).

10-4.2 Nitrogen Application

Small amounts of N may be recommended at seeding time to aid seedling establishment prior to the development of effective nodulation (Table 10-10) (85). Early plant growth can be limited by N deficiency when compared with plants that have received added N. Subsequently, the performance of plants with added N can be similar to those relying solely on symbiotically fixed N_2 (55). The time required to achieve effective nodulation may be characteristic of the cultivar (122). In a solution culture study, Heichel and Vance (76) found that addition of 50 mg L^{-1} of N resulted in fewer nodulated seedlings, smaller numbers of nodules per plant, and less nodule branching reflecting reduced meristematic area in very young seedlings as compared to 0 mg L^{-1} (Fig. 10-3). The source of N can affect plant performance. MacDowall (122) found nitrate N (NO_3^--N) to be more effective in promoting plant growth than ammonium N (NH_4^+-N). However, the NO_3^--N source seemed to suppress the onset of N_2 fixation more so than NH_4^+-N. Becana et al. (17) suggested that the level of nitrite N (NO_2^--N) in the nodules is more critical to plant per-

Table 10-10. Summary of state recommendations for N fertilizer at planting time for legume and grass-legume mixtures (85).†

N	Pure-legume	Grass-legume mixtures
kg ha⁻¹	——————— no. of responses ———————	
0	19	9
0-22.4	10	5
0-33.6	--	8
0-44.8	7	2
0-56.0	--	4
0-67.3	4	2
No recommendations listed	3	12
Indefinite recommendations	2	3

† Data from 45 states responding to inquiry.

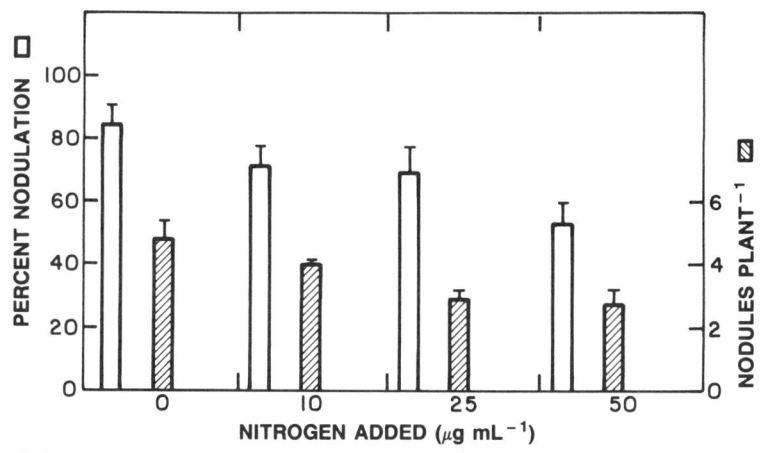

Fig. 10-3. Percent modulation and modules per plant for alfalfa grown with several concentrations of added N and inoculated with a mixed culture of five strains of *Rhizobium meliloti* (76).

formance than the actual N form that is supplied to plants. Nitrite is a strong and effective inhibitor of nitrogenase and inactivates leghemoglobin in the nodule. Nitrogen fixation may not be as sensitive for plants growing in soils as compared with plants grown in other culture media (85).

In field studies, N has been applied either in a band at planting or broadcast. Nitrogen in the band at planting can increase the uptake of P from the band, but probably only if N is limiting the growth of the crop. In band applications, NH_4^+-N can be more effective than NO_3^--N, but the difference is of limited practical value (182). If weeds are a potential problem at establishment and they are not controlled, N applications of 95 kg ha^{-1} have decreased the alfalfa component of the stand and increased the weed content (155). In Canada, Kunelius (111) observed a tendency in weedy stands for yield and number of alfalfa plants to decrease when 100 kg ha^{-1} N was applied postemergence. The relationship was less noticeable when few weeds were present. Nodulation was either not affected or decreased by the applications up to 100 kg^{-1} of N. Nodule weight was more sensitive than the number of nodules. The concentration of N in the plants after N application can be related to the preceding crop and its potential for using N (150). Following a small grain, the concentration increased when N was applied, while following a grass sod the concentration was not affected or decreased with added N. First cutting yields were increased by moderate N applications of 25 to 50 kg ha^{-1} N in two of three experiments by approximately 10 kg of alfalfa kg^{-1} of N, but the results were significant for the full establishment year in only 1 of 3 yr. In a 3-yr study in Oregon (Fig. 10-4) (49), yields and plant N concentrations increased with N applications only when the plants were ineffectively nodulated. All rates of N up to 225 kg ha^{-1} depressed nodulation, and in some comparisons N applications caused NO_3^--N accu-

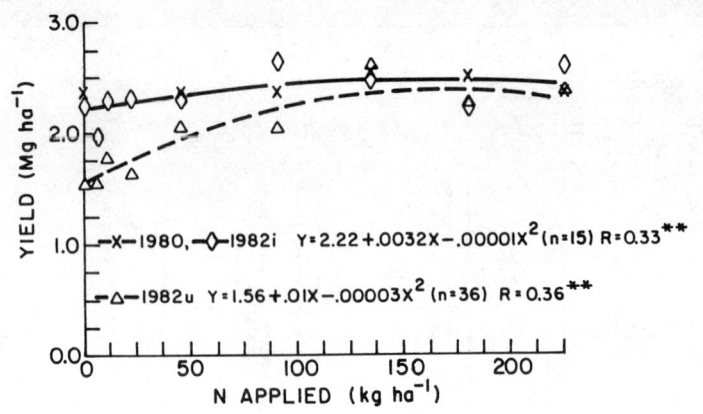

Fig. 10–4. Yield of inoculated (i) or uninoculated (u) 10-week-old alfalfa seedlings after application of N at establishment (49).

mulation in the plants. High levels of NO_3^--N accumulation could be detrimental to livestock performance. Nuttall (148) reported a yield response to 45 kg ha^{-1} of N, especially when seed had not been inoculated.

Many of the effects of N applied to alfalfa at planting are similar to those of N applied to established stands. Tesar (200) measured no effect of 0 to 225 kg ha^{-1} of N in obtaining alfalfa yields of 19 Mg ha^{-1}. Neither Rhykerd et al. (170) nor Lee and Smith (117) measured yield or consistent crude protein effects with applications up to 896 kg ha^{-1} of N. However, the plants grown by Rhykerd et al. (170) did appear taller and darker when N was applied. Lee and Smith (117) measured increased nonprotein N concentrations with N application. The NO_3^--N in plants equalled or exceeded 0.5 mg g^{-1} NO_3^--N at the 448 and 896 kg ha^{-1} of N rates. In a companion 2-yr study of applications up to 336 kg ha^{-1} of N, the alfalfa stand was not adversely affected. Nitrate-N also accumulated in alfalfa when N applications exceeded 224 kg ha^{-1} of N, and N_2 fixation as well as nodulation decreased with N application (178). In a Pennsylvania study (212), low rates of 22 or 45 kg ha^{-1} of N did not affect yield, crude protein, and/or total digestible nutrients. Nitrogen applications studied by Jenkins and Bottomley (95) increased late-season yield, uptake and concentration of N but, in general, N effects were not significant for total season production. This may be attributed to the changes in the soil environment as there was no N effect on any first harvest crops. The small contribution of third crop harvests to total forage yield reduced their potential impact on annual indexes of performance. In Israel (53), alfalfa yields were increased in the field when N was applied to a coarse-textured soil but not to a fine-textured soil.

Application of N to alfalfa-grass mixtures commonly will reduce the percentage of alfalfa in the mixture (149). However, if the alfalfa is not well nodulated, N application may increase the contribution of alfalfa to the stand productivity (165). A general fertilizer recommendation is to treat an alfalfa-grass stand with <30% alfalfa as a nonlegume stand (see Chapters 9 and 13 in this book).

Alfalfa can utilize residual soil N from the management associated with previous crops. The N can be as residual NO_3^--N or mineralized from organic additions. Mathers et al. (130) measured NO_3^--N depletion to 1.8 m by alfalfa after 1 yr of growth in soil that had received up to 45 t ha^{-1} yr^{-1} of manure for the preceding 3 yr. Nitrogen depletion was apparent to 3.6 m after 2 yr. They concluded that alfalfa could be used as a management option for the removal of excess NO_3^--N from soils and thus reduce the NO_3^--N pollution potential.

10-5 PHOSPHORUS

The concentration and quantity of P in harvested alfalfa forage are less than N or K. Melsted et al. (136) provide a critical value of 3.5 g kg^{-1} of P for the upper herbage at the early flowering stage. A sufficiency range of 2.6 to 7.0 g kg^{-1} of P for the upper 152 mm of the alfalfa plant has been suggested for first-cut alfalfa (103). These values are within the range reported from high-yield alfalfa research and production fields. Phosphorus concentrations, like other nutrients, generally decrease with maturity (64, 186).

10-5.1 Phosphorus in Plants

A specific function of P in alfalfa is its involvement in adenosine triphosphate (ATP) associated with nitrogenase activity. Vincent (210), Mortenson (139), and Bergersen (19) have associated the high ATP requirement for N_2 fixation and the essential role of P in the symbiotic process. The P effect on N metabolism has been demonstrated with a tropical alfalfa (61) (Table 10-11). Nodule number, size, and N_2 fixation levels all increased as the P level increased.

10-5.2 Phosphorus in Soils

Phosphorus is absorbed by alfalfa roots largely as orthophosphate from the soil solution. The P concentration in soil solution and the maintenance of this concentration are important for satisfactory alfalfa nutrition. Phosphorus moves through the soil solution to plant roots by

Table 10-11. Effect of P on nodule development of alfalfa 26 d after seeding (61).

	P levels, kg ha^{-1}				
	0	31	62	125	250
			mg		
Nodule dry wt	0.13	0.44	1.06	3.31	8.47
Weight/nodule	13	33	28	60	57
N content	0.01	0.03	0.07	0.15	0.65

diffusion and is readily converted to less insoluble organic and inorganic complexes. For these two reasons, P moves only short distances in soils and may be positionally unavailable or fixed in unavailable soil forms. Crocker et al. (40) observed that adequate lime increased the effectiveness of applied P by improving root growth, but that excessive lime applications would reduce P availability as it is precipitated with Ca.

Vesicular-arbuscular mycorrhiza (VAM) have been observed to promote the utilization of soil P by alfalfa. Vesicular-arbuscular mycorrhiza can increase yield and P concentration in the plant as compared to lime and P applications without VAM (110). However, the colonization by VAM decreased when P was added to the soil. Vesicular-arbuscular mycorrhiza increased the ratio of P uptake per unit of applied P as compared to the addition of P alone (147). When compared to uninoculated treatments, inoculation with the specific VAM *Glomus mosseare* in a calcaerous P-fixing soil in Spain increased alfalfa yield and plant concentrations of N, P, and K, especially when combined with rhizobium inoculation (6).

10-5.3 Phosphorus Application

Phosphorus application at seeding is an accepted practice in the USA (205) and Canada (183), particularly on soils testing low in P. Although direct effects on the alfalfa plants are temporary (Fig. 10-5), placement in a band below the seed when drill-seeding is recommended. The amounts

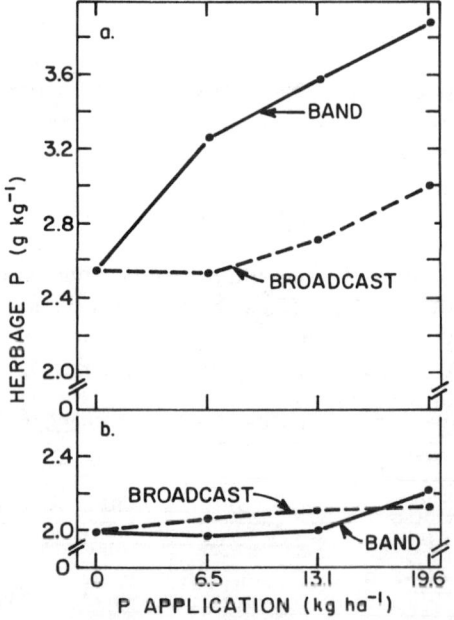

Fig. 10-3. Concentration of P in alfalfa seedlings 6 weeks (A) and 12 weeks (B) after P application in a band or as broadcast (183).

of P for fertilization at establishment are generally small and often in the 1-3-1 (N-P_2O_5-K_2O) ratio for conventional tillage. Adequate P is essential for the establishment of strong root systems (205), and benefits from supplementation are most apparent on infertile soils or when cool spring weather restricts P uptake (183, 205). Although P application is effective when soil tests are high (26), Sheard (182) recommended a reduction of P from 30 kg ha^{-1} to 15 kg ha^{-1}. Rates of 45 kg ha^{-1} applied at planting resulted in decreased seedling growth measured after 10 weeks in both low and high testing soils (182).

On established stands, annual broadcast applications according to soil tests have been satisfactory (8) (Table 10-12). Injecting P in established stands under arid and semiarid conditions has been evaluated. Injection of liquid ammonium polyphosphate at 10 cm depth on 30 cm centers was successful in increasing yields of the first harvest at two of three sites in western North Dakota. At the second harvest, the treatment was not different from surface application (62). In Canada, injected applications reduced yields when compared to surface applications and the effects persisted for 3 to 5 yr (118). Smith and Powell (190) concluded that 57 kg ha^{-1} Bray P1-P was an adequate soil level for yields of 7 to 8 Mg ha^{-1}. Mallarino et al. (125) reported that a response to top-dressed P on an alfalfa-smooth bromegrass-orchardgrass pasture was unlikely if the Bray P1-P exceeded 39 to 45 kg ha^{-1}. Alva et al. (3) and Tesar (204) measured no yield differences between annual rates of 45 or 48 and 90 or 96 kg ha^{-1} P. Tesar (199) reported no differences in 3-yr yields among annual rates of 0, 24, and 49 kg ha^{-1} P compared at a 12 Mg ha^{-1} yield level, when 39 kg ha^{-1} had been applied to alfalfa grown on a sandy loam soil with an available soil P of 24 kg ha^{-1}. On a silt loam with the same available P, there was no difference at a similar yield level in a 3-yr test of alfalfa that received 29 kg ha^{-1} of P at seeding and either 0 or 24 kg ha^{-1} of P annually. These data on two soil types, over a period of years, show no gain in yield or only minimal gains from annual topdressing of P if alfalfa receives adequate P at establishment.

10-5.4 Phosphorus Source

Most P is applied to alfalfa as ordinary (OSP) or triple superphosphate (TSP). Recent evidence suggests that other forms may be equally effective

Table 10-12. Effect of P fertilization on yield of alfalfa and soil test levels (9).

P fertilizer rate	Alfalfa yield, 3-yr avg	Soil test P levels		
		Initial	Second year	Third year
kg ha^{-1}	Mg ha^{-1}		kg ha^{-1}	
0	14.3	21.3	11.2	9.0
19.6	16.9	21.3	16.8	12.3
39.2	17.9	21.3	20.2	19.0
58.7	18.7	21.3	32.5	30.2

and may have advantages under specific conditions. In Australia, rock phosphate and lime applied to an acid soil was comparable to the production of alfalfa with OSP and lime (89). An advantage of the rock phosphate is the lack of soil acidification below the limed layer, a factor that may contribute to poor initial growth following fertilizer application in pasture settings (88). Once established, the alfalfa could grow satisfactorily, but initial seeding performance was undesirable (90). Also, additional studies suggested that gypsum ($CaSO_4 \cdot 2H_2O$) in OSP displaced Al into the soil solution which adversely affected alfalfa growth. In a calcareous clay soil, a mixture of rock phosphate with either 50 or 70% TSP was comparable to 100% TSP applications for alfalfa production (71).

Animal manure is another source of P that may be available where forage is utilized for livestock production. Goss and Stewart (63) obtained similar alfalfa yields when feedlot manure applications and TSP were compared in field and greenhouse studies. However, they suggested that TSP resulted in luxury consumption of P, based on the concentration in excess of the minimum P recommendation for alfalfa. On the other hand, manure provided adequate P, was more efficient in dry matter production per unit of P uptake, and resulted in a significant carryover effect in the following season. They suggest that while much of the manure P is inorganic as excreted, microbial activity utilizing the available energy sources in manure results in the conversion of the P to organic forms. These forms are then protected from soil adsorption reactions and become available slowly for plant use.

10-6 POTASSIUM

The K requirement of alfalfa is greater than that for any other nutrient. In high-yield alfalfa production systems, alfalfa is subject to more frequent harvests at immature growth stages, established on soils high in K, with large amounts of K applied annually (205). In comparison with past practices, these factors contribute to higher K concentrations in harvested herbage and greater K removal from soils. A decade ago the standard used for K removal was 17 kg Mg^{-1} which is much lower than the recent reports of removal in hay of from 22.5 to 25.0 kg Mg^{-1} (8, 113, 141, 180). Alfalfa can take up greater amounts of K than are sufficient for the level of dry matter produced. This is called *luxury consumption* (75). Results from most successful high-yield alfalfa research and producer fields reflect a degree of luxury consumption for K and that all K removed in the crop was not essentially for plant growth.

Grass invasion, weeds, and reduced stand persistence have been associated with inadequate K availability. The need to maintain adequate K is more critical with alfalfa-grass mixtures than pure alfalfa stands, because grasses can "out-compete" alfalfa for K (21).

A sufficient K concentration in the upper 152 mm of the alfalfa plant at early flower is about 25 g kg^{-1} (103, 136). The composition of the alfalfa plant is not uniform, however, and the impact of the proportion of various plant parts and their respective concentrations should be recognized when plant samples are used for diagnostic purposes. In general, the K concentration is in the order stems >leaves >roots (Table 10-6) (172, 173).

Sufficiency levels for adequate soil K are much less clear-cut. However, a recent survey of soil fertility and forage management specialists indicates that additional K is rarely recommended when the concentration of exchangeable K is greater than 300 kg ha^{-1} in the surface soil layer (115).

10-6.1 Potassium in Plants

Adequate K in the plant appears to be critical in early growth. At a leaf area index (LAI) of one, the carbon exchange rate (CER) was increased more with adequate K than at LAI 2, although adequate K improved the CER at both growth stages as compared to the rate at low K concentrations (218). Increased mesophyll resistance to CO_2 flow when K is limiting can be a factor in the reduced growth rate of shoots and roots under moderate K deficiency. Stomatal resistance to gas exchange may be a significant effect of K deficiency only under severely limiting conditions (154). Lower concentrations of reducing sugars in alfalfa leaves when the plants are supplied with adequate K probably reflect the role of K in the transport of photosynthate from source to sink (38).

The N concentration of alfalfa has been observed to increase with the availability of K (7). The rate of symbiotic fixation of N that is characteristic of well-nodulated alfalfa may be decreased by inadequate K availability because of reduced enzyme activity (38, 47, 121). The decrease in N fixation is possibly a secondary response of reduced photosynthesis, when shoot growth is restricted by low levels of K. Potassium also participates in the conversion of amino acids to proteins (146).

10-6.2 Potassium in Soils

Alfalfa roots absorb K from the soil solution. Potassium in the soil solution is in equilibrium with exchangeable K and other more slowly available K fractions. The amount of K in solution at any one time is influenced by many factors, including soil pH, wetting and drying cycles, temperature, level of other ions, and aeration. Potassium moves to plant roots by diffusion and with the transpiration stream as water is absorbed by alfalfa roots (mass flow). Diffusion accounts for the greatest soil-K movement (10). The amount of K reaching roots by diffusion increases in those crops that have high K requirements such as alfalfa (168). Diffusion rates are affected by moisture, temperature, root activity, and K concentrations in the soil. Actively growing alfalfa roots may create zones of K deficiency at the root-soil interface even in soils with high levels of available K. Soil and plant management factors that enhance good alfalfa root volume and distribution increase the efficient use of available K.

10-6.3 Potassium Application

Most fertilizer recommendations in the humid eastern USA call for annual applications of K based on soil test results (115). Split applications may be recommended when adding larger quantities of K. However, Tesar (204) found that splitting 372 kg ha^{-1} of K at rates of one-half each in the fall or spring, or one-fourth for each cutting, starting in early spring, did not increase average 3-yr yields on a loam soil at a yield level of 17 Mg ha^{-1}. Splitting a similar K rate between fall and spring on a sandy loam, likewise, did not increase yields. These recent data indicate that splitting K application is not necessary at rates up to 372 kg ha^{-1} which are adequate for yields of 17 Mg ha^{-1} on medium-textured soils. However, at K rates above 372 kg ha^{-1} split applications may be advantageous to avoid potential injury from chlorine in potassium chloride (KCl) fertilizer (189).

Surface-applied K is recovered by alfalfa more efficiently than that placed deeper in the soil. Peterson and Smith (156) estimated 41% recovery of K in 1 yr's growth after surface application and only 10 to 15% when the K was placed below 45 cm. Thus, topdressing is a simple, effective method for applying K.

Short-term recovery of applied K by alfalfa usually decreases as the rate of application increases. Over a 2-yr period alfalfa recovered 56% of the applied K from a medium-textured soil following the application of 448 kg ha^{-1}, but only 17% after 1792 kg ha^{-1} was applied (174). Nevertheless, the quantity of K recovered may be equivalent to that applied over a sustained period of cropping (39). The reduced recovery of K measured in some tests probably reflects utilization limitations of the cropping system rather than "loss" of applied K. Those soils which supply the greatest quantities of K to plants from the nonexchangeable sources will often be those which will temporarily "fix" the greatest quantities when K is applied after a period of depletion (134).

10-6.4 Potassium Source

In the USA, KCl is the most common K fertilizer for use on alfalfa. Rominger et al. (174) suggested that potassium sulfate (K$_2$SO$_4$) be used, in preference to KCl, at very high rates of application ($\geq$930 kg ha^{-1} of K) to reduce potential chloride toxicity. Also, when the application of secondary nutrients such as S or Mg are recommended, it may be beneficial to use sources of K other than KCl.

10-7 CALCIUM AND MAGNESIUM

Calcium and magnesium are associated closely with soil acidity and liming as mentioned. Kelling (103) gives sufficiency ranges in the top 152 mm of first-cut alfalfa as 5 to 30 g kg^{-1} for Ca and 3 to 10 g kg^{-1} for Mg.

10-7.1 Calcium and Magnesium in Plants

Calcium and Mg concentrations in alfalfa are greater than for the grasses at equivalent stages of maturity. For this reason, alfalfa is considered a superior source of these nutrients in animal rations than either forage grasses or corn silage. Actual concentrations can be quite variable as a result of several factors, including soil texture, pH, and date of harvesting. For example, Adams (1) found Ca content of alfalfa decreased from 14 g kg^{-1} at a 1 June harvest to 9.5 g kg^{-1} 4 weeks later. The Mg content dropped from 1.9 to 0.8 g kg^{-1} during the same period.

10-7.2 Calcium and Magnesium in Soils

Alfalfa produces good yields over a wide range of Ca/Mg ratios if the pH is within recommended levels for alfalfa (184). Magnesium concentrations within the plant, however, may be reduced below levels for good animal nutrition on some coarse-textured soils where either calcitic limestone is used exclusively or high levels of K exist. Potassium fertilization can depress both Ca and Mg concentrations in alfalfa (Table 10-13). Dionne (42) measured a greater yield response to Mg application on acid soil with 0.18 cmol kg^{-1} Mg than for another soil with 9.6 cmol kg^{-1} Mg (1% as compared to 24% saturation).

10-7.3 Calcium and Magnesium Application

Calcitic and dolomitic limestone are the principal sources for these nutrients. Low Mg levels can develop with continuous use of calcitic limestone. The prospect of encountering Mg deficiency can be reduced by using a dolomitic limestone source every third time that lime is applied to adjust pH level.

10-8 SULFUR

Changes in crop production and the environment have created the need to monitor the status of S in alfalfa over a broader array of conditions

Table 10-13. Nutrient concentration and yield of alfalfa with varying levels of K fertilizer (200).

K fertilizer rate	Dry matter yield	Nutrient concentration		
		K	Ca	Mg
kg ha^{-1}	Mg ha^{-1}		g kg^{-1}	
0	16.2	14.6	18.1	4.6
187	17.2	18.7	16.1	3.4
374	17.8	23.0	15.6	3.1

than in the past. Specifically, greater S removal in the higher yields of alfalfa, increased use of high analysis S-free fertilizer, reduced emission of S when fossil fuel is burned, and the removal of S from fungicides and crop protection chemicals have contributed to the increased need to monitor S availability and adjust fertilizer requirements accordingly.

The sufficiency range for the top 152 mm of first-cut alfalfa has been reported to be 3 to 5 g kg^{-1} (103).

10–8.1 Sulfur in Plants

Alfalfa yield loss from low S results, in part, from decreases in N_2 fixation (46). Plant N and S were correlated in field studies in Australia (5) and in growth chamber studies (126). Hoeft and Walsh (83) obtained increased uptake and higher concentration of S in Wisconsin field studies following the application of S, but there was no increase in N concentration. Protein-S levels are not as sensitive to changes in S status as other forms of plant S. However, plant morphological differences and some physiological differences in protein synthesis can influence the expression of S deficiency on S levels in the plant (214). Although alfalfa accumulates more S in roots and shoots than the cool-season grasses, such as tall fescue (*Festuca arundinacea* Schreb.) and orchardgrass, it apparently has fewer specific root absorption sites for S than the grasses. A smaller fraction of the S in the roots is transferred to shoots by alfalfa (12). In solution culture, Lopez-Jurado and Hannaway (120) measured maximum acetylene (C_2H_2) reduction and total plant N at 2.5 mg L^{-1} of S. The N/S ratio decreased below 11 as S concentration increased.

Several indicators of plant S status have been applied with success. Total S, sulfate S (SO$_4$-S), and N/S$_{total}$ gave satisfactory results in an Idaho field study. Maximum yield was achieved at 1.5 to 2.1 g kg^{-1} S and a N/S of 11. The yields did not differ, however, when N/S was <17 or 18 (214). The appropriate plant S concentration for maximum yield may be different than that for maximum N content. In an Australian field study, 2.0 g kg^{-1} S was adequate at maximum yield, but 2.8 g kg^{-1} of S was judged to be best when related to the maximum level of N (5). In this study, N/S was not as satisfactory as total plant S in the determination of adequate plant status because of the concurrent change in plant concentrations. Therefore, total plant S was the preferred indicator. Similarly, Sorensen et al. (194) did not find N/S of value in predicting plant response. Quigley and Jung (160) found 2.0 g kg^{-1} of S adequate for growth of alfalfa at yields of 11 t ha^{-1}, and that N/S decreased with S application while the N concentration remained constant. In a Canadian field study (69), there was little response to applied S when the plants contained 1.4 to 1.8 g kg^{-1} S. An every other year monitoring schedule was suggested to ensure that adequate S was available to the crop. Concentrations of S from 2.5 to 3.0 g kg^{-1} were associated with maximum yields in a Wisconsin field study, in which little yield increase was obtained when plants contained 2.0 to 2.2 g kg^{-1} S (83).

10-8.2 Sulfur in Soils

Much soil S in humid regions is in the organic form which when mineralized to sulfate (SO_4^{2-}) becomes available to alfalfa. In arid regions, sulfates of Ca, Mg, Na, and K can be common. Soils with marginal S levels for alfalfa production often are those that have not received manure or other S-containing amendments, are low in organic matter and clay content, and are well drained or highly leached through rainfall or irrigation.

Inorganic soil S is mobile and susceptible to leaching losses. Percolating SO_4^{2-} can accumulate in fine-textured subsoils and be available to deep-rooted crops, such as alfalfa. Although soil tests, usually for SO_4^{2-} content, can be useful, such tests alone may not prove to be satisfactory. Soil testing during the growing season did not provide Meyer and Marcum (137) with useful information. However, they felt that if 11 to 24 mg kg^{-1} of SO$_4$-S was measured before growth started in the spring little crop response would result from S applications. Gupta and Veinot (70) suggested 2 mg kg^{-1} of SO$_4$-S as the threshold for S response in their greenhouse studies. Nutall (148) considered the soil S test to be uncertain, but that a concentration of 4 mg kg^{-1} was adequate for crop growth. In a test of several extractants (84), calcium phosphate [Ca(H$_2$PO$_4$)$_2$] extractants with acetic acid solvent rather than water were found to give the best correlation with yield, S uptake by plants, and S concentration in control plants. If the soil S extracted with <6 mg kg^{-1}, plants grown in 66% of the soils responded to S applications, but only 14% responded if the soil test was >10 mg kg^{-1}. The addition of soil pH and the interaction of the extracted S with soil pH improved the predictive ability of the soil test (Fig. 10-6). Under greenhouse conditions, two groups of

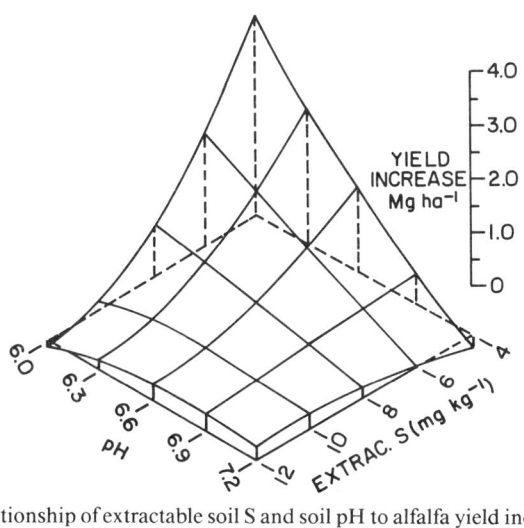

Fig. 10-6. Relationship of extractable soil S and soil pH to alfalfa yield increase in Wisconsin (84).

Kansas soils were found to differ in their relationship of extractable soil S to plant response (27). A coarse-textured group responded at the first harvest and extractable S was satisfactory in predicting the plant response. However, alfalfa grown in a finer-textured group of soils did not respond until the second and third harvests. For this group, organic matter content helped in predicting potential plant response. Hanson et al. (72) did not measure a yield response to S application to irrigated and nonirrigated fields of alfalfa in Missouri. Soil test levels of 6.7 mg kg^{-1} were apparently adequate for crop growth.

Subsoil S analysis may be recommended in combination with surface soil testing, plant analysis, and N/S determinations to derive S fertilizer requirements (159). Andrews (5) measured a greater yield response to spring-applied than fall-applied S, presumably as a result of changes in root morphology and the exploitation of additional soil volume during the growing season. In an Idaho study, S extracted by any of three extractants (0.1 M LiCl, 0.032 M KH$_2$PO$_4$ and water) for the 0 to 30 cm soil layer were correlated with yield, but inclusion of S from the 30 to 92 cm layer did not improve the relationship (213). Walker and Doornenbal (211) found several extractants based on 0.01 M CaCl$_2$ to be effective for their Canadian soils when samples were taken to 30 cm, but little advantage was obtained by sampling to 60 cm. Inclusion of the S content of increasing depth increments, with no weighting for depth increased the correlation of extractable soil S with plant response in an Australian study (216). A total of 3.5 mg kg^{-1} of SO$_4^{2-}$-S in the 0 to 80 cm layer was required for 90% of maximum yield. This relationship contrasts with the results of Probert and Jones (159) which were improved by weighting of the depth increment contributions. Alfalfa is less responsive to S applications in the field than under greenhouse conditions, and the potential of a field response is reduced in the presence of high subsoil S levels. Nevertheless, low subsoil levels do not ensure a crop response to applied S (98).

10-8.3 Sulfur Application

Coarse-textured soils low in organic matter have the greatest supplemental S requirement. Factors that contribute to low organic matter or that slow organic matter decomposition, such as low temperature, acidity or drought, contribute to low S availability. Where irrigation is used, the S content of irrigation water should be considered in developing S recommendations.

If S applications are required, they can be made in conjunction with the annual fertilizer program. An application of 28 kg ha^{-1} was optimum for alfalfa growth on responsive sites in a Wisconsin study, and in the 1st yr prilled S was less effective than SO$_4$-S (83). Recovery of S was greater at low rates of application. The SO$_4$-S was most effective when spring-applied and elemental S when fall-applied. Because of the differential availability of S from the various sources, a combination of SO$_4$-

S and elemental S was recommended for application once every 3 yr when S is required for crop growth. According to Meyer and Marcum (137), differences between SO_4-S and elemental S applications may decrease during the growing season with the potential increase in oxidation of the elemental form. A 56 kg ha^{-1} application was adequate for 2 yr of production in an Australian study (98). Crop response was greater from winter and early spring applications when the soil was cold and mineralization of organic sulfur was slow than when S applications were made in mid-summer.

10-9 BORON

Alfalfa is sensitive to B availability. Alfalfa yield was increased more by B application than the yields of red clover (*Trifolium pratense* L.) and timothy (*Phyleum pratense* L.) in a Canadian field study (68). The addition of this micronutrient to alfalfa fertilizers is accepted practice where known deficiencies occur. The sufficiency range for the top 152 mm of first cutting alfalfa is 30 to 80 mg kg^{-1} B (103). High yield systems remove from 0.50 to 0.70 kg ha^{-1} yr^{-1} B, which is above the levels that were considered normal in the past.

10-9.1 Boron in Plants

Boron deficiency may be associated with moisture stress. A condition called *alfalfa yellows* or *yellow top* that is caused by B deficiency is frequently mistaken for moisture stress because it may develop during short periods of summer drought. Gupta (68) measured some decrease in B concentration of field-grown alfalfa after a dry period in a Canadian study and leaf yellowing when the B concentration was <30 mg kg^{-1}. In a greenhouse study, alfalfa responded to B application more so than birdsfoot trefoil (*Lotus corniculatus* L.) and responded best under conditions of intermediate soil moisture treatment as compared to excessively dry or more optimal regimens (43). The visual symptoms of potato leafhopper [*Empoasca fabae* (Harris)] damage to alfalfa are similar to those of B deficiency.

Boron is relatively immobile in alfalfa tissue with the most recent growth first showing B deficiency symptoms. In an Illinois field study, B concentrations were in the order lower leaves>upper leaves>tips>upper stems ≥lowers stems for several B rates of application and plant maturities (138). Along with yellowing of the top leaves, there is a shortening of the upper internodes, which produces a rosette appearance in alfalfa. Mild deficiencies of B often go unrecognized because yields of any one cutting may be affected only slightly. However, alfalfa forage quality may be reduced by delayed maturity and leaf loss. Willet et al. (217) found B deficiency to be aggrevated by N application to greenhouse-grown alfalfa. They speculated this was a result of an interference with B nutrition in the plant. In a greenhouse study, Zn applications reduced B uptake on a low B soil possibly through interference with plant use or transport of B (25).

Boron concentration in plants is usually increased by B application (68, 138). Because of this uptake potential toxicity of B to plants is possible with excessive B applications (68).

10-9.2 Boron in Soils

Most available B in humid region soils is associated with soil organic matter. Boron is released to plants during organic matter decomposition. Organic matter decomposes readily when soils are moist, but decomposition essentially stops during dry periods. Thus, B deficiency on alfalfa may be observed during drought periods and not during periods of adequate rainfall. Boron availability is negatively correlated with pH, particularly at pH 6.0 or above. Therefore, the B supply can become limiting at high pH levels (Fig. 10-7) (40). Calcium may interfere with B uptake when soil B levels are high (58). Some B is absorbed by soil clay fractions, but is easily leached unless utilized by alfalfa or incorporated into soil organic matter. Leaching losses are greatest from coarse-textured soils, especially with high rainfall or irrigation. Gupta (68) found B toxicity of alfalfa growing in a fine sandy loam field soil to be short lived, reflecting the rapid potential decrease in B availability under these conditions.

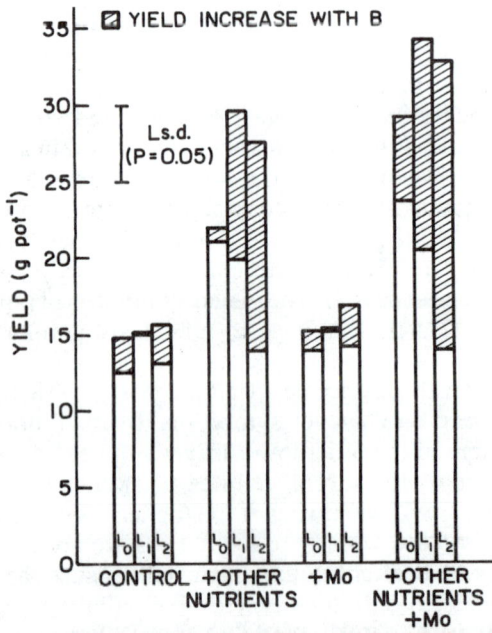

Fig. 10-7. Alfalfa yield response interactions of Mo and B with lime application (0, 1.25, or 2.5 t ha^{-1}) and a nutrient mixture (K, Ca, Mg, Fe, Mn, Cu, Zn, and Co) in an Australian greenhouse study (40).

10-9.3 Boron Application

Boron applications of from 1 to 3 kg ha^{-1} of B are generally recommended for alfalfa. This small amount of B can be applied with other nutrients in mixed fertilizers both in preplant and in maintenance applications. Fine-textured, slowly permeable soils may require only one application every 3 yr. Gupta (68) suggested alternate year applications of 2 kg ha^{-1} of B applications. Alfalfa intensively managed for production of high yields should receive annual B applications if B deficiencies have been noted in the past.

10-10 OTHER MICRONUTRIENTS

Other micronutrients such as Mo, Cu, Mn, Zn, Fe, and Co are not major problems in alfalfa production. Molybdenum deficiencies have occurred, especially on acid soils where the pH levels are below those recommended for high forage yields. Jones and Moschler (97) showed that alfalfa responds to Mo applications when pH levels ranged from 5.0 to 6.0. The response to Mo was negligible at pH 6.0 and above. They also noted that herbage N concentrations increased when Mo deficiency was corrected. Crocker et al. (40) found that Mo interacted positively with B applications as lime applications increased (Fig. 10-7). Molybdenum is known to be essential in the N_2-fixation process and in N assimilation into proteins. Application of lime can reduce the potential response to Mo if adequate soil reserves exist. The resulting concentration in forage can sometimes be in excess of that required for animal use (92, 96). However, if reserves are lacking, Mo in combination with lime is required for satisfactory production (67). Gupta (67) used total Mo (perchloric acid extraction) or oxalate-extractable to assess soil Mo status for potential plant response. In Kentucky, Thom (206) recommends the application of 43 to 86 g ha^{-1} Mo when soil pH<6.2 and the soil has not been limed within 16 to 26 weeks of alfalfa establishment or if the pH of an established stand is <6.2. However, it should be noted that Mo application does not substitute for a good liming program. Mortvedt (140) does not recommend Mo application if soil pH is >6.0 since levels of Mo toxic to livestock can accumulate in the plant. He suggested plant N status as an indicator of Mo sufficiency because of analytical difficulties with Mo. Under conditions of marginal Mo availability, the plant tops apparently are a more effective sink for Mo than the nodules (44).

Some yield responses have been reported from Cu additions where alfalfa is grown on high organic matter or muck soils and on highly weathered sandy soils. Copper is less available when the soil pH is high and more frequent deficiencies can be expected than at lower pH levels. Plant Cu and Zn levels were correlated with surface and subsoil EDTA extractable quantities in a field survey of New Zealand producer fields (132). Including soil pH did not improve the relationships between plant and soil levels. However, the correlation was improved when the data were evaluated within soil series. Forbes (56) found ethylenediaminetetraacetic acid (EDTA) extractable Cu comparable to Cu from perchloric acid digestion as a measure of soil Cu available for crop production even though the quantity removed by EDTA was less. A 5 mg kg^{-1} soil level with the EDTA extractant was adequate for 9 mg kg^{-1} in the plant. Plant levels of Cu can vary significantly from harvest to harvest, apparently as a result of differences in light intensity and temperature during the growth periods. Copper uptake by the plant decreases as the applications increase (57).

Forbes (56) suggests that both soil and plant analyses be used in the assessment of Cu status for alfalfa. On mineral soils, 11 to 17 kg ha^{-1} of copper sulfate (Cu$_2$SO$_4$) will be sufficient for alfalfa where deficiencies are known to occur. On organic soils, this rate should be doubled. There is residual effect of Cu applications to soil so that an application 1 yr in 3 may be adequate where the need exists (57).

Selenium in alfalfa is potentially toxic to animals which consume the feed. Soltanpour and Workman (193) evaluated the relationship of the diethylenetriaminepentaacetic acid (NH$_4$HCO$_3$-DTPA) extractable Se to alfalfa concentrations under greenhouse conditions. They found that Se concentration in alfalfa exceeded the recommended tolerance level of 5 mg kg^{-1} when the Se in the soil was $\geq$0.1 mg kg^{-1}.

In a field survey of potential contamination of alfalfa fields, Taylor and Allison (198) found only Pb to be related to rural, suburban, and industrial/highway settings. The concentration of Pb in the alfalfa collected increased from low to medium, with the range reported from 2.8 to 33.2 mg kg^{-1}. Cadmium, Ni, Cu, and Zn were generally within the expected range and were not related to location of the samples relative to cultural features.

10-11 ANIMAL MANURES AND MUNICIPAL WASTES

Animal manures can be applied prior to establishment of alfalfa or to the crop directly. Baylor and Waters (16) state that most of the top producers in the Pennsylvania Alfalfa Growers Program applied dairy cattle manure to the crop that preceded alfalfa in rotation.

The application of animal manures on established alfalfa does not result in efficient use of biologically available N from the manure. Additional production problems associated with manure spreading on alfalfa can be damage from traffic, fouling, and changes in stand composition. Hensler et al. (79) found a yield decrease or no yield change was measured when manure was applied to an established alfalfa-orchardgrass stand in the spring or after the first harvest. High rates of application, 90 or 135 t ha^{-1}, did reduce the amount of alfalfa in the stand. There was a tendency, but not significant, for the soil NO$_3$-N to increase after the high applications. Alfalfa (*M. media* Pers.) in a stand of alfalfa/smooth bromegrass (*Bromus inermis* Leyss.) decreased as the fertilizer applied or the beef cattle manure increased, but not when grown alone or with crested wheatgrass (*Agropyron cristatum* Schribn.) (87).

Even though some uncertainty in the resulting performance of the alfalfa exists, some farmers with high livestock/crop acreage ratios or limited storage capabilities may not be able to avoid spreading manure on established alfalfa stands. Where this practice is used, manure should be spread on stands that are in their last productive year when soils are firm and dry, preferably in the autumn. Manure should be applied thinly immediately after cutting if spreading must be done during the growing season.

Holt and Zenter (87) applied beef cattle manure (FYM) in the fall to established alfalfa, alfalfa mixtures with smooth bromegrass or crested wheatgrass and the pure grasses to approximate the rates of N and P supplied by inorganic fertilizer (IF) applications. The IF rates were 55 or 110 kg ha^{-1} of N and 12 or 24 kg ha^{-1} of P and the manure rates 11 or 22 t ha^{-1}. The yields with low IF and comparable FYM were not different for alfalfa, but FYM was less effective than IF for the alfalfa/smooth bromegrass mixture and the grasses. The utilization of N decreased less with the FYM than the IF. Net returns calculated for the forage operation were alfalfa>alfalfa/grass>grass and for IF were generally greater than FYM (Fig. 10–8). Net returns to alfalfa and the mixtures were more sensitive to forage value and nutrient cost than were the net returns to the grasses. If the value of FYM was changed to $0 t^{-1}, then the most favorable treatment for alfalfa or alfalfa/grass was the high rate of FYM. This was not the case for the grasses. The break-even price for FYM was $0 to 6 t^{-1}.

Land application of wastes from the processing of municipal wastewaters has become a possible option for disposal of the products which were previously discharged to waterways or buried in landfills. Mainte-

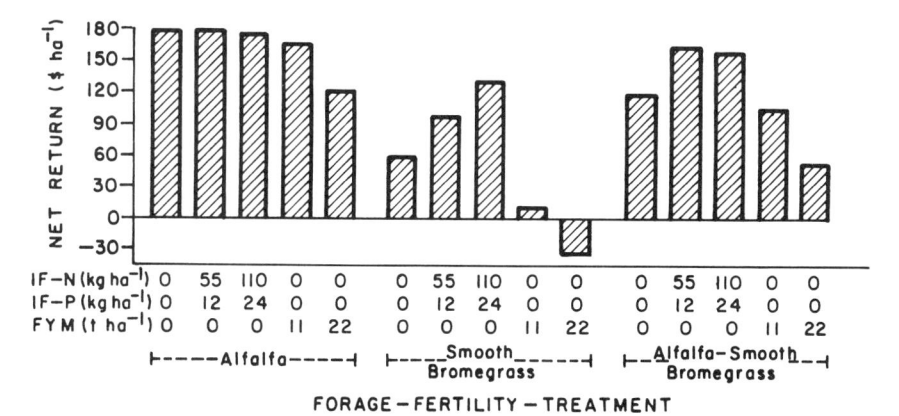

Fig. 10–8. Net returns ($ Canadian) of alfalfa, smooth bromegrass, and an alfalfa-smooth bromegrass mixture after applications of inorganic fertilizer (IF) and beef cattle manure (FYM) (87).

nance of a vegetative cover in the application areas is essential to the function of the renovation process. Species requirements include nutrient utilization, plant survival, and compatibility of general plant management demands with the application requirements.

In southwestern Saskatchewan, alfalfa survived applications of wastewater with high salt content when the wastewater was applied at rates which approximated the normal area rainfall amount plus a small excess for leaching. High rates of application in the complex landscape created areas with shallow water tables that lead to reduced alfalfa performance (91). Grasses can be more successful with high rates of water application, especially if a balanced fertility program is ensured. However, alfalfa is more flexible in supplemental fertility requirements than grasses, when the wastewater N concentration is low (23). Bole and Bell (22) found rates of N which approximated the N requirement did not result in potential groundwater contamination with NO_3^--N. The long-term effects on a site receiving waste water of good quality are probably not detrimental to continued crop growth (32).

In a greenhouse study, sewage sludge solids improved alfalfa yields when the crop was grown in an agricultural soil and in coal strip mine soil. The primary effect was amelioration of pH problems. Optimum rate of application was lower for alfalfa than tall fescue, probably due to differences in N utilization. Alfalfa was able to utilize larger amounts of nutrients than tall fescue as a result of its greater growth (196). This potential utilization is important for meeting the design requirements when sites will be receiving multiple applications.

ACKNOWLEDGMENT

Contribution from the Dep. of Agronomy, Pennsylvania Agric. Exp. Stn., The Pennsylvania State University, University Park, PA 16802 as no. 7729 in the Journal Series.

REFERENCES

1. Adams, R.S. 1980. Feeding forage to dairy cows. *In* Forage handbook for Pennsylvania. The Penn. State Univ. Coop. Ext. Serv., University Park, PA.
2. Alva, A.K., L.E. Lanyon, and K.T. Leath. 1985. Influence of P and K fertilization on phytophthora root rot or excess soil water injury of alfalfa cultivars. Commun. Soil Sci. Plant Anal. 16:229–243.
3. ----, ----, and ----. 1986. Production of alfalfa in Pennsylvania soils of differing wetness. Agron. J. 78:469–473.
4. Andrews, C.S. 1976. Effect of calcium, pH and nitrogen on the growth and chemical composition of some tropical and temperate pasture legumes. I. Nodulation and growth. Aust. J. Agric. Res. 27:611–623.
5. ----. 1977. The effect of sulphur on the growth, sulphur, and nitrogen concentrations of some tropical and temperate pasture legumes. Aust. J. Agric. Res. 28:807–820.
6. Azcon-Aguilar, C., and J.M. Barea. 1981. Field inoculation of medicago with V-A mycorrhiza and rhizobium in phosphate-fixing agricultural soil. Soil Biol. Biochem. 13:19–22.

7. Bailey, L.D. 1983. Effect of potassium fertilizer and fall harvests on alfalfa grown in the eastern Canadian prairies. Can. J. Soil Sci. 63:211-219.
8. Ball, J.A., and G. TenEyck. 1980. Top management of irrigated alfalfa produces top yields. Better Crops Plant Food 64 (Summer): 16-19.
9. ----, ----, and R.F. Nuttleman. 1980. Irrigated alfalfa research. Kansas State University, Ag Facts AF-51, Manhattan, KS.
10. Barber, S.A. 1968. Mechanisms of potassium absorption. p. 293-310. In V.J. Kilmer et al. (ed.) The role of potassium in agriculture. American Society of Agronomy, Madison, WI.
11. ----. 1984. Liming materials and practices. In F. Adams (ed.) Soil acidity and liming. 2nd ed. Agronomy 12:171-209.
12. Barney, R.F., Jr., L.P. Bush, and J.E. Leggett. 1984. Sulfur accumulation in tall fescue, orchardgrass, and alfalfa. Agron. J. 76:23-26.
13. Barta, A.L. 1983. Response of symbiotic N fixation and assimilate partitioning to K supply in alfalfa. Potash Review, Subject 7, International Potash Institute, Berne, Switzerland.
14. Baylor, J.E. 1974. Satisfying the nutritional requirements of grass legumes. p. 171-188. In D. A. Mays (ed.) Forage fertilization. American Society of Agronomy, Crop Science Society of America, and Soil Science Society of America, Madison, WI.
15. ----. 1983. Breaking alfalfa yield barriers: Research and farmer results. p. 1-10. In Proc. 29th Annual Farm Seed Conf., Kansas City, MO. 6 November. American Seed Trade Association, Washington, DC.
16. ----, and W.K. Waters. 1983. High alfalfa yields—inputs and returns. p. 106-111. In Proc. 1983 Forage Grassl. Conf., Madison, WI. 23-26 January. American Forage and Grassland Council, Lexington, KY.
17. Becana, M., P.M. Aparicio-Tejo, and M. Sanchez-Diaz. 1985. Levels of ammonia, nitrite, and nitrate in alfalfa root nodules supplied with nitrate. J. Plant Physiol. 119:359-367.
18. Bennett, D.L., and B.D. Doss. 1963. Effect of soil moisture regime on yield and evapotranspiration from cool season perennial forage species. Agron. J. 55:275-278.
19. Bergensen, F.J. 1971. Biochemistry of symbiotic N fixation in legumes. Ann. Rev. Plant Phys. 22:121.
20. Black, A.S., and K.C. Cameron. 1984. Effect of leaching on soil properties and lucerne growth following lime and gypsum amendments to a soil with an acid subsoil. N. Z. J. Agric. Res. 27:195-200.
21. Blaser, R.E., and E.L. Kimbrough. 1968. Potassium nutrition of forage crops with perennials. p. 423-445. In V.J. Kilmer et al. (ed.) The role of potassium in agriculture. American Society of Agronomy, Madison, WI.
22. Bole, J.B., and R.G. Bell. 1978. Land application of municipal sewage waste water: yield and chemical composition of forage crops. J. Environ. Qual. 7:222-226.
23. ----, W.D. Gould, and J.A. Carson. 1985. Yields of forage irrigated with waste water and the fate of added nitrogen-15 labeled fertilizer nitrogen. Agron. J. 77:715-719.
24. Bouma, D., E.S. Dowling, and D.J. David. 1981. Relation between aluminum content and the growth of lucerne and subterranean clover: their usefulness in the detection of aluminum toxicity. Aust. J. Exp. Agric. Anim. Husb. 21:311-317.
25. Brown, J.C., and J.H. Graham. 1978. Requirements and tolerance to elements by alfalfa. Agron. J. 70:367-373.
26. Bryant, 1983. How to establish alfalfa by no-till. Better Crops Plant Food 67 (Summer): 24-25.
27. Buchholz, D.D., and D.A. Whitney. 1985. Responsiveness of Kansas soil series to sulfur application under greenhouse conditions. Commun. Soil Sci. Plant Anal. 16:865-881.
28. Bula, R.J. 1972. Morphological characteristics of alfalfa plants grown at several temperatures. Crop Sci. 12:683-686.
29. Burns, J.C., C.L. Rhykerd, C.H. Noller, and K.R. Cummings. 1974. Influence of nitrogen, phosphorus, and potassium fertilization on the mineral concentrations in *Medicago sativa* L. I. Seasonal changes. Commun. Soil. Sci. Plant Anal. 5:247-259.
30. Buss, G.A., J.A. Lutz, Jr., and G.W. Hawkins. 1975a. Effect of soil pH and plant genotype on elemental concentration and uptake by alfalfa. Crop Sci. 15:614-617.
31. ----, ----, and ----. 1975b. Yield response of alfalfa cultivars and clones to several pH levels in Tatum subsoil. Agron. J. 67:331-334.
32. Campbell, W.F., R.W. Miller, J.H. Reynolds, and T.M. Schreeg. 1983. Alfalfa, sweetcorn, and wheat responses to long-term application of municipal waste water to cropland. J. Environ. Qual. 12:243-249.
33. Carlson, G.E. 1965. Photoperiodic control of adventitious stem initiation on roots. Crop Sci. 5:248-250.
34. Chamblee, D.S. 1958. The relative removal of soil moisture by alfalfa and orchardgrass. Agron. J. 50:587-589.

35. Coffindaffer, B.L., and D.J. Burger. 1958. Response of alfalfa varieties to daylength. Agron. J. 50:389–392.
36. Collins, M. 1983. Changes in composition of alfalfa, red clover, and birdsfoot trefoil during autumn. Agron. J. 75:287–290.
37. ----. 1985. Wilting and maturity effects on mineral concentrations in legume hay. Agron. J. 77:779–782.
38. ----, and S.H. Duke. 1981. Influence of potassium fertilization rate and form of photosynthesis and nitrogen fixation on alfalfa (*Medicago sativa*). Crop Sci. 21:481–485.
39. Cooke, G.W. 1978. Changing concepts on the use of potash. p. 361–405. *In* Potassium Research—Reviews and Trends. Proc. 11th Congr. Int. Potash Inst., Berne, Switzerland. 4–8 September. International Potash Institute, Berne.
40. Crocker, G.J., K.P. Sheridan, and I.C.R. Holford. 1985. Lucerne responses to lime and interactions with other nutrients on granitic soils. Aust. J. Exp. Agric. 25:337–346.
41. Devine, T.E., C.D. Foy, A.L. Fleming, C.H. Hanson, T.A. Campbell, J.E. McMurtry, and J.W. Schwarz. 1976. Development of alfalfa strains with differential tolerance to aluminum toxicity. Plant Soil. 44:73–79.
42. Dionne, J.L. 1980. Effets du magnesium et du pH du sol sur la lucerne cultivee en serre dans trois types de sol du Quebec. Can. J. Soil Sci. 60:274–284.
43. ----, and A.R. Pesant. 1978. Effects des doses de bore, des regimes hydriques et du pH des sols sur les rendements de la luzerne et du lotier et sur l'assimilabilite du bore. Can. J. Soil Sci. 58:369–379.
44. Doerge, T.A., P.J. Bottomley, and E.H. Gardner. 1985. Molybdenum limitations to alfalfa growth and nitrogen content on a moderately acid, high-phosphorus soil. Agron. J. 77:895–901.
45. Duell, R.W. 1974. Fertilizing forage for establishment. p. 67–93. *In* D. A. Mays (ed.) Forage fertilization. American Society of Agronomy, Crop Science Society of America, and Soil Science Society of America, Madison, WI.
46. Duke, S.H. 1983. Potassium and sulfur nutrition in alfalfa. p. 192–195. *In* Proc. Forage Grassl. Conf., Madison, WI. 23–26 January. American Forage and Grassland Council, Lexington, KY.
47. ----, M. Collins, and R.M. Soberalske. 1980. Effects of potassium fertilization on nitrogen fixation and nodule enzymes of nitrogen metabolism in alfalfa. Crop Sci. 20:213–219.
48. Duke, W.B., V.S. Rao, and J.F. Hunt. 1972. EPTC-atrazine residue interaction effect on seedling alfalfa varieties. Proc. N. E. Weed Sci. Soc. 26:258–262.
49. Eardly, B.D., D.B. Hannaway, and P.J. Bottomley. 1985. Nitrogen nutrition and yield of seedling alfalfa as affected by ammonium nitrate fertilization. Agron. J. 77:57–62.
50. Elliott, C.R., P.B. Hoyt, M. Nyborg, and B. Siemens. 1973. Sensitivity of several species of grasses and legumes to soil acidity. Can. J. Plant Sci. 53:113–117.
51. Evans, P.S. 1978. Plant root distribution and water use patterns of some pasture crop species. N. Z. J. Agric. Res. 21:261–265.
52. Fehrenbacher, J.B., B.W. Ray, and W.M. Edwards. 1965. Rooting volume of corn and alfalfa in shale-influenced soils in northwestern Illinois. Soil Sci. Soc. Am. Proc. 29:591–594.
53. Feigenbaum, S., and A. Hadas. 1980. Utilization of fertilizer nitrogen-nitrogen-15 by field grown alfalfa. Soil Sci. Soc. Am. J. 44:1006–1010.
54. ----, and K. Mengel. 1979. The effect of reduced light intensity and suboptimal potassium supply on N fixation and N turnover in Rhizobium infected lucerne. Buntehof Abstr. Buntehof Agric. Res. Stn., Hanover, Fed. Rep. of Germany, p. 29–30.
55. Fishbeck, K.A., and D.A. Phillips. 1981. Combined nitrogen and vegetative regrowth of symbiotically-grown alfalfa. Agron. J. 73:975–978.
56. Forbes, E.A. 1978a. Investigations into the availability to lucerne (*Medicago sativa*) of copper applied to yellow-brown pumice soils I. Farm survey of Cu in lucerne and soil. N. Z. J. Agric. Res. 21:629–636.
57. ----. 1978b. Investigations into the availability to lucerne (*Medicago sativa*) of copper applied to yellow-brown pumice soils II. Relationship between topdressing rate and Cu concentration in established lucerne. N. Z. J. Agric. Res. 21:637–642.
58. Fox, R.H. 1968. The effect of calcium and pH on boron uptake from high concentrations of boron by cotton and alfalfa. Soil Sci. 106:435–439.
59. ----, and W.P. Piekielek. 1983. Response of corn to nitrogen fertilizer and the prediction of soil nitrogen availability with chemical tests in Pennsylvania. Pennsylvania State Univ. Agric. Exp. Stn. Bull. 843, University Park, PA.
60. Foy, C.D. 1964. Toxic factors in acid soils of the southeastern United States as related to the response of alfalfa to lime. USDA Proc. Res. Rep. 80, U.S. Government Printing Office, Washington, DC.
61. Gates, C.T. 1974. Nodule and plant development in *Stylosanthes humilis*—A symbiotic response to P and S. Aust. J. Bot. 22:45–55.

62. Goos, R.J., B.E. Johnson, and C.A. Timm. 1984. Deep placement of phosphorus into established alfalfa. J. Fert. Issues 1:19–22.
63. Goss, D.W., and B.A. Stewart. 1979. Efficiency of phosphorus utilization by alfalfa from manure and superphosphate. Soil Sci. Soc. Am. J. 43:523–528.
64. Griffith, W.K. 1974. Satisfying the nutritional requirements of established legumes. p. 147–170. In D. Mays (ed.) Forage fertilization. American Society of Agronomy, Madison, WI.
65. ----. 1975. Factors affecting alfalfa stand persistence—the role of fertilizer. p. 75–87. In Fifth Annual Alfalfa Symp., Hershey, PA. 8 April. Certified Alfalfa Seed Council, Bakersfield, CA.
66. Gross, C.F., and G.A. Jung. 1981. Season, temperature, soil pH, and Mg fertilizer effects on heritage Ca and P levels and ratios of grasses and legumes. Agron. J. 73:629–634.
67. Gupta, U.C. 1969. Effect of interaction of molybdenum and limestone on growth and molybdenum content of cauliflower, alfalfa, and bromegrass on acid soils. Soil Sci. Soc. Am. Proc. 33:929–932.
68. ----. 1984. Boron nutrition of alfalfa, red clover and timothy grown on podzol soils of eastern Canada. Soil Sci. 137:16–22.
69. ----, and J.A. MacLeod. 1984. Effect of various sources of sulfur on yield and sulfur concentration of cereals and forages. Can. J. Soil Sci. 64:403–409.
70. ----, and R.C. Veinot. 1974. Response of corn to sulfur under greenhouse conditions. Soil Sci. Soc. Am. Proc. 38:785–788.
71. Hagin, J., and S. Katz. 1985. Effectiveness of partially acidulated phosphate rock as a science to plants in calcareous soils. Fert. Res. 8:117–128.
72. Hanson, R.G., N. Risner, and S.R. Maledy. 1984. Sulfur fertilization of two Aquic Hapludalf soils: I. Effect on alfalfa yield and quality. Commun. Soil Sci. Plant Anal. 15:227–237.
73. Harper, H.J. 1957. Effect of rainfall and fertilization on the yield and chemical composition of alfalfa over a 10 year period in northcentral Oklahoma. Soil Sci. Soc. Am. Proc. 21:47–51.
74. Harward, M.E., and H.M. Reisenauer. 1966. Reactions and movement of inorganic soil sulfur. Soil Sci. 101:326–335.
75. Havlin, J.L., D.G. Westfall, and H.M. Golus. 1984. Six years of phosphorus and potassium fertilization of irrigated alfalfa on calcareous soils. Soil Sci. Am. J. 48:331–336.
76. Heichel, G.H., and C.P. Vance. 1979. Nitrate-N and *Rhizobium* strain roles in alfalfa seedling nodulation and growth. Crop Sci. 19:512–518.
77. ----, ----, and D.K. Barnes. 1981. Symbiotic nitrogen fixation of alfalfa, Birdsfoot trefoil, and red clover. p. 336–338. In J.A. Smith and V.W. Hays (ed.) Proc. 14th Int. Grassl. Congr. Lexington, KY. 18–23 June. Westview Press, Boulder, CO.
78. Helyar, K.R., and A.J. Anderson. 1971. Effects of lime on growth of five species on aluminum toxicity and on phosphate availability. Aust. J. Agric. Res. 22:707–721.
79. Hensler, R.F., R.J. Olsen, S.A. Witzel, O.J. Atloe, W.H. Paulson, and R.F. Johannes. 1970. Effect of method of manure handling on crop yields, nutrient recovery, and runoff losses. Trans. ASAE 13:726–731.
80. Hill, R.R., Jr., and G.A. Jung. 1975. Genetic variability for chemical composition of alfalfa; mineral elements. Crop Sci. 15:652–657.
81. ----, and L.E. Lanyon. 1983. Phosphorus fertilizer response in experimental alfalfas selected for different phosphorus concentrations. Crop Sci. 23:973–976.
82. Hiltbold, A.E., and G.A. Buchanan. 1977. Influence of soil pH on persistence of atrazine in the field. Weed Sci. 25:515–520.
83. Hoeft, R.G., and L.M. Walsh. 1975. Effect of carrier, rate, and time of application of S on the yield, S and N content of alfalfa. Agron. J. 67:427–430.
84. ----, ----, and D.R. Keeney. 1973. Evaluation of various extracts for available soil sulfur. Soil Sci. Soc. Am. Proc. 37:401–404.
85. Hojjati, S.M., W.C. Templeton, Jr., and J.H. Taylor. 1978. Nitrogen fertilization in establishing forage legumes. Agron. J. 70:429–433.
86. Holland, C.W., and M.B. Tesar. 1981. Establishment of alfalfa in quackgrass sods with herbicides using conventional and sod seedling methods. Mich. Agric. Exp. Stn. Res. Rep. 420:9–18.
87. Holt, N.W., and R.P. Zenter. 1985. Effects of applying inorganic fertilizer and farmyard manure on forage production and economic returns in eastcentral Saskatchewan. Can. J. Plant Sci. 65:597–607.
88. Horsnell, L.J. 1985a. The growth of improved pastures on acid soils. I. The effect of superphosphate and lime on soil pH and on establishment of growth of Phalaris and lucerne. Aust. J. Exp. Agric. 25:149–156.
89. ----. 1985b. The growth of improved pastures on acid soils. II. The effect of lime

and phosphorus and their incorporation in acid soil on the growth of subterranean clover and lucerne pastures and on their response to superphosphate topdressing. Aust. J. Exp. Agric. 25:157–163.
90. ——. 1985c. The growth of improved pastures on acid soils. III. Response of lucerne to phosphate as effected by calcium and potassium sulfates and soil aluminum levels. Aust. J. Exp. Agric. 25:557–561.
91. Jame, Y.W., V.O. Biederbuck, W. Nicholaichuk, and H.C. Korven. 1984. Salinity and alfalfa yield under effluent irrigation in southwestern Saskatchewan. Can. J. Soil Sci. 64:323–332.
92. James, D.W., T.L. Jackson, and M.E. Howard. 1968. Effect of molybdenum and lime on the growth and molybdenum content of alfalfa grown on acid soils. Soil Sci. 105:397–402.
93. ——, W.H. Weaver, S. Roberts, and A.H. Hunter. 1975. Potassium in an arid loessial soil: Changes in availability as related to cropping and fertilization. Soil Sci. Soc. Am. Proc. 39:1111–1115.
94. Janghorbani, M., S. Roberts, and T.L. Jackson. 1975. Relationship of exchangeable acidity to yield and chemical composition of alfalfa. Agron. J. 67:350–354.
95. Jenkins, M.B., and P.J. Bottomley. 1984. Seasonal response of uninoculated alfalfa to nitrogen fertilizer. Agron. J. 76:959–963.
96. John, M.K., G.W. Eaton, V.W. Case, and H.H. Chuah. 1972. Liming of alfalfa (*Medicago sativa* L.) II. Effect on mineral composition. Plant Soil 37:363–374.
97. Jones, G.D., and W.W. Moschler. 1966. Alfalfa response to molybdenum on Tatum soil. Research Rep. 115. Virginia Polytechnic Institute, Blacksburg.
98. Jones, R.M. 1970. Sulphur deficiency of dryland lucerne in the eastern Darling Downs of Queensland. Aust. J. Exp. Agric. Anim. Husb. 10:749–754.
99. Jung, G.A., R.L. Reid, and J.A. Balasko. 1969. Studies on yield, management, persistence, and nutritive value of alfalfa in West Virginia. W. Va. Univ. Agric. Exp. Stn. Bull. 581T.
100. ——, and D. Smith. 1959. Influence of soil K and P content on cold resistance of alfalfa. Agron. J. 51:585–587.
101. Kamprath, E.J. 1971. Potentially detrimental effects from liming highly weathered soils to neutrality. Soil Crop Sci. Soc. Fla. Proc. 31:200–203.
102. Kapland, D.R., and G.O. Estes. 1985. Organic matter relationship to soil nutrient status and aluminum toxicity in alfalfa. Agron. J. 77:735–738.
103. Kelling, K.A. 1982. Alfalfa fertilization. Univ. of Wisconsin, Madison Ext. Pub. A2448.
104. ——, T. Erickson, and E.E. Schulte. 1981. DRIS-a new approach to alfalfa tissue analysis; results to date. p. 55–69. *In* Proc. Of the 1981 Alfalfa Forum, Annual Meeting of the Certified Alfalfa Seed Counc., Madison, WI. 30 March–1 April. Certified Alfalfa Seed Council, Woodland, CA.
105. ——, E.E. Schulte, and T. Erickson. 1986. Adapting DRIS for alfalfa: What are the diagnostic norms? Better Crops Plant Food 70 (Winter):18–10.
106. Kesmarki, I. 1981. A "blueprint" for maximizing yields of alfalfa. p. 261–272. *In* Proc. 16th Colloquium of the Int. Potash Inst. Berne, Switzerland.
107. Kilmer, V.J., O.L. Bennett, V.F. Stahly, and D.R. Timmons. 1960. Yield and mineral composition of eight forage species grown at four levels of soil moisture. Agron. J. 52:282–285.
108. Kimbrough, E.L., R.E. Blaser, and D.D. Wolf. 1971. Potassium effects on regrowth of alfalfa. Agron. J. 63:836–839.
109. Kohl, R.A., and J.J. Kolar. 1976. Soil water uptake by alfalfa. Agron. J. 68:536–538.
110. Kucey, R.M.N., and G.E.S. Diab. 1984. Effects of lime, phosphorus, and addition of vesicular-arbuscular (VA) mycorrhizal fungi on indigenous VA fungi and/or growth of alfalfa in a moderately acidic soil. New Phytol. 98:481–486.
111. Kunelius, H.T. 1974. Effects of weed control and N fertilization at establishment on the growth and nodulation of alfalfa. Agron. J. 66:806–809.
112. Lacefield, G.D., L.W. Murdock, and W.A. Talley. 1982. High yield alfalfa studies. p. 72–73. *In* 95th Annual Report. Univ. of Kentucky, Lexington, Agric. Exp. Stn.
113. Lanyon, L.E., J.E. Baylor, and W.K. Waters. 1983. Understanding alfalfa nutrient uptake. Better Crops Plant Food 67 (Summer):12–14.
114. ——, B. Naghshineh-Pour, and E.O. McLean. 1977. Effects of pH level on yields and composition of pearl millet and alfalfa in soils with differing degrees of weathering. Soil Sci. Soc. Am. J. 41:389–394.
115. ——, and F.W. Smith. 1985. Potassium fertilization of alfalfa and other forage legumes. p. 861–893. *In* R.D. Munson (ed.). Potassium in agriculture. 2nd ed. American Society of Agronomy, Madison, WI.
116. Lathwell, D.L., and M. Peech. 1964. Interpretation of chemical soil tests. Cornell Univ. Agric. Exp. Stn. Bull. 995.
117. Lee, C.-T., and D. Smith. 1972. Influence of nitrogen fertilizer on stands, yields of

herbage and protein, and nitrogenous fractions of field grown alfalfa. Agron. J. 64:527–530.
118. Leyshon, A.S. 1982. Deleterious effects on yield of drilling fertilizer into established alfalfa. Agron. J. 74:741–743.
119. Long, O.M., L.M. Sofley, J.A. Odom, and H. Morgan, Jr. 1972. Rates and depths of mixing lime and fertilizer for alfalfa. Tenn. Agric. Exp. Stn. Bull. 495.
120. Lopez-Jurado, G., and D.B. Hannaway. 1985. Sulfur nutrition effects on dinitrogen fixation of seedling alfalfa. J. Plant Nutr. 8:1103–1122.
121. Lynd, J.Q., G.V. Odell, Jr., and R.W. McNew. 1981. Soil potassium effects on nitrogenase activity with associated nodule components of hairy vetch at anthesis. J. Plant Nutr. 4:303–318.
122. MacDowall, F.D.H. 1985. Influence of nitrogen supply in developmental kinetics of symbiotic and nonsymbiotic alfalfa cv. Algonquin through first flowering. Can. J. Bot. 63:841–846.
123. MacLean, A.J., R.L. Halstead, and B.J. Finn. 1972. Effects of lime on extractable aluminum and other soil properties on barley and alfalfa grown in pot tests. Can. J. Soil Sci. 52:427–438.
124. Mahler, R.L. 1983. Influence of pH on yield and N and P nutrition of alfalfa grown on an andic Mission silt loam. Agron. J. 75:731–735.
125. Mallarino, A.P., W.F. Wedin, R.D. Voss, and C.P. West. 1983. Phosphorus requirements of alfalfa-smooth bromegrass-orchardgrass and reed canarygrass pastures under two grazing pressures. Agron. J. 75:291–294.
126. Martel, Y.A., and J. Zizka. 1977. Yield and quality of alfalfa as influenced by additions of S to P and K fertilization under greenhouse conditions. Agron. J. 69:531–535.
127. Marten, G.C. 1970. Temperature as a determinant of quality of alfalfa harvested by bloom stage or age criteria. p. 506–509. In Proc. Int. Grassl. Congr. 11th, New South Wales, Australia. American Forage and Grassland Council, Lexington, KY.
128. Martin, W.E., and J.E. Matocha. 1973. Plant analysis as an acid in the fertilization of forage crops. p. 393–426. In L.M. Walsh and J.D. Beaton (ed.) Soil testing and plant analysis, rev. ed. Soil Science Society of America, Madison, WI.
129. Matches, A.G., G.O. Mott, and R.J. Bula. 1962. Vegetative development of alfalfa seedlings under varying levels of shading and potassium fertilization. Agron. J. 54:541–543.
130. Mathers, A.C., B.A. Stewart, and R. Blair. 1975. Nitrate-nitrogen removal from soil profiles by alfalfa. J. Environ. Qual. 4:403–405.
131. McKenzie, R.C., and M. Nyborg. 1984. Influence of subsoil activity on root development and crop growth in soils of Alberta and northeastern British Columbia. Can. J. Soil Sci. 64:681–697.
132. McLaren, R.G., R.S. Swift, and B.F. Quin. 1984. EDTA-extractable copper, zinc and manganese in soils of the Canterbury Plains. N. Z. J. Agric. Res. 27:207–217.
133. McLean, E.O. 1971. Potentially beneficial effects from liming: chemical and physical. Soil Crop Sci. Soc. Fla. Proc. 31:189–196.
134. ----. 1976. Exchangeable K levels for maximum crop yields on soils of different cation exchange capacities. Commun. Soil Sci. Plant Anal. 7:823–838.
135. ----, and J.R. Brown. 1984. Crop response to lime in the midwestern United States. In F. Adams (ed.) Soil acidity and liming. 2nd ed. Agronomy 12:267–303.
136. Melsted, S.W., H.L. Motto, and T.R. Peck. 1969. Critical plant nutrient composition values useful in interpreting plant analysis data. Agron. J. 61:17–20.
137. Meyer, R.D., and D. Marcum. 1980. Alfalfa response to rate and source of sulfur in California. Sulphur Agric. 4:23–24.
138. Miller, D.A., and R.K. Smith. 1977. Influence of boron on other chemical elements in alfalfa. Commun. Soil Sci. Plant Anal. 8:465–478.
139. Mortenson, L.W. 1966. Components of cell-free extracts of clostridium pasteurianum required for ATP-dependent H_2 evolution from dithionite and for N fixation. Biochem. Biophys. Acta. 127:18.
140. Mortvedt, J.J. 1981. Nitrogen and molybdenum uptake and dry matter relationships of soybeans and forage legumes in response to applied molybdenum on acid soil. J. Plant Nutr. 3:245–256.
141. Mueller, J.P. 1981. Alfalfa yield potential in the Southeast. Better Crops Plant Food (Spring):3–5.
142. Munns, D.N. 1970. Modulation of *Medicago sativa* in solution culture. V. Calcium and pH requirements during infection. Plant Soil 32:90–102.
143. ----, and R.L. Fox. 1976. Depression of legume growth by liming. Plant Soil 45:701–705.
144. ----, and ----. 1977. Comparative lime requirements of tropical and temperate legumes. Plant Soil 46:533–548.
145. Nearpass, D.C. 1972. Hydrolysis of propazine by surface acidity of organic matter. Soil Sci. Soc. Am. Proc. 36:606–610.

146. Ngruyen, S.T., R. Paguin, L.J. O'Grady, and G. Ouelette. 1972. Influence de la fertilisation azote, phosphate et potassique sur l'incorporation des acides amines aux proteines et les rendements de la lucerne. Can. J. Plant Sci. 52:41–52.
147. Nielsen, J.P., and A. Jensen. 1983. Influence of vesicular-arbuscular mycorrhiza fungi on growth and uptake of various nutrients as well as uptake ratio of fertilizer P for lucerne (*Medicago sativa*). Plant Soil 70:165–172.
148. Nuttall, W.F. 1985. Effect of N, P, and S fertilizer on alfalfa grown on three soil types in northeastern Saskatchewan Canada. I. Yield and soil tests. Agron. J. 77:41–46.
149. ----, D.A. Cooke, J. Waddington, and J.A. Robertson. 1980. Effect of nitrogen and phosphorus fertilizer on a bromegrass and alfalfa mixture grown under two systems of pasture management. 1. Yield, percentage legume in sward and soil tests. Agron. J. 72:289–294.
150. Olsen, R.J., R.F. Hensler, O.J. Attoe, S.A. Witzel, and L.A. Peterson. 1970. Fertilizer nitrogen and crop rotation in relation to movement of nitrate nitrogen through soil profiles. Soil Sci. Soc. Am. Proc. 39:448–452.
151. Overdahl, C.J. 1972. Fertilizer experiments with alfalfa on a Brainard sandy loam. Univ. of Minn. Rep. 107.
152. Pearson, C.J., and L.A. Hunt. 1972. Effects of temperature on primary growth of alfalfa. Can. J. Plant Sci. 52:1007–1015.
153. Pearson, R.W., and C.S. Hoveland. 1974. Lime needs of forage crops. p. 301–322. *In* D.A. Mays (ed.) Forage fertilization. American Society of Agronomy, Madison, WI.
154. Peoples, T.R., and D.W. Koch. 1979. Role of potassium in carbon dioxide assimilation in *Medicago sativa* L. Plant Physiol. 63:878–881.
155. Peters, E.J., and J.F. Stritzke. 1970. Herbicides and nitrogen fertilizer for the establishment of three varieties of spring-sown alfalfa. Agron. J. 62:259–262.
156. Peterson, L.A., and D. Smith. 1973. Recovery of K$_2$SO$_4$ by alfalfa after placement at different depths in a low fertility soil. Agron. J. 65:769–772.
157. Porter, T.K., and J.H. Reynolds. 1975. Relationship of alfalfa cultivar yield to specific leaf weight, plant density, and chemical composition. Agron. J. 67:625–629.
158. Potash and Phosphate Institute. 1983. Alfalfa production for maximum profit. p. 97–106. *In* Maximum economic yield manual. Atlanta, GA.
159. Probert, M.E., and R.K. Jones. 1977. The use of soil analysis for predicting the response to sulphur of pasture legumes in the Australian tropics. Aust. J. Soil. Res. 15:137–146.
160. Quigley, E.H., and G.A. Jung. 1984. Alfalfa and corn response to sulfur fertilization on three Pennsylvania soils. Commun. Soil Sci. Plant Anal. 15:213–226.
161. Read, D.W.L., and C.A. Campbell. 1981. Bio-cycling of phosphorus in soil by plant roots. Can. J. Soil Sci. 61:587–589.
162. Rechcigl, J.E., and R.B. Reneau, Jr. 1984. Effect of subsurface acidity on alfalfa in a Tatum clay loam. Commun. Soil Sci. Plant Anal. 15:811–818.
163. ----, ----, and D.E. Starner. 1985a. Effect of subsurface amendments and irrigation on alfalfa growth. Agron. J. 77:72–75.
164. ----, D.D. Wolf, R.B. Reneau, Jr., and W. Kroantje. 1985b. Influence of surface liming on the yield and nutrient concentration of alfalfa established using no-tillage techniques. Agron. J. 77:956–959.
165. Rehm, G.W., J.T. Nichols, R.C. Sorensen, and W.J. Moline. 1975. Yield and botanical composition of an irrigated grass-legume pasture as influenced by fertilization. Agron. J. 67:64–68.
166. Reid, D.J., D.J. Lathwell, and M.J. Wright. 1965. Yield and carbohydrate responses of alfalfa seedlings grown at several levels of K fertilization. Agron. J. 57:434–437.
167. Reid, R.L., and G.A. Jung. 1974. Effects of elements other than nitrogen on the nutritive value of forages. p. 395–435. *In* D. Mays (ed.) Forage fertilization. American Society of Agronomy, Madison, WI.
168. Renger, M., and O. Strebel. 1979. Water and nutrient transport to plant roots as a function of depth and time under field conditions. p. 65–77. *In* 14th Colloquium. International Potash Institute, Berne, Switzerland.
169. Rhykerd, C.L., and C.J. Overdahl. 1972. Nutrition and fertilizer use. *In* C.H. Hanson (ed.) Alfalfa science and technology. Agronomy 15:437–465.
170. ----, K.L. Washburn, Jr., and C.H. Noller. 1970. Should N be applied to alfalfa? AY-184, Agronomy Guide, Purdue University, West Lafayette, IN.
171. Rice, W.A. 1975. Effect of CaCO$_3$ and inoculum level on nodulation and growth of alfalfa in an acid soil. Can. J. Soil Sci. 55:245–250.
172. Rominger, R.S., D. Smith, and L.A. Peterson. 1975a. Changes in elemental concentrations in alfalfa herbage at two soil fertility levels with advance in maturity. Commun. Soil Sci. Plant Anal. 6:163–180.
173. ----, ----, and ----. 1975b. Yields and elemental composition of alfalfa plant parts at late bud under two fertility level. Can. J. Plant Sci. 55:69–75.

174. ----, ----, and ----. 1977. Influence of high rates of topdressed KCl and K$_2$SO$_4$ on recovery of KCl and SO$_4$-S by alfalfa and residual amount in the soil. Commun. Soil Sci. Plant Anal. 8:489-507.
175. Roth, R.L., B.R. Gardner, G.K. Tritz, and E.A. Lakatos. 1983. Alfalfa response to water and nitrogen variables. 13th California Alfalfa Symposium. Calif. Coop. Ext. Serv., University of California-Davis.
176. Russelle, M.P., and C.C. Sheaffer. 1986. Use of the diagnosis and recommendation integrated system with alfalfa. Agron. J. 78:557-560.
177. Schultz, I., M.A. Turner, and J.G. Cooke. 1979. Effects of fertilizer potassium:sodium ratio on yield and concentration of sodium, potassium, and magnesium in lucerne and white clover. N. Z. J. Agric. Res. 22:303-308.
178. Schertz, D.L., and D.A. Miller. 1972. Nitrate N accumulation in soil profile under alfalfa. Agron. J. 64:660-664.
179. Seim, E.C., A.C. Caldwell, and G.W. Rehm. 1969. Sulfur response by alfalfa (*Medicago sativa* L.) on a sulfur-deficient soil. Agron. J. 61:368-371.
180. Sheaffer, C.C. 1984. Potash for irrigated alfalfa boosts yield and reduces stand loss. Better Crops Plant Food (Summer):8-9.
181. ----, M.P. Russells, D.B. Hestermann, and R.E. Stucker. 1986. Alfalfa response to potassium, irrigation, and harvest management. Agron. J. 78:464-468.
182. Sheard, R.W. 1980. Nitrogen in the P band for forage establishment. Agron. J. 72:89-99.
183. ----, G.J. Bradshaw, and D.L. Massey. 1971. Phosphorus placement for the establishment of alfalfa and bromegrass. Agron. J. 63:922-927.
184. Simson, C.R., R.B. Corey, and M.E. Sumner. 1979. Effect of varying Ca:Mg ratios on the yield and composition of corn and alfalfa. Commun. Soil Sci. Plant Anal. 10:153-162.
185. Simpson, J.R., and J. Lipsett. 1973. Effects of surface moisture supply on the subsoil nutritional requirements of lucerne (*Medicago sativa* L.). Aust. J. Agric. Res. 24:199-209.
186. Smith, D. 1968. The establishment and management of alfalfa. Univ. of Wisconsin Bull. 542. Madison, WI.
187. ----. 1970. Influence of temperature on the yield and chemical composition of five forage legume species. Agron. J. 62:520-523.
188. ----. 1971. Levels and sources of potassium for alfalfa as influenced by temperature. Agron. J. 63:497-500.
189. ----. 1975. Effects of topdressing in a low fertility silt loam soil on alfalfa herbage yields and composition and on soil K values. Agron. J. 67:60-64.
190. ----, and R.D. Powell. 1979. Yield of alfalfa as influenced by levels of P and K fertilization. Commun. Soil Sci. Plant Anal. 10:531-543.
191. ----, and A.K. Dobrenz. 1982. What are nutrient needs of high yield alfalfa. Better Crops Plant Food 66 (Summer):31-33.
192. Smith, G.S., D.R. Lauren, I.S. Cornforth, and M.P. Agnew. 1982. Evaluation of putrescine as biochemical indicator of the potassium requirements of lucerne. New Phytologist 91:419-428.
193. Soltanpour, P.N., and S.M. Workman. 1980. Use of NH$_4$HCO$_3$-DTPA soil test to assess availability and toxicity of selenium to alfalfa plants. Commun. Soil Sci. Plant Anal. 11:1147-1156.
194. Sorenson, R.C., E.P. Penas, and U.U. Alexander. 1968. Sulfur content and yield of alfalfa in relation to plant nitrogen and sulfur fertilization. Agron. J. 60:20-24.
195. Sparks, D.L., D.C. Martens, and L.W. Zelazny. 1980. Plant uptake and leaching of applied and indigenous potassium in Dothan soils. Agron. J. 72:551-555.
196. Stuckey, D.J., and T.S. Newman. 1977. Effect of dried anaerobically digested sewage sludge on yield and element accumulation in tall fescue and alfalfa. J. Environ. Qual. 6:271-274.
197. Sumner, M.E. 1979. Response of alfalfa and sorghum to lime and P on highly weathered soils. Agron. J. 71:763-766.
198. Taylor, R.W., and D.W. Allison. 1979. Cadmium, copper, lead, nickel, and zinc concentrations in alfalfa in Connecticut. Res. Rep. 55. Conn. Agric. Exp. Stn., The University of Connecticut, Storrs, CT.
199. Tesar, M.B. 1973. Harvest schedules to maintain alfalfa stands. p. 20-27. *In* Proc. 3rd Annual Alfalfa Symp., Certified Alfalfa Seed Council, Bakersfield, CA.
200. ----. 1981. High yield alfalfa research. p. 1-4. *In* Proc. Weed, Seed and Fertilizer Conf., Lansing, MI. 10 December. Michigan State University, Lansing.
201. ----. 1983a. Ten tons alfalfa without irrigation—a new record research yield. Better Crops Plant Food (Spring) 67:6-7.
202. ----. 1983b. Research leading to a yield of ten tons of alfalfa without irrigation. p. 1-10. *In* Proc. 19th Central Alfalfa Improvement Conf., Manhattan, KS. 8-10 June. Kansas State University, Manhattan.

203. ——. 1984. Ten tons alfalfa without irrigation in Michigan, a new world record. p. 59–65. *In* Proc. 14th Annu. Alfalfa Symp., East Lansing, MI. 21 March. Michigan State University, East Lansing.
204. ——. 1985. Fertilization and management for a yield of ten tons of alfalfa without irrigation. p. 327–333. *In* Proc. 1985 Forage Grassl. Conf., Hershey, PA. 3–6 March. American Forage and Grassland Council, Lexington, KY.
205. ——, K. Lawton, and B. Kawin. 1954. Comparison of band seeding and other methods of seeding legumes. Agron. J. 46:189–194.
206. Thom, W.D. 1984. Fertilizing alfalfa for optimum yields. Soil Sci. News & Views 5(10). Dep. of Argon., Univ. of Kentucky, Lexington, KY.
207. Upchurch, R.P., and R.L. Lovvorn. 1951. Gross morphological root habits of alfalfa in North Carolina. Agron. J. 43:493–498.
208. Van Riper, G.W. 1964. Influences of soil moisture on the herbage of two legumes and three grasses as related to dry matter yields, crude protein, and botanical composition. Agron. J. 56:45–50.
209. Viets, F.G. 1962. Fertilizers and the efficient use of water. Adv. Agron. 14:223–264.
210. Vincent, J.M. 1965. Environmental factors in the fixation of nitrogen by the legume. *In* W.V. Bartholomew and F.E. Clark (ed.) Soil nitrogen. Agronomy 7:384–435.
211. Walker, D.R., and G. Doornenbal. 1972. Soil sulphate. II. As an index of the sulphur available to legumes. Can. J. Soil Sci. 52:261–266.
212. Washko, J.B., and J.W. Price. 1970. Intensive management of alfalfa for forage production. p. 628–632. *In* Proc. 11th Int. Grassl. Congr., Queensland, Australia.
213. Westerman, D.T. 1974. Indexes of sulphur deficiency in alfalfa. I. Extractable soil SO_4-S. Agron. J. 66:576–581.
214. ——. 1975. Indexes of sulfur deficiency in alfalfa. II. Plant analysis. Agron. J. 67:265–268.
215. White, J.G.H. 1967. Establishment of lucerne on acid soils. p. 105–114. *In* R.H.M. Langer (ed.) The lucerne crop. Reed, Wellington, New Zealand.
216. White, P.J., M.J. Whitehouse, L.A. Warrell, and P.R. Burrill. 1981. Field calibration of a soil sulfate test on sward lucerne on the eastern Darling Downs, Queensland. Aust. J. Exp. Agric. Anim. Husb. 21:303–310.
217. Willett, J.R., P. Jakobsen, and B.A. Zarcinas. 1985. Nitrogen-induced boron deficiency in lucerne. Plant Soil 86:443–446.
218. Wolf, D.D., E.L. Kimbrough, and R.E. Blaser. 1976. Photosynthetic efficiency of alfalfa with increasing potassium nutrition. Crop Sci. 16:292–294.
219. ——, and H.E. White. 1984. No-till alfalfa establishment. p. 261–264. *In* Proc. Am. Forage Grassl. Counc. American Forage and Grassland Council, Lexington, KY.
220. Woodruff, C.M. 1967. Crop response to lime in the midwestern United States. *In* R.W. Pearson and F. Adams (ed.) Soils acidity and liming. Agronomy 12:207–231.
221. Woodhouse, W.W., and W.K. Griffith. 1973. Soil fertility and fertilization of forages. p. 403–415. *In* M.E. Heath (ed.) Forages. 3rd ed. The Iowa State University Press, Ames.

11 Alfalfa Water Relations and Irrigation

C. C. SHEAFFER

University of Minnesota
St. Paul, Minnesota

C. B. TANNER

University of Wisconsin
Madison, Wisconsin

M. B. KIRKHAM

Kansas State University
Manhattan, Kansas

Morphological and physiological features allow alfalfa (*Medicago sativa* L.) to adapt to a wide range of soil moisture conditions. In the humid regions of the USA, rainfall is usually sufficient for economic production; however, irrigation is often required for maximum production. In semi-arid and arid regions, irrigation is essential. Under moisture-limiting environments, alfalfa persists by extracting water from deep within the soil profile and by becoming semidormant. Despite its survival capabilities, growth is sensitive to water deficits in the upper soil profile.

Alfalfa water use is often considered extravagant, since seasonal evapotranspiration is large compared to other crops, primarily because of long periods of transpiration. Water requirements for production vary considerably depending mainly on climate and soil water supply and to a lesser extent on cultivar selection and management practices. Daily evapotranspiration rates approximate those of other crops with full ground cover. We will discuss the pathway of water movement, physiological and morphological effects of soil moisture deficits, water requirements for seed and forage production, and irrigation scheduling criteria.

11-1 PHYSIOLOGICAL EFFECTS OF WATER DEFICITS

11-1.1 Water Potential

The water status of plants, organs, and tissues is described either by the water potential (ψ_w) or less often by the relative water content (water content relative to that at full turgor). The components of the ψ_w include

---

Copyright 1988 © ASA-CSSA-SSSA, 677 South Segoe Road, Madison, WI 53711, USA.
Alfalfa and Alfalfa Improvement—Agronomy Monograph no. 29.

the solute contribution expressed as the osmotic potential (ψ_π) and the pressure potential ψ_p with $\psi_w = \psi_p + \psi_\pi$. In the protoplasm, ψ_p is the turgor potential and it usually is assumed that $\psi_p \geq 0$. In the absence of water flux across the plasmalemma, ψ_w of the wall and protoplasm are equal. Since ψ_π in the wall is much less than that in the protoplasm, in the wall $\psi_p \leq 0$ except during guttation. A third component, the matric potential (ψ_τ) often is added (e.g., $\psi_w = \psi_\pi + \psi_p + \psi_\tau$); however, the macroscopic measurements of water potential components include matric effects in ψ_π or ψ_p (138).

Increases in solute concentration are a mechanism for partial or complete turgor maintenance as ψ_w decreases. Potential benefits related to turgor pressure maintenance include delayed stomatal closure, which allows photosynthesis to continue despite reduced ψ_w, and in some instances continued leaf, stem, and root growth. Osmotic adjustment has been identified in many plant species (e.g., 208) and may occur in alfalfa (222). Research on osmotic adjustment as a mechanism for increased drought adaptation in alfalfa is warranted.

Brown and Tanner (23, 24) described methods specific to alfalfa for measurement of ψ_w and ψ_π. They found that when sap was extracted from frozen and thawed alfalfa leaves for ψ_π measurement, ψ_π increased if extraction was delayed after thawing (also 71). Subsequent unpublished work at Wisconsin has shown that when alfalfa leaflets are frozen in thin plastic bags (fingers of plastic gloves are suitable), thawed rapidly by placing the bagged leaflets between blocks of aluminum, and sap expressed immediately, no change of the sap ψ_π occurs with time. Additional methods of measuring ψ_w and its components in diverse plant species are reviewed by Baughn and Tanner (11), Campbell et al. (31), Neumann et al. (131), Richter (157), Richter et al. (158), Turner (205), and Tyree et al. (209).

11-1.2 Photosynthesis and Respiration

Although considerable research has been conducted on photosynthesis and respiration in alfalfa (see Chapter 6 in this book), few studies have evaluated the effects of moisture stress on these processes in alfalfa. However, since stomata regulate CO_2 exchange, water deficits that influence stomatal activity invariably affect photosynthesis. Also, the transfer and fixation of CO_2 internal to the stomata may be affected by water stress (206). Murata et al. (127) reported that photosynthesis and respiration of alfalfa were both reduced 40% due to plant stress induced by a soil matric potential (ψ_m) of -0.45 MPa. In another study, drought imposed by reducing soil moisture content to 37 to 40% of field capacity for 10 d at bud, flowering, and seed-filling stages decreased photosynthetic productivity by 22 to 35% and transpiration (T) by 20 to 28% (151). Begg and Turner (12) and Turner and Begg (206) summarized effects of moisture deficits on forage species (excluding alfalfa) and concluded that dark respiration and photorespiration were depressed whenever water deficit was sufficient to close stomata and decrease photosynthesis, but that the relative decrease in respiration was less than that of net photosynthesis. Although water stress may influence assimilate movement and distribution in plants because of changes in source and/or sink activity; specific data for alfalfa are lacking.

11-1.3 Symbiotic Relationships

Soil and plant moisture deficits depress legume symbiotic N_2 fixation. Survival, multiplication, and movement of rhizobia responsible for the development of the symbiotic relationship required for N_2 fixation (see Chapter 7 in this book) are reduced by soil water deficits (181). As a consequence, root hair infection and nodule initiation may be reduced or restricted to sites near the crown (65).

In plants with developed nodules, N_2 fixation is sensitive to moisture stress (36). Nitrogenase specific activity declined linearly until ψ_w reached -3.0 MPa when rates approached zero (Fig. 11-1). In general, nodules from plants under moisture stress have the same anatomy as those from well-watered plants, although nodule number per plant, nodule mass, and nodule size are reduced (36, 181). Nodules subjected to severe water stress resume activity when soil moisture content is restored. Under extreme moisture stress, nodule shedding may occur. Alfalfa nitrogenase activity decreased 85% when plants were subject to a water deficit (ψ_w of -2.0 MPa), but activity recovered to 70% of prewater deficit rates when turgor was restored (4). Engin and Sprent (52) reported that in white clover (*Trifolium repens* L.), which has an elongated nodule similar to that of alfalfa, recovery from stress was a two-phase process. The first phase involved rehydration of the nodule and the second involved renewed meristematic activity and production of new N_2-fixing tissue. Sprent (181) reviewed the effects of moisture stress on legume nodule structure and physiology; however, details relating specifically to alfalfa response are lacking.

Carter and Sheaffer (36) evaluated the relative effects of harvesting and plant water deficits on alfalfa N_2 fixation during two successive harvest/regrowth cycles. They concluded that nodule number and nitrogenase specific activity for water-stressed alfalfa were more closely associated with decreases in ψ_w and ψ_m than to either removal of herbage or concentration of root carbohydrates.

The principal cause of reduced nitrogenase activity under water stress may be the decline in photosynthesis that accompanies drought. A close relationship exists among photosynthesis, transpiration, and nitrogenase activity as ψ_w decreases (65). Photosynthesis, however, appears to recover rapidly following water stress, while there is a 1 to 2 d delay in the recovery of nitrogenase activity. This suggests that factors other than photosynthesis might be involved in reduced nitrogenase activity under water stress.

When subjected to drought, alfalfa cultivars vary in their ability to

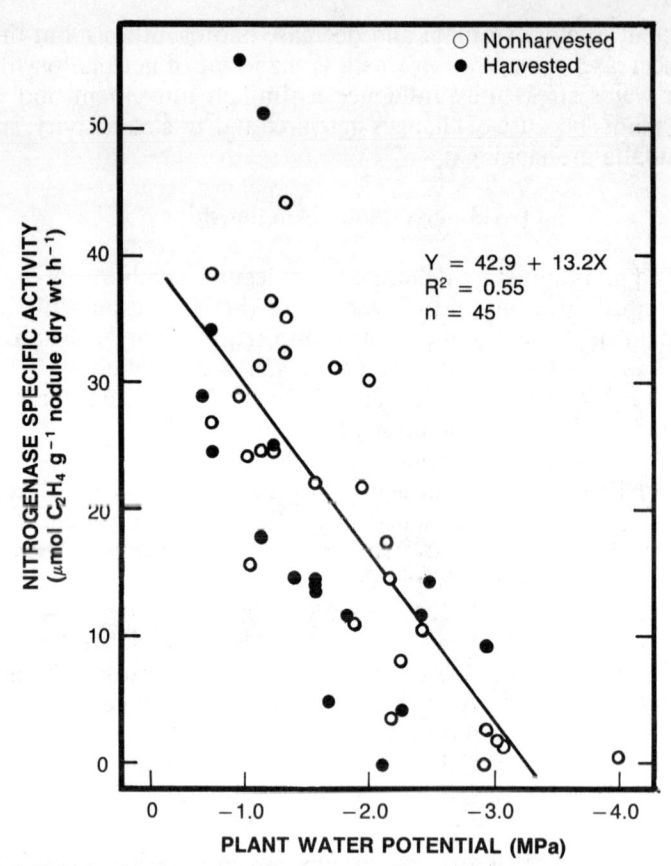

Fig. 11–1. Relationship between nitrogenase specific activity and plant water potential (ψ_w) for nonharvested and harvested alfalfa. Adapted from Carter and Sheaffer (36).

fix N_2. Cultivars adapted to dry conditions are likely to show smaller effects of water stress on N_2 fixation than those less well adapted. For example, 'Tierra de Campos,' normally grown under dryland conditions, was able to grow and fix N_2 at lower plant water potential than 'Aragon' (4). Recovery from stress was more rapid in the dryland-adapted cultivar than in Aragon (4, 169). Populations of alfalfa used in dryland hay production in subhumid environments have a higher N_2-fixation activity than those bred for forage production in humid or irrigated areas (163, 164). This suggests that drought-tolerant alfalfas can be selected for improved N_2 fixation under dry conditions.

Similarly, mycorrhizal associations with *Medicago* spp. roots are affected by soil water deficits. Hyphae of the fungi, which penetrate the soil and infect roots, enlarge the volume of soil from which a plant absorbs water and nutrients. Uptake of P is increased in the presence of mycorrhizae because P is relatively immobile in the soil (102, 109, 165). The number of mycorrhizal infections per unit root length of *M. truncatula* Gaertn. (barrel medic) at a soil water content of 0.22 kg kg^{-1} was more than four times greater than at 0.15 kg kg^{-1} ($\psi_m = -0.43$ MPa) and almost twice as great as at 0.28 kg kg^{-1} (saturation) (154).

11-2 PATHWAY OF WATER MOVEMENT

11-2.1 Root Systems

Root morphology of alfalfa grown on dryland (64, 217) and under irrigation (59, 104) is discussed in Chapter 4. We will concentrate on the growth and function of alfalfa roots in relation to water supply.

Rooting characteristics are dependent on stand age and soil characteristics. Rooting depths of 11 m were observed in an unirrigated 6-yr-old stand in Nebraska (101); however, the greatest depth that roots of alfalfa have been reported (39 m) occurred in a mine shaft under a field of alfalfa (121). Upchurch and Lovvorn (211) reported that by 304 d after seeding, roots had penetrated to 1.2 m and 1.8 m in a clay loam and sandy loam, respectively. In the 6th yr after seeding, roots had penetrated to 2.1 m in the sandy loam while maximum penetration in the clay loam was still only 1.2 m.

Generally, either excesses or deficits of soil moisture (wilting range) limit root growth and function (94, 105, 176). However, contrasting results have been reported for effects of soil moisture on alfalfa root characteristics. These may be attributable, in part, to variation in experimental procedure and soil characteristics. In addition, results from greenhouse studies where container size limits rooting may not be comparable to those from field studies (e.g. 19, 166).

Abdul-Jabbar et al. (2) reported that with line-source irrigation greatest total root mass and rooting depth of alfalfa in a clay loam occurred at the highest irrigation rate (Fig. 11-2). Root-length distribution was similar to that for root mass. They also reported a positive relationship between evapotranspiration (ET) and root mass (Fig. 11-3). In contrast, Carter and Sheaffer (36) reported that nonirrigated alfalfa (soil matric potential, ψ_m of -0.3 to -1.5 MPa) had greater root length to a 60 cm depth, and greater root mass to a 15 cm depth than irrigated alfalfa (ψ_m of -0.01 to -0.06 MPa). Jean and Weaver (90) and Weaver (217, 218) reported that roots of dryland alfalfa had greater primary and secondary branching than roots of irrigated alfalfa. Greenhouse experiments in small containers showed that with maintenance of soil water at -0.03 MPa, roots had greater branching and mass than under low soil moisture (maintained at 50% of the high moisture water requirement) (116).

Root growth response to soil moisture is influenced by cultivar characteristics. Root mass for moderately winterhardy 'Atlantic' was greater under low soil moisture, while root mass of the nonwinterhardy 'African' was greater under high soil moisture (14). Low and high soil moisture represented irrigation to field capacity when 80 and 30%, respectively, of the available water had been depleted from the root zone of a sandy loam.

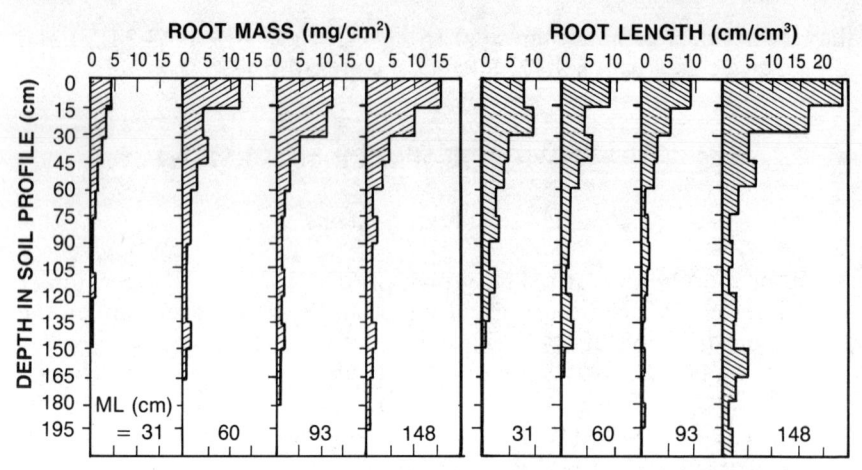

Fig. 11-2. Effect of moisture level (ML, rainfall + irrigation) on alfalfa root mass and root length distribution in the soil profile. Adapted from Abdul-Jabbar et al. (2).

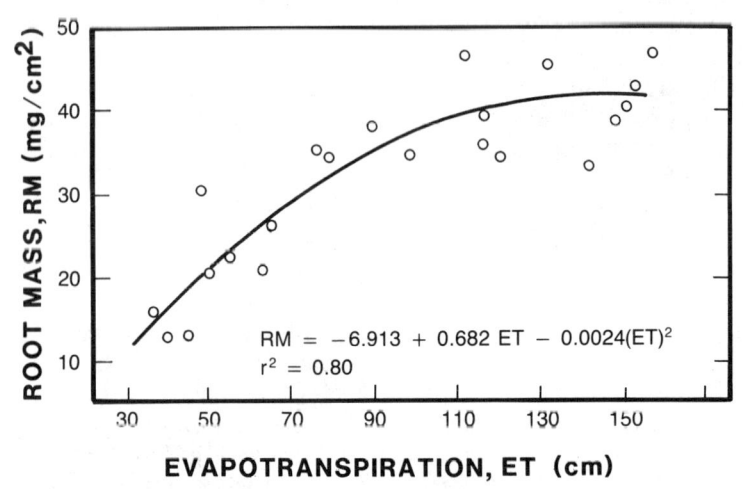

$$RM = -6.913 + 0.682\ ET - 0.0024(ET)^2$$
$$r^2 = 0.80$$

Fig. 11-3. The relationship between root mass and evapotranspiration of alfalfa. Adapted from Abdul-Jabbar et al. (2).

Root mass distribution of Atlantic was not altered by soil moisture level, but for African a greater proportion of the roots was in the upper 15 cm of the profile under high soil moisture. Carter et al. (37) determined differences in root mass and length among three alfalfa cultivars differing in winterhardiness and resistance to Phytophthora root rot (*Phytophthora megasperma* Drechs.) when they were grown in containers. At ψ_w of -0.8 MPa, 'WL318' had greater root mass than 'Agate' or 'Anchor', but at a ψ_w of -2.0 MPa root mass did not differ. Root length did not change as soil and plant water potential decreased, but over all moisture regimes Anchor and WL318 had greater root length than Agate. The existence of genetic variability in root characteristics in alfalfa could contribute to the development of cultivars with superior water procurement characteristics; however, no breeding work has been reported (8, 141).

Root distribution and growth are modified by harvest practices. In unharvested alfalfa, root dry matter yield increased throughout the growing season when adequate moisture was present; while there was little root mass increase with periodic harvesting at bud stage (69). Shoot removal resulted in a reduction in rate of root extension and depth of rooting of 302-d-old alfalfa (81). In a greenhouse study, nonharvested alfalfa had greater root mass with high (ψ_m of -0.03 MPa) than with low (ψ_m of -0.51 MPa) soil moisture; however, harvesting at first flower reduced differences in root mass between the two moisture regimes (36).

Inadequate fertility may limit subsoil root penetration and moisture extraction (19, 59). Bouton et al. (20) reported that alfalfa rooting depth and water extraction were improved by calcium carbonate ($CaCO_3$) addition to the subsoil. Subsoil application of P, B, and $CaCO_3$ increased root proliferation and yield of drought stressed alfalfa (176).

Availability of water and aeration limit root growth in compacted soil (100, 166). The often-cited characteristic of alfalfa roots to penetrate compacted soil (21) may result from their tendency to penetrate cracks to obtain water, rather than the characteristic for exerting more pressure than roots of other species. Fehrenbacher et al. (56) compared the penetration of maize (*Zea mays* L.) roots with those of alfalfa through soils derived from shale. Alfalfa roots penetrated the shale along cracks and cleavages while maize roots did not.

Because its roots can grow to considerable depth, alfalfa has been used to dry soil and prevent seepage of saline water to the surface (22, 26, 76). Even though alfalfa is not highly salt tolerant (156), it tolerates high salinity in the lower root zone without significant yield reduction, provided that the upper portion of the root zone is maintained relatively free of salinity (60).

11-2.2 Soil Water Depletion

When a drying cycle begins with the soil at field capacity, water is extracted mainly from the upper soil layers where root density is greatest and the flow path is shortest. As the upper soil dries, the zone of active extraction moves down, although it continues from the upper layers as long as the potential in the root xylem is lower than that of the soil water. If rainfall or irrigation wets the soil surface, this layer again becomes the zone of maximum extraction until the added water is depleted. There is evidence that when some crops become severely stressed, roots in the driest upper layer are shed or become nonfunctional and that time is required for roots to redevelop in the rewetted surface layer (87, 197, 221).

There is less water extracted from deeper soil layers even when low potentials exist in the upper roots (103, 185), presumably because of decreased root proliferation in the lower soil layers and because of decreases in water potential through the longer xylem pathway. As the more readily extracted water from the upper root zone is used, both the soil and plant water potentials decrease. When sufficient soil water is depleted so that stomata begin to close and reduce T, the proportion of total water depleted from "field capacity" in reasonably uniform soils is very close to the "rule-of-thumb" reported by Shockley (175): 40% of the depletion is from the upper fourth of the root zone, 70% from the upper half and 90% from the upper three-fourths. Six literature reports analyzed by Borg (18) confirm these depletion fractions (within 5%) on sandy loam, loam, and silt loam soils. In Wisconsin, similar results were found for a sand (193). In these seven experiments, the extraction depth varied from 1.2 to 1.6 m. Dense genetic horizons and compact pans will obviate this generalization.

11-2.3 Hydraulic Conductance

The anatomical and morphological characteristics of the vascular system of roots, stems, and leaves are described in Chapter 4 in this book and by Esau (54). Water movement occurs along a water potential or pressure gradient from the soil to the atmosphere through the roots, stems, and leaves. As water is transpired the leaf water potential decreases, creating a potential gradient from soil to leaf. Water flows from soil to leaf down this gradient. The pathway and resistances to water movement in plants were reviewed by Molz (123) and Turner and Burch (207), but little specific information is available for alfalfa. Water uptake is predominantly a passive process: a function of the difference between plant and soil water plant and of root density. Active absorption is negligible compared to that induced by transpiration (T). Although the major documented resistance to water flow is between the root periphery and xylem (104, 105), it is possible that a major resistance exists at the soil and root interface (55). At a high soil water content or in regions of high root density, there is little potential drop from soil to root (153). Even at low ψ_m (e.g., -1 MPa), flow through the soil is not likely to be limiting except for sands.

The potential drop between soil and leaf depends not only on T but also on the hydraulic conductance of the alfalfa plant and on the hydration/dehydration of foliage, crown and root (i.e., plant water capacitance) during transient conditions. Abdul-Jabbar et al. (1) related water conductance of alfalfa (leaf area index, LAI = 4.3) and ψ_w from 0730 to 1500 h. They found that hydraulic conductance (1.1–1.8×10^{-9} mm s^{-1} Pa^{-1}) was not correlated with ψ_w (-0.4 to -1.8 MPa) indicating little if any effect of stress on hydraulic conductance. Katerji et al. (98) also found nearly constant hydraulic conductance of alfalfa (LAI = 3) after correcting for capacitance under transient conditions; moreover, their measurements corresponded to the lower range of those measured by Abdul-Jabbar et al. (1), about in proportion to the LAI. Both studies indicate that over much of the day, ψ_w decreases approximately in proportion to T until the stomata exert control on T.

The effects of root and crown characteristics on hydraulic conductance have not been investigated. For example, during contractile growth of alfalfa roots in seedling development, the central xylem system, consisting of primary and secondary xylem, becomes undulated (54). This undulation should increase resistance to water flow and may effect plant response to moisture stress. Sharratt et al. (172) reported a lag in response of ψ_w to changes in evaporative demand for both irrigated and unirrigated alfalfa. This lag in ψ_w is greater for alfalfa than for other legumes, such as soybean (152), and may result from the water capacitance of the large tap root.

11-2.4 Stomatal Regulation

Stomata constitute the primary system for control of T. The stomatal density of alfalfa is greatest on the upper leaf surface, but its diffusive conductance is less than that of the lower surface (66, 78). This suggests that stomata on the upper leaf surface are less open or have shorter aperture lengths than those on the lower surface. Carlson et al. (33) measured stomatal density in field-grown alfalfa and found that leaflets from the uppermost node had a higher stomatal density than those from lower nodes. Means for the abaxial surface at node one and node three were 181 and 166 mm^{-1}, respectively. The adaxial surface had a higher stomatal density (mean of 217 mm^{-2}) than the abaxial surface (mean of 174 mm^{-2}). Cole and Dobrenz (42) reported similar stomatal density values although variation existed among cultivars.

Alfalfa stomata regulate water loss and CO_2 uptake during the diurnal cycle. Carter and Sheaffer (35) reported that conductance for well-watered alfalfa (predawn and midday ψ_w of -0.5 and -1.0 MPa, respectively) increased following sunrise, peaked at 1000 to 1200 h (maximum value of 3.3 cm s^{-1}) and declined through the midday and evening until stomatal closure at sunset. At moderate plant water deficits, leaf conductance was greatest during low evaporative demand periods (early morning and late afternoon) and least at midday. Under extreme plant moisture stress (predawn and midday ψ_w of -2.0 MPa and -4.5 MPa, respectively), conductance remained low (0.1 to 0.3 cm s^{-1}) throughout the day. Similar diurnal patterns of water regulation (expressed as stomatal resistance) in response to increased soil moisture deficits were reported by van Bavel (212) (Fig. 11-4). Baldocchi et al. (6) found that stomatal conductance of irrigated alfalfa decreased from 0900 to about 1400 h and then remained relatively constant until 1900 h.

Alfalfa stomata are sensitive to changes in ψ_w although the reported threshold ψ_w at which closure occurs varies. Carter and Sheaffer (35) reported that stomatal conductance began to decrease at ψ_w of -1.2 MPa and declined linearly until ψ_w reached -2.5 MPa (Fig. 11-5). The rate of decline was faster at higher than at cooler daily temperatures. With

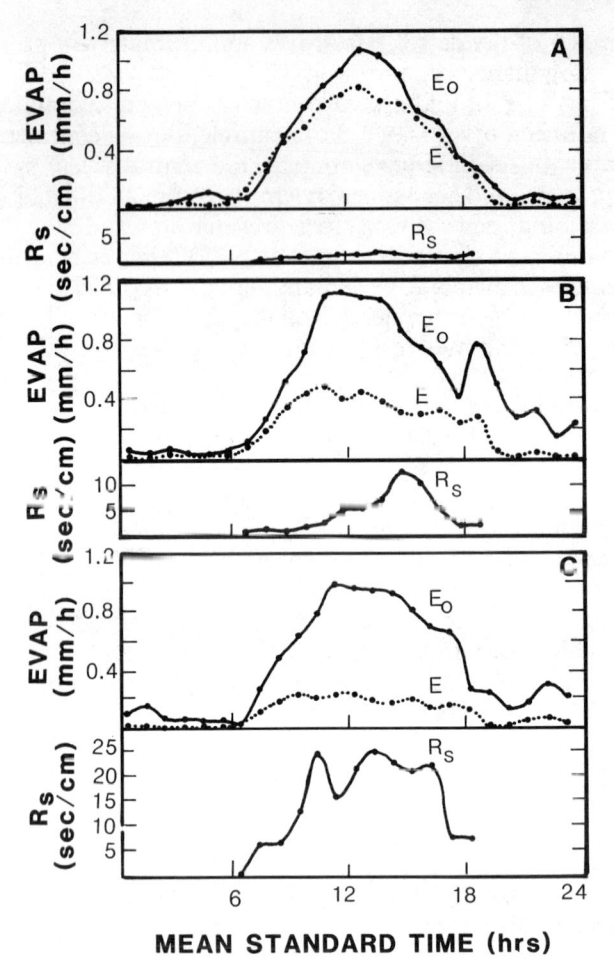

Fig. 11-4. Actual (E) and potential (E$_o$) hourly evaporation (EVAP) and canopy resistance (R$_s$) of alfalfa in Phoenix, AZ on 23(A), 27(B), and 31(C) d after irrigation. Adapted from van Bavel (212).

$\psi_w < -2.5$ MPa, (i.e., -2.5 MPa and drier), stomata were insensitive to changes in ψ_w, but T continued slowly indicating either incomplete stomatal closure or cuticular conductance. In other studies, conductance remained high until ψ_w was -1.5 MPa and reached a minimum at about -2.0 to -2.4 MPa (25). Since alfalfa herbage growth is minimal at $\psi_w < -1.0$ MPa to -1.5 MPa, it appears that considerable water loss may occur at water deficits limiting plant growth. In a greenhouse experiment, Hall and Larson (74) reported that decreases in leaf conductance in response to decreasing ψ_w (i.e., more negative ψ_w) were greater for full bloom than vegetative alfalfa. Following relief of plant moisture stress, stomatal activity recovered in 32 h.

Aparicio-Tejo et al. (4) compared the stomatal response to moisture deficits of Tierra de Campos, normally grown under dryland conditions

WATER RELATIONS AND IRRIGATION

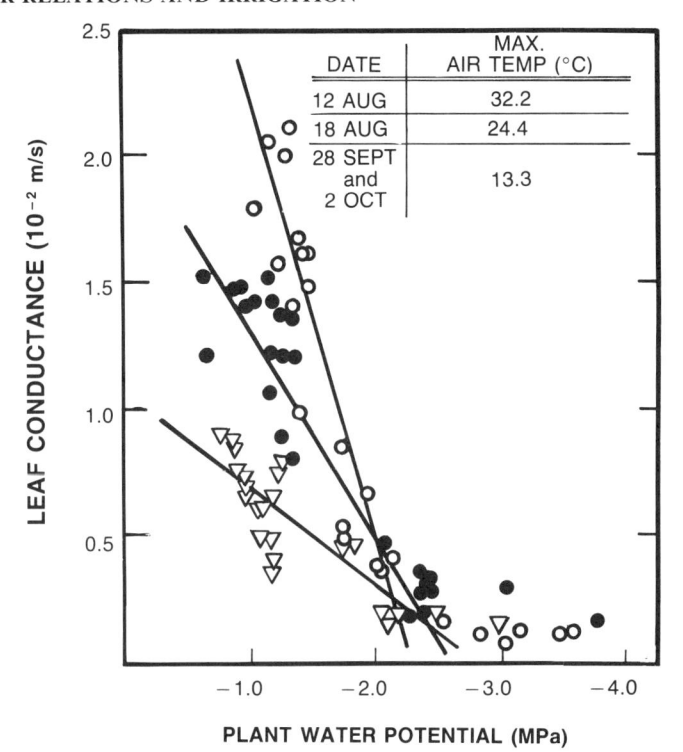

Fig. 11-5. Leaf conductance vs. plant water potential for alfalfa on three dates. Adapted from Carter and Sheaffer (35).

with that of Aragon normally grown under irrigation. Although Tierra de Campos (ψ_w of -1.7 MPa) had a similar conductance as Aragon (ψ_w of -1.4 MPa) at low soil moisture, the recovery of stomatal conductance was faster in Aragon with restoration of moisture. Hall and Larson (74) reported that stomatal conductance during moisture stress and stomatal recovery after rewatering did not differ for winterhardy 'Cody' and nonwinterhardy 'Sonora'. Sheehy et al. (174) reported a relationship between rate of photosynthesis (measured as CO_2 exchange rate, CER) and leaf conductance in genotypes of 'Vernal'. Genotypes with a greater CER (2280 mg m^{-2} s^{-1}) had a higher leaf conductance at CO_2 concentrations near the compensation concentration than genotypes with the lower CER (667 mg m^{-2} s^{-1}).

Ploidy level affects T. Pfeiffer et al. (147) determined that T was greater in an octoploid than in a tetraploid or diploid alfalfa. The increased T of the octoploid compared to the diploid and tetraploid could result from a greater leaf area. Setter et al. (170) studied CER, T, leaf diffusive resistance, stomatal size, stomatal density, and leaf area in diploid, tetraploid, and octoploid alfalfa and in parents and hybrids at diploid and tetraploid levels. They concluded that increased leaf area, CER, and T can be achieved easier with hybridization than by increasing ploidy level.

11-2.5 Transpiration

When alfalfa is grown on well-watered soil with LAI ≥ 3 ($\approx 80\%$ radiation absorption and T $\simeq$ ET) very high rates can be sustained; for example, ET of 14 mm d^{-1} with maximum daytime rates of 1.6 mm h^{-1} was measured in the field (162). The T from a full-cover, well-watered crop will be determined mainly by the meteorological heat supply, and will approach that of a wet surface with comparable albedo (solar radiance reflectance) and convective heat transfer characteristics. Some of the factors contributing to these high T rates are: (i) high stomatal conductance (1.5 to 3 cm s^{-1}); (ii) small leaves with high boundary layer conductances; (iii) high stem densities resulting in high parallel hydraulic conductance; and (iv) high root densities. When alfalfa is cut (LAI negligible), T is negligible and evaporation from wet soil may equal T from full canopy cover in humid regions (70–80% of full canopy cover T in advective regions), but may be only 20% of full canopy T if the surface is dry (223). As herbage regrowth occurs and LAI increases, T increases as a fraction of full cover in proportion to the increase in radiation intercepted by foliage; also the foliage receives a larger share of any convective heat supply. Accordingly, ET will approach a maximum at LAI <3; typically at a LAI of 1.5 to 2.0 when about two-thirds of the radiation is absorbed.

As soil water is depleted and the soil water potential in the upper root zone decreases, ψ_w decreases until stomata begin to close. Daily T then falls below the weather-driven T_{max}. With a further depletion of soil water, stomata close further and for longer periods (i.e., stomatal resistance increases) during the day, daily T or ET decreases below T_{max} or ET_{max}. This is shown in Fig. 11–6 and is supported by results of several experiments (25, 34, 133). Because the water potential drop through the plant increases with T, the "critical" depletion will vary somewhat with

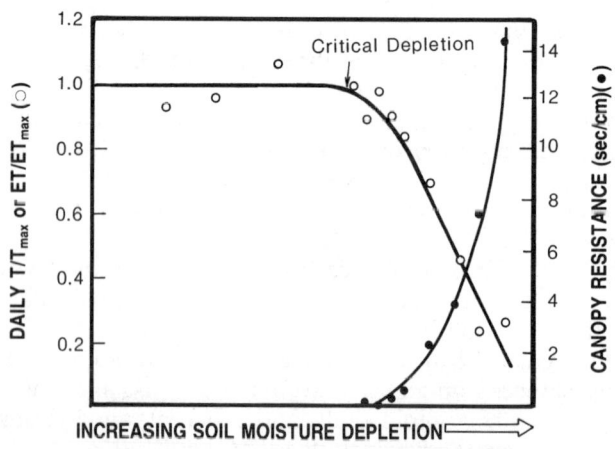

Fig. 11–6. Ratio of actual transpiration (T) or evapotranspiration (ET) to potential maximum T or ET and canopy resistance with increasing soil moisture depletion. Adapted from van Bavel (212).

T rate. Thus, wilting will be observed at a smaller depletion during a day of high T than when T is low.

11-2.6 Morphological Control of Water Loss

Christian (38) reviewed the changes in leaf morphology of alfalfa under arid conditions. Reduction of leaf and epidermal cell size reduces water loss. Effective leaf area reduction via cupping in response to a decline in plant water potential (203) is another mechanism for reducing energy load. Other morphological features such as leaf pubescence, which may increase plant resistance to insects (106, 134), and cuticle thickness have not been studied as features to increase alfalfa adaptation to water deficits. However, Baldocchi et al. (7) reported that ET was lower and water-use efficiency greater for a densely pubescent compared to standard, pubescent soybean [*Glycine max* (L.) Merr.]. Leaf pubescence facilitated penetration of solar radiation into the canopy.

Multiple layers of palisade cells in the leaf are typical of drought-resistant plants. *Medicago sativa* typically has only one layer of palisade cells, whereas *M. prostrata* Jacq. a more drought resistant species, has a double layer (180). Incorporation of this trait into commonly grown alfalfa types may increase drought tolerance.

11-3 CROP WATER REQUIREMENTS

11-3.1 Germination and Emergence

Under favorable moisture conditions, alfalfa imbibes all water required for germination within 4 to 8 h (119). Seed is capable of absorbing more than 100% of its dry mass in water due to high protein concentrations in the cotyledon and to a layer of cells in the testa which acts as a gelatinous sponge (32). McWilliam et al. (119) reported that when alfalfa was subjected to an osmotic "drought" created by polyethylene glycol, germination decreased as solution potential decreased to -1 MPa. Triplett and Tesar (204) reported that alfalfa emergence ceased when soil moisture tension was <-1 MPa in both a silt loam and a sandy loam.

Rate and extent of germination in sodium chloride (NaCl) and mannitol were related to alfalfa cultivar tolerance to drought and cold stress (51, 79, 210). Differences in cultivar germination have been observed with osmotic solutions with potentials of -0.04 to -0.09 MPa. These findings have been unreliable in predicting field results, and they are subject to criticism because toxic solutes in the osmotica may penetrate the seed coat (119).

Sowing alfalfa at or near the soil surface exposes the seed to moisture extremes. Although adequate moisture is present for germination, moisture stress during the later germination and early seedling growth stages may contribute to seedling mortality. For alfalfa germination, water equivalent to 125% of the seed dry mass must be imbibed (32). Moisture loss to below that amount before rupture of the seed coat will postpone germination. Following radical emergence, dehydration will drastically reduce seed viability. Firm soil, good soil-seed contact, increased planting depth, and mulch are helpful to ensure an adequate moisture supply for seedling establishment (126, 204) (see Chapter 19 in this book).

Irrigation can be used most effectively in promoting germination and emergence by wetting the soil profile to field capacity before seedbed preparation. Irrigation following seeding is sometimes utilized to promote germination and emergence (110); however, this practice can cause seed washing and crusting of some soils. The practice of withholding irrigation following emergence for the purpose of increasing root penetration is invalid because seedling root growth is suppressed more than shoot growth by moisture stress (67, 89).

11-3.2 Canopy Development and Biomass Production

Carter and Sheaffer (34) regressed mean daily plant water potential (ψ_w) on relative growth rate and found that at $\psi_w < -1.0$ MPa growth was slow. Under severe water stress (ψ_w of -2.5 to -3.0 MPa), leaf loss contributed to a negative growth rate. Brown and Tanner (25) also reported that little alfalfa leaf and stem growth occurred at $\psi_w < -1.0$ MPa. They reported that stems of irrigated alfalfa elongated faster during the day than at night because of higher daytime temperature, whereas stems of water stressed plants elongated faster at night despite lower night temperatures due to increased water potential and turgor. The sequence of leaf expansion followed by internode elongation was similar for well-watered and stressed alfalfa; however, rates of leaf and internode development were decreased by moisture stress.

Taylor (198) reported a linear decrease in herbage yield on a loam as soil matric potential, ψ_m, decreased from -0.1 to -0.4 MPa. Alfalfa canopy growth rate decreased 60 to 75% when ψ_m of a silty clay loam at 25 to 50 cm soil depths dropped below -0.25 MPa (100). However, since soil water content and conductivity at any given ψ_m varies with soil type, the relationship between alfalfa herbage yield and soil moisture content must be determined for each soil.

Water stress decreases stem number, stem diameter, internode number and length, and leaf size (45, 49, 66, 145, 213). Under monotonic drying, length of upper internodes was reduced more than length of basal internodes (25, 145). Brown and Tanner (25) reported that with monotonic drying the development of water stress 2 weeks after cutting reduced leaflet size and internode length of 'Saranac' alfalfa but not leaf and internode number or stem density (number m^{-2} soil) (see Fig. 11-4, Chapter 5 in this book). Perry and Larson (145) reported that effects of moisture stress on internode number and length were greater for nonwinterhardy than winterhardy cultivars. Although moderate moisture stress reduced leaf area and yield, the concomitant reduction of stem yield increased the proportion of leaves in the herbage (34). Alfalfa grown at ψ_m of -0.02 MPa partitioned 48% of its aerial dry matter into leaves, but at a ψ_m of -2.0 MPa, 62% of the aerial dry matter was leaves (213). With severe moisture stress, leaf loss may result in a reduced leaf/stem mass ratio (75).

Alfalfa recovers rapidly following release from moisture stress. Cowett and Sprague (45) reported that when moisture was supplied to drought-stressed plants, forage mass and stem numbers were comparable to those of nonstressed plants. Sheaffer and Barnes (173) reported that fall regrowth following a harvest and rainfall was greater for alfalfa which was previously unirrigated and under moisture stress than for irrigated alfalfa. Compensatory growth of other forage species occurred following release from moisture stress and was related to a build-up of water soluble carbohydrates during moisture stress (44, 82). Since alfalfa herbage production is reduced by a lower ψ_w than that required for initiating stomatal closure and restricting CER (25, 34, 35), fixed C is translocated to the roots and crown (41). Hall et al. (73) reported that moisture-stressed alfalfa (ψ_w of <-1.5 MPa) fixed radiolabeled CO_2. Stressed alfalfa exported 51% more radiolabel to the roots than did nonstressed alfalfa and had greater root starch concentrations. Andrew et al. (3) reported that moisture-stressed alfalfa tended to have higher root carbohydrate concentrations than well-watered alfalfa.

In seedlings, soil moisture deficits reduce root growth more than shoot growth (67), while the reverse occurs in mature plants (2, 19, 35, 94, 166, 206). This may be explained solely by an increase in root growth, or in part by decreased herbage production with increased partitioning of fixed C to roots, as described above.

Increased alfalfa leafiness, which is produced by moisture deficits, is a feature generally associated with increases in herbage quality (35, 75). Nitrogen concentrations of plants that do not fix N_2 generally increase under plant water stress (83); but effects of moisture deficits on alfalfa N concentrations have been inconsistent (35, 178, 213). Although increased herbage digestibilities of alfalfa grown under moisture stress have been reported, Carter and Sheaffer (34) did not observe increased herbage digestibilities until ψ_w decreased below -2.7 MPa. Increased digestibility has been associated with decreased concentrations of structural components of leaves and stems such as lignin and crude fiber (91, 178, 213). Rainfall and irrigation practices that promote plant diseases may decrease forage quality due to leaf loss.

11–3.3 Seed Production

Historically, water requirements have been regarded as lower for seed than for vegetative production (78). It was recommended that soil moisture be adequate to support crop growth through flowering, but restricted during seed formation to prevent vegetative regrowth (140, 178, 199). This management results in the production of a single large seed crop each season and restricts vegetative regrowth which often produces unripe seed at harvest. Contrary to established practice, Henderson and Yamada (80) reported that forage and seed yields were highly correlated ($R^2 = 0.81$) and that even moderate plant moisture stress (ψ_w of -1.0 to -1.2 MPa) was never beneficial to seed production. Although moderate moisture stress hastened the onset of flowering by a few days, plants not subjected to stress had more racemes and florets and greater seed yields. They concluded that with low pollination pressures provided by honeybees (*Apis mellifera*), seed setting is a prolonged continuous process and that new vegetative growth following flowering provides additional sites for seed production. In Israel (41, 68) and Canada (108), moderate rates of irrigation during flowering increased seed yields. Moisture stress during flowering reduced yields. Plant moisture status had no effect on flower pollination or pollinator activity, except when severe moisture stress (-2.5 to -3.0 MPa) prevented tripping (68, 80) (see Chapter 32 in this book).

11-3.4 Excess Water and Stand Persistence

Excessive moisture supplied by irrigation or rainfall is detrimental to alfalfa root and herbage growth and to stand persistence (100, 146, 214). Thompson and Fick (200) reported that alfalfa can endure flooding for 14 d at a soil temperature of 16°C, 10 d at 21°C, 7 to 8 d at 27°C and 6 d at 32°C. Near saturation of the soil in combination with high soil temperatures in the southwest and western USA causes "scalding" and plant death within 3 or 4 d (49, 70, 120). Effects of excess soil water are more severe immediately after cutting when little herbage exists. Flooding causes xylem necrosis and death of leaves beginning with the lowest on the stem. Initial effects of saturated soils are attributed to lack of O_2 in the root zone and to the formation of ethanol and other toxic substances in the roots (9, 38, 187). Effects of Phytophthora root rot appear secondary (9, 30, 200). Wet soils may enable Phytophthora oospores to germinate and release zoospores, or result in roots exuding ethanol which may attract zoospores (9).

Soil and plant moisture status influence alfalfa fall dormancy reaction and winter survival (see Chapter 8 in this book). In general, moisture stress increases and excess moisture decreases freezing tolerance (111). Excess soil moisture was associated with dehardening and decreased survival of alfalfa (29, 117, 187). Paquin and Mehuys (137) reported that drought stressing and freezing of alfalfa at 25% field capacity increased the cold tolerance of unhardened plants by 3.7°C compared to well-watered plants frozen at field capacity. Stout (186) reported that alfalfa, water-stressed during cold acclimation, contained less water and had a lower solute potential than nonstressed alfalfa.

11-4 CROP PRODUCTIVITY AND WATER USE

Under moisture limiting conditions, dry matter production decreases proportionally to the decrease in transpiration below the maximum that would occur if alfalfa were well supplied with water. This is shown in the relationship:

$$Y/Y_m = T/T_m, \qquad [1]$$

where Y_m is potential yield when T is equal to maximum, climate-driven transpiration (T_m). Because few T data are available and because at canopy closure, $T \simeq ET$, the relationship has been most often shown between ET and yield. A linear relationship between annual dry matter yield and ET (ET/ET max) was reported by de Wit (48) for the Great Plains and more recently by Bauder et al. (10) in North Dakota (Fig. 11-7), Retta and Hanks (155) in Utah, and Sammis (168) in New Mexico. Stewart and Hagan (184) reported a small deviation from this linear relationship due to higher productivity per unit of ET early in the season (attributed to translocation of stored dry matter from roots and crowns to herbage) and a lower productivity per unit of ET late in the season (from increased storage of photosynthate in roots and crowns). This linear relationship between ET and forage yield differs from that between root mass and ET (Fig. 11-3). Limitations in generalized assumptions governing the relationship between crop production and ET are discussed by Ritchie (160).

Sammis (168) summarized results of irrigation experiments from Nebraska (47), Nevada (202), and North Dakota (10) and concluded that 8.3 cm of water were required to produce 1 Mg ha^{-1} of alfalfa. Donovan and Meek (49) in California reported a similar relationship. Heichel (78) summarized research from diverse climates and reported that 5.6 to 7.3 cm of water are required to produced 1 Mg ha^{-1} of dry matter. However,

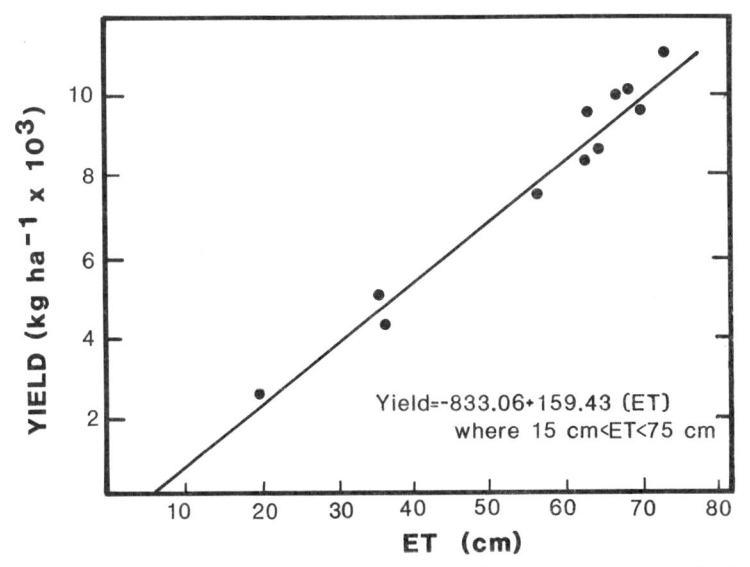

Fig. 11-7. Alfalfa dry matter yield as related to growing season evapotranspiration (ET). Adapted from Bauder et al. (10).

Table 11-1. Daily and seasonal evapotranspiration (ET) and yield from alfalfa.

Location	ET Daily Mean	ET Daily Max.	Seasonal	Forage yield	Measurement	Years	Reference
	mm			Mg/ha			
Kimberly, ID	5.2	11.0 (July)‡	958 (184 d)§	15.2	Lysimeter	1969-1975	224
Farmington, NM	†	†	1493 (231 d)	16.9	Lysimeter	1976-1977	168
Las Cruces, NM	†	†	1694 (332 d)	23.0	Lysimeter	1976-1977	168
Oakes, ND	†	†	685 (120 d)	10.9	Water balance	1973-1976	10
Scottsbluff, NE	5.2	7.1 (Aug.)	748 (150 d)	11.5	Water balance	1966-1968	47
Davis, CA	†	†	1514 (181 d)	20.0	Lysimeter	†	184
Phoenix, AZ	9.0	11.0 (May)	†	†	Lysimeter	1964	212
Mesa and Tempe, AZ	6.5	9.3 (June)	1890 (289 d)	†	Water balance	1946, 1950, 1962, 1963	53
Lethbridge, AB	5.1	9.1 (†)	648 (155 d)	11.7	Water balance	1950-1961	179
Reno, NV	8.5	†	1067 (126 d)	16.4	Lysimeter	1959-1961	202
Geneva, NY	3.3	5.3 (July)	407 (150 d)	†	Water balance	1953-1955	139

† Information not available.
‡ Month with highest maximum daily ET.
§ Average length of ET measurement.

the adoption of any dry matter yield/water input function is misleading since daily and seasonal water use and plant growth are dramatically influenced by climate and cultural practices.

Maximum daily water use or ET is typically 5 to 11 mm (Table 11-1), although extremes of 1.3 and 14 mm d^{-1} have been reported in Wisconsin (189) and Nebraska (162), respectively. Daily ET is primarily influenced by global radiation, stage of alfalfa development, temperature, and daylength (161, 182, 189). Water use is generally greatest during the warmest months of the year and with full canopy cover. Reduced growth or dormancy induced by temperature, daylength, or soil moisture deficits reduces daily potential ET. Water use during daytime hours usually constitutes the largest daily proportion (Fig. 11-4). Rosenberg (161), reported that nocturnal ET accounted for 21% of the total daily ET because of temperature inversion and stored soil heat. Seasonal ET rates are influenced primarily by length of growing season and temperature (78, 99, 107, 182). Seasonal rates range from 400 mm in the northeast to 1890 mm in the arid southwest (Table 11-1).

We define water-use efficiency (WUE) as the biomass (yield, Y) produced per unit area for a unit of ET, i.e., ET efficiency or Y/ET. When WUE is improved by management, the increase results from increased T as a fraction of ET; ET efficiency is increased although T efficiency is changed little, if any (196). Cultural practices such as weed control; timing of harvest in relation to irrigation; and minimization of surface runoff, soil evaporation, and deep percolation provide a higher proportion of applied water for use in T and thereby increase WUE. In addition, low fertility, nonoptimum growing temperatures, plant diseases, and insects that reduce yield, lower the leaf area, and reduce canopy closure increase soil evaporation, lower Y/ET and may lower Y/T. Generally the greatest WUE has been observed for spring growth (first harvest) of alfalfa (47, 182). This is associated with favorable growing temperatures and lower ET. Excessive summer temperatures (122) and reduced daylength in the fall (34) induce dormancy and reduce WUE.

Cultivar differences in yield response to soil moisture deficits have been reported (171). These differences have been associated with differences in herbage growth and regrowth rates (115), root characteristics (37, 116, 167), and T (43).

Numerous experiments based on various criteria for water application have demonstrated the positive effects of irrigation on alfalfa yield. Irrigation has been scheduled using soil moisture depletion (10, 34, 57, 58, 77, 96, 107, 112), pan evaporation (49, 177), and soil water budget models (15, 173) as criteria. Procedures for optimizing irrigation and WUE must be determined experimentally for different environments because of climatic and crop management effects on ET, variation in soil water-holding properties, cultivar response, and irrigation method efficiency. Thus in a semiarid area of Australia, greatest forage dry matter yield per unit of water applied occurred when irrigation approximated one-half Epan (177), while in subhumid North Dakota (10), yield per unit of water applied did not differ consistently among soil water depletion-based irrigation treatments which represented deficient, optimum, and excessive irrigation. On subirrigated soils, water table depth is critical for maximizing alfalfa production (202). Benz et al. (15) found that a water table depth between 1 and 2 m provided the greatest irrigated alfalfa yield, but production per unit water applied did not differ among irrigation treatments providing 0.5, 1.0, and 1.5 of the estimated soil water depletion.

11-5 IRRIGATION SCHEDULING

Most irrigation scheduling by various criteria is directed toward prevention of any yield-reducing stress with modifications for optimizing economical yields. Generally, it is assumed that production water use functions developed from experimental plots apply to field management. However, as Warrick and Gardner (216) and Gardner et al. (63) show, the spatial variability of soil properties and of irrigation applications will affect field-scale management. Greater irrigation efficiency and yield usually are obtained when some yield-reducing stress occurs in parts of the field before irrigation, than if irrigation is managed to prevent stress everywhere in the field. We do not address this problem in the following discussion.

The fact that leaf enlargement and stem elongation decrease at higher water potentials than stomatal closure creates some ambiguity in irrigation criteria. As long as stomata remain open, CO_2 assimilation continues along with T after shoot growth has decreased significantly. With little assimilate used to support shoot growth, increased partitioning of photosynthate to crown and roots relative to shoots occurs (73). The uncertainty in irrigation criteria hinges on whether either an appreciable or only a small fraction of this stored photosynthate is mobilized to shoot growth following relief of water stress. If only a small portion is mobilized, then greater yield will be obtained when the crop is irrigated to maintain maximum shoot growth. If almost all the stored photosynthate is mobilized to shoot growth, then the irrigation needs only to maintain uninterrupted CO_2 assimilation. Obviously, conditions between these extremes could occur, and the fraction of photosynthate mobilized could well depend on time of year for alfalfa cultivars that differ in fall growth (dormancy).

11-5.1 Plant-Based Criteria

Growth retardation (stem and leaf enlargement), stomatal conductance, T, and ψ_w are possible plant-based irrigation criteria. However, absolute values of these plant properties are impractical as irrigation criteria. For example, stomatal conductance is influenced by other environmental and plant parameters (e.g., light, soil fertility, and leaf age) in addition to plant water deficits. A comparison of stomatal conductance or other characteristics of moisture-stressed alfalfa with that of similarly treated well-watered alfalfa may allow an estimation of relative stress, but this comparison is often impossible in field situations. In addition, because of spatial variability of plants and soil, many leaves must be measured to provide an adequate sample, which may be impractical in routine scheduling operations. Canopy temperature, which depends on stomatal response and transpiration, is another option in determining plant moisture deficits. (See section 11-4.4 on crop temperature.)

Brown and Tanner (25) reported that stressed alfalfa (ψ_w of -1.5 MPa) was wilted and gray-green when a 50% decrease in relative growth rate occurred. Undoubtedly these and other visual features including leaf cupping and reduced leaf size provide a simple and quick indication of moisture stress (72), but they are difficult to quantify for use as irrigation criteria and may be influenced by cultivar, disease, and soil fertility (95, 220). In addition, yield reductions are frequent before visual symptoms appear.

11-5.2 Soil Water Criteria

As discussed earlier, the depletion of soil water below a "critical" level (Fig. 11-6) results in decreased growth rates, stomatal closure, and eventual loss of yield. The amount of depletion permissible for a specific cultivar and soil must be determined experimentally. Thereafter, depletion estimates are used to guide irrigation to prevent excessive depletion. The depletion can be estimated either from soil water measurements or estimates of ET, along with estimates of drainage (17). Commonly the "available soil water" in the effective root zone is determined as the difference between the $\psi_m = -1.5$ MPa retention and the "field capacity" or a substitute measurement, such as the -10 kPa (sands) or -33 kPa (loams, fine-textured soils) retention. Then a fractional depletion of the available water is permitted before irrigation (typically 40 to 65% depletion occurs before transpiration begins to decrease). Alternatively, "extractable soil water" is determined as the difference between the "field capacity" and the water remaining in the effective root zone when the crop is severely wilted; then a fractional depletion of the extractable water is permitted (typically 65-75% before T decreases). With the usual water extraction pattern in the root zone (see previous section 11-2.2 on soil water depletion), the "extractable water" must be less than, but correlated with, the "available water" for a given soil and T rate.

When soil water depletion is used as an irrigation criterion, monitoring of water depletion from the upper 1 m of the root zone is more convenient than determining extraction from the full root zone. The upper root zone is more dynamic and gives greater sensitivity. However, monitoring depletion in the lower half of the root zone provides information for potential storage of excessive rainfall and for control of leaching. In humid and subhumid regions, it is advisable to permit some water depletion in the lower root zone to provide storage for rainfall that may follow irrigation.

Since the effective rooting depth, soil water-holding capacity, and T demand (e.g., arid vs. humid region) vary with location and cultivar, site specific information on plant response to soil water content and depletion is required to develop irrigation criteria. As an example, Brown and Tanner (25) found that shoot growth (leaf enlargement and stem extension) of alfalfa (Saranac) grown on a sandy loam began to decrease at ψ_m (30-cm depth) of -35 kPa. This corresponded to a water depletion of about 3.5 cm and a midday ψ_w of -0.8 MPa (leaf enlargement and stem extension were negligible at $\psi_w < -1$ MPa). Measurable stomatal closure was obtained after about 5.5 cm of soil water had been depleted, 80% of which was from the top 1 m of soil. At that time, the midday ψ_w had decreased to about -1.5 MPa, and the ψ_m reading at 30 cm was < -80 kPa. In contrast, on finer-textured soils, $\psi_m < -30$ to -35 kPa would be permissible without decreasing shoot growth. Also, even when ψ_m at 30 cm was greater than -30 kPa, and nighttime ψ_w was high, growth was limited by low central Wisconsin nighttime temperatures. Accordingly, we might expect that if Saranac was grown on the same soil but farther south where nighttime temperatures were higher, the ψ_m could be < -30 to -35 kPa before shoot growth would be affected.

11-5.3 Estimating Soil Water Depletion

Moisture content and depletion in the root zone can be directly measured gravimetrically or indirectly using calibrated tensiometers, calibrated neutron meters (13, 72, 128) or time domain reflectometry (201). When soil water depletion in the effective root zone is assessed by neutron meter or time domain reflectometry measurements, the average T over a few days can be estimated provided drainage is accounted for (17), but daily T information may be essential.

Soil water depletion from "field capacity" also can be determined using ET assessments. Evapotranspiration can be measured by micrometeorological methods (e.g., energy balance/Bowen ratio and eddy correlation); these can provide ET over daily and even fractional hour periods (28, 97). Moreover, a measurement at a given site can provide a reasonable spatial average depending on height of measurement. Currently, the most convenient micrometeorological methods cannot be used for irrigation scheduling because of their cost and complexity.

"Climatonomic" methods requiring only atmospheric measurements and evaporation pans have been used to estimate the daily ET from full-cover, well-watered crops (ET_{max}) with variable success. In general, climatonomic and pan methods involve some assumptions that detract from their general applicability (28, 97, 125, 192). Many climatonomic methods have been proposed for estimating ET_{max} using two or three atmospheric parameters (e.g., air temperature, solar or net radiation, wind, and saturation deficit). While these proposed methods have provided ET_{max} estimates that often correlated well with measured ET averaged over periods of a week or more, they frequently were poorly correlated with daily ET. Even when ET estimates correlated well with daily ET, greatly different regression coefficients were found under different climates (e.g., humid and arid). We will discuss evaporation pans, the "combination" or "Penman" formula and two contractions of the combination formula (Priestley-Taylor and Makkink formulas) as methods to estimate ET.

11-5.3.1 Pans

The pan coefficient is the ratio of ET_{max} from a full-cover, well-watered crop to pan evaporation. Practice has shown that with experimental calibration pan evaporation can be used to estimate ET_{max}. There are two operational difficulties with pans. First, convective heat transfers at the pan are larger relative to radiation than for an extended alfalfa surface. Thus the pan coefficient can vary, depending on exposure (149). The variation of pan coefficient with exposure, which is more severe in semiarid than humid climates, requires that the coefficient be calibrated, although coefficients for a reference surface can be estimated (e.g., 50). In humid and subhumid climates, the pan coefficient will typically be 0.9 to 0.95, but in arid regions may be as low as 0.75 (49). The second problem is obtaining the pan evaporation from daily readings when there is rain. Because of splashout during rainfall (and overflow, rarely), the difference between daily measurements of pan level cannot be corrected reliably using gauged precipitation. A continuous recording of pan water levels is required for unambiguous measurement of pan evaporation.

11-5.3.2 Combination (Penman's) Formulas

Penman (142) combined the equations for convection transfer of sensible and latent heat with the energy balance to give an estimate of ET from planar, wet surfaces that required no measurement of surface temperature or vapor pressure. With this condition, the ET would be maximum (i.e., limited by heat available meteorologically):

$$\lambda ET_{max} = [s(R_n - G) + Ch_h(e^* - e)]/(s + \gamma) \qquad [2]$$

where ET is the evapotranspiration (λET is the equivalent latent heat), R_n is the net radiation (solar and thermal), G is the heat flow into or from the soil, λ is the latent heat of vaporization, $C = \rho c_p$ is the heat capacity of air (ρ = air density, c_p = specific heat), $\gamma = Pc_p/\epsilon\lambda$ is the psychrometer constant ($\epsilon = 0.622$ is the ratio of mole weight of water to air), and h_h is the transfer coefficient (conductance) for sensible heat from the surface to the height where the air temperature, water vapor pressure (e), and wind (to find h_h) are measured; e^* is the saturation vapor pressure at air temperature, and s is the slope of the saturation vapor pressure curve at a temperature midway between the surface and air temperatures (in practice taken at air temperature). Penman and Schofield (144) and Penman (143) subsequently modified Eq. [2] to account for the difference in the transfer of heat from the leaf surface and vapor through the stomata. Monteith (124) applied their concepts to a canopy, which essentially assumes that the canopy behaves as a big leaf in the following equation:

$$\lambda ET = [s(R_n - G) + Ch_h(e^* - e)]/[s + \gamma(1 + h_h/h_s)] \qquad [3]$$

where h_h is a lumped heat conductance (including canopy boundary layers) from individual leaves to the height at which air temperature, e, and the wind velocity are measured and h_s is a lumped "surface" conductance. Equations [2] and [3] represent the one-dimensional flux density from the surface and are not applicable to small plots where local advection (two-dimensional sensible and latent heat fluxes) occur. Monteith (125) and Kanemasu et al. (97) provide historical and technical reviews of these formulae. The surface conductance in Eq. [3] appears closely related to the mean leaf stomatal conductance in a column of foliage divided by the LAI; however, evaporation from a wet soil surface affects the surface conductance because the canopy is not a "big leaf" as assumed for Eq. [2] (132). For conditions where $h_s \gg h_h$, Eq. [3] reduces to Eq. [2]. Although Eq. [3] has been used to estimate ET (e.g., 188) using estimates of h_s (typically 1.7 to 3.0 cm s^{-1} for well-watered alfalfa), Eq. [2] has been used much more widely for irrigation scheduling.

Most of the parameters needed to compute ET in Eq. [2] and [3] can be measured with standard meteorological instruments combined with inexpensive, battery-powered dataloggers. Only R_n is difficult to measure routinely, but can be estimated from global solar radiation measurements and estimates of albedo and of long-wave, thermal radiation (27, 84) as indicated by Tanner and Jury (194). Collecting data above the field of interest avoids problems of extrapolating from central weather stations sited above surfaces not representative of the crop. Also, the mean (e*−e) can be measured and errors due to estimating from maximum/minimum temperature data can be avoided (150). Some caution is required in estimating h_h from wind speed. As discussed by Kanemasu et al. (97), the transfer coefficient for momentum (h_m) can be determined from wind speed and rational estimates of momentum roughness length (z_o) can be made from crop height (for alfalfa $z_o \approx 0.12$ times the alfalfa height). Although many workers have used h_m rather than h_h in Eq. [2] and [3], h_m is typically 50% larger than h_h. Kanemasu et al. (97) also indicate how to estimate the effect of temperature lapses and inversions on h_h, as does Brutsaert (28); this could be important at low winds, and the magnitude of the effect on ET should be estimated as described by these authors. It should be noted that in humid regions where the $Ch_h(e^* - e)$ term in Eq. [2] is small relative to $s(R_n - G)$, reasonable estimates can be obtained despite large error in h_h. In arid regions where (e*−e) is large, h_h is needed for greater accuracy. Also in arid regions, local advection can be important and McNaughton (118) describes a method to account for advection.

Importantly, Eq. [2] and [3] have been used to estimate the daily ET using 24-h means of R_n, G, V (wind speed for h_h), air temperature, and e. This is convenient in that 24-h data are most available and the 24-h G usually is negligible. The R_n is weighted by nighttime thermal radiation loss and more importantly, h_h and (e^*-e) are weighted heavily by nighttime values when T is negligible. There is considerable value in using mean parameters only for the daytime period of T to avoid ambiguities introduced by the very different night-time conditions when there is little T (150, 190, 195).

11–5.3.3 Priestley-Taylor and Makkink Formulas

Priestley and Taylor (148) found the ET from several wet natural surfaces, in absence of thermal inversions, to be correlated well with the radiation term in Eq. [2] as follows:

$$ET = \alpha_{pt}[s/(s+\gamma)](R_n - G) \qquad [4]$$

with $\alpha_{pt} \approx 1.3$. In humid regions α_{pt} for full-cover, well-watered crops will be within 10% of 1.3 (50). Others using 24-h R_n and G also have found good estimates of ET for full-cover, well-watered vegetation in humid regions where the convective heat supply to the surface is small (e.g., 28). This implies that the convective term in Eq. [2] is correlated with the radiative term despite varying temperature and wind conditions although a good correlation would not be expected. There is also no reason to expect the empirical α_{pt} to be constant, and indeed, α_{pt} has been found to vary seasonally for well-watered grass (129). The seasonal variation observed in α_{pt} was explained by the increased nighttime thermal radiation loss during shorter days, thus decreasing the 24-h R_n used in Eq. [4] relative to the daytime R_n, which drives T.

Earlier Makkink (114), recognizing that the albedo of many full-cover crop surfaces was essentially constant and that the net radiation often was proportional to the global (solar) radiation, proposed

$$\lambda ET = \alpha_m[s/(s+\gamma)]R_G \qquad [5]$$

where R_G is the global radiation. Pruitt and Doorenbos (150) found the Makkink α_m varied much less seasonally than α_{pt}. The Makkink formula neglects the soil heat flux which is much smaller than, and correlated with R_G.

In humid regions where the convective term in Eq. [2] is small, Eq. [4] and [5] offer operational convenience for ET estimates because measurements of vapor pressure and wind over the crop are not required. However, it is essential to calibrate these estimates and to establish α_{pt} or α_m by experiment. During drought years in humid regions, sufficient sensible heat may be carried into irrigated fields from drier adjacent areas so that ET estimates from these simpler formulae would be less satisfactory than those from Eq. [2].

11-5.3.4 Calibration

Any empirical ET estimate such as pans or the Penman, Priestley-Taylor and Makkink formulas should be calibrated for a particular crop and climate (150, 192). Comparison with lysimeter data is best, but neutron meter measurements over longer periods can be used if drainage is accounted for. While the Penman formula may apply to well-watered, full-cover crops over a greater range of climate regimes than either the Priestley-Taylor or Makkink formulas (or other even less physically based empiricisms), there still are sufficient ambiguities in representing the crop by the "big leaf" assumptions that calibration is advisable.

11-5.3.5 Crop Coefficients and Irrigation Scheduling

Because pan and climatonomic estimates are for maximum rates of ET (ET_{max}), ET is overestimated following cutting and until crop regrowth provides an LAI of 1.5 to 2. A standard practice is to use variable crop coefficients which account for the fractional reduction in ET below the ET_{max} (Fig. 11-8). The crop coefficient is defined as $K_c = ET/ET_{max}$, where K_c is determined by measurement of ET at different regrowth stages (50, 92, 183). Typically, the ET following cutting is measured for a dry soil surface, but with sufficient water in the root zone that T is not restricted. If the soil surface is wetted by rain then K_c is increased to account for the increased evaporation from the soil (223). Alternatively, evaporation from the soil can be estimated as described by Black et al. (17) and used by Ritchie (159). Procedures for estimating T as a function of regrowth were given by Ritchie and by Tanner and Jury (194) and can be modified to give ET/ET_{max} as a function of ground cover.

Irrigation has been scheduled using water budget models (93, 183, 219). These methods, some of which are computerized, sum soil water

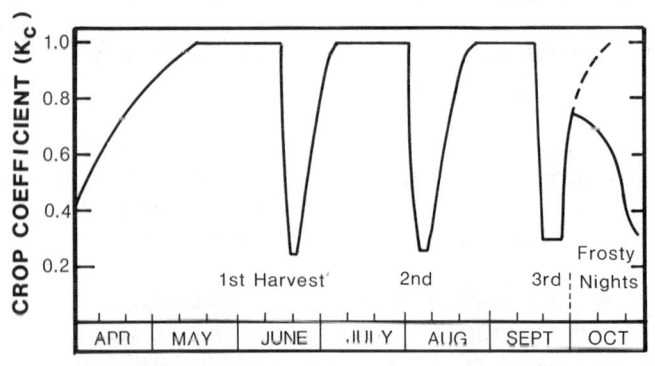

Fig. 11-8. Average seasonal basal crop coefficient curve for alfalfa in Kimberly, ID. Adapted from Wright (223).

depletions (determined by climatonomic estimates) and deduct water inputs (precipitation and irrigation) to provide net soil water status. For example, in the procedure described by Lundstrom and Stegman (113) potential ET is estimated using the Jensen-Haise equation (93) and a locally developed crop coefficient (K_c, e.g., Fig. 11-8) to correct for crop development. Potential ET is estimated as a function of the daily maximum air temperature and the average climatic solar radiation curve for a given location. Simplified versions of the above procedure require only input of maximum air temperature data to predict daily water use (219). However, the performance of such simplified models has been very variable and often unreliable. Tensiometer measurements are made to ensure that cumulative errors (from either ET estimates or water input estimates) in budgeting do not lead to excessive soil water depletion or excessive irrigation. Accuracy in using climatonomic methods to estimate ET will increase with access to improved meteorological data, and locally developed crop coefficient curves. This information may be more available in the future.

11-5.4 Crop Temperature

If sufficient water deficits develop in transpiring alfalfa that stomata begin to close and T decreases, the energy formerly used in T heats the leaves and is partitioned into sensible heat loss. The rise in leaf temperature accompanying a decrease in T can be used as a stress indicator for irrigation scheduling (191, 215). Infrared (IR) thermometry has been used to determine foliage temperatures since the early 1960s. Development of small, portable, handheld IR thermometers has greatly expedited the measurements, and considerable work has been done to quantify crop water stress via IR thermometer measurements. Jackson (86), O'Toole and Real (136), and Nielsen et al. (130) reviewed historical work, instrument calibration, and precautions for obtaining good measurements. Although IR thermometry provided convenient measurements of canopy temperature, leaf temperature is affected by other energy balance parameters so that the rise in leaf temperature must be referenced in some way.

Assuming that the canopy represents a big leaf, arguments similar to those used to derive Eq. [2] can be used to calculate the canopy foliage temperature (88). Using the same symbols as Eq. [2], but for measurement at a given time rather than time average values:

$$F_t - A_t = [(R_n - G)(\gamma/CH_h)][(1 + h_h/h_s) - (e^* - e)]/[s + \gamma(1 + h_h/h_s)] \quad [6]$$

where F_t and A_t are foliage and air temperatures, respectively. The foliage temperature is influenced by the stomata through h_s. Stomata respond to light and leaf/air vapor pressure gradients as well as water stress. Additionally, as Eq. [6] shows, F_t depends on air temperature, radiation (212), saturation deficit (85), and wind (135). Idso et al., (85) reported foliage—air temperature differences as a function of air vapor pressure deficit (Fig. 11-9). They also developed an index to aid in evaluation of water stress.

There are three approaches to determining the onset of stress with rise in foliage temperature. First, as suggested by Tanner (191) and Fuchs and Tanner (61), the foliage temperature of the test field can be compared with a well-watered reference area in the field. The main inconvenience in this approach is maintaining an irrigated reference area. Conversely, it has the advantage that environmental parameters are the same for the test and reference areas except those influenced by stress and decreasing T. Several reports indicate an increase in the temperature difference between unstressed and stressed alfalfa as moisture deficits increase (e.g., 35, 103, 172).

The second approach is to measure the spatial variability of foliage temperatures in a field, as suggested by Aston and van Bavel (5). Recognizing that there is spatial variability in infiltration and redistribution of soil water in the profile, they argued that plants in some areas would become stressed before plants in other areas. Accordingly, increased spatial variability of foliage temperatures could indicate the onset of stress. This approach has been used with maize by Gardner et al. (62) and Clawson and Blad (40).

The third approach, detailed by Jackson et al. (88), is to compare the measured foliage-air temperature difference with two extreme temperature differences calculated from Eq. [6] for prevailing atmospheric conditions: an upper limit found when h_s approaches zero, and a lower limit when h_s approaches infinity. The severity of the stress is indicated by the departure of the measured temperature from the lower limit. This approach should correct for emissivity of the canopy (0.97 to 0.98 for full-cover alfalfa) (86, 87).

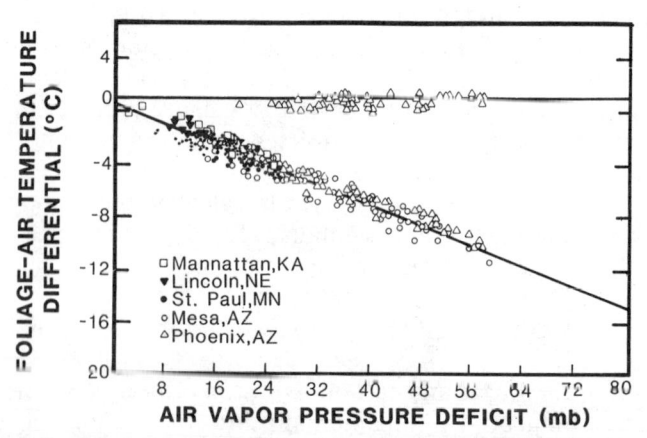

Fig. 11-9. Foliage-air temperature differential as determined by infrared thermometry vs. air vapor pressure deficit for well-watered alfalfa grown at several sites. Adapted from Idso et al. (85).

11-6 IRRIGATION APPLICATION SYSTEMS

Water is applied to alfalfa by three general systems; surface, sprinkler, and subirrigation. Examples and conditions for use of these are discussed by Bishop et al. (16); Christiansen and Davis (39); and Criddle and Kalisvaart (46). System choice is influenced by soil type, water supply, topography, and economics.

11-7 SUMMARY

Alfalfa water use or ET for forage production ranges from 1.3 to 14 mm d^{-1} and from 400 to 1900 mm per season. Daily and seasonal ET are influenced by climatic factors such as daily temperature and advection. Also, seasonal ET is influenced by length of growing season. There is a linear relationship between seasonal ET and forage dry matter yield under nonlimiting soil fertility. Approximately 5.6 to 8.3 cm of water are required to produce a Mg ha^{-1} of forage. However, these water requirements vary greatly with climate and crop and irrigation management. While forage yield is decreased by moisture deficits, forage quality is sometimes increased due primarily to changes in leaf/stem ratio and a decrease in forage structural components. Alfalfa is intolerant of saturated soil conditions, which are associated with excessive irrigation or poorly drained soils. Proper scheduling of irrigation and site selection may reduce this effect.

Alfalfa adaptation to stress from soil moisture deficits is enhanced by its extensive tap root system. Most root mass is in the upper soil profile, which consequently is the major zone of water extraction. In general, root mass is decreased by soil moisture deficits.

As in many crops, water loss via T is regulated primarily by stomata, which close at ψ_w of < -2.0 to -2.5 MPa. Moisture deficits affect processes such as CER, N$_2$ fixation, and C partitioning. Inadequate information is available on the effects of moisture stress on CER, osmotic adjustment, and water conductance.

Some researchers evaluating effects of irrigation on yield have failed to measure alfalfa response (i.e., ψ_w) to soil moisture treatments. As a result, it is difficult to compare results from experiments in which only water application is reported. While the difficulty of field research is appreciated, limitations of studies in containers with limited rooting volumes must be recognized.

Irrigation can be scheduled using plant- or soil-based criteria. Soil water depletion is the most widely used. Depletion can be measured directly (gravimetrically), indirectly (e.g., tensiometers) or based on pan or climatonomic models that estimate daily ET. Increased local monitoring of climatic conditions should increase the accuracy of scheduling models which employ ET estimates.

The study of alfalfa response to moisture deficits is directed at increasing efficiency of irrigation scheduling and forage production. Increased knowledge of physiological processes and crop development under moisture stress could contribute to the development of cultivars with greater drought resistance.

REFERENCES

1. Abdul-Jabbar, A.S., D.G. Lugg, T.W. Sammis, and L.W. Gay. 1984. A field study of plant resistance to water flow in alfalfa. Agron. J. 76:765-769.
2. ----, T.W. Sammis, and D.G. Lugg. 1982. Effect of moisture level on the root pattern of alfalfa. Irrig. Sci. 3:197-207.
3. Andrew, V.L., D.G. Lugg, A.S. Abdul-Jabbar, and T.W. Sammis. 1984. Carbohydrate levels and rooting patterns of alfalfa under different moisture regimes. Agron. Abstr. American Society of Agronomy, Madison, WI, p. 97.
4. Aparicio-Tejo, P.M., M.F. Sanchez-Diaz, and J.I. Pena. 1980. Nitrogen fixation, stomatal response and transpiration in *Medicago sativa*, *Trifolium repens* and *T. subterraneum* under water stress and recovery. Physiol. Plant. 48:1-4.
5. Aston, A.R., and C.H.M. van Bavel. 1972. Soil surface water depletion and leaf temperatures. Agron. J. 64:368-373.
6. Baldocchi, D.D., S.B. Verma, and N.J. Rosenberg. 1981. Seasonal and diurnal variation in CO_2 flux and CO_2-water flux ratio of alfalfa. Agric. Meteorol. 23:231-244.
7. ----, ----, ----, B.L. Blad, A. Garay, and J.E. Specht. 1983. Leaf pubescence effects on the mass and energy exchange between soybean canopies and the atmosphere. Agron. J. 75:537-543.
8. Barnes, D.K. 1983. Managing root systems for efficient water use: breeding plants for efficient water use. p. 127-136. *In* H.M. Taylor et al. (ed.) Limitations to efficient water use in crop production. American Society of Agronomy, Crop Science Society, and Soil Science Society of America, Madison, WI.
9. Barta, A.L. 1980. Regrowth and alcohol dehydrogenase activity in waterlogged alfalfa and birdsfoot trefoil. Agron. J. 72:1017-1020.
10. Bauder, J.W., A. Bauer, J.M. Ramirez, and D.K. Cassel. 1978. Alfalfa water use and production on dryland and irrigated sandy loam. Agron. J. 70:95-99.
11. Baughn, J.W., and C.B. Tanner. 1976. Excision effects on leaf water potential of five herbaceous species. Crop Sci. 16:184-190.
12. Begg, J.E., and N.C. Turner. 1976. Crop water deficits. Adv. Agron. 28:161-217.
13. Bell, J.R. 1976. Neutron probe practice. Inst. Hydrol. Rep. 9. Natural Environment Research Council, Wallingford, UK.
14. Bennett, O.L., and B.D. Doss. 1960. Effect of soil moisture level on root distribution of cool-season forage species. Agron. J. 52:204-207.
15. Benz, L.C., E.J. Doering, and G.A. Reichman. 1982. Water table and irrigation effects on alfalfa grown on sandy soils. Can. Agric. Eng. 24:71-74.
16. Bishop, A.A., M.E. Jensen, and W.A. Hall. 1967. Surface irrigation systems. *In* R.M. Hagan et al. (ed.) Irrigation of agricultural lands. Agronomy 11:865-884.
17. Black, T.A., W.R. Gardner, and C.B. Tanner. 1970. Water storage and drainage under a row crop on a sandy soil. Agron. J. 62:48-51.
18. Borg, H. 1980. Plant available water. M.S. thesis, Univ. of Wisconsin-Madison.
19. Bourget, S.J., and R.B. Carson. 1962. Effect of soil moisture stress on yield, water-use efficiency and mineral composition of oats and alfalfa grown at two fertility levels. Can. J. Soil Sci. 42:7-12.
20. Bouton, J.H., J.E. Hammel, and M.E. Sumner. 1982. Alfalfa, *Medicago sativa* L., in highly weathered, acid soils. IV. Root growth into acid subsoil of plants selected for acid tolerance. Plant Soil 65:187-192.
21. Bowen, H.D. 1981. Alleviating mechanical impedence. p. 21-57. *In* G.F. Arkin and H.M. Taylor (ed.) Modifying the root environment to reduce crop stress. American Society of Agricultural Engineers, St. Joseph, MI.
22. Brown, P.L., A.D. Halvorson, F.H. Siddoway, H.F. Mayland, and M.R. Miller. 1983. Saline-seep diagnosis, control, and reclamation. Conserv. Res. Rep. 30. USDA-ARS, Washington, DC.
23. Brown, P.W., and C.B. Tanner. 1981. Alfalfa water potential measurement: a comparison of the pressure chamber and leaf dew point hygrometer. Crop Sci. 21:240-244.
24. ----, and ----. 1983. Alfalfa osmotic potential: a comparison of water release curve and frozen tissue methods. Agron. J. 75:91-93.

25. ----, and ----. 1983. Alfalfa stem and leaf growth during water stress. Agron. J. 75:799-805.
26. Brun, L.J., and B.K. Worcester. 1975. Soil water extraction by alfalfa. Agron. J. 67:586-589.
27. Brutsaert, W.H. 1975. On a derivable formula for long-wave radiation from clear skies. Water Resour. Res. 11:742-744.
28. ----. 1982. Evaporation into the atmosphere. D. Reidel Publishing Co., Boston.
29. Calder, F.W., L.B. MacLeod, and L.P. Jackson. 1965. Effect of soil moisture content and stage of development on cold hardiness of the alfalfa plant. Can. J. Plant Sci. 45:211-218.
30. Cameron, D.G. 1973. Lucerne in wet soils—the effect of stage of regrowth, cultivar, air temperature, and root temperature. Aust. J. Agric. Res. 24:851-861.
31. Campbell, G.S., R.I. Papendick, E. Rabie, and A.J. Shayo-Ngowi. 1979. A comparison of osmotic potential, elastic modulus, and apoplastic water in leaves of dryland winter wheat. Agron. J. 71:31-36.
32. Cardwell, V.B. 1984. Germination and crop production. p. 53-92. *In* M.B. Tesar (ed.) Physiological basis of crop growth and development. American Society of Agronomy, Madison, WI.
33. Carlson, J.R., Jr., R.L. Ditterline, J.M. Martin, and R.E. Lund. 1981. Sampling stomatal density in alfalfa. Crop Sci. 21:467-469.
34. Carter, P.R., and C.C. Sheaffer. 1983a. Alfalfa response to soil water deficits. I. Growth, forage quality, yield, water use, and water-use efficiency. Crop Sci. 23:669-675.
35. ----, and ----. 1983b. Alfalfa response to soil water deficits. II. Plant water potential, leaf conductance and canopy temperature relationships. Crop Sci. 23:676-680.
36. ----, and ----. 1983c. Alfalfa response to soil water deficits. III. Nodulation and N_2 fixation. Crop Sci. 23:985-990.
37. ----, ----, and W.B. Voorhees. 1982. Root growth, herbage yield, and plant water status of alfalfa cultivars. Crop Sci. 22:425-427.
38. Christian, K.R. 1977. Effects of the environment on the growth of alfalfa. Adv. Agron. 29:183-227.
39. Christiansen, J.E., and J.R. Davis. 1967. Sprinkler irrigation systems. *In* R.M. Hagan et al. (ed.) Irrigation of agricultural lands. Agronomy 11:885-904.
40. Clawson, K.L., and B.L. Blad. 1982. Infrared thermometry for scheduling irrigation of corn. Agron. J. 74:311-316.
41. Cohen, Y., H. Bielorai, and A. Dovrat. 1972. Effect of timing of irrigation on total nonstructural carbohydrate level in roots and on seed yield of alfalfa (*Medicago sativa* L.). Crop Sci. 12:634-636.
42. Cole, D.F., and A.K. Dobrenz. 1970. Stomate density of alfalfa (*Medicago sativa* L.). Crop Sci. 10:61-63.
43. ----, ----, M.A. Massengale, and L.N. Wright. 1970. Water requirement and its association with growth components and protein content of alfalfa (*Medicago sativa* L.). Crop Sci. 10:237-239.
44. Corleto, A., and H.M. Laude. 1974. Evaluating growth potential after drought stress. Crop Sci. 14:224-227.
45. Cowett, E.R., and M.A. Sprague. 1962. Factors affecting tillering in alfalfa. Agron. J. 54:294-297.
46. Criddle, W.D., and C. Kalisvaart. 1967. Subirrigation systems. *In* R.M. Hagan et al. (ed.) Irrigation of agricultural lands. Agronomy 11:905-921.
47. Daigger, L.A., L.S. Axthelm, and C.L. Ashburn. 1970. Consumptive use of water by alfalfa in western Nebraska. Agron. J. 62:507-508.
48. de Wit, C.T. 1958. Transpiration and crop yields. Vers. Landbouwk. Onderz 64.6. Institute of Biological and Chemical Research on Field Crops and Herbage, Wageningen, Netherlands.
49. Donovan, T.J., and B.D. Meek. 1983. Alfalfa responses to irrigation treatment and environment. Agron. J. 75:461-464.
50. Doorenbos, J., and W.O. Pruitt. 1977. Crop water requirements. Irrigations and Drainage Paper 24 (revised). FAO, Rome.
51. Dotzenko, A.D., and J.G. Dean. 1959. Germination of six alfalfa varieties at three levels of osmotic pressure. Agron. J. 51:308.
52. Engin, M., and J.I. Sprent. 1973. Effects of water stress on growth and nitrogen-fixing activity of *Trifolium repens*. New Phytol. 72:117-126.
53. Erie, L.J., O.F. French, and Karl Harris. 1965. Consumptive use of water by crops in Arizona. Ariz. Agric. Exp. Stn. Tech. Bull. 169.
54. Esau, K. 1965. Plant anatomy. 2nd ed. John Wiley and Sons, New York.
55. Faiz, S.M.A., and P.E. Weatherley. 1978. Further investigations into the location and magnitude of the hydraulic resistances in the soil-plant system. New Phytol. 81:19-28.

56. Fehrenbacher, J.B., B.W. Ray, and W.M. Edwards. 1965. Rooting volume of corn and alfalfa in shale-influenced soils in northwestern Illinois. Soil Sci. Soc. Am. Proc. 29:591–594.
57. Finkel, H.J. 1983. Irrigation of alfalfa. p. 191–197. *In* H.J. Finkel (ed.) CRC handbook of irrigation technology. Vol. 2. CRC Press, Boca Raton, FL.
58. Fitzgerald, P.D., T.L. Knight, and G.G. Janson. 1976. Lucerne irrigation on light soils. N.Z. J. Exp. Agric. 5:23–27.
59. Fox, R.L., and R.C. Lipps. 1955. Subirrigation and plant nutrition. I. Alfalfa root distribution and soil properties. Soil Sci. Soc. Am. Proc. 19:468–473.
60. Francois, L.E. 1981. Alfalfa management under saline conditions with zero leaching. Agron. J. 73:1042–1046.
61. Fuchs, M., and C.B. Tanner. 1966. Infrared thermometry of vegetation. Agron. J. 58:597–601.
62. Gardner, B.R., B.L. Blad, and D.G. Watts. 1981. Plant and air temperature in differentially-irrigated corn. Agric. Meteorol. 25:207–217.
63. Gardner, W.R., A.W. Warrick, and A.D. Halderman. 1982. Soil variability and measures of irrigation efficiency. Paper 82-2106, Madison, WI. 27–30 June. American Society of Agricultural Engineers, St. Joseph, MI.
64. Garver, S. 1922. Alfalfa root studies. USDA Bull. 1087. U.S. Government Printing Office, Washington, DC.
65. Gibson, A.H., and D.C. Jordan. 1983. Ecophysiology of nitrogen-fixing systems. p. 301–390. *In* O.L. Lange et al. (ed.) Physiological plant ecology. Vol. III. Responses to the chemical and biological environment. Springer-Verlag, Berlin.
66. Gindel, I. 1968. Dynamic modifications in alfalfa leaves growing in subtropical conditions. Physiol. Plant. 21:1287–1295.
67. Gist, G.R., and G.O. Mott. 1957. Some effects of light intensity, temperature and soil moisture on the growth of alfalfa, red clover and birdsfoot trefoil seedlings. Agron. J. 49:33–36.
68. Goldman, A., and A. Dovrat. 1980. Irrigation regime and honeybee activity as related to seed yield in alfalfa. Agron. J. 72:961–965.
69. Graber, L.F., N.T. Nelson, W.A. Luekel, and W.B. Albert. 1927. Organic food reserves in relation to the growth of alfalfa and other perennial herbaceous plants. Univ. of Wisconsin Agric. Exp. Stn. Bull. 491.
70. Graham, H.J., F.I. Frosheiser, D.L. Stuteville, and D.C. Erwin. 1979. A compendium of alfalfa diseases. American Phytopathological Society, St. Paul.
71. Grange, R.E. 1983. Solute production during measurement of solute potential on disrupted tissue. J. Exp. Bot. 34:757–764.
72. Haise, H.R., and R.M. Hagan. 1967. Soil, plant, and evaporative measurements as criteria for scheduling irrigation. *In* R.M. Hagan et al. (ed.) Irrigation of agricultural lands. Agronomy 11:577–604.
73. Hall, M.H., C.C. Sheaffer, and G.H. Heichel. 1986. Patterns of photoassimilate partitioning in water stressed alfalfa. Agron. Abstr. American Society of Agronomy, Madison, WI, p. 95.
74. Hall, R.G., and K.L. Larson. 1982. Water stress of alfalfa during stress and recovery. Can. J. Plant Sci. 62:639–647.
75. Halim, R.A., and D.R. Buxton. 1985. Moisture stress effects on nutritive value of alfalfa. Iowa Academy of Sci. 92(2): abstr. 7.
76. Halvorson, A.D., and C.A. Reule. 1980. Alfalfa for hydrologic control of saline seeps. Soil Sci. Soc. Am. J. 44:370–374.
77. Hanson, E.G. 1967. Influence of irrigation practices on alfalfa yield and consumptive use. N. M. Agric. Exp. Stn. Bull. 514.
78. Heichel, G.H. 1983. Alfalfa. p. 127–155. *In* I.D. Teare and M.M. Peet (ed.) Crop-water relations. John Wiley and Sons, New York.
79. Heinrichs, D.H. 1959. Germination of alfalfa varieties in solutions of varying osmotic pressure and relationship to winterhardiness. Can. J. Plant Sci. 39:384–394.
80. Henderson, D.W., and H. Yamada. 1979. Irrigation management for alfalfa seed production. Univ. California West Side Field Stn. Annu. Rep.
81. Hodgkinson, K.C., and H.G. Baas Becking. 1977. Effect of defoliation on root growth of some arid zone perennial plants. Aust. J. Agric. Res. 29:31–42.
82. Horst, G.L., and C.J. Nelson. 1979. Compensatory growth of tall fescue following drought. Agron. J. 71:559–563.
83. Hsiao, T.C. 1973. Plant responses to water stress. Annu. Rev. Plant Physiol. 24:519–570.
84. Idso, S.B. 1981. A set of equations for full spectrum and 8 to 14 m and 10.5 to 12.5 m thermal radiation from cloudless skies. Water Resour. Res. 17:295–304.
85. ----, R.J. Reginato, D.C. Reicosky, and J.L. Hatfield. 1981. Determining soil induced plant water potential depressions in alfalfa by means of infrared thermometry. Agron. J. 73:826–830.

86. Jackson, R.D. 1982a. Canopy temperature and crop water stress. p. 43–85. In D. Hillel (ed.) Advances in irrigation, Vol. 1. Academic Press, New York.
87. ———. 1982b. Soil moisture inferences from thermal-infrared measurements of vegetation temperatures. IEEE Trans. Geosci. Remote Sens. GE-20:282–286.
88. ———, S.B. Idso, R.J. Reginato, and P.J. Pinter, Jr. 1981. Canopy temperature as a crop water stress indicator. Water Resour. Res. 17:1133–1138.
89. Janson, C.G. 1976. Lucerne irrigation: principles, practice and prospects. Span 19:37–39.
90. Jean, F.C., and J.E. Weaver. 1924. Root behavior and crop yield under irrigation. Pub. 357. Carnegie Institute, Washington, DC.
91. Jensen, E.H., M.A. Massengale, and D.O. Chilcote. 1967. Environmental effects on growth and quality of alfalfa. Nev. Agric. Exp. Stn. Western Regional Res. Pub. T9.
92. Jensen, M.E. 1974. Consumptive use of water and irrigation requirements. Report of the technical committee on irrigation water requirements, Irrig. Drain. Div., Am. Soc. Civ. Eng., New York.
93. ———, D.C.N. Robb, and C.E. Franzoy. 1970. Scheduling irrigations using climatic–crop-soil data. Irrig. Drain. Div. Am. Soc. Civ. Eng. J. 96 (IR1):25–38.
94. Jodari-Karimi, F., V. Watson, H. Hodges, and F. Whisler. 1983. Root distribution and water use efficiency of alfalfa as influenced by depth of irrigation. Agron. J. 75:207–211.
95. Jones, H.G. 1979. Visual estimation of plant water status of cereals. J. Agric. Sci. 92:83–89.
96. Joy, R.J., H.T. Poole, and A.K. Dobrenz. 1972. The effect of soil moisture regimes on water use efficiency and growth components of alfalfa. AZ Prog. Agric. 29:9–11.
97. Kanemasu, E.T., M.L. Wesely, B.B. Hicks, and J.L. Heilman. 1979. Techniques for calculating energy and mass fluxes. p. 156–182. In B.J. Barfield and J.F. Gerber (ed.) Modification of the aerial environment of plants. Monogr. 2. American Society of Agricultural Engineers, St. Joseph, MI.
98. Katerji, N., M. Hallaire, A. Perrier, and R. Durand. 1983. Transfert hydrique dans le vegetal. Acta Oecol., Oecol. Plant. 4:11–26.
99. Keller, W., and C.W. Carlson. 1967. Forage crops. In R.M. Hagan et al. (ed.) Irrigation of agricultural lands. Agronomy 11:607–621.
100. Kemper, W.D., and M. Amemiya. 1957. Alfalfa growth as affected by aeration and soil moisture stress under flood irrigation. Soil Sci. Soc. Am. Proc. 21:657–660.
101. Kiesselbach, T.A., J.C. Russel, and A. Anderson. 1929. The significance of subsoil moisture in alfalfa production. J. Am. Soc. Agron. 21:241–268.
102. Kirkham, M.B. 1983. Water resources research: potential contributions by plant and biological scientists. p. 157–170. In T.L. Napier et al. (ed.) Water resources research. Soil Conservation Society of America, Ankeny, IA.
103. ———, D.E. Johnson, Jr., E.T. Kanemasu, and L.R. Stone. 1983. Canopy temperature and growth of differentially irrigated alfalfa. Agric. Meteorol. 29:235–246.
104. Kohl, R.A., and J.J. Kolar. 1976. Soil water uptake by alfalfa. Agron. J. 68:536–538.
105. Kramer, P.J. 1983. Water relations of plants. Academic Press, New York.
106. Kreitner, G.L., and E.L. Sorensen. 1983. Erect glandular trichomes of *Medicago scutellata* (L.) Mill.: gland development and early secretion. Bot. Gaz. 144:165–174.
107. Krogman, K.K., and E.H. Hobbs. 1965. Evapotranspiration by irrigated alfalfa as related to season and growth stage. Can. J. Plant Sci. 45:309–313.
108. ———, and ———. 1977. Irrigation management of alfalfa for seed. Can. J. Plant Sci. 57:891–896.
109. Lambert, D.H., H. Cole, and D.E. Baker. 1980. Variation in the response of alfalfa clones and cultivars to mycorrhizae and phosphorus. Crop Sci. 29:615–618.
110. Lehman, W.F. 1979. Alfalfa production in the low desert valley areas of California. Univ. of California Leaflet 21097.
111. Levitt, J. 1972. Responses of plants to environmental stress. Academic Press, New York.
112. Lucey, R.F., and M.B. Tesar. 1965. Frequency and rate of irrigation as factors in forage growth and water absorption. Agron. J. 57:519–523.
113. Lundstrom, D.L., and E.C. Stegman. 1977. Checkbook method of irrigation scheduling. Paper NCR. 77-101. American Society of Agricultural Engineers, St. Joseph, MI.
114. Makkink, G.F. 1957. Ekzameno de la formula de Penman. Neth. J. Agric. Sci. 5:290–350.
115. McElgunn, J.D., and D.H. Heinrichs. 1975. Water use of alfalfa genotypes of diverse genetic origin at three soil temperatures. Can. J. Plant Sci. 55:705–708.
116. McIntosh, M.S., and D.A. Miller. 1981. Genetic and soil moisture effects on the branching-root trait in alfalfa. Crop Sci. 21:15–18.
117. McKenzie, J.S., and G.E. McLean. 1980. Changes in the cold hardiness of alfalfa

during five consecutive winters at Beaverlodge, Alberta. Can. J. Plant Sci. 60:703-712.
118. McNaughton, K.G. 1976. Evaporation and advection. J. R. Meteorol. Soc. 102:181-202.
119. McWilliam, J.R., R.J. Clements, and P.M. Dowling. 1970. Some factors influencing the germination and early seedling development of pasture plants. Aust. J. Agric. Res. 21:19-32.
120. Meek, B.D., T.J. Donovan, and L.E. Graham. 1980. Summertime flooding effects on alfalfa mortality, soil oxygen concentration, and matric potential in silty clay loam soil. Soil Sci. Soc. Am. J. 44:433-435.
121. Meinzer, O.E. 1927. Water supply. Paper A77. U.S. Government Printing Office, Washington, DC.
122. Metochis, Chr., and P.I. Orphanos. 1981. Alfalfa yield and water use when forced into dormancy by withholding water during the summer. Agron. J. 73:1048-1050.
123. Molz, F.J. 1981. Models of water transport in the soil-plant-atmosphere system. A review. Water Resour. Res. 17:1245-1260.
124. Monteith, J.L. 1965. Evaporation and environment. p. 205-234. *In* G.E. Fogg (ed.) The state and movement of water in living organisms. Academic Press, New York.
125. ----. 1981. Evaporation and surface temperature. J. R. Meteorol. Soc. 107:1-27.
126. Moore, R.P. 1943. Seedling emergence of small seeded legumes and grasses. J. Am. Soc. Agron. 35:370-381.
127. Murata, Y., J. Iyama, and T. Honma. 1966. Studies on the photosynthesis of forage crops. The influence of soil moisture content on the photosynthesis and respiration of seedlings in various forage crops. Proc. Crop Sci. Soc. Jpn. 34:385-390.
128. Nakayama, F.S., and R.J. Reginato. 1982. Simplifying neutron moisture meter calibration. Soil Sci. 133:48-52.
129. Nakayama, K., W.O. Pruitt, and B.A. Chandio. 1983. Some aspects of the Priestly and Taylor model to estimate evapotranspiration. Tech. Bull. Fac. Hortic., Chiba Univ. 32:25-30.
130. Nielsen, D.C., K.L. Clawson, and B.L. Blad. 1984. Effect of solar azimuth and infrared thermometer view direction on measured soybean canopy temperature. Agron. J. 76:607-610.
131. Neumann, H.H., G.W. Thurtell, K.R. Stevenson, and C.L. Beadle. 1974. Leaf water content and potential in corn, sorghum, soybean and sunflower. Can. J. Plant Sci. 54:185-195.
132. Norman, J.M., and G. Campbell. 1983. Application of a plant-environment model to problems of irrigation. p. 155-188. *In* D. Hillel (ed.) Advances in irrigation, Vol. 2. Academic Press, New York.
133. Ogata, G., L.A. Richards, and W.R. Gardner. 1960. Transpiration of alfalfa determined from soil water content changes. Soil Sci. 89:179-182.
134. Othman, R.B., E.L. Sorensen, G.H. Liang, and E.K. Horber. 1981. Density and distribution of erect glandular hairs on annual *Medicago* species. Bot. Gaz. 142:237-241.
135. O'Toole, J.C., and J.L. Hatfield. 1983. Effect of wind on the crop water stress index derived by infrared thermometry. Agron. J. 75:811-817.
136. ----, and J. Real. 1984. Canopy target dimensions for infrared thermometry. Agron. J. 76:863-865.
137. Paquin, R., and G.R. Mehuys. 1980. Influence of soil moisture on cold tolerance of alfalfa. Can. J. Plant Sci. 60:139-147.
138. Passioura, J.B. 1980. The meaning of matric potential. J. Exp. Bot. 31:1161-1169.
139. Peck, N.H., M.T. Vittum, and R.D. Miller. 1958. Evapotranspiration rates for alfalfa and vegetable crops in New York State. Agron. J. 50:109-112.
140. Pedersen, M.W., G.E. Bohart, V.L. Marble, and E.C. Klostermeyer. 1972. Seed production practices. *In* C.H. Hanson (ed.) Alfalfa science and technology. Agronomy 15:689-720.
141. Pederson, G.A., R.R. Hill, and W.A. Kendall. 1984. Genetic variability for root characters in alfalfa populations differing in winterhardiness. Crop Sci. 24:465-468.
142. Penman, H.L. 1948. Natural evaporation from open water, bare soil, and grass. Proc. R. Soc. London 193:120-146.
143. ----. 1953. The physical bases of irrigation control. Rep. 13th Int. Hortic. Congr. (1952) 2:913-924.
144. ----, and R.K. Schofield. 1951. Some physical aspects of assimilation and transpiration. Symp. Soc. Exp. Biol. 5:115-129.
145. Perry, L.J., and K.L. Larson. 1974. Influence of drought on tillering and internode number and length in alfalfa. Crop Sci. 14:693-696.
146. Peterschmidt, N.A., R.H. Delaney, and M.C. Greene. 1979. Effects of overirrigation on growth and quality of alfalfa. Agron. J. 71:752-754.
147. Pfeiffer, T., L.E. Schrader, and E.T. Bingham. 1980. Physiological comparisons of

isogenic diploid-tetraploid, tetraploid-octaploid alfalfa populations. Crop Sci. 20:299-303.
148. Priestley, C.H.B., and R.J. Taylor. 1972. On the assessment of surface heat flux and evaporation using large scale parameters. Mont. Weather Rev. 100:81-92.
149. Pruitt, W.O. 1966. Empirical method of estimating evapotranspiration using primarily evaporation pans. p. 57-61. *In* Conf. Evapotranspiration and its role in water resources management, Chicago. 5-6 December. American Society of Agricultural Engineers, St. Joseph, MI.
150. ----, and J. Doorenbos. 1977. Empirical calibration, a requisite for evapotranspiration formulae based on daily or longer mean climatic data. Int. Round Table Conf. on Evapotranspiration, Budapest, Hungary. 26-28 May. International Committee on Irrigation and Drainage, New Dehli, India.
151. Radeva, V., and A. Topchieva. 1979. Effect of periodic soil moisture tension on dry matter accumulation, seed yield and some physiological characters of lucerne. Commonwealth Agric. Bur. Record 1192037. Rasteniev'dni Nauki 16:13-25. (In Bulgarian, English sum.)
152. Reicosky, D.C., T.C. Kaspar, and H.M. Taylor. 1982. Diurnal relationships between evapotranspiration and leaf water potential of field grown soybeans. Agron. J. 74:667-673.
153. ----, and J.T. Richie. 1976. Relative importance of soil resistance and plant resistance in root water absorption. Soil Sci. Soc. Am. J. 40:293-297.
154. Reid, C.P., and G.D. Bowen. 1979. Effects of soil moisture on V/A mycorrhiza formation and root development in *Medicago*. p. 211-219. *In* J.L. Harley and R. Scott Russell (ed.) The soil-root interface. Academic Press, London.
155. Retta, A., and R.J. Hanks. 1980. Corn and alfalfa production as influenced by limited irrigation. Irrig. Sci. 1:135-147.
156. Richards, L.A. (ed.) 1954. Diagnosis and improvement of saline and alkali soils. USDA Agric. Handb. 60. U.S. Government Printing Office, Washington, DC.
157. Richter, H. 1978. Water relations of single drying leaves: evaluation with a dewpoint hygrometer. J. Exp. Bot. 29:277-280.
158. ----, F. Duhme, G. Glatzel, T.M. Hinckley, and H. Karlic. 1981. Some implications and applications of the pressure-volume curve technique in ecophysiological research. p. 263-272. *In* J. Grace et al. (ed.) Plants and their atmospheric environment. Blackwell Scientific Publishers, Oxford.
159. Ritchie, J.T. 1972. Model for predicting evaporation from a row crop with incomplete cover. Water Resour. Res. 8:1204-1213.
160. ----. 1983. Efficient water use in crop production: discussion on the generality of relations between biomass production and evapotranspiration. p. 29-44. *In* H.M. Taylor et al. (ed.) Limitations to efficient water use in crop production. American Society of Agronomy, Crop Science Society of America, and Soil Science Society of America, Madison, WI.
161. Rosenberg, N.J. 1969. Seasonal patterns in evapotranspiration by irrigated alfalfa in the Central Great Plains. Agron. J. 61:879-886.
162. ----, and S.B. Verma. 1978. Extreme evapotranspiration by irrigated alfalfa: A consequence of the 1976 midwestern drought. J. Appl. Meteorol. 17:934-941.
163. Rumbaugh, M.D., and D.A. Johnson. 1981. Screening alfalfa germplasm for seedling drought resistance. Crop Sci. 21:709-713.
164. ----, ----, and D.N. Rinehart. 1983. Stand density, shoot weight, and acetylene reduction activity on alfalfa populations subjected to field and greenhouse moisture gradients. Crop Sci. 23:784-789.
165. Safir, G.R., J.S. Boyer, and J.W. Gerdemann. 1972. Nutrient status and mycorrhizae enhancements of water transportation in soybean. Plant Physiol. 49:700-703.
166. Saini, G.R., and T.L. Chow. 1982. Effect of compact sub-soil and water stress on shoot and root activity of corn (*Zea mays* L.) and alfalfa (*Medicago sativa* L.) in a growth chamber. Plant Soil 66:291-298.
167. Salter, R., B. Melton, M. Wilson, and C. Currier. 1984. Selection in alfalfa for forage yield with three moisture levels in drought boxes. Crop Sci. 24:345-349.
168. Sammis, T.W. 1981. Yield of alfalfa and cotton as influenced by irrigation. Agron. J. 73:323-328.
169. Sanchez-Diaz, M.F., P.M. Aparicio-Tejo, C. Gonzalez-Murua, and J.I. Pena. 1982. The effect of NaCl salinity and water stress with polyethylene glycol on nitrogen fixation, stomatal response and transpiration of *Medicago sativa, Trifolium repens* and *Trifolium brachycalycinum*. Physiol. Plant. 54:361-366.
170. Setter, T.L., L.E. Schrader, and E.T. Bingham. 1978. Carbon dioxide exchange rates, transpiration, and leaf characters in genetically equivalent ploidy levels of alfalfa. Crop Sci. 18:327-332.
171. Shantz, H.L., and L.N. Piemeisel. 1927. The water requirements of plants at Akron, Colo. J. Agric. Res. 34:1093-1190.

172. Sharratt, B.S., D.C. Reicosky, S.B. Idso, and D.G. Baker. 1983. Relationships between leaf water potential, canopy temperature, and evapotranspiration in irrigated and non-irrigated alfalfa. Agron. J. 75:891–894.
173. Sheaffer, C.C., and D.K. Barnes. 1982. Potential for alfalfa production in Minnesota. p. 125–135. *In* Proc. 1982 Forage and Grassland Conf., Rochester, MN. American Forage and Grassland Council, Lexington, KY.
174. Sheehy, J.E., K.A. Fishbeck, and D.A. Phillips. 1980. Relationships between apparent nitrogen fixation and carbon exchange rate in alfalfa. Crop Sci. 20:491–495.
175. Shockley, D.R. 1955. Capacity of soil to hold moisture. Agric. Eng. 36:109–112.
176. Simpson, J.R., and J. Lipsett. 1973. Effects of surface moisture supply on the subsoil nutritional requirements of lucerne. Aust. J. Agric. Res. 24:199–209.
177. Snaydon, R.W. 1972a. The effect of total water supply, and of frequency of application, upon lucerne. I. Dry matter production. Aust. J. Agric. Res. 23:239–251.
178. ———. 1972b. The effect of total water supply, and of frequency of application upon lucerne. II. Chemical composition. Aust. J. Agric. Res. 23:253–256.
179. Sonmor, L.G. 1963. Seasonal consumptive use of water by crops grown in southern Alberta and its relationship to evaporation. Can. J. Soil Sci. 43:287–297.
180. Sorensen, E.L. 1985. Personal communication, Kansas State University, Manhattan.
181. Sprent, J.I. 1976. Water deficits and nitrogen-fixing root nodules. p. 291–315. *In* T.T. Kozlowski (ed.) Water deficits and plant growth, Vol. IV. Soil water measurement, plant responses, and breeding for drought resistance. Academic Press, New York.
182. Stanberry, C.O. 1955. Irrigation practices in growing alfalfa. p. 435–443. *In* Water. USDA Yearbook of Agriculture. U.S. Government Printing Office, Washington, DC.
183. Stegman, E.C., A. Bauer, J.C. Zubriski, and J. Bauder. 1977. Crop curves for water balance irrigation scheduling in S.E. North Dakota. N. Dak. Agric. Exp. Stn. Res. Rep. 66.
184. Stewart, J.I., and R.M. Hagan. 1969. Development of evapotranspiration–crop yield functions for managing limited water supplies. p. 23.505–23.530. Proc. 7th Int. Commission on Irrigation and Drainage, New Delhi, India. International Commission Conference, Mexico City, Mexico.
185. Stone, L.R., M.B. Kirkman, D.E. Johnson, and E.T. Kanemasu. 1982. Yield and water use of alfalfa. Keeping up with research. Kans. Agric. Exp. Stn. Bull. 58.
186. Stout, D.G. 1980. Alfalfa water status and cold hardiness as influenced by cold acclimation and water stress. Plant, Cell Environ. 3:237–241.
187. Suzuki, M. 1977. Effects of soil moisture on cold resistance of alfalfa. Can. J. Plant Sci. 57:315.
188. Szeicz, G., G. Endrodi, and S. Tajchman. 1969. Aerodynamic and surface factors in evaporation. Water Resour. Res. 5:380–394.
189. Tanner, C.B. 1957. Factors affecting evaporation from plants and soils. J. Soil Water Conserv. 12:221–227.
190. ———. 1960. Energy balance approach to evapotranspiration from crops. Soil Sci. Soc. Am. Proc. 24:1–9.
191. ———. 1963. Plant temperatures. Agron. J. 55:210–211.
192. ———. 1967. Measurement of evapotranspiration. *In* R.M. Hagan et al. (ed.) Irrigation of agricultural lands. Agronomy 11:534–574.
193. ———. 1986. Personal communication, University of Wisconsin, Madison, WI.
194. ———, and W.A. Jury. 1976. Estimating evaporation and transpiration from a row crop during incomplete cover. Agron. J. 68:239–243.
195. ———, and W.L. Pelton. 1960. Potential evapotranspiration estimates by the approximate energy balance method of Penman. J. Geophys. Res. 65:3391–3413.
196. ———, and T.R. Sinclair. 1983. Efficient water use in crop production: research or research. p. 1–27. *In* H.M. Taylor et al. (ed.) Limitations to efficient water use in crop production. American Society of Agronomy, Crop Science Society of America, and Soil Science Society of America, Madison, WI.
197. Taylor, H.M., and B. Klepper. 1974. Water relations of cotton: I. Root growth and water use as related to top growth and soil water content. Agron. J. 66:584–588.
198. Taylor, S.A. 1952. Use of mean soil moisture tension to evaluate the effect of soil moisture on crop yields. Soil Sci. 74:217–226.
199. ———, J.L. Haddock, and M.W. Pedersen. 1959. Alfalfa irrigation for maximum seed production. Agron. J. 51:357–360.
200. Thompson, T.E., and G.W. Fick. 1981. Growth response of alfalfa to duration of soil flooding and to temperature. Agron. J. 73:329–332.
201. Topp, G.C., J.L. Davis, and A.P. Annan. 1982. Electromagnetic determination of soil water content using TDR. Soil Sci. Soc. Am. J. 46:672–684.
202. Tovey, R. 1963. Consumptive use and yield of alfalfa grown in the presence of static water tables. Nev. Agric. Exp. Stn. Tech. Bull. 232.
203. Travis, R.L., and R. Reed. 1983. The solar tracking pattern in a closed alfalfa canopy. Crop Sci. 23:664–668.

204. Triplett, G.B., and M.B. Tesar. 1960. Effect of compaction, depth of planting, and soil moisture tension on seedling emergence of alfalfa. Agron. J. 52:681–684.
205. Turner, N.C. 1981. Techniques and experimental approaches for the measurement of plant water status. Plant Soil 58:339–366.
206. ----, and J.E. Begg. 1978. Responses of pasture plants to water deficits. p. 50–66. *In* J.R. Wilson (ed.) Plant relations in pastures. CSIRO, Melbourne, Australia.
207. ----, and G.J. Burch. 1983. The role of water in plants. p. 73–126. *In* I.D. Teare and M.M. Peet (ed.) Crop-water relations. John Wiley and Sons, New York.
208. ----, and M.M. Jones. 1980. Turgor maintenance by osmotic adjustment: a review and evaluation. p. 87–103. *In* N.C. Turner and P.J. Kramer (ed.) Adaptation of plants to water and high temperature stress. John Wiley and Sons, New York.
209. Tyree, M.T., M.E. MacGregor, A. Petrov, and M.I. Upenieks. 1978. A comparison of systematic errors between the Richards and Hammel methods of measuring tissue-water relations. Can. J. Bot. 56:2153–2161.
210. Uhvits, R. 1946. Effect of osmotic pressure on water absorption and germination of alfalfa seeds. Am. J. Bot. 33:278–285.
211. Upchurch, R.P., and R.L. Lovvorn. 1951. Gross morphological root habits of alfalfa in North Carolina. Agron. J. 43:493–498.
212. van Bavel, C.H.M. 1967. Changes in canopy resistance to water loss from alfalfa induced by soil water depletion. Agric. Meteorol. 4:165–176.
213. Vough, L.E., and G.C. Marten. 1971. Influence of soil moisture and ambient temperature on yield and quality of alfalfa forage. Agron. J. 63:40–42.
214. Wahab, H.A., and D.S. Chamblee. 1972. Influence of irrigation on the yield and persistence of forage legumes. Agron. J. 64:713–716.
215. Walker, G.K., and J.L. Hatfield. 1983. Stress management using foliage temperatures. Agron. J. 75:623–629.
216. Warrick, A.W., and W.R. Gardner. 1983. Crop yield as affected by spatial variations of soil and irrigation. Water Resour. Res. 19:181–186.
217. Weaver, J.E. 1926. Root development of field crops. McGraw-Hill Book Co., New York.
218. ----. 1968. Prairie plants and their environment. A fifty-year study in the midwest. Univ. of Nebraska Press, Lincoln.
219. Werner, H.D. 1978. Irrigation scheduling-checkbook method. Minn. Agric. Ext. Serv. M-160.
220. Wilde, S.A., and G.K. Voigt. 1952. The determination of color of plant tissues by the use of standard charts. Agron. J. 44:499–500.
221. Willatt, S.T. 1971. Model of water use by tea. Agric. Meteorol. 8:341–351.
222. Williams, R.J., and D.G. Stout. 1981. Evapotranspiration and leaf water status of alfalfa growing under advective conditions. Can. J. Plant Sci. 61:601–607.
223. Wright, J.L. 1982. New evapotranspiration crop coefficients. J. Irrig. Drain. Div., Am. Soc. Civ. Eng. 108:57–74.
224. ----, and M.E. Jensen. 1982. Peak water requirements of crops in southern Idaho. J. Irrig. Drain. Div., Am. Soc. Civ. Eng. 98:193–201.

12 Cutting Schedules and Stands

C.C. SHEAFFER

University of Minnesota
St. Paul, Minnesota

G.D. LACEFIELD

University of Kentucky
Princeton, Kentucky

V.L. MARBLE

University of California
Davis, California

Alfalfa (*Medicago sativa* L.) cutting management research and recommendations have changed dramatically since the early 1900s. Management systems that maximized stand persistence and herbage dry matter yields were once emphasized; however, more recently, systems that maximize forage nutrient yields and nutrient concentrations have increased in importance. These changes resulted from increased producer awareness of animal nutrition, hay marketing, and the value of forage nutrients. Newer cultivars that combine winterhardiness, rapid regrowth after cutting, and resistance to insects and stand-depleting diseases are adapted to frequent cutting.

Water availability and temperature greatly influence the productivity and survival of alfalfa and the suitability of a particular cutting schedule within a region. The total number of potential harvests can vary from 6 to 10 per season under irrigation in nonhumid areas, such as the southwest, to only one or two harvests per season in northern latitudes. Cutting schedules have been based on stage of growth, fixed time intervals, or crown shoot development. Although any criteria will provide a range in forage yield and quality, cutting according to stage of growth uses the plant as a harvest indicator and generally provides more consistent yield and quality among cultivars, and over years and locations. Producers frequently use a combination of criteria for scheduling haymaking with other farm operations.

Appropriate fall cutting is a major concern in many regions. Traditionally, fall cutting has been associated with risks to alfalfa stand persistence; and consequently with the adoption of conservative recommendations. Greater flexibility in fall cutting is now possible with disease resistant, adapted cultivars, and high levels of soil fertility. However, the

---

Copyright 1988 © ASA-CSSA-SSSA, 677 South Segoe Road, Madison, WI 53711, USA.
Alfalfa and Alfalfa Improvement—Agronomy Monograph no. 29.

effects of fall cutting are dramatically influenced by environmental conditions, such as air temperature and snowfall.

This chapter contains information originally presented in the earlier edition (143). We acknowledge the significant contributions to the scientific literature on alfalfa cutting management by Dale Smith. His research and writings form the basis for much of our present understanding of the subject.

12-1 THE IMPORTANCE OF CULTIVAR DEVELOPMENT

Most changes in alfalfa management systems can be attributed to the development of multiple-pest resistant alfalfa cultivars with increased persistence. Before development of cultivars with resistance to bacterial wilt [*Corynebacterium insidiosum* (McCull.) H.L. Jens.] (see Chapter 21 in this book), delaying cutting until full flowering was recommended for the long-term maintenance of stands (42, 65, 121, 160). Likewise, cultivars with resistance to other pests (diseases, insects, and nematodes), that stress the plant, can tolerate more frequent cutting or fall cutting with little effect on persistence or yield. Hill and Baylor (53) compared several cultivars under severe (bud stage) and normal (early flowering) cutting regimes. For both cutting treatments, alfalfa entries that had moderate or higher levels of resistance to anthracnose [*Colletotrichum trifolii* (Bain.)] and Fusarium wilt [*Fusarium oxysporum* Schl. f. *medicaginis* (Weimer) Sny. & Hans.] had superior long-term yields compared to less-resistant cultivars.

It is well documented that in northern regions alfalfa cultivars of *M. falcata* L. origin that have deep-set crowns, slow regrowth following harvest, and a strong dormancy reaction are more resistant to winter injury and more persistent under frequent cutting or grazing than *M. sativa* cultivars with similar pest resistance (51, 52). Also, *M. sativa* cultivars differ in winterhardiness and regrowth characteristics, which results in differences in persistence and yield under frequent cutting (22, 64, 126, 135, 150, 160). Twamley (159) in Ontario, Canada studied the performance of four cultivars over a 4-yr period with two, three, and four cuttings per season (two summer harvests and one or two autumn harvests). The four cultivars, Vernal, Ranger, Grimm, and DuPuits, provided all combinations of high and low winterhardiness and of high and low bacterial wilt resistance. Only the cv. Vernal that combined both disease resistance and winterhardiness performed well under all cutting schedules. With the hardy cultivars, Vernal and Ranger, it was possible to compensate for untimely cutting by addition of fertilizer. Jung et al. (61) in West Virginia and Tesar and Yager (156) in Michigan also found that cultivar winterhardiness and disease resistance were important for alfalfa tolerance to fall cutting.

Alfalfa cultivars with moderate resistance to several pathogens and with sufficient winterhardiness for adaptation to specific climatic regions generally have superior yields and persistence in recommended cutting systems. It may be possible to develop high-yielding cultivars with increased tolerance to frequent defoliation (16, 133, 162). This may be critical with grazing (22), or with frequent cutting to provide nutritious forage (i.e., for "dehy" or for forage used in dairy industries). Alfalfa yield and persistence might be improved by selecting genotypes with increased root and crown carbohydrate reserves during initiation of regrowth and earlier crown shoot development (preflowering) in the regrowth cycle (17). This agrees with Smith's (143) proposal that the ideal cultivar for long-term production of high-quality forage should have large numbers of new shoots already growing below the normal mowing height at harvest.

12-2 CROP DEVELOPMENT AND CUTTING SYSTEMS

The effect of a particular harvest system on seasonal alfalfa forage yield, forage quality, and stand persistence is related to the morphological development of the crop at each harvest within the system.

12-2.1 Canopy Development

The morphological development of herbage has been classified in several ways (41, 63, 65). Developmental stages are broadly categorized as vegetative, bud, flower (bloom), and seed pod. Smith (139, 143) popularized the use of first flower (10% bloom) as a maturity guide to stage of harvesting by producers, but for research a more refined system is needed that is capable of quantifying changes in maturity to relate plant development to yield and changes in quality. Kalu and Fick (63) proposed a 10-stage numerical system based on development and mean weight of individual stems.

Total dry matter accumulation within an alfalfa growth cycle was most rapid and nearly linear until early flowering and decreased thereafter (66, 103). In humid areas, maximum dry matter yield for any single growth period usually occurs when plants are cut from early to late flowering; however, leaf loss from lower portions of the canopy may reduce herbage yields of alfalfa harvested at late flowering (14, 34, 36). Marble (76) in California reported that for the intermediate dormant 'Caliverde,' dry matter accumulation was not linear until after the bud stage. After full flowering, dry matter accumulation declined rapidly due to leaf loss. Although leaf loss in mature canopies is accentuated by leaf diseases, this can be reduced by using disease-resistant cultivars. Basal regrowth from crown and axillary buds may increase herbage yields if plants are not cut until flowering or later stages especially in open or lodged canopies.

Kilcher and Heinrichs (66) reported that stem yield increased linearly from early vegetative to late-flowering stages, while leaf yield increased only until early flowering. Yields of stems and leaves were equal at early flowering, but by late flowering stems made up 60% and leaves only 40%

of the total yield. Fick and Holthausen (34) reported that at early to midflowering, alfalfa leaf blades constituted approximately 20% of the herbage while stems comprised 75%. Bula et al. (11) and Fuess and Tesar (36) reported that the decrease in leaf/stem ratio with canopy maturity was greater for the initial growth compared to regrowths.

Yields at a particular harvest are influenced by environmental conditions and cultivar selection (see Chapter 5 in this book). In the northern USA, leaf area indices (LAIs) and dry matter yields of "dormant-type" alfalfa are usually greatest for the first growth cycle of the season (11, 30, 33, 36, 83). In southern and western areas, "nondormant-type" alfalfa may yield less during the earliest growth cycle than at subsequent harvests and may be lower yielding than dormant alfalfa in the first growth cycle (50, 75, 77). Regrowth of dormant alfalfa generally has a greater leaf/stem ratio and smaller stem diameter than the initial growth (11, 99, 108). In California, Marble (77) reported that the leaf/stem ratio of nondormant alfalfa varied throughout the year. Leaf/stem ratio was high for the first cutting, declined until the fourth cutting in mid-June, and then rose again at the sixth and seventh cutting in the fall.

12-2.2 Forage Quality

The decline in alfalfa forage quality with advancing maturity is well documented (see Chapter 14 in this book). This decline in quality is associated primarily with a decrease in LAI, in the leaf/stem ratio, and to an increase in fibrous constituents of the stem (1, 14, 75, 108, 166). Saving leaves is of prime importance in haymaking because leaves contain more nutrients than stems (14, 108, 121). Smith (140, 141) and Rominger et al. (120) found that leaves from early bloom alfalfa had higher concentrations than stems of total digestible nutrients (TDN), crude protein (CP), fat, total ash, starch, total nonstructural carbohydrates, P, Ca, Mg, Al, Fe, Sr, B, Cu, Zn, and Mn. Stems had higher reducing and total sugars, crude fiber, K, and Cl. Digestibility and CP concentration of stems declined at a faster rate than those of leaves with increased maturity (14, 34).

Highest concentrations of nutrients are usually obtained with forage harvested at immature stages with highest yields of nutrients per land area obtained at 10% flowering (4, 6, 38, 40, 61, 96, 97, 137). In humid regions, yields of TDN, CP, and minerals per land area decrease after early flowering with the loss of lower leaves from shading, aging, and disease (1, 14, 36, 169). Conversely, in less humid areas, where leaf disease is less severe, harvesting at 50% flowering results in maximum nutrient yields (75, 77).

12-2.3 Carboyhydrate Reserves

Energy for alfalfa regrowth originates from stored nonstructural carbohydrate (NSC) reserves in roots and crowns or from residual leaves and stems (see Chapter 6). In general, storage and utilization of NSC root reserves follow a cyclic pattern of decrease during early vegetative regrowth following dormancy or cutting until plants are 20- to 30-cm tall (33, 39, 45, 104, 115, 147). Nonstructural carbohydrates in the roots and crowns then increase with increasing plant maturity until full flowering but are generally considered adequate by first flower.

Carbohydrates accumulate in dormant cultivars during autumn in response to decreasing temperatures and daylength. These stored NSC are the main source of energy during overwintering (62). As much as 50% of the NSC stored in roots during the autumn were used in respiration during a Wisconsin winter (12). Kust and Smith (67) and Meyer and Nelson (95) reported a positive correlation between the quantity of stored root NSC in the fall and hay yields the following year. Carbohydrate reserves fluctuate less and are often low in summer in the subtropics (73).

Frequent harvests of immature alfalfa (vegetative to bud) or fall harvesting which prevents vegetative regrowth from developing enough to replenish reserves can result in reduced root NSC concentrations. This may be associated with stand decline and yield loss (115, 118, 136). Cutting at full flower is advantageous to alfalfa persistence because it results in maximum NSC concentrations in the root and crown. However, in cultivars with multiple-pest resistance, NSC levels associated with cutting at full flower are usually in excess of that required for persistence, and losses in herbage nutrient concentration and yield often offset the increased persistence from infrequent harvest. The deleterious effects of frequent cutting (e.g., low NSC reserves and predisposition to winter injury) may be reduced if alfalfa flowers at least once annually before cutting to permit adequate storage of NSC reserves.

Flowering is not a prerequisite to obtain high NSC storage in roots. During the autumn in northern latitudes, fall dormant cultivars can attain adequate NSC storage with 20 to 25 cm of vegetative growth and flowers need not be present (45). Furthermore, alfalfa may not flower at high altitudes in the tropics or during the fall and winter in mild areas, but root reserves are stored as top growth accumulates. Soil moisture deficits can induce accumulation of root NSC reserves (see Chapter 11 in this book).

12–3 HARVESTING ON A FIXED INTERVAL

Cutting on a fixed interval, or using a fixed number of cuts per season with no particular attention paid to stage of development, is a common practice. It has been studied extensively (Table 12–1). Since fixed interval harvesting does not account for the effect of environmental conditions and dormancy differences among cultivars, the most satisfactory interval between cuttings will vary with location, climate, and season of the year.

Table 12-1. Interval between harvests at which highest seasonal yields per acre were obtained.

Reference	Location	Years	Criteria	Harvest interval, d	Harvests/season	Cultivar
			Dormant cultivars			
168	OH	Several	Herbage	38–45	3–4	Common† and Grimm
3	CT	Several	Herbage and protein	41–50	3	Common† and Grimm
55, 56	WA	1947–1948	Nitrogen	41	3	Ladak
57	WA	1950–1952	Herbage	37	3–4	Ladak, Turkestan, Ranger, and Buffalo
				40	3	
110	OH	1957–1958	Herbage and protein	45	3	Vernal
67	WI	1956–1957	Herbage and protein	48	3	Vernal
135	WI	1954–1956	Herbage and protein	45	3	Vernal and Narragansett
117	TN	1965–1966	Herbage and TDN‡	42	4	Buffalo
36	MI	1961–1962	Herbage	45	3	Vernal
150	MI	1968–1969	Herbage	40	4	Saranac
147	MO, IA	1963–1964	Herbage, protein, and TDN	45	3–4	Vernal and DuPuits
35	WI, MN					
131	PA	1967–1968	Herbage, protein, and TDN	35	4	Saranac, Cayuga, Vernal, and Buffalo
105	MO	1970–1972	Herbage	35	4	Saranac
91	MD	1974–1977	Herbage, Digestible DM	40	4	WL311, Williamsburg
5	MN	1981–1983	Prime grade herbage	45	4	Saranac AR, Ramsey
				30		
			Nondormant cultivars			
111	CA	1949–1951	Herbage	35	7	§
24	Columbia, SA	§	Herbage	63	5–6	Peruvian
166	CA	1954	Protein and TDN	42	5	Common†
				35	6	
106	AR	1956–1962, 1961–1964	Herbage	41	5	Rhizoma, Sevelra, Vernal, Buffalo, and Lahontan
60	SA, Australia	1965–1966	Herbage, protein, and TDN	42	6	Hunter River
77	CA	1975–1978	Herbage, TDN	37–42	5	Dawson, WL318, Lahontan WL512, Moapa 69, U.C. Salton
		1975–1978	Protein	28	7	Dawson, WL318, Lahontan

† Common strains of alfalfa are no longer recognized as cultivars.   ‡ TDN = total digestible nutrients.   § Not stated by authors.

harvesting does not account for the effect of environmental conditions and dormancy differences among cultivars, the most satisfactory interval between cuttings will vary with location, climate, and season of the year.

12-3.1 Northern Regions

The preponderance of research in the 1960s using persistent cultivars in the northern Lake States showed that three cuttings before 1 September gave higher seasonal nutrient yields, and usually higher herbage yields than cutting at more or less-frequent intervals (15, 26, 36, 40, 47, 67, 85, 110, 136, 138, 139, 145, 146, 155, 159). In this region, three cuttings at fixed intervals are most frequently obtained by dividing the growing season (late May – early September) into three segments with harvests about 1 June, 15 July, and 30 August. Smith (139) showed that seasonal yields of herbage, CP, and TDN in Wisconsin from three cuttings with the above schedule were 20, 46, and 31% higher, respectively, than with two cuttings on about 20 June and 25 August.

Fuess and Tesar (36) in Michigan studied the reasons for higher herbage yields from three cuttings at early flowering when compared with two cuttings at full flowering. They found that three-cut alfalfa yielded 17% more herbage primarily as the result of reduced leaf loss. Leaf loss accounted for two-thirds of the difference in yield and the remaining difference was attributed to higher net photosynthesis by the less mature leaves of the three-cut alfalfa.

Since the mid-1960s, the use of persistent cultivars with rapid recovery after cutting in combination with high soil fertility and insect control has permitted more frequent cutting, with increased yields of herbage and nutrients without sacrificing persistence. These schedules generally involve four cuttings per year and include mid-October cutting after which there is little vegetative regrowth. Tesar (149), in southern Michigan, reported that a system with four harvests: late May to early June (bud); 5 to 10 July (first flower); 20 to 25 August (first flower); and 15 to 30 October resulted in 10% greater herbage yield and greater forage quality and nutrient yield than the standard three-cut system (1 June, 15 July, and 30 August). Herbage DM yields of the highest-yielding cultivars subjected to this four-cut system averaged 17.6 Mg ha^{-1} in the 10th yr of a long-term study. In Michigan, Savoie et al. (122) showed that a four-cut system (15 May, 30 June, 15 August, and 15 October) was more profitable than a three-cut system (15 May, 30 June, and 15 August) for 90% of the time throughout a 26-yr period. In southern Minnesota, Brink and Marten (6) reported that for two locations, a similar four-cut system with the last cut on 1 October had greater short-term herbage, in vitro digestible dry matter (IVDDM), and CP yields than seven other systems which included three or four cuts per year. However, a system with harvests on 25 May, 29 June, 31 August, and 15 October resulted in the highest yield of top-grade alfalfa. The persistence and high yield of alfalfa under a four-cut system in the northern states of Michigan and Minnesota is likely dependent on the last date of cutting, 1 October or later. Little regrowth, which would reduce NSC reserves, occurs after this date in these or other North Central States resulting in good winter survival and persistence.

12-3.2 West and Southwest Regions

Marble (75,77) and Hagemann and Marble (48) studied the interactions between cultivars of different dormancies and fixed-interval cutting frequencies at three locations extending over 970 km from north-central California (Davis) to the Mexican border (El Centro). The nondormant cultivars used in these 3- and 4-yr studies were highest yielding at all locations when harvest frequency was 28 d or less. Dormant and semidormant cultivars were more productive than nondormant cultivars when cut at 33- and 42-d intervals, persisted longer (especially at the longer cutting intervals), and had less weed invasion except when they were harvested at intervals of 28 d and less. In California's Imperial Valley adjacent to the Mexican border, nondormant cultivars yielded more than semidormant and dormant cultivars over a 3-yr period at cutting intervals of 3, 4, 5, and 6 weeks. By the end of the 2nd year, stands of the dormant and semidormant cultivars were depleted and weeds were invading particularly with the frequent 3-week cutting interval. Herbage yields increased linearly when intervals were extended from 3 through 6 weeks (75).

The same linear increase in herbage yields was found with longer cutting intervals over a 4-yr period at two locations in central California, which has a slightly milder summer climate and a colder dormant period when growth virtually ceases during the winter (75, 77). Weed invasion in the late summer, winter, and early spring increased sharply when cutting intervals were <33 to 35 d. Harvest intervals of 6 weeks for dormant and semidormant cultivars reduced weed invasion and Egyptian alfalfa weevil [*Hypera brunneipennis* (Boheman)] damage (75, 77). Harvest intervals had a significant effect on total yield during each year. By the first cut of the 2nd yr, schedules with cuttings at intervals of less than 4 weeks during the 1st yr yielded 0.8 to 1.2 Mg ha^{-1} less than those with intervals greater than 4 weeks.

12-3.3 Advantages of Fixed Interval Harvesting

A fixed system of cutting based on calendar date or time intervals may allow easier scheduling of harvesting with other field activities, such as irrigation. Thus, many western producers harvesting 6 to 10 times per season use a fixed time interval to schedule cutting during warmer months and rely on crop development to schedule cuttings during cooler months.

In northern states, cool, cloudy weather often delays floral development in the first spring regrowth of alfalfa. For this reason, producers in southern Minnesota frequently cut alfalfa before 1 June regardless of stage of maturity to avoid the high probability of inclement weather in early June and potential delays in subsequent cutting. Subsequent harvests of the four-cut system are based on maturity. It is therefore evident that a combination of harvesting by stage of growth and calendar date may be advantageous in the production of alfalfa.

12-4 HARVESTING BY STAGE OF GROWTH

Harvesting at specific phenological stages of development has been the subject of much research (7, 10, 13, 25, 32, 61, 65, 69, 85, 96, 97, 102, 121, 139, 166, 170, 172). A harvest schedule based on plant maturity depends on the stage of plant development to indicate the proper time to cut and the number of cuttings possible in a season.

Fixed-interval harvesting does not consider the effects of environmental conditions (see Chapter 5 in this book) and cultivar characteristics (see Chapters 26 and 27 in this book) on morphological and physiological development. It also disregards the fact that plant maturity on a particular date will vary among years, locations, and cultivars. Temperature and cultivar characteristics greatly influence the rate of maturation (19). For example, Jensen et al. (58) grew nondormant 'Moapa' and dormant 'Nebraska C-614' at warm (33/17°C day/night) and cool (24/4°C day/night) temperatures. At 28 d following harvest, the nondormant alfalfa grown at the warm and cool temperatures had 66 and 31% flowering, respectively, while the dormant alfalfa had 3 and 0% flowering, respectively. Similar effects of temperature on rate of alfalfa development have been reported by others (79, 105, 139, 142).

Cutting according to stage of development is superior to cutting at fixed intervals in obtaining consistent forage yield and quality (48, 58, 75, 79, 119, 166). In three studies from southern to northern Wisconsin, Smith et al. (145) showed the advantage of cutting whenever the first flowers appeared (10% flowering) until early September as compared with three cuttings by date. With the first flower cutting regime, yields of CP and IVDDM from Vernal and DuPuits alfalfa were increasingly higher from south to north because of the progressively later maturity of the first crop. First flower in the first crop of Vernal occurred about 4 June in the south, about 16 June in the center, and about 22 June in the northern part of the state. In contrast, calendar-date and plant-maturity cutting of Vernal and DuPuits showed no clear difference in CP and IVDDM yields in a cooperative study involving Missouri, Iowa, Wisconsin, and Minnesota (85, 146). Even so, it was concluded that stage of maturity was the best criterion on which to base time of harvesting and ensure high-quality herbage over a broad latitude.

In northern areas where three to four harvests per season are common, most investigations have shown that cutting winter-dormant cultivars at first flower (10% flowering) is the best compromise to optimize herbage and nutrient yields and stand persistence (143). First flower is a stage: (i) that approximates 10% flowering when a near maximum yield of nutrients is attained; (ii) that is easier to recognize than 10% flowering; (iii) when root reserves have been restored to a reasonably high level; and (iv) when crop growth rate has begun to decrease markedly (36, 103, 104, 105, 145).

In contrast, in southwestern and western USA, where winters are milder and nonwinter-dormant cultivars are grown, cutting at 25 to 50% flowering ensures high herbage yield, reasonable forage quality and good persistence (32, 74, 75, 77, 78, 84, 119). Feltner and Massengale (32) obtained the highest herbage yields of Moapa and 'Lahontan' by harvesting at 10% flowering for 3 yr, but later, Robison and Massengale (119) and Massengale (84) reported that cutting at 50% bud to 10% flowering did not provide sufficient time for roots to accumulate NSC reserves resulting in retarded regrowth, reduced yield, and loss of stands by the summer of the 2nd yr. Rapid stand loss occurred when temperatures, especially at night, were extremely high in July and August (32, 48, 119). Other workers (28, 31, 157) from warm, low-latitude areas have shown greatest persistence, regrowth potential, and root NSC concentrations when harvest occurred at 25 to 50% flowering. Under tropical conditions where plants had no dormant period, the highest herbage yields occurred at 10% flowering compared with harvesting by three other stages of plant development during 4 yr (30 cuttings) (13).

In low-latitude areas of the northern and southern hemisphere, growth continues slowly throughout the winter months when daylength is inadequate to promote flowering. Under these conditions, flowering cannot be used as a guide to cutting and fixed-interval harvesting is preferred (84). In areas where insects such as lygus (*Lygus* spp.) blast flower buds and prevent recognition of floral development, alternative guides to cutting are used, such as fixed dates and crown bud development.

12-5 HARVESTING BY CROWN SHOOT DEVELOPMENT

12-5.1 Alfalfa Regrowth from Buds

Alfalfa regrowth occurs from crown buds and axillary buds on stems. Crown buds are responsible for spring regrowth and are primarily formed during the previous fall (46, 101). Additional crown buds also develop in spring before initiation of regrowth (103). Management systems that stress alfalfa plants and reduce NSC reserves will decrease the number and size of fall crown buds (46, 131, 168).

If alfalfa is not cut, one or more additional regrowths may originate from the crown during the growing season (103). If cut, regrowth occurs from crown buds and from axillary buds located on the stubble of the previous crop (70, 94, 134). Bud origin and rate of regrowth are influenced by cultivar, stage of maturity at harvest, amount of stubble remaining following cutting, and cutting frequency within a system. If alfalfa is cut at full flowering or later stages of maturity when NSC reserves are high, buds on the stubble are more basal, more numerous, and more developed (5, 71, 134). Meyer and Larson (94) in North Dakota reported that when Vernal alfalfa was harvested to a 7.5-cm stubble height at full flower, 92% of the new stems originated from the crown, but for a first-flower harvest only 80% of the stems originated from the crown. With a 2.5-cm stubble, all new stems originated from the crowns of alfalfa cut at first and full flower. Taller stubble provides more sites for regrowth, but may suppress crown bud development (70, 72). Langer and Steinke (68) reported that in systems with frequent cuttings, there were fewer crown buds and regrowth occurred mainly from axillary buds on the cut stems.

12-5.2 New Crown Shoots and Cutting

Because of the frequent association between basal shoot development, alfalfa maturity, and NSC reserves, basal shoot elongation has served as an indicator that the crop is ready to cut. In northern areas where winter-dormant cultivars are grown, this method is not superior to cutting according to stage of growth or a fixed schedule because environmental factors influence new shoot elongation. Shoots often develop from the crown when prolonged dormancy induced by drought is broken or when canopy lodging exposes the crown to light (46, 167). In desert areas of southern Arizona, Massengale (84) reported that length of shoots of Lahontan varied from 2.5 to 10 cm throughout the year when plants were cut repeatedly at 10% bloom under high temperature. Marble (75), in California, also reported that crown shoot regrowth varied seasonally and noted a cultivar interaction at comparable stages of maturity.

In southern Arizona and southern California, alfalfa does not flower during the winter and early spring because of cool temperatures and short daylengths; thus Massengale (84) recommended that 5-cm of regrowth from crown buds be used as a guide for cutting. However, he found that cutting on a fixed-interval basis was as satisfactory if not superior. At high altitudes near the equator where flowering is sparse and sporadic, Crowder et al. (24) recommended cutting when new crown shoots reach 5 cm. Marble et al. (78) also reported that crown bud development was a more reliable guide for cutting than floral development. They indicated that 10% flowering was equivalent to 60% of the alfalfa plants with crown bud regrowth 1.2 to 1.9 cm long.

New shoot development (below the normal mowing height) at the time of herbage cutting (i.e. first flower) is desirable in increasing the rate of herbage regrowth (143). If the existing interval of 5 to 7 d between cutting and production of a LAI adequate for significant photosynthesis (36) could be shortened by genetic manipulation, yields might be increased, particularly in intensive management systems. Chatterton et al. (17) found that cultivars with advanced crown bud development at cutting recovered more rapidly than those with less bud development. Additional study is warranted on crown bud characteristics of modern cultivars and use of this characteristic as a guide to cutting.

12-6 HEIGHT OF CUTTING

Cutting height can influence yield and survival of alfalfa when carbohydrate root reserves are reduced by frequent cutting (98, 143). A tall stubble provides more photosynthetic area that in turn, furnishes additional energy for initial regrowth after cutting (70, 71). Langer and Steinke (68) reported that in frequent cutting systems where the number of crown buds was inadequate, a tall stubble provided more sites for axillary bud development.

For cutting systems that do not deplete NSC reserves, higher herbage and nutrient yields are obtained with short vs. tall cutting heights (80, 98, 143, 161). In a 2-yr Wisconsin study, Vernal alfalfa produced higher herbage yields with a 7.6-cm than with a 2.5-cm stubble with frequent (five and six times), but not with less frequent cutting (three and four times) (67). In another Wisconsin study (147), Vernal alfalfa was cut three, four, five, or six times before early September for 2 yr, leaving stubbles of 2.5, 5, 7.6, and 15 cm. Yields of herbage decreased as cutting frequency or stubble height increased. Averaged over all cutting frequencies, herbage yields decreased as stubble height increased. A tall stubble was needed to maintain high yields only with the most frequent cutting schedule. Meyer and Larson (94) reported that for two- and three-cut systems in North Dakota, total season herbage yield decreased approximately 1.1 Mg ha^{-1} for each additional 5 cm of stubble above 2.5 cm to a height of 13 cm. Although cutting at increased heights generally increased forage CP and IVDDM concentrations, yields of CP and IVDDM were greatest for 2.5- or 7.5-cm stubbles.

In Arizona, Robison and Massengale (119) found that declines in herbage yield and stand of Moapa alfalfa over a 2-yr period were less when plants were cut to an 8-cm rather than a 2.5-cm stubble height. In all probability, photosynthesis and root reserves were low because of high summer temperature.

Leach (70) suggested that the advantage of leaving a stubble (5 cm with or without leaves as compared to no stubble) is that there are more sites available for regrowth, and that this may be more important than root NSC reserve level or residual leaf area. Watters and Henderlong (165) reported that cutting at 13 and 25 cm increased axillary bud de velopment and resulted in 8 to 76% of the new shoots originating above 5 cm on the old stubble. Plants cut at a 25-cm height at 3- or 4-week intervals maintained NSC levels comparable to a control harvested every 5 weeks to a 3- to 5-cm stubble height. In a greenhouse study (23), cutting to a 7.5-cm stubble height increased alfalfa axillary bud and stem formation and increased plant dry weight yield compared to cutting at a 2.5-cm stubble height. In contrast, Wolf and Blaser (173) in a field experiment, reported that stems originating from axillary buds on old stems contributed less to herbage regrowth than those originating from crown buds.

The value of residual leaves has been questioned (9). Older and less efficient basal leaves may be more of a handicap than a benefit to regrowth, especially if they photosynthesize slowly, shade the plant base, and prevent new shoot development. Similarly, Leach (70) suggested that the rapidity with which new leaves are formed after cutting may be more important than the amount of leaf area left on the stubble.

Since bases of mature alfalfa stems are of lower quality and usually contain fewer leaves than stem tops (14), harvesting the upper portion of a canopy results in higher-quality forage. In Nebraska (107), the top half of a full-bloom alfalfa canopy yielded 51% of the total herbage, 60% of the IVDDM, 64% of the CP, 77% of the carotene, and only 39% of the total fiber. Producers might obtain higher-quality forage from mature alfalfa canopies by increasing the cutting height, but this may be difficult to achieve with modern farm equipment.

Increased cutting heights are warranted for fall cuttings. Stubble (at least 15 cm) catches snow and provides a longer duration of snow cover. It also reduces fluctuations in soil temperature in the fall and early spring.

12-7 FALL HARVEST MANAGEMENT

12-7.1 Critical Period Concept

Fall cutting has been associated with reduced persistence and yield since the work of Silkett et al. (131) and Rather and Harrison (116) in Michigan. They stated that "two cuttings (with the second in mid-August) are safest in Michigan." Their recommendation for the northern states was sound because they found that a third cutting in mid-September sometimes reduced yield and persistence. In addition, alfalfa persistence in that era was generally limited by the lack of bacterial wilt resistant cultivars and lack of information on the importance of soil fertility for stand maintenance. Their work led to the accepted recommendation in northern states that alfalfa should not be cut from 4 to 6 weeks before the first killing frost ($-2.2°C$) (2, 143). This interval, often referred to as the fall "critical period" (143), occurs from early September through mid-October in most northern states but may begin in August in Canada.

Recent research in Michigan (151, 156, 175) and in Minnesota (80) showed that the recommendation not to cut during the fall critical period can be liberalized to afford greater management flexibility. Furthermore, the concept of a critical period is not always valid, particularly with use of a winter-dormant, disease-resistant cultivar and high soil fertility. In Michigan (156, 175), three-cut harvest systems with final cuts on 15 September, 1 or 15 October had yields and persistence similar to or greater than those of the standard three-cut system with the final cut on 1 September. Marten (80), in southern Minnesota, also reported similar 4-yr yields and stands for a three-cut system with final cuts on 1 or 15 September. In another Minnesota study (128), at three southern locations, a three-cut system with a final cut on 15 September had greater 4-yr yields than a standard three-cut system with the final cut on 1 September. However, a four-cut system with a final cut on 15 September frequently resulted in yield and stand loss. Cutting schedules which include a fourth cut following a killing frost are currently recommended in northern lake states (6, 128, 149, 152, 153). These schedules utilize fall forage regrowth and provide a minimum risk to long-term yield or persistence.

McKenzie et al. (90) evaluated the effect of fall cutting during critical periods in Alberta, Canada. The critical period includes August to early September in central and northern Alberta; and September in southern Alberta. In southern Alberta, herbage yield the subsequent year was reduced by a final cut in mid-September in a three-cut system but not in a two-cut system. In central Alberta, a two-cut system with a final cut during the critical period consistently reduced subsequent yields compared to a two-cut system with a final cut in October. However, in northern Alberta, the detrimental effects of a second cut during the critical period were not observed unless followed by a severe winter.

In addition to the aforementioned reports, research in Maryland (25), Ontario (35), Kansas (46), Saskatchewan (54), Washington (57), Nevada (59), West Virginia (61), Alabama (86), North Dakota (95), and Oklahoma (130) indicated that fall cutting was not always detrimental to alfalfa yield and stand persistence.

12-7.2 Soil Fertility and Cultivars

High levels of soil fertility, particularly K, reduce the stress induced by fall-cutting practices (43, 61, 110, 127, 144, 153, 159). In eastern Canada, Bailey (3) reported that fall (October) cutting had no detrimental effects on stand productivity when alfalfa was adequately fertilized with K. Smith (144) reported that a fourth cut in mid-October of unfertilized (no K) alfalfa reduced final stands by 50% compared to no fall cutting; however, at annual K rates of 448 kg ha^{-1} or greater, stand reductions were <13%. Unfortunately, many reports on the effects of fall cutting practices on alfalfa yield and persistence do not provide an adequate description of soil fertility levels under which the experiment was conducted (7, 37, 54, 57, 116). Use of inadequate soil fertility may lead to false conclusions as to the effects of fall cutting.

Cultivar characteristics and stand age interact with soil fertility and climate to modify the effects of fall cutting. Winterhardiness and disease resistance (especially bacterial wilt) have been identified as important cultivar characteristics in moderating the risks associated with fall harvesting (61, 86, 87, 88, 89). Tesar and Yager (156) reported that K fertilization reduced winter injury which resulted from the combined effects of bacterial wilt and a third cut in mid-September. However, they concluded that in northern states both bacterial wilt resistance and K fertilization were essential for long-term persistence of fall-cut alfalfa.

Alfalfa stands in the 1st yr of production are less susceptible to the negative effects of fall harvesting than older stands (27, 87, 125). This may be explained by the greater overall stress resistance of younger plants which is related to a lower incidence of disease infestation. Also, established alfalfa stands may be more susceptible to leaf injury from early frosts than seedling stands (87).

12-7.3 Cutting Schedules and Plant Maturity

Conflicting results from fall cutting studies can be partially explained by the effects of management and environmental factors on plant health (163). High concentrations of NSC in roots and crowns have been associated with increased persistence and winter survival by alfalfa (8, 21, 45, 67, 87, 95). The primary argument for "no fall cutting" is that NSC reserves required for winter survival are reduced by fall cutting (143). However, seasonal cutting frequency, and stage of plant development at fall cutting influence plant NSC status and can offset the potential negative effects of fall cutting. In West Virginia (61), fall cutting of plants after flowering (high concentration of stored NSC) greatly reduced the effects of fall cutting on subsequent yield and stand persistence. Minnesota studies (7) showed that the interval between cuttings (i.e., stage of maturity at harvest) was as important as time of fall cutting in predisposing alfalfa to winter injury. Furthermore, in Michigan, Tesar and Yager (156) found that alfalfa root NSC reserves in December were similar following a third cutting on 15 September or 1 or 15 October. They concluded that fall cutting is not harmful as long as there is adequate time for replenishment of carbohydrate reserves (indicated by at least 10% flowering) between the second (mid-July) and third cutting.

In several studies, fall cutting had no effect on yield or stand persistence when root NSC were not affected (86, 117, 130, 153). Brink and Marten (6) suggested that there may be a root NSC threshold above which alfalfa persistence is not detrimentally affected regardless of cutting time. This may explain why persistence was similar for alfalfa subjected to three-cut systems with final cuts on either 1 or 15 September or 15 October (80, 128, 130, 156).

Frequently, erroneous conclusions regarding fall cutting have been drawn from comparisons of cutting schedules with different numbers of total cuts per season (37, 67). For example, Kust and Smith (67) compared the effects of a three-cut schedule with a final cut before the critical period to those of a four-cut schedule with a final cut during the critical period and concluded that fall cutting was detrimental to persistence of alfalfa. Unfortunately, the effects of fall cutting were confounded with those due to number of cuts per season. To determine the effects of fall cutting, evaluations should be based on cutting schedules with similar numbers of cuts per season.

12-7.4 Environment

Detrimental effects of fall cutting are substantially reduced with lower levels of environmental stress. In the southern USA, mild winter temperatures and the presence of viable basal leaves on plants insure adequate NSC reserves and minimize effects of fall harvesting (117, 130). In Alabama, 'Williamsburg' was tolerant of a fourth cutting in September or early October (86). In Maryland, systems with a third cutting in early (20 August–15 September), middle (10 September–1 October) and late 10–25 October) fall had no effect on alfalfa persistence (25).

In the northern USA and Canada, snow cover protects plants from the harmful effects of low winter air temperature (87, 88, 109). In Minnesota, at snow depths >10 cm, air temperatures to $-40°C$ did not significantly influence soil temperature (124). The deleterious effects of fall cutting occur more frequently at locations without dependable winter snow cover (87). McKenzie et al. (90) and McKenzie and McLean (88) in Alberta, Canada identified several other environmental factors which may influence winterhardiness and interact with fall cutting to decrease alfalfa persistence. Periods of warm weather during the fall can stimulate the development of crown buds and the production of new shoots, delay hardening, and reduce maximum cold hardiness potential. Saturated soils during hardening in the fall cause plants to deharden and to become more susceptible to environmental stresses which are not normally lethal (see Chapter 8 in this book). Winter injury of fall-cut alfalfa is enhanced by excessive irrigation (128, 152).

Fall cutting removes stubble that may catch and hold snow during winter and early spring (44, 139). Strips of unmown alfalfa can be left to catch snow or alfalfa can be grazed to leave 15 cm of top growth. On poorly drained soils in the lower Corn Belt where heaving is a problem, it is desirable to have 15 to 25 cm top growth to protect the soil during winter and reduce soil temperature fluctuation (102, 114).

12–7.5 Forage Yield and Quality

Yield and quality of fall-harvested alfalfa vary greatly depending upon crop maturity and environmental conditions. However, cool temperatures and short daylengths during the autumn delay maturation and reduce the usual decline in alfalfa quality (163). Late fall cuttings should not be delayed long after the first killing frost because of rapid leaf loss. Collins (20) in Wisconsin and Collins and Taylor (21) in Kentucky reported that following a killing frost, forage yield of alfalfa previously cut in early September declined more rapidly than forage quality. The decline in yield was probably associated with loss of older leaves at the bottom of the canopy.

Even in situations of minimum risk to alfalfa persistence, fall weather in northern regions may provide poor drying conditions and haymaking may be difficult. Alternative forage harvesting/storage systems which require less field drying or grazing should be considered (see Chapter 19 in this book).

12–7.6 Recommendation

It is impossible to guarantee that fall cutting will have no adverse effects on long-term yield and persistence of alfalfa. Contrary to recent results from Oklahoma (130), Michigan (156), and Minnesota (80, 128), fall cutting did reduce alfalfa vigor in Missouri (102), California (123), and Kentucky (21).

Recommendations on fall cutting can be made only after considering all factors that influence alfalfa productivity. Obviously, plants that are stressed by frequent cutting, disease, low soil fertility, or extreme climatic conditions would be affected more by fall cutting than disease-resistant plants with high NSC reserves grown on high-fertility soil. A recommendation to avoid fall cutting would provide the least risk to long-term alfalfa productivity, but this recommendation is often too conservative for producers who: (i) maintain stands for short durations (2–3 yr), (ii) attempt to maximize short–term nutrient or dry matter production; (iii) use modern disease resistant cultivars and high levels of soil fertility; and (iv) require management flexibility.

12–8 SPRING MANAGEMENT

Timing the first cut of alfalfa in the spring is important because it usually dictates the total number of harvests per growing season, and it may influence the recovery of stands that have been damaged during winter. Smith (143) reviewed the literature and concluded that the first harvest of winter-injured plants (e.g., slow to recover in spring, chlorotic, and few shoots per plant) should be made at full flower. This allows accumulation of NSC and healing of winter-injured tissues. Cutting the first crop of a winter-injured stand at bud (9 June) reduced second crop yield in August by 57% and increased weed invasion compared with a late-flower first-crop harvest (139).

On soils subject to freezing and thawing, alfalfa crowns may be heaved aboveground at the time of the first spring harvest (113). In addition to delaying the first spring harvest, cutting at above-normal stubble heights may be beneficial in promoting recovery.

Cutting healthy stands of alfalfa at first flower allows near maximum dry matter and nutrient yield and restoration of high levels of root reserves (143). However, flexibility in spring management is required because (i) in much of the USA an initial harvest at first flower is complicated by unfavorable weather conditions, (ii) early spring feed shortages often necessitate harvesting at immature stages, (iii) producers often are willing to sacrifice forage yield to obtain higher forage nutrient concentrations, (iv) timing of the initial spring harvest may be a method of insect control, and (v) late spring frosts may destroy the growing points and stunt plant growth. Smith (139) reported that early spring clipping of plants (15- to 20-cm tall) followed by two full flower cuttings reduced forage dry matter, CP, and TDN yields compared with three harvests at early flowering. In contrast, Wolf and Blaser (173) cut the initial spring growth of alfalfa when it was 15, 30, 45, and 65 cm (first flower) tall and harvested the regrowth at first flower. Total season and subsequent year herbage yields were similar among cutting treatments. However, they recommended that for the 15-cm cutting height about one-half of the leaf area (LAI of 0.5) be left to insure fast regrowth. This system provides high-quality forage early in the growing season and may delay the subsequent cutting until a period of more favorable haymaking weather. In North Dakota, Dodds et al. (27) recommended early (before flowering) cutting if one-third to one-half of the new spring regrowth is wilted by frost. The second cut of frosted stands which have been mowed early should be delayed until mid-flowering to allow recovery of NSC reserves.

12-9 INSECT RELATIONS

Harvest scheduling and harvest procedures can influence populations of insect pests (see Chapter 22 in this book). In some northern regions, timing of the initial spring harvest aids in control of alfalfa weevil [*Hypera postica* (Gyllenhal)]. Early spring cutting (first flower or earlier) removes many alfalfa weevil eggs and larvae (49). In later cutting, weevils pupate, fall to the ground and are not killed by harvesting. With severe alfalfa weevil infestation, use of insecticides after the first cutting may be needed to avoid substantial yield and quality losses in the second cutting (171).

For control of potato leafhopper [*Empoasca fabae* (Harris)], cutting should be timed to remove eggs and nymphs because adults migrate from fields during cutting (112). Nymphal populations were reduced 95% when alfalfa was cut to a 2- to 5-cm stubble height (132). Harvesting practices that left excess stubble or uncut borders on fields provided habitat for nymphs and adults and allowed for reestablishment of harmful populations. Tesar (154) found less leafhopper damage to second and third growth alfalfa in a four- than in a three-cut harvest system because of the shorter interval between harvests. Environmental and geographical factors will influence the efficacy of using harvest management as a method to control insects, and often time of insect development and migration is not synchronized with alfalfa growth. In many situations, particularly under intensive management, it may be preferable to use insecticides rather than suffering yield and quality losses from untimely harvest.

12-10 STAND DECLINE

Plant population and herbage yields usually decline with increasing stand age (92, 93, 126, 148). Stand decline is associated with competition among plants and the effects of disease, environmental conditions, and harvest management. Severe reductions in plant populations are possible before significant yield reduction occurs because decreases in plant populations are often compensated for by increased stem numbers and dry weight yields per plant (18, 23, 72, 100).

Sund and Barrington (148) reported no effect on seeding and 1-yr-old stand yields as alfalfa populations decreased from 160 to 46 plants m^{-2}. For an alfalfa-bromegrass (*Bromus* spp.) mixture, Marten et al. (82) reported no differences in dry matter yield among stands containing from 18 to 108 plants m^{-2}. In Michigan, 10-yr-old alfalfa stands with a plant density of 19 plants m^{-2} had forage dry matter yields of 19 Mg ha^{-1}, which were similar to yields of a 1-yr-old stand with a density of 210 plant m^{-2} (150). The above results generally support the idea that for economical hay production in non moisture-limiting northern environments, minimum alfalfa populations of from 20 to 40 plants m^{-2} are required (93, 158). For subtropical regions, maximum yields are obtained with populations of 30 and 40 plant m^{-2} for rain-grown and irrigated alfalfa, respectively (73). On dryland sites or when rhizomatous cultivars are utilized, populations of <9 plants m^{-2} may be acceptable.

Decline in alfalfa stand density has little direct effect on total forage quality. Meyer and Bolger (93) reported no effect of populations from 11 to 484 plants m^{-2} on CP or acid detergent fiber concentration. Similar effects of plant population on leaf percent and forage quality were reported by Sund and Barrington (148). In a space-plant study, stem diameter and lignin concentration decreased and stem digestibility increased as plant density increased from 9 to 144 plants m^{-2} (18). Plant density had no effect on total forage and leaf digestibility.

Ingress of grass and broadleaf weeds (see Chapter 23 in this book) into alfalfa stands is associated not only with stand aging or injury but also with environmental conditions following a harvest that are favorable for weed germination. Normally, alfalfa competes intensely for light and moisture with invaders, but untimely or frequent harvesting can result in ingress of annual weeds. Many weeds do not reduce yield or total forage quality; while others may have a dramatic effect on alfalfa yield, persistence, and quality (29, 81, 129, 158, 174).

12-11 SUMMARY

Cutting of alfalfa may be scheduled using stage of plant development, fixed time intervals, crown bud development or a combination of these criteria. Harvesting by stage of plant development provides the most consistent yield and quality over cultivars, years, and locations. Often a combination of the three criteria are used for scheduling, because flowering does not always occur in all environments and because alternative criteria may improve scheduling of alfalfa harvesting with other field operations.

Cutting schedules can be adjusted to obtain variable levels of forage yield, quality, or stand persistence. With increasing alfalfa maturity in a regrowth cycle, forage nutrient concentrations decrease while forage dry matter yield and root carbohydrates generally increase to about midflowering. Cutting at first flower (10% bloom) has generally resulted in the best combination of seasonal herbage and nutrient (energy and protein) yield and stand persistence. The development of cultivars with increased persistence has added flexibility in adjusting cutting schedules to produce both high yields and high concentrations of nutrients. With the current emphasis on forage quality, more research is warranted on cutting schedules to produce forage for specific classes of livestock.

The effects of cutting management variables, such as cutting frequency, cutting height, and fall cutting on forage yield, forage quality, and stand persistence are dramatically influenced by environment and plant factors. Important environmental factors include precipitation (including snow cover), air, temperature, soil fertility, and soil drainage (see Chapters 5, 8, and 10 in this book). Plant factors include cultivar winterhardiness, disease resistance, and morphology (see Chapter 4 in this book). Evaluation of the effects of traditional and new cutting practices on yield, quality, and persistence of modern cultivars is needed in many regions. Documentation of environmental conditions will improve the interpretation of results.

REFERENCES

1. Allinson, D.W., M.B. Tesar, and J.W. Thomas. 1969. Influence of cutting frequency, species, and nitrogen fertilization on forage nutritional value. Crop Sci. 9:504–508.

2. Ahlgren, G.H. 1956. Forage crops. 2nd ed. McGraw-Hill Book Co., New York.
3. Bailey, L.D. 1983. Effects of potassium fertilizer and fall harvests on alfalfa grown on the eastern Canadian prairies. Can. J. Soil Sci. 63:211–219.
4. Baumgardt, B.R., and Dale Smith. 1962. Changes in the estimated nutritive value of the herbage of alfalfa, medium red clover, ladino clover and bromegrass due to stage of maturity and year. Wis. Agric. Exp. Stn. Res. Rep. 10.
5. Bibbey, R.O. 1960. Shoot dominance and other factors affecting the crown of Vernal alfalfa. Proc. Can. Soc. Agron. 6:109–113.
6. Brink, G.E., and G.C. Marten. 1983. The effect of selected cutting management systems on grade one alfalfa yield, total nutrient yield, and stand persistence of alfalfa. p. 31–36. In Proc. Am. Forage Grass. Counc., Eau Claire, WI. 23–26 January. American Forage and Grassland Council, Lexington, KY.
7. Brown, B.A. 1963. Alfalfa varieties and their management. Conn. Agric. Exp. Stn. Bull. 376.
8. ----, and R.I. Munsell. 1942. The effect of cutting systems on alfalfa. Conn. Agric. Exp. Stn. Bull. 242.
9. Brown, R.H., R.B. Cooper, and R.E. Blaser. 1966. Effects of leaf age on efficiency. Crop Sci. 6:206–209.
10. Bryant, H.T., and R.E. Blaser. 1963. Effect of defoliation of four alfalfas and one birdsfoot trefoil variety on tops and roots. Va. Agric. Exp. Stn. Bull. 548.
11. Bula, R.J., D.A. Holt, R.G. May, and M.M. Schreiber. 1975. Environmental physiology, modeling and simulation of alfalfa growth. II. Biomass accumulation characteristics of an alfalfa canopy. Purdue Agric. Exp. Stn. Bull. 76.
12. ----, and Dale Smith. 1954. Cold resistance and chemical composition in overwintering alfalfa, red clover and sweetclover. Agron. J. 46:397–401.
13. Buller, R.E., and A. Sanchez. 1960. Effect of the maturity of alfalfa at harvest on forage production and stand in the Valley of Mexico. Agron. Abstr. American Society of Agronomy, Madison, WI, p 62.
14. Buxton, D.R., J.S. Hornstein, W.F. Wedin, and G.C. Marten. 1985. Forage quality in stratified canopies of alfalfa, birdsfoot trefoil, and red clover. Crop Sci. 25:273–279.
15. Chance, C.M., W.L. Griffeth, C.W. Loomis, A.A. Johnson, and C.S. Winkelblech. 1961. Forages: production, harvesting, utilization. New York Agric. Exp. Stn., Cornell University, Misc. Bull. 39:14–15.
16. Chatterton, N.J., S. Akao, G.E. Carlson, and W.E. Hungerford. 1977. Physiological components of yield and tolerance to frequent harvests in alfalfa. Crop Sci: 17:918–923.
17. ----, G.E. Carlson, R.H. Hart, and W.E. Hungerford. 1974. Tillering, non-structural carbohydrates, and survival relationships in alfalfa. Crop Sci. 14:783–787.
18. Cherney, J.H., J.J. Volenec, and K.D. Johnson. 1986. Forage quality of alfalfa as influenced by plant density. p. 127–131. In Proc. Am. Forage Grass. Counc., Athens, GA. 15–17 April. American Forage and Grassland Council, Lexington, KY.
19. Christian, K.R. 1977. Effects of the environment on the growth of alfalfa. Adv. Agron. 29:183–227.
20. Collins, M. 1983. Changes in composition of alfalfa, red clover, and birdsfoot trefoil during autumn. Agron. J. 75:287–291.
21. ----, and T.H. Taylor. 1980. Yield and quality of alfalfa harvested during autumn and winter and harvest effects on the spring crop. Agron. J. 72:839–844.
22. Counce, P.A., J.H. Bouton, and R.H. Brown. 1984. Screening and characterizing alfalfa for persistence under mowing and continuous grazing. Crop Sci. 24:282–285.
23. Cowett, E.R., and M.A. Sprague. 1962. Factors affecting tillering in alfalfa. Agron. J. 54:294–297.
24. Crowder, L.V., J. Vanegas, and J. Silva. 1960. The influence of cutting interval on alfalfa production in the high Andes. Agron. J. 52:128–130.
25. Decker, A.M., H.A. McDonald, R.C. Wakefield, and G.A. Jung. 1960. Cutting management of alfalfa and ladino clover in the northeast. R. I. Agric. Exp. Stn. Bull. 356.
26. Dexter, S.T. 1964. Alternative three-cutting systems for alfalfa. Agron. J. 56:386–388.
27. Dodds, D.L., D. Meyer, and B. Rice. 1981. Alfalfa management in North Dakota. North Dak. Ext. Serv. Circular R–571 (Rev.).
28. Droushiotis, D.N. 1980. The effects of the stage of growth and height on the yield of irrigated lucerne. Nicocia Agric. Res. Inst. Tech. Bull. 36.
29. Dutt, T.E., R.G. Harvey, R.S. Fawcett, N.A. Jorgensen, H.J. Larsen, and D.A. Schlough. 1979. Forage quality and animal performance as influenced by quackgrass (*Agropyron repens*) control in alfalfa (*Medicago sativa*) with pronamide. Weed Sci. 27:127–132.
30. Evans, D.W., and R.N. Peaden. 1984. Seasonal forage growth rate and solar energy conversion of irrigated Vernal alfalfa. Crop Sci. 24:981–984.

31. Farnworth, J. 1973. The effect of cutting frequency on Hasawi alfalfa. Ministry Agric. and Water, Saudi Arabia. Pub. 16.
32. Feltner, K.C., and M.A., Massengale, 1965. Influence of temperature and harvest management on growth, level of carbohydrates in the roots and survival of alfalfa (*Medicago sativa* L.). Crop Sci. 5:585–588.
33. Fick, G.W. 1984. Simple simulation models for yield prediction applied to alfalfa in the Northeast. Agron. J. 76:235–239.
34. ----, and R.S. Holthausen. 1975. Significance of parts other than blades and stems in leaf-stem separations of alfalfa herbage. Crop Sci. 15:259–262.
35. Folkins, L.P., J.E.R. Greenshields, and F.S. Nowosad. 1961. Effect of date and frequency of defoliation on yield and quality of alfalfa. Can. J. Plant Sci. 41:188–194.
36. Fuess, F.W., and M.B. Tesar. 1968. Photosynthetic efficiency, yields, and leaf loss in alfalfa. Crop Sci. 8:159–163.
37. Fulkerson, R.S. 1970. Location and fall harvest effects in Ontario on food reserve storage in alfalfa (*Medicago sativa* L.) p. 555–559. *In* M.J.T. Norman (ed.) Proc. 11th Int. Grassl. Congr., Queensland, Australia.
38. ----, D.N. Mowat, W.E. Tossell, and J.E. Winch. 1967. Yield of dry matter, in vitro-digestible dry matter and crude protein of forage. Can. J. Plant Sci. 47:683–690.
39. Gabrielson, B.C., D.H. Smith, and C.E. Townsend. 1985. Cicer milkvetch and alfalfa as influenced by two cuttings schedules. Agron. J. 77:416–422.
40. Gasser, H., and L. Lachance. 1969. Effect of dates of cutting on dry matter production and chemical content of alfalfa and birdsfoot trefoil. Can. J. Plant Sci. 49:339–349.
41. Gengenbach, B.G., and D.A. Miller, 1972. Variation and heritability of protein concentrations in various alfalfa plant parts. Crop Sci. 12:767–769.
42. Graber, L.F. 1953. A half century of alfalfa in Wisconsin. Wis. Agric. Exp. Stn. Bull. 502.
43. ----, and V.G. Sprague. 1938. The productivity of alfalfa as related to management. J. Am. Soc. Agron. 30:38–54.
44. Grandfield, C.O. 1934. The effect of the time of cutting and of winter protection on the reduction of stands in Kansas common, Grimm, and Turkestan alfalfas. J. Am. Soc. Agron. 26:179–188.
45. ----. 1935. The trend of organic food reserves in alfalfa roots as affected by cutting practices. J. Agric. Res. 50:697–709.
46. ----. 1943. Food reserves and their translocation to the crown buds as related to cold and drought resistance in alfalfa. J. Agric. Res. 67:33–47.
47. Gross, H.D., C.P. Wilsie, and J. Pesek. 1958. Some responses of alfalfa varieties to fertilization and cutting treatments. Agron. J. 50:161–164.
48. Hagemann, R.W., and V.L. Marble. 1983. Variety response to cutting schedules in Imperial Valley. p. 6–15. *In* Proc. 13th California Alfalfa Symp., Holtville, CA.
49. Hamlin, J.C., W.C. McDuffie, F.V. Lieberman, and R.W. Bunn. 1943. Prevention and control of alfalfa weevil damage. USDA Farmers Bull. 1930.
50. Heichel, G.H. 1983. Alfalfa. p. 127–155. *In* I.D. Teare and M.M. Peet (ed.) Crop-water relations. John Wiley and Sons, New York.
51. Heinrichs, D.H. 1973. Winterhardiness of alfalfa cultivars in southern Saskatchewan. Can. J. Plant Sci. 53:773–777.
52. ----, J.E. Troelsen and K.W. Clark. 1960. Winterhardiness evaluation in an alfalfa. Can. J. Plant Sci. 40:638–644.
53. Hill, R.R., Jr. and J.E. Baylor. 1983. Genotype × environment analysis for yield in alfalfa. Crop Sci. 23:811–815.
54. Irvine, R.B., and J.D. McElgunn. 1982. Effect of eight three-cut harvesting schedules on production of alfalfa forage under irrigation in southwestern Saskatchewan. Can. J. Plant Sci: 62:107–110.
55. Jackobs, J.A. 1950. The influence of spring-clipping; interval between cuttings, and date of last cutting on alfalfa yields in the Yakima Valley. Agron. J. 42:594–597.
56. ----. 1952. The influence of cutting practices on the nitrogen content of Ladak alfalfa in the Yakima Valley. Agron. J. 44:132–135.
57. ----, and D.L. Oldemeyer. 1955. The response of four varieties of alfalfa to spring clipping, intervals between clippings, and fall clipping in the Yakima Valley. Agron. J. 47:169–170.
58. Jensen, E.H., M.A. Massengale, and D.O. Chilcote. 1967. Environmental effects on growth and quality of alfalfa. Nev. Agric. Exp. Stn. Bull. T9.
59. ----, R.R. Skivington, and V.R. Bohman. 1981. Dormant season grazing of alfalfa. Nev. Agric. Exp. Stn. R142.
60. Judd, P., and J.C. Radcliffe. 1970. The influence of cutting frequency on the yield, composition and persistence of irrigated lucerne. Aust. J. Exp. Agric. Anim. Husb. 10.48–52.

61. Jung, G.A., R.L. Reid, and J.A. Balasko. 1969. Studies on yield, management, persistence and nutritive value of alfalfa in West Virginia. West Va. Agric. Exp. Stn. Bull. 581T.
62. ----, and D. Smith. 1961. Trends of cold resistance and chemical changes over winter in the roots and crowns of alfalfa and medium red clover. I. Changes in certain nitrogen and carbohydrate fractions. Agron. J. 53:359–364.
63. Kalu, B.A., and G.W. Fick. 1981. Quantifying morphological development of alfalfa for studies of herbage quality. Crop Sci. 21:267–271.
64. Kehr, W.R., E.C. Conrad, M.A. Alexander, and F.G. Owen. 1963. Performance of alfalfas under five management systems. Nebr. Agric. Exp. Stn. Res. Bull. 211.
65. Kiesselbach, T.A. 1918. Forage crops. Nebr. Agric. Exp. Stn. Bull. 169.
66. Kilcher, M.R., and D.H. Heinrichs. 1974. Contribution of stems and leaves to the yield and nutrient level of irrigated alfalfa at different stages of development. Can. J. Plant Sci. 54:739–742.
67. Kust, C.A., and Dale Smith. 1961. Influence of harvest management on levels of carbohydrate reserves, longevity of stands and yields of hay and protein from Vernal alfalfa. Crop Sci. 1:267–269.
68. Langer, R.H.M., and T.D. Steinke. 1965. Growth of lucerne in response to height and frequency of defoliation. J. Agric. Sci. 64:291–294.
69. Langille, J.E., L.B. MacLeod, and F.S. Warren. 1965. Influence of harvesting management on yield, carbohydrate reserves, etiolated regrowth, and potassium utilization of alfalfa. Can. J. Plant Sci. 45:383–388.
70. Leach, G.J. 1968. The growth of the lucerne plant after cutting: the effect of cutting at different stages of maturity and at different intensities. Aust. J. Agric. Res 19:517–530.
71. ----. 1970. Shoot growth on lucerne plants cut at different heights. Aust. J. Agric. Res. 21:583–591.
72. ----. 1979. Regrowth characteristics of lucerne under different systems of grazing management. Aust. J. Agric. Res. 30:445–465.
73. ----, and R.J. Clements. 1984. Ecology and grazing management of alfalfa pastures in the subtropics. Adv. Agron. 37:127–154.
74. Marble, V.L. 1971. Relating alfalfa yields, quality and stand persistence to harvest frequency. p. 19–28. In Proc. 2nd California Alfalfa Symp., Fresno, CA. 7–8 December. Agriculture Extension, University of California, Davis.
75. ----, 1974. How cutting schedules and varieties affect yield, quality, and stand life. p. 47–57. In Proc. 4th California Alfalfa Symp., Fresno, CA. 4–5 December. Cooperative Extension, University of California, Davis.
76. ----. 1976. Weevil damage influenced by alfalfa varieties and cutting schedules. p. 79–87. In Proc. 6th California Alfalfa Symp., Fresno, CA. 8–9 December. Cooperative Extension, University of California, Davis.
77. ----. 1980. Affect of harvest frequencies and variety on yield, quality and stand life. p. 22–38. In Proc. 10th California Alfalfa Symp., Visalia, CA.
78. ----, K.G. Bagget, R.W. Benton, R.H. Gripp, and P.D. Smith. 1985. High-quality hay production in the mountain counties of northern California. Calif. Agric. 39:27–30.
79. Marten, G.C. 1970. Temperature as a determinant of quality of alfalfa harvested by bloom stage or age criteria. p. 506–509. In M.J.T. Norman (ed.) Proc. 11th Int. Grassl. Congr., Queensland, Australia.
80. ----. 1980. Late autumn harvest of third crop alfalfa can allow long stand persistence in the north. Forage Grassl. Progress 21:3–4.
81. ----, and R.N. Andersen. 1975. Forage nutritive value and palatability of 12 common annual weeds. Crop Sci. 15:821–827.
82. ----, W.F. Wedin, and W.F. Hueg. 1963. Density of alfalfa plants as a criterion for estimating productivity of an alfalfa bromegrass mixture on fertile soil. Agron. J. 55:343–344.
83. Martin, N.P., and D.A. Schriever. 1986. Alfalfa management practices used by Minnesota alfalfa growers. p. 151–156. In Proc. Am. Forage Grassl. Counc., Athens, GA. 15–17 April. American Forage and Grassland Council, Lexington, KY.
84. Massengale, M.A. 1974. Management of alfalfa for increased forage production. p. 69–71. In Proc. California and Arizona Low Desert Alfalfa Symp., El Centro, CA. 16–17 January. Agricultural Extension, University of California, Davis, CA.
85. Matches, A.G., W.F. Wedin, G.C. Marten, Dale Smith, and B.R. Baumgardt. 1970. Forage quality of Vernal and DuPuits alfalfa harvested by calendar date and plant maturity schedules in Missouri, Iowa, Wisconsin and Minnesota. Wis. Agric. Exp. Stn. Res. Rep. 73.

86. Mays, D.A., and E.M. Evans. 1973. Autumn cutting effects on alfalfa yield and persistence in Alabama. Agron. J. 65:290-292.
87. McKenzie, J.S., and G.E. McLean. 1980. Some factors associated with injury to alfalfa during the 1977-78 winter at Beaverlodge, Alberta. Can. J. Plant Sci. 60:103-112.
88. ----, and ----. 1980. Changes in the cold hardiness of alfalfa during five consecutive winters at Beaverlodge, Alberta. Can. J. Plant Sci. 60:703-712.
89. ----, P. Pankiw, and B. Siemens. 1981. Peace alfalfa. Can. J. Plant Sci. 61:473-474.
90. ----, W.A. Rice, L. Folkins, M. Hanna, and S. Freyman. 1980. Some aspects of establishment and cutting management of forage crops. NRG 80-3 Agric. Canada, Beaverlodge, Res. Stn.
91. McNemar, J.H. 1978. The response of two alfalfa cultivars to fertility, irrigation, and cutting management. M.S. thesis, Univ. of Maryland.
92. Meyer, D.W. 1984. Stand age effects on alfalfa forage yields. p. 120-121. 1985 Crop Production Guide, North Dakota Agric. Assoc., Bismarck.
93. ----, and T.P. Bolger. 1983. Influence of plant density on alfalfa yield and quality. p. 37-41. In Proc. Am. Forage Grassl. Counc., Eau Claire, WI. 23-26 January. American Forage Grassland Council, Lexington, KY.
94. ----, and K.L. Larson. 1975. Alfalfa management in North Dakota. North Dak. Farm Res. 32:3-9.
95. ----, and B.L. Nelson. 1983. Fall harvest effects on forage yield, forage quality and root TNC reserves of irrigated alfalfa. Agron. Abstr., American Society of Agronomy, Madison, WI, p. 110.
96. Meyer, J.H., and L.G. Jones. 1962. Controlling alfalfa quality. Calif. Agric. Exp. Stn. Bull 784.
97. Moline, W.J., and W.F. Wedin. 1969. Yield and quality evaluations of first-cutting Vernal alfalfa. Iowa State J. Sci. 43:261-273.
98. Monson, W.G. 1966. Effect of sequential defoliation, frequency of harvest and stubble height on alfalfa (*Medicago sativa* L.). Agron. J. 58:635.
99. Mowat, D.N., R.S. Fulkerson, and E.E. Gamble. 1967. Relationship between stem diameter and *in vitro* digestibility of forages. Can. J. Plant Sci. 47:423-426.
100. Mullen, R.E., J.J. Vorst, H.E. Laborde, and C.L. Rhykerd. 1977. Yield and stand dynamics of *Medicago sativa* (alfalfa) as influenced by seeding management. p. 789-801. In E. Wojahn and H. Thöns (ed.) Proc. 13th Int. Grassl. Congr. Leipzig, German Democratic Republic. 18-27 May. Akademie-Verlag, Berlin.
101. Musgrave, D.J., and R.H.M. Langer. 1977. Crown development of two diverse genotypes of lucerne. N.Z.J. Agric. Res. 20:453-458.
102. Nelson, C.J., and M. Mitchell. 1974. Alfalfa cutting management. Research Rep., southwest Missouri Center. Mo. Agric. Res. Stn., Columbia, p. 22-27.
103. ----, and Dale Smith. 1968. Growth of birdsfoot trefoil and alfalfa. II. Morphological development and dry matter contribution. Crop Sci. 8:21-25.
104. ----, and ----. 1968. Growth of birdsfoot trefoil and alfalfa. III. Changes in carbohydrate reserves and growth analysis under field conditions. Crop Sci. 8:25-28.
105. ----, and ----. 1969. Growth of birdsfoot trefoil and alfalfa. IV. Carbohydrate reserve levels and growth analysis under two temperature regimes. Crop Sci. 9:589-591.
106. Offutt, M.S. 1967. Effect of cutting interval on the relative performance of pasture- and hay-type alfalfa varieties. Ark. Agric. Exp. Stn. Bull. 719.
107. Ogden, R.L., and W.R. Kehr. 1968. Field management for dehydration and hay production. p. 23-27. In Proc. 10th Tech. Alfalfa Conf., USDA-ARS.
108. Onstad, D.W., and G.W. Fick. 1983. Predicting crude protein, in vitro true digestibility, and leaf proportion in alfalfa herbage. Crop Sci. 23:961-964.
109. Ouellet, C.E. 1977. Monthly climatic contribution to winter injury of alfalfa. Can. J. Plant Sci. 57:419-426.
110. Parsons, J.L., and R.R. Davis. 1960. Forage production of Vernal alfalfa under differential cutting and phosphorous fertilization. Agron. J. 52:441-443.
111. Peterson, M.L., and R.M. Hagan. 1953. Production and quality of irrigated pasture mixtures as influenced by clipping frequency. Agron. J. 45:283-287.
112. Pienkowski, R.L., and J.T. Medler. 1962. Effect of alfalfa cutting on the potato leaf hopper, *Empoasca fabae*. J. Econ. Entomol. 55:973-978.
113. Pike, D.R., and J.F. Stritzke. 1984. Alfalfa (*Medicago sativa*)-Cheat (*Bromus secalenus*) competition. Weed Sci. 32:751-756.
114. Portz, H.L. 1967. Frost heaving of soil and plants. I. Incidence of frost heaving of forage plants and meteorological relationships. Agron. J. 59:341-344.
115. Rai, S.D., D.A. Miller, and C.N. Hittle. 1973. Influence of cutting frequency on drymatter yield and root reserves of alfalfa. Indian J. Agric. Sci. 43:388-392.
116. Rather, H.C., and C.M. Harrison. 1938. Alfalfa management with special reference to fall treatment. Mich. Agric. Exp. Stn. Spec. Bull. 292.

117. Reynolds, J.H. 1971. Carbohydrate trends in alfalfa (*Medicago sativa* L.) roots under several forage harvest schedules. Crop Sci. 11:103–106.
118. ----, and Dale Smith. 1962. Trend of carbohydrate reserves in alfalfa, smooth bromegrass, and timothy grown under various cutting schedules. Crop Sci. 2:333–336.
119. Robison, G.D., and M.A. Massengale. 1968. Effect of harvest management and temperature on forage yield, root carbohydrates, plant density, and leaf area relationships on alfalfa. Crop Sci. 8:147–151.
120. Rominger, R.S., D. Smith, and L.A. Peterson. 1975. Yields and elemental composition of alfalfa plant parts at late bud under two fertility levels. Can. J. Plant Sci. 55:69–75.
121. Salmon, S.C., C.O. Swanson, and C.W. McCampbell. 1925. Experiments relating to time of cutting alfalfa. Kans. Agric. Exp. Stn. Tech. Bull. 15.
122. Savoie, P., L.D. Parsch, C.A. Rotz, R.C. Brook, and J.R. Black. 1985. Simulation of forage harvest and conservation on dairy farms. Agric. Systems 17:117–131.
123. Schoner, C.A., Jr. 1982. The Effect of late fall harvest on alfalfa stand and yield. p. 15–21. *In* Proc. 12th California Alfalfa Symp., Fresno, CA. 8–9 December. Cooperative Extension, University of California, Davis.
124. Sharratt, B.S. 1984. Influence of the winter and growing season environments in the production of alfalfa. Ph.D. diss. Univ. Minnesota, St. Paul (Diss. Abstr. DA8424740).
125. Sheaffer, C.C. 1983. Seeding year harvest management of alfalfa. Agron. J. 75:115–119.
126. ----, and D.K. Barnes. 1982. Potential for alfalfa production in Minnesota. p. 125–136. *In* Proc. Am. Forage Grassl. Counc., Rochester, MN. 22–24 February. American Forage and Grassland Council, Lexington, KY.
127. ----, M.P. Russelle, O.B. Hesterman, and R.E. Stucker. 1986. Alfalfa response to potassium, irrigation, and harvest management. Agron. J. 78:464–468.
128. ----, J.V. Wiersma, D.A. Warnes, D.L. Rabas, W.E. Lueschen, and J.H. Ford. 1986. Fall harvesting and alfalfa yield, persistence and quality. Can. J. Plant Sci. 66:329–338.
129. ----, and D.L. Wyse. 1982. Common dandelion (*Taraxacum officinale*) control in alfalfa (*Medicago sativa*). Weed Sci. 30:216–220.
130. Sholar, J.R., J.L. Caddel, J.F. Stritzke, and R.C. Berberet. 1983. Fall harvest management of alfalfa in the southern plains. Agron. J. 75:619–622.
131. Silkett, V.W., C.R. Megee, and H.C. Rather. 1937. The effect of late summer and early fall cutting on crown bud formation and winterhardiness of alfalfa. J. Am. Soc. Agron. 29:53–62.
132. Simonet, D.E., and R.L. Pienkowski. 1979. Impact of alfalfa harvest on potato leafhopper populations with emphasis on nymphal survival. J. Econ. Entomol. 72:428–431.
133. Simons, R.G. 1984. Genotypic differences in alfalfa regrowth. p. 58. *In* Proc. 29th Alfalfa Improve. Conf., Lethbridge, AB. 15–20 July. North American Alfalfa Improvement Conference, Beltsville, MD.
134. Singh, Y., and J.E. Winch. 1974. Morphological development of two alfalfa cultivars under various harvesting schedules. Can. J. Plant Sci. 54:79–87.
135. Smith, Dale. 1961. Association of fall growth habit and winter survival in alfalfa. Can. J. Plant Sci. 41:244–251.
136.----. 1962. Carbohydrate root reserves in alfalfa, red clover, and birdsfoot trefoil under several management schedules. Crop Sci. 2:75–78.
137.----. 1964. Chemical composition of herbage with advance in maturity of alfalfa, medium red clover, ladino clover, and birdsfoot trefoil. Wis. Agric. Exp. Stn. Res. Rep. 16.
138. ----. 1965. Forage production of red clover and alfalfa under differential cutting. Agron. J. 57:463–465.
139. ----. 1968. The establishment and management of alfalfa. Wis. Agric. Exp. Stn. Bull. 542.
140. ----. 1969. Influence of temperature on the yield and chemical composition of Vernal alfalfa at first flower. Agorn. J. 61:470–473.
141. ----. 1970. Yield and chemical composition of leaves and stems of alfalfa at intervals up the shoots. J. Agric. Food Chem. 18:652–656.
142. ----. 1970. Influence of temperature on the yield and chemical composition of five forage legume species. Agron. J. 62:520–523.
143. ----. 1972. Cutting schedules and maintaining pure stands. *In* C.H. Hanson (ed.) Alfalfa science and technology. Agronomy 15:481–496.
144. ----. 1975. Effects of potassium topdressing a low fertility silt loam soil on alfalfa herbage yields and composition and on soil K values. Agron. J. 67:60–64.
145. ----, M.L. Jones, R.F. Johannes, and B.R. Baumgardt. 1966. The performance of

Vernal and DuPuits alfalfa harvested at first flower or three times by date. Wis. Agric. Exp. Stn. Res. Rep. 23.
146. ----, G.C. Marten, A.G. Matches, and W.F. Wedin. 1968. Dry matter yields of Vernal and DuPuits alfalfa harvested by calendar date and plant maturity schedules in Missouri, Iowa, Wisconsin, and Minnesota. Wis. Agric. Exp. Stn. Res. Rep. 37.
147. ----, and C.J. Nelson. 1967. Growth of birdsfoot trefoil and alfalfa. I. Responses to height and frequency of cutting. Crop Sci. 7:130–133.
148. Sund, J.M., and G.P. Barrington. 1976. Alfalfa seeding rates: their influence on dry matter yield, stand density and survival root size and forage quality. Wis. Agric. Exp. Stn. Res. Bull. R2786.
149. Tesar, M.B. 1970. Response of alfalfa to frequent cutting and variable dates of fall cutting. Agron. Abstr. American Society of Agronomy, Madison, WI, p. 57.
150. ----. 1979. Alfalfa variety trials p. 49-56. In Central Alfalfa Improve. Conf., 1979 Rep. Lincoln, NE.
151. ----. 1981. Fall cutting of alfalfa under 3- and 4-cutting systems in Michigan. p. 14–16. In Proc. 17th Central Alfalfa Improve. Conf., East Lansing, MI. 30 June–1 July. Central Alfalfa Improvement Conference, East Lansing.
152. ----. 1984. 10 Tons alfalfa without irrigation in Michigan, a new world record. p. 59–65. In Proc. National Alfalfa Symposium, East Lansing, MI. 21 March. Certified Alfalfa Seed Council, Inc., Woodland, CA.
153. ----. 1985. Fertilization and management for a yield of ten tons of alfalfa without irrigation. p. 326–332. In Proc. American Forage Grassl. Counc., Hershey, PA. 3–6 March. American Forage and Grassland Council, Lexington, KY.
154. ----. 1986. Personal communication. Michigan State University, East Lansing, MI.
155. ----, E.C. Doll, and K. Lawton. 1963. Effect of P and K on alfalfa cut two and three times annually. Agron. Abstr. American Society of Agronomy, Madison, WI, p. 112–113.
156. ----, and J.L. Yager. 1985. Fall cutting of alfalfa in the North Central USA. Agron. J. 77:774–778.
157. Tiharuhondi, E.R. 1975. Influence of management on herbage yield, persistence, carbohydrate reserves and nutritive value of "African" and "Florida 66" alfalfa. Ph.D. diss. Univ. of Florida, Gainsville (Diss. Abstr. 36:529B).
158. Triplett, G.B., R.W. Van Keuren, and J.D. Walker. 1977. Influence of 2,4-D, pronamide, and simazine on dry matter production and botanical composition of an alfalfa–grass sward. Crop Sci. 17:61–65.
159. Twamley, B.E. 1960. Variety, fertilizer, management interactions in alfalfa. Can. J. Plant Sci. 40:130–138.
160. Tysdal, H.M., and T.A. Kiesselbach. 1939. The differential response of alfalfa varieties to time of cutting. J. Am. Soc. Agron. 31:513–519.
161. Van Riper, G.E., and F.G. Owen. 1964. Effect of cutting height on alfalfa and two grasses as related to production, persistence and available soil moisture. Agron. J. 56:291–295.
162. Veronesi, F., A. Mariani, M. Fulcinelli, and S. Arcioni. 1986. Selection for tolerance to frequent cutting regimes in alfalfa. Crop Sci. 26:58–61.
163. Walgenbach, R.P. 1983. Autumn management of alfalfa affects forage yield, quality and persistence. p. 112–117. In Proc. Am. Forage Grassl. Counc., Eau Claire, WI. 23–26 January. American Forage and Grassland Council, Lexington, KY.
164. Washko, J.B., and J.W. Price. 1970. Intensive management of alfalfa for forage production. p. 628–632. In M.J.T. Norman (ed.) Proc. 11th Int. Grassl. Congr., Queensland, Australia. University of Queensland Press, St. Lucia.
165. Watters, V.L., and P.R. Henderlong. 1978. Alfalfa regrowth morphology and TNC changes with increased defoliation height and frequency. Agron. Abstr. American Society of Agronomy, Madison, WI, p. 106.
166. Weir, W.C., L.G. Jones, and J.H. Meyer. 1960. Effect of cutting interval and stage of maturity on the digestibility and yield of alfalfa. J. Anim. Sci. 19:5–19.
167. Willard, C.J. 1951. The management of alfalfa meadows. Adv. Agron. 3:93–112.
168. ----, L.E. Thatcher, and J.S. Cutler. 1934. Alfalfa in Ohio. Ohio Agric. Exp. Stn. Bull. 540.
169. Willis, W.G., D.L. Stuteville, and E.L. Sorensen. 1969. Effects of leaf and stem diseases on yield and quality of alfalfa forage. Crop Sci. 9:637–640.
170. Wilsie, C.P., and Takahashi, M. 1937. The effect of frequency of cutting on the yield of alfalfa under Hawaiian conditions. J. Am. Soc. Agron. 29:236–241.
171. Wilson, M.C. 1984. Manage the alfalfa weevil to improve alfalfa yield and quality. Certified Alfalfa Seed Council, Woodland, CA.
172. Winch, J.E., R.W. Sheard, and D.N. Mowat. 1970. Determining cutting schedule for

maximum yield and quality of bromegrass, timothy, lucerne and lucerne/grass mixture. J. Br. Grassl. Soc. 25:44–52.
173. Wolf, D.D., and R.E. Blaser. 1981. Flexible alfalfa management: early spring utilization. Crop Sci. 21:90–93.
174. ----, and C.L. Foy. 1984. Alfalfa yield response to a between-cutting contact herbicide. Crop Sci. 24:645–648.
175. Yager, J., and M.B. Tesar. 1968. Effect of fall cutting on subsequent production and carbohydrate reserves of alfalfa (*Medicago sativa* L.) Agron. Abstr. American Society of Agronomy, Madison, WI, p. 53.

13 Relationships with Other Species in a Mixture

DOUGLAS S. CHAMBLEE

North Carolina State University
Raleigh, North Carolina

MICHAEL COLLINS

University of Kentucky
Lexington, Kentucky

Alfalfa (*Medicago sativa* L.) is frequently sown in mixtures with grasses. In order to maintain a desirable balance between the legume and grass component, we need to understand the principles of competition and individual plant response to microclimate.

In legume-grass mixtures the grass component is usually the aggressor and dominates the legume after a few growing seasons. Several years ago, difficulty was encountered in many areas of the world in maintaining a desirable balance of alfalfa in alfalfa-grass mixtures. In many areas we now have difficulty maintaining a perennial grass with alfalfa. This is particularly true in the northern Great Lake States (78, 80) with mixtures of alfalfa and either smooth bromegrass (*Bromus inermis* Leyss.) or timothy (*Phleum pratense* L.) under the widely used three-cut system. The development of more vigorous, persistent alfalfa cultivars and the adaptation of improved soil fertility and management practices have contributed to this change. In many areas of the northern USA, improved alfalfa cultivars may be cut an extra time and this extra cut has proven disadvantageous to some grasses.

The introduction of alfalfa into grass dominant swards, such as tall fescue (*Festuca arundinacea* Schreb.), has been demonstrated to provide the equivalent in dry matter production of that obtained by applying from about 200 (88) to 300 kg ha^{-1} of N (86) to pure stands of tall fescue.

13-1 PRINCIPLES OF COMPETITION

Mixtures of species and varieties have been utilized because environmental conditions could not be accurately predicted (69). Furthermore, several species or cultivars have been included in a mixture to ensure against failure and to combine in one sward the different seasonal

---

Copyright 1988 © ASA-CSSA-SSSA, 677 South Segoe Road, Madison, WI 53711, USA.
Alfalfa and Alfalfa Improvement—Agronomy Monograph no. 29.

growth patterns of the constituents. The introduction of cool-season grasses, such as ryegrass (*Lolium* spp., annual and perennial) and the cereals, into alfalfa swards illustrates the latter. In addition to the above reasons for utilizing mixtures, a complex mixture might be more efficient in utilizing available environmental resources (69). Donald (27) has stated that, "it is reasonable to suggest that two species of contrasting habit, with respect to branching, leaf distribution, height, root distribution, mineral uptake, or other morphological or physiological character, will together be able to exploit the total environment more effectively than a monoculture, and will, thereby, give increased yield."

Competition is strongest among individual plants that are most similar (36). The demands of the individual plants of like species appear at the same place, same time, and with the same intensity above- and belowground. In discussing the general nature of dominance, Donald (27), stated that "when two species grow together, one can envisage in terms of the general concept of the nature of competition by Clements et al. (18) that one species may be more successful than the other in securing an undue "share" of the light, the water, or the nutrients, and that as a consequence, its yield per plant will be increased while the yield of the other species will be decreased." After an extensive review of the literature regarding trials in which two species were grown together, Donald (27) concluded that the yield of the mixture will usually be less than that of the higher yielding of the pure cultures, and that the yield of the mixture may be greater or less than the lower-yielding pure culture. As discussed above, there are theoretical considerations that suggest that a mixture might be superior to a pure stand of either a grass or a legume, and yield advantages have been observed for some alfalfa-grass mixtures (see section 13–3.3). Only a few tests conducted with alfalfa-grass mixtures or other legume-grass mixtures prove conclusively the yield advantages for mixtures.

Density of seeding is a critical factor in the comparisons of mixtures vs. monocultures (69). Both species (assuming that alfalfa is seeded with one grass) must be grown at their optimum density, and this condition has not always been fulfilled. Level of soil fertility is another factor critical to proper evaluation and interpretation of mixtures vs. pure stands. Frequently, grasses grown in pure stand do not receive added N, or only low levels of N that are insufficient for maximum growth. Similarly, alfalfa may not receive adequate fertilizer for maximum growth. Consequently, it may be difficult to evaluate the merits of the mixtures vs. pure stands. In most comparisons of pure cultures of alfalfa vs. pure cultures of grass vs. mixture of the two, N is not applied to the mixture. The use of N fertilizer at relatively low levels will increase the yields of the alfalfa-grass mixture under certain environmental conditions (see section 13–2.3).

The high yield potential of alfalfa relative to many perennial grasses was calculated by Cooper (20). In summarizing some maximum growth rates from various studies in the world, he noted that alfalfa produced a maximum of 23 g m^{-2} d^{-1} in the western USA. Orchardgrass (*Dactylis glomerata* L.) in the United Kingdom and a cross between perennial ryegrass (*L. perenne* L.) × Italian ryegrass (*L. multiflorum* Lam.) in New Zealand produced 18.9, perennial ryegrass in the Netherlands 20.0, and bermudagrass [*Cynodon dactylon* (L.) Pers.] 21.2 g m^{-2} d^{-1} in the southern USA.

Several techniques have been developed to introduce and control interspecific competition in alfalfa-grass mixtures. A desired balance of alfalfa and grass may be achieved in part by choice of species, date of seeding, density of seeding, planting pattern, soil fertility, and time and height of defoliation.

13-2 SPECIFIC COMPETITIVE EFFECTS

Rhodes (69) noted that the term *competition* can be used to describe those events that lead to the modification of the growth and development of a plant resulting from its association with other plants. According to Guy (36), when plants such as alfalfa are grown in pure stands, intraspecific competition is necessary for effective utilization of the environment. Following similar logic, we can conclude that in a mixture, interspecific competition is necessary for effective utilization of the environment. Generally, the maximum yield per plant of an individual forage grass or legume in pure stand is obtained at densities too low to produce maximum production per unit area of land surface. An exception to this premise has been observed by Champness (17) who noted that small seedlings of timothy growing close to larger ones or among white clover plants (*Trifolium repens* L.) often remained healthy, whereas others surrounded by bare ground died. Close proximity to other seedlings led to lower moisture loss and smaller fluctuations in temperature because of the vegetative cover.

13-2.1 Competition for Light

A forage mixture, such as alfalfa-grass, may develop sufficient foliage within a few weeks after planting to intercept all measurable light. Thus, competition for light may begin in a very few days after seedlings emerge.

Maximal absorption by the leaves is necessary for optimal utilization of solar radiation and this is related to leaf-area index (6). The spatial arrangement of foliage, as well as total leaf area is important in the interception of light (99). Grasses generally have more inclined leaves than clovers, and thus require more foliage to intercept a given amount of light (11). Consequently, this suggests the possibility of increasing yield by using species or cultivars with more erect leaves.

The net photosynthesis of alfalfa does not appear to be maximized until light intensity reaches 1012 μmol photon m^{-2} s^{-1}(6). Pritchett and Nelson (68) placed alfalfa and bromegrass under light intensities ranging from 34 to 610 μmol photon m^{-2} s^{-1}. Dry weights of both species were reduced as light intensity was reduced. Alfalfa nodulation was decreased as light intensity was decreased and was completely inhibited at 55 μmol photon m^{-2} s^{-1}. Brown et al. (12) found the light saturation value at 1227 μmol photon m^{-2} s^{-1} for bermudagrass and 431 μmol photon m^{-2} s^{-1} for orchardgrass with alfalfa being intermediate between the two grasses. Light intensity data were originally reported in foot-candles and later in lux units. Lux units were divided by 50 to obtain an approximation of SI units μmol photon m^{-2} s^{-1}. Low photosynthetic photon flux density reduced bermudagrass yields from 13.36 g pot^{-1} at 1050 μmol photon m^{-2} s^{-1} to 5.10 at 420 μmol photon m^{-2} s^{-1}, both at 30°C day temperature (39). Smith (78) concluded that orchardgrass was more shade tolerant than most common forage grasses, and, therefore, less suppressed by the associated legume. One may infer from these data that light can become a critical factor in the growth of alfalfa as well as the associated grasses.

Grasses may offer competition at various stages of growth, particularly as new growth is arising from the crown buds. Chamblee and Lovvorn (16) postulated that crown buds of alfalfa were completely shaded by the close association of tall fescue. Many buds failed to develop in the dense, vigorous grass sod.

Interactions exist between competition factors such as light and nutrients (37). Shading, i.e., competition for light, could reduce root size and thereby make a plant less competitive for soil nutrients. Other research indicates that the availability of K affects photosynthetic efficiency of alfalfa (103). Alfalfa C exchange (μmol m^{-2} s^{-1}) was increased by increasing soil K level as was total C fixed.

Various studies have shown that legumes with high light requirements may be eliminated completely from the sward when grown in association with grasses (with root systems partitioned) in competition only for light (27). On the other hand, investigators, have postulated that legumes, such as alfalfa, may produce a favorable "light" environment for the growth of such shade-tolerant grasses as orchardgrass (see section 13-3.3).

13-2.2 Competition for Moisture

Grasses commonly grown with alfalfa have a higher proportion of the root system in the upper 30 cm than alfalfa. Alfalfa roots also penetrate the soil to a greater depth than those of most grasses grown with alfalfa. Thus, once it is established, alfalfa competes favorably for available soil moisture.

Under humid conditions, Ward et al. (95) found that the orchardgrass component of a mixture was increased more by irrigation than was the alfalfa component. The alfalfa component in an alfalfa-bromegrass mixture in North Dakota (52) by the fall of the 3rd yr had decreased to 25% under low moisture conditions (<25 cm annually during the growing season). At high moisture levels the alfalfa component comprised 98% of the mixture. Under the humid conditions referred to above, moisture was probably a more limiting factor for the growth of orchardgrass than alfalfa, because most orchardgrass roots are confined to the upper levels. Yields of N-fertilized timothy were reduced more by drought than were yields of orchardgrass and bromegrass, indicating significant species differences among the cool-season grasses (33).

Powell and Kardos (67) found the alfalfa-grass mixtures differed only slightly from pure alfalfa in efficiency of water use. When each was grown in pure stands in the Pennsylvania study, alfalfa was more efficient than either bromegrass or orchardgrass. In Nebraska, the amount of water left in the soil after the 1st yr was the same for alfalfa-grass mixtures (bromegrass and orchardgrass) and alfalfa (91). In the 2nd yr, alfalfa-orchardgrass removed significantly more water than alfalfa alone in the 30- to 60-cm zone. There were no differences at other depths.

Chamblee (15) concluded that the favorable competitive performance of alfalfa when grown with grasses in certain environments results not only from alfalfa obtaining water at lower depths than grasses, but also from the equal competition of alfalfa with some grasses (orchardgrass in this study) for available soil moisture in the upper soil levels. Wolf (102) found that mixtures containing alfalfa extracted less soil moisture at 30 cm, but more at 122 cm, than mixtures containing ladino clover (*T. repens* L.) or birdsfoot trefoil (*Lotus corniculatus* L.). Utilizing radioactive P to characterize the root activity of subirrigated alfalfa, Fox et al. (30) found that alfalfa roots were highly active in the surface depths when moisture from rainfall was adequate. These data and those of Chamblee (15) support the thesis that alfalfa may be very competitive in the horizon near the surface for both moisture and nutrients.

13-2.3 Competition for Nutrients

The nature of the competition between alfalfa and grasses for nutrients has been partially explained by Drake et al. (29). They found that alfalfa roots had a cation-exchange capacity (CEC) nearly double that of the roots of bromegrass and other common perennial grasses. A plant root surface having a high CEC would absorb relatively more divalent cations, such as Ca, than a plant root, such as grasses, with a low CEC. Their investigation showed that the grasses were less capable of divalent cation absorption, but competed better for K than associated legumes in a soil low in available K.

Investigations have shown that alfalfa is placed at a disadvantage in association with grasses, for available soil S and K and even for P during establishment (61). However, alfalfa competes favorably when P is limited since its root system extends deeper than associated grasses. Competition for K may account, in part, for the suppressive effect of grasses on legumes (7). Competition indices for yield of alfalfa taproots grown with each grass increased with K and decreased with N fertilization (54).

In studies of alfalfa-grass mixtures in North Carolina (16), tall fescue suppressed the growth of alfalfa much more than orchardgrass. Alfalfa frequently was lower in K content when grown with tall fescue than when grown with orchardgrass (16). Also it has been reported (40, 55) that alfalfa dominated when K was sufficient for both components of an alfalfa-grass mixture but grass dominated where K was limited (40, 55). The relative ability of alfalfa to take up K placed at various depths in the soil or surface applied has been evaluated by Peterson and Smith (65). They found that alfalfa recovered 11% of the K added as potassium sulfate (K_2SO_4) to the soil at a depth 82.5 cm, and the recovery was greatest for surface-applied K.

Under greenhouse conditions competition between alfalfa and grass species was observed for both N and K (53). An adequate K supply alone was not effective in preventing the reduction of legume yield normally experienced in alfalfa-grass mixtures with heavy applications of N. Studies in Canada indicated that applications of N were necessary to maintain a balance of grass with alfalfa (48). Nitrogen applications of 90 kg ha^{-1} with or without added P increased yields and maintained 40% grass in the mixture. Where N was omitted or where P only was added the grass component was about 12%.

13-2.4 Temperature Effects

Temperature affects competition indirectly as it affects growth and evapotranspiration. In Maine, Stafford (84) found that alfalfa was superior to grasses in emergence at lower soil temperatures (Table 13-1). However, subsequent vegetative development of alfalfa was retarded more by cold soil than that of bromegrass and orchardgrass. Seven days after seeding in pots at a soil temperature of 9°C, no orchardgrass and only three seedlings of alfalfa had emerged; at 19°C orchardgrass did not emerge in 7 d, while 77% of the alfalfa seedlings emerged. High soil temperatures (29°C) were particularly detrimental to the seedling vegetative development of the grasses. Wilsie (98) reported the cardinal temperatures for germination: alfalfa- minimum 1°, optimum 30°; and timothy- minimum 3°, optimum 26°C.

Yields of alfalfa roots and foliage increased with temperature in-

Table 13-1. Effect of soil temperature on yields† of forage species 36 d after seeding, and percentage seedling emergence 14 d after seeding in a greenhouse experiment. Adapted from Stafford (84).

Temperature	Alfalfa		Bromegrass		Timothy		Orchardgrass	
°C	g	%	g	%	g	%	g	%
9	0.96	72	2.58	58	0.48	39	1.22	17
19	7.43	84	8.66	73	6.50	78	6.50	73
29	5.54	84	4.33	76	1.56	64	1.19	24

† Yields in grams of dry matter/pot.

creases up to at least 19.4°C under greenhouse conditions (60). Optimum temperatures for growth for alfalfa and for orchardgrass have been reported as 21 and 17°C, respectively. However, Brown (10) found that 21°C was optimum for orchardgrass growth under the conditions of his experiments. Dinitrogen fixation, an important factor in legume growth, is influenced by temperature. In a greenhouse study increasing the root temperature from 16 to 30°C reduced alfalfa N_2 fixation rate (acetylene reduction) by approximately 50% (4).

Alfalfa grown at 32/24°C day/night temperatures reached first flower after 21 d of regrowth, and had leaves that were significantly thinner and lower in starch concentration than those from alfalfa grown to first flower at 21/12°C (82). Plants grown at 21/12°C required 45 d to reach first flower. In another controlled environment study, Knievel and Smith (49) found that growth at 32/24°C day/night temperatures reduced the time required to reach late anthesis by 35, 25, and 28% for timothy, orchardgrass, and tall fescue, respectively compared with growth at 18/10°C. At the warm temperature, the height of the shoot at early head was reduced for timothy (39%) and tall fescue (35%) but not for orchardgrass. The same increase in growth temperature decreased by 50% the time required for bromegrass to reach early anthesis (77). Results from several studies of temperature effects on shoot growth of alfalfa and several grasses are summarized in Table 13-2. Compared with warmer (32/24°C) and cooler (21/12°C) temperatures, alfalfa shoot growth rate was highest at 27/18°C (34). At that temperature, alfalfa reached first flower in 30 d and shoot height was greater than at the other temperatures.

Growth of 'Vernal' alfalfa (76) and timothy, and bromegrass (77) was evaluated under similar experimental conditions in growth chambers maintained at temperatures of 32/24°C and 18/10°C. Shoot yield of alfalfa at first flower under warm conditions was reduced 69% below yields

Table 13-2. Summary of temperature effects on shoot growth of alfalfa and several grasses.

Species	Maturity stage	18/10	21/12	27/18	32/24	Reference
			g pot^{-1}			
Alfalfa	First flower	6.40			1.31	76
Alfalfa	First flower	3.82			1.82	76
Alfalfa	First flower		4.27	4.06	2.49	34
Timothy	Early anthesis	30.30			7.80	77
Bromegrass	Early anthesis	12.40			6.20	77
Timothy	Stems and leaf blades between early heading and late anthesis	7.82			1.23	49
Orchardgrass†		2.52			0.70	49
Tall fescue†		5.21			0.80	49

† Same growth stage as timothy.

under the cool-temperature regime. This value is similar to the reduction in shoot yield for timothy (74%) and higher than that for bromegrass (50% reduction), when they were harvested at early anthesis. Similar reductions in growth with increasing temperature were reported for timothy (84%), tall fescue (73%), and orchardgrass (61%) between early heading and late anthesis (49).

13-3 MUTUAL EFFECTS OF ABOVE- AND BELOW-GROUND ASSOCIATIONS OF ALFALFA-GRASS MIXTURES

13-3.1 Morphological and Physiological Differences Affecting Competition

Basic morphological and physiological differences between alfalfa and the various grasses materially affect the nature of competition that develops between species in various mixtures. As previously noted, the CEC of alfalfa and grass roots affects their ability to absorb and compete for various nutrients. Also, the root systems of alfalfa and grasses are distinctly different in morphology. Lamba et al. (50) noted that more than one-half of the weight of roots of alfalfa, bromegrass, and timothy was in the upper 20 cm of soil and that alfalfa penetrated to a greater depth than the grasses. In some soils grasses will have a higher percentage of their root systems in the upper 20 cm than alfalfa. Grass roots are smaller and more fibrous than alfalfa and will occupy the soil more thoroughly near the soil surface.

After cutting or grazing, alfalfa must regrow principally from new basal buds. On the other hand, grasses can regrow from the expansion of cut leaves because of meristematic activity at the base of the leaf (3), as well as from the development of tillers at the stem base. The nature of regrowth explains why the regrowth of alfalfa is often delayed more after defoliation than grasses. Grasses vary in growth habit, time of flowering, and in manner of regrowth and location of shoot apices. Consequently, cutting systems must be devised in terms of these differences among species.

Rate of emergence and growth are important factors in compounding forage seed mixtures. Blaser et al. (9) have shown that alfalfa, relative to many other legumes and grasses, is an aggressive species in the seedling stage. Assigning a seedling vigor score of 100 to alfalfa, they reported scores for red clover (*T. pratense* L.) and ladino clover of 63 and 38, respectively, whereas orchardgrass, tall fescue, bromegrass, and timothy had values of 42, 45, 52, and 17, respectively. The more rapid seedling growth of alfalfa may be explained by a higher relative rate of CO_2 uptake (70).

13-3.2 Below-ground Effects

In addition to below-ground competitive effects, such as competition for moisture and nutrients, beneficial effects may be obtained from an alfalfa-grass mixture. Mutual advantages might arise from the effective exploitation of the total environment by the different species (alfalfa, grass). Also, under certain conditions, alfalfa supplies N for growth of the associated grass by excretion, sloughing of nodules, or decay of the root system. Wilson (100) determined that a long day-length and relatively low temperatures induced the type of metabolism necessary for N excretion.

In a series of experiments, Wilson and Wyss (101) obtained evidence of N excretion by alfalfa grown with perennial ryegrass. Dilz and Mulder (26) reported that alfalfa supplied 8% of the N in the tops of ryegrass during the main growing period. In Australia, excretion accounted for most of the N transference from alfalfa to grass rather than sloughing of nodules or decay of root systems (75). Moncada de la Fuente (58) observed that growth and N content of fescue increased slightly when it was grown in a common nutrient supply with alfalfa. In Wisconsin, certain grasses grown with alfalfa were appreciably higher in percentage of N than grasses grown alone, with or without supplemental N (90).

13-3.3 Above- and Below-ground Effects

Alfalfa and grass grown in combination not only compete above- and below-ground for various growth factors, but also may have a beneficial effect on the environment for growth, e.g., N may be made available by the alfalfa component. Chamblee (14) separated the root systems of alfalfa and orchardgrass by sheet metal partitions that prevented intermingling of root systems, but permitted intermingling of aboveground parts. Comparisons were made to the two species grown without root partitions. In this study (Table 13-3) both alfalfa and orchardgrass were benefitted at various times by mixing as compared with their performance in pure stand (Orchardgrass received 22 kg ha^{-1} of N at seeding only.) Orchardgrass produced more growth when grown between two rows of alfalfa than between two rows of orchardgrass, both with and without root partitions. It was postulated that aboveground benefits to orchardgrass included the shading effect of alfalfa, which influenced air and soil temperatures (indirectly affecting evaporation of soil moisture) (Table 13-3).

Alfalfa was benefitted in a more indirect manner in that orchardgrass offered less above- and below-ground competition in this experiment (Table 13-3). At certain times (first spring harvest, 1949, nonpartitioned block), the yield of both orchardgrass and alfalfa was increased by their association with one another, to the extent that total yield of the mixture was greater than alfalfa or orchardgrass grown in pure stands. To make this comparison on a comparable basis, one must double the yield of a single row of alfalfa (A)A(A) or orchardgrass (O)O(O) and compare these yields with the addition of the single-row yield of orchardgrass (A)O(A) plus the single-row yield of alfalfa (O)A(O). In late summer of the 2nd yr of these trials, orchardgrass was damaged by *Rhizoctonia solani* Kuehn;

Table 13-3. Seasonal yields of orchardgrass and alfalfa grown in 15-cm spaced rows in pure stands, and alternate rows, as influenced by row combinations and partitioning of the root systems in a field in North Carolina. Adapted from Chamblee (14).

Combinations of rows†	First spring harvest Part.‡	First spring harvest No part.	Avg of four annual harvests Part.	Avg of four annual harvests No part.
		g row⁻¹		
		1948		
(O)O(O)	57	74	25	32
(A)O(A)	82	79	37	37
LSD (0.05)	12		5	
		1949		
(O)O(O)	6	10	6	9
(A)O(A)	29	45	13	17
LSD (0.05)	5		2	
		1948		
(A)A(A)	48	54	46	47
(O)A(O)	50	64	52	67
LSD (0.05)	NS		5	
		1949		
(A)A(A)	46	45	52	48
(O)A(O)	62	73	75	86
LSD (0.05)	9		7	

† O = orchardgrass; A = alfalfa. Yields are for the center row of the three-row combinations.
‡ Part. indicates that the root systems of individual rows were spaced 15 cm from adjacent rows and partitioned by metal partitions to a depth of 76 cm.

the damage appeared to be more severe in mixture with alfalfa than in pure stand. Velich and Charvat (94) conducted a somewhat similar study in which alfalfa and orchardgrass were grown alone in rows (25 cm apart) and in alternating rows (12.5 cm apart) with and without below-ground and aboveground competition. With unrestricted competition above- and below-ground, alfalfa yield was reduced 25%. The competition aboveground accounted for 3% of the reduction.

Aberg et al. (1) did not report a significant gain or loss in forage or root yields for both members of an alfalfa-grass association. The significant gains in a crop combination usually were made by the most vigorous species and losses by the least vigorous. Roberts and Olson (73) found that the largest gains from association occurred when a legume with a vigorous growth habit, such as alfalfa, was associated with a grass with a weaker growth habit, such as Kentucky bluegrass (*Poa pratensis* L.) (Table 13-4). There were no comparisons in which both legume and grass were either benefitted or injured by associated growth vs. growth in pure stand. The weight of the legume component (tops and roots) for the

Table 13-4. Influence of associated growth of alfalfa and grass on total weight and total N content as compared with their performance in pure stand in a greenhouse trial. Adapted from Roberts and Olson (73).

Alfalfa with:	Differences in favor of the mixture between yield in pure stands† and in mixed stands‡					
	Tops	Roots	Total	Tops	Roots	Total
	—————wt§, g—————			——— total N, mg ———		
Kentucky bluegrass	+6.01	−0.40	+5.61	+175	+41	+216
Redtop	+2.37	−0.72	+1.65	+93	+13	+106

† 15 grass and 15 legume plants each in pure stand on one-half unit area.
‡ 15 grass and 15 legume plants in mixed stand on unit area.
§ No N applied.

Table 13-5. Yield of the alfalfa component of alfalfa-grass mixtures when grown with various grasses as a percentage of its production when grown alone in a field trial near Lafayette, IN. Adapted from McCloud and Mott (56).

Grass	1946	1947	1948
	——— % of alfalfa grown alone ———		
Bromegrass	42	22	66
Kentucky bluegrass	81	90	117
Timothy	53	46	85

Table 13-6. Yield of various alfalfa-grass mixtures expressed as a percentage of the yield of alfalfa grown in pure stand. Adapted from McCloud and Mott (56).

Alfalfa with:	1946	1947	1948
	——— % of alfalfa grown alone ———		
Bromegrass	179	175	144
Kentucky bluegrass	86	116	146
Timothy	74	105	137

alfalfa-Kentucky bluegrass and alfalfa-redtop (*Agrostis alba* L.) mixture was 26.7 and 20.3 g, respectively; the weight of the grasses was 7.2 and 12.6 g, respectively.

McCloud and Mott (56) reported that the performance of different legume-grass mixtures, not fertilized with N, varied from mutually depressive, no interaction, to beneficial and mutually beneficial. Their results (Tables 13-5 and 13-6) illustrate the wide diversity of competitive and associational effects that different grasses may have in different years when grown with alfalfa. Smooth bromegrass was aggressive in all years, and on the average, reduced the alfalfa component more than other grasses. Yet, the mixture of alfalfa-bromegrass was the most productive of all mixtures compared with pure alfalfa.

13-4 MANAGEMENT EFFECTS ON ALFALFA-GRASS MIXTURES

13-4.1 Seeding Time and Density

Time of seeding of an alfalfa-grass mixture markedly influences the emergence and seedling growth of individual species. Density of seeding is a critical factor in the comparison of mixtures vs. single species. Black (5) found in studies with mixtures of red clover and alfalfa that the growth of one species was depressed when that species was present at high densities and the other at low densities, and conversely, was increased when that species was present at low densities and the other at high densities. His data confirmed previous proposals that the most severe competition will occur among individuals that are the most similar. Yamada and Horiuchi (106) concluded that the time when plants begin to compete with each other is determined not only by spacing but also by their growth rate.

Dense seedings of grass with alfalfa may reduce the number of alfalfa plants at establishment (16). High seeding rates of grass (17 kg ha^{-1}) mixed with alfalfa reduced total yields of forage, although similar amounts of grass were produced each year regardless of seeding rate. Evidently, the excessive number of grass seedlings reduced the number of alfalfa seedlings to less than the optimum population.

Seedling growth of associated grasses may be seriously suppressed by high densities of alfalfa (84). Density of seeding affects both weed invasion and microclimate. A dense canopy of alfalfa and/or grass will reduce soil and air temperatures and will conserve soil moisture when the moisture level is relatively high (22).

13-4.2 Alternate Rows, Row Widths, and Broadcasting

Many researchers have seeded alfalfa and grass in alternate rows rather than broadcasting or drilling them together in the row. This represents an attempt to reduce interspecies competition and thereby maintain a better balance between legumes and grass (Fig. 13-1). In most of these studies alternate rows have been inferior to other methods, particularly with wider rows. The inferiority of alternate rows may be attributed partially to lack of optimum spacing of individual plants, particularly alfalfa, or in some tests to the more aggressive growth of grass in alternate rows (16). In Minnesota, Tewari and Schmid (89) reported that alfalfa and grass mixed in the row yielded more and had a higher percentage of legumes than alternate rows. Similar results were noted by Fyfe and Rogers (32). Chamblee and Lovvorn (16) obtained higher alfalfa yields from broadcast alfalfa-grass than from alternate rows spaced 15 cm apart.

Wide spacing of alternate rows of alfalfa-grass mixtures often has resulted in lower yields. In England, yields of alfalfa decreased as row spacing increased from 9 to 71 cm (38). Although yields of orchardgrass

Table 13-7. Total dry matter yield and N content of bromegrass at four distances from the nearest alfalfa row in the plot, St. Paul, MN. Adapted from Tewari and Schmid (89).

Distance of grass from alfalfa	Yield	N content
cm	g/61-cm row	g kg^{-1}
15	65	24.3
30	39	22.9
46	35	22.2
61	34	21.1

Fig. 13-1. Alfalfa seeded with grass. *Left*, mixed in the row; *center*, alternate rows; *right*, broadcast.

increased with wider row spacing in the first 2 harvest years, they were greatest at 18-cm spacing in the 3rd and 4th yr. Fyfe and Rogers (32) obtained lower dry matter yields from alfalfa-grass in 30- than in 15-cm row widths. In alternate row plantings, alfalfa may have a substantial influence on the yield and N content of a grass at a distance of 15 cm, but has little apparent effect at distances of 30 cm or more (Table 13-7).

Results from a semiarid area in Canada (51), where the long-term rainfall mean for the 4-mo period April-July totaled 189 mm, are different from those reported in the USA and England. Alfalfa and Russian wild ryegrass (*Elymus junceus* Fisch.) or crested wheatgrass [*Agropyron desertorum* (Fisch.) Schult.] were seeded alone or in alternate rows spaced 30, 60, or 90 cm apart. Six seeding rates were used ranging from 17 to 100 seed/m row. Narrow row spacing (30 cm) initially produced the highest yields. By the 5th yr, there had been a transition in yield advantage from the 30-cm through the 60-cm to the 90-cm rows, and alternate grass-alfalfa stands were producing the highest dry matter yields.

13-4.3 Species Effects—Pure vs. Mixtures

Mixtures of alfalfa-grass may or may not yield more than pure stands of either species. Increased yields are obtained frequently from mixtures relative to the production of alfalfa alone, although these increases are not of great magnitude (Tables 13-3 and 13-4). In a single season, an increased yield of up to 80% may be realized from mixtures (Table 13-6). Over a period of years, however, typical maximum increases will average approximately 10 to 15%. In Canada, the influence of alfalfa on total yield was such that no significant total yield differences could be assigned to associated grasses (47).

Many investigators have reported no yield advantage for alfalfa-grass mixtures over alfalfa alone. In Iowa, Wilsie (97) found that over a 5-yr period alfalfa-grass did not yield significantly more total forage than that produced by alfalfa alone, except toward the end of the experiment when bacterial wilt [*Corynebacterium insidiosum* (McCull.) H. L. Jens] had depleted the stand of alfalfa. Similar results have been obtained by others (13, 21, 25). Carter and Scholl (13) observed little difference in yield between grasses (orchardgrass or bromegrass) receiving 269 kg ha^{-1} of N and mixtures of alfalfa-grass receiving no N.

Researchers reporting increased yields by the addition of a grass as compared with alfalfa alone include Plancquaert (66). He observed, in a 5-yr study in France, that yields were generally, but not always higher from mixtures. In a 2-yr study in Belgium, mixtures of alfalfa with orchardgrass, Italian ryegrass, and perennial ryegrass outyielded pure alfalfa by 11.9, 6.7, and 4.5%, respectively, while other grasses in mixtures had little effect on total yield (93). In Wales, some mixtures produced about 6% more than pure alfalfa and some about 10% less (alfalfa-orchardgrass) (24). Williams (96) made the observation in Britain that self-sown grasses became prevalent and competitive with the alfalfa crop without adding much to yield. Therefore, they add small quantities of a non-aggressive grass, such as meadow fescue (*Festuca pratensis* L.) or timothy, to control unsown grasses. In New Zealand (31), a mixture of *Bromus willdenowii* and alfalfa produced from 3 (1st yr) to 15% (2nd yr) more dry matter than a pure stand of alfalfa. McCloud and Mott (56) observed average increases over 3 yr ranging from 5 to 66% (Table 13-6). Miller (57) reported from Illinois that mixtures of alfalfa with either orchardgrass, smooth bromegrass, or reed canarygrass (*Phalaris arundinacea* L.) produced a maximum of 4.5% more dry matter yield than alfalfa grown alone over a 3-yr period. Sprague et al. (83) observed that in New Jersey alfalfa yielded from a third to a half more with bromegrass than with orchardgrass.

In recent years, there have been an increasing number of reports of new alfalfa cultivars dominating associated grasses. Davis and Tyler (24) reported that an outstanding feature of their trial with alfalfa-grass mixtures (included orchardgrass) was the steady increase in the proportion of alfalfa during the trial (Table 13-8). Also, their data show the variation that exists among grasses in competitive ability.

According to Ridgeman et al. (72), the greatest yields of alfalfa-grass mixtures were realized with the use of high- yielding cultivars of alfalfa. They also reported higher grass yields when using low-yielding alfalfa cultivars, but grass yields were never sufficient to compensate for low alfalfa yields. Conversely, Jackobs (41) concluded that the productivity of the alfalfa-grass mixture generally increased with the productivity of the grass grown alone.

Grasses with the same potential yield capacity may differentially affect the associated alfalfa (16, 41). These investigators report that tall fescue had a greater depressive effect on the associated alfalfa than orchardgrass even though the two grasses were about equally productive. In North Carolina, Chamblee and Lovvorn (16) noted at the end of the 4th yr that twice as many alfalfa plants were present with orchardgrass as with tall fescue. In Washington, however, orchardgrass was very competitive with alfalfa (42). When it was added to a mixture with alfalfa there was a reduction of 0.49 kg of legume for each 0.45 kg of increase in the grass.

In Kentucky, the yields of orchardgrass or tall fescue with alfalfa were very low in the 1st yr and increased in later years (87). In some years, tall fescue was severely damaged by *R. solani*, and it was surmised that alfalfa created an ideal environment near the ground surface for the development of this disease (see section 13-3.3).

The addition of short-lived species to alfalfa and alfalfa-perennial grass species may affect establishment and growth of the mixture. Jackobs (43) conducted a study to determine the effect of the addition of short-lived legumes, such as nonwinter-hardy alfalfas or red clover, or short-lived perennial ryegrass to a mixture that contained a long-lived alfalfa and/or long-lived orchardgrass. The addition of short-lived alfalfa to mixtures with long-lived alfalfa did not influence performance in the 1st yr of grazing but increased yield in the second. Red clover, while present, increased both the legume component and total yield. Perennial ryegrass (short-lived) increased the grass component, but depressed the legume component and total yield in the 1st yr.

In grazing trials in Pennsylvnia, Jung et al. (45) showed that orchardgrass in mixture with alfalfa contributed 50 and 80% of the dry matter

Table 13-8. Yields of alfalfa in pure stand and alfalfa component of alfalfa-grass mixtures,† Aberystwyth, Wales. Adapted from Davis and Tyler (24).

Year	Alfalfa pure stand	Alfalfa + meadow fescue	Alfalfa + Italian ryegrass	Alfalfa + orchardgrass
1957	100†	55	37	47
1959	100	92	86	67

† 100 represents value given to alfalfa in pure stand; others are percent yields of alfalfa component relative to alfalfa in pure stand.

in May of the 1st and 3rd yr, respectively, whereas perennial ryegrass contributed 35 to 40% at the same period. In most trials, alfalfa seeded with grass has shown more dominance in midsummer than in early spring when the grasses generally produce their maximum growth. Kalton and Wilsie (46) reported that the first cuttings of alfalfa-grass contained 45 to 73% alfalfa, whereas second cuttings contained 80 to 90% alfalfa, with grass as the balance.

Sod seeding of legumes such as alfalfa, into grass sods has become an accepted practice. The degree of grass suppression has not been determined for all conditions. In North Carolina, paraquat (1,1'-dimethyl-4,4'-bipyridinium ion) applied to a tall fescue sod increased first season dry matter yields of the alfalfa component from 4615 to 7813 kg ha^{-1} after a September seeding and from 2175 to 4176 kg ha^{-1} after an October seeding (74). Similarly in Illinois, the alfalfa component was increased in each of 3 yr by use of herbicides at time of planting (62). In the 3rd yr, the alfalfa-tall fescue mixture produced 11.0 and 8.2 Mg ha^{-1} with and without herbicide use. Groya and Sheaffer (35) showed that alfalfa dry matter yield and root weight were generally greater when seeded into Kentucky bluegrass than into smooth bromegrass.

13-4.4 Cutting Practices

Alfalfa and grasses respond differently to various management practices because they differ markedly both morphologically and physiologically. Likewise, differences exist in date of maturity among species of grasses and cultivars of alfalfa. These differences require the use of appropriate management practices for each combination of species.

Dotzenko and Ahlgren (28) found that alfalfa grown in mixture with grass responds essentially the same to cutting frequency as when grown in pure stand. In their studies, the yield of the alfalfa component in alfalfa-bromegrass mixtures was only 10% as much when cut at an immature stage (25 cm high) as when harvested at the seedpod stage. Yet, yield of the grass component was not significantly reduced by more frequent harvest.

In pure stand, the productivity of most grass species decreases with increased frequency of clipping (19). In mixture with alfalfa the relative percentage of grasses may be increased by early or frequent cutting (19, 63, 83) because alfalfa growth and stands are reduced under these conditions and the alfalfa offers less competition. Barker et al. (2) in England observed that late-autumn defoliation led to a greater yield and proportion of alfalfa and a lesser yield and proportion of grass (orchardgrass).

Frequent cutting (every 30-35 d, four cuts annually) of alfalfa-grass mixtures in Indiana removed the apices of smooth bromegrass and timothy before the basal buds were fully developed and capable of growth (71). Alfalfa usually had begun to bloom in 30 to 35 d. Orchardgrass was able to persist better than bromegrass and timothy because the stem did not elongate after the first cutting. Also, as observed by Blaser et al. (8), orchardgrass recovers rapidly after cutting as a result of the initiation of numerous lateral shoots at the base of each culm even before the first spring growth is removed. In the trials in Indiana (71), smooth bromegrass and timothy essentially disappeared by the end of the third season and orchardgrass continued to persist with alfalfa in approximately a 50:50 ratio.

The general adoption of a three-cut alfalfa management system in the North Central region has created a problem in the maintenance of grasses with alfalfa as compared with the traditional two-cut system (81). If the first harvest is delayed until late June, as it would under a two-cut system, orchardgrass, bromegrass, and timothy are at or beyond the heading stage when cut. However, a first cut in early June, as is recommended for alfalfa today, comes when bromegrass is in the late boot stage, timothy is in the stem elongation stage and only orchardgrass has headed. Poor regrowth of grasses when cut during early reproductive growth is related to low numbers and/or size of tillers present at that time (44). In Wisconsin, the percentage of total herbage represented by bromegrass in an alfalfa-grass mixture cut three times each season decreased from 57% in the year after seeding to 32% 2-yr later (59). In another experiment, (105) bromegrass and timothy were eliminated from alfalfa stands after 2 yr of cutting three times each season. More cuts (five each season) maintained nearly balanced mixtures of alfalfa-bromegrass and alfalfa-timothy. In another 2-yr study in Wisconsin (80) using two stubble heights (4 and 10 cm), stands of both timothy and bromegrass were reduced more by three than by four cuts annually. Cutting at the jointing and boot stages of maturity resulted in lowest grass yield in an alfalfa-bromegrass mixture (64). The same general pattern was apparent for bromegrass grown with N added and no alfalfa and for plants shaded to reduce light by 60%. At the same maturity stage, however, the number of tillers present per plant was lower for bromegrass with alfalfa than for grass with N or with N and shading.

Davis and Parsons (23) conducted studies in the greenhouse of rest and harvest periods with alfalfa-grass mixtures. Their rest periods of 7, 21, and 42 d followed either complete defoliation in 1 d or partial defoliation over intervals of 7 and 21 d (Table 13-9). The yields of the alfalfa-grass mixture were greatly reduced by frequent defoliation. With a 7-d rest period, the yield of the alfalfa-grass was approximately doubled as the harvest period was increased from 1 to 21 d. The yield was approximately doubled again under the 21-d harvest period by increasing the rest period from 7 to 42 d (Table 13-9). The percent stands of alfalfa and timothy were sharply reduced by frequent defoliation, while the bromegrass stand was not affected.

Height, as well as frequency of cutting, may affect the growth of the individual grass species or alfalfa cultivars and, consequently, the ratio of grass to legume. Wolf et al. (104) observed in greenhouse studies that the proportion of grass in mixtures of alfalfa with bromegrass, orchard-

Table 13-9. Effect of rest and harvest periods† on dry-matter production per plot‡, and percentage of initial legume and grass stand remaining after a 224-d differential cutting period in a greenhouse trial. Adapted from Davis and Parsons (23).

Dry matter and percent stand	Rest period 7 d			Rest period 21 d			Rest period 42 d		
	Harvest period in days			Harvest period in days			Harvest period in days		
	1	7	21	1	7	21	1	7	21
				Alfalfa-bromegrass					
Dry matter, g	88	123	170	260	206	236	373	349	325
Initial legume stand, %	51	56	91	100	93	99	93	99	100
Initial grass stand, %	93	94	93	92	99	100	87	99	94
				Alfalfa-timothy					
Dry matter, g	94	114	175	229	209	252	429	422	341
Initial legume stand, %	70	84	99	100	100	99	100	100	94
Initial grass stand, %	40	59	80	59	54	84	70	87	89

† One-day harvest period—cut all growth at once to a 3.8-cm stubble. Seven-day harvest period—cut four times over a 7-d period, removing from top of plant one-fourth of growth first cut, and progressively additional increments. Twenty-one-day harvest period—cut eight times over a 21-d period removing one-eighth of total growth first cut and progressively additional increments.
‡ Each plot consisted of 14 grass and 14 legume plants seeded in a check design. Each plot was 35.6 by 55.9 cm in dimension.

grass, or timothy increased with increases from 2.5 to 12.7 cm in stubble height. Van Riper and Owen (92) reported 2-yr average yields of 8.3 and 5.5 Mg ha^{-1} for Vernal cut three times each season at stubble heights of 5.1 and 12.7 cm, respectively. The same increase in stubble height reduced by a similar amount the yields of alfalfa-bromegrass and alfalfa-orchardgrass mixtures cut on the same schedule. Smith and Jacques (79) left a 4- or 10-cm stubble on orchardgrass, reed canarygrass, tall fescue, bromegrass, and timothy grown in alternating 15-cm rows with Vernal alfalfa. After 2-yr of cutting three times each year, tall fescue, bromegrass, and early-, medium-, and late-maturing timothy yielded more at the 10-cm stubble. Orchardgrass and reed canarygrass were unaffected by stubble height.

The grass in the mixture and frequency of harvest may greatly influence weed invasion. During a 3-yr period in Kentucky, alfalfa grown alone and cut three times annually contained 15% weeds, while alfalfa-orchardgrass contained 4% weeds (85). Alfalfa alone cut five times contained 28% weeds and alfalfa-orchardgrass, 13% weeds.

13-5 SUMMARY

Alfalfa is frequently sown in mixture with grasses. Mixtures of alfalfa-grass may or may not yield more than pure stands of alfalfa. The more typical increases will average approximately 10 to 15%. A desired balance of alfalfa with grass may be achieved in part by adjustments in the choice of species, date of seeding, density of seeding, fertilizers, and time and height of defoliation.

Alfalfa competes strongly for various above- and below-ground growth factors. In many areas, difficulties have been encountered in maintaining a grass with alfalfa. Improved alfalfa cultivars that are better adapted to more frequent harvest have contributed to the problem of maintaining grass with alfalfa under more-frequent cutting regimes.

REFERENCES

1. Aberg, E., I.J. Johnson, and C.P. Wilsie. 1943. Association between species of grasses and legumes. J. Am. Soc. Agron. 35:357–369.
2. Barker, M.G., F. Hanley, and W.J. Ridgman. 1955. Studies on lucerne and lucerne-grass leys. I. Summer and autumn managment of a lucerne-grass mixture grown on heavy land. J. Agric. Sci. 46:362–376.
3. Barnard, C. 1964. Grasses and grassland. Macmillan Publishing Co., New York.
4. Barta, A.L. 1978. Effect of root temperature on dry matter distribution, carbohydrate accumulation, and acetylene reduction activity in alfalfa and birdsfoot trefoil. Crop Sci. 18:637–640.
5. Black, J.M. 1960. An assessment of the role of planting density in competition between red clover and lucerne in the early vegetative stage. Okios 11(1):26–42.
6. Blackman, G.E., and J.N. Black. 1959. Physiological and ecological studies in the analysis of plant environment. 12. The role of the light factor in limiting growth. Ann. Bot. 23:131–145.
7. Blaser, R.E., and N.C. Brady. 1950. Nutrient competition in plant associations. Agron. J. 42:128–135.
8. ----, W.H. Skrdla, and T.H. Taylor. 1952. Ecological and physiological factors in compounding forage seed mixtures. Adv. Agron. 4:179–216.
9. ----, T.H. Taylor, W. Griffeth, and W. Skrdla. 1956. Seedling competition in establishing forage plants. Agron. J. 48:1–6.
10. Brown, E.M. 1939. Some effects of temperature on the growth and chemical composition of certain pasture grasses. Mo. Agric. Exp. Stn. Res. Bull. 299.
11. Brown, R.H., and R.E. Blaser. 1968. Leaf area index in pasture growth. Herb. Abstr. 38:1–9.
12. ----, ----, H.L. Dunton. 1966. Leaf-area index and apparent photosynthesis under various microclimates for different pasture species. p. 108–113. In A.G.G. Hill (ed.) Proc. 10th Int. Grassl. Congr., Helsinki, Finland. 7–16 July. Finnish Grassland Association, Helsinki.
13. Carter, L.P., and J.M. Scholl. 1962. Effectiveness of inorganic nitrogen as a replacement for legumes grown in association with forage grasses. I. Dry matter production and botanical composition. Agron. J. 54:161–163.
14. Chamblee, D.S. 1958a. Some above- and below-ground relationships of an alfalfa-orchardgrass mixture. Agron. J. 50:434–437.
15. ----. 1958b. The relative removal of soil moisture by alfalfa and orchardgrass. Agron. J. 50:587–589.
16. ----, and R.L. Lovvorn. 1953. The effects of rate and method of seeding on the yield and botanical composition of alfalfa-orchardgrass and alfalfa-tall fescue. Agron. J. 45:192–196.
17. Champness, S. 1950. Effect of microclimate on the establishment of timothy grass. Nature (London) 165:325.
18. Clements, F.E., J.E. Weaver, and H.C. Hanson. 1929. Plant Competition: An analysis of community functions. Carnegie Institution of Washington, Washington, DC.
19. Comstock, V.E., and A.G. Law. 1948. The effect of clipping on the yield, botanical composition, and protein content of alfalfa-grass mixtures. Agron. J. 40:1074–1083.
20. Cooper, J.P. 1970. Potential production and energy conversion in temperate and tropical grasses. Herb. Abstr. 40:1–15.
21. Cords, H.P. 1967. Hay yields of orchardgrass in three maturity classes. Nev. Agric. Exp. Stn. T6.

22. Cowett, E.R., and M.A. Sprague. 1963. Effect of stand density and light intensity on the microenvironment and stem production of alfalfa. Agron. J. 55:432–434.
23. Davis, R.R., and J.L. Parsons. 1961. The effect of length of rest period and length of harvest period on yield and survival of forage crops. Ohio Agric. Exp. Stn. Circ. 99.
24. Davis, W.E., and B.F. Tyler. 1962. The yield and composition of lucerne, grass and clover under different systems of management J. Br. Grassl. Soc. 17:306–314.
25. Decker, A.M., H.A. Macdonald, R.C. Wakefield, and G.A. Jung. 1960. Cutting management of alfalfa and ladino clover in the northeast. Northeast Reg. Pub. R. I. Agric. Exp. Stn. Bull. 356.
26. Dilz, K., and E.G. Mulder. 1962. Effect of associated growth on yield and nitrogen content of legume and grass plants. Plant Soil 16–17:229–237.
27. Donald, C.M. 1963. Competition among crop and pasture plants. Adv. Agron. 15:1–118.
28. Dotzenko, A., and G.H. Ahlgren. 1950. Response of alfalfa in an alfalfa-bromegrass mixture to various cutting treatments. Agron. J. 42:246–247.
29. Drake, M., J. Vengris, and W.G. Colby. 1951. Cation-exchange capacity of plant roots. Soil Sci. 72:139–147.
30. Fox, R.L., R.C. Lipps, A.W. Moore, and H.F. Rhoades. 1958. Soil fertility practices for alfalfa production in the Central Platte Valley. Neb. Agric. Exp. Stn. Bull. 444.
31. Fraser, T.J. 1982. Evaluation of 'Grassland matua' prairie grass and 'Grasslands maru' *Phalaris* with and without lucerne in Canterbury. N. Z. J. Exp. Agric. 10(3):235–237.
32. Fyfe, J.L., and H.H. Rogers. 1965. Effects of varying variety and spacing on yields and composition of mixtures of lucerne and tall fescue. J. Agric. Sci. 64:351–359.
33. George, J.R., C.L. Rhykerd, C.H. Noller, J.E. Dillon, and J.C. Burns. 1973. Effect of N fertilization on dry matter yield, total-N, N recovery, and nitrate-N concentration of three cool-season forage grass species. Agron. J. 65:211–216.
34. Greenfield, P.L., and Dale Smith. 1973. Influence of temperature change at bud on composition of alfalfa at first flower. Agron. J. 65:871–874.
35. Groya, F.L., and C.C. Sheaffer. 1980. Effect of moisture and competition on sod-seeded alfalfa. Agron. Abstr. Americn Society of Agronomy, Madison, WI, p. 99.
36. Guy, P. 1966. Intraspecific competition in forage plants. Vol. 1. p. 183–189. *In* Proc. 9th Int. Grassl. Congr., Sao Paulo, Brazil. 7–20 Jan. 1965. Alarico Limitada, Sao Paulo.
37. Hall, R.L. 1974. Analysis of the nature of interference between plants of different species. I. Concepts and extension of the de Wit analysis to examine effects. Aust. J. Agric. Res. 25:739–747.
38. Hanley, F., R.H. Jarvis, and W.J. Ridgman. 1964. The effects of nitrogenous manuring, inter-row distance and method of sowing on the yields of a lucerne-cocksfoot ley. J. Agric. Sci. 62:425–431.
39. Henderson, M.S., and D.L. Robinson. 1982. Environmental influences on yield and *in vitro* true digestibility of warm-season perennial grasses and the relationships to fiber components. Agron. J. 74:943–946.
40. Hunt, O.J., and R.E. Wagner. 1963. Effects of phosphorus and potassium fertilizers on legume composition of seven grass-legume mixtures. Agron. J. 55:16–19.
41. Jackobs, J.A. 1952. The performance of six grasses growing alone and in combination with legumes with differential nitrogen and phosphate fertilization in a Yakima Valley pasture. Agron. J. 44:573–578.
42. ----. 1963. A measurement of the contribution of ten species to pasture mixtures. Agron. J. 55:127–131.
43. ----. 1966. The role of short-lived species in seeding mixtures for grasslands. Vol. 1. p. 414–416. *In* Proc. 9th Int. Grassl. Congr., Sao Paulo, Brazil. 7–20 Jan. 1965. Alarico Limitada, Sao Paulo.
44. Jewiss, O.R. 1972. Tillering in grasses-its significance and control. J. Br. Grassl. Soc. 27:65–82.
45. Jung, G.A., L.L. Wilson, P.J. LeVan, R.E. Kocher, and R.F. Todd. 1982. Herbage and beef production from ryegrass-alfalfa and orchardgrass-alfalfa pastures. Agron. J. 74:937–942.
46. Kalton, R.R., and C.P. Wilsie. 1953. Effect of bromegrass variety on yield and composition of a brome-alfalfa mixture. Agron J. 45:308–311.
47. Kilcher, M.R. 1959. Grass-alfalfa seeding ratios and control of alfalfa domination in mixtures. J. Br. Grassl. Soc. 14:29–35.
48. ----. 1966. Fertilizers and seed ratios for controlling lucerne domination in mixtures. J. Br. Grassl. Soc. 21:135–139.
49. Knievel, D.P., and Dale Smith. 1973. Influence of cool and warm temperatures and temperature reversal at inflorescence emergence on growth of timothy, orchardgrass, and tall fescue. Agron. J. 65:378–383.
50. Lamba, P.S., H.L. Ahlgren, and R.J. Muckenhirn. 1949. Root growth of alfalfa, medium red clover, bromegrass and timothy under various soil conditions. Agron. J. 41:451–458.

51. Leyshon, A.J., M.R. Kilcher, and J.D. McElgunn. 1981. Seeding rates and row spacings for three forage crops grown alone or in alternate grass-alfalfa rows in southwestern Saskatchewan. Can. J. Plant Sci. 61(3):711–717.
52. Lorenz, R.J., C.W. Carlson, G.A. Rogler, and H. Holmen. 1961. Bromegrass and bromegrass-alfalfa yields as influenced by moisture level, fertilizer rates and harvest frequency. Agron. J. 53:49–52.
53. MacLeod, L.B. 1965a. Effect of nitrogen and potassium on the yield, botanical composition, and competition for nutrients in three alfalfa-grass associations. Agron. J. 57:129–134.
54. ----. 1965b. Effect of nitrogen and potassium fertilization on the yield, regrowth, and carbohydrate content of the storage organs of alfalfa and grasses. Agron. J. 57:345–350.
55. ----, and R. Bradfield. 1963. Effect of liming and potassium fertilization on the yield and composition of an alfalfa association. Agron. J. 55:435–439.
56. McCloud, D.E., and G.O. Mott. 1953. Influence of association upon the forage yield of legume-grass mixtures. Agron. J. 45:61–65.
57. Miller, D.A. 1984. Forage crops. McGraw-Hill Incorporated Book Co., New York.
58. Moncada de la Fuente, J.M. 1967. Influence of leguminous plants on the growth and nitrogen nutrition of associated grass. Ph.D. diss. Univ. of North Carolina, Raleigh (Diss. Abstr. 27, no. 9, 2953B).
59. Newman, R.C., and Dale Smith. 1972. Influence of two seeding patterns, nitrogen fertilization and three alfalfa varieties on dry matter and protein yields and persistence of alfalfa-grass mixtures. Univ. of Wisconsin, College of Agric. and Life Sci. Res. Rep. 2377.
60. Nielsen, K.F., R.L. Halstead, A.J. Maclean, R.M. Holmes, and S. J. Bourget. 1961. Effects of soil temperature on the growth and chemical composition of lucerne. p. 287–292. In C.L. Skidmore (ed.) Proc. 8th Int. Grassl. Congr., Reading, UK. 11–21 July 1960. Alden Press, Oxford, UK.
61. O'Connor, K.F. 1967. Sociability of lucerne. In R.H.M. Langer (ed.) The lucerne crop. Reed Wellington, Auckland, Sydney.
62. Olsen, F.J., J.H. Jones, and J.J. Patterson. 1981. Sod-seeding forage legumes into a tall fescue sward. Agron. J. 73:1032–1036.
63. Paulsen, G.M., and Dale Smith. 1968. Influences of several management practices on growth characteristics and available carbohydrate content of smooth bromegrass. Agron. J. 60:375–379.
64. ----, and ----. 1969. Organic reserves, axillary bud activity and herbage yields of smooth bromegrass as influenced by time of cutting, nitrogen fertilization, and shading. Crop Sci. 9:529–534.
65. Peterson, L.A., and Dale Smith. 1973. Recovery of K_2SO_4 by alfalfa after placement at different depths in a low fertility soil. Agron. J. 65:769–772.
66. Plancquaert, P. 1967. Etude sur la production des associations luzerne-graminees. (Study on the yield of lucerne-grass mixtures.) Institut Technique Cereales et Fourrages, Paris, France.
67. Powell, R.D., and L.T. Kardos. 1968. Effect of moisture regimes and harvests on efficiency of water use by ten forage crops. Soil Sci. Soc. Am. Proc. 32:871–874.
68. Pritchett, W.L., and L.B. Nelson. 1951. The effect of light intensity on the growth characteristics of alfalfa and bromegrass. Agron. J. 43:172–177.
69. Rhodes, I. 1970. Competition between herbage grasses. Herb. Abstr. 40:115–121.
70. Rhykerd, C.L., R. Langston, and J.B. Peterson. 1959. Effect of light treatment on the relative uptake of labeled carbon dioxide by legume seedlings. Agron. J. 51:7–9.
71. ----, C.H. Noller, J.E. Dillon, J.B. Ragland, B.W. Crowl, G.C. Naderman, and D.L. Hill. 1967. Managing alfalfa-grass mixtures for yield and protein. Ind. Agric. Exp. Stn. Bull. 839.
72. Ridgman, W.J., F. Hanley, and M.G. Barker. 1956. Studies on lucerne and lucerne-grass leys. J. Agric. Sci. 47:50–58.
73. Roberts, J.L., and F.R. Olson. 1942. Interrelationships of legumes and grasses grown in association. J. Am. Soc. Agron. 34:695–701.
74. Rogers, D.D., D.S. Chamblee, J.P. Mueller, and W.V. Campbell. 1985. Fall no-till seeding of alfalfa into tall fescue as influenced by time of seeding and grass and insect suppression. Agron. J. 77:150–157.
75. Simpson, J.R. 1965. The transference of nitrogen from pasture legumes to an associated grass under several systems of management in pot culture. Aust. J. Agric. Res. 16:915–926.
76. Smith, Dale. 1969. Influence of temperature on the yield and chemical composition of 'Vernal' alfalfa at first flower. Agron. J. 61:470–473.
77. ----. 1970. Influence of cool and warm temperatures and temperature reversal at inflorescence emergence on yield and chemical composition of timothy and brome-

grass at anthesis. p. 510–514. *In* M.J.T. Norman (ed.) Proc. 11th Int. Grassl. Congr., Queensland, Australia. 13–23 April. University of Queensland Press, St. Lucia, Queensland.
78. ----. 1981. Forage management in the North. Kendall/Hunt Publishing Co., Dubuque, IA.
79. ----, and A.V.A. Jacques. 1973. Influence of alfalfa stand patterns and nitrogen fertilization on the yield and persistence of grasses grown with alfalfa. Wis. Agric. Exp. Stn. Res. Rep. R2480.
80. ----, ----, and J.A. Balasko. 1973. Persistence of several temperate grasses grown with alfalfa and harvested two, three, or four times annually at two stubble heights. Crop Sci. 13:553–556.
81. ----, and D.A. Rohweder. 1977. Establishing and managing alfalfa. Wis. College of Agric. and Life. Sci. Res. Rep. 1741.
82. ----, and B.E. Struckmeyer. 1974. Grass morphology and starch accumulation in leaves of alfalfa plants grown at high and low temperatures. Crop Sci. 14:433–436.
83. Sprague, M.A., E.R. Cowett, and M.V. Adams. 1964. Early and deferred cutting management of alfalfa, ladino white clover, bromegrass, and orchardgrass. Crop Sci. 4:35–38.
84. Stafford, R.F. 1969. Stand density effects of alfalfa on associated grass seedlings. M.S. thesis, University of Maine, Orono.
85. Taylor, T.H. and W.C. Templeton. Unpublished data.
86. ----. 1982. 'Fergus' birdsfoot trefoil—An adapted variety for Kentucky. Ky. Agric. Ext. Serv. AGR-104.
87. Templeton, W.C., Jr., T.H. Taylor, and J.R. Todd. 1965. Comparative ecological and agronomic behavior of orchardgrass and tall fescue. Ky. Agric. Exp. Stn. Bull. 699.
88. Tesar, M.B. 1974. Nitrogen on grasses compared to alfalfa-grass mixtures in northern Michigan. Mich. State Univ. Agric. Exp. Stn. Res. Rep. 256.
89. Tewari, G.P., and A.R. Schmid. 1960. The production and botanical composition of alfalfa-grass combinations and the influence of the legume on associated grasses. Agron. J. 52:267–269.
90. Van Riper, G.E. 1960. Protein difference among varieties of alfalfa and red clover and grass combinations. Agron. J. 52:549–550.
91. ----. 1964. Influence of soil moisture on the herbage of two legumes and three grasses as related to dry matter yields, crude protein, and botanical composition. Agron. J. 56:45–50.
92. ----, and F.G. Owen. 1964. Effect of cutting height on alfalfa and two grasses as related to production, persistence, and available soil moisture. Agron. J. 56:291–295.
93. Van Slijcken, A., and A. Andries. 1963. Lucerne or lucerne/grass? Report on a comparison between pure lucerne and some lucerne/grass mixtures. Rev. Agric. Brux. 16:1411–1427.
94. Velich, J., and V. Charvat. 1976. Explanation of the interspecific relationships between lucerne and cocksfoot in a mixed sward. (Czechoslovakia) Rostlinna Vyroba 22:763–769.
95. Ward, C.Y., J.N. Jones, J.H. Lillard, J.E. Moody, R.H. Brown, and R.E. Blaser. 1966. Effects of irrigation and cutting management on yield and botanical composition of selected legume-grass mixtures. Agron. J. 58:181–184.
96. Williams, T.E. 1980. Grass, its production and utilization. p. 6–69. *In* W. Holmes (ed.) Herbage production: Grasses and leguminous forage crops. Blackwell Scientific Publications, Oxford, UK.
97. Wilsie, C.P. 1949. Evaluation of grass-legume associations with emphasis on the yields of bromegrass varieties. Agron. J. 41:412–420.
98. ----. 1952. Crop adaptation and distribution. W.H. Freeman and Co., New York.
99. Wilson, J.W. 1961. Influence of spatial arrangement of foliage area on light interception and pasture growth. p. 275–279. *In* C.L. Skidmore (ed.) Proc. 8th Int. Grassl. Congr., Reading, UK. 11–21 July 1960. Alden Press, Oxford, UK.
100. Wilson, P.W. 1940. The biochemistry of symbiotic nitrogen fixation. The University of Wisconsin Press, Madison, WI.
101. ----, and O. Wyss. 1937. Mixed cropping and the excretion of nitrogen by leguminous plants. Soil Sci. Soc. Am. Proc. 2:289–297.
102. Wolf, D.D. 1964. Soil moisture extraction trends of several legume-grass mixtures as affected by cutting frequency and nitrogen fertilization. Agron. J. 56:467–469.
103. ----, E.L. Kimbrough, and R.E. Blaser. 1976. Photosynthetic efficiency of alfalfa with increasing potassium nutrition. Crop Sci. 16:292–294.
104. ----, K.L. Larson, and Dale Smith. 1962. Grass-alfalfa yields and food storage of associated alfalfa as influenced by height and frequency of cutting. Crop Sci. 2:363–364.
105. ----, and Dale Smith. 1964. Yield and persistence of several legume-grass mixtures as affected by cutting frequency and nitrogen fertilization. Agron. J. 56:130–133.

106. Yamada, T., and S. Horiuchi. 1961. On the bias of quantitative characters and the change of their distribution in a population due to inter-plant competition. p. 297–302. *In* C.L. Skidmore (ed.) Proc. 8th Int. Grassl. Congr., Reading, UK. 11–21 July 1960. Alden Press, Oxford, UK.

14 Feeding Value (Forage Quality)

G. C. MARTEN

USDA-ARS
St. Paul, Minnesota

D. R. BUXTON

USDA-ARS
Ames, Iowa

R. F BARNES

American Society of Agronomy
Madison, Wisconsin

Potential forage feeding value does not necessarily indicate attainable animal performance. Animal performance is influenced by many nonforage factors such as animal genetic, physiological, and environmental factors as well as interactions within the plant-animal complex (Fig. 14-1). Direct expression of alfalfa-feeding value as units of gain or animal performance attained would, therefore, be of limited worth. *Feeding value* as used in this chapter means potential feeding value or *forage quality*. These terms encompass potential nutritive value (type and amount of digestible nutrients and the efficiency of their use), extent of occurrence of antiquality factors, and potential intake.

We emphasize expressions of available energy because alfalfa rations are most frequently limited by concentration of energy. We also emphasize crude protein because alfalfa is frequently a rich source of protein that is especially complementary to low-protein, high-energy concentrates in complete rations. However, a balance of nutrients is required for optimum animal production, and digestibility and intake of digestible energy and protein are not complete estimates of the potential feeding value of alfalfa. Other nutrients such as minerals and vitamins, as well as antiquality constituents in alfalfa, are discussed in Chapter 15 in this book.

This chapter primarily describes the general and specific characteristics of alfalfa feeding value and preharvest factors that influence feeding value. Post harvest factors and on-farm feeding of dehy, hay, silage, greenchop, and dewatered (wet-fractionated) alfalfa are discussed in Chapters 17 and 18, respectively. Also, genetic and cultivar differences in alfalfa feeding value are described in Chapter 26.

Copyright 1988 © ASA-CSSA-SSSA, 677 South Segoe Road, Madison, WI 53711, USA. *Alfalfa and Alfalfa Improvement*—Agronomy Monograph no. 29.

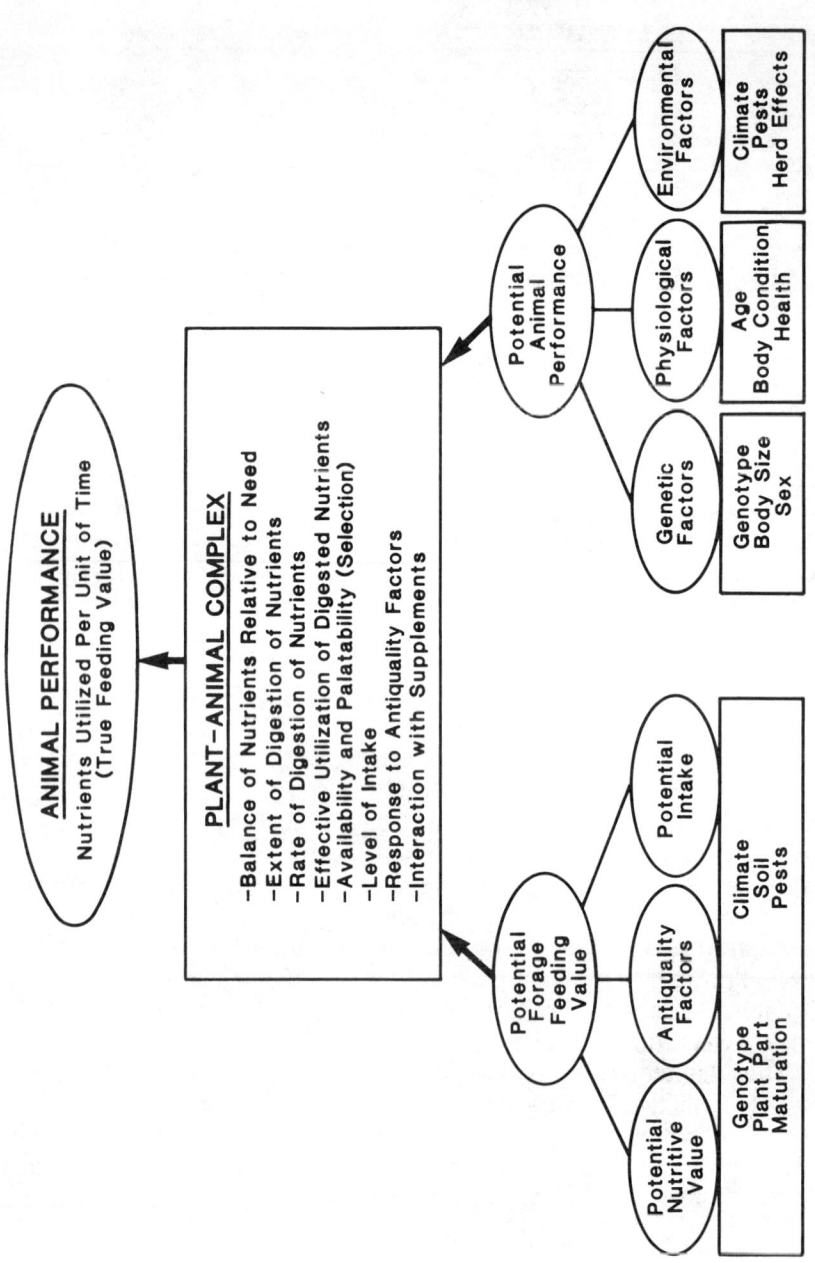

Fig. 14–1. Factors associated with true forage feeding value.

14-1 CHARACTERISTICS OF ALFALFA-FEEDING VALUE

14-1.1 Digestibility, Intake, and Efficiency of Utilization

The voluntary intake of digestible nutrients of alfalfa herbage by ruminants is greater than that of grasses because much more alfalfa dry matter (and that of other legumes) is in the form of cell solubles that are readily available for absorption in the digestive system (156, 190, 193). Although the levels of cell-wall components (fibrous material) in alfalfa are lower than those in grasses, the cell walls of alfalfa are highly lignified and less available than those of grasses (187).

Cell-wall concentration of forage diets is accepted as the best single chemical predictor of intake potential (195, 198). This is because intake of pure forage diets in growing ruminants is regulated by the undigested and retained rumen residues that constitute physical fill. The effect of these variables is apparent in comparisons of animal intake and performance from alfalfa and grass diets with essentially the same level of digestible dry matter (DDM) (15, 17, 53, 54, 188, 201). Despite nearly equal digestibility in the studies cited, the intake of the grasses and animal daily gains achieved were lower than those of alfalfa. The intake, passage, and digestion of alfalfa occurred at faster rates, thereby increasing the consumption of digestible nutrients per day from alfalfa over grasses of equal digestibility.

Animal performance can be estimated on the basis of digestible energy intake because available energy is often the first limiting factor in high forage rations. It is recognized, however, that some forms of digested energy may be utilized with greater efficiency than others. Digestible energy obtained from more fibrous materials is not used as efficiently for fattening as that obtained from material of higher quality (121, 141). Thus, the total advantage of high-quality alfalfa extends beyond that indicated by digestible nutrient concentration. Alfalfa's advantage includes the potential for increased consumption, faster rate of digestibility, and perhaps more efficient conversion of digested energy to productive energy. In addition, the inorganic fraction in alfalfa may have a beneficial effect on animal performance (31, 94).

14-1.1 Hay-Grading Standards

Commercial hay-grading standards, originally proposed by the American Forage and Grassland Council (AFGC), depended upon use of acid detergent fiber (ADF) for estimating digestibility by ruminants and neutral detergent fiber (NDF) for estimating intake of forages, including alfalfa. The term NDF is synonymous to cell-wall components. However, in vitro rumen fermentation is the universally preferred procedure for estimation of digestibility in research, and it is often expressed as in vitro digestible dry matter (IVDDM) (127, 130). The IVDDM procedure is not recommended for routine, commercial hay-quality testing because it is difficult to standardize and expensive to conduct in commercial laboratories.

In 1984, the U.S. Alfalfa Hay Quality Committee, organized by AFGC and the National Hay Association, decided to develop and test improved uniform standards for expressing the feeding value of alfalfa hay (182). The committee agreed that: energy values should be expressed as DDM; DDM should be calculated from ADF; laboratory determinations should include percentages of crude protein (CP) and ADF reported *on a dry wt. basis*; as well as percentage of dry matter (DM) reported on an *as received basis*; and estimated digestible dry matter (EDDM) should be calculated from the ADF analysis with an accepted equation (e.g., that of Rohweder; 159). The committee further agreed that additional optional organoleptic and chemical factors could be added (including NDF to estimate intake) and that CP, ADF, and DM could be determined by any method that gives results within the acceptable range established by a certifying association. In late 1984, the National Alfalfa Hay Testing Association issued a certification manual that incorporated the above standards. This manual was reprinted in 1986. A rapid, accurate, and precise method recognized by the association that holds considerable promise for the routine analysis of CP, ADF, and DM, as well as other quality components, is near infrared reflectance spectroscopy (128, 149).

14-1.2 Significance of Cell-Wall Composition to Digestibility and Intake

Cell walls contain most of the plant portion that is resistant to degradation by enzymes of microorganisms present in mammalian gastrointestinal tracts. Cell contents, including protein, lipid, sugar, and starch, are readily available and essentially completely digested as is cell-wall pectin. Conversely, the remaining cell-wall components are rarely completely digested. Consequently, most of the variability in digestion and voluntary animal intake of alfalfa herbage is associated with variation in cell-wall concentration and the intrinsic characteristics of cell-wall components. Alfalfa cell walls may contain cellulose, hemicellulose, protein, lignin, waxes, cutin, and minerals as well as pectin. The portion of alfalfa herbage represented by cell walls ranges from about 200 g/kg of DM in leaves to more than 700 g/kg in stem bases of nearly mature plants (7, 107). Plant cell walls can be divided into a thin primary wall and a thick secondary wall.

14-1.2.1 Composition of Primary Walls

A cell wall is usually recognized as primary when formed during the time of increasing cell-surface area. The primary cell wall is the only wall of parenchyma cells in the cortex and pith of stems and the mesophyll of leaves (59). The composition of primary walls is about 100, 250, 350, and 300 g/kg of protein, cellulose, pectin, and hemicellulose, respectively

(6, 14, 76, 179). Primary walls have little or no secondary thickening and limited lignification; thus they are readily degraded by microorganisms.

14-1.2.2 Composition of Secondary Walls

After cell growth stops, cell walls thicken and the secondary wall is formed. In contrast to primary walls, secondary walls do not contain protein and may vary significantly in composition and structure among cell types (179). Secondary walls consist of a network of cellulose fibrils embedded in an amorphous matrix of hemicellulose, pectin, and lignin (143, 180, 184). Generally, young cell walls are richer in pectin and lower in cellulose than older cell walls (76).

14-1.2.3 Cellulose

Cellulose, the major structural polysaccharide in plants, is composed of molecules that have more than 10 000 glucosyl units (76, 143). Hydrogen bonding among neighboring molecules results in threadlike structures called *microfibrils*. The microfibril organization adds resistance to microbial and chemical digestion (76).

Accessibility of the cellulose surface to cellulytic enzymes may be the primary physical feature influencing enzymatic hydrolysis (45, 178). Hydrolysis requires direct contact between the substrate cellulose and the enzyme (46). Thus, specific surface area and lignin concentration are the most important structural features influencing hydrolysis of cellulose (69). Cellulose concentration usually correlates with digestibility of DM only to the extent that its availability is determined by lignification or other limiting factors (191).

Cellulose in herbage is usually measured as the insoluble residue remaining after treatment with strong alkali. This residue contains some impurities such as arabinose and xylose (193). Cell walls in alfalfa stems can be composed of up to 700 g of cellulose/kg and those in leaves up to 600 g of cellulose/kg (7, 124).

14-1.2.4 Hemicellulose

Hemicellulose has no unique chemical definition. It is the cell-wall polysaccharides solubilized by aqueous alkali after removal of water soluble and pectic carbohydrates (184). In herbage, starch, pectin, and hemicelluloses are often difficult to separate because of their overlapping solubility characteristics (193). Hemicellulose has many side chains and a much shorter chain length than cellulose (143). Alfalfa herbage has a lower ratio of hemicellulose to cellulose than that of grasses. Concentration of hemicellulose in alfalfa ranges from about 150 to more than 300 g/kg of cell wall (7, 124).

In addition to forming a matrix among the cellulose fibrils, some hemicellulose is located in the middle lamella. The amount and composition of hemicellulose probably varies with tissue type and maturity (210).

Because of the wide variety of sugar residues and glycosidic linkages in hemicellulose, a wide assortment of enzymes is required for its degradation in mammalian gastrointestinal tracts. Hemicellulases are thought to be endoenzymes produced by both bacteria and protozoa that randomly attack the glycosidic chain (193).

14-1.2.5 Neutral Sugars

Cellulose and hemicellulose are composed primarily of neutral sugars. Glucose is the most common sugar unit found in plant cell walls and represents 60 to 70% of total neutral sugars in alfalfa herbage (7). Xylose is frequently the second most numerous neutral sugar. Albrecht (7) found that the proportion of glucose and xylose increased during alfalfa maturation while arabinose and galactose decreased. He also observed that glucose, arabinose, galactose, mannose, and rhamnose were present in higher concentrations in cell walls of alfalfa leaves than in cell walls of stems. Conversely, xylose was more concentrated in stem than in leaf cell walls. The xylose/arabinose ratio in stems increased from about 2.0 to 3.8 during advancing maturity, while the ratio in leaves was less than unity, and did not change with alfalfa maturity.

Insofar as the xylose/arabinose ratio gives an indication of arabinose branching from the main xylose core (137), immature alfalfa stems apparently have more arabinose branching than mature stems. Brice and Morrison (26) demonstrated a positive relationship between xylan digestibility and xylose/arabinose ratios, perhaps because arabinose side chains must be removed before the xylose core can be hydrolyzed. Others have concluded that arabinose side chains linked to xylose units are hydrolyzed more readily than in-chain xylose units (72). Albrecht (7) digested alfalfa stems in rumen fluid and observed that galactose and arabinose were degraded more rapidly than xylose.

14-1.2.6 Pectic Substances

Pectic substances are polysaccharides composed largely of a galacturonic acid chain substituted with arabinan and galactan side chains. They are located in the middle lamella as well as the primary cell wall (6, 14, 184). Hot neutral detergent dissolves most pectins, which suggests that they probably are not bonded to the cell-wall matrix (193).

Pectic substances occur in higher concentrations in alfalfa and other legumes than in grasses. In legume herbage, pectin concentrations range from about 40 to 120 g/kg of DM (14, 39, 185). Digestion coefficients of pectins in alfalfa range from 850 to 980 g/kg (39).

14-1.2.7 Cell Wall Protein

Extensin, a hydroxyproline- rich protein, can be an important constituent of primary cell walls that is not found in secondary walls (179).

It consists of four-unit arabinose chains and single galactose molecules linked to the protein core (6). These glycoproteins are more resistant to digestion than cellular protein because of their close association with cell-wall polysaccharides (193). The importance of cell-wall protein to total herbage digestion and animal nutrition is small, however, because of low concentration of protein in the total cell wall.

14-1.2.8 Lignin

Lignin is a condensed phenylpropanoid polymer of high molecular weight composed mostly of three monomers: *p*-coumaryl, coniferyl, and sinaphyl alcohols (81). The monomers are interlinked to stable C by C and ether bonds. These bonds are not susceptible to simple hydrolysis that increases the difficulty of lignin structural analysis. Lignification occurs mostly in maturing cells that have specialized functions for water conduction and/or mechanical support (59, 81). Lignin and hemicellulose form a matrix among the cellulose fibrils and impart a rigid characteristic to cell walls.

Crude lignin, as determined in forages, usually contains substances other than true lignin, such as condensed tannins and cutin that are also highly indigestible (193). The crude lignin concentration in alfalfa cell walls is much greater than that of grass species at equivalent maturity (199). Within legume species, cell walls of alfalfa herbage have a greater crude lignin concentration than cell walls of white clover (*Trifolium repens* L.) and red clover (*T. pratense* L.), but less than cell walls of birdsfoot trefoil (*Lotus corniculatus* L.) (32). In contrast to grass species, herbage of alfalfa and other legumes has only a small variation in the degree of lignification per unit of cell wall tissue with advancing maturity. The portion of crude lignin in alfalfa cell walls is about 150 g/kg for both stems and leaves (7).

Lignin is usually the major factor limiting digestibility of cell walls. In addition to its low digestibility (102), lignin inhibits the digestion of cell-wall polysaccharides (192). Furthermore, the undigested lignin-carbohydrate residue acts as a ballast in the rumen and reduces forage intake (199). Some researchers think that lignin concentration determines the extent of digestion, but not the rate of digestion (117, 200).

The manner in which lignin limits digestion of cell-wall components suggests involvement of chemical, physical, and nutritional factors. First, covalent bonds between lignin and hemicellulose have been documented and strongly implicated as inhibitors of hemicellulose digestion (38, 62, 67, 143, 146). The lignin core may be linked to hemicellulose via ester bonds through *p*-coumaric and ferulic acids (85, 101). Also, phenolic acid and acetyl substances have been implicated as limiting hemicellulose digestion by binding to xylan (13, 86, 142, 186).

Second, although there is no evidence of chemical bonds between lignin and cellulose (144), a negative relationship usually exists between lignin concentration and cellulose digestibility (81). Morrison (143) described a "cage" theory for ligno-hemicellulosic complex protection of cellulose from rumen microorganisms and their enzymes. In this hypothesis, "bars of the cage" of young cell walls are thought to be far enough apart to allow enzyme access. With maturation, the bars grow closer together to restrict the entry of enzymes.

Finally, in addition to chemical bonding and physical restriction of nutrients by lignin, simple phenolic monomers from lignin and cell solubles can exert toxic effects on bacteria involved in nutritional systems (3, 50, 61, 102, 103, 219). Removal of soluble phenolic compounds from alfalfa increases the digestion of cellulose and protein (100). Paracoumaric and ferulic acids are the primary phenolics in forages. The concentration of these phenolics increases with advancing maturity in the herbage of some plants but not in alfalfa (102, 186). Additionally, alkyl groups on the benzene ring of phenolics increase their toxicity (83, 101, 186).

In contrast to grasses, alfalfa herbage has little or no free phenolic monomers. Furthermore, alkali-labile phenolic monomers, which are relatively loosely bound compounds, are present in alfalfa and other legume cell walls at concentrations below those found in grass cell walls (86, 104, 105, 148).

Cell walls are not uniformly lignified. Lignin concentration is greatest in the corners of cell walls, followed by the middle lamella and the remaining cell wall (1). The chemical composition of lignin in secondary walls has been shown to differ from that located in the middle lamella (1, 208, 209).

Lignin in alfalfa and other legumes is probably more highly cross-linked than grass lignin which results in fewer reactive sites being available for combination with other molecules, such as hemicellulose (71). Legume lignins are also less soluble and less easily removed from cell walls than grass lignins (86). Furthermore, at equal digestibility, alfalfa and other legumes have a greater lignin concentration than grasses. Also, the association between the amount of lignin in the cell wall and rate of cell-wall digestion is closer in grasses than in legumes (71). Thus, the type and extent of lignin-polysaccharide bonding may affect digestion more than the amount of lignin (18). Jung et al. (105) speculated that the lower phenolic monomer concentration in alfalfa than in grasses may account for the smaller inhibitory effect of alfalfa lignin on cell-wall digestibility.

Additional evidence of differences in alfalfa and grass lignin is found in their response to alkali. Alkali treatment has been used to increase digestibility of grasses, probably as a result of saponification of lignin-hemicellulose bonds. The lignin-carbohydrate bond of alfalfa is less susceptible to hydrolysis by alkali, and alkali treatment has resulted in less improvement in digestibility of alfalfa than of grasses (84, 192).

14–1.2.9 Potentially Digestible Cell Wall

Wilkins (211) first noted that cell walls can be separated into two fractions; an indigestible fraction and a potentially digestible fraction.

The indigestible fraction is not digested even when exposed to rumen microorganisms for an infinite amount of time. The ratio of cell wall to lignin mass in the undigestible plant fraction has been reported to be similar for several species (175, 176). This observation has led to the general conclusion that lignin protects from digestion about three times its mass in cell wall (199). Thus, although lignin concentration is an important factor in determining the size of the indigestible fraction, the evidence suggests that structural features of both cell-wall polysaccharides and lignin are contributing factors.

As illustrated in Fig. 14-2, the cell-wall concentration of alfalfa herbage is less than that of most grasses at equivalent maturity, but the concentration of indigestible cell wall is greater in alfalfa because of a higher lignin concentration. Digestion occurs at a faster rate in alfalfa than in grasses, but it is potentially more complete in grass species (196, 199).

The digestibility of cell walls is negatively correlated with both the concentration of cell wall and the concentration of lignin within the cell wall. Based on the inverse relationship between cell-wall concentration and cell-wall digestibility, Belyea et al. (21) concluded that digestible cell wall per unit of organic matter (cell-wall concentration × digestible cell-wall fraction) is relatively constant at about 200 g/kg for legumes and 400 g/kg for grasses.

14-1.3 Significance of Cell and Tissue Type

Although alfalfa total herbage is usually analyzed for quality, feeding quality varies greatly among plant parts, tissues, and cell types (74). Leaves

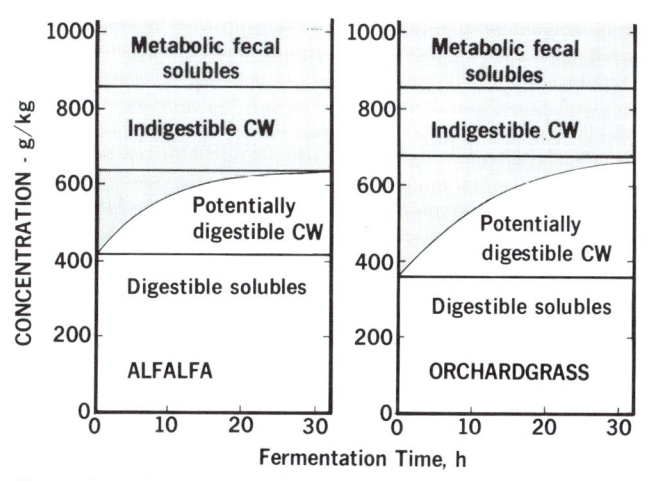

Fig. 14-2. Comparison of the sources of DDM and digestion rates for potentially digestible cell walls in alfalfa and orchardgrass and their effect on digestibility as fermentation time changes. A rate for solubles is not shown because their digestion is so fast that it is unaffected by the amount of forages fed alone. CW = cell walls. Adapted from Waldo and Jorgensen (199).

and stems, which represent about 95% of total herbage (65), show marked differences in digestibility. Additionally, tissues and cell types within leaves and stems have a wide range in digestibility.

14-1.3.1 Digestion of Cell Types

Generally mesophyll and phloem cells are degraded rapidly, epidermis and parenchyma bundle sheath cells are degraded slowly, and sclerenchyma and lignified vascular tissues are resistant to degradation (4, 5, 25). Brazle and Harbers (25) reported that the cuticle, lower epidermis, and epidermal pubescence of alfalfa leaves were resistant to digestion. They also observed that the stem epidermis was only partially degraded after digesting in the rumen for 24 h. After 72 h of digestion, the lignified vascular tissue and the cuticle of stems remained undigested.

14-1.3.2 Nutritive Value of Leaves and Stems

Alfalfa leaves are more digestible and have more nutrients than stems, even when plants are in the prebud stage (7, 33, 107, 110, 169, 213), although extremely young stem tips may be equal to leaves in digestibility (183). The nutritive value of leaves deteriorates much more slowly than that of stems between early vegetative and late bloom stages (110). Leaf digestibility varies only slightly with maturity and position within the plant canopy (7, 33, 151, 183). Leaves contain two to three times the concentration of CP as stems (145). Albrecht (7) and Buxton et al. (33) found that CP concentration of alfalfa leaves declined continuously from the vegetative to the early seed stage, with lower canopy leaves having less CP than leaves near the canopy top. They found a range in values of nearly 100 g/kg of DM. Conversely, Kalu and Fick (107) reported that alfalfa leaf CP increased by up to 90 g/kg during progressive growth of the vegetative stages before leveling off or decreasing during later stages.

In contrast to leaves, alfalfa stems decline in digestibility with advancing maturity at a rate of 4 to 5 g/(kg/d) and decline in CP at a rate of about 2 g/(kg/d) (7, 33, 183). Stem bases are less digestible than stem tops with a rate of change along the stem of about 20 g/(kg/stem) node (33), as illustrated in Fig. 14-3. Digestibility of alfalfa stem segments is related closely to their cell-wall concentration and lignin concentration of cell walls. Buxton and Hornstein (32) showed that variation in cell-wall concentration and lignin concentration of cell walls accounted for 95% of the observed variation in IVDDM concentration among stem segments of three legume species, including alfalfa.

14-2 PREHARVEST FACTORS THAT INFLUENCE FEEDING VALUE

14-2.1 Stage of Plant Development

The cumulative effect of environment is integrated through plant growth and development, and expressed in a large part by the morphological stage of development (see Chapter 4 in this book). Decreases in

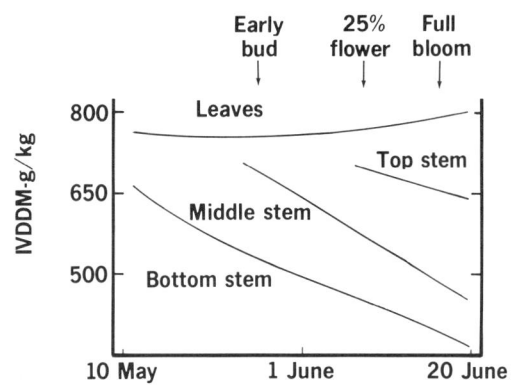

Fig. 14-3. Changes in IVDDM of alfalfa leaves and stem segments during maturation of spring growth. The bottom stem segment constitutes the lowest six main stem internodes (293 mm when fully elongated). The middle stem segment constitutes the next six internodes (295 mm). Adapted from Buxton et al. (33).

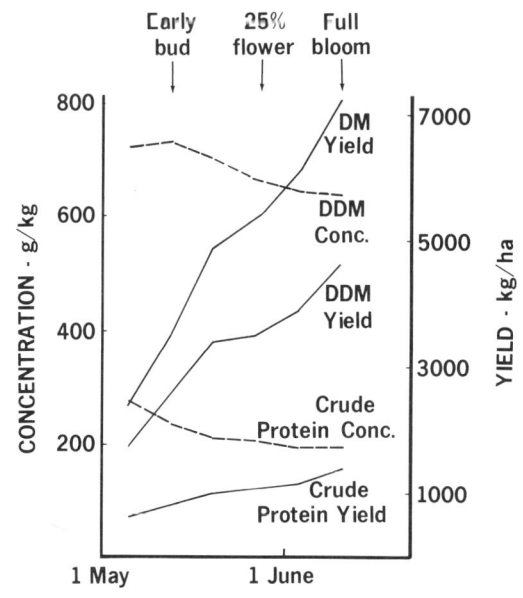

Fig. 14-4. Changes in DDM concentration and yield, crude protein concentration and yield, and total DM yield of alfalfa forage during maturation of spring growth. Adapted from R.F Barnes, (16).

nutritive value, voluntary intake, and animal performance potential are associated with stage of development (64, 65, 115, 199). The effect of development stage of alfalfa at time of harvest on in vivo digestibility, yields of digestible nutrients and DM, and concentration of CP is illustrated in Fig. 14-4.

14-2.1.1 Methods for Determining Development Stage

In the past, classification systems have lacked the uniformity necessary to make comparisons among studies and the precision to predict alfalfa-feeding value. Kalu and Fick (106, 107) described a 10-stage, numerical system which consists of three vegetative, two bud, two flower, and three seed-pod stages. They showed that this system could adequately predict nutritive value of alfalfa herbage grown in several environments in New York.

14-2.1.2 Leaf/Stem Ratio

As alfalfa advances in maturity, the stem mass increases and the leaf-to-stem mass ratio (L/S) decreases. This is a major factor contributing to the low quality of mature herbage (7, 151). Albrecht (7) found that L/S decreased from 1.4 in the late vegetative stage to 0.7 or lower by the early pod stage. The most rapid change in L/S occurred before mid-bud. Changes in alfalfa L/S were closely related to the heat sum above a 5°C base temperature in a study by Onstad and Fick (151). The decrease in L/S is most pronounced in the shaded, lower one-half of the plant canopy because of leaf drop and increase in stem mass. Although there is a net loss in leaf mass in the lower portion of the canopy, total canopy leaf mass normally increases until plants are into the flowering stage (33, 218).

The L/S of spring-growth alfalfa is frequently less than that of regrowth herbage at the same morphological stage (151). This may be attributed to differences in origin of stems, which come from crown buds in spring growth compared to mostly axillary basal buds in regrowth (147); differences in temperature and daylength patterns (194); or effects of a larger water deficit during summer compared to spring (197) (see Chapter 12).

14-2.1.3 Rate of Change of Quality Factors with Advancing Maturity

Kalu and Fick (106, 107) reported that IVDDM, CP, NDF, and lignin concentration in total herbage were closely related to morphological stage of development of alfalfa. The IVDDM decreased about 43 g/kg of DM with each unit change in their 10-stage maturity system. In Montana and Utah, Anderson et al. (11) observed that alfalfa DDM decreased 2.8 g/(kg/d) and CP declined 2.0 g/(kg/d) during spring growth. The rate of decline in digestiblity was similar to that of reports by Jung et al., (99) and Richards et al. (158), but not as rapid as frequently reported for grasses (119, 189). Both the rate of change in morphological stage and digestibility are strongly influenced by temperature (126, 151). Consequently, the rate of decline in digestibility with time is faster in summer when temperature is high than in spring or autumn when temperature is low (42, 151). Alfalfa herbage continues to decline in feeding value after being exposed to an autumn frost (43); frost damage may cause leaf loss which increases the rate of decline in herbage quality.

Advancing maturity reduces animal production, partly because of a lowered concentration of digestible energy in the herbage and partly because of lowered voluntary intake (189). Anderson et al. (11) reported that intake of first-crop alfalfa was reduced 0.21 g/kg (body wt.)$^{0.75}$ for each day in delay of harvest after vegetative stage in May. Variation in voluntary intake is more important than variation in digestibility in determining animal performance (11, 199).

The feeding value for dairy cows (*Bos* spp.) of first-flower, half-bloom, and full-bloom alfalfa used as the sole ration for entire lactations was determined by Dawson et al. (51). After 3 yr of testing with Holstein cows, they reported production of 5035, 4428, and 4075 kg of milk per cow for these three growth stages, respectively. Average DMD by sheep (*Ovis aries*) were 630, 610, and 580 g/kg of DM, respectively.

Scientists in Minnesota (55, 140) evaluated late-bud and full-bloom alfalfa hays for the growth of heifers and milk cows. Heifers starting at about 272 kg liveweight made average daily gains of 0.74 kg/d on the late-bud hay compared to 0.63 kg/d on the full-bloom hay. Milk production of first-calf heifers fed late-bud hay as the sole energy source was 4000 kg fat-corrected milk per 280-d period compared to 3000 kg fat-corrected milk for heifers fed full-bloom hay. The superior performance was consistent with a 14% greater daily intake of late-bud hay and a greater digestibility. Thus, the differential in alfalfa-feeding value for milk cows was greater than it was for growing heifers. This likely was related to much greater energy needs of milk cows.

14-2.1.4 Spring Growth vs. Aftermath Growth

First-cutting alfalfa has been characterized by a greater extent of decline in digestibility with increasing maturity than occurs in subsequent cuttings (94, 155). In general, later cuttings are not as high in initial feeding value as is first-cutting alfalfa. Also, the decline in quality, while not as extensive, occurs more rapidly in the summer than in the spring because high temperatures speed rate of development as well as respiratory losses of nonstructural carbohydrates (68, 96, 126, 167, 194).

14-2.1.5 Optimum Harvest Schedule for High-Quality Forage

Cutting alfalfa for maximum forage quality, yield of digestible nutrients per unit land area, and stand survival may involve some compromises (see Chapter 12). Although the best quality feed is produced from immature alfalfa, the greatest financial return may at times be realized by harvesting at more mature stages depending on the need to maximize yield, reduce costs of harvest, ensure stand survival or ensure rapid drying. The nutrient intake needs of different classes of livestock may also dictate the financial returns from feeding of alfalfa cut at various growth stages.

Harvesting by a growth stage criterion has generally been a more reliable indicator of alfalfa feeding value than calendar date (126, 134, 170, 171, 189, 216). Sheaffer (161) concluded that when established early in the spring without a companion crop in Minnesota, newly seeded alfalfa could be harvested initially at the late-bud to early flower stage. In the year of seeding in highly fertile soils, he also found that the highest yield of top-quality forage could be obtained without stand loss by harvesting three to four times.

Brink and Marten (28) reported that alfalfa harvest schedules in Minnesota based on early first harvests (prebloom stages) with three additional harvests ending at various times in the autumn provided the greatest yields of top grade (maximum quality) alfalfa, even though these schedules did not provide the greatest total yields of IVDDM and CP. They concluded that a producer's decision to obtain either the greatest yield of most consumable (maximum quality) alfalfa or the greatest yield of nutrients per se should depend on the needs of the livestock being fed.

14–2.2 Climatic and Edaphic Factors

14–2.2.1 Geographic Location

Harvesting on specific calendar dates at widely separated geographic locations, particularly locations at different latitudes, will result in alfalfa harvested at several maturities (171). Environmental differences at various locations influence yield and quality (96, 134, 139, 171). Growing alfalfa at different elevations results in maturity differences similar to shifts in latitude. Anderson and Thacker (12) reported that on the same date, alfalfa was most mature at the lower elevations, but the rate of daily decline in DMD was greatest at the higher elevation. The greatest uniformity in quality can be obtained by cutting at a uniform stage of growth (134, 171).

14–2.2.2 Seasonal and Yearly Variation

Environmental factors that vary from season to season and year to year are responsible for many unexplained differences in alfalfa quality. Meyer and Jones (139) described three distinct seasonal periods that differentially affected lignin, protein, and estimated total digestible nutrients (TDN) concentration within growth stages of California-grown alfalfa. Chemical composition and estimates of feeding value can be influenced by daylength and illuminance (41, 68, 202).

According to Christian (40), leaf development of alfalfa may be more rapid at long daylengths, but leaf size is apparently not affected. The increase in stem diameter and length during long days decreases the L/S ratio, with a concomitant reduction in feeding value. Specific leaf weight declines with reduced illuminance. Shaded stems are more succulent and pliable. However, L/S ratio remains fairly constant when shading occurs (202). Shading to exclude 47 to 73% of sunlight did not influence CP concentration of the upper herbage of alfalfa during early or late summer; but shading increased CP during mid-summer. Although total nonstructural carbohydrate concentration of the upper herbage decreased as shading increased in all three periods, rates of cell wall digestion were not influenced by shading.

Large year-to-year variations in DMD of alfalfa herbage have been reported (11, 19, 166). Seasonal weather patterns influence the development of plants differently each year, so that more or less time is required for the plant to reach a specific stage of growth. The effect of these variables supports the use of a morphological measure of growth stage, such as flowering or regrowth from crown buds, rather than a calendar date or time interval to determine time of cutting (134, 139, 170).

14-2.2.3 Diurnal Variation

Curtis (49) reported an increase in carbohydrate levels in alfalfa herbage from 45 to 61 g/kg of DM from morning to afternoon. He further reported that alfalfa plots cut in the evening yielded more total DM than similar plots cut in the morning. However, Adolph et al. (2) subsequently reported no significant differences in DM yield between alfalfa cut in the evening vs. the following morning. A digestibility experiment using rabbits (*Oryctolagus cuniculus*) also showed no significant differences between alfalfa cut at the different times.

Diurnal variation was observed in the concentrations of reducing sugars, ash, and P in the upper 10 cm of 'Ranger' alfalfa (9, 10). The maximum amounts of reducing sugars were found between 1000 and 1400 h. Water-soluble carbohydrate concentration in alfalfa was shown by Holt and Hilst (93) to follow a curvilinear diurnal trend, from a low at 0600 to maximum levels at 1200 h and a slight decrease by 1800 h. These observations confirm results reported by Kivimae (114). Holt and Hilst (93) also reported that nonstructural polysaccharide concentration followed a nonlinear daily trend, with the most rapid increase occurring in the afternoon. This suggests that a significant portion of water-soluble sugars was converted rapidly into polysaccharides during the afternoon. The diurnal shift in organic acid level reported by Burns et al. (30) supports their conclusion. Lechtenberg et al. (118) found that IVDDM values of second-growth alfalfa were 16 g/kg greater at 1800 h compared to 0600 h; starch concentration of leaves increases by 100 g/kg during the day while that in stems did not change, which accounted for most of a concurrent change in L/S ratio from 1.1 to 1.5.

14-2.2.4 Ambient Temperature

It is difficult to assess the effects of field temperature on alfalfa growth, composition, and quality. Controlled environment studies showed, however, that the rate of maturation (negatively associated with quality) increased with increased temperature in the range of about 20 to 30°C (68, 96, 126, 153, 168, 173, 197).

The L/S ratio at first and 50% bloom of alfalfa was lower at 16/10°C than at 27/21°C, and stem diameter was greater in the cool environment (126, 197). On the other hand, alfalfa grown at 18/10°C had a higher L/

S ratio at the late-bud stage than that grown at either 26/18 or 34/26°C (203). Regardless of L/S ratios, alfalfa harvested at specific growth stages had greater CP concentration in herbage when grown in warm compared to cool temperature regimes (126, 167, 197, 203). However, warm temperatures often decreased IVDDM and nonstructural carbohydrate concentrations (126, 167, 197) without consistently affecting crude fiber, ADF, or lignin concentrations.

In field studies, Hidiroglou et al. (90), working with alfalfa and timothy, obtained highest levels of DDM and digestible organic matter during the cooler periods of the growing season. Jensen et al. (96) reported similar observations under field conditions in Arizona. Conversely, Meyer and Jones (139) indicated that temperature had no apparent effect on alfalfa quality in California. This type of inconsistency is to be expected, because differences in ambient temperature are seldom independent of other associated variables.

14-2.2.5 Soil Type

Soil type per se is less important in determining forage quality than its indirect effect through water-holding capacity, soil aeration, and nutrient availability. Additionally, compositional differences noted on soil types at widely separated locations are difficult to evaluate because of confounding with other environmental factors (171). Meyer and Jones (139) reported that lignin was lower and protein higher in alfalfa produced on heavy clay loams than on sandy soils. These patterns are probably associated with a higher L/S ratio and shorter plant height for alfalfa grown on heavy clay soils, compared to loam or sandy soils (217).

14-2.2.6 Soil Moisture

Moisture levels and soil compaction have had variable influence on alfalfa CP concentration (22, 24, 36, 37, 70, 73, 111, 197, 203). Carter and Sheaffer (37) found that water stress will increase N concentration in plants that do not fix N_2. Therefore, some of the inconsistent CP responses to alfalfa water deficits may be explained by variations in N_2-fixation capabilities. In California, Donovan and Meek (56) found that a moisture deficit increased alfalfa CP in cool months, but not in warm months.

The reduced vigor associated with drought often causes a stunted, leafier plant which has finer stems, reduced fiber concentration, and increased IVDDM (29, 37, 56, 96, 197). Carter and Sheaffer (37) found that the IVDDM concentration of alfalfa was not affected by plant water potential until it fell below -2.7 MPa, at which time the increased digestibility was associated with an increased L/S. Moderate to severe water stress slows maturity of alfalfa and increases the IVDDM concentration of stems (75, 177).

Excess soil moisture can reduce yields and lower the concentration of feed nutrients. Irrigation at a 50% level of minimum available soil moisture produced higher yields of alfalfa, and a higher concentration of CP and in vitro digestible cellulose, than either higher or lower levels of irrigation (22) (see Chapter 11 in this book).

14-2.2.7 Soil Fertility

The composition and the nutritional value of forages are affected by the availability of several essential elements in the soil (8, 20, 89, 136, 157, 165). Hill and Guss (91) reviewed literature that revealed that alfalfa provides most of the 10 mineral elements required by dairy cattle, although deficiencies occur for P, Na, Cu, and Zn. Beeson (20) pointed out that response to P application is largely dependent on the native fertility of the soil. Large differences in alfalfa forage P concentration response to P fertilizer application occurred in Pennsylvania soils (92). Hanson and MacGregor (80), summarizing a 10-yr study, concluded that the concentration of P and K in 10-yr-old alfalfa was related to the quantities of these elements applied by annual topdressings. The accumulation of Ca and K by alfalfa is more related to the ratios of these elements than to their absolute content in the soil (40). Heinemann et al. (89) fed domestic rabbits alfalfa hay produced on soil with low-available P (1.2 g of P/kg of DM) and high-available P (2.6 g of P/kg). They observed retarded growth, lower mature body weights, impaired breeding efficiency, abnormal bone structure, and greater bone fragility in rabbits fed alfalfa hay produced on the soil with low-available P. They concluded that increased skeletal strength, improved breeding efficiency, and increased rate of growth could be obtained by proper P fertilization of the soil.

Nutritional factors other than the concentration of chemical elements in plants may be affected by the availability and uptake of elements in the soil (8, 30, 34, 60, 125, 157). Smith and Albrecht (174) reported that excessive use of a single fertilizer element, or unbalanced soil fertilization, may increase maximum yields but reduce biological value. When Wedin et al. (206) fed alfalfa to guinea pigs (*Cavia* spp.), their average daily gains reflected the fertility ratings of the soil types on which the alfalfa was grown. Moderate fertilization of 182 kg/ha of P and K resulted in increased daily gains in several instances. However, fertilization with 363 kg/ha of P and K tended to decrease weight gains. Higher L/S ratios may occur in plants when growth is restricted by a P or K deficiency, resulting in greater CP concentrations (162). The need to balance soil fertility to avoid mineral imbalances and digestive upsets in livestock has been investigated (125, 133). Martz et al. (133) suggested that dairy heifers consumed more alfalfa-bromegrass (*Bromus inermis* Leyss.) hay which had been fertilized with a balanced fertilizer mixture than hay that had received only N.

Calder and MacLeod (34) found that the IVDDM of first-cut, but not second-cut, alfalfa was increased by applications of K fertilizer. In Minnesota, Sheaffer et al. (162) reported that IVDDM concentrations of alfalfa grown in soils having varying moisture status were not affected by K fertilization. Fertilizer application may increase the yield of digestible nutrients per unit area by increasing DM yield with no influence on digestibility (205). Potential changes in the relative contribution of alfalfa and grasses in mixtures must be considered when determining the effect of mineral fertilizers on forage-feeding value.

The application of S-containing fertilizer to low S soils resulted in improved animal performance of lambs fed alfalfa hay grown on the soils (157). The S concentration of the alfalfa was increased, but other accompanying changes in composition appeared to have a greater effect on feeding value than changes in S. Methionine supplementation did not improve the feeding value of hay grown on low-S soils. The S intake of sheep fed alfalfa hay from low-S soils was reflected in blood serum inorganic sulfate levels and appeared to be influenced by the N/S ratios (157, 207).

The concentration of organic acids (30) and relative proportions of free amino acids (125) in alfalfa will be altered by soil fertilization.

14-2.3 Diseases and Insects

Diseases and insects limit alfalfa production (see Chapters 22 and 23 in this book). Although producers may be concerned primarily with yield reduction and survival of the stand, feeding value may be affected by pests. Any condition reducing the L/S ratio, increasing the fiber concentration, or reducing the protein or carotene concentration can be expected to lower feeding value. These relationships explain the major effects of diseases and insects on feeding value.

A seven-state study showed that accumulation of the estrogenic component, *coumestrol*, was closely associated with the incidence of foliar diseases (79). Infection with each of four fungus pathogens increased the coumestrol level of alfalfa forage (79, 123). Coumestrol concentration ranged from 2 mg/kg for healthy tissue grown in controlled growth rooms to 429 mg/kg for forage infected with foliar diseases in the field. Severe potato leafhopper [*Empoasca fabae* (Harris)] yellowing was associated with higher levels of coumestrol in alfalfa plants, but the higher levels may have been caused by secondary invasion by saprophytes (79). Loper (122) also found that a pea aphid [*Acyrthosiphon pisum* (Harris)]-susceptible cultivar of alfalfa had higher coumestrol concentration than did two aphid-resistant cultivars when all were exposed to aphid attack. Sherwood et al. (164) reported that coumestrol increased only within the alfalfa tissue that was infected with specific diseases. Three-foliar pathogens increased coumestrol in leaves but not roots, whereas a root pathogen increased coumestrol in roots but not leaves or stems.

The effects of estrogenic compounds in alfalfa on ruminants have been studied extensively. However, a documented report of reduced fertility in livestock consuming alfalfa is unknown to the authors. Other less important estrogenic responses have been reported for sheep consuming alfalfa (88). Oldfield et al. (150) reported a stilbestrol-like stimulation of growth in sheep fed alfalfa. A similar growth response has not been observed in beef cattle.

Bickoff et al. (23) demonstrated an increase in phenolic compounds other than coumestrol in alfalfa leaf tissue as a result of disease infection. Minimal information is available on the possible effects of such compounds on the responses of animals. Hanson (78) emphasized the need to reassess the relation of foliar diseases to forage quality and animal performance.

Brigham (27) found that alfalfa leaflets uninfected with Cercospora leaf spot (*Cercospora medicaginis* Ell. & Ev.) contained considerably more CP (330 g/kg of DM) than those that had from 25 to 50% of the surface covered with lesions (CP of 180 g/kg of DM). Also, infected leaves had less fat and more crude fiber and ash than uninfected leaves. In Minnesota (131), alfalfa plants inoculated with *Phoma medicaginis* Malbr. & Roum. 3 weeks before harvest produced forage that contained significantly less IVDDM, and more ADF and NDF than did disease-free forage. These changes were attributed partly to reduced quality of infected leaves and partly to reduced leaf percentage of infected forage. Leath et al. (116) compared alfalfa leaves that were either disease free or moderately infected with *P. medicaginis*. Diseased leaves had less IVDDM, CP, and total nonstructural carbohydrates; more ADF, NDF, lignin, and cellulose; less palatability to adult meadow voles; and less digestibility for weanling meadow voles. Willis et al. (212) reported an 18% reduction in alfalfa forage carotene but no change in CP when plants were not sprayed vs. sprayed weekly with a fungicide.

Damage from pest insects will affect the feeding value of alfalfa. Reduced CP and carotene concentrations in alfalfa have been associated with infestation by potato leafhopper (48, 77, 95, 97, 98, 108, 109, 113, 138, 154, 160, 170, 172, 215), alfalfa weevil [*Hypera postica* (Gyllenhal)] (66, 204), and adult meadow spittlebug (*Philaenus spumarius* (L.)] (135). Protein losses caused by potato leafhopper are especially severe when alfalfa nears maturity (160). However, Parman and Wilson (152) reported that fourth-stage nymphs of meadow spittlebug increased L/S ratio and CP concentration in alfalfa herbage, although DM yields were unaffected. They developed a regression equation to predict increases in CP concentration associated with the number of nymphs per unit area (214); the R^2 of the equation was 0.73.

Meyer and Jones (139) and Cuperus et al. (47) found that pea aphid damage did not affect CP concentration of alfalfa. These results contrast with those of Harper and Lilly (82); Harvey et al. (87); and Kindler et al. (112) who found pea aphid infestation of alfalfa reduced both CP and carotene concentrations. However, Meyer and Jones (139) reported a trend toward higher lignin concentration when aphid damage became severe, and Cuperus et al. (47) found that aphid days per net sweep were negatively correlated with IVDDM of second regrowth alfalfa. Kindler et al. (112) also reported reduced digestibility of alfalfa damaged by pea aphid.

Liu and Fick (120) reported that digestibility and CP concentrations of alfalfa leaves in a three-cut system were reduced during peak larval feeding by large populations of alfalfa weevil. However, weevil defoliation did not influence leaf percentage or total herbage quality because stem growth was retarded. In a two-cut system, quality was greater when weevils were not controlled because of reduced lodging and renewed growth of high-quality herbage when larval activity subsided. Livestock poisoning, especially of horses (*Equus caballus*), may occur when animals consume dead blister beetles (*Epicauta* spp.) that have previously infested alfalfa that is cut for hay (35).

To date, other possible effects of insects upon quality have not been assessed. For example, aphids leave a deposit of honeydew on plant tissue, which could contribute to mold growth and reduced feeding value.

14-2.4 Weeds

Invading weeds may or may not decrease the feeding value of alfalfa pasture and hay crops depending on species, maturation, and soil fertility. In Nevada, Cords (44) found a negative correlation between CP concentration of alfalfa hay and concentration of nonlegume winter-annual weeds that were quite mature by the time of alfalfa harvest. The principal weeds that infested the hay were flixweed [*Descurainia sophia* (L.) Webb], downy brome (*Bromus tectorum* L.), and wild barley (*Hordeum leporinum* Link.). On the other hand, Marten and Andersen (129) and Temme et al. (181) found that three common annual invaders of new alfalfa seedings had CP, ADF, NDF, and IVDDM concentrations essentially equivalent to that of high-quality alfalfa when they were harvested in mixture with the alfalfa or at the same time as alfalfa in late June to mid-July on fertile soils in Minnesota and Wisconsin. The weeds were redroot pigweed (*Amaranthus retroflexus* L.), common lambsquarters (*Chenopodium album* L.), and common ragweed (*Ambrosia artemisiifolia* L.). However, several weeds had either less feed nutrients or less digestibility than alfalfa. These included giant foxtail (*Setaria faberii* Herrm.), Pennsylvania smartweed (*Polygonum pensylvanicum* L.), yellow foxtail [*Setaria glauca* (L.) P. Beauv.], barnyardgrass [*Echinochloa crus-galli* (L.) P. Beauv.], and shepherd's purse [*Capsella bursa-pastoris* (L.) Medic].

Perennial and biennial weeds also have variable forage quality. Fawcett et al. (63), Dutt et al. (58), and Marten et al. (132) found quackgrass [*Agropyron repens* (L.) Beauv.] to have the typical perennial grass deficiency of reduced CP or an undesirable excess of NDF compared to alfalfa. Dutt et al. (57) reported that animal intake, in vivo digestibility, and nutritive value index of alfalfa were reduced by infestation with yellow rocket (*Barbarea vulgaris* R. Br.). However, several investigators (52, 57, 132, 163) found that some perennial weeds may have equivalent or greater forage quality than the associated alfalfa. Among these high-quality weeds are common dandelion (*Taraxacum officinale* Weber), white cockle (*Lychnis alba* Mill.), and immature Jerusalem artichoke (*Helian-*

thus tuberosus L.). Marten et al. (132) found that swamp smartweed (*Polygonum coccineum* Muhl.) had especially low IVDDM compared to alfalfa, and that curly dock (*Rumex crispus* L.), Canada thistle [*Cirsium arvense* (L.) Scop.], hoary alyssum [*Berteroa incana* (L.) DC.], swamp smartweed, and Jerusalem artichoke were all much less palatable than alfalfa when they were grazed by lambs. Marten et al. (132) also reported that white cockle, quackgrass, and dandelion were often as palatable to grazing sheep as was alfalfa. They concluded that the intrinsic forage quality significance of common weeds that invade alfalfa must be decided in terms of individual species and local situations. Although some weeds have acceptable or even excellent forage quality, weed species are not desirable in alfalfa stands because they are indicative of either less than optimum management or of depleted alfalfa plant vigor. Weeds also are sources of unwanted seeds or vegetative propagules that invade surrounding or subsequent crops, and some weeds are toxic to livestock (see Chapter 23 in this book).

14-3 SUMMARY

The potential feeding value of alfalfa exceeds that of perennial grasses at similar stages of maturation primarily because alfalfa has greater intake potential associated with faster digestibility. Alfalfa can provide adequate nutrition as the sole ingredient in many livestock feeding programs when supplemented with the proper minerals. Its primary deficiency is insufficient available energy for the production needs of high-producing animals.

An improved, uniform procedure for expressing the feeding value of alfalfa hay has been developed for commercial use. Available energy is expressed as estimated DDM calculated via equation from assay for ADF, and the hay is analyzed for CP and dry matter (moisture). Additional optional organoleptic and chemical factors may be measured, including those that are useful for estimating intake potential.

Many factors determine the influence of alfalfa cell-wall composition on forage digestibility and intake. Among these are the specific composition of primary and secondary cell walls and the presence of different specific cell and tissue types in leaves compared to stems. Alfalfa leaves almost always have higher and more uniform feeding value than stems. Lignin is usually the major factor limiting digestibility of cell walls because of its involvement in chemical, physical, and nutritional phenomena.

Major factors that influence the potential feeding value of alfalfa during growth and development include: growth stage at the time of cutting; L/S ratio; climatic and edaphic factors, such as geographic location, seasonal and yearly variation, illuminance-associated diurnal variation, ambient and soil temperature, soil type, soil moisture, and soil fertility; disease and insect damage; and weed infestation. Growth stage at the time of cutting is often the most easily controlled factor that determines the worth of alfalfa as a feedstuff.

REFERENCES

1. Adler, Erich. 1977. Lignin chemistry—past, present and future. Wood Sci. Technol. 11:169-218.
2. Adolph, W.H., H.A. MacDonald, Hui-Lan Yeh, and G.P. Lofgren. 1947. Content and digestibility of morning and evening cuttings of alfalfa. J. Anim. Sci. 6:347-351.
3. Akin, D.E. 1982. Forage cell wall degradation and p-coumaric, ferulic, and sinapic acids. Agron. J. 74:424-428.
4. ----. 1983. Structural characteristics limiting digestion of forage fiber. p. 511-514. In J.A. Smith and V.W. Hays (ed.) Proc. 14th Int. Grassl. Congr., Lexington, KY. 15-24 June 1981. Westview Press, Boulder, CO.
5. ----, D. Burdick, and G.E. Michaels. 1974. Rumen bacterial interrelationships with plant tissue during degradation revealed by transmission electron microscopy. Appl. Microbiol. 276:1149-1156.
6. Albersheim, P. 1976. The primary cell wall. p. 225-274. In J. Bonner and J.E. Varner (ed.) Plant biochemistry. Academic Press, New York.
7. Albrecht, K.A. 1983. Studies on nitrogen accumulation, fiber chemistry, and in vitro digestibility of alfalfa. Ph D. diss. Iowa State Univ., Ames (Diss. Abstr. 8407047).
8. Albrecht, W.A., and G.E. Smith. 1941. Biological assays of soil fertility. Soil Sci. Soc. Am. Proc. 6:252-258.
9. Allen, R.S., P.G. Homeyer, N.L. Jacobson, N.R. Gould, R.E. Worthington, and R.J. Johnson. 1957. Effect of irrigation and time of day on composition of young alfalfa. J. Anim. Sci. 16:1055 (Abstr.).
10. ----, R.E. Worthington, N.R. Gould, M.L. Jackson, and A.E. Freeman. 1961. Diurnal variation in composition of alfalfa. J. Agric. Food Chem. 9:406-408.
11. Anderson, M.J., G.F. Fries, D.V. Kopland, and D.R. Waldo. 1973. Effect of cutting date on digestibility and intake of irrigated first-crop alfalfa hay. Agron. J. 65:357-360.
12. ----, and D.R. Thacker. 1970. Elevation effects on nutritive characteristics of alfalfa. J. Dairy Sci. 53:676 (Abstr.)
13. Bacon, J.S.D., A.H. Gordon, and E.J. Morris. 1975. Acetyle groups in cell-wall preparations from higher plants. Biochem. J. 149:485-487.
14. Bailey, R.W. 1973. Structural carbohydrates. p. 157-211. In G.W. Butler and R.W. Bailey (ed.) Chemistry and biochemistry of herbage. Academic Press, London.
15. Balwani, T.L., R.R. Johnson, K.E. McClure, and B.A. Dehority. 1969. Evaluation of green chop and ensiled sorghums, corn silage and perennial forages using digestion trials and VFA (volatile fatty acids) production in sheep. J. Anim. Sci. 28:90-97.
16. Barnes, R.F 1970. Unpublished data. USDA-ARS, Purdue Univ.
17. ----, and G.O. Mott. 1970. Evaluation of selected clones of Phalaris arundinacea L. I. In vivo digestibility and intake. Agron. J. 62:719-722.
18. Barton, F.E., and D.E. Akin. 1977. Digestibility of delignified forage cell walls. J. Agric. Food Chem. 25:1299-1303.
19. Baumgardt, B.R., and D. Smith. 1962. Changes in estimated nutritive value of the herbage of alfalfa, medium red clover, ladino clover, and bromegrass due to stage of maturity and year. Wis. Agric. Exp. Stn. Rep. 10.
20. Beeson, K.C. 1946. The effect of mineral supply on the mineral concentration and nutritional quality of plants. Bot. Rev. 12:424-455.
21. Belyea, R.L., M.B. Foster, and G.M. Zinn. 1983. Effect of delignification on in vitro digestion of alfalfa cellulose. J. Dairy Sci. 66:1277-1281.
22. Bezeau, L.M., and L.G. Sonmor. 1964. The influence of levels of irrigation on the nutritive value of alfalfa. Can. J. Plant Sci. 44:505-508.
23. Bickoff, E.M., G.M. Loper, C.H. Hanson, J.H. Graham, S.C. Witt, and R.R. Spencer. 1967. Effect of common leafspot on coumestans and flavones in alfalfa. Crop Sci. 7:259-261.
24. Bourget, S.J., and R.B. Carson. 1962. Effect of soil moisture stress on yield, water-use efficiency and mineral composition of oats and alfalfa grown at two fertility levels. Can. J. Soil Sci. 42:7-12.
25. Brazle, F.K., and L.H. Harbers. 1977. Digestion of alfalfa hay observed by scanning electron microscopy. J. Anim. Sci. 46:506-512.
26. Brice, R.E., and I.M. Morrison. 1982. The degradation of isolated hemicellulose and

27. Brigham, R.D. 1959. Effect of Cercospora disease on forage quality of alfalfa. Agron. J. 51:365.
28. Brink, G.E., and G.C. Marten. 1983. Contemporary vs. traditional harvest management of alfalfa—effects on yield, quality, and persistence. Agron. Abstr. American Society of Agronomy, Madison, WI, p. 103.
29. Brown, P.W., and C.B. Tanner. 1983. Alfalfa stem and leaf growth during water stress. Agron. J. 75:799–805.
30. Burns, J.C., C.H. Noller, C.L. Rhykerd, and T.S. Rumsay. 1968. Influence of fertilization on some organic acids in alfalfa, *Medicago sativa* L. Crop Sci. 8:1–2.
31. Burroughs, Wise, Paul Gerlaugh, and R.M. Bethke. 1950. The influence of alfalfa hay and fractions of alfalfa hay upon the digestion of ground corncobs. J. Anim. Sci. 9:207–213.
32. Buxton, D.R. and J.S. Hornstein. 1984. Unpublished data, USDA-ARS, Iowa State Univ.
33. ----, ----, W.F. Wedin, and G.C. Marten. 1985. Forage quality in stratified canopies of alfalfa, birdsfoot trefoil, and red clover. Crop Sci. 25:273–279.
34. Calder, F.W., and L.B. MacLeod. 1968. In vitro digestibility of forage species as affected by fertilizer application, stage of development and harvest date. Can. J. Plant Sci. 48:17–24.
35. Capinera, J.L. 1986. Will one blister beetle kill a horse? p. 143–149. *In* Proc. 16th Natl. Alfalfa Symp., Fort Wayne, IN. 5–6 March. The Certified Alfalfa Seed Council, Davis, CA.
36. Carlton, A.E., C.S. Cooper, R.H. Delaney, A.L. Dubbs, and R.F. Eslick. 1968. Growth and forage quality comparisons of sainfoin (*Onobrychis viciaefolia* Scop.) and alfalfa (*Medicago sativa* L.). Agron. J. 60:630–632.
37. Carter, P.R., and C.C. Sheaffer. 1983. Alfalfa response to soil water deficits. I. Growth, forage quality, yield, water use, and water-use efficiency. Crop Sci. 23:669–675.
38. Chesson, A., A.H. Gordon, and J.A. Lomax. 1983. Substituent groups linked by alkali-labile bonds to arabinose and xylose residues of legume, grass and cereal straw cell walls and their fate during digestion by rumen microorganisms. J. Sci. Food Agric. 34:1330–1340.
39. ----, and J.A. Monro. 1982. Legume pectic substances and their degradation in the ovine rumen. J. Sci. Food Agric. 33:852–859.
40. Christian, K.R. 1977. Effects of the environment on the growth of alfalfa. Adv. Agron. 29:183–227.
41. Coffindaffer, B.L., and O.J. Burger. 1958. Response of alfalfa varieties to daylength. Agron. J. 50:389–392.
42. Collins, Michael. 1983. Changes in composition of alfalfa, red clover, and birdsfoot trefoil during autumn. Agron. J. 75:287–291.
43. ----, and T.H. Taylor. 1984. Quality changes of late summer and autumn produced alfalfa and red clover. Agron. J. 76:409–415.
44. Cords, H.P. 1973. Weeds and alfalfa hay quality. Weed Sci. 21:400–401.
45. Cowling, E.B. 1975. Physical and chemical constraints in the hydrolysis of cellulose and lignocellulose materials. Biotechnol. Bioeng. Symp. 5:163–181.
46. ----, and Wynford Brown. 1969. Structural features of cellulosic materials in relation to enzymatic hydrolysis. *In* E.T. Reese and G. Hajny (ed.) Cellulases and their application. Adv. Chem. Series 94:152–186.
47. Cuperus, G.W., E.B. Radcliffe, D.K. Barnes, and G.C. Marten. 1982. Economic injury levels and economic thresholds for pea aphid, *Acyrthosiphon pisum* (Harris), on alfalfa. Crop Prot. 1:453–463.
48. ----, ----, ----, and ----. 1983. Economic injury levels and economic thresholds for potato leafhopper (*Homoptera: Cicadellidae*) on alfalfa in Minnesota. J. Econ. Entomol. 76:1341–1349.
49. Curtis, O.F. 1944. The food content of forage crops as influenced by the time of day at which they are cut. J. Am. Soc. Agron. 36:401–416.
50. Davidson, P.M., and A.L. Branden. 1981. Antimicrobial activity of nonhalogenated phenolic compounds. J. Food Prot. 44:623–632.
51. Dawson, J.R., D.V. Kopland, and R.R. Graves. 1940. Yield, chemical composition, and feeding value for milk production of alfalfa hay cut at three stages of maturity. USDA Tech. Bull. 739. U.S. Government Printing Office, Washington, DC.
52. Doll, J.D. 1981. Dandelions in alfalfa don't affect forage quality. Hoard's Dairyman 125:671.
53. Donker, J.D., G.C. Marten, and P.K. Bhargava. 1982. A comparison of grass and legume hays fed to Holstein heifers and lambs. J. Anim. Sci. 55, Suppl. 1:307 (Abstr.).
54. ----, ----, R.M. Jordan, and P.K. Bhargava. 1976. Effects of drying on forage quality of alfalfa and reed canarygrass fed to lambs. J. Anim. Sci. 42:180–184.

55. ----, Harbans Singh, and H.W. Mohrenweiser. 1968. Forage evaluation. I. Performance of Holstein heifers fed only early-cut or late-cut alfalfa hay on a free-choice basis. J. Dairy Sci. 51:362-366.
56. Donovan, T.J., and B.D. Meek. 1983. Alfalfa responses to irrigation treatment and environment. Agron. J. 75:461-464.
57. Dutt, T.E., R.G. Harvey, and R.S. Fawcett. 1982. Feed quality of hay containing perennial broadleaf weeds. Agron. J. 74:673-676.
58. ----, ----, ----, N.A. Jorgensen, H.J. Larsen, and D.A. Schlough. 1979. Forage quality and animal performance as influenced by quackgrass control in alfalfa with pronamide. Weed Sci. 27:127-132.
59. Esau, K. 1977. Anatomy of seed plants. 2nd ed. John Wiley and Sons, New York.
60. Evans, J.L., J. Arroyo-Aguilu, M.W. Taylor, and C.H. Ramage. 1965. Date of harvest of New Jersey forages as related to the nutrition of ruminant animals. N.J. Agric. Exp. Stn. Bull. 814.
61. Fahey, G.C., Jr., S.Y. Al-Haydari, F.C. Hinds, and D.E. Short. 1980. Phenolic compounds in roughages and their fate in the digestive system of sheep. J. Anim. Sci. 50:1165-1172.
62. ----, Jr., G.A. McLaren, and J.E. Williams. 1978. Hemicellulose digestion by lambs as affected by activation energy between lignin and hemicellulose. Nutr. Rep. Int. 18:281-288.
63. Fawcett, R.S., R.G. Harvey, D.A. Schlough, and I.R. Block. 1978. Quackgrass control in established alfalfa with pronamide. Weed Sci. 26:193-198.
64. Ferebee, D.B., D.O. Frickson, C.N. Haugse, K.L. Larson, and M.L. Buchanan. 1972. Digestibility and chemical composition of bromic and alfalfa throughout the growing season. N.D. Farm Res. 30:3-7.
65. Fick, G.W., and R.S. Holthausen. 1975. Significance of parts other than blades and stems in leaf-stem separations of alfalfa herbage. Crop Sci. 15:259-262.
66. Flessel, J.K., and H.D. Niemczyk. 1971. Theoretical values of fully grown first-cutting alfalfa lost to alfalfa weevil larvae. [*Hypera postica*]. J. Econ. Entomol. 64:328-329.
67. Gaillard, B.D.E., and G.N. Richards. 1975. Presence of soluble lignin-carbohydrate complexes in the bovine rumen. Carbohydr. Res. 42:135-145.
68. Garza, R.T., R.F Barnes, G.O. Mott, and C.L. Rhykerd. 1965. Influence of light intensity, temperature and growing period on the growth, chemical composition and digestibility of Culver and Tanverde alfalfa seedlings. Agron. J. 57:417-420.
69. Gharpuray, M.M., Y.H. Lee, and L.T. Fan. 1983. Structural modification of lignocellulosics by pretreatments to enhance enzymatic hydrolysis. Biotechnol. Bioeng. 25:157-172.
70. Gifford, R.O., and E.H. Jensen. 1967. Some effects of soil moisture regimes and bulk density on forage quality in the greenhouse. Agron. J. 59:75-77.
71. Gordon, A.J. 1975. A comparison of some chemical and physical properties of alkali lignins from grass and lucerne hays before and after digestion by sheep. J. Sci. Food Agric. 26:1551-1559.
72. Gordon, A.H., J.A. Lomax, and A. Chesson. 1983. Glycosidic linkages of legume, grass and cereal straw cell walls before and after extensive degradation by rumen microorganisms. J. Sci. Food Agric. 34:1341-1350.
73. Groskopp, M.D., J.M. Sund, and J.T. Murdock. 1963. Irrigated alfalfa in central Wisconsin. Wis. Agric. Exp. Stn. Bull. 558.
74. Hacker, J.B., and D.J. Minson. 1981. The digestibility of plant parts. Herb. Abstr. 51:459-482.
75. Halim, R.A., and D.R. Buxton. 1983. Unpublished data. USDA-ARS, Iowa State Univ.
76. Hall, J.L., T.J. Flowers, and R.M. Roberts. 1982. Plant cell structure and metabolism. 2nd ed. Longman, New York.
77. Ham, W.E., and H.M. Tysdal. 1946. The carotene content of alfalfa strains and hybrids with different degrees of resistance to leafhopper injury. J. Am. Soc. Agron. 38:68-74.
78. Hanson, C.H. 1966. Foliar diseases and forage quality. p. 1209-1213. *In* Proc. 9th Int. Grassl. Congr., Sao Paulo, Brazil. 7-20 Jan. 1965. Edicoes Harico Limitada, Sao Paulo.
79. ----, G.M. Loper, G.O. Kohler, E.M. Bickoff, K.W. Taylor, W.R. Kehr, E.H. Stanford, J.W. Dudley, M.W. Pedersen, E.L. Sorensen, H.L. Carnahan, and C.P. Wilsie. 1965. Variation in coumestrol content of alfalfa as related to location, variety, cutting, year, stage of growth, and disease. USDA Tech. Bull. 1333.
80. Hanson, R.C., and J.M. MacGregor. 1966. Soil and alfalfa plant characteristics as affected by a decade of fertilization. Agron. J. 58:3-5.
81. Harkin, J.M. 1973. Lignin. p. 323-373. *In* G.W. Butler and R.W. Bailey (ed.) Chemistry and biochemistry of herbage. Academic Press, New York.
82. Harper, A.M., and C.E. Lilly. 1966. Effects of pea aphid on alfalfa in southern Alberta. J. Econ. Entomol. 59:1426-1427.

83. Harris, P.J., R.D. Hartley, and K.H. Lowry. 1980. Phenolic constituents of mesophyll and non-mesophyll cell walls from leaf laminae of *Lolium perenne*. J. Sci. Food Agric. 31:959-962.
84. Hartley, R.D. 1983. Degradation of cell walls of forages by sequential treatment with sodium hydroxide and a commercial cellulose preparation. J. Sci. Food Agric. 34:29-36.
85. ----, and J. Haverkamp. 1984. Pyrolysis-mass spectrometry of the phenolic constituents of plant cell walls. J. Sci. Food Agric. 35:14-20.
86. ----, and E.C. Jones. 1977. Phenolic components and degradability of cell walls of grass and legume species. Phytochemistry 16:1531-1534.
87. Harvey, T.L., H.L. Hackerott, and E.L. Sorenson. 1971. Pea aphid injury to resistant and susceptible alfalfa in the field. J. Econ. Entomol. 64:635-636.
88. Hawk, H.W., H.F. Rigther, C.H. Gordon, C.H. Hanson, T.H. Brinsfield, and J.C. Derbyshire. 1967. Stimulatory effects of diseased and disease-free alfalfa hay on the uterus of ovariectomized ewes. J. Anim. Sci. 26:567-570.
89. Heinemann, W.W., M.E. Ensminger, W.E. Hammand, and J.E. Oldfield. 1957. The effects of phosphate fertilization of alfalfa on growth, reproduction, and body composition of domestic rabbits. J. Anim. Sci. 16:467-475.
90. Hidiroglou, M., P. Dermine, H.A. Hamilton, and J.E. Troelsen. 1966. Chemical compostion and in vitro digestibility of forage as affected by season in northern Ontario. Can. J. Plant Sci. 46:101-109.
91. Hill, R.R., Jr., and S.B. Guss. 1976. Genetic variability for mineral concentration in plants related to mineral requirements of cattle. Crop Sci. 16:680-685.
92. ----, and L.E. Lanyon. 1983. Phosphorus fertilizer response in experimental alfalfas selected for different phosphorus concentrations. Crop Sci. 23:973-976.
93. Holt, D.A., and A.R. Hilst. 1969. Daily variation in carbohydrate content of selected forage crops. Agron. J. 61:239-242.
94. Horn, G.W., and W.M. Beeson. 1969. Effects of corn distillers dried grains with solubles and dehydrated alfalfa meal on the utilization of urea nitrogen in beef cattle. J. Anim. Sci. 28:412-417.
95. Hower, A.A., and R.A. Byers. 1977. Potato leafhoppers reduce alfalfa quality. Sci. Agric. 24:10-11.
96. Jensen, E.H., M.A. Massengale, and D.O. Chilcote. 1967. Environmental effects on growth and quality of alfalfa. Nev. Agric. Exp. Stn. Western Regional Res. Pub. T9.
97. Johnson, H.W. 1934. Nature of injury to forage legumes by the potato leafhopper. J. Agric. Res. 49:379-406.
98. ----. 1938. Further determinations of the carbohydrate-nitrogen relationship and carotene in leaf-hopper-yellowed and green alfalfa. Phytopathology 28:273-377.
99. Jung, G.A., R.L. Reid, and J.A. Balasko. 1969. Studies on yield, management, persistence, and nutritive value of alfalfa in West Virginia. West Va. Agric. Exp. Stn. Bull. 581T.
100. Jung, H.G., and G.C. Fahey, Jr. 1981. Effect of phenolic compound removal on in vitro forage digestibility. J. Agric. Food Chem. 29:817-820.
101. ----, and ----. 1983a. Nutritional implications of phenolic monomers and lignin: a review. J. Anim. Sci. 57:206-219.
102. ----, and ----. 1983b. Effects of phenolic monomers on rat performance and metabolism. J. Nutr. 113:546-556.
103. ----, and ----. 1983c. Interactions among phenolic monomers and in vitro fermentation. J. Dairy Sci. 66:1235-1263.
104. ----, ----, and J.E. Garst. 1983a. Simple phenolic monomers of forages and effects of in vitro fermentation on cell wall phenolics. J. Anim. Sci. 57:1294-1305.
105. ----, ----, and N.R. Merchen. 1983b. Effects of ruminant digestion and metabolism on phenolic monomers of forages. Br. J. Nutr. 50:637-651.
106. Kalu, B.A., and G.W. Fick. 1981. Quantifying morphological development of alfalfa for studies of herbage quality. Crop Sci. 21:267-271.
107. ----, and ----. 1983. Morphological stages of development as a predictor of alfalfa herbage quality. Crop Sci. 23:1167-1172.
108. Kehr, W.R., R.L. Ogden, and S.D. Kindler. 1970. Diallel analyses of potato leafhopper injury to alfalfa. Crop Sci. 10:584-586.
109. ----, ----, and ----. 1975. Management of four alfalfa varieties and control of damage from potato leafhopper. Univ. Nebr. Agric. Exp. Stn. Res. Bull. 275.
110. Kilcher, M.R., and D.H. Heinrichs. 1974. Contribution of stems and leaves to the yield and nutrient level of irrigated alfalfa at different stages of development. Can. J. Plant Sci. 54:739-742.
111. Kilmer, V.J., O.L. Bennett, V.F. Stahly, and D.R. Timmons. 1960. Yield and mineral composition of eight forage species grown at four levels of soil moisture. Agron. J. 52:282-285.

112. Kindler, S.D., W.R. Kehr, and R.L. Ogden. 1971. Influence of pea aphids and spotted alfalfa aphids on the stand, yield of dry matter and chemical composition of resistant and susceptible varieties of alfalfa. J. Econ. Entomol. 64:653–657.
113. Kindler, S.D., W.R. Kehr, R.L. Ogden, and J.M. Schalk. 1973. Effect of potato leafhopper injury on yield and quality of resistant and susceptible alfalfa clones. J. Econ. Entomol. 66:1298–1302.
114. Kivimae, Arnold. 1959. Chemical composition and digestibility of some grassland crops, with particular reference to changes caused by growth, season and diurnal variation. Acta Agric. Scand. Suppl. 5.
115. Kühbauch, W., and L. Pletl. 1981. Estimation of feed quality of white clover, red clover and lucerne using morphological criteria and/or chemical constituents of plants. I. Report: estimation of feed quality from stem length or from cell wall content of the stem. 2. Acker Pflanzenbase 150:271–280.
116. Leath, K.T., W.A. Kendall, and J.S. Shenk. 1977. *Phoma* leafspot and alfalfa quality. p. 46–47. *In* Rep. Alfalfa Improvement Conf., 26th, South Dakota State Univ., Brookings. 6–8 June. ARM-NC-7, Agricultural Reviews and Manuals, USDA-SEA, Washington, DC.
117. Lechtenberg, V.L., V.F. Colenbrander, L.F. Bauman, and C.L. Rhykerd. 1974. Effect of lignin on rate of *in vitro* cell wall and cellulose disappearance in corn. J. Anim. Sci. 39:1165–1169.
118. ----, D.A. Holt, and H.W. Youngberg. 1971. Diurnal variation in nonstructural carbohydrates, *in vitro* digestibility, and leaf to stem ratio of alfalfa. Agron. J. 63:719–724.
119. Leilla, F., N.A. Jorgensen, J.M. Scholl, H,W, Ream, and E.L. Jensen. 1977. In vivo and in vitro measures as predictors of the nutritive value of alfalfa and bromegrass hays. Wis. Agric. Exp. Stn. Res. Bull. R2903.
120. Liu, B.W.Y., and G.W. Fick. 1975. Yield and quality losses due to alfalfa weevil. Agron. J. 67:828–832.
121. Lofgreen, G.P., and W.N. Garrett. 1968. A system for expressing net energy requirements and feed values for growing and finishing beef cattle. J. Anim. Sci. 27:793–806.
122. Loper, G.M. 1968. Effect of aphid infestation on the coumestrol content of alfalfa varieties differing in aphid resistance. Crop Sci. 8:104–106.
123. ----, C.H. Hanson, and J.H. Graham. 1967. Coumestrol content of alfalfa as affected by selection for resistance to foliar disease. Crop Sci. 7:189–192.
124. Luckett, C.R., R.L. Ogden, T.J. Klopfenstein, and W.R. Kerr. 1967. Composition and digestibility of alfalfa leaf and stem fractions. J. Anim. Sci. 26:936.
125. MacLeod, L.B., and Michio Suzuki. 1967. Effect of potassium on the content of amino acids in alfalfa and orchardgrass grown with NO_3^- and NH_4^- nitrogen in nutrient solution culture. Crop Sci. 7:599–605.
126. Marten, G.C. 1970. Temperature as a determinant of quality of alfalfa harvested by bloom or age criteria. p. 506–509. *In* M.J.T. Norman (ed.) Proc. 11th Int. Grassl. Congr., Surfers Paradise, Queensland, Australia. 13–23 April. University of Queensland Press, St. Lucia.
127. ----. 1981. Chemical, in vitro, and nylon bag procedures for evaluating forage in the USA. p. 39–55. *In* J.L. Wheeler and R.D. Mochrie (ed.) Forage evaluation: Concepts and techniques. Proc. Workshop, Armidale, NSW, Australia. 27–31 Oct. 1980. American Forage Grassland Council, Lexington, KY.
128. ----. 1984. Analysis of alfalfa hay by near infrared reflectance spectroscopy. p. 27–30. *In* V.L. Marble (ed.) Proc. Natl. Alfalfa Hay Quality Testing Workshop, Chicago. 22-23 March. Agronomy Extension, Univ. of California, Davis.
129. ----, and R.N. Andersen. 1975. Forage nutritive value and palatability of 12 common annual weeds. Crop Sci. 15:821–827.
130. ----, and R.F Barnes. 1980. Prediction of energy digestibility of forages with in vitro rumen fermentation and fungal enzyme systems. p. 61–71. *In* W.J. Pigden et al. (ed.) Standardization Anal. Method. Feeds. Proc. Int. Workshop, Ottawa, Canada. 12–14 Mar. 1979. International Development Research Centre, Ottawa, Ontario.
131. ----, and R.D. Wilcoxson. 1974. Unpublished data, USDA-ARS, Univ. of Minnesota.
132. ----, D.L. Wyse, and C.C. Sheaffer. 1983. Nutritive value and palatability of alfalfa and smooth bromegrass compared to common perennial invaders. p. 170–172. *In* Proc. 1983 Forage and Grassl. Conf., Eau Claire, WI. 23–26 January. American Forage and Grassland Council, Lexington, KY.
133. Martz, F.A., M.R. Teel, C.L. Hill, and D.L. Hill. 1962. Plant N-K balance means more and better forage. Crops Soils 14(6):13–14.
134. Matches, A.G., W.F. Wedin, G.C. Marten, D. Smith, and B.R. Baumgardt. 1970. Forage quality of Vernal and DuPuits alfalfa harvested by calendar date and plant maturity schedules in Missouri, Iowa, Wisconsin, and Minnesota. Wis. Agric. Exp. Stn. Res. Rep. 73.

135. Mathur, R.B., and R.L. Peinkowski. 1967. Influence of adult meadow spittlebug [Philaenus spumarius] feeding on forage quality. J. Econ. Entomol. 60:207–209.
136. Matrone, G., R.L. Lovvorn, W.J. Peterson, F.H. Smith, and J.A. Weybrew. 1949. Studies of the effect of phosphate fertilization on the composition and nutritive value of certain forages for sheep. J. Anim. Sci. 8:41–51.
137. McNeil, M., P. Albersheim, L. Taiz, and R.L. Jones. 1975. The structure of plant cell walls. VII. Barley aleurone cells. Plant Physiol. 55:64–68.
138. Medler, J.T., and E.H. Fisher. 1953. Leafhopper [*Empoasca fabae*] control with methoxychlor and parathion to increase alfalfa hay production. J. Econ. Entomol. 46:511–513.
139. Meyer, J.H., and L.G. Jones. 1962. Controlling alfalfa quality. Calif. Agric. Exp. Stn. Bull. 784.
140. Mohrenweiser, H.W., and J.D. Donker. 1968. Forage evaluation. III. Comparison of several methods of evaluating two alfalfa hays fed to lactating cows. J. Dairy Sci. 51:373–377.
141. Moore, L.A., H.M. Irvin, and J.C. Shaw. 1953. Relationship between TDN [Total Digestible Nutrient content] and energy value of feeds. J. Dairy Sci. 36:93–97.
142. Morrison, I.M. 1974. Structural investigations on the lignin-carbohydrate complexes of *Lolium perenne*. Biochem. J. 139:197–204.
143. ----. 1979. Carbohydrate chemistry and rumen digestion. Proc. Nutr. Soc. 38:269–274.
144. ----. 1980. Changes in the lignin and hemicellulose concentrations of ten varieties of temperate grasses with increasing maturity. Grass Forage Sci. 35:287–293.
145. Mowat, D.N., R.S. Fulkerson, W.E. Tossell, and J.E. Winch. 1965. The in vitro digestibility and protein content of leaf and stem portions of forages. Can. J. Plant Sci. 45:321–331.
146. Neilson, M.J., and G.N. Richards. 1982. Chemical structures in a lignin-carbohydrate complex isolated from the bovine rumen. Carbohydr. Res. 104:121–138.
147. Nelson, C.J., and Dale Smith. 1968. Growth of birdsfoot trefoil and alfalfa. II. Morphological development and dry matter distribution. Crop Sci. 8:21–25.
148. Newby, V.K., R.M. Sablon, R.L.M. Synge, K.V. Casteele, and C.F. Van Sumere. 1980. Free and bound phenolic acids of lucerne (*Medicago sativa* cv. Europe). Phytochemistry 19:651–657.
149. Norris, K.H., R.F Barnes, J.E. Moore, and J.S. Shenk. 1976. Predicting forage quality by infrared reflectance spectroscopy. J. Anim. Sci. 43:889–897.
150. Oldfield, J.E., C.W. Fox, A.V. Bahn, E.M. Bickoff, and G.O. Kohler. 1966. Coumestrol in alfalfa as a factor in growth and carcass quality in lambs. Anim. Sci. 25:167–174.
151. Onstad, D.W., and G.W. Fick. 1983. Predicting crude protein, in vitro true digestibility, and leaf proportion in alfalfa herbage. Crop Sci. 23:961–964.
152. Parman, V.R., and M.C. Wilson. 1982. Alfalfa crop responses to feeding by the meadow spittlebug. J. Econ. Entomol. 75:481–486.
153. Pearson, C.J., and L.A. Hunt. 1972. Effects of temperature on primary growth of alfalfa. Can. J. Plant Sci. 52:1007–1015.
154. Poos, F.W., and H.W. Johnson. 1936. Injury to alfalfa and red clover by the potato leafhopper. J. Econ. Entomol. 29:325–331.
155. Reid, J.T., W.K. Kennedy, K.L. Turk, S.T. Slack, G.W. Trimberger, and R.P. Murphy. 1959. Symposium on Forage Evaluation: I. What is forage quality from the animal standpoint? Agron. J. 51:213–216.
156. Reid, R.L., and G.A. Jung. 1966. Factors affecting the intake and palatability of forages for sheep. p. 863–869. *In* Proc. 9th Int. Grassl. Congr., Sao Paulo, Brazil. 7–20 Jan. 1965. Edicoes Harico Limitada. Sao Paulo, Brazil.
157. Rendig, V.V., and W.C. Weir. 1957. Evaluation by lamb feeding tests of alfalfa hay grown on a low-sulfur soil. J. Anim. Sci. 16:451–461.
158. Richards, C.R., G.F.W. Haenlein, M.C. Calhoun, J.D. Connolly, and H.G. Weaver. 1962. Date of cut vs. the combination of crude fiber and crude protein as estimators of forage quality. J. Anim. Sci. 21:844–847.
159. Rohweder, D.A. 1984. Estimating forage hay quality. p. 31–37. *In* V.L. Marble (ed.) Proc. Natl. Alfalfa Hay Quality Testing Workshop, Chicago. 22–23 March. Agronomy Extension, Univ. of California, Davis.
160. Shaw, M.C., and M.C. Wilson. 1986. The potato leafhopper: scourge of leaf protein—and root carbohydrates too? p. 152–160. *In* Proc. 16th Natl. Alfalfa Symp., Fort Wayne, IN. 5–6 March. The Certified Alfalfa Seed Council, Davis, CA.
161. Sheaffer, C.C. 1983. Seeding year harvest management of alfalfa. Agron. J. 75:115–119.
162. ----, M.P. Russelle, O.B. Hesterman, and R.E. Stucker. 1986. Alfalfa response to potassium, irrigation, and harvest management. Agron. J. 78:464–468.
163. ----, and D.L. Wyse. 1982. Common dandelion (*Taraxacum officinale*) control in alfalfa (*Medicago sativa*). Weed Sci. 30:216–220.

164. Sherwood, R.T., A.F. Olah, W.H. Oleson, and E.E. Jones. 1970. Effect of disease and injury on accumulation of a flavonoid estrogen, coumestrol, in alfalfa. Phytopathology 60:684–688.
165. Singleton, P.C., and L.I. Pointer. 1961. Phosphate improves yield and nutritive value of alfalfa. Wyo. Agric. Exp. Stn. Bull. 380.
166. Smith, Dale. 1964. Chemical composition of herbage with advance in maturity of alfalfa, medium red clover, ladino clover, and birdsfoot trefoil. Wis. Agric. Exp. Stn. Res. Rep. 16.
167. ----. 1969. Influence of temperature on the yield and chemical composition of 'Vernal' alfalfa at first flower. Agron. J. 61:470–472.
168. ----. 1970a. Influence of temperature on the yield and chemical composition of five forage legume species. Agron. J. 62:520–523.
169. ----. 1970b. Yield and chemical composition of leaves and stems of alfalfa at intervals up the shoots. J. Agric. Food Chem. 18:652–656.
170. ----, M.L. Jones, R.F. Johannes, and B.R. Baumgardt. 1966. The performance of Vernal and DuPuits alfalfa harvested at first flower or three times by date. Wis. Agric. Exp. Stn. Res. Rep. 23.
171. ----, G.C. Marten, A.G. Matches, and W.F. Wedin. 1968. Dry matter yields of Vernal and DuPuits alfalfa harvested by calendar date and plant maturity schedules in Missouri, Iowa, Wisconsin, and Minnesota. Wis. Agric. Exp. Stn. Res. Rep. 37.
172. ----, and J.T. Medler. 1959. Influence of leafhoppers on the yield and chemical composition of alfalfa hay. Agron. J. 51:118–119.
173. ----, and B.E. Struckmeyer. 1974. Gross morphology and starch accumulation in leaves of alfalfa plants grown at high and low temperatures. Crop Sci. 14:433–436.
174. Smith, D.E., and W.A. Albrecht. 1942. Feed efficiency in terms of biological assays of soil treatments. Soil Sci. Soc. Am. Proc. 7:322–330.
175. Smith, L.W., H.K. Goering, and C.H. Gordon. 1972. Relationships of forage compositions with rates of cell wall digestion and indigestibility of cell walls. J. Dairy Sci. 55:1140–1147.
176. ----, ----, D.R. Waldo, and C.H. Gordon. 1971. In vitro digestion rate of forage cell wall components. J. Dairy Sci. 54:71–76.
177. Snaydon, R.W. 1972. The effect of total water supply, and frequency of application, upon lucerne. II. Chemical composition. Aust. J. Agric. Res. 23:253–256.
178. Stone, J.E., A.M. Scallan, E. Donefer, and E. Ahlgren. 1969. Digestibility as a simple function of a molecule of similar size to a cellulase enzyme. Adv. Chem. Ser. 95:219–241.
179. Talmadge, K.W., K. Keegstra, W.D. Bauer, and P. Albersheim. 1973. The structure of plant cell walls. I. The macromolecular components of the walls of suspension-cultured sycamore cells with a detailed analysis of the pectic polysaccharides. Plant Physiol. 51:158–173.
180. Tanner, G.R., and I.M. Morrison. 1983. The effect of saponification, reduction, and mild acid hydrolysis on the cell walls and cellulose treated cell walls of *Lolium perenne*. J. Sci. Food Agric. 34:137–144.
181. Temme, D.G., R.G. Harvey, R.S. Fawcett, and A.W. Young. 1979. Effects of annual weed control on alfalfa forage quality. Agron. J. 71:51–54.
182. Templeton, W.C., Jr. 1984. Background and recommendations of the National Alfalfa Hay Quality Committee for expressing nutritive value of alfalfa hay. p. 3–7. *In* Proc. Natl. Alfalfa Hay Quality Testing Workshop, Chicago. Agronomy Extension, Univ. of California, Davis.
183. Terry, R.A., and J.M.A. Tilley. 1964. The digestibility of the leaves and stems of perennial ryegrass, cocksfoot, timothy, tall fescue, lucerne, and sainfoin as measured by an *in vitro* procedure. J. Br. Grassl. Soc. 19:363–372.
184. Theander, O., and P. Åman. 1979. The chemistry, morphology, and analysis of dietary fiber components. *In* G.E. Inglett and S.I. Falkenhag (ed.) Dietary fibers: Chemistry and nutrition. Academic Press, New York.
185. ----, and ----. 1980. Chemical composition of some forages and various residues from feeding value determinations. J. Sci. Food Agric. 31:31–37.
186. ----, P. Uden, and P. Åman. 1981. Acetyl and phenolic acid substituents in timothy of different maturity and after digestion with rumen microorganisms or a commercial cellulase. Agric. Environ. 5:127–133.
187. Tomlin, D.C., R.R. Johnson, and B.A. Dehority. 1965. Relationship of lignification to in vitro cellulose digestibility of grasses and legumes. J. Anim. Sci. 24:161–165.
188. Troelsen, J.E., and J.B. Campbell. 1959. Nutritional quality of forage crops adapted to southwestern Saskatchewan as determined by their digestibility and dry matter intake when fed to sheep. Can. J. Plant Sci. 39:417–430.
189. ----, and ----. 1969. The effect of maturity and leafiness on the intake and digestibility of alfalfas and grasses fed to sheep. J. Agric. Sci. 73:145–154.

190. Van Soest, P.J. 1964. Symposium on nutrition and forage and pastures: new chemical procedures for evaluating forages. J. Anim. Sci. 23:838–845.
191. ----. 1973. The uniformity and nutritive availability of cellulose. Fed. Proc. 32:1804–1808.
192. ----. 1981. Limiting factors in plant residues of low biodegradability. Agric. Environ. 6:135–143.
193. ----. 1982. Nutritional ecology of the ruminant. O & B Books, Corvallis, OR.
194. ----, D.R. Mertens, and B. Deinum. 1978. Preharvest factors influencing quality of conserved forage. J. Anim. Sci. 47:712–720.
195. ----, and J.B. Robertson. 1980. Systems of analysis for evaluating fibrous feeds. p. 49–60. In W.J. Pigden et al. (ed.) Standardization Anal. Method. Feeds. Proc. Int. Workshop, Ottawa, ON, Canada. 12–14 Mar. 1979. International Development Research Centre, Ottawa, ON.
196. Varga, G.A., and W.H. Hoover. 1983. Rate and extent of neutral detergent fiber degradation of feedstuffs in situ. J. Dairy Sci. 66:2109–2115.
197. Vough, L.R., and G.C. Marten. 1971. Influence of soil moisture and ambient temperature on yield and quality of alfalfa forage. Agron. J. 63:40–42.
198. Waldo, D.R. 1985. Regulation of forage intake in ruminants. p. 233–237. In R.F Barnes et al. (ed.) Proc. (Trilateral Workshop) Forage Legumes for Energy Efficient Animal Production, Palmerston North, New Zealand. 30 April–4 May 1984. USDA-ARS, Washington, DC.
199. ----, and N.A. Jorgensen. 1981. Forages for high animal production: Nutritional factors and effects of conservation. J. Dairy Sci. 64:1207–1229.
200. ----, L.W. Smith, and E.L. Cox. 1972. Model of cellulose disappearance from the rumen. J. Dairy Sci. 55:125–129.
201. ----, D.J. Thomson, H.K. Goering, and H.F. Tyrrell. 1982. The voluntary intake, growth rate, and tissue retention of cattle fed grass or legume silage. Proc. Nutr. Soc. 41:24A.
202. Walgenbach, R.P., and G.C. Marten. 1981. Release of soluble protein and nitrogen in alfalfa. II. Influence of shading. Crop Sci. 21:859–862.
203. ----, ----, and G.R. Blake. 1981. Release of soluble protein and nitrogen in alfalfa. I. Influence of growth temperature and soil moisture. Crop Sci. 21:843–849.
204. Walstrom, R.J., P.A. Jones, and G.F. Gastler. 1970. Effect of phorate for partial control of alfalfa weevil on nutritional values of alfalfa hay. J. Econ. Entomol. 63:1374–1375.
205. Washko, J.B., and J.W. Price. 1970. p. 628–632. In Intensive management of alfalfa for forage production. Proc. 11th Int. Grassl. Congr., Surfers Paradise, Queensland, Australia. University of Queensland Press, St. Lucia.
206. Wedin, W.F., A.W. Burger, and H.L. Ahlgren. 1956. Effect of soil type, fertilization, and stage of growth on yield, chemical composition, and biological value of ladino clover (*Trifolium repens* L.) and alfalfa (*Medicago sativa*). Agron. J. 48:147–152.
207. Weir, W.C., and V.V. Rendig. 1952. Studies on the nutritive value for lambs of alfalfa hay grown on a low sulfur soil. J. Anim. Sci. 11:780 (Abstr.).
208. Whiting, P., and D.A. I. Goring. 1982a. Chemical characterization of tissue for the middle lamella and secondary wall of black spruce tracheids. Wood Sci. Technol. 16:261–267.
209. ----, and ----. 1982b. Relative reactivities of middle lamella and secondary wall lignin of black spruce wood. Holzforschung 36:303–306.
210. Wilkie, K.C.B. 1979. The hemicelluloses of grasses and cereals. Adv. Carbohydr. Chem. Biochem. 36:215–264.
211. Wilkins, R.J. 1969. The potential digestibility of cellulose in forage and faeces. J. Agric. Sci. 73:57–64.
212. Willis, W.G., D.L. Stuteville, and E.L. Sorensen. 1969. Effects of leaf and stem diseases on yield and quality of alfalfa forage. Crop Sci. 9:637–640.
213. Wilman, David, and M.A.K. Altimimi. 1984. The in-vitro digestibility and chemical composition of plant parts in white clover, red clover, and lucerne during primary growth. J. Sci. Food Agric. 35:133–138.
214. Wilson, M.C., and V.R. Parman. 1982. Economic thresholds for the meadow spittlebug on alfalfa. Forage Grassl. Progress 23:2–4.
215. ----, J.K. Stewart, and H.D. Vail. 1979. Full season impact of the alfalfa weevil, meadow spittlebug, and potato leafhopper. J. Econ. Entomol. 72:830–834.
216. Winch, J.E., R.W. Sheard, and D.N. Mowat. 1970. Determining cutting schedules for maximum yield and quality of bromegrass, timothy, lucerne, and lucerne/grass mixtures. J. Br. Grassl. Soc. 25:44–52.
217. Woodman, H.E., and R.E. Evans. 1935. Nutritive value of lucerne. IV. The leaf-stem ratio. J. Agric. Sci. 25:578–597.
218. Woodward, F.I., and J.E. Sheehy. 1979. Microclimate, photosynthesis and growth of lucerne (*Medicago sativa* L.). II. Canopy structure and growth. Ann. Bot. 44:709–719.
219. Zemek, J., B. Kosikova, J. Augustin, and D. Joniak. 1979. Antibiotic properties of lignin components. Folia Microbiol. 24:483–486.

15 Antiquality Factors and Nonnutritive Chemical Components

R.E. HOWARTH

Agriculture Canada, Research Station
Saskatoon, Canada

Among legume and nonlegume forage crops, alfalfa (*Medicago sativa* L.) ranks near the top in nutritive value, particularly for ruminant animals. Nevertheless, antiquality factors in alfalfa (Table 15-1) have received considerable attention because of the economic importance and wide geographic adaptation of the crop. The antiquality factors have stimulated study of the nonnutritive chemical constituents of alfalfa in order to identify undesirable substances, and to develop methods for removing them by plant breeding, crop management, animal management, as well as by secondary processing of the crop. From 1950 to 1960, much attention was devoted to the saponins and estrogenic constituents. In recent

Table 15-1. The antiquality constituents of alfalfa herbage.

Antiquality constituents	Major effects	References
Calcium/P ratio	High Ca/P ratio impairs P metabolism.	--
Chlorphyll	Degradation products in leaf protein concentrate may cause photosensitization.	68, 108
o-Dihydroxyphenols	Oxidation to o-quinones may impair the quality of leaf protein concentrate.	62, 87, 88
Isoflavonoids	Estrogenic activity occasionally causes infertility in sheep and cattle.	6, 86, 103
Lignin	Reduced feed intake and digestibility as herbage matures.	32, 33, 38
Nitrate	Levels may be above 0.2% but toxicity is not normally a problem.	5
Proteins	Frothy bloat in ruminants.	44, 74, 76, 77
	Protease inhibitors may inhibit growth of monogastric animals.	22, 23, 94
Saponins	Reduced feed intake and growth in monogastric animals, especially poultry. Reduced egg production.	9, 10, 15, 93
	Proposed role in ruminant bloat is in doubt.	65, 75
Unknown	Allergic reactions.	5

Copyright 1988 © ASA-CSSA-SSSA, 677 South Segoe Road, Madison, WI 53711, USA.
Alfalfa and Alfalfa Improvement—Agronomy Monograph no. 29.

years, the proteins and phenolic constituents have received attention because of their significance in ruminant bloat and leaf-protein extraction.

In a review of toxic constituents in legume forage crops, Smolenski et al. (102) listed 14 toxic constituents and livestock diseases. Several of these well-known toxic constituents, namely, lathyrogenic amino acids, cyanogens, nitro compounds, and Se compounds, will not be considered in this chapter because they do not occur at toxic levels in alfalfa. Similarly, livestock diseases caused by mycotoxin contaminants, aflatoxicosis, lupinosis, the slobber factor, and sweetclover disease, do not occur in alfalfa.

15-1 HERBAGE PROTEINS

Chemical characterization of herbage proteins normally begins with homogenization of fresh tissue in neutral buffer solutions, and subsequent separation into soluble and insoluble fractions by centrifugation. In herbage from immature alfalfa, containing about 20% crude protein, the soluble fraction contains 45 to 50% of the total N, and 50 to 55% of the total N is insoluble. Approximately, 45 to 50% of soluble N is protein N, and 40 to 45% is nonprotein N.

The insoluble fraction contains cell walls and membranes. The chloroplast membrane fragments of herbage contain approximately 50% protein (73). Although these alfalfa proteins have not been well characterized, they are known to be hydrophobic because they have a structural association with the membrane lipids. The membrane proteins can be extracted with either phenol-acetic acid-water or with detergent solutions. For example, they are dissolved in the neutral detergent fiber procedure, which is used frequently in the analysis of forages (95). After dissolution of the membrane proteins, the residual protein is that associated with the cell wall. Cell-wall protein represents approximately 5% of total herbage N.

Soluble herbage proteins are classified into two groups, fraction 1 protein and fraction 2 protein, on the basis of molecular weight. Traditionally, these proteins were called *cytoplasmic proteins*, but in the intact cell, the fraction 1 protein, and probably some of the fraction 2 proteins, are located in the stroma region of chloroplasts.

Fraction 1 protein is a single large protein with molecular weight in excess of 500 000 Da (79). It is the principal soluble protein in chloroplasts and constitutes 20 to 50% of the soluble herbage proteins (46, 73). Fraction 1 protein is the enzyme, ribulose-5-diphosphate carboxylase, which is responsible for CO_2 fixation in photosynthesis. Noguchi et al. (89) have described the physical characteristics of alfalfa fraction 1 protein. McArthur et al. (79) identified alfalfa fraction 1 protein as 18S protein, in reference to its sedimentation coefficient in the ultracentrifuge. This terminology has received wide use in the literature on ruminant bloat (85), but in recent years the fraction 1 terminology has been preferred.

Fraction 2 proteins are a mixture of soluble proteins. The alfalfa fraction 2 proteins have been characterized by electrophoresis (47). Up to 17 discrete bands are easily identifiable, and many other very faint bands are present.

Interest in commercial production of alfalfa leaf protein concentrates and ruminant bloat has prompted a number of investigations into the solubility and other characteristics of alfalfa fraction 1 and 2 proteins (3, 4, 28, 42, 53, 59, 72, 99, 116). They have isoelectric points and minimum solubilities in the range of pH 3 to 6.5. The effects of several environmental variables on soluble protein concentrations in alfalfa have been described (112, 113, 114). Maximum soluble protein concentration occurred in alfalfa grown at day/night temperatures of 26/18°C. Nitrogen fertilizer increased soluble protein while K fertilizer and moisture stress (at 26/18°C) had no effect, and shading decreased soluble protein.

15-2 RUMINANT BLOAT

Ruminant bloat is usually associated with the consumption of fresh alfalfa. It is encountered less frequently in feeding alfalfa hay, and rarely in feeding silage and haylage. Processed alfalfa pellets, that are normally used as supplemental feedstuffs, are not a significant factor in causing bloat.

Bloat is an acute, digestive disorder of ruminant animals. The rumen is distended by gas retained within the rumen contents. Distension of the rumen exerts pressure on the circulatory and pulmonary systems causing severe stress and sometimes death of the animal. The gas, which is produced by microbial fermentation in the rumen, remains dispersed throughout the rumen contents and cannot be eliminated in the usual manner.

The bloat-causing potential of alfalfa is greatest during the vegetative growth stage, and subsequently decreases to low or insignificant levels at the early bloom stage of development. Outbreaks of bloat are associated with lush rapid growth, cool night temperatures, and moderate day temperatures. Although livestock producers frequently associate bloat with the consumption of herbage that is wet from rain or dew, bloat can and does occur at other times.

In cattle (*Bos* spp.), average death losses due to pasture bloat are approximately 1% of the animals grazing legume pastures. It is the risk of more severe losses that is the major deterrent to the use of pastures containing a large proportion of alfalfa. The costs of bloat include management time, cash expenditures for either prevention or treatment, and reduced feed intake in nonlethal cases of bloat. Also, graziers depending on grass pastures have the added cost of N fertilizer to maintain herbage production.

In seeded pastures, a mixture of alfalfa and grass prevents bloat and provides some of the benefits of a legume. Mixtures are important in the production of beef cattle, particularly under the extensive grazing conditions of the Great Plains region. When pastures contain a high proportion of alfalfa, the daily administration of antibloat substances are necessary to prevent bloat, especially when the forage is lush and immature. Mineral oil, vegetable oils, and certain synthetic detergents will prevent bloat. Poloxalene, an example of an antibloat detergent, may be administered in supplemental grain, molasses licks, or salt-molasses blocks. In general, the use of antibloat substances is restricted to irrigated or to small, intensively managed pastures.

15-2.1 Foaming Agents

Following the recognition that pasture bloat is caused by frothiness of the rumen contents (52), much research has been devoted to the identification of the foaming agents responsible for this condition. Emphasis has been given to surface-active foaming agents that include saponins, soluble-forage proteins, saliva proteins, and proteins from rumen microbes (18).

15-2.2 Saponins

Alfalfa saponins were the objective of a major investigation by Lindahl et al. (66) who concluded that alfalfa saponins might contribute to alfalfa pasture bloat. Later Cheeke (15) reviewed the evidence for saponins as a cause of bloat. However, when high- and low-saponin strains of alfalfa were fed to cattle, the two alfalfa strains produced similar frequencies of bloat (76) indicating that alfalfa saponins were not the cause of pasture bloat. Although bloat and frothy rumen contents can be produced by the administration of saponins to experimental animals, convincing evidence that they cause bloat in commercial herds is lacking.

15-2.3 Soluble Proteins

Several research groups have implicated the soluble herbage proteins in pasture bloat (18). McArthur et al. (79), identified 18S or fraction 1 protein as the principal foaming agent and bloat-causing factor in alfalfa. Subsequently, Howarth et al. (44) demonstrated that alfalfa fraction 2 proteins also will stabilize foams. They concluded that the entire soluble protein fraction rather than fraction 1 protein alone is a potential cause of rumen frothiness and bloat. In rumen fluid from cattle fed fresh alfalfa, Majak et al. (77) found that soluble protein concentrations were adequate to stabilize foams, but they did not find any relationship with the occurrence of bloat. The data suggest that other substances must be involved with the immediate onset of bloat. Jones and Lyttleton (54) implicated complexes of herbage lipids, soluble forage proteins, and modified salivary proteins in legume pasture bloat. Recently, Majak et al. (74, 75, 77) have suggested that chloroplast membrane fragments, which contain approximately 50% protein and 50% lipid, contribute to bloat. Because the frothy condition of ruminal contents is a complex mixture of surface-active substances, as well as particulate material, it is impossible, and perhaps unwise to identify a single substance as the foaming agent responsible for bloat. The soluble alfalfa proteins together with the protein in chloroplast membrane fragments, constitute a major portion of the total protein in alfalfa herbage. Therefore, the protein constituents of the herbage provide the best simplified explanation for the bloat-causing potential of alfalfa herbage.

15-2.4 Rate of Digestion

Comparisons of bloat-inducing and nonbloating legumes have added to our understanding of alfalfa bloat. Because bloat occurs within a few hours after feeding, the initial digestion of the ingested herbage may be a factor in the occurrence of bloat. Rumen microorganisms digest bloat-inducing legumes, including alfalfa, more rapidly than nonbloating legumes (45). This may result, in part, from a thinner epidermis and mesophyll cell walls in leaves of alfalfa and other bloat-inducing legumes (64). In addition, condensed tannins, which are very effective inhibitors of rumen microorganisms, are found in a number of nonbloating legumes, e.g., birdsfoot trefoil (*Lotus corniculatus* L.), sainfoin (*Onobrychis viciifolia* Scop.), crownvetch (*Coronilla varia* L.), and sericea lespedeza [*Lespedeza cuneata* (Dum.) G. Don.], but not in alfalfa.

15-2.5 Development of Bloat-safe Alfalfa

The excellent nutritive value, high yield, N_2 fixation and other desirable agronomic traits of alfalfa, make it the preferred legume forage crop over wide areas of the temperate crop zones of the world. However, the risk of bloat remains a problem, restricting the use of alfalfa in grazing systems.

The prospect of breeding a bloat-safe alfalfa cultivar has been considered by several alfalfa breeders. In the early 1970s, Agriculture Canada undertook a major commitment to the development of a bloat-safe alfalfa cultivar. The absence of an effective, inexpensive screening method for isolating bloat-safe plants was a major obstacle that had to be resolved. Goplen et al. (29) considered several available options. These included selection for fraction 1 (18S) protein, total soluble proteins, foam volume, tannins, initial rate of digestion, mechanical strength of mesophyll cell walls, enzyme digestion, rates of gas production during in vitro fermentation, and dry matter loss by leaching.

The initial rate of digestion using a modified in situ bag digestion procedure was identified as the most promising approach to breeding a bloat-safe alfalfa strain. In this method, selection would be practiced for a slower initial rate of digestion (43). After one cycle of selection, alfalfa strains selected for slow and fast initial rates of digestion showed these same traits when fed to sheep (*Ovis aries*) (60). Continued divergence in subsequent cycles of selection for slow initial rate of digestion has improved the prospects of developing a bloat-safe cultivar.

The expense and time required for repeated assays are disadvantages of the nylon bag digestion procedure. From a genetic viewpoint, the preferred approach to a bloat-safe alfalfa would involve the introduction of the gene or genes for the biosynthesis of condensed tannins in alfalfa herbage (20). However, this would require an intergeneric transfer, using modern biotechnology methods because all *Medicago* spp. lack condensed tannins in their herbage (30). Several biotechnology research laboratories have shown interest in this problem, but to date, the intergeneric transfer of tannins has not been achieved.

15-3 PHENOLIC CONSTITUENTS

A large number of aromatic plant constituents can be grouped into a single class based on the presence of a C_6C_3 unit, phenylpropane, or a derivative of this structure (37). They range from a single aromatic ring structure, e.g., fumaric acid to complex polymeric structures formed by condensation of the simple structures. The polymeric structures include tannins and lignin, both of which have important effects on the nutritive value of forages. The phenolic compounds also play important roles in protection against plant diseases and pest insects.

15-3.1 Benzoic Acids

Three benzoic acids, *p*-hydroxybenoic, vanillic, and salicylic, occur in alfalfa herbage (88). Vanillic and *p*-hydroxybenzoic acids are present both free and bound to other constituents. The bonds to other constituents are both alkali- and acid-labile linkages indicating ester and amide bonds. Salicylic acid occurs only in the bound form. Newby et al. (88) also reported three cinnamic acids, *p*-coumaric acid, ferulic acid, and sinapic acid in alfalfa herbage. Ferulic and *p*-coumaric acids were both free and bound. Sinapic acid was present only in the alkali-labile bound form. *N*-feruloylglycylphenylalanine has been isolated from alfalfa leaf protein concentrate by partial acid hydrolysis (111).

15-3.2 Chlorogenic Acid

Chlorogenic acid, an ester of caffeic acid with quinic acid, was found in alfalfa protein concentrates (28, 61, 62). Monties and Rambourg (87) detected only trace amounts of chlorogenic acid in their leaf protein preparations. Newby et al. (88) in an extensive fractionation of alfalfa herbage, did not detect either chlorogenic acid nor caffeic acid in alfalfa leaves. Hudson and Mahgoub (48) did not detect caffeic acid. Quinic and shikimic acids have been reported in alfalfa (5). It appears that alfalfa may contain relatively low concentrations of chlorogenic acid and that the levels of this acid varies among cultivars.

15-3.3 Coumarins

Coumarins have an aromatic ring structure formed by an internal lactone of the phenylpropane (C_6C_3) structure. These volatile substances occur widely in forage crops, contributing to the characteristic odor of freshly cut hay. Umbelliferone, (7-hydroxycoumarin) is the most widely distributed coumarin. There do not appear to be any reports of coumarins in alfalfa.

15-3.4 Flavonoids

Flavonoids, an important class of phenolic compounds, are characterized by a $C_6C_3C_6$ structure that is formed by a condensation of the phenylpropane structure with an aromatic ring structure derived from the shikimic acid biosynthetic pathway. Flavonoid compounds include anthocyanidins, flavones, flavonols, flavanols, and others.

15-3.5 Anthocyanidins

The anthocyanidins are the plant pigments responsible for flower color. Three anthocyanidins, i.e., delphinidin, petunidin, and malvidin, occur as glycosides in alfalfa (5). A variable number of glycosides have been reported, which may be a consequence of partial hydrolysis during isolations. Sarkar and Howarth (98) identified three anthocyanidins, i.e., cyanidin, pelargonidin, and peonidin, in alfalfa herbage. These are probably responsible for the red color occasionally observed in alfalfa petioles, stems, and roots.

15-3.6 Flavones

Five flavones have been identified in alfalfa. They are tricin, apigen, crysoeriol, 4',7-dihydroxyflavone, and 3',4',7-trihydroxyflavone (5, 97). Alfalfa flavones occur as 7-glucosides and 7-glucuronides (97). The concentrations of 4',7-dihydroxyflavone in alfalfa herbage increased in response to fungal infection or exposure to ozone (49, 101). Tricin caused smooth muscle relaxation when applied to guinea pig (*Cavia* spp.) intestinal strips (5).

15-3.7 Favonols

Hudson and Mahgoub (48) identified two flavonols, quercetin and myricetin, in alfalfa leaves. Conversely, Saleh et al. (97) failed to detect flavonols in alfalfa herbage.

15-3.8 Flavanols

Catechins (flavan-3-ols) and flavan-3,4-diols frequently occur together. Because flavanols are the monomeric units of condensed tannins they have been investigated in relation to the occurrence of bloat. Milić (81) reported the occurrence of catechin in alfalfa herbage, while other investigators obtained negative tests for catechins (30, 54, 96, 98). Flavanols produce a cherry-red color upon reaction with vanillin-HCl, but a false positive reaction can be obtained from the red color of anthocyanidins in HCl (98). Thus is would appear that alfalfa herbage does not contain flavolans, although they are found in the seed coat of alfalfa (30).

15-3.9 Isoflavonoids

The isoflavonoid compounds are related closely to the flavonoids but differ in having a branched $C_6C_3C_6$ skeleton. Alfalfa contains two isoflavonoid groups, the isoflavones and coumestans, that are known for their estrogenic activity in farm animals. Four isoflavones have been found in alfalfa, namely, daidzein, formononetin, genistein, and biochanin A (5). Nine coumestans have been identified in alfalfa, namely, 7,12-dihydroxycoumestan (coumestrol), 4'-0-methylcoumestrol, 3'- methoxycoumestrol, lucernol, medicagol, sativol, trifoliol, and 11,12-dimethoxy-7-hydroxycoumestan (5), and more recently, a 3-hydroxy-7-9-dimethoxycoumestan, named *wairol* (7).

Two other isoflavonoid groups, pterocarpans and isoflavans, occur in alfalfa herbage in response to fungal infection or they can be induced by treatment with cupric chloride. These include pterocarpan and medicarpin, as well as the two isoflavans sativan and vestitol (24, 25, 50, 51, 58). Pterocarpans and isoflavans represent the most abundant classes of isoflavonoid phytoalexins produced by leguminous plants. They are thought to play an important role in disease resistance.

15-4 TANNINS

The term *tannin* must be defined because various definitions have been used in the literature on chemical composition of agricultural products. The preferred definition is that of Swain and Bate-Smith (107), who defined tannins as water-soluble phenolic compounds that have molecular weights from 500 to 3000 Da and the capability to precipitate proteins. Two groups of natural products have these properties, (i) the condensed tannins which are polymeric forms of flavanols (catechin and flavan-3,4-diol), and (ii) the hydrolyzable tannins which are esters of a sugar (usually glucose) and polyhydroxyphenolic acids. Upon hydrolysis, which is relatively easy, the latter group yields the sugar, gallic acid, and one or more derivatives of gallic acid (e.g., ellagic acid and chebulic acid).

The significance of tannins in animal feeding has provided an incentive to find tannins in alfalfa. Direct tests for protein precipitants in alfalfa herbage have been negative (34, 53). In addition, most investigations have failed to detect flavanols in alfalfa herbage (1, 30, 48, 53, 87, 96, 98). On the other hand, scientists in Yugoslavia found catechins, gallic acid, and gallotannins in the alfalfa cv. Panonia (21, 22, 81, 82, 83). These five reports used the same methods for identification of the plant constituents. In our laboratory, we were unable to confirm the presence of condensed tannins in Panonia alfalfa.

On the basis of these reports, one must conclude that alfalfa herbage does not contain tannins as defined above. Some authors have used the term tannin with reference to total phenolic substances, including the low-molecular-weight constituents that do not precipitate proteins. These constituents represent 2 to 3% of alfalfa herbage dry matter (5, 84).

The plant phenolic compounds, including those found in alfalfa, have a strong affinity for proteins (110). Both the low-molecular-weight monomeric substances (often called *polyphenols*) and the polymeric tannins, bind to the proteins by hydrogen bonding between the phenolic hydroxyl groups and the amide groups in the peptide linkages of proteins. Tannins precipitate proteins because they are large enough to form cross-linkages between protein molecules. The low-molecular-weight phenols bind to proteins without inducing precipitation. Although both tannins and low-molecular-weight phenols are classified as nonspecific enzyme inhibitors, the protein precipitating activity of tannins makes them particularly effective. The plant phenols serve as plant defense mechanisms against invasion by microorganisms and against feeding by herbivores (106). In addition, they may reduce digestibility of plant foodstuffs by inactivation of enzymes in the digestive tracts of herbivores and other animals.

15-4.1 Oxidation of Tannins and Phenols

Phenolic substances are readily susceptible to oxidation and the oxidation products form covalent linkages to amino acid side chains in proteins. Lysine and methionine side chains are particularly vulnerable. In this manner, phenols reduce the availability of essential amino acids to monogastric animals and so reduce the biological value of proteins. Covalent cross-links caused by tannins also reduce the digestibility of proteins.

Oxidation of phenols may occur either by enzyme activity or by nonenzymatic mechanisms. Enzymatic oxidation is catalyzed by phenolase (phenol oxidase), for which chlorogenic acid is a preferred substrate. *o*-Quinones produced by the oxidation of *o*-dihydroxyphenols react rapidly and nonenzymatically with proteins. This explains the interest (62, 87, 88) in the amount of phenolics bound to alfalfa leaf protein concentrates. The finding that chlorogenic acid and other *o*-dihydroxyphenols were absent or present in relatively low amounts indicates that enzymatic oxidation of phenols is unlikely to be important in impairing the nutritive quality of alfalfa leaf protein (88). Similarly, the absence of tannins in alfalfa contributes to the relatively high yields of alfalfa leaf protein concentrate.

Nonenzymatic oxidation of phenols may occur during high-temperature drying of alfalfa products but it does not appear to be of practical importance, perhaps because of the relatively low level of phenols in alfalfa (2-3%). Compared to high-temperature drying, freeze-drying of alfalfa herbage did not improve the protein digestibility of alfalfa meal fed to rats (*Rattus* spp.) (16) and to turkey poults (*Meleagris gallopavo*) (41).

15-4.2 Potential Benefits of Tannins

Plant tannins are usually considered to be antiquality factors because of their negative effects on protein quality and nutrient digestibility. Nevertheless, forage tannins have potential beneficial value in ruminant nutrition (109). They are effective antibloat substances, possibly associated with a reduction in the initial rate of digestion or through the precipitation of foaming agents in rumen fluid. All tannin-containing legume forages are nonbloating. In addition, tannins may contribute to rumen by-pass of protein thereby enhancing the efficiency of protein utilization in ruminant animals. In this situation, the tannin-protein complex would by-pass or escape microbial digestion in the rumen, with protein becoming available for digestion in the lower intestinal tract where an alkaline pH disrupts the tannin-protein complex. For these reasons, and possibly to aid in resistance to insect infestations, a low to moderate content of tannins in alfalfa would be a desirable trait. Several investigators are studying the possibilities of introducing tannins into the alfalfa plant.

15-5 LIGNIN

Lignin is an insoluble, amorphous macromolecule. The basic structural unit of lignin is the phenylpropane (C_6C_3) unit. In the biosynthesis of lignin, the phenylpropane structure is provided by three cinnamic acids, i.e., p-coumaric, ferulic, and sinapic acids. These monomeric units are joined together by C-to-C and strong ether linkages that are highly resistant to chemical hydrolysis. The proportions of the three monomeric acids vary among species, and perhaps even among different cell structures within a species. Cross-linkages are formed with carbohydrates, proteins, and other plant constituents so that lignin encrusts and strengthens the more fragile polysaccharide constituents of the plant cell wall.

Typical lignin concentrations in alfalfa herbage are 5 to 14% with a range of 3 to 8% in leaves; and 6 to 15% in stems. Lignin concentrations in alfalfa tend to be higher than in the temperate grasses (38). The structure of alfalfa lignin has not been characterized. Herbage lignins contain N, perhaps as a consequence of cross-linkages with protein. Nitrogen contamination of herbage lignin preparations is common because of high-protein levels in herbage, compared to most other ligno-cellulosic materials. The original purpose of the acid-detergent fiber procedure (95) was the removal of protein and other interfering substances, in the preparation of forages for lignin analysis.

Freudenburg and Harkin used a combination of ball milling, solvent extraction, and acid hydrolysis to obtain alfalfa lignin (cited in 38). Compared to grass lignin, alfalfa lignin was more highly condensed, with fewer side chains for interaction with other plant constituents (31, 32). Compared to wheat straw lignin, alfalfa lignin is less soluble in 10% KOH and the alkali soluble lignin contains fewer linkages to hemicellulose (33).

Plant fiber digestibility decreases with increasing lignin content. Lignin prevents fiber digestion by protection of cellulose from enzyme activity, and by the prevention of swelling of the fiber to a condition suitable for penetration by microbial polysaccharidases (38). In ruminant digestion the unfermentable cellulose and hemicellulose amounts to about 2.5 times the lignin content of the forage. These carbohydrates become available to microbial fermentation if the lignin carbohydrate bond is broken by pretreatment of fibers with alkali (95). Acid labile lignin-carbohydrate linkages may be cleaved in silage (38). Gordon (32, 33) suggested that alfalfa lignin reduces the digestibility of plant fiber to a lesser degree (per unit weight of lignin) than grass lignin.

15-6 SAPONINS

Saponins are plant glycosides which, on hydrolysis, yield sapogenins and sugars. The diverse biochemical and biological activities of saponins have encouraged the characterization of alfalfa saponins, and several excellent reviews have been published (9, 10, 15, 36). The alfalfa sapogenins have pentacyclic triterpenoid structures. The sapogenins reported in alfalfa include the following: soyasapogenols A, B, C, D, and E, medicagenic acid, hederogenin (78), and lucernic acid (2). The various sugars released by hydrolysis of mixed alfalfa saponins include glucose, arabinose, xylose, galactose, rhamnose, glucuronic acid, and galacturonic acid (78). Alfalfa may contain up to 33 different saponins (2), but only a few have received detailed structural analysis (78).

Alfalfa herbage contains 2 to 3% saponins. The concentration is low in spring and fall and peaks in mid-summer (91) when temperatures are high and soil moisture low. Stems contain about 0.5% saponins; flowers and roots 2.5 to 3.5%; and seeds 1.2 to 1.5% (115). Qualitative differences in chemical structure occur among saponins found in herbage, flowers, and roots (78). Leaves of 'Du Puits' had nearly twice the saponin concentration of 'Lahontan' (36).

Saponins have a wide range of biological and physiological effects that result from the bipolar nature of saponin molecules. Their interactions with both hydrophyllic and hydrophobic substances results from the combination of both sapogenin (hydrophobic) and carbohydrate (hydrophyllic) moities in the saponin molecule. Saponins also form complexes with cholesterol and other hydroxy steroids.

15-6.1 Responses of Animals

Cheeke (15) reviewed the physiological effects of saponins in animals, most of which are detrimental. Saponins are highly toxic to fish and amphibians. Fish toxicity and the characteristic lysis of red blood cells (hemolytic activity) have provided useful bioassays for the isolation and characterization of alfalfa saponins. Inhibition of growth, enzyme activities, smooth muscle activity, and absorption of nutrients have been attributed to saponins. Medicagenic acid is the primary toxic sapongenin in alfalfa saponins (36). Formation of a cholesterol-saponin complex in the gut may reduce levels of blood and tissue cholesterol in monogastric animals. This effect has been investigated for possible value in the prevention of atherosclerosis.

Poultry are more sensitive than other farm animals to the toxic effect of alfalfa saponins. Inclusion of alfalfa meal at levels of around 10% in poultry rations causes pronounced inhibition of chick (*Gallus gallus domesticus*) growth (10) and reduces egg production in layers (9). The effect can be prevented by feeding 1% cholesterol. Growth of chicks and rats was greater when fed low-saponin alfalfa meal compared to normal or high saponin alfalfa meal (92). The depressed egg production in layers fed 20% alfalfa meal disappeared soon after alfalfa was withdrawn from the ration (40).

Levels of 20% alfalfa have been incorporated into swine (*Sus scrofa domesticus*) growing rations with no adverse effects on growth (15). Alfalfa saponins are not known to affect growth of ruminants. No detrimental effect was observed when alfalfa hay containing either 2.62 or 1.72% saponin was fed to calves (8).

The biochemical mechanism for the growth-depressing effects of alfalfa saponins are not known. Cheeke (15) concluded that the adverse effects in poultry fed 20% alfalfa meal are almost entirely a response to alfalfa saponins, rather than high fiber levels. Inhibition of cellular enzymes or digestive enzymes or inhibition of feed intake are possible mechanisms.

Lindahl et al. (65) produced experimental bloat in sheep by oral and by intravenous administration of a composite mixture of alfalfa saponins, but the levels administered were high compared to normal intakes of alfalfa saponins. For many years, saponins have been considered as a possible cause of bloat from alfalfa. Several mechanisms have been proposed to explain the effects of saponins, namely, as a foaming agent in the production of frothy rumen contents, as a toxic inhibitor of rumen motility and eructation, and as a source of viscous slime from the degradation of saponins by rumen microorganisms. High- and low-saponin, near-isogenic strains of alfalfa, containing 1.94 and 0.82% saponin, respectively, produced similar frequencies of bloat in cattle (76). On the basis of these data and other negative evidence, Majak et al. (76) concluded that saponins are not responsible for bloat in cattle fed fresh alfalfa.

15-6.2 Response of Plant Pests

Alfalfa saponins inhibit the growth of several fungal species, including *Sclerotium rolfsii* Sacc. (10, 93), *Pythium* spp. (10, 63), *Trichoderma viride* Pers. ex Fr. (63), and *Rhizoctonia solani* DC ex Fr. (63). Medicagenic acid saponins have the principal antifungal activity. These particular saponins bind cholesterol, and their toxic effect is exerted only on those fungi which contain cholesterol in their membranes (10).

Alfalfa saponins showed no effect on growth of *Stemphyllium botryosum* Wallr., *Ascochyta imperfecta* Pk., *Leptosphaerulina briosiana* (Poll.) Graham & Luttrell, *Colletotrichum destructivum* O'Gara, and *Mycoplasma* spp. (10, 63). Growth of *Fusarium roseum* Lk. ex Fr. emend. Sny., & Hans. and *Phytophthora megasperma* Drechs. was stimulated by alfalfa saponins (63). Hanson et al. (36) indicated that the foliar pathogens of alfalfa are relatively insensitive to saponins. At present, there is no evidence that breeding for resistance to disease has increased the saponin content of alfalfa.

Some insect pests are sensitive to alfalfa saponins. Growth of the stored product pest red flour beetle [*Tribolium castaneum* (Herbst)] was inhibited by alfalfa saponins (10). The aglycones medicagenic acid and hederagenin were responsible for the toxicity, which could be prevented by concurrent feeding of chloresterol or plant sterols.

Resistance of alfalfa to the pea aphid [*Acyrthosiphon pisum* (Harris)] was decreased by selection for low saponin alfalfa, and, conversely, selection for high-saponin alfalfa increased resistance to the pea aphid (93). However, there is no indication that selection for resistance to insects has increased the saponin content of alfalfa cultivars.

15-7 ESTROGENIC ACTIVITY

Natural constituents of plants which exhibit estrogenic activity in higher animals are commonly referred to as *phytoestrogens*. A large number of natural products are estrogenic. They occur in many plants or plant products used for animal feed and for human foods (103). Some specific compounds that cause the estrogenic effect(s) have been identified and related to the physiological consequences of consuming estrogenic plants. Many isoflavones and coumestans are protoestrogens and must be converted into active estrogens in the rumen or liver (102). For example, formononetin is demethylated in the rumen to produce daidzein (67). Typical estrogenic responses in animals include hypertrophy of the uterus, vagina, vulva, and mammae of female mammals; hypertrophy of the mammae and accessory glands of male mammals; and inhibition of the gonads of both sexes. These effects are usually temporary and disappear with a change to a diet free of phytoestrogens (103).

Estrogenic activity in alfalfa may cause infertility and nymphomania in cattle (86), although infertility in sheep and cattle grazing subterranean clover (*Trifolium subterraneum* L.) or red clover (*T. pratense* L.) are of much greater economic significance. Typical estrogenic responses are observed occasionally in sheep and cattle consuming alfalfa (6, 103). Several reports suggest that estrogens cause an increase in the incidence of mastitis in cattle (13, 27). Compared to cattle, sheep are more sensitive to phytoestrogens found in alfalfa (86).

The stimulating effects of estrogens on growth and improvement of feed efficiency in cattle have been applied widely in commercial production programs. It has been suggested that phytoestrogens in pastures may produce similar beneficial effects in farm animals. Alfalfa containing large amounts of coumestrol was reported to stimulate growth and improve carcass quality in lambs (90). It is doubtful, however, that estrogens occur in alfalfa in either sufficient quantities or consistently over time, to produce a beneficial effect. Coumestrol, an estrogen from alfalfa, did not stimulate growth when fed to cattle (104). The effect of estrogens on animal growth is species specific; estrogens cause inhibition of growth in rodents (103).

Most phytoestrogens fall into three groups of compounds: the isoflavones, the coumestans, and the resorcyclic acid lactones. The estrogenic coumestans found in alfalfa are coumestrol and 4'-0-methyl coumestrol (5). The alfalfa isoflavones with estrogenic activity are biochanin A, daidzein, formononetin, and genistein (5). The estrogenic activity of the coumestans is about one order of magnitude higher than that of the isoflavones (5). In sheep, coumestrol is 15 times more active than the isoflavones when administered intraruminally. The resorcyclic acid lactones have not been reported in alfalfa.

The occurrence of estrogenic activity in plants has been associated with various plant diseases. The coumestrol content of alfalfa is increased by the fungi *Phoma medicaginis* Malbr. & Roum. var. *medicaginis* Broerema, *Cylindrocladium scoparium* Morgan, *Colletotrichum trifolii* Bain. & Essary, *Uromyces striatus* Schroet., *L. briosana* (Poll.) Graham & Luttrell, and *Pseudopeziza medicaginis* (Lib.) Sacc. (70, 100). Sherwood et al. (100) reported the following concentrations of coumestrol in alfalfa plants inoculated with *U. striatus*: none in disease-free leaves, 115 ppm in infected leaves, 168 ppm in the pustules caused by the organisms, and 746 ppm in the urediospores.

Stuthman et al. (105) reported coumestrol in alfalfa in the absence of foliar pathogens. Also extremely high concentrations of coumestrol were found in spotted, barrel medic (*Medicago littoralis* Rohde ex Loisel.) leaves which were free of bacteria or fungi (103). Aphid-damaged portions of alfalfa contain increased levels of coumestrol (69). This observation was confirmed in Australia (56) where infestations of pea aphid and blue-green alfalfa aphid (*Acyrthosiphon kondi shinji*) caused increased coumestrol levels in alfalfa herbage.

An extensive study has been conducted in the USA on variation in the coumestrol content of alfalfa (35). Estrogenic activity of alfalfa will vary with location, cutting, year of harvest, and growth stage. There is genetic variation in the estrogenic activity of alfalfa (35). Selection of alfalfa for resistance to foliar disease affected the coumestrol content (71).

15-8 VOLATILE COMPONENTS

Host-plant volatile constituents often mediate insect feeding and reproductive behavior. The volatile constituents of alfalfa leaves, stems, flowers, and pods have been described (11, 12). In these studies a Temax trap was used to collect volatiles, with (Z)-3-hexenyl acetate and (Z)-3-hexenol, the principal volatiles found in alfalfa herbage and seed pods. The principal volatile from flowers was (E)-β-ocimene, with smaller amounts of methylbutanol, (Z)-3-hexenyl acetate, decanyl acctate, (Z)-3-hexenol, and dodecanylacetate. A total of 19 volatiles were identified from leaves, 17 from pods, and 16 from herbage. Steam distillation of the volatiles in fresh herbage released a larger number of compounds (11, 12, 57) some of which may be degradation products.

In earlier studies (5), 2-hexanal, 3-hexanals, and 3-hexanols were identified as the volatile constituents responsible for the characteristic grassy aroma of alfalfa and other temperate forages. The introduction of 3-hexanol into the rumen or lungs of the cow produced a typical grass flavor in milk.

15-9 MINERALS

A layer of mineral material covers the exterior surface of plant cell walls (80). McManus et al. (80) suggest that these insoluble mineral substances might significantly inhibit cell-wall digestion by rumen microorganisms. The levels of silica in alfalfa, 0.68 to 0.95%, are much lower than in grasses [2.2–5.9%, in smooth bromegrass (*Bromis inermis* Leyss.) (26)]. The alfalfa cell-wall fraction (neutral detergent fiber residue) contains Ca and P as the predominant minerals. They occur as hydroxylpatite [$Ca_5(PO_4)_3OH$] and whitelockite [$(CaMg)_5(PO_4)_2$]. The Ca and P are removed by acid detergent leaving quartz as the dominant mineral on alfalfa acid detergent fiber residue. In alfalfa, no adverse effects have been associated with cell-wall mineralization.

Forage nitrate (NO_3^-) concentrations in excess of 0.2% present a risk of NO_3^- toxicity in ruminant livestock. At the prebud stage the NO_3^- content of alfalfa ranges from 0.18 to 0.32%, which decreases to 0.12 to 0.17% at the green pod stage (5). Thus, NO_3^- concentrations are not excessive, and toxicity is not normally a problem. Caution might be exercised in utilization of prebud alfalfa grown on soils that are high in N.

The Ca/P ratio in alfalfa (6:1–13:1) is higher than the 2:1 ratio recommended for livestock feeds. Thus, alfalfa has the potential to produce a Ca/P imbalance. In practice, this potential imbalance is avoided by feeding low Ca salt-mineral mixtures in conjunction with alfalfa.

15-10 ALKALOIDS

Stachydrine, the simplest of the pyrrolidine alkaloids, is the principal quaternary N base in alfalfa. It is present at 0.64 to 0.93% and represents 1.9 to 2.5% of the total N in herbage. Other minor bases are choline, trimethylamine, and betaine. In addition to stachydrine, alfalfa seed contains both homostachydrine and trigonelline (5). In a chick-feeding study, stachydrine did not affect growth. It decreased the incidence of perosis by an unknown mechanism (19).

15-11 PURINES AND PYRIMIDINES

The purine and pyrimidine constituents of alfalfa have been observed to stimulate the growth of the rumen microorganism *Bacillus subtilus*. Adenine, adenosine, and guanosine are the most abundant in alfalfa juice. Other purines and pyrimidines that have been identified are guanine, xanthine, hypoxanthine, isocytosine, inosine, and cytidine (5).

15-12 PHOTOSENSITIZATION

Photosensitization of rats fed alfalfa-protein concentrate was caused by degradation products of chlorophyll, phaeophobide, and chlorophyllide (68, 108).

15-13 PROTEASE INHIBITORS

Water-soluble protease inhibitors have been isolated from alfalfa herbage. Hazelwood et al. (39) obtained two heat-labile inhibitors, with isoelectric points at pH 9.0 and 9.5, and with inhibitory activity against trypsin and pepsin. These inhibitors were associated with fraction 1 protein during isolation of soluble proteins from alfalfa. Dehydrated alfalfa meal (14, 17, 94) contains a water-soluble trypsin inhibitor that is slowly inactivated by heat. Inhibitory activity was two- to threefold greater in leaves than in stems, and the concentration in the leaves increased after the prebloom stage (14). Delić (21) and Delić et al. (22) isolated a trypsin inhibitor and separated two components by electrophoresis.

The nutritional significance of these protease inhibitors is not well established. Ramirez and Mitchell (94) suggested the trypsin inhibitor could be partially responsible for growth depression when alfalfa meal is added to broiler rations at levels of 10% or more. When fed at a level of 40% in the diet, the trypsin inhibitor reduced growth of mice (*Mus musculus*) and rats (22, 23). It also caused agglutination of red blood cells. Hazlewood et al. (39) suggested a role for protease inhibitors in either the etiology of ruminant bloat, or as a beneficial contributor to efficiency of protein utilization by means of rumen by-pass.

15-14 ALLERGENIC COMPOUNDS

Two organic macromolecular substances have been isolated from alfalfa pollen and partially identified. These substances have been found to cause allergic reactions on persons subject to bronchial asthma and hay fever (5). Some laboratory technicians working with freshly chopped alfalfa report mild headaches and irritation of eyes and nasal passages.

15-15 SUMMARY

For ruminant animals, which consume most alfalfa forage, lignin and bloat are the principal antiquality factors. The increase in lignin content with increasing maturity of forage, and its negative effect on the potential digestibility of cell-wall fiber, is responsible for immense economic losses. With further advances in our knowledge of lignin chemistry, it may be possible to reduce the negative impact of alfalfa lignin through breeding. The risk of ruminant bloat prevents greater utilization of alfalfa in pastures, especially for beef cattle production. Current research on breeding a bloat-safe alfalfa cultivar offers promise.

A calcium/phosphate imbalance is corrected easily with mineral supplements. Examples of economic losses from estrogenic activity and NO_3^- toxicity are rare. Well-known antiquality factors in other legume forages, which do not occur in alfalfa, are tannins, coumarin, alkaloids, toxic amino acids (lathyrogens), cyanogens, nitro compounds, and Se.

The adverse effect of alfalfa meal on growth of monogastric animals appears to be primarily associated with saponins. The possible contributions of polyphenols, trypsin inhibitors, lipid oxidation products, and a reported hemaglutinin remain to be established. The concentrations of monomeric polyphenols, particularly o-dihydroxyphenols, are relatively low. This is a positive factor with respect to alfalfa protein digestibility and quality. Digestibilities of alfalfa leaf-protein concentrate may be reduced by either lipid oxidation products or polyphenols.

REFERENCES

1. Bate-Smith, E.C., and N.H. Lerner. 1954. Leuco-anthocyanins. 2. Systematic distribution of leuco-anthocyanins in leaves. Biochem. J. 58:126-132.
2. Berrang, B., K.H. Davis, Jr., M.E. Wall, C.H. Hanson, and M.W. Pedersen. 1974. Saponins of two alfalfa cultivars. Phytochemistry 13:2252-2260.
3. Betschart, A.A. 1974. Nitrogen solubility of alfalfa protein concentrate as influenced by various factors. J. Food Sci. 39:1110-1115.
4. ----, and J.E. Kinsella. 1973. Extractability and solubility of leaf protein. J. Agric. Food Chem. 21:60-65.
5. Bickoff, E.M., G.O. Kohler, and Dale Smith. 1972. Chemical composition of herbage. In C.H. Hanson (ed.) Alfalfa science and technology. Agronomy 15:247-282.
6. ----, R.R. Spencer, S.C. Witt, and B.E. Knuckles. 1969. Studies on the chemical and biological properties of coumestrol and related compounds. USDA Tech. Bull. 1408. U.S. Government Printing Office, Washington, DC.
7. Biggs, D.R., and G.J. Shaw. 1980. Wairol, a new coumestan from *Medicago sativa*. Phytochemistry 19:2801-2802.
8. Binns, W., and M.W. Pedersen. 1964. Effect of feeding high saponin hay to four-month old Holstein-Friesian bulls. Proc. 7th Conf. Rumen Function, Chicago, IL.
9. Birk, Y. 1969. Saponins. p. 169-210. In I.E. Liener (ed.) Toxic constituents of plant foodstuffs. Academic Press, New York.
10. Bondi, A., Y. Birk, and B. Gestetner. 1973. Forage saponins. p. 511-528. In G.W. Butler and R.W. Bailey (ed.) Chemistry and biochemistry of herbage. Academic Press, New York.
11. Buttery, R.G., and J.A. Kamm. 1980. Volatile components of alfalfa: Possible insect host plant attractants. J. Agric. Food. Chem. 28:978-981.
12. ----, and ----, and L.C. Ling. 1982. Volatile components of alfalfa flowers and pods. J. Agric. Food. Chem. 30:739-742.
13. Brookbanks, E.O., R.A.S. Welch, and M.R. Coup. 1969. Oestrogens in pasture and a possible relationship with mastitis. N.Z. Vet. J. 17:159-160.
14. Chang, H.-Y., G.R. Reeck, and H.L. Mitchell. 1978. Alfalfa trypsin inhibitor. J. Agric. Food Chem. 26:1463-1464.
15. Cheeke, P.R. 1971. Nutritional and physiological implications of saponins: A review. Can. J. Anim. Sci. 51:621-632.
16. ----, and R.O. Meyer. 1975. Protein digestibility and lysine availability in alfalfa meal and alfalfa protein concentrate. Nutr. Rep. Int. 12:337-344.
17. Chien, T.F., and H.L. Mitchell. 1970. Purification of a trypsin inhibitor of alfalfa. Phytochemistry 9:717-720.
18. Clarke, R.T.J., and C.S. W. Reid. 1973. Foamy bloat of cattle: A review. J. Dairy Sci. 57:753-785.
19. Conner, M.A., J.B. Stark, J.C. Fritz, and G.O. Kohler. 1973. Stachydrine: Content in alfalfa and biological activity in chicks. Agric. Food Chem. 21:195-198.
20. Dalrymple, E.J., B.P. Goplen, and R.E. Howarth. 1984. Inheritance of tannins in birdsfoot trefoil. Crop. Sci. 24:921-923.
21. Delić, I., 1972. A contribution to investigation of chemical and biological characteristics and way of action of alfalfa inhibitory matters. Ph.D. diss., Univ. of Zagreb, Yugoslavia.
22. ----, T. Stojisavljević, and S. Stojanović. 1974. An investigation of trypsin inhibitor inactivation in mice under in vivo conditions. Acta Vet. 24:1-5.
23. ----, S. Stojanović, N. Vucurević, and T. Stojsavljević. 1975. An investigation of the possiblity for improving the nutritive value of alfalfa meal with supplements of cholesterol, methyl donors, and $Fe(NO_3)_3 \cdot 9H_2O$. Acta Vet. 25:121-125.
24. Dewick, P.M., and M. Martin. 1979a. Biosynthesis of pterocarpan and isoflavan phytoalexins in *Medicago sativa*: The biochemical interconversions of pterocarpans and 2'-hydroxyisoflavans. Phytochemistry. 18:591-596.
25. ----, and ----. 1979b. Biosynthesis of pterocarpan, isoflavan and coumestan metabolites of *Medicago sativa*: chalcone, isoflavone and isoflavanone precursors. Phytochemistry 18:597-602.
26. Ferebee, D.B., D.O. Erickson, C.N. Haugse, K.L. Larson, and M.L. Buchanan. 1972. Digestibility and chemical composition of brome and alfalfa throughout the growing season. N. Dak. Farm Res. 30:3-7.
27. Frank, N.A., V.L. Sanger, W.D. Pounden, A.D. Pratt, and R.Van Keuren. 1967. Forage estrogens and their possible influence on bovine mastitis. J. Am. Vet. Med. Assoc. 150:503-507.
28. Free, B.L., and L.D. Satterlee. 1975. Biochemical properties of alfalfa protein concentrate. J. Food Sci. 40:85-89.

29. Goplen, B.P., R.E. Howarth, G.L. Lees, W. Majak, J.P. Fay, and K.-J. Cheng. 1983. Evolution of selection techniques in breeding for bloat-safe alfalfa. p. 221-223. *In* J.A. Smith and V.W. Hays (ed.) Proc. 14th Int. Grassl. Congr., Westview Press, Boulder, CO.
30. ----, ----, S.K. Sarkar, and K. Lesins. 1980. A search for condensed tannins in annual and perennial species of *Medicago, Trigonella*, and *Onobrychis*. Crop Sci. 20:801-804.
31. Gordon, A.J. 1974. Effect of plant genotype and digestion on the chemical and physical properties of forage lignins. Ph.D. diss., University of Guelph, Guelph, Canada.
32. ----. 1975. A comparison of some chemical and physical properties of alkali lignins from grass and lucerne hays before and after digestion by sheep. J. Sci. Food Agric. 26:1551-1559.
33. ----, and B.D.E. Gaillard. 1976. The relationship between lignin and carbohydrate in the hemicellulose A, B, and C fractions extracted from lucerne and wheat straw with alkali. p. 55-65. *In* Carbohydrate research in plants and animals. Misc. papers 12. Landbouwhogeschool, Wageningen, Netherlands.
34. Gutek, L.H., B.P. Goplen, R.E. Howarth, and J.M. McArthur. 1974. Variation of soluble proteins in alfalfa, sainfoin, and birdsfoot trefoil. Crop Sci. 14:495-499.
35. Hanson, C.H., G.M. Loper, G.O. Kohler, E.M. Bickoff, K.W. Taylor, W.R. Kehr, E.H. Stanford, J.W. Dudley, M.W. Pedersen, E.L. Sorensen, H.L. Carnahan, and C.P. Wilsie. 1965. Variation in coumestrol content of alfalfa as related to location, variety, cutting year, stage of growth, and disease. USDA Tech. Bull. 1333. U.S. Government Printing Office, Washington, DC.
36. ----, M.W. Pedersen, Bertold Berrang, M.E. Wall, and K.H. Davis, Jr. 1973. The saponins in alfalfa cultivars. p. 33-52. *In* A.G. Matches (ed.) Anti-quality components of forages. Spec. Pub. 4. Crop Science Society of America, Madison, WI.
37. Harborne, J.B. 1980. Plant phenolics. p. 329-402. *In* E.A. Bell and B.V. Charlwood (ed.) Secondary plant products. Vol. 8. Encyclopedia of plant physiology. Springer-Verlag New York, New York.
38. Harkin, J.M. 1973. Lignin. p. 323-373. *In* G.W. Butler and R.W. Bailey (ed.) Chemistry and biochemistry of herbage, Vol. 1. Academic Press, New York.
39. Hazelwood, G.P., J.M. Horsnell, and J.L. Mangan. 1983. Trypsin inhibitors of lucerne association with leaf fraction 1 protein. Phytochemistry 22:1107-1111.
40. Heywang, B.W., C.R. Thompson, and A.R. Kemmer. 1959. Effect of alfalfa saponin on laying hens. Poult. Sci. 38:968-971.
41. Holder, D.P., T.W. Sullivan, G.O. Kohler, and A.L. Livingston. 1975. Freeze-dried alfalfa protein as a protein source for turkeys, 0-4 weeks of age. Poult. Sci. 54:86-90.
42. Hood, L.L., and J.R. Brunner. 1975. Compositional and solubility characteristics of alfalfa protein fractions. J. Food Sci. 40:1152-1154.
43. Howarth, R.E., B.P. Goplen, S.A. Brandt, and K.-J. Cheng. 1982. Disruption of leaf tissues by rumen microorganisms: An approach to breeding bloat-safe forage legumes. Crop Sci. 22:564-568.
44. ----, J.M. McArthur, M. Hikichi, and S.K. Sarkar. 1973. Bloat investigations: denaturation of alfalfa fraction II proteins by foaming. Can. J. Anim. Sci. 53:439-443.
45. ----, K.-J. Cheng, J.P. Fay, W. Majak, G.L. Lees, B.P. Goplen, and J.W. Costerton. 1982. Initial rate of digestion and legume pasture bloat. p. 719-722. *In* J.A. Smith and V.W. Hays (ed.) Proc. 14th Int. Grassl. Congr., Westview Press, Boulder, CO.
46. ----, S.K. Sarkar, A.C. Fesser, and G.W. Schnarr. 1977a. Some properties of soluble protein from alfalfa (*Medicago sativa*) herbage and their possible relationship to ruminant bloat. J. Agric. Food Chem. 25:175-179.
47. ----, W. Majak, D.E. Waldern, S.A. Brandt, A.C. Fesser, B.P. Goplen, and D.T. Spurr. 1977b. Relationships between ruminant bloat and the chemical compostion of alfalfa herbage. I. Nitrogen and protein fractions. Can. J. Anim. Sci. 57:345-357.
48. Hudson, B.J.F., and S.E.O. Mahgoub. 1980. Naturally occurring antioxidants in leaf lipids. J. Sci. Food Agric. 31:646-650.
49. Hurowitz, B., E.J. Pell, and R.T. Sherwood. 1979. Status of coumestrol and 4',7-dihydroxyflavone in alfalfa foliage exposed to ozone. Phytopathology 69:810-813.
50. Ingham, J.L. 1979. Isoflavonoid phytoalexins of the genus *Medicago*. Biochem. System. Ecol. 7:29-34.
51. ----, and R.L. Miller. 1973. Sativin: an induced isoflavan from the leaves of *Medicago sativa* L. Nature (London) 242:125-126.
52. Johns, A.T. 1954. Bloat in cattle on red clover. N.Z. J. Sci. Technol. 35 (A):289-320.
53. Jones, W.T., L.B. Anderson, and M.D. Ross. 1973. Bloat in cattle. XXXIX. Detection of protein precipitants (flavolans) in legumes. N.Z. J. Agric. Res. 16:441-446.
54. ----, and J.W. Lyttleton. 1973. Bloat in cattle. XXXVIII. The foaming properties of rumen liquor. N.Z. J. Agric. Res. 16:161-168.
55. ----, and J.L. Mangan. 1976. Large-scale isolation of fraction I leaf protein from lucerne (*Medicago sativa* L.) J. Agric. Sci. 86:495-501.

56. Kain, W.M., and D.R. Biggs. 1980. Effect of pea aphid and bluegreen lucerne aphid (*Acyrthosiphon* spp.) on coumestrol levels in herbage of lucerne (*Medicago sativa*). N.Z. J. Agric. Res. 23:563–568.
57. Kami, T. 1983. Composition of the essential oil of alfalfa. J. Agric. Food Chem. 31:38–41.
58. Khan, F.Z., and J. Milton. 1975. Phytoalexin production by lucerne (*Medicago sativa* L.) in response to infection by *Verticillium*. Physiol. Plant Pathol. 7:179–187.
59. Knuckles, B.E., D. deFremery, E.M. Bickoff, and G.O. Kohler. 1975. Soluble protein from alfalfa juice by membrane filtration. J. Agric. Food Chem. 23:209–212.
60. Kudo, H., K.-J. Cheng, M.R. Hanna, R.E. Howarth, B.P. Goplen, and J.W. Costerton. 1985. Ruminal digestion of alfalfa strains selected for slow and fast initial rates of digestion. Can. J. Anim. Sci. 65:157–161.
61. Lahiry, N.L., and L.D. Satterlee. 1975. Release and estimation of chlorogenic acid in leaf protein concentrate. J. Food Sci. 40:13–26.
62. ----, ----, H.W. Hsu, and G.W. Wallace. 1977. Characterization of the chlorogenic acid binding fraction in leaf protein concentrates. J. Food Sci. 42:83–85.
63. Leath, K.T., K.H. Davis, Jr., M.E. Wall, and C.H. Hanson. 1972. Vegetative growth responses of alfalfa pathogens to saponin and other extracts from alfalfa. Crop Sci. 12:851–856.
64. Lees, G.L. 1984. Cuticle and cell wall thickness: relation to mechanical strength of whole leaves and isolated cells from some forage legumes. Crop Sci. 24:1077–1081.
65. Lindahl, I.L., R.E. Davis, and R.T. Tertell. 1957. Production of bloat and other symptoms in intact sheep by alfalfa saponin administration. p. 2–15. USDA Tech. Bull. 1161. U.S. Government Printing Office, Washington, DC.
66. ----, ----, ----, G.E. Whitmore, W.T. Shalkop, R.W. Dougherty, C.R. Thompson, G.R. Van Atta, E.M. Bickoff, E.D. Walter, A.G. Livingston, J. Guggolz, R.H. Wilson, M.B. Sideman, and F. DeEds. 1957. Alfalfa saponins, studies on their chemical, pharmacological and physiological properties in relation to ruminant bloat. USDA Tech. Bull. 1161. U.S. Government Printing Office, Washington, DC.
67. Livingston, A.L. 1978. Forage plant estrogens. J. Toxicol. Environ. Health 4:301–324.
68. Lohrey, E., B. Tapper, and E.L. Hove. 1974. Photosensitization of albino rats fed on lucerne-protein concentrate. Br. J. Nutr. 31:159–166.
69. Loper, G.M. 1968. Effect of aphid infestation on the coumestrol content of alfalfa varieties differing in aphid resistance. Crop Sci. 8:104–106.
70. ----, and C.H. Hanson. 1964. Influence of controlled environmental factors and two foliar pathogens on coumestrol in alfalfa. Crop Sci. 4:480–482.
71. ----, ----, and J.H. Graham. 1967. Coumestrol content of alfalfa as affected by selection for resistance to foliar disease. Crop Sci. 7:1189–1192.
72. Lu, P.-S., and J.E. Kinsella. 1972. Extractability and properties of protein from alfalfa leaf meal. J. Food Sci. 37:94–99.
73. Lyttleton, J.W. 1973. Proteins and nucleic acids. p. 63–103. *In* G.W. Butler and R.W. Bailey (ed.) Chemistry and biochemistry of herbage. Academic Press, New York.
74. Majak, W., J.W. Hall, and R.E. Howarth. 1986. The distribution of chlorophyll in rumen contents and the onset of bloat in cattle. Can. J. Anim. Sci. 66:97–102.
75. ----, R.E. Howarth, K.-J. Cheng, and J.W. Hall. 1983. Rumen conditions that predispose cattle to pasture bloat. J. Dairy Sci. 66:1683–1688.
76. ----, ----, A.C. Fesser, B.P. Goplen, and M.W. Pedersen. 1980. Relationships between ruminant bloat and compostion of alfalfa herbage. II. Saponins. Can. J. Anim. Sci. 60:699–708.
77. ----, ----, and P. Narasimhalu. 1985. Chlorophyll and protein levels in bovine rumen fluid in relation to alfalfa pasture bloat. Can. J. Anim. Sci. 65:147–156.
78. Malinow, M.R. 1984. Triterpenoid saponins in mammals: Effects on cholesterol metabolism and atherosclerosis. p. 229–246. *In* W.D. Nes et al. (ed.) Isopentenoids in plants. Marcel Dekker, New York.
79. McArthur, J.M., J.E. Miltimore, and M.J. Pratt. 1964. Bloat investigations: the foam stabilizing protein of alfalfa. Can. J. Anim. Sci. 44:200–206.
80. McManus, W.R., R.G. Anthony, L.L. Grout, A.S. Malin, and V.N.E. Robinson. 1979. Biocrystallization of mineral material on forage plant cell walls. Aust. J. Agric. Res. 30:635–649.
81. Milić, B.L. 1972. Lucerne tannins. I. Content and composition during growth. J. Sci. Food. Agric. 23:1151–1156.
82. ----, and S. Stojanović. 1972. Lucerne tannins. III. Metabolic fate of lucerne tannins in mice. J. Sci. Food Agric. 23:1163–1167.
83. ----, ----, and N. Vucurević. 1972. Lucerne tannins. II. Isolation of tannins from lucerne, their nature and influence on the digestive enzymes in vitro. J. Sci. Food Agric. 23:1157–1162.
84. Miltimore, J.E., J.M. McArthur, B.P. Goplen, W. Majak, and R.E. Howarth. 1974.

Variability of fraction 1 protein and heritability estimates for fraction 1 protein and total phenolic constituents in alfalfa. Agron. J. 66:384–386.
85. ----, ----, J.L. Mason, and D.L. Ashby. 1970. Bloat investigations. The threshold fraction I (18S) protein concentration for bloat and relationships between bloat and lipid, tannin, Ca, Mg, Ni, and Zn concentrations in alfalfa. Can. J. Anim. Sci. 50:61–68.
86. Moule, G.R., A.W.H. Braden, and D.R. Lamond. 1963. The significance of oestrogens in pasture plants in relation to animal production. Anim. Breed. Abstr. 31:139–157.
87. Monties, B., and J.C. Rambourg. 1978. Occurrence of flavonoids (flavones and coumestans) in alfalfa (*Medicago sativa* var. Europe) leaf proteins. Ann. Technol. Agric. 27:629–654.
88. Newby, V.K., R.M. Sablon, R.L.M. Synge, K.V. Casteele, and C.F. Van Sumere. 1980. Free and bound phenolic acids of lucerne (*Medicago sativa* cv. Europe). Phytochemistry 19:651–657.
89. Noguchi, H., T. Maekawa, S. Fujimoto, I. Satake, and M. Sakakibara. 1978. Physicochemical studies on fraction I protein from alfalfa. Agric. Biol. Chem. 42:1553–1558.
90. Oldfield, J.E., C.W. Fox, A.V. Bahn, E.M. Bickoff, and G.O. Kohler. 1966. Coumestrol in alfalfa as a factor in growth and carcass quality in lambs. J. Anim. Sci. 25:167–174.
91. Pedersen, M.W. 1978. Low saponin alfalfa for non-ruminants. p. 148–155. *In* R.E. Howarth (ed.) Proc. 2nd Int. Green Crop Drying Congress, Univ. of Saskatchewan, SK, Canada.
92. ----, J.O. Anderson, J.C. Street, L.C. Wang, and R. Baker. 1972. Growth response of chicks and rats fed alfalfa with saponin content modified by selection. Poul. Sci. 51:458–464.
93. ----, D.K. Barnes, E.L. Sorensen, G.D. Griffin, M.W. Nielson, R.R. Hill, Jr., F.I. Frosheiser, R.M. Sonoda, C.H. Hanson, O.J. Hunt, R.N. Peaden, J.H. Elgin, Jr., T.E. Devine, M.J. Anderson, B.P. Goplen, L.J. Elling, and R.E. Howarth. 1976. Effects of low and high saponin selection in alfalfa on agronomic and pest resistance traits and the interrelationship of these traits. Crop Sci. 16:193–198.
94. Ramirez, J.S., and H.L. Mitchell. 1960. The trypsin inhibitor of alfalfa. J. Agric. Food Chem. 8:393–395.
95. Robertson, J.B., and P.J. Van Soest. 1981. The detergent system of analysis and its application to human foods. p. 123–158. *In* W.P.T. James and O. Theander (ed.) The analysis of dietary fiber in food. Marcel Dekker, New York.
96. Rumbaugh, M.D. 1979. The search for condensed tannins in the genus *Medicago*. Agron. Abstr. American Society of Agronomy, Madison, WI, p. 75.
97. Saleh, N.A., L. Boulos, S.I. El-Negoumy, and M.F. Abdalla. 1982. A comparative study of the flavonoids of *Medicago radiata* with other *Medicago* and related *Trigonella* species. Biochem. Systematics Ecol. 10:33–36.
98. Sarkar, S.K., and R.E. Howarth. 1976. Specificity of the vanillin test for flavanols. J. Agric. Food Chem. 24:317–320.
99. ----, ----, M. Hikichi, and J.M. McArthur. 1975. Soluble proteins of alfalfa (*Medicago sativa*) herbage. Fractionation by ammonium sulfate and gel chromatography. J. Agric. Food Chem. 23:626–630.
100. Sherwood, R.T., A.F. Olah, W.H. Oleson, and E.E. Jones. 1970. Effect of disease and injury on accumulation of a flavonoid estrogen, coumestrol, in alfalfa. Phytopathology 60:684–688.
101. Skarby, L., and E.J. Pell. 1979. Concentrations of coumestrol and 4′,7-dihydroxyflavone in four alfalfa cultivars after exposure to ozone. J. Environ. Qual. 8:285–286.
102. Smolenski, S.J., A.D. Kinghorn, and M. Balandrin. 1981. Toxic constituents of legume forage plants. Econ. Bot. 35:321–355.
103. Stob, M. 1983. Naturally occurring food toxicants estrogens. p. 81–100. *In* M. Rechcigl, Jr. (ed.) CRC handbook of naturally occurring food toxicants. CRC Press, Boca Raton, FL.
104. ----, W.M. Beeson, T.W. Perry, and M.T. Mohler. 1968. Effects of coumestrol in combination with implanted and orally administered diethylstilbestrol on gains and tissue residues in cattle. J. Anim. Sci. 27:1638–1642.
105. Stuthman, D.D., E.M. Bickoff, R.L. Davis, and M. Stob. 1966. Coumestrol differences in *Medicago sativa* L. free of foliar disease symptoms. Crop Sci. 6:333–334.
106. Swain, T. 1977. Secondary compounds as protective agents. Ann. Rev. Plant Physiol. 28:479–501.
107. ----, and E.C. Bate-Smith. 1962. Flavonoid compounds. p. 755–809. *In* A.M. Florkin and H.S. Mason (ed.) Comparative biochemistry, Vol. 3. Academic Press, New York.
108. Tapper, B.A., E. Lohrey, E.L. Hove, and R.M. Allison. 1975. Photosensitivity from chloroform derived pigments. J. Sci. Food Agric. 26:277–284.
109. Ulyatt, M.J., J.A. Lancashire, and W.J. Jones. 1977. The nutritive value of legumes. Proc. N.Z. Grassl. Assoc. 38:107–118.

110. Van Sumere, C.F., J. Albrecht, A. Dedonder, H. de Pooter, and I. Pé. 1975. Plant proteins and phenolics. p. 211–264. *In* J.B. Harborne and C.F. Van Sumere (ed.) The chemistry and biochemistry of plant proteins. Academic Press, New York.
111. ----, H.J. Houpeline-de Cock, Y.C. Vindevaghel-DeBacquer, and G.E. Fockenier. 1980. N-Feruloylglycyl-L-phenylalanine isolated by partial hydrolysis of bulk leaf protein of lucerne, *Medicago sativa*, cv. Europe. Phytochemistry 19:704–705.
112. Walgenbach, R.P., and G.C. Marten. 1981a. Release of soluble protein and nitrogen in alfalfa. II. Influence of potassium and nitrogen fertilizer on three cultivars. Crop Sci. 21:852–855.
113. ----, and ----. 1981b. Release of soluble protein and nitrogen in alfalfa. III. Influence of shading. Crop. Sci. 21:859–862.
114. ----, ----, and G.R. Blake. 1981. Release of soluble protein and nitrogen in alfalfa. I. Influence of growth temperature and soil moisture. Crop Sci. 21:843–849.
115. Wall, M. 1971. Toxins of animal and plant origin. Gordon and Breach Science Pub., London.
116. Wang, J., and J.E. Kinsella. 1975. Composition of alfalfa leaf protein extracts. J. Food Sci. 40:1156–1160.

16 Pasture Production and Utilization

R. W. VAN KEUREN

Ohio Agricultural Research and Development Center
Wooster, Ohio

A. G. MATCHES

Texas Tech University
Lubbock, Texas

Alfalfa (*Medicago sativa* L.) is a superior pasture legume for many classes of livestock because of its high yield, forage quality, and wide climatic and soil adaptation. Alfalfa is a dependable and economical protein source for the grazing animal because it is independent of soil N. The protein is of excellent quality, with a good amino acid profile especially important to nonruminants on pasture, such as swine (*Sus* spp.), poultry (*Gallus* spp.), and horses (*Equus* spp.). Alfalfa is an excellent source of Ca, Mg, P, and pro-vitamin A (carotene) and vitamin D. Intake of alfalfa is usually greater than that of grasses of equal digestibility.

Alfalfa provides greater flexibility for the livestock producer than many other pasture legumes. Because of upright growth habit and rapid recovery, it lends itself to harvest as hay, silage, or soilage, as well as pasture. This flexibility is a desirable characteristic in livestock areas where harvested feed is utilized for portions of the year. Its drought resistance also provides a more uniform seasonal growth pattern than that of most legumes and grasses.

16-1 ADAPTATION AND DISTRIBUTION OF ALFALFA FOR PASTURE

In the Corn Belt region of the USA, where alfalfa is especially important to dairying, it is grown as a pasture and hay crop in rotation with maize (*Zea mays* L.). In that region, no other legume provides the combination of high yield, stand longevity, and feed quality of alfalfa. In the northern part of the Corn Belt, alfalfa for pasture is grown most commonly with smooth bromegrass (*Bromus inermis* Leyss.). In the southern part, it is used most frequently with orchardgrass (*Dactylis glomerata* L.).

Alfalfa is grown in the rangelands of the western USA and Canada for beef cattle (*Bos* spp.) and sheep (*Ovis aries*), often with cool-season grasses such as crested wheatgrass [*Agropyron desertorum* (Fisch.) Schult.],

---

Copyright 1988 © ASA-CSSA-SSSA, 677 South Segoe Road, Madison, WI 53711, USA.
Alfalfa and Alfalfa Improvement—Agronomy Monograph no. 29.

intermediate wheatgrass [*A. intermedium* (Host) Beauv.], and Russian wildrye (*Elymus junceus* Fisch.). It is useful for improving deteriorated dryland pasture.

In Utah, Rumbaugh and Pederson (143) showed that in a semiarid environment alfalfa can survive and remain productive when grazed for 23 yr. However, they contended that moisture stress may prohibit use of alfalfa as range or permanent pasture in locations with <280 mm annual precipitation.

Alfalfa is now important in dryland pastures of South Africa, Argentina, southeastern Australia, and New Zealand (142,175). Alder and Minson (2) defined the potential value of alfalfa pasture for summer dry periods in England.

Alfalfa is grown widely as a highly productive irrigated hay and pasture crop in western and southwestern USA, western Canada, Chile, South Africa, and southern Australia. Also, it is excellent for supplementing dryland pastures. Its value in subtropical regions is recognized (30,53,110,111,179). The development of rhizomatous and creeping-rooted cultivars of alfalfa (80) has contributed significantly to expanded pasture use in semiarid areas.

16-2 ALFALFA PASTURE MANAGEMENT

Alfalfa subjected to grazing does not differ appreciably in defoliation management from that harvested for hay. Thus, physiological factors discussed in Chapters 5 and 12 are applicable when alfalfa is grazed.

In northern USA, harvesting alfalfa for hay at the late-bud to early flower stage with four cuts annually, or to one-tenth bloom with three cuts is commonly recommended (see Chapter 12 in this book). Typically, these schedules will have a 35- to 42-d recovery period between harvests. An autumn recovery growth must also occur before killing frost to maintain stands.

Most evidence indicates that the present adapted cultivars of alfalfa require rotational grazing with a recovery period similar to a hay schedule to maintain good stands for more than 2 or 3 yr. In practice, grazing is commonly initiated when the alfalfa is in late vegetation stages, followed by a 7- to 10-d grazing period and a 30- to 40-d recovery period. Grazing the first paddock is generally initiated earlier than for hay harvest because the subsequent paddocks are advancing in maturity. Grazing also differs from mechanical harvesting in that there is a gradual removal of the photosynthetic tissue over a period of time, nor is there as complete a removal. Mechanical harvesting results in immediate removal of the entire plant almost to the soil surface, leaving virtually no remaining photosynthetic material. Thus, theoretically it should be possible to initiate grazing at an earlier stage of maturity for each grazing period compared with hay harvest. Counce (40) also suggests that factors contributing to persistence under grazing are different from those contributing to persistence under mowing.

Limited research has been done on the effect of early grazing of alfalfa. Temperature appears to have a major effect on alfalfa growth and development (58) and alfalfa may be more tolerant of grazing during spring when temperatures are normally lower (57,98). Soil temperature and moisture may also be more favorable for alfalfa stand persistence in spring than in summer (81). In Virginia, Allen et al. (3) found that early spring grazing for 2 consecutive yr with lambs had minimal effect on alfalfa stand longevity and productivity. These results confirm earlier research showing that early spring clipping had little or no effect on seasonal yield during the current or subsequent years (89).

The rapid decline of alfalfa stands with continuous or short-period rotational grazing has been recognized for many years in the humid pasture region of North Central USA. This pasture decline has been demonstrated with sheep (61,62) and dairy cows (6,135,149). Cooke et al. (39) reported stand losses under continuous grazing by steers in the northern parklands of Saskatchewan, as did Heinemann (77) with irrigated alfalfa.

The rapid decline in alfalfa productivity under continuous grazing also occurred in Australia (21,59,119,131,152,155). In a 4-yr Australian trial, continuous sheep grazing of alfalfa-phalaris (*Phalaris stenoptera* Hack.)-subterranean clover (*Trifolium subterraneum* L.) practically eliminated the alfalfa (123). Rotational grazing every 4 weeks reduced alfalfa considerably, while grazing every 8 weeks maintained a productive stand. Similar results were obtained with dryland alfalfa in New South Wales (131). McKinney (119) reported lowest stands of alfalfa with continuous grazing with sheep. In other research in New South Wales conducted by Southwood and Robards (155), alfalfa stands under continuous grazing with sheep at a high stocking rate were eliminated in 17 weeks, but persisted under a low stocking rate for 3 yr, at which time the stand was lost from a prolonged drought.

The effect of rotational grazing on alfalfa persistence has been shown by a number of studies. Bateman and Keller (15) maintained 'Ranger' alfalfa through the ninth grazing season under irrigation in Utah, when pastures were grazed by dairy cows three times each season for the first 3 yr (42-d recovery) and four times each season thereafter (35-d recovery). Alfalfa did not persist beyond the 3rd yr in irrigated pastures in Alberta, Canada, with a rotation of 1 week of grazing by sheep and 3 weeks recovery (177). The best alfalfa management for New Zealand is a 4-d grazing period with close grazing to bare ground followed by a 36-d recovery period (88).

In England, alfalfa did not persist well under biweekly close defoliation by sheep, but persisted under monthly defoliation (48). In southeastern England, grazing at 8-week intervals maintained the composition of an alfalfa-orchardgrass pasture (173). An 8-week recovery early or late in the season did not overcome the adverse effect of more frequent grazing (11).

In Alberta, Canada, the percentage of alfalfa in smooth bromegrass-alfalfa-red fescue (*Festuca rubra* L.) pastures increased under rotational grazing from 23 to 47%, and herbage was of higher quality than that from continuously grazed pasture (169). Over the 4 yr, mean annual precipitation was 406 mm. Average stocking rate was higher with rotational than with continuous grazing.

Leach (103) lists several factors that influence alfalfa persistence under rotational or intermittent grazing, including the duration of the resting and grazing periods, their timing in relation to growth, development, and weather, and the severity of grazing. Of these the most critical is the interval between grazings (21,90,119). Janson (90) suggests that the duration of the resting and grazing periods interacts with climate, allowing greater flexibility in drier regions and seasons.

Leach (104) in Queensland, Australia compared resting alfalfa for 32 or 44 d combined with grazing with sheep for 4 or 16 d; resting for 56 d with grazing for 4 d, and resting for 40 d with grazing for 8 d. Irrigation was used to avoid prolonged droughts and to ensure that moisture supply approached the average seasonal pattern. Annual rainfall for this area averages 775 mm. Survival of alfalfa was poorer with 4 d of grazing than with either 8 or 16 d of grazing. Competition from grass invasion was greatest with the 4-d grazing treatment and may have been responsible for the greater decline in alfalfa stands. Over all treatments, stand declines averaged about five plants/m^2.

Persistence of the alfalfa was satisfactory under rotational grazing with sheep for 4 yr in New South Wales (155). The researchers concluded that each grazing period should be short (5-10 d), and that this was of greater importance than number of rotational paddocks (5 vs. 7). McKinney (119) found with rotational grazing that the minimum rest period to maintain alfalfa stands ranged from 5½ weeks in summer to 8 weeks in winter. He suggested that if the rest period is too short, there will be a loss of alfalfa plants; if too long, the decreasing leaf-stem ratios will substantially reduce the quality per unit yield of alfalfa.

Nuttall et al. (125) in western Canada grazed smooth bromegrass-alfalfa pastures in a four paddock rotation. A higher percentage of alfalfa was maintained when cattle numbers were adjusted periodically to utilize the forage available than where animal numbers were maintained at 3.7 head/ha and fed supplementary barley.

Research of Thompson et al. (161) in New South Wales showed that the decline in alfalfa density was related to stocking rate. In a comparison of stocking rates ranging from 8 to 21 ewes/ha, greatest decline in alfalfa stands occurred with rates above 12 ewes/ha. Similarly, research in Victoria, Australia (139) showed that alfalfa stands were maintained at 7.4 and 9.9 ewes/ha, but declined at 12.4 and 14.8 ewes/ha.

The effect of alfalfa grazing is influenced by weather conditions. Grazing treatments applied in a wet year had a greater negative effect on subsequent alfalfa vigor than similar treatments applied in a dry year (173). In Oregon, Gross and Matheson (66) found an interaction between irrigated pasture production and weather conditions. Alfalfa-tall fescue gave higher beef production per ha during hot summers than during cool summers. In contrast, Kentucky bluegrass (*Poa pratense* L.)-white clover (*T. repens* L.) gave higher production of beef in cool summers.

Wilman (176) in the United Kingdom observed inconsistent results when an alfalfa-orchardgrass mixture was defoliated four, five, or six times by sheep. Alfalfa stands were maintained with four grazings per year with a 54-d recovery period. During a wet season, six grazings were detrimental to alfalfa stands with the alfalfa declining to 2% of total herbage yield as compared to 66% alfalfa with four grazings. During a drier year there were no differences in stand decline among the three frequencies of grazing. Apparently, the wetter season favored orchardgrass that dominated the alfalfa when frequently defoliated. There results show that when alfalfa is grown in a mixture with grass, the optimum grazing regime for maintaining alfalfa stands may vary over years depending on weather conditions. Leach and Ratcliff (106) have also reported on the detrimental effect of grass competition on alfalfa persistence under subtropical conditions.

In the North Central region of the USA, which generally experiences severe winters, an autumn recovery period for alfalfa is required to assure winter survival. Grazing alfalfa in September and October in Michigan severely damaged the stand (137,138). In a review paper, Washko (170) cited work in West Virginia showing slight damage to alfalfa grazed during early September or early October, but no damage to that grazed in late October. Also, continuous grazing at a light stocking rate during September and October was not harmful.

In Germany, an autumn rest period of 45 to 55 d is considered optimum to assure successful overwintering of alfalfa (175). September grazing in southeastern England reduced alfalfa yield and persistence in subsequent years compared with grazing in October (10). In other studies, winter grazing of alfalfa in late February and late March reduced yields the following years (70).

In Tennessee the effect of all-season, three-paddock rotational grazing (20 d recovery) was compared with the same grazing procedure following removal of a hay crop in May (162). By the fifth grazing season the pastures grazed all season were 75% orchardgrass-25% alfalfa, while those grazed following hay harvest were 80% alfalfa. These results, together with the results from Virginia studies, suggest that a combination of grazing and hay harvests would result in better alfalfa persistence than under grazing only. Early spring grazing (3) on some paddocks, with other paddocks harvested first for hay followed by late-summer grazing (4) would also allow better control of stage of growth for grazing. Paddocks not needed for grazing in spring would be set aside for first harvest hay.

Swanson et al. (160) observed slower recovery of alfalfa-grass after harvesting as soilage than after daily strip grazing, as well as better stand maintenance with strip grazing. Alfalfa that was strip grazed daily was injured less by summer drought than the chopped alfalfa. Davis and Pratt (49) found that rotationally grazed alfalfa was less seriously affected by drought than was continuously grazed alfalfa.

Broad-crowned types of alfalfa, such as 'Vernal', 'Nomad', and 'Rhizoma', were more persistent under a wide range of pasture management in Nebraska than narrow-crowned types such as 'Buffalo', 'DuPuits', and 'Grimm' (95). Some cultivars show good persistence when grazed continuously under dryland conditions. Under humid conditions (44), Vernal grazed by sheep outyielded pasture-type 'Teton'. Recent research in Wyoming (75,76) showed that creeping and tap-rooted alfalfa cultivars responded similarly to defoliation pattern. In a comparison of erect, spreading, and creeping-rooted alfalfa grown in Queensland, Australia under dryland grazing (65,105), survival of creeping-rooted lines was poor. Several spreading lines (105) showed good survival for up to 6 yr with yields similar to the local cv. Hunter River. They postulated that spreading genotypes may be useful in the development of improved cultivars for prolonged dryland grazing. Because of their rhizomatous spreading habit, they may have a deeper crown and a longer period of bud production that may contribute to better survival under grazing (102).

The rangelands of the North American Great Plains are predominantly grasses. Legumes are not used in most range seedings. However, alfalfa can be interseeded successfully into range vegetation (145), and can make a valuable contribution to the production from dryland pastures seedings in range areas (20,28,46,56).

For many years, 'Ladak' has been used as a rangeland alfalfa in the northern Great Plains. Rhizoma, with a rhizomatous spreading habit, persisted better than Ladak under range grazing (20). Alfalfa with a creeping-rooted habit appears to be the best type for dryland range conditions (80). 'Rambler', one of the first creeping-rooted cultivars, persisted better than Ladak, Grimm, Vernal, and Rhizoma under drought conditions in Saskatchewan (31). Rambler had superior persistence and yield in dryland pastures (7). 'Travois', developed as a rangeland alfalfa in South Dakota, had superior winterhardiness and grazing persistence (144). In Australian studies near Canberra, creeping-rooted types were markedly superior to local strains under continuous grazing by sheep (45).

Alfalfa can persist very well under some range conditions because of intermittent grazing (depending on the seasonality of the companion grass) and moderate or deferred grazing designed to leave a high stubble for reserve feed and best grass survival. Moore (122) described using a combination of pastures for rangeland conditions in South Dakota, such as alternating pasture-type alfalfa-bromegrass-intermediate wheatgrass with native range.

The value of alfalfa for feed, the development of grazing types for dryland regions, and the possibility of breeding strains low in bloat potential show promise for the expanded use of alfalfa in range areas.

16-3 ALFALFA AS PASTURE FOR DAIRY COWS

16-3.1 Milk Production from Alfalfa-Grass Pasture

Excellent milk production per cow and per hectare is obtainable on alfalfa-grass pastures (Table 16-1). Per ha production as high as 6948 kg of 4% fat-corrected milk (FCM) has been reported for a 24-week grazing period without supplemental feed (25). Hill and Lundquist (82) reported that an individual cow produced a daily average of 24.7 kg of 4% FCM for 84 d on alfalfa-grass pasture without supplemental feed on a daily intake of 101 kg of green forage (20 kg of dry matter). They concluded that about 50% of the milk for the lactation could be expected during the 17-week pasture period and that alfalfa-grass pasture alone supported milk production greater than 15.8 kg per d without body weight loss. Rumery et al. (146) reported that the 4-yr average feed replacement value of a hectare of irrigated alfalfa-smooth bromegrass pasture was 459 kg of grain, 1206 kg of alfalfa hay, and 5022 kg of corn silage on an as-fed basis.

Orchardgrass-alfalfa produced more milk per ha but less milk per cow than smooth bromegrass-alfalfa (127,149). The use of alfalfa with smooth bromegrass was preferable to N fertilization of smooth bromegrass for dairy pasture (54,171). Alfalfa was superior in milk production per ha to sudangrass [*Sorghum sudanense* (Piper) Stapf.] (126,136) but similar to annual pastures in daily production of 4% FCM (156). Blaser et al. (19) reported an average daily production by 15.2 kg of 4% FCM by "top-grazers" (cows allowed to graze first in a pasture) on alfalfa-orchardgrass compared with 13.9 kg produced by "bottom-grazers".

16-3.2 Pasture Utilization Methods

Rotational grazing was similar to continuous grazing in individual animal performance but superior in forage utilization and higher milk production per hectare (25,49). Likewise, similar milk production per cow was obtained from daily strip and continuous grazing of alfalfa-smooth bromegrass (22) and alfalfa-orchardgrass (109). Surplus feed mechanically harvested during peak growth from strip-grazed pastures increases the production of total digestible nutrient (TDN) per ha. Brundage and Sweetman (23) reported that daily milk production was more uniform with strip grazing than with rotational grazing. Bryant et al. (24) suggested the top and bottom grazer technique for obtaining maximum output per ha and per animal.

In recent years, there has been a strong trend in dairy areas of the USA and Canada toward year-around drylot feeding. Mechanical har-

Table 16-1. Representative animal production per hectare and per animal for dairy cattle, feeder steers, and lambs from pasture in humid, irrigated, and dryland pasture regions.

Species	Period	Performance			Reference and remarks
Dairy cattle		4% FCM, kg/ha	Avg daily 4% FCM (kg)	Grazing d/ha	
1. Alfalfa-orchardgrass	20 weeks	6948	12.8	585	25 Rot. grazing, 2-yr avg
Ladino clover-orchardgrass		4493	10.5	415	No suppl. feed
White clover-Kentucky bluegrass		5396	11.5	432	
2. 3-paddock rotation grazing	17 weeks	6347	18.1	350	167 Alf.-brome, 3-yr avg
Daily strip grazing		6680	16.9	400	Limited grain suppl.
Soilage		9640	17.3	535	
Feeder steers		Liveweight, kg/ha	ADG, kg	Grazing d/ha	
1. Alfalfa-smooth bromegrass		314	0.78	402	115 Humid region, 5-yr avg
Orchardgrass		274	0.78	352	
2. Alfalfa-orchardgrass	22 weeks	1049	0.95	1118	165 Irrigated, 3-yr avg
Orchardgrass		595	0.78	768	
3. Alfalfa-smooth bromegrass	17 weeks	222	0.99	222	39 Parklands, 6-yr avg
Alfalfa-intermediate wheatgrass		222	0.94	230	
				ha/yearling	
4. Alfalfa-intermediate wheatgrass	113 d	49	0.66	0.56	46 Range, 10-yr avg
Intermediate wheatgrass	110 d	29	0.59	0.75	
Lambs				Grazing d/ha	
1. Alfalfa-smooth bromegrass	14 weeks	643	0.13	4892	93 Humid region, 2-yr avg
Bromegrass		487	0.11	4545	Early weaned lambs
2. Alfalfa	108 d	493	0.15	--	85 Irr., sudan hay fed
Birdsfoot trefoil-orchardgrass		378	0.14	--	Feeder lambs
				Mature animal d/ha (including ewes)	
3. Alfalfa	22 weeks	957	0.22	3336	78 Irr., 2 yr-avg
Alfalfa-orchardgrass		868	0.19	3684	Suckling lambs
Orchardgrass		522	0.17	3005	

vesting and feeding of alfalfa silage and hay can result in greater milk production per area than grazing. However, animal performance has usually been similar if the forage is utilized at comparable growth stages (74,97,99,100,101,167). Although soilage increases forage utilization compared to grazing, the equipment and labor costs of the farmer are considerably higher (134).

Using the chromogen-chromic oxide technique, Bryant et al. (25) estimated daily dry matter (DM) intakes of 2.25 kg and 1.48 kg/45 kg liveweight and dry matter digestibilities (DMD) of 69.0 and 56.9% by cattle grazing alfalfa-orchardgrass during early season (May-June) and late season (August-September), respectively. Marten et al. (114) reported season average organic matter digestibilities (OMD) of 69.0 and 64.0% for alfalfa-smooth bromegrass pasture for 2 yr using the chromogen technique. Larsen et al. (101) reported a daily intake of 0.91 kg of forage dry matter/45 kg liveweight by dairy cows strip grazing alfalfa-ladino (white) clover-smooth bromegrass.

Despite the high productivity of alfalfa-grass pastures for dairy cows, grazing may be at least as expensive as mechanical harvesting, especially with larger herds. This is true because grazing requires more land, fencing, pasture clipping, labor for moving animals to and from pasture, and often poorer utilization of forage compared with mechanical harvesting (168). These trials were conducted in the north central dairy region of the USA where agricultural land values are high and where dairymen must provide housing, feed storage, and handling facilities, and forage harvesting equipment for a winter-feeding period of 26 to 30 weeks.

16-3.3 Grain Supplementation on Pasture

Although alfalfa-grass pasture provides adequate protein for high-producing dairy cows (135), milk production is increased by feeding an energy supplement such as ground corn and oat (*Avena sativa* L.) (49). Holstein cows producing 22.5 kg of milk daily can be supported on alfalfa-smooth bromegrass pasture alone (Fig. 16-1), but above that level, concentrates should be provided (51). Total protein requirements of cows producing up to 47 kg of milk daily may be met with alfalfa supplemented with grain as an energy source (37). In the major dairy areas of the USA and Canada, grain supplements are relatively inexpensive and commonly fed to high-producing dairy cows on pasture.

16-4 ALFALFA AS PASTURE FOR BEEF CATTLE

16-4.1 Production on Pasture

In many regions of the world, beef cattle are finished on pasture only. Alfalfa pasture is widely used for finishing cattle in Argentina, where alfalfa production rivals that in the USA (17). In the USA and Canada,

Fig. 16-1. More than 23 kg of milk per cow per day is obtainable on alfalfa-grass pasture alone (51). Total protein requirements of cows producing more than 45 kg of milk daily may be met by alfalfa pasture.

feeder calves are commonly wintered on hay and silage, then put on pasture, followed by finishing in drylot. Alfalfa-grass pastures can contribute greatly to these feeding programs.

Often alfalfa or alfalfa-grass pastures have yielded higher average daily gains by cattle, more liveweight gain per area, and greater carrying capacity than grass alone both with and without irrigation. Liveweight gains of more than 896 to 1120 kg/ha and average daily gains (ADG) of more than 0.9 kg season-long were reported with yearling steers on irrigated alfalfa-grass pasture without supplemental feed (Table 16-1) (83,165). Steers attained a grade of "U.S. Standard" after 22 weeks on pasture. Steers grazing irrigated alfalfa produced 501 kg of liveweight gain per ha, compared with 402 kg for those on birdsfoot trefoil (*Lotus corniculatus* L.)-orchardgrass. Average daily gains were similar for the two kinds of pasture: 0.75 and 0.79 kg, respectively (85). Cattle on alfalfa-grass attained a higher degree of finish than those on grass pasture (165).

In limited rainfall areas, including alfalfa with grasses markedly increased beef production and carrying capacity, compared to grasses alone (20,46,56,174). In eastern Colorado, steers on alfalfa-intermediate wheatgrass produced a 4-yr average liveweight gain per ha of 62 kg, compared to only 36 kg on intermediate wheatgrass; cattle on alfalfa-grass gained slightly better than those on grass alone (20). Alfalfa-smooth bromegrass and alfalfa-intermediate wheatgrass pastures in northern Saskatchewan parklands provided about 224 kg liveweight gain per ha and slightly >0.9 kg of ADG (Table 16-1) (39).

In the humid central USA, McVey et al. (120) obtained 5-yr averages of 370, 324, and 298 kg of beef/ha from alfalfa-smooth bromegrass, alfalfa-orchardgrass, and alfalfa-tall fescue (*Festuca arundinacea* Schreb.), respectively. They also reported ADG of 0.55, 0.34, and 0.38 kg/d, respectively. Matches (115) reported that alfalfa-smooth bromegrass pasture gave higher beef production per ha in a 5-yr study in Missouri than did tall fescue, orchardgrass, or orchardgrass-birdsfoot trefoil pastures. Average daily gains were comparable (slightly <0.9 kg) except for tall fescue, which gave slightly more than 0.45 kg. Scholl et al. (148) found that alfalfa-orchardgrass produced higher cattle gain per ha, slightly more grazing days, and similar ADG when compared with smooth bromegrass.

16-4.2 Pasture Utilization Methods

In general, continuous, rotational, and daily strip grazing plus mechanical harvesting of pasture have resulted in comparable animal daily gains, but usually the more intensive methods have usually increased animal gains per ha (77,85,87,121). For example, Ittner et al. (87) reported respective ADG of 0.81, 0.86, and 0.86 kg accompanied by gains per ha of 467, 650, and 788 kg with rotational grazing, strip grazing, and soilage. The greater animal product per ha from soilage resulted largely from more complete usage of the forage. Steers on irrigated alfalfa pasture utilized only about 65% of the available forage by grazing compared to soiling (132).

Meyer et al. (121), using the chromogen technique and total fecal collection, found that steers initially weighing about 248 kg and rotationally grazed on alfalfa pasture consumed 6.2 kg of DM daily. This included a limited amount of oat hay fed for bloat control. The steers required 389 kg of pasture DM/45 kg of gain. Steers on daily strip-grazed alfalfa pasture consumed 5.76 kg of DM daily and required 406 kg of DM per 45 kg gain. Soilage-fed steers consumed slightly more daily DM (6.75 kg) and DM per 45 kg gain (483 kg DM daily) than steers on rotational pasture. Average daily gains of steers was similar among the three methods. Feeding alfalfa as hay resulted in significantly (0.05) lower ADG compared with grazing and soilage. Forage utilized as pasture was 52% of that utilized as soilage, while beef production per ha from pasture was 64% of that from soilage, indicating forage selectivity by grazing steers. In a similar study, slightly lower values for all methods were obtained by Lofgreen et al. (108). Lofgreen and Meyer (107) reported DMD of 59.4% for alfalfa pasture in bud to early bloom stage, using the chromogen technique.

16-4.3 Grain Supplementation on Pasture

Steers usually will not reach a good to choice market grade on alfalfa pasture alone unless supplemental grain is fed (79). Energy is often the first limiting factor in the feeding value of forages (43,60,136,140). Baker and co-workers (8,9,12), Heinemann and Van Keuren (79), and Bryant et al. (26) confirmed this for alfalfa by feeding grain to pastured cattle. Baker (12) reported an ADG of 1.13 kg for steers full-fed grain for part of the alfalfa pasture period, compared with 1.07 kg for those fed in drylot. Baker and Baker (8) obtained similar results. Heinemann and Van Keuren (79) reported that steers given ground ear corn as half of their total feed (approximately 4.5 kg daily per head) for the last 60 d of a 165-d grazing period on alfalfa-orchardgrass attained U.S. Good to Choice grade. Steers that were not fed grain required drylot feeding for an additional 60 d. In England, Alder et al. (1) in England obtained only a small increase in steer gains by feeding barley (*Hordeum vulgare* L.) grain or beet pulp on alfalfa pasture compared with alfalfa pasture alone.

16-4.4 Cow-Calf Production

Although alfalfa is used to some extent in pastures for beef cow herds, only limited data are available on the value of this legume for beef cows and calves. Gross (67) conducted a 9-yr study in Manitoba comparing alfalfa-smooth bromegrass to smooth bromegrass alone for cow-calf production. He found the addition of alfalfa to the grass increased carrying capacity and calf production per ha. Generally, similar gains from beef cows and steers were obtained from alfalfa-perennial ryegrass (*Lolium perenne* L.) and alfalfa-orchardgrass pastures in Pennsylvania, with slightly better intake and gains from the alfalfa-perennial ryegrass mixtures (94). Also, perennial ryegrass was also reported to be less competitive with alfalfa than orchardgrass. The excellent gains of steers and lambs on alfalfa pasture suggest that equally good responses will be obtained by using alfalfa in pastures for beef cow and calf production and that the result will be higher weaning weights than from grass alone.

Alfalfa-tall fescue utilized as hay (two to three harvests) followed by late fall-early winter grazing (October to December) has been shown to be a useful system in Ohio for providing winter forage for beef cow herds (163,164). The alfalfa persisted well under this method of utilization. The alfalfa-tall fescue utilized in this way extended the grazing season for up to 8 weeks, and provided high-quality pasture for lactating, fall-calving beef cows or for conditioning spring-calving cows in preparation for the winter period.

16-5 ALFALFA AS PASTURE FOR SHEEP

16-5.1 Production on Pasture

Alfalfa is excellent sheep pasture, either alone or seeded with grass (Table 16-1). It is unexcelled for fattening lambs on pasture, with average daily gains as high as 0.38 kg obtained from suckling lambs (166). Weaned lambs grazing alfalfa pasture (Fig. 16-2) approached lambs in drylot on an all-concentrate ration in ADG, 0.24 and 0.28 kg/d, respectively (117), and in carcass characteristics, with less fat accumulation for the lambs

Fig. 16-2. Alfalfa is unexcelled as pasture for fattening lambs, with average daily gains as high as 0.24 kg (78).

grazing alfalfa (118). Fattening lambs have produced gains per ha as high as 896 to 1008 kg on irrigated alfalfa pasture (166). Lambs grading low U.S. Choice can be marketed directly off pasture in June at 140 d of age with no supplemental feed. Lambs also can be weaned and marketed directly off mixtures of alfalfa-smooth bromegrass (73) and alfalfa-orchardgrass (112). Alfalfa is used extensively in Australia and New Zealand as finishing pasture for fat lamb production.

Alfalfa or alfalfa with smooth bromegrass, orchardgrass, or timothy (*Phleum pratense* L.) has yielded higher sheep liveweight gain and more grazing days per ha than grasses alone (61,69,72,78,93). Jordan and Marten (92) found no difference in ADG of lambs grazing alfalfa or alfalfa-smooth bromegrass pastures until the grass became more than half of the mixture. They suggested that the inclusion of 50% or more high-quality grass in alfalfa pastures will reduce lamb performance because of reduced intake associated with high cell wall concentrations in grasses. Under range conditions common to much of the northern Great Plains of North America, adding alfalfa to crested wheatgrass, intermediate wheatgrass, and Russian wildrye increased liveweight gains per ewe and per ha, increased carrying capacity, and reduced forage consumption per pound of liveweight gain (28). Clark and Wilson (32) found that lambs grazing alfalfa with orchardgrass, tall fescue, or reed canarygrass (*Phalaris arundinacea* L.) produced better gains than did those grazing grasses alone. Alfalfa carried four dry ewes/ha compared with < one ewe/ha on native pastures in the low rainfall area of New South Wales (131,141).

16-5.2 Pasture Utilization Methods

In the humid region, rotational grazing of alfalfa by sheep with three paddocks resulted in more animal product per ha than with two paddocks (93). Average daily gains was not affected by grazing method. Harrison et al. (73) found little difference in animal gains per ha when comparing continuous and rotational grazing. They used the same stocking rate for both systems. As the forage in the continuously grazed pastures matured, the ewes dropped in milk production and their lambs tended to gain less than those on rotational grazing. Fuellemen et al. (62) obtained more animal product per ha from moderate alternate grazing than from continuous or heavy alternate grazing.

Under dryland conditions, continuous grazing of alfalfa-Russian wildrye and alfalfa-crested wheatgrass was as satisfactory for sheep production as a seasonal grazing system (spring or fall grazed) (27).

Feeder lambs on irrigated alfalfa pasture gained faster than those on alfalfa soilage (0.15 and 0.09 kg/d, respectively) and reached choice slaughter grade sooner (85). The pastured lambs consumed more forage, and utilized it more efficiently. Soiling increased lamb production 5%/ha over grazing. These researchers concluded that soiling did not appear economically promising compared with grazing for fattening lambs.

Alfalfa at nearly all stages of growth is eaten readily by sheep, although they exhibit increasing selectivity of leaves over stems with increasing plant maturity (5). Hull et al. (85), using the chromogen technique and total fecal collection, found that feeder lambs with an average initial weight of 29.7 kg consumed 1.06 kg of alfalfa pasture DM and 63 g of sudangrass hay DM (bloat preventative) daily and required 355 kg of forage DM to produce 45 kg of gain. Gain per day was 0.14 kg. Marten and Jordan (113) reported that alfalfa pasture grazed by yearling wethers had an OMD of 70.5% as estimated by the chromogen technique. The meat from lambs and yearlings grazing alfalfa had a different and more intense flavor than that from sheep which had grazed grass pastures (128).

16-5.3 Grain Supplementation on Pasture

Grain feeding generally has not proved necessary for suckling lambs on legume-grass pasture (73,78,166). Feeding grain to early weaned lambs, however, on alfalfa-smooth bromegrass pasture gave ADG of 0.20 kg, compared with 0.13 for the control (93). Feeding grain improved carcass grade and increased pasture carrying capacity. Also, it alleviated the problems of inadequate pasture during late summer and of light, under-finished lambs at the end of the grazing season. Grain can be fed all season or started when pasture production begins to decline.

Jordan and Marten (91) reported that in 2 of 3 yr ad libitum grain feeding of both suckling and weaned lambs increased their ADG on alfalfa-smooth bromegrass pasture. Grain supplementation increased the 3-yr ADG of suckling lambs from 0.18 to 0.22 kg, and of weaned lambs from 0.13 to 0.19 kg. Jordan and Marten (92) suggested more efficient utilization of the highly soluble alfalfa protein occurs when fed in conjunction with a more concentrated form of energy, such as grain. However, as pointed out by Garrett et al. (64), a feeder must determine if the increased daily gains and/or improved carcass grades would be sufficient to pay for the cost of the supplement. These authors, in a study on irrigated alfalfa pastures, found that pastures alone were more profitable.

16-6 EFFECT OF ALFALFA ESTROGENS

The adverse effects of plant estrogens on sheep have been documented for subterranean, red (*Trifolium pratense L.*), and white clovers and birdsfoot trefoil. The effects are primarily on reproductive efficiency of the ewe, often resulting in delayed lambing. Using the mouse uterine weight technique, estrogenic activity of freshly harvested alfalfa was shown highest in early bud stage (133). Stob et al. (157) reported a wide variation between estrogenic activity of alfalfa genotypes. Later studies by Hanson et al. (71) showed that the estrogenicity of alfalfa forage associated with accumulation of coumestrol resulted primarily from infection with foliar diseases caused by fungal pathogens. Alfalfa that is poor in quality because of foliar diseases is likely to be high in coumestrol.

In general, grazing alfalfa pasture has not been reported to have an adverse effect on the reproductive performance of sheep. Matthews and Foote (116) found that ewes grazing alfalfa prior to and during the breeding period were not different from ewes on wheatgrass pastures in average date of conception, percentage of ewes lambing, or body weight gain. The alfalfa-pastured ewes had a 20% higher lambing rate (lambs born per ewe). They suggested that the beneficial flushing effect of alfalfa apparently resulted in more ovulations, a higher rate of fertilization of the eggs, or a combination of both. Parker (129) showed that young ewes bred while grazing alfalfa pastures had a significantly higher lambing rate than those bred while grazing Kentucky bluegrass or during drylot feeding. Mature ewes had similar lambing rates among the three feeding regimes. All ewes on alfalfa pasture had similar lambing dates compared with those on Kentucky bluegrass pasture or in drylot. However, New Zealand studies have shown a reduction in lambing percentages primarily resulting from reduced twinning by mating ewes on grass-alfalfa pastures (147).

16-7 ALFALFA AS PASTURE FOR SWINE

Swine can utilize pasture for a portion of their protein, vitamin, and mineral needs. Pastures were needed originally to balance the rations commonly fed hogs (153,154), but comparable results can now be obtained in drylot. Although pasture is used for all classes of swine, it is considered more desirable for breeding animals. Older animals are able to utilize pasture to a much greater extent than are young growing pigs (38). Sows have a large digestive capacity, and are capable of consuming 4.5 to 6.3 kg of a bulk ration during pregnancy and 7.2 to 8.1 kg during lactation. Snyder (153) reported that mature hogs thin in flesh gained slightly <0.22 kg/d on alfalfa pasture alone. Becker et al. (16) reported that farrowing and weaning records for sows and gilts fed on alfalfa pasture were similar to those fed in drylot, with some saving of feed on pasture. Whatley et al. (172) and Self et al. (150) reported that breeding swine required less feed during pregnancy and lactation when pastured on alfalfa-grass.

Growing and finishing pigs fed on alfalfa pastures generally required less feed than those fed in drylot (29,47,84,86). Danielson et al. (47) reported ADG of 0.76 and 0.80 kg for young finishing swine on alfalfa pasture and drylot, respectively. They estimated that 5 to 12% of the total DM intake on pasture was in the form of alfalfa. Pasture-fed hogs had slightly less back fat and less estimated carcass fat in a study by Hudman and Peo (84). Dean and Thompson (50) suggested alfalfa and ladino clover as the best hog pasture forages. They listed exercise, parasite reduction, and reduced feed cost as advantages for pasture over drylot.

16-8 ALFALFA AS PASTURE FOR POULTRY

Pasture has long been depended upon to supplement poultry rations, providing a good source of protein, vitamins, and minerals. In many parts of the world, poultry still forage for all or some of their feed. In these areas, pasture is still the cheapest source of quality protein and vitamins. Poultry need a pasture that is young, leafy, nutritious, and easily grazed if they are to obtain a significant part of their total feed requirements from the herbage.

Early attempts at confinement of poultry were not successful until the discovery of nutritional factors that were supplied by pasture were discovered (96). These factors can now be supplied by dehydrated alfalfa and other sources, and pastures have become uneconomical for large-scale egg factories.

However, pasture is still of considerable importance for smaller laying flocks and for range rearing of turkeys (*Meleagris gallopavo*) and other poultry. The nutritive value of poultry pasture was reviewed by Cowlishaw and Eyles (41). Pasture may comprise up to 20% of the feed intake of laying birds, and it can provide much of the needed protein and vitamins. Two legumes, alfalfa and ladino clover, have wide acceptance for poultry pasture in the USA (178). Alfalfa is the most satisfactory perennial pasture for poultry (68,130).

Barnett et al. (13) found that the use of pasture resulted in a higher percentage of broad breasts in turkeys and provided more sanitary conditions than confinement rearing. Sunde et al. (159) reduced feed costs one-third by using good-quality ladino clover for turkeys on a low-protein ration compared with high-protein rations on poor pasture or no pasture. Turkey gains, however, were lower on the low-protein ration. Similar results were obtained with alfalfa pasture (Sunde, 158).

Berry (18) compared alfalfa range with grass [winter wheat (*Triticum aestivum* L.) in winter-spring, sudangrass in summer-fall], and with bare soil. He reported the largest return over feed cost when the laying flock was provided with alfalfa range. Limiting hens to the alfalfa range for 2 hr each day returned a larger profit than allowing continuous free range on alfalfa. Cowlishaw et al. (42) found better persistence of the pasture stand with a stocking rate of 150 laying hens/ha vs. a stocking rate of 500 or 750 birds/ha. Shoulders and Bragg (151) pointed to the need for rotational grazing of alfalfa pasture with poultry to maintain the stand. They recommended restricting each grazing period to 7 to 10 d.

16-9 ALFALFA AS PASTURE FOR HORSES

Although not widely used, alfalfa is recognized as an excellent horse pasture for many regions (52,55). Morrison (124) listed alfalfa-smooth bromegrass as an ideal pasture for horses.

In a grazing experiment in eastern Washington, Galgan et al. (63) found that alfalfa improved both yield and protein content of pasture when mixed with grass. Alfalfa also made a superior horse feed if it became necessary to cut surplus pasture for hay.

16-10 BLOAT IN ANIMALS ON ALFALFA PASTURE

A large annual livestock loss occurs from bloat. Even greater economic loss results by limiting use of high-yielding legumes in pastures because of the fear of bloat (14). However, many researchers working with alfalfa pastures reported few, if any, problems with bloat in dairy cattle (7,95,126), sheep (80,91,95,166), and beef cattle (7,46,56,95). Van Keuren and Heinemann (165) reported an average death loss from bloat of only 1.7% after fattening steers on legume-grass pastures for 3 yr. They concluded that despite bloat loss of 5% the 1st yr, it was more profitable to include legumes in the pastures than to use grasses alone. Repetitive bloat appears most frequently with the same animals. Brundage and Petersen (22) noted that bloat only occurred within sets of animals. Cole and Boda (33) reviewed the evidence for animal differences to bloat susceptibility.

The following precautions are commonly cited to help prevent bloat: (i) avoid grazing immature stands; (ii) use a mixture that has at least 50% grass; (iii) never turn hungry animals into lush stands of alfalfa but rather let them prefill with dry forage; (iv) where bloat has been a problem, give overnight feedings of grass hay, or graze the alfalfa pasture alternately with grass pasture; and (v) with fattening animals, move chronic bloaters to the feedlot. Other practices, such as pasturing alfalfa only when free of dew, restricting the area and length of grazing period, and providing a bunker of hay on pasture have not proved consistently useful. Of these suggested practices, only supplemental feeding of oat hay on alfalfa pasture (36) and joint feeding of alfalfa pasture with sudangrass or sudangrass pasture (34,35) have been tested experimentally and reported effective.

Cole and Boda (34) summarized the numerous control treatments that provide bloat protection for varying time lengths. All of these control measures—feeding antibiotics (such as penicillin), detergents, methyl silicone, vegetable oils or animal fats, or spraying pastures with emulsified peanut oil or tarrow—provide relatively short-term bloat control. The most recent control measure, the use of poloxalene, also requires daily ingestion for bloat prevention (14). The ultimate bloat control in alfalfa grazing will depend upon determining the substance(s) or characteristics of the plant that cause bloat. This may enable plant breeders to reduce or eliminate bloat potential of alfalfa (see Chapter 15 in this book).

16-11 SUMMARY

Beginning with its Asian origin as a range plant, alfalfa has had a long history as a superior grazing legume for many classes of livestock. The advantages of alfalfa are determined by DM production, high protein yield potential, and excellent animal intake. High daily milk production by dairy cows and excellent daily gains by fattening cattle and sheep are obtainable from grazing alfalfa or alfalfa-grass mixtures. Digestible energy is the limiting factor in alfalfa pastures. Supplemental feeding of grain to dairy cows, sheep, and fattening cattle balances alfalfa's high-protein level and extends its usefulness as a pasture. Alfalfa interseeded into dryland pastures in range areas improves animal performance and carrying capacity. Likewise, alfalfa pasture is useful for poultry, swine, and horses. Mature breeding sows can be maintained on alfalfa pasture alone. When ruminant bloat is a concern, a number of animal management and prophylactic measures are available. Recent reports indicate that it may be possible to develop alfalfa cultivars low in bloat potential. Because alfalfa requires a period of regrowth following grazing in order to regenerate new energy reserves, rotational grazing is recommended. The proper length of grazing and rest periods will differ according to stocking rate and environmental conditions for regrowth.

REFERENCES

1. Alder, F.E., M.J. Head, and J.F.R. Berting. 1956. Carbohydrate supplements for beef cattle on grass/clover and grass/lucerne mixtures. Proc. Br. Soc. Anim. Prod. 1956:78–89.
2. ———, and D.J. Minson. 1960. Lucerne as a grazing crop. Agriculture 67:448–451.
3. Allen, V.G., D.D. Wolf, J.P. Fontenot, J. Cardina, and D.R. Notter. 1986a. Yield and

regrowth characteristics of alfalfa grazed with sheep: I. Spring grazing. Agron. J. 78: 974-979.
4. ----, ----, ----, ----, and ----. 1986b. Yield and regrowth characteristics of alfalfa grazed with sheep: II. Summer grazing. Agron. J. 78: 979-985.
5. Arnold, G.W. 1960. Selective grazing by sheep of two forage species at different stages of growth. Aust. J. Agric. Res. 11:1026-1033.
6. Arny, A.C., and A.R. Schmid. 1958. Rotation pasture studies, 1936-1947. Minn. Agric. Exp. Stn. Tech. Bull. 223.
7. Ashford, R., and D.H. Heinrichs. 1967. Grazing of alfalfa varieties and observations on bloat. J. Range Manage. 20:152-153.
8. Baker, G.N., and M.L. Baker. 1952. The use of various pastures in producing finished yearling steers. Nebr. Agric. Exp. Stn. Bull. 414.
9. ----, ----, and J. Jackson. 1956. Concentrates for yearling steers on alfalfa pasture and in drylot. Nebr. Agric. Exp. Stn. Bull. 435.
10. Barker, M.G., F. Hanley, and W.J. Ridgman. 1955. Studies on lucerne and lucerne-grass leys. I. Summer and autumn management of a lucerne-grass mixture grown on heavy land. J. Agron. Sci. 46:362-376.
11. ----, ----, and ----. 1957. Studies on lucerne and lucerne-grass leys. IV. The effect of systems of grazing management on the persistence of a lucerne-cocksfoot ley. J. Agric. Sci. 48:361-365.
12. Baker, M.L. 1938. The use of alfalfa and native grass pasture in producing finished cattle. Nebr. Agric. Exp. Stn. Bull. 315.
13. Barnett, B.D., E.C. Naber, J.B. Cooper, and C.L. Morgan. 1958. Influence of range and confinement rearing of turkeys on growth, feed consumption and body conformation. Poult. Sci. 37:1304-1308.
14. Bartley, E.E. 1967. Progress in bloat prevention. Agric. Sci. Rev. 5(1):5-12.
15. Bateman, G.Q., and W. Keller. 1956. Grass-legume mixtures for irrigated pastures for dairy cows. Utah Agric. Exp. Stn. Bull. 382.
16. Becker, D.E., B.G. Harmon, A.H. Jensen, and W.F. Nickelson. 1965. Levels of feed for gestating sows and gilts on pasture. Ill. Agric. Exp. Stn. AS-620.
17. Beetle, A.A. 1954. The Argentine literature on range management. J. Range Manage. 7:125-127.
18. Berry, L.N. 1938. Ranges for the laying flock. N. Mex. Agric. Exp. Stn. Bull. 255.
19. Blaser, R.E., R.C. Hammes, Jr., H.T. Bryant, W.A. Hardison, J.P. Fontenot, and R.W. Engel. 1960. The effect of selective grazing on animal output. p. 601-606. In Proc. 8th Int. Grassl. Congr., Hurley, UK. 11-21 July. British Grassland Society, Grassland Research Institute, Hurley, UK.
20. Bonham, C.D., and D.F. Hervey. 1963. Grass versus grass-legume mixtures. Colo. Agric. Exp. Stn. Prog. Rep. PR74-A.
21. Brownlee, H. 1973. Effects of four grazing management systems on the production and persistence of dryland lucerne in central western New South Wales. Aust. J. Exp. Agric. Anim. Husb. 13:259-262.
22. Brundage, A.L., and W.E. Petersen. 1952. A comparison between daily rotational grazing and continuous grazing. J. Dairy Sci. 35:623-630.
23. ----, and W.J. Sweetman. 1958. Comparative utilization of alfalfa-bromegrass pasture under rotational and daily strip grazing. J. Dairy Sci. 41:1777-1780.
24. Bryant, H.T., R.E. Blaser, R.C. Hammes, Jr., and W.A. Hardison. 1961. Method for increased milk production with rotational grazing. J. Dairy Sci. 44:1733-1741.
25. ----, ----, ----, and ----. 1961. Comparison of continuous and rotational grazing of three forage mixtures by dairy cows. J. Dairy Sci. 44:1742-1750.
26. ----, R.C. Hammes, Jr., R.E. Blaser, and J.P. Fontenot. 1965. Effects of feeding grain to grazing steers to be fattened in drylot. J. Anim. Sci. 24:676-680.
27. Campbell, J.B. 1961. Continued versus repeated-seasonal grazing of grass-alfalfa mixtures at Swift Current, Saskatchewan. J. Range Manage. 14:72-77.
28. ----. 1963. Grass-alfalfa versus grass-alone pastures grazed in a repeated-seasonal pattern. J. Range Manage. 16:78-81.
29. Carlisle, G.R., D.E. Becker, B.C. Breidenstein, and H.W. Norton. 1966. Effect of different protein levels of rations fed to growing-finishing pigs on pasture and drylot. Ill. Agric. Exp. Stn. AS-633a.
30. Christian, C.S., and N.H. Shaw. 1952. A study of two strains of Rhodes grass (*Chloris gayana* Kunth.) and of lucerne (*Medicago sativa* L.) as components of a mixed pasture at Lawes in south-east Queensland. Aust. J. Agric. Res. 3:277-299.
31. Clark, R.D. 1960. Personal communication. Research Station, Canada Department of Agriculture, Lethbridge, AB.
32. ----, and D.B. Wilson. 1966. Sheep production on irrigated pasture as influenced by forage mixture and fertilization. Can. J. Anim. Sci: 46:97-106.
33. Cole, H.H., and J.M. Boda. 1960. Continued progress toward controlling bloat. A review. J. Dairy Sci. 43:1585-1614.

34. ——, and M. Kleiber. 1945. Bloat in cows on alfalfa pasture. Am. J. Vet. Res. 6:188–193.
35. ——, S.W. Meade, and M. Kleiber. 1942. Bloat in cattle. Calif. Agric. Exp. Stn. Bull. 662.
36. Colvin, H.W., Jr., P.T. Cupps, and H.H. Cole. 1958. Efficacy of oat hay as a legume bloat preventative in cattle. J. Dairy Sci. 41:1557–1564.
37. Conrad, H.R., and J.W. Hibbs. 1965. Dairy ration for high productivity. Ohio Agric. Res. Dev. Center Ohio Rep. 50(3):35.
38. Conrad, J.H., and W.R. Beeson. 1957. A comparison of drylot and pasture for producing pork rapidly and economically. Purdue Univ. Agric. Exp. Stn. Mimeo A.H. 213.
39. Cooke, D.A., S.E. Beacom, and W.K. Dawley. 1965. Pasture productivity of two grass-alfalfa mixtures in northeastern Saskatchewan. Can. J. Plant Sci. 45:162–168.
40. Counce, P.A., J.H. Bouton, and R.H. Brown. 1984. Screening and characterizing alfalfa for persistence under mowing and continuous grazing. Crop Sci. 24:282–285.
41. Cowlishaw, S.J., and D.E. Eyles. 1957. The nutritive value of herbage for poultry. Nutr. Abstr. Rev. 27:983–996.
42. ——, D.E. Eyles, and J.G. Astbury. 1956. The effect of two rates of stocking on contrasting swards for poultry. J. Br. Grassl. Soc. 11:155–161.
43. Crampton, E.W. 1957. Interrelationship between digestible nutrient and energy content, voluntary dry matter intake, and the over-all feeding value of forages. J. Anim. Sci. 16:546–552.
44. Cuykendall, C.H., and G.C. Marten. 1968. Defoliation by sheep-grazing versus mower-clipping for evaluation of pasture. Agron. J. 60:404–408.
45. Daday, H. 1968. Heritability and genotypic and environmental correlations of creeping root and persistency in *Medicago sativa* L. Aust. J. Agric. Res. 19:27–34.
46. Dahl, B.E., A.C. Everson, J.J. Norris, and A.H. Denham. 1967. Grass-alfalfa mixtures for grazing in eastern Colorado. Colo. Agric. Exp. Stn. Bull. 529-S.
47. Danielson, D.M., J.E. Butcher, and J.C. Street. 1969. Estimation of alfalfa pasture intake and nutrient utilization by growing-finishing swine. J. Anim. Sci. 28:6–12.
48. Davis, A.G. 1947. Lucerne growing. Emp. J. Exp. Agric. 15:113–118.
49. Davis, R.R., and A.D. Pratt. 1956. Rotational vs. continuous grazing with dairy cows. Ohio Agric. Exp. Stn. Bull. 778.
50. Dean, B., and W.C. Thompson. 1968. Pastures for hogs. Ky. Agric. Exp. Stn. Leaflet 312.
51. Donker, J.D., G.C. Marten, and W.F. Wedin. 1968. Effect of concentrate level on milk production of cattle grazing high-quality pasture. J. Dairy Sci. 51:67–73.
52. Duell, R.W. 1961. Better pastures for horses and ponies. N.J. Agric. Ext. Serv. Bull. 350 A.
53. Edye, L.A., and K.P. Haydock. 1967. Breeding creeping-rooted lucerne for the subtropics. Aust. J. Agric. Res. 18:891–901.
54. Edgerly, C.G.M., and J.F. Carter. 1960. Pastures for dairy cows. N. Dak. Exp. Stn. Farm Res. 21(5):5–10.
55. Ensminger, M.E. 1959. The stockman's handbook. The Interstate Publishers, Danville, IL.
56. ——, H.G. McDonald, A.G. Law, E.J. Warwick, E.J. Kreizinger, and V.B. Hawk. 1944. Grass and grass-alfalfa mixtures for beef production in eastern Washington. Wash. Agric. Exp. Stn. Bull. 444.
57. Feltner, K.C., and M.A. Massengale. 1965. Influence of temperature and harvest management on growth level of carbohydrates in the roots, and survival of alfalfa (*Medicago sativa* L.). Crop Sci. 5:585–589.
58. Field, T.R.O., C.J. Pearson, and L.A. Hunt. 1976. Effects of temperature on the growth and development of alfalfa. Herb. Abstr. 46:145–150.
59. FitzGerald, R.D. 1974. The effect of intensity of rotational grazing on lucerne density and ewe performance at Wagga Wagga, Australia. p. 127–133. *In* Proc. 12th Int. Grassl. Congr., Moscow, USSR. 11–20 June.
60. Forbes, R.M., and W.P. Garrigus. 1950. Some relationships between chemical composition, nutritive value, and intake of forages grazed by steers and wethers. J. Anim. Sci. 9:354–362.
61. Fuelleman, R.F., W.L. Burlison, and W.G. Kammlade. 1944. A comparison of brome-grass and orchardgrass pastures. J. Am. Soc. Agron. 36:849–858.
62. ——, ——, and ——. 1948. Methods of management of a bromegrass-alfalfa mixture. J. Anim. Sci. 7:99–109.
63. Galgan, M.W., W.C. Green, M.E. Ensminger, J.K. Patterson, A.G. Law, and J.L. Schwendiman. 1956. Pasture for horses. Wash. Agric. Exp. Stn. Circ. 289.
64. Garrett, W.N., W.C. Weir, J.H. Meyer, and G.P. Lofgreen. 1960. Effect of various energy supplements on gains, yields and carcass grades of lambs grazing alfalfa pasture. J. Anim. Sci. 19:773–779.

65. Gramshaw, D., H.G. Bishop, and D.H. Ludke. 1982. Performance of grazed creeping-rooted lucernes on two soils in central Queensland. Aust. J. Exp. Agric. Anim. Husb. 22:177–181.
66. Gross, A.E., and K. Matheson. 1964. Irrigated pastures produce high beef gains. Crops Soils 17(3):22–23.
67. Gross, A.T.H. 1970. Personal communication. Research Station, Research Branch, Agriculture Canada, Brandon, MB.
68. Gutteridge, H.S., and F.S. Nowosad. 1955. Pasture for poultry. Can. Dep. Agric. Pub. 771.
69. Hamilton, R.I., J.M. Scholl, and A.L. Pope. 1969. Performance of three grass species grown alone and with alfalfa under intensive pasture management. Agron. J. 61:357–361.
70. Hanley, F., W.J. Ridgman, and J.D. Whitear. 1964. The effect of date of winter grazing on the yield of a lucerne-grass ley. J. Agric. Sci. 62:281–284.
71. Hanson, C.H., G.M. Loper, G.O. Kohler, E.M. Bickoff, K.W. Taylor, W.R. Kehr, E.H. Stanford, J.W. Dudley, M.W. Pedersen. E.L. Sorensen, H.L. Carnahan, and C.P. Wilsie. 1965. Variation in coumestrol content of alfalfa. USDA Tech. Bull. 1333.
72. Harrison, C.M., H.M. Brown, and H.C. Rather. 1947. The production of forage crop mixtures under different systems of management, the consequent effect on corn yields, and the re-establishment of alfalfa. J. Am. Soc. Agron. 39:214–223.
73. ----, C.L. Cole, and H.C. Rather. 1941. Continuous and rotation grazing of alfalfa-bromegrass pastured with ewes and lambs. Soil Sci. Soc. Am. Proc. 6:303–308.
74. Harshberger, K.E., E.E. Ormiston, J.R. Staubus, and R.V. Johnson. 1965. A nutritional assessment of methods of harvesting summer forage for dairy cows. Ill. Agric. Exp. Stn. Bull. 709.
75. Hart, R.H. 1984. Personal communication, USDA and Colorado State University, Ft. Collins, CO.
76. ----, and A.O. Gdara. 1982. Response of alfalfa cultivars to defoliation patterns. Agron. Abstr. American Society of Agronomy, Madison, WI, p. 121.
77. Heinemann, W.W. 1970. Continuous and rotation grazing by steers on irrigated pastures. Wash. Agric. Exp. Stn. Bull. 724.
78. ----, and R.W. Van Keuren. 1958. A comparison of grass-legume mixtures, legumes, and grass under irrigation as pasture for sheep. Agron. J. 50:189–192.
79. ----, and ----. 1958. Irrigated pastures and a limited grain ration for fattening steers. Wash. Agric. Exp. Stn. Bull. 585.
80. Heinrichs, D.H. 1963. Creeping alfalfas. Adv. Agron. 15:317–337.
81. ----, and K.F. Nielsen. 1966. Growth response of alfalfa varieties of diverse genetic origin to different root zone temperatures. Can. J. Plant Sci. 46:291–298.
82. Hill, D.L., and N.S. Lundquist. 1952. Milk production from pasture. Purdue Univ. Agric. Exp. Stn. Circ. 386.
83. Hubbard, W.A., and H.H. Nicholson. 1964. Irrigated grass-legume mixtures as summer pasture for yearling steers. Can. J. Plant Sci. 44:332–336.
84. Hudman, D.B., and E.R. Peo, Jr. 1960. Carcass characteristics of swine as influenced by levels of protein fed on pasture and in drylot. J. Anim. Sci. 19:943–947.
85. Hull, J.L., J.H. Meyer, G.P. Lofgreen, and A. Strother. 1957. Studies on forage utilization by steers and sheep. J. Anim. Sci. 16:757–765.
86. Hutchinson, H.D., A.H. Jensen, S.W. Terrill, and D.E. Becker. 1956. Comparison of free-choice and complete rations on pasture and drylot. Ill. Agric. Exp. Stn. AS435.
87. Ittner, N.R., G.P. Lofgreen, and J.H. Meyer. 1954. A study of pasturing and soiling alfalfa with beef steers. J. Anim. Sci. 13:37–43.
88. Iversen, C.E. 1967. Grazing management of lucerne. p. 129–133. In R.H. Langer (ed.) The lucerne crop. Reed, Wellington, New Zealand.
89. Jacobs, J.A. 1950. The influence of spring-clipping, interval between cuttings, and date of last cutting on alfalfa yields in the Yakima Valley. Agron. J. 42:594–597.
90. Janson, C.G. 1982. Lucerne grazing management research. p. 85–90. In R.B. Wynn-Williams (ed.) Lucerne for the 80's. Spec. Pub. Agronomic Society of New Zealand, Palmerston North, New Zealand.
91. Jordan, R.M., and G.C. Marten. 1968. Effect of weaning, age at weaning and grain feeding on the performance and production of grazing lambs. J. Anim. Sci. 27:174–177.
92. ----, and ----. 1983. Lamb production on pasture, as affected by forage and grain variables. p. 137–140. In Proc. 1983 Forage and Grassl. Conf., Eau Claire, WI. 23–26 January. American Forage and Grassland Council, Lexington, KY.
93. ----, and W.F. Wedin. 1961. Lamb production as affected by forage mixture and grazing management of perennial pastures. J. Anim. Sci. 20:898–902.
94. Jung, G.A., L.L. Wilson, P.J. LeVan, R.E. Kocher, and R.F. Todd. 1982. Herbage and beef production from ryegrass-alfalfa and orchardgrass-alfalfa pastures. Agron. J. 74:937–942.

95. Kehr, W.R., E.C. Conrad, M.A. Alexander, and F.G. Owen. 1963. Performance of alfalfas under five management systems. Nebr. Agric. Exp. Stn. Res. Bull. 211.
96. Kennard, D.C., and V.D. Chamberlin. 1934. Shall the layers be ranged or confined? Ohio Agric. Exp. Stn. Bimonthly Bull. 171:193–198.
97. Kennedy, W.K., J.T. Reid, and M.J. Anderson. 1959. Evaluation of animal production under different systems of grazing. J. Dairy Sci. 42:679–685.
98. Langer, R.H.M. 1973. Lucerne. p. 357–360. In R.H.M. Langer (ed.) Pastures and pasture plants. Reed, Wellington, New Zealand.
99. Larsen, H.J. 1959. Methods of forage utilization in the Midwest. J. Dairy Sci. 42:574–578.
100. ----, and R.F. Johannes. 1965. Summer forage: stored feeding, green feeding, and strip grazing. Wis. Agric. Exp. Stn. Res. Bull. 257.
101. ----, ----, J.M. Sund, and M.F. Finner. 1965. Systems of summer forage utilization. p. 615–617. In Proc. 9th Int. Grassl. Congr., Sao Paulo, Brazil, 7–20 January. Edicoes Harico Limitada, Sao Paulo.
102. Leach, G.F. 1977. The survival of some erect and spreading lucerne lines at Lawes, south-east Queensland. Aust. J. Exp. Agric. Anim. Husb. 17:412–416.
103. ----. 1978. The ecology of lucerne pastures. p. 290–308. In J.R. Williams (ed.) Plant relations in pastures. CSIRO, Melbourne.
104. ----. 1979. Lucerne survival in south-east Queensland in relation to grazing management systems. Aust. J. Exp. Agric. Anim. Husb. 19:208–215.
105. ----, D. Gramshaw, and F.H. Kleinschmidt. 1984. The survival of erect and spreading lucerne, Medicago sativa, under grazing at Lawes and Bioela, southern Queensland. Trop. Grassl. 16:206–213.
106. ----, and D. Ratcliff. 1979. Lucerne survival in relation to grass management on a brigalow land in south-east Queensland. Aust. J. Exp. Agric. Anim. Husb. 19:198–207.
107. Lofgreen, G.P., and J.H. Meyer. 1956. A method for determining total digestible nutrients in grazed forage. J. Dairy Sci. 39:268–273.
108. ----, ----, and M.L. Peterson. 1956. Nutrient consumption and utilization from alfalfa pasture, soilage, and hay. J. Anim. Sci. 15:1158–1165.
109. Logan, V.S., and V. Miles. 1958. Pasture management studies. I. Daily strip grazing versus free range grazing of dairy cattle on cultivated pastures. Can. J. Anim. Sci. 38:133–144.
110. t'Mannetje, L. 1967. Pasture improvement in the Eshdale district of south-eastern Queensland. Trop. Grassl. 1:9–19.
111. Marshall, S.P., and J.M. Myers. 1963. Unirrigated and irrigated alfalfa-oat-clover pasture for dairy cows. Fla. Agric. Exp. Stn. Bull. 659.
112. Marten, G.C., and R.M. Jordan. 1970. Personal communication, USDA and University of Minnesota, St. Paul.
113. ----, and ----. 1967. Pasture quality for sheep as estimated by chromogen vs. nitrogen indicators. J. Anim. Sci. 26:1165–1168.
114. ----, W.F. Wedin, and J.D. Donker. 1963. A comparison of two established fecal index systems for estimating the digestibility and consumption of forages by grazing dairy cattle. Agron. J. 55:265–268
115. Matches, A.G. 1970. Personal communication, Texas Tech. University, Lubbock.
116. Matthews, D.H., and W.C. Foote. 1961. Reproductive performance of ewes pastured on different sequences of alfalfa and wheatgrass during the breeding season. J. Anim. Sci. 20:677 (Abstr.).
117. McClure, K.E., and R.W. Van Keuren. 1985. Effect of grazed forages and all concentrate in drylot on lamb performance. J. Anim. Sci. 61 (Suppl. 1): 340 (Abstr. 326).
118. ----, ----, and P.G. Althouse. 1985. Effect of grazed forages and concentrate in drylot on lamb carcass characteristics. J. Anim. Sci. 61 (Suppl. 1): 341 (Abstr. 327).
119. McKinney, G.T. 1974. Management of lucerne for sheep grazing on the southern Tablelands of New South Wales. Aust. J. Exp. Agric. Anim. Husb. 14:726–734.
120. McVey, W.M., H.N. Wheaton, and G.O. Mott. Undated. Alfalfa-grass pastures for beef cattle. Purdue Univ. Agric. Ext. Serv. Mimeo AY-148.
121. Meyer, J.H., G.P. Lofgreen, and N.R. Ittner. 1956. Further studies on the utilization of alfalfa by beef steers. J. Anim. Sci. 15:64–75.
122. Moore, R.A. 1970. Symposium on pasture methods for maximum production in beef cattle: Pasture systems for a cow-calf operation. J. Anim. Sci. 30:133–137.
123. Moore, R.M., N. Barrie, and E.H. Kipps. 1946. Grazing management: continuous and rotational grazing by Merino sheep. Aust. CSIRO Bull. 201.
124. Morrison, F.B. 1956. Feeds and feeding. The Morrison Publishing Co., Ithaca, NY.
125. Nuttall, W.F., D.A. Cooke, J. Waddingston, and J.A. Robertson. 1980. Effect of nitrogen and phosphorous fertilizers on a bromegrass and alfalfa mixture grown under two systems of pasture management. I. Yield, percentage legume in sward, and soil tests. Agron. J. 72:289–294.

126. Olson, T.M., and T.A. Evans. 1938. Ten years of experimental results on cultivated pastures. S. Dak. Agric. Exp. Stn. Bull. 324.
127. Owen, F.G., R.G. Hinders, P.E. Schleusener, and G.E. Van Riper. 1963. Value of irrigated bromegrass-alfalfa and orchardgrass-alfalfa pastures for lactating dairy cows. J. Dairy Sci. 46:830–834.
128. Park, R.J., A. Ford, D.J. Minson, and R.I. Baxter. 1975. Lucerne-derived flavour in sheep meat as affected by season and duration of grazing. J. Agric. Sci. 84:209–213.
129. Parker, C.F. 1968. Reproductive performance of ewes mated on bluegrass, alfalfa, and ladino pastures and under drylot confinement. Ohio Agric. Res. Dev. Center Res. Summary 28:13–14.
130. Payne, L.F., and C.L. Gish. 1943. Grass and alfalfa as silage, forage, and meal for poultry. Kans. Agric. Exp. Stn. Bull. 320.
131. Peart, G.R. 1968. A comparison of rotational grazing and set stocking of dryland lucerne. Proc. Aust. Soc. Anim. Prod. 7:110–113.
132. Petersen, M.L., G.P. Lofgreen, and J.H. Meyer. 1956. A comparison of the chromogen and clipping methods for determining the consumption of dry matter and total digestible nutrients by beef steers on alfalfa pasture. Agron. J. 48:560–563.
133. Pieterse, P.J.S., and F.N. Andrews. 1956. The estrogenic activity of alfalfa and other feedstuffs. J. Anim. Sci. 15:25–36.
134. Porter, R.M., and S.R. Skaggs. 1958. Forage and milk yields from alfalfa under three different harvesting systems. N. Mex. Agric. Exp. Stn. Bull. 421.
135. Pratt, A.D., and R.R. Davis. 1956. High- vs. low-protein grain mixtures as supplements to legume-grass pastures. J. Dairy Sci. 39:1304–1308.
136. ----, ----, and R.W. Van Keuren. 1956. Sudangrass vs. alfalfa-grass for dairy pasture and silage in northeastern Ohio. Ohio Agric. Res. Dev. Center Res. Bull. 990.
137. Rather, H.C., and A.B. Dorrance. 1935. Pasturing alfalfa in Michigan. J. Am. Soc. Agron. 27:57–65.
138. ----, and ----. 1938. A study of the time of pasturing alfalfa. J. Am. Soc. Agron. 30:130–134.
139. Reeve, F.L., and M.F. Sharkey. 1980. Effect of stocking rate, time of lambing and inclusion of lucerne on prime lamb production in north-east Victoria. Aust. J. Exp. Agric. Anim. Husb. 20:637–653.
140. Reid, J.T., K.L. Turk, W.A. Hardison, C.M. Martin, and P.G. Woolfolk. 1955. The adequacy of some pastures as the sole source of nutrients for growing cattle. J. Dairy Sci. 38:20–28.
141. Robards, G.E., and G.R. Peart. 1967. Outstanding results from rotationally grazed lucerne. Agric. Gaz. N.S.W. 78:14–19.
142. Rogers, V.E. 1967. Adaptability of lucerne to soil and climate. p. 36–46. In R.H.M. Langer (ed.) The lucerne crop. Reed, Wellington, New Zealand.
143. Rumbaugh, M.D., and M.W. Pedersen. 1979. Survival of alfalfa in five semiarid range seedings. J. Range Manage. 32:48–51.
144. ----, G. Semeniuk, R. Moore, and J.D. Colburn. 1965. Travois—an alfalfa for grazing. S. Dak. Agric. Exp. Stn. Bull. 525.
145. ----, and T. Thorn. 1965. Initial stands of interseeded alfalfa. J. Range Manage. 18:258–261.
146. Rumery, M.G.A., E.C. Conard, and J.C. Adams. 1956. Irrigated pastures for milk production. Nebr. Agric. Exp. Stn. Bull. 436.
147. Scales, G.H., and R.A. Moss. 1976. Mating ewes on lucerne. N.Z. Agric. 132(2):21–22.
148. Scholl, J.M., M.D. Groskopp, and J.M. Sund. 1966. p. 487–490. In Proc. 10th Int. Grassl. Congr., Helsinki, Finland. 7–16 July. Finnish Grassland Society, Helsinki.
149. Seath, D.M., W.C. Templeton, Jr., D.R. Jacobson, W.M. Miller, and T.H. Taylor. 1962. Grazing comparisons of two alfalfa-grass-ladino clover mixtures for dairy cows. Ky. Agric. Exp. Stn. Bull. 676.
150. Self, H.L., R.H. Grummer, O.E. Hays, and H.G. Spies. 1960. Influence of three different feeding levels during growth and gestation on reproduction, weight gains and carcass quality in swine. J. Anim. Sci. 19:274–282.
151. Shoulders, J.F., and D.D. Bragg. 1959. Poultry pastures for Virginia. Va. Agric. Ext. Serv. Circ. 822.
152. Smith, M.Y. 1970. Effects of stocking rate and grazing management on the persistence and production of dryland lucerne on deep sands. p. 624–628. In Proc. 11th Int. Grassl. Congr., Surfers Paradise, Queensland, Australia. 13–n23 April. University of Queensland Press, St. Lucia.
153. Snyder, W.P. 1907. Growing hogs in western Nebraska. Nebr. Agric. Exp. Stn. Bull. 99.
154. ----. 1930. Pork production at the North Platte substation. Nebr. Agric. Exp. Stn. Bull. 243.

155. Southwood, O.R., and G.E. Robards. 1975. Lucerne persistence and the productivity of ewes and lambs grazed at two stocking rates within different management systems. Aust. J. Exp. Agric. Anim. Husb. 15:747–752.
156. Spahr, S.L., E.E. Ormiston, and R.G. Peterson. 1967. Sorghum-sudan hybrid SX-11, Piper sudangrass, and alfalfa-orchardgrass for dairy pastures. J. Dairy Sci. 50:1925–1934.
157. Stob, M., R.L. Davis, and F.N. Andrews. 1957. Strain differences in the estrogenicity of alfalfa. J. Anim. Sci. 16:850–853.
158. Sunde, M.L. 1974. Personal communication, Univ. of Wisconsin, Madison, WI.
159. ----, H.R. Bird, J.M. Sund, and M.J. Wright. 1956. The use of low protein diets for turkeys on range. Poult. Sci. 35:1106–1116.
160. Swanson, E.W., J.B. MacLaren, and E.J. Chapman. 1959. A comparison of milk production and forage crop utilization from strip grazing versus green chop feeding. Tenn. Agric. Exp. Stn. Bull. 292.
161. Thompson, J.A., K.P. Sheridan, and B.A. Hamilton. 1976. The effects of rates of stocking with rotational grazing on the productivity of dryland lucerne at Tamworth, New South Wales. Aust. J. Exp. Agric. Anim. Husb. 16:845–853.
162. Van Horn, A.G., W.M. Whitaker, and R.H. Lush. 1956. Effects of early and delayed grazing on orchardgrass-alfalfa-ladino clover pastures. Tenn. Agric. Exp. Stn. Bull. 249.
163. Van Keuren, R.W. 1979. Alfalfa in winter pastures for beef cow/calf programs. p. 39. In 26th Alfalfa Improve. Conf. Rep. USDA-SEA ARM-NC-7.
164. ----. 1985. All-season forage systems for beef cow herds in the humid temperate regions of the United States. p. 1050–1052. In Proc. 3rd AAAP Animal Sci. Congr., Seoul, Korea. 6–10 May. Organizing Committee, the 3rd AAAP Animal Science Congress, Seoul, Korea.
165. ----, and W.W. Heinemann. 1958. A comparison of grass-legume mixtures and grass under irrigation as pasture for yearling steers. Agron. J. 50:85–88.
166. ----, and ----. 1962. Annual and perennial irrigated pastures and progesterone-estradiol implants for lamb production. Wash. Agric. Exp. Stn. Bull. 641.
167. ----, A.D. Pratt, H.R. Conrad, and R.R. David. 1966. Utilization of alfalfa-bromegrass as soilage, strip-grazing, and rotational grazing for dairy cattle. Ohio Agric. Res. and Dev. Center Res. Bull. 989.
168. ----, E.T. Shaudys, R.H. Baker, and A.D. Pratt. 1966. Economy of grazing, soilage, and stored feeding methods for summer feeding of dairy cattle. p. 505–510. In Proc. 10th Int. Grassl. Congr., Helsinki, Finland. 7–16 July 1960. Finnish Grassland Society, Helsinki.
169. Walton, P.D., R. Martinez, and A.W. Bailey. 1981. A comparison of continuous and rotational grazing. J. Range Manage. 34:19–21.
170. Washko, J.B. 1970. Fall harvest management of alfalfa. Forage Grassl. Progress 11(3):3–4.
171. Wedin, W.F., J.D. Donker, and G.C. Marten. 1965. An evaluation of nitrogen fertilization in legume-grass and all-grass pasture. Agron. J. 57:185–188.
172. Whatley, J.A., Jr., I.T. Omtvedt, J.B. Palmer, and D.F. Stephens. 1959. A comparison of pasture and confinement systems for raising hogs. Okla. Agric. Exp. Stn. Misc. Pub. MP/55, p. 8–14.
173. Whitear, J.D., F. Hanley, and W.J. Ridgman. 1962. Studies on lucerne and lucerne-grass leys. VI. Further studies on the effect of systems of grazing management on the persistence of a lucerne-cocksfoot ley. J. Agric. Sci. 59:415–428.
174. Whitman, W.C., L. Langford, R.J. Douglas, and T.J. Conlon. 1963. Crested wheatgrass and crested wheatgrass-alfalfa pastures for early-season grazing. N. Dak. Agric. Exp. Stn. Bull. 442.
175. Whyte, R.O., G. Nilsson-Lessner, and H.C. Trumble. 1953. Legumes in agriculture. FAO Agricultural Studies 21.
176. Wilman, D. 1977. The effect of grazing compared with cutting, at different frequencies, on a lucerne-cocksfoot ley. J. Agric. Sci. 88:483–492.
177. Wilson, D.B., and R.D. Clark. 1961. Performance of four irrigated pasture mixtures under grazing by sheep. Can. J. Plant Sci. 41:533–543.
178. Winter, A.R., and E.M. Funk. 1960. Poultry science and practice. J.B. Lippincott Co., New York.

ADDITIONAL REFERENCE

179. Van Keuren, R.W. 1986. Alfalfa hay production in tropical climates. In Proc. I Simposio do Colegio Brasileiro de Nutricao Animal, Campinas, SP, Brazil. 1–3 October. Technical Committee, Colegio Brasileiro de Nutricao Animal, Campinas, SP, Brazil.

17 Role in Livestock Feeding—Greenchop, Silage, Hay, and Dehy

H. R. CONRAD

Ohio Agricultural Research and Development Center
Wooster, Ohio

T. J. KLOPFENSTEIN

University of Nebraska
Lincoln, Nebraska

In the USA, alfalfa (*Medicago sativa* L.) is a basic forage for maximizing milk production and provides an important source of nutrients for dairy and beef cattle (*Bos* spp.). Its usage for livestock is expected to increase because of the projected demand for production at low costs in the next decade, and the probable competition for grains among animals, the human population, and industries producing alcohol fuels. Moreover, dairy farms in the USA have the capacity to double the current production of alfalfa forage.

Before reviewing the many factors that contribute to the gain or loss of nutritive value of alfalfa forage, it is important to consider why alfalfa production is often minimal on many dairy farms. In the past third of a century, cereal grains have become important sources of feed for ruminants in the USA. During this period, application of sophisticated technology to genetically superior cultivars of cereal crops increased the production of grain beyond the needs of man and nonruminant animals. Mounting surpluses depressed prices for cereal crops, making them attractive and valuable sources of feed for ruminants. Therefore, much grain was marketed in the form of meat and dairy products. Feed grains are rich energy sources well suited to the mechanized and automated feeding that accompanied their increased use in the livestock industry (20). Only the superior combination of all nutrients in alfalfa has maintained the production of alfalfa as a competitive enterprise on many dairy farms. Computations of the relative dollar value of poor-quality forages, on the basis of relative nutritive content and cost of harvest, show why maize (*Zea mays* L.) silage is selected over low-quality forage on many dairy- and beef-producing units (19).

Alfalfa is preferred to other forages in feeding ruminants (13). The demand arises because its primary nutritive value is based on: rapid passage through the gastrointestinal tract; the large amount of soluble

---

Copyright 1988 © ASA-CSSA-SSSA, 677 South Segoe Road, Madison, WI 53711, USA.
Alfalfa and Alfalfa Improvement—Agronomy Monograph no. 29.

protein provided for rumen microorganisms for resynthesis of protein, synthesis of B vitamins, and stimulation of cellulose digestion; the value of vitamins A, E, and K or their precursors, all of which are vital protective nutrients; the significant amount and supplementary value of all essential nutrients when alfalfa forages are fed to dairy cattle (11); and the fact that alfalfa has relatively large amounts of cell solubles and lowest amount of cell walls in comparison with other forages (44).

Alfalfa in combination with the bacteria (*Rhizobium meliloti* Dangead) is a highly effective symbiosis for biological N_2 fixation. Alfalfa is the feedstuff by which that N and other nutrients are carried to an equally powerful symbiotic mechanism in the rumen of cattle and other ruminants. There is a possibility, however, that large amounts of N can be lost by ruminal proteolysis and absorption of ammonia through the rumen wall in energy-deficient diets (40). This problem can be ameliorated by allocating cereal grain as 20% of the dry matter fed with alfalfa greenchop, hay, and dehydrated (dehy) and 33% of the dry matter fed with alfalfa silage. With appropriate levels of grain in the ration, N losses in the urine are trivial (9).

It is noteworthy that the milk production per cow has essentially doubled during the last 33 yr in the USA (44). Consequently, requirements for this increased milk production have been met by an increase in the total amount of feed consumed and changes in the protein to energy ratio. As the milk production has increased, so has the concentration of total N and Ca in the feed supply. Alfalfa at different stages of maturity differs greatly in feeding value and storage properties. The process of harvesting and storing plants of different chemical content introduces losses that differentially change their feeding value.

17-1 POSTHARVEST NUTRITIONAL CHANGES

Alfalfa is harvested and stored primarily as hay or silage for use on the farm. Alfalfa hay and silage together with greenchop and commercially dehydrated alfalfa account for all the harvested alfalfa fed. The feeding value of harvested alfalfa may be changed by postharvest factors as much as by the precutting environment and history of the plant. Conservation and storage systems are designed to minimize the loss and deterioration of nutrients.

Preservation and storage as hay or silage depend primarily on the reduction of moisture content and/or atmospheric oxygen to levels that will yield minimum metabolic changes induced either by chemical deterioration of plant cells or microorganisms. There are many associated variables within these two main effects that influence feeding value. Physiological maturity and growth periods often differ for the different types of alfalfa storage. Thus, the nutritional merits of different systems of harvesting are not easily compared because of different nutritional quality at harvest time. Immature alfalfa is the most suitable for grazing. Alfalfa from bud to maturity is selected for greenchopping, whereas silage is often harvested at full-bud stage, and hay may represent all degrees of maturity depending on weather patterns at harvest. Because protein is a major qualitative parameter for pricing dehydrated alfalfa, dehy is cut usually at the late bud stage or early bloom.

The intake of dry matter may vary differently among various preservation methods. A summary of effects on voluntary intake and digestibility by the method of hay harvesting is shown in Table 17-1. Artificial drying caused no change in intake or digestibility in sheep. The greater loss in digestible energy of hay was associated with increased leaf loss in rain-damaged hay. Reid (38) reported that the net percentage of the dry matter in harvested forages that can be preserved for feeding were most in barn dried with heat.

The greater the drying under field conditions, the greater the harvesting loss. For direct cut silage containing 17 to 20% dry matter, the field losses were only 2 to 4%. For field-cured hay containing 80 to 90% dry matter, the field and harvesting losses were 21 to 28% (38). Thus, the field losses of alfalfa nutrients are inversely proportional to the losses in storage. More than 95% of the alfalfa fed on farms is harvested as hay and silage. Although there is a substantial international market for dehydrated products, greenchop and dehydrated alfalfa are only 5% of the total fed.

Table 17-1. The range in percentage of dry matter available to cattle, with estimated percentage intake and digestibility in comparison with freshly cut materials as greenchop or freeze-dried materials. Sources: Conrad et al. (10), Demarquilly and Jarrige (12), Pratt and Conrad (36), Pratt et al. (37), and Reid (38).

Harvesting method	Dry matter available	Dry matter intake	Dry matter digestibility
	%	%	%
Artificially dried			
Minimum	88	100	100
Maximum	100	100	100
Silage, 55-65% moisture			
Minimum	85	95	95
Maximum	88	95	100
Silage, 40-50% moisture in sealed structure			
Minimum	85	100	90
Maximum	89	101	100
Silage, direct cut, 75-82% moisture			
Minimum	65	67	80
Maximum	84	90	94
Hay, without rain			
Minimum	68	78	85
Maximum	76	100	100
Hay, rain damaged			
Minimum	50	66	77
Maximum	68	78	85

17-2 NUTRITIONAL QUALITY OF ALFALFA HAY

Alfalfa hay is produced by drying the green forage crop to critical levels of moisture to interrupt plant respiration, and prevent mold growth and excessive heating. The critical level of moisture in forage varies with storage conditions, the ambient temperature in the storage area, density of the hay, and air circulation. Although alfalfa hay is stabile at 20 to 22% moisture, green crops must be dried to <15% moisture for storage without danger of heat damage and spoilage (17). The greater the density from stacking or piling hay into mows and hay barns, the greater the likelihood of mold. Molding is accompanied by reduced digestibility and poor acceptance by cattle (33). It is estimated that one-third of the baled hay in the eastern USA molds from excessive moisture in bales at harvest time.

The most important aspect of hay quality is the amount of usable or metabolizable energy consumed by the animal (44). Thus the amount of animal production becomes the critical measurement of hay quality. Many factors affect hay quality for cattle so that no one characteristic can serve to predict animal production. Some of the important factors that determine alfalfa quality for cattle are stage of maturity, chemical composition, leaf-stem ratio, physical form, foreign material (particularly weeds and dust), damage or deterioration during harvest and storage, and the presence of antiquality substances such as estrogens, thyrotoxic factors, and toxic amines and their condensation products (30).

17-2.1 Field and Barn Curing

Hay should be dried to a safe moisture percentage for storage, approximately 15%, by field curing or barn drying with heated or unheated air. Each drying system has a different effect on nutritive value. During the past decade, field curing has been used almost entirely because of the high cost of fuel for heating air for drying. In general, the feeding value depression in alfalfa hay increases with the length of field exposure before baling or stacking. Often this greater nutritional loss results from rain leaching, leaf shatter, and mechanical losses in handling. The major effects and interactions of hay-handling and conditioning methods have been reported (29). Conditioning by crusher, crimper, or flail harvester has reduced field drying time by as much as 32% and reduced alfalfa leaf loss. Uniform crushing generally is more effective than crimping for increasing stem-drying rates and conservation of nutrients, but it is more difficult to achieve. In general, nutritional value is enhanced by crushing or crimping because hay may be baled 1 d earlier. Reduced field exposure time and particularly reduced leaf loss has been shown to improve feeding value (30). Dairy cows consumed 500 to 900 g more field-cured hay/d and produced slightly more milk when fed crushed alfalfa compared to uncrushed.

17-2.2 Dessicants

Many problems related to hay making could be solved if hay dried quickly in the field or if it could be stored at a high-moisture content without spoilage. Recently, chemical drying aids have been used successfully to reduce drying time. The technique of spraying with carbonates to hasten drying of alfalfa has been described (22, 43, 48). The chemical treatments disrupt the cuticle and cell membranes increasing the rate of water loss. Available results to date indicate that the nutritional value of such forages is improved and, with harvesting completed within a day or a day and one-half, the loss of nutrients from weathering is diminished. Rotz (39) studied the economic feasibility of chemically conditioning alfalfa with potassium carbonate (KCO_3) treatments in whole farm models using base information from dairy farms. The reduced dry matter and protein losses obtained with chemical conditioning were more economical in hay production than in haylage production. Increased dry matter and protein obtained from the chemical treatment justified the cost of application on cuttings other than the initial spring cutting for hay or haylage. The effectiveness and economics of the chemical treatment were very sensitive to swath structure, such as double windrowing. The value of the treatment depends on increasing its effectiveness for first cuttings of alfalfa made when weather conditions are often unfavorable.

17-2.3 High-Moisture Hay

Damage from heating, molding, and respiratory losses of soluble carbohydrates in high-moisture alfalfa hay (above 20% moisture) can be limited by controlling plant and microbial metabolism. This can be accomplished with ammonia (NH_3) or organic acids as preservatives (23). Heat generated by metabolic activity of microorganisms, particularly fungal microorganisms, increases the hay temperature. Heat-resistant fungi may remain active at temperatures up to 65°C (18, 32).

Losses begin at about 22% moisture and total nutrient loss can be expected with 35 and 45% moisture in the hay crop (14). When hay is stored at 20 to 25% moisture, and sometimes at <20% moisture, weight losses will occur from respiratory utilization of soluble nutrients. The loss in dry-baled hay is about 1% of dry matter for each 1% of moisture above 18%. Also, molding and heating of wet hay cause serious losses in protein digestibility (16). The microbial growth in moist hay can be prevented by treatment with preservatives, particularly the organic acids and NH_3 (14, 22, 27, 41, 46).

Hay containing 18% moisture or less may be protected from overheating by small amounts of propionic acid. Knapp et al. (27) successfully used propionic acid to reduce nutrient loss from heating in alfalfa hay. They reported minimal storage losses and maintenance of quality in high-moisture hay treated with propionic acid at a rate of 1% of wet hay. Lower application rates were ineffective. Alfalfa requires relatively large amounts of propionic acid for preservation because of its relatively high surface area and hollow stems. Consequently, a preservative with rapid volatilization characteristics, such as anhydrous NH_3, has advantages for preservation.

Anhydrous NH_3 added at the rate of 1 to 2% of the dry matter inhibits mold growth if effectively contained in high-moisture hay previous to storage (30). Alfalfa hay treated in this manner is an improved feed for cattle and often contains more crude protein than untreated hay. The N of NH_3 is available for utilization in the rumen when treated hay is fed with an adequate amount of starch from cereal grains. Cows readily consume ammoniated hay and produce well (46). The use of 3% or more of NH_3 on wet hay may cause toxic compounds to form and pass into milk (47).

17-2.4 Unique Characteristic

A unique contribution of alfalfa hay arises from insalivation during bolus formation before swallowing. A meal of alfalfa hay generally lasts from 27 to 45 min (3). Long hay has a greater meal duration and is more thoroughly insalivated with the buffers of saliva. This makes a positive contribution to the overall well-being of ruminants and ensures continuance of a normal acid-base balance. Conditions for rumen synthesis are optimized. Many modern feeding systems keep cattle at the stress point of acidosis. This is often true with the low pH in the rumen of cattle consuming typical diets of corn silage and corn-based concentrates. It is also interesting that the best performance by tested cows in the USA are in the arid western states where large amounts of high-protein alfalfa are produced and fed to the dairy cattle.

17-3 NUTRITIONAL VALUE OF ALFALFA SILAGE

Alfalfa silages may be separated into three groups on the basis of moisture level (which is related to the amount of proteolysis). Thus, silage intake, as determined by utilization of N, is highly variable. High-moisture or direct-cut silage has 70% or more moisture. Wilted silage has 60 to 70% moisture. Low-moisture silage ranges between 40 and 60% moisture.

Four major advantages exist to support ensiling alfalfa rather than making hay. First, reduced field losses occur with the use of improved harvesting and handling equipment. Field losses in making silage may be only 20% of those with hay making. Thus, more nutrients are preserved for feeding. Second, less delays at harvest because of unfavorable weather increases the time for regrowth of high-protein alfalfa. Third, silage is better adapted to mechanization, flexibiliity in the choice of a feeding program, and computerized feeding for maximum utilization of nutrients. Fourth, the period for utilization may be extended over months and years with minimum loss of nutrients after the stability of the silage has been reached (34).

17-3.1 Direct-Cut Silage

Direct-cut alfalfa stored without field drying is likely to contain more than 20% moisture. At first cut in the spring, direct-cut alfalfa silage may contain as much as 80% moisture. On average, one can expect to harvest direct-cut alfalfa at about 76% moisture. Noller and Thomas (34) point out that the higher the moisture content, the more critical is the need for a low pH to obtain good preservation. There is the added problem of the massive effusion of water-containing solubles and nutrients from the stored silage. This causes offensive odors that can make the area socially unacceptable. In some alfalfa silages harvested directly, dry matter losses exceeded 30% (36). In general, the forage dry matter intake when direct-cut silages are used is only half that of wilted silage, greenchop, or hay. Thus, a major disadvantage is encountered in the depression of forage intake or direct-cut silages (31). The intake depression arises from the formation of NH_3 and amines in the silage (2, 36, 42).

17-3.2 Optimizing Nutritive Value

The reduced digestibility with advancing maturity of alfalfa involves a reduction both in digestible energy and digestible protein, and consequently, a reduction in net energy and monetary value. The losses in harvesting and storage of alfalfa silages have a significant monetary value. As the protein content of alfalfa is decreased from 20 to 13%, irrespective of cause, the value of lost protein is approximately $30/t of silage dry matter when soybean meal is worth $200/t. The value of the lost net energy not contained in the protein is an additional $12/t of forage dry matter (4). Changes in N fractions in alfalfa silages through harvest and storage processes largely result from proteolysis and deamination in wet alfalfa (<50% dry matter) or denaturation in high dry matter alfalfa (>50% dry matter). Because milking cows fed high-protein alfalfa as their principal forage may receive as much as 60% of their total protein from the alfalfa, it is important to minimize protein degradation during harvest and storage. In alfalfa forages, protein degradation is estimated by the retention of insoluble N. Degradation is greatest in the direct-cut silages but wilted alfalfa silages may have as much as 20% ammoniacal N (7). The treatment of direct-cut alfalfa silage with formic acid improves the recovery of insoluble N; formic acid, coupled with formaldehyde, may retain as much as 97% of the insoluble N for nutrient utilization. The retention of undegraded protein in treated alfalfa silages increases ruminant growth and shifts energy deposition from fat to protein (44). Wilted silages undergo measurable protein degradation in the wilting process causing probable decreases in feeding value (45).

Overheating of high-dry matter silage and subsequent denaturation and formation of heat-damaged protein, which is indigestible, is a continuing problem. Prange et al. (35) found alfalfa silage (48% dry matter) and alfalfa hay, in square bales, provided similar amounts of amino acids for absorption by lactating cows. Damage in wilted silages has been found in 35% of the samples surveyed (44). Normally, this heat damage is accompanied by the formation of large amounts of acid-detergent-insoluble N, but its formation is not always predictable from linear regression analyses.

17-4 NUTRITIVE VALUE OF GREENCHOPPED ALFALFA

In the recent past, greenchop was used in many dairy-feeding systems during the summer months. However, its high labor requirement confines greenchop to emergency forage programs, to providing supplemental forage in late summer, and to use at silage harvest.

Losses of greenchopped alfalfa are minimal. Expected losses of nutrients are <2% (21). Although alfalfa greenchop provides a high-protein forage which is used efficiently, protein declines with approaching maturity. In addition, digestibility declines progressively over time because the digestion of the stem diminishes by 0.6%/d during maturation (21). With continuous day-by-day harvesting, digestibility of the total forage may vary from a high of 72% of the dry matter to a low of 55% of the dry matter during a 6-week growth cycle (10).

Proteins in greenchop alfalfa are consumed as intact plant parts. Most of the plant cells remain undisturbed until after they enter the rumen. Relative to ensiled alfalfa, ammoniacal N released from greenchop alfalfa in the rumen is less. Also, direct ruminal losses are lower. Since greenchop is much higher in water content, the rumen passthrough time is decreased and delivery of passthrough (bypass proteins) to the intestines for direct digestion is greater. A net saving is realized in energy and total protein, but protein quality may be slightly lower. The net effect is that at the same digestibility coefficient, the requirement for cereal grain supplements with greenchopped alfalfa is one-half that required with ensiled forage (9, 37).

17-5 DEHY FOR PROTEIN IN RUMINANTS

Maximizing the use of various alfalfa products when they are available for use in beef cattle rations depends on their defined nutrient content. At the farm level, alfalfa can produce 2.5 times as much protein per hectare as soybeans [*Glycine max* (L.) Merr.], 2 times as much as corn silage, and 3 times as much as corn grain. Alfalfa is an excellent source of Ca, K, and trace minerals for beef cattle. As an energy source, alfalfa is inferior to maize but superior to most other hay crops and crop residues. Protein is often the key nutrient in beef cattle production. The effective use of alfalfa is dependent on the class of beef cattle to be fed. Protein is wasted if the brood cow is fed alfalfa containing 15 to 20% protein when her requirement is 7 to 9% protein.

17-5.1 Finishing Beef Cattle

In finishing cattle, use of alfalfa or other forages is primarily a management decision, but dehy alfalfa may be the best forage. In Nebraska (24), two trials were conducted using dehydrated alfalfa in two pellet sizes and corn silage as three forage sources in high-grain finishing rations. Both 0.63-cm, finely ground dehy pellets and 0.95-cm, coarsely ground dehy pellets maintained intake and gain as well as corn silage, and the dehy produced more efficient gains. Liver abscesses were not controlled quite as well with dehy as with corn silage. These results, as well as considerable data with alfalfa hay, indicated that alfalfa was probably not surpassed as a forage in finishing rations. It now appears that dehy has desirable fibrous characteristics as well as high-nutrient content for use in finishing rations. Performances in calves weighing from 150 to 350 kg is determined primarily by forage quality. Therefore, the best-quality alfalfa should be fed to growing calves and the poorest quality alfalfa to brood cows or finishing cattle.

Klopfenstein (25) developed a system of using dehydrated alfalfa as a source of ruminally undegraded protein for growing and finishing beef cattle. Dehydrated alfalfa was shown to be slowly degraded in the rumen because of heat applied during drying. More recent research with alfalfa showed that the heat treatment increased the value of alfalfa protein in ruminant diets (26, 28). The research of Britton et al. (1) emphasizes that alfalfa proteins respond to heating with decreased rumen degradability and increased efficiency of protein use in ruminants because more of the heated proteins are absorbed in the lower part of the digestive tract. Leaf protein responds more to heating effects than stem protein, so an increase in leaves will increase the effect of dehydration.

17-5.2 Heat Damage

A warning is in order regarding heating of alfalfa protein. Protein that is indigestible in the small intestine because of excessive heat damage is of no use to the animal. Hay crop silages, dehydrated alfalfa, dried brewers grains, and dried distillers grains are feeds that occasionally may be subjected to severe heat damage during drying (15). However, carefully controlling the heating process results in only small decreases in digestibility. Conrad and Hibbs (6) found that protein digestibility was reduced from 67% before dehydration to 64% after dehydration. Similar effects were reported by Britton et al. (1). A modest decrease of 2 to 4% N protein digestibility represents a trade-off for the direct losses in ammoniacal N encountered when protein digests in the rumen in the absence of heat treatment.

17-5.3 Solubility of Protein

In general, the extent of degradation of feed protein in the rumen is dependent on the solubility of the protein. As with dehydrated alfalfa, a part of the dietary protein almost always escapes rumen degradation, with highly variable estimates among feedstuffs in the amount escaping. Also, the proportion of dietary protein reaching the small intestine for digestion may depend on the level of feed intake. Thus, the amount of protein degraded in the rumen is a function of physical properties of the protein, retention time in the rumen, and protein solubility. By heating, the proteins of dehydrated alfalfa are made less soluble and the time in the rumen is decreased because of smaller particle size. Consequently, the amount of protein escaping ruminal degradation and bypassing the rumen may reach 50% or more of the digestible crude protein. When the amount of bypassed protein is approximately 50%, the feeding program maximizes the use of protein for both purposes; that is, bypass protein and the need to maximize protein synthesis in the rumen. Alfalfa is the best of all the stimuli tested for increasing microbial growth in the rumen. Fresh alfalfa sustained a higher concentration of rumen bacteria than the common dairy rations based on grain, protein concentrates, and silages (6). The rumen is capable of synthesizing from 1.5 to 1.8 kg of protein daily in cows at maximum feed intake (8). This is about the same amount of protein as found in 45 to 50 kg of milk. Thus, subsystems providing this segment of protein to dairy cows should not be jeopardized by making the proteins completely insoluble. Dehy appears to be a valuable protein in supplying both N and amino acids of the needed proteins for dairy cattle.

17-5.4 Formulating Balanced Rations

The distribution of amino acids in proteins used for bypassing rumen digestion should be given attention in formulating balanced rations. One method is to compare the profile of essential amino acids of feedstuffs with milk. An example is shown in Table 17-2 where the amino acid contents of rumen bacteria and protozoa, dehydrated alfalfa, soybean meal, and dried brewers grains are compared as coefficients of the amino acid content of milk. It is noteworthy that all the feedstuffs are relatively low in one or more of the essential amino acids. In dehy, methionine is in relatively short supply. Methionine in rumen bacteria is needed to complement the essential amino acids of alfalfa. With respect to histidine, the situation is reversed. Thus, the pitfall of a single source of protein needs to be avoided when depending on bypass protein in high-producing cows.

Generally, the amount of protein escaping ruminal degradation increases with increased feed intake and the subsequent increase in rate of passage from the rumen. This may be an added value of large dehydrated alfalfa pellets (that is, 0.95-cm cut, 0.95-cm dia) when fed to dairy cows

Table 17-2. The ratios† of nine essential amino acids in rumen microorganisms, dehydrated alfalfa, brewers grains, and soybean meal to the amino acids of milk protein.

Essential amino acid	Rumen bacteria	Rumen protozoa	Dehy. alfalfa	Brewers grains	Soybean meal
Histidine	0.67	0.71	1.15	0.88	1.25
Methionine	1.00	0.62	0.52	0.76	0.35
Phenylalanine	1.55	1.18	1.30	1.47	1.30
Lysine	1.05	1.26	0.70	0.40	0.79
Leucine	0.75	0.87	1.00	1.23	0.90
Isoleucine	0.67	0.81	0.73	1.06	0.69
Threonine	1.36	1.22	1.41	0.88	1.01
Valine	0.92	0.83	1.09	1.04	0.86
Arginine	1.41	1.31	1.47	1.38	3.18

† Ratio = Essential amino acid in the feed or rumen microorganisms to amino acid content of milk (e.g., ratio of histidine in rumen bacteria = 1:0.67 or 67% of the amount of histidine in milk).

In a two-lactation switchover trial, Conrad and Hibbs (5) showed that the 0.95-cm pellets of dehydrated alfalfa were equal in nutrient value for lactating cows to long-cut alfalfa hay as 25% of the total dry matter in the ration of dairy cattle. A significant observation was that the mean milk fat percentage was 3.9% with the alfalfa pellets and 3.7% with the alfalfa hay. Milk production and milk protein were slightly increased, again indicating the dehy pellets have fiber value as well as nutrient value when properly used in dairy cattle diets.

17-6 SUMMARY

In general, alfalfa is a superior feed for cattle, in the form of both good quality hay and silage. Alfalfa is quickly digested, relatively high in protein, high in cell solubles, and therefore, low in cell walls or neutral detergent fiber. Alfalfa should be managed to maintain digestibility. Other nutritional benefits are its carotene, vitamin E, and mineral content. Because it is a leafy plant, losses of dry leaves may be large at harvest. If ensiled directly from the field without wilting, large losses may occur with the leaching of nutrients. Carefully wilting alfalfa will result in the optimum nutritional responses of cattle. Protein losses or denaturation may be a costly factor if heating results from continuing metabolism in newly harvested alfalfa. Losses may be minimized in alfalfa containing more than 20% moisture by applying 1 to 1.5% ammonia or propionic acid. Both of these inhibit growth of molds. Dehydrated alfalfa meal is used widely in pelleted form to provide carotene, fiber, and partly soluble protein that will pass through the rumen. Coarsely ground pelleted dehy, about 1 mm in diameter, is a useful product to supply up to 25% of the cow's diet.

REFERENCES

1. Britton, R., D. Dorn, T. Klopfenstein, J. Ward, and J. Merril. 1983. Harvesting and processing effects on alfalfa protein utilization by ruminants. P. 653-657. *In* Proc. 14th Int. Grassl. Congr., Lexington, KY. 15-24 June 1981. Westview Press, Boulder CO.
2. Clancy, M., P.G. Wangsness, and B.R. Baumgardt. 1977. Effect of silage extract on voluntary intake, rumen fluid constituents and rumen motility. J. Dairy Sci. 60:580-590.
3. Conrad, H.R. 1971. Limits on Voluntary Feed Intake in Dairy Cattle. Proc. Distillers Feed Res. Conf. 26:18-26.
4. ----, and J.W. Hibbs. 1968. Early cut forage is worth more. Ohio Rep. 53(6):90-91.
5. ----, and ----. 1973. Coarse alfalfa pellets substitute for hay. Ohio Rep. 58(1):21-22.
6. ----, and ----. 1981. Utilization of alfalfa protein in dairy cows. Feedstuffs 53(6):28-30.
7. ----, ----, and A.D. Pratt. 1960. Nitrogen metabolism in dairy cattle. Ohio Agric. Exp. Stn. Bull. 861.
8. ----, ----, and ----. 1967. Effect of plane of nutrition and source of nitrogen and methionine synthesis in cows. J. Nutr. 91:343-350.
9. ----, ----, ----, and R.R. Davis. 1961. Nitrogen metabolism in dairy cattle. I. The influence of grain and meadow crops harvested as hay, silage, or soilage on efficiency of nitrogen utilization. J. Dairy Sci. 44:85-95.
10. ----, ----, ----, and ----. 1962. Relationships between forage growth stage, digestibility, nutrient intake and milk production in dairy cows. Ohio Agric. Exp. Stn. Bull. 914.
11. ----, and F.A. Martz. 1985. Forages for dairy cattle. p. 550-559. *In* M.E. Heath et al. (ed.) Forages—the Science of Grassland Agriculture. 4th ed. Iowa State University Press, Ames.
12. Demarquilly, C., and R. Jarrige. 1970. The effect of method of forage conservation on digestibility and voluntary intake. p. 733-737. *In* Proc. 11th Int. Grassl. Congr., Surfers Paradise, Queensland, Australia. 13-23 April. University of Queensland Press, St. Lucia.
13. Dornfield, D., D.A. Rohweder, and W.T. Howard. 1983. Income, costs and profits associated with feeding selected forages to dairy cows. p. 825-827. *In* Proc. 14th Int. Grassl. Congr., Lexington, KY. 15-24 June 1981. Westview Press, Boulder, CO.
14. Drew, L.D., H.M. Keener, R.W. VanKeuren, H.R. Conrad, and W.E. Gill. 1974. Preservatives for hay. Ohio Rep. 59(2):38-39.
15. Goering, H.K. 1976. A laboratory assessment on the frequency of overheating in commercial dehydrated alfalfa samples. J. Animal Sci. 43:869-872.
16. ----, P.J. VanSoest, and R.W. Hemken. 1973. Relative susceptibility of forages to heat damage as affected by moisture, temperature and pH. J. Dairy Sci. 56:137-143.
17. Gordon, C.H. 1974. Preserving grassland products for ruminant feeding. p. 68-85. *In* H.B. Sprague (ed.) Grasslands of the United States. Iowa State University Press, Ames.
18. Gregory, P.H., and M.E. Lacey. 1963. Mycological examination of dust from moldy hay associated with Farmer's Lung disease. J. Gen. Microbiol. 30:75-88.
19. Hibbs, J.W., and H.R. Conrad. 1973. Alfalfa—its value in the dairy ration. p. 53-60. *In* Proc. 3rd Annual Alfalfa Symp. 29 March Ohio Agricultural Research & Development Center, Wooster.
20. Hodgson, R.E. 1974. The place of forages in animal production. p. 57-67. *In* H.B. Sprague (ed.) Grasslands of the United States. Iowa State University Press, Ames.
21. Huffman, C.F. 1959. Summer feeding of dairy cattle. J. Dairy Sci. 42:1495-1561.
22. Johnson, T.R., J.W. Thomas, and C.A. Rotz. 1983. Quality of alfalfa hay chemically treated at cutting to hasten field drying. J. Dairy Sci. 66:1052-1056.
23. Kjelgaard, W.L., P.M. Anderson, L.D. Hoffman, L.L. Wilson, and W.W. Harpster. 1983. Round baling from field practices through storage and feeding. p. 657-660. *In* Proc. 14th Int. Grassl. Congr., Lexington, KY. 15-24 June 1981. Westview Press, Boulder, CO.
24. Klopfenstein, T. 1973. Maximizing alfalfa in beef rations. p. 71-80. *In* Proc. 3rd Annual Alfalfa Symp. Ohio Agricultural Research & Development Center, Wooster.
25. ----. 1974. New concepts in protein nutrition for ruminants. p. 81. *In* Proc. Nebraska Feed Nutr. Conf., Grand Island. 23-24 April.
26. ----, C. Dorn, R.L. Ogden, W.R. Kehr, and T.L. Hanson. 1978. Field wilted and direct cut dehydrated alfalfa as protein sources for growing beef cattle. J. Animal Sci. 46:1780-1788.
27. Knapp, W.R., D.A. Holt, and V.L. Lechtenberg. 1976. Propionic acid as a hay preservative. Agron. J. 68:120-123.
28. Prudlo, V., and T. Klopfenstein. 1978. In vitro studies on dried alfalfa and comple-

mentary effects of dehydrated alfalfa and urea in ruminant rations. J. Anim. Sci. 46:499–507.
29. Larsen, W.E., and A.R. Rider. 1985. Mechanization of forage harvesting and handling. p. 452–459. In M.E. Heath et al. (ed.) Forages—The science of grassland agriculture. 4th ed. Iowa State University Press, Ames.
30. Lechtenberg, V.L., and R.W. Hemken. 1985. Hay quality. p. 460–469. In M.E. Heath et al. (ed.) Forages—The science of grassland agriculture. 4th ed. Iowa State University Press, Ames.
31. McCullough, M.E. 1962. Some factors influencing intake of direct-cut silage by dairy cows. J. Dairy Sci. 45:116–119.
32. Miller, L.G., D.C. Clanton, L.F. Nelson, and O.E. Hoehne. 1967. The nutritive value of hay baled at various moisture contents. J. Anim. Sci. 26:1369–1373.
33. Mohanty, C.P., N.A. Jorgensen, M.J. Owens, and H.H. Voelker. 1967. Effect of molding on the feeding value and digestibility of alfalfa hay. J. Dairy Sci. 50:990.
34. Noller, H.C., and J.W. Thomas. 1985. Hay-crop silage. p. 517–527. In M.E. Heath et al. (ed.) Forages—The science of grassland agriculture. 4th ed. Iowa State University Press, Ames.
35. Prange, R.W., M.D. Stern, N.A. Jorgensen, and L.D. Satter. 1980. Sites and extent of nutrient digestion in lactating cows fed ensiled or baled alfalfa. J. Dairy Sci. 63 (Suppl. 1):135.
36. Pratt, A.D., and H.R. Conrad. 1965. The need for unfermented grain or forage with high-moisture grass-legume silage for dairy cattle. Ohio Agric. Exp. Stn. Res. Bull. 979.
37. ----, R.R. Davis, H.R. Conrad, and J.H. Vandersall. 1961. Soilage and silage for milk production. Ohio Agric. Exp. Stn. Bull. 871.
38. Reid, J.T. 1973. Forages for dairy cattle. p. 664–676. In M.E. Heath et al. (ed.) Forages—The science of grassland agriculture 3rd ed. Iowa State University Press, Ames.
39. Rotz, C.A. 1984. Economics of chemical conditioning of alfalfa. p. 278–284. In Proc. 1984 Forage Grassl. Conf., Houston, TX. 23–26 January. American Forage and Grassland Council, Lexington, KY.
40. Satter, L.D., and R.E. Roffler. 1975. Nitrogen requirement and utilization in dairy cattle. J. Dairy Sci. 58:1219–1237.
41. Sheaffer, C.C., and N.A. Clark. 1975. Effects of organic preservative on the quality of aerobically stored high moisture hay. Agron. J. 67:660–662.
42. Thomas, J.W., L.A. Moore, and J.F. Sykes. 1961. Further comparisons of alfalfa hay and alfalfa silage for growing dairy heifers. J. Dairy Sci. 44:862–873.
43. Tullberg, J.N., and D.J. Minson. 1978. The effect of potassium carbonate solutions on the drying of lucerne. J. Agric. Sci. 91:557–561.
44. Waldo, D.R., and N.A. Jorgensen. 1981. Forages for high animal production: Nutritional factors and effects of conservation. J. Dairy Sci. 64:1207–1229.
45. ----, J.E. Keys, Jr., and C.H. Gordon. 1973. Preservation efficiency and dairy heifer response from unwilted formic and wilted untreated silages. J. Dairy Sci. 56:129–136.
46. Weiss, W.P., V.F. Collenbrander, and V.L. Lectenberg. 1982. Feeding dairy cows high moisture alfalfa hay preserved with anhydrous ammonia. J. Dairy Sci. 65:1212–1218.
47. ----, H.R. Conrad, C.M. Martin, R.F. Cross, and W.L. Shockey. 1986. Etiology of ammoniated hay toxicosis. J. Anim. Sci. 63:525–532.
48. Wieghart, M., J.W. Thomas, and M.B. Tesar. 1980. Hastening drying rate of cut alfalfa with chemical treatment. J. Anim. Sci. 51:1–9.

18 Wet Fractionation Processes and Products

NEAL A. JORGENSEN

University of Wisconsin
Madison, Wisconsin

RICHARD G. KOEGEL

USDA-ARS
Dairy Forage Research Center
Madison, Wisconsin

Wet fractionation is a process in which juice is expressed from freshly cut herbage which is normally at a relatively early stage of maturity (e.g., late bud to early bloom in the case of alfalfa *Medicago sativa* L.). The liquid fraction thus obtained is frequently slightly more than half of the fresh crop weight and contains 7 to 10% dry matter. This amounts to 20 to 25% of the total crop dry matter and might contain up to 33% of the crude protein in the crop. The resulting solid or fibrous fraction usually varies in moisture content from 65 to 75% and has an average particle size considerably smaller than that of the initial crop. The liquid fraction is occasionally used in its entirety, either fresh or preserved, as part of animal rations (including those of nonruminants such as swine, *Sus scrofa domesticus*). More frequently, however, it is further divided into a high-protein fraction containing 40 to 60% crude protein on a dry matter basis and a deproteinized juice fraction with 7% or less dry matter and negligible true protein content. The fibrous fraction is almost always used for ruminants and may be fed fresh, ensiled, or dehydrated.

18–1 REASONS FOR USING THE SYSTEM

There are several advantages associated with the wet-fractionation process. These include: (i) reduction in field losses made possible by immediate removal of the crop from the field, (ii) increased weather independence of the harvesting process allowing the plants to be cut at the exact stage of maturity desired, and (iii) increased flexibility of prodcut utilization made possible by a high protein-low fiber fraction, including feeding to nonruminants.

Copyright 1988 © ASA-CSSA-SSSA, 677 South Segoe Road, Madison, WI 53711, USA.
Alfalfa and Alfalfa Improvement—Agronomy Monograph no. 29.

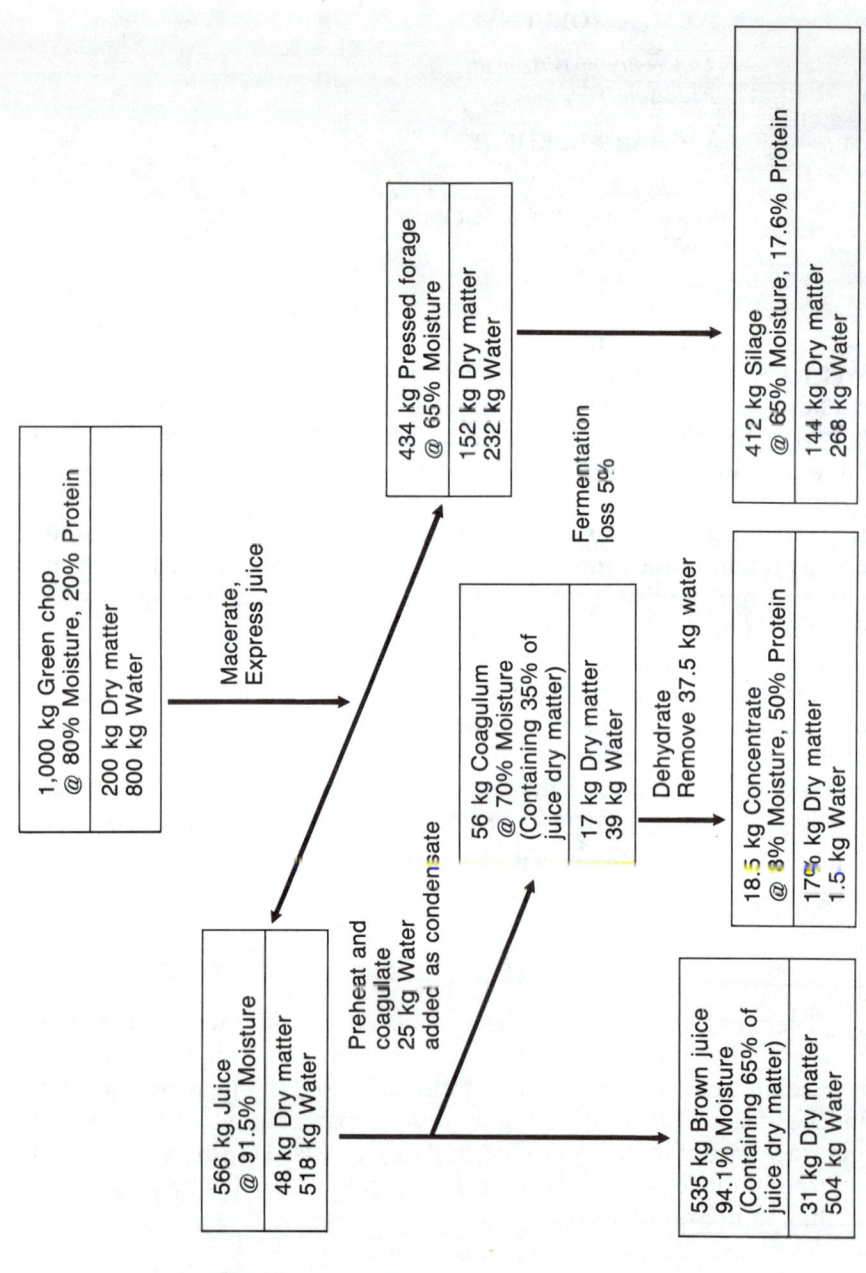

Fig. 18–1. Quantities of products formed at various steps in the processing of 1000 kg of alfalfa. Courtesy, H. D. Bruhn.

18-2 PRODUCTION

Since the reasons for carrying out the wet fractionation process vary considerably, there are a number of variations in the process including:
1. Plant species.
2. Scale of operation, from <50 kg/h to >40 t/h.
3. Location of operation, including field, farmstead, or centralized processing plant.
4. Types of equipment and methods used for the various subprocesses.
5. Priorities, including whether to maximize protein concentrate production or maximize forage quality.
6. Use and form of end products and method of preservation.

One processing pathway and the approximate weights of the various forage fractions are shown in Fig. 18-1. Other pathways are described by Pirie (26) and Telek and Graham (32).

Despite the diversity in the wet fractionation process, the freshly cut herbage generally undergoes four major steps:
1. Cell rupture, also referred to as "maceration" or "pulping".
2. Juice expression.
3. Separation of protein concentrate from the juice fraction.
4. Preservation and storage of products.

18-3 CELL RUPTURE

While cell rupture and juice expression are sometimes carried out concurrently in such devices as roll presses and screwpresses, the dissimilarity in requirements of the two processes leads to less efficient use of equipment and energy.

Cell walls can be ruptured by subjecting them to a sufficiently high level of force imbalance across individual cells. These can be shear forces, in which case cell walls fail by angular deformation, or unbalanced normal forces which cause the semirigid cell wall to transmit pressure to the cell contents at the high force location. This pressure in turn causes the wall to burst outwardly at a low force location. Uniform or hydrostatic forces, no matter how great, have little effect since in this case the cell wall is supported on the incompressible cell contents.

Application of the required unbalanced forces across individual cells in large masses of material is made difficult by the extremely low coefficient of friction of the forage materials with machine elements, especially after a small amount of juice is released by a few randomly ruptured cells. The same phenomenon causes uniaxial forces to be transformed into ineffective hydrostatic pressure in masses of forage.

Most frequently cell rupture has been carried out making use of the inertia of the forage by impacting it with relatively high velocity rotating machine elements. This is the case in various hammer mills, attrition mills, and rotating knife devices. The visco-elastic nature of the fresh forage material and the randomness of these processes cause them to have a high energy requirement. Cell rupture can also be carried out in steel crushing rolls. However, the high rate of slippage of the forage relative to the rolls caused by the low coefficient of friction results in reduced throughput and high energy requirements. Pirie (26) cites energy requirements of 10 to 40 kW-h/t. for a rotary pulper operating at 800 to 1700 revolutions/min.

A cell-rupture device was developed at the University of Wisconsin which creates the necessary force imbalance by extruding the herbage through orifices (24). This concept was developed into a continuous rotary machine by the use of a cylindrical rotating orifice ring with a cylindrical roller running on its internal surface to extrude herbage radially outward through the orifices. This design was built in various sizes, the largest of which had a crop throughput in excess of 20 t/h while generally requiring <1.6 kW- h/t.

Evaluating the actual percentage of cells ruptured in a large quantity of herbage by a given process is a desirable, but challenging task. Emetarom and Barrington (7) proposed washing a representative sample of processed herbage under prescribed conditions. This wash water is then filtered and the electrical conductivity of the filtrate is determined. This filtrate conductivity is then compared with that from the same herbage which was subjected to a standard treatment, i.e., 2 min. in a Waring blender. While this procedure does not determine the percentage of cells ruptured, it does yield an index to indicate the relative effectiveness of a given process.

18-4 EXPRESSION OF JUICE

A mass of herbage which has undergone a cell rupture process contains a certain volume of solids and a certain volume of voids that is occupied by both liquid and gas. The solids and the liquid are essentially incompressible. Expression of juice requires that the mass be compressed until the volume of voids is reduced to less than the volume of the liquid contained in them. A rough requirement for the expression of enough juice to reduce the herbage from a moisture content of 80 to 67% wet basis (w.b.) (liquid/solid ratio reduced from 4:1 to 2:1) is that a pressure of at least 700 kPa be maintained for more than 1 min. A lower pressure can be partially compensated for by a longer holding time, or a shorter holding time can be partially compensated for by a higher pressure. The time rate of juice expression can be increased somewhat by reorientation of the material while it is compressed. This is done at the cost of an increased energy input.

Juice expression can be carried out as a batch process, but frequently it is desired to accomplish it as a steady-flow process. A number of presses are available for this purpose. The most commonly used press for forage fractionation is the screw press. This press consists of a helix or screw rotating within a perforated cylindrical barrel. A restriction device at the outlet end of the barrel causes a pressure build-up in the material entering at the opposite end and traversing the barrel under the action of the rotating screw. Double screw presses also have been used. These have one right-hand and one left-hand screw which counterrotate and whose axes are parallel. The screw flightings overlap slightly and the barrel has "Figure 8" cross section.

Screw presses have functioned quite satisfactorily. Their energy requirements and initial costs tend to be high relative to their throughput. The rubbing action of the herbage on the barrel and screw and the great degree of material reorientation result in a higher fiber content in the juice, equipment wear, and a high energy requirement.

Roll and belt presses have both been used for forage fractionation. The former type has been developed for the crushing and pressing of sugarcane (*Saccharum officinarum* L.) and similar applications. The latter consists of a tensioned flat belt running around two or more pullies. The herbage is loaded on the belt in a uniform layer and is pressed between the belt and pulley, one of which is perforated to allow the juice to exit.

Neither type of press is capable of reducing the moisture content of macerated alfalfa herbage for 80 to 65% (w.b.) in a single pass. The roll press can provide adequate pressure, but not adequate holding time. When additional passes are made, the expressed juice rewets the incoming herbage thus decreasing the effect. The belt press normally provides neither adequate pressure nor hold time, the maximum pressure available being T/R, where T is the maximum tension per unit width which the belt can withstand and R the pulley radius. The hold time is approximately the time during which the pulley rotates 180°. Increased pressure is sometimes provided by forcing the belt against the pulley with additional rolls. Additional hold time can be provided by placing the herbage between two perforated belts and running these around multiple pullies in a "serpentine" fashion.

A "cone" or "vee" press for forage fractionation was designed, constructed, and evaluated at the University of Wisconsin (31). This rotary press has two opposing pressing surfaces in the form of rather flat cones whose axes intersect at the cone vertices, but are slightly nonparallel. The nonparallel axes cause the cone surface to have a minimum spacing and a maximum spacing between them at locations 180° apart. The herbage is packed between the cones at the maximum spacing, travels and is compressed between the slowly rotating cones for 180° of rotation until the minimum spacing or "nip" is reached, and is then removed from between the cones by a stationary plow. The cone press evaluated could reduce the moisture in 12 to 15 t/h of alfalfa herbage from 80 to 65% (w.b.) when rotated at 0.3 revolutions/min. The energy requirement did not exceed 1.5 kW-h/t.

18-5 PROTEIN SEPARATION

The protein fraction of the juice is usually separated by means of coagulation. Membrane technology has also been used experimentally. Coagulation may be accomplished by the addition of heat or by lowering the pH to below 4.5. The latter may be accomplished either by the addition of acid or by the production of acid during anaerobic fermentation of the juice.

Frequently, heat coagulation is accomplished by the injection of steam into the juice. The juice can, however, also be heated using a scraped surface heat exchanger. When fresh alfalfa juice is heated, the coagulated protein floats to the surface and may be separated by skimming or straining. Since the protein molecules have a greater density than the juice, the floatation appears to be made possible by naturally occurring gas bubbles adhering to the coagulated protein. The protein concentrate obtained by skimming or straining usually has a moisture content between 85 and 90%, w.b. More than half of this liquid can be removed by pressing (29). Kohler et al. (13) and others have used a decanter centrifuge to separate heat coagulated protein from juice. They have achieved solids contents as high as 50% by this method. When too much time elapses between cutting and processing, the effectiveness of the heat coagulation/floatation process suffers. This may be due to proteolysis and other related changes in the juice. Two hours is probably the maximum allowable time between cutting and processing.

The quantities of various acids required to lower the pH of alfalfa juice sufficiently to precipitate the protein was studied by Ajibola et al. (1). Ajibola et al. (2) also studied the autofermentation of alfalfa juice as a protein precipitation technique. They found that the variable-length lag phase preceding fermentation could be eliminated by adding 5% (v/v) of the supernatant from the previous day's fermentation. In so doing, they were consistently able to lower the pH to the desired level of 4.5 within 24 h at ambient temperature. Increasing the temperature further reduced this time. The final pH achieved was determined by the juice content of nonstructural carbohydrates. If carbohydrate content was insufficient to lower the pH to the level desired, this could be accomplished by addition of a carbohydrate source to the juice.

After precipitation, it is possible to decant approximately 67% of the juice volume. The precipitate thus obtained is much finer grained than heat coagulated protein and is thus more difficult to further dewater by pressing or centrifugation. It can, however, be resolubilized by pH adjustment if this is desired, whereas heat coagulated protein is irreversibly denatured.

18-6 PROTEIN PRESERVATION

Preservation of the protein concentrate has been carried out by dehydration, chemical preservatives, or by mixing it with grain to create a high energy-high protein supplement which was stored under anaerobic conditions. Dehydration has been carried out in spray dryers, drum dryers, or by forming the partially dewatered concentrate into pellets and passing heated air through it. Propionic acid in concentrations above 1.5% was found to be an effective preservative (30). This method may be particularly attractive where protein concentrate must be held for several months until it can be mixed with grain for final storage under anaerobic conditions.

18-7 DEPROTEINIZED JUICE

The deproteinized juice has been used as a liquid fertilizer, as a culture medium for single cell organisms, and as a feed when concentrated or dried along with the pressed forage fraction.

18-8 UTILIZATION

18-8.1 Pressed Forage

Pressed forage, the high fiber fraction after pressure fractionation, may be fed and/or stored as fresh, ensiled, or dried material (32).

Concentration of the fiber components in the forage fraction increases as a result of the wet fractionation process (Table 18-1). Thus, it is reasonable to assume dry matter intake and digestibility of pressed forage

Table 18-1. Mean composition of wilted vs. pressed alfalfa silages.†
Source: Russell et al. (28) JAS, with approval of the American Society of Animal Science; Lu et al. (15) JDS, with approval of the American Dairy Science Association.

	Russell et al. (28)		Lu et al. (15)	
	Wilted	Pressed	Wilted	FPS
Dry matter, %	42.1	30.5	46.4	30.2
Percentage of dry matter:				
Crude protein	18.5	15.9	23.3	16.8
Neutral detergent fiber	47.2	57.8	40.4	54.6
Acid detergent fiber	37.6	47.4	30.7	39.0
Organic acids, percentage of dry matter:				
Acetic	2.1	3.8	1.5	1.0
Propionic	0.2	0.5	0.1	0.1
Butyric	0.1	1.2	0.1	0.1
Lactic	3.8	2.7	4.6	3.5
Total	6.2	8.2	6.2	4.6
Percentage of total N:				
Nonprotein N	20.4	27.2	45.0	34.0
Ammonia N	8.6	3.2	7.2	3.8

† Wilted = total herbage cut and field-wilted; FPS = pressed forage treated with 0.5% formic acid (w/w).

would be lower than the original plant material. However, they were not reduced to the extent expected (17). This is attributed to perturbation caused by the mechanical force during the wet fractionation process. Since application of mechanical force causes disruption of the cellular structure of plant tissue, the degree of crystallinity or rigidity of physical barriers is reduced. Consequently, part of the fiber fraction not available for bacterial degradation becomes available (Table 18-2). This counterbalances the depression in dry matter intake and digestibility caused by removal of cell solubles. Digestibility of fiber components increased by both cell maceration and pressure fractionation (17). Compared to direct cut hay, hay made from pressed forage is about equal in support of milk production and weight gain in ruminants (5).

Fermentation pattern is critical for the utilization of silage made from pressed forage (hereafter called *pressed silage*) by ruminants. Reduction in moisture content, removal of nonstructural carbohydrate or cell contents, and disruption of plant cell walls are the three major factors affecting fermentation of ensiled pressed forages. The extent of fermentation is expected to be reduced as a result of moisture reduction and the removal of cell contents. On the other hand, the extent of fermentation is enhanced by the disruption of plant cell walls. Therefore, the extent and pattern of fermentation in pressed silage is determined by the extent of reduction and enhancement which reflects the amount of moisture and cell contents being removed and the number of cells being disrupted. This partially explains the inconsistency of results regarding quality and acceptability of pressed silage among research groups (6, 25, 27, 28, 35). It has been demonstrated that cell maceration enhances the extent of fermentation, and pressure fractionation reduces the extent of fermentation (28). Compared to silage made from direct-cut forage, pressed silage appears to be of better quality due to a more desirable fermentation pattern and a higher intake and digestibility by ruminants (35). However, a lower dry matter intake was reported in animals fed pressed silage as

Table 18-2. Mean apparent digestibility of wilted vs. pressed alfalfa silages.†
Source: Russell et al. (28) JAS, with approval of the American Society of Animal Science; Lu et al. (15) JDS, with approval of the American Dairy Science Association.

	Digestibility, %			
	Russell et al. (28)		Lu et al. (15)	
Measurement	Wilted	Pressed	Wilted	FPS
Dry matter	62.8	61.9	67.1	6.3
Crude protein	71.1	71.2	76.5	71.9
Neutral detergent fiber (NDF)	58.2	59.4	59.3	62.8
Acid detergent fiber (ADF)	56.4	55.4	57.2	58.6
Hemicellulose (NDF-ADF)	72.1	75.7	57.4	69.4
Cellulose	60.7	62.3	68.1	69.8

† Silages: Wilted = total herbage field-wilted; FPS = formic acid-treated pressed forage.

compared to wilted silage (6, 28). Dairy cows (*Bos* spp.) fed pressed silage produced slightly less or the same level of milk as those fed wilted silage. The undesirable fermentation pattern in pressed silage as evidenced by high concentrations of ammonia nitrogen (NH_3N) and volatile organic acids was related to the lower dry matter intake (17, 28). Higher moisture content than conventional wilted silage and/or the disruption of cell walls are the two main factors contributing to undesirable fermentation in pressed silage (Table 18-1). Extracting more juice from green plants in order to reduce moisture and consequently improve the condition for ensiling has not been successful. Reducing moisture by double pressing depressed dry matter intake and digestibility. This was due to removal of large quantities of readily digestible nutrients (17).

Milk cow response to pressed silage treated with formic acid was equal to that of conventionally wilted silage. (Table 18-3). Through the addition of formic acid, both preservation characteristics and animal productive performance were improved compared to untreated pressed silage (16). Formic acid treatment depressed concentrations of NH_3N, nonprotein N, total volatile fatty acids, and acetic acid, and increased lactic acid content of the pressed silage (16). Chemical modification to improve ensiling condition has provided a method for making pressed silage competitive with high-quality conventionally wilted silage.

18-8.2 Protein Concentrate

Chloroplastic and cytoplasmic proteins represent two major extractable fractions from green plants (Table 18-4). The proteins can be coagulated and separated from deproteinized juice. Chloroplastic protein is precipitated from raw juice when heated at 50 to 55°C. Cytoplasmic protein is the fraction obtained from the reprecipitation of the supernatant fraction by heating at 80°C or higher. The procedure for fractionating the whole protein into cytoplasmic and chloroplastic proteins

Table 18-3. Mean response of cows fed wilted vs. pressed alfalfa silages.[†] Source: Lu et al. (15) JDS, with approval of the American Dairy Science Association.

Measure	Control	FPS
Dry matter intake:		
Silage, kg/d	10.5	13.3
Concentration, kg/d	7.1	7.1
Total, kg/d	17.6	20.4
Production:		
Milk, kg/d	21.9	22.8
Fat, %	4.1	4.0
4% FCM, kg/d[‡]	22.7	22.8
Protein, %	3.4	3.5

[†] Silage: wilted = total herbage field wilted to 46.4% dry matter; FPS = 30.2% dry matter, treated with 0.5% formic acid (w/w).
[‡] 4% FCM = 4% fat-corrected milk.

Table 18-4. Composition of alfalfa protein concentrates. Source: University of Wisconsin Plant Juice Protein Team.

Protein concentrate	Crude protein	Ether extract	Crude fiber	Ash	N-free extract
Whole	46-62	6-9	1 3	10-12	14-17
Chloroplastic (green)	38-46	11-14	4-6	14-16	20-22
Cytoplasmic (white)	82-91	0.4-0.7	0.5-1	0.3-0.5	8-10

has not been standardized among research groups, and the terminology is loosely used. Protein concentrate will be referred to as *whole protein* in this section.

Processing conditions are likely to be the most important factor affecting quality of protein concentrates. Numerous chemical reactions, both enzymatic and nonenzymatic, can occur after the green plant is harvested. Composition of protein concentrate which ultimately dictates quality is influenced by the end products of these reactions. Under conditions found in raw juice, which provides a favorable condition for proteolysis to occur, true protein can be converted to nonprotein N.

Consequently, both the extractability and quality of protein are reduced with time. Nonenzymatic reactions such as the reactions between phenolic compounds and amino acids in extracted protein will reduce availability of amino acids (8, 34).

18-8.2.1 Ruminants

The utilization of protein concentrate by ruminants is important if the wet fractionation process is to be considered as an on-the-farm process. Production and utilization of protein concentrate would allow farmers to redistribute forage proteins and thus reduce cost of purchased protein supplements. Extent of protein degradation is important in determining the value of protein sources for ruminants. A protein source that provides optimal ruminal microbial protein synthesis while allowing maximum escape from ruminal degradation followed by intestinal absorption may be considered as an ideal protein source for ruminants if what escapes the rumen is of high biological value. Substantial evidence suggests that the rate and extent of ruminal degradation of alfalfa protein concentrate can be altered by processing conditions (14, 15, 18, 21). Alfalfa protein concentrate coagulated by heat at 80°C is more resistant to microbial degradation in the rumen than that coagulated at 60°C. Depending upon the processing condition, ruminal degradability of alfalfa protein concentrate ranges from 51 to 74% (17). Using regression analysis, degradability of alfalfa protein concentrate was found to be 56% (15). Apparent absorption of N in the small intestine ranged from 66 to 70% for alfalfa protein concentrate. Absorption of daily amino acid intake in the intestine ranged from 71 to 81% (17, 19). In a comparison to soybean [*Glycine max* (L.) Merr.] meal, alfalfa protein concentrate appeared to be more resistant to ruminal degradation and it was well utilized by lactating dairy cows (20).

18-8.2.2 Nonruminants

High concentrations of saponin in the diet affect microbial fermentation (22). Evidence supports the speculation that ruminants are less sensitive than nonruminants to undesirable compounds in leaf protein concentrate, such as saponins and polyphenols (10). Quality of protein and the association of antiquality substances with leaf protein concentrates represent two major concerns in the utilization of such proteins by nonruminants and poultry. Addition of methionine and lysine improved the utilization of leaf protein concentrates by rats (*Rattus* spp.) and chicks (*Gallus gallus domesticus*). Alfalfa protein concentrate is comparable in amino acid composition to the major plant protein sources, such as soybean meal. Substantial evidence suggests that availability rather than composition is the major factor affecting quality of leaf proteins. Reduction in lysine availability is related to undesirable processing conditions during coagulation and dehydration (36, 37). A time delay between extraction and coagulation can result in oxidation of methionine, thereby reducing availability of methionine. Availability of S-containing amino acids and lysine is reduced as a result of processing conditions which favor reactions between phenolic compounds and amino acids (8, 23).

Antiquality substances such as saponins, and estrogenic flavonoid compounds are critical to the utilization of leaf protein by nonruminants. Feeding alfalfa protein concentrate prepared from high saponin strains retards growth of poultry, swine, rabbits (*Oryctolagus cuniculus*), and rats (4,9,33). Coumestrol, one of the estrogenic flavonoid compounds, precipitates with leaf protein concentrate during coagulation (11). Estrogenic favonoid compounds can interact competitively with natural estrogens in animals. Precautions need to be taken to prevent the contamination of leaf protein concentrates with antiquality substances. Protein concentrate prepared by anaerobic fermentation may contain less toxic material than that prepared by heat coagulation. It is possible that leaf protein concentrate prepared by anaerobic fermentation may be better utilized by nonruminants. However, the loss of protein due to the conversion of protein to nonprotein N during fermentation should not be overlooked (17).

18-8.2.3 Humans

Consumption of leaf protein concentrate by human subjects was reviewed by Pirie (26). It is clear that leaf protein concentrate cannot substitute completely for milk. However, leaf protein has a potential to improve the health of malnourished children, especially in underdeveloped areas. When compared to values suggested by FAO, amino acid composition of leaf protein is good. Cytoplasmic protein prepared from alfalfa is comparable to casein in essential amino acid composition (14). Recent studies suggested that soluble alfalfa protein can be prepared to serve as an ingredient for human food (12). Consumption or leaf protein of fractions of leaf protein will be expanded provided continuing research efforts are undertaken.

18-8.3 Deproteinized Juice

Deproteinized juice contains nonstructural carbohydrates, nonprotein N, K, P, and other minerals. Deproteinized juice may be used as a medium for single cell protein production or as a fertilizer (32). It may not be feasible to feed deproteinized juice to animals due to the existence of antiquality substances and low dry matter content. However, high moisture and nonstructural carbohydrates make it a potential medium for production of microorganisms. Substantial amounts of organisms can be produced from anaerobic fermentation of deproteinized juice (3). Yield of single cell protein can be increased by the addition of N and P to deproteinized juice. Deproteinized juice can be used as a fertilizer due to its nutrient content. Application of deproteinized juice to the field stimulates alfalfa regrowth. Using deproteinized juice as a fertilizer is important if the concept of an on-the-farm process is adopted.

18-9 SUMMARY

Wet fractionation of alfalfa is a process which divides plant tissue into three major fractions: (i) pressed forage, (ii) protein concentrate, and (iii) deproteinized juice. The process involves direct harvest followed by cell rupture or maceration and juice expression. This yields a pressed forage and a raw juice. The pressed forage can be fed fresh, ensiled, or dehydrated. Separation of protein concentrate from the juice is usually by heat coagulation or acid treatment. The protein concentrate can be preserved in wet or dried forms and can be an excellent feed or food for all classes of animals. The deproteinized juice may be used as a fertilizer, medium for single cell production, or added as a nutrient source to the pressed forage before dehydration. The process can increase utilization efficiency of forage nutrients.

REFERENCES

1. Ajibola, O.O., R.G. Koegel, R.J. Straub, and H.D. Bruhn. 1982. Acid precipitation of alfalfa juice protein. ASAE Paper 82-1538. American Society of Agricultural Engineers, St. Joseph, MI.
2. ———, R.J. Straub, H.D. Bruhn, and R.G. Koegel. 1981. Fermentation of plant juice as a protein separation technique. ASAE Paper 81-1528. American Society of Agricultural Engineers, St. Joseph, MI.
3. Beker, M.J., A.A. Upitis, and I.J. Krauze. 1979. Fractionation of plant green mass and microbiological transformation of its components. Izv. Acad. Sci. Latrian 5:61-65.

4. Cheeke, P.R., J.H. Kinzell, and M.W. Pedersen. 1977. Influence of saponins of alfalfa utilization by rats, rabbits and swine. J. Anim. Sci. 45:476.
5. Connell, J., and D.G. Cramp. 1975. The nutritive value for dairy cows of artificially dried lucerne wafers made from either the fresh crop or after mechanical dewatering. Proc. Br. Soc. Anim. Prod. 4:112–113.
6. Derbyshire, J.C., C.H. Gordon, R. D. Holdren, and J.R. Menear. 1969. Evaluation of dewatering and wilting as moisture reduction methods for hay crop silage. Agron. J. 61:928–931.
7. Emetarom, C., and G.P. Barrington. 1977. Electrical conductivity as a means of estimating effectiveness of maceration of green plant material. ASAE Paper 77–1062. American Society of Agricultural Engineers, St. Joseph, MI.
8. Fatunso, M., and M. Byers. 1977. Effect of pre-press treatments of vegetation on the quality of the extracted leaf protein. J. Sci. Food Agric. 28:375–380.
9. Hegsted, M., and H.M. Linkswiller. 1980. Protein quality of high and low saponin alfalfa protein concentrate. J. Sci. Food Agric. 31:377–381.
10. Jayasuriya, M.C.N., D. Wijeyatune, and H.G.D. Perera. 1982. Rumen and post-rumen fermentation of spent tea leaf protein and other protein sources studied by the nylon bag method. Anim. Feed Sci. Technol. 7:221–730.
11. Knuckles, B.E., D. de Fremery, and G.O. Kohler. 1976. Coumestrol content of fractions obtained during wet processing of alfalfa. J. Agric. Food Chem. 24:1177–1188.
12. ----, and G.O. Kohler. 1982. Functional properties of edible protein concentrates from alfalfa. J. Agric. Food Chem. 30:748–752.
13. Kohler, G.O., R.H. Edwards, and D. de Fremery. 1983. LPC for feeds and foods. p. 508–524. *In* L. Telek and H.D. Graham (ed.) Leaf protein concentrates. AVI Publishing Co., Westport, CT.
14. Lu, C.D. 1981. Wet fractionation of alfalfa: utilization of pressed forage and protein concentrate by ruminants. Ph.D. thesis. University of Wisconsin, Madison (Diss. Abstr. 8117521).
15. ----, N.A. Jorgensen, and C.H. Amundson. 1982a. Ruminal degradation and intestinal absorption of alfalfa protein concentrate by sheep. J. Anim. Sci. 54:1251–1262.
16. ----, ----, and G.P. Barrington. 1979. Wet fractionation process: preservation and utilization of pressed alfalfa forage. J. Dairy Sci. 62:1399–1407.
17. ----, ----, and G.P. Barrington. 1982b. Intake, digestibility and rate of passage of silages and hays from wet fractionation of alfalfa. J. Dairy Sci. 63:2051–2059.
18. ----, ----, and G.P. Barrington. 1983a. Quantitative studies of amino acid flow in the digestive tract of sheep fed alfalfa protein concentrate. J. Nutr. 113: 2390–2402.
19. ----, ----, A.L. Pope, and R.J. Straub. 1983b. Digestion and nutrient flow in the gastrointestinal tract of sheep fed alfalfa protein concentrate prepared by various methods. J. Anim. Sci. 55:690–699.
20. ----, ----, and L.D. Satter. 1983c. Comparative study of site and extent of nutrient digestion in lactating dairy cows fed alfalfa protein concentrate or soybean meal. J. Dairy Sci. (Suppl. 1) 66:187.
21. ----, ----, R.J. Straub, and R.G. Koegel. 1981. Quality of alfalfa protein concentrate with changes in processing conditions during coagulation. J. Dairy Sci. 64:1561–1570.
22. ----, L.S. Tsai, L.M. Rode, and N.A. Jorgensen. 1983b. Effect of alfalfa saponins on fermentation in continuous culture of mixed rumen bacteria. J. Anim. Sci. (Suppl. 1) 57:305.
23. Mcleod, M.N. 1973. Plant toxins—their role in forage quality. Nutr. Abstr. Rev. 44:804–808.
24. Nelson, F.W., G.P. Barrington, R.J. Straub, and H.D. Bruhn. 1983. Design parameters for rotary extrusion macerators. Trans. ASAE 26(4):1011–1015.
25. Oelshlegel, F.J., Jr., J.R. Schroeder, and M.A. Stahmann. 1969. Protein concentrates: use of residues as silage. J. Agric. Food Chem. 17:796–798.
26. Pirie, N.W. 1978. Leaf protein and other aspects of fodder fractionation. Cambridge University Press, Cambridge.
27. Raymond, W.F., and C.E. Harris. 1957. The value of the fibrous residue from leaf protein extraction as a feeding-stuff for ruminants. Br. Grassl. Soc. J. 12:166–170.
28. Russell, J.R., J.P. Hurst, N.A. Jorgensen, and G.P. Barrington. 1978. Wet plant fractionation: utilization of pressed alfalfa silage. J. Anim. Sci. 46:278–287.
29. Straub, R.J., and H.D. Bruhn. 1978. Mechanical dewatering of alfalfa protein concentrate. Trans. ASAE 21(3):414–418.
30. ----, ----, G.P. Barrington, and R.G. Koegel. 1982. Chemical preservation of alfalfa juice protein. ASAE Paper 82-1539. American Society of Agricultural Engineers, St. Joseph, Mi.
31. ----, and R.G. Koegel. 1983. Evaluation of a cone press for forage fractionation. Trans. ASAE 26(4):1016–1021.
32. Telek, L., and G.D. Graham. 1983. Leaf protein concentrates. AVI Publishing Co., Westport, CT.

33. Tung, J.Y., R.J. Straub, J.M. Scholl, and M.I. Sunde. 1977. Methods used to evaluate biological protein quality and saponin concentration of various alfalfa juice protein. ASAE Paper 77-1010. American Society of Agricultural Engineers, St. Joseph, MI.
34. Van Sumere, C.F., J. Albrecht, A. Dedonder, H. de Pooter, and I. Pe. 1975. Plant proteins and phenolics. p. 211-213. *In* J. Harborne, and C.F. Van Sumere (ed.) The chemistry and biochemistry of plant proteins. Academic Press, London.
35. Vartha, E.W., R.M. Allison, and L.R. Fletcher. 1973. Protein extracted herbage for sheep feeding. N.Z.J. Exp. Agric. 1:171-174.
36. Walker, A.F. 1979a. Determination of protein and reactive lysine in leaf-protein concentrates by dye-binding. Br. J. Nutr. 42:445-454.
37. ----. 1979b. A Comparison of the Dye-Binding and Fluorodinitrobenzene methods for determining reactive lysine in leaf-protein concentrate. Br. J. Nutr. 42:455-465.

19 Equipment for Harvesting, Transporting, Storing, and Feeding

A. W. PAULI

Deere & Company
Moline, Illinois

V. L. LECHTENBERG

Purdue University
West Lafayette, Indiana

W. F. WEDIN

Iowa State University
Ames, Iowa

Alfalfa (*Medicago sativa* L.) and alfalfa-based forage account for almost 60% of the total tonnage of forages harvested in the USA. Area per farm, climatic conditions, type of terrain, and end use all have a direct effect on the type and size of equipment and on the harvest, storage, and feeding system used by farmers. Equipment and storage needs must be viewed as part of a total system. Typically, alfalfa harvest, storage, and feeding systems have become more dependent on mechanical power as machines have become more efficient and laborsaving.

In 1982, about 80% of forage with alfalfa as a component was baled (Table 19-1). Alfalfa haylage (low-moisture silage) has become popular in the Corn Belt and Lake and Northern States and now accounts for

Table 19-1. Estimated alfalfa and alfalfa-based forage harvest methods in the USA.

Harvest methods	Tg (million t),[†] 1982	Percentage of total
Bales:		
Rectangular	45.3	55.0
Large round	20.8	25.3
Total hay (baled)	66.1	80.3
Haylage[‡]	11.6	14.1
Loose (stacks)	2.5	3.0
Meal and pellets	1.6	2.0
Green chop	0.5	0.6
Total	82.3	100.0

[†] Tonnage corrected to moisture equivalent of baled hay.
[‡] Wilted before storage to reach moisture level as low as 40%.

Copyright 1988 © ASA-CSSA-SSSA, 677 South Segoe Road, Madison, WI 53711, USA. *Alfalfa and Alfalfa Improvement*—Agronomy Monograph no. 29.

14% of harvested tonnage. Loose stacks, most popular in the northern Great Plains States, account for 3%; meal and pellets, 2%; and green chop only 0.6% of the total tonnage.

The increasing cost of labor and high-seasonal labor requirements for storing and feeding alfalfa have forced continuing improvements in haymaking methods and equipment. Innovations that result in labor saving and lower costs per unit of feed value are the goal of modern harvest and handling systems.

19-1 HARVESTING

19-1.1 Mowers and Rakes

The mower has long been the standard machine for cutting alfalfa (Fig. 19-1 and Fig. 19-2). Its use has declined in recent years with adoption of mower-conditioners that both cut and condition the crop in one operation. Even though the harvested area of alfalfa has remained rather constant, sales of mowers in the USA declined from 25 000 in 1970 to 11 000 in 1982. However, the mower likely will be important in alfalfa harvesting for many years.

Tractor-operated mowers may be side-mounted, three-point hitch-mounted or pull type. Most cutter bar mowers have either a 2.1- or a 2.7-m cutting width. The reciprocating, sickle-bar principle embodied in

Fig. 19-1. Sickle-bar mower. Courtesy, New Holland, New Holland, PA.

EQUIPMENT FOR ALFALFA PRODUCTION 569

Fig. 19-2. Rotary-cut mowers: (*top*) covered in operation and (*bottom*) uncovered. Courtesy, New Holland, New Holland, PA.

mowers has been in common use for more than 100 yr (20). Its efficiency was improved with development of the balanced pitmanless drive. This development permitted higher knife speeds and faster travel through the field. While it is still the dominant mower type in the USA, sales of sicklebar mowers are declining relative to rotary-cutting types.

Rotary cutting, the major method for cutting grasses in Europe for many years, was introduced in the USA during the mid-1970s (Fig. 19-1 and Fig. 19-2). In 1982, rotary mowers accounted for 40% of all cutterbar mowers sold in the USA. Rotary cutting is accomplished by a series of rapidly rotating horizontal knives. Cutting assemblies may be drum type (driven from the top through a vertical shaft) or disc type (belt driven below the knife mechanism). Initial cost of a rotary mower is greater than that of the sickle-bar type. Rotary mowers are attractive to producers, however, because of their capability for higher field speeds without plugging under difficult conditions. This includes mowing fields of lodged high-yielding alfalfa, those with rodent mounds, and those with uneven terrain. Rotary mowers are available in cutting widths from 1.8 to 2.7 m.

Rakes are used to form windrows when alfalfa is left in the swath for drying, to turn windrows prior to baling to speed drying, and to consolidate windrows to match baler or forage harvester capacity. The side-delivery rake, either the wheel or parallel-bar type, is most common.

In 1982, the parallel-bar rake accounted for more than 80% of the 11 000 side-delivery rakes sold in the USA. This type of side-delivery rake has a series of parallel bars that form an oblique reel. Hay is moved forward and to the side by teeth attached to the bars. Ground-driven, semi-mounted, side-delivery rakes are most common. They are used as single units or as two units coupled together with a tandem hitch to make a single windrow from two or more swaths.

The hydraulically driven, side-delivery rake is a recent development that is gaining popularity among growers with large areas of alfalfa (Fig. 19-3). It consists of two independently controlled cylinders arranged to discharge the windrow between the cylinders. The rake cylinders are driven by hydraulic motors that provide constant raking velocity and the capability for controlling windrow size from the tractor. Current models can form windrows from swaths up to 9.75-m wide.

Fig. 19-3. Hydraulically driven tandem side-delivery rake used to form large windrows for baling or chopping. Courtesy, Vermeer Manufacturing Co., Pella, IA

The wheel rake consists of several diagonally positioned individual wheels each of which is equipped with spring teeth. The wheels revolve from ground contact to form a windrow behind the last wheel. Because wheels are mounted individually, wheel rakes are particularly useful on uneven or rough terrain.

Tedders were in common use several years ago to fluff swaths or windrows for faster drying. Their use in alfalfa harvesting and drying declined with the introduction of conditioners that caused less loss of leaves.

19-1.2 Mower-Conditioners and Windrowers

The use of pull-type mower-conditioners, developed to combine the functions of mowers and pull-type conditioners, has increased in the USA since the late 1960s. Both sickle-bar and rotary-cutting types are used (Fig. 19-4). Typical sickle-bar models have cutting widths ranging from 2.2 to 4.3 m. Some are pulled rigidly behind a tractor; others have a pivot hitch that allows cutting on either side of the tractor. Rotary-cutting mower-conditioners currently available have cutting widths from 2.2 to 3.0 m.

Self-propelled (SP) windrowers were first used during the 1950s in irrigated alfalfa growing areas of western USA (Fig. 19-5). Their use has expanded to all areas of the country where harvested tonnage or areas of alfalfa justifies their purchase. They are typically equipped with air-conditioned operator stations, hydrostatic transmissions, engines ranging in power from 52 to 71 kW, and cutting widths of 3.7 to 4.9 m. Both the mower-conditioner and SP windrower have the cutting mechanism mounted on the front of a platform with a conditioning device at the rear. As hay leaves, the conditioning rolls, adjustable deflector shields are used to form a windrow. The shields can be used also to adjust height and width of windrows.

19-1.2.1 Enhanced Drying and Preservatives

Drying is the greatest impediment to making good-quality alfalfa hay in much of the USA. Under natural field conditions, hay drying often requires 2 to 3 d and is slowed by high-relative humidity. Frequently, rainfall occurs before hay is completely dried, which further slows the drying process and contributes to yield and nutrient losses.

Risk of rain during drying is greater in the midwestern and eastern states than in the west. Fast drying is critical in these states, because alfalfa harvested for hay must be dried to <25% moisture or chemically treated before it can be stored without spoilage (13). Hay dries slowly because of plant resistance to evaporation caused by the cuticle layer and by the presence of membranes and chemical constituents within plant cells. Overcoming this resistance to evaporation greatly speeds hay drying.

Conditioning has become a standard practice to speed field drying of alfalfa. Conditioning is accomplished by one of several mechanical devices, by application of chemical agents, or both.

Fig. 19–4. (*Top*) Nonpivot and (*bottom*) center-pivot pull-type mower conditioners will cut, condition, and windrow alfalfa or leave it in the swath. Courtesy, Deere & Co., Moline, IL; New Holland, New Holland, PA.

19–1.2.1.1 Mechanical Conditioning. Mechanical conditioning involves crushing, crimping, or abrasion of stems and leaves to permit faster evaporation of moisture from the interior of the plant.

Fig. 19-5. Self-propelled windrower, most often used in larger alfalfa operations, cuts, conditions, and windrows alfalfa. Courtesy, Deere & Company, Moline, IL.

Conditioning is very effective in reducing the drying time of stems. Early tests indicated that mechanical conditioning reduced drying time as much as 30% (26). In crushing or crimping, cut alfalfa is passed between two rolls rotating in opposite directions. Roll clearance is adjustable to adequately crush stems and leaves without unduly increasing harvest loss. Smooth or intermeshing corrugated rolls are made of steel, rubber, or synthetic polymer. Most manufacturers offer options of steel, rubber, or one steel and one rubber roll.

Abrasion, accomplished by passing cut alfalfa through a series of rapidly rotating tines, is another approach to mechanical conditioning. These machines can be adjusted for maximum conditioning effect on stems with minimal leaf loss.

19-1.2.1.2 Chemical Conditioning. This is a comparatively new development for which specific field application methods remain to be developed. In some studies, drying rates were increased 40% above those of mechanically conditioned alfalfa (22). In other situations, chemical conditioning has not been shown to increase drying rates. The objective of chemical conditioning is to remove or alter the evaporation-limiting cuticle on the stem surface, thus speeding water loss and minimizing leaf loss. It is accomplished by spraying a solution of chemicals on the crop during cutting (29). The solution often includes potassium (K_2CO_3) or sodium (Na_2CO_3) carbonates. Research indicates that a combination of both chemical and mechanical conditioning is more effective than either method alone (22).

19-1.2.1.3 Preservatives.
Storing hay at >20% moisture results in mold growth. These mold organisms, through respiration, generate heat and reduce the carbohydrates present in hay. Some molds grow actively at temperatures up to 65°C. If hay temperature rises to 65°C, chemical reactions can lead to spontaneous combustion. Hay stored at 30 to 40% moisture is most likely to undergo these reactions.

The heat generated in hay by mold growth causes serious losses in protein digestibility (6). Also, weight losses can be severe when hay is stored at high moisture. Studies show that hay stored <20% moisture loses from 4 to 10% of its dry weight during storage (8). Other studies (9,13) have shown the relationship between dry weight loss and storage moisture percentage, with losses averaging approximately one percentage unit for each percentage of moisture above 10.

Chemical preservatives can prevent mold growth in moist hay. Organic acids and related compounds, when applied to alfalfa hay at 10 kg/t, have reduced heating and storage losses and preserved quality in hay stored at 20 to 30% moisture. Application rates of <10 kg/t are less effective.

Organic acids are applied usually at baling, using a baler-mounted sprayer that delivers the chemical onto the hay as it enters the bale chamber. While organic acids, especially propionic acid (11,24), are effective in preventing mold in high-moisture hay, these acids are quite corrosive and can damage equipment.

Anhydrous ammonia has been shown to kill mold spores (2) and prevent mold growth in high moisture hay when applied correctly. Alfalfa hay treated with anhydrous ammonia is excellent livestock feed and higher in crude protein percentage than untreated hay (10). Dairy animals readily consume ammoniated hay and produce well (32). Equipment to apply anhydrous ammonia in a field operation is not commercially available.

19-1.3 Balers

19-1.3.1 Small Rectangular Balers

The self-tying baler (Fig. 19-6) has been the most popular hay-packaging machine since the early 1940s. Currently, it is used to bale more than 50% of alfalfa harvested in the USA (Table 19-1). Sales have declined, however, from 30 000 in 1970 to about 8000 in 1984. The growing acceptance of haylage and development of the large round baler are major factors in this decline.

The most common bale size is 35 × 45 cm in cross section with length varying from 80 to 90 cm. Larger bales up to about 45 × 55 cm in cross section and up to 115 cm in length are popular in the western USA. Bale knotters are available for tying with wire or with synthetic or sisal twine. Twine tying is used with 85% of these balers.

Most balers are pull-type and driven by the tractor power take-off.

EQUIPMENT FOR ALFALFA PRODUCTION

Fig. 19-6. Small rectangular baler. This type baler is used with at least 50% of alfalfa harvested in the USA. Courtesy, New Holland, Inc., New Holland, PA.

Other pull-type models are powered by baler-mounted engines. Some SP balers are used in western states, but sales are low.

Baler productivity depends on size of the window, tractor power, baler capacity, and uniformity of terrain. Capacities of 10 to 12 t/h are typical. In the western USA, balers may average 600 to 700 t per season, compared with about 100 t in the East. In some areas, custom operators may bale 2000 to 3000 t/yr.

19-1.3.2 Large Rectangular Balers

In 1978, the first commercially available large rectangular baler was introduced in the USA (Fig. 19-7). It forms a 1.2 × 1.2 × 2.4 m bale weighing up to 900 kg that is tied with six heavy-duty, plastic twines. Typical baler capacity is 20 t/h. Equipment for accumulating, loading, and processing these bales is available. Other manufacturers now sell large rectangular balers in the USA and Europe.

Large rectangular balers are used most frequently in the West where there are large areas in high-yielding irrigated alfalfa. Much of the crop in this region is sold as hay and transported long distances by truck. The large rectangular bale provides a low-labor alternative to loading many small bales on a truck. Bale dimensions and density provide for loads of maximum truck weight with the lowest unit transport cost.

Moisture evaporation from large bales is slow because of their size and density. Thus, to prevent spoilage, hay must be drier at baling than

Fig. 19–7. Large rectangular baler system includes (*top*) baler, bale accumulator; (*center*) loader-retriever; and (*bottom*) processor. Courtesy, Hesston Corp., Hesston, KS.

with small rectangular bales. Relatively high temperature and low humidity weather typical in the western USA makes this region ideal for the use of large rectangular bales. High investment costs make these systems best suited to growers or custom operators who bale 750 t or more of alfalfa per season.

19-1.3.3 Large Round Balers

The large round baler, introduced in the early 1970s, has been adopted rapidly by USA hay producers (Fig. 19-8). By 1979, sales of these balers were greater than sales of the small rectangular baler. In 1982, they accounted for about 60% of all balers sold in the USA.

Fixed and variable chamber types of large round balers are available. With the fixed chamber type, bale dimension is determined by chamber size, i.e., only one bale size is possible. Hay at the center of the bale is compressed less than hay on the outside of the bale. With variable chamber balers, all hay entering the chamber is compressed continuously, and density is more uniform throughout the bale. Bale size can be varied by the operator.

Several bale sizes and weights can be obtained from currently available large round balers. Most bale diameters range from 1.2 to 1.8 m, widths from 1.2 to 1.5 m, and bale weight from 300 to 900 kg.

Fig. 19-8. Large round baler forms, automatically ties, and unloads bales. Courtesy, Deere & Co., Moline, IL.

19-1.3.4 Small Round Bales

Balers for making small round bales about 45 cm in diameter and 90-cm long were popular during the late 1950s and early 1960s. While no longer manufactured, several of these balers are in use. These bales were well adapted for low-labor feeding systems in which bales could be left unprotected in the field. The system was particularly popular in cow-calf operations in the Great Plains. However, low baler capacity, difficulties in mechanized handling, and development of large round balers contributed to their decline in popularity.

19-1.4 Forage Harvesters

Forage harvesters are used to chop alfalfa for haylage (field dried to attain 40–60% moisture) and green-chop feeding. Forage harvesters use either precision-cut or flail-type cutting devices. Precision-cut, the most popular, is either pull-type or SP (Fig. 19–9). Self-propelled models are used primarily by large dairy farmers, dehydrators, custom operators, and other operations where large quantities of alfalfa are harvested. Pull-type harvesters are used in smaller operations; however, the largest of the pull-type harvesters now have the same harvesting capacity as a SP harvester.

Precision-cut harvesters may be equipped with a windrow attachment for picking up wilted alfalfa for haylage or a cutter-bar attachment for green-chopping. Precision-cut harvesters have various numbers of knives on a rotating cutterhead. Two knife configurations are available: (i) several full-length knives positioned at an angle across the width of the cutterhead or (ii) many narrow knives about 8.5-cm wide positioned perpendicular to the outside edge of the cutterhead. Both types of knives revolve against a fixed shear bar, chopping forage into lengths as short as 3 mm.

Forage harvesters are now available with devices that detect ferrous metal in the windrow as it is fed into the harvester. Chopping action is thus quickly stopped before the metal reaches the cutterhead. This not only prevents damage to the machine, but also reduces the possibility of livestock deaths from "hardware disease" caused by ingestion of small pieces of wire or other metal. Other devices automatically keep the harvester on the row and automatically move the discharge spout to ensure delivery of chopped material into the wagon or trailer during turns or on contoured or uneven land. They reduce operator fatigue and increase harvester productivity.

Cutting with flail-type forage harvesters is accomplished by slinging knives attached to a revolving shaft. These harvesters are used most often to cut alfalfa for green chop feeding where precision cutting is not required.

EQUIPMENT FOR ALFALFA PRODUCTION

Fig. 19-9. (*Top*) Pull-type and (*bottom*) SP forage harvesters with windrow-pickup attachment. Haylage is discharged into trailing covered forage wagon. Courtesy, Gehl Co., West Bend, WI; Deere & Co., Moline, IL.

19-1.5 Other Methods

19-1.5.1 Cubers

Field cubers were popular in the western USA from the mid-1960s to the mid-1970s for compressing alfalfa into very dense cubes easily handled mechanically from field through feeding. However, cubes with the required durability for handling can be formed only at moistures of 12% or less without additional particle-size reduction, chemical additives, or both. This limits cubing primarily to the lower humidity areas of the West. The geographic limitation, high-energy requirement, and low machine capacity contributed to the decline in cubing. Although field cubing is practiced in some local areas, cubing is restricted largely to stationary operations for specialty uses.

19-1.5.2 Stackers and Stackformers

Stackers are used primarily in the Northern Great Plains. Sweep attachments with long teeth are mounted on tractor front-end loaders to collect the hay from the windrow and transport it to the stacking area. The sweep is either tripped to dump the hay on the stack or it is pushed off the sweep mechanically. Sometimes, cage-like frames are used to aid in stackforming. As a stack is completed, the frame is removed and pulled to a new stacking area.

Many loose-hay stacking operations have been replaced by stackformers consisting of large mobile trailers with sides and a hydraulically movable top used to compress the hay. A flail or retractable finger conveyor picks alfalfa up from the windrow as the trailer is pulled through the field. The hay is deposited in the trailer until full, then compressed. This cycle is repeated several times until the stack is completed and unloaded on the ground. Stacks should have a rounded top to shed rain.

With the development of less labor-intensive and more efficient large round and rectangular balers, the use of both loose stacking and stackformers has declined. They now account for <5% of alfalfa harvested in the USA.

19-2 TRANSPORTING

Large rectangular and large round balers and increased use of haylage have necessitated changes in methods for transporting alfalfa. With these changes, mechanical handling has been substituted for hand labor. Transporting methods vary widely depending on locality and end use.

19-2.1 Small Rectangular Balers

As bales leave the baler, they can be thrown mechanically into a trailing wagon (Fig. 19-10), pushed over an extended slide onto a wagon or accumulator, or dropped to the ground. Bales loaded on trailers or wagons typically are moved to barns or other storage facilities and unloaded. This system is most popular in the Corn Belt, Lake States, and Northeastern region.

Accumulators automatically place six or eight bales in a group that is dumped as the baler travels through the field (Fig. 19-10). The groups are picked up with front-end loaders on tractors and stacked at the edge

EQUIPMENT FOR ALFALFA PRODUCTION 581

Fig. 19-10. Methods for handling small rectangular bales include (A) accumulator that collects bales for pickup by front-end loader, (B) SP bale wagon that picks up and stacks bales and places stack on ground, and (C) bale thrower that places bales into wagon pulled behind baler. Courtesy, Farmhand, Inc., Excelsior, MN; New Holland, Inc., New Holland, PA; Deere & Co., Moline, IL.

of the field or on transport vehicles. This method is used most often in the Great Plains.

Bales dropped on the ground are usually picked up by an automatic bale wagon (Fig. 19-10) or by a side loader on trucks or trailers. Bale wagons are available to mechanically pick up 70 to 160 bales, depending on bale size, and stack them on the wagon. Large SP and pull-type models place the entire stack on the ground by tilting the wagon bed. Also, smaller pull-type bale wagons will unload one bale at a time.

Side-mounted bale-pickup loaders move bales from the ground to truck or trailer. They are relatively inexpensive and eliminate considerable hand labor.

19-2.2 Large Rectangular Bales

The weight and physical size of large rectangular bales require special handling equipment and large tractors (Fig. 19-7). The bales usually are picked up with a combination grapple-spike attachment on a front-end farm or industrial loader and stacked on the flatbed truck for long distance transport or movement to local storage. The same equipment is used to load bales from temporary storage and to unload at the point of feeding.

Tractors and loaders used with these bales should have sufficient strength, power, and weight or wheel ballast for stability in handling 1135-kg loads.

19-2.3 Large Round Bales

Several methods have been developed for handling large round bales, ranging from simple and inexpensive to more complex and costly equipment. The methods are of three general types—equipment for handling single bales, for handling more than one bale at a time without stacking capability, and multi-bale stackers and retrievers (Fig. 19-11).

Equipment for moving individual bales usually consists of simple probes or "spears" inserted into the bale, forks placed under the bale, or a grapple or other arrangement that clamps the bale. They may be used with a tractor three-point hitch, a front-end loader, a towed single-bale mover, or combinations of these. Single-bale movers can be used to move bales to and from outside storage without stacking. Some models also have the capability for unrolling bales for feeding. Single-bale equipment is particularly adapted to smaller dairy and cow-calf operations in the Midwest, South, and East.

Multi-bale handling equipment consists of stacking and nonstacking type. Nonstacking types are towed with tractor or truck to pick up two to six bales in the field for transport to storage. They may be used for retrieving bales from storage, but usually do not have unrolling or processing capabilities.

Other types of multi-bale equipment can pick up four to eight bales in the field for stacking up to three bales high in storage. Some models

EQUIPMENT FOR ALFALFA PRODUCTION

Fig. 19-11. Large round bales may be picked up and transported individually with (A) a three-point hitch bale handler, (B) front-end loader, or (C) with a multi-bale stacker-retriever. Courtesy, New Holland, Inc., New Holland, PA; Deere & Co., Moline, IL; Olds Ag-Tech Industries, Olds, AB, Canada

stand the first row of bales on end with succeeding layers placed on edge. This type can be used to retrieve bales, and a chopper option may be available to process bales for feeding. Higher cost of multi-bale loader-stackers limits their use to large farms or ranches, many of which are cow-calf operations in the western USA.

19-2.4 Chopped Haylage or Silage

Forage is typically discharged by the forage harvester into a forage wagon towed behind the harvester or into a truck that is driven alongside. Other models discharge into harvester-mounted containers and unload onto trucks for transport to storage.

Forage wagons should have covered tops to prevent loss of material in the field and during hauling. Most wagons have a capacity of 17 to 20 m^3 and are designed to self-unload. They are used on many dairy farms in the eastern half of the USA. Trucks with high sides are preferred in large operations where hauling distances are great and speed in transport is important. They usually are equipped with a hydraulic-tilt bed to facilitate unloading.

19-3 STORAGE AND FEEDING

Timely handling of forage into, during, and out of storage is essential for high-livestock production efficiency. Forage is a "perishable product". Thus, handling systems that save labor and time are needed to minimize losses in the field, from weather, during storage, and in feeding. The level of mechanization achieved in alfalfa harvesting, storage, and feeding depends on the type of harvest and storage system used and on the feeding needs of the producer. Haylage systems are usually highly mechanized throughout the total system while hay systems range from highly mechanized to relatively unmechanized, especially with regard to storage and feeding.

19-3.1 Rectangular Bales

Fifty-five percent of all alfalfa in the USA is harvested, stored, and fed using the small, rectangular-bale system. An advantage of hay baled in this manner is that it may be stored in the open or in many different types of inexpensive structures. It can be reloaded and transported as desired to distant feeding areas and markets. Mechanized methods of handling bales were adopted by many producers as farm labor became scarce and expensive. While alternative methods of harvest, i.e., haylage and large round bales, are available, advantages often cited for a small, rectangular-bale system include: adaptability to small production areas; adaptability to existing storage structures; relative ease in marketing; ease in control of feeding; and high cost of alternative machines and storage structures.

Historically, the small, rectangular-bale system is labor intensive. Handling of bales in and out of storage is difficult to mechanize with low-cost machines. The type of machine and amount of handling are influenced by the type of storage; open stack, shed, pole barn, or mow. In western USA with low rainfall, open storage is common. In the Great Plains and the southwestern USA, pole barns and sheds are popular. In the Corn Belt, Lake States, and Northeast region most rectangular bales are stored in mows. The mow, once used for loose hay storage, is not ideal for bale storage but is available at low cost on many farms.

Elevators of various lengths and types are used for conveying bales from trucks into storage or from storage to trucks for transport to feedlots or market. Small electric motors, tractor power take-offs, or gasoline engines are used to power elevators and conveyors.

Conventional double-chain elevators are in common use for moving bales from wagon to mow. Special bale elevators have a single chain in the center with spikes for catching the bale and pushing it along. One type of bale wagon automatically unloads bales into the bale elevator leading to the mow.

Movable knock-off shields on bale conveyors can be adjusted to discharge the bales for random storage or hand stacking in the barn. Random stacking requires little additional storage space, but bale removal is more difficult than when bales are stacked uniformly.

Feeding of baled hay remains one of the most difficult of all chores to mechanize economically. Before the hay is fed, the wire or twine must be removed. In open range feeding, the wire or twine is cut and the bale pushed off the truck. Feeding from a mow requires moving bales by hand to the nearest opening and dropping them directly into a feed bunk or transport vehicle. Feeding a small number of bales daily prohibits a large expenditure on mechanization if labor is available. For large-volume feedlot operations, it is economical to mechanize bale handling as much as possible.

Small, pull-type, automatic bale wagons can be equipped to retrieve bales from a stack, or retrievers can be mounted on separate vehicles.

However, retrieving is somewhat difficult with inside storage. Small bale wagons can unload one bale at a time. Front-end loaders are often used in batch feeding.

19-3.2 Large Round Bales

Large round balers greatly reduce the labor requirements for hay harvest. However, large bales are frequently stored outside (Fig. 19-12) and, therefore, subjected to quality deterioration and weathering losses. Environmental conditions during storage and the manner in which bales are fed to livestock determines the magnitude of storage and feeding losses (9,12). Feeding losses with large round bales can be as low as 4 to 5%,

Fig. 19-12. (*Left*) Large round bales, commonly stored outside, should be placed on a well-drained site. (*Right*) Open cages are typically used in feeding these bales. Courtesy, Univ. of Kentucky, Lexington.

comparable to those from small rectangular bales, when the feeding system is carefully managed (Fig. 19-12). Conversely, losses can exceed 50% when feed is mismanaged.

When large round bales are stored outside, weather damage reduces hay quality. Generally, this damage is confined to the outer portion of the bale, and, under southern Corn Belt conditions, represents approximately 13% of the total bale weight (12).

The greatest advantages of large, round-bale harvesting systems are reduced labor and faster harvest. Some or even all of these benefits can be offset by additional storage and feeding losses if bales are stored outside. However, losses are not greater than with small rectangular bales if large bales are stored inside and if the feeding system is managed to limit animal access to hay, thus reducing feeding waste.

Plastics may be used to cover large round bales stored as hay. There is little question that a cover for each round bale, or for groups of bales, will reduce dry matter and nutrient losses. Adoption of the practice has been limited, however, by unfavorable economics.

Use of plastic covers has increased more for storage of large round bales at high moisture than for dry hay. Large, round-bale silage (haylage, balage) is finding acceptance where silo capacity is inadequate, where curing conditions for hay are unsatisfactory or unpredictable, or where there is interest in feeding a higher moisture content forage. Plastic covers have been used extensively in Europe for several years.

In Pennsylvania (9), alfalfa was baled, bagged, and ensiled at 50% moisture. In this test there was a loss of about 10% dry matter whereas hay round-baled at 18% moisture sustained a 24% loss. In Kentucky (5), alfalfa bagged at 65% moisture and stored was consumed readily by mature dry ewes. In general, research indicates a substantial reduction in losses. However, cost for plastic covers may be $5 or more per tonne.

Though the use of large round bales at a higher-moisture content is only an evolving practice, salient points to consider in choosing and using the system are: bales should be made from small windrows to increase bale density; moisture should be 65% or less, with the optimum about 50%; a silage inoculant is advisable; plastic covers need protection and must be checked periodically for leaks; and bales must be bagged as soon as possible to prevent heating.

Popularity of the large round bale has been especially great among beef cow-calf producers. Farmers must be prepared to adapt and modify storage and feeding practices, however, to minimize storage losses and feeding waste. One disadvantage of large round bales is their limited marketability. It is difficult to obtain dense loads and to transport large round bales. Thus, sales of large round bales are usually confined to small marketing regions near the hay source.

19-3.3 Hay Grinding

Animals consuming hay derive nutrients by digesting the fibrous components. The amount of nutrients obtained is determined by the digestibility of the hay, the rate of digestion, and the rate of passage or length of time feed is retained in the digestive tract. As a rule, high-quality hay is digested more rapidly than low-quality hay that stays in the digestive tract for a longer time. Therefore, animals consume less low-quality hay.

Grinding to reduce particle size speeds passage, increases hay consumption, and improves animal production. The benefits from grinding are usually greater with low-quality hays, presumably because high-quality hays are lower in fiber, digest faster, and pass through the animal more quickly. Also, hay grinding may help to mechanize feeding and reduce feeding waste.

The most common machine for grinding hay is a large tub grinder. Grinders are available in several sizes and can accommodate loose hay, large round bales, and small rectangular bales.

19-3.4 Haylage

Haylage is harvested forage that has been dried before storage to at least 65, and as low as 40% moisture. The proportion of alfalfa and alfalfa-grass forage handled as haylage has increased significantly since the 1940s (16). Second to baling, it is the most important method for processing, storing, and feeding of alfalfa and alfalfa-grass forage. Reasons for its continued popularity are: reduced risk of weather damage; ease of mechanization; and lower field loss of nutrients (31).

The significance of plant moisture in harvesting, processing, storage, and feeding of haylage should not be underestimated. Forty to 60% moisture is most desirable for good fermentation. Serious difficulties (Fig. 19-13) may be encountered if moisture content is not optimal (3).

Alfalfa is low in available carbohydrates, precursors of organic acids, formed during fermentation. Cations and protein, high in alfalfa, buffer changes in pH (27). If a sufficiently low pH is not attained during fermentation, proteolysis results (17). Storing haylage at low moisture reduces proteolysis and results in low-soluble N content (7,18). Hawkins

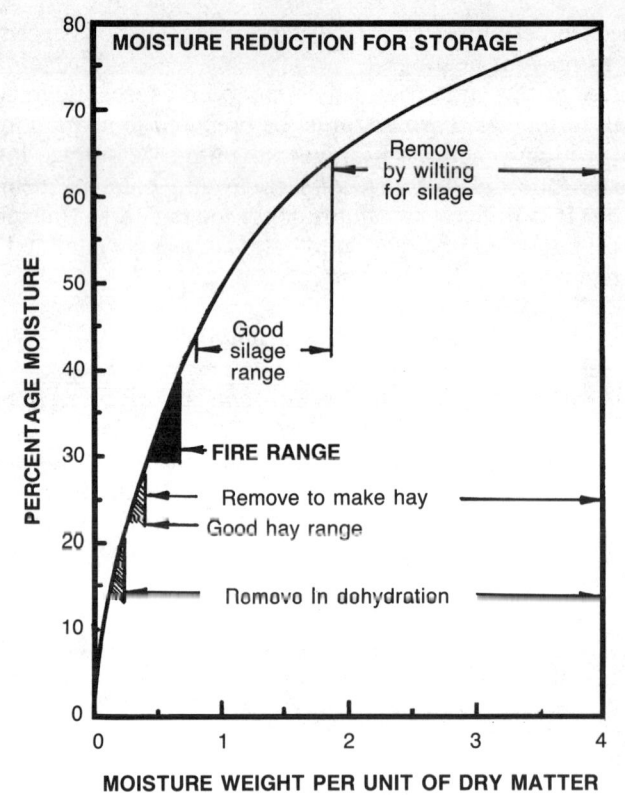

Fig. 19-13. Moisture reduction for storage. From Bruhn (3).

et al. (7) reported that dry matter consumption by sheep increased as silage moisture percentage decreased from 78 to 20%. Proteolysis was inversely related to silage consumption. Others have observed similar relationships (1,28). In a comparison of 45% moisture haylage with sun-cured alfalfa hay, quality differences could be attributed to shattering during baling and chopping (19).

Estimates of storage losses in haylage range from 2 to 12% from surface spoilage and fermentation. Storage losses an be reduced by minimizing air leaks and covering the top of the silage.

19-3.5 Upright Silos

Upright silos, or sealed-storage structures, are most commonly of concrete stave or metal construction. Typical capacities are given in Table 19-2.

Sealed storage structures have distinct advantages over conventional silos, but are higher in cost than concrete-stave silos. Losses from properly sealed and capped upright concrete silos are not appreciably different from those with O_2 limited (metal) structures.

Table 19-2. Approximate capacity of vertical silos for haylage (assumed 500 g kg^{-1} moisture concentration. Adapted from Van Fossen (30).

Settled haylage,[†] depth	Silo diam, m				
	3.7	4.9	6.1	7.3	8.5
m			t		
6.1	22	38	60	85	118
7.3	27	49	78	111	151
8.5	34	64	96	138	189
9.8	42	74	118	169	230
11.0	51	87	138	199	272
12.2	58	103	161	230	314
13.4	67	118	185	265	363
14.6	76	134	209	301	410
15.9		151	234	337	461
17.1		169	261	376	512
18.3		185	288	414	560
19.5			316	454	615
20.7			345	492	669
22.0				532	726
23.2				573	782
24.4				615	838

[†] As much as 3.1 m of extra height is often added to the silo to provide space for silage to settle and for silo-unloader space at the top.

With upright storage, the final link from field to storage is the forage blower. The hopper-type blower (Fig. 19–14) can accommodate rapid unloading and a wide range of forage moisture.

Top and bottom unloaders are available for upright storage structures. The basic design of the top unloader includes chipper and auger knives that loosen the silage, a gathering mechanism (usually an auger) that conveys the forage to the center of the silo, and a blower to remove the silage.

Most bottom unloaders operate on the chain-saw principle (Fig. 19–15). Some bottom unloaders operate with a sweep-auger arm and discharge auger which can be serviced without removal from the silo.

Complete mechanical feeding from upright silos is most popular on dairy farms. As a rule, haylage is dropped onto a conveyor which discharges into a feed bunk (Fig. 19–15). For dairy, new free-stall housing units can be designed for either bunk or fence-line feeding. General guidelines and examples of commercially available haylage feeders for dairies are available from public and private agencies. Batch movement of haylage from upright storage with an unloading wagon, common in beef cattle feeding, is adaptable to fence-line feeding in dairy cattle.

19–3.6 Horizontal Silos

Because they are difficult to seal and compact adequately for haylage, horizontal silos are used primarily for direct-cut or high-moisture silage

Fig. 19-14. The forage blower is used to move forages in to upright storage. Courtesy, Artsway, Armstrong, IA.

(60–68%). They are best suited for large volume horizontal systems using tractor-mounted loaders and feed wagons. Horizontal silos have a low investment cost per tonne of storage, but high-spoilage losses in the absence of adequate sealing and air exclusion. They should be on a well-drained site to expedite water run-off. Sealing with plastic has become a common practice. Storage capacities are given in Table 19-3.

19-3.7 Green Chop

Green-chopped alfalfa is harvested and fed directly one or more times daily. If hauled in a forage wagon, forage is conveyed or dumped into self-feeders. Alternatively, wagons or trailers designed specifically for both transport and feeding may be used.

19-3.8 Stacked Hay

Stacking of hay evolved as a convenient way to store forage in earlier days when labor was available, and transportation was not as convenient. Even then, stacking was mechanized in many innovative ways, and followed with the development of stack movers. Electric fences or cages are restraining methods that reduce losses in feeding cattle. Without restriction, cattle will spoil well over 50% of a forage stack, whether of loose

EQUIPMENT FOR ALFALFA PRODUCTION

Fig. 19-15. Example of (*top*) unloading and (*bottom*) bunk feeding of haylage. Courtesy, A. O. Smith Harvestore, DeKalb, IL.

Table 19-3. Approximate capacities of horizontal silos. Adapted from Van Fossen (30).

Depth	Silo floor width, m								
	6.1	9.2	12.2	15.2	18.3	21.4	24.4	27.4	30.5
m	t								
3.1	41	60	77	95	113	132	150	168	186
3.7	50	72	93	115	137	159	180	202	224
4.3	60	85	111	136	161	187	212	238	263
4.9	70	99	128	157	186	215	244	273	211
5.5	80	112	145	178	210	243	276	308	250
6.1	91	127	163	200	236	272	308	345	381

and long form, or from flail-chopped forage in a machine-formed stack moved to the point of feeding. In Alabama, waste ranged from 35 to 46% when large haystacks were fed to cattle (21). At Purdue, hay waste was reduced to 3.7% when large hay packages were fed using a feeding rack (14). Management is the key to reducing wastage from stacked hay, as with large round bales.

19-3.9 Cubes

Storage and feeding methods for cubes have been developed in the western USA. Reduced storage space, labor savings in handling, increased consumption by animals, and less wastage when feeding are among the reasons for the acceptance of cubed alfalfa by livestock feeders.

Bulk density of cubes is about 400 kg m^{-3} compared to 200 for bales. Transport costs are reduced and storage space requirements cut by one-half. Storage structures should have roofs for protection from rain and smooth concrete or asphalt floors for easy cube removal with a loader.

Belt, auger, and chain-and-flight conveyors are suitable for conveying cubes. Cubes can be handled mechanically with tractor front-end loaders, dump trucks, and bottom-unloading trucks. There are some fines (loose hay particles) in most cubes, especially after being handled two or three times. Although not desirable, fines up to 10% by weight are usually acceptable if kept distributed throughout the mass. Fines concentrated in one place impede natural air flow and may cause heating if moisture is near or above the critical content of 14%.

Storage structures should be designed for ease of filling and removal for feeding. Mechanization includes self-feeding barns, mechanical conveyor feeding, and forage wagon feeding. Cubes are fed whole to dairy cows, and often crushed or ground and mixed with concentrates for beef cows. Uncrushed cubes mix fairly well with concentrates and are consumed well from conventional feed bunks.

19-4 SYSTEMS ANALYSIS

The potential productivity of forages in the USA is generally estimated to be two to three times greater than that currently realized (4). Although the technology exists to achieve this potential, farmers may be reluctant to adopt new forage production practices.

This reluctance is a response to the complexity of the total production system. An improved forage management practice may appear desirable agronomically, but it may contribute little or have a detrimental effect on a farmer's operation when evaluated as a component of the total production system. In some situations, a rather drastic change in animal management may be necessary to capitalize on a change in forage production. These complications make it extremely difficult for farmers to assess the economic value of specific changes in management or technology. The economic payoff may be many steps removed from the improved practice.

The development of large-scale computer simulation models has made it possible to accurately assess the consequences of a change in one component of a complex system and to evaluate the effect of this change on the economic output of the total system (15,23,25). Recent advances in computer capabilities have placed such systems analysis within reach of individual farmers and farm managers on home and office computers. As appropriate simulation and decision-making models are developed, capability will exist on farms to inexpensively evaluate alternative alfalfa production and management techniques in combination with various animal-feeding schemes and management strategies. These models will evaluate such factors as alfalfa quality, ration requirements, marketing strategies, and machinery needs.

This capability will help alfalfa producers to improve the management of their complex forage and livestock system. Thus, importance and payoff of new technology will become more apparent and the gap between current and potential productivity will narrow.

REFERENCES

1. Barry, T.N., D.C. Mundell, R.J. Wilkins, and D.E. Deever. 1970. The influence of formic acid and formaldehyde additives and type of harvesting machine on the utilization of nitrogen in lucerne silages. 2. Changes in amino acid composition during ensiling and their influence on nutritive value. J. Agric. Sci. 91:717–725.
2. Bothast, R.J., E.B. Lancaster, and C.W. Hesseltine. 1973. Ammonia kills spoilage molds in corn. J. Dairy Sci. 56:241–245.
3. Bruhn, H.D. 1978. Barn and silo fires. Hoard's Dairyman 123(9):607.
4. Bula, R.J., V.L. Lechtenberg, and D.A. Holt. 1981. Potential of temperate zone cultivated forages for ruminant animal production. p. 7–28. *In* Winrock Int. Rep. Potential of the world's forages for ruminant animal production. Winrock International Livestock Research and Training Center, Morrilton, AR.
5. Evans, J.K., and C.T. Dougherty. 1984. Bagged silage works well in trials. Hoard's Dairyman 129(9):593.
6. Goering, H.K., C.H. Gordon, R.W. Hemken, D.R. Waldo, P.J. Van Soest, and L.W.

Smith. 1972. Analytical estimate of nitrogen disgestible in heat-damaged forages. J. Dairy Sci. 55:1275-1290.
7. Hawkins, D.R., H.E. Henderson, and D.B. Purser. 1970. Effect of dry matter levels of alfalfa silage on intake and metabolism in the ruminant. J. Anim. Sci. 31:617-625.
8. Hodgson, R.E., J.B. Shepherd, L.G. Schoenleber, H.N. Tysdal, and W.H. Hosterm. 1946. Progress report on comparing the efficiency of three methods of harvesting and preserving forage crops. Agric. Eng. 27:219-222.
9. Kjelgaard, W.L., P.M. Anderson, L.D. Hoffman, L.L. Wilson, and H.W. Harpster. 1983. Round baling from field practices through storage and feeding. p. 657-660. In Proc. 14th Int. Grassl. Congr., Lexington, KY. June 1981. Westview Press, Boulder, CO.
10. Knapp, W.R., D.A. Holt, and V.L. Lechtenberg. 1975. Hay preservation and quality improvement by anhydrous ammonia treatment. Agron. J. 67:766-769.
11. ----, ----, and ----. 1976. Propionic acid as a hay preservative. Agron. J. 68:120-123.
12. Lechtenberg, V.L. 1978. Preserving and increasing the quality in harvested forages. p. 85-91. In Proc. Am. Forage Grassl. Counc., Raleigh, NC. American Forage and Grassland Council, Lexington, KY.
13. ----, and D.A. Holt. 1982. Innovations in hay harvesting and storing. p. 38-47, In Proc. Natl. Alfalfa Symp., Lexington, KY. 6-8 April. Certified Alfalfa Seed Council, Woodland, CA.
14. ----, W.H. Smith, S.D. Parsons, and D.C. Petritz. 1974. Storing and feeding large hay packages for beef cows. J. Anim. Sci. 39:1011-1015.
15. Loewer, O.J., E.M. Smith, G. Benock, T.C. Bridges, L. Wells, N. Gay, S. Burgess, L. Springate, and D. Debertin. 1981. A simulation model for assessing alternate strategies for beef production with land, energy, and economic restraints. Trans. ASAE 24:164-173.
16. McCullough, M.F. 1978. Silage—some general considerations. In M.C. McCullough (ed.) Fermentation of silage—a review. National Feed Ingredients Association, Des Moines, IA.
17. McDonald, P., and R.A. Edwards. 1976. The influence of conservation methods on digestion and utilization of forages by ruminants. Proc. Nutr. Soc. 35:201-211.
18. ----, and R. Whittenbury. 1973. The ensilage process. In G.W. Butler and R.W. Bailey (ed.) Chemistry and biochemistry of herbage, Vol. 3. Academic Press, London.
19. Merchen, N.R., and L.D. Satter. 1983. Digestion of nitrogen by lambs fed alfalfa conserved as baled hay or as low moisture silage. J. Anim. Sci. 56:943-951.
20. Miller, H.F., Jr. 1960. Power in the harvest. In Power to produce. USDA Yearbook of Agriculture. U.S. Government Printing Office, Washington, DC.
21. Renoll, F.S., W.G. Anthony, L.A. Smith, and J.L. Stallings. 1971. Comparison of baled and stacked systems for handling and feeding hay. Auburn Univ. Agric. Exp. Stn. Progress Rep. 97.
22. Rotz, C.A., J.W. Thomas, T.R. Johnson, and D.A. Herrington. 1982. Mechanical and chemical conditioning to speed alfalfa drying. Am. Soc. Agric. Eng. Paper 82-1036. American Society of Agricultural Engineering, St. Joseph, MI.
23. Savoie, P.L., D. Parsch, C.A. Rotz, R.C. Brook, and J.R. Black. 1985. Simulation of forage harvest and conservation of dairy farms. Agric. Systems 17:117-131.
24. Sheaffer, C.C., and N.A. Clark. 1975. Effects of organic preservatives on the quality of aerobically stored high-moisture baled hay. Agron. J. 67:660-662.
25. Smith, E.M., and O.J. Loewer. 1981. A non-specific crop growth model. Am. Soc. Agric. Eng. Paper 81-4013. American Society of Agricultural Engineering, St. Joseph, MI.
26. Sorenson, J.W., Jr., and N.K. Person, Jr. 1967. Harvesting and drying selected forage crops. Texas A & M Univ. Bull. 1071
27. Stallings, C.C., R. Townes, B.W. Jesse, and J.W. Thomas. 1981. Changes in alfalfa haylage during wilting and ensiling with and without additives. J. Anim. Sci. 53:765-773.
28. Thomas, J.W. 1978. Preservative for conserved forage crops. J. Anim. Sci. 47:721-735.
29. Tullberg, J.N., and D.J. Minson. 1978. Effect of potassium carbonate solution on the drying of lucerne. J. Agric. Sci. 91:557-561.
30. Van Fossen, Larry. 1978. Selecting and locating silage storage. Iowa State Univ. Coop. Ext. Serv. Pm-417b.
31. Waldo, D.R. 1977. Potential of chemical preservation and improvement of forages. J. Dairy Sci. 60:306-326.
32. Weiss, W.P., V.F. Colenbrander, and V.L. Lechtenberg. 1982. Feeding dairy cows high moisture alfalfa hay preserved with anhydrous ammonia. J. Dairy Sci. 654:1212-1218.

20 Geographic Adaptation and Cultivar Selection

BILL MELTON

New Mexico State University
Las Cruces, New Mexico

JIM B. MOUTRAY

AgriPro
Ames, Iowa

JOE H. BOUTON

University of Georgia
Athens, Georgia

In the USA, the area planted to alfalfa (*Medicago sativa* L.) increased from approximately 0.8 million ha (2 million acres) in 1900, to a peak of approximately 12.1 million ha (30 million acres) in 1958 (Fig. 20 1). The area involved in alfalfa production decreased gradually after 1962 to 10.2 million ha (25.2 million acres) in 1983 but stabilized between 10.4 and 10.9 million ha (25.7–26.8 million acres) by 1985. Expansion of alfalfa in the USA has followed regional trends. Early alfalfa introductions into the Eastern States failed because of soil acidity except for a few locations near Syracuse, NY and on limestone soils in Virginia (34). By the late 1930s, alfalfa was established as a major crop in the Northeastern region. After the late 1940s, this region accounted for approximately 8% of the total alfalfa area in the USA (5, 16, 41, 56). Successful introductions of Chilean, Peruvian, and Spanish alfalfa types into the West in the mid-1800s established the Western region as the primary alfalfa-producing area (34, 41, 56). Lack of winterhardiness in these alfalfas limited their use in other regions. Identification and development of winter-hardy cultivars from introduced strains stimulated the expansion of alfalfa in the North Central region. Alfalfa production was concentrated in the North Central region by the late 1930s. The Western and North Central States have remained the major growing regions in the USA with 25 and 57% of the total, respectively. The South Central, Mid-Atlantic region contains 6 to 7% and the Southern region <1% of the total alfalfa area in the USA.

Several factors contributed to the regional growth patterns of the alfalfa industry. These have been divided into sequential events: (i) *Prior*

Copyright 1988 © ASA-CSSA-SSSA, 677 South Segoe Road, Madison, WI 53711, USA.
Alfalfa and Alfalfa Improvement Agronomy Monograph no. 29.

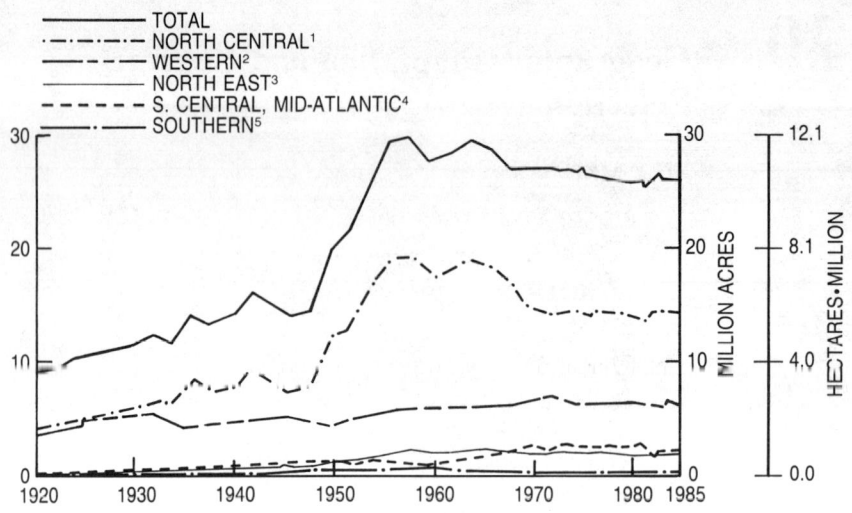

Fig. 20-1. Alfalfa acreage harvested in different regions of the USA. 1 = North Central: North Dakota, South Dakota, Nebraska, Minnesota, Iowa, Michigan, Wisconsin, Illinois, Ohio, and Kansas. 2 = Western: Arizona, New Mexico, California, Colorado, Wyoming, Montana, Utah, Nevada, Idaho, Washington, and Oregon. 3 = Northeast: Pennsylvania, New York, Connecticut, Delaware, Maine, Massachusetts, New Hampshire, New Jersey, Rhode Island, and Vermont. 4 = South Central Mid-Atlantic: Oklahoma, Missouri, Arkansas, Kentucky, Tennessee, North Carolina, Maryland, Virginia, and West Virginia. 5 = Southern: Texas, Louisiana, Alabama, Georgia, South Carolina, Florida, and Mississippi.

to 1925—identification and selection of winter-hardy cultivars; (ii) *1925 to 1955*—development of cultivars combining winterhardiness with resistance to bacterial wilt [caused by *Corynebacterium insidiosum* (McCull.) H.L. Jens.] and increased development of cultural practices associated with liming, soil fertility, and *Rhizobium* inoculation; and (iii) *1955 to 1980s*—development of multiple-pest resistance in cultivars varying in winterhardiness (61).

20-1 ADAPTATION AND CULTIVAR DEVELOPMENT

Winter temperature is a primary factor governing adaptation of alfalfa cultivars from north to south on the basis of winterhardiness. Lowe et al. (56) separated alfalfa into three types—hardy, medium hardy, and nonhardy to reflect adaptation to regions with severe, medium, or mild winter climates. Recognition of more detailed winter-hardiness zones (Fig. 20-2) and better definition of levels of winterhardiness has resulted in the development of cultivars with more specific areas of adaptation. The system currently in use in Minnesota, California, New Mexico, and elsewhere, identifies nine indices based on fall dormancy: very dormant [index 1], dormant [2], moderately dormant [3], three categories of semidormant or intermediate dormancy [4,5,6], moderately nondormant [7], nondormant [8], very nondormant [9] (12,58,61,70,75,76,77). The fall dormancy indices can be fitted into the plant-hardiness zone map (Fig. 20–2A). Nondormant cultivars are best adapted to zones 8 and 9; semidormants to zones 5, 6, and 7, and dormants into zones 2, 3, 4, and 5.

Rainfall is a primary factor affecting cultivar adaptation in an east-to-west direction because of its influence on soil moisture, soil pH, and prevalence of disease and insect problems. Acid soils are usually associated with high rainfall and alkaline soils with low rainfall. Alfalfa grows best on deep, well-drained, slightly alkaline soils. The alfalfa weevil [*Hypera postica* (Glyllenhal)], potato leafhopper [*Empoasca fabae* (Harris)], and pea aphid [*Acyrthosiphon pisum* (Harris)] have become problems in the humid zones, while all aphids are recognized as primary problems under low rainfall. Foliar diseases are more prevalent with high rainfall while root and crown diseases are of concern in regions with either high rainfall or irrigation. In 1974, Barnes et al. (10) listed 21 diseases, 10 insects, and three nematodes as major production hazards in alfalfa. Between 1975 and 1985, these numbers have increased with the addition of Verticillium wilt (caused by *Verticillium albo-atrum* Reinke & Berth.) blue alfalfa aphid (*Acyrthosiphon kondoi* Shinji) and various species of nematodes (19,25,35,38,40,54,62).

Increased precision in defining levels of winterhardiness and need for improved levels of multiple-pest resistance have contributed to a proliferation in the number of alfalfa cultivars. Prior to 1955, about 33 recognized cultivars were grown in the USA and Canada. Between 1956 and 1975, the number of cultivars increased to approximately 160 (11). In 1983, Miller and Melton (61) listed more than 400 cultivars or brand names as having been offered in the USA and/or Canadian seed markets. The development of the private seed industry has also added to the number of available cultivars. The proportion of privately developed cultivars in the periods 1956 to 1960, 1961 to 1965, 1966 to 1970, 1971 to 1975, 1976 to 1980, and 1981 to 1985 increased from 20, 59, 66, 72, 88, and 92%, respectively. (Appendix Table 1.) In 1986, the Certified Alfalfa Seed Council listed and described 125 certified alfalfa cultivars of which only 22% were developed by public agencies (6).

20–2 CULTIVAR RELEASE PROCEDURES

Procedures for cultivar release have been refined to encourage germplasm conservation, accurate representation of the cultivar to the consumer, and protection of rights of the developing organization. Most plant-breeding organizations, either public or private, have Varietal Release Committees that review experimental data on potential new cultivars to determine if release is warranted and to assist in selecting an appropriate name. If release is recommended, descriptive data are submitted to the National Certified Alfalfa Variety Review Board (NCAVRB). This board was organized, and evaluation criteria developed, as a co-

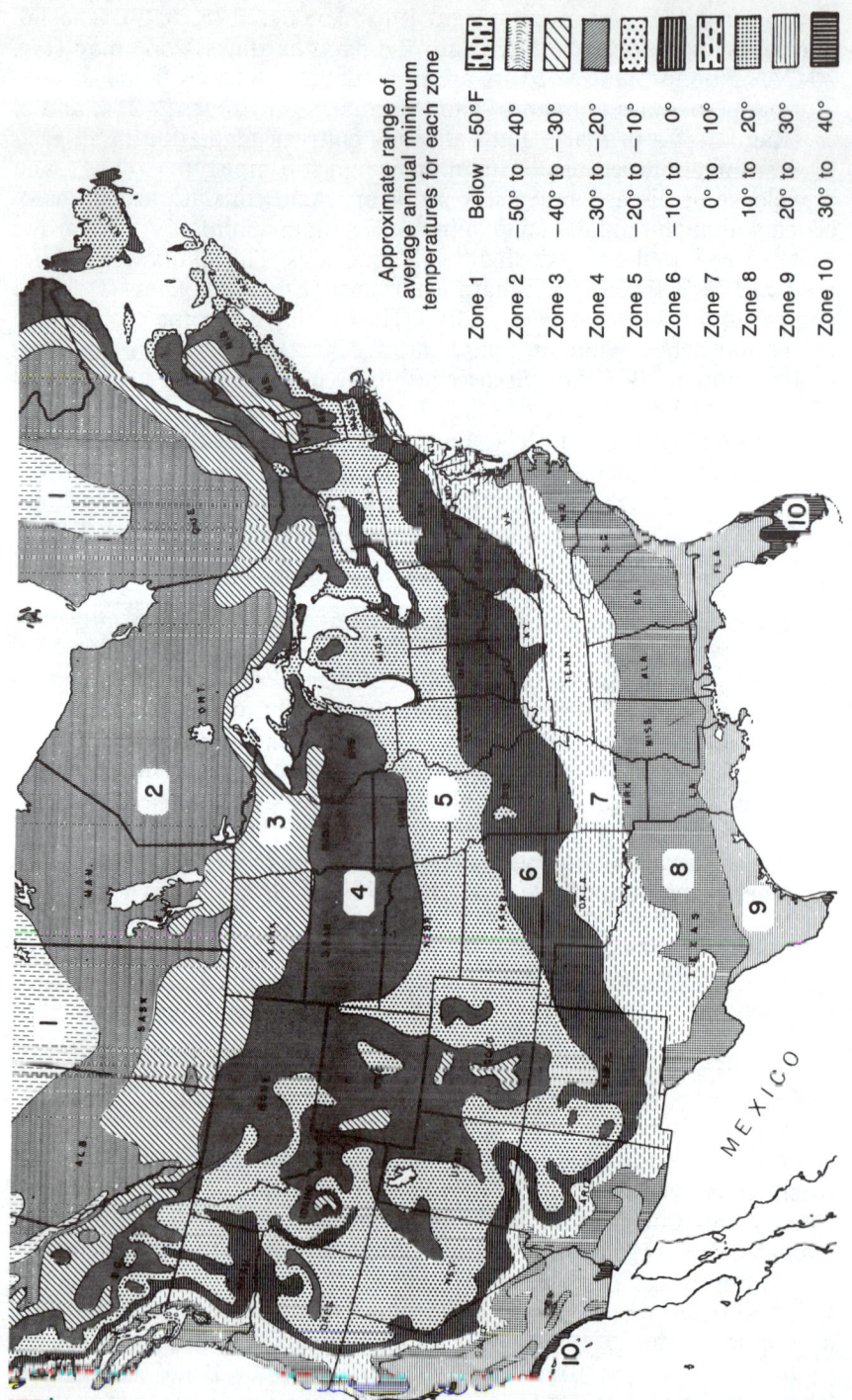

Fig. 20-2. Plant-hardiness zone map. Source: USDA-ARS Misc. Pub. 814.

operative effort by the Joint Alfalfa Work Conference and state seed-certifying agencies (29) (see Chapter 33 in this book). The NCAVRB reviews and evaluates information submitted by public or private organizations and publishes a summary of their findings. Evaluation criteria include: cultivar name and experimental designation, genetic origin, breeding procedure, descriptive data on morphological traits, disease and insect resistance, fall dormancy, forage and seed yield, probable area of adaptation, intended usage, and procedures for maintenance of stock seed classes. Also, the application requests a date when seed will be available and whether or not Plant Variety Protection will be requested. The NCAVRB assists in avoiding duplication of cultivar names and/or infringement on trademarks. The NCAVRB may grant a favorable review or request additional information to support application of claims. The annual report, that includes a brief description of those cultivars receiving a favorable review, is distributed to research workers, seed-certifying and trade organizations, Crop Science Society of America, Plant Variety Protection Office, and Federal Extension Service. Upon receiving a favorable review, the cultivar is eligible for seed certification and developers are urged to publish a registration article in *Crop Science* (Appendix Table 2). Registration provides a permanent record of germplasm types, cultivar characteristics, and ownership. A seed sample is requested for storage in the National Seed Storage Laboratory, Ft. Collins, Colorado, as part of the registration procedure. This step is considered vital to germplasm conservation. The plant breeder, or developing organization, may apply for Plant Variety Protection (2). The Plant Variety Protection Act, 1970, offers legal recourse for developers of new cultivars of plants which reproduce through seed if there is improper exploitation by others. The developer furnishes a complete description of the cultivar including defined procedures for seed multiplication to the Plant Variety Protection Office, Hyattsville, MD. This agency processes the application and may or may not grant a certificate of protection. Approval is based on evidence of distinctness, uniformity, and stability. The breeder, or anyone assigned breeder's rights, is entitled to legal recourse under federal law if anyone sells, exports, imports, or multiplies seed of the cultivar as a step toward marketing it for growing purposes without authority of the owner. The owner of a cultivar can specify that seed is to be sold by cultivar name only as a class of certified seed. Sale of such seed by cultivar name as uncertified seed constitutes a violation of the Federal Seed Act.

Foundation seed of several major public cultivars, e.g., Ranger and Vernal, was produced in large quantities following acceptance by the National Foundation Seed Project. This project declined in importance with the shift from public to private cultivars. Currently, it is being phased-out and new cultivars are not accepted (see Chapter 33 in this book).

20-3 ADAPTATION IN NONHUMID REGIONS

The nonhumid region consists of the 11 western states and the western parts of North Dakota, South Dakota, Nebraska, Kansas, Oklahoma, and Texas (Fig. 20–3). Alfalfa production is limited by low rainfall and the availability and cost of irrigation water. The most important recent changes in alfalfa culture in this region is the development of alfalfa as a primary cash crop with sales in foreign and interstate commerce and markets. Thus, the future status of alfalfa is dependent on demand generated by the livestock industry and competition from other crops. Long-term drought conditions in the Southern and South Central, Mid-Atlantic regions have recently reduced numbers of livestock and reduced demand and prices for western hay.

The alfalfa area increased between 1970 and 1980 in 7 of the 11 western states with decreases only in Arizona and California (5,59). The decrease in Arizona was attributed to cost and availability of irrigation

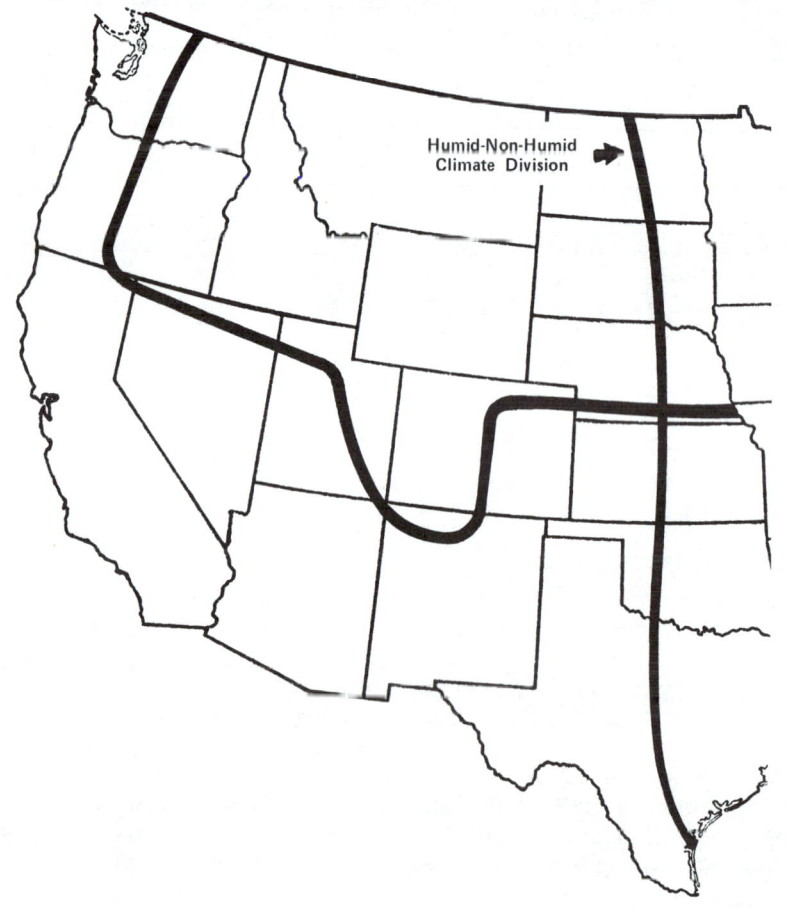

Fig. 20–3. Alfalfa-growing areas in the nonhumid region of the USA. The heavy lines separate: (1) the nonhumid from the humid region, and (2) the approximate dividing line between growing areas for dormant and the combined areas for semidormant and nondormant cultivars. Adapted from Lowe et al. (56).

water while the decrease in California resulted, in part, to competition from other crops. The largest increases in alfalfa through 1983 were in New Mexico (38%) and Nevada (27%) (5,31,59). Between 1980 and 1985, the alfalfa production area in this region decreased by 158 000 ha (390 000 acres). Some decreases occurred in 6 of the 11 Western States with increases in Colorado, Nevada, Oregon, and California. California, Idaho, Montana, and Colorado are the largest production areas in this region. Twenty-four to 25% of the alfalfa area in the USA is in the Western States.

The nonhumid region is very diverse and strongly influenced by the Rocky Mountains. Cultivar adaptation is dependent largely on elevation coupled with latitude and longitude. For example, New Mexico, one of the more southerly states in the region, recommends some cultivars that are adapted to the North Central region as well as semidormant and nondormant cultivars (58). Lowe et al. (56) separated the nonhumid region into three production areas: The Great Plains, the Intermountain region, and the Southwest. However, they recognized the limitations of these descriptive areas of adaptation. In 1973, the Western Alfalfa Improvement Conference appointed the Committee to Establish Ecological Zones for Alfalfa Production in the Western United States (28). Fourteen climatic zones were identified with some zones represented in as many as six states (Table 20-1). It was assumed that similar cultivars should be adapted within each climatic zone.

In the nonhumid region, alfalfa is grown under two major cultural systems; dryland and irrigated. Dryland culture relies on natural rainfall, while irrigated production depends on water from rivers, wells, and/or irrigation projects. Of less importance is what has been referred to as limited irrigation, in which irrigation waters from runoff are available on a limited and unscheduled basis.

Table 20-1. Characteristics of 14 climatic zones in the western USA identified by cluster analysis procedure, using frost-free days, mean temperature, and latitude as parameters. From Elgin et al. (28).

Climatic zone	Frost-free days	Mean temperature (°F)	Latitude
1	50 + 5.98	40 + 0.85	N 42° 44' + 30'
2	94 + 3.52	45 + 0.54	N 42° 05' + 19'
3	105 + 4.47	45 + 0.38	N 44° 20' + 13'
4	115 + 4.86	43 + 0.41	N 48° 20' + 4'
5	115 + 3.40	48 + 0.43	N 39° 26' + 16'
6	125 + 3.19	47 + 0.41	N 46° 27' + 18'
7	136 + 5.85	49 + 0.40	N 43° 06' + 23'
8	142 + 2.13	50 + 0.33	N 40° 26' + 15'
9	144 + 10.08	49 + 1.34	N 36° 04' + 34'
10	160 + 0.74	53 + 0.67	N 36° 55' + 23'
11	181 + 5.87	51 + 0.54	N 45° 45' + 27'
12	209 + 5.44	61 + 0.56	N 34° 41' + 35'
13	256 + 6.15	62 + 0.75	N 36° 16' + 31'
14	278 + 8.60	74 + 0.57	N 33° 08' + 12'

20–3.1 Adaptation Under Dryland Conditions

Dryland alfalfa growing areas are primarily in North Dakota, South Dakota, Montana, and Wyoming with lesser amounts in Utah, Colorado, and northern New Mexico. In Montana and Wyoming, dryland alfalfa accounts for about 50 and 20% of total area, respectively. Dryland production areas vary considerably in winter temperatures, but have an average rainfall from 25 to 50 cm (10–20 inches). Alfalfa is grown in pure stands for hay production, interseeded on rangelands, or sown in mixtures with pasture grasses. In all cultural systems, alfalfa is subject to grazing during periods of drought or during winter months even when planted in pure stands. Forage yields from pure stands vary from near 0 to >7 t/ha (3 tons/acre) (18,30,37,49,58,79).

Problems in dryland production areas are a function of inadequate or inconsistent supplies of soil moisture. Stand establishment has been of primary importance and has been studied from the standpoint of both management and plant improvement (26,27,37,49,69,68,81). In dryland areas, alfalfa stands may be maintained beyond their period of maximum yield because of problems encountered in establishing new stands. Fields in production for 50 to 75 yr have been reported in many dryland areas. Peak productivity usually occurs in the 2nd or 3rd yr of growth and then declines because of exhaustion of subsoil moisture (13,30,34,37,49,50,66,78). Persistence has also been identified as a primary selection criteria in breeding programs because of its importance to production (55,56). Research in Canada and South Dakota has resulted in several released cultivars described as root proliferating and/or rhizomatous (33,66,67). Lowe et al. (56) described these "pasture-type" cultivars as having low crowns, a procumbent growth habit, drought resistance, marked fall dormancy, slow recovery after cutting, and a high degree of winterhardiness. These types yield less than "hay-type" cultivars when grown in environments conducive to high productivity. They are frequently low in seed production.

Drought tolerance is an obvious survival trait under these conditions. Carter (18) reported that, under severe drought conditions, 'Teton' and 'Travois' survived best. Both of these cultivars are pasture types. Rumbaugh (67) described Travois as a root-proliferating type and Teton as rhizomatous. However, other morphological or physiological factors related to drought tolerance may be present in different cultivars. Cultivar differences in water requirement and yield with less than optimum amounts of moisture have been reported (20,79). 'Ladak' was the most widely grown dryland cultivar in Montana and Wyoming while 'Dawson' was used in New Mexico. Garver (30) found that Ladak roots had greater soil depth penetration than those of 'Grimm'. In southwestern Saskatchewan, Kilcher and Heinrichs (50) found three creeping rooted cultivars persisted better than Ladak in grass mixtures. Lowe et al. (56) listed 'Nomad', 'Rambler', 'Rhizoma', 'Sevelra', Teton, and Travois as pasture-type alfalfa. Rumbaugh (67) added 'Roamer', 'Drylander', 'Kane', 'Roverde', 'Spredor', 'Cancreep', and 'Victoria'. More recently 'Maverick' and 'Spredor II' have been released for pasture use. C-3 and C-6 germplasm sources were developed to provide genetic materials for dryland or pasture-type alfalfas (61). C-3 consists of an intercross of plant material tracing to 63 cultivars thought to possess adaptive characteristics for dryland production. C-6 traces to 12 sources of *Medicago falcata, M. media*, 'Cossack', 'Canadian Variegated', 'Cherna', and 'Sibturk' origins. In some areas, Ladak types (Ladak, 'Ladak 65', and 'Ladak 75') are preferred because of its winterhardiness, high forage yield at the first cutting, and the ability to become dormant during periods of summer drought (59).

Most disease and insect problems encountered in humid or irrigated areas may occur under dryland conditions. Damage may be less pronounced because of relatively thin stands, and comparatively low forage yields. Bacterial wilt is apparently important in stand longevity as indicated by the release of the wilt-tolerant cv. Ladak 65. Townsend (78) describes the various species of grasshoppers (*Melanoplus, Camnula* spp.) as being very destructive. Control is by application of pesticides because little is known about resistance mechanisms. Pocket gophers (*Thomomys, Geomys*, and *Cratogeomys* spp.) were considered a greater production hazard on dryland alfalfa than any other pest (78). Alfalfa plants with creeping roots were less easily damaged than plants with taproots.

Because of low yield potential, research in developing new cultivars is limited for the dryland areas.

20–3.2 Adaptation Under Irrigation

Alfalfa ranked as the leading cash crop in five western states and was second or third in five other states in this region (5,21,59). In many areas, alfalfa is the primary source of income for farmers. Development of alfalfa as a cash crop has forced producers to examine production practices closely in attempts to maximize yield and quality. In some areas, production has changed from a rotational sequence to almost a monoculture of continuous alfalfa. This cultural system has led to extensive studies of allelopathy and autotoxicity. Autotoxicity and allelopathy effects have been reported from Wyoming and Nevada, but results from other areas have been inconsistent as to extent, severity, and other crop involvement (14,45,47,51). Autotoxicity and/or allelopathic effects were attributed to plant exudates, by-products of decomposition, buildup of diseases, depletion of soil minerals, and depletion of soil moisture (47,49,51).

Widespread adoption of sprinkler irrigation systems, in an effort to increase applied water-use efficiency and decrease costs associated with irrigation, has resulted in disease problems not previously encountered in this region. Anthracnose (caused by *Collelotrichum trifolii* Bain.) is now a most important disease problem in New Mexico (59). Foliar diseases, such as downy mildew (caused by *Peronospora trifoliorum* De Bary), common leafspot [caused by *Pseudopeziza medicaginis* (Lib.) Sacc.], Stemphylium leafspot [caused by *Pleospora herbarum* (Pers. ex. Fr.) Rab. or *Stemphylium botryosum* Wallr.], and Stagonospora leafspot [caused by *Stagonospora meliloti* (Lasch) Petr.] have increased in frequency and severity of occurrence. Southwestern alfalfa cultivars are generally susceptible to most foliar diseases.

Phytophthora root rot (caused by *Phytophthora megasperma* Drechs. f. sp. *medicaginis*), first identified in California, is listed as the first or second most important disease problem of alfalfa in seven of the western states (59). This disease is extremely prevalent in areas with heavy clay soils and flood irrigation. Bacterial wilt [caused by *Corynebacterium insidiosum* (McCull.) H. L. Jens.] and Fusarium wilt [caused by *Fusarium oxysporum* Schl. f. sp. *medicaginis* (Weimer) Sny. and Hans.] are equally as important in the western states as in the other alfalfa-growing areas. 'Ranger' and 'Vernal' were grown extensively in the western states because of their resistance to bacterial wilt. Nondormant cultivars are usually resistant to Fusarium wilt but extremely susceptible to bacterial wilt (61). Newer semidormant cultivars, usually have at least moderate levels of resistance to both bacterial and Fusarium wilts (61). In southwestern areas, Rhizoctonia stem, crown, and root canker (caused by *Rhizoctonia solani* Kuehn) causes extensive summer stand losses under high temperatures (36,46,56,59). Spring blackstem (caused by *Phoma medicaginis* Malbr. and Roum.) was listed as a significant pathogen in Nevada, Wyoming, and New Mexico. Damping-off (caused by *Pythium* spp. and *Rhizoctonia* spp.) can be severe in seedling alfalfa during periods of high temperature shortly after planting. Phymatotrichum root rot (i.e., Texas root rot or cotton root rot) [caused by *Phymatotrichum omnivorum* (Shear) Dug.] affects specific areas in southern California, Nevada, Arizona, New Mexico, and Texas. Significant areas of alfalfa were removed from production in the plains area of west Texas because of this disease. No sources of resistance are available in alfalfa. Verticillium wilt resistance is now a major factor in cultivar adaptation in affected areas. Resistance to Verticillium wilt is now available in dormant and semidormant cultivars.

Nematodes, although a factor in alfalfa production for many years, are now considered to be more important and widespread than previously noted. For many years, nematodes have been responsible for serious stand losses in parts of Nevada, Washington, Idaho, and Utah. More recently, they have been reported as a primary cause of stand loss in Arizona and southern California (54). Early work was concerned primarily with the stem nematode [*Ditylenchus dipsaci* (Kuhn) Filipjev]. Recently, the rootknot nematodes (*Meloidogyne hapla* Chitwood, *M. incognita*, and *M. javanica*) have assumed greater importance in some areas. 'Lahontan' was developed specifically for resistance to the stem nematode (72). 'Lew' was developed as a nondormant with resistance to the stem nematode (61). More recently, 'New Syn XX' and 'New Syn YY' germplasms were developed with extremely high levels of resistance to both the stem and root-knot nematodes (61). 'UC Cibola' was selected from a nematode-infested field primarily for resistance to the root-knot nematode, but *M. arenaria*, stubby root nematode (*Trichodorus* spp.), root lesion nematode (*Pratylenchus* spp.), stunt nematode (*Tylenchorhynchus* spp.), dagger nematode (*Xiphinema* spp.), and ring nematode (*Criconemoides* spp.) were present (54). All of these nematodes are detrimental to alfalfa growth but damage has not been quantified. Assigning damage to a specific nematode species is often confounded by the presence of several species in the soil.

The various aphids rank as the most important insect problem in the western region. The pea aphid is widespread over the entire region and periodically reduces yield and can damage stands. 'New Mexico 11-1' was one of the first cultivars developed in the west for resistance to this pest (61). Emphasis on the pea aphid was removed when the spotted alfalfa aphid [*Therioaphis maculata* (Buckton)] was discovered in New Mexico in 1954. The spotted alfalfa aphid seriously threatened the entire alfalfa industry. Lahontan was found to be resistant to the spotted alfalfa aphid and contributed valuable germplasm to subsequent cultivar developments (61). 'Zia', a semidormant, and 'Moapa', a nondormant, were the first cultivars developed specifically for resistance to this pest (56,61,81). These cultivars were followed by such releases as 'Washoe' and 'Mesilla' which combined resistance to the spotted alfalfa aphid and the pea aphid (56,61). Most newer cultivars are resistant to both of these aphids. In 1975, the blue alfalfa aphid was found in Arizona and southern California. By 1983, it had spread to New Mexico, Nevada, Utah, and as far east as Kansas (59,62). 'CUF 101', a nondormant, was developed specifically for resistance to the blue alfalfa aphid and was later shown to have resistance to all three aphids (62).

The alfalfa weevil [*Hypera postica* (Gyllenhal)] was first identified as a serious pest in alfalfa in the intermountain zone of the Western region. In 1972, it migrated through Oklahoma into New Mexico. In New Mexico, the alfalfa weevil exhibited a different reproductive pattern in that all growth stages can be found in the field the year around (32). A degree of resistance has been achieved in some improved cultivars (see Chapter 28) but has not been a factor in cultivar adaptation or selection in the Western region. The Egyptian alfalfa weevil [*Hypera brunneipennis* (Boheman)] was identified in the Yuma Valley of Arizona in the 1930s (22). By the late 1970s, it had spread into most alfalfa-producing areas of southern and central California and to other small, isolated areas, such as the Four-Corners Region of New Mexico, Arizona, Colorado, and Utah. No cultivars with tolerance to the Egyptian alfalfa weevil are known.

In the western region, level of fall dormancy (winterhardiness) is a primary consideration in determining areas of cultivar adaptation (Table 20-2). The north to south temperature gradient is highly confounded by elevation so that dormant cultivars are grown as far south as New Mexico and Arizona. The nondormant adaptive area starts in the southern one-third of New Mexico and extends across most of Arizona, southern Ne-

Table 20-2. Fall dormancy classification of alfalfa cultivars grown in western USA and estimated percent of the total area devoted to each grouping.†

State	Nondormant 9 8 7	Semidormant Dormancy index 6 5 4	Dormant 3 2 1
Montana			100%‡
Wyoming			100%
Colorado		60%	40%
Utah§		40%	60%
Idaho		10%	80% 10%
Oregon		50%	50%
Washington		60%	40%
Nevada	5%	55%	40%
New Mexico	10%	80%	10%
California	70%	20%	10%
Arizona	95%	5%	

† From a survey of alfalfa workers in western states (59).
‡ Estimated percent of area of cultivars of various dormancy classifications.
§ Some nondormant cultivars grown in southwestern part of state.

vada into the Central Valley of California. Nondormant cultivars usually have been characterized as nonpersistent. This may be a function of intensive use associated with a long growing season and many seasonal harvests. Improvements in nondormant cultivars have focused on factors associated with increased stand persistancy without sacrificing rapid regrowth and high yields associated with this type of cultivar. 'UC Salton' and 'UC Cargo' were selected for persistence under conditions conducive to scald and stand loss. 'Florida 66' was selected for persistence under conditions conducive to stand loss, and the improved version 'Florida 77' has resistance to the spotted alfalfa aphid (61). Florida 77 has more recently been shown to be resistant to three root-knot nematodes (*M. arenaira, M. incognita,* and *M. javanica*) which could contribute to its increased persistency (9,43,44). Following the release of Moapa in 1957, most nondormant cultivars have had resistance to the spotted alfalfa aphid. Pea aphid resistance was later added to a number of cv. e.g., WL 515, Maxidor, Rincon, and Granada. CUF 101 was developed for resistance to the blue alfalfa aphid and combined resistance to all three aphid species (62). CUF 101 is widely grown and has contributed germplasm to several blue alfalfa aphid resistant nondormant proprietary cultivars.

Stem nematode damage was partially alleviated through the development of Lew, then WL 515, with several nondormant, stem nematode resistant cultivars expected now in the evaluation stage. UC Cibola was released in 1982 primarily for resistance to several genera of root-knot nematodes (54).

'Sonora' and 'Sonora 70' were selected as extremely nondormant cultivars providing some resistance to alfalfa mosaic virus. Levels of downy mildew resistance have been described in several nondormant cultivars, such as 'WL 501R' and 'WL 512' (46). 'Pierce' is typical of newer nondormant cultivars with resistance to all three aphids, stem nematode, Phytophthora root rot, and Fusarium wilt plus moderate levels of resistance to downy mildew, common leafspot, and a low level of resistance to bacterial wilt (57). Nondormant germplasm sources have been developed with high levels of resistance to both race 1 and race 2 anthracnose (61).

There is a group of cultivars often described as being intermediate in fall dormancy, between Mesilla and Moapa. Recent releases include cultivars in the WL 400 series, 'Baron', and 'AS-13R+'. Baron is resistant to all three aphid species, Phytophthora root rot, bacterial wilt, Fusarium wilt, and anthracnose. AS-13R+ has resistance to the spotted alfalfa aphid, Phytophthora root rot, bacterial wilt, Fusarium wilt, common leafspot, root-knot nematode, stem nematode, and has low levels of resistance to the pea aphid and blue alfalfa aphid. These types of cultivars are generally recommended for the high desert valleys in the boundary areas between true nondormant and semidormant areas of adaptation.

Semidormant cultivars are those in the fall-dormancy range between Lahontan and Mesilla. These types usually have resistance to the pea aphid and spotted alfalfa aphid with recent efforts adding resistance to the blue alfalfa aphid. Semidormant cultivars usually have moderate levels of resistance to bacterial and Fusarium wilts. Some of the earliest multiple-pest resistance breeding efforts were in semidormant cultivars. Lahontan was developed to combine resistance to bacterial wilt and the stem nematode and was later found to have resistance to the spotted alfalfa aphid (73). Zia, Washoe, Mesilla, and 'Dona Ana' utilized some Lahontan germplasm. These cultivars possess moderate to high levels of resistance to pea aphid, spotted alfalfa aphid, bacterial wilt, Fusarium wilt, and stem nematode. Dona Ana also has a moderate level of resistance to Phytophthora root rot (60). Recent plant-breeding efforts in this dormancy group have been toward anthracnose resistance and enhanced levels of resistance to stem and root knot nematodes. 'Deseret' was released by Utah on the basis of frost tolerance and resistance to the stem nematode. A number of new, multiple-pest resistant proprietary cultivars are adapted for use in the semidormant adaptive area with representative types being the WL 300 series, Pioneer Hi-Bred Int., "5" series, Northrup King and Co. 'Pike', and Ferry-Morse 'GT-55'. These cultivars are grown widely in the boundary areas between dormants and nondormants including southwestern Oklahoma, the panhandle of Texas, New Mexico, northern Arizona, central Nevada, southern Colorado, south-central Utah, central area of California, Oregon, and Washington.

In general, dormant cultivars are grown north of an irregular line extending from central Colorado, through central Utah, Nevada, and the northern one-third of California. Dormant and semidormant cultivars are both grown in some areas along the boundary. Newer dormant cultivars reflect the trend toward multiple-pest resistance with moderate to high levels of resistance to pea aphid, spotted alfalfa aphid, bacterial wilt, Fusarium wilt, Phytophthora root rot, nematodes, and anthracnose. Vernal and Ranger are grown in these areas but are being replaced with newer cultivars that have higher levels of multiple-pest resistance and increased yield potential.

A recent change among both dormant and semidormant cultivars resulted from the discovery of Verticillium wilt in the Pacific Northwest. This led to the rapid development and release of resistant cv., i.e., Vernema, Apollo II, Trumpetor, and WL 316. The European cv. Vertus was used as an early source of resistance. The early Verticillium wilt resistant cultivars were closer to the semidormant types but resistant, dormant cultivars are now available. The rapid development and multiplication of the public and proprietary cultivars in response to the Verticillium wilt threat illustrates the responsiveness of public and private breeders to the needs of alfalfa producers. The early Verticillium wilt resistant cultivars are rapidly being replaced with newer types combining Verticillium wilt resistance with multiple-pest resistance in several dormancy types.

20-4 ADAPTATION IN HUMID REGIONS

The humid region includes all of the area east of a line which approximately divides North Dakota in the north and Texas in the south (Fig. 20-4). Rainfall in the region varies from 41 cm (16 inches) on the western edge to more than 122 cm (48 inches) in several southern states (56). Factors limiting production in the region vary and include timeliness and amount of rainfall, diseases, insects, and soil acidity. Lowe et al. (56), using Trewartha's (77) climatic classifications and alfalfa hardiness zones by Aamodt (1), divided the humid region into the North Central; Northeast; South-Central, Mid-Atlantic; and Southern production areas (Fig. 20-4). They were careful to point out that these boundaries are only approximate divisions among climatic areas that uniquely affect alfalfa production.

20-4.1 North Central Region

The North Central region contains 53 to 58% of the alfalfa-growing area of USA and grows more than 70% of the alfalfa in the humid region. Seven of the 10 states in this region grow more than 0.4 million ha (1 million acres) with Wisconsin being the largest alfalfa-growing state in the USA with more than 1.2 million ha (3 million acres). Production

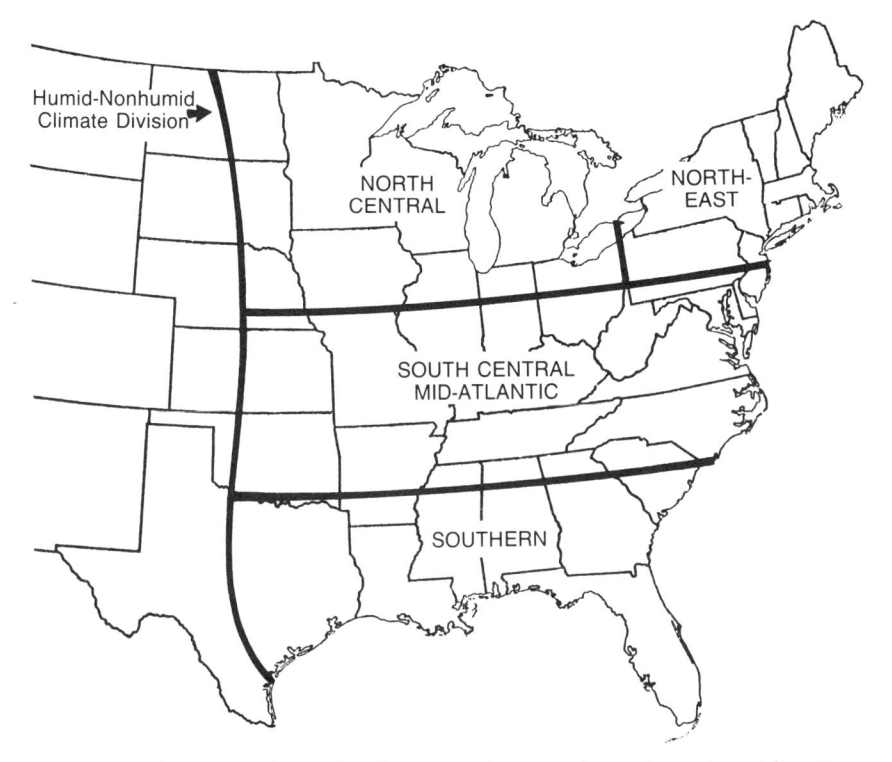

Fig. 20-4. Alfalfa adaptation regions in the humid areas of the USA. Adapted from Lowe et al. (56).

trends since 1970 indicate a decrease in the alfalfa area in Nebraska, Minnesota, and Iowa with increases in Michigan, Wisconsin, and Ohio. South Dakota decreased significantly in alfalfa production in 1985. Ohio has shown the largest percentage increase in alfalfa production area since 1970.

Annual precipitation varies from 61 to 102 cm (24–40 inches) over much of the region but is reduced to 41 cm (16 inches) on the western edge (1,3,8,56). In most years, distribution of precipitation is favorable for high yields of alfalfa. In the transition zone between humid and nonhumid regions, use of sprinkler irrigation increased considerably in the 1970s. In the northern half of the region, snowfall of 102 to 152 cm (40–60 inches) usually provides good insulation against low winter temperatures.

A freeze-free period of 120 to 170 d is common to most of the area. Small parts of the northern sections of Minnesota, Wisconsin, and Michigan have only 90 freeze-free d. Average January minimum temperatures range from $-20°C$ ($-5°F$) in the northwest to $-4°C$ ($25°F$) in the southeast. Average July maximum temperatures are in the 27 to 32°C (80–90°F) range. These temperature extremes are among the greatest in the continental USA and present a challenge to the development of adaptive, productive cultivars. Cultivars must not only withstand extremely cold winter temperatures but also respond well to favorable temperatures during the growing season.

Much of the North Central region has excellent soils for alfalfa production. Many are fertile, deep, adequately drained, with high moisture-holding capacity. Liming is required for maximum yields. Poor drainage may reduce production and persistence in some areas. However, the availability of improved cultivars with resistance to Phytophthora root rot has helped to reduce stand failures.

Other diseases that can have a major impact on persistence are bacterial wilt, anthracnose, Fusarium wilt, and Verticillium wilt (36,59). The effects of bacterial wilt in the North Central region have been known for more than half a century. In the early 1970s, Phytophthora root rot was shown to limit persistence. Since the 1970s, the effects of anthracnose have become more widespread and recognized in the area. In 1980, Verticillium wilt was found in Wisconsin and by 1986, it had been identified in seven North Central States (19,25,35,38). In the past 5 yr, cultivars incorporating resistance to Verticillium wilt resistance have increased in importance.

Leaf diseases often mentioned as important are spring blackstem, summer blackstem [caused by *Cercospora medicaginis* Ell. and Ev.], common leafspot [caused by *Pseudopeziza medicaginis* (Lib.) Sacc.], and lepto leaf spot [caused by *Leptosphaerulina briosiana* (Poll.) Graham and Luttrell].

Potato leafhopper and alfalfa weevil rank first and second as the most important insect pests of the North Central region (10,23,48,52). In some areas, alfalfa may be damaged by pea aphid, several species of grasshoppers, and meadow spittlebug (*Philaenus spumarius* L.) (59). Occasional damage by the spotted alfalfa aphid occurs in Nebraska.

Winterhardiness was the primary factor in adaptation of early cultivars to this region. Many of these, such as 'Grimm', were introductions or derived from introductions (56). Bacterial wilt became the primary factor in stand persistence in the 1930s and 1940s. Ranger and Vernal were developed as the first winter-hardy, wilt-resistant cultivars. In the 35 years following the release of Vernal, there has been a shift toward faster growing, less fall-dormant cultivars primarily through the incorporation of Flemish germplasm. In the first decade after the release of Vernal, two additional cultivars with Vernal dormancy were released compared to three each with Ranger and 'Saranac' dormancy. The next decade saw the approval of 10 new Vernal types, 19 cultivars of Ranger dormancy, and 16 of Saranac dormancy by the NCAVRB. During this period, proprietary cultivars increased to 15 to 20% of the total usage. From 1980 to 1985, nine new cultivars of Vernal dormancy, 25 of Ranger dormancy, and 56 of Saranac dormancy were approved by NCAVRB. (Appendix Table 3.) Ranger dormancy, and 56 of Saranac dormancy were approved by the NCAVRB. Adaptation of these less-dormant types to the North Central region has been improved through breeding for disease resistance, winter survival, and adoption of improved management practices. Nevertheless, these less-dormant types may decrease the margin of safety for winter survival. Useage of proprietary cultivars presently accounts for 70 to 75% of the total.

20-4.2 Northeast Region

The alfalfa-producing area of the Northeast region has remained relatively stable since 1970 with slightly more than 0.8 million ha (2 million acres). New York and Pennsylvania are the leading alfalfa-producing states in the region (5).

Climate in the Northeast region is influenced by large bodies of water and elevation differences. Precipitation ranges from 76 to 132 cm (30–52 inches) with more than one-half coming during the freeze-free period (1,3,9). However, as compared to the North Central region, a larger percentage occurs as snow, which gives greater protection from low winter temperatures. A freeze-free period of 120 to 150 d is most common in the region but varies from 80 d at some higher elevations to 200 d along the Atlantic coast. Average minimum January temperatures range from $-18°C$ (0°F) in northern Maine to $-4°C$ (25°F) in southern Pennsylvania. Average maximum July temperatures over much of the region are in the 27 to 30°C (80–85°F) range. The cooler summer temperatures result in slower growth and longer harvest intervals than found in the North Central region.

Soils in the Northeast region are highly variable. Steep slopes, poor drainage, shallow soils, and stoniness limit alfalfa production on most land. Where alfalfa is grown, liming is normally essential for good production and persistence (3,56,71). Adequate drainage is an important consideration in choosing land for alfalfa production.

Bacterial wilt has been recognized as an important factor in premature stand depletion for more than 30 yr (56). Since about 1970, anthracnose has been recognized as a major factor in yield losses and persistence in the southern part of the region. More recently, Phytophthora root rot has been recognized as an important disease especially on imperfectly drained soils. Fusarium wilt and crown rots were frequently mentioned as problems by alfalfa workers in the region (59). In 1981, Verticillium wilt was found in New York and Pennsylvania and in several additional sites in the next 3 yr (53).

The most prevalent leaf diseases are spring blackstem, summer blackstem, common leafspot, lepto leafspot, and Stemphylium leafspot (59). These diseases lower yield, hay quality, and affect adaptability of cultivars.

During the past decade, the alfalfa weevil has declined in severity. It is no longer considered to be the most important insect pest in the Northeast region. This is because of several factors, including biological control, better management technique, and improved insecticides. Potato leafhopper now ranks as the most important insect pest in the region (52,59). In recent years, significant production losses from this insect have been documented (52). Damage is less severe on some cultivars, but breeding for resistance has been very slow. Also, alfalfa blotch leafminer (*Liriomyza* spp.) and clover root curculio (*Sitona hispidulus* F.) are important pests in the area (59). Meaningful cultivar differences have not been reported for these two pests.

Cultivars grown in the Northeast region are generally moderately dormant. 'Narragansett,' released in 1945 from Rhode Island, proved to be adapted to many Northeast environments (34,56). Narragansett's success was partly attributed to a high level of winterhardiness. In the early 1950s, several Flemish strains were introduced into the Northeast. These high-yielding, faster growing alfalfas lacked the winterhardiness of Narragansett, but demonstrated satisfactory winter survival. 'DuPuits' was widely grown on the better drained soils until the mid-1960s (56). Bacterial wilt was a primary problem with both Narragansett and DuPuits. In 1963 and 1966, the cv. Saranac, developed from DuPuits, and Iroquois, developed from Narragansett, were released. These bacterial wilt-resistant cultivars soon replaced older types in the Northeast region. Vernal was used in the region but not extensively. During the 1970s, Iroquois was the most widely used cultivar in the area (63). Saranac continued to be strongly represented until 1978, when the anthracnose-resistant cv. Saranac AR increased in importance. In recent years, emphasis has been given to adding resistance to Phytophthora root rot and Verticillium wilt, in the development of new multiple-pest resistant cultivars adapted to this region (63,64). A number of proprietary cultivars with resistance to bacterial wilt, Fusarium wilt, Phytophthora root rot, anthracnose, and some resistance to pea aphid and Verticillium wilt, have achieved substantial acceptance. Proprietary cultivars now represent and estimated 60 to 70% of the total area in the Northeast region and a higher percent in southern portion.

20–4.3 South Central, Mid-Atlantic Region

The South Central, Mid-Atlantic region contains approximately 10% of the alfalfa production area of the USA (Fig. 20–1). Most of this area is in Kansas with approximately 0.4 million ha (1 million acres). Alfalfa production in this region has followed national trends with a decrease in area between 1970 and 1980, but a slight increase in 1985. Production areas decreased in Oklahoma, Missouri, and Arkansas; and increased in Kentucky, Tennessee, North Carolina, Maryland, and West Virginia. The region is very diverse. The South Central, Mid-Atlantic region lies between 35 and 40° N Lat and is bordered by the Atlantic Ocean on the east and the eastern halves of Oklahoma and Kansas on the West (Fig. 20–4). The climate is varied, with an expectation of a mean annual temperature of 10 to 15°C (50–60°F) with 24 weeks of freeze-free conditions (9,24). Thee is a definite north to south gradient of cool to warm temperatures and winter average minimums can reach −10°C (14°F). At the higher elevations, temperatures are cooler. Therefore, alfalfa cultivars are usually moderately dormant to semidormant types.

Annual rainfall ranges from 71 cm (28 inches) in the western part to above 109 cm (43 inches) in the eastern part of the region. Although total rainfall is sufficient for alfalfa growth in much of the region, its distribution is poor resulting in both long- and short-term periods of drought (8,24).

The majority of the soils are acidic, often infertile, and possess a low base saturation. Generally, soils are highly weathered. To successfully grow alfalfa, recommended liming and fertility practices should be followed.

Major diseases listed by alfalfa workers in the region were Phytophthora root rot, Fusarium wilt, anthracnose, Sclerotinia crown rot (caused by *Sclerotinia trifoliarum* Eriks.), and common leaf spot (59). Rhizoctonia crown rot and root canker (caused by *Rhizoctonia solani* Kuehn) were considered to be of minor importance.

The major insect pests were alfalfa weevil and potato leafhopper. The threecornered alfalfa hopper [*Spissistilus festinus* (Say)], spotted alfalfa aphid, pea aphid, blue alfalfa aphid, and armyworm (*Prodenia* spp.) can cause serious losses. The aphids are more prevalent in the western section of this region.

Throughout the early 1900s, 'Common' alfalfas were used extensively in this region (34,56,59). Commons are still used, especially in the western areas and occupy up to 65% of the production area in Oklahoma. 'Kansas Common' was a superior strain during the early periods of alfalfa production, but its importance now lies in its contribution of germplasm to later cv. such as Buffalo, Williamsburg, Cody, Dawson, and Kanza (56). Buffalo was developed for resistance to bacterial wilt; Cody for resistance to the spotted alfalfa aphid, and Kanza for resistance to the pea aphid, spotted alfalfa aphid, and bacterial wilt. This group of cultivars is often referred to as the "Plains Type" and represent an unique germplasm that has been added to a number of proprietary cv., including Apollo and Cimarron (59). The eastern part of this region has been subjected to the alfalfa weevil, and foliar diseases, especially anthracnose. Cultivar usage in this region is now in a state of change. Multiple-pest resistant proprietary cultivars are increasing in importance in the western area and resistance to anthracnose and Phytophthora root rot are assuming greater significance in the east.

20–4.4 Southern Region

Although increasing in importance, production in the Southern region represents <1% of that in the USA (Fig. 20–1). The Southern region produces a lot of forage but alfalfa production is limited by prevalent insect, nematode, and disease problems leading to poor stand persistency, and poor haymaking weather conditions affecting hay quality.

The Southern region has long growing seasons and mild winters. An average mean annual temperature of 15 to 21°C (59–70°F), and a freeze-free period of 34 to 43 weeks is expected, depending on north to south location and elevation (1,9,24). Summers are hot and humid with many days above 32°C (90°F). Rainfall is high with yearly averages exceeding 119 cm (47 inches)/yr, but distribution can be poor leading to both long- and short-term droughts. Alfalfa yields are usually high in the spring, low in the summer, with a smaller peak in production in the fall (17).

Soils in the region are both acidic and infertile (24,65,73). Liming and fertilization are necessary for acceptable alfalfa yields. Soils of the Coastal Plains [i.e., the land extending up to 400 km (250 miles) inland from the Atlantic Ocean and Gulf of Mexico] are young, sandy, and well drained with the exception of "flatwoods" which are poorly drained (24). Although alfalfa production is restricted because of poor stand persistence, the Coastal Plain has potential for expanded use. The northern portion of the Coastal Plain extends into the Piedmont, Ridge and Valley, Appalachian, and Interior Low Plateaus (24,65,73). Soils in these areas are older, heavier, and more erosive. Most alfalfa in the region is grown in these areas. The Mississippi Alluvial Valley is an area with fertile, but variable soils. Very little alfalfa is produced in this area because of limited livestock production and competition from row crops.

A number of diseases occur in the region with Phytophthora root rot, Fusarium wilt, Sclerotinia crown rot, anthracnose, common leaf spot, summer, blackstem, and Rhizoctonia stem blight the most prevalent (59).

The alfalfa weevil and potato leafhopper are the predominant insect pests with fall armyworms, threecornered alfalfa hopper, and various aphids causing considerable damage. Nematodes are destructive in the Coastal Plain (39).

Semidormant cultivars are generally grown in the northern portion of the region with nondormants grown in the southern areas. Cultivar development has not received emphasis from either public or private agencies, although multiple-pest resistant cultivars adapted to the region should perform well. Florida 77 was developed from Florida 66 for persistence and resistance to the spotted alfalfa aphid. Florida 77 was later shown to be resistant to the root-knot nematode and possibly resistant to other root nematodes (9). This could account for its improved persistency.

20–5 ADAPTATION IN CANADA

Alfalfa was introduced originally into the province of Ontario in 1871 (15,16). Production remained largely in Ontario until 1908 when the cv. Grimm and Baltic were introduced from Minnesota into Alberta (15). From these points of introduction, alfalfa culture spread into all Canadian provinces (Fig. 20–5). Bolton (15) reported that alfalfa production increased from 0.2 million ha (0.5 million acres) in 1910 to 0.6 million ha

Fig. 20-5. Alfalfa-growing areas of Canada based primarily on winter temperatures and humidity.

(1.5 million acres) in 1951. In 1982, Goplen et al. (33) estimated that 4 to 6 million ha (10–12 million acres) of alfalfa was grown in Canada for hay and pasture purposes (Appendix Table 4). In addition, some alfalfa is grown for green manure, dehydration, and seed production.

20-5.1 Eastern Canada

Eastern Canada, which includes Ontario, Quebec, and the Atlantic provinces grows approximately one-third of the alfalfa in Canada (4). In Ontario, alfalfa production is concentrated near the Great Lakes. Annual precipitation ranges from 76 to 91 cm (30–36 inches) with high humidity during the growing season. Average July maximum temperatures range from 24 to 32°C (75–90°F). Minimum winter temperatures of −23 to −34°C (−10–−30°F) are common in the important alfalfa-producing areas (33). Snowfall is usually adequate to provide insulation against winter injury. Established stands produce three cuttings. Drainage may limit stand longevity on many soils. Liming is usually required for high yields.

Bacterial wilt is found in Ontario, but rarely in the other eastern provinces (33). Phytophthora root rot is an important disease in most areas (33). Verticillium wilt has recently become widespread in southern Ontario (7,40). It was identified in Quebec in 1962, but was apparently eradicated (7). Root and crown rots (caused by *Fusarium* spp.) are severe in Quebec and the Atlantic provinces (33,42). The most important leaf and stem diseases are common leaf spot, Lepto leaf spit, Stemphylium leaf spot, spring blackstem, and yellow leafblotch [caused by *Leptotrochila medicaginis* (Fckl.) Schuepp.].

Insects causing significant crop losses in eastern Canada are the potato leafhopper, alfalfa blotch leaf miner, alfalfa weevil, and pea aphid. Damage from potato leafhopper and alfalfa weevil is most common in southern areas (33,42). The alfalfa blotch leaf miner is a relatively new pest occurring throughout the area (59). Pea aphid are found in all areas and occasionally build up to damaging levels.

The northern root-knot and root lesion nematodes are reported to be widespread (33). In addition to direct effects, nematode damage predisposes plants to infection by various root and crown diseases.

Cultivars grown in eastern Canada are very dormant to moderately dormant. The original alfalfa introduction into Ontario was from Lorraine, France. The strain originating from this seedstock, known as 'Ontario Variegated,' was widely grown throughout eastern Canada for several decades (15). In the 1960s, Ontario Variegated was replaced gradually by Ranger, Rhizoma, Vernal, and DuPuits (15). New cultivars are characterized by increased levels of resistance to bacterial wilt. In 1983, Vernal accounted for nine percent of the alfalfa seed used in eastern Canada (5). Saranac and Iroquois combined accounted for 28% of the seed with 'Algonquin' and 'Angus' accounting for 9%. The use of proprietary cultivars developed in the USA has increased steadily over the past 15 years and now accounts for approximately 40% of seed usage.

20–5.2 Western Canada

Western Canada consists of the prairie provinces of Alberta, Saskatchewan, and Manitoba and the coastal province of British Columbia (Fig. 20–4). In western Canada, the area in alfalfa roughly doubled each decade from 1910 until 1940 and then tripled during the next 10-yr period. By 1961, there were 0.9 million ha (2.3 million acres) of alfalfa and alfalfa-grass mixtures grown for hay. In 1981, there were 1.8 million ha (4.4 million acres) of alfalfa in this region (4,15).

Bolton (15) divided alfalfa production in the prairie provinces into two general zones. One is a grassland area of brown and dark brown soils in southern Saskatchewan and southern Alberta. This area has an annual precipitation of about 36 cm (14 inches), much of which falls during the growing season, and low humidity. Winter temperatures may fall to $-45°C$ ($-50°F$), and rise to $40°C$ ($105°F$) in July. Summer droughts are common. Production of both dryland and irrigated alfalfa is substantial. One cutting is made under dryland conditions with two or three cuttings in irrigated areas. Much of the dryland alfalfa is planted in grass mixtures and grazed.

The second area described by Bolton (15) is a belt of black soils with parkland vegetation located north and east of the grassland area. In this zone, alfalfa production is under dryland conditions. Precipitation ranges from 41 to 51 cm (16–20 inches). Typically, alfalfa is cut once for hay or grazed. In British Columbia, the heaviest concentration of alfalfa is found in the southern interior valleys and in the Peace river area.

Bacterial wilt is most common in irrigated areas (33). Crown bud rot caused by a group of organisms including *Fusarium* spp., *Rhizoctonia* spp., is widespread and sometimes severe. Winter crown rot, or snow mold caused by a low-temperature *Basidiomycete*, is destructive some years (33). Serious losses have been attributed to Fusarium wilt, root, and crown rots caused by *Fusarium* spp. In recent years, Verticillium wilt has been widespread in British Columbia and parts of the prairie provinces (7,40). Phytophthora root rot is present but not recognized as a major problem (33,59). Leaf and stem diseases of importance are spring blackstem, yellow leaf blotch, common leafspot, and Lepto leafspot. Severity of these diseases is much less than in eastern Canada.

Insect pests in western Canada include pea aphid, alfalfa weevil, and grasshoppers (33). Stem nematodes cause significant losses in irrigated areas.

In the prairie provinces of western Canada, winterhardiness is essential for persistence. Grimm was the first winter-hardy cultivar grown in this area. It was responsible for the large expansion in production during the next several decades (15). Selections from Minnesota Grimm known as '19A', 'Grimm Sask. 145', and 'Grimm Sask. 666' represented further improvements in winterhardiness (15). Later, the introductions, Ladak and 'Cossack', were used but did not replace Grimm. In areas where bacterial wilt was a problem, Ladak's and Cossack's slight resistance resulted in an extra year of production (15). Ranger and Vernal were used where extreme winterhardiness was not required.

Branching, or creeping rooted, cultivars are preferred for dryland production. Current cv. are Rambler, Roamer, Kane, Drylander, and Rangelander. These cultivars, with the exception of Rangelander, have resistance to bacterial wilt (42). 'Beaver' is adapted for dryland production except in the very dry areas (59).

In the southern irrigated areas where winterhardiness is not as critical, less-dormant, multiple-pest resistant cultivars perform satisfactory. Resistance to bacterial wilt and the stem nematode is important and there is increasing interest in resistance to Verticillium wilt.

20-5.3 Licensing Procedure in Canada

In order for seed of an alfalfa cultivar to be sold by name in Canada, it must be licensed by the Food Production and Inspection Branch of Agriculture Canada. A cultivar will be placed on the recommended list for a specific area or province if it is (i) a licensed cultivar, and (ii) if it meets standards set forth by testing authorities in these areas. Standards are based primarily on forage yield and persistence over a minimum number of years at several locations. To be considered for licensing, the yield of a cultivar must at least equal that of current check cultivars in provincial trials. Resistance to bacterial wilt is a prerequisite in most provinces. In Ontario, resistance to Phytophthora root rot and Verticillium wilt are added requirements for licensing. In recent years, data from private trials have been considered in the licensing process.

REFERENCES

1. Aamodt, O.S. 1941. Climate and forage crops. *In* G. Hambridge and M.J. Drown (ed.) Climate and man. USDA Yearbook of Agriculture. U.S. Government Printing Office, Washington, DC.
2. Anonymous. 1971. The plant variety protection act. USDA Consumer and Marketing Service. C+MS-89. U.S. Government Printing Office, Washington, DC.
3. ----. 1981. Land resource regions and major land resource areas of the United States. CSC, USDA Agric. Handb. 296. U.S. Government Printing Office, Washington, DC.
4. ----. 1983. Forage seed useage survey. Statistics Canada. Agriculture Statistics Division. Canada Department of Agriculture, Ottawa, Canada.
5. ----. 1986. Annual crop summary. USDA, Agric. Marketing Service, Crop Reporting Board, Washington, DC.
6. ----. 1986. Certified alfalfa varieties. Certified Alfalfa Seed Council, Woodland, CA.
7. Aube, C., and W.E. Sackston. 1964. Verticillium wilt of forage legumes in Canada. Can. J. Plant Sci. 44:427–432.
8. Baldwin, J.L. 1973. Climates of the United States. U.S. Dep. of Commerce. UDS 551.582(73). U.S. Government Printing Office, Washington, DC.
9. Baltensperger, D.D., K.H. Quesenberry, R.A. Dunn, and M.M. Abd-Elgawed. 1985. Root-knot nematode interaction with berseem clover and other temperate forage legumes. Crop Sci. 25:848–851.
10. Barnes, D.K., F.I. Frosheiser, E.I. Sorensen, M.W. Nielsen, W.F. Lehman, K.T. Leath, R.H. Ratcliffe, and R.J. Baker. 1974. Standard tests to characterize pest resistance in alfalfa varieties. USDA-ARS ARS-NC-19.
11. ----, E.T. Bingham, R.P. Murphy, O.J. Hunt, D.F. Beard, W.H. Skrdla, and L.R. Teuber. 1977. Alfalfa germplasm in the United States. Genetic vulnerability, use improvement, and maintenance. USDA-ARS Tech. Bull. 1571.
12. ----, D.M. Smith, R.E. Stucker, and L.J. Elling. 1978. Fall dormancy in alfalfa: a valuable predictive tool. p. 34. *In* Proc. of 26th Alfalfa Improve. Conf., Brookings, SD. 6–8 June. U.S. Government Printing Office, Washington, DC.
13. Blum, L.J., and B.K. Worcester. 1975. Soil water extraction by alfalfa. Agron. J. 67:586–589.
14. Bohnenblust, K.E. 1982. The effect of allelopathy or autotoxicity on alfalfa seedling establishment. p. 25–26. *In* Proc. 28th Alfalfa Improve. Conf., Davis, CA. 13–16 July. Curlie Printing Co., Minneapolis.
15. Bolton, J.L. 1962. Alfalfa: botany, cultivation and utilization. Interscience Publishers, New York.
16. ----, B.P. Goplen, and H. Baenziger. 1972. World distribution and historical developments. *In* C.H. Hanson (ed.) Alfalfa science and technology. Agronomy 15:1–34.
17. Bouton, J.H., D.T. Wood, G.V. Calvert, B.J. Deal, R.B. Moss, E.E. Worley, and P.C. Worley. 1982. Performance of alfalfa varieties in Georgia, 1978–81. Univ. Georgia Agric. Exp. Stn. Res. Rep. 408.
18. Carter, J.F. 1964. Alfalfa production in North Dakota. N. Dak. Agric. Exp. Stn. Bull. 448.
19. Christen, A.A., and R.N. Peaden. 1981. Verticillium wilt in alfalfa. Plant Dis. 65:310–321.
20. Cole, D.F., A.K. Dobrenz, M.A. Massengale, and L.N. Wright. 1970. Water requirement and its association with growth components and protein content of alfalfa (*Medicago sativa* L.). Crop Sci. 10:237–240.
21. Cothern, J.H. 1981. Marketing alfalfa in an uncertain economic environment. p. 103–107. *In* Proc. 11th California Alfalfa Symp., Fresno, CA. 9–10 December. California Extension Service, Davis.
22. Cothran, W.R., and C.G. Summers. 1971. Biology and control of the Egyptian alfalfa weevil, *Hypera brunneipennis* (Boh.) in California. p. 59–62. *In* Proc. California Alfalfa Prod. Symp., Fresno, CA. 7–8 December. California Agriculture Extension Service, Davis, CA.
23. Cusperus, G., E.B. Radcliff, D.K. Barnes, and G.C. Marten. 1983. Proper potato leafhopper management. p. 21. *In* Proc. 18th Central Alfalfa Improve. Conf., Manhattan, KS. 8–10 June.
24. Daniels, R.B., B.L. Allen, H.H. Bailey, and F.H. Beinroth. 1983. Physiography. p. 3–16. *In* S.W. Buol (ed.) Soils of the southern states and Puerto Rico. Southern Coop. Series Bull. 174. North Carolina State Univ., Raleigh.
25. Delwiche, P.A., C.R. Grau, and D.C. Arny. 1981. Host range of the alfalfa strain of *Verticillium albo-atrum*. p. 4. *In* Proc. 17th Central Alfalfa Improve. Conf., East Lansing, MI. 30 June–1 July.
26. Dotzenko, A.K., and J.G. Dean. 1959. Germination of six alfalfa varieties at three levels of osmotic pressure. Agron. J. 51:308–309.

27. ----, and T.E. Haus. 1960. Selection of alfalfa lines for their ability to germinate under high osmotic pressure. Agron. J. 52:200-201.
28. Elgin, J.H., Jr., et al. 1982. Alfalfa growing areas of the western United States. Their climates and similarities. USDA-ARS Tech. Bull. 1651.
29. Garrison, R.H. 1978. Personal communication to members of the National Certified Alfalfa Variety Review Board.
30. Garver, Samuel. 1946. Alfalfa in South Dakota: Twenty-one years of research at the Redfield station. S. Dak. Agric. Exp. Stn. Bull. 383.
31. Gerhard, D.G., and C.M. Hayes. 1982. New Mexico Agricultural Statistics. USDA, N. Mex. Dep. of Agric.
32. Gholson, L., J. Lopez, J. Arledge, and B. Melton. 1982. Alfalfa weevil management for New Mexico—1981-1982. N. Mex. Coop. Ext. Ser. 400J-31.
33. Goplen, B.P., H. Baenziger, L.D. Bailey, A.T.H. Gross, M.R. Hanna, R. Michaud, K.W. Richards, and J. Waddington. 1982. Growng and managing alfalfa in Canada. Agriculture Canada Pub. 1705/E.
34. Graber, L.F. 1950. A century of alfalfa culture in America. Agron. J. 42:525-535.
35. Graham, J.H., R.N. Peaden, and D.W. Evans. 1977. Verticillium wilt of alfalfa found in the United States. Plant Dis. Rep. 61:337-340.
36. ----, D.L. Stuteville, F.I. Frosheiser, and D.C. Erwin. 1979. A compendium of alfalfa diseases. American Phytopathological Society, St. Paul.
37. Grandfield, C.O., and W.H. Metzger. 1936. Relation of fallow to restoration of subsoil moisture in an old alfalfa field and subsequent depletion after reseeding. J. Am. Soc. Agron. 28:115-123.
38. Grau, C.R., P.A. Delwiche, and R.L. Norgren. 1981. Verticilliumwilt of alfalfa in Wisconsin. Plant Dis. 65:843-844.
39. Haaland, R.L., C.S. Hoveland, F. Grey, E. Clark, and R. Rodriguez-Kabana. 1979. Rhizosphere problems limiting alfalfa production in the deep south. p. 30. *In* Proc. 26th Alfalfa Improvement Conf., Brookings, SD. 6-8 June. U.S. Government Printing Office, Washington, DC.
40. Hanna, M.R. 1979. Verticillium wilt disease of alfalfa. Forage Notes, Canada. 24:19-20.
41. Hanson, C.H., and R.L. Davis. 1972. Highlights in the United States. *In* C.H. Hanson (ed.) Alfalfa science and technology. Agronomy 15:35-51.
42. Heinrichs, D.H. 1969. Alfalfa in Canada. Pub. 1377. Canada Department of Agriculture, Ottawa, Canada.
43. Horner, E.S. 1979. Plant breeding progress on solving alfalfa problems in the deep south. p. 91-92. *In* Proc. 36th Southern Pasture and Forage Crop Improvement Conf., Beltsville, MD. USDA-SEA, New Orleans.
44. Hoveland, C.S., R.L. Haaland, and M.W. Alison, Jr. 1981. Performance of alfalfa varieties in Alabama. Agronomy and Soils Dep. Series 68. Auburn Univ., Alabama.
45. Jensen, E.H., and B.J. Hartman. 1981. Does alfalfa following alfalfa affect seedling growth? p. 37-40. *In* Proc. 11th California Alfalfa Symp., Fresno, CA. 9-10 December. California Agriculture Extension Service, Davis.
46. Kawaguchi, I.I. 1971. Varietal characteristics in alfalfa, their classification and use in choosing varieties or brands for specific uses and locations. p. 88-94. *In* Proc. California Alfalfa Prod. Symp., Fresno, CA. California Agriculture Extension Service, Davis, CA.
47. Kehr, W.R. 1982. Introduction to problems in continuous alfalfa. p. 20-23. *In* Rep. 28th Alfalfa Improve. Conf., Davis, CA. 13-16 July. Curlie Printing Co., Minneapolis.
48. ----, R.L. Ogden, and J.D. Kindler. 1975. Management of four alfalfa varieties to control damage from potato leafhoppers. Nebr. Agric. Exp. Stn. Res. Bull. 275.
49. Kiesselbach, T.A., J.C. Russell, and A. Anderson. 1929. The significance of subsoil moisture in alfalfa production. J. Am. Soc. Agron. 21:241-268.
50. Kilcher, M.R., and D.H. Heinrichs. 1966. Persistence of alfalfas in mixture with grasses in a semiarid region. Can. J. Plant Sci. 46:163-167.
51. Klein, R.R., and D.A. Miller. 1980. Allelopathy and its role in agriculture. Soil Sci. Plant Anal. 11:43-56.
52. Lange, R.T. 1980. Yield and nutritional losses attributable to potato leafhopper on alfalfa. M.A.G. Paper. Department of Entomology, Pennsylvania State University, University Park.
53. Leath, K.T. 1984. Personal communication, University Park, PA.
54. Lehman, W.F., V.L. Marble, Les Ede, Bill Thyr, J. Radewald, L. Teuber, M. Campbell, and J. Davidson. 1981. Progress in breeding alfalfas with resistance to nematodes. p. 66-69. *In* Proc. 11th California Alfalfa Symp., Fresno, CA. 9-10 December. California Agriculture Extension Service, Davis.
55. Lorenz, R.J. 1982. Alfalfa in western grazing management systems. *In* A.C. Wilton (chair) Alfalfa for dryland grazing. USDA-ARS. Agric. Inf. Bull. 44. U.S. Government Printing Office, Washington, DC.

56. Lowe, C.C., V.L. Marble, and M.D. Rumbaugh. 1972. Adaptation, varieties and usage. *In* C.H. Hansen (ed.) Alfalfa science and technology. Agronomy 15:391–413.
57. Marble, Vern. 1981. Characteristics of alfalfa varieties and brands used in California. p. 73–89. *In* Proc. 11th California Alfalfa Symp., Fresno, CA. 9–10 December. California Agriculture Extension Service, Davis.
58. Melton, B., J. Arledge, F. Smith, F. Matta, J. Gregory, and R. Hooks. 1983. Alfalfa variety trials in New Mexico. N. Mex. Agric. Exp. Stn. Bull. 700.
59. ----, J.H. Bouton, and J.B. Moutray. 1984. Survey of alfalfa workers. Unpublished.
60. ----, D. Miller, L. Teuber, and M. Walton. 1983. Dona Ana alfalfa. N. Mex. Agric. Exp. Stn. Release Notice.
61. Miller, D., and B. Melton. 1983. Description of alfalfa cultivars and germplasm sources. N. Mex. Agric. Exp. Stn. Spec. Rep. 53.
62. Nielsen, M.W., and W.F. Lehman. 1977. Multiple aphid resistance in CUF-101 alfalfa. J. Econ. Entomol. 70:13–14.
63. Pardee, W.D. 1980. 1980 Report northeast seed use survey small seeded grasses and legumes. Cornell Univ. PB80-2.
64. ----. 1983. 1983 Report northeast seed use survey small seeded grasses and legumes. Cornell Univ. PB83-2.
65. Perkins, H.F., H.J. Byrd, and F.T. Richie, Jr. 1983. Ultisols-light-colored soils of the warm temperate forest lands. p. 73–86. *In* S.W. Buol (ed.) Soils of the southern states and Puerto Rico. Southern Coop. Ser. Bull. 174. North Carolina State University, Raleigh.
66. Ries, R.E. 1982. Environmental factors and alfalfa persistence in dryland pastures and rangeland. *In* A.C. Wilton (chair.) Alfalfa for dryland grazing. USDA-ARS. Agric. Inf. Bull. 444. U.S. Government Printing Office, Washington, DC.
67. Rumbaugh, M.D. 1982. Origins of alfalfa cultivars used for dryland grazing. *In* A.C. Wilton (chair.) Alfalfa for dryland grazing. USDA-ARS Agric. Inf. Bull. 444. U.S. Government Printing Office, Washington, DC.
68. ----, and D.A. Johnson. 1981. Screening alfalfa germplasm for seedling drought tolerance. Crop Sci. 21:709–713.
69. ----, and T. Thorn. 1965. Initial stands of interseeded alfalfa. J. Range Manage. 48:258–261.
70. Smith, Dale. 1961. Association of fall growth habit and winter survival in alfalfa. Can. J. Plant Sci. 41:244–251.
71. ----. 1981. Forage management in the north. 4th ed. Kendall/Hunt Publishing Co., Dubuque, IA.
72. Smith, O.F. 1958. Registration of Lahontan alfalfa. Agron. J. 50:684–685.
73. Slusher, D.F., and S.A. Lytle. 1983. Alfisols-light-colored soils of the humid temperate areas. p. 61–72. *In* S.W. Buol (ed.) Soils of the southern states and Puerto Rico. Southern Coop. Ser. Bull. 174. North Carolina State University, Raleigh.
74. Teuber, L.R., B.J. Hartman, and W.L. Green. 1980. Insights after one year of fall dormancy determinations at several locations. p. 67. *In* Proc. 27th Alfalfa Improve. Conf., Madison, WI. 8–10 July. U.S. Government Printing Office, Washington, DC.
75. ----, and V.L. Marble. 1981. Use of climatic and fall dormancy data to describe varieties for California. p. 116–117. *In* Proc. 11th California Alfalfa Symp., Fresno, CA. 9–10 December. California Agriculture Extension Service, Davis.
76. ----, ----, W.F. Lehman, I.I. Kagauchi, M.K. Miller, B.J. Hartman, O.J. Hunt, D.K. Barnes, B. Burrows, D.L. Lancaster, and R.H. Gripp. 1984. Use of climatic and fall dormancy data to allocate alfalfa varieties to regional variety trials. California Agriculture. March-April.
77. Townsend, C.E. 1982. Diseases, insects, and other pests of rangeland alfalfa. *In* A.C. Wilton (chair.) Alfalfa for dryland grazing. USDA-ARS Agric. Inf. Bull. 444. U.S. Government Printing Office, Washington, DC.
78. Trewartha, G.T. 1968. An introduction to climate. 4th ed. McGraw-Hill Book Co., NY.
79. Wilson, M., B. Melton, J. Arledge, D. Baltensperger, R. Salter, and C. Edminster. 1983. Performance of alfalfa cultivars under less than optimum moisture conditions. N. Mex. Agric. Exp. Stn. Bull. 702.
80. ----, ----, and C.E. Watson. 1957. Registration of Zia alfalfa. Agron. J. 52:405–406.
81. Younis, M.A., F.C. Stickler, and E.L. Sorenson. 1963. Reaction of seven alfalfa varieties under simulated moisture stress in the seedling stage. Agron. J. 55:177–181.

21 Diseases and Nematodes

KENNETH T. LEATH

USDA-ARS
University Park, Pennsylvania

DONALD C. ERWIN

University of California
Riverside, California

GERALD D. GRIFFIN

USDA-ARS
Logan, Utah

Alfalfa (*Medicago sativa* L.) diseases are many and varied. They are usually caused by fungi, bacteria, nematodes, and viruses or mycoplasma-like organisms. Also, diseases can arise from imbalances within the plant, i.e., physiogenic, in the absence of any pathogenic agent, and serious injury can result from air pollutants or other toxic sources. This chapter is limited to a discussion of diseases that result from causal agents, injury caused by air pollutants, as well as selected physiogenic disorders. Disorders that result from climatic or nutritional stresses are dealt with in chapters devoted to environmental conditions and to soil fertility.

It has been estimated that about one-fourth of the U.S. alfalfa hay crop and one-tenth of the seed crop are lost annually to disease (72), amounting to nearly $400 million in 1972. These estimates are conservative because they reflect neither inflation nor losses resulting from reduced quality, shortened stand life, predisposition to other stresses, and costs involved in developing resistant cultivars.

Many of the diseases that attack alfalfa can limit production. Although disease is usually singled out as a separate entity for purposes of discussion, the reader should be aware that a disease is not a singular event. Rather, diseases are only one component of the total stress load on a plant, and they interact with each other, with other pests, and with numerous abiotic factors in what may be considered "a cumulative stress load." Because a disease is apparent for a brief period at one harvest does not mean that the effects of the disease are limited to the same time span. The effects from one disease outbreak can adversely affect subsequent production, even into the next crop year.

Alfalfa is consumed mainly on farms where it is produced, with only specialized hay growers offering a product for sale. The impression that

---

Copyright 1988 © ASA-CSSA-SSSA, 677 South Segoe Road, Madison, WI 53711, USA.
Alfalfa and Alfalfa Improvement—Agronomy Monograph no. 29.

alfalfa is not a high value per hectare crop has limited research on disease epidemiology and crop-loss assessment. Fungicides have had only limited use on alfalfa, primarily in seed production or as seed treatments. Consequently, economic thresholds for disease losses have not been defined as in those crops that are subject to heavy fungicide and nematicide use, and the related development of sophisticated disease forecasting and pest management has not occurred.

All of these factors influence the approaches taken to manage diseases. There are few actual control measures that can be implemented once a disease epidemic is underway (136). Therefore, it is essential to avoid severe disease losses by growing disease-resistant cultivars—our main line of defense. Furthermore, epidemics can be delayed by applying management strategies that favor vigorous growth of the crop and/or reduce opportunities for the increase of inoculum (138).

Application of sound management practices reduces disease losses and results in greater economic returns from each crop. The use of resistant germplasm is most important in reducing disease losses, but care must be taken to ensure that new resistant cultivars are not susceptible to other pathogens. The increased use of Flemish germplasm during the 1950s and 1960s may, indeed, have resulted in the elevation of Lepto leaf spot [*Leptosphaerulina briosiana* (Poll.) Graham & Luttrel.] and anthracnose (*Colletotrichum trifolii* Bain.) from minor to major disease status (72). Rotation of alfalfa with a dissimilar crop, such as maize (*Zea mays* L.), reduces pathogen inoculum and the overall disease levels. The maintenance of high soil fertility, especially P and K, together with a soil pH of 6.5 to 7, contributes to the expression of maximum disease resistance potential in alfalfa (144). The use of certified seed, recommended irrigation and harvest practices, and sound sanitation procedures all contribute to minimizing disease losses in alfalfa (136).

Diseases are discussed in this chapter based on the primary plant organ or process affected, i.e., leaf and stem, crown and root diseases, etc. Detailed information on pathogenic organisms has been kept to a minimum, with referrals to literature where such information is available. The selection of host-plant resistance to disease is covered in chapter 27. Established methods for screening alfalfa for pest resistance have been summarized by Elgin et al. (48).

21-1 DISEASES OF LEAVES AND STEMS

Many pathogens cause diseases of alfalfa leaves and stems. In addition to the biotic pathogens, air pollutants injure leaf tissue and reduce the plant's efficiency (72,73). Alfalfa often exhibits various leaf spotting that cannot be traced to an infectious agent. These disorders are probably physiogenic or environmental in origin.

Foliar diseases reduce the overall energy efficiency of the plant, which results in a major loss of yield and quality. When diseased leaves are retained on the plant, there is still a loss of quality constituents (154) and a less favorable ratio of desirable to undesirable constituents. Estrogenic activity may increase in diseased alfalfa with a concomitant reduction in palatability and digestibility. Leaf and stem diseases reduce both the photosynthetic capability of a plant as well as the translocation of carbohydrates. Foliar pathogens, such as *Phoma medicaginis*, which causes spring black stem, occasionally kill entire stems and invade crown and root tissues. Severe foliar disease, especially in the fall, may predispose plants to other stresses and contribute to subsequent death of plants during the winter.

If a particular leaf or stem disease becomes severe, attempts are made to develop resistant cultivars. Currently available, improved cultivars are subject to significant losses from foliar diseases in the humid production regions as well as in some irrigated areas. Based on field symptoms, it could be concluded that currently grown cultivars have little resistance to foliar diseases. However, when newer cultivars are grown side-by-side with older cultivars, like Ranger and Buffalo, a difference is readily apparent. Obviously, higher levels of foliar disease resistance are needed. Losses remain unacceptably high.

Fungicides are not used to control foliar diseases. Although fungicides can control most foliar diseases, potentially undesirable cost/benefit ratios and residues have prevented their use on alfalfa.

Early harvest of heavily diseased fields may prevent a potential epidemic of a foliar disease. In general this practice reduces forage yield, increases forage quality, and removes much fungal inoculum from the field. Foliar disease losses can be expected to decline with a frequent harvesting schedule, because the disease has a shorter time and less favorable conditions to increase in severity.

Viruses are mostly systemic in the plant and, although not strict foliar pathogens, they do reduce the efficiency of leaf and stem functions. Most symptoms of virus infections are visible in alfalfa leaves, hence viruses cannot be ignored in a discussion of leaf diseases. The total impact of viruses on alfalfa is not well documented.

21-1.1 Physiogenic Leaf Spots

Under certain conditions, alfalfa may exhibit leaf spots that cannot be attributed to infectious agents nor to chemical injury. A minute black spotting on leaves and stems was described by Kreitlow and Kilpatrick (131). This spotting, which is favored by high light and cool temperatures in greenhouses, is of more concern to researchers than to farmers. White spotting and necrosis on leaves have been reported as physiogenic disorders (9), with the vestigial flower and branched raceme characteristics described by Dudley and Wilsie (47). Although these disorders are characterized as physiogenic, their actual causes remain unknown and might involve unidentified pathogens.

21–1.2 Air Pollution Injury

Sulfur dioxide (SO_2), hydrogen fluoride (HF), ozone (O_3), and peroxyacetyl nitrate (PAN) are the most common air pollutants affecting alfalfa (73). Sulfur dioxide and HF are products of industrial processes or fossil fuel combustion, whereas O_3 and PAN arise in large part from the action of sunlight on automobile exhaust. Thus, injury from the first two pollutants usually occurs downwind from point sources, such as steel mills, aluminum smelters, or coal-fired power plants. In contrast, leaf injury from O_3 and PAN can occur hundreds of miles from urban centers as the photochemical haze moves with air masses.

Air pollution injury symptoms can be confused with nutritional disorders, pesticide injury, and physiogenic disorders. Ozone, the most important phytotoxic air pollutant, causes small, isolated, interveinal necrotic flecks on the leaflets and occasional chlorosis (Plate 21–1) (20,101). Damage can be serious enough to cause leaf death and abscission, especially of older leaves. Ozone in California has lowered yield of a susceptible cultivar up to 13% (21). Sluggish air movement and bright sunlight enhance O_3 injury.

Injury caused by PAN may resemble that caused by O_3 and SO_2. The silvery/copper sheen that frequently follows PAN injury aids diagnosis, but can also result from frost or other agents. Sulfur dioxide causes chlorosis and cream to light-brown bands between veins, frequently at the leaf margins, and occasionally defoliation. In one study, experimental exposure to SO_2 reduced alfalfa yields by 9% (21). Injury caused by different air pollutants can be similar and can even be attributed to damage from herbicides (103). Possible nearby sources should be sought whenever HF injury is suspected. Plant tissue analysis can detect HF accumulation in leaves.

Strains of alfalfa resistant or tolerant to air pollution have been selected (107), and cultivars vary in sensitivity (220). Although not of major concern at present, air pollutants, such as O_3, pose a future threat to successful alfalfa production in specific areas. This could warrant efforts to increase resistance to injury.

21–1.3 Bacterial Leaf Spot

This leaf spot, caused by *Xanthomonas alfalfae* (Riker, Jones & Davis) Dows., was first reported from Wisconsin in 1935. The disease is widespread but seldom of economic importance.

Symptoms begin as small, diffuse, chlorotic areas within which circular water-soaked areas develop (22, 194). Water-soaked areas are most apparent on the undersides of leaves. As lesions dry, they often appear shiny and later become translucent. Lesions also occur on stems (166) and can elongate over more than one internode. Defoliation and stunting of seedlings (210) are the main effects of this disease.

Infection occurs through stomates or wounds (34), and warm, moist weather favors disease development. This pathogen overwinters in debris or in soil and may survive for several years in hay or in debris in stored seed. The bacteria are spread from plant to plant by rain and wind.

Cultivars resistant to bacterial leaf spot could be developed based on observed differences among plants in response to infection. Losses to bacterial leaf spot are less in spring-seeded alfalfa than in fall or late-summer seedlings. Spring seeding should be encouraged in areas where this disease is a problem.

21-1.4 Common Leaf Spot

Common or Pseudopeziza leaf spot [caused by the fungus *Pseudopeziza medicaginis* (Lib.) Sacc.] is a serious disease of alfalfa worldwide. The disease causes economic losses in yield and hay quality (167, 218). Leaflets are lost as a direct result of infection and the shattering of diseased leaves during harvest. The disease can develop during any harvest period if rain or dew is frequent (119).

The fungus causes small, circular, brown or black spots with dentate margins. Lesions can be up to 3 mm in diameter and usually do not coalesce. In older lesions, a central raised fruiting structure, an apothecium, is often apparent. The tan or gray apothecia are readily visible with a hand lens. Apothecia form mainly on the upper surface of the leaf. Severely infected leaflets turn yellow and drop, whereas young infected leaflets may be reduced in size (Fig. 21-1).

Fig. 21-1. Severe common leaf spot, some showing an apothecium in the center.

Ascospores are produced by the fungus in apothecia on leaflets. Spores are released in the spring and spread to new growth by wind and rain, with subsequent infections occurring when cool, wet conditions prevail (202). The fungus overwinters in dead leaves.

Some cultivars are moderately resistant, and high levels of resistance have been identified for incorporation into future cultivars. Early harvest of severely diseased stands is beneficial because leaf retention is enhanced and inoculum removed from the field.

21-1.5 Yellow Leaf Blotch

This disease [caused by the fungus *Leptotrochila medicaginis* (Fckl.) Schüepp] occurs worldwide where climatic conditions are favorable. It is most severe in the Great Plains states of the USA, and occurs in most northern states as well as in Canada and Europe. It is more severe on leaves than on stems or petioles.

Symptoms first appear as small, chlorotic spots on the leaf's upper surface (118). These enlarge, often as yellow streaks, and later become yellow to orange blotches. The blotches are fan-shaped, expanding toward the leaf margin. There can be several blotches on a leaflet, and small, black pycnidia form in the blotches. Spores are produced in the pycnidia, but these spores apparently are not infectious to the plant. The central area of the blotch turns brown or black as leaflets die, dessicate, and curl downward. Many leaflets fall from the plant.

Apothecia form on both leaf surfaces adjacent to the darkened central area of the blotch. Ascospores produced in the apothecia can be produced during the growing season; however, they might not occur until the following spring on fallen leaves (118). Spore discharge can continue until July from overwintered apothecia. Spores germinate over a temperature range of 2 to 31°C and can infect leaves within 4 h from start of germination.

Losses can be reduced by harvesting early, and spring burning of crop residues and crop rotation reduce disease incidence (155). Cultivars with *Medicago falcata* germplasm have some resistance.

21-1.6 Stemphylium Leaf Spot

This is a common foliar disease of alfalfa (caused by the fungus *Stemphylium botryosum* Wallr.) that occurs across much of the humid alfalfa growing region during warm, moist weather. It is most important in the eastern and southern USA. Stemphylium leaf spot has been reported from Europe, New Zealand, Australia, and Egypt (73, 114, 201).

Typical symptoms begin as round to oval, slightly sunken, dark-brown spots. Often the center portion is light brown or tan, and spots are surrounded by a chlorotic ring. Fan-shaped bleached areas expanding outward toward the tip of the leaflet are common. Some lesions develop a target-like appearance with concentric rings. Usually a large single lesion will cause a leaflet to abscise (206); defoliation can begin as early as 5 d after infection. Stems and petioles are also attacked (Fig. 21-2).

A cool-temperature strain of the fungus occurs in the western USA (42), which produces lesions that are distinct from those described for the eastern strain. The disease is most prevalent in early spring in interior valleys but persists all year in coastal valleys. Lesions in these areas are more elongate and irregular with tan centers, seldom exceed 3 to 4 mm in length, and do not have concentric rings.

The conidial stage of the fungus is critical in the disease cycle. The fungus overwinters primarily in plant tissue and produces conidia when warm temperatures (ca. 25°C) and moist conditions prevail. The fungus can be seed-borne.

Fig. 21-2. Leaflets infected with large or small developing Stemphylium leafspots. The smaller lesions show a yellow halo; the large lesions are concentrically ringed. Such lesions are typical of those caused by the Eastern strain.

Early harvest during severe disease outbreaks can reduce the loss of leaves as well as the amount of inoculum in the field. Some available cultivars have low to moderate levels of resistance to this disease.

21-1.7 Leptosphaerulina Leaf Spot

This disease [caused by the fungus *Leptosphaerulina briosiana* (Poll.) Graham & Luttrell] is a serious problem of alfalfa in eastern and central USA (73). It also defoliates alfalfa in eastern Canada, Europe, Asia, and Africa (49, 219).

Although this leaf spot attacks petioles, it is most damaging on young leaves. Spots are initially small and black and may remain as "pepper spots" on slow-growing or mature leaves (114). On young, fast-growing leaves, the spots enlarge and develop tan or bleached centers. Borders of spots are generally brown with chlorotic "halos" usually surrounding each lesion. Such lesions can enlarge, coalesce, and kill leaflets. Dead leaflets may or may not remain attached. Early infection of spring growth can cause stunting.

The fungus produces ascospores in dead leaves, and ascospores are forcibly ejected, which aids their dissemination by wind. Cool, moist conditions favor spore production and infection. Spring and early summer are peak seasons for this disease. However, fall epidemics are also common. The disease occurs in midwinter in the southern states. The fungus overwinters as ascospores and mycelium in dead leaves.

Early harvest is the only available management practice for reducing losses.

21-1.8 Stagonospora Leaf Spot

This disease [caused by the fungus *Stagonospora meliloti* (Lasch) Petr.] occurs in the USA and Europe (73). The leaf spot has a diffuse margin and a bleached center, and leaflets drop rapidly after infection (128). Spots, similar to those on leaves, also occur on stems, and infected stems may exhibit internal red-orange flecks, much like those caused by the fungus in root tissue.

Spores are produced in pycnidia on lower stems and leaves and are distributed in irrigation and rainwater. A role for the ascospores in the disease cycle has not been determined. The root rot phase of the disease is the most important and may represent a continuation of an earlier leaf or stem disease phase (52).

No control for the disease is available.

21-1.9 Downy Mildew

This disease (caused by the obligate parasitic fungus, *Peronospora trifoliorum* dBy.) occurs in many temperate areas of the USA, Canada, Europe, and at higher elevations of Central and South America (73).

DISEASES AND NEMATODES

Spring and fall are seasons of highest disease activity, because the fungus requires cool, moist conditions to flourish. In fall seedings, the disease can kill seedlings and cause complete stand failures. Generally, most losses in established stands occur during spring growth.

The first symptom of disease is chlorosis, usually localized on leaflets. Occasionally, systemic infections cause symptoms on entire leaves or shoots and the death of young plants (126). Infected leaves become twisted with downward-curled margins. Heavily infected stands develop a gray cast. The fungus produces external conidiophores and conidia, which are conspicuous as a gray to pale violet covering of the lower leaf surface. Oospores are produced in diseased tissue (Fig. 21-3).

Fig. 21-3. (A) Downy mildew on upper and lower leaf surfaces; (B) deformed leaves at tip of stem.

The fungus overwinters in systemically infected crown buds and shoots of alfalfa. Conidia from surviving shoots and systemically infected early spring growth serve as primary inoculum. Conidia are produced in darkness during periods of high humidity. Conidia, which usually are not long lived, are spread by wind and rain. The spore germinates and the germtube penetrates the leaf either via a stomate or directly through the epidermis. Secondary infection cycles can occur every 5 d. Seed also can become infected (130).

Selection methods for resistance are available (215), and control can be achieved by growing resistant cultivars. Seeding in the spring is recommended if downy mildew has been a serious problem. Alfalfa should be cut on schedule or early, because this removes inoculum from the field and allows the stand to dry out. Higher temperatures in subsequent growth periods inhibit disease development (19, 155).

21-1.10 Powdery Mildew

This minor disease (caused by the fungus *Erysiphe polygoni* D.C.) has been reported in the USA (229). Although this disease does not occur anywhere on a regular basis, it was reported on alfalfa in Egypt during a recent survey (201).

No studies have been made of the host-pathogen relationships, and no control measures have been attempted.

21-1.11 Rust

Rust [caused by the fungus *Uromyces striatus* Schroet. var. *medicaginis* (Pass.) Arth.] occurs throughout the world. Rust is more of a problem in the southern USA, where it may remain active throughout the winter, than in the North where activity is limited to a short period in the fall before frost. Losses can be most severe with delayed harvests, especially during seed production. Rust has been reported from South Africa, Egypt, USSR, Israel, and the Sudan (72, 73, 201).

Reddish brown masses of uredospores rupture the epidermis of leaflets, petioles, and stems, causing leaflets to shrivel and fall. The fungus overwinters in southern states as uredospores and mycelium in plant tissues. As weather conditions improve, new growth is infected. Uredospores blow north during the summer, and secondary spread occurs within a field during fall. Teliospores form in old uredial sites, but the aecial stage on *Euphorbia cyparissias* L. is rare.

Resistant cultivars are available and should be grown in problem areas. Harvesting before an epidemic becomes severe conserves leaves and reduces inoculum for secondary infection.

21-1.12 Bacterial Stem Blight

This disease (caused by *Pseudomonas syringae* van Hall) occurs in many areas of the northern USA and in Europe. Most disease damage occurs in the first cutting and often is associated with late-spring frost injury. Leaf injury can also be severe. This disease causes severe damage only in the high, cool valleys of the western USA where it can reduce first-crop hay harvests by 40 to 50% (73).

Early symptoms (196) appear as water-soaked, yellowish to olive-green lesions, usually starting at a node and extending down from one to three internodes. Lower internodes are most affected. Lesions turn amber then become black and glistening. The bacteria do not invade the vascular system but cause stunted, weak, brittle stems that break easily. Infected leaf bases and petiolules are common.

Cold, wet conditions in the spring favor the disease, which generally follows frost injury. The bacteria survive in plant debris in soil and enter stems through frost cracks. Secondary spread usually does not occur and losses are largely restricted to first-cut hay.

Resistant cultivars are not available; however, frost-tolerant cultivars should incur less loss to this disease than less tolerant cultivars.

21–1.13 Summer Black Stem and Leaf Spot

This disease (caused by the fungus *Cercospora medicaginis* Ell. & Ev.) causes losses in the USA and many other countries (73). Damage occurs during unseasonably warm growing periods, particularly in the central states, and the first cutting is seldom affected. This disease usually increases in severity in subsequent cuttings (72).

Early leaf symptoms (14) appear as small, brown spots on both surfaces that enlarge into irregular roundish brown lesions up to 6 mm in diameter. At high humidity spores are produced in the lesions, which become gray or silvery. Defoliation accompanies heavy infection. Elongate, dark-brown lesions form on stems and coalesce to produce a black stem appearance. Infected stems often are stunted, chlorotic, and may die prematurely. Leaves on infected stems often are small and may drop (Fig. 21–4).

The fungus overwinters in infected stems and produces spores when climatic conditions are favorable. Disease severity relates directly to duration of moisture on plant surfaces. Inoculum buildup parallels plant growth and canopy formation with the attendant increase in relative humidity under the canopy. The fungus can be seed-borne, and wind (15) and rain move spores from plant to plant. Secondary inoculum comes from lesion sporulation.

Harvest should not be delayed if the disease is present, because mature stands favor rapid disease development. Early harvest should minimize losses and reduce inoculum. High levels of resistance are not available in current cultivars, but some cultivars appear to be affected less than others.

21–1.14 Spring Black Stem and Leaf Spot

The fungus *Phoma medicaginis* Malbr. & Roum. var. *medicaginis* Boerema causes this stem and leaf disease. It is widespread throughout

Fig. 21-4. Large Cercospora lesions on trifoliolate leaf.

Fig. 21-5. Spots on leaflets and petiole caused by *Phoma medicaginis*.

most of the alfalfa growing areas of the USA, Canada, and Europe (17, 117, 162). The disease reduces yields of forage and seed. Peak disease periods are usually in the spring and fall, but the pathogen can remain active throughout a cool growing season (Fig. 21-5).

The fungus attacks leaves, petioles, and stems and is believed to contribute to crown and root rot problems and seedling diseases. In early spring, small dark spots appear on lower leaves and stem areas. Young stems can be girdled and killed. Irregularly shaped lesions on leaves increase in size and coalesce. Severely infected leaflets turn yellow, wither, and drop. Pods and seeds are susceptible to attack. Stem lesions enlarge and coalesce, often covering much of the lower stem surface. Infections vary in severity with some sites restricted to epidermal tissue, whereas other infections penetrate inner stem tissues and cause stem dysfunction. The disease is most severe on injured stems (10).

The fungus survives in crowns of plants, stem lesions, dead stubble, and debris on or in the soil (73). Generally, the fungus does not survive over 2 yr in the soil, but can survive up to 8 yr in infected seed (155). In the spring, spores ooze from pycnidia to be spread by splashing rain drops, wind, and insects. New shoots are infected as they grow through crop residues and past infected stubble. Free moisture on plant surfaces is necessary for infection. Damage is most severe in the first harvest.

Disease losses can be reduced by early harvest, as well as by using certified seed produced in arid regions. Spring burning of crop residues (2), crop rotation, and seed treatment have been suggested to help reduce inoculum (72). No highly resistant cultivars are available, but those selected in the field for "leaf spot" resistance may have some resistance to spring black stem (189).

21-1.15 Anthracnose

Anthracnose [caused by the fungus *Colletotrichum trifolii* (Bain.)] has become a major disease of alfalfa in the USA and other parts of the world (72). The increased use of Flemish germplasm is generally considered to be responsible for the current importance and prevalence of anthracnose (195).

The fungus can attack leaves, although this symptom is rarely encountered in the field. The most characteristic symptoms occur on stems as irregularly shaped, blackened areas or as large sunken, oval, or diamond-shaped lesions. Stem lesions usually occur on the lower third of the stem and often at the base of the stem. Large lesions are straw-colored with brown or black borders and may girdle stems (Fig. 21-6). Characteristically, girdled shoots wilt and droop in a shepherd's-crook form, and wither, die, and bleach out in this crooked form. Fungus fruiting bodies (acervuli) are common in stem lesions.

Frequently stems die without apparent stem lesions. This results from crown infections by *C. trifolii*, a disease stage that nearly always follows stem infections and causes death of the plant (11). Although stem lesions occur only rarely in many dry areas of the western USA, the crown-rot stage is still prevalent. Diseased crown tissue has a blue-black appearance not typical of other common rots. Symptoms are less definitive however, when other pathogens are also active in the crowns.

The fungus overwinters in the field in some southern states, but at

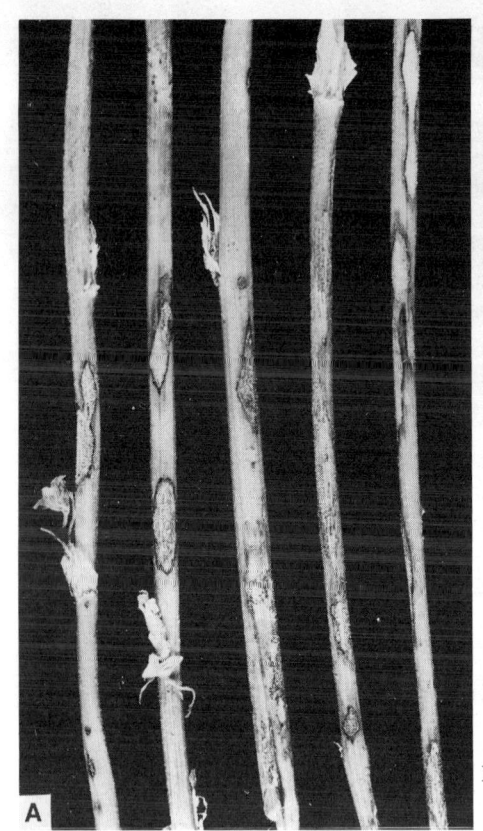

Fig. 21-6. (A) Stem lesions typical of anthracnose disease, and (B) dead stem showing diagnostic "shepherd's crook."

northern latitudes it survives mainly in debris sheltered in barns on farm equipment (27, 149). Spores are spread by splashing rain and are carried down into crowns by rain and heavy dew. Also, the fungus grows down from stem infections into the crowns. Maximum severity of the disease occurs on late harvests during warm, moist periods.

Cultivars are available that have high levels of resistance to this disease and should be grown wherever anthracnose has been a problem. Races of the fungus occur (176, 241), but only race 1 has been widespread and of economic importance. Expression of resistance in some germplasm lines may be influenced by temperature (243). Cleaning debris from cutters and wagons before first harvest in the spring and during the growing season whenever possible is good protection against introduction of the pathogen into new fields. Mowing young stands before old stands also helps to reduce its spread. Other forage legumes are hosts for *C. trifolii* (239).

21-1.16 Sclerotinia Crown and Stem Rot

This disease is caused mainly by the fungus *Sclerotinia trifoliorum* Eriks. and to a lesser degree by *S. sclerotiorum* (Lib.) dBy. It was first reported on alfalfa in 1915 (73). Disease severity varies greatly among years and fields. Usually, scattered plants or plants in small areas are affected, but large-scale plant losses are possible. Plants of all ages become diseased, with the worst outbreaks often occurring in seedling stands. In cold regions, e.g., northern USA, the fungus is most active in early spring, especially with late snow cover. Further south the fungus may be active all winter. Diagnosis can be difficult because affected plants may not die until later in the season when temperatures rise.

The first symptoms of Sclerotinia rot are usually yellowed leaves and stems that become flaccid and collapse (69). Stem areas appear bleached, with white, fluffy mycelium growing over the dead plant surfaces as well as on the soil surface close to the plant. The mycelium is very evident when temperatures are cool and free water prevails inside the crop canopy. Mycelial growth also occurs under basal snow cover on stems. Dry mycelium appears brown or gray. Hard black sclerotia form on and within diseased stems and crowns when environmental and host conditions no longer favor disease development. Sclerotia can be found free in the soil when plant tissues decay (Fig. 21–7).

The fungus, in its sclerotial form, persists for many years in the soil. Sclerotia vary in shape and range from minute to 10-mm long. Sclerotia serve as the primary inoculum source by producing cup-shaped apothecia, borne on short stalks, which in turn produce ascospores, the primary infectious propagules. Sclerotia produced in the spring remain inactive until fall when, under cool, moist conditions, they germinate; ascospores are produced and disseminated causing primary infection (183). Timing of apothecia development and spore release varies with latitude and elevation. In mild climates, infection may occur repeatedly throughout the winter. Free water is necessary for infection, and the disease is soon arrested with the advent of warm, dry conditions.

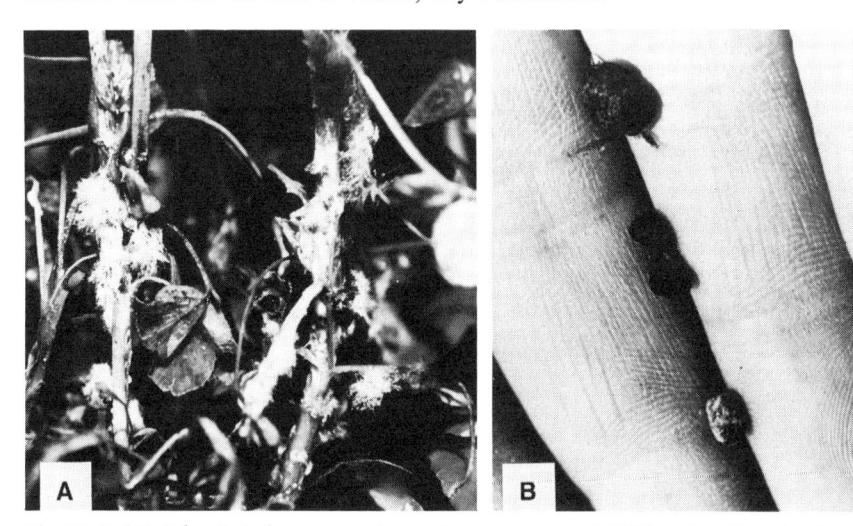

Fig. 21–7. (A) *Sclerotinia* fungus growing on lower stems, and (B) fungal resting structures (sclerotia) found in stems and on the soil surface adjacent to dead plants.

Genetic resistance to *S. trifoliorum* is not available, therefore management must be used to minimize crop losses. Rotations of at least 3 to 4 yr between legume crops, deep plowing to bury sclerotia, and planting clean seed are the recommended strategies. Fungicides have shown some promise especially if applied when apothecia are present in the field (242).

21-1.17 Alfalfa Stem Nematode

Ditylenchus dipsaci (Kühn) Filipjev is the most important nematode pathogen on alfalfa and is found in practically every corner of the globe where alfalfa is grown. It was first reported on alfalfa in Germany in 1881 and in the USA in 1923. It is most frequently reported as a serious pest in areas of heavy soils, high rainfall, and heavy spring rains, or in irrigated fields.

Ditylenchus dipsaci contains several distinct biological races. Green (80) found that several species of weeds and wild plants are hosts of *D. dipsaci* in England. These may be a source of different biological races. An alfalfa race from the USA caused characteristic symptoms on several nonhost seedlings but reproduced only on sainfoin (*Onobrychis viciifolia* Scop.).

All stages of the nematode, except the first larval stage, which molts in the egg, can attack the alfalfa plant. The nematode enters primordial bud tissue and migrates into developing buds. Infected stems enlarge and usually discolor; the nodes swell; and the internodes become shortened. The disease symptoms are induced in reaction to enzymes secreted by the nematode and also by a physiological imbalance of auxins produced in the plant. Growing stems may succumb to infection or overcome the swelling and develop normally. Stem necrosis results after long periods of parasitism accompanied by moderate temperatures and high humidity. Blackening of the stem can be observed up to 0.3 m or more above the ground. Great numbers of nematodes can be found inside the blackened stem. The numbers of stems per crown become fewer as the alfalfa crown is destroyed, and eventually the entire plant dies. The stand of alfalfa is thinned, leaving space for weeds and grasses. In addition, Boelter et al. (18) have shown that stem nematode reduces cold tolerance in plants resulting in greater winter kill.

Soil temperature plays an important role in the host-parasite relationships of *D. dipsaci* on alfalfa (81) (Plate 21-2 and Fig. 21-8). When heavy nematode infection is combined with warm, humid weather, the nematode migrates into the leaf tissue, causing a curling and distortion of the leaves. Plants show "white flagging," a symptom attributed to nematode infection of leaf tissue and destruction of chloroplasts. *Ditylenchus dipsaci* can also infect the alfalfa seed if the inflorescences are infected. Roots are occasionally infected, resulting in internal cavities. There may also be gall-like outgrowths that girdle the crown.

Large numbers of nematodes can survive periods of environmental stress in the alfalfa crown tissue, although only small numbers have been

DISEASES AND NEMATODES

Fig. 21-8. Stems with shortened internodes caused by the alfalfa stem nematode. Normal stem is at left.

recovered from alfalfa field soil at any time of the year. *Ditylenchus dipsaci* is unable to survive >2 yr under a nonhost cropping of grain under heavy rainfall, or when standard irrigation practices are followed. *Ditylenchus dipsaci* has been known to survive for 20 yr in a dormant stage. Differences in longevity of this nematode appear to depend on availability of host tissue in the soil, soil moisture relationships, and possible physiological differences within races and nematode populations.

Ditylenchus dipsaci is able to parasitize plants over a wide tissue range and also interacts with other plant pathogenic organisms. It can predispose plant tissue to bacterial wilt, and Hawn and Hanna (96) found that *D. dipsaci* could also reduce bacterial wilt resistance in alfalfa.

Seed was once considered the major means of dissemination of *D. dipsaci*; however, anything that moves nematode-infested soil and alfalfa tissue will move nematodes. They are spread readily by machinery during harvest season and by rain and irrigation water from infected alfalfa debris left in fields. The reuse of waste irrigation water is probably the most common method of nematode dissemination (60).

Resistance is the only practical method of nematode control. Resistance to *D. dipsaci* was first seen in a Turkistan alfalfa selection planted in a nematode-infested field in northern Utah. Resistant cultivars have been developed from this and other selections.

A 2- or 3-yr rotation, using nonhost crops such as grain, bean, or sugar beet (*Beta vulgaris* L.), can reduce the nematode population below the level of detection. However, recontamination via machinery, animals, or waste irrigation water can quickly negate the beneficial effects of crop rotation.

Certain agronomic practices can partially alleviate the parasitism of *D. dipsaci* on alfalfa. In irrigated areas of the arid and semiarid western USA, the alfalfa stem nematode is usually a problem only in the first cutting of alfalfa. Inasmuch as each cutting requires reinfection primarily from the soil, infection can be reduced if the alfalfa is cut when the top 5 to 8 cm of the soil is dry. This results in little or no nematode invasion of subsequent plant growth. Nematode infection in spring plant growth is significantly reduced when fall burning is used for weed control. Spring burning, however, promotes new growth and increased infection and is therefore not recommended.

Normally, the decline in yield varies with the source of nematode infestation. If the nematode source is from irrigation water or if the number of nematodes in the soil at the time of planting is high, serious decline occurs in 2 to 3 yr. If the source of infestation is from seed, machinery, or from a low population in the soil, the decline is more gradual.

21-1.18 Alfalfa Mosaic

Alfalfa mosaic is caused by various strains of alfalfa mosaic virus (AMV) that vary in virulence and symptomatology. The disease, first described by Weimer (237), occurs worldwide and causes economic losses. In Canada a 30% loss in forage yield has been reported (155), along with reduced nodulation and poor winter survival (226). In Minnesota, Frosheiser (62), reported that a 53% incidence of infection reduced yield by 11% but did not alter winter survival. The incidence of disease sometimes approaches 100% (155, 168, 221).

Because of strain diversity, plant response will vary from a very mild mosaic to death (943, 62). The most common symptom is an interveinal light-green or yellow mottle, which may or may not be accompanied by stunting. Other symptoms include various types of contorted leaves and petioles. Rapid root necrosis and plant death can occur following infection with specific strains of AMV on some host genotypes. Often infection may cause no symptoms or only a slight yield loss and gradual debilitation of the plant over several growing seasons. These stunted plants may be hidden under the crop canopy and not detected. Also, high summer temperatures and long daylength prevent the expression of leaf symptoms. Serological tests are probably the best way to determine the presence of AMV. However, direct assay of suspected plants on cowpea [*Vigna sinensis* Torner) Savi] is simple and quite reliable (221).

Strains of AMV attack over 73 plant genera (73), with most strains occurring in alfalfa. Virus particles are transmitted in sap, pollen, and seed. However, the aphid *Acyrthosiphon pisum* (Harris), along with other aphids (155), are the most effective means of transmission in the field. Virus particles remain infective for at least 5 yr in seed, and seed-borne virus probably constitutes the primary inoculum in most fields. Seed transmission of up to 31% has been reported (155), but 3 to 8% is more common (63). Aphids distribute the virus from plant to plant and field to field, with limited distribution by sap on mower blades. Dissemination of the virus can be extremely rapid with the incidence increasing from 3% shortly after seeding to nearly 100% in the fall of the second year (221).

The use of virus-free seed along with aphid control and a sound weed control program will help to delay spread of the virus. Mowing young stands first might delay the introduction of virus from older, infected fields.

All alfalfa cultivars are susceptible but some plants are resistant. The numerous strains of AMV portend that a resistant cultivar might be short-lived in the field.

21–1.19 Witches' Broom

This disease is probably caused by a mycoplasma-like organism rather than a virus and occurs in western USA, Canada, Australia, USSR, and probably in other semiarid regions (73). The disease is of minor importance in North America; however, it causes more damage in the drier areas of New South Wales and Queensland in Australia (98, 99) and also in Saudi Arabia (36).

Short spindly shoots from the crown and axillary buds on the stems are characteristic of this disease (163). A proliferation of fine stems are common on infected plants. Affected plants assume a yellowish cast because of the marginal chlorosis of leaflets and the presence of light-green stems. Leaflets are often small and puckered. Symptoms are most pronounced under high summer temperatures and moisture stress with remission common during winter. Greening of flowers is another characteristic of this disease.

Mycoplasma-like particles are found in the phloem of plants exhibiting witches' broom symptoms, but pathogenicity according to Koch's postulates has not been demonstrated. Leafhoppers are considered to be the primary vectors of this pathogen. No direct control measures are available, but area-wide removal of old diseased plantings, reseeding, and maintenance of highly productive stands will reduce spread to new plantings. No resistant cultivars are available.

21-1.20 Alfalfa Enation

This disease, which is serious in France and other parts of Europe (73), is caused by a bacilliform virus (3, 4).

Infected plants exhibit enations on the undersides of leaves, mostly along the midveins, which are crinkled. In general, plants are not stunted but often appear bushy.

The virus is not transmitted mechanically but can be transmitted by grafting and by the cowpea aphid (*Aphis craccivora* Koch.) (147). Symptoms appear about 9 d after inoculation. Transmission by either the pea aphids or seeds is not known to occur.

No control measures are available.

21-1.21 Transient Streak

This disease occurs in Australia (16) and may be fairly widespread in Canada (155, 177). The disease symptoms are manifested as yellow streaks along main lateral veins of leaflets. These streaks may be little more than small spots or can be up to 2-mm wide and extend to the leaf margin. Some stunting and reduced branching may occur. Symptoms can be masked completely in warm weather. As much as 30% infection in 2- to 3-yr-old stands have been reported in Canada, although symptoms are often visible only on new leaves in the spring.

The virus has been identified in several other plant species, but little is known about its spread. Sap transmission has been demonstrated, but neither seed nor insect transmission have been reported.

21-1.22 Alfalfa Latent

This virus was found in alfalfa in Nebraska (230) but causes no symptoms. The virus is sap and aphid transmitted but probably not seed transmitted. The potential damage by this virus is not known, nor has any control been developed.

21-2 DISEASES OF ROOTS

Most of the root diseases of alfalfa are caused by fungi or nematodes, but physiogenic disorders are known and bacteria are beginning to be implicated in root dysfunctions (151). Generally, fungi pathogenic to roots kill root tissues directly and reduce the plant's absorptive, N_2 fixing, storage, and anchoring capabilities. Symptoms of root diseases often manifest themselves in leaves and stems.

Some root diseases are caused by single specific pathogens; others are caused by a complex of organisms that is not always the same. The perennial association of alfalfa roots with soil microflora and fauna affords great opportunity for interactions among organisms to produce disease. Time and exposure enable slow or weak pathogens to cause serious disease problems. The rate of development of root disease is often dependent upon the environment and the type of management to which a crop is subjected. Stresses of many kinds, biotic and abiotic, enhance development of certain root diseases.

Although it may be possible to select resistance to some root diseases involving a single pathogen, it could be difficult or impossible to identify resistance to pathogen complexes (139). Without resistance, management becomes the major strategy in slowing the rate of disease development. Good management prolongs stand profitability.

21–2.1 Seedling Diseases

Seedling diseases affect the plant during preemergence and postemergence stages. At the preemergence stage, the fungus penetrates the seed before emergence. Because the time period during which infection occurs is so short, accurate diagnosis of preemergence seedling disease is difficult. *Pythium* spp. are especially active at this stage. After the plant emerges, the hypocotyl tissue may be penetrated and infected. The hypocotyl shrinks and the plant lodges and dies. The immature root may also be affected by one or more of the pathogens. Plants generally become resistant to seedling pathogens at 5 to 10 d of age; however, *P. megasperma* f. sp. *medicaginis* is an exception, because this fungus also causes root rot on adult plants.

Several fungi have been associated with seedling diseases in different areas. *Pythium ultimum* (86, 212), *P. debaryanum* (25), *P. sylvaticum*, *P. paroecandrum*, *P. hypogynum*, *P. torulosum* (212), *P. megasperma* f. sp. *medicaginis* (198), *Aphanomyces euteiches* (104, 161, 198), *Fusarium culmorum*, *F. acuminatum* (86), *Cylindrocladium* spp. (173), and *Rhizoctonia solani* (86), have been implicated as seedling pathogens and potentially could be involved in most areas where alfalfa is grown. It has been shown that seedling mortality was higher when *Fusarium oxysporum* (84) or *Pythium ultimum* (222) and root knot nematode occurred together. The etiology of seedling diseases varies from one region to another probably because of different environmental conditions.

The incidence of damping off of several crops is considered to be inversely proportional to the velocity of emergence (134); low temperature crops were more susceptible at high temperatures and vice versa (134).

Pythium spp. appear to be the most frequent cause of seedling disease. A "forked root condition" due to killing of taproot at the seedling stage by *P. ultimum* (86) is a secondary effect that extends beyond the seedling stage.

Rhizoctonia solani is generally pathogenic during periods of high temperature and causes postemergence damping off.

Control of seedling diseases should include management of soil water and planting good quality seed at optimal depth. Seed treatment with fungicides did not increase the stand enough to increase yield (228). However, currently available fungicides have significantly increased stand establishment under experimental conditions (58, 221), and metalaxyl is currently being used as a seed treatment against losses caused by *Pythium* and *Phytophthora* spp.

21-2.2 Phytophthora Root Rot

Phytophthora megaspera (Drechs.) f. sp. *medicaginis* Kuan & Erwin (Pmm), the causal agent of Phytophthora root rot, was described in 1954 as *P. cryptogea* (53), changed in 1965 to *P. megasperma* (54), and in 1982 to *P. megasperma* f. sp. *medicaginis* (Pmm) (132) because of host specificity and to differentiate it from *P. megasperma* on other hosts. The fungus is relatively specific to alfalfa, although it will attack a few other legumes (56, 76, 115). Phytophthora root rot occurs in most of the alfalfa producing areas of North America (26, 29, 61, 76, 184, 198), in Mexico (1), and in many other regions of the world (113, 156, 185).

Plants infected by Pmm become stunted and leaves become yellow or purple. Rotted tap root tissue is firm, but brown with diffuse margins. Infections often start where lateral roots emerge. Nodules on seedling plants are also readily infected (78). In a field, rotting of taproots of surviving plants at similar depths, plus the brown to yellow color of diseased tissue extending through the cortex into the xylem are useful diagnostic tools. Severely diseased plants wilt and die; often the taproot is girdled (Fig. 21-9).

Root rot occurs in wet soils especially when free water persists for an extended period, and activity of the fungus is greatest within a range of soil temperatures of 18 to 27°C. Irrigation water can disseminate zoospores within and between fields. The fungus may survive as oospores (211), either in soil or in diseased plant tissue. Various levels of virulence of the pathogen exist among field isolates (55, 59). A high temperature strain has been reported (max. growth at 39°C) in the southwest USA (190).

Moderate to high resistance is now available in a large number of cultivars from both public and private breeding programs (55, 64, 65). However, even resistant cultivars can become diseased if inoculum is abundant and the soil remains wet for extended periods. Therefore, use of resistant cultivars should also be accompanied by optimum water management.

Irrigation water should not stand for extended periods on soil. The severity of Phytophthora root rot in the midwestern USA was increased markedly by long periods of flooding (65, 184), and the effects of Phytophthora root rot, excess soil-water stress, and other stresses are difficult to separate (5). The success of water management depends on the drainage characteristics of the soil and on the prevention of flooding for extended times. Where high rainfall occurs, low areas in fields should be drained. The greater lability of *P. megasperma* zoospore cysts in drier soils com-

DISEASES AND NEMATODES 643

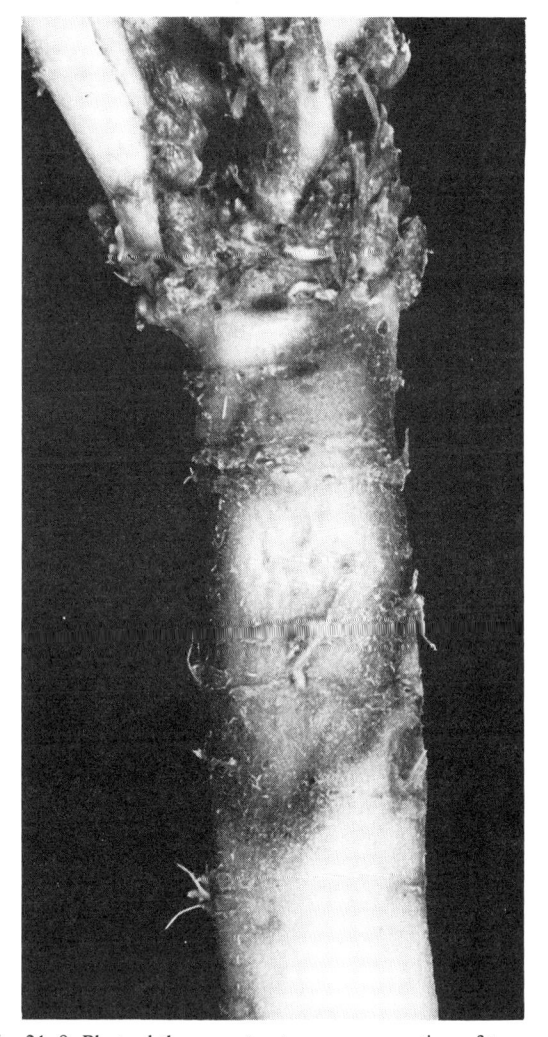

Fig. 21-9. Phytophthora root rot on upper portion of taproot.

pared to those of *P. cinnamomi* (70) supports the validity of management of soil moisture as a primary control measure.

21-2.3 Stagonospora Root Rot

Stagonospora root rot and leaf spot [caused by *Stagonospora meliloti* (Lasch.) Petr.] was described near Riverside, CA, and Madison, WI, by Jones and Weimer (128). The disease occurs in nearly all alfalfa growing areas in California. It has been reported in the eastern USA (105) and in some areas of Australia (186) and New Zealand (35, 213).

Severely infected plants recover slowly after mowing. The tissue on the surface of infected crown branches and tap roots is often rough and cracked. The most definitive symptom is a reddish-brown flecking that occurs within the woody tissue (52, 73). The disease progresses slowly, and usually 12 to 26 weeks elapse before the root rots completely (Plate 21-3).

Infection of alfalfa in California appears to occur in the spring of the year when low soil temperatures prevail, and probably begins with infected leaves and stems (see Stagonospora leaf spot) and progresses to the crown and tap roots. The disease persists throughout the year except in the hot southwest U.S. desert areas. Root rot was most severe on wound-inoculated plants at 24°C (128).

Resistant cultivars are not currently available, but promising germplasm has been identified.

21-2.4 Rhizoctonia Root Canker

Rhizoctonia root canker of mature plants (caused by *Rhizoctonia solani* Kuhn (207)) occurs mainly in the southwestern USA (207) and in western Queensland, Australia (186), irrigated areas of Egypt (200), and in New Zealand (35) during the summer season. Dark sunken lesions with a yellow-brown border occur on large tap roots, especially in the region where lateral roots emerge. When lesions coalesce, the entire tap root rots off, which may cause death of the plant. In winter, the lesions heal, become dark brown to black, and the disease becomes inactive (Fig. 21-10).

The optimum temperature for growth of *R. solani* (25-30°C) is similar to the optimum soil temperature for root canker formation (30°C) (209). These data correlate well with the soil temperature (7.6-cm depth) during summer at Bard, CA. For the most part, only specific *R. solani* isolates from root lesions were capable of causing root canker. Isolates from alfalfa seedlings and from other crops, such as sugar beet, potato (*Solanum tuberosum* L.), and cotton (*Gossypium hirsutum* L.) did not cause root canker (208).

Root canker plays an important part in the stand loss complex of alfalfa in the lower desert areas of southwestern USA. The high-temperature, flooding injury commonly called "scald" occurs under similar conditions. Root canker also seems to be associated with excessive irrigation, which may play a role in plant susceptibility. Why Rhizoctonia root canker is important only in the desert areas is not understood. Although high soil temperature is correlated with the incidence and severity of the disease, soil temperatures in other areas, such as the San Joaquin Valley in California, where the disease is not prevalent, are also well within the range of temperatures reported to be optimum for *R. solani* (209).

No resistant cultivars are known.

21-2.5 Crown Bud Rot

Crown bud rot is caused by *Rhizoctonia solani* (95, 52). In contrast to the high temperature requirement for root canker caused by *R. solani*,

Fig. 21-10. Cankers on taproots caused by *Rhizoctonia solani*.

the maximum severity of crown bud rot occurred when soil temperatures (2.45-cm depth) were about 16°C (89). *Rhizoctonia solani* spread throughout the cortical tissue and xylem vessels (90, 52).

No control measures are known.

21-2.6 Violet Root Rot

Violet root rot is caused by *Rhizoctonia crocorum* DC ex. Fr. (Syn. *R. violaceae* Tul.; *Helicobasidium purpureum* Pat.) but is rarely seen or reported. In a field in Wisconsin in which all plants in a large area were killed, roots were covered by a mantle of violet to purplish brown hyphae (87). In Australia, the disease has only been destructive in small areas. Foliage of affected plants became yellow and wilted. Plants were covered with mycelium and "tiny black bodies known as sclerotes, which are able to survive for long periods in the soil." The disease occurred in discrete patches that gradually expanded (185).

No control measures are known.

21-2.7 Crown and Root Rot Complex

Crown rot occurs in varying degrees in all areas in which alfalfa is grown. The causes of crown rot in a broad sense have not been definitively

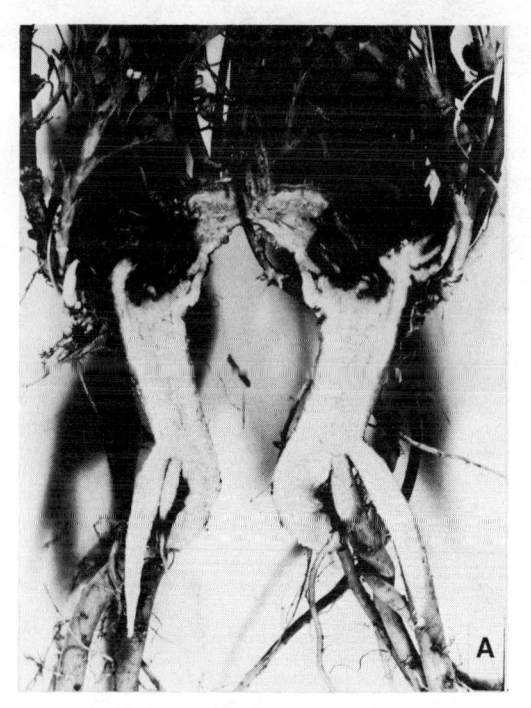

Fig. 21-11. (A) Early rot in split crown showing portion of crown no longer producing stems. (B) Upright growth of healthy plant on right is contrasted with asymmetrical, spreading growth of plant with crown rot on left.

and unequivocally determined. There are numerous causal agent-crown disease relationships that can be readily diagnosed, such as Stagonospora crown rot, anthracnose crown rot, and Rhizoctonia crown rot; however, a definitive explanation of the causes of chronic crown necrosis (52), collar rot, heart rot, and hollow crown (231, 234) has eluded researchers for decades.

There are many pathogens that contribute to the crown rot causal complex, and it is unlikely that the complex is the same in different areas or under different environmental conditions (Fig. 21-11). The etiological complex in the northeastern USA (143) included *Fusarium oxysporum*, *F. solani*, and *F. roseum*, which were the most frequently isolated fungi associated with root rot, and *F. tricinctum* and *F. moniliforme*, which were isolated less frequently. All of the isolates tested colonized either red clover (*Trifolium pratense* L.) or alfalfa seedlings and maintained a parasitic relationship with the plant without causing any apparent damage. It is generally accepted that stress factors govern the degree of damage that these fungi cause (143, 174). Many other fungi and a few bacteria that are weakly virulent on alfalfa (67, 141, 150, 159, 227), and *Myrothecium* spp. (142) have been identified as components of the root rot complex.

Wounding has been associated with the root and crown rot complex, and early reports credited wounds as being the main points of entry for crown rot fungi. Root feeding insects have been implicated in the wound-pathogen-disease syndrome (46, 133, 143). Many cortical-rotting Fusaria, however, penetrate roots directly (30, 141), and stubble provides ready-made entry sites for many crown pathogens (191). Although wounds enable some pathogens to bridge the alfalfa root integument, they also result in increased penetration of root tissues adjacent to the wound and increased the severity of disease caused by some Fusaria (217).

Control measures are not well defined; however, preventing stresses such as drought, insect feeding, excess soil water, and nutrient deficiencies, as well as adjusting soil acidity by liming, and following recommended mowing schedules, all contribute to slowing disease development. Direct damage to crowns by heavy equipment could also favor crown rot development. Because many organisms contribute to the crown and root rot complex, the prospect of selecting resistant germplasm is poorer than with many other diseases. Not all the organisms associated with crown and root rots are primary pathogens, however, and some may not be pathogens at all. Actually, only a small percentage of Fusaria isolated from diseased roots had the ability to penetrate intact roots or cause disease (141). This is encouraging, and recent reports (139, 164, 165, 191, 244) raise cautious optimism over the prospect of selection for a decreased rate of root and crown rot development. Certain cultivars, not selected for root rot resistance per se, develop less rot than other cultivars under certain conditions (165, 213). *Medicago falcata* has shown resistance to a complex of crown rot pathogens in Canada (94), and despite slower regrowth may yield as well as some *M. sativa* cultivars.

21-2.8 Scald

Scald is a physiological disease that is particularly serious in the hot irrigated areas in the southwestern USA and in other areas of the world where high temperatures prevail in the summer. The actual cause is most likely the interaction of high soil temperature and the duration that the soil is saturated at a high soil temperature (57). Oxygen deficiency is likely involved. "High temperature flooding injury" is more descriptive, but "scald" has gained wide acceptance. The disease also occurs in areas of more moderate temperatures.

Plants in high temperature areas are often killed within a week after the flooding episode. The soil temperatures at which the disease is most likely to occur range from 35 to 42°C. At these temperatures flooding of the soil for periods >30 h can cause death of an entire stand. Scald can occur at lower temperatures following a long period of soil saturation. Plants that have been recently clipped are much more susceptible than plants with foliage.

Affected plants become yellow in color and wilt within a week. The vascular tissue of the roots either becomes necrotic or the root collapses completely. A putrid odor is often associated with scald. Brown vascular discoloration associated with scald could be confused with bacterial wilt or dwarf. However, the foliar symptoms of these diseases differ from scald.

Water management is the most important factor for control. Irrigation for short periods of time is most effective (148) and should be delayed after mowing until plants have regrown to at least 10 cm in length. There is field evidence that some cultivars are more tolerant of scald than others (148). In Ohio some cultivars that were resistant to *Phytophthora megasperma* f. sp. *medicaginis* were more tolerant to long periods of flooding at moderate temperatures (~25°C). Barta (12, 13) proposed that ethanol produced in the flooded roots might be the toxic agent.

21-2.9 Winter Crown Rot or Coprinus Snow Mold

Winter crown rot, caused by a basidiomycete, *Coprinus psychromorbidus* Redhead & Traquair (187), occurs mainly in Canada under snow cover at 1 to 2°C (23, 224, 225). The fungus, called the "low temperature Basidiomycete" because the basidial stage had not been found, was readily isolated from diseased crown tissue and proved to be pathogenic (39). As a result of hydrocyanic acid produced by the fungus under snow cover, crown and crown bud tissue becomes black and soft and, if the affected area is extensive, the plant is killed (146).

The use of sodium tetraborate as a foliar spray in the autumn has been reported as a control measure (145).

21-2.10 Phymatotrichum Root Rot

Phymatotrichum root rot is caused by *Phymatotrichum omnivorum* (Shear) Dug. (syn. *Ozonium omnivorum* Shear). This disease has also been called cotton root rot, ozonium root rot, and Texas root rot. The distribution of the disease is limited to the southwestern part of the USA (Texas, New Mexico, Arizona, and southern California) and northern and central Mexico where the fungus was endemic. The disease affects cotton and up to 2300 other species of dicotyledonous plants during the summer months (152, 214).

The disease often occurs in a "fairy ring" pattern. Leaves of plants with advanced stages of root rot become bronzed and eventually wilt and the plants die. The lesions on roots are sunken and could be confused with the round to elliptical lesions of Rhizoctonia root canker in the absence of careful examination. The disease occurs only in the summer months when soil temperatures are high (153). Even susceptible plants are not affected during the winter months. The fungus persists in the soil at depths up to 2 m as long-lived sclerotia (152).

Extensive research on control of the disease has not provided highly successful control measures because of the persistence of the fungus at great depths in soil. The broad host range of the organism limits the value of crop rotation. Heavy applications of animal manures and incorporation of large amounts of green manure from winter cover crops have partially suppressed the fungus in soil.

There is no known source of genetic resistance to the disease; however, vigorous rooting types favor recovery of alfalfa following root rot in the summer. Although there has been little evidence of spread from field to field, movement of soil from affected counties is prohibited.

21-2.11 Crown Wart

Crown wart [caused by the fungus *Urophylicitis alfalfae* (Lagerh.) Magn. (210) (Syn. *Physoderma alfalfae*] was first described by Jones and Drechsler (124).

It has occurred sporadically on the Pacific slope of the USA (106, 124), in Pennsylvania (135), in Europe (72), in New Zealand (35), and in Australia. The disease has not had a marked economic impact on alfalfa production.

Urophylicitis alfalfae infects the crown buds in the spring, as swimming zoospores emanate from sporangia produced by the germination of resting spores from disintegrated warts (199) (Fig. 21-12). After infection the crown buds become swollen and the wart becomes marbled on the surface. The presence of globose resting spores (40-59 μm in diameter) in the wart is a definitive sign of this obligate parasite. The presence of resting spores distinguishes this disease from crown gall [caused by *Agrobacterium tumefaciens* (Smith & Townsend) Conn.], which occurs less frequently. The crown warts change from white to brown and eventually disintegrate during the summer. Presumably, the resting spores germinate during wet weather in the following spring. Although no control measures are known, rotation to nonsusceptible crops should be effective in reducing disease severity.

Fig. 21-12. Wart-like growths at base of stems caused by the fungus *Urophlyctis alfalfae*.

21-2.12 Brown Root Rot (Plenodomus Root Rot)

Brown root rot (caused by *Plenodomus meliloti*, D. and S.) occurs mainly in Canada during winter or early spring after the soil has thawed. The pathogen is more virulent on sweet clover (*Melilotus alba* Sesr.) roots than on alfalfa (197). *Plenodomus meliloti*, as well as *Sclerotinia sclerotiorum*, is capable of penetration of the natural layer of secondary tissue (37) following winter dormancy. The low optimum temperature for growth (15°C) is associated with lack of disease progress during the summer. The dark-brown lesions are sunken and usually separated from the surrounding healthy tissue by a darker brown ring. The firmness of the rotted tissue distinguishes it from the soft rot caused by *Sclerotinia* spp., which also attack alfalfa in early spring. Brown to black pycnidia of *P. meliloti* form on and in the affected tissues. Resistance to *P. meliloti* has not been found.

21-2.13 Cylindrocarpon Root Rot

The causal agent, *Cylindrocarpon ehrenbergi* Wr., is part of a complex of fungi that attacks roots of alfalfa and sweet clover in Canada during the early spring after thawing of frozen soil (38). The disease is more severe in sweet clover than in alfalfa and occurs in virgin soils as well as on alfalfa soils. Plants are affected in the early spring but not during the growing season. Freezing during dormancy predisposes plants to infection. *Cylindrocarpon obtusisporum* is also lightly to moderately pathogenic; *C. radicicola* is weakly pathogenic; and *C. olidum* is nonpathogenic. *Cylindrocarpon ehrenbergi* isolates, which grew optimally at 19°C, were the most damaging in early spring, and isolates that grew best at 24°C were most virulent in summer. *Cylindrocarpon ehrenbergi* penetrates roots directly through lenticels, or through tissues at the bases of lateral roots. Infected tissue initially appears water-soaked and light brown but becomes dark brown with age. Often salmon- to orange-colored sclerotia-like stromata of the fungus form on the infected root. *Medicago falcata* may be a source of resistant germplasm (38).

21-2.14 Southern Blight

Sclerotium rolfsii Sacc. [sexual stage *Pellicularia rolfsii* (Cruzi) West] causes a root and crown rot of alfalfa and on many other crops where warm weather prevails, soils are wet, and high levels of inoculum (sclerotia) occur in the soil. The disease is readily diagnosed by the presence of brown, mustard seed-like sclerotia in diseased tissue. The sclerotia are the principal means of survival. The affected tissue becomes tan to white, and the plant dies rapidly when severely attacked. There are few reports on southern blight of alfalfa (77). Aycock (8) presents a comprehensive, general review of the pathogen and the diseases it causes.

21-2.15 Charcoal Rot

Macrophomina phaseolina (Tassi) G. Goid. causes a root and crown rot following drought stress. The disease is also known as charcoal rot on *Sorghum* spp. and ashy stem blight on bean. There are few reports on this disease. The pathogen has a wide host range that includes alfalfa. The disease was reported in Alabama by Gray et al. (77) in a field where the plants were under severe water stress. The disease can be identified by small sclerotia that form in infected host tissue giving it a "charcoal" appearance. The fungus may penetrate plant tissue, but pathogenicity and symptoms of the disease require the predisposing effect of drought stress.

21-2.16 Mycoleptodiscus Crown and Root Rot

Mycoleptodiscus terrestris (Gerd.) Ostazeski (syn. *Leptodiscus terrestris* Gerd) (68, 175) causes a dark-colored root rot of alfalfa and occurs within a complex of *Fusarium* spp., *Phoma medicaginis* and others. The fungus is more virulent on red clover than on alfalfa. *Mycoleptodiscus terrestris* is pathogenic to several other leguminous hosts as well as alfalfa and causes preemergence and postemergence damping off as well as a black rot of lateral and tap roots. Affected crowns and roots are dark and necrotic, and numerous black, spherical sclerotia form within the tissue. Apparently these bodies perpetuate the fungus.

No control measures have been reported.

21-2.17 Northern Root-Knot Nematode

Meloidogyne hapla Chitwood, referred to as the northern root-knot nematode, is actually widely distributed. It is the most common species in the northern hemisphere, where rhizosphere soil temperatures drop to 0°C or below, and summer soil temperatures seldom exceed 25 to 30°C. The predominance of *M. hapla*, however, has been challenged in the northern USA with the discovery of the Columbia root-knot nematode, *M. chitwoodi*, Golden et al. (71).

Meloidogyne hapla is considered a mild parasite on alfalfa, but severe invasion of alfalfa seedlings from highly infested soil is capable of causing a high mortality in young alfalfa plants (170). Under low soil infestations, however, alfalfa plants may appear normal, resulting in satisfactory plant growth and yields. As is true with most root-knot species, *M. hapla* is most severe in sandy to sandy-loam soils.

Nematode numbers (eggs) are greatly reduced by freezing temperatures, and it is practically nonpathogenic at soil temperatures below 15°C. It is often difficult to find spring populations in cultivated soil after a severe winter when the frostline has extended below the rhizosphere (Fig. 21-13).

Probably the greatest importance of *M. hapla* on alfalfa is the associated economic problem created when highly susceptible crops are grown in rotation with alfalfa. *Meloidogyne hapla* is also involved in relationships between alfalfa and other alfalfa pathogens. *Meloidogyne hapla* increases the incidence of bacterial wilt in wilt susceptible and resistant alfalfa cultivars (83, 114). Griffin and Thyr (84) found that the combination of *M. hapla* and *Fusarium oxysporum* increased the incidence of Fusarium wilt and reduced plant growth. There is also a synergistic interaction between *M. hapla* and *D. dipsaci* (82). *Ditylenchus dipsaci* predisposes root-knot resistant alfalfa to infection by *M. hapla*. *Meloidogyne hapla* or *M. incognita* interact with Pmm to increase the severity of root rot symptoms and overcome Pmm resistance (240).

A unique characteristic of *M. hapla* is its symptomology on host plant tissue. *Meloidogyne hapla* galls are usually readily identified by the lateral root growth from galls. Gall size is variable, however, depending on the physiological response of the plant.

There are few options for control of *M. hapla* on alfalfa. Crop rotation, where a susceptible alfalfa cultivar is used, is impractical, because large populations of nematodes are produced over the life of the alfalfa

DISEASES AND NEMATODES 653

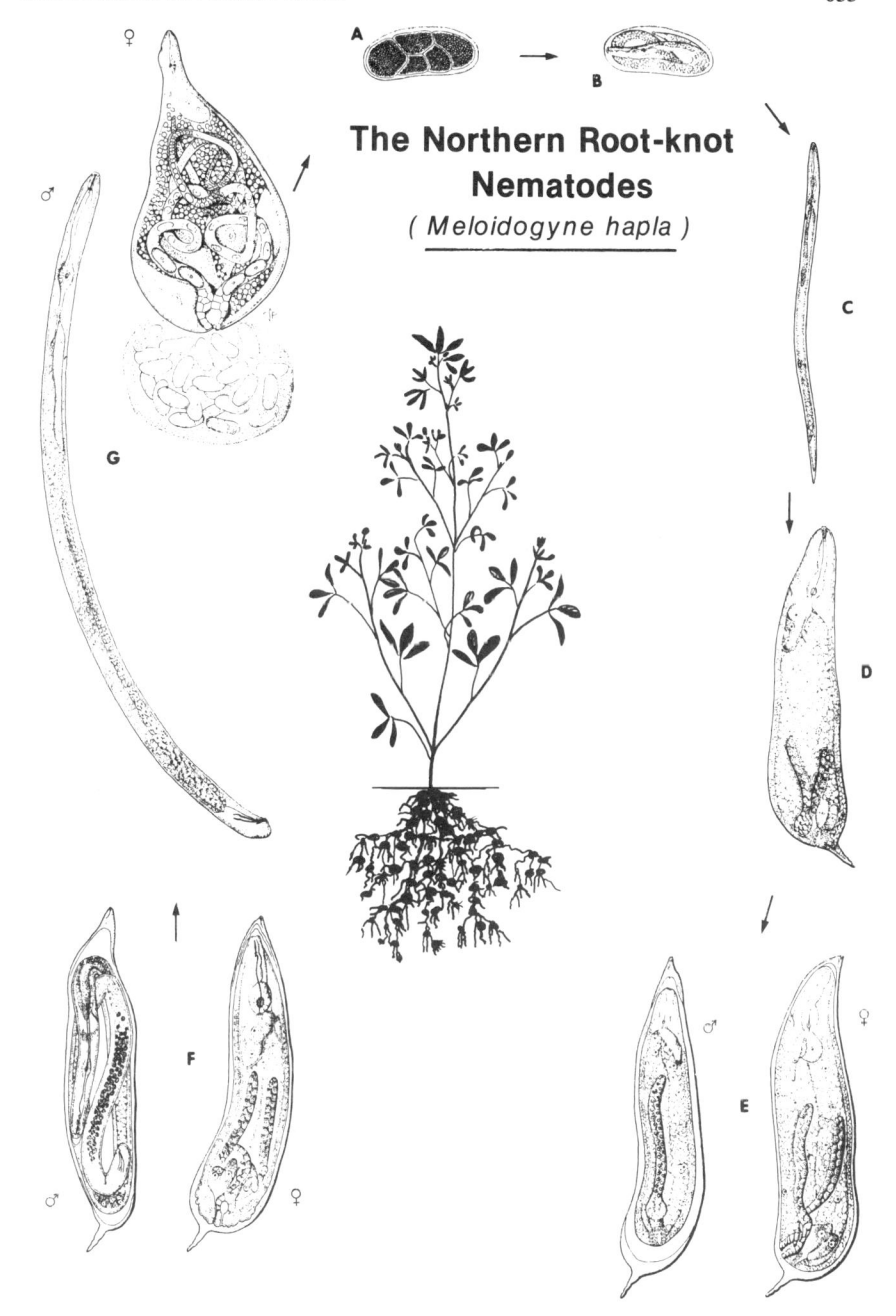

Fig. 21-13. Life cycle of the northern root-knot nematode. A = egg, B-C = second stage larvae, D-F = third and fourth stage larvae, G = adults. Note female with gelatinous matrix enclosing eggs. Courtesy of the Dep. of Nematology, Univ. of California, Riverside.

plant, depending on climatic zone. *Meloidogyne hapla* also has a wide host range and will reproduce on most plants other than those of the grain and grass family. The most practical control is the use of resistant cultivars. Resistance to *M. hapla* was first selected from Vernal alfalfa, and now resistant germplasm is available to alfalfa breeders.

21–2.18 Southern Root-Knot Nematode

Meloidogyne incognita [(Kofoid and White) Chitwood] is probably the most important root-knot species in relation to damage and economic loss to cultivated plants. This species consists of four races of nematodes, distinguished by host specificity. *Meloidogyne incognita* is unable to survive the harsh cold temperatures of temperate climates and is found only in areas with optimum temperature ranges of 25 to 30°C. In the USA, *M. incognita* is not found above geographical zones where the average January temperature is below -1°C. Because <10% of the acreages of alfalfa and alfalfa mixtures is found in areas climatically suited to *M. incognita*, it does not rank in importance with *M. hapla*. *Meloidogyne incognita* also is less virulent on alfalfa than is *M. hapla*.

Meloidogyne incognita has been associated with disease complexes in alfalfa. McGuire et al. (160) found that Fusarium wilt (*F. oxysporum*) was more severe on alfalfa in the presence of *M. incognita* than when associated with *M. arenaria* or *M. javanica*.

The primary method of control is the use of resistant cultivars. Resistance to *M. incognita* has been identified in nondormant African and Chilean cultivars.

21–2.19 Javanese Root-Knot Nematode

The javanese root-knot nematode, *Meloidogyne javanica* (Treub.) Chitwood, is found under warmer conditions than *M. hapla*, and the northern limit is near the 7.2°C isotherm for average January temperatures.

Meloidogyne javanica commonly occurs on alfalfa in the arid southwestern USA, but invasion and virulence is dependent on the alfalfa cultivar, crop rotation history, and soil type. Like *M. hapla*, *M. javanica* is a problem on alfalfa, mainly because alfalfa is included in rotational programs with other nematode-susceptible crops, such as cotton and melons.

Reynolds and O'Bannon (188) found that *M. javanica* was more virulent than *M. incognita* when tested on susceptible cultivars.

Certain breeding lines with African and Sirsa parentage that were selected for resistance to the spotted alfalfa aphid, *Therioaphis maculata* Buckton, were resistant to *M. javanica*.

Another root-knot nematode species, *M. arenaria* Chitwood, has approximately the same northern limitation, geographically, as *M. incognita*. Although important on alfalfa in some areas of California, *M. ar-*

enaria should be considered of minor importance generally because of its limited geographical distribution.

21-2.20 Root Lesion Nematodes

Root lesion nematodes, *Pratylenchus* spp., are found throughout the world from the tropics to the most northern temperate areas. They attack a wide range of plants and play an important role in damaging both cultivated and noncultivated plants, including alfalfa.

Pratylenchus spp. are among the most important plant-parasitic nematodes, and stunted plant growth is usually associated with this nematode. Damage caused by root lesion nematodes is difficult to evaluate accurately, because other soil microorganisms usually invade the nematode-infected roots. Aboveground symptoms are also difficult to assess, because other pathogens and stresses can produce similar symptoms. No symptoms develop when nematode numbers are low, but when numbers are high and environmental conditions ideal, infected plants become stunted.

Females deposit eggs in root tissue or in the soil. Both juvenile and adult forms enter roots by forcing their way between or through the epidermal and cortical cells and then feed on the cell contents as they migrate through the root tissue. The points of entrance into roots also provide access for other soil microorganisms. Townshend and Stobbs (223) found that root lesions first appeared as water-soaked areas, but later became discolored, coalesced, and the coloration then intensified until the entire root became blackened.

Chapman (28) found that *Pratylenchus* spp. severely damaged alfalfa that had been spring sown in winter wheat (*Triticum aestivum* L.) fields. Hence, a high nematode population density plus competition with wheat under moist conditions resulted in the severe infection and destruction of alfalfa root tissue. Conversely, after a summer fallow, fall-sown alfalfa made good growth and the roots were only slightly damaged. *Pratylenchus penetrans*, however, is apparently the most important lesion nematode on alfalfa. It is found throughout the USA, Canada, and Europe and has a wide host range. Mauza and Webster (157) found that *P. penetrans* interacted with *Fusarium* spp. to stunt alfalfa. *Pratylenchus penetrans* also interacted with *M. incognita* to reduce alfalfa root growth.

Because many crops serve as hosts for root lesion nematodes, crop rotation is not as good a control measure as for other nematodes. There are no alfalfa cultivars resistant to root lesion nematodes.

21-2.21 Ectoparasitic Nematodes

Several genera of nematodes have been associated with alfalfa. These include *Criconemella, Helicotylenchus, Hoplolaimus, Paratylenchus, Paratrichodorus, Tylenchorhynchus,* and *Xiphinema*. Noel and Lownsbery (169) found that *T. clarus* Allen suppressed the growth of alfalfa and caused a reduction in feeder root growth, but no lesions or other abnormal root growth were observed.

Norton (171) reported an inverse relationship between alfalfa yields and *Xiphinema americanum* Cobb in Iowa, and stunted plants were associated with high populations of nematodes. Artificial infestation of soil with *X. americanum* reduced plant growth.

Several viruses found on alfalfa are vectored by nematodes; alfalfa is a host for the pea early browning virus, which is transmitted by a *Paratrichodorus* sp.

21-3 DISEASES OF THE VASCULAR SYSTEM

Vascular diseases, or "wilts" as they are commonly called, result primarily from the internal disruption of movement in the main water transport system of the plant—the xylem. Although infection by wilt pathogens usually occurs in the roots, the symptoms and actual pathogenesis usually occur in the stems, petioles, and leaves. The pathogens become established systemically in the water-conducting system of the plant. Symptoms of wilts vary considerably and are therefore discussed specifically for each disease. Wilt diseases have the potential to limit production of alfalfa cultivars lacking specific resistance.

21-3.1 Fusarium Wilt

Fusarium wilt of alfalfa was first described in 1927 (232, 233) in the USA and has since been found in many other areas of the world. The causal agent is *Fusarium oxysporum* Schl. f. sp. *medicaginis* (Weimer) Syn. & Hans.; however, race 1 and race 2 of *F. oxysporum* f. sp. *vasinfectum* (Atk.) Snyd. & Hans. and *F. oxysporum* f. sp. *cassia* Armst. & Armst. have also been reported to be pathogenic to alfalfa (6).

Shoots become yellow to reddish in color and wilt, and often only one side of the plant is affected. Cross sections of roots show a reddish-brown discoloration of the outer tissue in the xylem. The disease is favored by warm temperatures, and infected plants usually die within the season. Soil moisture was not a critical factor in disease development (50).

Frosheiser and Barnes (66) reported large differences in resistance among cultivars in field evaluations. Many current cultivars, especially nondormant types, have high levels of resistance (106), and resistance appears broadly effective and stable (51). The fungus survives in organic matter and persists in soil as chlamydospores for as long as 10 yr. Because the fungus persists for long periods in soil, control by rotation of <5 yr is not successful.

21-3.2 Verticillium Wilt

The type of Verticillium wilt, which is most important on alfalfa, is caused by the dark-mycelial fungus, *Verticillium albo-atrum* Reinke et

Berth. Because there has been controversy in the literature on nomenclature of *V. albo-atrum* and of *V. dahliae* Kleb., it should be emphasized that the alfalfa strain found in the USA (74) produces dark mycelium at the bases of conidiophores and does not produce microsclerotia, which are characteristic of *V. dahliae* (88, 205). Verticillium wilt caused by *V. dahliae* has been reported but is not usually of economic importance.

Verticillium wilt of alfalfa was described in Sweden in 1918 (97) and in Germany in 1938 (192). It has been found in other parts of Europe, including Great Britain (116). The disease was reported in Canada in 1964 (7) but not until 1976 in the USA (74). The disease was first found in the states of Washington and Oregon (74) and subsequently in many other states and Canada (7, 75, 79, 137, 203). Important information on this disease has been summarized by Peaden and Christen (178) and Huang and Atkinson (108).

Symptoms of Verticillium wilt are summarized by Christen and Peaden (33). Early symptoms include chlorotic streaks or blotches on leaflets. Often chlorosis occurs in a V-shaped area at the tip of the leaflet. This later becomes dried and bleached, taking on a pinkish to orange-brown hue. Young leaflets tend to curl upward or more often twist to form a spiral along the midvein. Less curling occurs with older leaflets. Leaf symptoms occur at first on one stem of a plant, occurring later on additional stems. Stems remain green and erect, often issuing new healthy leaves in close proximity to diseased leaves. Symptom development is explained by the xylem vessel distribution in alfalfa stems (85) and the disruption of individual vessel function by the pathogen (180). Symptoms of Verticillium wilt appear more frequently in the second or third year than in the first year. The xylem tissue in tap roots becomes orange or brown in color. However, this symptom alone does not serve to distinguish this disease from other vascular wilt diseases. Isolation of *V. albo-atrum* from stems is the most confirmatory type of evidence (Fig. 21–14).

The alfalfa strain of *V. albo-atrum* is carried on plant debris associated with seed (116), and within seeds from infected plants (31, 32, 204). *Verticillium albo-atrum* persists on the surface of seeds for as long as 36 weeks (204). Spores that are produced on stems of infected plants during moist weather can be spread by aphids (110). Inoculum within plant debris can be distributed by wind, machines, and insects (109, 111). Conidia enter via cut stems and the fungus invades the xylem tissue and proliferates by mycelial growth and the production of new conidia (179). Symptomless plants in resistant cultivars can also harbor the fungus (181).

Although Verticillium wilt is fairly widespread in the USA and Canada, there are many locations where it has not been reported. Therefore, efforts to delay or prevent its spread are justified. Cleaning of debris from machinery and wagons and power washing or disinfesting cutter bars before moving to clean fields are recommended. The mowing of youngest stands first reduces carry-over of pathogens from older, infested fields.

Fig. 21–14. Top of stem afflicted with Verticillium wilt. (A) Note rolling of leaflets along longitudinal axis. (B) Note initiation of new leaf while older leaves are dying. (C) Spores (conidia) of *Verticillium albo-atrum* produced on stems of diseased alfalfa. × 1350. Courtesy, H.C. Huang.

DISEASES AND NEMATODES

Protectant fungicide treatment of seed should reduce the amount of inoculum in a seed lot but will not be effective against internally borne fungus. Rotation of 2 or more years to a nonhost crop (e.g., corn, grasses) will also help to reduce disease losses (108, 129). The pathogen can survive in weed species, therefore a sound program of weed control reduces inoculum sources.

The disease has been controlled in Europe by resistant cultivars. Most cultivars in the USA are susceptible; however, newer cultivars are available with usable levels of resistance.

21-3.3 Bacterial Wilt

Bacterial wilt, caused by the bacterium *Clavibacter michiganense* subsp. *insidiosum* [*Corynebacterium insidiosum* (McCulloch) Jensen] (syn. *Aplanobacter insidiosum* McCull.) (158), was first described in Wisconsin (120, 125). The disease now occurs in most alfalfa growing regions of the world.

Diseased plants are stunted and recover slowly after mowing. The stems proliferate and become spindly. Leaves are slightly cupped with marginal chlorosis. Even when distinct foliar symptoms are not apparent, pale yellow to brown discoloration of the xylem tissue in the outermost part of the stele is a characteristic symptom of the disease. In many infected plants, this discoloration also occurs in spots on the inner side of the bark. This symptom distinguishes bacterial wilt from dwarf in areas such as the San Joaquin Valley of California where both diseases often occur together (106). (Fig. 21-15).

The bacterium survives in infected plant material as long as 8 to 10 yr (40) and in seed from severely infected plants (41). The bacteria enter plants through wounds such as those made by mowing or freezing and thawing (127). The disease usually does not become evident or severe until the second or third year.

Jones (122) described the plugging of xylem vessels by the bacterium and believed that during periods of rapid growth and new vessel development, a plant was more tolerant of disease effects. The plugging phenomenon was considered by Ries and Strobel (193) to be due to a glycopeptide produced by the bacterium, which acted as a phytotoxin. Dey and van Alfen (45) reported that water conductance was not blocked in stems, but in petioles and leaflet veins, was decreased 60-fold by bacterial wilt infection.

Because growth of the pathogen is limited to temperatures of <31°C, the progress of the disease is greatest at relatively low to moderate temperatures. In general, the disease is most economically important in areas with high rainfall or where the land is irrigated.

The severity and incidence of bacterial wilt is increased by other root pathogens or pests. The northern root-knot nematode, *Meloidogyne hapla*, increased the incidence of bacterial wilt in Iowa (172) and in Nevada and Utah (112). *Ditylenchus dipsaci*, the stem and bulb nematode, is a

Fig. 21-15. Plant stunted by bacterial wilt.

vector of *C. insidiosum* (91, 92, 93) and is capable of lowering the resistance of the cv. Beaver to bacterial wilt (96). The larvae of the clover root curculio (*Sitona hispidulus* F.) increased the incidence of bacterial wilt (102).

The most successful control measure is the use of resistant cultivars. Soon after the first description of the disease (120, 121) research to find resistance was initiated. Since Jones' (123) identification of resistant plants, numerous cultivars with high levels of stable resistance have been developed.

Nonleguminous hosts to *C. insidiosum* have not been identified. Therefore, control by rotation, about 3 yr out of alfalfa, reduces the inoculum in the soil. Care must be taken to destroy all infected plants and to incorporate infected debris into the soil for maximum effectiveness.

21-3.4 Dwarf

The dwarf disease was first reported as a disease of unknown etiology (235). Stunted plants occur in a random pattern that often starts at the edges of fields. Infected plants are blue-green in color and have smaller than normal leaves on fine stems, which distinguishes the disease from bacterial wilt, which induces yellowing. Dwarf causes a yellow-brown vascular necrosis in the outer wood of the tap root but does not cause yellow spots on the inner bark as often occurs with bacterial wilt. Infected plants die within 24 to 32 weeks. The disease was less severe in dry soil than when soil was frequently irrigated (236). Weimer (238) reported that the causal agent could be transmitted by grafting but not by mechanical means and considered it to be a virus. Exhaustive tests indicated that a fungus was not involved, and, though many bacteria were associated with the disease, their pathogenicity could not be proved. He concluded: "The possibility that dwarf is due to a bacterium may not be entirely excluded, however, all the evidence obtained argues against such a conclusion." Thus, he concluded, that the causal agent was a virus, which was transmissible by grafting, but not mechanically with plant juice.

Hewitt et al. (100) proved that the causal agent of dwarf (considered to be a virus) was the same as the causal agent of Pierce's disease of grape (*Vitis vinifera* L.), which had been described in 1892 (182) as the "Anaheim disease" responsible for the destruction of much of the grape industry near Anaheim, CA. The causal agent was transmitted between alfalfa and grape by sharpshooter leafhoppers (100).

Recently, the causal agent has been determined to be a gram-negative, fastidious, xylem-limited bacterium, which has not yet been named (44).

The epidemiology of the disease is undoubtedly associated with the ecology of the sharpshooter leafhopper vectors. Many plants other than grape and alfalfa are hosts of the dwarf pathogen. In the 1930s and 1940s, the disease was especially severe in southern California, but since then the disease has been found only occasionally. Currently it is not considered a serious factor in stand decline. The reason for this change in the incidence of disease is unknown, but the dwarf disease could become serious again.

Tolerance to dwarf has been identified (106), although the tolerance or resistance of current cultivars is not known.

21-4 THE ALFALFA DISEASE SITUATION—AN OVERVIEW

Losses due to diseases are a major problem for alfalfa growers worldwide, and as new areas in developing countries come into alfalfa production, disease problems can be expected to increase. Crop management has and will continue to have a key role in reducing losses from disease. Undesirable cost-benefit ratios and concern over potential residue problems have discouraged the use of fungicides on alfalfa, although the economics of their use may be changing (24). New methods of genetic manipulation could hold the answer to improving resistance to diseases that have not responded to conventional breeding methods.

Disease-resistant cultivars will continue to be the main strategy for minimizing disease losses in alfalfa, but we must be careful to continue to increase research efforts in cultural, biological, and chemical control development. Advances in these areas will enhance the benefits we derive from genetic resistance. The value of the alfalfa crop will greatly influence the interest of private industry in fungicide development. The importance of alfalfa as a cash crop is increasing, and every year more alfalfa is grown for direct sale. When alfalfa is priced according to its nutritional quality, and this day is coming, there will be more interest in controlling diseases.

Alfalfa is still the number one forage legume and a mainstay of the dairy industry, and this is not likely to change. Too much of the nation's and the world's crop is lost each year to disease. Major progress has been made in reducing disease losses over the past 60 yr. Indeed, without resistance to bacterial and Verticillium wilts, anthracnose, and Phytophthora root rot, alfalfa could not be grown in many areas around the world.

A coordinated and sustained research program at federal, state, and commercial levels is the only solution to today's disease problems in alfalfa and to those that are yet to come.

REFERENCES

1. Aguirre, R.J., R.B. Hine, and M.H. Schonhorst. 1983. Distribution of Phytophthora root rot of alfalfa in central Mexico and development of disease resistance in Mexican cultivars of alfalfa. Plant Dis. 67:91–94.
2. Alberta Agriculture. 1981. Alberta forage manual. Edmonton, AB, Canada.
3. Alliot, B., J. Giannotti, and P.A. Signoret. 1972. Mise en évidence de particules bacilliformes de virus associées à la maladie à énations de la Luzerne. C. R. Acad. Sci. Sér. D. 274:1974–1976.
4. ----, and P.A. Signoret. 1972. La maladie à énations de la Luzerne, une maladie nouvelle pour la France. Phytopathol. Z. 74:69–73.
5. Alva, A.K., L.E. Lanyon, and K.T. Leath. 1985. Excess soil water and Phytophthora root rot stresses of Phytophthora root rot sensitive and resistant alfalfa cultivars. Agron. J. 77:437–442.
6. Armstrong, G.M., and J.K. Armstrong. 1965. Further studies on the pathogenicity of three forms of *Fusarium oxysporum* causing wilt of alfalfa. Plant Dis. Rep. 49:412–416.
7. Aubé, C., and W.E. Sackston. 1964. Verticillium wilt of forage legumes in Canada. Can. J. Plant Sci. 44:427–432.
8. Aycock, R. 1966. Stem rot and other diseases caused by *Sclerotium rolfsii*. N. C. Agric. Exp. Stn. Tech. Bull. 174.
9. Azizi, M.R., and D.K. Barnes. 1977. Characterization and inheritance of a spotted leaf trait in alfalfa. Crop Sci. 17:126–132.
10. Banttari, E.E., and R.D. Wilcoxson. 1964. Effect of pea aphids on spring black stem of alfalfa. Phytopathology 54:1415–1417.
11. Barnes, D.K., S.A. Ostazeski, J.A. Schillinger, and C.H. Hanson. 1969. Effect of anthracnose (*Colletotrichum trifolii*) infection on yield, stand, and vigor of alfalfa. Crop Sci. 9:344–346.
12. Barta, A.L. 1980. Regrowth and alcohol dehydrogenase activity in water logged alfalfa and birdsfoot trefoil. Agron. J. 72:1017–1020.
13. ----. 1984. Ethanol synthesis and loss from flooded roots of *Medicago sativa* L. and *Lotus corniculatus* L. Plant, Cell Environ. 7:187–191.
14. Baxter, J.W. 1956. Cercospora black stem of alfalfa. Phytopathology 46:398–400.
15. Berger, R.D., and E.W. Hanson. 1963. Pathogenicity, host-parasite relationships, and morphology of some forage legume Cercosporae, and factors related to disease development. Phytopathology 53:500–508.
16. Blackstock, J. McK. 1978. Lucerne transient streak and lucerne latent, two new viruses of lucerne. Aust. J. Agric. Res. 29:291–304.
17. Boerema, G.H., M.M.J. Dorenbosch, and L. Leffring. 1965. A comparative study of black stem fungi on lucerne and red clover and the footrot fungus on pea. Neth. J. Plant Pathol. 71:79–89.
18. Boelter, R.H., F.A. Gray, and R.H. Delaney. 1985. Effect of *Ditylenchus dipsaci* on alfalfa mortality, winterkill, and yield. J. Nematol. 17:140–144.

19. Bolton, J.L. 1962. Alfalfa, botany cultivation, and utilization. Interscience Publishers, New York.
20. Brennan, E., I.A. Leone, and P.M. Halisky. 1969. Response of forage legumes to ozone fumigations. Phytopathology 59:1458–1459.
21. Brewer, R.F. 1982. The effects of ozone and sulfur dioxide on alfalfa yields and hay quality. University of California Rep. ARB Agreement A1-038-33. Parlier, CA.
22. Brigham, R.D. 1957. Stem lesions associated with *Xanthomonas alfalfae*. Phytopathology 47:309–310.
23. Broadfoot, W.C., and M.W. Cormack. 1941. A low-temperature basidiomycete causing early spring killing of grasses and legumes in Alberta. Phytopathology 31:1058–1061.
24. Broscious, S.C., and H.W. Kirby. 1985. Fungicides and application timing in relation to leaf spot disease and yield of alfalfa. Phytopathology 75:1346.
25. Buchholtz, W.F. 1942. Influence of cultural factors on alfalfa seedling infection by *Pythium debaryanum* Hesse. Iowa Agric. Exp. Stn. Res. Bull. 296:571–592.
26. Bushong, J.W., and J.W. Gerdemann. 1959. Root rot of alfalfa caused by *Phytophthora cryptogea* in Illinois. Plant Dis. Rep. 43:1178–1183.
27. Carroll, R.B., E.R. Jones, and R.H. Swain. 1977. Winter survival of *Colletotrichum trifolii* in Delaware. Plant Dis. Rep. 61:12–15.
28. Chapman, R.A. 1954. Meadow nematodes associated with failure of spring-sown alfalfa. Phytopathology 44:542–545.
29. Chi, C.C. 1966. Phytophthora root rot of alfalfa in Canada. Plant Dis. Rep. 50:451–453.
30. ----, W.R. Childers, and E.W. Hanson. 1964. Penetration and subsequent development of three *Fusarium* species in alfalfa and red clover. Phytopathology 54:434–437.
31. Christen, A.A. 1982. Demonstration of *Verticillium albo-atrum* within alfalfa seed. Phytopathology 72:412–414.
32. ----. 1983. Incidence of external seedborne *Verticillium albo-atrum* in commercial seed lots of alfalfa. Plant Dis. 67:17–18.
33. ----, and R.N. Peaden. 1981. Verticillium wilt in alfalfa. Plant Dis. 65:319–321.
34. Claflin, L.E., D.L. Stuteville, and D.V. Armbrust. 1973. Wind-blown soil in the epidemiology of bacterial leaf spot of alfalfa and common blight of bean. Phytopathology 63:1417–1419.
35. Close, R.C. 1967. Diseases of lucerne in New Zealand. p. 248–256. *In* R.H.M. Langer (ed.) The lucerne crop. Reed, Wellington, New Zealand.
36. Cook, A.A., and A.C. Wilton. 1985. Witches' broom disease of alfalfa in Saudi Arabia. Plant Dis. 69:83.
37. Cormack, M.W. 1934. On the invasion of roots of *Medicago* and *Melilotus* by *Sclerotina* sp. and *Plenodomus meliloti* D. and S. Can. J. Res. 11:474–480.
38. ----. 1937. *Cylindrocarpon ehrenbergi* Wr. and other species, as root parasites of alfalfa and sweetclover in Alberta. Can. J. Res. Sect. C: 15:403–424.
39. ----. 1948. Winter crown rot or snow mold of alfalfa, clovers, and grasses in Alberta. I. Occurrence, parasitism, and spread of the pathogen. Can. J. Res. Sect. C: 26:71–85.
40. ----. 1961. Longevity of the bacterial wilt organism in alfalfa hay, pod debris, and seed. Phytopathology 51:260–261.
41. ----, and J.F. Moffatt. 1956. Occurrence of the bacterial wilt organism in alfalfa seed. Phytopathology 46:407–409.
42. Cowling, W.A., D.G. Gilchrist, and J.H. Graham. 1981. Biotypes of *Stemphylium botryosum* on alfalfa in North America. Phytopathology 71:679–684.
43. Crill, P., D.J. Hagedorn, and E.W. Hanson. 1970. Alfalfa mosaic, the disease and its virus incitant. University of Wisconsin Res. Bull. 280.
44. Davis, M.J., A.H. Purcell, and S.V. Thomson. 1978. Pierce's disease of grapevines: Isolation of the causal bacterium. Science 199:75–77.
45. Dey, R., and N.K. van Alfen. 1979. Influence of *Corynebacterium insidiosum* on water relations of alfalfa. Phytopathology 69:942–946.
46. Dickason, E.A., C.M. Leach, and A.E. Gross. 1968. Clover root curculio injury and vascular decay of alfalfa roots. J. Econ. Entomol. 61:1163–1168.
47. Dudley, J.W., and C.P. Wilsie. 1956. Inheritance of branched raceme and vestigial flower in alfalfa. Agron. J. 49:320–323.
48. Elgin, J.H. et al. (ed.). 1984. Standard tests to characterize pest resistance in alfalfa cultivars. USDA Misc. Pub. 1434. U.S. Government Printing Office, Washington, DC.
49. Elliot, A.M., and R.D. Wilcoxson. 1964. Effect of temperature and moisture on formation and ejection of ascospores and on survival of *Leptosphaerulina briosiana*. Phytopathology 54:1443–1447.
50. Emberger, G., and R.E. Weltry. 1983. Effect of soil water matric potential on resistance to *Fusarium oxysporum* f. sp. *medicaginis* in alfalfa. Phytopathology 73:208–212.
51. ----, and ----. 1983. Evaluation of virulence of *Fusarium oxysporum* f. sp. *medicaginis* and Fusarium wilt resistance in alfalfa. Plant Dis. 67:94–98.

52. Erwin, D.C. 1954. Relation of *Stagonospora, Rhizoctonia*, and associated fungi to crown rot of alfalfa. Phytopathology 44:137–144.
53. ———. 1954. Root rot of alfalfa caused by *Phytophthora cryptogea*. Phytopathology 44:700–704.
54. ———. 1965. Reclassification of the causal agent of root rot of alfalfa from *Phytophthora cryptogea* to *P. megasperma*. Phytopathology 55:1139–1143.
55. ———. 1966. Varietal reaction of alfalfa to *Phytophthora megasperma* and variation in virulence of the causal fungus. Phytopathology 56:653–657.
56. ———, and B.W. Kennedy. 1957. Studies on Phytophthora root rot of alfalfa. Phytopathology 47:520.
57. ———, ———, and W.F. Lehman. 1959. Xylem necrosis and root rot of alfalfa associated with excessive irrigation and high temperatures. Phytopathology 49:572–578.
58. Falloon, R.E., and R.A. Skipp. 1982. Fungicide seed treatments improve lucerne establishment. Proc. N. Z. Weed Pest Contr. Conf. 35:127–129.
59. Faris, M.A. 1985. Variability and interaction between alfalfa cultivars and isolates of *Phytophthora megasperma*. Phytopathology 75:390–394.
60. Faulkner, L.R., and O.J. Bolander. 1970. Acquisition and distribution of nematodes in waterways of the Columbia Basin in eastern Washington. J. Nematol. 2:362–367.
61. Frosheiser, F.I. 1969. Phytophthora root rot of alfalfa in the upper Midwest. Plant Dis. Rep. 53:595–597.
62. ———. 1969. Variable influence of alfalfa mosaic virus strains on growth and survival of alfalfa and on mechanical and aphid transmission. Phytopathology 59:857–862.
63. ———. 1970. Virus-infected seeds in alfalfa seed lots. Plant Dis. Rep. 54:591–594.
64. ———. 1980. Conquering Phytophthora root rot with resistant alfalfa cultivars. Plant Dis. 64:909–912.
65. ———, and D.K. Barnes. 1973. Field and greenhouse selection for Phytophthora root rot resistance in alfalfa. Crop Sci. 13:735–738.
66. ———, and ———. 1978. Field reaction of artificially inoculated alfalfa populations to the Fusarium and bacterial wilt pathogens alone and in combination. Phytopathology 68:943–946.
67. Gaudet, D.A., D.C. Sands, D.C. Mathre, and R.L. Ditterline. 1980. The role of bacteria in the root and crown rot complex of irrigated sanfoin in Montana. Phytopathology 70:161–167.
68. Gerdemann, J.W. 1954. Pathogenicity of *Leptodiscus terrestris* on red clover and other Leguminosae. Phytopathology 44:451–455.
69. Gilbert, A.H., and C.W. Bennett. 1917. *Sclerotinia trifoliorum*, the cause of stem rot of clovers and alfalfa. Phytopathology 7:432–442.
70. Gisi, U. 1983. Biophysical aspects of the development of Phytophthora. p. 109–119. *In* D.C. Erwin et al. (ed.) Phytophthora: Its biology, taxonomy, ecology, and pathology. American Phytopathological Society, St. Paul.
71. Golden, A.M., J.H. O'Bannon, G.S. Santo, and A.M. Finley. 1980. Description and SEM observation of *Meloidogyne chitwoodi* n. sp. (Meloidogynidae) a root-knot nematode on potato in the Pacific Northwest. J. Nematol. 12:319–327.
72. Graham, J.H., K.W. Kreitlow, and L.R. Faulkner. 1972. Diseases. *In* C.H. Hanson (ed.) Alfalfa science and technology. Agronomy 15:497–526.
73. ———, F.I. Frosheiser, D.L. Stuteville, and D.C. Erwin. 1979. A compendium of alfalfa diseases. American Phytopathological Society, St. Paul.
74. ———, R.N. Peaden, and D.W. Evans. 1977. Verticillium wilt of alfalfa found in the United States. Plant Dis. Rep. 61:337–340.
75. Grau, C.R., P.A. Delwiche, R.L. Norgren, T.E. O'Connell, and D.P. Maxwell. 1981. Verticillium wilt of alfalfa in Wisconsin. Plant Dis. 65:843–844.
76. Gray, F.A., W.H. Bohl, and R.H. Abernethy. 1983. Phytophthora root rot of alfalfa in Wyoming. Plant Dis. 67:291–294.
77. ———, R.L. Haaland, E.M. Clark, and D.M. Ball. 1980. Diseases of alfalfa in Alabama. Plant Dis. 64:1015–1017.
78. ———, and R.B. Hine. 1976. Development of Phytophthora root rot of alfalfa in the field and the association of *Rhizobium* nodules with early root infections. Phytopathology 66:1413–1417.
79. ———, and D.A. Roth. 1982. Verticillium wilt of alfalfa in Wyoming. Plant Dis. 66:1080.
80. Green, C.D. 1981. The effect of weeds and wild plants on the reinfestation of land by *Ditylenchus dipsaci*. p. 217–224. *In* J.M. Thresh (ed.) Pests, pathogens, and vegetation. Pitman, London.
81. Griffin, G.D. 1968. The pathogenicity of *Ditylenchus dipsaci* to alfalfa and the relationship of temperature to plant infection and susceptibility. Phytopathology 50:929–932.
82. ———. 1980. The interrelationships of *Meloidogyne hapla* and *Ditylenchus dipsaci* on resistant and susceptible alfalfa. J. Nematol. 27:287–293.

83. ----, and O.J. Hunt. 1972. Effect of temperature and inoculation timing on the *Meloidogyne hapla/Cornybacterium insidiosum* complex in alfalfa. J. Nematol. 4:70-71.
84. ----, and B.D. Thyr. 1978. Interaction of *Meloidogyne hapla* and *Fusarium oxysporum* on alfalfa. J. Nematol. 10:289.
85. Grove, A.R., and G.E. Carlson. 1972. Morphology and anatomy. *In* C.H. Hanson (ed.) Alfalfa science and technology. Agronomy 15:103-122.
86. Hancock, J.G. 1983. Seedling diseases of alfalfa in California. Plant Dis. 67:1203-1208.
87. Hansen, E.W. 1963. Violet root rot of alfalfa in Wisconsin. Plant Dis. Rep.47:306-307.
88. Hawksworth, D.L., and P.W. Talboys. 1970. *Verticillium albo-atrum*. Commonw. Mycol. Inst. Descr. Pathog. Fungi Bact. no. 255. Commonwealth Mycological Institute, Kew, Surrey, UK.
89. Hawn, E.J. 1958. Studies on the epidemiology of crown bud rot of alfalfa in southern Alberta. Can. J. Bot. 36:239-250.
90. ----. 1959. Histological study on crown bud rot of alfalfa. Can. J. Bot. 37:1247-1249.
91. ----. 1963. Transmission of bacterial wilt of alfalfa by *Ditylenchus dipsaci* (Kühn). Nematologica 9:65-68.
92. ----. 1970. New technique for studying a bacterium/nematode interaction in alfalfa. J. Nematol. 2:272-273.
93. ----. 1971. Mode of transmission of *Corynebacterium insidiosum* by *Ditylenchus dipsaci*. J. Nematol. 3:420-421.
94. ----, W.B. Berkenkamp, and J.B. LaBeau. 1981. Evaluation of losses in alfalfa hay production caused by crown rot. Can. J. Plant Pathol. 3:103-105.
95. ----, and M.W. Cormack. 1952. Crown bud rot of alfalfa. Phytopathology 42:510-511.
96. ----, and M.R. Hanna. 1967. Influence of stem nematode infestation on bacterial wilt reaction and forage yield of alfalfa varieties. Can. J. Plant Sci. 47:203-208.
97. Hedlund, T. 1923. Om nagra sjukdomar och skador pa vara lantbruksvaxtee, 1922. Sver. Allm. Jordbrukstidskr. 5:166-168.
98. Helms, K. 1957. Witches' broom disease of lucerne. II. Field studies on factors influencing symptom expression, disease incidence, and mortality rate. Aust. J. Agric. Res. 8:148-161.
99. ----. 1958. Witches' broom disease of lucerne. III. A greenhouse study on symptom expression and growth in relation to frequency of watering and cutting. Aust. J. Agric. Res. 9:319-327.
100. Hewitt, W.B., B.R. Houston, N.W. Frazier, and J.H. Freitag. 1946. Leafhopper transmission of the virus causing Pierce's disease of grape and dwarf of alfalfa. Phytopathology 36:117-128.
101. Hill, A.C., M.R. Pack, M. Treshow, R.J. Downs, and L.G. Transtrum. 1961. Plant injury induced by ozone. Phytopathology 51:356-363.
102. Hill, R.R., Jr., J.J. Murray, and K.E. Zeiders. 1971. Relationship between clover root curculio injury and severity of bacterial wilt in alfalfa. Crop Sci. 11:306-307.
103. Hitchcock, A.E., D.C. McClune, L.H. Weinstein, D.C. MacLean, J.S. Jacobson, and R.H. Mandl. 1971. Effect of hydrogen fluoride fumigation on alfalfa and orchardgrass: A summary of experiments from 1952 through 1965. Contrib. Boyce Thompson Inst. 24:363-386.
104. Holub, E.B., and C.R. Grau. 1985. Studies on *Aphanomyces euteiches* as a pathogen of alfalfa. Cent. Alf. Improve. Conf. Proc. 19:36.
105. Horsfall, J.G. 1940. A study of meadow-crop diseases in New York. Cornell Univ. Memo. 130.
106. Houston, B.R., D.C. Erwin, E.H. Stanford, M.W. Allen, D.H. Hall, and A.O. Paulus. 1960. Diseases of alfalfa in California. Calif. Agric. Exp. Stn. Cir. 485.
107. Howell, R.K., T.E. Devine, and C.H. Hanson. 1971. Resistance of selected alfalfa strains to ozone. Crop Sci. 11:114-115.
108. Huang, H.C., and T.G. Atkinson (ed.). 1983. Verticillium wilt of alfalfa. Agriculture Canada Research Branch Contribution 1982-8E Revised. Agriculture Canada, Ottawa, Canada.
109. ----, and A.M. Harper. 1985. Survival of *Verticillium albo-atrum* from alfalfa in feces of leaf chewing insects. Phytopathology 75:206-208.
110. ----, ----, E.G. Kokko, and R.J. Howard. 1983. Aphid transmission of *Verticillium albo-atrum* to alfalfa. Can. J. Plant Pathol. 5:141-147.
111. ----, and K.W. Richards. 1983. *Verticillium albo-atrum* contamination of leaf pieces forming cells for the alfalfa leaf-cutter bee. Can. J. Plant Pathol. 5:248-250.
112. Hunt, O.J., G.D. Griffin, J.J. Murray, M.W. Pedersen, and R.N. Peaden. 1971. The effects of root knot nematodes on bacterial wilt in alfalfa. Phytopathology 61:256-259.

113. Irwin, J.A.G. 1974. Reaction of lucerne cultivars to *Phytophthora megasperma* the cause of a root rot in Queensland. Aust. J. Exp. Agric. Anim. Husb. 14:561–565.
114. ——. 1984. Etiology of a new *Stemphylium*-incited leaf disease of alfalfa in Australia. Plant Dis. 68:531–532.
115. ——, and J.L. Dale. 1982. Relationships between *Phytophthora megasperma* isolates from chickpea, lucerne and soybean. Aust. J. Bot. 30:199–210.
116. Isaac, I. 1957. Wilt of lucerne caused by species of *Verticillium*. Ann. Appl. Biol. 45:550–558.
117. Johnson, E.M., and W.D. Valleau. 1933. Black stem of alfalfa, red clover, and sweet clover. Ky. Agric. Exp. Stn. Bull. 339.
118. Jones, F.R. 1918. Yellow-leafblotch of alfalfa caused by the fungus *Pyrenopeziza medicaginis*. J. Agric. Res. 13:307–330.
119. ——. 1919. The leaf-spot diseases of alfalfa and red clover caused by the fungi *Pseudopeziza medicaginis* and *Pseudopeziza trifolii*, respectively. USDA Bull. 759. U.S. Government Printing Office, Washington, DC.
120. ——. 1925a. A new bacterial disease of alfalfa. Phytopathology 15:243–244.
121. ——. 1925b. A new bacterial disease of alfalfa. Plant Dis. Rep. 9:28–29.
122. ——. 1928. Development of the bacteria causing wilt in the alfalfa plant as influenced by growth and winter injury. J. Agric. Res. 37:545–569.
123. ——. 1934. Testing alfalfa for resistance to bacterial wilt. J. Agric. Res. 48:1085–1098.
124. ——, and C. Drechsler. 1920. Crown wart of alfalfa caused by *Urophlyctis alfalfae*. Phytopathology 10:295–323.
125. ——, and L. McCulloch. 1926. A bacterial wilt and root rot of alfalfa caused by *Aplanobacter insidiosum* L. McC. J. Agric. Res. 33:493–521.
126. ——, and J.H. Torrie. 1946. Systemic infection of downy mildew in soybean and alfalfa. Phytopathology 36:1057–1059.
127. ——, and J.L. Weimer. 1928. Bacterial wilt and winter injury of alfalfa. USDA Circ. 39. U.S. Government Printing Office, Washington, DC.
128. ——, and ——. 1938. Stagonospora leaf spot and root rot of forage legumes. J. Agric. Res. 57:791–812.
129. Keinath, A.P., and R.L. Millar. 1986. Persistence of an alfalfa strain of *Verticillium albo-atrum* in soil. Phytopathology 76:576–581.
130. Kreitlow, K.W., J.H. Graham, and R.J. Garber. 1953. Diseases of forage grasses and legumes in the northeastern States. Pa. Agric. Exp. Stn. Bull. 573.
131. ——, and R.A. Kilpatrick. 1967. A physiogenic leaf and stem spot of some forage legumes. Plant Dis. Rep. 51:619–622.
132. Kuan, T.-L., and D.C. Erwin. 1980. Formae speciales differentiation of *Phytophthora megasperma* isolates from soybean and alfalfa. Phytopathology 70:333–338.
133. Leach, C.M. 1959. A survey of root deterioration of *Medicago sativa* in Oregon. Plant Dis. Rep. 43:622–625.
134. Leach, L.D. 1947. Growth rates of host and pathogen as factors determining the severity of preemergence damping-off. J. Agric. Res. 75:161–179.
135. Leath, K.T. 1978. Crown wart of alfalfa in Pennsylvania. Plant Dis. Rep. 62:621–623.
136. ——. 1981. Pest management systems for alfalfa diseases. p. 293–302. *In* D. Pimentel (ed.) Handbook of pest management in agriculture, Vol. 3. CRC Press, Boca Raton, FL.
137. ——. 1982. Verticillium wilt in Pennsylvania. p. 35. *In* Proc. 28th Improve. Conf., Davis, CA. 12–16 July. Alfalfa Improvement Conference, Davis, CA.
138. ——. 1983. Minimizing disease losses in forage crops through management. p. 579–581. *In* J.A. Smith and V.W. Hays (ed.) Proc. 14th Int. Grassl. Congr., Lexington, KY. 15–24 June 1981. Westview Press, Boulder, CO.
139. ——. 1985. The crown and root rot complex of forage legumes—problems and progress. Proc. Am. Forage Grassl. Conf. p. 109–112.
140. ——, and R.R. Hill, Jr. 1974. *Leptosphaerulina briosiana* on alfalfa: Relation of lesion size to leaf age and light intensity. Phytopathology 64:243–245.
141. ——, and W.A. Kendall. 1978. Fusarium root rot of forage species: Pathogenicity and host range. Phytopathology 68:826–831.
142. ——, and ——. 1983. *Myrothecium roridum* and *M. verrucaria* pathogenic to roots of red clover and alfalfa. Plant Dis. 67:1154–1155.
143. ——, F.L. Lukezic, H.W. Crittenden, E.S. Elliott, P.M. Halisky, F.L. Howard, and S.A. Ostazeski. 1971. The fusarium root rot complex of selected forage legumes in the northeast. Pa. State Univ. Bull. 777.
144. ——, and R.H. Ratcliffe. 1974. The effect of fertilization on disease and insect resistance. p. 481–503. *In* D.A. Mays (ed.) Forage fertilization. American Society of Agronomy, Crop Science Society of America, and Soil Science Society of America, Madison, WI.
145. LeBeau, J.B., and J.G. Atkinson. 1967. Borax inhibition of cyanogenesis in snow mold of alfalfa. Phytopathology 57:863–865.

146. ----, and J.G. Dickson. 1955. Physiology and nature of disease development in winter crown rot of alfalfa. Phytopathology 45:667–673.
147. LeClant, F., B. Alliot, and P.A. Signoret. 1973. Transmission et épidémiologie de la maladie à énations de la Luzerne (Lev.) Premiers résultats. Ann. Phytopathol. 5:441–445.
148. Lehman, W.F., S.J. Richards, D.C. Erwin, and A.W. Marsh. 1968. Effect of irrigation treatments on alfalfa (*Medicago sativa* L.) production, persistence, and soil salinity in southern California. Hilgardia 39:277–295.
149. Lukezic, F.L. 1974. Dissemination and survival of *Colletotrichum trifolii* under field conditions. Phytopathology 64:57–59.
150. ----, D.C. Hildebrand, M.N. Schroth, and P.A. Shinde. 1982. Association of *Serratia marcenscens* with crown rot of alfalfa in Pennsylvania. Phytopathology 72:714–718.
151. ----, K.T. Leath, and R.G. Levine. 1983. *Pseudomonas viridiflava* associated with root and crown rot of alfalfa and wilt of birdsfoot trefoil. Plant Dis. 67:808–811.
152. Lyda, S.D. 1978. Ecology of *Phymatotrichum omnivorum*. Annu. Rev. Phytopathol. 16:193–209.
153. ----, and E. Burnett. 1971. Influence of temperature on *Phymatotrichum* sclerotial formation and disease development. Phytopathology 61:728–730.
154. Mainer, A., and K.T. Leath. 1978. Foliar diseases alter carbohydrate and protein levels in leaves of alfalfa and orchardgrass. Phytopathology 68:1252–1255.
155. Matens, J.W., W.L. Seaman, and T.G. Atkinson (ed.) 1984. Diseases of field crops in Canada. Canadian Phytopathological Society, Harrow, ON, Canada.
156. Matsumoto, N., and T. Araki. 1978. Alfalfa root rot caused by *Phytophthora megasperma* Drechsler in Japan. Nippon Skokubutsu Byori Gakkaiho 44:214–217.
157. Mauza, B.E., and J.M. Webster. 1982. Suppression of alfalfa growth by concomitant populations of *Pratylenchus penetrans* and two *Fusarium* species. J. Nematol. 14:364–367.
158. McCulloch, L. 1925. *Aplanobacter insidiosum* N. Sp. The cause of an alfalfa disease. Phytopathology 15:496–497.
159. McDonald, W.C. 1955. The distribution and pathogenicity of the fungi associated with crown and root rotting of alfalfa in Manitoba. Can. J. Agric. Sci. 35:309–321.
160. McGuire, J.M., H.J. Walters, and D.A. Slack 1958. The relationship of root-knot nematodes to the development of Fusarium wilt in alfalfa. Phytopathology 48:344.
161. McKeen, W.E., and J.A. Traquair. 1980. *Aphanomyces* sp., and alfalfa pathogen in Ontario. Can. J. Plant Pathol. 2:42–44.
162. Mead, H.W. 1964. Resumé of data on black stem of alfalfa caused by *Ascochyta imperfecta* Peck. Can. Plant Dis. Surv. 44:134–141.
163. Menzies, J.D. 1946. Witches' broom of alfalfa in North America. Phytopathology 36:762–774.
164. Michaud, R., and C. Richard. 1985. Evaluation of alfalfa cultivars for reaction to crown and root rot. Can. J. Plant Sci. 65:95–98.
165. ----, ----, and C. Gagnon. 1984. Progress in breeding for resistance to crown and root rot in alfalfa. p. 26. *In* Proc. 29th Alfalfa Improve. Conf., Lethbridge, Alberta, Canada. 16–20 July. Alfalfa Improvement Conference, Lethbridge.
166. Moffett, M.L., and J.A.G. Irwin. 1975. Bacterial leaf and stem spot (*Xanthomonas alfalfae*) of lucerne in Queensland. Aust. J. Exp. Agric. Anim. Husb. 15:223–226.
167. Morgan, W.C., and D.G. Parberry. 1977. Effects of Pseudopeziza leaf spot disease on growth and yield in lucerne. Aust. J. Agric. Res. 28:1029–1040.
168. Mueller, W.C. 1965. Progressive incidence of alfalfa mosaic virus in alfalfa fields. Phytopathology 55:1069.
169. Noel, G.R., and B.F. Lownsbery. 1978. Effect of temperature on the pathogenicity of *Tylenchorhynchus clarus* to alfalfa and observations on feeding. J. Nematol. 10:195–198.
170. Noling, J.W., and H. Ferris. 1985. Influence of *Meloidogyne hapla* on alfalfa yield and host population dynamics. J. Nematol. 17:415–421.
171. Norton, D.C. 1963. Population fluctuations of *Xiphinema americanum* in Iowa. Phytopathology 53:66–68.
172. ----. 1969. *Meloidogyne hapla* as a factor in alfalfa decline in Iowa. Phytopathology 59:1824–1828.
173. Ooka, J.J., and J.Y. Uchida. 1982. Cylindrocladium root and crown rot of alfalfa in Hawaii. Plant Dis. 66:947–948.
174. O'Rourke, C.J., and R.L. Millar. 1966. Root rot and root microflora of alfalfa as affected by potassium nutrition, frequency of cutting, and leaf infection. Phytopathology 56:1040–1046.
175. Ostazeski, S.A. 1967. An undescribed fungus associated with a root and crown rot of birdsfoot trefoil (*Lotus corniculatus*). Mycologia 59:970–975.
176. ----, J.E. Elgin, and J.E. McMurtrey. 1979. Occurrence of anthracnose on formerly anthracnose-resistant 'Arc' alfalfa. Plant Dis. Rep. 63:734–736.

177. Paliwal, Y.C. 1982. Lucerne transient streak—a virus of alfalfa newly recognized in North America. Phytopathology 72:989.
178. Peaden, R.N., and A.A. Christen. 1984. A guide for identification of Verticillium wilt in alfalfa. USDA Inform. Bull. 456. U.S. Government Printing Office, Washington, DC.
179. Pennypacker, B.W., and K.T. Leath. 1983. Dispersal of *Verticillium albo-atrum* in the xylem of alfalfa. Plant Dis. 67:1226-1229.
180. ———, and ———. 1986. Anatomical response of a susceptible alfalfa clone infected with *Verticillium albo-atrum*. Phytopathology 76:522-527.
181. ———, ———, and R.R. Hill, Jr. 1984. Resistant alfalfa plants as symptomless carriers of *Verticillium albo-atrum*. Phytopathology 74:855.
182. Pierce, N.B. 1892. The California vine disease. USDA Div. Veg. Pathol. Bull. 2. U.S. Government Printing Office, Washington, DC.
183. Prior, G.D., and J.H. Owen. 1964. Pathological anatomy of *Sclerotinia trifoliorum* on clover and alfalfa. Phytopathology 54:784-787.
184. Pulli, S.K., and M.B. Tesar. 1975. Phytophthora root rot in seedling-year alfalfa as affected by management practices inducing stress. Crop Sci. 15:861-864.
185. Purss, G.S. 1959. Root rot of lucerne. Queensl. Agric. J. 85:767-770.
186. ———. 1965. Diseases of lucerne. Queensl. Div. Plant Indus. Leafl. 812.
187. Redhead, S.A., and J.A. Traquair. 1981. *Coprinus* sect. herbicolae from Canada, notes on extra limital taxa, and the taxonomic position of a low temperature basisiomycete forage crop pathogen from Western Canada. Mycotaxon 13:374-404.
188. Reynolds, H.W., and J.H. O'Bannon. 1960. Reaction of sixteen varieties of alfalfa to two species of root-knot nematodes. Plant Dis. Rep. 44:441-443.
189. Rhodes, L.H. 1986. Severity of spring black stem on alfalfa cultivars in Ohio. Plant Dis. 70:746-748.
190. Ribeiro, O.K., D.C. Erwin, and R.A. Khan. 1978. A new high temperature *Phytophthora* pathogenic to roots of alfalfa. Phytopathology 68:155-161.
191. Richard, C., R. Michaud, A. Frève, and C. Gagnon. 1980. Selection for root and crown rot resistance in alfalfa. Crop Sci. 20:691-695.
192. Richter, H., and M. Klinkowski. 1938. Wirtelpilz welkkrankheit an Luzerne und Espanette (Erreger: *Verticillium albo-atrum* R&B). Nachrichtenbl. Dtsch. Pflanzenschutzdienst (Berlin) 18:57-58.
193. Ries, S.M., and G.A. Strobel. 1972. Biological properties and pathological role of a phytotoxic glycopeptide from *Corynebacterium insidiosum*. Physiol. Plant Pathol. 2:133-142.
194. Riker, A.J., F.R. Jones, and M.C. Davis. 1935. Bacterial leaf spot of alfalfa. J. Agric. Res. 51:177-182.
195. Roberts, D.A., R.E. Ford, C.H. Ward, and D.T. Smith. 1959. Colletotrichum anthracnose of alfalfa in New York. Plant Dis. Rep. 43:352.
196. Sackett, W.G. 1910. A bacterial disease of alfalfa. Colo. Agric. Exp. Stn. Bull. 158.
197. Sanford, G.B. 1933. A root rot of sweet clover and related crops caused by *Plenodomus meliloti* Dearness and Sanford. Can. J. Res. 8:337-348.
198. Schmitthenner, A.F. 1964. Prevalence and virulence of *Phytophthora*, *Aphanomyces*, *Pythium*, *Rhizoctonia*, and *Fusarium* isolated from diseased alfalfa seedlings. Phytopathology 54:1012-1018.
199. Scott, C.E. 1920. A preliminary note on the germination of *Urophylctis alfalfae*. Science N.S. 52:225-226.
200. Seif El-Nasr, H.I., and K.T. Leath. 1983. Crown and root fungal diseases of alfalfa in Egypt. Plant Dis. 67:509-511.
201. ———, and ———. 1983. Foliar diseases of alfalfa in Egypt. Plant Dis. 67:691-695.
202. Semeniuk, G. 1984. Common leafspot of alfalfa. Ascospore discharge and plant infection in the field. Phytopathol. Z. 110:290-300.
203. Sheppard, J.W. 1979. Verticillium wilt, a potentially dangerous disease of alfalfa in Canada. Can. Plant Dis. Surv. 59:3.
204. ———, and S.N. Needham. 1980. Verticillium wilt of alfalfa in Canada: Occurrence of seed-borne inoculum. Can. J. Plant Pathol. 2:159-162.
205. Smith, H.C. 1965. The morphology of *Verticillium albo-atrum*, *V. dahliae*, and *V. tricorpus*. N. Z. J. Agric. Res. 8:450-478.
206. Smith, O.F. 1940. Stemphylium leaf spot of red clover and alfalfa. J. Agric. Res. 61:831-846.
207. ———. 1943. Rhizoctonia root canker of alfalfa (*Medicago sativa*). Phytopathology 33:1081-1085.
208. ———. 1945. Parasitism of *Rhizoctonia solani* from alfalfa. Phytopathology 35:832-837.
209. ———. 1946. Effect of soil temperature on the development of Rhizoctonia root canker of alfalfa. Phytopathology 36:638-642.

210. Sparrow, F.K. 1973. Chytridiomycetes, Hyphochytridiomycetes. p. 85–110. *In* G.C. Ainsworth et al. (ed.) The fungi: An advanced treatise. Vol. IVB. A taxonomic review with keys: Basidiomycetes and lower fungi. Academic Press, New York.
211. Stack, J.P., and R.L. Millar. 1985. Relative survival potential of propagules of *Phytophthora megasperma* f. sp. *medicaginis*. Phytopathology 75:1398–1404.
212. Stelfox, D., and J.R. Williams. 1980. *Pythium* species in alfalfa in central Alberta. Can. Plant Dis. Surv. 60:4–35, 36.
213. Stephen, R.C., D.J. Saville, I.C. Harvey, and J. Hedley. 1982. Herbage yields and persistence of lucerne (*Medicago sativa* L.) cultivars and the incidence of crown and root disease. N. Z. J. Exp. Agric. 10:323–332.
214. Streets, R.B., and H.E. Bloss. 1973. Phymatotrichum root rot. Monogr. no. 8. American Phytopathological Society, St. Paul.
215. Stuteville, D.L. 1984. Downy mildew resistance. Laboratory methods. p. 23–25. *In* J.E. Elgin et al. (ed.) Standard tests to characterize pest resistance in alfalfa cultivars. USDA Misc. Publ. 1434. U.S. Government Printing Office, Washington, DC.
216. ----, and E.L. Sorensen. 1966. Distribution of leaf spot and damping-off (*Xanthomonas alfalfae*) of alfalfa in Kansas, and new hosts. Plant Dis. Rep. 50:731–734.
217. Stutz, J.C., K.T. Leath, and W.A. Kendall. 1985. Wound-related modifications of *Fusarium roseum* penetration, development and root rot in forage legumes. Phytopathology 75:920–924.
218. Summers, C.G., and W.D. McClellan. 1975. Effect of common leaf spot on yield and quality of alfalfa in the San Joaquin Valley of California. Plant Dis. Rep. 59:504–506.
219. Sundheim, L., and R.D. Wilcoxson. 1965. *Leptosphaerulina briosiana* on alfalfa: Infection and disease development, host-parasite relationships, ascospore germination and dissemination. Phytopathology 55:546–553.
220. Thompson, C.R., G. Kats, E.L. Pippen, and W.H. Isom. 1976. Effect of photochemical air pollution on two varieties of alfalfa. Environ. Sci. Technol. 10:1237–1241.
221. Thyr, B.C. 1985. Personal communciation. USDA-ARS. College of Agriculture, Reno, NV.
222. Townshend, J.L. 1984. Inoculum densities of 5 plant parasitic nematodes in relation to alfalfa seedling growth. Can. J. Plant Pathol. 6:309–312.
223. ----, and L. Stobbs. 1981. Histopathology and histochemistry of lesions caused by *Pratylenchus penetrans* in roots of forage legumes. Can. J. Plant Pathol. 3:123–128.
224. Traquair, J.A. 1980. Conspecificity of an unidentified snow mold basidiomycete and a *Coprinus* species in the section *Herbicolae*. Can. J. Plant Pathol. 2:105–184.
225. ----, and E.J. Hawn. 1982. Pathogenicity of *Coprinus psychromorbidus* on alfalfa. Can. J. Plant Pathol. 4:106–108.
226. Tu, J.C., and T.M. Holmes. 1980. Effect of alfalfa mosaic virus infection on nodulation, forage yield, forage protein, and overwintering of alfalfa. Phytopathol. Z. 97:1–9.
227. Turner, V., and N.K. van Alfen. 1983. Crown rot of alfalfa in Utah. Phytopathology 73:1333–1337.
228. Tyler, L.J., R.P. Murphy, and H.A. MacDonald. 1956. Effect of seed treatment on seedling stands and on hay yields of forage legumes and grasses. Phytopathology 46:37–44.
229. United States Department of Agriculture (USDA). 1960. Index of plant diseases in the U.S. USDA Agric. Handb. 165. U.S. Government Printing Office, Washington, DC.
230. Veerisetty, V., and M.K. Brakke. 1977. Alfalfa latent virus, a naturally occurring carlavirus in alfalfa. Phytopathology 67:1202–1206.
231. Weimer, J.L. 1927. Observations on some alfalfa root troubles. USDA Circ. 425. U.S. Government Printing Offices, Washington, DC.
232. ----. 1927. A wilt disease of alfalfa caused by *Fusarium* sp. Phytopathology 17:337–338.
233. ----. 1928. A wilt disease of alfalfa caused by *Fusarium oxysporum* var. *medicaginis*, n. var. J. Agric. Res. 37:419–433.
234. ----. 1930. Alfalfa root injuries resulting from freezing. J. Agric. Res. 40:121–143.
235. ----. 1931. Alfalfa dwarf, a hitherto unreported disease. Phytopathology 21:71–75.
236. ----. 1933. Effect of environmental and cultural factors on the dwarf disease of alfalfa. J. Agric. Res. 47:351–368.
237. ----. 1934. Studies on alfalfa mosaic. Phytopathology 24:239–247.
238. ----. 1936. Alfalfa dwarf, a virus disease transmissible by grafting. J. Agric. Res. 53:333–347.
239. Welty, R.E. 1982. Forage legume hosts of races 1 and 2 of *Colletotrichum trifolii*. Plant Dis. 66:653–655.
240. ----, K.R. Barker, and D.L. Lindsey. 1980. Effects of *Meloidogyne hapla* and *M. incognita* on Phytophthora root rot of alfalfa. Plant Dis. 64:1097–1099.
241. ----, and J.P. Mueller. 1979. Occurrence of a highly virulent isolate of *Colletotrichum trifolii* on alfalfa in North Carolina. Plant Dis. Rep. 63:666–670.

242. ----, and J.O. Rawlings. 1984. Effect of benomyl on Sclerotinia crown and stem rot of alfalfa. Plant Dis. 68:294–296.
243. ----, and ----. 1985. Effects of inoculum concentration and temperature on anthracnose severity in alfalfa. Phytopathology 75:593–598.
244. Wilcoxson, R.D., D.K. Barnes, F.I. Frosheiser, and D.M. Smith. 1977. Evaluating and selecting alfalfa for reaction to crown rot. Crop Sci. 17:93–96.

22 Insects and Mites

GEORGE R. MANGLITZ

USDA-ARS
Lincoln, Nebraska

ROGER H. RATCLIFFE

USDA-ARS
Beltsville, Maryland

The numbers and kinds of insects that inhabit alfalfa (*Medicago sativa* L.) are incredible. Thus in New York, 591 species of insects were collected from alfalfa grown near Ithaca (151). Most of these insects are incidental visitors, others parasitize or prey on the relatively few harmful species, and still others spend most of their time visiting flowers, sharing in the important job of pollination (pollinators are discussed in Chapter 32 in this book). However, >100 species damage alfalfa with a conservative annual cost estimate of U.S. $260 million (6). Fortunately, damaging species are not omnipresent and those that do cohabit a single geographical area generally reach damaging numbers at different times of the year or in different years.

Gyrisco (59) pointed out that research on insects that damage forage crops was revitalized immediately following World War II. This resulted partly from the development of new, effective, and inexpensive insecticides. Economic factors and the importance of forage quality also were partly responsible. The use of insecticides has continued, but may be restricted due to environmental concerns, rising production costs, and realization that the supply of fossil fuels is limited. In response to these concerns, research efforts in biological control, host plant resistance, and management systems that integrate control methods have been strengthened.

This chapter briefly discusses the insects that feed on alfalfa, their damage and methods of control. For discussion purposes, insects are grouped according to the nature of damage to alfalfa: (i) insects that consume foliage; (ii) insects that suck sap; (iii) insects that feed on roots; and (iv) insects that affect seed production.

22-1 INSECTS THAT CONSUME FOLIAGE

22-1.1 Alfalfa Weevil Complex

The greatest insect damage to alfalfa in the USA is caused by the alfalfa weevil complex. This weevil was introduced accidentally into the

USA on three different occasions: *Hypera postica* (Gyllenhal) about 1904 near Salt Lake City, UT (213); Egyptian alfalfa weevil [*H. brunneipennis* (Boheman)] near Yuma, AZ, about 1939 (230); and *H. postica* near Baltimore, MD about 1951 (157). The last introduction was referred to as the eastern strain, to distinguish it from the population originally found in Utah, which became known as the western strain. The faster moving eastern strain met the western strain along a north-south line that passed through central Nebraska (103, 128). The Egyptian alfalfa weevil remained confined to the southwestern USA. Although the three weevils are nearly identical in appearance, they appear to be separated by biological differences, with partial cross incompatibility reported between the eastern and western strains (20, 21, 232). Recent genetic studies reveal no chromosome differences among the three groups of weevils (90). Allozyme analysis has shown the Egyptian and eastern strains to be quite similar, with neither much different from the western strain (93). Furthermore, partial cross incompatibility of the western strain with the other two strains was associated with the presence of a rickettsia in the western weevil (91, 92). These recent findings strongly suggest that there is only one species of alfalfa weevil in the USA.

The principal damage to alfalfa is caused by the feeding of larvae on interveinal tissue, sometimes leaving only the leaf veins (Fig. 22-1A). Heavily infested fields may have a whitish appearance. The greatest abundance of the weevil, and period of greatest crop damage, occurs during the first growth of alfalfa in the spring. Under some conditions (particularly when many newly formed adults are present) the second growth may be retarded. Crop damage late in the season is rare. In general the crop is most sensitive to weevil feeding shortly after it begins growing in the spring. Populations of insects that would seriously damage the crop in early spring cause minor damage if hatched shortly before the first harvest when the alfalfa is taller and more tolerant to feeding.

When the weevil larvae attain their full size, approximately 9.5 mm, they spin lace-like cocoons, either on lower portions of the plants or on the surface of the soil. In 7 to 14 d the adult emerges from the cocoon as a brown snout beetle about 4.8 mm in length with a dark stripe centered on its back.

In most areas the alfalfa weevil produces one generation per year and overwinters as an adult. In southern areas they may produce eggs before winter, but in northern areas eggs are generally produced only in the spring. The eggs are deposited inside stems of alfalfa plants (Fig. 22-1B). The resulting larvae mature (Fig. 22-1C), pupate and become adults by early summer (Fig. 22-1D). Many of these new adults then leave the fields and remain inactive in protected places during the summer (124, 214). Eggs, larvae, and adults overwinter in some southern locations (152) and some oviposition may occur on warm days in midwinter (10).

Cultivars that have a degree of resistance to the alfalfa weevil may reduce field losses (see Chapter 28 in this book). In addition, biological

INSECTS AND MITES

Fig. 22-1. Stages of the alfalfa weevil and its damage to alfalfa: (A) damage to alfalfa by larvae; (B) alfalfa stem split to reveal eggs that were deposited inside; (C) larvae; (D) adult.

controls have proven helpful to growers in areas of the eastern USA, where establishment of species of weevil parasites are responsible for a decline in weevil populations (44, 175). In Nebraska one species of parasite in combination with a late spring freeze reduced weevil populations below economic levels for several years (128). A fungal disease that effectively controlled alfalfa weevil in Canada (68) has spread across the USA (162). However, there is little a grower can do to initiate or encourage biological control agents. Growers face annual decisions as to the need to apply an insecticide and, if so, when. A number of states have developed systems to help in these decisions. These systems are based on temperature accumulations, number and growth stage of weevils, and growth stage of the crop (236).

A bibliography of the alfalfa weevil was compiled by Wood et al. (239).

22-1.2 Clover Leaf Weevil

The clover leaf weevil, *Hypera punctata* (F.), is very similar in appearance to the alfalfa weevil but is distinctively larger in size. The clover leaf weevil adult is about 6.4 mm in length, although its larvae attain lengths of 12.7 mm. The clover leaf weevil was introduced accidentally into New York about 1880 and now occurs throughout the clover (*Trifolium* spp.) and alfalfa growing regions of the USA (135).

Adult females insert eggs into alfalfa and clover stems in the fall of the year, although egg laying also may occur in the spring or continue throughout the winter on warm days, depending on latitude. Regardless of when the eggs are laid and whether hatching occurs in spring or fall, most damage to alfalfa occurs by the larvae feeding in early spring. Larvae feed primarily on leaves, making various sized, irregularly shaped holes (4). Although this insect has the potential to cause serious losses, it is usually controlled by a naturally occurring fungus (171).

22-1.3 Caterpillars

Several kinds of caterpillars, the larvae of butterflies or moths, feed on alfalfa.

The alfalfa caterpillar, *Colias eurytheme* Boisduval, is widely distributed throughout the USA but is rarely a serious problem outside the southwest (5). Hovanitz (88) reported two races of the alfalfa caterpillar; one preferred alfalfa and the other red clover (*Trifolium pratense* L.). In areas where the two crops were grown in close proximity the races hybridized and many offspring were functionally sterile, providing a unique control. There are a great number of adult color forms (188).

The alfalfa looper, *Autographa californica* (Speyer), is usually found in alfalfa fields, particularly in the western states. It is rarely a serious problem. The caterpillars, about 25.4-mm long when mature, are dark olive-green with darker stripes down the back and sides (42). Insect parasites, bacterial, fungal, and viral diseases generally hold the insect in check (160, 42, 35).

The forage looper *Caenurgina erechtea* (Cramer), is a minor pest of alfalfa in Kansas and the surrounding Plains region. There may be three to four generations per year, with the later generations often causing the most damage (45).

The green cloverworm, *Plathypena scabra* (F), is common on alfalfa, clover, and soybean [*Glycine max* (L.) Merr.] in the eastern half of the USA. When mature, the light green larvae are about 32-mm long with faint white stripes on the sides. The larvae are quite active and wiggle when disturbed. The adults are dark-brown moths with a wing span of about 31.7 mm. There are two to four generations per year. Although quite common, these insects are influenced by a number of natural controls and seldom damage alfalfa. Lentz and Pedigo (113) reported that the larvae were attacked by 10 species of insect parasites, a nematode, and a granulosis virus.

Several species of cutworms may attack alfalfa. The adult stages of these insects are dull-colored moths, often called "millers," that are commonly attracted to lights. The larvae are somewhat similar, vary from 25.4- to 38.1-mm long, and are smooth and brown, gray, or blackish, often with stripes or spots. Most species of cutworms overwinter as immature larvae that continue feeding in the spring. They feed mostly at night with some cutting off whole stems and others feeding on foliage. During the day they remain active under debris or in the soil. The army cutworm, *Euxoa auxiliaris* (Grote) is the most important cutworm on alfalfa, and responsible for serious damage to a number of other crops. It has a unique life history. Moths of the single annual generation mature on the plains on either side of the Continental Divide in May, migrate shortly thereafter to high elevations in the Rocky Mountains, and then return to the plains in the fall to lay eggs (159). Eggs hatch in the same fall. The small larvae overwinter and damage alfalfa early in the spring just as the plant begins to grow. In general, established stands of alfalfa recover from damage after larvae mature, and moths leave the fields. In fields seeded the previous fall, heavy infestations of the army cutworm can result in severe plant mortality (131, 127). The feasibility of rearing the army cutworm on an artificial diet for 10 generations has been reported by Sutter and Miller (205). A sex attractant (a mixture of Z-5-tetradecen-1-yl acetate and E-7-tetradecen-1-yl acetate), which could be used as a detection tool, was found for adult males of the army cutworm (204). The variegated cutworm, *Peridroma saucia* (Hübner), is widely distributed over the USA and at times is very destructive to alfalfa. This insect has several generations and most frequently damages alfalfa growth following the first cutting. The variegated cutworm damages a number of crops. Berry et al. (14) found that larvae reared on peppermint (*Mentha piperita* L.) had a higher level of midgut microsomal aldrin epoxidase, and consequently were more resistant to several insecticides than were those reared on alfalfa, bush snap beans (*Phaseolus vulgaris* L.), garden beets (*Beta vulgaris* L.), or curly dock (*Rumex crispus* L.). In the Pacific Northwest the redbacked cutworm *Euxoa ochrogaster* (Guenée), often damages alfalfa in the spring. The granulate cutworm, *Feltia subterranea* (F.); the pale western cutworm, *Agrotis orthognia* Morrison; the dingy cutworm, *Feltia ducens* Walker; bristly cutworm, *Lacinipolia renigera* (Stephens); and *Tathorlynehus angustiorata* (Grote) occasionally inhabit alfalfa (224).

The yellowstriped armyworm, *Spodoptera ornithogalli* (Guenée), is widely distributed in the USA, though most common in the South. It feeds on many crops, including alfalfa. Mature larvae are about 38-mm long and pale gray to black with a pair of black triangular spots on the back of most segments and a yellowish orange stripe along the side. There are three to four generations per year (224). A closely related species, the western yellow striped armyworm, *S. praefica* Grote, is found in the far West. It also feeds on a number of crops but seldom damages the field in which it originates. The major difficulty arises when an alfalfa field is cut and the armyworms migrate to adjacent fields of alfalfa or other crops. When these caterpillars, from large fields of alfalfa, converge upon an adjacent field they often completely defoliate plants in a 12- to 15-m wide area nearest the cut alfalfa (42).

Alfalfa is commonly attacked by three species of webworm caterpillars (*Loxostege* spp.): the alfalfa webworm, *Loxostege commixtalis* (Walker), garden webworm, *Achyra rantalis* (Guenée), and beet webworm *L. sticticalis* (L.). They are general feeders that attack a wide range of plants and occasionally cause serious damage to alfalfa, particularly during July and August (45, 187). The larvae of all three species web the tops of the plants together and feed within the webs until nothing is left but skeletons of leaves and stems. All overwinter as larvae in the soil, pupate, and emerge as adults in the spring.

22–1.4 Blister Beetles

More than a dozen species of blister beetles (*Epicauta* spp.) have been recorded as feeding on alfalfa. Smith and Franklin (188) reported that a small gray blister beetle, *Epicauta murina* (LeConte), was the most abundant of eight species studied in alfalfa; the black blister beetle, *E. pennsylvanica* (DeGeer), and the three-striped blister beetle, *E. leminscata* (F.), were the next most abundant. The beetles range from 12- to >25-mm long and are black, gray, brown, or striped. All are long, narrow, cylindrical, comparatively soft-bodied beetles with heads rather distinct from the rest of their bodies. Adults feeding on the foliage and flowering parts of alfalfa and many other plants cause ragged, stunted plants, and masses of the beetles may strip leaves from plants in a short time. A secondary problem that may arise from blister beetle feeding in alfalfa is accidental poisoning of livestock that were fed hay containing dead beetles (29). Poisoning results from the presence of the toxin cantharidin in dead beetles. Capinera et al. (29) reported six species of blister beetles in alfalfa in sufficient numbers to indicate a threat to livestock. The six species are: *E. sabricii* (LeConte), *E. immaculata* (Say), *E. maculata* (Say), *E. murina*, *E. pennylvanica*, and *E. sericans* (LeConte).

Blister beetles overwinter as larvae in the soil, transform to pupae, and emerge as adults in the spring. Females deposit from 50 to several hundred eggs, usually in small cavities in the soil. Eggs hatch into very active predatory larvae. Larvae of some species are beneficial, feeding on eggs of grasshoppers and crickets.

22–1.5 Grasshoppers

Many species of grasshoppers frequent alfalfa fields, but four are important pests, particularly in the midwestern and western USA. These are the migratory grasshopper, *Melanoplus sanguinipes* (F.), the two-striped grasshopper, *M. bivittatus* (Say), the differential grasshopper, *M. differentialis* (Thomas), and the red-legged grasshopper, *M. femurrubrum* (DeGeer). In New York, Pimentel and Wheeler (151) reported these species, except the differential grasshopper, as secondary herbivores in alfalfa. During severe outbreaks, grasshoppers can destroy an entire crop. Also, they attack flowers and reduce seed set. Weed control in alfalfa may reduce grasshopper damage, because grasshoppers feed on many common weeds found in alfalfa fields. The differential and migratory grasshoppers probably cannot complete development solely on alfalfa (9). There is indication of some resistance to grasshoppers in alfalfa (74, 83).

Grasshoppers deposit their eggs in pods in the soil during the fall. Eggs hatch in the spring, and young nymphs mature to adults in 40 to 60 d. Many adults persist until cold weather. Both young and mature grasshoppers are voracious feeders. All four species mentioned have a single generation per year, except for the migratory grasshopper, which has two or three generations in the southern part of its range (147).

22–1.6 Leafminers

Alfalfa leaves may be disfigured by the larvae of small flies called leafminers. Until recently, only larvae of *Liriomyza* spp. were reported to cause minor feeding damage on alfalfa in the USA, although heavy damage has been noted (136). Larvae of an *Agromyza* spp., *A. frontella* (Rondani), were observed feeding on alfalfa in Massachusetts in 1968 (136). This species, named the alfalfa blotch leafminer (ABL), is native to Europe and currently is distributed throughout the northeastern USA and adjoining Canadian provinces (58). It is the only leafminer of general importance on alfalfa, although the economic impact of damage has not been quantified. Most of the description of leafminer biology and damage, which follows, refers to *A. frontella*.

The adults of both *Agromyza* and *Liriomyza* spp. lay their eggs in leaf tissue, and the larvae feed between the upper and lower surfaces of the leaf (78). Adults of ABL are small, dull-black, hump-backed flies (Plate 22–1A). The feeding pattern of ABL larvae (referred to as mines) differs from that caused by larvae of *Liriomyza* spp. by being "blotch-shaped" in appearance rather than "serpentine." Adults of the ABL usually oviposit near the base of the leaflet by inserting the egg between the upper and lower leaf surfaces through a perforation made with the ovipositor. There is usually one egg deposited per leaflet, but there may be more, and it is not uncommon to find up to three larvae developing in some leaflets. Newly emerged larvae move from the point of oviposition to the leaf border, and then to the apex, and then turn distally. They gradually broaden the mine on each side of the midvein, forming a large central blotch. The completed mine often has the appearance of a comma (Plate 22-1B). Mature larvae cut holes in the upper leaf surface, emerge from the mine (Plate 22-1C), and drop to the soil where they pupate. The ABL typically develop through three to four generations per year (3, 8, 153), but additional generations may occur if favorable temperature conditions exist. The ABL overwinters as a partially developed pupa that completes development the following spring. Pupal development may be affected both by photoperiod and temperature (138). Early harvest may decrease pupal survival by exposing them to higher soil temperatures and/or lower humidity conditions associated with the removal of the crop canopy (207).

Most damage occurs from larval feeding and subsequent loss of injured leaves during harvesting operations. Adult females cause additional damage when feeding by boring small pinholes in the underside of the leaflets with the ovipositor (Plate 22-1D), from which they imbibe plant juices. Punctures vary from a few to over 100 per leaflet and can be observed by holding leaves up to the light. High numbers of adult feeding punctures may adversely affect larval survival and development by reducing the quantity and quality of leaf tissue available (163). Losses in yield and quality of alfalfa have been reported (25, 206), but reports differ as to the significance of the damage on both yield and quality (120). Richard and Guibord (169) reported an increase in the incidence of spring black stem (*Phoma medicaginis* Malbr. & Roum.) on leaflets attributed to feeding by adult ABL. Leafminers can be controlled effectively with insecticides, but because of the questionable economic impact of feeding injury on alfalfa, chemical control generally is not recommended (13, 120, 209). Methods of population assessment for life stages of the ABL and pest management strategies have been developed to aid growers in the selection and timing of control measure (3, 69, 134, 241). Biological control of the ABL appears very promising (51). A number of native and European parasites have been studied as potential biological control agents for the ABL (50, 79, 80). The ABL populations in the northeastern USA were held well below suggested economic injury levels by the combined effects of native and imported parasites (81). Efforts to select for ABL resistance in alfalfa have been relatively unsuccessful (85). No significant differences were found in ABL feeding injury to or survival on the alfalfa cv. Anchor, Iroquois, Saranac, or Vernal (52). However, differences were demonstrated among nonglandular and glandular-haired *Medicago* spp. in oviposition preferences by ABL (121) (see Chapter 28 in this book).

22-2 INSECTS THAT SUCK SAP

22-2.1 Aphids

The similarities among the several species of aphids (*Therioaphis* spp.) that damage alfalfa are: relatively small soft bodies; piercing mouth

Alfalfa and Alfalfa Improvement

© 1988 ASA-CSSA-SSSA

Chapter 21—Diseases and Nematodes
   Kenneth T. Leath, Donald C. Erwin,
   and Gerald D. Griffin

Plates 21-1, 21-2, 21-3, and 21-4

Chapter 22—Insects and Mites
   George R. Manglitz and Roger H. Ratcliffe

Plate 22-1

Plate 21-1. Typical symptoms of ozone injury.

Plate 21-2. White flagging induced by infection of the alfalfa stem nematode.

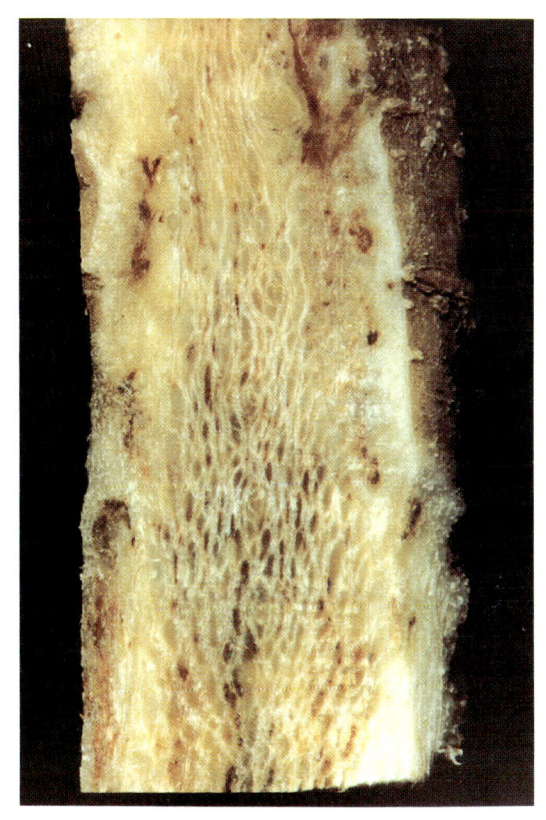

Plate 21-3. Taproot showing reddish brown flecking characteristic of Stagnospora root rot.

Plate 21-4. Typical appearance of field with Verticillium wilt (note chlorotic appearance of diseased plants).

Plate 22-1. Stages of the alfalfa blotch leafminer and its damage to alfalfa leaves: (A) adult alfalfa blotch leafminer; (B) pinhole leaf punctures made by adult female; (C) fully grown larva emerging from leaf; (D) characteristic "comma" shaped blotches made by larvae. Photos courtesy of R.M. Hendrickson, Jr.

parts, which prevent consumption of plant tissue, relegating them to liquid diets; and the production of multiple generations restricted to parthenogenetic females present during the summer months.

22-2.1.1 Spotted Alfalfa Aphid

The spotted alfalfa aphid, *Therioaphis maculata* (Buckton), is very destructive to alfalfa, particularly from Nebraska, south to Texas, and west to the Pacific Ocean. Confusion over its identity has persisted following the discovery of the insect in New Mexico, USA (217). The spotted alfalfa aphid (SAA) now is considered, by many taxonomists, to be a recognizable segregate of *Therioaphis trifolii* [Monell that should be named *T. trifolii* form *maculata* (30)]. In Europe, *T. trifolii* feeds on both red clover and alfalfa (87). Two separate introductions into the USA were composed of segregates of the European population that not only had distinct and exclusive host requirements (red clover *or* alfalfa) but also were morphologically distinguishable from each other. The yellow clover aphid, (*T. trifoli* form *trifoli*) is the clover feeding form. There appears to be no natural interbreeding of the two forms in the USA, and they function as separate species because of isolation by host requirements (129). Of the one or two plant species that serve as common hosts for both the yellow clover aphid and the spotted alfalfa aphid none are commonly grown inthe USA (126). Although SAA may not be a valid species it is a distinct and important entity.

Feeding by SAA (Fig. 22-2) causes leaves to turn yellow and abscise.

Fig. 22-2. Nymphs and adults of the spotted alfalfa aphid on an alfalfa leaflet.

Seedling plants can be killed by very light infestations. Older plants can be stunted, and heavily infested plants defoliated. In addition, the aphids secrete a sticky honeydew on which a black mold grows that lowers the quality of the hay. In Arizona there may be 20 to 40 generations per year (216). In the northern USA, egg laying forms appear in the fall in response to short day lengths (174), and overwintering occurs in the egg stage (125). A bibliography of the spotted alfalfa aphid was published in 1974 by Davis et al. (43).

22-2.1.2 Pea Aphid

The most common aphid in alfalfa, the pea aphid (PA), *Acyrthosiphon pisum* (Harris), is widely distributed over the world and probably introduced into the USA from Europe (Fig. 22-3). It attracted attention as a pest of pea (*Pisum sativum* L.) in 1899 (32). The first reported outbreak on alfalfa occurred in 1921 in Kansas and adjacent states (186). This species varies with respect to host plant preference (139, 57). Pea aphid color normally varies from light to dark green. Recently, a red form, long known in Europe (132) has been found in the USA were it differs from the green form in its reaction to resistant alfalfa (108). Klostermeyer (102)

Fig. 22-3. Pea aphids on an alfalfa stem.

observed that large populations of PA feeding on stem terminals stopped apical growth, prevented flowering, and reduced seed yield. Harper and Lilly (72) found that alfalfa heavily infested with PA contained less carotene than uninfested alfalfa. Cold-hardiness was also reduced by PA infestation (71). Leath and Byers (111) demonstrated a positive association between pea aphid and fusarium (*Fusarium* spp.) root rot damage. A Nebraska study, under field cages, suggests an economic injury level of 79 PA per day per stem (99), whereas in Minnesota, in open field conditions, a slightly higher injury level, 114 PA per day per stem was determined (39). Several sampling methods were used in Minnesota (sweep net, vacuum sampler, and stem sampling) with the conclusion that stem sampling appeared to be more precise and time efficient (39). Recently constructed life tables on the pea aphid in Wisconsin indicate emigration by adult alatae and fungal disease to be a major source of PA mortality (95). In northen latitudes, true sexual forms develop in the fall and overwintering eggs are produced. In some areas the pea aphid overwinters chiefly as an egg but summer forms may survive mild winters (37). A recent bibliography of the pea aphid has been published (71).

22-2.1.3 Blue Alfalfa Aphid

The blue alfalfa aphid, *Acyrthosiphon kondoi* Shinji, a recent immigrant to the USA, made its first appearance near Bakersfield, CA, in April 1974 (181). This insect, apparently native to the Far East, was little known prior to 1974. In addition to the USA this aphid damages alfalfa in Argentina (118), New Zealand (98), and Australia (15). The blue alfalfa aphid damages alfalfa by stunting plants, causing leaf curl, chlorosis, and eventual leaf drop (198). In the laboratory, the blue aphid feeds on a number of other legumes (53). Survival and total fecundity is higher at low (10-15°C) than at high (20-25°C) temperatures (107), which may explain why the greatest activity of this insect is in the early spring (12). The areas of principal damage in the USA have been from Oklahoma, westward to the Pacific. The insect, however, has been in Nebraska since 1977 and recently reported in Georgia and Kentucky (201). Kodet and Nielson (106) failed to produce sexual forms of this aphid in the laboratory following treatments involving 15 combinations of temperature and photoperiod. Thus, the inability of eggs to overwinter may limit the northern distribution of this species.

22-2.2 Leafhoppers

The potato leafhopper, *Empoasca fabae* (Harris), is the most important of several leafhoppers that attack alfalfa. It is the most serious alfalfa pest throughout much of the midwestern and eastern states (40). Both adults and nymphs pierce leaves and stems and suck the juices (Fig. 22-4). Feeding interferes with carbohydrate translocation; the foliage turns yellow and often various shades of red and purple (97), a condition com-

Fig. 22-4. The potato leafhopper on an alfalfa leaflet (adult above left, nymph below right).

monly called hopperburn. Leafhopper feeding reduces yield, delays maturity, and lowers quality of forage by reducing protein and carotene (100, 101, 238). In addition, leafhopper infested plants are more severely damaged by certain root pathogens than noninfested plants (111). Most of the damage is to the second and third crops. Adult potato leafhoppers are wedged-shaped pale green insects about 3-mm long. The nymphs resemble adults, except that they are smaller and lack wings; they are very active and run backward, forward, or sideways with equal agility. Potato leafhoppers overwinter in the Gulf States (46), and they migrate northward each year in the warm air currents of spring. Time of arrival in northern states will vary with weather patterns (149). Females deposit eggs in small stems and leaf veins of alfalfa and other crops. Eggs hatch in 6 to 9 d into whitish nymphs that soon turn yellowish green. There are several generations per season, with a 3-week interval in the development of adults from eggs. Timing and thoroughness of alfalfa harvest affects survival of potato leafhopper nymphs and is important in control (182). Management strategies and methods for sampling leafhopper populations and assessing economic damage levels are available to aid growers in making control decisions (31, 40, 55, 56, 109, 143, 183). A bibliography of the potato leafhopper is available (60).

Many other species of leafhoppers are found on alfalfa. In Arizona, Nielson and Currie (142) listed *Aceratagallia curvata* Oman, *Acinopterus angulatus* Lawson, and the western potato leafhopper, *Empoasca abrupta* DeLong as economically important species. Species commonly found but of minor importance are the aster leafhopper, *Macrosteles fascifrons* (Stal); the painted leafhopper, *Endria inimica* (Say); the clover leafhopper, *Aceratagallia sanguinolenta* (Provancher); *Agallia constricta* Van Duzee; *Draeculacephala antica* (Walker); *Graminella nigrifrons* (Forbes); and *Deltocephalus flavocostatus* Van Duzee.

22-2.3 Meadow Spittlebug

The meadow spittlebug, *Philaenus spumarius* (L.), is a pest on the first harvest of alfalfa in the northeastern and northcentral regions of the USA (148). Nymphs of these insects secrete a liquid and then force air into it to form masses of white froth or spittle in which they live. They suck plant sap thereby stunting plants, delaying maturity, and reducing yield (148). Also alfalfa regrowth after the first cut may be reduced (238). Feeding from heavy infestations may cause rosetting of terminal growth.

The meadow spittlebug overwinters as an egg deposited in late summer or early fall on grain stubble or old alfalfa and other plant stems. The white eggs are deposited in a row of 1 to 30 (227) and hatch in early spring as pinkish nymphs that soon turn yellowish green and become adults in late May or early June. They are about 6-mm long, somewhat triangular in shape, and resemble leafhoppers. They are more robust, however, and jump quickly when disturbed. The predominant color ranges from gray to brown, but some may be striped or mottled. Many adults leave alfalfa fields during the summer and scatter widely. In August and September they return to alfalfa and other legumes for egg laying.

22-2.4 Three-cornered Alfalfa Hopper

The three-cornered alfalfa hopper, *Spissistilus festinus* (Say), attacks many crops, but alfalfa is one of the most important (133). This insect occurs over the eastern and southern USA, but is more abundant in southern areas. The most serious losses to alfalfa occur from Arkansas, west to California. Adults and nymphs suck plant sap by puncturing stems either randomly or in a regular line that completely girdles the stem. A gall usually forms on weakened stems and translocation ceases to the upper part of the plant. Many stems become yellow, wilt, break off, or die, with the possibility that entire fields exhibit an unthrifty appearance.

The adult is triangular, light green, and about 6-mm long. The nymphs are similar in shape but are either straw colored or have a greenish hue. They are covered with spines and hairs.

The three-cornered alfalfa hopper overwinters as an adult, and may be active throughout the year during mild winters in the Southwest (234). In Louisiana it overwinters as an adult in a state of reproductive diapause (140). The white, rather oblong eggs are deposited beneath the epidermis of alfalfa stems in long slits cut by the female (133). Females may shred the stem in the process of oviposition, further damaging the plant. Eggs hatch in 12 to 17 d during the summer. The nymphal period lasts from 22 to 37 d, and the entire period from egg to adult averages about 50 d (longer at lower temperatures). There are about four generations per year in the southern part of the insect's range. Alfalfa cultivars differ in susceptibility to the three-cornered hopper, but none are classified as resistant (167).

22–2.5 Mites

Mites belong to the class Arachnida, which also includes scorpions, ticks, and spiders. They are small, usually eight-legged animals that are very difficult to see because of their small size. Several species of *Tetranychus* occur on alfalfa: the two-spotted spider mite, *Tetranychus urticae* Koch, the strawberry spider mite, *T. turkestani* Ugarov and Nikolski, the Pacific spider mite, *T. pacificus* McGregor, and the carmine spider mite, *T. cinnabarinus* (Boisduval) (26).

Mites damage alfalfa by sucking juice from leaves and seed pods. They are usually found on the undersides of the leaves, where they spin a delicate protective webbing. Infested leaves become mottled and brown, whereas seed pods turn brown and dry and may fail to fill. Adults winter in trash or litter on the ground (114) and migrate to growing plants in the spring.

An Eriophyid mite, often called the alfalfa or lucerne bud mite, *Eriophyes medicaqinis* Keifer, may cause damage to alfalfa by reducing the growth of seedlings (112, 170).

22–3 INSECTS THAT FEED ON ROOTS

22–3.1 Clover Root Curculio

The clover root curculio, *Sitona hispidulus* (F.), probably introduced from Europe, was found in New Jersey, USA, in 1876 (233). The adult is a grayish black snout beetle about 4-mm long. The adults (Fig. 22–5) feed on the leaves by chewing irregular holes in the margins, but most damage is caused by larvae feeding on roots. Larvae begin feeding on root nodules and tender fibrous roots; subsequently they chew large cav-

Fig. 22–5. Adult clover root curculio on alfalfa stem and leaflet. Photo courtesy of A.A. Hower, Jr.

ities along the sides of the main root. Fourth and fifth instar larvae are responsible for the major damage to tap roots (48). Byers and Kendall (24) reported that survival of larvae was higher on alfalfa roots with nodules than on those without nodules. The effect of feeding damage on alfalfa production differs, possibly as a result of location and severity of infestations. Underhill et al. (218) indicated that damage to young alfalfa was surprisingly severe, and that larval injury contributed to winter heaving in alfalfa stands. Dickason et al. (47) found that larvae caused much injury to alfalfa roots, both directly and indirectly by enhancing the entrance of pathogens, but no yield loss was observed. Hill et al. (86) indicated that feeding damage caused by clover root curculio larvae facilitates the entrance of the bacterial wilt organism, *Corynebacterium insidiosum* (McCull.) H.L. Jens.; however, such damage is not essential to the spread of the disease. Neal and Ratcliffe (137) found that insecticides applied to control *S. hispidulus* larvae in a year of heavy infestation resulted in significant increases in alfalfa yield at first harvest and in regrowth following first harvest. No such advantages could be demonstrated during years of lower larval infestations. No resistance to *S. hispidulus* larvae has been found in alfalfa (24).

The clover root curculio overwinters as an adult under trash and debris on the surface of the soil but may be somewhat active on warm days during the winter. Females begin depositing eggs on leaves or ground early in the spring. The eggs are minute and white when first laid but soon become jet black. Eggs hatch in about 2 weeks, with a larval period of about 3 weeks. When fully grown, the larvae are about 6-mm long, white, and legless with a light chocolate-colored head. A pupal period of from 8 to 10 d is passed in an earthen cell in the soil. Several other species of *Sitona* inhabit alfalfa; all are similar in size and appearance to the clover root curculio. *Sitona flavescens* Marsham and *S. sissifrons* Say are found in the more northern areas of the USA. The larvae cause root damage similar to that caused by the clover root curculio. *Sitona sissifrons* feeds on alfalfa in Alaska (226). Also, in the Pacific Northwest, adult pea leaf weevils, *S. lineata* (L.) feed heavily on alfalfa foliage; however, they are not recorded as damaging to the roots of alfalfa (158).

22–3.2 Alfalfa Snout Beetle

The alfalfa snout beetle, *Otiorhynchus ligustici* (L.) was first reported as a pest of alfalfa in 1933 (82) after accidental introduction into New York, USA, from Europe, possibly in the late 1800s (243). The main damage is caused by larvae that gnaw deep channels in roots. The foliage in heavily infested fields may turn yellowish, with damage accentuated during periods of drought. This beetle has been reported as a serious pest of alfalfa in Europe (146), where it is widespread. The alfalfa snout beetle could pose a serious threat to alfalfa and other legumes in the USA if it should spread to other areas (242). However, it has spread very slowly, increasing in distribution from the original two counties in New York to only four counties by 1978 (242).

The adult is rather large, 9- to 12-mm long, and black with irregular patches of pearl gray scales (146). The wings are fused together, rendering the insect incapable of flight. Only females are known, so parthenogenesis is the assumed mode of reproduction.

The eggs of the alfalfa snout beetle are deposited singly in the soil from May to September. An individual female deposits from 125 to 500 eggs. The eggs hatch in about 2 weeks, and the larvae work their way down through the soil and feed first on the side roots of alfalfa plants and subsequently on the taproot. Full-grown larvae are about 12-mm long, legless, and yellowish white with a brown head. Most of the larvae are full grown by November and pass the winter in this stage. Pupation occurs in June or July of the following year. The pupal period lasts from 22 to 25 d, but newly formed adults remain in the pupal cell until the following spring.

Lincoln and Palm (116) obtained fair to good control of the alfalfa snout beetle by rotating infested legume fields with row crops, because the larvae develop poorly on such crops. The failure of this approach can be attributed to the lack of a coordinated community effort (243).

22-3.3 White-fringed Beetles

White-fringed beetles, *Graphognathus* spp., are general feeders that attack >170 species of plants (244). Three species are found in the USA. The beetles, first found in Alabama and Florida in 1936, now occur in all states south of Maryland, Kentucky, and Missouri, and west to Texas.

Since the major distribution of white-fringed beetles and alfalfa rarely coincide, these insects have caused little damage to the crop. There are examples, however, where these species have seriously damaged alfalfa. The larvae feed on roots, and if the population is heavy, plants are killed. Damage is usually confined to small areas ranging from a few square feet to a few acres.

22-4 INSECTS THAT AFFECT SEED PRODUCTION

In a broad sense, any insect that damages alfalfa and reduces plant vigor affects the yield of seed. We have noted that blister beetles, grasshoppers, pea aphids, and mites may attack flowering parts, pods, and seeds. However, there are several insects whose damage is principally to flowering parts and seed.

22-4.1 Alfalfa Seed Chalcid

The alfalfa seed chalcid, *Bruchophagus roddi* (Gussakovsky), was first reported from Kansas, USA, in 1904 (212). The larvae, which feed in the developing alfalfa pod, can destroy from 2 to 80% of the alfalfa seed in certain localities and during certain years (219). The female chalcid forces her ovipositor through the pod and into the developing seed where she deposits an egg, usually only one per seed. Urbahns (219) found that chalcid flies favor green, half-grown pods for oviposition. Females are attracted to young (12 to 14-d-old) seed pods by a volatile olfactory attractant (210). As the larva develops, it devours the contents of the seed, leaving only the shell (Fig. 22-6A). Many of these seeds are crushed and blown out with the chaff during harvesting.

The adult chalcid is a small wasp about 2-mm long. It is jet black in color except for parts of the legs, which are yellowish brown. The small, almost colorless eggs cannot be seen without a microscope. The larvae are legless and white except for the brown mandibles. The white pupae gradually turn black.

Eggs hatch in 4 to 10 d, with larvae completing their development in 12 to 30 d, depending upon temperature. The larvae pupate within the seeds. The pupal period lasts from 6 to 10 d in the summer and longer during cool weather. The adults cut a circular hole in the seed coat and emerge. There are several generations per year, depending on location and temperature. Nielson (141) found that the chalcid developed through two generations in Arizona from about 1 June through September. The pupae of the second generation entered diapause and remained in this condition until the following spring. Induction and termination of diapause was controlled by photoperiod.

In some areas alfalfa seed chalcid injury can be reduced by use of cultural methods such as timing crop maturity to avoid seasonal peaks of the insects and destroying the previous year's crop residue (1, 2, 70). Progress in identifying chalcid resistance in alfalfa is discussed in chapter 28.

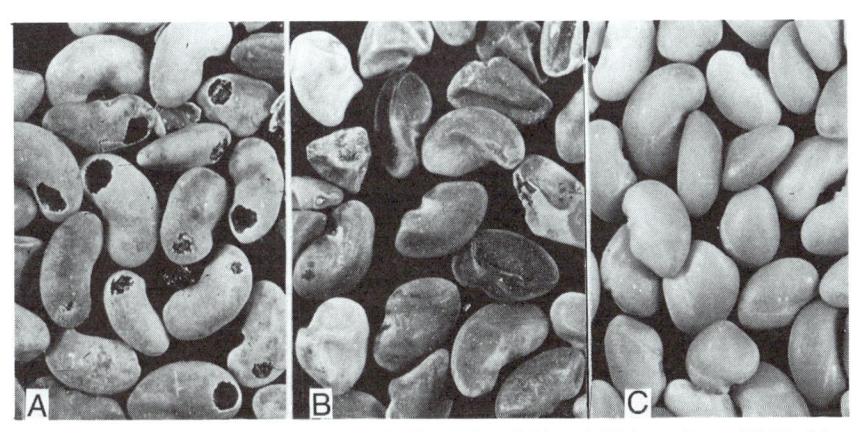

Fig. 22-6. Alfalfa seed damaged by (A) alfalfa seed chalcid and (B) lygus bugs. (C) Healthy, plump seeds are shown on the right.

22-4.2 Plant Bugs

The plant bugs feeding on alfalfa are comprised of several species from the family Miridae, commonly called mirids. The most common and economically important plant bugs include species of *Lygus* and *Adelphocoris*, but species of *Plagiognathus*, *Lopidea*, *Poecilocapsus*, and *Neurocolpus* have been recorded on alfalfa (231). There are three major species of lygus bug that damage alfalfa seed, namely: *Lygus hesperus* Knight; the pale legume bug *L. elisus* Van Duzee; and the tarnished plant bug *L. lineolaris* (Palisot de Beauvois). *Lygus hesperus* and *L. elisus* are important in the West, especially in Arizona and southern California (200, 203). *Lygus lineolaris* is the most common species in the eastern half of the USA (231) and is found as a pest of alfalfa in many western states (165, 225). A fourth species, *L. desertus* Knight, reported as present throughout Arizona, is much less abundant than either *L. hesperus* or *L. elisus* (23). These *Lygus* spp. are similar in size, shape, and color. They are rather flat, slightly <6-mm long, and about half as wide. They vary from pale green to reddish brown, and most have a V-shaped yellowish mark on their backs.

Lygus bugs pierce alfalfa tissues and suck the juices. This can cause distorted growth and rosetting of plants. Strong (202) found that injury caused by *L. hesperus* is due principally to digestion of plant tissues by enzymes secreted during feeding. Although feeding by lygus bug (and other mirids) can reduce yield and quality of alfalfa forage (225), the most important damage is to flowers and seeds (Fig. 22-6B). In Utah, lygus bugs were the most abundant insects in alfalfa seed fields, and their feeding caused blasted buds, flower drop, and shriveled seed (194). *Lygus* spp. are attracted to alfalfa (179) and especially to alfalfa infested with aphids. They feed directly on aphids (117), aphid honeydew (22), and on aphid mummies containing pupae of parasitic wasps (231).

Lygus bugs overwinter in the adult stage in grass clumps and under litter. In warm areas they are somewhat active throughout the year. The eggs, laid in alfalfa stems and leaf petioles, hatch in about 15 d. There are four to seven generations per year, depending on location.

Three *Adelphocoris* spp. cause damage to alfalfa seed similar to that caused by lygus feeding. These are the alfalfa plant bug, *A. lineolatus* (Goeze); the rapid plant bug, *A. rapidus* (Say); and the superb plant bug, *A. superbus* (Uhler). The alfalfa plant bug was first found in North America in Nova Scotia, Canada, in 1917 (104) and in the USA in Iowa in 1929 (105). It probably was introduced from Europe, where it is an important pest of alfalfa seed crops. It has spread over much of the eastern alfalfa growing areas and as far west as Utah. Hughes (94) indicated that the spread of this plant bug throughout Minnesota coincided with a decline in yields of alfalfa seed. Radcliffe and Barnes (164) reported that this plant bug also caused severe leaf yellowing and plant stunting. The rapid plant bug is found in the eastern two-thirds of the USA. It is usually less abundant and, as a result, less damaging than the alfalfa plant bug. The superb plant bug probably is native to the western USA (195) and has been found from Iowa, west to Colorado and Utah, and south to Arizona.

The development of all three species is similar. They overwinter as eggs deposited in alfalfa stems, although some eggs may be laid in leaves. The eggs hatch in 10 to 20 d, depending upon the species and the temperature. The nymphs become full grown in 16 to 28 d. There are usually two generations per year.

Predators and parasites of *Lygus* spp. have been investigated (166, 223) with no success, to date, in developing effective biological control measures.

22-4.3 Thrips

Thrips, mostly *Frankliniella* spp., are minute insects that are sometimes very abundant on alfalfa blooms. They range from yellowish to dark brown to black. The insects damage flowering parts by puncturing and sucking juice, which, in turn, may cause some discoloring and withering of the flowers.

Thrips lay their eggs within the floral parts of plants or on leaves. When weather is favorable, they can complete development from egg to adult in about 2 weeks (193).

22-4.4 Other Pests

Several species of stink bugs damage alfalfa seed, particularly in the Southwest (173), by inserting their mouthparts through the pods and into the seeds and sucking out the contents. The damaged seeds then collapse into flattened shells that soon shrivel, dry, and turn brown. The stink bugs that attack alfalfa are rather large shield-shaped bugs. Several species that are similar in appearance are involved in damaging seed crops. They range from about 11 to 16 mm in length and vary in color from green to yellowish green, and from dull olive to gray. The Say stink bug, *Chlorochroa sayi* Stal, is probably the most important. Stink bugs overwinter as adults in protected places. In the spring the females deposit their eggs on the foliage. Hatching occurs approximately in 7 d with nymphs maturing in about 30 d. There are several generations per year.

Common field crickets, *Gryllus* spp., found in most areas of the USA, are pests in the northern portion of the Central States. Although crickets feed on many parts of the plant, they prefer flowers and developing seeds, especially when seeds are in the milk and early dough stages (180). They gnaw on the edges of the seed pods and pull out the seeds. In some areas entire seed crops have been destroyed.

Adult field crickets are 20-mm long, black in color, with wings shorter than the body. Each female lays from 150 to 400 single eggs in the soil. The winter is usually passed in the egg stage, although a small percentage may overwinter as nymphs. Eggs hatch in the spring, and the young become adults during July and August. Most adults die by mid-September, but a few persist until killed by freezing weather. There is only one genertion per year.

A leaf roller, *Aphelia pallorana* (Robinson), is a minor pest of alfalfa seed. The insect ties racemes of alfalfa flowers together, thus preventing pollination. Therefore, when the insect is abundant, it can reduce seed set. Less damage to seed production is caused when the insect rolls leaves together and feeds within the rolls (191). The species overwinters as small larvae on the dormant growth with pupation within the tied leaves or racemes. The adults are golden yellow moths. There are two generations per year in Utah.

22-5 CONTROL

22-5.1 General Discussion

With the rather sudden availability of synthetic organic insecticides in the period immediately after World War II, the control of forage crops took on new dimensions (59). In time, however, research to control insects damaging alfalfa changed in response to problems associated with overuse of insecticides and to the obvious need to reduce the cost of production. This shift in emphasis is illustrated in a report by Armbrust et al. (7) summarizing results from a 5-y multistate "Integrated Pest Management" research program for the control of alfalfa insects. Much of the information gathered concerned sampling techniques and details of pest insects and parasite life histories, all of which are essential to the development of model systems. In addition to accurate detection and prediction of populations, there is a critical need for reliable information on economic injury levels (the level at which the insect population is causing crop damage that equals or exceeds the cost of controlling the insects) and economic thresholds (that point in the development of an insect population where it can be predicted that the population will reach the economic injury level if not controlled).

22-5.2 Tactics

The tactics for insect control on alfalfa can be grouped under four general categories: (i) cultural control involving the use of regular or slightly altered farm practices to adversely affect pest insects; (ii) resistant cultivars; (iii) biological control based on the introduction of exotic parasites and/or predators as well as the artificial increase of native parasites, predators, and/or pathogens; and (iv) the use of chemical control.

22-5.2.1 Cultural Control

Manipulation of cultural practices are among the oldest forms of insect control. In a forage crop such as alfalfa, timing harvests is the most common manipulation (early vs. delayed harvests). For example, delayed harvesting of alfalfa helps to control potato leafhoppers (177, 178) because eggs deposited in plant growth are removed from the field together with the delayed first cutting (96); however, a delay in harvest will reduce crop quality. In Wisconsin, an early second cutting was often more effective than a delayed first cutting (150). Also, earlier than normal harvesting, which is commonly done for general insect control, prevents further damage to the crop and appears to reduce insect populations, perhaps because they are exposed to much higher temperatures when the hay is cut (185). Early harvesting will help to control alfalfa caterpillars if practiced on a community-wide basis (5). Community action is essential because adult butterflies move to nearby fields when their food sources (alfalfa flowers) are removed (88, 189). Other changes in the timing of harvests can be beneficial in reducing populations of various pest insects. Barnes (8) reported that rotating alfalfa with cotton (*Gossypium* spp.) or other cultivated crops reduced populations of grasshoppers. Wildermuth (235) found that a community-wide program of cultivating alfalfa seed fields in the fall and winter, following certain crop sanitation measures, and producing seed on the same cutting helped to control the alfalfa seed chalcid.

Another method of control, flaming devices, used earlier to control alfalfa weevils (228) and pea aphid (18), has been reevaluated (16, 19). This practice has not gained wide acceptance, possibly because of the added cost of fuel for flaming, and the need to supplement flaming with an application of insecticide to obtain effective control of the weevil (11, 49). The burning of old growth during the winter or early spring, however, has provided effective control of several alfalfa insects (94, 115, 211).

Other cultural practices used to reduce populations of pest insects include short crop rotations, seeding on uninfested ground, protection of fields with baited furrows, clean culture, cultivation, dragging, and dust mulching, use of gathering machines, and pasturing (54, 59).

22–5.2.2 Resistant Cultivars

The general subject of insect resistance is covered in chapter 28; however, it would be remiss not to mention resistant cultivars in a discussion of insect control. Resistant cultivars provide long-lasting control at minimal cost to the grower. For example, the modest investment in the development of 'Moapa' provided estimated annual savings in insect control of more than U.S. $3 million (67).

The study of resistance in alfalfa to the pea aphid and potato leafhopper was initiated nearly 40 yr ago (17, 145, 176); however, only two resistant cultivars were available before 1954 (144). A short time later, Lahontan alfalfa was found to have resistance to the newly introduced spotted alfalfa aphid (89, 196). In 1979, Ratcliffe (168) published a listing of 26 alfalfa cultivars, each of which resists one or more of the following insects: alfalfa weevil, pea aphid, spotted alfalfa aphid, blue alfalfa aphid, meadow spittlebug, and potato leafhopper.

22-5.2.3 Biological Control

Releases of parasites to control the alfalfa weevil in the western states were summarized by Clausen (36). Subsequently, similar efforts in the eastern states were successful (44, 175). The eastern weevil population may have a mechanism of normal immunity (egg encapsulation) to *Bathyplectes curculionis* Thomson, one of the most effective parasites (161). Nevertheless, this parasite has provided an effective control of the eastern weevil (128). Also, parasites of plant bugs that inhabit alfalfa also have been studied (34, 35). The performance of parasites is a major advantage. Once established, they maintain themselves and spread naturally (61, 130, 237). Three parasites of the spotted alfalfa aphid have been established successfully (220, 221), and recent attempts to establish a parasite of the pea aphid were successful (27, 28, 119, 192, 222).

The value of predators of aphids, including a great number of insect species and even birds, has been recognized for a long time. One of the most important predators of aphids is the convergent lady beetle, *Hippodamia convergens* Guérin-Méneville (184). Alfalfa can be seriously damaged, however, before the beetles can bring the aphids under control. Attempts to move the beetles from their hibernation quarters to fields when control is needed have not produced the desired results (37, 41, 75). Some conservation of predators of alfalfa insects can be achieved by the selection of appropriate insecticides and by scheduling application on the basis of need (190, 199). The insect predators commonly found in alfalfa fields occasionally feed on other pest insects, but their preferred diet appears to be aphids (110, 240).

A fungus disease of pea aphids has provided another important approach in the control of this pest (76, 122). In addition, a number of fungi kill spotted alfalfa aphids (63, 64, 65). The use of other pathogens to control alfalfa caterpillars (polyhedrosis virus and *Bacillus thuringiensis* Berliner) was reviewed by Clausen (36). Subsequently, Hall and Stern (66) and Stern et al. (197) showed that commercial preparations of *B. thuringiensis* were as effective as certain insecticides. Hill (84) reported natural control of green cloverworms (*Plathypena scabra* Fabricius) by a fungus, *Botrytis rileyi* Farl. Also, several pathogenic organisms are reported to attack the alfalfa weevil. These include two fungi, *Beauveria bassiana* (Balsamo) (38, 77); and *Erynia phytonomi* (68); a microsporidian, *Nosema* spp. (123); and a nematode, *Hexamermis arvalis* Poinar and Gyrisco (154, 155, 156). Methods have not been developed to use these organisms to control the alfalfa weevil. A fungus disease will furnish effective control of the larvae of the clover leaf weevil. Contagion is so rapid and thorough that other control measures may not be needed (215). Hall (62) reported that two virus diseases were important in holding the alfalfa looper in check, and Thompson (208) indicated that field applications of a bacteria-virus combination produced mortality of the alfalfa caterpillar more rapidly than the virus alone. The control obtained by Thompson, however, was neither as long lasting nor as complete as when only the virus was applied.

These selected examples document the potential of biological control. It must be recognized, however, that certain disadvantages are characteristic of many approaches to biological control; certain organisms, particularly pathogens, are very dependent upon temperature and humidity; other organisms, particularly parasites and predators, tend to build up in numbers and control the host insect after the crop has been damaged, and, frequently, there is little an individual grower can do to initiate or enhance biological control.

22-5.2.4 Chemical Control

Although chemicals have been used to control insects for >100 yr they were not used extensively on alfalfa until after World War II. However, when new insecticides were found to furnish excellent control of some insects and produce increased yields of alfalfa forage and seed, the use of selected insecticides on alfalfa became widely accepted.

Several types of organic synthetic insecticides are used on alfalfa: chlorinated hydrocarbons, organophosphates, and carbamates. The chlorinated hydrocarbons give excellent control of many pests of alfalfa. They have been phased out of use because they tend to leave residues on the harvested crop and in the products from animals that feed on the forage. In contrast, the organophosphates, such as diazinon [O,O,-diethyl$O$-(2-isopropyl-6-methyl-4-pyrimidinyl)phosphorothioate], mevinphos [methyl(E)-3-hydroxycrotonate dimethyl phosphate], methyl parathion [O-O-diemthyl O-(p-nitrophenyl)phosphorothioate], parathion [O-O-diethyl O-(p-nitrophenyl)phosphorothioate], malathion [O,O-diethyl phosphorodithioate of diethylmercapto=succinate], demeton [O,O-diethyl O(and S)-[2-(ethylthio)ethyl] phosphorothioates], and naled [1,2-dibromo-2,2-dichloroethyl dimethyl phosphate], are widely used and, if properly applied, these insecticides do not leave harmful residues on the crop at harvest. The carbamates, carbaryl (1-naphthyl methylcarbamate) and carbofuran (2,3-dihydro-2,2-dimethyl-7-benzofuranyl methylcarbamate) are valuable insecticides.

Insecticides are available as emulsifiable concentrates, wettable powders, granules, dusts, and oil solutions. The first two are most important on alfalfa with emulsifiable concentrates receiving wide acceptance. Dusts are not popular because of problems associated with drift, whereas oil solutions can produce phytotoxic effects.

The correct insecticide must be applied properly at the right time and in the amount indicated on the registered label. Use of more than necessary is wasteful, will not increase control, and may be hazardous to humans, livestock, and wildlife. Use of too little is equally wasteful because of unsatisfactory control. Researchers are constantly searching for

Table 22-1. Economic thresholds for alfalfa weevil pest management decision-making. From Wilson (236).

Heat units[†]	Plant height	Stem tips with feeding	Decision
	cm (inches)		
300	<15.2 (<6)	25%[‡]	Reevaluation in 7 d. If the number of weevil larvae average at least one per stem and damage is increasing, spray with a long residual insecticide.
400	22.9 (9)	50%	Spray with a long residual insecticide if weevil larvae average one or more per stem.
500	30.5 (12)	75%	Spray with a short residual insecticide. If field is cut at this time, reevaluate field after cutting and treat within 7 d if weevils are still active.
600	38.0+ (15+) or bud stage	75-100%	Best to cut and remove crop; spray stubble within 7 d if weevils are still active.
750	Short or no regrowth	50% on regrowth	If no regrowth within 4 to 5 d of cutting and weevils are present, feeding on "bark" of old stems, spray immediately.
800			Beyond need for control measures. Weevil population gone or declining rapidly.

[†] Heat unit accumulation above a base temperature of 48°F from 1 January.
[‡] Counts of larvae in addition to feeding are advised since mortality of winter hatching larvae frequently occurs and treatment at this stage may be too early.

new and improved insecticides for use on alfalfa, and recommendations are subject to frequent change. Therefore, insecticides are not listed for control of specific insects. Current information may be obtained from local county extension offices, state agricultural experiment stations, and the USDA.

22-6 CONTROL SYSTEMS

The alfalfa crop, the insects feeding on it, and the natural enemies of those insects interact with each other, with temperature, moisture, and other variables, and these complex interactions ultimately determine the extent of insect damage to the crop. When decisions are made relative to the application of chemicals to control insects, one must consider the amount and value of damage that may occur without control, and the cost of controlling a particular insect, or insects. It is difficult for an individual to deal effectively with all of the variables involved in this decision. Current trends in managing alfalfa, and other crops, are based on systems that integrate and analyze numerous variables and aid in decision making.

A report summarizing 5 yr of modeling and systems analysis research on the alfalfa crop, the alfalfa weevil, and alfalfa weevil parasites was published by Ruesink et al. (172). Descriptive simulation models and management models were developed.

Although the research referred to above is fairly sophisticated, it can be translated into simple steps that alfalfa growers can follow with relative ease, provided they have access to degree day accumulations and periodically sample the weevil population (Table 22-1). An additional example is provided by Wedberg et al. (229).

22-7 SUMMARY

A great number of insect species feed on alfalfa, and some are capable of inflicting serious damage to the crop. The alfalfa weevil is the most important species consuming foliage. Several species of caterpillars and grasshoppers are, at times, important consumers of foliage. The potato leafhopper is the most serious pest that sucks sap from plants. Several species of aphids also cause serious damage by removing sap from plants. Alfalfa roots are damaged by the clover root curculio, related species of *Sitona*, and the alfalfa snout beetle, although the latter has a restricted geographical range. Insects can adversely affect seed production by feeding on and damaging flowers (i.e., Lygus bugs) or by feeding directly on the developing seed (i.e., alfalfa seed chalcids).

Damage to alfalfa can be reduced by using resistant cultivars, cultural, chemical, and biological control methods, or a combination of methods.

REFERENCES

1. Ahring, R.M., J.O. Moffett, and R.D. Morrison. 1984. Date of pod set and chalcid fly infestation in alfalfa seed crops in the southern Great Plains. Agron. J. 76:137-140.
2. ----, T.L. Springer, O.R. Jones, J.F. Stritzke, and J.O. Moffett. 1985. Alfalfa seed production—Southern Great Plains. Okla. Agric. Exp. Stn. Bull. B-776.
3. Andaloro, J.T., T.M. Peters, and A.J. Alicandro. 1983. Population dynamics of the alfalfa blotch leafminer, *Agromyza frontella*, and its influence on alfalfa yield in Massachusetts. Environ. Entomol. 12:510-514.
4. Anonymous. 1956. The clover leaf weevil and its control. USDA Farmer's Bull. 1484. U.S. Government Printing Office, Washington, DC.
5. ----. 1963. Control of the alfalfa caterpillar. USDA Leaflet 325. U.S. Government Printing Office, Washington, DC.
6. App, B.A., and G.R. Manglitz. 1972. Insects and related pests. *In* Alfalfa science and technology. Agronomy 15:527-554.
7. Armbrust, E.J., B.C. Pass, D.W. Davis, R.G. Helgesen, G.R. Manglitz, R.L. Pienkowski, and C.G. Summers. 1980. General accomplishments toward better insect control in alfalfa. p. 187-216. *In* C. Huffaker (ed.) New technology of pest control. John Wiley and Sons, New York.
8. Barnes, O.L. 1959. Effect of cultural practices on grasshopper populations in alfalfa and cotton. J. Econ. Entomol. 52:336-337.
9. ----. 1963. Food-plant tests with the differential grasshopper. J. Econ. Entomol. 56:396-399.
10. Bass, M.H. 1967. Notes of the biology of the alfalfa weevil, *Hypera postica*, in Alabama. Ann. Entomol. Soc. Am. 60:295-298.

11. Bennett, S.E. 1968. A decade with the alfalfa weevil in Tennessee. Tenn. Agric. Exp. Stn. Bull. 446.
12. Berberet, R.C., D.C. Arnold, and K.M Soteres. 1983. Geographical occurrence of *Acyrthosiphon kondoi* Shinji in Oklahoma and its seasonal incidence in relation to *Acyrthosiphon pisum* (Harris), and *Therioaphis maculata* (Buckton) (Homoptera: Aphididae). J. Econ. Entomol. 76:1064–1068.
13. Bereza, K. 1979. Alfalfa blotch leafminer. Ontario Ministry of Agriculture and Food Factsheet 79-132. Ontario Ministry of Agriculture, Toronto, ON, Canada.
14. Berry, R.E., S.J. Yu, and L.C. Terriere. 1980. Influence of host plants on insecticide metabolism and management of variegated cutworm. J. Econ. Entomol. 73:771–774.
15. Bishop, A.L., P.J. Walters, R.H. Holtkamp, and B.C. Dominiak. 1982. Relationships between *Acyrthosiphon kondoi* and damage in three varieties of alfala. J. Econ. Entomol. 75:118–122.
16. Bishop, J.L., and R.L. Pienkowski.1967. Early season control of the alfalfa weevil. J. Econ. Entomol. 27:262–264.
17. Blanchard, R.A., and J.E. Dudley. 1934. Alfalfa plants resistant to the pea aphid. J. Econ. Entomol. 27:262–264.
18. ———, H.B. Walker, and O.K. Hedden. 1933. Burning for the control of aphids on alfalfa in the Antelope Valley of California. USDA Circ. 287.
19. Blickenstaff, C.C. 1965. Flaming—a new approach to alfalfa weevil control. p. 12–13 *In* Proc. 2nd Annu. Symp. Use of Flame in Agric. Natural Gas Processors Association and National LP Gas Association, Oak Brook, IL.
20. ———. 1965. Partial intersterility of eastern and western U.S. strains of the alfalfa weevil. Ann. Entomol. Soc. Am. 58:523–526.
21. ———. 1969. Mating competition between eastern and western strains of the alfalfa weevil. *Hypera postica*. Ann Entomol. Soc. of Amer. 62:956–958.
22. Butler, G.D., Jr. 1968. Sugar for the survival of *Lygus hesperus* on alfalfa. J. Econ. Entomol. 61:854–855.
23. ———. 1970. Temperature and the development of egg and nymphal stages of *Lygus desertus*. J. Econ. Entomol. 63:1994–1995.
24. Byers, R.A., and W.A. Kendall. 1982. Effects of plant genotypes and root nodulation on growth and survival of *Sitona* spp. larvae. Environ. Entomol. 11:440–443.
25. ———, and K. Valley. 1981. Losses in digestible dry matter and crude protein in alfalfa caused by the alfalfa blotch leafminer. Melsheimer Entomol. Ser. 31:8–13.
26. California, State of. 1969. California Agric. Exp. Stn. Ext. Serv. Bull. (unnumbered).
27. Cameron, P.J., G.P. Walker, and D.J. Allan. 1981. Establishment and dispersal of the introduced parasite *Aphidius eadyi* (Hymenoptera: Aphidiidae) in the North Island of New Zealand and its initial effect on the pea aphid, *Acyrothosiphon pisum*. N.Z. J. Zool. 8:105–112.
28. Campbell, A., and M. Mackauer. 1975. The effect of parasitism by *Aphidius smithi* (Hymenoptera: Aphidiidae) on reproduction and population and population growth of the pea aphid. Can. Entomol. 107:919–926.
29. Capinera, J.L., D.R. Gardner, and F.R. Stermitz. 1985. Canthardin levels in blister beetles (Coleoptera: Meloidae) associated with alfalfa in Colorado. J. Econ. Entomol. 78:1052–1055.
30. Carver, M. 1978. The scientific nomenclature of the spotted alfalfa aphid (Homoptera: Aphididae). J. Aust. Entomol. Soc. 17:287–288.
31. Cherry, R.H., K.A. Wood, and W.G. Ruesink. 1977. Emergence trap and sweep net sampling for adults of the potato leafhopper from alfalfa. J. Econ. Entomol. 70:279–282.
32. Chittenden, F.H. 1909. The pea aphis. USDA Circ. 43. U.S. Government Printing Office, Washington, DC.
33. Clancy, D.W. 1968. Distribution and parasitization of some *Lygus* spp. in western United States and Central Mexico. J. Econ. Entomol. 61:443–445.
34. ———. 1969. Parasitization of cabbage and alfalfa loopers in southern California. J. Econ. Entomol. 62:1078–1083.
35. ———, and H.D. Pierce. 1966. Natural enemies of some *Lygus* Bugs. J. Econ. Entomol. 59:853–858.
36. Clausen, C.P. 1956. Biological control of insect pests in the continental United States. USDA Tech. Bull. 1139. U.S. Government Printing Office, Washington, DC.
37. Cook, W.C. 1963. Ecology of the pea aphid in the blue mountain area of eastern Washington and Oregon. USDA Tech. Bull. 1287. U.S. Government Printing Office, Washington, DC.
38. Cothran, W.R., and G.G. Gyrisco. 1966. An entomogenous fungus observed attacking alfalfa weevil adults in New York. J. Econ. Entomol. 59:243–244.
39. Cuperus, G.W., E.B. Radcliffe, D.K. Barnes, and G.C. Marten. 1982. Economic injury levels and economic thresholds for pea aphid, *Acyrthosiphon pisum* (Harris), on alfalfa. Crop Prot. 1:453–463.

40. ----, ----, ----, and ----. 1983. Economic injury levels and economic thresholds for potato leafhopper (Homoptera: Cicadellidae) on alfalfa in Minnesota. J. Econ. Entomol. 76:1341-1349.
41. Davidson, W.M. 1924. Observations and experiments on the dispersion of the convergent lady beetle in California. Trans. Am. Entomol. Soc. 50:163-175.
42. Davis, D.W., and G.F. Knowlton. 1976. Foliage pests of alfalfa p. 3-12. *In* D.W. Davis (ed.) Insects and nematodes associated with alfalfa in Utah. Utah Agric. Exp. Stn. Bull. 494.
43. ----, M.P. Nichols, and E.J. Armbrust. 1974. The literature of arthropods associated with alfalfa: I. A bibliography of the spotted alfalfa aphid, *Therioaphis maculata* (Buckton) (Homoptera: Aphidae). Biol. Notes (Ill. Nat. Hist. Surv.) 87.
44. Day, W.H. 1981. Biological control of the alfalfa weevil in the northeastern United States. *In* G.C. Papavizas (ed.) Biological control in crop production. BARC Symp. 1-3 May 1979. Allanheld, Osmum, Totowa, NJ.
45. Dean, G., and R.C. Smith. 1936. Insects injurious to alfalfa in Kansas. 29th Biennial Report of the Kansas State Board of Agriculture, p. 202-249.
46. Decker, G.C., and H.B. Cunningham. 1968. Winter survival and overwintering areas of the potato leafhopper. J. Econ. Entomol. 61:154-161.
47. Dickason, E.A., C.M. Leach, and A.E. Gross. 1968. Clover root curculio injury and vascular decay of alfalfa roots. J. Econ. Entomol. 61:1163-1168.
48. Dintenfass, L.P., and G.C. Brown. 1986. Feeding rate of larval clover root curculio, *Sitona hispidulus* (Coleoptera: Curculionidae), on alfalfa taproots. J. Econ. Entomol. 79:506-510.
49. Dorsey, C.K., and L.P. Stevens. 1969. Experiments to control the alfalfa weevil with low-volume and ultra-low-volume spray treatments. Bull. 579T-May 1969. W. Va. Univ., Agric. Exp. Stn.
50. Drea, J.J., D. Jeandel, and F. Gruber. 1982. Parasites of Agromyzid leafminers (Diptera: Agromyzidae) on alfalfa in Europe. Ann. Entomol. Soc. Am. 75:297-310.
51. Drea, J.J., Jr., and R.M. Hendrickson, Jr. 1986. Analysis of a successful classical biological control project: The alfalfa blotch leafminer (Diptera: Agromyzidae) in the northeastern United States. Environ. Entomol. 15:448-455.
52. Drolet, J., and J.N. McNeil. 1984. Performance of the alfalfa blotch leafminer, *Agromyza frontella* (Diptera: Agaromyzidae), on four alfalfa varieties. Can. Entomol. 116:795-800.
53. Ellsbury, M.M., and M.W. Nielson. 1981. Comparative host plant range studies of the blue alfalfa aphid, *Acyrthosiphon kondoi* Shinji and the pea aphid, *Acyrthosiphon pisum* (Harris) (Homoptera: Aphididae). USDA Teach. Bull. 1639. U.S. Government Printing Office, Washington, DC.
54. Essig, E.O., and A.E. Michelbacher. 1933. The alfalfa weevil. Calif. Agric. Exp. Stn. Bull. 567.
55. Fleischer, S.J., W.A. Allen, J.M. Luna, and R.L. Pienkowski. 1982. Absolute-density estimation from sweep sampling, with a comparison of absolute-density sampling techniques for adult potato leafhopper in alfalfa. J. Econ. Entomol. 75:425-430.
56. Flinn, P.W., and A.A. Hower. 1984. Effects of density, stage, and sex of the potato leafhopper, *Empoasca fabae* (Homoptera: Cicadelidae), on seedling alfalfa growth. Can. Entomol. 116:1543-1548.
57. Frazer, B.D. 1972. Life tables and intrinsic rates of increase of apterous black bean aphids and pea aphids on broad bean (Homoptera: Aphididae). Can. Entomol. 104:1717-1722.
58. Guppy, J.C. 1981. Bionomics of the alfalfa blotch leafminer, *Agromyza frontella* (Diptera: Agromyzidae), in eastern Ontario. Can. Entomol. 113:593-600.
59. Gyrisco, G.G. 1958. Forage insects and their control. Ann. Rev. Entomol. 3:421-448.
60. ----, D. Landman, A.C. York, B.J. Irwin, and E.J. Armbrust. 1978. The literature of arthropods associated with alfalfa: IV. A bibliography of the potato leafhopper. Univ. of Ill. Spec. Pub. 51.
61. Hagen, A.F., and G.R. Manglitz. 1967. Parasitism of the alfalfa weevil in the western plains states from 1963-1966. J. Econ. Entomol. 60:1663-1666.
62. Hall, I.M. 1953. The role virus diseases in the control of the alfalfa looper. J. Econ. Entomol. 46:1110-1111.
63. ----, and P.H. Dunn. 1957a. Fungi on spotted alfalfa aphid. Calif. Agr. 11:5, 14.
64. ----, and ----. 1957b. Entomophthorus fungi parasitic on the spotted alfalfa aphid. Hilgardia 27:159-165.
65. ----, and ----. 1958. Artificial dissemination of entomophthorus fungi pathogenic to the spotted alfalfa aphid in California. J. Econ. Entomol. 51:341-344.
66. ----, and V.M. Stern. 1962. Comparison of *Bacillus thuringiensis* Berliner var. *thuringiensis* and chemical insecticides for control of the alfalfa caterpillar. J. Econ. Entomol. 55:862-865.

67. Hanson, C.H. 1961. Moapa alfalfa pays off. Crops Soils 13:11-12.
68. Harcourt, D.G., J.C. Guppy, D.M. Macleod, and D. Tyrrell. 1974. The fungus *Entomophthora phytonomi* pathogenic to the alfalfa weevil, *Hypera postica*. Can. Entomol. 106:1295-1300.
69. ----, J.M. Yee, and J.C. Guppy. 1983. Two models for predicting the seasonal occurrence of *Agromyza frontella* (Diptera: Agromyzidae) in eastern Ontario. Environ. Entomol. 12:1455-1458.
70. Harpaz, I. 1978. Cultural control of the alfalfa seed chalcid, *Bruchophagus roddi*, in Israel. FAO Plant Prot. Bull. 26:158-162.
71. Harper, A.M., and S. Freyman. 1983. Cold-hardiness of 1-, 2-, and 3-year old alfalfa infested with the pea aphid (Homoptera: Aphididae). Can. Entomol. 115:1243-1244.
72. ----, and C.E. Lilly. 1966. Effects of the pea aphid on alfalfa in southern Alberta. J. Econ. Entomol. 59:1426-1427.
73. ----, J.P. Miska, G.R. Manglitz, B.J. Irwin, and E.J. Armbrust. 1978. The literature of arthropods associated with alfalfa: III. A bibliography of the pea aphid. Ill. Agric. Exp. Stn. Spec. Pub. 50.
74. Harvey, T.L., and H.L. Hackerott. 1976. Grasshopper-resistant alfalfa selected in the field. Environ. Entomol. 5:572-574.
75. Hatch, H.M., and C. Tanasee. 1948. The liberation of *Hippodamia convergens* in the Yakima Valley of Washington 1943-1946. J. Econ. Entomol. 41:993.
76. Hawkins, J.H. 1937. Pea aphids. Maine Agric. Exp. Stn. Bull. 387.
77. Hedlund, R.C., and B.C. Pass, 1968. Infection of the alfalfa weevil, *Hypera postica* by the fungus *Beauveria bassiana*. J. Invertebr. Pathol. 11:25-34.
78. Hendrickson, R.M., Jr., and S.E. Barth. 1978. Biology of the alfalfa blotch leafminer. Ann. Entomol. Soc. Am. 71:295-298.
79. ----, and ----. 1979a. Effectiveness of native parasites against *Agromyza frontella* (Rondani) (Diptera: Agromyzidae), an introduced pest of alfalfa. J. N.Y. Entomol. Soc. 87:85-90.
80. ----, and ----. 1979b. Introduced parasites of *Agromyza frontella* (Rondani) in the USA. J. N.Y. Entomol. Soc. 87:167-174.
81. ----, and J. A. Plummer. 1983. Biological control of alfalfa blotch leafminer (Diptera: Agromyzidae) in Delaware. J. Econ. Entomol. 76:757-761.
82. Herrick, G.W. 1933. *Otiorhynchus ligustici*, L., a European snout beetle new to this country. J. Econ. Entomol. 26:731-732.
83. Hewitt, G.B. and J.D. Berdahl. 1984. Grasshopper food preferences among alfalfa cultivars and experimental strains adapted for rangeland interseeding. Environ. Entomol. 13:828-831.
84. Hill, C.C. 1925. Biological studies of the green cloverworm. USDA Bull. 1336. U.S. Government Printing Office, Washington, DC.
85. Hill, R.R., Jr., and R.A. Byers. 1979. Allocation of resources in selection for resistance to alfalfa blotch leafminer. Crop Sci. 19:253-257.
86. ----, J.J. Murray, and K.E. Zeiders. 1971. Relationships between clover root curculio injury and severity of bacterial wilt in alfala. Crop Sci. 11:306-307.
87. Hille Ris Lambers, D., and R. van den Bosch. 1964. On the genus *Therioaphis* Walker, 1870, with descriptions of new species Homoptera: Aphididae). Zooligische Verhandelingen Uitgegeven Door Het Rijksmuseuum van Natuurlijke Historie Te Leiden 68. Rijksmuseum van Natuurlijke Historie, Leiden, Netherlands.
88. Hovanitz, W.J. 1944. Physiological behavior and geography in control of the alfalfa butterfly. J. Econ. Entomol. 37:740-745.
89. Howe, W.L., and Smith, O.F. 1957. Resistance to the spotted alfalfa aphid in Lahontan alfalfa. J. Econ. Entomol. 50:320-324.
90. Hsiao, C., and T.H. Hsiao. 1984. Cytogenetic studies of alfalfa weevil (*Hypera postica*) strains (Coleoptera: Curculionidae). Can. J. Genet. Cytol. 26:348-353.
91. ----, and ----. 1985. Rickettsia as the cause of cytoplasmic incompatibility in the alfalfa weevil, *Hypera postica* (Gyllenhal). J. Invertebr. Pathol. 45:244-246.
92. ----, and ----. 1985. Hybridization and cytoplasmic incompatibility among alfalfa weevil strains. Entomol. Exp. Appl. 37:155-159.
93. ----, and J.M. Stutz. 1985. Discrimination of alfalfa weevil strains by allozyme analysis. Entomol. Exp. Appl. 37:113-121.
94. Hughes, J.H. 1943. The alfalfa plant bugs *Adelphocoris lineolatus* (Goeze) and other Miridae (Hemiptera) in relation to alfalfa-seed production in Minnesota. Minn. Agric. Exp. Stn. Techn. Bull. 161.
95. Hutchinson, W.D., and D.B. Hogg. 1985. Time-specific life tables for the pea aphid *Acythosiphon pisum* (Harris) on alfalfa. Res. Popul. Ecol. 27(2):231-253.
96. Jewett, H.H. 1934. The relation of time of cutting to leafhopper injury to alfalfa. Ky. Agric. Exp. Stn. Bull. 348:51-59.
97. Johnson, H.W., 1938. Further determinations of the carbohydrate-nitrogen relation-

ship and carotene in leaf-hopper-yellowed and green alfalfa. Phytopathology 28:273-277.
98. Kain, W.M., D.S. Atkinson, R.S. Marsden, M.J. Oliver, and T.V. Holland. 1977. Bluegreen lucerne aphid damage in lucerne crops within southern North Island. Proc. N.Z. Weed Pest Control Conf. 30th, 177-181.
99. Karner, M.A., G.R. Manglitz, W.R. Kehr, and R.L. Ogden. 1984. Manipulation of pea aphid populations to determine economic injury levels on alfalfa in Nebraska. (Unpublished data).
100. Kehr, W.R., R.L. Ogden, and S.D. Kindler. 1975. Management of four alfalfa varieties to control damage from potato leafhoppers. Univ. Nebr. Exp. Stn. Res. Bull. 275.
101. Kindler, S.D., W.R. Kehr, R.L. Ogden, and J.M. Schalk. 1973. Effect of potato leafhopper injury on yield and quality of resistant and susceptible alfalfa clones. J. Econ. Entomol. 66:1298-1302.
102. Klostermeyer, E.C. 1962. The relationship among pea aphids, Lygus bugs and alfalfa seed yields. J. Econ. Entomol. 55:462-465.
103. Klostermeyer, L.E., and G.R. Manglitz. 1979. Distribution of eastern and western alfalfa weevil in Nebraska determined by cross-matings (Coleoptera: Curculionidae). J. Kans. Entomol. Soc. 52:209-214.
104. Knight, H.H. 1922. Nearctic records for species of Miridae known heretofore only from the Palaeartic region (Heteroptera). Can. Entomol. 53:287-288.
105. ———. 1930. An European plant-bug (*Adelphocoris lineolatus* (Groeze) found in Iowa (Hemp. Miridae). Entomol. News 41:4-6.
106. Kodet, R.T., and M.W. Nielson. 1980. Effect of temperature and photoperiod on polymorphisms of the blue alfalfa aphid, *Acyrthosiphon kondoi*. Environ. Entomol. 9:94-96.
107. Kodet, R.T., M.W. Nielson, and R.O. Kuehl. 1982. Effect of temperature and photoperiod on the biology of the blue alfalfa aphid, *Acyrthosiphon kondoi* Shinji. USDA Tech. Bull. 1660. U.S. Government Printing Office, Washington, DC.
108. Kugler, J.L., and R.H. Ratcliffe. 1983. Resistance in alfalfa to a red form of the pea aphid (Homoptera: Aphididae). J. Econ. Entomol. 76:74-76.
109. Lamp, W.O., R.J. Barney, E.J. Armbrust, and G. Kapusta. 1984. Selective weed control in spring-planted alfalfa: Effect on leafhoppers and plant hoppers (Homoptera: Auchenorrhyncha), with emphasis on potato leafhopper. Environ. Entomol. 13:207-213.
110. Lavallee, A.G., and F.R. Shaw. 1969. Preferences of golden-eye lacewing larvae for pea aphids, leafhopper and plant bug nymphs, and alfalfa weevil larvae. J. Econ. Entomol. 62:1228-1229.
111. Leath, K.T., and R.A. Byers. 1977. Interaction of fusarium root rot with pea aphid and potato leafhopper feeding on forage legumes. Phytopathology 67:221-229.
112. Lehman, W.F., and R.A. Flock. 1970. Two microscopic mites, *Aceria medicaginis* and *Tarsonemus setifer*, found on alfalfa in the desert area of the southwestern United States. J. Econ. Entomol. 63:293-294.
113. Lentz, G.L., and L.P. Pedigo. 1975. Population ecology of parasites of the green cloverworm in Iowa. J. Econ. Entomol. 68:301-304.
114. Lieberman, F.V., and G.F. Knowlton. 1955. Section II. Injurious insects and mites and their control. Utah State Univ. Agric. Exp. Stn. Circ. 135:23-32.
115. Lilly, C.E., and C.A. Hobbs. 1962. Effects of spring burning and insecticides on the superb plant bug *Adelphocoris superbus* (Uhl.) and associated fauna in alfalfa seed fields. Can. J. Plant Sci. 42:53-61.
116. Lincoln, C., and C.E. Palm. 1941. Biology and ecology of the alfalfa snout beetle. N.Y. Agric. Exp. Stn. Ithaca Mem. 236.
117. Lindquist, R.K., and E.L. Sorensen. 1970. Interrelationships among aphids, tarnished plant bugs, and alfalfas. J. Econ. Entomol. 63:192-195.
118. Luna, A. 1977. Diferenciacion de las dos especies principales de pulgones que dañan a la alfalfa en La Argentina. Programa Alfalfa INTA/FAO ARG 75/006, Hoja Informativa, Ano 2, 2:1-11. INTA, Castelar, Argentina.
119. MacKauer, M. 1983. Determination of parasite preference by choice tests: The *Aphidius smithi* (Hymenoptera: Aphidiidae)—pea aphid (Homoptera: Aphididae) model. Ann. Entomol. Soc. Am. 76:256-261.
120. MacCollom, G.B., G.L. Baumann, N.L. Gilroy, and J.G. Welch. 1982. Alfalfa blotch leafminer, *Agromyza frontella* (Diptera: Agromyzidae), effects on alfalfa in Vermont. Can. Entomol. 114:673-680.
121. MacLean, P.S., and R.A. Byers. 1983. Ovipositional preferences of the alfalfa blotch leafminer (Diptera: Agromyzidae) among some simple and glandular-haired *Medicago* species. Environ. Entomol. 12:1083-1086.
122. MacLeod, D.M. 1955. A fungus enemy of the pea aphid, *Macrosiphum pisi* (Kaltenbrock). Can. Entomol. 87:503-505.
123. Maddox, J.V., and W.H. Luckmann. 1966. A microsporidian disease of the alfalfa weevil. J. Invertebr. Pathol. 8:543-544.

124. Manglitz, G.R. 1958. Aestivation of the alfalfa weevil. J. Econ. Entolmol. 51:506–508.
125. Manglitz, G.R., C.O. Calkins, R.J. Walstrom, S.D. Hintz, S.D. Kindler, and L.L. Peters. 1966. Holocyclic strain of the spotted alfalfa aphid in Nebraska and adjacent states. J. Econ. Entomol. 59:636–639.
130. ———, and J.M. Schalk. 1970. Occurrence and hosts of *Aphelinus semiflavus* Howard (Hymenoptera: Eulophidae) in Nebraska. J. Kans. Entomol. Soc. 43:309–314.
131. ———, ———, L.W. Andersen, and K.P. Pruess. 1973. Control of the army cutworm on alfalfa in Nebraska. J. Econ. Entomol. 66:299.
126. ———, and H.J. Gorz. 1974. Additional hosts of the "Yellow Clover Aphid Complex". J. Econ. Entomol. 67:453–454.
129. ———, and L.M. Russell. 1974. Cross matings between *Therioaphis maculata* (Buckton) and *T. trifolii* (Monell) (Hemptera: Homoptera: Aphididae) and their implications in regard to the taxonomic status of the insects. Proc. Entomol. Soc. Wash. 76:290–296.
127. ———, W.R. Kehr, D.L. Keith, J.M. Mueke, J.B. Campbell, R.L. Ogden, and T.P. Miller. 1980. Alfalfa insect management studies 1971–1977. Nebr. Agric. Exp. Stn. Res. Bull. 293.
128. ———, L.E. Klostermeyer, and D.L. Keith. 1981. Comparisons of eastern and western strains of the alfalfa weevil in Nebraska. J. Econ. Entomol. 74:581–588.
132. Markkula, M. 1963. Studies on the pea aphid, *Acyrthosiphon pisum* Harris (Hom., Aphididae), with special reference to the differences in the biology of the green and red forms. Ann. Agric. Fenn. 2. 1–30.
133. Meisch, M.V., and N.M. Randolph. 1965. Life-history studies and rearing techniques for the three-cornered alfalfa hopper. J. Econ. Entomol. 58:1057–1059.
134. Mellors, W.K., and R.G. Helgesen. 1980. Life table analysis for the alfalfa blotch leafminer, *Agromyza frontella* in central New York. Environ. Entomol. 9:738–742.
135. Metcalf, C.L., and W.P. Flint. 1939. Destructive and useful insects. 2nd ed. McGraw-Hill Book Co., New York.
136. Miller, D.E., and G.L. Jensen. 1970. Agromyzid alfalfa leaf miners and their parasites in Massachusetts. J. Econ. Entomol. 63:1337–1338.
137. Neal, J.W., Jr., and R.H. Ratcliffe. 1975. Clover root curculio: Control with granular carbofuran as measured by alfalfa regrowth, yield, and root damage. J. Econ. Entomol. 68:829–831.
138. Nechols, J.R., M.J. Tauber, C.A. Tauber, and R.G. Helgesen. 1983. Environmental regulation of dormancy in the alfalfa blotch leafminer, *Agromyza frontella* (Diptera: Agromyzidae). Ann. Entomol. Soc. Am. 76:116–119.
139. Neiman, E.L. 1971. The identification and characterization of pea aphid biotypes. Ph.D. diss. Univ. of Nebraska-Lincoln (Diss. Abst. Int. B. 32:6454).
140. Newsom, L.D., P.L. Mitchell, and N.N. Troxclair, Jr. 1983. Overwintering of the threecornered alfalfa hopper in Louisiana. J. Econ. Entomol. 76:1298–1302.
141. Nielson, M.W. 1976. Dispause in the alfalfa seed chalcid, *Bruchophagus roddi* (Gussakovsky) in relation to natural photoperiod. Environ. Entomol. 5:123–127.
142. ———, and W.E. Currie. 1962. Leafhoppers attacking alfalfa in the Salt River Valley of Arizona. J. Econ. Entomol. 55:803–804.
143. Onstad, D.W., C.A. Shoemaker, and B.C. Hansen. 1984. Management of potato leafhopper, *Empoasca fabae* (Homoptera: Cicadellidae), on alfalfa with the aid of systems analysis. Environ. Entomol. 13;1046–1058.
144. Painter, R.H. 1958. Resistance of plants to insects. Annu. Rev. Entomol. 3:267–290.
145. Painter, R.H. and C.O. Grandfield. 1935. Preliminary report on resistance of alfalfa varieties to the pea aphid. Agron. J. 27:71–674.
146. Palm, C.E. 1935. The alfalfa snout beetle, *Brachyrhinus ligustici* L. N.Y. Agric. Exp. Stn. Ithaca Bull. 629.
147. Parker, J.R., and R.V. Connin. 1964. Grasshoppers—their habits and damage. USDA Agric. Info. Bull. 287. U.S. Government Printing Office, Washington, DC.
148. Parman, V.R., and M.C. Wilson. 1982. Alfalfa crop responses to feeding by the meadow spittlebug (Homoptera: Cercopidae). J. Econ. Entomol. 75:481–486.
149. Peterson, A.G., J.D. Bates, and R.S. Saini. 1969. Spring dispersal of some leafhoppers and aphids. J. Minn. Acad. Sci. 35:98–102.
150. Pienkowski, R.L., and J.T. Medler. 1962. Effects of alfalfa cuttings on the potato leafhopper, *Empoasca fabae*. J. Econ. Entomol. 55:973–978.
151. Pimental, D., and A.G. Wheeler, Jr. 1973. Species and diversity of arthropods in the alfalfa community. Environ. Entomol. 2:659–668.
152. Pitre, H.N. 1969. Field studies on the biology of the alfalfa weevil, *Hypera postica*, in northeast Mississippi. Ann. Entomol. Soc. Am. 62:1485–1489.
153. Plummer, J.A., and R.A. Byers. 1981. Seasonal abundance and parasites of the alfalfa blotch leafminer, *Agromyza frontella*, in central Pennsylvania. Environ. Entomol. 10:105–110.
154. Poinar, G.O., and G.G. Gyrisco. 1960. A nematode parasite of the alfalfa weevil (*Hypera postica* (Gyll) J. Econ. Entomol. 53:178–179.

155. ----, and ----. 1962. A new mermithid parasite of the alfalfa weevil, *Hypera postica* (Gyllenhal). J. Insect Pathol. 4:201–206.
156. ----, and ----. 1962. Studies on the bionomics of *Hexamermis arvalis* Poinar and Gyrisco, a mermithid parasite of the alfalfa weevil, *Hypera postica* (Gyllenhal). J. Insect Pathol. 4:469–483.
157. Poos, F.W., and T.L. Bissell. 1953. The alfalfa weevil in Maryland. J. Econ. Entomol. 46:178–179.
158. Prescott, H.W., and M.M. Reeher. 1961. The pea leaf weevil—an introduced pest of legumes in the Pacific Northwest. USDA Tech. Bull. 1233. U.S. Government Printing Office, Washington, DC.
159. Pruess, K.W. 1967. Migration of the army cutworm, *Chorizagrotis auxiliaris* (Lepidoptera: Noctuidae). I. Evidence for a migration. Ann. Entomol. Soc. Am. 60:910–920.
160. Puttarudriah, M. 1953. The natural control of the alfalfa looper in central California. J. Econ. Entomol. 46:723.
161. Puttler, B. 1967. Interrelationship of *Hypera postica* (Coleoptera: Curculionidae) and *Bathyplectes curculionis* (Hymenoptera: Ichneumonidae) with particular reference to encapsulation of the parasite eggs by the weevil larvae. Ann. Entomol. Soc. Am. 60:1031–1038.
162. ----, D.L. Hostetter, S.H. Long, and R.E. Pinnell. 1978. *Entomophthora phytonomi*, a fungal pathogen of the alfalfa weevil in the mid-Great Plains. Environ. Entomol. 7:670–671.
163. Quiring, D.T., and J.N. McNeil. 1984. Adult-larval intraspecific competition in *Agromyza frontella* (Diptera: Agromyzidae). Can. Entomol. 116:1385–1391.
164. Radcliffe, E.B., and D.K. Barnes. 1970. Alfalfa plant bug injury and evidence of plant resistance in alfalfa. J. Econ. Entomol. 63:1995–1996.
165. ----, R.W. Weires, R.E. Stucker, and D.K. Barnes. 1976. Influence of cultivars and pesticides on pea aphid, spotted alfalfa aphid, and associated arthropod taxa in a Minnesota alfalfa ecosystem. Environ. Entomol. 5:1195–1207.
166. Rakickas, R.J., and T.F. Watson. 1974. Population trends of *Lygus* spp. and selected predators in strip-cut alfalfa. Environ. Entomol. 3:781–784.
167. Randolph, N.M., and M.V. Meisch. 1970. Evaluation of chemicals and resistant alfalfa varieties for control of the three-cornered alfalfa hopper. J. Econ. Entomol. 63:979–981.
168. Ratcliffe, R.H. 1979. Insect resistance in alfalfa: Present status and future possibilities. p. 64–69. *In* H.D. Wells (ed.) Proc. 36th Southern Pasture and Forage Crop Improv. Conf., Beltsville, MD. 1–3 May. SEA-AR-Southern Region, New Orleans.
169. Richard, C., and M.O'C. Guibord. 1980. Relationship of alfalfa blotch leafminer with spring black stem. Can. J. Plant Sci. 60:265–266.
170. Ridland, P.M., and G.M. Halloran 1980. Influence of alfalfa bud mite on growth of alfalfa under different temperatures. Crop Sci. 20:790–792.
171. Rockwood, L.P. 1950. Entomogenous fungi of the family Entomophthoraceae in the Pacific northwest. J. Econ. Entomol. 43:704–707.
172. Ruesink, W.G., C.A. Shoemaker, A.P. Gutierrez, and G.W. Fick. 1980. The systems approach to research and decision making for alfalfa insect control. p. 217–246. *In* C. Huffaker (ed.) New technology of pest control. John Wiley and Sons, New York.
173. Russell, E.E. 1952. Stink bugs on seed alfalfa in southern Arizona. USDA Cir. 903. U.S. Government Printing Office, Washington, DC.
174. Schalk, J.M., and G.R. Manglitz. 1972. Influence of light and temperature on the production of sexuales of *Therioaphis maculata* and *T. riehmi*. Environ. Entomol. 1:209–213.
175. Schroder, R.F.W., and W.W. Metterhouse. 1980. Population trends of the alfalfa weevil (Coleoptera: Curculionidae) and its associated parasites in Maryland and New Jersey, 1966–1970. N.Y. Entomol. Soc. 88:151–163.
176. Searls, E.M. 1932. A preliminary report of the resistance of certain legumes to certain homopterous insects. J. Econ. Entomol. 25:46–49.
177. ----. 1934. The effect of alfalfa cutting schedules upon the occurrence of the potato leafhopper (*Empoasca fabae* Harris) in Wisconsin. J. Econ. Entomol. 27:80–88.
178. ----. 1935. Further studies on the effect of controlling the potato leafhopper (*Empoasca fabae* Harris) in alfalfa by designed cutting. J. Econ. Entomol. 28:831–833.
179. Sevacherian, V., and V.M. Stern. 1975. Movements of lygus bugs between alfalfa and cotton. Environ. Entomol. 4:163–165.
180. Severin, H.C. 1935. The common black field cricket a serious pest in South Dakota. S.D. Agric. Exp. Stn. Bull. 295.
181. Sharma, R., and V. Stern. 1980. Blue alfalfa aphid: Economic threshold levels in southern California. Calif. Agric. 34:16–17.
182. Simonet, D.E., and R.L. Pienkowski. 1979. Impact of alfalfa harvest on potato leaf-

hopper populations with emphasis on nymphal survival. J. Econ. Entomol. 72:428–431.
183. ——, ——, D.G. Martinez, and R.D. Blakesless. 1979. Evaluation of sampling techniques and development of a sampling program for potato leafhopper adults on alfalfa. Environ. Entomol. 8:397–399.
184. Simpson, R.G., and C.C. Burkhardt. 1960. Biology and evaluation of certain predators of *Therioaphis maculata* (Buckton). J. Econ. Entomol. 53:89–94.
185. Smith, R.C. 1956. Effects of high soil and air temperatures on certain alfalfa insects. J. Kans. Entomol. Soc. 29:1–19.
186. ——, and E.W. Davis. 1926. The pea aphid as an alfalfa pest in Kansas. J. Agric. Res. 33:47–57.
187. ——, and W.W. Franklin. 1954. The garden webworm—*Loxostege similalis* Guen. as an alfalfa pest in Kansas. J. Kans. Entomol. Soc. 27:27–38.
188. ——, and ——. 1961. Research notes on certain species of alfalfa insects at Manhattan (1904–1956) and at Fort Hays, Kansas (1948–1953). Kans. Agric. Exp. Stn. Report of Progress 54.
189. Smith, R.F., D.E. Bryan, and W.W. Allen. 1949. The relation of flights of *Colias* to larval population density. Ecology 30:288–297.
190. ——, and K.S. Hagen. 1959. The integration of chemical and biological control of the spotted alfalfa aphid. Part III. Impact of commercial insecticide treatments. Hilgardia 29:131–154.
191. Snow, S.J., and S. McClellan. 1951. *Tortrix pallorana*, a pest of seed alfalfa in Utah. J. Econ. Entomol. 44:1023.
192. Soldan, T., and P. Stary. 1981. Parasitogenic effects of *Aphidius smithi* (Hymenoptera: Aphidiidae) on the reproductive organs of the pea aphid, *Acyrthosiphon pisum* (Homoptera: Aphididae). Acta Entomol. Bohemoslov. 78:243–253.
193. Sorenson, C.J. 1932. Insects in relation to alfalfa-seed production. Utah State Univ. Agric. Exp. Stn. Circ. 98.
194. ——. 1939. Lygus bugs in relation to alfalfa seed production. Utah Agric. Exp. Stn. Bull. 284.
195. ——. 1946. Mirid-bug injury as a factor in declining alfalfa yields. *In* Fifth Annual Faculty Research Lecture. Utah State Agric. Coll.
196. Stanford, E.H. 1956. Aphid resistant alfalfa plants. Calif. Agric. 10:3.
197. ——, V. Sevacherian, A. Mueller, and J. Ryan. 1968. Effect of Naled, Trichlorofon, and *Bacillus thuringiensis* on three species of lepidopterous larvae attacking alfalfa in California. J. Econ. Entomol. 61:1324–1327.
198. ——, R.J. Sharma, and C. Summers. 1980. Alfalfa damage from *Acyrthosiphon kondoi* and economic threshold studies in southern California. J. Econ. Entomol. 73:145–148.
199. Stern, V.M., and R. van den Bosch. 1959. Field experiments on the effects of insecticides. Hilgardia 29:103–130.
200. Stitt, L.L. 1940. Three species of the genus *Lygus* and their relation to alfalfa seed production in southern Arizona and California. USDA Tech. Bull. 741. U.S. Government Printing Office, Washington, DC.
201. Stoetzel, M. 1985. Personal communication, USDA-ARS. Insect Identification and Beneficial Insect Introduction Institute, Beltsville, MD.
202. Strong, F.E. 1970. Physiology of injury caused by *Lygus hesperus*. J. Econ. Entomol. 63:808–814.
203. ——, J.A. Sheldahl, P.R. Hughes, and Esmat M.K. Hussein. 1970. Reproductive biology of *Lygus hesperus* Knight. Hilgardia 40:105–147.
204. Struble, D.L., and G.E. Swailes. 1977. A sex attractant for the adult male of the army cutworm, *Euxoa auxiliaris*: A mixture of Z-5-tetradecen-1-yl acetate and E-7-tetradecen-1-yl acetate. Environ. Entomol. 6:719–724.
205. Sutter, G.R., and E. Miller. 1972. Rearing the army cutworm on an artificial diet. J. Econ. Entomol. 63:717–718.
206. Suzuki, M., and L.S. Thompson. 1981. Effects of alfalfa blotch leafminer on chemical components of alfalfa. Can. J. Plant Sci. 61:595–600.
207. Therrien, P., and J.N. McNeil. 1985. Effects of removing the plant canopy on pupal survival and adult emergence of the alfalfa blotch leafminer, *Agromyza frontella* (Diptera: Agromyzidae). Can. Entomol. 117:167–170.
208. Thompson, C.G. 1956. The use of certain microorganisms to control the alfalfa caterpillar. Proc. Int. Congr. Entomol., 10th 4:693.
209. Thompson, L.S. 1981. Field evaluation of insecticides for control of the alfalfa blotch leafminer and its effect on alfalfa yield in Prince Edward Island. J. Econ. Entomol. 74:363–365.
210. Tingey, W.M., and M.W. Nielson. 1974. Alfalfa seed chalcid: Nonpreference resistance in alfalfa. J. Econ. Entomol. 67:219–221.
211. Tippins, H.H. 1964. Effect of winter burning on some pests of alfalfa. J. Econ. Entomol. 57:1003–1004.

INSECTS AND MITES

212. Titus, E.G. 1904. Some preliminary notes on the clover-seed chalcis-fly. Preliminary Notes. USDA Div. of Entomol. Bull. 44.
213. ----. 1910. The alfalfa leaf weevil. Utah State Coll. Agric. Exp. Stn. Bull. 110.
214. ----. 1913. The control of the alfalfa weevil. Utah Agric. Exp. Stn. Circ. 10:105-120.
215. Tower, D.C., and F.A. Fenton. 1920. Clover leaf weevil. USDA Bull. 922. U.S. Government Printing Office, Washington, DC.
216. Tuttle, D.M., O.L. Barnes, M.W. Nielson, F.D. Roth, and M.H. Schornhorst. 1958. The spotted alfalfa aphid in Arizona. Ariz. Agric. Exp. Stn. Bull. 294.
217. ----, and G.D. Butler. 1954. The yellow clover aphid—a new alfalfa pest in the southwest. J. Econ. Entomol. 47:1157.
218. Underhill, G.W., E.C. Turner, Jr., and R.G. Henderson. 1955. Control of the clover root curculio on alfalfa with notes on life history and habits. J. Econ. Entomol. 48:184-187.
219. Urbahns, T.D. 1920. The clover and alfalfa seed chalsis-fly. USDA Bur. Entomol. Bull. 812.
220. van den Bosch, R., E.I. Schlinger, and E.J. Dietrick. 1957. Imported parasites established. Calif. Agric. 11:11-12.
221. ----, ----, ----, J.C. Hall, and B. Puttler. 1964. Studies on succession, distribution, and phenology of imported parasites of *Therioaphis trifolii* (Monell) in southern California. Ecology 45:602-621.
222. ----, ----, C.E. Lagace, and J.C. Hall. 1966. Parasitism of *Acyrthosiphon pisum* by *Aphidius smithi* a density dependent process in nature (Homoptera: Aphididae) (Hymenoptera: Aphidiidae). Ecology 47:1049-1054.
223. Van Steenwyk, R.A., and V.M. Stern. 1976. The biology of *Peristenus stygicus* (Hymenoptera: Braconidae), a newly imported parasite of lygus bugs. Environ. Entomol. 5:931-934.
224. Walkden, H.H. 1950. Cutworms, armyworms and related species attacking cereal and forage crops in the central great plains. USDA Circ. 849. U.S. Government Printing Office, Washington, DC.
225. Walstrom, R.J. 1983. Plant bug (Heteroptera: Miridae) damage to first-crop alfalfa in South Dakota. J. Econ. Entomol. 76:1309-1311.
226. Washburn, R.H., and L.J. Klebesadel. 1964. *Sitona scissifrons* (Coleoptera: Curculionidae), a potential hazard to alfalfa production in Alaska. J. Econ. Entomol. 57:995.
227. Weaver, C.R., and D.R. King. 1954. Meadow spittlebug. Ohio Agric. Exp. Stn. Res. Bull. 741.
228. Webster, F.M. 1912. Preliminary report on the alfalfa weevil. USDA Bur. Entomol. Bull. 112. U.S. Government Printing Office, Washington, DC.
229. Wedberg, J.L., W.G. Ruesink, E.J. Armbrust, and D.P. Bartell. 1977. Alfalfa weevil pest management program. Ill. Ext. Circ. 1136.
230. Wehrle, L.P. 1939. A new insect introduction. Bull. Brooklyn Entomol. Soc. 34:170.
231. Wheeler, A.G., Jr. 1974. Studies on the arthropod fauna of alfalfa. VI. Plant bugs (Miridae). Can. Entomol. 106:1267-1275.
232. White, C.E., E.J. Armbrust, and J. Ashley. 1972. Cross-mating studies of eastern and western strains of alfalfa weevil. J. Econ. Entomol. 65:85-89.
233. Wildermuth, V.L. 1910. The clover root curculio. USDA Bur. Entomol. Bull. 85, Part III: 29-38. U.S. Government Printing Office, Washington, DC
234. ----. 1915. Three-cornered alfalfa hopper. J. Agric. Res. 343-362.
235. ----. 1931. Chalcid control in alfalfa seed production. USDA Farmer's Bull. 1642. U.S. Government Printing Office, Washington, DC.
236. Wilson, M.C. 1984. Manage the alfalfa weevil to improve alfalfa yield and quality. Certified Alfalfa Seed Council, Woodland, CA.
237. ----, R.T. Huber, J.F. Gerhold, and T.R. Hintz. 1969. Buildup of the alfalfa weevil parasite *Bathyplectes curculionis* in Indiana. J. Econ. Entomol. 62:1517-1518.
238. ----, J.K. Stewart, and H.D. Vail. 1979. Full season impact of the alfalfa weevil, meadow spittlebug and potato leafhopper in an alfalfa field. J. Econ. Entomol. 72:830-834.
239. Wood, K.A., E.J. Armbrust, D.P. Bartell, and B.J. Irwin. 1978. The literature of arthropods associated with alfalfa: V. A bibliography of the alfalfa weevil, *Hypera postica* (Gyllenhal), and the Egyptian alfalfa weevil, *Hypera brunneipennis* (Boheman) (Coleoptera: Curculionidae). Ill. Agric. Exp. Stn. Spec, Pub. 54.
240. Yadava, C.P., and F.R. Shaw. 1968. The preferences of certain coccinellids for pea aphids, leafhoppers and alfalfa weevil larvae. J. Econ. Entomol. 61:1104-1105.
241. Yee, J.M., and D.G. Harcourt. 1983. SIM WEEVIL/SIM ABL, an IPM monitoring system for the alfalfa weevil and alfalfa blotch leafminer. Agric. Can. Res. Br. Contrib. 1983-5E. Research Program Service, Agriculture Canada, Ottawa.
242. York, A.C., and G.G. Gyrisco. 1978. Dosage-mortality response and field control of adult alfalfa snout beetle. J. Econ. Entomol. 71:783-784.

243. ----, ----, and C.M. Edmonds. 1971. The status of the alfalfa snout beetle in New York State. J. Econ. Entomol. 64:1332–1333.
244. Young, H.C., B.A. App, J.B. Gill, and H.S. Hollingsworth. 1950. White-fringed beetles and how to combat them. USDA Cir. 850. U.S. Government Printing Office, Washington, DC.

23 Weeds and Weed Control

ELROY J. PETERS

University of Missouri
Columbia, Missouri

DEAN L. LINSCOTT

USDA-ARS
Ithaca, New York

Before the advent of selective herbicides, weeds in alfalfa (*Medicago sativa* L.) were controlled mainly by management and cultural practices applied before and after the crop was planted. Selective herbicides came into use in the mid-1940s, but were used very little in alfalfa because they were expensive, not entirely effective, and the hazard from herbicidal injury was high. Interest in herbicides increased following the introduction of more selective herbicides in the mid-1950s. The most effective herbicides that were safe on seedling alfalfa were the preplant incorporated herbicides that were effective for controlling weed grasses and some broadleaf weeds and for which alfalfa had considerable tolerance. Prior to 1982, postemergence herbicides that controlled emerged weed grasses caused too much injury to alfalfa, but recently developed herbicides are tolerated by alfalfa and have potential for controlling weed grasses in seedling as well as in mature alfalfa.

In the past, the main objective during the seedling year was to get the stand established. It was soon discovered, however, that spring seedlings could be established successfully with herbicides, and with competition from weeds eliminated, high yields of forage could be obtained during the seedling year.

Much of the early research with herbicides was devoted to seedling stands. Research since about 1970 has demonstrated the efficacy of herbicides for weed control in mature alfalfa and a number of them have been labeled for this use.

Herbicides are valuable tools that may be substituted for some cultural practices in controlling unwanted vegetation, for example, in limited tillage systems. In general, they represent an additional approach to weed control that should be integrated with other good cultural practices.

23–1 LOSSES CAUSED BY WEEDS

23–1.1 Competition for Water, Nutrients, and Light

Weeds reduce yields of alfalfa through competition for water, nutrients, light, and space; and they lower the quality of forage and seed (19, 33, 38, 142).

Copyright 1988 © ASA-CSSA-SSSA, 677 South Segoe Road, Madison, WI 53711, USA.
Alfalfa and Alfalfa Improvement—Agronomy Monograph no. 29.

Because alfalfa seedlings grow slowly, they are especially susceptible to weed competition and often fail to become established because of heavy weed infestations in the absence of adequate control measures. Seeding failures are costly because of the lost investment in seed, time, and tillage. Furthermore, seeding failures may interfere with rotational sequences and cause much inconvenience in adjusting management practices. When seedings fail, farmers often are forced to make costly investments in the establishment of emergency feed crops, or to purchase feed.

23–1.2 Production of Inhibitors and Toxins

In addition to competition for water, nutrients, and light, some weeds may reduce yield through the production of toxins or growth inhibiting materials (83). It is often difficult to establish alfalfa on newly plowed land that had been infested with quackgrass [*Agropyron repens* (L.) Beauv.], and alfalfa growing on such land is low in vigor and yield (66). Investigations attribute the low vigor of alfalfa to toxins from substances originating from quackgrass (97, 145). Similarly, decomposition products from plant tissue may be toxic to other plants (98). Wilson (170) showed that ground Canada thistle [*Cirsium arvense* (L.) Scop.] tissue reduced alfalfa growth. The parasitic dodders (*Cuscuta* spp.) have been shown to cause production of substances toxic to host plants (60, 61) and also reduce content of alfalfa (130).

23–1.3 Reduced Quality of Forage and Seed

Weeds are objectionable because they lower the nutritional value of alfalfa forage. Many weeds are not eaten by livestock, and others, when eaten, contribute less protein and minerals than alfalfa (19, 38, 142). Marten and Anderson (81) showed that redroot pigweed (*Amaranthus retroflexus* L.) had a nutritional composition and digestibility equal to that of high quality alfalfa when harvested in mid-July. However, weeds are often beyond their optimum nutritional stage when harvested with alfalfa. Weeds such as bromes (*Bromus* spp.), quackgrass, and johnsongrass [*Sorghum halepense* (L.) Pers.] even at optimum stages are less nutritious than alfalfa (19, 38). Although young weeds may be eaten by livestock, they become unpalatable with advancing maturity and subject to refusal.

Weeds such as curly dock (*Rumex crispus* L.) and horseweed [*Conyza canadensis* (L.) Cronq.] have coarse stems that livestock reject. Other weeds have awns, spines, and thorns that cause cattle (*Bovine* spp.) to avoid them. When eaten, awns of grass species, such as foxtail barley (*Hordeum jubatum* L.) and weedy bromegrasses (*Bromus* spp.), have been known to injure the mouths of animals (43, 94, 113). The awns may become embedded in the tissues and cause mouth infections. Spines and thorns may cause similar physical damage if eaten.

Some weeds have objectional odors or flavors and will not be eaten by cattle (94). Other weeds, such as wild garlic (*Allium vineale* L.) and ragweed (*Ambrosia artemisiifolia* L.), may be eaten and impart a undesirable flavor to milk or meat of cattle (94, 112).

Another objection to weedy hay is that it may contain weed seeds, and when transported or fed, introduces weed seeds to new areas. Because of its undesirable characteristics, weedy hay usually will sell at a reduced price when marketed (144), and in some situations, it may prove unacceptable in the market (119). Thus, hay infested with noxious weed seeds, such as dodder (*Cuscuta* spp.), may have no market value.

Where alfalfa is grown for processing in dehydrating plants, weedy alfalfa may not be accepted. Dehydrated alfalfa must have high protein and other nutritive qualities (129). Weeds frequently reduce the protein and nutritive value of products made from dehydrated alfalfa.

Large monetary losses occur when weed-infested alfalfa is grown for seed. Those losses result from reduced production as well as from increased costs involved with growing, transporting, and conditioning of combined seed to provide weed-free seed. When seedling alfalfa was kept weed free, it yielded 820 kg/ha seed compared with 45 kg/ha from weedy alfalfa (33). It is estimated that in the USA, weeds reduce production of alfalfa seed by 12% (152). An additional 4% of the seed is lost in conditioning, and a 2% loss results from lower prices because of quality. These losses during seed production amounted to about $9.5 million in the USA in 1957.

A high proportion of many types of weed seeds are dormant and germinate over periods of years; therefore, once a field becomes infested, eradication is difficult. For this reason, alfalfa growers attempt to prevent infestation by planting weed-free seed. Dodder is an especially troublesome weed in the alfalfa seed growing areas of the western USA. Many weed seeds, such as dodder, are similar in size to alfalfa seed and difficult to separate once mixed. Consequently, seed conditioning plants require expensive equipment and considerable labor to remove weed seeds. The presence of weed seed that are difficult to remove from alfalfa during seed conditioning result in heavy clean-outs and large losses to the seed grower.

23-2 WEED PROBLEMS

23-2.1 Spring Seedlings

Weeds may be a more serious problem in spring-planted alfalfa than in alfalfa planted at other times. In spring, abundant moisture and variable temperature contribute to the germination of large numbers of weed seeds. Furthermore, the winter environment breaks the dormancy of many weed seeds to allow germination in spring. Many weed seeds have already begun to germinate when the alfalfa is planted and emerge ahead of the alfalfa. By the time the alfalfa seedlings emerge from the soil, weed seedlings frequently have made some growth and have a competitive advantage.

Annual weeds are the most serious weed problem in spring seedings of alfalfa. Although seedlings of perennial weeds may be present, they usually grow slowly and seldom cause as much competition for seedling alfalfa; however, seedling quackgrass and yellow nutsedge (*Cyperus esculentus* L.) can be very competitive. If the soil is heavily infested with rhizomes of johnsongrass or quackgrass, and their growth is not reduced by either tillage or herbicides, then regrowth of these perennials will suppress the development of alfalfa seedlings. Sometimes fleshy roots of weeds such as curly dock and dandelion (*Taraxacum officinale* Weber), present in a previous crop, will survive tillage and produce new growth in seedling alfalfa.

23–2.2 Summer and Fall Seedings

Weeds in summer and fall seedings may be less of a problem than in spring seedings. Favorable conditions in the spring cause weeds to germinate abundantly and reduce the supply of nondormant weed seeds by late summer and fall, and relatively few of the remaining seeds are ready to germinate. Daylength and other environmental conditions cause many weeds found in alfalfa seedings to produce more growth in spring than in late summer. As the daylength shortens in late summer, most annual weeds that emerge in summer go into the reproductive stage in a shorter time than those that germinate in the spring, and consequently offer less competition to new seedings. Furthermore, most summer annual weeds are frost sensitive and subject to killing at a time when alfalfa can be expected to continue growth.

Winter annual weeds, such as common chickweed [*Stellaria media* (L.) Cyrillo], shepherdspurse [*Capsella bursa-pastoris* (L.) Medic.], knawel (*Scleranthus annus* L.), and *Bromus* spp., are serious problems in some summer seedings. These winter annual species germinate from late summer through fall and winter. They continue growth after the alfalfa becomes dormant and frequently become so dense that they seriously reduce stands and yields (15, 160).

23–2.3 Established Alfalfa Stands

Summer annual weeds seldom infest mature alfalfa stands providing the stands are dense and growing vigorously. Shading from the alfalfa prevents weed seedlings from becoming established (22, 26). In thin alfalfa stands weeds can increase in open areas where sunlight reaches the ground. Winter annual weeds that germinate and grow while the alfalfa is dormant can become a serious problem even in vigorous stands of alfalfa (3, 7).

Under intensive alfalfa production in irrigated areas of California and in areas where the water table is high enough for subirrigation, annual weeds are a common problem. Here, nondormant or intermediate-dormant types of alfalfa are harvested five to six times during the year following fall seeding. Excessive traffic from harvesting machinery and frequent cutting may cause alfalfa stands to decline soon after establishment, and annual weeds invade these thin stands.

Dodder species are annual weeds that are troublesome in both seedling and mature stands of alfalfa (146). These parasitic weeds germinate in the soil and twine around the alfalfa plants (32, 172). On contacting the alfalfa plant, dodder produces haustoria that penetrate the cells of the host plant. The dodder then loses its connection with the soil and lives entirely from nutrients extracted through the haustoria from the host plant. In addition to draining nutrients from its host, dodder also causes an increase in toxic compounds in the host plant (60, 61). Dodder plants grow rapidly and often produce dense mats that cover the alfalfa plants and reduce the amount of sunlight reaching the alfalfa.

Dodder is especially troublesome in alfalfa seed fields, where it not only reduces yields but contaminates the alfalfa seed. Lee and Timmons (71) showed that when heavy infestations of dodder were controlled, seed yields were increased 500%. Dodder produces abundant seed, and a high percentage of the seed is dormant. Dawson (25) planted dodder seed and found that only 10% of the seed germinated during the initial 2-yr period. Thus, once a field is contaminated with dodder, seed will germinate for many years and continue to reinfest the field. Because dodder is such a tenacious weed, alfalfa seed containing dodder should not be planted. Dawson (27) stated that methods used in seed fields that control <100% of the dodder plants are unsatisfactory. Conversely, dodder observed in alfalfa fields grown for forage production may not become a serious problem unless the infestation is heavy and harvesting is delayed. Early cutting when alfalfa is in the bud stage will help to reduce damage.

Three species of dodder infest alfalfa fields (32, 172). Largeseed dodder (*Cuscuta indecora* Choisy) and field dodder (*C. campestris* Yunck.) seed are similar in size to alfalfa seed and very difficult to remove from harvested alfalfa seed. Smallseed dodder (*C. planiflora* Tenore) seeds are smaller and easily separated from those of alfalfa. Smallseed dodder is troublesome mainly in the western USA, whereas other dodders are widespread.

Dense stands of mature alfalfa reduce the changes of biennial and perennial weed seedlings from becoming established (136). Most biennial and perennial weeds present in mature stands of alfalfa were present in the seedling alfalfa. However, perennial weed seedlings may come into established alfalfa stands, especially as they become thin (69).

Weeds such as quackgrass, johnsongrass, and field bindweed (*Convolvulus arvensis* L.) may become established as seedlings in alfalfa, but most often these weeds originate from rhizomes and root stocks that survived tillage before the alfalfa was sown. Quackgrass increases in density in older alfalfa stands, and is especially troublesome in thin stands of alfalfa.

23-2.4 No-till Planting

One of the primary purposes of tillage is to control weeds. With the advent of herbicides, opportunities increased for reducing tillage in alfalfa production. Management for weed control is of primary importance, however, because of shifts in weed populations in response to changes in tillage and herbicide practices, and the inherent weaknesses of presently available herbicides (79). In theory, annual weed species are most readily handled in no-tillage systems. Since fields are not plowed, seeds from annuals escaping control remain on or close to the soil surface, and hence remain susceptible to unfavorable environmental effects. Over time, annual weed populations should decline in limited-tillage systems with good management (147).

The technology for controlling most annual and perennial grass species in no-till alfalfa production is known, but the needed herbicides are not labeled for this purpose. Further, those presently available are subject to improper use so that grass problems, particularly perennial grasses, remain significant at least in the near term. Perennial broadleaf weeds represent a significant concern both now and in the future. Selective herbicides available for many perennial broadleaf species are marginal in effectiveness. Without the benefit of tillage, perennial broadleaf species normally associated with hayfields and sods have become an increasingly difficult problem. These weeds include: plantain species (*Plantago* spp.), dandelion (*Taraxacum officinale* Weber) (127), cinquefoils (*Potentilla* spp.), goldenrods (*Solidago* spp.), hawkweeds (*Hieraicum* spp.), docks and sorrels (*Rumex* spp.), several species of thistles (Cardus and *Cirsium* spp.), milkweeds (*Asclepias* spp.), and others (73).

Problems with perennial species are reduced significantly with a combination of conventional and no-tillage management practices. Thus the impact of perennial species would be diminished if a crop or crops tilled and planted using conventional tillage preceded no-tillage alfalfa. Build up of weeds is most easily prevented with the rotation of crops, tillage practices, and herbicide control measures.

23-3 CULTURAL WEED CONTROL METHODS: ALFALFA SEEDINGS

23-3.1 Preventative Weed Control

The best way to limit weeds in alfalfa fields is to prevent their initial introduction. Acquiring weed-free seed is the first step.

Federal and state seed laws aid in protecting the farmer from buying weed seeds through usual commercial channels. The laws do not prevent a grower from using his own seed. A grower who plans to use his own seed for planting can obtain a seed purity analysis from a state or private seed laboratory. Appropriate steps based on the results can then be taken.

Cleaning of tillage and harvesting implements when moving from field to field is essential to prevent transport of seeds and plant parts. Combines, in particular, will carry weed seed lodged in various places on the machines. Plows and harrows can transport vegetative parts of perennial weeds, e.g., the rhizomes of quackgrass. Flood and irrigation water frequently move weed seeds from one area to another.

23-3.2 Time of Planting

Weed problems in alfalfa are frequently avoided by planting at a time of year when weed infestations are expected to be low and growing conditions are most advantageous to the alfalfa. Where alfalfa is grown under natural rainfall, spring seedings have the advantage of more dependable moisture supply than do late summer seedings. In northern areas where the growing season is short, spring seedings will give the plants sufficient time to attain reasonable size before cold weather. However, weed competition may be severe in spring plantings.

In some situations, especially in southern areas where the growing season is long, alfalfa is often sown in summer or fall to avoid the spring weed problem. Also, alfalfa planted in late summer may grow more rapidly and thus be able to compete more successfully with weeds. The lower temperatures in the spring result in slow germination and development of alfalfa seedlings, whereas the summer environment may be more favorable, particularly if there is adequate rainfall. Blaser et al. (13) compared the dry weight alfalfa in Virginia and found 38% greater gain in dry weight from August seedings than from March seedings during the same time span following planting.

There are other problems, however, that make late summer or fall seedings hazardous. Rainfall is frequently limited during late summer and early fall, and seedings frequently fail due to drought. Insect populations, such as grasshoppers (*Melanoplus* spp.), are usually greater in late summer or fall than in spring so that alfalfa seedlings may be destroyed. If alfalfa seedlings do not become well established, they may heave out of the soil because of frequent thawing and freezing. Late summer seedings may not produce enough top growth to cover the soil and reduce or prevent erosion on erosive sites. It may be desirable to plant a winter grain companion crop with alfalfa to reduce the prospects of erosion as well as the deleterious consequences of frequent thawing and freezing. There is some evidence that no-tillage plantings in late summer may be less susceptible to heaving and other winter related injury because of the vegetative cover.

23-3.3 Tillage and Cropping Practices

Major factors in avoiding weed problems in seedling alfalfa are the use of tillage practices and cropping sequences that will deplete the supply of weed seeds and vegetative parts in the soil. The cultural practices must begin before alfalfa is planted (137). Cultivation is sometimes used either alone or with smother crops on fields infested with perennial weeds such as quackgrass, bindweed, and leafy spurge (*Euphorbia esula* L.). Cultivation destroys the top growth of perennial weeds, disrupts the underground system, and gradually depletes food reserves so that the weed infestation is reduced or eliminated.

Herbicides can be substituted for mechanical fallowing. Herbicides such as amitrole [3-amino-*s*-triazole], 2,4-D [2,4-dichlorophenoxy) acetic acid], dalapon [2,2-dichloropropionic acid], and glyphosate [*N*-(phosphonomethyl) glycine] can be used for killing annual and perennial vegetation in advance of seeding. With some herbicides, there is a waiting period before new seedings can be made following herbicide application, because time is required for detoxification. Restrictions, as indicated on the label, must be followed carefully.

A short period of fallow before planting alfalfa also may be beneficial for controlling weeds. Sometimes spring seeding may be delayed until several crops of weeds are destroyed by tillage or herbicides. With summer seeding, tillage may be done during most of the growing season until alfalfa is planted. In other programs, early maturing crops such as small grains may be grown and several crops of weeds destroyed between harvest of the small grain and planting of alfalfa in late summer. If few or no weeds are present after the last tillage, alfalfa may be sown with as little disturbance of the seedbed as possible. This has been called the *stale seedbed technique*; tillage is avoided so that new weed seeds are not brought to the surface.

Herbicides may be more effective than mechanical fallowing for stale seedbeds because each tillage brings another supply of weed seeds to the surface. Lee (70) used paraquat (1,1′dimethyl-4,4′-bipyridinium salt), diquat (6,7-dihydrodipyrido[1,2-*a*:2′,1′c]pyrazinediium salt), propham (isopropyl carbanilate), plus 2,4-D and a combination of amitrole + ammonium thiocyanate in January to kill broadleaf weeds and volunteer grasses. Grasses for seed production were seeded in March without seedbed disturbance. Paraquat poses no danger to alfalfa if applied prior to planting because it is inactivated by the soil upon contact.

Linscott et al. (75) studied the use of paraquat on a stale seedbed that was plowed in spring, but not seeded until midsummer. Paraquat plus a surfactant was applied immediately prior to planting alfalfa. Good control of seedling weeds was obtained at rates as low as 0.33 kg/ha of paraquat. At 1.1 kg/ha of paraquat, sufficient suppression of quackgrass and yellow nutsedge was obtained to ensure excellent establishment of alfalfa.

23-3.4 Influence of Soil Fertility Levels on Weed Control

Any practice that promotes greater seedling vigor will render alfalfa more competitive with weeds. Alfalfa grows best on soils of neutral pH and uses large quantities of soil nutrients; therefore, adequate soil fertility is important to promote vigor. Banding of fertilizer, especially P, with the seed while planting promoted greater early growth of alfalfa and reduced yields of weeds (4). Tesar et al. (143) obtained greater seedling vigor, and 20% more alfalfa plants in a stand seeded directly over fertilizer drilled 3.75-cm deep in 17.5-cm rows, compared with broadcasting seed over banded fertilizer.

In the northeastern USA, several states reported greater seedling growth when P and K were banded 2.5 to 5 cm below seed sown 0.6- to 1.2-cm deep (4). In the majority of seedings, greater seeding growth from banded fertilizer occurred early in the season, with no differences noted later in the season. Stimulation of growth early in the season is desirable because at this stage alfalfa is most susceptible to weed competition.

Pearson and Wakefield (100) and Peters and Stritzke (107) found that additions of NH_4NO_3 increased yields of annual weeds and decreased yield of alfalfa. Combining N with herbicides decreased effectiveness of the herbicides (100). The high level of N (55 kg/ha) combined with 2,4-DB [4-(2,4-dichlorophenoxy) butyric acid], 2,4-DB + dalapon or EPTC (S-ethyl dipropylthiocarbamate) caused a significant loss in alfalfa stand (100). However, a combination of NH_4NO_3 fertilizer and EPTC + 2,4-DB increased yields of seedling 'DuPuits' and 'Vernal' alfalfa but not 'Ranger' (107). In a Canadian study, the application of N fertilizer in spring at seedling emergence increased yields of the first cutting in 2 out of 3 yr (67).

23-3.5 Companion Crops

The widespread practice of seeding alfalfa with a companion crop modified the spring weed problem. The earlier term "nurse crop" has given way to "companion crop," because the associated crop is strongly competitive with alfalfa. The use of a companion crop essentially substitutes a less competitive population for another. Small grains and early canning pea (*Pisum sativum* L.) used as companion crops with alfalfa are more desirable than weeds because they are removed as pasture, hay, or seed and do not compete for the entire season.

In addition to reducing growth of weeds, the companion crop will yield an economic return of forage or grain and straw while alfalfa is becoming established. When weeds are controlled with herbicides, yields of pure seedings of alfalfa will be increased, but the total forage yield for the season will usually be greater with a companion crop (109, 122, 131, 161).

Companion crops are used very little in the southern Corn Belt where oat (*Avena sativa* L.) is no longer grown extensively and wheat (*Triticum aestivum* L.) is heavily fertilized to produce grain and, thus, is too competitive for seedling alfalfa (102). In the humid East and the northern Corn Belt, the use of a companion crop continues as a common but declining cropping practice. In the northeast, >60% of new alfalfa plantings are made without companion crops.

23–3.6 Clipping New Alfalfa Stands

Clipping done at the proper time is an effective method for controlling annual broadleaf weeds. The objective is to remove most or all of the leaf surface and lateral buds from which new growth arises. Weed grasses cannot be controlled effectively by mowing because growth is regenerated from crown buds near the soil surface below the point of mowing. Timeliness of clipping in relation to seedling alfalfa development is important. Clipping arrests annual broadleaf competition, but damage to the alfalfa seedlings will result if it is done too soon.

In a 14-yr, six state study comparing clipping on four different dates, a single clipping of stands with or without oats, about the time the oats were ripe (60–90 d after seeding), was beneficial in terms of hay yields the following year (5). Mazzoni and Scholl (85) found a 1 August mowing of a stand seeded 24 May, without a companion crop, four times as effective as an 8 July mowing. Second year yields from the early cutting were depressed compared with the 1 August cutting. Peters (101) found that weedy plots mowed in June of the year of seeding, in a year with ample rainfall, yielded only one-third as much alfalfa as herbicide-treated plots, but yielded second and third cuttings the same year, and all cuttings in the year following seeding were the same regardless of early management. In a very dry year, however, all plots treated with herbicide yielded more hay throughout the seeding year compared with plots mowed in June. However, the treatments had no effect on yield during the second year of production. Wakefield and Pearson (161) found that delayed clipping favored seasonal control of weeds, especially grass weeds. In the mixed weed stands, shade from broadleaved plants severely suppresses weed grasses so that less regrowth occurs as clipping is delayed.

Clipping after harvest of a small grain crop is advised, if necessary to remove heavy weed growth and/or to remove grain stubble that would adversely affect the quality of the subsequent hay crop (169). In Ohio and Wisconsin, clipping prior to 1 September did not reduce yields the following year (63, 169). In the northern and central USA, clipping as low as possible is advisable because alfalfa recovers primarily from the crown regardless of cutting height and the growing points of many weeds are removed by low cutting. In southern regions, a stubble height of 7.5 cm or more may be accepted as standard practice. Thus the cutting height should coincide with local or regional recommendations. Clippings should be removed if there is any indication that they will smother the alfalfa.

23-4 CHEMICAL WEED CONTROL METHODS: ALFALFA SEEDINGS

23-4.1 General Considerations

Herbicides often have been used to supplement cultural weed control measures. Herbicides used in alfalfa may be broadly classified as primarily for either broadleaf species or grass species, although there are many exceptions where a specific herbicide will control certain species of both classifications. Herbicides active on weed grasses also will be active on forage grasses and any small grains used as companion crops. At present, there are no herbicides registered that will remove weed grasses from forage grasses.

23-4.2 Benefits of Herbicides During Establishment

Although herbicides are applied on seedlings grown with and without companion crops, the principal benefit is from applications made in the absence of companion crops. Because of generally favorable moisture conditions, spring-sown alfalfa seedlings growing without competition from weeds or companion crops seldom fail and frequently produce one to three cuttings of hay during the seedling year. The principal advantage in yield is realized during the year of seeding, although improved yields in subsequent years and greater longevity of stands have been reported (80).

Yields of >6700 kg/ha of oven-dry alfalfa from two or three cuttings during the seedling year have been obtained (101, 109). Other first-year hay yields (in kg/ha) reported include 6540 in Wisconsin (41), 4500 in New York (80), and 5000 in Maine (14).

In the first cutting following seeding, higher yields of alfalfa are obtained from a pure alfalfa seeding treated with herbicides over a seeding in a companion crop. Yields of >2240 kg/ha dry matter within 12 weeks of a May seeding have been reported (58, 68, 85, 123, 161).

The degree of carry-over effect from weed control at the time of seeding is varied. Peters (101) and Kust (68) found that herbicides had little effect on yields of second cutting alfalfa in years with normal rainfall. In a year of deficient rainfall, the depressing effect of weeds in the first cutting carried over into the second cutting.

In some situations, yield differences can be expected in the second and subsequent years of growth following the use of herbicides for weed control during establishment. In studies in six northeastern states, three forage mixtures, with alfalfa seeded in each, with and without oat, each with four clipping dates, there were no significant differences in yields the second year (5). At some locations in some years, there were significant management effects. It was concluded that poor distribution of rainfall affected alfalfa seedling vigor and lowered second-year yields more consistently than any other variable studied.

In contrast, in a long-term study in New York, Lucey et al. (80) compared yields during the seeding year and for the following 2 yr of alfalfa seeded alone and alfalfa seedings made in oat or barley (*Hordeum vulgare* L.) companion crops. Alfalfa yields alone were 4256, 12 768, and 10 952 kg/ha in the seeding, second and third years, respectively, as compared to 582, 9408 and 9408 kg/ha, respectively, when seeded in grain harvested at the mature stage. However, if the cereals were harvested at the silage stage, the alfalfa yields were 1680, 10 304 and 8960 kg/ha, respectively. The greatest total herbage production during the seedling year was obtained from the seeding in cereal cut at the silage stage, 6384 kg/ha of 15% moisture.

The use of herbicides on summer seedings may or may not influence overwintering. Wakefield and Pearson (162), in Rhode Island, reported serious winter heaving of alfalfa in an August seeding when herbicides were used to kill common chickweed and shepherdspurse. The authors assumed that the mulching effect of the living weeds reduced heaving. On the other hand, Kerkin and Peters (59) obtained over twice as much alfalfa in June from a seeding made the previous August (3470 vs. 1450 kg/ha) by using a herbicide that controlled a heavy fall growth of black mustard (*Brassica nigra* L.). The gain was attributed to better establishment of alfalfa plants not subjected to competition from mustard.

23-4.3 Herbicides for Pure Seedings of Alfalfa

The information on herbicides given here and later in the text is intended to give the reader some general information on characteristics and uses of herbicides that either are, have been, or may be registered for use in the USA. The reader is urged to consult herbicide specialists and container labels for specific recommendations and precautions for using herbicides.

EPTC is highly effective on annual weed grasses and nutsedge (*Cyperus* sp.) and only marginally effective on many broadleaf weeds. Effectiveness of EPTC is increased if it is incorporated into the top 5 to 7.5 cm of soil prior to planting. If surface applied, it is largely lost through volatilization. Incorporation by cross cultivation with a disk, harrow, or other implement soon after application reduces loss of the herbicide into the atmosphere. If a disk is used, it should not be set deep. Temporary stunting and malformation of a few alfalfa leaves may occur, but normal growth will follow. Injury to alfalfa can occur when the soil is dry and temperatures are high at planting time. Sometimes EPTC is incorporated into the soil by overhead irrigation soon after herbicide application, or the herbicide can be injected into the irrigation water.

Subsurface placement of EPTC has been studied by Linscott et al. (79). This technique involves placement of nozzles on the rear of a split-boot assembly on a grain drill, so that EPTC is applied at seeding in a 5-cm band under an 18-cm drill row. In Linscott's experiments, adequate annual grass control was obtained with EPTC applied to only two-sevenths of the total area. Broadleaf weed control, however, was incomplete, requiring a postemergence application of 2,4-DB. This technique allows EPTC application during seeding, gives immediate coverage with soil, and thus reduces loss by volatility. Dawson (29, 31) demonstrated that EPTC impregnated in a coating applied to alfalfa seed, controlled weeds in a narrow band near the seed. He later reported (30) that broadcast applications of coated seed controlled weeds at a planting rate of 20 kg/ha and greater. The high seeding rate resulted in sufficient treated seeds to compensate for injury from EPTC.

Dinoseb (2-sec-butyl-4,6-dinitrophenol), formerly called DNBP in the USA, was one of the earlier selective herbicides used on seedling alfalfa (108, 119). Dinoseb is a contact herbicide, effective only within prescribed conditions. If the temperature is in excess of 27°C, alfalfa may be injured or killed. If the temperature is <10°C, alfalfa injury is slight, but likewise, activity on most weed species is marginal. Dinoseb has been used to some extent in the northern USA, but less so in the southern Corn Belt and other areas where high temperatures are apt to occur at application time. Under ideal conditions, weeds are killed with dinoseb when <2.5- to 5-cm tall. Weed grasses, in particular, become increasingly resistant with age. Dinoseb is relatively toxic to humans and animals and care must be exercised to prevent inhalation and body contact. When properly used, it is effective in killing weeds at low cost. However, dinoseb is no longer labeled for use.

Benefin (N-butyl-N-ethyl a,a,a-trifluoro-2,5-dinitro-p-toluidine) is highly effective for controlling seedling grasses. It is incorporated into the seedbed similar to EPTC before alfalfa is planted. It does not provide consistent control of broadleaf weeds, and is especially poor for the control of common ragweed. Alfalfa is seldom injured with benefin even at rates higher than needed for weed control

Profluralin [N-(cyclopropylamino)-6-(methylthio)-s-triazine] was available for use for a number of years as a preplant herbicide. It controlled weed grasses and some broadleaf weeds. It is no longer manufactured.

The amines and esters of 2,4-DB have been shown to be effective for postemergence control of broadleaf weeds in alfalfa (10, 53, 58, 77, 104, 125). The herbicide is usually applied after the legumes have two or three true leaves and are 5 to 8 cm in height. Broadleaf weeds are readily killed when 8 cm or less in height. This herbicide has virtually no activity on emerged grasses.

Lambsquarters (*Chenopodium album* L.), common ragweed, and red root pigweed are controlled with 2,4-DB. These three species are found most frequently in spring seedings. Other species controlled are curly dock seedlings, henbit (*Lamium amplexicaule* L.), pennycress (*Thlaspi arvense* L.), shepherdspurse, prickly lettuce (*Lactuca serriola* L.), Russian thistle (*Salsola kali* L.), kochia (*Kochia scoparia* (L.) Roth), and smartweed (*Polygonum* spp.). *Brassica* spp. vary considerably in their sensitivity to this chemical. A related species, wild radish (*Raphanus raphanistrum* L.), is quite resistant, as is common chickweed. In contrast, wild radish and common chickweed are very sensitive to dinoseb. Temporary injury to alfalfa from 2,4-DB may occur as a narrowing and crinkling of the leaflets. Injury from 2,4-DB may be noted when seedling alfalfa is not growing rapidly, especially under cold and dry or hot and dry conditions. The ester formulation is more likely to cause these formative effects than is the amine. However, the ester formulations usually give better weed control, and it should be selected when growing conditions are good, when species difficult to control are involved, or when weeds are beyond the optimum stage for control.

Differences in metabolism of alfalfa and broadleaf weeds apparently permit alfalfa to tolerate 2,4-DB when many weeds do not. Wain and Wightman (159) report that 2,4-DB is toxic only to plants containing an enzyme system capable of beta-oxidation, which converts the chemical into the phytotoxic acetic acid analog. They suggest that much less beta-oxidation occurs in alfalfa than in many broadleaf weed species; thus, 2,4-DB could be used as a selective postemergence treatment. However, Linscott et al. (78) report that alfalfa possesses an efficient beta-oxidation system and may be tolerant because it is capable of synthesizing inactive chlorophenoxy compounds with longer side chains than the parent compound.

Propham, formerly known as IPC, has limited use for preemergence weed control during alfalfa establishment. It has been used for controlling winter annual weeds in fall seedlings on the west coast of the USA where winter annual grasses predominate (15). Its use is confined to the cooler periods of the year because of rapid loss by volatility under high temperatures. Propham must move into the soil and contact plant roots to be effective, therefore, application on dry soil is likely to be ineffective unless irrigation or rain follows soon after treatment. An analog, chlorpropham (1-methylethyl 3-chlorophenlcarbamate), has a similar limited use in alfalfa establishment, primarily in later stages of establishment. It is used in established stands.

23–4.4 Grass Herbicides

Several new herbicides that selectively control grasses in alfalfa are under development. These include: sethoxydim [2-[1-(ethoxyimino)-butyl]-5-[2-(ethylthio)propyl]-3-hydroxy-2-cyclohexen-1-one], fluazifop [(±)-butyl-2-[4-[[5(trifluoromethyl)-2-pyridinyl]]oxy]phenoxy]propanoic acid], haloxyfop [2-[4-[[3-chloro-5-(trifluoromethyl)-2-pyridinyl]oxy]phenoxy]propanoic acid], and others (1, 8, 11, 64, 111, 163). These herbicides are most effective when applied postemergence to rapidly growing grass in seedling alfalfa. Of the three mentioned, haloxyfop is most effective per unit of active ingredients. Rates of 0.025 kg/ha will control grasses, with the lower ranges effective on annuals and the higher rates required for perennials. Reaction of fluazifop, a close analog of haloxyfop, is similar but about four times the active ingredient is required (11). Fluazifop is effective on perennial grasses at 0.4 kg/ha, with lower rates effective on annual grasses. Sethoxydim controls annual grasses at the 0.1 to 0.4 kg/ha range. It is not as effective on perennial grass as either haloxyfop or fluazifop (1). None of these grass herbicides have a significant effect on annual and perennial broadleaf species. Likely, they will be used in combination with broadleaf herbicides for controlling weeds in seedling alfalfa. Oil from various crop plants at about 1 L/ha increases the effectiveness of these herbicides.

At least a dozen new herbicides are under new development for weed control in soybean [*Glycine max* (L.) Merr.]. Some of these herbicides may find use for weed control problems in alfalfa.

23-4.5 Combinations and Mixtures of Herbicides

Herbicides, tolerated by seedling alfalfa, are not equally effective on weed grasses and broadleaf weeds. Grasses tend to increase in number and vigor when only broadleaf weeds are controlled. This can lead to competition even more serious than in untreated alfalfa because weed grasses tend to be more competitive to seedling alfalfa than broadleaf weeds (86). Two herbicides are sometimes used to control both broadleaf weeds and weed grasses. The herbicides are mixed together prior to application or are applied separately.

23-4.5.1 EPTC and 2,4-DB

Incorporation of EPTC into soil before planting has been used, followed by postemergence application of 2,4-DB when broadleaf weeds were not satisfactorily controlled with EPTC. Linscott et al. (79) found this combination to be particularly effective when placing EPTC in subsurface bands. Control of grasses in and between the bands was satisfactory with EPTC, but a postemergence application of 2,4-DB was needed to control broadleaf weeds. In several states, a broadcast incorporation of EPTC followed by 2,4-DB postemergence is a standard recommendation for weed control in seedling alfalfa.

23-4.5.2 EPTC + Dinoseb

Preplant applications of EPTC followed by an early postemergence treatment of dinoseb controls a wide range of grass and broadleaf weeds, but some injury to alfalfa may occur (76). As indicated previously, particular attention must be given to temperature at the time of dinoseb application. Additionally, reduced rates of dinoseb should be considered when following EPTC.

23-4.5.3 Fluazifop + 2,4-DB

Postemergence tank mixtures of fluazifop and 2,4-DB provide a spectrum of control of grass and broadleaf weeds without injury to alfalfa.

Some antagonistic effect has been noted in control of grasses (155), which suggests the need for an upward adjustment in the fluazifop rate. This effect is masked at rates >0.4 kg/ha.

23-4.5.5 Sethoxydim + 2,4-DB

The spectrum of control is similar to that of the fluazifop and 2,4-DB mixture. It differs in the absence of apparent antagonistic effects at field dosages.

23-4.5.6 Other Combinations

Limited research has been conducted on fluazifop and sethoxydim tank mixes with dinoseb. Although the combination provided effective weed control, the results were inconclusive in terms of damage to alfalfa with dinoseb alone compared with the mixture.

23-4.6 Herbicides on Alfalfa Sown with Companion Crops

Weed growth is reduced by small grain companion crops; nevertheless, at times weed populations justify the use of herbicides. Effective weed control will benefit both the small grain and the alfalfa.

The phenoxy herbicides have been used to control broadleaf weeds in small grains sown with alfalfa. Alfalfa has a low tolerance to 2,4-D and MCPA [(4-chloro-2-methylphenoxy)acetic acid], but the oat crop forms a canopy that protects alfalfa from severe injury. The herbicide 2,4-DB, a close relative of 2,4-D and MCPA, is less toxic to alfalfa. In general, it has replaced 2,4-D and MCPA for use on alfalfa sown with small grains.

Dinoseb has been relatively safe on both small grains and alfalfa if used as prescribed. It is particularly effective on *Brassica* spp. commonly associated with spring grains, but has little effect on weed grasses. Applications are safer on the crop and more effective on weeds if made within a temperature range of 16 to 27°C before the small grain is >16-cm tall. The degree of dinoseb is linked closely with temperatures, with selectivity decreasing rapidly above 27°C.

23-5 NO-TILL AND OTHER METHODS OF ESTABLISHMENT OF ALFALFA

The concept of establishing legumes by no-tillage or limited tillage methods is not new. Graber (45) was one of the early researchers to attempt renovation of pastures with legumes without tillage. He recognized the importance of weeds, and later increased the success rate of seedings by a limited-tillage discing operation prior to planting. Sprague (133, 134) recognized the potential for substituting herbicides for tillage when planting alfalfa in pastures. Subsequently, numerous studies ensued throughout the USA and the world (40, 135, 148, 149, 153, 154, 164). Using herbicides for vegetation control, Taylor et al. (140, 141) and Decker and Dudley (34) were successful in combining herbicides and no-tillage planting techniques as were Triplett et al. (149) and others who followed (46, 89, 103, 150, 155, 167). The research of Decker and co-workers (34, 35) and a Kentucky group (139, 140, 141) on no-tillage equipment for forage plantings laid the groundwork for present day implements. However, it was the combination of the best weed and pest control practices with appropriate planting equipment that increased the reliability of no-tillage alfalfa plantings. These advances have made no-tillage planting of alfalfa a viabile alternative to conventional systems. Available practices have application in sod-plantings and pasture renovations, and afford an attractive option where alfalfa is a component of a cropland rotation.

Linscott (74), summarizing experiments in New York over a 9-yr period, reported a 80% success rate when planting alfalfa directly into grain crop stubbles and after silage corn. This rate was less than the 93% reported earlier (72). Plantings were made in mid- or late-summer after harvest or early in the following spring. In these experiments, paraquat or glyphosate was applied to emerged vegetation before planting. If perennials as well as annual species were the problem, glyphosate was the herbicide of choice because it was translocated and had greater efficacy. If annuals, particularly annual broadleaf weeds, were of primary concern, then paraquat applied prior to no-till alfalfa planting may suffice. Other investigators report similar findings of successful no-tillage planting of alfalfa in crop stubble (155). Red clover (*Trifolium pratense* L.) and birdsfoot trefoil (*Lotus corniculatus* L.) have particularly high establishment percentages when planted into stubble by no-tillage methods (72, 74).

Volunteer grains can compete very seriously with no-till planted alfalfa, if there is inordinate loss of grain from combines, and especially if crop residues are not spread. Strips of volunteer grain that are not controlled will prevent the establishment of alfalfa. In these situations, postemergence treatments with sethoxydim, fluazifop or haloxyfop will control volunteer grain as well as weed grasses that have not emerged at time of preplanting treatments (155). In some fields, the volunteer small grains might be controlled by delaying preplanting treatments with paraquat or glyphosate until the grain has germinated. However, to do so could result in an undesirable delay in planting. The selective postemergence herbicide option for volunteer grain and grasses is attractive.

Herbicide options, applied preplanting and postemergence, are similar for alfalfa no-till planted in spring into crop stubbles from the previous year. However, if significant perennial species are present, they can be controlled with fall applications of glyphosate, or glyphosate and 2,4-D, or dicamba. This procedure allows for earlier no-till planting in the spring. If needed, a paraquat application can be made prior to planting. In the absence of a fall treatment, a spring preplanting treatment with paraquat or glyphosate delayed until the proper stage of weed development would be necessary. As with summer plantings, selective control of grasses post-planted with fluazifop, haloxyfop, or sethoxydim and annual broadleaf weeds with 2,4-DB or dinoseb is feasible.

Weed control in no-till seeding of alfalfa in sods presents more problems than seeding in crop stubbles. In sod, control of competing perennial species, particularly the grasses, prior to and after planting alfalfa is critical and, of course, more difficult. In addition, pests, diseases, allelopaths, and/or products of decaying vegetation can inhibit the seeding growth and development of alfalfa. Linscott (74) over a 9-yr period reported a no-till planting success rate of 60% for alfalfa planted in sods. He and others report significant problems with molluscs (*Mollusci* spp.) diseases (56, 166), and other pests (16, 39). Careful attention to proper vegetation control, pest control, timing of the control measures (24, 52, 65), and planting can increase the probabilities of successful no-till alfalfa establishment in sod (74, 82, 90, 126). Adequate control of the sod species is vital to obtaining a high percentage of successful stands, with consideration given to the total spectrum of pests, disease, and allelopaths.

Mueller-Warrant and co-workers (91, 92, 93), Nichols and Peters (95), Nichols et al. (96), and others successfully established no-till alfalfa in spring after fall applications of glyphosate or pronamide [3,5-dichloro(N-1,1-dimethyl-2-propynyl) benzamide]. Other scientists have used these herbicides in combination with 2,4-D or dicamba (3,6-dichloro-2-methoxybenzoic acid) to kill sod species in the fall. Other herbicides such as dalapon (116) and amitrole also have been used (35, 135). Fall treatments present several advantages. Plantings in spring can be made early; the killed vegetation will have time to decay and present a less favorable habitat for pests of seedling alfalfa; and allelopaths produced from dying and dead vegetation will dissipate to a considerable extent (72, 74). Further, if some species escape the fall treatments, it is possible to add an appropriate preplanting treatment in spring.

The nature and efficacy of the herbicides used dictate the procedures necessary for spring treated sod control. Glyphosate applied after sods have made significant growth is effective. However, planting should be delayed for 3 weeks (17, 82, 89, 90) or more following treatment until the sod is brown and decaying and the field has received a significant rain. This procedure will prevent some potential residue problems, and will help to negate deleterious effects from allelopaths of certain weed species. Delay in planting after paraquat application usually is not necessary, particularly if the sods are short and weak. However, in view of paraquat's marginal effectiveness on some species, split applications before planting are usually most effective, particularly on sods that have not been weakened by intensive grazing or hay harvest. A 3- to 4-week period should separate split applications of paraquat. The interval can be reduced if there is sufficient regrowth to intercept the herbicide. The combination of 2,4-D and paraquat will give better broadleaf control, provided planting is delayed for at least 3 weeks to allow 2,4-D to decompose. In split treatments of paraquat, 2,4-D can be added to the tank mix at the first treatment.

Problem grasses that emerge after alfalfa planting can be controlled with the grass herbicides mentioned previously, fluazifop, haloxyfop, and others.

Seedling dandelions and other broadleaf species can be serious in no-till seedings or alfalfa in sod, especially in the absence of fall treatments with either 2,4-D or dicamba. Postemergence treatments with 2,4-DB to very young broadleaf weed seedlings can be effective. Conversely, if the seedling broadleaf weeds are allowed to grow to >5 cm, 2,4-DB is not very effective. Broad spectrum broadleaf weed control in alfalfa-seeded sods remains a problem (115).

As indicated previously, control of the sod species is improved if they are in a physiologically weakened conditioned at the time of herbicide treatment. Frequent grazing or clipping can weaken sods to the extent that they are more susceptible to herbicides (115, 117). Grazing or harvesting the forage removes the surface vegetation. Failures of no-till alfalfa are common because of excessive surface plant residues, resulting in planting difficulties, pest havens, a favorable microclimate for disease, and the production of allelopaths (18). The less surface vegetation the better is a sound general rule in alfalfa sod seedings (115). If surface vegetation is dense, small alfalfa seedlings may not emerge.

The practice of burning killed vegetation followed by alfalfa planting would increase the probability of successful plantings (103). If sods are severely weakened by grazing, some investigators have established alfalfa and other legumes successfully in sod without herbicides (139). However, the probabilities are greater for success if grazing is augmented by herbicides before and/or after planting (115).

23-6 WEED CONTROL METHODS: ESTABLISHED STANDS

Competition is the main factor involved in keeping weeds out of established alfalfa. Mature alfalfa that is fertilized properly and managed to maintain a rapidly growing vigorous stand will provide enough shade to kill weed seedlings during the growing season (20). However, winter annual weeds may come into dense stands of alfalfa when the alfalfa is dormant.

Weed control practices that remove perennial and biennial weed seedlings from seedling alfalfa will prevent these plants from becoming problems in mature stands. For example, alfalfa treated with 2,4-DB in the seedling year was free of curly dock the following year (101). Perennial weeds in established stands are difficult to control with mowing or tillage, but progress has been made in the development of herbicidal practices.

23-6.1 Tillage

Although tillage for control of specific weeds in mature alfalfa has been advocated and practiced to some extent (32), the value of tillage is often questioned. In Nebraska, tillage with several implements had no effect on yields of alfalfa (62). Bruns and Heinemann (15) thought that cultivation with a spring tooth cultivator caused excessive damage to alfalfa crowns, and clumps of downy brome (*Bromus tectorum* L.) in moist soil tended to survive tillage. Tillage is objectionable in humid areas and in some irrigated arid areas (20) because alfalfa roots sustain some damage from the tillage tools, and wounds in the roots are easily invaded by disease organisms, such as those causing bacterial wilt (*Corynebacterium insidiosum*) (132).

Various tillage implements have been used during the dormant season or in stubble after hay has been cut in attempts to remove weeds (32, 62). Most of the tough fleshy taproots of alfalfa slide by tillage tools as they move through the soil. To minimize damage to alfalfa roots, spring toothed harrows or field cultivators should be equipped with straight chisel type teeth. Weeds with taproots similar to alfalfa will escape the tillage operation. The spike tooth harrow and disk may be useful for controlling annual or shallow rooted weeds. The disk, if set too deep, will damage alfalfa (32). The finger weeder and flextine harrow are useful tools for controlling germinating weeds where the soil surface is loose. In irrigated areas where soil moisture can be regulated, tillage may be used in alfalfa stubble to create a dust mulch at the soil surface (32). Frequent tillage is used until the soil dries and a dust mulch is formed 5-cm deep. The dust mulch prevents weed seed germination. Alfalfa will grow through the dust mulch and irrigation then can be resumed. Growth of alfalfa will then be rapid enough to suppress weeds by shading, provided that the alfalfa stand is dense. Tillage may injure some crown buds and delay growth of alfalfa.

23-6.2 Mowing

Mowing can be effective for controlling certain weeds in alfalfa. Mowing reduces food reserves and reduces the vigor of weeds (2, 9). Hodgson (51) mowed alfalfa twice per year and reduced the infestation of Canada thistle to <1% after 3 yr. Alfalfa emerged early in the spring and recovered quickly after mowing to offer severe competition to Canada thistle. Schreiber (124), in 4 yr, nearly eliminated Canada thistle from alfalfa pastured rotationally with sheep (*Ovis* spp.). He mowed at the end of each grazing period. The alfalfa was grazed twice the first year, three times annually during the second and third years, and four times the fourth year.

Johnsongrass can be reduced greatly and nearly eliminated in alfalfa mowed three or more times per year for hay (99). Similarly, mowing and competition from alfalfa can reduce bindweed infestations, but may not eliminate them (137).

Mowing is not effective for controlling such weeds as quackgrass,

Russian knapweed (*Centaurea repens* L.), and leafy spurge. Quackgrass begins growth early in the spring, grows little in summer, and continues growth late in the fall so that mowing during the summer while harvesting hay is not effective (168). Russian knapweed and leafy spurge emerge before or at the time alfalfa emerges in the spring, and produce rank growth that stays ahead of the alfalfa (36, 37).

Mowing, in addition to decreasing the infestation of perennial weeds, is also helpful in preventing seed production and further infestations. Sometimes alfalfa needs to be mowed early to prevent weeds from producing seed.

23–6.3 Miscellaneous Methods of Control

If small patches of weeds occur, they are sometimes removed by hand. Where noxious weeds begin to invade an area and are present in small patches, it may be advantageous to cut or dig the weeds and burn them.

In some instances fire can be used to control weeds. Propane or butane torches are used to control patches of dodder or other annual weeds, thus preventing seed production and subsequent spread of the infestation (32). Flaming is a two-step operation. The first flaming kills the weeds; and when the dead weeds have dried, they are burned with a second flaming. Flaming also may be employed to destroy dodder that remains attached to alfalfa stubble after cutting. If dodder is to be controlled, flaming should be intense enough to burn alfalfa stubble to which the dodder is attached.

Growing perennial forage grasses, such as smooth bromegrass (*Bromus inermis* Leyss.), with alfalfa is effective in keeping weeds out (168). Grasses provide cover between alfalfa plants, which is especially helpful in reducing invasion by winter annual weeds. Under some circumstances during alfalfa establishment and in mature alfalfa hay fields and meadows, rope wick or rolling brush applicators can be used to apply glyphosate, dicamba, and other herbicides to weeds above the alfalfa canopy. The procedure works particularly well where alfalfa containing meadows have been selectively grazed by livestock leaving tall weeds.

23–6.4 Chemical Control Methods: Established Stands

Controlling weeds in mature stands with herbicides is different enough from controlling weeds in seedlings that separate discussions are warranted. The deep rooting habit of established alfalfa permits the use of some herbicides that are unsuitable for seedling stands. Some herbicides can be used in seedlings as well as established stands, but often application techniques vary.

Chlorpropham, known also as CIPC, is a carbamate herbicide that is especially effective for controlling winter annual weeds during the dormant season (3, 7, 156). It is most effective when applied to the soil before or when weed seeds germinate in late fall or early winter (21). Chloropropham is also effective on chickweed that has made considerable growth (3, 12). Chlorpropham will injure forage grasses and should not be used on alfalfa-grass mixtures. Granular formulations have often been used instead of sprays to avoid injury when alfalfa foliage is present (21).

Propham, also known as IPC, is a carbamate herbicide especially active on grasses. It is related to chlorpropham and has been used as a preemergence or early postemergence treatment. Propham is broken down rapidly by microorganisms in the soil (44). Breakdown is hastened by high temperatures, so it is most useful in the fall when temperatures are lower (21).

Propham has been used to control winter annual grasses and chickweeds, but its use has declined with the discovery of chlorpropham. Chlorpropham has longer life in the soil and higher toxicity to grass seedlings than propham.

Dimethyl tetrachloroterephthalate (DCPA) is effective on weed grasses if applied prior to emergence. The utility of DCPA is limited because it is not effective on a number of broadleaf weeds (21). It has been used to some extent to control dodder in alfalfa seed fields, and is effective if applied before dodder seedlings become wrapped around the alfalfa stems (32, 87). DCPA is more effective if applied when little alfalfa growth is present so that a minimum amount of spray is intercepted by the alfalfa foliage and a maximum amount of herbicide reaches the soil surface.

Dichlobenil (2,4-dichlorobenzonitrile) has been used to some extent in alfalfa seed fields in the Pacific Northwest (28). It may cause some injury to alfalfa and reduce forage yield, but has not affected seed yield. It kills dodder and suppresses established perennial weeds such as Canada thistle, quackgrass, and dandelion (28). Dichlobenil injures seedling alfalfa. Alfalfa >1-yr-old presumably escapes serious injury from dichlobenil because it is deep-rooted and the herbicide generally does not move readily into the soil (84, 88). Incorporation increases not only the herbicidal activity of dichlobenil, but also injury to alfalfa (28). Dichlobenil is considered safe on alfalfa if applied to established plants during the dormant season or after the hay crop has been harvested.

Dinoseb, known also as DNBP, is a dinitro phenol that has been used in various formulations. The amine salts are most commonly used with water and are effective for controlling germinating and newly emerged broadleaf weeds (54, 119). Dinoseb is considered safe on the crop only if applied when alfalfa is dormant, or shortly after cutting, before the alfalfa has grown much (55). Although dinoseb injures plant tissue on contact, it is considered a selective herbicide because the contact action on some plants is greater than on others. If dinoseb is applied in oil, the contact action is enhanced and seedling weed grasses can be controlled as well as broadleaf weeds. When applied in oil, dinoseb will injure alfalfa unless applied before regrowth begins. Dinoseb in oil has been used frequently to spot-treat small areas of infestation. Dinoseb is effective on chickweed, but two applications or more may be necessary if chickweed growth is heavy (114). Unfortunately, this herbicide is no longer available for use.

Diuron [N-(3,4-dichlorophenyl)-N,N-dimethylurea] is a substituted urea herbicide that has been applied as a preemergence or early postemergence treatment to control seedlings of both grass and broadleaf weeds in alfalfa (6). It is low in solubility and, therefore, is not readily leached in heavy soils (165). Diuron is leached more readily on soils low in organic matter or low in cation exchange capacity (151) and can damage alfalfa on such soils when applied at heavy rates. Diuron is safe on alfalfa only if applied during the dormant season and will injure alfalfa severely if applied on first-year growth. It is not labeled for use in many states in humid areas.

Simazine [6-chloro-N,N-diethyl-1,3,5-triazine-2,4-diamine] is a triazine herbicide that has been used to control both grass and broadleaf weeds in established alfalfa (47, 48, 49, 57, 105, 120, 127, 138, 157). Simazine is most effective when applied as fall or winter treatments before weeds emerge or when they are very small. Simazine is not easily leached in heavy soils and tends to remain near the surface (118, 121, 128). Alfalfa <1-yr-old is injured by simazine. Also, on sandy soils or in situations where simazine is leached into the soil, established alfalfa will be injured. The prospect of injury to established stands increases at rates >1 kg/ha (69).

Simazine, although most effective when applied before seeds germinate, is also effective on shallow-rooted weeds. Hastings and Kust (49) and Kust (69) were able to control hoary alyssum (*Berteroa incana* (L.) DC), yellow rocket (*Barbarea vulgaris* R. Br.), and white cockle (*Lychnis alba* Mill.) when simazine was applied to rosettes of these weeds in mid-September. The weeds did not die but were injured so severely that competition from alfalfa the following spring eliminated them. Peters and O'Leary (110) reported that simazine applied in the dormant season caused severe chlorosis and prevented jointing and flowering of white cockle. Simazine at 3.3 kg/ha applied to alfalfa stubble controlled yellow rocket but injured alfalfa (23). Simazine is also effective for controlling quackgrass, a perennial (1, 42).

Terbacil [5-chloro-3-(1,1-dimethylethel)-6-methyl-2,4(1H,3H)-pyrimidinedione] is a urea herbicide that is applied to alfalfa that is >1-yr-old after plants become dormant in the fall, or in the spring before new growth starts. It is applied in the fall or winter for semidormant or nondormant alfalfa. Terbacil is effective on many broadleaf weeds and weed grasses before and after emergence (57, 105, 120, 157). Among weeds controlled are dandelion, perennial sow thistle (*Sonchus arvensis* L.) (157), and annual bromes (57, 105).

Pronamide (3.5 dichloro(*N*-1,1-dimethyl-2-propynl)benzamide can be used on new or established alfalfa stands preemergence or postemergence to weeds. It can be applied in the fall after the last cutting to control winter annual as well as many perennial weeds. It controls or suppresses quackgrass (38, 42) and is effective on downy brome (57).

Metribuzin [4-amino-6-(1,1-dimethylethyl)-3-(methylthio)-1,2,4-triazin-5(4H)-one] is used prior to spring growth in dormant alfalfa established 1 yr or longer. Injury may occur if spring growth has begun at time of application. The herbicide controls a number of broadleaf weeds and weed grasses both preemergence and postemergence (105, 127).

Paraquat is a contact herbicide that burns foliage on contact. It kills the tops of broadleaf weeds and weed grasses. Many weeds, especially perennials, will recover from paraquat treatment. Paraquat will burn the foliage of actively growing alfalfa and reduce yields of the subsequent cutting (50). It can be applied when alfalfa is dormant or after cutting before regrowth. Ideally, applications should be made before the appearance of basal stem regeneration in alfalfa. Thus, alfalfa cutting schedule adjustments may assist in preventing damage. Paraquat is effective on downy brome (57) and several other annual grass species as well as chickweed which is common in established alfalfa.

Hexazinone [3-cyclohexyl-6-(dimethylamino)-1-methyl-1,3,5-triazine-2,4(1H, 3H)-dione] is applied to alfalfa that has been established 1 yr or more. It is applied in the fall or winter when alfalfa is dormant or in the spring before new growth begins. It is effective preemergence or early postemergence on a number of winter and summer annual weeds (105), as well as on bluegrass (*Poa pratensis* L.) and dandelion. Also, hexazinone can be impregnated on dry bulk fertilizer and applied to alfalfa before 5 cm of regrowth has occurred. Injury is avoided because the fertilizer falls to the ground and contact with alfalfa leaves is minimized.

Where established alfalfa is irrigated, EPTC can be metered into the irrigation water to control many weed grasses and broadleaves. When applied with irrigation water, EPTC moves into the soil with the water and is effective for controlling weeds. In regular practice, EPTC is usually shallowly incorporated into the soil to place the chemical for best effect and to reduce herbicide loss by volatilization. Where the soil surface can be tilled to incorporate the herbicide, trifluralin [2,6-dinitro-N,N-dipropyl-4-(trifluoromethyl)benzenamine] is used on established stands. Some injury to alfalfa will occur from tillage, but trifluralin is especially helpful for controlling weed grasses.

At the time of this writing, a number of new postemergence grass killing herbicides, including sethoxydim, fluazifop, haloxyfop, and others, have been reported as killing annual and perennial grasses in alfalfa without injury to alfalfa (1, 8, 50, 64). These herbicides will eliminate weedy grasses in alfalfa throughout the year. There are several other grass herbicides under development for use on legumes which may find a use for alfalfa. The future looks particularly bright for control of weed grasses.

The amines or esters of 2,4-DB have been used on alfalfa before flowering and during the dormant season (158). They are effective for controlling many broadleaf weeds in alfalfa, including yellow rocket, when applied early in spring at the proper stage of weed development (106); however, at rates safe for alfalfa, many species of perennial weeds cannot be controlled. Not withstanding the control measures presently available, effective and safe procedures for a great number of perennial broadleaf weed species are still lacking. Selective control of broadleaf species remains a significant challenge for weed scientists now and in the future.

23-7 SUMMARY

Competition of weeds with alfalfa for water, nutrients, and light causes serious reductions in yields of alfalfa forage and seed. Alfalfa is most vulnerable to weed competition when it is in the seedling stage. Cultural practices such as using clean seed fallowing, selecting proper planting dates, clipping weeds, and using companion crops continue to make valuable contributions to reducing weed competition in alfalfa, but herbicides have become increasingly important. Herbicides, with spring plantings, have greatly increased the chances of successful establishment and have made it possible to harvest up to three cuttings of hay in the first year. No-tillage plantings of alfalfa into crop stubbles residue and into sods are now feasible because of advances in new herbicide development.

Herbicides have also controlled winter annual weeds in summer or fall plantings and thus assure more frequent success in establishment. Summer and winter annual weeds in mature stands of alfalfa also have been successfully controlled with herbicides. Herbicides have been used to control some perennial weed species in alfalfa, but other perennial species are more difficult to control, and their control remains a significant problem.

Major failings of presently available herbicides include; injury to alfalfa with improper use, absence of control over a significant range of weed species, and inability to remove weed grasses from alfalfa-forage grass mixtures.

REFERENCES

1. Ahrens, T.M., and R.G. Harvey. 1983. Established alfalfa-quackgrass control study. Res. Rep. North Cent. Weed Control Conf. 40:22–28.
2. Aldous, A.E. 1929. The eradication of brush and weeds from pasture lands. J. Am. Soc. Agron. 21:660–666.
3. Aldrich, R.J. 1957. Chickweed control in alfalfa. USDA Tech. Bull. 1174. U.S. Government Printing Office, Washington, DC.
4. Anonomyous. 1960. Band and broadcast seeding of alfalfa-bromegrass in the Northeast. Northeast Reg. Pub. 41. Md. Agric. Exp. Stn. Bull. A-108.
5. ----. 1963. Seedling management of grass-legume associations in the northeast. Northeast Reg. Pub. 42. N.J. Agric. Exp. Stn. Bull. 804.
6. Appleby, A.P., W.R. Furtick, and D.G. Swan. 1966. Weed control in established alfalfa with soil-applied herbicides. Abstr. Weed Science Society America, section 1, vol. 6, p. 1.
7. Arnold. J.L., and P.W. Santelmann. 1970. Herbicides for winter annual weeds in established alfalfa. Proc. South Weed Sci. Soc. 23:115–119.

8. Arnold, W.E., M.A. Wrucke, and S.R. Gylling. 1983. Screening for weed control in seedling alfalfa. Res. Rep. North Cent. Weed Control Conf. 40:36.
9. Arny, A.C. 1932. Variations in organic reserves in underground parts of five perennial weeds from late April to November. Minn. Agric. Exp. Stn. Tech. Bull. 84.
10. Bayer, G.H. 1968. A three-year summary of weed control for establishment of legume seedings. Proc. Northeast. Weed Control Conf. 22:306-312.
11. Beardmore, R.A., M.A. Trimmer, and D.L. Linscott. 1984. Effects of fluazifop, haloxyfop and sethoxydim on water uptake in oats. Proc. Annu. Northeast. Weed Sci. Soc. 38:9.
12. Berggren, F. W., and R.A. Peters. 1953. Chemical chickweed control and its influence on yields of alfalfa-grass mixtures. Proc. Northeast. Weed Control Conf. 7:205-212.
13. Blaser, R.E., T.Taylor, W. Griffeth, and W. Skrdla. 1956. Seedling competition in establishing forage plants. Agron. J. 48:1-6.
14. Brown, C.W., and R.F. Stafford. 1970. Get top yields from alfalfa seedings. Better Crops Plant Food 54(1):16-18.
15. Bruns, V.F., and W.W. Heinemann. 1959. Control of downy brome in alfalfa. USDA Tech. Bull. 1197 U.S. Government Printing Office, Washington, DC.
16. Byers, R.A. 1979. Arthropod and mollusc pests of no-till forages. Symposium on Minimum- and No-Tillage Systems for Agronomic Crops in the Northeast. NE Branch Am. Soc. Agron., Rutgers Univ., p. 13-15.
17. Campbell, M.H. 1974. Effects of glyphosate on the germination and establishment of surface-sown pasture species. Aust. J. Exp. Agric. Anim. Husb. 14:557-560.
18. Cardina, J., and N.L. Hartwig. 1983. Glyphosate and/or quackgrass effect on no-tillage alfalfa seeding in quackgrass sod (*Agropyron repens*). Proc. Annu. Meet. Northeast. Weed Sci. Soc. 37:63-67.
19. Cords, H.P. 1973. Weeds and alfalfa hay quality. Weed Sci. 21:400-401.
20. ----, and L.M. Stockton. 1979. Control of summer annual weeds in seed alfalfa. Nev. Agric. Exp. Stn.
21. Crafts, A.S. 1961. The chemistry and mode of action of herbicides. Interscience Publishers, New York.
22. ----, and W.W. Robbins. 1962. Weed control. McGraw-Hill Book Co., New York.
23. Currey, W.L., and R.A. Peters. 1968. Control of yellow rocket (*Barbarea vulgaris*) and other broadleaf weeds associated with established alfalfa. Proc. Northeast. Weed Control Conf. 22:455-458.
24. Davis, H.E., R.S. Fawcett, and R.G. Harvey. 1978. Effect of fall frost on the activity of glyphosate on alfalfa (*Medicago sativa*) and quackgrass (*Agropyron repens*). Weed Sci. 26:41-45.
25. Dawson, J.H. 1965. Prolonged emergence of field dodder. Weeds 13:373-374.
26. ----. 1966. Response of field dodder to shade. Weeds 14:4-5.
27. ----. 1966. Factors affecting dodder control with granular CIPC. Weeds 14:255-259.
28. ----. 1970. Dodder control in alfalfa with dichlobenil. Weed Sci. 18:225-230.
29. ----. 1980. Selective weed control from EPTC applied with seed of alfalfa (*Medicago sativa*). Weed Sci. 28:607-611.
30. ----. 1983. Tolerance of alfalfa to EPTC. Weed Sci. 31:103-108.
31. ----, and R.G. Harvey. 1981. Management systems for weeds in alfalfa. *In* D. Pimentel (ed.) Handbook of pest management in agriculture. CRC Press, Boca Raton, FL.
32. ----, W.O. Lee, and F.L. Timmons. 1969. Controlling dodder in alfalfa. USDA Farmers Bull. 2211. U.S. Government Printing Office, Washington, DC.
33. ----, and C.M. Rincker. 1982. Weeds in new seedings of alfalfa (*Medicago sativa*) for seed production: Competition and control. Weed Sci. 30:20-25.
34. Decker, A.M., and R.F. Dudley. 1976. Minimum tillage establishment of five forage species using five sod-seeding units and two herbicides, p. 140-145. *In* J. Luchok et al. (ed.) Hill lands. Proc. Int. Symp., Morgantown, WV. 3-9 October, West Virginia University Books, Morgantown.
35. ----, M.J. Retzer, M.L. Sarna, and H. Dicker. 1969. Permanent pastures improved with sod-seeding and fertilization. Agron. J. 61:243-247.
36. Derscheid L.A., K.E. Wallace, and R.L. Nash. 1960. Leafy spurge control with cultivation, cropping and chemicals. Weeds 8:115-127.
37. ----, ----, and ----. 1960. Russian knapweed control with cultivation, cropping and chemicals. Weeds 8:268-278.
38. Dutt, T.E., R.G. Harvey, R.S. Fawcett, N.A. Jorgenson, H.J. Larsen, and D.A. Schlough. 1979. Forage quality and animal performance as influenced by quackgrass (*Andropyron repens*) control in alfalfa (*Medicago sativa*) with pronamide. Weed Sci. 27:127-132.
39. Faix, J.J., C.J. Kaiser, and M.E. Farris. 1980. Insect and weed control in no-till alfalfa establishment. Ill. Agric. Exp. Stn., Dixon Springs Agric. Ctr. Rep. 8:34-38.
40. ----, G.E. McKibben, and C.J. Kaiser. 1977. Sod suppressants and preemergence herbicides for sod-seeded alfalfa. Proc. North Cent. Weed Control Conf. 32:59-61.

41. Fawcett, R.A., and R.G. Harvey. 1978. Field comparison of seven dinitroanaline herbicides for alfalfa seedling establishment. Weed Sci. 26:123-127.
42. Fawcett, R.W., R.G. Harvey, D.A. Schlough, and T.R. Block. 1978. Quackgrass (*Andropogon repens*) control in established alfalfa (*Medicago sativa*) with pronamide. Weed Sci. 26:193-198.
43. Fleming, C.E., and N.F. Peterson. 1919. Don't feed fox-tail hay to lambing ewes. Nev. Agric. Exp. Stn. Bull. 97.
44. Freed, V.H. 1951. Some factors influencing the herbicidal efficacy of isopropyl N phenol carbamate. Weeds 1:48-60.
45. Graber, L.F. 1927. Improvement of permanent bluegrass pastures with sweet clover. J. Am. Soc. Agron. 19:994-1006.
46. Groya, F.L., and C.C. Sheaffer. 1981. Establishment of sod-seeded alfalfa at various level of soil moisture and grass competition. Agron. J. 73:560-565.
47. Guenthner, H.R., and H.P. Cords. 1975. Winter annual weeds in alfalfa in western and northern Nevada. Univ. of Nev., Reno Coop. Ext. Serv. C169.
48. Harvey, R.G., D.A. Rohweder, and R.S. Fawcett. 1976. Susceptibility of alfalfa cultivars to triazine herbicides. Agron. J. 68:632-634.
49. Hastings, R.E., and C.A. Kust. 1970. Control of yellow rocket and white cockle in established alfalfa. Weed Sci. 18:239-333.
50. Himmelstein, F.J., and R.A. Peters. 1983. Crabgrass-alfalfa competition as influenced by application of postemergence herbicides. Proc. Annu. Meet. Northeast. Weed Sci. Soc. 37:74-79.
51. Hodgson, J.M. 1958. Canada thistle (*Cirsium arvense* Scop.) control with cultivation, cropping and chemical sprays. Weeds 6:1-11.
52. Holland, C., and M.B. Tesar. 1981. Establishment of alfalfa (*Medicago sativa*) in quackgrass (*Agropyron repens*) sods with herbicides using conventional and sod-seeding methods. Res. Rep. 470, Mich. State Univ. Agric. Exp. Stn., p. 9-18.
53. Hull, R.J., and R.C. Wakefield. 1959. A comparison of pre- and post-emergence herbicidal treatments for the establishment of legume seedings. Proc. Northeast. Weed Control Conf. 13:178-187.
54. Johnson, H.W., R.B. Carr, and O.A. Leonard. 1949. Herbicide applications to seedling and established alfalfa. Proc. South. Weed Conf. 2:43-47.
55. Jones, L.G. and W.A. Harvey. 1952. Weed control in perennial legumes. Proc. California Weed Conf. 4:109-114.
56. Kalmbacker, R.S., D.R. Minnick, and F.G. Martin. 1979. Destruction of sod-seeded legume seedlings by the snail (*Polygra cereolus*). Agron. J. 71:365-368.
57. Kapusta, G., and C.F. Strieker. 1975. Selective control of downy brome in alfalfa. Weed Sci. 23:202-206.
58. Kerkin, A.J., and R.A. Peters. 1957. Herbicidal effectiveness of 2,4-D, MCPB, neburon and other materials as measured by weed control and yields of seedling alfalfa and birdsfoot trefoil. Proc. Northeast. Weed Control Conf. 11:128-138.
59. ----, and ----. 1958. Effect of herbicidal treatment on the winter heaving of late summer seeded alfalfa. Proc. Northeast. Weed Control Conf. 12:159-167.
60. Khamna, S.K., R.K. Khamna, and G.G. Sanwal. 1975. Studies on alfalfa grown after cutting dodder (*cuscuta*) infection: Changes in content and nature of phenolics. Indian J. Exp. Biol. 13:407-409.
61. ----, ----, and ----. 1976. Influence of shading and dodder (*cuscuta*) infection on the content and nature of alfalfa phenolics. Indian J. Exp. Biol. 14:375-376.
62. Kiesselback, T.A., and A. Anderson. 1927. Alfalfa in Nebraska. Nebr. Exp. Stn. Bull. 222.
63. Klebesadel, L.J., and D. Smith. 1958. The influence of oat stubble management on the establishment of alfalfa and red clover. Agron. J. 680-683.
64. Knake, E.L. 1983. Herbicides for legume establishment. Proc. No. Cent. Weed Cont. Conf. 40:32-33.
65. Koch, D.W., G.W. Mueller-Warrant, and J.R. Mitchell. 1983. Sod-seeding of forages. I. Alternative to conventional establishment. Station Bull. 525. N.H. Agric. Exp. Stn.
66. Kommedahl, T., J.B. Kotheimer, and J.V. Bernardini. 1959. The effects of quackgrass on germination and seedling development of certain crop plants. Weeds 7:1-12.
67. Kunelius, H.T. 1974. Effects of weed control and N fertilization at establishment on the growth and modulation of alfalfa. Agron. J. 66:806-809.
68. Kust, C.A. 1968. Herbicides or oat companion crops for alfalfa establishment and forage yields. Agron. J. 60:151-154.
69. ----. 1969. Selective control of hoary alyssum in alfalfa. Weed Sci. 17:99-101.
70. Lee, W.O. 1965. Herbicides in seedbed preparation for the establishment of grass and seed fields. Weeds 13:293-297.
71. ----, and F.L. Timmons. 1956. Evaluation of pre-emergence and stubble treatments for control of dodder in alfalfa seed crops. Agron. J. 48:6-10.

72. Linscott, D.L. 1979. Projections for no-tillage forage systems for the Northeast. p. 10–12. In Symposium on minimum- and no-tillage systems for agronomic crops in the Northeast. Proc. NE Branch Am. Soc. Agron., New Brunswick, N.J. June. Rutgers University, New Brunswick.
73. ———. 1979. A report on economic losses due to weeds-forages. Proc. Annu. Meet. Northeast. Weed Sci. Soc. Suppl. 33:58–62.
74. ———. Forage crop establishment with limited-tillage systems. Symposium: Weed control in no-tillage agriculture in the Northeast. Proc. Annu. Meet. Northeast. Weed Sci. Soc. Suppl. 37:68–72.
75. ———, A.A. Akhavein, and R.D. Hagin. 1969. Paraquat for weed control prior to establishing legumes. Weed Sci. 17:428–431.
76. ———, and R.D. Hagin. 1968. Interaction of EPTC and DNBP on seedlings of alfalfa and birdsfoot trefoil. Weed Sci. 16:182–184.
77. ———, ———, and A.A. Akhavein. 1967. Weed control during establishment of alfalfa and birdsfoot trefoil. Proc. Northeast. Weed Control Conf. 21:270–277.
78. ———, ———, and J.E. Dawson. 1968. Synthesis of homologs to 2,4-DB by alfalfa: A mechanism of resistance to this herbicide. J. Agric. Food Chem. 16:844–847.
79. ———, R.J. Seaney, and R.D. Hagin. 1967. Subsurface placement of EPTC for weed control in seedling legumes. Weed Sci. 15:259–264.
80. Lucey, R.F., R.R. Seaney, and R.F. Burt. 1970. Alfalfa establishment production and botanical composition as affected by competition and management of spring grain companion crops. Agron. Abstr. American Society of Agronomy, Madison, WI, p. 53.
81. Marten, G.C., and R.N. Anderson. 1975. Forage nutritive values and palatability of 12 common annual weeds. Crop Sci. 15:821–827.
82. Martin, N.P., C.C. Sheaffer, D.L. Wyse, and D.A. Schreiver. 1983. Herbicide and planting date influence establishment of sod-seeded alfalfa. Agron. J. 75:951–955.
83. Martin, P, and B. Rademacher. 1960. Studies on the mutual influences of weeds and crops. p. 143–152. In J.L. Harper (ed.) The biology of weeds. Blackwell Scientific Publications, Oxford, UK.
84. Massoni, P. 1961. Movement of 2,6-dichlorobenzonitrile in soil and plants in relation to its physical properties. Weed Res. 1:142–146.
85. Mazzoni, L.E., and J.M. Scholl. 1964. Effect of chemical and mechanical weed control in spring-seeded legumes on establishment of interseeded grasses. Agron. J. 56:403–405.
86. McCarty, M.K., and P.F. Sand. 1958. Chemical weed control in seedling alfalfa. I. Control of weedy grasses. Weeds 6:152–160.
87. McNeely, G.H., E.C. Hoffman, E. Bayer, and C.L. Foy. 1966. Control of dodder in alfalfa with DCPA. Calif. Agric. 20(3):14–16.
88. Miller, C.W., I.E. Demoranville, and A.J. Charig. 1967. Effect of water on the persistence of dichlobenil. Weed Res. 7:164–167.
89. Moshier, L., and D. Penner. Use of glyphosate in sod-seeding alfalfa (*Medicago sativa*) establishment. Weed Sci. 26:163–166.
90. Mueller, J.P., and D.S. Chamblee. 1984. Sod-seeding of Ladino clover and alfalfa as influenced by seed placement, seedling date and grass supression. Agron. J. 76:284–289.
91. Mueller-Warrant, G.W., and D.W. Koch. 1980. Establishment of alfalfa by conventional and minimum-tillage techniques in quackgrass dominant sward. Agron. J. 72:884–889.
92. ———, and ———. 1983. Fall and spring herbicide treatment for minimum-tillage seeding of alfalfa (*Medicago sativa*). Weed Sci. 31:391–395.
93. ———, ———, and J.R. Mitchell. 1983. Sod-seeding of forages. II. Vegetation control. N.H. Agric Exp. Stn. Stn. Bull. 526.
94. Muenscher, W.C. 1949. Weeds. Macmillan Publishing Co., New York.
95. Nichols, R.L., and R.A. Peters. 1980. Effect of timing and herbicides on the no-tillage establishment of red clover, alfalfa and birdsfoot trefoil. Proc. Annu. Meet. Northeast. Weed Sci. Soc. 34:91.
96. ———, ———, and B.G. Mullinix, Jr. 1983. Effects of herbicides and treatment dates on the establishment of sod-seeded red clover, birdsfoot trefoil, and alfalfa. Res. Rep. 78, Storrs Agric. Exp. Stn., University of Connecticut, Storrs.
97. Ohman, J.H., and T. Kommedahl. 1960. Relative toxicity of extracts from vegetative organs of quackgrass to alfalfa. Weeds 8:666–670.
98. Patrick, Z.A., T.A. Toussoun, and L.W. Koch. 1964. Effect of crop residue decomposition products on plant roots. Annu. Rev. Phytopathol. 2:267–292.
99. Paulling, J.R. 1945. Johnsongrass as a weed in Missouri. University of Missouri, Coll. of Agric., Agric. Ext. Serv. Circ. 517.
100. Pearson, J.O., and R.C. Wakefield. 1965. Effects of herbicides and nitrogen levels on spring seedings of alfalfa. Proc. Northeast. Weed Control Conf. 19:180–184.

101. Peters, E.J. 1964. Pre-emergence, preplanting and postemergence herbicides for alfalfa and birdsfoot trefoil. Agron. J. 56:415–419.
102. ——. 1968. Herbicides for establishing legumes in the southern corn belt. Proc. North Cent. Weed Control Conf. 23:55.
103. ——. 1976. Aerial application of herbicides, seed and fertilizer improves forage on Ozark Hill Lands. p. 167–171. In J. Luchok et al. (ed.) Hill lands. In Proc. Int. Symp., Morgantown, WV. 3–9 October. West Virginia University Books, Morgantown.
104. ——, and F.S. Davis. 1960. Control of weeds in legume seedings with 4-2(2,4-DB), dalapon and TCA. Weeds 8:349–367.
105. ——, R.A. McKelvey and Richard Mattas. 1984. Controlling weeds in dormant and nondormant alfalfa (Medicago sativa). Weed Sci. 32:154–157.
106. ——, and J.F. Stritzke. 1964. Control of yellow rocket in alfalfa. Res. Rep. North Cent. Weed Control Conf. 18:84–85.
107. ——, and J.F. Stritzke. 1970. Herbicides and nitrogen fertilizer for the establishment of three varieties of spring-sown alfalfa. Agron. J. 62:259–262.
108. Peters, R.A. 1954. Preliminary report of the effect of weed control chemicals on new forage seedings. Proc. Northeast. Weed Control Conf. 8:331–339.
109. Peters, R.A. 1961. Legume establishment as related to the presence or absence of an oat companion crop. Agron. J. 53:195–198.
110. ——, and R.M. O'Leary. 1967. Herbicidal response of white cockle (Lycnis alba) and other winter weeds associated with alfalfa. Proc. Northeast. Weed Control Conf. 21:299–302.
111. ——, and A.J. Kerkin. 1959. Combinations of chemicals for weed control in new alfalfa seedlings. Proc. Northeast. Control Conf. 13:175–177.
112. Pipal, F.J. 1914. Wild garlic and its eradication. Ind. Agric. Exp. Stn. Bull. 176.
113. Platt, K., and E.R. Jackman. 1946. The cheatgrass problem in Oregon. Oreg. State Coll. Ext. Bull. 668.
114. Ralcigh, S.M. 1952. Control of chickweed in alfalfa. Proc. Northeast. Weed Control Conf. 6:257–258.
115. Rayburn, E.B., J.F. Hunt, and D.L. Linscott. 1981. Three year summary of no-till establishment research on New York farm sites. Proc. Annu. Meet. Northeast. Weed Sci. Soc. 35.65-66.
116. ——, ——, and ——. 1981. Herbicide and tillage effects on legume establishment in bromegrass sod. Proc. Annu. Meet. Northeast. Weed Sci. Soc. 35:67–68.
117. ——, D.L. Linscott, and J.F. Hunt. 1980. Influence of management prior to direct-planting on the establishment of legumes. Proc. Annu. Meet. Northeast. Weed Sci. Soc. 34:97–98.
118. Roadhouse, F.E.B., and L.A. Birk. 1961. Penetration of and persistence in soil of the herbicide 2-chloro-4,6-bis(ethylamino)-S-triazine (simazine). Can. J. Plant Sci. 41:252–260.
119. Robbins, W.W., A.S. Crafts, and R.N. Raynor. 1952. Weed control. 2nd ed. McGraw-Hill Book Co., New York.
120. Robinson, L.R., C.F. Williams, and W.D. Laus. 1978. Weed control in established alfalfa. Weed Sci. 26:37–40.
121. Rodgers, E.G. 1968. Leaching of Seven S-triazines. Weed Sci. 67:117–120.
122. Schmid, A.P., and R. Behrens. 1972. Herbicide versus oat companion crop for alfalfa establishment. Agron. J. 64:151–159.
123. Schreiber, M.M. 1960. Pre-emergence herbicides on alfalfa and birdshoot trefoil. Weeds 8:291–299.
124. Schreiber, M.M. 1967. Effect of density and control of Canada thistle on production and utilization of alfalfa pasture. Weeds 15:138–142.
125. Shaw, W.C., and W.A. Gentner. 1957. The selective herbicidal properties of several variously substituted phenoxyalkycarboxolic acids. Weeds 5:75–92.
126. Sheaffer, C.C., and D.R. Swanson. 1982. Seeding rates and grass suppression for sod-seeded red clover and alfalfa. Agron. J. 74:355–358.
127. Sheaffer, C.C., and D.L. Wyse. 1982. Common dandelion control in alfalfa (Medicago sativa). Weed Sci. 30:216–220.
128. Sheets, T.J. 1959. The comparative toxicities of monuron and simazine in soil. Weeds 7:189–194.
129. Silker, R.E. 1962. Dehydration of forage crops. Forages. 2nd ed. Iowa State Univ. Press, Ames.
130. Singh, D.V. 1972. Studies on alfalfa (Medicago sativa) grown after cutting dodder (Cuscuta) infection: starch and nucleic acid metabolism. Indian J. Exp. Biol. 10(2):140–142.
131. Smith, D. 1960. Yield and chemical composition of oats for forage with advance in maturity. Agron. J. 52:637–639.
132. ——. 1962. Forage management in the North. W.C. Brown Book Co., Dubuque, IA.

133. Sprague, M.A. 1952. Substitution of chemicals for tillage in pasture renovation. Agron. J. 44:405–409.
134. ----. 1960. Seedbed preparation and improvement of unplowable pastures using herbicides. Proc. Int. Grassl. Congr., 8th, 264–266.
135. ----, R.D. Ilnicki, R.J. Aldrich, A.H. Kates, T.P. Evard, and R.W. Chase. 1962. Pasture improvement and seedbed preparation with herbicides. N.J. Agric. Exp. Stn. Bull. 803.
136. Stahler, L.M. 1948. Shade and moisture factors in competition between selected crops and field bindweed (*Convolvulous arvensis*). J. Am. Soc. Agron. 40:490.
137. ----, and L.A. Derscheid. 1948. Cultural methods of noxious weed control in South Dakota. S.D. Agric. Exp. Stn. Circ. 75.
138. Swan, D.G. 1978. Effects of repeated herbicide applications on alfalfa (*Medicago sativa*). Weed Sci. 26:151–153.
139. Taylor, R.W., and D.W. Alinson. 1983. Legume establishment in grass sods using minimum-tillage seeding techniques without herbicide application: forage yield and quality. Agron. J. 75:167–172.
140. Taylor, T.H., J.S. Foote, J.H. Snyder, E.M. Smith, and W.C. Templeton, Jr. 1972. Legume seedling stands resulting from winter and spring sowings in Kentucky bluegrass (*Poa pratensis* L.) sod. Agron. J. 64:535–538.
141. ----, E.M. Smith, and W.C. Templeton, Jr. 1969. Use of minimum-tillage and herbicide for establishing legumes in Kentucky bluegrass (*Poa pratensis* L.) swards. Agron. J. 61:761–766.
142. Temme, D.G., R.G. Harvey, R.S. Fawcett, and A.W. Young. 1979. Effects of annual weed control on alfalfa forage quality. Agron. J. 71:51–54.
143. Tesar, M.B., K. Lawton, and B. Kawin. 1954. Comparison of band seeding and other methods of seeding legumes. Agron J. 46:189–194.
144. Thomas, R. 1976. Weed control and its effect on quality (alfalfa) hay [Market prices]. p. 59–61. *In* Proc. Calif. Alfalfa Symp., Davis. December. University of California, Davis.
145. Toai, T.V., and D.L. Linscott. 1979. Phytotoxic effect of residues. Weed Sci. 27:595–598.
146. Torell, P.J. 1968. Ten treatments for controlling dodder in alfalfa seed fields. University of Idaho Coll. of Agric. Current Info. Series 75. University of Idaho, Moscow.
147. Tripplet, G.B., J.R. Abernathy, C.R. Fenster, W. Flinchum, D.L. Linscott, E.L. Robinson, L. Standifer and J.D. Walker. 1984. Weed control for reduced tillage systems. USDA Farmers Bull. AD-fo-2279. U.S. Government Printing Office, Washington, DC.
148. ----, Jr., R.W. Van Keuren, and D. Walker. 1977. Influence of 2,4-D pronamide and simazine on dry matter production and botanical composition of an alfalfa-grass sward. Crop Sci. 17:61–65.
149. ----, ----, and V.H. Watson. 1975. The role of herbicides in pasture renovation. p. 29–41. *In* Proc. No-Tillage Forage Symp., Columbus. October. Ohio State University, Columbus.
150. Underwood, J.F., and J.C. Clay. 1976. Forage improvements in Appalachian Ohio through sod-seeding. p. 316–319. *In* J. Luchok et al. (ed.) Hill lands. Proc. Int. Symp., Morgantown, WV. 3–9 October. West Virginia University Books, Morgantown.
151. Upchurch, R.P. 1958. The influence of soil factors on the phytotoxicity and plant selectivity of diuron. Weeds 6:161–171.
152. USDA. 1965. Losses in agriculture. USDA Handb. 291. U.S. Government Printing Office, Washington, DC.
153. VanKeuren, R.W. 1976. Hill land improvement in eastern U.S.A. p. 77–90. *In* J. Luchok et al. (ed) Hill lands. Proc. of an Int. Symp., Morgantown, WV. West Virginia Books, Morgantown.
154. ----, and G.B. Tripplett. 1970. Seeding legumes into established grass swards. Proc. Int. Grassl. Cong., 11th, 131–134.
155. Vaughan, R.H., and D.L. Linscott. 1983. Herbicide performance in no-tillage legume establishment in grain stubble. Proc. Annu. Meet. Northeast. Weed Sci. Soc. 37:68–72.
156. Vengris, J. 1957. Downy chess control in alfalfa. Proc. Northeast. Weed Control Conf. p. 139–142.
157. Waddington, J. 1980. Chemical control of dandelion (*Taraxacum officinale*) and perennial sowthistle (*Sonchus arvensis*) in alfalfa (*Medicago sativa*) grown for seed. Weed Sci. 28:164–167.
158. ----, J. Gebhardt, and D.A. Pulkinen. 1976. Forage yield and quality alfalfa following late fall applications of 2,4-D or 2,4-DB [(2,4-dichlorophenoxy) acetic acid, 4-(2,4-dichlorophenoxy) butyric acid] herbicides. Can. J. Plant Sci. 56:929–934.
159. Wain, R.L., and F. Wightman. 1954. The growth regulating activity of certain w-substituted alkylcarboxylic acids in relation to their B-oxidation within the plant. Proc. R. Soc. London, B. 142:525–536.

160. Wakefield, R.C., and R.J. Hull. 1960. Chickweed control in alfalfa seedings. Proc. Northeast. Weed Control Conf. 18:320–326.
161. ----, and J.O. Pearson. 1964. Effects of herbicides and management factors on establishment of alfalfa seedings. Proc. Northeast. Weed Control Conf. 18:319–324.
162. ----, and ----. 1965. Control of winter annual weeds in alfalfa seedings. Proc. Northeast. Weed Control Conf. 19:199–203.
163. ----, and C.D. Sawyer. 1983. Control of annual weedy grasses in no-till alfalfa seedings. Proc. Annu. Meet. Northeast. Weed Sci. Soc. 37:80.
164. Watkins, E.M., and J.E. Winch. 1970. Assessment and improvement of roughland pastures in Ontario. Ontario ARDA Projects 25021 and 6011. Ontario Agricultural College, Guelph, ON, Canada.
165. Weldon, L.W., and F.L. Timmons. 1962. Penetration and persistence of diuron in soil. Weeds 9:195–203.
166. Welty, L.E., R.L. Anderson, R.H. Delaney, and P.F. Hensleigh. 1981. Glyphosate timing effects on establishment of sod-seeded legumes and grasses. Agron. J. 73:813–817.
167. White, J.G.H. 1970. Establishment of Lucerne (*Medicago sativa* L.) in uncultivated country by sod-seeding and over-seeding. Proc. Int. Grassl. Congr. 11th, 134–138.
168. Willard, C.J. 1962. Weed control in forages. p. 382–388. *In* H.D. Hughs et al. (ed.) Forages. Iowa State University Press, Ames.
169. ----, L.E. Thatcher, and S.S. Cutler. 1934. Alfalfa in Ohio. Ohio Agric. Exp. Stn. Bull. 540.
170. Wilson, R.G., Jr. 1981. Effects of Canada thistle (*Cirsium arvense*) residue on growth of some crops. Weed Sci. 29:159–164.
171. ----. 1981. Weed control in established dryland alfalfa (*Medicago sativa*). Weed Sci. 29:615–618.
172. Yuncker, T.G. 1932. The genus Cuscuta. Mem. Torrey Bot. Club 18:113–300.

24 Cytology and Cytogenetics of Alfalfa

T. J. McCOY

University of Arizona
Tucson, Arizona

E. T. BINGHAM

University of Wisconsin
Madison, Wisconsin

This chapter is a comprehensive review of the cytology and cytogenetics of cultivated alfalfa (*Medicago sativa* L.) and its hybrids with wild species. A detailed account of cytogenetics of the entire *Medicago* genus was compiled by Lesins and Lesins (100). The somatic karyotypes of *Medicago* spp. published previously (97) are not included, and the discussion of interspecific hybridization is restricted to interspecific hybrids with alfalfa.

Recently a number of important research results have been documented. These include: highly repeatable and comparatively efficient methods of haploid isolation, development of diploid alfalfa comprised entirely of cultivated germplasm, and identification of meiotic and gametophytic mutants that have proven useful in ploidy manipulations and in the recovery of new interspecific hybrids. These results have broadened the spectrum of basic research approaches and potential germplasm donors for alfalfa improvement. These findings have added considerably to the body of information broadly defined as alfalfa reproductive biology, and they have been useful in basic genetics and breeding research of alfalfa.

24-1 MICROSPOROGENESIS, MEGASPOROGENESIS, AND GAMETOPHYTE DEVELOPMENT

Microsporogenesis in alfalfa was first described by Reeves (129). A better understanding of microsporogenesis and megasporogenesis in alfalfa has been achieved by studies of mutants of the reproductive process (105, 108, 122, 150). These recent studies help to clarify the likely pathway of polyploid evolution in the genus *Medicago*.

Normal microsporogenesis follows a developmental sequence typical of most dicotyledonous plants. A number of sporogenous cells develop in each anther, and follow characteristic pollen mother cell (PMC) de-

Copyright 1988 © ASA-CSSA-SSSA, 677 South Segoe Road, Madison, WI 53711, USA.
Alfalfa and Alfalfa Improvement—Agronomy Monograph no. 29.

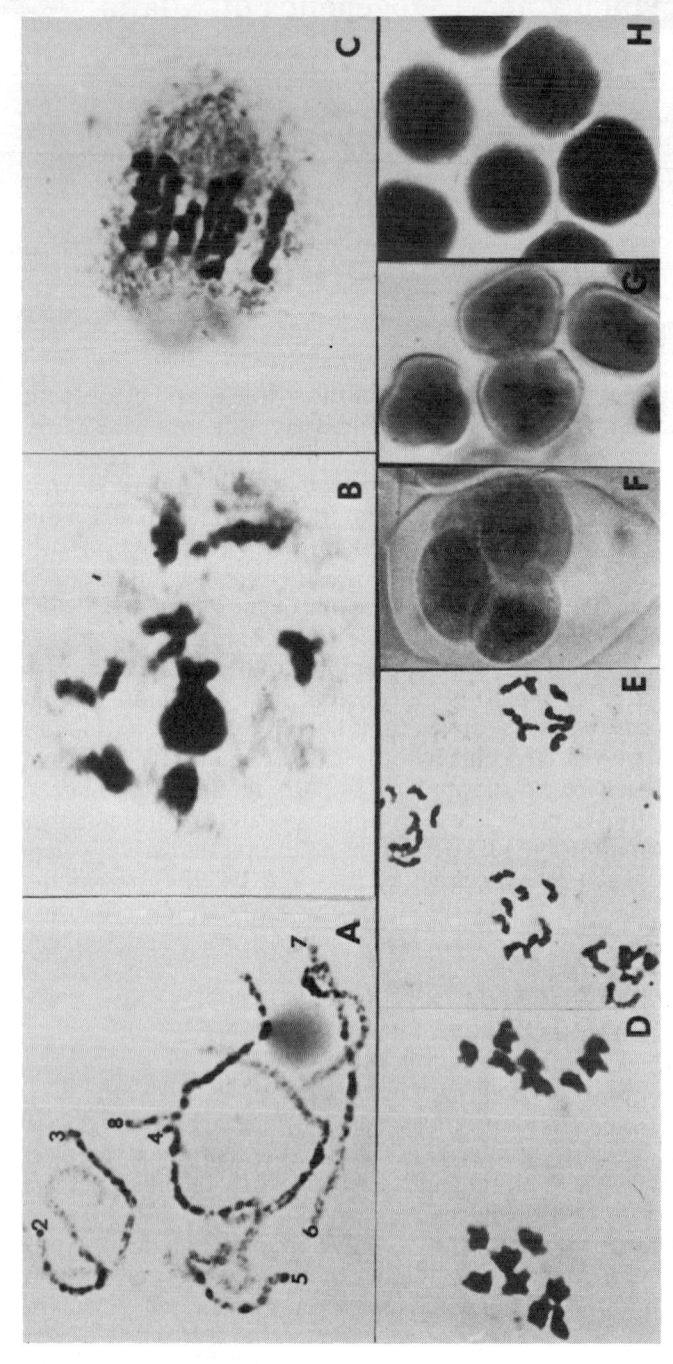

Fig. 24-1. Stages of microsporogenesis in a normal diploid plant of *M. sativa*. A. Pachytene showing eight bivalents with chromosomes numbered at short arm telomeres (68). B. Diakinesis with eight bivalents (112). C. Metaphase I with eight bivalents (112). D. Metaphase II with eight chromosomes in each nucleus (148). E. Telophase II (148). F. Quartet stage (112). G. Individual $n = x = 8$ microspores (112). H. Pollen grains with dark-staining generative nucleus (112).

velopment. The various stages of meiosis can be analyzed (Fig. 24–1A–1F), and numerous studies on chromosome pairing have been conducted in recent years. Although pachytene analysis is possible in alfalfa, particularly at the diploid level, diakinesis and metaphase pairing relationships are easier to study and have been extensively utilized to determine chromosome affinities. As discussed later in this chapter, chromosomes pair mainly as bivalents at all ploidy levels studied.

Simultaneous cytokinesis occurs after the second meiotic division such that four haploid microspores (Fig. 24–1F) are formed from one PMC. The first postmeiotic division occurs in the developing microspore to produce a binucleate mature pollen grain consisting of a generative cell and a tube cell. The second postmeiotic mitotic division does not occur until several hours postpollination, while the tube is growing down the style. Division of the generative nucleus at this time produces two sperm nuclei, resulting in the mature male gametophyte. Treatment of the germinating pollen tube with colchicine in vitro enables one to arrest division of the generative nucleus and count the chromosome number of the male gametophyte (10).

In addition to the classical work on megasporogenesis and female gametophyte development in alfalfa (27, 28, 29, 43, 56, 128), several recent contributions have clarified and expanded on the original discussions. These include female reproductive development studies, which have been conducted recently with the "cleared pistil" technique (122, 153).

Development of the female gametophyte follows the classic polygonum type (Fig. 24–2A–2L). Two or more primary sporogenous cells develop into megaspore mother cells (MMC's) in most ovules (43, 56, 122, 128). Further development of the MMC's in a given ovule frequently proceeds at different rates. After the first meiotic division, a cell plate is formed (Fig. 24–2C), dividing the MMC into a micropylar and chalazal cell. A linear tetrad of megaspores (Fig. 24–2E) results from cytokinesis following the second meiotic division. The three micropylar megaspores degenerate (Fig. 24–2F). The chalazal megaspore proceeds through three mitotic divisions (Fig. 24–2H, 2I, 2J), resulting in a seven celled, eight-nucleate, mature gametophyte (the embryo sac). The embryo sac consists of an egg cell flanked by two synergid cells, a large binucleate central cell and three antipodal cells (Fig. 24–2J).

24–2 FERTILIZATION AND EMBRYO DEVELOPMENT

Depending on genotype, species, and environment, fertilization in *Medicago* occurs between 24 and 44 h postpollination (43, 46, 47, 58, 133). Pollen tubes enter the female gametophyte between the egg and the synergids. The sperm nuclei are discharged near the egg. One sperm nucleus and the egg nucleus ultimately fuse to form a zygote. The other sperm nucleus fuses with the polar nuclei of the central cell to produce

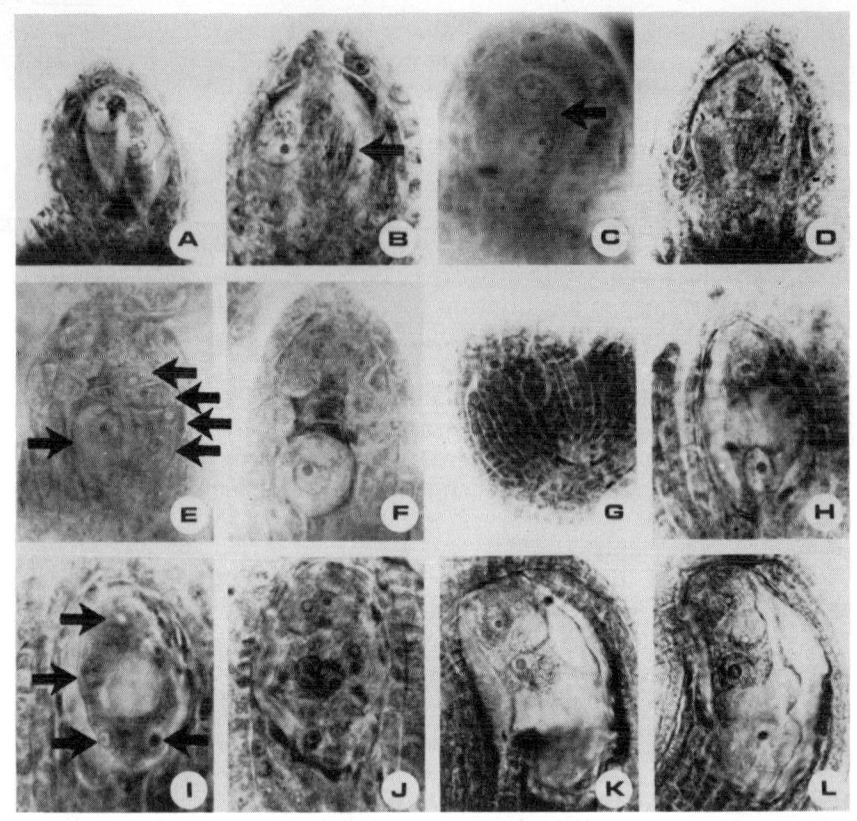

Fig. 24-2. Megasporogenesis in alfalfa, normal (A–L taken from 122). A. Two sporogenous cells, both in prophase. B. Two sporogenous cells, one in prophase, one at MI (*arrow*). C. Dyad, cell plate forming (*arrow*). D. MII, nearly linear. E. *Right side*—tetrad of megaspores still visible (*arrows*). *Left side*—chalazal cell enlarging (*arrow*), three micropylar cells degenerating. F. Chalazal cell is primary megaspore, three micropylar cells degenerating. G. *Entire ovule*—integuments have grown to surround nucellus. H. Two-nucleate stage of embryo sac. I. Four nucleate stage of embryo sac (*arrows*). J. Eight-nucleate, seven cell embryo sac. K. Egg apparatus and polar bodies, antipodals degenerated. L. Egg apparatus and fused polar bodies.

the endosperm mother cell. (See Chapter 30 in this book for details of fertilization in self-fertile and self-sterile material.)

Transverse division of the zygote at about 30 to 46 h results in a two-celled proembryo (47, 58, 133) with a small apical cell and an enlarged basal cell. At this time, four endosperm nuclei are present, demonstrating that endosperm development occurs at a faster rate than embryo development in the early stages. Further divisions of the apical cell result in a six-celled proembryo (43). It is from the apical cell of this six-cell proembryo that the embryo develops. The other cells form the suspensor, these divide, and the suspensor increases in size. Day-by-day embryo developmental sequence has been presented by Fridriksson and Bolton (58) and Sangduen et al. (133) for alfalfa and by Sangduen et al.

(134) for interspecific hybrids between alfalfa and *M. scutellata*. The globular stage of development occurs at 7 to 8 d postpollination (DPP). At the globular stage the embryo consists of eight to 16 cells in the embryo proper, and one or two chalazal suspensor cells with two basal suspensor cells. The endosperm is coenocytic. The heart stage occurs at 10 to 13 DPP. At this stage the initiation of cotyledon growth is evident and the primordia of the epicotyl and hypocotyl are initiated. The endosperm is still coenocytic although very dense and undergoing rapid development. At 13 DPP, Sangduen et al. (133) report that the suspensor is at its maximum size with 16 to 20 chalazal cells and two basal cells; the endosperm has become cellular. The embryo reaches the torpedo stage at 14 to 16 DPP, followed by a period of rapid apical cell divisions and curvature of the cotyledons. The embryo attains full size at about 25 DPP, and seeds are mature at about 35 DPP. In addition to the light microscope studies, electron microscopy (133, 134) has provided an ultrastructural description of embryogenesis following intra- and interspecific crosses.

Several investigations have followed the development of interspecific hybrid embryos (45, 58, 110, 134). Fridriksson and Bolton (58) studied embryo development following crosses of alfalfa and the following species: *M. arborea, M. blancheana, M. lupulina, M. marina, M. platycarpa, M. rigidula, M. ruthenica,* and *M. scutellata.* Only *M. lupulina* showed no evidence of fertilization. Sangduen et al. (136) in their study of pollen germination and pollen tube growth also found good evidence of frequent fertilization following selected interspecific crosses. Failure to produce hybrids was attributed to postfertilization barriers, involving either too vigorous embryo growth or lagging endosperm development. Cooper and Brink (45) also found interspecific hybrid failure associated with abnormal endosperm development. The important point is that a postfertilization barrier should be amenable to embryo rescue techniques. McCoy and Smith (110) have shown that preculturing the ovule prior to removal of the embryo for culturing resulted in the recovery of several new interspecific combinations (described later). At the ultrastructural level, Sangduen et al. (134) found that late heart stage hybrid embryos of *M. sativa* × *M. scutellata* had inactive dictyosomes and endoplasmic reticulum in the ovules. They believed embryo abortion resulted from failure of nutrient metabolism and transport in all tissues. Specific deficiencies included failure to metabolize starch and lipid out of maternal tissue at 5 to 7 DPP, failure of the nucleus to utilize food reserves, and failure of the embryonic suspensor to transport nutrients because of abnormalities in the endoplasmic reticulum, microbodies, and transfer wall surfaces.

24-3 CYTOLOGY OF MEIOTIC AND GAMETOPHYTIC ABNORMALITIES

Meiotic and gametophytic abnormalities under simple genetic control have been important in plant genetics and breeding. Thus, a single mutant in corn, *ig* for indeterminate gametophyte (82), results in male sterility and paternal as well as maternal haploids; all of which are of potential value in maize (*Zea mays* L.) improvement. Frequently meiotic abnormalities lead to male sterility (see Chapter 30 in this book).

Gametes with the sporophytic chromosome number are referred to as $2n$ gametes. Therefore, in alfalfa a $2n$ gamete from a diploid has 16 chromosomes, a $2n$ gamete from a tetraploid has 32 chromosomes. The occurrence of $2n$ gametes is not new, either to plant genetics or to alfalfa genetics and breeding. In 1935, a mechanism for $2n$ gamete production in *Datura* was reported to be under simple genetic control (137). Since then, several examples of single gene control of $2n$ gamete formation have been documented in several species including: maize (131), potato (*Solanum tuberosum* L.) (117), and alfalfa (105). In potato, an autopolyploid species like alfalfa, intensive research has been directed toward utilization of a reproductive mutant, *ps* (for parallel spindles), that results in the production of $2n$ male gametes by a first division restitution (FDR) mechanism (117). The resultant heterozygosity transferred in tetraploid-diploid matings has made this a valuable mutant in potato improvement (114).

The random occurrence of $2n$ gametes has had an important role in germplasm transfer across ploidy levels, especially in the transfer of useful agronomic traits from diploid *M. falcata* to tetraploid cultivated germplasm (7, 12). In addition, it is now generally accepted that $2n$ gametes are the major mechanism for movement of genetic material from lower to higher ploidy levels during evolution (50, 70). When $2n$ gametes are formed by a type of first division restitution mechanism, e.g., parallel spindles in potato (117) and alfalfa (150), or a desynaptic mutant in potato (127), then most if not all of the heterozygosity present in the sporophyte is transferred in the $2n$ gamete.

The first diploid alfalfa clone that produced a high frequency of $2n$ pollen, designated W31, was identified in 1975 (18). Several additional clones were identified as a result of studies on reproductive stability of diploid populations containing various levels of cultivated and wild germplasm. Clone W31 was used in cytological and genetic studies to determine the mechanism of $2n$ pollen formation and the inheritance of the mechanism. Vorsa and Bingham (150) concluded that $2n$ pollen resulted from a disorientation of spindles in metaphase II of microsporogenesis (Fig. 24–3A–3D). The mechanism was similar to the parallel spindle mechanism in potato (117). Therefore, the resulting $2n$ pollen was genetically equivalent to that formed after first division restitution.

The FDR mechanism of $2n$ pollen formation is controlled by a single recessive gene, designated *rp* (105). Plants homozygous for *rp* produce variable levels of $2n$ pollen depending on the environment; however, *rprp* plants produce a minimum of 4% $2n$ pollen regardless of the environment in which they are grown.

Plants that produce a high frequency of $2n$ eggs have been identified

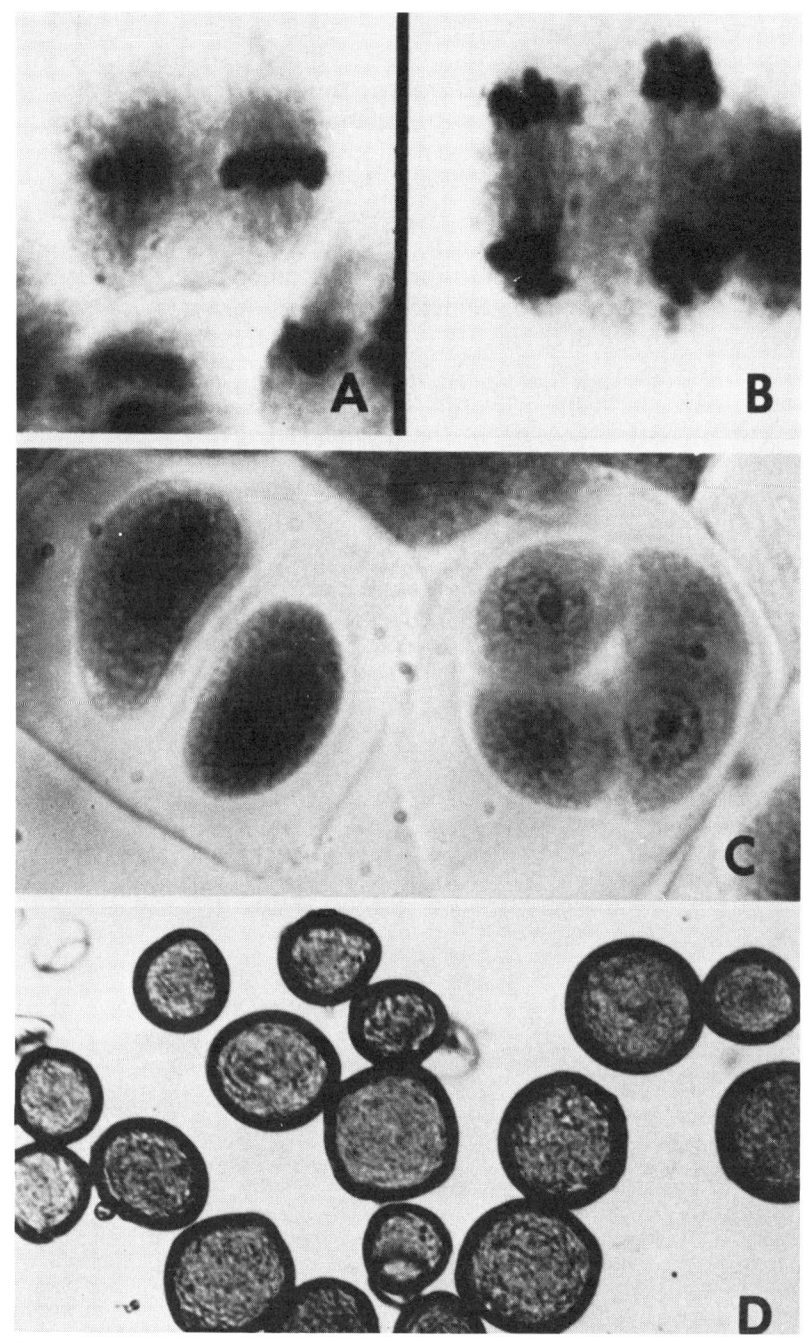

Fig. 24–3. Stages of microsporogenesis in a plant homozygous for *rp*, resulting in 2*n*-pollen formation by a mechanism genetically equivalent to first division restitution. A. Metaphase II showing disorientation of spindles. Spindles are basically parallel to each other in one plane as opposed to the normal 60° orientation to form a tetrahedron (150). B. Anaphase II (150). C. Dyad resulting from spindle disorientation forming two 2*n* microspores with a quartet from normal microsporogenesis that results in four *n* microspores (112). D. *n* and 2*n* pollen grains (112).

in alfalfa (122). As indicated earlier, a cell plate is formed following the first meiotic division, and a linear tetrad of megaspores results from cytokinesis following the second meiotic division is normal female gametophyte development. In $2n$ egg gametophyte development, the alteration occurs after anaphase II. The micropylar cell undergoes cytokinesis following the second meiotic division; however, the chalazal cell lacks cytokinesis (Fig. 24–4A, 4B). The two micropylar megaspores degenerate (Fig. 24–4C, 4D), and the two nuclei in the chalazal cell fuse prior to the mitotic divisions (Fig. 24–4E) to form the gametophyte. An alternative sequence that occurs in some gametophytes is for fusion to occur during the first two mitotic divisions (Fig. 24–4F, 4G, 4H). The final result is an embryo sac with a $2n$ egg. In the first sequence all nuclei of the embryo sac are $2n$; in the second sequence the embryo sac would be composed of n and $2n$ nuclei. Because of the mechanism involved, $2n$ eggs are genetically equivalent to second division restitution (SDR). Therefore, all loci between the centromere and the first crossover are homozygous in the $2n$ egg. Heterozygosity present in the sporophyte is not transferred through SDR $2n$ gametes.

Two reports documenting the lack of postmeiotic cytokinesis during microsporogenesis have been published. Pfeiffer and Bingham (122) found

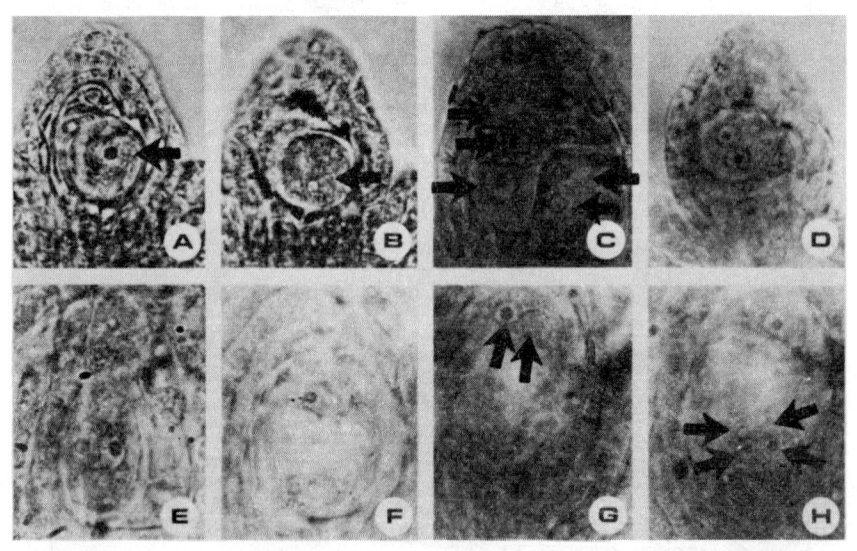

Fig. 24–4. Altered sequence of megasporogenesis in $2n$ egg producers and mechanism of $4n$ pollen formation in diploid alfalfa (A–H taken from 122). A and B. Two planes of focus through the same ovule. Two micropylar megaspores degenerating. Two nuclei and no cell plate in the chalazal cell. Only one chalazal nucleus is visible in each plane of focus (*arrows*). C. *Right side*—chalazal cell with two nuclei (*arrows*), two micropylar megaspores degenerating. *Left side*—normal tetrad, three megaspores visible (*arrows*). D. Unreduced primary megaspore. E. Nuclear fusion prior to first mitosis. F. Nuclear fusion following first mitosis. G. and H. Nuclear fusion following second mitosis. Two planes of focus. Note size of nucleoli in the two micropylar nuclei (*arrows G*) as compared with the four chalazal nuclei (*arrows H*).

that several plants that produced a high level of $2n$ eggs had variable expression ranging from occasional to the complete absence of the postmeiotic cytokinesis. Similarly, McCoy and Smith (108) have identified diploid clones with the complete lack of cytokinesis. An inheritance study demonstrated that a single recessive gene, designated *jp*, controlled the lack of cytokinesis. Cytological studies confirmed that a single four-nucleate microspore developed from one PMC instead of the normal, four, single-nucleate microspores from one PMC (Fig. 24–5A). Fusion of the

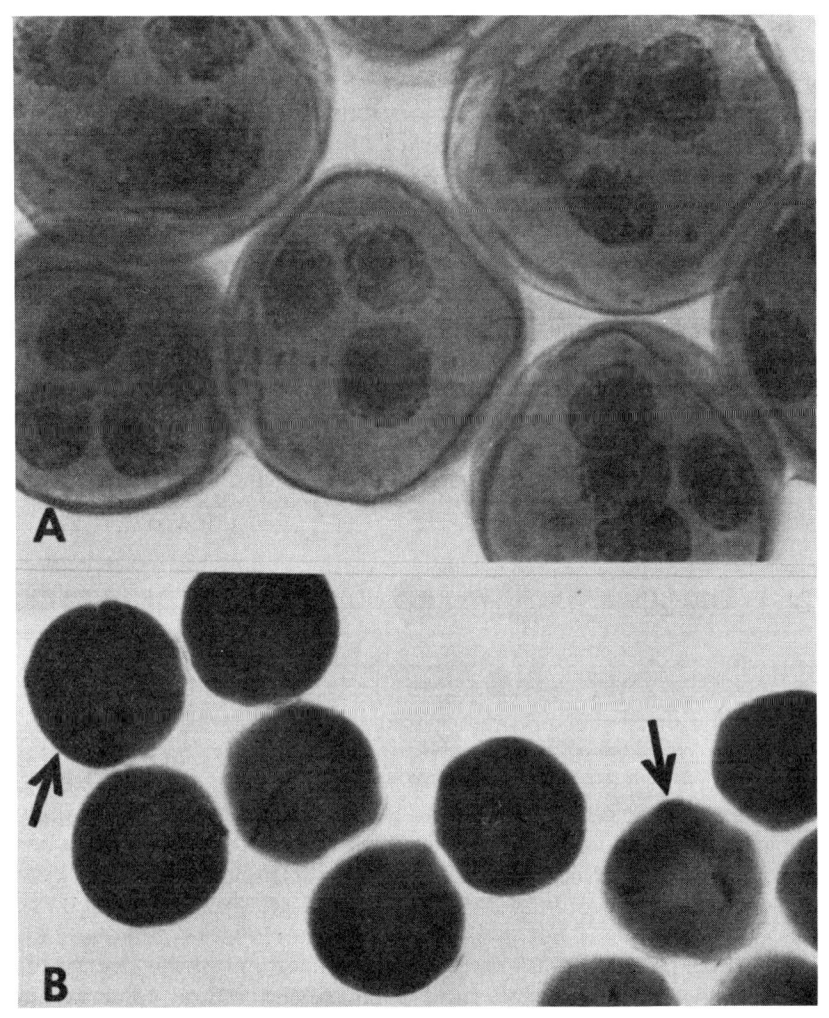

Fig. 24–5. Microsporogenesis in a diploid plant homozygous for *jp*. A Four-nucleate microspores (the products or meiosis in plants that lack the postmeiotic cytokinesis) (108). B. $4n$ pollen grains from plants lacking the postmeiotic cytokinesis. Most pollen grains have one generative nucleus due to fusion of all four nuclei before the first postmeiotic mitosis; however, some pollen grains (*arrows*) have more than one "generative" nucleus due to partial fusion (108).

four nuclei into a single nucleus occurred in 80 to 100% of the microspores depending on genotype (Fig. 24-5B). This results in the formation of a $4n$ gametophyte from a $2n$ sporophyte.

McCoy and Smith (108) found that plants homozygous for *jp* produced $4n$ pollen that was essentially nonfunctional. Although some $4n$ pollen germinated, only five progeny were recovered from thousands of $8x$-$2x$ (*jpjp*) crosses. This may be a consequence of the severe inbreeding in the $4n$ pollen from the diploid genotypes used. Plants producing $4n$ pollen were identified by Pfeiffer and Bingham (122), due to the production of octoploid progeny, following $4x$-$2x$ crosses, indicating the presence of functional $4n$ pollen. The octoploids probably arose from fusion of a $2n$ egg ($2n = 4x = 32$) with a $4n$ pollen ($4n = 4x = 32$). Research with more vigorous and more heterozygous diploid plants has shown that $4n$ pollen is capable of effecting fertilization, and several hundred octoploids have been produced from $8x$-$2x$ and $4x$-$2x$ matings (21, 112).

Female gametophyte alterations also are present in clones homozygous for *jp* (108). At the diploid level, *jpjp* clones tend to have a greater frequency of $2n$ eggs, although this trend is inconsistent. At the tetraploid level, *jpjpjpjp* clones may produce an elevated frequency of haploids in $4x$-$2x$ crosses. Preliminary cytological observations demonstrate the presence of supernumerary nuclei in embryo sacs of *jpjp* clones (112), which are similar to the supernumerary nuclei in embryo sacs of soybean [*Glycine max* (L.) Merr.] homozygous for ms_1 (48). Abnormalities in the female gametophyte may be partially responsible for several successful interspecific hybrids produced betwen *M. sativa* and other *Medicago* species (106, 109, 110).

24-4 ORIGIN AND CYTOLOGY OF EUPLOIDS OF ALFALFA

24-4.1 Diploids

The majority of the 56 recognized species of *Medicago* are diploids with most $2n = 2x = 16$; however, four species have $2n = 14$ and one species is comprised of $2n = 16$ and $2n = 14$ types. In addition to diploid forms, tetraploid (both $2n = 32$ and $2n = 30$) species and two hexaploid species (*M. cancellata* and *M. saxatillis*) are known. Several species occur at more than one ploidy level. Diploid and tetraploid forms are known for *M. falcata*, *M. lupulina*, *M. papillosa*, and *M. prostrata*, as well as *M. sativa*. Tetraploid and hexaploid forms are known for *M. arborea*. Lesins and Lesins (100) compiled an excellent monograph of the chromosome number, species relationships, and potential agricultural value of *Medicago* spp.

Genetic transfer across ploidy levels in alfalfa has been documented (12, 20, 148), and Bingham and Saunders (20) manipulated the chromosomes of a single clone to seven ploidy levels, and demonstrated the freedom to transfer germplasm across ploidy levels. A euploid series from diploid to octoploid has been established and manipulated using the $4x$ cultivated genome. The alfalfa ploidy series has been used in physiological studies of polyploids, in analytic breeding research, research on the optimum ploidy level, and transfer of useful agronomic traits both up and down the ploidy series.

Much chromosome manipulation work has been devoted to the identification of an effective and highly reproducible method for isolating haploids ($2n = 2x = 16$) in cultivated alfalfa (13). A workable method utilizes the $4x$-$2x$ cross to recover parthenogenetic eggs. It is hypothesized that one sperm fertilizes the central cell to initiate endosperm development, and the egg develops into a haploid plant. In this system the diploid male parent does not contribute germplasm. Therefore, the male parent can be another unadapted *Medicago* spp. Although there are obvious genotypic differences in haploid frequency (13, 108), it appears that this method can be used to extract haploids from any cultivated tetraploid alfalfa plant. Haploid frequencies up to 10 haploids per 1000 pollinations are common (13). McCoy and Smith (108) produced up to 28 haploids per 1000 pollinations using tetraploid plants homozygous for *jp*. Haploids are important in interploidy gene transfer, especially in attempts to utilize the analytic breeding method proposed by Chase (35, 36).

Chromosome pairing studies on haploids have demonstrated predominately bivalent pairing (17, 39, 68, 147). Eight bivalents are generally observed at pachytene (Fig. 24–1A), diakinesis (Fig. 24–1B), and metaphase I (Fig. 24–1C). A haploid that was an apparent synaptic mutant with many univalents and some unpaired cells has been reported (17), although univalents occur at a low frequency in most haploids. Other chromosomal alterations, e.g., duplications or interchanges, have been observed infrequently. One haploid with an interchange was described by Stanford and Clement (147) and one haploid was found (39) with a tandem duplication of a major portion of the satellite region of the nucleolus organizer chromosome.

Haploid crossing behavior has been studied by several investigators (17, 39, 88, 147). In general, the frequency of $2n$ gametes is relatively higher in haploids than in normal diploids. In $2x$-$4x$ crosses involving haploids as the $2x$ parent, up to 25 seeds per 100 pollinations are produced; most of the progeny are tetraploids, with a few triploids.

The first haploids had reduced female fertility and were male sterile. The lack of functional pollen (though some of it is stainable) precluded their use in haploid × haploid matings. However, female fertility is adequate, and Bingham and Gillies (17) found that among 54 haploids, two had normal fertility in haploid × diploid matings, 48 had reduced fertility to varying degrees, and four were female sterile. Several reports have demonstrated that haploids are slightly more fertile in crosses with diploid *M. falcata* than in crosses with diploid alfalfa. Recent haploids of improved tetraploids have been fertile as seed and pollen parents (21).

The inability to produce cultivated diploid alfalfa directly from haploid × haploid matings required development of an alternative approach. Thus, haploids were crossed with *M. falcata* diploids, which resulted in male-fertile F_1 hybrids. The F_1 hybrids were then backcrossed to different raw haploids for six generations to obtain essentially 100% *Cultivated Alfalfa at the Diploid Level,* designated CADL (18). CADL has been basic to studies on the physiology of polyploids, breeding theory, and analytic breeding in *Medicago,* and genetic analyses at the diploid level. The herbage yield, seed production, and crossing behavior of CADL have been described (18). Herbage yield is reduced from the tetraploid level, averaging about 72% of the tetraploid. Self and cross fertility of CADL is comparable to tetraploids in terms of seed number; however, the seeds are smaller and seed yield by weight is less than tetraploids. The populations are stable at the diploid level although some plants produce a high frequency of $2n$ gametes. New CADL populations are being developed. Each year additional germplasm from tetraploid cultivars is scaled down to the diploid level by haploidy and triploid bridge-crosses. Individual populations of CADL have been developed with germplasm from 'Vernal', 'Saranac', and 'Iroquois', respectively (21).

24–4.2 Triploids

Triploids occur in the progeny of $4x$-$2x$ or $2x$-$4x$ crosses (10, 20, 42, 81, 85). In addition, triploids have been recovered from $2x$-$2x$ matings where either a $2n$ egg or a $2n$ pollen has effected fertilization. Given the chromosomal constitution of the gametes, one would expect a high frequency of triploids following $4x$-$2x$ or $2x$-$4x$ crosses; however, the frequency of triploids from these matings is low (about one triploid per 1000 pollinations). This has been referred to as the triploid block. Several explanations for the triploid block have been advanced, including the hypothesis that embryo vs. endosperm chromosome sets must be in a ratio of 2:3, or that in the endosperm the ratio must be two sets of maternal chromosomes to one set of paternal chromosomes. Neither situation prevails in $2x$-$4x$ or $4x$-$2x$ matings, although more triploids are usually recovered from the latter cross.

Cytological studies on alfalfa (*M. sativa sensu lato*) triploids have documented a range of three to four trivalents (Fig. 24–6) at metaphase I (10, 42, 81). Trivalent frequency is thus greater than multivalent frequency in tetraploids. This phenomenon has been observed in several other plant species. Further meiotic studies of chromosome behavior at anaphase I show frequent deviations from the expected 12-12 disjunction.

Triploids reported prior to 1984 were very low in female fertility and essentially male sterile. Although stainable pollen frequency ranged from 0 to >50%, pollen germination was usually low and the pollen was not functional in crosses. Most early triploids were from diploid *M. falcata* × tetraploid alfalfa crosses. Many triploids produced at Wisconsin were frequently from crossing a cytoplasmic male sterile tetraploid *M. sativa* × diploid *M. falcata.* Tetraploid hybrids from these $4x$-$2x$ matings in-

CYTOLOGY AND CYTOGENETICS 749

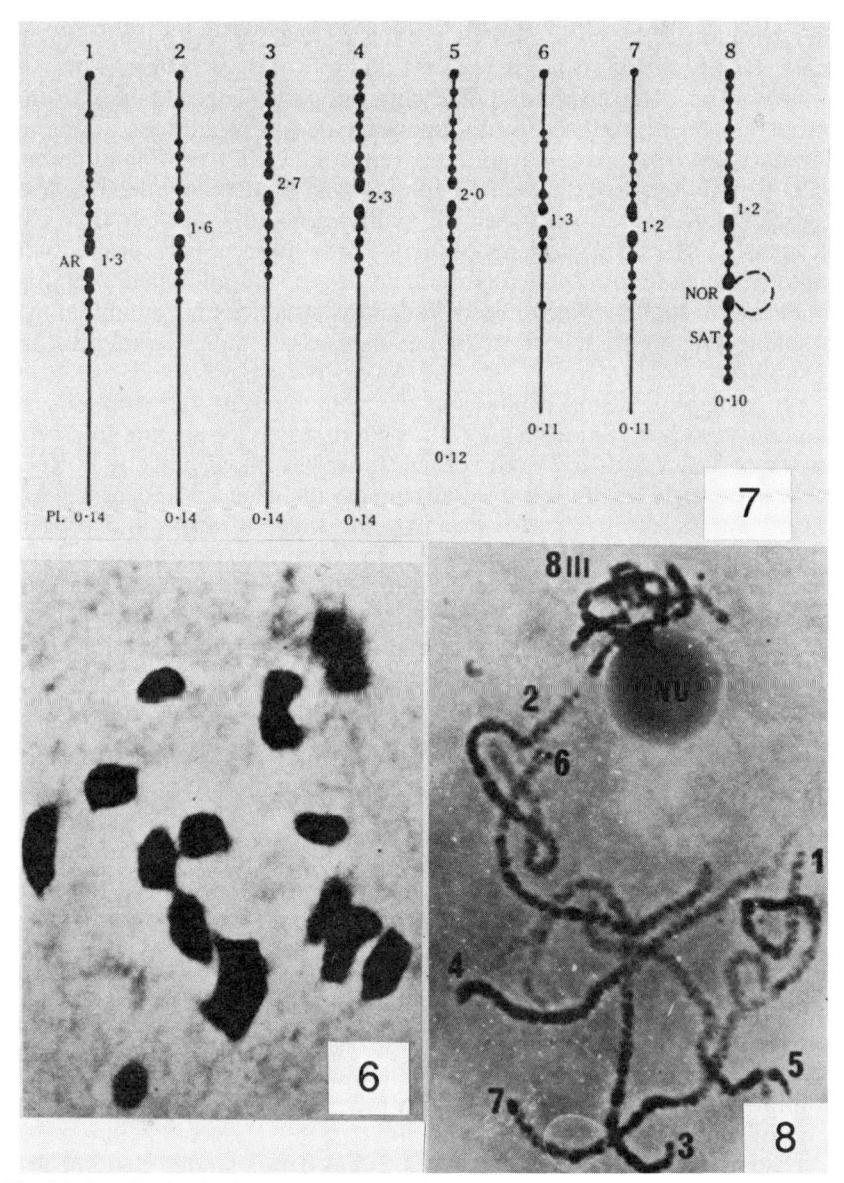

Fig. 24-6 to 24-8. Fig. 24-6. Metaphase I in a triploid alfalfa plant with bivalent and trivalent associations near equatorial plate and univalents near the poles (10). Fig. 24-7. Standard haploid idiogram of pachytene chromosomes of *M. sativa* showing centromeres and arm ratios (AR), proportional lengths (PL), positions of major chromomeres, and position of nucleolus organizer region (NOR) and distal satellite (SAT) (67). Fig. 24-8. Pachytene stage of meiosis in trisomic alfalfa plant ($2n = 2x + 1 = 17$), trisomic for chromosome 8 (67).

dicated that the *M. falcata* plants used were restorers (12); therefore, the male sterility of the triploids was not due to a lack of restoration.

An array of 10 triploids possessing only cultivated alfalfa chromosomes has been produced using highly fertile partial inbreds of Vernal, Saranac, and Iroquois as tetraploid parents and CADL as diploid parents (22). Female fertility of the 10 triploids was similar to other triploids (between zero and 25 seeds per 100 pollinations in 3x-2x, 3x-4x, or 3x-6x crosses). However, pollen production was noticeably improved with some triploids used as pollen parents and even self-pollinated. Although much of the pollen was aborted and nonstainable, the remainder was stainable ranging in size from that typical of diploids, tetraploids, and hexaploids.

In general, crossing triploids as the pollen parent with triploids, tetraploids and hexaploids resulted in predominately near-tetraploid progeny from 3x-3x and 4x-3x crosses, and near hexaploid progeny in 6x-3x crosses, because of the selective advantage of balanced and nearly balanced gametic numbers. However, a wide range of aneuploid progeny (chromosome numbers of $2n$ = 29, 31, 35, 39 and 47) were also recovered from 4x-3x and 6x-3x crosses. In a sample of seven self-progeny of one triploid, each plant had a different chromosome number ($2n$ = 31, 32, 36, 45, 46, 48, and 52). The fact that triploids produce euploids and aneuploids, from the diploid to the hexaploid level, makes triploids extremely useful for both bridge crosses in germplasm transfer across ploidy levels and for isolating aneuploids.

24-4.3 Tetraploids

The tetraploid level ($2n$ = $4x$ = 32) is the ploidy level of most commercially grown alfalfa. Although genetic studies confirm the autopolyploid structure of alfalfa, the species has an extremely low level of multivalent formation. The majority of cytological studies have demonstrated 16 bivalents or at least 15 bivalents at metaphase I of meiosis (4, 42, 44, 62, 69, 79, 87, 113). Various levels of multivalent formation have been observed ranging from an average of 0.62 per cell (69) to 1.68 per cell (42). Even in colchicine-doubled autotetraploids the frequency of multivalent formation has been low (42, 119). The reasons for the low level of multivalent pairing are unknown.

Detailed meiotic analysis of diploids has demonstrated that one arm of most chromosomes is more heterochromatic and physically shorter than the other arm. In addition, chromosomes frequently pair as rod bivalents, indicating the lack of a cross-over event in one arm. Given that multivalents require a crossover in both chromosome arms, this low level of genetic exchange would result in a low frequency of multivalents. Natural selection for bivalent pairing may have an important role in pairing affinities. In maize, colchicine-doubled autotetraploids had relatively high levels of multivalent formation; however, selection for increased fertility over a 10 yr period resulted in a population with significantly lower levels of multivalent formation (60). From a cytogenetic viewpoint, multivalent associations are less reliable than bivalent associations for producing correct chromosomal disjunction. Multivalents can contribute to aneuploidy and reduce fertility, because of improper alignment and disjunction at anaphase I. Thus, a reduction in multivalent frequency is beneficial for reproductive efficiency.

Tetraploid alfalfa evidently carries a large number of recessive mutations for various types of reproductive and morphological abnormalities, as well as recessive lethal genes. Evidence is provided by the cultivated tetraploid germplasm that served as the source of the abnormalities discovered in CADL, i.e., meiotic and gametophytic mutants (discussed earlier) and female sterility (see chapter 30). The tetraploid level is, in effect, protected by tetrasomic ratios such that homozygotes expressing abnormal traits segregate rarely.

24-4.4 Pentaploids

Pentaploids result from a variety of crosses. The most direct is $6x$-$4x$ or the reciprocal crosses, but they have been produced from $3x$-$4x$ and $2x$-$4x$ crosses (20), and possibly other interploidy level crosses. Pentaploids have been obtained from $2x$-$4x$ crosses using haploids as the $2x$ parent (17, 40). Presumably these pentaploids result from the formation of $3n$ nuclei in the $2x$ parent due to abnormal cytokinesis. Pentaploids can also be produced from $3x$-$4x$ crosses, although the majority of progeny from this cross were at or near the tetraploid level (10). Pentaploids were the most frequent progeny in backcrosses between triploid *M. sativa* × *M. dzhawakhetica* F_1 hybrids and $4x$ *M. sativa*. This is significantly different from the frequency of pentaploids in the progeny of $3x$-$4x$ crosses within *M. sativa* (109).

Chromosome behavior in pentaploids has been inferred from the results of crossing the pentaploid with other ploidy levels, especially $4x$ and $6x$, and not from direct analysis. The fertility of pentaploids is reduced compared to that of tetraploids and hexaploids, because of the high frequency of univalents and subsequent formation of unbalanced gametes. Although information on crossing behavior is limited, it is known that pentaploids make a good bridge ploidy level between $6x$ and $4x$ levels. The majority of the univalents are either lost or grossly unbalanced gametes fail to function because the majority of $5x$-$4x$ progeny have 32 or 33 chromosomes. Plants derived from crossing pentaploid *M. sativa* and *M. dzhawakhetica* BC_1 plants back to tetraploid *M. sativa* were predominantly tetraploid, or 33- or 34-chromosome number plants (112).

24-4.5 Hexaploids

There are two naturally occuring hexaploid ($2n = 6x = 48$) species in the genus, which are *M. cancellata* and *M. saxatilis*. In addition, *M. arborea* has both tetraploid and hexaploid types (100). The cytogenetic structure of the hexaploid species has been described as alloautoploid. Lesins (93) felt that *M. saxatilis* had a genomic constitution consisting of two genomes closely related to *M. sativa* and four genomes from a wild diploid species, *M. rhodopea*. Likewise, the genomic constitution of *M. cancellata* was proposed as including two genomes of a species closely related to *M. sativa*, and four genomes of another diploid species, *M. rupestris* M.B. (100).

The first reported *M. sativa* hexaploids were observed in the progeny of 8x-4x crosses, where the octoploid was produced by colchicine doubling (3, 79). Hexaploids can also be produced from 4x-4x, 3x-6x, and 5x-6x crosses (20, 104). Interest in hexaploids was stimulated when Bingham and Binek (16) reported the occurrence of spontaneous hexaploids in the cv. Saranac. Spontaneous hexaploids most likely result from the union of a normal, n gamete ($n = x = 16$) with a $2n$ gamete ($2n = 2x = 32$). A spontaneous hexaploid has been found in Narragansett (138), and the authors (21, 112) have documented spontaneous hexaploids in six different tetraploid lines and populations. In addition, hexaploids have been regenerated from tissue culture of a diploid (130), a tetraploid (72), and after treatment of tetraploid cuttings with colchicine (21). Under certain conditions, hexaploid cells appear to have a selective advantage.

Lesins et al. (102) produced three separate sets of hexaploid alfalfa. One set that should have been vigorous with a high level of heterozygosity was discarded on the basis of seed production. Another set, derived from 8x-4x crosses, had good seed production but poor forage production. However, one (the third set) performed well, with the authors concluding that it should be possible to develop hexaploid cultivars. However, chromosome instability is a major problem in breeding hexaploids (143, 152), with low seed set observed in many hexaploid populations (102, 152). It appears that hexaploids of spontaneous origin may be more fertile and more chromosomally stable (16), than hexaploids derived from 8x-4x or 3x-6x crosses (143). Fertility comparable to tetraploid plants was observed under greenhouse conditions in spontaneous hexaploids of Saranac (16). In other hexaploids, seed production has been low, ranging from 0 to 1.5 seeds per pollination (102, 152).

Smith et al. (143) found that seed production was increased 52 and 123% after one and two cycles of selection for hexaploidy. In two studies (143, 152), selection for hexaploidy showed no significant effect on chromosomal stability. In another study, Lesins et al. (102) developed chromosomally stable hexaploid populations after four cycles of selection for both fertility and hexaploidy. Smith et al. (143) postulated that heterozygosity is important to the chromosomal stability of hexaploids and discussed the problem of maximizing hexaploid heterozygosity and chromosome stablity. It should be possible to develop stable hexaploids because two species are true-breeding natural hexaploids and one species (*M. arborea*) includes both tetraploid and hexaploid types.

24-4.6 Heptaploids

The first reported heptaploid ($2n = 7x = 56$) was observed in the progeny of a $4x$-$6x$ cross, presumably the result of a $2n$ egg ($2n = 4x = 32$) from the female parent and an n gamete ($n = 3x = 24$) from the male parent (16). The most direct methods for producing heptaploids would be from a $6x$-$8x$ cross or its reciprocal. Bingham and Saunders (20) produced a heptaploid plant in the development of the diploid through octoploid series from a single tetraploid clone. It was produced by crossing the hexaploid with an octoploid. The octoploid used was from chromosome doubling in tissue culture. In addition, near heptaploid plants with $2n = 55$ to 58 have been produced by $3x$-$3x$, $3x$-$6x$, and $4x$-$3x$ crosses (22). The heptaploid is of little value other than academic interest.

24-4.7 Octoploids

Octoploids ($2n = 8x = 64$) were derived originally from chromosome doubling of cultivated tetraploids with colchicine (3, 79). More recently, a number of octoploids have been recovered by spontaneous doubling in tissue culture (19, 20, 72, 78). Also, octoploids have been produced by sexual polyploidization via $4x$-$2x$ crosses, whereby a $2n$ egg ($2n = 4x = 32$) was fertilized by $4n$ ($4n = 4x = 32$) pollen (122). Several diploid mutants now have been identified which produce viable $4n$ pollen, which results in octoploid progeny in matings with 32-chromosome megagametophytes (21, 112).

In general, octoploid plants do not breed true, although no detailed studies have been made of chromosome numbers in later generations of octoploid populations. Root tip chromosome counts of plants in two Wisconsin octoploid populations, following two generations of seed increase, indicated chromosome numbers ranged from near $2n = 32$ to $2n = 64$ with the modal class near $2n = 48$. As discussed under hexaploids, when the tetraploid genome of cultivated *M. sativa* is elevated, there appears to be a selective advantage near the hexaploid level.

Cytological studies of octoploid alfalfa plants are limited. Julen (79) found that bivalents were the most common configuration in octoploid plants, with some univalents and multivalents. Somatic cells of octoploids exhibited a great deal of chromosome instability in root tips of plants produced from intercrossing colchicine induced octoploids. Sadasivaiah and Lesins (132) observed octoploid, tetraploid, and aneuploid chromosome numers in cells of the same root tip. Chromosome reduction was associated with meiosis-like mitotic divisions.

The octoploid level is an important ploidy level of research on physiological aspects of polyploidy. Several studies including the effects of ploidy on agronomic, biochemical, and physiological traits have been conducted with diploid, tetraploid, and octoploid clones composed of similar genes.

24-5 ORIGIN, CYTOLOGY, AND USES OF ANEUPLOIDS

Trisomics at the diploid level ($2n = 2x + 1 = 17$) are easily recovered following $3x$-$2x$ crosses (10, 33, 81). Although several types of trisomics are recognized in plant cytogenetics (83), only primary trisomics (where the extra chromosome is one of the normal chromosomes of the complement) have been identified in alfalfa. Monosomics ($2n = 2x - 1 = 15$) have not been reported in alfalfa.

The first trisomics reported were derived from crosses of triploid *M. sativa* × *M. falcata* hybrids with diploid *M. falcata*. Subsequently, Kasha and McLennan (81) produced a number of trisomics from crossing triploids with diploid alfalfa and diploid *M. falcata*.. Thirty-nine of 93 progeny (40%) from the $3x$-$2x$ crosses were trisomic. Several double trisomics ($2n = 18$) and triple trisomics were recovered.

Cytological identification of trisomics by pachytene analysis resulted in positive confirmation of five of the eight chromosomes (67). Based on the pachytene idiogram of alfalfa (68, 97; and Fig. 24-7), trisomics of chromosomes 1, 4, 6, 7, and 8 (Fig. 24-8) have been confirmed. Chromosome banding techniques such as G- or C-banding for the identification of trisomics in mitotic cells have not been successful. In some species, morphological criteria have been useful for trisomic identification (83); however, they have not been conclusive in alfalfa (33, 67, 81).

To be useful for assigning genes to chromosomes and for determining linkage groups, the trisomic must be transmitted at an efficient rate. In alfalfa, trisomic transmission is sufficiently high to be useful in genetic studies (33, 81). Buss and Cleveland (33) used the frequency of trivalents in pollen mother cells (PMC's) to predict the frequency of $n + 1$ gametes. They made the assumption that PMC's with trivalents would produce $n + 1$ gametes, whereas PMC's with a univalent would lose the extra chromosome. They predicted that 0 to 23% of the microspores would be $n + 1$, depending on the trisomic used. Kasha and McLennan (81) found that 5.5 to 24.7% of the progeny were trisomic when the trisomic was the female parent, and 0.6 to 17.4% of the progeny were trisomic when the trisomic was the male parent.

The sample of trisomics thus far analyzed is not adequate to determine whether certain primary trisomics occur more frequently than others. Gillies (67) speculated that if chiasma frequency is reduced in the shorter chromatic arms of alfalfa, as indicated by Clement and Stanford (40), then chromosomes with chromatic short arms (e.g., 3, 4, and 5) would produce a lower frequency of trivalents in the triploid and therefore a lower frequency of gametes carrying extra chromosomes 3, 4 or 5 (67).

Only limited gene-chromosome mapping has been accomplished in alfalfa. A dominant purple flower color gene is found on chromosome 8, the nucleous organizer chromosome (80). Chromosome 8 is the most easily identified trisomic because of the presence of three chromosomes with satellites. Trivalents were frequent in trisomics for chromosome 8 (67) and female and male transmissions were high, averaging 20% through the female and 11% through the male. This is in agreement with the assumption of Buss and Cleveland (33), that trisomic transmission would be high in plants with high trivalent frequency. Other genes linked to specific chromosomes include a male-sterile gene linked with chromosome 6, and a dominant calyx spot gene and a recessive yellow cotyledon gene on one of the chromosomes with a short chromatic arm (chromosome 3, 4, or 5). Trisomics for chromosome 7 have been characterized as having dark green leaves, but inclusion of short stem internodes in the original description for trisomic 7 (80) was not confirmed (67).

Aneuploids near the tetraploid level have been produced in several ways including: $3x$-$4x$ crosses (10, 146), screening of varietal populations for spontaneous aneuploids (11), progeny of asynaptic plants (17, 146), haploid × tetraploid matings (17, 147), and polyembryonic seeds (146). The most efficient method is $3x$-$4x$ crossing. Stanford (146) found two plants with 31 chromosomes, two with 32 chromosomes, and one with 33 chromosomes. Binek and Bingham (10) produced a high frequency of aneuploids from $3x$-$4x$ crosses. Chromosome numbers ranged from 28 to 40 with 109 out of 229 (48%) having 28, 30, or 31 chromosomes. If efficient procedures could be identified, a large number of aneuploids could be recovered from screening varietal populations. Bingham (11) found 6% aneuploids in the variety Vernal and 3% in Saranac. Progeny of an asynaptic plant were found to have chromosome numbers ranging from 25 to 34 with nine plants having 31 chromosomes (146). Although trisomic diploids ($2n = 17$) trisomic tetraploids ($2n = 31$) do occur in haploid × tetraploid crosses, they occur at a low frequency.

Stanford (146) crossed 30-chromosome plants, obtained from polyembryonic seeds, with a normal tetraploid to obtain 31-chromosome progeny. This illustrates how the 28- or 30-chromosome progeny of $3x$-$4x$ crosses could be used to derive additional 31-chromosome, trisomic tetraploids.

Stanford (146) proposed using trisomic tetraploids for gene-chromosome mapping, and was successful in associating a tetrasomic gene "sticky leaf" with a specific trisomic. However, chromosome mapping with 31-chromosome plants requires large family sizes, and it is recognized that gene-chromosome mapping studies can be done much more efficiently with trisomic diploids. In addition, identification of the extra chromosomes in trisomic diploids is feasible with pachytene analysis (67). Given that alfalfa is characterized by tetrasomic inheritance, gene-chromosome mapping at the diploid level will be applicable to the tetraploid. Establishment of a complete trisomic series and the assignment of genes, such as isozyme markers (125, 126), should be feasible over the next several years. Alfalfa plants regenerated from tissue culture are an additional source of aneuploids (see Chapter 29 in this book).

24-6 USES OF HAPLOIDS

The isolation of haploids from $4x$-$2x$ crosses (13) formed the basis for research that has demonstrated the autopolyploid genetic structure of alfalfa (17, 68) and the importance of maximizing heterozygosity in alfalfa production (51). In addition, haploids have made it possible to develop diploid populations of cultivated alfalfa (CADL), composed of germplasm from the cultivated tetraploid (18).

The advantage of haploidy and the subsequent development of CADL is that diploids can be doubled to produce tetraploids with a defined genotypic structure. A tetraploid derived by doubling a diploid can have a maximum of two alleles at any locus. This provides for the establishment of defined populations with a maximum of two alleles at any locus that can be used for breeding theory research. Dunbier and Bingham (51) scaled tetraploid alfalfa down to the diploid level by haploidy, then produced diploid hybrids that were chromosome-doubled by colchicine to produce diallelic duplex parents. Subsequent crossing established a single cross (SC) and then a double cross (DC) generation, each with a certain theoretical genetic structure. The theoretical structure indicated a greater frequency of tri- and tetra-allelic loci in the DC generation than in the SC population from 100% diallelic parents. Yield and fertility studies indicated the DC population was superior to the SC, and the SC population was superior to the parents, providing evidence that maximum heterozygosity is important for superior plant performance. This conclusion has been substantiated in additional studies (15, 107).

Pfeiffer and Bingham (123) demonstrated improvements in both forage yield and fertility by selecting within two-allele populations. They concluded that such improvements came from recombination and accumulation of the most favorable of two additive alleles present in the original plant. The results strongly suggest that linkage blocks or linkats, as defined by Demarly (49), containing favorable alleles at different linked loci contribute to vigor and yield in alfalfa.

The prospect of extracting haploid plants from essentially any tetraploid makes alfalfa amenable to the analytic breeding scheme proposed by Chase (35) for potato, and later extended to several other polyploid species (36). Basically, this scheme involves reduction of the autotetraploid to the diploid level where selection and breeding are conducted, followed by returning to the tetraploid via chromosome doubling or unreduced gametes.

In addition to the above applications, the haploid method has been used to transfer the capacity to regenerate plants (from callus and suspension cultures) to the diploid level (See chapter 29).

24-7 CYTOGENETIC AND GENETIC CONFIRMATION OF AUTOPOLYPLOIDY

Chromosome pairing and pachytene analysis studies of haploids provide support for an autotetraploid interpretation. Bingham and Gillies

(17) found that metaphase chromosomes paired mainly as bivalents in 15 haploids from 10 different sources of cultivated alfalfa. This demonstration of the high degree of chromosome homology, plus the normal pachytene pairing observed in haploids and the similarity in arm ratio between the haploid complement and diploid alfalfa and diploid *M. falcata* (68) confirm the conclusion that cultivated alfalfa is an autotetraploid.

Previous reports have summarized genetic studies in alfalfa (8, 30). Historically, there was disagreement whether alfalfa is an allotetraploid with disomic inheritance or an autotetraploid with tetrasomic inheritance. During the first half of the 20th century, most genetic studies assumed disomic inheritance. In 1951 Stanford (145) presented unequivocal proof of tetrasomic inheritance in alfalfa. Stanford (145) stressed the importance of the critical F_3 generation to determine mode of inheritance. Since 1951, definitive genetic studies have proven the autopolyploid nature of alfalfa, because most genes studied exhibit tetrasomic inheritance.

The genetics of polysomic polyploids, like alfalfa, are complex. The salient features of this complexity have been described by Bingham (15). Considering a locus with two alleles (*A* and *a*) five combinations are possible in an autotetraploid. These are: *AAAA* (referred to as quadruplex), *AAAa* (triplex), *AAaa* (duplex), *Aaaa* (simplex), and *aaaa* (nulliplex). However, a polysomic locus has the potential of having more than two alleles. The existence of multiple alleles has been inferred by various breeding studies (15, 30, 51, 107), and genetic analysis of peroxidase and leucine aminopeptidase isozymes has provided direct evidence for multiple alleles (125, 126). An additional complexity of polysomic polyploids is that a locus is indistinguishable from a chromosome segment of potentially linked genes (15, 49). Also, the buffering capacity of the autotetraploid genome allows the retention of deleterious or lethal genes that can result in aberrant segregation patterns.

A number of genes controlling morphological and anatomical traits have been identified (8, 30; see Chapter 4 in this book). For example, the inheritance of genes that affect root morphology and anatomy (24, 25, 54), leaf morphology and anatomy (5, 6, 26, 71), chlorophyll deficiency (31), and flower color (14) have been described. In addition, several genes controlling either male sterility or female sterility have been described (see Chapter 30 in this book).

The inheritance of resistance to several diseases has been determined (see Chapter 21 in this book). Genetic control of the following diseases has been described: anthracnose (*Colletotrichum trifolii* Bain.) (34), bacterial wilt [*Corynebacterium insidiosum* (McCull.) H. L. Jens.] (149), Fusarium wilt [*Fusarium oxysporum* Schl. f. sp. *medicagnis*] (73), and Phytophthora root rot (*Phytophthora megasperma* Drechs.) (76, 77, 103). In addition, the inheritance of stem nematode [*Ditylenchus dipsaci* (Kühn) Filipjev] resistance has been determined (53). Genetic analysis of N_2

fixation parameters has identified several genes that affect nodulation (121; see chapter 7).

Tissue culture and the accompanying somaclonal variation has been a source of additional mutants in alfalfa. The genetics of several novel mutants identified in regenerated plants have recently been determined (19, 130; see chapter 29).

24-8 USES OF THE PLOIDY SERIES

24-8.1 Ploidy Level Effects on Physiological and Agronomic Parameters

The development of repeatable techniques for establishing a ploidy series (20) has stimulated research on the physiology of polyploids. Isogenic diploid-tetraploid and isogenic tetraploid-octoploid populations developed by Pfeiffer et al. (124) were used to study ploidy effects on several parameters including: inbreeding depression (57), ribulose-1, 5-bisphosphate carboxylase oxygenase (115, 116), photosynthesis (118, 124), and N_2 fixation (86, 124). In addition, agronomic characters (1, 2, 52), CO_2 exchange rates, and transpiration (140) have been studied in genetically equivalent ploidy levels of alfalfa. Results of these experiments relative to ploidy level effects are presented in Table 24-1.

24-8.2 Optimum Ploidy Level for Alfalfa

Although cultivated alfalfa is an autotetraploid species, it is not known if this is the optimum ploidy level from the perspective of agronomic performance. Given that hexaploid and octoploid plants have generally larger leaves, consideration has been given to developing alfalfa with elevated ploidy levels (16, 102, 143, 152). The major problem associated with elevated ploidy levels of *M. sativa* is their chromosomal instability (143, 152). However, the instability problem may be solvable, because three species exist naturally at the hexaploid level and are chromosomally stable.

Producing a population of elevated ploidy level for farm use could be accomplished by several methods: (i) use a $2n$ gamete-producing tetraploid as one parent and a normal tetraploid for the other parent, (ii) generate self-sterile octoploids by chromosome doubling with colchicine or through tissue culture and use these as females in crosses with normal tetraploids, or (iii) produce triploids (preferably ones that exclusively produce $2n$ eggs) that when crossed with hexaploid males result in hexaploid progeny. Hexaploid and/or octoploid populations could be maintained and increased separately and blended with tetraploid seed at the time of commercial release. Although these methods are speculative, they illustrate approaches whereby cytogenetic manipulations may contribute to increased alfalfa production.

Table 24-1. Results of experiments testing the effects of ploidy level on various agronomic and physiological parameters.

Character	Ploidy comparisons	References
DNA concentration	$2x = 4x$	52
Water soluble protein concentration	$2x = 4x$	52
Percentage total N	$2x = 4x$	1, 2, 52
Dry matter yield	$4x$ two times $>2x$	52
Fresh weight per trifoliate	$4x$ two times $>2x$	52
DNA per trifoliate	$4x$ two times $>2x$	52
Glucose-6-phosphate dehydrogenase activity	$4x$ two times $>2x$	52
Percentage digestible dry matter	$2x = 4x$	1, 2
Stomatal density	$4x > 2x$ and $8x > 4x$	140
CO_2 exchange rates	$4x = 2x$ and $8x = 4x$	124, 140
N_2 fixation	$4x > 2x$, for first 10 d	86
	$4x = 2x$, after 10 d	86
	$4x = 8x$	86
	$4x = 2x$ and $8x = 4x$	124
Transpiration	$4x = 2x, 8x > 4x$	124
Inbreeding depression	$2x$ inbreeding depression $> 4x$	57
Concentration of DNA in protoplasts		
Concentration of ribulose-1,5-bisphosphate carboxylase in protoplasts		
Chlorophyll and chloroplasts per cell	$4x$ two times $>2x$, $8x$ generally two times $>4x$	118
Concentration of ribulose-1,5-bisphosphate carboxylase/oxygenase in leaves	$4x = 2x, 8x = 4x$	116
Kinetic properties of ribulose-1,5-bisphosphate carboxylase/oxygenase	$4x = 2x, 8x = 4x$	115

24-8.3 Uses of Meiotic and Gametophytic Abnormalities

The ploidy series may be used in the analytic breeding scheme proposed by Chase (35) utilizing the meiotic mutant, *rp*. Analytic breeding applied to polyploid crops requires a three-step process, namely: (i) reduction of a tetraploid to the diploid level; (ii) breeding and selection at the diploid level; and (iii) transfer of diploid germplasm back to the tetraploid level. The *rp* gene conditions $2n$ pollen formation by a mechanism genetically equivalent to first division restitution (FDR). Therefore, these $2n$ gametes transfer an estimated 80% (15) of the heterozygosity present in the diploid. All of the heterozygosity from the centromere to the first crossover and half of the heterozygosity beyond the first crossover is transferred.

The effects of *rp* on heterosis at the population level were recently investigated experimentally and theoretically (107). The experimental test used three diploid *rp* clones, designated 2xrprp, and their tetraploid counterparts, designed 4xCD, produced by colchicine doubling. The 2xrprp plants and 4xCD plants were then crossed to the same tetraploid cytoplasmic male-sterile clone, designated 4xCMS. Progenies from the crosses were space planted in the field. Shoot biomass data from six harvests showed a highly significant yield difference between 4xCMS × 2xrprp families vs. their counterpart 4xCMS × 4xCD families. The biomass of 2xrprp crosses was 12% to 32% greater than the yields of 4xCD crosses with an average gain of 19%. Assuming that crossover frequencies are not affected by ploidy level and that only bivalent pairing occurs in the tetraploid, the theoretical model indicated that $2n$ FDR pollen from 2xrprp plants should have 12.5 to 50% more heterozygous loci, on average, than the n pollen from the 4xCD plants. The biomass data suggests that heterozygosity is important, and adds further support to the importance of maximizing heterozygosity to achieve maximum yield (15, 51).

McCoy and Walker (111) have discussed a method of forming highly heterozygous tetraploid families by union of $2n$ FDR eggs (from a diploid producing a high frequency of $2n$ eggs) with $2n$ FDR pollen (from a diploid producing a high frequency of $2n$ pollen). In addition, $2n$ FDR gametes may eventually be utilized in forming highly heterozygous hexaploid populations. Other applications include the possible use of $2n$ gametes in identifying superior genotypes, and as an important experimental tool for identification of intra- and inter-locus effects in alfalfa.

24-9 CYTOGENETICS OF THE *MEDICAGO* GENUS

Extensive pachytene idiograms and karyotype data from somatic chromosomes of some *Medicago* spp. have been published (32, 41, 61, 62, 63, 64, 65, 66, 74, 97). This section includes only the standard pachytene idiogram of diploid alfalfa (Fig. 24-7). An excellent description of the *Medicago* genus has been published (100). This chapter summarizes relationships among species in the genus and emphasizes efforts to utilize other species of the genus. *Medicago* species and their $2n$ chromosome numbers are given in Table 24-2.

The most common chromosome number in the *Medicago* genus is $2n = 2x = 16$; however, four species of diploids have $2n = 14$ chromosomes and one species (*M. murex*) has both $2n = 14$ and $2n = 16$ types. The $2n = 14$ type originated by chromosome rearrangement, whereby two chromosomes united and the centromere of one of the chromosomes was lost (101). At pachytene the segments of both chromosomes are recognizable. According to Lesins and Lesins (100), three of the other $2n = 14$ species (*M. polymorpha, M. praecox,* and *M. rigidula*) probably arose in a similar manner, because all three species have one exceptionally long chromosome that could have arisen by fusion of two chromosomes. *Medicago constricta* has no exceptionally long chromosomes. It may have arisen by secondary chromosome rearrangement.

There are only two annual species at the tetraploid level, *M. rugosa* and *M. scutellata*. Both of these species have 30 rather than 32 chromosomes as reported previously (9). *Medicago scutellata* was found to have two pairs of chromosomes with a nucleolus organizer region (NOR),

Table 24-2. Chromosome number of species of the genus *Medicago*.

Perennials		Annuals	
Medicago species	2n chromosome no.	*Medicago* species	2n chromosome no.
arborea	32, 48	arabica	16
cancellata	48	bonarotiana	16
carstiensis	16	ciliaris	16
cretacea	16	constricta	14
daghestanica	16	coronata	16
dzhawakhetica	32	disciformis	16
falcata	16, 32	doliata	16
glomerata	16	granadensis	16
glutinosa	32	heyniana	16
hybrida	16	intertexta	16
ignatzii	16	laciniata	16
marina	16	lanigera	16
papillosa	16, 32	littoralis	16
pironae	16	lupulina	16, 32
platycarpa	16	minima	16
prostrata	16, 32	murex	14, 16
rhodopea	16	muricoleptis	16
rupestris	16	noeana	16
ruthenica	16	orbicularis	16
sativa	16, 32	polymorpha	14
saxatilis	48	praecox	14
suffruticosa	16	radiata	16
		rigidula	14
		rotata	10
		rugosa	30
		sauvagei	16
		scutellata	30
		secundiflora	16
		shepardii	16
		soleirolii	16
		tenoreana	16
		tornata	16
		truncatula	16
		turbinata	16

whereas *M. rugosa* has only one pair of chromosomes with an NOR. The two species could have arisen by hybridization of a $2n = 14$ and a $2n = 16$ species, followed by polyploidization. Interestingly, phenolic taxometric studies have demonstrated that both $2n = 14$ and $2n = 16$ species show relationship to the $2n = 30$ species (37, 142).

Outside of the *M. falcata-sativa-glutinosa* complex, only four perennial species (*M. arborea, M. dzhawakhetica, M. papillosa,* and *M. prostrata*) are reported as $2n = 4x = 32$. In addition, two hexaploid perennial species are known (*M. cancellata* and *M. saxatilis*), and *M. arborea* has both tetraploid and hexaploid types. *Medicago cancellata* and *M. saxatilis* are thought to be alloautoploids (93, 95, 100) with two genomes from *M. sativa* and four genomes from *M. rhodopea* (*M. saxatilis*) or *M. rupestris* (*M. cancellata*).

Chromosome morphology has been observed on the basis of mitotic tissue (97). In general, idiograms prepared from somatic chromosomes are not accurate enough to distinguish homologous chromosomes. It is difficult to make accurate comparisons because of differences in pretreatments and methods of preparation of the chromosomes (97, 100).

Attempts to utilize chromosome banding technniques have been unsuccessful in *Medicago*. Given that *Medicago* spp. generally have small and similar chromosomes (2.3–5.0 μ) at mitotic metaphase, detailed somatic karyotypes are difficult in the absence of banding techniques. However, Johnson et al. (78) have documented chromosomal aberrations including interchanges in alfalfa protoclones (plants regenerated from protoplasts; see Chapter 29 in this book), and Schlarbaum et al. (139) have recently clarified species relationships in section intertextae by somatic karyotype analysis.

Definitive karyotype analysis can be conducted at the pachytene stage of meiosis, particularly at the diploid level (32, 41, 61, 62, 63, 64, 65, 66, 68, 74). All species closely related to *M. sativa* (species included in the *M. sativa-falcata-glutinosa* complex) had similar pachytene karyotypes (64, 97), and these karyotypes were the same as haploids of cultivated alfalfa (68). In addition, Gillies (65) found some species that were distantly related to *M. sativa* (*M. daghestanica, M. pironae, M. rhodopea,* and *M. rupestris* had pachytene karyotypes similar to *M. sativa*. In contrast, some distantly related species (e.g., *M. hybrida* and *M. suffruticosa*) had pachytene karotypes distinctly different from *M. sativa* (66). Hybrids between diploid *M. sativa* and diploid *M. rhodopea* and diploid *M. rupestris* have been produced (106, 110). Meiotic analysis of the diploid F_1 hybrids indicated high chromosome pairing affinity (see next section).

24-10 CYTOGENETICS OF INTERSPECIFIC HYBRIDS BETWEEN *M. SATIVA* AND OTHER *MEDICAGO* SPECIES

Although interest has long been expressed in *Medicago* interspecific hybridization, and several useful traits have been identified in wild *Medicago* spp. (7, 23, 75, 84, 141, 144), there has been little success in crossing outside of the *M. sativa* complex (100). As discussed in chapter 3, species of the *M. sativa* complex belong to one biological species. All members freely intercross (especially when adjusted to the same ploidy level), have excellent chromosome pairing in F_1 hybrids, and have no hybrid sterility in the F_1 and later generations. Germplasm from all members of the complex can be readily incorporated into cultivated germplasm.

Attention is given here to the production, cytology, and crossing behavior of successful interspecific hybrids between *M. sativa* and other *Medicago* spp. Species of the *Medicago* genus according to subgenera and section are presented in Table 24–3. Alfalfa has now been successfully hybridized with *M. cancellata, M. daghestanica, M. dzhawakhetica, M. glomerata, M. hybrida, M. marina, M. papillosa, M. pironae, M. prostrata, M. rhodopea, M. rupestris, M. saxatilis,* and *M. scutellata.* Hybrids

Table 24-3. The genus *Medicago*.

I. Subgenus *Lupularia*
    (a) *M. lupulina*   (b) *M. secundiflora*
II. Subgenus *Orbicularia*
    A. Section Carstiensae
        (a) *M. carstiensis*
    B. Section Platycarpae
        (a) *M. platycarpa*   (b) *M. ruthenica*
    C. Section Orbiculares
        (a) *M. orbicularis*
    D. Section Hymenocarpos
        (a) *M. radiata*
    E. Section Heyniana
        (a) *M. heyniana*
    F. Section Cretaceae
        (a) *M. cretacea*
III. Subgenus *Medicago*
    A. Section Falcago
        1. Subsection Falcatae
            (a) *M. falcata*   (b) *M. glomerata*   (c) *M. glutinosa*
            (d) *M. prostrata*   (e) *M. sativa*
        2. Subsection Rupestres
            (a) *M. cancellata*   (b) *M. rhodopea*   (c) *M. rupestris*
            (d) *M. saxatilis*
        3. Subsection Daghestanicae
            (a) *M. daghestanica*   (b) *M. pironae*
        4. Subsection Papillosae
            (a) *M. dzhawakhetica*   (b) *M. papillosa*
    B. Section Arboreae
        (a) *M. arborea*
    C. Section Marinae
        (a) *M. marina*
    D. Section Suffruticosae
        (a) *M. suffruticosa*   (b) *M. hybrida*
IV. Subgenus *Spirocarpos*
    A. Section Rotatae
        (a) *M. bonarotiana*   (b) *M. noeana*   (c) *M. rotata*
        (d) *M. rugosa*   (e) *M. shepardii*   (f) *M. scutellata*
    B. Section Pachyspirae
        (a) *M. constricta*   (b) *M. doliata*   (c) *M. littoralis*
        (d) *M. murex*   (e) *M. rigidula*   (f) *M. soleirolii*
        (g) *M. tornata*   (h) *M. truncatula*   (i) *M. turbinata*
    C. Section Leptospirae
        (a) *M. arabica*   (b) *M. coronata*   (c) *M. disciformis*
        (d) *M. laciniata*   (e) *M. lanigera*   (f) *M. minima*
        (g) *M. poymorpha*   (h) *M. praecox*   (i) *M. sauvagei*
        (j) *M. tenoreana*
    D. Section Intertextae
        (a) *M. ciliaris*   (b) *M. granadensis*   (c) *M. intertexta*
        (d) *M. muricoleptis*

have been produced from seed in all of these examples except *M. daghestanica*, *M. hybrida*, *M. marina*, *M. pironae*, and *M. rupestris*. Only two hybrids have been produced from seed with *M. rhodopea*, and only one hybrid was produced from seed with *M. scutellata*.

Sangduen et al. (135) have recovered the only hybrid between *M.*

sativa and the annual *M. scutellata,* by using gibberellic acid (GA) treatment of peduncles and pedicels immediately following pollination. However, the main effect of the GA treatment was to increase self-fertility of the normally, self-sterile alfalfa clone used as the female parent.

Pollination and fertilization studies have demonstrated that zygotes are formed and embryogenesis is initiated for a number of interspecific combinations (58, 110, 120, 134). These results indicate that the major barrier to recovery of *Medicago* interspecific hybrid is postfertilization, which should make the production of hybrids amenable to embryo culture. In addition, the rapidly expanding technology of somatic hybridization via protoplast fusion should be applicable to *M. sativa,* given that alfalfa can be regenerated from protoplasts (see chapter 29).

24–10.1 Embryo Culture Rescue of Interspecific Hybrids

Previous efforts to obtain *Medicago* interspecific hybrids via embryo culture were unsuccessful (55, 59). McCoy and Smith (110) attempted various embryo rescue techniques, e.g., direct culture of hybrid embryos, nurse endosperm techniques, and the embryo transplant technique of Williams and deLautour (151). A modification of these techniques, designated the ovule-embryo culture method (110), has resulted in successful recovery of interspecific hybrids between alfalfa and several other *Medicago* spp.

Basically, the ovule-embryo culture technique consists of culturing the fertilized ovule containing a developing hybrid embryo for 5 to 7 d. The hybrid embryo is then excised and placed directly on the appropriate medium where it develops into a plant. This technique resulted in the successful hybridization of alfalfa with *M. daghestanica (2x), M. hybrida* (2x), *M. marina (2x), M. papillosa* (2x), *M. pironae* (2x), *M. rupestris* (2x), and *M. rhodopea* (2x). In addition, several trispecies hybrids were produced. These include: *M. sativa-M. dzhawakhetica* 3x × *M. cancellata* (6x), and *M. sativa-M. dzhawakhetica* (3x) × *M. saxatilis* (6x). The technique has proven useful in recovering backcross progeny, which frequently cannot be recovered from seed (106, 110).

24–10.2 Perennial Interspecific Hybrids

Results of hybridizing *M. sativa* with other species of the subgenus *Medicago* are summarized in Table 24–4. Successful hybridization has occurred with species of all sections except Arborae. A detailed description of all successful hybridizations is as follows.

24–10.2.1 *Medicago sativa* L. × Species of Subsection Falcatae

24–10.2.1.1 *Medicago sativa* L. × *M. glomerata* Balb. *Medicago glomerata* is a diploid ($2n = 2x = 16$) species with upright stems, yellow flowers, and coiled pods covered with glandular hairs. In addition, the

Table 24-4. Results of hybridizing *M. sativa* with all other species of the subgenus *Medicago*, genus *Medicago*.

Subgenus *medicago*	Results of hybridization with *Medicago sativa*
Section Falcago:	
Subsection Falcatae	
M. glomerata	Easily hybridized with *M. sativa*; >90% bivalent pairing. However, F_1 had low fertility (92).
M. prostrata	Easily hybridized with *M. sativa*; >90% bivalent pairing. Hybridization barrier when *M. prostrata* was the female parent (91).
Subsection Rupestres	
M. cancellata	Hybridizes with $4x$ and $6x$ *M. sativa* (90, 143, 152).
M. rhodopea	Two hybrids recovered from thousands of pollinations (95). Hybrids readily recovered from ovule-embryo culture (110). >75% bivalent pairing in F_1 (112).
M. rupestris	Hybrids readily recovered from ovule-embryo culture (110). Chromosomes pair mainly as bivalents in F_1 (106). Ovule-embryo culture required to produce backcross also.
M. saxatilis	Readily hybridizes at the $4x$ and especially at the $6x$ level (93, 143, 152).
Subsection Daghestanicae	
M. daghestanica- *M. pironae*	Two trispecies hybrids (*M. daghestanica-M. pironae-M. sativa*) have been produced by crossing doubled *M. daghestanica-pironae* hybrid with *M. sativa* (94).
M. daghestanica	Several diploid hybrids recovered from ovule-embryo culture (112).
M. pironae	Several diploid hybrids recovered from ovule-embryo culture (112).
Subsection Papillosae	
M. dzhawakhetica	Triploid hybrids comprised of two genomes of *M. dzhawakhetica* and one genome of *M. sativa* are readily recovered (109). Backcross (BC) progeny difficult to produce, but subsequent BC generations readily recovered.
M. papillosa	Triploid hybrids comprised of two genomes *M. papillosa* and one genome of *M. sativa* are easily recovered.† Diploid hybrids containing one genome of *M. papillosa* and one genome of *M. sativa* have been recovered from ovule-embryo culture (110).
Section Arboreae	
M. arborea	No hybrids recovered.
Section Marinae	
M. marina	Two diploid hybrids from crossing $2x$ *M. sativa* × $2x$ *M. marina* recovered from ovule-embryo culture (110).
Section Suffruiticosae	
M. suffruticosa	No hybrids recovered.
M. hybrida	One diploid hybrid from crossing $2x$ *M. sativa* × $2x$ *M. hybrida* recovered from ovule-embryo culture (110).

† Lesins (89) published as *M. dzhawakhetica* but later taxonomic data (100) indicate it was *M. papillosa*.

pods have a parenchymatous wall separating seeds in the pod, similar to species of subsection Papillosae. Hybrids between *M. sativa* and *M. glomerata* were obtained readily, averaging 80 seeds per 100 flowers pollinated for the crosses *M. sativa* × *M. glomerata*, *M. falcata* × *M. glomerata*, and the reciprocal (92). However, based on a greater number of meiotic irregularities and relatively low seed set in F_1 hybrids of *M. sativa* × *M. glomerata*, Lesins (92) concluded that *M. glomerata* should be considered a species separate from the members of the *M. sativa-falcata* complex. McCoy and Smith (108) found that certain clones of *M. glomerata* may be superior pollinators in $4x$-$2x$ crosses to produce haploids.

24–10.2.1.2. *Medicago sativa* **L.** × *M. prostrata* **Jacq.** *Medicago prostrata* is a yellow-flowered species with tightly coiled pods having both diploid ($2n = 2x = 16$) and tetraploid ($2n = 4x = 32$) representatives (98). The pachytene karyotype of *M. prostrata* is similar in chromosome arm ratios and proportional lengths to diploid *M. sativa* (64). The presence of erect glandular hairs (84), which offer potential resistance to the alfalfa weevil (*Hypera postica*), has generated interest in *M. prostrata* (75).

Hybrids can be produced between *M. prostrata* and *M. sativa* at both the diploid and tetraploid levels (91). Chromosome pairing is excellent in the diploid hybrids (> 90% of the cells with eight bivalents) with few instances of aberrant chromosome behavior. Sorensen et al. (144) produced a number of tetraploid hybrids from *M. prostrata* ($2x$) × *M. sativa* ($4x$) and reciprocal crosses. However, Lesins (91) observed low seed set and shrunken seeds and distinctly smaller seed when *M. prostrata* was the female parent. Therefore, *M. prostrata* was identified as a separate species more closely related to *M. glomerata* than to *M. sativa* (92, 100).

24–10.2.2 *Medicago sativa* × **Species of Subsection Rupestres**

24–10.2.2.1 *Medicago sativa* × *M. rhodopea* **Velen.** Mature *M. rhodopea* plants are semiprostrate, branch profusely from the crown, have narrow oblanceolate leaflets, yellow flowers, and coiled pods, either with or without spines. The chromosome number is diploid ($2n = 2x = 16$). The pachytene karyotype of *M. rhodopea* is similar to alfalfa (65). A large, dark-stained body was observed in the nucleolus by Gillies (65). The body, which contained RNA, was described as an accessory nucleolus. Two hybrids between alfalfa and *M. rhodopea* were produced by Lesins (95). One was a triploid ($2n = 3x = 24$) recovered from crossing diploid alfalfa with a chromosome doubled tetraploid *M. rhodopea*, and the other was an aneuploid ($2n = 31$) produced by crossing tetraploid alfalfa with chromosomally doubled tetraploid *M. rhodopea*. Since no euploid plants were recovered, Lesins (95) hypothesized that some deleterious factor was removed, or its effect was reduced by changing the genomic ratio due to unbalanced chromosome sets, or by the loss of one chromosome.

The ovule-embryo culture method (110) has been successfully applied to *M. sativa* (2*x*) × *M. rhodopea* (2*x*) hybrids. Twenty-eight F_1 hybrids have been recovered, and 24 were found to be diploid ($2n = 16$), two were triploid ($2n = 24$), and two were tetraploid ($2n = 32$). Cytological observations indicated hybrid embryo abortion at the late heart stage of development. The fact that all hybrids are euploid and most contain balanced chromosome sets when embryo rescue is used, may indicate that the interspecific incompatibility factor results in endosperm breakdown, and the artificial ovule-embryo culture medium counteracts the effects of any incompatibility factor.

Chromosome pairing was excellent in the diploid F_1 hybrids (112). The majority of the diakinesis and metaphase I (MI) cells had eight bivalents. Although some hybrids had extremely low or zero pollen germination, other hybrids had up to 62% germination. Cross fertility of most hybrids averaged 0.3 seeds per pollination as the maternal parent. In the two hybrids produced by Lesins (95), chromosome pairing at metaphase I indicated a high level of homology. The triploid hybrid had one to four trivalents in most MI cells, and the 31-chromosome plant had five bivalents plus a univalent.

24-10.2.2.2 *Medicago sativa* L. × *M. rupestris* M.B. *Medicago rupestris* M.B. is a diploid ($2n = 2x = 16$) species, closely related to *M. rhodopea*. It is characterized by small, wedge-shaped leaflets, yellow flowers, and coiled pods with prominent veins. Leaves and stems are covered with simple, appressed hairs. The pachytene karyotype of *M. rupestris* is almost identical to *M. rhodopea*, and excellent chromosome pairing was observed in two *M. rhodopea* × *M. rupestris* hybrids (65).

The first interspecific hybrids between alfalfa and *M. rupestris* were produced with the ovule-embryo culture method (106). Although no hybrids were recovered from seed, over 50 F_1 hybrids have now been recovered by the ovule-embryo culture method. Chromosome pairing in the F_1 hybrids has been excellent with most cells having eight bivalents. However, pollen germination in all hybrids is <5%, and no first backcross progeny to alfalfa have been recovered, regardless of whether the F_1 was used as the male or the female parent. First backcross progeny (BC_1) were easily recovered with the ovule-embryo culture method (106). Male and female cross fertility of BC_1 plants has been adequate. *Medicao rupestris* can be added to the list of potential germplasm sources for alfalfa improvement.

24-10.2.2.3 *Medicago sativa* L. × *M. cancellata* M.B. *Medicago cancellata* M.B. ($2n = 6x = 48$) is characterized by obovate to narrowly wedge-shaped leaflets, yellow flowers, and coiled pods with reticulate veins (100). Borges et al. (23) indicated that *M. cancellata* is resistant to Stemphyllium leafspot. This hexaploid has been used in crosses with recently synthesized hexaploid alfalfa and the naturally occurring hexaploid species (143, 152).

The first hybrids between alfalfa and *M. cancellata* were produced by Lesins (90). Pentaploid hybrids were obtained following crosses of alfalfa (4x) and *M. cancellata* (6x). However, only one hybrid resulted from the cross of alfalfa (6x) and *M. cancellata* (6x). This difficulty in crossing at the same ploidy level must relate to the particular genotype(s), because other research indicates a significantly greater level of crossability between *M. cancellata* and alfalfa (143, 152). Chromosome pairing in the F_1 hybrids was good although some abnormal behavior was observed (90). Genetic recombination between the two species does occur.

Medicago cancellata is easily hybridized with *M. saxatilis* (93), producing 0.6 seeds per pollination. The genetic structure of *M. cancellata* has been described as an alloautoploid (100). It has been proposed that *M. cancellata* consists of two genomes of alfalfa and four genomes of *M. rupestris,* a species morphologically similar to *M. cancellata.* Further cytogenetic studies are necessary to support this hypothesis.

24-10.2.2.4 *Medicago sativa* L. × *M. saxatilis* M.B.

Medicago saxatilis is another naturally occurring hexaploid species ($2n = 6x = 48$) closely related to *M. cancellata. Medicago saxatilis* has a more robust plant type than *M. cancellata.* Leaflets are obovate with serrated margins, the yellow flowers are much larger than either *M. cancellata* or *M. sativa,* and pods have either spines or corrugated edges.

Interspecific hybrids of *M. sativa-M. saxatilis* are obtained readily whether 4x or 6x *M. sativa* plants are used, but especially when *M. sativa* is the maternal parent (93). Although chromosome pairing has not been studied, few abnormalities were observed at anaphase (21% of A1 cells with lagging univalents) indicating fairly good pairing. Yen and Murphy (152) found that chromosome instability was similar whether hexaploid alfalfa was intercrossed or crossed with *M. cancellata* or *M. saxatilis.* *Medicago cancellata* × *M. saxatilis* produced chromosomally stable populations.

The genomic structure of *M. saxatilis* is similar to *M. cancellata.* Lesins (93) hypothesized that *M. saxatilis* is an alloautoploid. Two genomes were assumed to come from *M. sativa* and four genomes from *M. rhodopea,* based on morphological characteristics.

24-10.2.3. *Medicago sativa* L. × Species of Subsection Daghestanicae

The two species of subsection Daghestanicae are *M. daghestanica* Rupr. and *M. pironae* Vis. Lesins (99) demonstrated that both perennial species are diploids ($2n = 2x = 16$). *Medicago daghestanica* is unique in that it is the only *Medicago* spp., other than *M. sativa,* with purple flowers. Plants of *M. daghestanica* are prostrate with thin, wiry stems, bluish-green and obovate leaflets, and coiled pods with short, conical spines. *Medicago pironae* is characterized by semi-upright to upright growth habit, wiry stems, small, obcordate leaflets, yellow flowers, and coiled spined pods covered with glandular, articulate hairs.

Viable F_1 hybrids between *M. daghestanica* and *M. pironae* were easily obtained by Lesins and Gillies (96). All hybrids were diploid, and completely male and female sterile. Pachytene analysis identified different chromosome patterns, with structural dissimilarities indicated by nonpairing of some distal euchromatic segments and uneven lengths of chromosomes. The most frequent chromosome configuration was seven bivalents and two univalents. More than half of the anaphase I cells had bridges, and only 10% of the anaphase II cells had normal separation, confirming the structural differences between the two genomes (96).

No hybrids were obtained from seed when diploid alfalfa was crossed with either diploid *M. daghestanica* or diploid *M. pironae* (96). However, two trispecies hybrids were produced from crossing 4x *M. sativa* and the 4x *M. daghestanica-M. pironae* hybrid produced via colchicine treatment (94). One hybrid had 30 chromosomes, the other 34. Multivalents were observed at metaphase I, indicating the potential for genetic exchange between *M. sativa* and *M. daghestanica-M. pironae*. Interspecific hybrids between diploid *M. sativa* and diploid *M. daghestanica*, and between diploid *M. sativa* and diploid *M. pironae* have been produced with the ovule-embryo culture method (112).

24-10.2.4 *Medicago sativa* × Species of Subsection Papillosae

Lesins and Lesins (100) have identified three species in the subsection Papillosae, namely *M. dzhawakhetica* Bordz., *M. ignatzii* L. & L., and *M. papillosa* Boiss. First, we will present current identifying characteristics of the three species and then clarify the species designation used in previous reports.

Medicago dzhawakhetica exists only at the tetraploid level ($2n = 4x = 32$) (100). It is a prostrate to ascending species with obovate leaflets, yellow flowers, and glabrous pods. *Medicao ignatzii* and *M. papillosa* differ from *M. dzhawakhetica* in that diploid ($2n = 2x = 16$) and tetraploid ($2n = 4x = 32$) types are known, and both species have pods covered with articulate, glandular hairs. The distinction between *M. ignatzii* and *M. papillosa* is that *M. ignatzii* has distinctly larger flowers, larger pods, and larger leaves. The two species were given species rank because of the complete sterility of F_1 hybrids of *M. ignatzii* × *M. papillosa* (100).

Previous reports (38, 89) of interspecific hybrids between *M. sativa* and *M. dzhawakhetica* were actually based on *M. sativa* × *M. papillosa*. In both studies, the pods of the diploid parent had articulate glandular hairs [Clement (38) used a diploid directly and Lesins (89) used a chromosomally doubled diploid.] The seed pod characters exclude *M. dzhawakhetica* and implicate *M. papillosa*. Therefore, Lesins (89) and Clement (38) were reporting on the hybridization of *M. sativa* × *M. papillosa*. Lesins (89) found that successful hybridization only occurred with uneven ploidy levels. Triploid hybrids resulted from crossing (2x) *M. sativa* × (4x) *M. papillosa*. These hybrids thus contained two genomes of *M. pap-*

illosa (89). In contrast, Clement (38) studied a single, diploid hybrid between *M. sativa* and *M. papillosa* (as *M. dzhawakhetica*), and found excellent chromosome pairing (predominantly eight bivalents), and had a workable level of fertility when backcrossing the F_1 hybrid to *M. sativa*. Clement (38) concluded that there was considerable homology between genomes.

The first hybrids between *M. sativa* and *M. dzhawakhetica* were produced by McCoy and Smith (109). In common with *M. papillosa*, unequal ploidy levels were required. In addition, the reproductive mutant, *jp* was essential for efficient hybridization. When diploid alfalfa clones homozygous for *jp* were crossed with tetraploid *M. dzhawakhetica*, hybrids were produced at a level equal to intraspecific matings of diploid alfalfa. Almost all hybrids were triploid ($2n = 3x = 24$), comprised of two genomes of *M. dzhawakhetica* and one genome of alfalfa. In contrast to Lesins' triploid hybrids of *M. sativa-M. papillosa* (89), McCoy and Smith (109) found trivalent frequencies in some hybrids equal to trivalent frequencies observed in alfalfa triploids (10, 42, 113). The level of pairing in triploid hybrids suggests excellent potential for genetic exchange between *M. sativa* and *M. dzhawakhetica* genomes.

The triploid F_1 hybrids of *M. sativa* × *M. dzhawakhetica* were completely male sterile. No pollen germination was observed in the F_1's. Backcrossing the F_1 hybrid as female parent to diploid alfalfa resulted in only two BC_1 progeny (109). The backcross to tetraploid alfalfa was difficult but feasible. Cytological studies demonstrated that ovules in many of the F_1 hybrids degenerated early and an embryo sac was never formed (112). Concentrating on F_1 hybrids with normal ovule development and using the ovule-embryo culture method to recover BC_1 progeny resulted in a significant increase in the number of BC_1 progeny. The majority of the BC_1 progeny had 40 chromosomes although chromosome number ranged from 28 to 40. Male and female fertility was restored in most of the BC_2 plants, particularly in those plants at or near the tetraploid level (112).

24-10.2.5 *M. sativa* × Species of Section Marinae

Medicago marina, a diploid ($2n = 2x = 16$), is the only species of Section Marinae. Plants have a characteristic grayish appearance, resulting from a dense cover of simple hairs. *Medicago marina* is the only *Medicago* spp. with a seashore habitat.

Two hybrids between diploid *M. sativa* and diploid *M. marina* have been recovered by ovule-embryo culture (110). Both hybrids are diploid and extremely weak, and neither hybrid has produced flowers.

24-10.2.6 *M. sativa* × Species of Section Suffruticosae

Two diploid ($2n = 2x = 16$) species, *M. suffruticosa* and *M. hybrida*, comprise Section Suffruticosae. *Medicago hybrida* is a yellow-flowered, prostrate species. Recently a diploid hybrid was recovered via the ovule-embryo culture method after crossing diploid *M. sativa* and diploid *M. hybrida* (110).

24-10.3 *Medicago sativa* L. × Annual Species

24-10.3.1. *Medicago sativa* L. × *M. scutellata* Mill.

Medicago scutellata is one of two *Medicago* spp. with $2n = 30$ chromosomes (9). The first and only perennial × annual interspecific hybrid in the *Medicago* genus has been produced from *M. sativa* × *M. scutellata* (135). Somatic chromosome number of the hybrid was unstable, ranging from 30 to 64 chromosomes in root tip cells. The primary shoot was hexaploid ($2n = 48$) and a later shoot was tetraploid ($2n = 32$), both unexpected because *M. scutellata* has 30 chromosomes. The hybrid was perennial, and intermediate for most other morphological characteristics. Chromosome pairing studies have not been conducted, and no progeny have been obtained.

24-10.4 Summary Comments on Interspecific Hybridization

Many new hybrids have been produced and many more are likely to be produced in the future, either by sexual means coupled with embryo rescue or via somatic means using protoplast fusion (see Chapter 29 in this book). These hybrids and future hybrids have expanded the opportunities for gene transfer in alfalfa. Interspecific hybridization research holds potential for producing and testing many new allo- and "allo-auto" polyploid combinations. Future research may produce new genome combinations with a genetic structure superior to the autotetraploid genetic structure of current alfalfa.

REFERENCES

1. Arbi, N., D. Smith, and E.T. Bingham. 1979. Dry matter and morphological responses to temperature of alfalfa strains with differing ploidy levels. Agron. J. 71:537–577.
2. ----, ----, ----, and R.M. Soberalske. 1978. Herbage yields and levels of N and IVDDM from five alfalfa strains of different ploidy levels. Agron. J. 70:873–875.
3. Armstrong, J.M. 1954. Cytological studies in alfalfa polyploids. Can. J. Bot. 332:531–542.
4. Atwood, S.S., and P. Grun. 1951. Cytogenetics of alfalfa. Bibliogr. Genet. 14:133–188.
5. Azizi, M.R., and D.K. Barnes. 1977. Characterization and inheritance of a spotted leaf trait in alfalfa. Crop Sci. 17:126–132.
6. Baenziger, H. 1977. Inheritance of torn-leaf mutant in alfalfa. Can. J. Plant Sci. 57:47–50.
7. Barnes, D.K., E.T. Bingham, R.P. Murphy, O.J. Hunt, D.F. Beard, W.H. Skrala, and L.R. Teuber. 1977. Alfalfa germplasm in the United States: Genetic vulnerability, use, improvement and maintenance. USDA-ARS Bull. 1571. U.S. Government Printing Office, Washington, DC.
8. ----, and C.H. Hanson. 1967. An illustrated summary of genetic traits in tetraploid and diploid alfalfa. USDA-ARS Tech. Bull. 1370. U.S. Government Printing Office, Washington, DC.
9. Bauchan, G.R., and J.H. Elgin, Jr. 1984. A new chromosome number for the genus *Medicago*. Crop Sci. 24:193–195.
10. Binek, A., and E.T. Bingham. 1970. Cytology and crossing behavior of triploid alfalfa. Crop Sci. 10:303–306.

11. Bingham, E.T. 1968. Aneuploids in seedling populations of tetraploid alfalfa, *Medicago sativa* L. Crop Sci. 8:571–574.
12. ——. Transfer of diploid *Medicago* spp. germplasm to tetraploid *M. sativa* L. in 4x-2x crosses. Crop Sci. 8:760–762.
13. ——. 1971. Isolation of haploids of tetraploid alfalfa. Crop Sci. 11:433–435.
14. ——. 1973. Interaction of two basic color factor genes in alfalfa. Crop Sci. 13:393–394.
15. ——. 1980. Maximizing heterozygosisty in autopolyploids. p. 471–489. *In:* W.H. Lewis (ed.) Polyploidy: Biological relevance. Plenum Press, New York.
16. ——, and A. Binek. 1969. Hexaploid alfalfa, *Medicago sativa* L.: Origin, fertility, and cytology. Can. J. Genet. Cytol. 11:359–366.
17. ——, and C.B. Gillies. 1971. Chromosome pairing, fertility, and crossing behavior of haploids of tetraploid alfalfa, *Medicago sativa* L. Can. J. Genet. Cytol. 13:195–202.
18. ——, and T.J. McCoy. 1979. Cultivated alfalfa at the diploid level: Origin, reproductive stability, and yield of seed and forage. Crop Sci. 19:97–100.
19. ——, and ——. 1986. Somaclonal variation in alfalfa. Plant Breed. Rev. 4:123–152.
20. ——, and J.W. Saunders. 1974. Chromosome manipulations in alfalfa: Scaling the cultivated tetraploid to seven ploidy levels. Crop. Sci. 14:474–477.
21. ——. Unpublished results.
22. Blake, N.K., and E.T. Bingham. 1986. Alfalfa triploids with functional male and female fertility. Crop Sci. 26:643–645.
23. Borges, O.L., E.H. Stanford, and R.K. Webster. 1976. Sources and inheritance of resistance to Stemphyllium leafspot of alfalfa. Crop Sci. 16:458–461.
24. Brick, M.A., and D.K. Barnes. 1981. Inheritance and ontogeny of a lobed-cambium trait in alfalfa roots. J. Hered. 72:419–422.
25. ——, and ——. 1982. Inheritance and anatomy of root bark area in alfalfa. Crop Sci. 22:747–752.
26. ——, ——, and A. K. Dobrenz. 1984. Inheritance and anatomy of a narrow leaflet trait in alfalfa. Crop Sci. 24:787–790.
27. Brink, R.A., and D.C. Cooper. 1940. Double fertilization and development of the seed in angiosperms. Bot. Gaz. (Chicago) 102:1–25.
28. ——, and ——. 1941. Incomplete seed failure as a result of somatoplastic sterility. Genetics 26:487–505.
29. ——, and ——. 1947. The endosperm in seed development. Bot. Rev. 13:423–541.
30. Busbice, T.H., R.R. Hill, Jr., and H.L. Carnahan. 1972. Genetics and breeding procedures. *In:* C.H. Hanson (ed.) Alfalfa science and technology. Agronomy 15:283–318.
31. Buss, G.R., and D.K. Barnes. 1975. Inheritance of a chlorotic seedling trait that affects zygote survival in diploid alfalfa. Crop Sci. 15:185–186.
32. ——, and R.W. Cleveland. 1968. Pachytene chromosomes of diploid *Medicago sativa* L. Crop Sci. 8:744–747.
33. ——, and ——. 1971. Meiosis of trisomics of diploid alfalfa. Crop Sci. 11:808–810.
34. Campbell, T.A., J.A. Schillinger, and C.H. Hanson. 1974. Inheritance of resistance to anthracnose in alfalfa. Crop Sci. 14:667–668.
35. Chase, S.S. 1963. Analytic breeding in *Solanum tuberosum* L.: A scheme utilizing parthenotes and other diploid stocks. Can. J. Genet. Cytol. 5:359–363.
36. ——. 1964. Analytic breeding of amphipolyploid plant varieties. Crop Sci. 4:334–337.
37. Classen, D., C. Nozzolillo, and E. Small. 1982. A phenolic-taxometric study of *Medicago* (Leguminosae). Can. J. Bot. 60:2477–2495.
38. Clement, W.M., Jr. 1963. Chromosome relationships in a diploid hybrid between *Medicago sativa* L. and *M. dzhawakhetica* Bordz. Can. J. Genet. Cytol. 5:427–432.
39. ——, and W.F. Lehman. 1962. Fertility and cytological studies of a dihaploid plant of alfalfa, *Medicago sativa* L. Crop Sci. 2:451–453.
40. ——, and E.H. Stanford. 1961. A mechanism for the production of tetraploid and pentaploid progeny from diploid × tetraploid crosses of alfalfa. Crop Sci. 1:11–14.
41. ——, and ——. 1963. Pachytene studies at the diploid level in *Medicago*. Crop Sci. 3:147–150.
42. Cleveland, R.W., and E.H. Stanford. 1959. Chromosome pairing in hybrids between tetraploid *Medicago sativa* L. and diploid *Medicago falcata* L. Agron. J. 51:488–492.
43. Cooper, D.C. 1935. Macrosporogenesis and embryology of *Medicago*. J. Agric. Res. (Washington, DC) 51:471–477.
44. ——. 1939. Artificial induction of polyploidy in alfalfa. Am. J. Bot. 26:65–67.
45. ——, and R.A. Brink. 1940. Somatoplastic sterility as a cause of seed failure after interspecific hybridization. Genetics 25:593–617.
46. ——, and ——. 1940. Partial self-compatibility and the collapse of fertile ovules as factors affecting seed formation in alfalfa. J. Agric. Res. (Washington, DC) 60:453–472.
47. ——, ——, and H.R. Albrecht. 1937. Embryo mortality in relation to seed formation in alfalfa (*Medicago sativa*) Am. J. Bot. 24:203–213.

48. Cutter, G.L., and E.T. Bingham. 1977. Effect of soybean male-sterile gene ms_1 on organization and function of the female gametophyte. Crop Sci. 17:760–764.
49. Demarley, Y. 1979. The concept of linkat. p. 257–265. *In:* A.C. Zeven and A.M. vanHarten (ed.) Broadening the genetic base of crops. Centre for Agricultural Publishing and Documentation (PUDOC), Wageningen, Netherlands.
50. deWet, J.M.J. 1980. Origins of polyploidy. p. 3–15. *In* H.L. Lewis (ed.) Polyploidy: Biological relevance. Plenum Press, New York.
51. Dunbier, M.W., and E.T. Bingham. 1975. Maximum heterozygosity in alfalfa: Results using haploid-derived autotetraploids. Crop Sci. 15:527–531.
52. ----, D.L. Eskew, E.T. Bingham, and L.E. Schrader. 1975. performance of genetically comparable diploid and tetraploid alfalfa: Agronomic and physiological parameters. Crop Sci. 15:211–214.
53. Elgin, J.H., Jr. 1979. Inheritance of stem-nematode resistance in alfalfa. Crop Sci. 19:352–354.
54. ----, D.K. Barnes, and K.W. Kreitlow. 1971. Description and inheritance of a tumor trait on stem cuttings of alfalfa. J. Hered. 62:189–192.
55. ----, J.E. McMurtrey, and G.W. Schaeffer. 1977. Attempted interspecific hybridization of *Medicago scutellata* and *M. sativa.* Agron. Abstr. American Society of Agronomy, Madison, WI, p. 54.
56. Farley, H.M., and A.H. Hutchinson. 1941. Seed development in *Medicago* hybrids. I. The normal ovule. Can. J. Res. Sect. C. 19:421–437.
57. Fox, C.C. 1981. A comparison of the effects of inbreeding on isogenic diploid-tetraploid alfalfa (*Medicago sativa* L.). Ph.D. thesis. Univ. of Wisconsin, Madison (Diss. Abstr. 82-03162).
58. Fridriksson, S., and J.L. Bolton. 1963. Development of the embryo of *Medicago sativa* L. after normal fertilization and after pollination by other species of *Medicago.* Can. J. Bot. 41:23–33.
59. ----, and ----. 1963. Preliminary report on the culture of alfalfa embryos. Can. J. Bot. 41:439–440.
60. Gilles, A., and L.F. Randolph. 1951. Reduction of quadrivalent frequency in auto tetraploid maize during a period of 10 years. Am. J. Bot. 38:12–17.
61. Gillies, C.B. 1968. The pachytene chromosomes of a diploid *Medicago sativa.* Can. J. Genet. Cytol. 10:788–793.
62. ----. 1970. Alfalfa chromosomes. I. Pachytene karyotype of a diploid *Medicago falcata* L. and its relationship to *M. sativa* L. Crop Sci. 10:169–171.
63. ----. 1971. Pachytene studies in $2n = 14$ species of *Medicago.* Genetica 42:278–298.
64. ----. 1972. Pachytene chromosomes of perennial *Medicago* species. I. Species closely related to *M. sativa.* Hereditas 72:277–288.
65. ----. 1972. Pachytene chromosomes of perennial *Medicago* species. II. Distantly related species whose karyotypes resemble *M. sativa.* Hereditas 72:289–302.
66. ----. 1972. Pachytene chromosomes of perennial *Medicago* species. III. Unique karyotypes of *M. hybrida* Trautv. and *M. suffruticosa* Ramond. Hereditas 72:303–310.
67. ----. 1977. Identification of trisomics in diploid lucerne. Aust. J. Agric. Res. 28:309–317.
68. ----, and E.T. Bingham. 1971. Pachytene karyotypes of $2x$ haploids derived from tetraploid alfalfa (*Medicago sativa*)—evidence for autotetraploidy. Can. J. Genet. Cytol. 13:397–403.
69. Grun, P. 1951. Variations in the meiosis of alfalfa. Am. J. Bot. 38:475–482.
70. Harlan, J.R., and J.M.J. deWet. 1975. On O. Winge and a prayer: The origins of polyploidy. Bot. Rev. 41:361–390.
71. Hartman, B.J., O.J. Hunt, R.N. Peaden, and B.D. Thyr. 1978. Inheritance of stippled leaf in alfalfa. Crop Sci. 17:517–518.
72. Hartman, C.L., T.J. McCoy, and T.R. Knous. 1984. Selection of alfalfa (*Medicago sativa*) cell lines and regeneration of plants resistant to the toxin(s) produced by *Fusarium oxysporum* f. sp. *medicaginis.* Plant Sci. Lett. 34:183–194.
73. Hijano, E.H., D.K. Barnes, and F.I. Frosheiser. 1983. Inheritance of resistance to Fusarium wilt in alfalfa. Crop Sci. 13:31–34.
74. Ho, K.M., and K.J. Kasha. 1972. Chromosome homology at pachytene in diploid *Medicago sativa, M. falcata* and their hybrids. Can. J. Genet. Cytol. 14:829–838.
75. Horber, E.K., E.L. Sorensen, and K.J.R. Johnson. 1980. Resistance to alfalfa weevil and potato leafhopper increased by glandular and simple hairs. p. 51. *In:* Rep. of the 27th Alfalfa Improve. Conf., Madison, WI. 8–10 July. AR, North Central Region, USDA-SEA, Peoria, IL.
76. Irwin, J.A.G., D.P. Maxwell, and E.T. Bingham. 1981. Inheritance of resistance to *Phytophthora megasperma* in diploid alfalfa. Crop Sci. 21:271–276.
77. ----, ----, and ----. 1981. Inheritance of resistance to *Phytophthora megasperma* in tetraploid alfalfa. Crop Sci. 21:277–283.

78. Johnson, L.B., D.L. Stuteville, S.E. Schlarbaum, and D.Z. Skinner. 1984. Variation in phenotype and chromosome number in alfalfa protoclones regenerated from non-mutagenized calli. Crop Sci. 24:948–951.
79. Julen, B. 1944. Investigations on diploid, triploid, and tetraploid lucerne. Hereditas 30:567–582.
80. Kasha, K.J. 1968. Studies with trisomics in alfalfa. Forage Notes 14:7–8.
81. ----, and H.A. McLennan. 1967. Trisomics in diploid alfalfa. I. Production, fertility, and transmission. Chromosoma 21:232–242.
82. Kermicle, J.L. 1969. Androgenesis conditioned by a mutation in maize. Science 166:1422–1424.
83. Khush, G.S. 1973. Cytogenetics of aneuploids. Academic Press, New York
84. Kreitner, G.L., and E.L. Sorensen. 1979. Glandular trichomes on *Medicago* species. Crop Sci. 19:380–384.
85. Ledingham, G.F. 1940. Cytological and development studies of hybrids between *Medicago sativa* and a diploid form of *M. falcata*. Genetics 25:1–15.
86. Leps, W.T., W.J. Brill, and E.T. Bingham. 1980. Effect of alfalfa ploidy on nitrogen fixation. Crop Sci. 20:427–430.
87. Lesins, K. 1952. Some data on the cytogenetics of alfalfa. J. Hered. 43:287–291.
88. ----. 1957. Cytogenetic study on a tetraploid plant at the diploid chromosome level. Can. J. Bot. 35:181–196.
89. ----. 1961. Interspecific crosses involving alfalfa. I. *Medicago dzhawakhetica* (Bordz.) Vass. × *M. sativa* L. and its peculiarities. Can. J. Genet. Cytol. 3:135–152.
90. ----. 1961. Interspecific crosses involving alfalfa. II. *Medicago cancellata* M. B. × *M. sativa* L. Can. J. Genet. Cytol. 3:316–324.
91. ----. 1962. Interspecific crosses involving alfalfa. III. *Medicago sativa* L. × *M. prostrata* Jacq. Can. J. Genet. Cytol. 4:14–23.
92. ----. 1968. Interspecific crosses involving alfalfa. IV. *Medicago glomerata* × *M. sativa* with reference to *M. prostrata*. Can. J. Genet. Cytol. 10:536–544.
93. ----. 1970. Interspecific crosses involving alfalfa. V. *Medicago saxatilis* × *M. sativa* with reference to *M. cancellata* and *M. rhodopea*. Can. J. Genet. Cytol. 12:80–86.
94. ----. 1971. Interspecific hybrids involving alfalfa. VI. Ineffectiveness of alloploidy in induction of fertility in *Medicago pironae* × *M. daghestanica* hybrids. Can. J. Genet. Cytol. 13:437–442.
95. ----. 1972. Interspecific hybrids involving alfalfa. VII. *Medicago sativa* × *M. rhodopea*. Can. J. Genet. Cytol. 14:221–226.
96. ----, and C.B. Gillies. 1968. Relationships of taxa in genus *Medicago* as revealed by hybridization. II. *M. pironae* × *M. daghestanica* with reference to *M. sativa*. Can. J. Genet. Cytol. 10:454–459.
97. ----, and ----. 1972. Taxonomy and cytogenetics of *Medicago*. In: C.H. Hanson (ed.) Alfalfa science and technology. Agronomy 15:53–86.
98. ----, and I. Lesins. 1960. Sibling species in *Medicago prostrata* Jacq. Can. J. Genet. Cytol. 2:416–417.
99. ----, and ----. 1961. Some little-known *Medicago* species and their chromosome complements. Can. J. Genet. Cytol. 3:7–9.
100. ----, and ----. 1979. Genus Medicago (Leguminosae): A taxogenetic study. Dr. W. Junk Publishers, The Hague, Netherlands, p. 228.
101. ----, ----, and C.B. Gillies. 1970. *Medicago murex* with $2n = 16$ and $2n = 14$ chromosome complements. Chromosoma 30:109–122.
102. ----, S.M. Singh, I. Baysal, and R.S. Sadasivaiah. 1975. An attempt to breed hexaploid alfalfa (*Medicago* spp.). Z. Pflanzenzuecht. 75:192–204.
103. Lu, N.S., D.K. Barnes, and F.I. Frosheiser. 1973. Inheritance of *Phytophthora* root rot resistance in alfalfa. Crop Sci. 13:714–717.
104. Mariani, A. 1975. Cytogenetic research on hexaploid alfalfa. *Medicago sativa* L. Caryologia 28:359–373.
105. McCoy, T.J. 1982. The inheritance of $2n$ pollen formation in diploid alfalfa *Medicago sativa*. Can. J. Genet. Cytol. 24:315–323.
106. ----. 1985. Interspecific hybridization of *Medicago sativa* L. and *M. rupestris* M.B. using ovule-embryo culture. Can. J. Genet. Cytol. 27:238–245.
107. ----, and D.E. Rowe. 1986. Single cross alfalfa (*Medicago sativa* L.) hybrids produced via $2n$ gametes and somatic chromosome doubling: Experimental and theoretical comparisons. Theor. Appl. Genet. 72:80–83.
108. ----, and L.Y. Smith. 1983. Genetics, cytology, and crossing behavior of an alfalfa (*Medicago sativa*) mutant resulting in failure of the post-meiotic cytokinesis. Can. J. Genet. Cytol. 25:390–397.
109. ----, and ----. 1984. Uneven ploidy levels and a reproductive mutant required for interspecific hybridization of *Medicago sativa* L. and *M. dzhawakhetica* Bordz. Can. J. Genet. Cytol. 26:511–518.

110. ----, and ----. 1986. Interspecific hybridization of perennial *Medicago* species using ovule-embryo culture. Theor. Appl. Genet. 71:772-783.
111. ----, and K.A. Walker. 1984. Alfalfa. p. 171-192. *In:* P.V. Ammirato et al. (ed.) Handbook of plant cell culture, Vol. 3. Macmillan Book Co., New York.
112. ----. Unpublished results.
113. McLennan, H.A., J.M. Armstrong, and K.J. Kasha. 1966. Cytogenetic behavior of alfalfa hybrids from tetraploid by diploid crosses. Can. J. Genet. Cytol. 8:544-555.
114. Mendiburu, A.O., and S.J. Peloquin. 1977. The significance of $2n$ gametes in potato breeding. Theor. Appl. Genet. 49:53-61.
115. Meyers, S.P., C. Briaegar, L.E. Schrader, D.B. Jordan, and W.L. Ogren. 1983. Ploidy effects in isogenic populations of alfalfa (*Medicao sativa* L.). IV. Similarity in physical and kinetic properties of ribulose-I, 5-bisphosphate carboxylase/oxygenase. Plant Physiol. 71:966-968.
116. ----, S.L. Nichols, G.R. Baer, W.T. Molin, and L.E. Schrader. 1982. Ploidy effects in isogenic populations of alfalfa. I. Ribulose-1,5-bisphosphate carboxylse, soluble protein, chlorophyll, and DNA in leaves. Plant Physiol. 70:1704-1709.
117. Mok, D.W.S., and S.J. Peloquin. 1975. The inheritance of three mechanisms of diplandroid ($2n$ pollen) formation in diploid potatoes. Heredity 335:295-302.
118. Molin, W.T., S.P. Meyers, G.R. Baer, and L.E. Schrader. 1982. Ploidy effects in isogenic populations of alfalfa. II. Photosynthesis, chloroplast number, ribulose-1,5-bisphosphate carboxylase, chlorophyll, and DNA in protoplasts. Plant Physiol. 70:1710-1714.
119. Obajami, A.O., and E.T. Bingham. 1973. Inbreeding cultivated alfalfa in one tetraploid-haploid-tetraploid cycle: Effects on morphology, fertility, and cytology. Crop Sci. 13:36-39.
120. Oldemeyer, R.K. 1956. Interspecific hybridization in *Medicago*. Agron. J. 48:584-585.
121. Peterson, M.A., and D.K. Barnes. 1981. Inheritance of ineffective nodulation and non-nodulation traits in alfalfa. Crop Sci. 21:611-616.
122. Pfeiffer, T.W., and E.T. Bingham. 1983. Abnormal meiosis in alfalfa, *Medicago sativa:* Cytology of $2n$ egg and $4n$ pollen formation. Can. J. Genet. Cytol. 25:107-112.
123. ----, and ----. 1983. Improvement of fertility and herbage yield by selection within two-allele populations of tetraploid alfalfa. Crop Sci. 23:633-636.
124. ----, L.E. Schrader, and E.T. Bingham. 1980. Physiological comparisons of isogenic diploid-tetraploid, tetraploid-octoploid alfalfa populations. Crop Sci. 20:299-303.
125. Quiros, C.F. 1982. Tetrasomic segregation for multiple alleles in alfalfa. Genetics 101:117-127.
126. ----, and K. Morgan. 1981. Peroxidase and leucine-aminopeptidase in diploid *Medicago* species closely related to alfalfa: Multiple gene loci, multiple allelism and linkage. Theor. Appl. Genet. 60:221-228.
127. Ramanna, M.S. 1983. First division restitution gametes through fertile desynaptic mutants of potato. Euphytica 32:337-350.
128. Reeves, R.G. 1930. Development of the ovule and embryo sac of alfalfa. Am. J. Bot. 17:239-246.
129. ----. 1930. Nuclear and cytoplasmic division in the microsporogenesis of alfalfa. Am. J. Bot. 17:29-40.
130. Reisch, B., and E.T. Bingham. 1981. Plants from ethionine-resistant alfalfa tissue cultures: Variation in growth and morphological characters. Crop Sci. 21:783-788.
131. Rhoades, M.M., and E. Dempsey. 1966. Induction of chromosome doubling at meiosis by the elongate gene in maize. Genetics 54:505-522.
132. Sadasivaiah, R.S., and K. Lesins. 1974. Reduction of chromosome number in root tip cells of *Medicago*. Can. J. Genet. Cytol. 16:219-227.
133. Sangduen, N., G.L. Kreitner, and E.L. Sorensen. 1983. Light and electron microscopy of embryo development in perennial and annual *Medicago* species. Can. J. Bot. 61:837-849.
134. ----, ----, and ----. 1983. Light and electron microscopy of embryo development in an annual × perennial *Medicago* species cross. Can. J. Bot. 61:1241-1257.
135. ----, E.L. Sorensen, and G.H. Liang. 1982. A perennial × annual *Medicago* cross. Can. J. Genet. Cytol. 24:361-365.
136. ----, ----, and ----. 1983. Pollen germination and pollen tube growth following self-pollination and intra- and interspecific pollination of *Medicago* species. Euphytica 32:527-534.
137. Satina, S., and A.F. Blakeslee. 1935. Cytological effects of a gene in *Datura* which causes dyad formation in sporogenesis. Bot. Gaz. 96:521-532.
138. Sayers, E.R. 1961. The relationship of chromosome number and pollen stainability to seed set in *Medicago sativa* L., variety Narragansett. M.S. thesis. Cornell Univ., Ithaca, New York.
139. Schlarbaum, S.E., E. Small, and L.B. Johnson. 1984. Karyotypic evolution, morphological variability and phylogeny in *Medicago* sect. Intertextae. Plant Syst. Evol. 145:203-222.

140. Setter, T.L., L.E. Schrader, and E.T. Bingham. 1978. Carbon dioxide exchange rates, transpiration, and leaf characters in genetically equivalent ploidy levels of alfalfa. Crop Sci. 18:327-332.
141. Shade, R.E., T.E. Thompson, and W.R. Campbell. 1975. An alfalfa weevil larval resistance mechanism detected in *Medicago*. J. Econ. Entomol. 63:399-404.
142. Simon, J.P. 1976. Relationship in annual species of *Medicago*. V. Analysis of phenolics by means of one-dimensional chromatographic techniques. Aust. J. Bot. 15:83-93.
143. Smith, S.E., R.P. Murphy, and D.R. Viands. 1984. Reproductive characteristics of hexaploid alfalfa derived from $3x$-$6x$ crosses. Crop Sci. 24:169-172.
144. Sorensen, E.L., N. Sangduen, and G.H. Liang. 1980. Transfer of glandular hairs from diploid *Medicago prostrata* to tetraploid *M. sativa*. p. 33. *In:* Rep. of the 27th Alfalfa Improve. Conf., Madison, WI. 8-10 July. AR, North Central Region, USDA-SEA, Peoria, IL.
145. Stanford, E.H. 1951. Tetrasomic inheritance in alfalfa. Agron. J. 43:222-225.
146. ----. 1959. The use of chromosome deficient plants in cytogenetic analyses of alfalfa. Agron. J. 51:470-472.
147. ----, and W.M. Clement, Jr. 1958. Cytology and crossing behavior of a haploid alfalfa plant. Agron. J. 50:589-592.
148. ----, ----, and E.T. Bingham. 1972. Cytology and evolution of the *Medicago sativa-falcata* complex. *In:* C.H. Hanson (ed.) Alfalfa science and technology. Agronomy 15:87-101.
149. Viands, D.R., and D.K. Barnes. 1980. Inheritance of resistance to bacterial wilt in two alfalfa gene pools: Qualitative analysis. Crop Sci. 20:48-54.
150. Vorza, N., and E.T. Bingham. 1979. Cytology of $2n$ pollen formation in diploid alfalfa *Medicago sativa*. Can. J. Genet. Cytol. 21:525-530.
151. Williams, E.G., and G. deLautour. 1980. The use of embryo culture with transplanted nurse endosperm for the production of interspecific hybrids in pasture legumes. Bot. Gaz. 141:252-257.
152. Yen, S.T., and R.P. Murphy. 1979. Cytology and breeding of hexaploid alfalfa. I. Stability of chromosome number. Crop Sci. 19:389-393.
153. Young, B.A., R.T. Sherwood, and E.C. Bashaw. 1979. Cleared-pistil and thick sectioning techniques for detecting aposporous apomixis in grasses. Can. J. Bot. 57:1668-1672.

25 Breeding and Quantitative Genetics

M. D. RUMBAUGH

USDA-ARS
Logan, Utah

J. L. CADDEL

Oklahoma State University
Stillwater, Oklahoma

D. E. ROWE

USDA-ARS
Reno, Nevada

Alfalfa (*Medicago sativa* L.) is a polymorphic species, adapted to many soils and climates. Inherent variation is immense, the introgression of *M. falcata* L. into *M. sativa* has increased genetic variation and range of adaptation. Alfalfa is grown extensively in the temperate climates of all continents.

As with all crops, the methods and procedures used in breeding alfalfa depend upon the botanical, physiological, genetic, and reproductive characteristics of the species. Alfalfa, a perennial with perfect flowers, is naturally crosspollinated by bees, tolerates comparatively little inbreeding, and can be vegetatively propagated by stem cuttings. It has a chromosome complement of $X = 8$ with both diploid and tetraploid forms (65). All cultivars are autotetraploids and the inheritance of economic traits is therefore quite complex.

25-1 BREEDING

25-1.1 Constraints and Objectives

At the initiation of an improvement program, the breeder makes several important decisions that will determine the success or failure of the program. Decisions must be made concerning the traits to be improved, the extent of improvement desired, the parental germplasm, the unit of selection, and the mating system to be used. Additional decisions involve population sizes, selection intensities, selection techniques, field test procedures, and the time frame in which the breeding objectives are

---

Copyright 1988 © ASA-CSSA-SSSA, 677 South Segoe Road, Madison, WI 53711, USA.
Alfalfa and Alfalfa Improvement—Agronomy Monograph no. 29.

to be attained within the constraints of the experimental resources available to the program. It is essential that breeding objectives be well defined, succinctly stated, and attainable. If this is not done, the program could become too broad and suffer from the absence of focus on major objectives. This detracts from the effectiveness of the researcher and is wasteful of program resources. As in any scientific endeavor, time and resources must be used efficiently in order to maximize progress.

It is a truism that genetic improvement of a population is not possible beyond the limits established by the genes present in that population. Therefore, choice of the germplasm to be included in a breeding program is critical. Most often the breeder should use sources of the desired genes that are known to be adapted and to possess desirable agronomic attributes. Plant introductions and similar exotic germplasms are valuable sources of genes that are absent or infrequent within improved populations but they should not be routinely incorporated into breeding plans.

The breeder also must be cognizant of the methods required to increase seeds of the improved cultivar and still maintain its desirable characteristics. For the past 25 yr, most breeders have stressed the development of cultivars with multiple pest resistance. Large numbers of parental clones are selected from well adapted and productive source populations by appropriate screening techniques. Selected clones are then incorporated into genetically broad-based synthetics from which commercial seed is produced under limited generation schemes (see Chapter 32 in this book). Methods used in producing seed of hybrid cultivars will be quite different and the breeder must be aware of such differences.

25-1.2 Units of Selection and Mating Systems

25-1.2.1 Ecotype Selection

Alfalfa grows from inside the arctic circle to the southern hemisphere and from sea level to >2500-m elevation. It is cultivated without irrigation in regions where there is <200-mm annual precipitation as well as in regions that receive 2500 mm of precipitation. The first level of breeding alfalfa, identification of types generally adapted to a given region, is ecotype selection. In many regions the identification of adapted ecotypes is based primarily on the relative need for winter dormancy and pest resistance. More fall dormant types are generally more resistant to being killed by low winter temperatures. Also, more dormant alfalfas have the ability to withstand drought conditions by becoming dormant upon the onset of dry conditions even during warm parts of the growing season. Less dormant types tend to grow later in the fall and earlier in the spring, thus, they tend to yield more where high levels of winter hardiness are not important.

In general, alfalfas that evolved in, or were selected for, humid regions have more foliar disease resistance, whereas types from arid regions tend to have higher levels of resistance to crown and root diseases because of the likelihood of irrigation during their development.

Alfalfa ecotypes also differ in their adaptation to grazing. Dormant types of *M. sativa* and *M. falcata* have broad crowns, and some have the ability to propagate themselves with underground stems and creeping roots. Plants with low-growing, broad crowns are more resistant than erect types to trampling by livestock.

Ecotype nurseries are useful in the identification of forage cultivars, experimental strains and populations, or introductions that could prove to be well adapted to a particular environment (37). Nurseries are usually set up as single row plots separated by 1 m with spacings of 0.3 to 1.0 m between plants within rows. A minimum of 50 plants per entry should be observed.

Entries in replicated or unreplicated ecotype nurseries can be scored for fall growth, foliar pest resistance, crown height, etc., during 2 or 3 yr and then uprooted to determine relative damage from crown and root pests. These observations, together with observations related to reaction to environmental stress (cold, heat, flooding, drought, frost, pollution, etc.) can be beneficial, especially in the initial stages of a breeding program.

Bolton (8), Lowe et al. (69), Barnes et al. (4), and Lorenz et al. (68) listed ecotypes and cultivars of hay and grazing alfalfas. Barnes et al. (4) classified alfalfa in the USA into nine different germplasm sources that evolved in comparatively distinct regions of the world. Their classification is based primarily on winter hardiness, plant morphology, and geographic origin. The nine germplasm sources are: *M. falcata*, Ladak, *M. varia*, Turkistan, Flemish, Chilean, Peruvian, Indian, and African.

25-1.2.2 Interpopulation Breeding

Interpopulation improvement utilizes the concept of open breeding populations; allowing genes to flow from one population to another. Plants in one population (cultivar, strain, ecotype, or any germplasm source) are crossed either by hand or by bees (commonly *Apis mellifera* L. or *Megachile rotundata* F.) with plants from other populations. Selection is applied to improve one (or sometimes both) population(s) by isolating plants with desirable traits from both sources. Rowe and Hill (83) suggested that if nonadditive gene action was important for forage yield, then interpopulation improvement with reciprocal recurrent selection and topcross selection should be superior to most intrapopulation improvement methods.

25-1.2.2.1 Strain Building. The preservation of germplasm and the development of new cultivars that have a broad genetic base is founded on the concept of strain building (19, 38, 58, 95). Tysdal et al. (95) explained that strain building is a general term covering all forms of strain improvement rather than a specific procedure. The objective is to develop a superior strain characterized by having higher frequencies of genes for the expression of a particular character or characters while retaining a sufficiently broad gene base so that inbreeding effects will be minimal in advanced generations. As an interpopulation improvement method, this entails crossing plants from two different sources and selecting in subsequent generations for improved qualities. Any of the mating systems described below can be employed to achieve this objective.

Busbice et al. (17) listed several cultivars and germplasm sources developed on the basis of this concept. Mass selection or phenotypic recurrent selection can be used in strain building to improve any genetically controlled characteristic that has a relatively high heritability (25, 38, 95). Some type of progeny test should be included when traits with low heritabilities are to be improved (83, 94).

25–1.2.2.2 Synthetics. The term *synthetic variety* (cultivar) has been defined in several different ways. Hayes and Garber (42) first used it with maize (*Zea mays* L.) as a cross of several inbred strains with subsequent selection in the F_1 and advanced generations to improve the cultivar. According to them, farmers could use their own seed year after year without special conditions for pollen control. They also indicated that a synthetic cultivar was developed from selected germplasm and could be improved in advanced generations by selection.

Tysdal and Crandall (94) defined an alfalfa synthetic "variety" (cultivar) as a cultivar developed by crossing, compositing, or planting together, two or more strains or clones, with bulk seed being harvested and replanted in successive generations. Natural intercrossing and successive generations of seed increase were important considerations in the development of their concept. Allard (1) emphasized that the parents of a synthetic must be selected on the basis of combining ability. Busbice (14) defined it as a cultivar produced by random mating of several parents so that all possible matings between parents have an equal probability of occurring. He excluded the requirement of selecting for combining ability and improvement in advanced generations. Simmonds (88) used the term synthetic to describe an open-pollinated cultivar but acknowledged that the term often applies to various generations of experimental strains derived from several selected parents. He differentiated synthetics from open-pollinated cultivars in that synthetics can be reconstituted from the original selected parents. Wood (100) defined a synthetic cultivar as an advanced generation of an open-pollinated population composed of a group of selected inbreds, clones, or hybrids.

Regardless of the definition, synthetics in alfalfa are used in advanced generations as commercial cultivars. The parents are always selected for some particular trait or traits but seldom for combining ability per se. Synthetic cultivars permit the expression of heterosis to a degree, usually less than hybrids, while providing a practical method for seed multiplication. It is possible to fix specific traits in synthetic cultivars of alfalfa, and they can provide a source of improved germplasm because they represent an efficient approach to conserving genes (38).

Parents for synthetic cultivars in alfalfa are selected by many different methods. In an open breeding system the parents can be selected from such diverse sources as ecotypes, cultivars, and experimental strains. Although production of a synthetic cultivar is relatively simple, a wise choice of parents for the Syn 0 generation is crucial, for this will determine the performance of the synthetic. Decisions as to which and how many parents to include, fix the minimum degree of inbreeding that the eventual cultivar will sustain in subsequent generations. Busbice (14) demonstrated a method of computing inbreeding in advanced generations based on the relationship among parental plants, degree of selfing, and ploidy level.

Kehr et al. (60) utilized narrow based synthetics with varying numbers of parental clones (two–six) to show that forage yields generally decreased over generations of synthesis, especially from Syn 1 to Syn 2. Some particular strains maintained stable relative yields during advanced generations much better than others. Seed yields for the breeder and foundation seed generations were compared for eight synthetic cultivars by Kehr et al. (59). They found yields from breeder seed were lower than yields from foundation seed. They concluded that plant breeders should use at least one commercial seed generation of their experimental cultivars in seed yield trials of new synthetics. Broad based synthetics formed from 100 or more clones would be expected to be more stable during seed multiplication than narrow based synthetics.

Busbice (15) developed formulae for predicting the yield of synthetic cultivars. Effects of the following factors were considered: (i) ploidy level, (ii) amount of selfing, (iii) number of parents, (iv) inbreeding in parents, (v) combining ability of parents, (vi) relationship among parents, and (vii) generation of multiplication. Breeders should concentrate on selection of parents rather than on combinations of parents because little additional gain would be expected when more than four parents are involved (45).

Neither polycross- nor self-progeny tests were effective in increasing yield because of the confounding effects of genotype × environment interactions (18). Busbice and Gurgis (16) showed that clonal progeny tests were more effective than topcross or S_1 progeny tests in predicting first-cross yields. They also presented prediction formulae for the performance of synthetic cultivars that should be helpful in determining the number of clones to include, which parents to use, and yield of the cultivar in advanced generations. Based on their calculations, 16 unrelated, noninbred parents should adequately minimize inbreeding at equilibrium.

25–1.2.2.3 Backcross Breeding. Backcrossing is one of the most conservative breeding methods. It is especially appropriate when a particularly well adapted population has only a few genetic deficiencies. Backcrossing narrows the germplasm base and increases the probability that the parents of the subsequent synthetics will be related. Backcross breeding, as defined for self-pollinated crops, must be modified to minimize inbreeding effects when applied to alfalfa. In self-pollinated plants the donor line or plant can be crossed to only a few or even one plant from the recurrent parent, and only a few individuals need be selected and backcrossed each generation. Stanford and Houston (93) described a backcross method for use with alfalfa that involved many (100–300) relatively unrelated plants in the recurrent and donor populations. 'Caliverde' (91), 'Saranac' (73), 'Iroquois' (74), and 'Apalachee' (87) are examples of cultivars developed by modified backcrossing procedures. Although improved cultivars have been developed using backcross procedures, breeders tend to utilize other methods to attain their objectives.

25–1.2.2.4 Cultivar Crosses. Complementary strain crossing, strain crossing, or cultivar crossing are terms applied to breeding procedures designed to incorporate positive dominant traits from two or more sources into a single strain. Resulting strains can be used as cultivars or as source populations for further improvement (17, 27).

Hanson et al. (39) and Elgin et al. (27) showed that yields of strain crosses were usually equal to or slightly greater than the mean of the parental strains. Pest resistance of resulting strains was, in most situations, at least equal to the midparent level, and the level of resistance and yielding ability of strain crosses seldom decreased from Syn 1 through Syn 3 or 4.

Busbice et al. (17) illustrated mathematically that strain crossing should be an important concept in breeding autotetraploid species. If two strains possess favorable dominant genes at different loci at a frequency of 0.5, the strain cross between the two would have a gene frequency of 0.25 for each. At equilibrium, 46.7% of the resulting population would have both dominant genes, 43.3% would have one dominant gene, and only 10.0% would carry neither. In comparison, the population distribution in a diploid species is 19.1, 49.2, and 31.6%, respectively. Strain crosses provide a useful approach to the development of multiple pest resistance in alfalfa because most genes for resistance are dominant.

Hill (48) reported an appreciable amount of heterosis from crosses among unrelated, noninbred parental strains, whereas forage yield was reduced when parental cultivars were partially inbred. Inbred autotetraploid plants transmit part of their inbreeding to the progeny. Bingham (5, 6, 7) proposed that the use of at least four highly selected, diverse strains in a type of double cross would maximize heterosis in autotetraploids. He suggested that breeder seed be produced by hand crossing the strains in pairs and then mixing the seed in a field planting to produce foundation seed. By chance about 50% of the plants in the next generation (certified seed) would be double crosses and represent the highest state of heterozygosity.

Strain crosses may be made in several different ways. Plants from each strain can be paired at random and reciprocally hand crossed, or plants from both strains can be grown in alternate rows in cages and crossed with bees. Even though a certain amount of intrastrain pollination and selfing would take place in the latter, no practical difference was observed in the resulting population examined by Elgin et al. (27).

Another method of strain crossing that will decrease the amount of intrastrain crossing and selfing is to sow two strains in a grid so that the nearest neighbors for every plant are plants of the other strain. One strain can be cut back several days before the other on the flush of growth prior to seed production. When the second strain comes into bloom, an abundant supply of pollen from the first strain will be present. With intensive pollinator activity, the vast majority of flowers on the second strain will be pollinated with pollen from the complementary strain. High levels of self incompatability will further increase the proportion of inter-strain crosses.

25-1.2.2.5 Hybrid Alfalfa. The high level of inbreeding depression and hybrid vigor manifested by alfalfa led to an interest in hybrid cultivars early in alfalfa breeding. The advantages of hybrids over synthetics listed by Busbice et al. (17) include: (i) greater avoidance of inbreeding, (ii) reduction of natural selection when seed production is outside the area of adaptation, and (iii) full utilization of nonadditive gene action. In addition to fixing desirable traits and optimizing the expression of heterosis, hybrid cutivars have a certain mystic hold in the eyes of alfalfa producers because of the demonstrated success of hybrid corn and sorghum [*Sorghum bicolor* (L.) Moench].

Tysdal et al. (95) envisioned hybrids as one approach to overcoming the apparent ceiling on yield associated with synthetics. They proposed the vegetative propagation of two self-sterile inbred plants in fields of about 12 ha and allowing them to intercross. Seed harvested from two such fields (two different hybrids) would be mixed in equal proportions and sown in a third field. From the latter, double cross, chance-hybrid seed would be harvested. The four original clones would have been selected, not only for combining ability, but also for a high degree of self-sterility in order to reduce the amount of self pollination.

Childers and Barnes (21) described the events that led to the production and marketing of the first hybrid alfalfa utilizing cytoplasmic male sterility in 1968. The hybrid production process was patented by the L. Teweles Seed Co., Milwaukee, WI. The authors also summarized the problems associated with production of hybrid alfalfa seed. The fact that alfalfa is insect pollinated and that the bees effective in pollination frequently show a preference for certain plants increase the problems greatly. Both strains in a hybrid alfalfa seed production field must be attractive to bees to ensure pollen transfer from one strain to the other. The tetraploid inheritance of alfalfa also complicates the development of hybrids if nuclear genes control fertility. Conversely, alfalfa does have some traits that facilitate hybrid seed production. Alfalfa does not have to be reestablished annually and flower fertility is of little consequence in commercial forage production fields.

Barnes et al. (3) concluded that the self-incompatibility in alfalfa is not reliable enough for pollen control in hybrid seed production unless gene markers were used to determine the extent of selfing and sibbing. The amount of pollen produced by male-sterile parents also may vary and affect the percentage of crossing (76). Hybrids may be distinguished from selfs in field crosses by the use of electrophoresis or genetic markers (72). Pedersen and Barnes (75) considered the use of seed size as an indication of hybridity, since selfed seed was smaller than sibbed seed and sibbed seed was smaller than outcrossed seed. Hunt et al. (56) suggested using a black seed color trait to help distinguish hybrid seed on male steriles from pollenizer (male-fertile plants) seed when the two types were planted in mixed stands. They also suggested using the small seed size associated with the black seeds to separate hybrid seed from pollenizer seed.

Bingham (6, 7) recently showed that maximum heterozygosity and hybrid vigor would be expressed only in a tetra-allelic condition. This could be accomplished by using four parental strains instead of the two proposed for the production of most diploid hybrid cultivars. In this way up to 75% of the plants in commercial plantings could be double crosses and potentially tetra-allelic.

25–1.2.3 Intrapopulation Selection

Alfalfa breeders have utilized many different forms of selection within single populations to improve an array of characteristics. Intrapopulation breeding procedures have been most notably successful in increasing pest resistance. Genes for resistance to a specific pest may occur at low frequencies in a population(s) available for improvement. Consequently, breeders have had to screen large numbers of plants to identify those with resistance. By intercrossing resistant plants and reselecting within the resulting populations, the frequency of genes for resistance has been increased to usable levels.

The choice of unit of selection varies with the particular trait of interest and the precision of the selection test. Units of selection for alfalfa improvement may be individual plants or families of plants. Families are usually half sibs or full sibs. We have divided the following discussion of intrapopulation selection into individual plant selection, selection among families, selection within families, and combinations of these as selection units.

25–1.2.3.1 Individual Selection

25–1.2.3.1.1 Mass Selection

Using mass selection, desirable plants from a source population are chosen, from which seed is harvested and composited, without progeny tests, to produce the subsequent generation. Selection based on the phenotype of plants within a source population without pollen control is probably the simplest form of breeding cross-pollinated plants. It can be effective for highly heritable traits. The main weaknesses in simple mass selection are: (i) uncertain identification of superior genotypes, (ii) the source of pollen is unknown, and (iii) inbreeding depression arises when small populations result from very intensive selection. Natural selection can play an important role in mass selection for traits such as resistance to pests or environmental stress. Most other intrapopulation breeding procedures for alfalfa are basically refinements of mass selection.

Division of the source population into grids as proposed by Gardner (32, 33), Verhalen et al. (96), and Burton (11, 12) for corn, cotton (*Gossypium hirsutum* L.), and Pensacola bahiagrass (*Paspalum notatum* Flügge), respectively, should reduce the environmental effects within grids and improve the effectiveness of mass selection in alfalfa for traits with medium and low heritability. Grid selection can be applied to plants in solid seeded field stands, spaced plants in rows, or plants grown in flats for greenhouse or growth chamber screening trials. With this method, the breeder applies a uniform selection intensity to plants within each grid, saving the same percentage from each that he would otherwise apply to the whole population. The use of grids to reduce the environmental variation among units of selection can be imposed on progeny test nurseries when seed supplies and/or space limits the use of replications or experimental blocks.

25–1.2.3.1.2 Maternal Selection

Selection based only on maternal phenotype is sometimes referred to as maternal selection. When progeny tests are not utilized it should be more correctly called mass selection. When clones are used, then "clonal line selection" would be the preferred term. When sexually produced progeny are used, the preferred term should indicate the type of progeny test

25–1.2.3.1.3 Clonal Line Selection

A simple modification of mass selection, clonal line selection, can assist the breeder in identifying superior genotypes. Cloning selected plants and observing the clonal lines in replicated tests is one approach to eliminating escapes for certain traits. Two obvious disadvantages of clonal line selection include: (i) the extra time required for making stem cuttings and transplanting in the same generation, and (ii) the failure of plants from stem cuttings to develop normal tap roots.

25–1.2.3.1.4 Progeny Tests

Progeny tests can be effective in the identification of superior genotypes. Normally plants are selected from a source population, their progeny are observed, and the original plants saved or discarded on the basis of average progeny performance. At this point, the original plants are intercrossed to form subsequent generations. The progeny do not contribute to the synthesis of the next population.

At least five different progeny tests have been described for breeding alfalfa:

1. Open-pollinated
2. Selfed
3. Topcross
4. Polycross
5. Diallel.

The choice of progeny test depends primarily on the inheritance of the trait or traits of interest to the breeder. The major disadvantage to all progeny tests is the requirement of an extra generation and the time (frequently >1 yr) involved in each selection cycle. Selfed progeny tests and particularly topcross and diallel progeny tests also require a significant investment in seed production. Consequently, progeny tests are not used for traits with high heritabilities and little genotype × environment interaction. It is generally more efficient to employ repeated cycles of mass selection for such traits.

1. Open-pollinated progeny tests consist of harvesting seed from selected plants and testing the progeny from each selected plant in separate plots. Remnant seed from plants selected on the basis of progeny performance is composited to form the next generation, or maternal plants can be intercrossed to provide seed for subsequent generations. When remnant seed is used, it should be kept in mind that half the genes in the next generation come from unknown paternal sources. When only a small proportion of the individuals in the source are chosen, a high proportion of pollen comes from inferior (phenotypically unselected) plants.

2. Selfed progeny tests are planted with self-pollinated seed obtained by tripping flowers on selected plants. Selfed seed from each plant is harvested, maintained separately, and sown in individual selfed progeny rows for evaluation. The family is referred to as S_1. Disadvantages associated with selfed progeny tests include: (i) self incompatability reduces seed set on some plants and requires tripping of many flowers, and (ii) varying amounts of inbreeding depression can complicate comparisons among the progeny of the original plants.

3. Topcross progeny performance can be evaluated when selected plants are each crossed with a common tester (single clone or composite of pollen from a strain). Tysdal and Crandall (94) found that topcross progeny tests were effective in identifying parents for synthetics. Specific combining ability with the tester can be evaluated using this procedure. Plants in different families produced with a common tester are half sibs and those within a family are full sibs (if a single clone is used as the tester).

4. Polycross progeny tests, first described for alfalfa by Tysdal et al.

(95), have been used more frequently than other tests because it represents a compromise in the amount of work required to obtain seed of half-sib families and the information derived from other types of progeny tests. It requires more work than open-pollinated progeny tests, but the pollen source is restricted to a known population. Much less work is required than for topcross and diallels.

For polycross progeny breeding, selected plants are cloned, transplanted to an isolated polycross seed production nursery, and allowed to intermate at random. Seed harvested from all propagules within a clone is composited and sown in progeny rows, maintaining the identity of each clone. Plants within a polycross family are half sibs and the pollen parent is one of the selected plants from the original population. This avoids the problem that occurs in mass selection where pollen can be provided by inferior (nonselected) plants.

Individual clones should be replicated to ensure random mating among selected plants and minimize the effect of pollination between neighboring plants. Clones may have to be cut back at different times and early flowers may have to be removed to ensure synchronization of flowering. Selected plants may be intercrossed either by hand pollination or with bee pollinators, although bee preference for certain clones has been cited as a source of a bias in resulting progenies. If bees are used as pollinators, they must be highly active during flowering. Variable quantities of selfed seed can have a negative effect on the forage or seed yield of the progenies.

5. Diallel crosses require much more work to produce seed, but provide more information to the breeder because both male and female parents are known for each family. In this system each selected clone is crossed with every other selected clone, and the seed is kept separate. In general, the number of clones in a diallel cross is held to a minimum because $[n(n - 1)]/2$ represents the number of progeny rows required for a complete diallel, excluding reciprocals and selfs. If reciprocal differences are thought to be of importance $n(n - 1)$ families are required, and n^2 families are required when selfed progenies are included. With only 10 selected clones, 45, 90, and 100 different crosses would be required, respectively. Considering the number of individual flowers to be emasculated and pollinated to produce seed for such a study, 50- or even 20-plant diallels represent more effort than most plant breeders are willing to expend for routine progeny tests. On the other hand, diallel cross progeny tests provide valuable information about genetic control of important traits that have low heritabilities. Diallel families represent variation within and among half-sib and full-sib families and among unrelated families that can reveal general and specific combining abilities and information about additive, dominance, and perhaps epistatic gene action. The level of inbreeding of the parental clones will affect the performance of their progenies (46), and the genetic interpretation of the analysis. Partial diallel crosses have been suggested by some investigators as one approach to testing a large number of parent plants (66).

25-1.2.3.2 Family Selection Family selection differs from mass selection in that progeny tests are used to evaluate each plant's genotype. It is generally more effective than mass selection for traits with low heritability and high genotype × environment interaction. Family selection differs from individual plant selection methods with progeny tests in that progenies from the originally selected plants are intercrossed to produce subsequent generations rather than using the original plants for crossing and recombination.

25-1.2.3.2.1 Half-sib Family Selection

Half-sib selection can be used in intrapopulation improvement. Seed is harvested from selected plants in an open-pollinated or polycross source population. Progeny from each selected plant is kept separate and sown in replicated progeny tests. The worst families are eliminated, and the best families are intercrossed to permit recombination and to produce subsequent generations. This procedure is similar to the "modified ear-to-row" method outlined by Lonnquist (67) for corn.

Compton and Comstock (22) proposed identification of the best families, based on half-sib corn progenies, and the use of remnant seed from the selected original plants for intercrossing and recombination. This modification allows selection of both male and female parents used to produce subsequent generations. Hill and Byers (49) proposed the use of a modified ear-to-row procedure for alfalfa that depends on selecting plants and isolating them for intermating.

25-1.2.3.2.2 Full-sib Family Selection

This differs from half-sib family selection in that plants in a family are full sibs produced by selfing, topcrosses, or diallels. The main deterrent to widespread use of this method in alfalfa is the large amount of work required to make the prescribed pollinations.

25-1.2.3.3 Within Family Selection. Progenies for within family selection are developed as described above for family selection methods. In this situation the best individual or several of the better individuals within each family are used for intercrossing and recombination. This method enables the breeder to maintain a broad gene base in the new source population. Selection pressure will be lower than when only plants from the best families are saved.

25-1.2.3.4 Combined Selection. Methods that employ selection both within and among families can be used in various ways. Weighting selection within or among families allows breeders to adjust selection pressure. The breeder may want to use only the very best plants in the upper 50% of the progeny rows. Thus, much more selection pressure will be exerted within families than among families. Progress in selection will be coupled with the maintenance of a wide gene base. This approach is particularly desirable when selecting for resistance to some pest or environmental factor that is not sufficiently severe to furnish an effective screen for the selection of individual plants. Consider, as an example, the problem of breeding for cold tolerance in a year when temperatures were not sufficiently low for satisfactory discrimination. Selection for cold tolerance could be based on family units, whereas intensive selection within families could be based primarily on other traits such as yield or rate of regrowth.

25-1.2.3.4.1 Maternal-line Selection

Maternal-line selection as described by Fryer (30) should be considered combined family selection. He improved alfalfa seed yield using progeny tests and selecting within and among half-sib families. Fryer eliminated the poorest plants from all rows and kept the best one, two, or three plants in the best rows and intercrossed them to produce an improved source population for further selection.

Hanson and Carnahan (37) used the term maternal-line selection for improving perennial forage grasses; however, their methods do not seem to differ from individual plant selection methods combined with progeny tests, as described above.

25-1.2.3.5 Recurrent Selection. Repeated cycles of mass selection in alfalfa differ only slightly from simple recurrent selection and phenotypic recurrent selection. In corn breeding, selected plants are frequently selfed and selfed progenies intercrossed to form the next generation. Conversely, alfalfa breeders tend to use simple recurrent selection in which selected plants or clones are intercrossed to form the next generation. They rely on the perennial growth habit of alfalfa, and the capacity to propagate plants vegetatively, to preserve genes in much the same way that corn breeders rely on selfing.

Simple recurrent selection, recurrent selection for general combining ability, recurrent selection for specific combining ability, and reciprocal recurrent selection can be applied to alfalfa as well as to corn for intrapopulation improvement. It should be noted, however, that pollen control in corn is much simpler, and simultaneous crossing and selfing are facilitated by monoecy and multiple ears with many seeds. Simple recurrent selection is the only one of the four methods practical in applied alfalfa breeding. It is used with or without modification to take advantage of alfalfa's perennial nature. The other three methods, that are designed specifically for improvement of combining ability, are seldom used in alfalfa except for genetic studies. Recurrent selection in alfalfa can be practiced in combination with the various progeny testing methods described above.

25-1.3 Selection for Multiple Traits

Plant breeders often face situations where it is essential to modify more than one trait. Usually several attributes of a population should be improved prior to its release as a cultivar. These improvements can be achieved by any one of three fundamental breeding approaches: (i) tandem selection, (ii) independent culling levels or successive elimination, or (iii) index selection.

25–1.3.1 Tandem Selection

In tandem selection programs, the breeder improves a desirable trait for one or more generations until it is fixed in the population at an acceptable level. Then additional cycles of selection are initiated for a second desirable trait, etc. The number of cycles executed per trait and the sequence of traits for which improvement is sought may be varied, but only one trait is selected at any one time in a tandem breeding program. Assuming that the traits are inherited independently, i.e., the genetic correlations are zero, that each of n traits is to be improved for one generation, and that the traits are of equal importance, the genetic improvement will be equal to the sum of the products of the heritabilities (h_j^2) and selection differentials ($P_j - \overline{P}_j$):

$$h_1^2(P_1 - \overline{P}_1) + h^2(P_2 - \overline{P}_2) + \ldots + h_n^2(P_n - \overline{P}_n) = \sum_j^n h_j^2(P_j - \overline{P}_j).$$

The average change in economic value per generation is proportional to $ih\sigma_g$ or $ih^2\sigma_p$ where i is the selection differential in standard measure, and σ_g^2 and σ_p^2 are the additive genetic and phenotypic variances, respectively (77).

25–1.3.2 Independent Culling

In independent culling, a separate selection threshold is chosen for each trait and selection units are retained for breeding only if they are above the threshold for each trait simultaneously. A variant of this scheme has been called "successive elimination." Successive elimination is used primarily when selecting for multiple disease and insect resistance. Survivors from screening trials with pest A are tested for pest B, survivors of pest B will be selected for resistance to pest C, and so on.

The method of independent culling decreases the selection intensity for individual traits when the size of the source population and the number of individuals to be retained for breeding are held constant. The more traits considered, the larger the decrease (77). If selection is for one trait and the better 25% of the population is retained for breeding, the selection differential is 1.30 standard deviations. When two traits are selected and 25% of the population is retained, selection can only be among the best 50% of individuals for each trait, and the selection differential for a single trait decreases to 0.80. When three traits are considered, the proportion selected is 0.63 for each trait and the selection differential decreases to 0.58 standard deviations for each trait. However, if the traits are of equal importance and genetically independent, the total improvement will be greater than selection for any single trait. With n traits undergoing simultaneous selection at equal intensities, the selection differential reduces from $i=z/b$ to $i'=z'/b^{1/n}$, where z' is the ordinate of the point of truncation and $b^{1/n}$ is the proportion of plants retained. If the heritabilities and variances are equal in all n traits, the expected increase in breeding value per generation reduces to $ni'h\sigma_g$. Genetic progress relative to tandem selection equals ni'/i under the assumptions of equal importance, equal variance, and equal heritability of all traits. This ratio will always exceed unity, and, hence, independent culling is more effective than tandem selection provided that the stated assumptions are valid.

25-1.3.3 Index Selection

In index selection breeding, plants are evaluated for several traits and these values are weighted according to economic and genetic relationships prior to summing values to obtain an index or total score. Individuals or progenies with superior weighted scores are used to advance the population in the breeding program. Smith (89) first proposed index selection based on discriminant functions for selecting cultivars of wheat (*Triticum aestivum* L.). Hazel (43) extended the concept to interbreeding populations and Hazel and Lush (44) compared the efficiency of index selection to the methods of tandem selection and independent culling. Index selection proved to be superior to independent culling, which, in turn, is superior to tandem selection. This conclusion was extended to cases in which the traits were correlated and of varying economic value and variability (101). Harris (40, 41) and Sales and Hill (86) have shown that there may be an appreciable error in predicting the gain from index selection unless the sample size is rather large, on the order of 1000 individuals. Index selection will be most successful when heritabilities and genetic correlations are comparatively high and environmental correlations comparatively low.

The superiority of index selection as a breeding method increases with an increase in the number of traits under selection and decreases when the traits differ greatly in importance, or when selection intensity is increased. Some scientists have suggested that a selection index is best suited to crops in which the value of the marketable crop is determined by readily measurable attributes rather than with crops, such as alfalfa, where the harvested portion is used directly as livestock feed (90). It should also be pointed out that a second important application of selection index theory is in predicting the breeding value of a single trait from several different sources of information (10).

Alfalfa breeders frequently select for multiple traits in an informal way without defining precisely the selection intensities or assigned economic and genetic weights. Experience and a general understanding of relationships among traits can enable breeders to realize significant progress. Also, procedures may be modified at any point in the improvement program. In selecting for multiple pest resistance, Hanson et al. (38) used independent culling levels during the first 10 generations of selection conducted in the field and practiced tandem selection in subsequent generations conducted in the greenhouse. When valid estimates of genetic correlations among the traits are not available, a modified index, termed the base index, in which the traits are weighted only by relative economic values may be used (97). Elgin et al. (26) found a base index method to be more effective than independent culling or tandem selection for improvement of pest resistance in alfalfa.

25-1.4 Selection for Threshold Traits

There are many multifactorial characters of biological interest that exhibit a discontinuous distribution. Examples include the presence or absence of a plant organ or structure, immunity to a disease, germination of a seed, and survival or death of a plant following exposure to a given agent. Such characters have been called "quasi-continuous variations" because the mode of inheritance is like that of a continuously varying character. However, the character is not expressed until a certain threshold or level of genotype and environment is attained.

Individual plants can have a phenotypic value of + or −, or of 0 or 1, indicating the presence or absence of a threshold trait. Groups of individuals, however, can have any value in the form of the proportion, percentage, or incidence of the individuals in one or the other of the two classes. This value may be transformed from the visible to the underlying genetic scale by reference to a table of probabilities of the normal curve. It is then possible to state the phenotypic value of a population in terms of its standard deviation and to compare means of populations if they have the same standard deviation. Although differences among the means of families can be analyzed and realized heritability computed, the breeder cannot discriminate between high and low individuals on the underlying scale (29). Response to selection depends in the usual way on the selection differential, but the selection differential does not depend primarily on the proportion selected as with an observable, continuously varying trait. Selected individuals are a random sample from the desired class, and the mean of the selected individuals is the mean of the desired class irrespective of the numbers retained. The selection differential is greatest when the proportion retained for breeding is exactly equal to the incidence.

In some situations both a threshold and a ceiling may exist. There then will be a trichotomy of phenotypes but the analysis is similar to those cases just described. The breeder must be aware, however, that not all dichotomous traits depend on genetic thresholds. Environmental factors may determine completely the presence or absence of a character. Although alfalfa breeders have studied dichotomous traits, such as the ability of seeds to germinate in osmotic solutions, we are not aware of any detailed genetic analysis of selection for such traits.

25-1.5 Cultivar Synthesis and Multiplication

Superior characteristics developed by the breeders must be maintained during seed production if the cultivar is to be successful. The plan for seed increase must reflect the genetic system controlling the improved traits and become a part of the cultivar description and release statement or registration. In general, the number and severity of the constraints on seed multiplication will increase when improvements are conditioned by specific genetic configurations or specific combining abilities. When large populations are reselected by procedures based on additive gene action and from which comparatively large numbers of individuals are retained for breeding, there need be less concern about genetic shift, genetic drift, inbreeding, or reconfiguration of the genetic structure of the population. In addition, the potential for change in agronomic performance is reduced when a cultivar is multiplied within its region of adaptation, where differences in plant survival and relative seed production per plant are minimized. This approach to breeding ignores the importance of testing specific combining ability, and the resulting population is genetically stable during the seed increase phase of the release program, provided isolation requirements are met (9).

As the number of parents in a synthetic cultivar is decreased, the potential for inbreeding during multiplication of the seed from the syn 0 to syn 1 and advanced generations becomes more of a risk (14). The degree of inbreeding changes from generation to generation until equilibrium is reached with the greatest change exhibited in the Syn 2 generation. However, Syn 2 yield performance could be predicted from clonal F_1 performance quite accurately, whereas Syn 1 performance could not (82). Advanced generations of two-clone synthetics (components of double-cross synthetics) were consistently lower yielding than any four-clone synthetic that had the same two clones in its parentage (70). This agreed with theoretical expectations on inbreeding. Hill and Elgin (50) found that the greatest expected gain from selection within a large population of experimental synthetic cultivars tracing to one parental population occurred when the synthetics had more than four but <16 parents each.

The performance of synthetics in which the parental clones had been selected intensively for specific combining ability in the Syn 1 would decline as the seed was increased to Syn 2 or Syn 3. However, it is more common to find that specific combining ability effects are so small compared to general effects that breeding procedures dependent on general effects are more efficient than those that capitalize on specific effects (52).

If the breeding program is directed toward development of hybrid cultivars, an exceptional level of control must be maintained during seed multiplication. The limited number of parental clones and their selection based on nonadditive genetic effects render such cultivars especially sensitive to improper field isolation, breakdown in cytoplasmic male sterility, and unauthorized increase to advanced generations.

25-2 QUALITATIVE GENETICS

There have been numerous studies of the genetic segregation of qualitative trait differences among alfalfa clones. The results of these investigations are summarized in Chapter 24 in this book.

25-3 QUANTITATIVE GENETICS

The development of genetic theory for cross-pollinated autotetraploid species, such as alfalfa, has been minimal in comparison with genetic theory developed for diploids. It might be argued that autotetraploid genetic theory is 15 to 25 yr behind that of diploid genetics. However, this may be misleading because autotetraploid theory does not appear to parallel that of diploid genetics.

The paucity of published work on autotetraploid genetics is partially attributable to several characteristics of the diploid gamete and its formation. In autotetraploids, gamete frequency is not equated to gene frequency. Also, even when there are only two or three genes at a locus, there are a relatively large number of possible genotypes and, thus, there are a large number of possible intra- and interlocus interactions. The third restriction on autotetraploid genetic theory development is the high dimensionality of the algebraic expressions, which are often complex and difficult to solve, and even when solved they may not have meaningful interpretations.

The development of a mathematical representation of autotetraploid genetics has been nearly exclusive for the single locus. When the effects of many loci in one individual or of one locus in many individuals are considered, the usual assumption is that the effects of loci are strictly additive.

For all genetic model development, simplifying assumptions are necessary and the extrapolation of results as descriptions or predictions must be made on the basis of those assumptions.

25-3.1 Autotetraploid Genetic Model

A useful and interpretable assumption for modeling is that a given population is in random mating equilibrium (RME). Under this assumption the frequency of every genotype in a population is equivalent to the products of the frequencies of the genes in the population. For a population to stay in RME it cannot be subjected to any selection pressure, all clones must contribute gametes equally to the next generation, and mating must be completely random. Thus a population of size n will produce the next generation as the products of $(n^2 - n)$ crosses and n selfs, each of which will occur at a frequency of $(1/n^2)$.

A population that is not in RME (thus is in disequilibrium) and not subjected to selection pressure will approach asymptotically the equilibrium state. Two-thirds of the disequilibrium is lost with each generation of random mating, and the equilibrium is approximated in four or five generations of random mating (35).

Since multivalent pairing is rather rare in alfalfa (92) we make the simplifying assumption of random chromosome inheritance. The alternative is called random chromatid inheritance and provides a mechanism for increased homozygosity in the progeny via the mechanism of double reduction (24).

The notation and terminology for autotetraploids is indicated in the subsequent description of a very general genetic model of the effects at a single locus in an autotetraploid model developed by Kempthorne (63). In this model the four alleles at the A locus are indicated by subscripts i, j, k, and l, $(A_iA_jA_kA_l)$. The genotypic value is expressed as effects of individual genes and all of their possible interactions. Then the genotypic values of genotype $A_iA_jA_kA_l$ is

$$\alpha_i + \alpha_j + \alpha_k + \alpha_l + \beta_{ij} + \beta_{ik} + \beta_{il} + \beta_{jk} + \beta_{jl} + \beta_{kl} + \gamma_{ijk} + \gamma_{ijl} + \gamma_{ikl} + \gamma_{jkl} + \delta_{ijkl}.$$

In this model individual or main effects (α's) are called additive effects and are analogous to the effects of the same name in diploid genetics. The effects associated with an interaction of two alleles (β's) are called digene effects and are analogous to heterotic effects at a single locus in diploid genetics. Interactions involving three alleles (γ's) or four alleles (δ) do not have any analogous effects in diploids.

An autotetraploid genotype can be indicated by different terms depending on the reference point. When the number of different alleles at a given locus is important, genotypes composed of one, two, three, and four different alleles are called, respectively, monoallelic, diallelic, triallelic, and tetraallelic or equivalently monogenic, digenic, trigenic, and quadragenic.

When interest is in the frequency of only the allele B in the genotypes, terminology is altered. If we let all alleles other than B at the locus be indicated by a lower case of the same letter, then genotypes *BBBB*, *BBBb*, *BBbb*, *Bbbb*, and *bbbb* are named, respectively, quadraplex, triplex, duplex, monoplex (simplex), and nulliplex.

A second, more recent model of genetics at a single locus of an autotetraploid was developed by Hill (45), who made the assumption that there never were more than two genes at a locus. This enormously simplified the mathematics. In his model all effects are a reflection of the frequency of the B alleles in the five genotypes (Table 25-1). Since the differences in genotypic values among the five genotypes provide the only information about genetic factors at a locus, the genotypic value of the nulliplex is subtracted from the values of all genotypes, resulting in simpler mathematical expressions without loss of information. The genotypic value of the nulliplex could also be interpreted as the "background" effect.

The genotypic values of the five genotypes are describable in four

Table 25-1. The mathematical model of genotypic values of genotypes in population P and the equations for populations parameters and their variances† (45).

Genotypes	Genotypic value
BBBB	$4A + 6D + 4T + F$
BBBb	$3A + 3D + T$
BBbb	$2A + D$
Bbbb	A
bbbb	0
Genetic effects	Genotypic variances‡
Additive $(\alpha) = A + 3pD + 3p^2 T + p^3 F$	Additive variance $(\sigma_A^2) = 4pq\alpha^2$
Digenic $(\beta) = D + 2pT + p^2 F$	Digenic variance $(\sigma_D^2) = 6p^2 q^2 \beta^2$
Trigenic $(\gamma) = T + pF$	Trigenic variance $(\sigma_T^2) = 4p^3 q^3 \gamma^2$
Quadragenic $(\delta) = F$	Quadragenic variance $(\sigma_F^2) = p^4 q^4 \delta^2$

† A, D, T, and F are genetic effects associated with an individual allele and the interaction of two, three, and four alleles, respectively.
‡ Total genetic variance (σ_G^2) is $\sigma_A^2 + \sigma_D^2 + \sigma_T^2 + \sigma_F^2$.

terms (A, D, T, and F) with genetic interpretations. The effect associated with the A parameter is that of individual B alleles, the additive effects. The coefficient for this parameter indicates the dosage or number of B alleles in a genotype. The D parameter indicates the interaction of two B alleles and its coefficient is found with combinatorics as C_2^i where i is the number of B alleles in the genotype. The T parameter is the effect associated with the interaction of three B alleles and its coefficient is found via combinatorics as C_3^i. The F parameter is the interaction of all four B alleles and it has a coefficient of C_4^4 or 1.

The frequency of the B allele in some designated population is indicated by p and that of the b allele by q with the restriction that $p + q = 1.0$. When the two allele population is in RME, the frequencies of genotypes are given by the binomial expansion $(p + q)^4$ and the frequency of gamete genotypes is given by the binomial expansion $(p + q)^2$ (Table 25-2).

For the population in RME the genotype parameters A, D, T, and F do not have a one-to-one correspondence to the population effects called additive, digenic, trigenic, and quadragenic (Table 25-1). Except for the quadragenic effect, the population effect and their variances are functions of two or more genotypic effects, unless some or all of these effects have a value of zero.

For some hypothetical populations, genotypic values can be assigned to the genotypes in Table 25-1 to reflect different types of genic action at the B locus. In Table 25-3 four such models are indicated where genotypic values are expressed as a proportion of the arbitrary genotypic value (h) for some trait. These models can then be solved for the genotype parameters, A, D, T, and F.

In the additive model the genotypic value is proportional to the frequency of the B allele in the genotype. The monoplex dominance model in autotetraploids, which is similar to complete dominance in diploids,

BREEDING AND QUANTITATIVE GENETICS

Table 25-2. Frequencies of genotypes in populations that are or are not in RME and where the frequency of the B allele is $p = x + y$ and the frequency of the b allele is $q = y + z$. Then $(p + q) = (x + 2y + z) = 1.0$.

Genotypes	Equilibrium population	Nonequilibrium population
	Sporophyte	
BBBB	p^4	x^2
BBBb	$4p^3 q$	$4xy$
BBbb	$6p^2 q^2$	$4y^2 + 2xz$
Bbbb	$4pq^3$	$4yz$
bbbb	q^4	z^2
	Gamete	
BB	p^2	x
Bb	$2pq$	$2y$
bb	q^2	z

Table 25-3. Genotypic vlaues associated with four types of genic action at the B locus and solutions for A, D, T, and F for the population.

		Genic action		
Genotype	Additive	Monoplex dominance	Duplex dominance	Over dominance
		h		
BBBB	1	1.0	1.0	0
BBBb	0.75	1.0	1.0	0.75
BBbb	0.50	1.0	1.0	1.0
Bbbb	0.25	1.0	0.0	0.75
bbbb	0.0	0.0	0.0	0
Genotype effect		Solutions for genotype effects		
A	0.25	1.0	0.0	0.75
D	0.0	−1.0	1.0	−0.5
T	0.0	1.0	−2.0	0.0
F	0.0	−1.0	3.0	0.0

has a genotypic value of h for every genotype with at least one B allele. With duplex dominance the genotypic value is h when there are two or more B alleles in the genotype. For the overdominance case, genotypic values are maximized and symmetric about the duplex genotype. Though the possible types of dominance and overdominance are infinite, a few selected models can provide useful descriptive information about the dynamics of autotetraploid populations.

Components of genetic variance can be estimated by using mating designs developed for diploid genetics. Any design used for diploids that does not require inbreeding can be used for autotetraploids (66). Mating designs have been outlined in texts such as Chapter 4 of Hallauer and Miranda Fo (36). Differences between the analysis of autotetraploids and diploids lie in the interpretation of the covariances between relatives. The methodology for determining the covariance among relatives for a population in RME was determined by Kempthorne (61). For the single locus some of the more commonly used covariances are those of parent and offspring ($\sigma_A^2/2 + \sigma_D^2/6$), half-sibs ($\sigma_A^2/4 + \sigma_D^2/36$), and of full-sibs ($\sigma_A^2/2 + 2\sigma_D^2/9 + \sigma_T^2/12 + \sigma_F^2/36$). When two loci are considered, the genetic variance is further decomposed to reflect interactions between loci after the procedure of Kempthorne (62)

$$\sigma_G^2 = \sigma_A^2 + \sigma_D^2 + \sigma_T^2 + \sigma_F^2 + \sigma_{AA}^2 + \sigma_{AD}^2 + \sigma_{AT}^2 + \sigma_{AF}^2 + \sigma_{DD}^2 + \sigma_{DT}^2 + \sigma_{DF}^2 + \sigma_{TF}^2 + \sigma_{TT}^2 + \sigma_{FF}^2.$$

The covariance of parent and offspring involving two loci is

$$\text{COV}_{\text{PO}} = \sigma_A^2/2 + \sigma_D^2/6 + \sigma_{AA}^2/4 + \sigma_{AD}^2/24 + \sigma_{DD}^2/36.$$

The usual methods of solving for variance components is to equate the mean squares in an analysis of variance to its expected variance components and then find solutions for the variance components. Levings and Dudley (66) suggest that estimation of trigenic, quadragenic, and interaction effects is prohibited by their small coefficients.

25-3.1.1 Heritability

Heritability is a sometimes elusive concept that is useful in predicting and measuring response to selection and relationships of genotypic value to phenotypic value. This term can have several different technical interpretations (57) and, unless properly understood, it may connote more than it measures.

Narrow sense heritability (h^2) is defined as the ratio of additive genetic variance to phenotypic variance in a genetically stable large population in RME. The assumption of a large population in RME is necessary to remove the effects of random genetic drift (24). Biases in means introduced by disequilibrium are discussed later in this chapter.

For the defined population an estimate of h^2 would seem to be possible, directly from the methods of variance components analysis considered earlier. An inspection of the components of covariances indicates no exact estimate of σ_A^2 exists for autotetraploids unlike that for diploids, where σ_A^2 can be estimated as twice the covariance of parent and offspring.

If genic action at all loci is strictly additive, then digenic, trigenic, and quadragenic variances will be zero, and the covariance of offspring and parent would be only a function of additive variance. Levings and Dudley (66) proposed using twice the estimate of the covariance of parent and offspring as an estimate of h^2.

Likewise, the broad-sense heritability measure (H) is a ratio of variances: total genetic variance divided by total phenotypic variance. The same reference population as that used with the h^2 interpretation is appropriate, i.e., a large genetically stable population in RME.

The determination of H may be no less difficult than that of h^2.

Consider the following expression for the phenotype (X) of the ith individual in the jth environment.

$$X_{ij} = \mu + \alpha_i + \beta_j + \gamma_{ij}.$$

Where μ is the mean value of the trait measured in a nearly infinite population, α is the genetic value of the ith genotype, β is the average effect of the jth environment on all genotypes, and γ_{ij} is interaction or nonadditivity of genetic and environmental effects on phenotype. The phenotypic variance for the trait (X) in the arbitrary population is (29)

$$\sigma_P^2 = \sigma_G^2 + \sigma_E^2 + \sigma_I^2 + \sigma_{GE}.$$

Where σ_G^2 is genetic variance, σ_E^2 is environmental variance, σ_I^2 is interaction variance, and σ_{GE} is the covariance of genotypes and environments. The variances are necessarily nonnegative but there is no such restriction on the covariance. If there is a random-association of genotypes and environments, then $\sigma_{GE} = 0$. This assumption might not be realistic in field plantings where stands might be selectively thinned by factors unique to the particular environment. If the assumption of no interaction of genotypes and environments is valid; that is, the ranking of all genotypes is constant for all environments, then σ_I^2 will be zero.

With the interaction of genotype and environment and the covariances of genotype and environment equated to 0.0, the expression of phenotypic variance is simplified to $\sigma_P^2 = \sigma_G^2 + \sigma_E^2$, which facilitates estimation of H. Sometimes the estimate of heritability is considered to be a broad-sense measure when the additive variance cannot be separated from other genetic variance or interactions.

Realized heritability estimated in the absence of a reference population is useful in selection experiments. This is a descriptive measure of a particular selection situation that has limited usefulness for extrapolation to other populations. Though several symbols may be used for realized heritability, H is probably most common. For this measure $H = R/S$ where R is the deviation of the mean of the progeny of selections as deviations from the population mean, and S is the theoretical mean of the selections as deviations from population means in standardized units. It is necessary and sufficient that both the progeny of selections and the original population be normally distributed. The value of S is usually found by knowing the percentage of the original population selected (36).

When there are multiple cycles of selection for one trait in a population, realized heritability can be estimated in either of two ways. The gains may be regressed on their respective selection differentials (S) or expressed as the ratio of total gains to total selection differential (29). Extensive theoretical investigations have been made by W. G. Hill (53, 54, 55) into the interactions of selection pressure on heritability estimation, the variation in estimates of H, and the value of divergent selection or selection in one direction.

25-3.1.2 Selection

One of the most useful results of research in quantitative genetics has been the development of methodologies and expressions for predicting the change in gene frequencies and the mean response of selected populations. This research has facilitated comparisons of different selection schemes as nested in different types of genic action. Although the development of the prediction equations is made for restrictive and often simplistic hypothetical situations, they can be useful in comparing relative efficiencies. Even when some of the restrictions are not fulfilled research has suggested that comparisons are relatively accurate (80).

Lush (71) first indicated that expected gain or response to selection was predicted by multiplying the selection differential (S) by the heritability (h^2). This expression was more rigorously developed by Griffing (34), who indicated that the prediction was linear over cycles, whereas the response would necessarily by asymptotic to fixation of favorable alleles. In any event, since h^2 is a function of gene frequencies and the effect of selection is to change gene frequency, the initial prediction equation is exact for one cycle of selection and approximate for additional cycles only when the change in gene frequency is very small.

Prediction equations for different selection schemes in an autotetraploid population have been developed using a two gene model (45) for one theoretical population. This is expanded as needed to two or three theoretical populations.

With a single cycle of selection, change in gene frequency (dp) is predicted with the following equation (28):

$$dp = \frac{S\sigma_{xy}}{\sigma_p^2}$$

where S is the selection differential, σ_{xy} is the covariance between frequency of desirable allele in selected units (x), and the genotypic value of the corresponding units (y) upon which selection of x is based, and σ_p^2 is the phenotypic variance of the x units.

The predictive equations are developed in two stages. First the ratio (i/σ_p^2) is assigned a constant value K and expressions for covariance are developed. This covariance is referred to as the covariance of selection. Second, equations for K are developed to reflect the breeding procedure.

For every change in gene frequency, a change in genotypic mean value is expected. It is necessary to define the objectives of selection, however, before expressions can be developed for the change in population mean or means with the change in gene frequency.

Consider that there are at least two different objectives. First there is intrapopulation improvement where a single population is subjected to selection for improvement of one or more characteristics. Predictive equations for this objective have been rather completely developed and summarized by Rowe and Hill (85). The second objective is to improve the mean value from a cross of two populations measured either in the Syn 1 generation or when the cross is in RME. Predictive equations for the second objective are fragmentary because of mathematical complexities. Crosses of three or more populations have not been considered.

A brief outline of the major steps in modeling intrapopulation improvement procedures with reference to one locus can be informative. The mean ($\bar{X}$) of a population in RME with p as the frequency of B is

$$\bar{X} = 4pA + 6p^2D + 4p^3T + p^4F.$$

With selection, gene frequency will change by a quantity dp and the selected population in RME will have a mean different from the original population, indicated by $d\bar{X}$ which is

$$d\bar{X} = 4 \alpha dp + 6 \beta(dp)^2 + 4 \gamma(dp)^3 + \delta(dp)^4$$

(85).

When the assumption is made that the change in gene frequency is small, then the square and higher exponents of dp can be expected to be near zero and the above expression is simplified by ignoring the nonadditive population effects

$$d\bar{X} = 4 \alpha dp \text{ or } 4\alpha K\sigma_{xy}$$

where K equals the ratio of the selection differential to the phenotypic variance (S/σ_p^2).

The covariance of selection is given for five intrapopulation selection schemes that can be used on alfalfa (Table 25-4). The term Mass[1] is given to simple recurrent phenotypic selection where there is control of both pollen and seed parents in production of the selected progenies.

Table 25-4. Theoretical expected changes of five intrapopulation breeding schemes and their phenotypic variances.

Method	Expected change $d\bar{X}$ in population P mean	Generations per cycle	Phenotypic† variance σ_p^2
Mass[1]	$4kpq\sigma_{(p)}^2$	1	$\sigma_w^2 + \sigma_G^2$
Mass[2]	$2kpq\alpha_{(p)}^2$	1	$\sigma_w^2 + \sigma_G^2$
S1PT	$4kpq\alpha_{(p)} [\alpha_{(p)} + C\ddagger]$	3	$(\sigma_w^2 + \sigma_{S1}^2 - \sigma_{S1F}^2)/rn + \sigma_e^2/r + \sigma_{S1F}^2$
HSPT	$2kpq\alpha_{(p)}^2$	2 or 3	$(\sigma_w^2 + \sigma_G^2 - \sigma_{HS}^2)/rn + \sigma_e^2/r + \sigma_{HS}^2$
TX§	$2kpq\alpha_{(p)} [\alpha_{(u)} + (p - u)\beta_{(u)}]$	3	$(\sigma_w^2 + \sigma_{GTX}^2 - \sigma_{TX}^2)/rn + \sigma_e^2/r + \sigma_{TX}^2$

† σ_w^2, σ_G^2, σ_{S1}^2, σ_{S1F}^2, σ_{HS}^2, σ_e^2, σ_{GTX}^2, σ_{TX}^2, r, n represent environmental variance, total genetic variance, total genetic variance in an S_1 population, variance due to S_1 family means, variance due to half-sib family means, error variance, total genetic variance of a set of topcross family means, variance due to topcross family means, replications of a progeny, and number of plants in a progeny plot, respectively.
‡ $C = [(1 - 2p)/4]D - [(3p^2 - 2p)/2]T - [(10p^3 - 6p^2 - p)/12]F$.
§ Frequency of B is u in testor population (U).

When there is recurrent phenotypic selection with no control of the pollen parent, i.e., it is a random sample of the whole population, it is called Mass[2] selection.

For half-sib progeny test selection (HSPT), individuals are selected on the basis of the performance of half-sib families that were produced by crossing parents with pollen representative of the whole population. With the S_1 progeny test selection scheme (S1PT), individuals are selected on the basis of their progenies produced by selfing and the selections are then intercrossed. When individuals of one population are pollinated by a random sample of pollen from a second population (U) where the frequency of the B allele is u, and the progenies from these crosses are used to make selections in the original population, the procedure is called a topcross progeny test or (TX). As in all previous schemes, selections are intercrossed to produce the selected population.

From Table 25-4 it is apparent that Mass[1], Mass[2], and HSPT procedures have covariances that are functions of gene frequency and additive gene effects (pq) and the ranking of the responses is obvious. With the S1PT scheme, prediction of gain is frustrated by an effect (C) that may or may not be positive. In the TX procedure, additive and digenic components of the second population are important as well as gene frequencies in both populations and additive effects of the first population. The probable time required to complete a single cycle of selection is not constant for all these procedures and is shown in the column "Generations per cycle" of Table 25-4.

The components of σ_p^2 are presented in Table 25-4 for each selection procedure. This variance is affected by whether evaluations were made on individual plants or on their progenies. The size and source of progenies are important considerations. The variances are reduced by divisors rn and r when progenies are used. The use of a progeny test is expected to improve the precision of selection through replication but may not be of much value unless heritability is very low (45).

Development of analytical expressions for comparing interpopulation selection experiments consists of a single publication (83). This was a comparison of three procedures with equal numbers of generations assuming several types of genic action and a constant selection pressure. One method involved improving the parent populations of a cross via the intrapopulation selection procedure HSPT and then crossing. The second involved crossing both parent populations to a third population (the TX scheme) and then intercrossing the selections. The third procedure was reciprocal recurrent selection (23) where each of the two parent populations was used as a tester population for the other.

The predication equations developed for the procedure involving HSPT included additive and trigenic population parameters, whereas the equations for the other two procedures included expressions of additive, digenic, and trigenic population effects. Results of the study (83) indicated that crossing to a third population was the best procedure if that third population was inferior to both tested populations with respect to the selected trait.

25-3.1.3 Population Response Models

A model of the autotetraploid population is needed to describe genetic factors affecting genotypic values. Such a model has been developed for unique situations of populations in RME (61), but is of limited utility because populations are rarely, if ever, in RME.

Since inbreeding has such a catastrophic effect on alfalfa, the modeling of performance based on the inbreeding coefficient has been tested. The inadequacies of such a model to describe the observed reduction in forage and seed yields have been well documented (2, 13, 64, 95, 98, 99).

Busbice and Wilsie (20) developed a model to describe inbreeding responses. Their model weights the genotypes by the frequency of heterogenic interactions. Only the initial population is assumed to be in RME and noninbred. The difficulty with this model is that equation means are nonlinear and a minimum of eight generations is required to fit the model.

Gallais (31) also modeled the inbreeding population, which at generation zero was in RME and noninbred, but used a slightly different rationale. His model was developed by a weighting of genotypes based on the frequency of homogenic interactions and requires at least four inbred generations when initial frequencies of model parameters are known.

Hill (47) developed a model for inbreeding studies of self-pollinated generations where parental genotypes and origins are not specified. This model was developed from two-allele single locus theory and is suitable for analysis of small families.

A more recent population model (51) has been developed from Hill's single locus population model (45). Initial populations are assumed to be in RME and from the results of population crosses estimates were made of additive, digenic, and quadragenic parameters.

All of these models were developed with assumptions on the dynamics of the genetic factors in the autotetraploid population, and, hence, require further testing to establish the accuracy, correctness, and limitations of the models.

The testing of models is not only very expensive in time and facilities but also inconclusive. Testing of models can indicate the inaccuracy of a model as a lack of fit of some data set, but there is no test for the correctness of a model, per se. Comparisons of models are reduced to examining the lack of fit for each model, none of which is significant (78, 79). Though the model in essence is correct, the assumptions made for fitting a model, such as random mating and equal fertility, may not be attained in any population.

25-3.1.4 Gametic and Linkage Disequilibriums

Gametic products of each heterozygous genotype are not simply products of gene frequencies in the genotypes as seen in Table 25-2. Thus for genotype $BBBb$ the frequency of B is $P = 0.75$ and expected gametic products would be $(p + q)^2$ or p^2 for BB, $2pq$ for Bb, and q^2 for bb where $p + q = 1.0$. But the actual frequencies with random chromosome segregation at meiosis is $(1/2)BB$ and $(1/2)Bb$, thus there is a disequilibrium in the frequencies of gametes at one locus. The disequilibrium for products of one locus is called gametic disequilibrium, whereas that involving two or more loci on the same or different chromosomes is called linkage disequilibrium.

As indicated earlier, with random mating the gametic disequilibrium is dissipated with two-thirds of the disequilibrium lost per generation. A gametic disequilibrium can affect genotypic means of a population and responses to selection.

A study by Rowe and Hill (84) indicated that the means of the first generation of synthetic varieties could be greatly increased or decreased depending upon the gene frequencies of the parents of the synthetic and the presence of nonadditive genic action. When genic action was strictly additive at a locus, the population means were not affected by the gametic disequilibrium, but the genetic variance available for selection went to zero in some cases.

It has been demonstrated that rankings of test cross progenies can be affected by gametic disequilibrium (85). If populations are crossed to a single tester population and if the tested populations are in RME, the progenies of each test cross will have rankings reflecting the relative frequency of desirable genes, though the means may be biased. When the tested population is not in RME, such as a single plant, the rankings of the progenies may not reflect the relative frequencies of desirable genes. Also, advanced generations of synthetics could show an increase or decrease in mean genotypic values depending on the type of genic action and gene frequencies in synthetic parents.

Gametic disequilibrium can increase or decrease the covariance of selection from that found with the population in RME (80). Also, the relationships between different selection procedures is also affected. Half-sib progeny test selection is expected to have a covariance of one-half that of Mass[1] selection, but with the gametic disequilibrium this ratio can be changed by a plus or minus 20%.

The effects and evolution of linkage disequilibrium in a population are not understood at this time. Development of recursive relationships showing the approach to equilibrium and the evolution of a population mean and its constituent genotypes have not been forthcoming.

25-3.1.5 Inbreeding

The detrimental effects of inbreeding are well documented. Selfing a population for only two or three generations leads to near extinction of the population through loss of vigor and increased sterility. Because the effect of inbreeding is so consistent, it has been associated with changes in the mean performance of synthetic varieties over generations. Considerable effort has been expended to model this dependency.

One difficulty in measuring the level of inbreeding depends on the array of genotypes found at a locus. Consider that different levels and types of inbreeding exist for the following genotypes where two genes with the same subscript are identical by descent: $A_i A_i A_i A_i$, $A_i A_i A_i A_j$, $A_i A_i A_j A_k$, and $A_i A_i A_j A_j$. This problem of inbreeding level is partially alleviated by relying on a coefficient of kinship rather than a coefficient of inbreeding.

Busbice (14) developed a theory of the evolution of inbreeding of autotetraploids and used it to formulate prediction equations for yield of synthetic varieties in advancing generations (15, 16). Prediction of yields of synthetics was based on a weighting of the average yield in selfed progenies of parent clones and yields of all possible crosses in a diallel involving all parents. The yields were determined over multiple years and locations.

Tests of Busbice's predictive formulae (16, 82) indicated that they were surprisingly accurate, but costs restricted the measure of accuracy to the nine clones of 'Cherokee'. The accuracy of Busbice's predictions are dependent upon equal gametic production and complete random intercrossing of all parents and progenies in each generation, a very unrealistic assumption (81). Also, the $n^2 - n$ crosses of the diallel must be representative of the near infinite combinations of crosses occurring in the synthetic approaching RME. This is a difficult requirement to attain when parents of synthetics are selected for genetic diversity in an attempt to maximize heterozygosity.

REFERENCES

1. Allard, R.W. 1960. Principles of plant breeding. John Wiley and Sons, New York.
2. Aycock, M.K., and C.P. Wilsie. 1968. Inbreeding Medicago sativa L. by sib-mating. II. Agronomic traits. Crop Sci. 8:481–485.
3. Barnes, D.K., E.T. Bingham, J.D. Axtell, and W.H. Davis. 1972. The flower, sterility mechanisms, and pollination control. Agronomy 15:123–141.
4. ----, ----, R.P. Murphy, O.J. Hunt, D.F. Beard, W.H. Skrdla, and L.R. Teuber. 1977. Alfalfa germplasm in the United States: Genetic vulnerability, use, improvement, and maintenance. USDA Tech. Bull. 1571. U.S. Government Printing Office, Washington, DC.
5. Bingham, E.T. 1979. Maximizing heterozygosity in autopolyploids. p. 471–489. In W.H. Lewis (ed.) Polyploidy: Biological relevance. Plenum Press, New York.
6. ----. 1983. Maximizing hybrid vigor in autotetraploid alfalfa. p. 130–143. In Better Crops for food. CIBA Found. Symp. 97. Pitman Books, London.
7. ----. 1983. Molecular genetic engineering vs. plant breeding. Plant Mol. Biol. 2:221–228.
8. Bolton, J.L. 1962. Alfalfa botany, cultivation and utilization. World Crops Books, Leonard Hill, London.
9. Brown, D.E., W.R. Kehr, and G.R. Manglitz. 1981. Isolation requirements for foundation alfalfa seed fields. Crop Sci. 21:628–629.
10. Bulmer, M.G. 1980. The mathematical theory of quantitative genetics. Oxford University Press, Oxford, UK.
11. Burton, G.W. 1974. Recurrent restricted phenotypic selection increases forage yields of Pensacola bahiagrass. Crop Sci. 14:831–835.

12. ----. 1982. Improved recurrent restricted phenotypic selection increases bahiagrass forage yields. Crop Sci. 22:1058-1061.
13. Busbice, T.H. 1968. Effects of inbreeding on fertility in *Medicago sativa* L. Crop Sci. 8:231-234.
14. ----. 1969. Inbreeding in synthetic varieties. Crop Sci. 9:601-604.
15. ----. 1970. Predicting yield of synthetic varieties. Crop Sci. 10:265-269.
16. ----, and R.Y. Gurgis. 1976. Evaluating parents and predicting performance of synthetic alfalfa varieties. USDA ARS-S-130. U.S. Government Printing Office, Washington, DC.
17. ----, R.R. Hill, Jr., and H.L. Carnahan. 1972. Genetics and breeding procedures. *In* C.H. Hanson (ed.) Alfalfa science and technology. Agronomy 15:283-318.
18. ----, O.J. Hunt, J.H. Elgin, Jr., and R.N. Peaden. 1974. Evaluation of the effectiveness of polycross- and self-progeny tests in increasing the yield of alfalfa synthetic varieties. Crop Sci. 14:8-11.
19. ----, and J.O. Rawlings. 1974. Combining ability in crosses within and between diverse groups of alfalfa introductions. Euphytica 23:86-94.
20. ----, and C.P. Wilsie. 1966. Inbreeding depression and heterosis in autotetraploids with application to *Medicago sativa* L. Euphytica 15:52-67.
21. Childers, W.R., and D.K. Barnes. 1972. Evolution of hybrid alfalfa. Agric. Sci. Rev. 10:11-18.
22. Compton, W.A., and R.E. Comstock. 1976. More on modified ear-to-row selection in corn. Crop Sci. 16:122.
23. Comstock, R.E., H.F. Robinson, and P.H. Harvey. 1949. A breeding procedure designed to make maximum use of both general and specific combining ability. Agron. J. 41:360-367.
24. Crow, J.F., and M. Kimura. 1970. An introduction to population genetics theory. Harper and Row, New York.
25. Dudley, J.W., R.R. Hill, Jr., and C.H. Hanson. 1963. Effects of seven cycles of recurrent phenotypic selection on means and genetic variances of several characters in two pools of alfalfa germplasm. Crop Sci. 3:543-546.
26. Elgin, J.H., Jr., R.R. Hill, Jr., and K.E. Zeiders. 1970. Comparison of four methods of multiple trait selection for five traits in alfalfa. Crop Sci. 10:190-193.
27. ----, J.E. McMurtrey, III, B.J. Hartman, B.D. Thyr, E.L. Sorensen, D.K. Barnes, F.I. Frosheiser, R.N. Peaden, R.R. Hill, Jr., and K.T. Leath. 1983. Use of strain crosses in the development of multiple pest resistant alfalfa with improved field performance. Crop Sci. 23:57-64.
28. Empig, L.T., C.O. Gardner, and W.A. Compton. 1972. Theoretical gains for different population improvement procedures Nebr. Agric. Exp. Stn. Bull. M26. Revised ed.
29. Falconer, D.S. 1982. Introduction to quantitative genetics. Longman, New York.
30. Fryer, J.R. 1939. The maternal-line selection method of breeding for increased seed-setting in alfalfa. Sci. Agric. (Ottawa) 20:131-139.
31. Gallais, A. 1967. Moyennes des populations tetraploids. Ann. Amelior. Plant. 17:215-227.
32. Gardner, C.O. 1961. An evaluation of effects of mass selection and seed irradiation with thermal neutrons on yield of corn. Crop Sci. 241-245.
33. ----. 1969. The role of mass selection and mutagenic treatment in modern corn breeding. *In* H.D. Loden and D. Wilkinson (ed.) Proc. 24th Annu. Corn Sorghum Res. Conf., 24th, Chicago. December. American Seed Trade Association, Washington, DC.
34. Griffing, B. 1960. Theoretical consequences of truncation selection based on the individual phenotype. Aust. J. Biol. Sci. 13:307-343.
35. Haldane, J.B.S. 1930. Theoretical genetics of autotetraploids. J. Genet. 22:349-372.
36. Hallauer, A.R., and J.B. Miranda Fo. 1981. Quantitative genetics in maize breeding. Iowa State University Press, Ames.
37. Hanson, A.A., and H.L. Carnahan. 1956. Breeding perennial forage grasses. USDA Tech. Bull. 1145. U.S. Government Printing Office, Washington, DC.
38. Hanson, C.H., T.H. Busbice, R.R. Hill, Jr., O.J. Hunt, and A.J. Oakes. 1972. Directed mass selection for developing multiple pest resistance and conserving germplasm of alfalfa. J. Environ. Qual. 1:106-111.
39. ----, H.O. Graumann, W.R. Kehr, H.L. Carnahan, R.L. Davis, J.W. Dudley, L.J. Elling, C.C. Lowe, D. Smith, E.L. Sorensen, and C.P. Wilsie. 1964. Relative performance of alfalfa varieties, variety crosses, and mixtures. USDA-ARS Production Res. Rep. 83. U.S. Government Printing Office, Washington, DC.
40. Harris, D.L. 1963. The influence of errors of parameter estimation upon index selection. p. 491-500. *In* W.D. Hanson and H.F. Robinson (ed.) Statistical genetics and plant breeding. NRC Pub. 982. National Academy of Science, Washington, DC.
41. ----. 1964. Expected progress from index selection involving estimates of population parameters. Biometrics 20:46-72.

42. Hayes, H.K., and R.J. Garber. 1919. Synthetic production of high protein corn in relation to breeding. J. Am. Soc. Agron. 11:309-318.
43. Hazel, L.N. 1943. The genetic basis for constructing selection indexes. Genetics 28:476-490.
44. ----, and J.L. Lush. 1942. The efficiency of three methods of selection. J. Hered. 33:393-399.
45. Hill, R.R., Jr. 1971. Effect of the number of parents on the mean and variance of synthetic varieties. Crop Sci. 11:283-286.
46. ----. 1975. Parental inbreeding and performance of alfalfa single-crosses. Crop Sci. 15:373-375.
47. ----. 1976. Response to inbreeding in alfalfa populations derived from single clones. Crop Sci. 16:237-241.
48. ----. 1983. Heterosis in population crosses of alfalfa. Crop Sci. 23:48-50.
49. ----, and R.A. Byers. 1979. Allocation of resources in selection for resistance to alfalfa blotch leafminer in alfalfa. Crop Sci. 19:253-257.
50. ----, and J.H. Elgin, Jr. 1981. Effect of number of parents on performance of alfalfa synthetics. Crop Sci. 21:298-300.
51. ----, and ----. 1985. Analysis of pest resistance in alfalfa with a new autotetraploid genetic model. Can. J. Genet. Cytol. 27:39-46.
52. ----, K.T. Leath, and K.E. Zeiders. 1972. Combining ability among four-clone alfalfa synthetics. Crop Sci. 12:627-630.
53. Hill, W.G. 1971. Design and efficiency of selection experiments for estimating genetic parameters. Biometrics 27:293-311.
54. ----. 1972. Estimation of realized heritabilities from selection experiments. I. Divergent selection. Biometrics 28:747-765.
55. ----. 1972. Estimation of realized heritabilities from selection experiments. II. Selection in one direction. Biometrics 28:767-780.
56. Hunt, O.J., D.K. Barnes, R.N. Peaden, and B.J. Hartman. 1976. Inheritance of black seed character and its use in alfalfa improvement. p. 33. In D.K. Barnes (sec.) Rep. 25th Alfalfa Improve. Conf. Cornell Univ., Ithaca, NY. 13-15 July. USDA-ARS-NC-52, USDA-SEA, Peoria, IL
57. Jacquard, A. 1983. Heritability: One word, three concepts. Biometrics 39:465-477.
58. Jenkin, T.J. 1931. The method and techniques of selection, breeding and strain-building in grasses. Imp. Bur. Plant Genet.: Herb. Plant Bull. 3:5-34.
59. Kehr, W.R., D.K. Barnes, D.E. Brown, J.H. Elgin, Jr., and E.L. Sorensen. 1983. Seed yields from breeder and foundation seed of eight alfalfa cultivars. Crop Sci. 23:256-258.
60. ----, H.O. Gaumann, C.C. Lowe, and C.O. Gardner. 1961. The performance of alfalfa synthetics in the first and advanced generations. Nebr. Agric. Exp. Stn. Bull. 200.
61. Kempthorne, O. 1955. The correlation between relatives in a simple autotetraploid population. Genetics 40:168-174.
62. ----. 1955. The theoretical values of correlations between relatives in random mating populations. Genetics 40:153-167.
63. ----. 1957. An introduction to genetic statistics. John Wiley and Sons, New York.
64. Kirk, L.E. 1932. Further contributions on the technique employed in the breeding of herbage and forage plants. Imp. Bur. Plant Genet., Herb. Plants Bull. 7:1-32.
65. Lesins, K.A., and I. Lesins. 1979. Genus *Medicago* (Leguminosae): A taxogenetic study. W. Junk Publ., The Hague.
66. Levings, D.S., III, and J.W. Dudley. 1963. Evaluation of certain mating designs for estimation of genetic variance in autotetraploid alfalfa. Crop Sci. 3:532-535.
67. Lonnquist, J.H. 1964. A modification of the ear-to-row procedures for the improvement of maize populations. Crop Sci. 4:227-228.
68. Lorenz, R.J., R.E. Ries, C.S. Cooper, C.E. Townsend, and M.D. Rumbaugh. 1982. Alfalfa for dryland grazing. USDA Agric. Info. Bull. 444. U.S. Government Printing Office, Washington, DC.
69. Lowe, C.C., V.L. Marble, and M.D. Rumbaugh. 1972. Adaptation, varieties, and uses. In C.H. Hanson (ed.) Alfalfa science and technology. Agronomy 15:391-413.
70. ----, R.W. Cleveland, and R.R. Hill, Jr. 1974. Variety synthesis in alfalfa. Crop Sci. 14:321-325.
71. Lush, J.L. 1945. Animal breeding plans. Iowa State University Press, Ames.
72. Miller, M.K., M.H. Schonhorst, and R.G. McDaniel. 1972. Identification of hybrids from alfalfa crosses by electrophoresis of single seed proteins. Crop Sci. 12:535-537.
73. Murphy, R.P., and C.C. Lowe. 1966. Registration of Saranac alfalfa. Crop Sci. 6:611.
74. ----, and ----. 1968. Registration of 'Iroquois' alfalfa. Crop Sci. 8:396.
75. Pedersen, M.W., and D.K. Barnes. 1973. Alfalfa seed size as an indicator of hybridity. Crop Sci. 13:72-75.
76. ----, and ----. 1973. Alfalfa pollen production in relation to percentage of hybrid seed produced. Crop Sci. 13:652-656.

77. Pirchner, F. 1969. Population genetics in animal breeding. W.H. Freeman and Co., San Francisco.
78. Rice, J.S., and J.W. Dudley. 1974. Gene effects responsible for inbreeding depression in autotetraploid maize. Crop Sci. 14:390–393.
79. ----, and ----. 1983. Comparison of genetic models for inbreeding in autotetraploids using maize data. Crop Sci. 23:651–654.
80. Rowe, D.E. 1982. Effects of gametic disequilibrium on selection in an autotetraploid population. Theor. Appl. Genet. 64:69–74.
81. ----. 1985. Variability among alfalfa clones in seed production. I. Effective population size. Crop Sci. 25:611–614.
82. ----, and R.Y. Gurgis. 1982. Evaluation of alfalfa synthetic varieties: Prediction of yield in advanced generations and average clone effects. Crop Sci. 22:868–871.
83. ----, and R.R. Hill, Jr. 1981. Inter-population improvement procedures for alfalfa. Crop Sci. 21:392–397.
84. ----, and ----. 1984. Effect of gametic disequilibrium on means and on genetic variances of autotetraploid synthetic varieties. Theor. Appl. Genet. 68:69–74.
85. ----, and ----. 1984. Theoretical improvement of autotetraploid crops: Interpopulation and intrapopulation selection. USDA Tech. Bull. 1689. U.S. Government Printing Office, Washington, DC.
86. Sales, J., and W.G. Hill. 1976. Effect of sampling error on efficiency of selection indices. Anim. Prod. 22:1–17; 23:1–14.
87. Sherwood, R.T., J.W. Dudley, T.H. Busbice, and C.H. Hanson. 1967. Breeding alfalfa for resistance to the stem nematode, *Ditylenchus dipsaci*. Crop Sci. 7:382–384.
88. Simmonds, N.W. 1981. Principles of crop improvement. Longman Group Limited, London.
89. Smith, H.F. 1936. A discriminant function for plant selection. Ann. Eugen. (London) 7:240–250.
90. Sprague, G.F. 1966. Quantitative genetics in plant improvement. p. 315–354. *In* K.J. Frey (ed.) Plant breeding. Iowa State University Press, Ames.
91. Standord, E.H. 1952. Transfer of disease resistance to standard varieties. Proc. Int. Grassl. Congr., 6th 2:1585–1590.
92. ----, W.M. Clement, Jr., and E.T. Bingham. 1972. Cytology and evolution of the *Medicago sativa-falcata* complex. Agronomy 15:87–102.
93. ----, and E.R. Houston. 1954. The backcross technic as a method of breeding alfalfa. p. 44–45. *In* Rep. 14th Alfalfa Improve. Conf., Davis, CA. 3–7 August. U.S. Department of Agriculture, Washington, DC.
94. Tysdal, H.M., and B.H. Crandall. 1948. The polycross progeny performance as an index of the combining ability of alfalfa clones. J. Am. Soc. Agron. 40:293–306.
95. ----, T.A. Kiesselbach, and H.L. Westover. 1942. Alfalfa breeding. Nebr. Agric. Exp. Stn. Res. Bull. 124.
96. Verhalen, L.M., J.L. Baker, and R.W. McNew. 1975. Gardner's grid system and plant selection efficiency in cotton. Crop Sci. 15:588–591.
97. Williams, J.S. 1962. The evaluation of a selection index. Biometrics 18:375–393.
98. Williams, R.D. 1931. Self-fertility in lucerne. Welsh Plant Breed. Stn. Bull. Ser. H. 12:217–220.
99. Wilsie, C.P. 1958. Effect of inbreeding on fertility and vigor of alfalfa. Agron. J. 50:182–185.
100. Wood, D.R. (ed.) 1983. Glossary of terms. p. 271–279. *In* Crop breeding. American Society of Agronomy and Crop Science Society of America, Madison, WI.
101. Young, S.S.Y. 1961. A further examination of the relative efficiency of three methods of selection for genetic gains under less-restricted conditions. Genet. Res. 2:106–121.

26 Breeding for Yield and Quality

R. R. HILL, JR.

USDA-ARS
University Park, Pennsylvania
University Park, Pennsylvania

J. S. SHENK

Pennsylvania State University
University Park, Pennsylvania

R. F BARNES

American Society of Agronomy
Madison, Wisconsin

Alfalfa (*Medicago sativa* L.) has the capacity to produce high yields of high quality forage. Considerable scientific effort has been and is being devoted annually to improvement of both yield and quality of alfalfa. The approaches are varied, depending on location, resources available to the breeder, and the specific objectives of the breeding program.

Yield and quality can be influenced by genetic and nongenetic factors. The effects of harvest frequency, soil fertility, method of harvest and preservation, and the presence or absence of disease and insect pests, and other environmental factors on alfalfa yield and quality are covered elsewhere in this monograph. We will restrict our discussion to those aspects of yield and quality that are related to breeding.

26–1 YIELD

Alfalfa forage yields have increased dramatically since 1940, and breeding has been directly responsible for an important portion of this increase. Progress in improving resistance to disease and insect pests, adaptation to specific environments, and tolerance to frequent harvests, have made substantial contributions to the yield performance of alfalfa cultivars. This discussion will be restricted to breeding for increased yield per se.

26–1.1 Recent Trends in Breeding for Yield

The genetic increases in alfalfa yields have been small compared with those realized in most grain crops. Hill and Kalton (38) estimated that

---

Copyright 1988 © ASA CSSA-SSSA, 677 South Segoe Road, Madison, WI 53711, USA.
Alfalfa and Alfalfa Improvement—Agronomy Monograph no. 29.

total genetic yield improvement in alfalfa between 1956 and 1974 was about 3%. A similar rate of improvement was estimated by Elliott et al. (14). We used data from trials reported by the Central Alfalfa Improvement Conference to update the estimated progress in breeding for improved yield (Table 26-1). These results indicate that progress in increasing alfalfa yields in the central region have been greater since 1971 than during the time period included in the survey of Hill and Kalton (38). Data from Hill et al. (40) were used to plot yields in percent of 'Vernal' vs. year of release of new cultivars (Fig. 26-1). The increase in yield between 1971 and 1981 were erratic and no method of estimating genetic increases in yield showed more than a fraction of a percentage gain each year. The increases are cumulative, however, and over a period of time, the increases result in significant improvements in yield.

There are a number of reasons for the slower rate of progress in increasing yields with alfalfa than with the grain crops. One of the most obvious is the perennial growth habit of alfalfa. Factors that affect winter survival and the storage of photosynthates for production of the next crop must be considered in the breeding program. In addition, experimental strains must be evaluated for several years before decisions can be made in selection programs. Thus, equivalent gains per cycle of se-

Table 26-1. Yields of newer alfalfa cultivars in percent of Vernal.

Region, USA	Year	
	1975†	1983
North Central	103	109
Northeastern	103	--

† Adapted from Hill and Kalton (38).

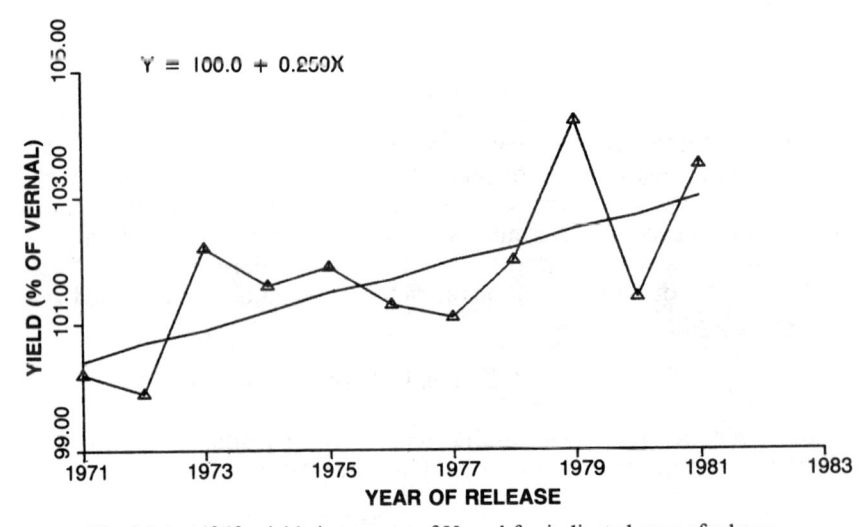

Fig. 26-1. Alfalfa yields in percent of Vernal for indicated year of release.

lection for alfalfa and an annual crop could translate into significantly lower gains per year for alfalfa.

Although forages were not mentioned explicitly, Evans (15) offered another reason why genetic gains for forage yield may be lower than those for grain yield. He reported that a large portion of the yield increase in many crops had been achieved by altering regulatory processes rather than by altering assimilatory processes. Thus, breeders have increased the proportion of plant assimilates going into the desired plant organs (seed in many crops) without increasing total plant growth. Because the total plant is used for forage, alfalfa breeders have not had the opportunity or the need to extensively modify regulatory processes, and yield increases must be achieved by altering assimilatory processes.

Perhaps another reason for slower progress in breeding for yield in alfalfa is the diversion of breeding effort from yield to pest resistance. Dramatic increases in yield can be observed when resistant and susceptible cultivars are grown in environments that favor pest development. Pest resistance does not result in an increase in yield, however, when the crop is grown in the absence of the pests for which resistance has been obtained. Breeding for pest resistance is an essential and effective aspect of alfalfa germplasm improvement, but it does not necessarily increase the frequency of genes for yield per se in populations under improvement.

Breeding methods for pest resistance, discussed in chapters 27 and 28, are not easily adapted to include selection for yield. More than half the alfalfa cultivars released since 1973 have had more than 40 parents (Table 26-2), and the association between number of parents and time period is not independent of other factors. In most situations, the broadly

Table 26-2. Number of parents per alfalfa synthetic by time period.†

Time period	1-3	4-8	9-16	17-40	>40	Total
Before 1963	1	1	2	3	0	7
	0.1‡	1.2	1.5	1.5	2.7	
1963-1978	1	14	8	6	4	33
	0.6	5.6	7.2	7.0	12.6	
1969-1972	1	2	7	8	8	26
	0.6	4.3	5.5	5.5	10.2	
1973-1978	0	10	14	11	23	58
	1.0	9.8	12.6	12.2	22.4	
1979-1982	0	1	5	7	29	42
	0.8	7.1	9.1	8.9	16.1	
Totals	3	28	36	35	64	166

† Data obtained by summarizing past reports of the Certified Alfalfa Variety Review Board and includes only those varieties that had descriptions that gave the number of parents per variety.
‡ Numbers on second line indicate expected number of synthetics in the cell if number of parents per synthetic and time periods were independent. Chi-square for independence = 54.6*, significant at the 0.05 probability level.

based synthetics released in recent years represent the end product of a selection program for pest resistance in populations known to have acceptable yield potential. A large number of parents are used to minimize changes in gene frequencies for traits other than the one considered in the pest resistance selection program. At first glance, it might seem reasonable to suggest that narrow based synthetics should be produced by selection for yield within pest resistant germplasm. Obviously, the task of finding 10 or fewer clones with superior yield potential would be easier than the task of finding 40 or more superior parents. As the discussion below will show, however, there are very good reasons for the use of broad based synthetics in alfalfa breeding, and a suitable alternative to their use has not yet been found.

Although increases in yield of alfalfa have been less than those in the grain crops, progress is being made in genetic improvements in alfalfa yield. There is abundant genetic variability within and among alfalfa cultivars, and there is no indication that we are close to the limits that can be achieved in breeding for increased yield. There is every reason to believe that alfalfa yields will continue to increase, and that the rate of increase may be greater in the future than it has been in the past.

26–1.2 Gene Action for Yield

Gene action for yield in alfalfa is not as well understood as it is for maize (*Zea mays* L.), although many of the breeding methods used with alfalfa were adapted from maize breeding methodology. Bingham (5) recently reviewed theoretical aspects of heterosis in autotetraploids such as alfalfa. One of the more popular theories for an explanation of heterosis in alfalfa can be traced back to the work of Demarly (12) and Busbice and Wilsie (9). The theory requires that multiple alleles exist and that the relative vigor ranking from the most to the least vigorous is tetraallelic ($a_i a_j a_k a_l$) > triallelic ($a_i a_i a_j a_k$) > diallelic duplex ($a_i a_i a_j a_j$) > diallelic simplex ($a_i a_i a_i a_j$) > monoallelic ($a_i a_i a_i a_i$), where a_i represents an allele and different subscripts indicate different alleles of a multiple allele series. The assumed relative vigor of the genotypes is inversely proportional to the level of inbreeding of the individual genotypes (6).

The above model for heterosis agrees with many of the observations that have been made in alfalfa breeding. The approach to homozygosity in autotetraploids is much slower than in diploids (21), but the observed inbreeding depression for yield follows a curve similar to that for maize (60). Tetra- and triallelic types disappear rapidly with inbreeding in autotetraploids, and if these were the most vigorous types, the inbreeding depression would be much greater than would be expected on the basis of a simple approach to homozygosity (6, 9). Synthetics with a smaller number of parents would have a greater proportion of di- and monoallelic genotypes as random mating equilibrium is approached, and would thus have poorer performance than synthetics with larger numbers of parents; a prediction that has been observed experimentally (7, 35). Inbred parents would transmit a higher frequency of homozygous gametes and would thus have a greater frequency of di- and monoallelic types than progeny of noninbred parents. This is substantiated by the poor performance of crosses between inbred parents that has been observed in alfalfa (6, 28, 32).

The model for heterosis in alfalfa is similar in many aspects to the early models for maize that assumed large levels of overdominance. It has been shown that heterosis in maize can be explained by linkage and gene action no more complicated than partial to compelte dominance (56). The true model for alfalfa may eventually be shown to be due to linked, dominant, or partially dominant alleles. Demarly (12) proposed that the "alleles" may actually be linkage blocks that he called "linkats." Also, Bingham (5) suggested that the association bewteen heterosis and heterozygosity in alfalfa may be the result of linked alleles. The multiple allele model can be formulated with only two linked alleles (Table 26-3). The number of alleles becomes limitless when the entire genome of alfalfa is considered.

Even if gene action for yield can eventually be shown to be the result of linked, dominant, or partially dominant alleles, the implications for alfalfa breeders would be very different from that for maize breeders. The ideal cultivar in a diploid such as maize would be one that contained a specific pair, $a_i a_j$, of alleles or chromosome segments in each individual in the cultivar. An analogous situation for an autotetraploid such as alfalfa would be a cultivar in which each individual was tetraallelic, $(a_i a_j a_k a_l$, for a specific set of four alleles or chromosome segments. The easiest way to obtain the ideal cultivar in diploids is the production of hybrids between inbred lines. Production of the ideal cultivar for an autotetraploid

Table 26-3. Relationships between multiple alleles and chromosome segments for two linked alleles in an autotetraploid.

Genotype	Genotype symbol	Chromosome arrangement
Tetraallelic	$a_1 a_2 a_3 a_4$	A | A | a | a B | b | B | b
Triallelic	$a_1 a_1 a_2 a_3$	A | A | A | a B | B | b | B
Diallelic duplex	$a_1 a_1 a_2 a_2$	A | A | A | A B | B | b | b
Diallelic simplex	$a_1 a_1 a_1 a_2$	A | A | A | A B | B | B | b
Monoallelic	$a_1 a_1 a_1 a_1$	A | A | A | A B | B | B | B

such as alfalfa is impossible by sexual means with current genetic tecynology (17, 25). Broad based synthetics produce a high level of heterozygosity because they contain a high proportion of a very large number of tetra- and triallelic types (25). We know from variability that exists within most alfalfa cltivars that a subsample of plants with yield greater than the average of the original population must exist. The subsample of better performing plants would probably contain a specific set of tetra- and triallelic genotypes, and that subsample cannot be produced sexually without also producing a high frequency of di- and monoallelic genotypes. With current genetic technologies, alfalfa breeders are faced with a dilemma; breeding methods that produce low frequencies of di- and monoallelic genotypes make no use of any advantage in yield gains that potentially could come from specific individual or small sets of tetra- and triallelic types; and breeding methods that produce specific individual or small sets of tetra- and triallelic types produce high frequencies of di- and monoallelic types. Current breeding methods for yield in alfalfa make practically no use of specific combining ability that results from nonadditive gene interaction between specific sets of alleles (26, 29).

Current theory relative to the expression of heterosis in alfalfa synthetics is based on assumptions that random mating occurs during generations of seed increase, and that all genotypes have an equal probability of survival in the field. In actual practice, it is likely that neither of these assumptions hold. It is known that alfalfa genotypes differ in their seed producing ability and the actual genotypic array obtained in the field differs substantially from theoretical expectations. In addition, only a small portion of the seed sown in an alfalfa field survive beyond the first year of the stand, and it is unlikely that the elimination of seedlings is completely random. Research to gain a better understanding of the effects of "natural" selection that occurs during seed increase and stand establishment may provide important insights to breeding for increased yield.

Trends since 1963 indicate that most alfalfa breeders have made a conscious decision to limit their efforts to broad based synthetics (Table 26-2). There are few acceptable alternatives. Both theoretical (6, 26) and experimental (7, 32, 35) results show that even a small amount of inbreeding cannot be tolerated in commercial alfalfa cultivars. If heterosis in alfalfa, like that in maize, is due to chromosome segments of linked genes, then considerable progress could probably be made through selection to increase the frequencies of the more desirable chromosome segments. Recent research by Pfeiffer and Bingham (51) indicate that increases in yield can be obtained by selection for rearrangements of genes on chromosomes, and that heterosis in alfalfa may indeed involve the effects of linked genes. More research is needed to develop effective methods of increasing frequencies of chromosome segments (as opposed to increasing frequencies of favorable alleles) in alfalfa populations.

Unconventional reproductive methods potentially offer a means of producing alfalfa populations heterozygous for a single or small set of tetraallelic genotypes. This subject has been discssed briefly by Bingham (5). Details are provided in Chapter 29 in this book.

26-1.3 Genotype × Environment Interactions

Most alfalfa cultivars are developed for use over broad geographic areas, where they are subject of a wide array of environmental conditions. Consequently, genotype × environment interactions pose problems in breeding and evalation. This is further complicated by the fact that the cultural conditions and generation tested by the breeder may differ from the cultural practices and generation used commercially (progeny tests or early generation synthetics in space plant or drilled row plots vs. broadcast stands).

Busbice et al. (8) compared yields of experimental synthetics with parents selected on the basis of polycross- and selfed-progeny test performance. Progeny test evaluations were conducted at a single location, and experimental synthetics derived from selected parents were tested in a wide range of environments. Neither method of selecting parents produced an experimental synthetic with consistently superior performance over a range of environments. Busbice et al. (8) attributed the poor response of the two selection procedures to genotype × environment interactions. The interactions could have resulted from two sources: (i) different cultural conditions for the progeny (single row plots) and synthetics (broadcast plots), and (ii) interactions with location effects that could not be detected in the evaluation of progeny at a single location. Selection based on progeny tests conducted at several locations could possibly reduce the impact of the second type of interaction. Production of enough seed for a multilocation progeny test would be difficult or impossible if progeny evaluations were conducted in broadcast plots, and drastic departures from broadcast plots could accentuate discrepancies associated with the first type of interaction mentioned above. Second generation progeny tests (31) could be conducted at several locations, but this method of breeding for yield has not been fully evaluated.

Methods for analyzing the stability of cultivars over a range of environments have been developed for annual crop species (13, 16). Although these methods have been applied to perennial crops, they may not be completely appropriate because year effects are often confounded with age of stand effects (34). A method of predicting environmental responses for one generation (the certified seed generation) on the basis of measurements made on an earlier generation (progeny tests or early generation synthetics) is needed for alfalfa breeding.

Breeding for multiple pest resistance may be a method for increasing stability of afalfa cultivars over a range of environments just as it is an indirect method of increasing yield. Hill and Baylor (34) found that moderate or higher levels of resistance to anthracnose (*Colletotrichum trifolii* Bain.), and Fusarium wilt [*Fusarium oxysporum* Schl. f. *medicaginis* (Wcimer) Sny. & Hans.] was needed for stable performance over two locations, 3 yr, and two management programs. Cultivars with no resistance to a particular pathogen would eventually show poor performance in those environments that favored the pathogen. Multiple pest resistance is not, however, the complete solution to cultivar stability over a range of environments. Hill and Baylor (34) observed that some cultivars with multiple pest resistance were not stable over the environments included in their study.

The fact that maize double crosses had smaller genotype × environment interactions than did single crosses (57) indicates that broad based cultivars are more stable than narrow based ones over a range of environments. The same phenomenon has been observed with alfalfa synthetics (35, 48). The greater stability of broad based synthetics over a range of environments is another reason for their popularity in breeding for yield in alfalfa.

26–2 QUALITY

Although alfalfa is one of the highest quality forages available to livestock feeders, it is not a perfect forage and breeding is devoted to improvement of quality. The ultimate goal of breeding to enhance forage quality is the improvement of animal performance. Breeding programs are based on alteration of chemical constituents known to affect animal performance because animal evaluation on a large number of experimental lines or families would be prohibitively expensive.

26–2.1 Selection of Chemical Traits

A large number of chemical constituents in forages are known to affect animal performance (3, 61). Although not all of these chemical constituents are important in alfalfa, a very large number of them are potential candidates in programs devoted to improving alfalfa quality. The constituent chosen for study should have the following properties:
1. Be relatively easy to measure so that large numbers of genotypes, lines, or families can be evaluated.
2. Be heritable enough to permit response to selection.
3. Be strongly correlated with animal performance so that successful change in the chemical constituent will result in improved animal performance. This strong relationship between the chemical constituent and animal performance should exist within alfalfa. A strong relationship over different species is not necessarily an indication that successful selection within a given species will result in improved animal performance.

Before the advent of near infrared reflectance spectroscopy (NIRS) for forage analysis (55), alfalfa breeders were limited in their choice of constituents on which to base a forage quality improvement program. The labor and technical support required for wet-laboratory analyses often dictated that only one chemical constituent could be considered for inclusion in the selection program. Conversely, estimates of the concentration of several chemical constituents can be obtained with NIRS analysis for less effort than that required for a single constituent in the wet-laboratory (53). The two major disadvantages of NIRS analysis are, cost of equipment and difficulties in instrument calibration. Despite these disadvantages, NIRS adds many forage constituents to the list of those that meet the first criterion given above. The greater accuracy of NIRS may increase the list of constituents that meet the second criterion. The capacity to increase the number of replications and to use several consstituents in a selection index would be improved with access to NIRS forage analysis.

Regardless of the method of forage analysis, all constituents included in a selection program should be identified with care. Cooper (11) divided plant breeding objectives for improving forage quality into the four broad groups that are discussed in the following sections.

26-2.1.1 Constituents Related to Energy Content and Digestibility

The most popular traits in this category have been in vitro dry matter disappearance (IVDMD) and any of several mesures of fiber. The objectives in an alfalfa breeding program would involve increasing IVDMD or decresing fiber concentration. Alfalfa has a greater lignin concentration than most forage grasses, and reduction of lignin concentration has often been expressed as a valid goal for breeders interested in improving alfalfa forage quality.

More breeding effort has been devoted to alteration of IVDMD than other traits in this category. Both high (10, 54) and low (27, 33, 45) estimates of heritability for IVDMD in alfalfa have been reported. The conflicting results may reflect different germplasm pools or differences in laboratories, and they offer little in the way of specific guidelines on the prospects of altering IVDMD through breeding. However, results obtained by Thomas et al. (59) would suggest that IVDMD is a valid trait for breeders to use in improving alfalfa forage quality. In these studies, selection for high and low IVDMD in alfalfa did result in increased and decreased digestibility, respectively, when forage from geneticaly altered populations was fed to sheep (*Ovis aries*).

Shenk (53) reported that NIRS analysis was considerably more accurate than wet-laboratory methods for measuring IVDMD, and heritability would be expected to be higher with the use of NIRS. Alfalfa breeders should ensure that factors responsible for low heritability have been corrected before initiating a large selection program to improve digestibility with IVDMD. Heritability can be increased by restricting the environment in which genetic materials are evaluated. This could not be recommended until it is demonstrated that selections made in the restricted environment maintain their superiority over a range of environments.

Heritability for acid detergent fiber (ADF) and lignin concentration in two alfalfa populations was three to five times greater than that for IVDMD (33). Other researchers have reported genetic variability for ADF, lignin, or one of the other meausres of forage fiber in alfalfa (2, 10, 45, 54).

A selection program for high and low lignin concentration in alfalfa resulted in a negative correlated response for forage yield (30). High lignin selections had a significantly greater forage yield than did low lignin selections. Further evaluation of lines from this selection program indicated that much of the observed response was due to a genetic alteration of the leaf-to-stem ratio (46). The high lignin lines had a lower leaf-to-stem ratio that gave greater yields and greater lignin concentration. Similar results were reported for lignin in alfalfa by Shenk and Elliott (54).

Sumberg et al. (58) successfully selected for high and low ADF in alfalfa. Experimental synthetics with significantly different ADF concentration were produced from parents selected on the basis of a weighted average ADF concentration over the three harvests of a growing season. Only those plants with high or low ADF concentration, but which did not differ from the population mean for yield, were selected. Yields of the resulting synthetics were not significantly different, a result that was different from that reported by Hill (30). The results to date indicate that selection for higher IVDMD or for lower fiber concentration should not be initiated in the absence of procedures that protect against a correlated decrease in forage yield.

26–2.1.2 Toxic or Harmful Constituents

Elliott et al. (14) discussed a number of alfalfa forage constituents in this category. The use of the forage must be known before an accurate assessment of toxic or harmful constituents can be made. Many of the "harmful" constituents are not serious problems when forage is fed to ruminants, but might be toxic if fed to monogastric animals.

Bloat can be a serious problem in animals on pastures that contain alfalfa and certain other legumes. Foam production associated with soluble proteins in forage is thought to be the cause of bloat (41), although many researchers recognize that other factors may be involved. Several approaches to solving the bloat problem have been proposed and/or attempted.

Bloat-safe legumes have tannins that combine with proteins to form insoluble complexes, whereas bloat-causing legumes have little or no tannin (43, 44). An exception to this generalization is cicer milkvetch (*Astragalus cicer* L.), a bloat-safe legume, and one in which condensed tannins could not be found (52). A search of a number of *Medicagos* revealed no entries with condensed tannins (20). The apparent absence of genetic variability for tannin concentration makes breeding for altered tannin concentration impossible. If variability for tannin concentration can be found or created, the effect on bloat incidence is uncertain. Alfalfa with a high tannin concentration would probably be less palatable than existing germplasm. Although lower palatability is generally considered undesirable, a less palatable alfalfa may be preferred in afalfa-grass pastures.

Resistance to cell rupture is greater in bloat-safe than in bloat-causing legumes (41). A glass bead method of measuring susceptibility to cell rupture demonstrated that cell walls are stronger in bloat-safe legumes (47). A nylon bag technique was used to demonstrate that cell wall destruction rates by rumen microorganisms was greater for alfalfa than for the bloat-safe legumes (42). No differences were found in cell wall destruction rate for six alfalfa cultivars, or for the high and low saponin strains from these six cultivars, but differences were found among the 11 parental clones of 'Beaver' alfalfa.

Saponins are consdiered a harmful or undesirable constituent in afalfa. They were once thought to be an important factor in the incidence of bloat (14). However, Majak et al. (49) found no significant difference between high- and low-saponin strains of alfalfa developed by Pedersen et al. (50) for incidence of frothy rumen ingesta or for bloat. Negative relationships between saponin concentration and growth of chickens (*Gallus gallus*) and other monogastrics were reviewed by Elliott et al. (14).

26–2.1.3 Essential Constituents that Can be Supplemented

It is within this category of constituents that alfalfa breeders are likely to have the most trouble making and defending a decision on breeding for improved quality. The costs of initiating a breeding program vs. the cost of animal diet supplementation must be considered, and accurate cost comparisons are difficult. Changing economic conditions can eliminate the need to breed for a constituent before the breeding program is completed.

Alfalfa forage generally has more Ca and K and less P than required for the diet of a moderately producing dairy animal (37). The excess Ca is typical of many forage legumes, and the excess K is characteristic of most forages. Most forages contain less P than recommended for dairy animals (62).

Many mineral related problems in forages are caused by imbalances rather than the concentration of a single mineral element. Alfalfa forage has been characterized as containing an excess of Ca and a deficiency of P. The milk-fever downer-calf syndrome may be related to excessive Ca-to-P ratios in diets of dairy cattle (*Bos botaurus*) fed a large proportion of high quality legume forage (36). Thus, breeding to decrease the Ca concentration, increase the P concentration, or both would be a valid goal of breeders interested in improving alfalfa forage quality.

Genetic variability for Ca and P concentration, the Ca-to-P ratio, and the concentration of other mineral elements have been reported for alfalfa (24, 37). Hill and Jung (37) compared expected gains from selection with target values that met animal needs and concluded that it should be possible through breeding to increase the P concentration to the level recommended for the diet of a moderately producing dairy cow. Their data suggest, however, that it would be extremely difficult to reduce the Ca-to-P ratio to the recommended levels.

Full-sib families selected for high P concentration had lower forage yield than did families selected for low P concentration (30). The results would be as expected if selection for P concentration altered the leaf-stem ratio as did selection for lignin concentration (46). Experimental synthetics from high and low selections were evaluated at two P fertility levels on two soil types (39). The high selections had a greater P concentration than did the low selections at the two soil fertility levels, but differences between the high and low selections were small at the low P soil fertility level. Neither type of synthetic had more than half the P concentration recommended for dairy cattle at the low P soil fertility levels. Hill and Lanyon (39) concluded that alfalfa forage with genetically altered P concentration would contribute little to improved animal health unless it was grown at recommended soil fertility levels.

An indirect approach to reducing animal health problems related to mineral imbalances in forages would be to grow alfalfa in mixtures with a grass. Berg and Hill (4) evaluated breeding goals for timothy, *Phleum pratense* L., when grown alone and when grown in a mixture with alfalfa. The degree of mineral imbalance was reduced for the mixture compared to either forage alone. Production of alfalfa grass mixtures to reduce mineral imbalance problems is particularly appropriate for situations, such as pastures, where diet supplementation may be difficult.

Protein, like many of the minerals, can be supplemented in the diets of animals that are fed in confinement. A stronger justification can be made for breeding to increase protein concentration, however, than can be made for altering mineral concentration in alfalfa. Alfalfa forage is produced on many animal farms as a protein source. Forages with high protein concentration generally have better overall quality than those with lower protein concentration, and breeding for increased protein concentration can be an indirect way of improving other forage quality constituents (11). Alfalfa was the most energy efficient of the species discussed by Heichel (22) for production of protein per unit of cultural energy. Thus, breeding efforts to further improve one of the strongest assets of alfalfa can be justified.

Heritability for protein concentration was the highest of all the quality traits investigated by Hill and Barnes (33). Genetic variability for protein concentration was reported by Heinrichs (23), who found that protein concentration was greater in the leaves than in the stems. The results of Heinrichs (23) and those of Hill (30) and Kephart et al. (46) indicate that selection for protein concentration without consideration of forage yield might decrease total forage yield by increasing the leaf-to-stem ratio. Sumberg et al. (58) found that selection for high and low crude protein concentration was successful in afalfa, but syntheticcs from parents selected for high or low crude protein were significantly different in the second and third harvests only. Their high protein synthetics had lower ADF concentrations than did the low protein synthetics, which confirms the suggestion by Cooper (11) that selection for protein concentration may indirectly improve other quality constituents.

26-2.1.4 Physical and Chemical Factors that Influence Storage of Forages as Hay or Silage

Little has been done by breeders to alter characteristics that affect storage of alfalfa forage. Many of the reactions are not well enough understood to permit breeders to identify a constituent or small group of constituents that could be changed through breeding.

Storage problems seem to be more frequent with silage and haylage than with dry hay. Reduced N digestibility occurs when haylage is stored with too little moisture and overheats (19). Methods of determining heat damaged protein have been developed (18), but the chemical reactions that occur during heat damage have not been established. If heat damage occurs only with specific types of proteins, breeders could possibly change the protein form to one that is less susceptible to heat damage. A specific constituent or small group of constituents would need to be identified, however, before breeders could initiate effective selection programs.

Chemical preservatives are used as aids in forage harvesting. These include compounds that aid in fermentation, speed up drying in the field, or reduce the time of exposure to weather conditions at harvest. Interest in chemicals that aid in harvest operations is high because much low quality alfalfa hay results from exposure to unfavorable weather during harvest. It is not known at present if the effectiveness of forage preservatives could be enhanced by genetic alteration of certain chemical constituents in alfalfa.

26-2.2 Philosophies on Breeding for Quality

The ultimate goal of a breeding program to improve forage quality is improved animal performance. Measurement of animal performance on all, or even a small subset, of experimental lines is clearly impractical and impossible. Forage breeders must attempt to improve animal performance by breeding for chemical constituents that affect animal performance. Much of the information on relationships between chemical constituents and animal performance comes from experiments that involved two or more forage species. Strong correlations between chemical constituents and animal performance across forage species do not necessarily indicate that successful genetic alteration of the constituent within a single species will result in improved animal performance. An exmaple of the problem was demonstrated with research on saponin. Bloat-safe legumes have low saponin contents, but differences between high- and low-saponin alfalfa lines were not significant (49). All assumed relationships are not negative, because selection for high and low IVDMD by Thomas et al. (59) did alter digestibility in feeding trials with sheep. The final merit of a chemical constituent as a trait for use in breeding programs to improve forage quality cannot be made until genetically altered forage is evaluated in comparative animal feeding trials. Research that reports heritability of one or a few chemical constituents addresses only a small part of the problem.

Alfalfa is such a high quality forage that some may argue that breeding efforts should be placed on increasing yield and improving those characteristics that influence yield. We ran the feed rationing program of Adams (1) to determine reductions in feed costs from genetic changes in protein, total digestible nutrients (TDN), minerals, and yield (Table 26-4). There were four runs of the program for each line in Table 26-4; (i) the "before selection" concentration for a ration of alfalfa alone, (ii) the before selection concentration for a ration of alfalfa and corn silage, (iii) the "after selection" concentration for a ration of alfalfa alone, and (iv) the after selection concentration for a ration of alfalfa and corn silage. The cost figures in Table 26-4 are reductions in total feed costs that would be expected from the indicated change in protein (from 170–180 g kg^{-1}), TDN (from 560–580 g kg^{-1}), mineral (a 10% increase), or yield (from 11.27–11.88 mg ha^{-1}) in alfalfa through breeding.

When alfalfa was fed alone, greatest projected cost reductions were realized from a modest increase in forage yield, followed by increases in protein and TDN concentrations (Table 26-4). Projected cost reductions from increasing alfalfa yield or TDN were less for rations that included corn silage, but did not change for protein. The results of our projections indicate that alfalfa breeders could easily justify devoting part of their resources to increasing protein concentration and yield in alfalfa. Also, they support Cooper's (11) suggestion that efforts by forage breeders to increase protein concentration are warranted.

When used as a feed for ruminants, alfalfa contains few harmful or toxic constituents. Bloat on pastures that contain alfalfa is probably one of the more serious problems related to toxic or harmful constituents. Breeding to reduce bloat incidence when alfalfa is grazed is a valid goal,

Table 26-4. Cost reductions in dairy cow feed costs from changes in alfalfa traits that might result from alfalfa breeding.

Trait	Genetic change		Dairy cow ration	
	Before selection	After selection	Alfalfa alone	Alfalfa plus corn silage
			— U.S. $ cow^{-1} d^{-1} —	
Protein, g kg^{-1}	170	180	0.04	0.04
TDN, g kg^{-1}	560	580	0.03	0.01
Minerals	10% increase		0.00	0.01
Yield, mg ha^{-1}	11.27	11.88	0.12	0.04

but it appears that we are still in the "constituent defining" stage of this problem.

26-3 SUMMARY

Moderate genetic gains in alfalfa yields can be documented, and progress continues to be made. Genetic increases in yield of alfalfa are less than those for most grain crops. The lower rate of gain in alfalfa yield can be attributed to greater emphasis on pest resistance vs. yield, the perennial nature of the crop, and the complex genetics of autotetraploids.

Gene action for yield in alfalfa is a type that exhibits heterosis and a severe inbreeding depression. The inbreeding depression is greater than would be expected for autotetraploids on the basis of approach to homozygosity alone. Although conclusive proof has not yet been obtained, heterosis in alfalfa is probably due to effects of linked, dominant, or partially dominant genes. Autotetraploidy and the mode of reproduction in alfalfa make it impossible to utilize heterosis that results from specific combinations of genes or chromosome segments, as can be done with single crosses between diploid, inbred lines. Broad based synthetics provide high levels of heterozygosity with many different heterozygous genotypes, and are probably less sensitive to genotype × environment interactions than narrow based synthetics.

Alfalfa is an extremely high quality forage for ruminant animals. Experimental strains with genetically altered concentrations of some quality constitutents have been developed, but we know of no commercial cultivar developed specifically for enhanced forage quality. Selection for quality constituents without regard for yield often results in high-quality, low-yielding experimental populations. Concentrated efforts on breeding for quality alone is not recommended.

REFERENCES

1. Adams, R.S. 1984. Feeding programming for dairy cattle. Pennsylvania State Univ. Ext. Memo. DSE-84-56.
2. Allinson, D.W., F.C. Elliott, and M.B. Tesar. 1969. Variations in nutritive value among species of *Medicago* genus as measured by laboratory techniques. Crop Sci. 9:634–637.
3. Barnes, R.F. 1973. Laboratory methods of evaluating feeding value of herbage. p. 179–214. In G.W. Butler and R.W. Bailey (ed.) Chemistry and biochemistry of herbage, Vol. 3. Academic Press, New York.
4. Berg, C.C., and R.R. Hill, Jr. 1983. Quantitative inheritance and correlations among forage yield and quality components in timothy. Crop Sci. 23:380–384.
5. Bingham, E.T. 1979. Maximizing heterozygosity in autotetraploids. p. 471–489. In W.H. Lewis (ed.) Polyploidy, biological relevance. Plenum Press, New York.
6. Busbice, T.H. 1969. Inbreeding in synthetic varieties. Crop Sci. 9:601–604.
7. ----, and R.Y. Gurgis. 1976. Evaluating parents and predicting performance of synthetic alfalfa varieties. USDA Bull. ARS-S-130. USDA-ARS, Washington, DC.
8. ----, O.J. Hunt, J.H. Elgin, Jr., and R.N. Peaden. 1974. Evaluation of the effectiveness of polycross and self-progeny tests in increasing the yield of alfalfa synthetic varieties. Crop Sci. 14:8–11.
9. ----, and C.P. Wilsie. 1966. Inbreeding depression and heterosis in autotetraploids with application to *Medicago sativa* L. Euphytica 15:52–67.

10. Chaverra-Gil, H., R.L. Davis, and R.F. Barnes. 1967. Inheritance of in vitro digestibility and associated characteristics in *Medicago sativa*. Crop Sci. 7:19-21.
11. Cooper, J.P. 1973. Genetic variation in herbage constituents. p. 379-417. *In* G.W. Butler and R.W. Bailey (ed.) Chemistry and biochemistry of herbage, Vol. 2. Academic Press, New York.
12. Demarly, Y. 1963. Genetique des tetraploids et amelioration des plantes. Ann. Amelior. Plant. 13:307-400.
13. Eberhart, S.A., and W.A. Russell. 1966. Stability parameters for comparing varieties. Crop Sci. 6:36-40.
14. Elliott, F.C., I.J. Johnson, and M.H. Schonhorst. 1972. Breeding for forage yield and quality. *In* C.H. Hanson (ed.) Alfalfa science and technology. Agronomy 15:319-333.
15. Evans, L.T. 1980. The natural history of crop yield. Am. Sci. 68:388-409.
16. Freeman, G.H. 1973. Statistical methods for the analysis of genotype-environment interactions. Heredity 31:339-354.
17. Gallais, A. 1968. Interactions between alleles and their variability in autotetraploid cross-fertilized plants. Consequences for selection. Genetica Agrar. 23:312-323.
18. Goering, H.K. 1976. A laboratory assessment on the frequency of overheating in commercial dehydrated alfalfa samples. J. Anim. Sci. 43:869-872.
19. ----, D.R. Waldo, and R.S. Adams. 1974. Nitrogen digestibility of wilted hay-crop silage. Proc. Int. Grassl. Congr., 13th 13:625-631.
20. Goplen, B.P., R.E. Howarth, S.K. Sarkar, and K. Lesins. 1980. A search for condensed tannins in annual and perennial species of *Medicago*. Crop Sci. 20:801-804.
21. Haldane, J.B.S. 1930. Theoretical genetics of autotetraploids. J. Genet. 22:349-372.
22. Heichel, G.H. 1976. Agricultural production and energy resources. Am. Sci. 64:64-72.
23. Heinrichs, D.H. 1970. Variation of chemical constituents within and between alfalfa populations. p. 267-270. *In* M.J.T. Norman (ed.) Proc. Int. Grassl. Congr., 11th, Queensland, Australia. 12-23 April. University of Queensland Press, St. Lucia, Queensland, Australia.
24. ----, J.E. Torelsen, and F.G. Warder. 1969. Variation of chemical constituents and morphological characters within and between alfalfa populations. Can. J. Plant Sci. 49:293-305.
25. Hill, R.R., Jr. 1968. Some practical considerations of the effects of autotetraploidy on the choice of an alfalfa breeding program. p. 33-37. *In* C.H. Hanson (sec.) Rep. 21st Alfalfa Improve. Conf., (Reno, NV). 9-11 July. USDA-ARS, CR66-68, Beltsville, MD.
26. ----. 1971. Effect of the number of parents on the mean and variance of synthetic varieties. Crop Sci. 11:283-286.
27. ----. 1974. Heritability and genetic interrelationships for quality components in alfalfa. p. 12. *In* D.K. Barnes (sec.) Rep. P. 24th Alfalfa Improve. Conf., Tucson, AZ. 8-10 October. USDA-ARS, St. Paul.
28. ----. 1975. Parental inbreeding and performance of alfalfa single crosses. Crop Sci. 15:373-375.
29. ----. 1977. Quantitative genetics of forages—potentials and pitfalls. Agron. Abstr. American Society of Agronomy, Madison, WI, p. 58.
30. ----. 1980. Selection for phosphorus and lignin content in alfalfa. p. 56. *In* D.K. Barnes (sec.) Rep. 27th Alfalfa Improve. Conf., Madison, WI. 8-10 July. USDA-SEA, Peoria, IL.
31. ----. 1981. Second-generation progeny test for forage breeding. *In* J.A. Smith and V.W. Hays (ed.), Proc. Int. Grassl. Congr., 14th, Lexington, KY. 15-24 June. Westview Press, Boulder, CO.
32. ----. 1983. Heterosis in population crosses of alfalfa. Crop Sci. 23:48-50.
33. ----, and R.F. Barnes. 1977. Genetic variability for chemical composition of alfalfa. II. Yield and traits associated with digestibility. Crop Sci. 17:948-952.
34. ----, and J.E. Baylor. 1983. Genotype × environment interaction analysis for yield in alfalfa. Crop Sci. 23:811-815.
35. ----, and J.H. Elgin, Jr. 1981. Effect of the number of parents on performance of alfalfa synthetics. Crop Sci. 21:298-300.
36. ----, and S.B. Guss. 1976. Genetic variability for mineral concentration in plants related to mineral requirements of cattle. Crop Sci. 16:680-685.
37. ----, and G.A. Jung. 1975. Genetic variability for chemical composition of alfalfa. I. Mineral elements. Crop Sci. 15:652-657.
38. ----, and R.R. Kalton. 1976. Current philosophies in breeding for yield. p. 51. *In* D.K. Barnes (sec.) Rep. 25th Alfalfa Improve. Conf., Ithaca, NY. 13-15 July. USDA-SEA, Peoria, IL.
39. ----, and L.E. Lanyon. 1983. Phosphorus fertilizer response in experimental alfalfas selected for different phosphorus concentrations. Crop Sci. 23:973-976.
40. ----, and J.L. Rosenberger. 1985. Methods of combining data from germplasm evaluation trials. Crop Sci. 25:467-470.

41. Howarth, R.E., B.P. Goplen, A.C. Fesser, and S.A. Brandt. 1978. A possible role for leaf cell rupture in legume pasture bloat. Crop Sci. 18:129-133.
42. ----, ----, S.A. Brandt, and K.J. Cheng. 1982. Disruption of leaf tissues by rumen microorganisms: An approach to breeding bloat-safe forage legumes. Crop Sci. 22:564-568.
43. Jones, W.T., and J.W. Lyttleton. 1971. Bloat in cattle. XXXIV. A survey of legume forages that do and do not produce bloat. N. Z. J. Agric. Res. 14:101-104.
44. ----, ----, and R.T.J. Clarke. 1970. Bloat in cattle. XXXIII. The soluble proteins of legume forage in New Zealand, and their relationship to bloat. N. Z. J. Agric. Res. 13:149-156.
45. Kellogg, D.W., B.A. Melton, C.E. Watson, Jr., and D.D. Miller. 1976. Genetic and environmental effects on nutritive content of alfalfa. N. M. Agric. Exp. Stn. Res. Rep. 308.
46. Kephart, K.D., D.R. Buxton, and R.R. Hill, Jr. 1984. Leaf to stem ratio and stem cell wall components of alfalfa selected for divergent herbage lignin concentration. Agron. Abstr. American Society of Agronomy, Madison, WI, p. 159.
47. Lees, G.L., R.W. Howarth, B.P. Goplen, and A.C. Fesser. 1981. Mechanical disruption of leaf tissue and cells in some bloat-causing and bloat-safe forage legumes. Crop Sci. 21:444-448.
48. Lowe, C.C., R.W. Cleveland, and R.R. Hill, Jr. 1974. Variety synthesis in alfalfa. Crop Sci. 14:321-325.
49. Majak, W., R.E. Howarth, A.C. Fesser, B.P. Goplen, and M.W. Pedersen. 1980. Relationships between ruminant bloat and the composition of alfalfa herbage. II. Saponins. Can. J. Anim. Sci. 60:699-708.
50. Pedersen, M.W., B. Berrang, M.E. Wall, and K.H. Davis, Jr. 1973. Modification of saponin characteristics of alfalfa by selection. Crop Sci. 13:731-735.
51. Pfeiffer, T.W., and E.T. Bingham. 1983. Improvement of fertility and herbage yield by selection within two-allele populations of tetraploid alfalfa. Crop Sci. 23:633-636.
52. Sarkar, S.K., R.W. Howarth, and B.P. Goplen. 1976. Condensed tannins in herbacious legumes. Crop Sci. 16:543-546.
53. Shenk, J.S. 1981. How NIR can help in measuring forage quality for breeding and utilization programs. p. 3-7. *In* S.A. Ostazaski (sec.) Proc./Summaries 4th Eastern Forage Improve. Conf., Beltsville, MD. 7-9 July. USDA-ARS, Beltsville.
54. ----, and F.C. Elliott. 1970. Two cycles in directional selection for improved nutritive value of alfalfa. Crop Sci. 10:710-712.
55. ----, I. Landa, M.R. Hoover, and M.O. Westerhaus. 1981. Description and evaluation of a near infrared reflectance spectro-computer for forage and grain analysis. Crop Sci. 21:355-358.
56. Sprague, G.F., and S.A. Eberhart. 1977. Corn breeding. *In* G.F. Sprague (ed.) Corn and corn improvement. Agronomy 18:305-362.
57. ----, and W.T. Federer. 1951. A comparison of variance components in corn yield trials. II. Error, year × variety, location × variety, and variety components. Agron. J. 43:535-541.
58. Sumberg, J.E., R.P. Murphy, and C.C. Lowe. 1983. Selection for fiber and protein concentration in a diverse alfalfa population. Crop Sci. 23:11-14.
59. Thomas, J.W., J.L. McCampbell, M.B. Tesar, and F.C. Elliott. 1968. Improved nutritive value of alfalfa by crossing and selection based on in vitro methods. J. Anim. Sci. 27:1783-1784.
60. Tysdal, H.M., T.A. Kiesselbach, and H.L. Westover. 1942. Alfalfa breeding. Nebr. Agric. Exp. Stn. Bull. 124.
61. Ulyatt, M.J. 1973. The feeding value of herbage. p. 131-178. *In* G.W. Butler and R.W. Bailey (ed.) Chemistry and biochemistry of herbage, Vol. 3. Academic Press, New York.
62. Underwood, E.J. 1966. The mineral nutrition of livestock. Commonw. Agric. Bur. The Central Press (Aberdeen) Ltd., United Kingdom.

27 Breeding for Disease and Nematode Resistance

J. H. ELGIN, JR.

USDA-ARS
Beltsville, Maryland

R. E. WELTY

USDA-ARS
Corvallis, Oregon

D. B. GILCHRIST

University of California
Davis, California

Plant diseases and nematodes are often major limitations to alfalfa (*Medicago sativa* L.) production. They contribute to failure in stand establishment, reduction in stand longevity, and significant yield losses. Annual losses to U.S. alfalfa production from diseases and nematodes are estimated at 10% or approximately $500 million dollars. These estimates do not include either the cost of reestablishing stands or the loss of production during the establishment period. The perennial nature of alfalfa reduces the effectiveness of cultural practices, such as crop rotation, and increases the cost of pesticides in the control of both diseases and nematodes. Plant resistance is the most practical means of controlling many of these pests. Development of cultivars resistant to economically important diseases and nematodes has made a significant contribution to the successful production of alfalfa. Also, the use of multiple pest resistant cultivars has reduced the prospect of water and soil pollution from pesticides and the accumulation of pesticide residues in milk or meat. The goal of most current alfalfa breeding programs is to incorporate resistance to as many economically important pests as possible while improving the desirable agronomic traits needed in high-yielding cultivars.

This chapter focuses on the principles used to develop disease and nematode resistant cultivars and highlights progress made in developing resistance to specific diseases and nematodes. (See Chapter 21 for diseases and nematodes and Chapter 25 for breeding procedures) The authors acknowledge the efforts of W. R. Kehr, F. I. Frosheiser, R. D. Wilcoxson, D. K. Barnes, O. J. Hunt, L. R. Faulkner, and R. N. Peaden in their development of chapters 15 and 16 of *Alfalfa Science and Technology* (ASA

Copyright 1988 © ASA-CSSA-SSSA, 677 South Segoe Road, Madison, WI 53711, USA.
Alfalfa and Alfalfa Improvement—Agronomy Monograph no. 29.

Monograph 15) published in 1972. This chapter serves as an updated revision of their work and incorporates much of the information assembled by these previous authors.

27-1 PRINCIPLES OF DISEASE AND NEMATODE RESISTANCE BREEDING

27-1.1 Understanding Host-Pathogen Interaction

Breeding for disease and nematode resistance is similar to breeding for other characteristics. There is a need, however, to understand the pathogen, host-pathogen interactions, and environmental effects on host and pathogens. Alfalfa is an autotetraploid cross-pollinated species that exists naturally as a heterogeneous population of plants. An alfalfa cultivar is a population of plants improved for one or more specific characteristics that can be retained through subsequent generations of seed multiplication. Although individual plants within a cultivar have similarities, there is considerable variability among individuals for numerous characteristics. Disease and nematode resistance in alfalfa, therefore, is characterized according to the frequency of resistant plants in a population. This may be expressed either as a percentage of resistant plants or as a disease severity index (DSI), the calculated average numerical disease severity rating for the individual plants tested.

In breeding for disease and nematode resistance, the ultimate objective is to identify and maintain individual plants in the population that carry genes for resistance. These plants are interpollinated to produce seeds that give rise to populations of plants that have both a higher level of pest resistance and a higher frequency of resistant plants. Large numbers of plants must be included in the breeding program to minimize inbreeding and to conserve genetic variability for other desirable traits.

Breeding for disease and nematode resistance requires understanding the pathogen, including its host range, growth requirements, life cycle, geographic distribution, and tendency to develop races or biotypes. Factors affecting the host-pathogen relationship include the portion of plant affected, symptoms, sources of inoculum, mode of pathogenesis, and how inoculum concentration influences disease. Knowledge of how environmental factors affect the host-plant interaction is essential when the breeder wishes to create epidemics for the purpose of selecting resistant plants. Environmental factors include temperature, moisture, light, soil fertility, soil moisture, pH, stage of plant growth, and interactions with other microorganisms that may affect disease development.

For greatest success, disease screening techniques need to be developed that are dependable and that predict plant resistance in both greenhouse and field experiments. When this has been accomplished, resistance can be evaluated in either the field or greenhouse or both. Evaluations confined to the greenhouse sometimes preclude identifying plants that possess a useable level of field resistance or that possess other useful agronomic traits. Information about the causal organism and standardized techniques for disease screening are available for many alfalfa pathogens (68, 110).

27–1.2 Sources of Resistance

Usually the best source of resistance to a disease or nematode will be found in areas of the world where the pathogen and host coevolved. For example, the first known resistance to the stem nematode [*Ditylenchus dipsaci* (Kühn) Filipjev] was reported in an introduction from Turkestan (FC 19304) (280), where alfalfa has been grown for thousands of years, often in association with nematodes. Fortunately, it is not always necessary to search for pest resistance in unadapted sources. Recently developed cultivars and gene pools with parentage tracing to several resistance sources of diverse origin may contain sufficient resistance for the breeder to select plants with a combination of both pest resistance and desirable agronomic traits.

27–1.3 Breeding Procedures Employed

Recurrent phenotypic selection has been the most successful method employed in developing disease-resistant alfalfa populations. It has proven to be an extremely effective method for increasing the frequency of desirable genes in alfalfa populations. The procedure involves mass screening followed by intercrossing 100 or more resistant plants under isolation to develop a new population. This population is available for testing and for repeating the cycle as needed. The method has been successful in increasing resistance to bacterial wilt [*Corynebacterium insidiosum* (McCull.) H. L. Jens] (14), anthracnose (*Colletotrichum trifolii* Bain.) (60), common leaf spot [*Pseudopeziza medicaginis* (Lib.) Sacc.] (109), downy mildew (*Peronospora trifoliorum* dBy.) (266), Fusarium wilt (*Fusarium oxysporum* Schl. f. sp. *medicaginis* (Weimer) Sny. & Hans.) (96), Phytophthora root rot (*Phytophthora megasperma* Drechs. f. sp. *medicaginis*) (97), Verticillium wilt (*Verticillium albo-atrum* Reinke & Berth.) (219), rust (*Uromyces striatus* Schroet.) (141), Stemphylium leaf spot (*Stemphylium botryosum* Wallr.) (277), and stem nematode (75).

In some programs backcrossing has been used to increase or transfer resistance. For example, the modified backcross method was used to transfer bacterial wilt resistance into 'Saranac' (196), a Flemish type alfalfa, and root-knot nematode (*Meloidogyne* spp.) resistance into 'Nevada Synthetic XX' germplasm (220). A more widely used technique relies on strain crosses to combine dominantly inherited pest resistance into populations with multiple pest resistance (83). This method represents a modified hybrid approach to alfalfa breeding; increasing heterozygosity and the potential for increasing yield in the resultant population. A prerequisite for the development of successful strain crosses with multiple pest resistance is the availability of alfalfa populations with high levels of resistance to different pests, and complementary effects for yield and agronomic performance.

Hybrid alfalfa using a cytoplasmic male-sterile line also has been proposed (59, 42). To date it has not been practical because of difficulties in bee pollination and seed production on the male-sterile parents. If the hybrid system becomes feasible it would be useful for developing multiple pest-resistant cultivars.

27-1.4 Inheritance of Resistance

Mode of inheritance of resistance to some of the major disease and nematode pests of alfalfa has been studied. In most reports, resistance has been conditioned by one or a few genes with varying degrees of dominance. There is little indication of linkage between resistance to one disease and resistance to another. However, there are a few exceptions. For example, linkage occurs between the separate genes controlling resistance to the two root-knot nematodes, *Meloidogyne hapla* and *M. javanica* (107). Physiologic (or pathogenic) races capable of attacking previously resistant cultivars of alfalfa are not important for most pathogens. However, the discovery of a second physiologic race of *Colletotrichum trifolii*, which causes anthracnose, and races of *Peronospora trifoliorum*, which cause downy mildew, illustrate that this can occur (210, 272, 308).

27-2 BREEDING FOR RESISTANCE TO SPECIFIC DISEASES

27-2.1 Bacterial Diseases

27-2.1.1 Bacterial Wilt [*Corynebacterium insidiosum* (McCull.) H. L. Jens.]

The development of bacterial wilt-resistant cultivars has been one of the most notable accomplishments in breeding for disease resistance in alfalfa. Resistance was first identified in Ladak and certain Turkestan introductions in the 1930s. 'Ranger' and 'Buffalo' were the first cultivars bred for wilt resistance (293). This marked the beginning of breeding for alfalfa disease resistance. Essentially all winter hardy and moderately winterhardy cultivars developed since 1965 have had resistance to bacterial wilt. 'Vernal' with about 42% resistant plants is the accepted standard used for comparing bacterial wilt resistance of cultivars (Fig. 27-1) (5).

Several inoculation methods have proven satisfactory, either alone or in combination in identifying resistant plants. These include introducing inoculum by scraping roots in a bacterial suspension (155, 225), injecting roots or crowns (53, 193), puncturing the leaf petiole (108), wounding cotyledons (172), and the root-soak method (53, 84). The latter is the most effective method. A bacterial suspension is prepared by soak-

Fig. 27-1. Stand depletion from bacterial wilt in the susceptible cv. Narragansett at Madison, WI. Vernal was bred for resistance.

ing 50 g of finely ground fresh, frozen, or dried infected roots in 1 L of tap water for 20 to 30 min before inoculation. Fresh or frozen roots are usually more satisfactory than dried roots as a source of inoculum (93). Seven- to 10-week-old plants are dug, and the roots are washed and immersed in this bacterial suspension for 20 to 30 min. Often, the ends of the immersed roots will be clipped to facilitate entry of the organism and subsequent transplanting operations. The tops are cut off and the plants transplanted in the field. Surviving plants are dug 12 to 16 weeks after inoculation and taproots are cross-sectioned to identify those free of vascular discoloration. A combination of the cotyledon wounding and the root-soak method has been effective for screening large populations of seedlings (14, 93, 172).

Although pure cultures of the bacteria can be used for inoculum, they may be less virulent than inoculum from diseased host tissue (168). Infected plant material is easily handled and stored, and provides a mixture of sources that may vary in virulence (51, 53).

Disease severity can be influenced by many factors, including strains of the pathogen (99, 198), crop management (166, 260, 291), soil fertility (170, 291), soil temperature and moisture (169, 214), winter injury (157, 226, 312), nematodes (130, 131, 147, 204, 264, 230, 231, 232, 298), other pathogens (193, 316, 317), and insects (139).

The inheritance and nature of resistance has been investigated by many scientists (31, 64, 132, 156, 221, 224, 226, 254, 278, 315). Most investigators concluded that inheritance is complex, and some report a comparatively simple pattern of inheritance. Viands and Barnes (294) studied the inheritance of bacterial wilt resistance in two diverse germplasm pools. In one germplasm pool a single dominant tetrasomic gene coded for resistance; however, two other genes with smaller, additive gene effects were noted. In the second germplasm pool only the two genes with smaller additive gene effects were found. They speculated that the dominant gene in the first germplasm pool originated from Turkestan germplasm.

Resistance to bacterial wilt is highly heritable and rapid progress with selection can be made in developing resistant populations (64, 99, 221, 292). An interaction was observed between *C. insidiosum* and *Rhizobium meliloti*, in which a degree of protection against *C. insidiosum* was obtained in plants inoculated with *R. meliloti* (24). Other studies have shown a small negative correlation between resistance to bacterial wilt and N_2-fixation rate (295).

27–2.1.2 Bacterial Leaf Spot [*Xanthomonas alfalfae* (Riker, Jones, & Davis) Dows.]

No resistant cultivars have been developed, but plants with apparent resistance were selected from 'Cody' (273) and 'Kanza" (271). Progenies derived from intercrossing selected Kanza plants were equal or superior in resistance to their parents.

The bacterium can be readily isolated from diseased alfalfa, maintained, and increased on potato-dextrose agar. A bacterial suspension sprayed on plants, containing silicone carbide dust, has been effective for inoculation. Plants are kept in a growth chamber at 24 to 27°C for 2 weeks, and the survivors reinoculated in the same manner or by puncturing each internode with a dissecting needle dipped in inoculum (30, 271). The inheritance of resistance has not been identified.

27–2.1.3 Other Bacterial Diseases

Two diseases of lesser importance are bacterial stem blight caused by *Pseudomonas syringae* van Hall, and alfalfa dwarf caused by a Gram-negative bacterium (109).

No resistant varieties are available for bacterial stem blight. However, differences in bacterial stem blight resistance were noted among plants in Colorado (249). In Utah, Richards (245) reported that 'Ladak' and 'Grimm' were less injured than other cultivars. Whitehead and Pinnell (311) observed Ladak and Vernal to be more resistant than other cultivars in Missouri. There appears to be some association between the disease and low temperatures; with frost-resistant cultivars generally injured less by the disease than those that are frost-susceptible.

The alfalfa dwarf bacterium is carried to susceptible plants by leafhopper vectors. Although the disease is now relatively rare in the USA, it was important about 45 yr ago in southern California. Dwarf-resistant California 'Common 49' was selected from California 'Common' plants that produced normal growth in dwarf-infected fields (143, 144). Alfalfas of diverse origin were equally susceptible in some controlled tests (143, 303).

27-2.2 Fungal Diseases

27-2.2.1 Anthracnose (*Colletotrichum trifolii* Bain.)

Conscious selection for anthracnose resistance in alfalfa began in about 1968, when it was observed that several alfalfa cultivars, developed by recurrent selection in the field in the southern and mid-Atlantic states, had measurable levels of resistance (15, 207). 'Arc', the first cultivar with anthracnose resistance was released in 1974 (62). In areas where the disease occurs, anthracnose resistance has the potential to increase alfalfa forage yields by an average of 10% (69).

The fungus grows on many laboratory media and produces abundant inoculum on oatmeal agar, V-8® juice agar, or lima bean agar (68). Inoculum should be harvested from 7- to 10-d-old cultures because conidia lose viability with continued incubation. The effectiveness of the inoculum will decrease with culture age with a concomitant increase in disease escapes (306). Inoculum also can be prepared by drying and grinding anthracnose infected stems. Both dry and wet inoculum are effective (34) when sprayed on 2-week-old seedlings kept in a moist chamber at 20 to 25°C for 48 h (68). Plants are ready to score for disease reaction 14 d after removal from the moist chamber (207), with the disease developing in susceptible plants at temperatures of 10 to 30°C (310). Hypodermic needle inoculations have also been used (209).

Only one race of the anthracnose fungus was known to occur until 1977 when the disease developed on a previously anthracnose-resistant cultivar (308). Since then race 2 has been discovered in North Carolina, Maryland, and Virginia (208, 210, 307, 309). The cultivars Arc and Saranac serve to differentiate the two races. Race 1 is virulent on Saranac only, whereas race 2 is equally virulent on both cultivars. Saranac AR is highly resistant to both races. Most commercial cultivars and breeding lines resistant to race 1 are susceptible to race 2 (79, 307). Many breeding lines and cultivars resistant to both races trace to a single germplasm source, Beltsville 2-An4 (61). Some annual and perennial *Medicago* spp. have high resistance to both races.

Anthracnose resistance is highly heritable. One dominant tetrasomically inherited gene, as occurs in Arc, conditions for resistance to race 1 only, whereas a second dominant gene, as in Saranac AR, conditions for resistance to both race 1 and race 2 (80). There is an excellent correlation between resistance measured in the greenhouse and that recorded under field conditions. However, in some cases other genes may modify the behavior of a cultivar that displays susceptible ratings in greenhouse tests, so that a high degree of field survival is observed in spite of the presence of anthracnose.

In the development of Arc one cycle of recurrent phenotypic selection in 'Team' increased the frequency of highly resistant plants (race 1) from 18 to 75% (Fig. 27-2). Recurrent phenotypic selection for race 1 anthracnose resistance also was effective in Vernal, Saranac, and 'Glacier' (60).

Fig. 27–2. Resistance to anthracnose (race 1) demonstrated in the cultivar Arc, which was bred by selecting race 1 anthracnose resistant plants from the var. Team. Saranac, 'Weevilchek', and Williamsburg are susceptible.

A similar response to selection for resistance to race 2 has been reported (81, 82). Most dormant and semidormant cultivars appear to contain a low frequency of genes for anthracnose resistance. Growth of the fungus is inhibited in tissue of resistant plants, whereas it enters and grows rapidly in epidermal, cortical, and vascular tissue of susceptible plants (256).

27–2.2.2 Spring Black Stem (*Phoma medicaginis* Malbr. & Roum.)

No highly resistant cultivars are available for this disease. However, plants with varying degrees of resistance have been found in many cultivars and *Medicago* spp. (10, 145, 228, 237, 238, 286, 300). Black stem resistance is heritable (248, 276) and can be increased by selection (78, 237). 'Teton' has moderate resistance.

Inoculum can be provided by growing the fungus on barley (*Hordeum vulgare* L.) and wheat (*Triticum aestivum* L.) grain, followed by storage under refrigeration (187, 239, 240). The concentration of inoculum is very important in distinguishing resistant from susceptible plants.

Disease develops on plants held in a moist chamber at 16 to 24°C (66, 191) with 48 h usually sufficient for leaf infection, whereas 3 to 4 d may be necessary for stem infection. Disease ratings are made about 10 to 14 d following inoculation (18, 239, 240). Nutritive composition of the inoculum tends to increase disease severity (10). Also, the environment influences the severity of spring black stem, because plants resistant in the field are frequently susceptible in the greenhouse (248, 276). There is some evidence of physiological races of the pathogen (192).

27-2.2.3 Summer Black Stem (*Cercospora medicaginis* Ell. & Ev.)

Alfalfa cultivars and clones differ in resistance to summer black stem (29, 124, 195, 314). Buffalo, Cody, and 'Williamsburg' have been reported to be less affected by summer black stem than most cultivars. The pathogen will grow on many laboratory media, but it sporulates best on V-8 juice (19) or carrot decoction medium (17). Infected stems may serve as a source of inoculum (29). Methods for inoculating plants are essentially the same as those described for spring black stem. Physiologic races have been reported (20).

The heritability of resistance appears high. Greenhouse and field reactions of clones were significantly correlated (29). In an eight-clone diallel both dominant and recessive genes appeared to influence the expression of resistance (276).

27-2.2.4 Common Leaf Spot [*Pseudopeziza medicaginis* (Lib.) Sacc.]

Resistant plants have been found in many *Medicago* spp, populations, and cultivars (160, 238). Cultivars of Flemish origin are the most widely used source of resistance; however, resistance has been found in winter-hardy cultivars, such as Teton, Ladak, 'Travois', and others. Recurrent phenotypic selection has been effective in increasing resistance to common leaf spot in a number of populations (78, 109, 136, 257), with several releases of resistant germplasm (121, 282, 283, 284, 285). Even though few breeding programs have been selecting specifically for common leaf spot resistance, most new winterhardy and moderately winterhardy cultivars have more resistance than nondormant or older dormant cultivars, such as Ranger.

The pathogen grows slowly in culture. Inoculum is prepared from stock cultures that are mixed in sterile water and poured over the surface of oatmeal agar (68). Apothecia are produced on the agar surface but recovery of pathogenic ascospore suspensions have not been reported. Plants are inoculated by placing them beneath inverted sporulating cultures in a moist chamber at 20°C for 24 to 48 h (68). Infected leaves have been used as a source of inoculum. Although plants can be inoculated as soon as the unifoliolate leaves unfold, infection will be more uniform if several of the trifoliolate leaves have developed. Lesions appear on the leaves of susceptible plants in 14 to 20 d. Results from greenhouse and field tests were consistent (92, 134, 159). There are no reports of either differences in virulence or discrete biotypes or races. In many areas natural epidemics of common leaf spot are frequent and severe enough, especially during the year of seeding, to permit field selection for resistance (2, 3, 4, 35, 109, 221).

Heritability of resistance was relatively high in several studies (4, 35, 58, 109, 221). General combining ability effects were more important than specific combining effects in determining resistance (2, 35, 221). Realized gain in selection for resistance was equal or better than that predicted (3, 4, 221).

27–2.2.5 Downy Mildew (*Peronospora trifoliorum* dBy.)

Cultivars vary markedly in their frequency of resistant plants. Comparisons among 36 cultivars under a severe downy mildew epidemic in a space-planted nursery at Rosemount, MN, indicated that Saranac and 'Narragansett" had >90% resistant plants, whereas Ranger and Vernal had about 55% resistant plants (12). Most cultivars in the study fell somewhere between these extremes. Evaluations of downy mildew resistance have been conducted primarily in the field (124, 126, 157, 163, 181, 182, 199, 270), but laboratory evaluation methods are available (68).

The fungus is an obligate parasite that must be maintained on alfalfa plants in the greenhouse or stored at $-20°C$ until used. Disease symptoms develop at 15 to 20°C (218). Considerable variability in virulence exists among isolates (68) and different races have been identified (272).

Backcross and self-pollination data obtained in the development of 'Caliverde' suggested a degree of dominance for resistance (267). Pedersen and Barnes (222) indicated that downy mildew resistance was conditioned by one tetrasomically inherited gene with incomplete dominance. In contrast, Jones and Torrie (163) suggested that susceptibility was dominant.

Resistance was incorporated into Caliverde by a modified backcross method, whereas clonal selection has been used in achieving a measure of resistance in other cultivars, such as 'Mesa-Sirsa', 'Sonora', 'Uinta', and several proprietary releases.

27–2.2.6 Fusarium Root Rot [*Fusarium* spp., including *F. solani* (Mart.) Appel & Wr., *F. oxysporum* Schl., *F. roseum* Lk. ex Fr. emend. Sny. & Hans., and *F. tricinctum* (Corda) Saccardo.]

Fusarium crown and root rot, recognized as one of the most widespread diseases of alfalfa, has been subjected to extensive studies in the USA (85, 176) and Canada (189). In field plantings, genetic differences in disease resistance were detected among and within cultivars (313). Furthermore, progress was reported in selecting for increased resistance to crown and root rot pathogens by applying inoculum to transverse cuts in the taproot or by a bare-root-soak method (244). Management practices that reduce plant stress and the use of cultivars less affected by various plant stress factors may provide the best approach to control of root rot losses from these pathogens.

Stress factors that favor development of disease include temperature (46, 302), soil moisture (46, 302), soil nutrition (38, 205, 270), insects (63, 85, 140, 178), nematodes, and other pathogens (188, 299, 316), and cutting frequency.

27-2.2.7 Fusarium Wilt (*Fusarium oxysporum* Schl. f. sp. *medicaginis* (Weimer) Sny. & Hans.

Alfalfa populations have been evaluated for resistance to Fusarium wilt, in the field and greenhouse (Fig. 27-3) (86, 98). Recurrent selection has been effective in increasing resistance (316), and today, a number of public and proprietary cultivars have acceptable levels of resistance. Although gene frequencies for resistance were low, substantial progress in selection for resistance was possible after only one generation. Inheritance studies have shown resistance in winterhardy and nonhardy germplasm to be controlled by one dominant gene and one incompletely dominant gene (135).

The method of Fusarium wilt inoculation (root dip) is similar to that used in screening for resistance to bacterial wilt (68). It is important, however, to avoid mixing inoculum of the two pathogens because of an unexplained interaction between the fungus and bacterium in the same plant (98). Until more is known about variation in pathogenicity of wilt-causing strains of *Fusarium*, it is desirable to establish wilt nurseries and to screen germplasm for use in a specific area by using isolates from that area.

27-2.2.8 Leptosphaerulina Leaf Spot [*Leptosphaerulina briosiana* (Poll.) Graham & Luttrell]

Resistant plants have been reported in many *Medicago* spp. and cultivars (124, 165, 186, 212, 238, 297). Most attempts to increase re-

Fig. 27-3. Effect of Fusarium wilt on plant survival in the cv. Apalachee. Row on right, inoculated with *Fusarium oxysporum*; row on left uninoculated check.

sistance have been unsuccessful (234, 138, 243); however, some recently released cultivars are described as having resistance.

The fungus grows well in culture and produces ascospores on V-8 juice or oatmeal agar when provided with light that includes near-ultraviolet wavelengths (174). Plants are generally inoculated by suspending inverted sporulating cultures over the plant tops (68, 234, 138, 275, 297), and kept moist at 20°C for at least 36 h. It is not necessary to provide light during the infection period. Typical symptoms develop in about 7 d on plants grown w/370 μmol m^{-2} s^{-1} light for 12 h daily. Disease severity is evaluated by the number and the type of spots on upper leaves (212, 275). Evidence of races was obtained in greenhouse (174) and field studies (186, 211, 212). These results may be of little consequence because most isolates are highly aggressive and disease scores for most cultivars exceed 3.5 (on a scale of 1 to 5, with 5 = most susceptible).

Little is known about inheritance of resistance. Narrow-sense heritability estimates of 3 to 80% were reported from a recurrent selection program in which little response to selection was observed (78). Pathogenicity of isolates was significantly different across cultivars, and the cultivar × isolate interaction was significant (186). This is a very difficult disease to control by breeding, in part, because of problems encountered in duplicating results from field and greenhouse screening tests. These inconsistencies may result from the presence of races, the absence of finite resistance, and the lack of adequate information on critical environmental factors influencing the expression of reproducible levels of disease.

27–2.2.9 Phytophthora Root Rot (*Phytophthora megasperma* Drechs. f. sp. *medicaginis*)

Phytophthora root rot occurs in nearly every area of the world where alfalfa is grown. It is especially severe in regions of high rainfall, poor drainage, or where alfalfa is irrigated by flooding. Water management and deep tillage to break compacted soil layers will help reduce the severity of the disease. Resistant cultivars represent the only practical method of disease control. A number of resistant cultivars are available.

Phytophthora megasperma is isolated by placing newly infected roots on solid media containing antibiotics (289). Inoculum can be increased on V-8 juice agar or broth (88). Inoculum is incorporated into soil or sand in which alfalfa plants are to be grown for screening in the greenhouse or the growth chamber at 20 to 24°C (68). Others have reported success in inoculating seedlings 12 d after planting, by adding inoculum between rows, watering, and flooding for 3 d (142). The organism requires abundant soil moisture for sporulation, inoculum dissemination, infection, and disease development (89, 94). In Minnesota, daily irrigation of a field nursery by overhead sprinklers provides adequate soil moisture for optimum disease development (68). Older plants can be inoculated individually by wounding the taproot, placing infested agar next to the wound, and replacing the soil, which is then kept saturated for 2 to 4 weeks. Lesion development at the wound site is the criterion for evaluating host reaction.

Lu et al. (184) reported that resistance in 'Lahontan' and Vernal clones was conditioned by one tetrasomic recessive gene with incomplete dominance for susceptibility. The nulliplex and simplex conditions confered high and intermediate levels of resistance, respectively. Duplex, triplex, and quadruplex genotypes were susceptible. Resistance in Lahontan- and Vernal-type cultivars appeared to be conditioned by the same gene.

Irwin et al. (151) working with clones from NAPB 0310 germplasm and 'Hunter River' found resistance to be a dominant trait conditioned by two complementary genes with incomplete dominance. The expression of resistance required at least a duplex genotype at one locus and a simplex genotype at the other locus. Crosses between a resistant NAPB 0310 plant and a resistant Hunter River plant indicated that different genes conditioned disease reaction in the two sources. Irwin et al. (150) found a similar inheritance scheme in diploid alfalfa with resistance conditioned by two dominant complementary genes.

Recurrent phenotypic selection was effective in increasing resistance in Minnesota (Fig. 27-4) (97, 131, 145). Two cycles of intensive field selection followed by reinoculation of individual parent plants in the greenhouse to eliminate escapes prior to intercrossing resulted in an increase of resistant plants from <5% in the original source to about 65%.

Northern root-knot (*Meloidogyne hapla* Chitwood) and southern root-knot [*M. incognita* (Kofoid & White) Chitwood] nematodes were found

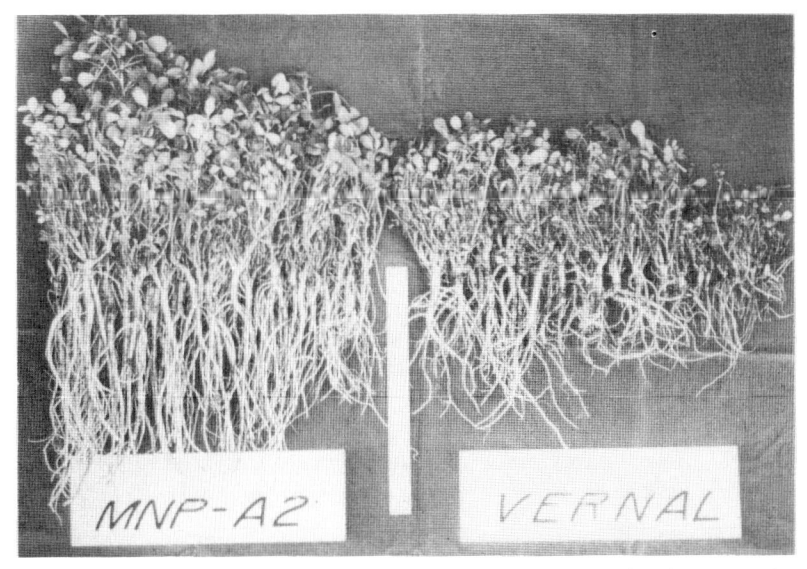

Fig. 27-4. Resistance to Phytophthora root rot in MNP-A2 resulting from selection of resistant plants in a field nursery at St. Paul, MN. MNP-A2 was the experimental designation for 'Agate'.

to interact with and reduce resistance to *P. megasperma*. Combined inoculations with either of these nematodes and *P. megasperma* were detrimental to alfalfa in all criteria of disease severity, as well as plant growth and vigor (304).

27–2.2.10 Rhizoctonia Crown and Root Rot and Seedling Blight (*Rhizoctonia solani* Kuehn)

Rhizoctonia blight is a sporadic disease with losses ranging from slight to severe. Most damage occurs during warm and wet periods. The fungus is isolated easily from infected tissue. No economic control measures are commonly available. However, seedling resistance has been reported in three resistant germplasm lines (16).

The fungus grows optimally at 25 to 30°C (262) and isolates vary in pathogenicity, environmental requirements, and host range (40, 87, 261). Inoculum may be increased in several media and may be mixed with soil for disease evaluations (39, 87, 167).

27–2.2.11 Rust (*Uromyces striatus* Schroet.)

Rust causes only sporadic damage in North America. Losses occur primarily in summer and fall in the warmer alfalfa-growing regions. Resistance has been reported in *Medicago* spp. and in various cultivars and experimental populations (65, 125, 136, 141, 170, 247). 'Cherokee' and Teton, as well as several germplasm releases, have high levels of resistance (121, 165).

The fungus is an obligate parasite, and hence inoculum must be increased on alfalfa plants or on detached leaves (78, 141, 170). Inoculum is prepared by diluting urediospores with talcum powder (68, 78) and dusting leaves of test plants. Inoculated plants should be kept moist for at least 24 h to ensure uniform infection. Rust severity is evaluated by number and size of uredia. Also, resistance may be evaluated in the field, where a mixture of races may exist (141, 301). Greenhouse, laboratory, and field reactions have been similar (141, 170).

Cope (43) reported that resistance was controlled by three or four major genes and one or two genes with small effects. Two of the genes for resistance were partially dominant. Heritability estimates of 2 to 80% have been reported (43, 65, 78, 190). Recurrent phenotypic selection was effective in increasing the level of rust resistance in two broad-based germplasm pools (65, 122, 141).

27–2.2.12 Sclerotinia Crown and Stem Rot (*Sclerotinia trifoliorum* Eriks.)

Damage to alfalfa from sclerotinia crown and stem rot varies from season to season. It may involve entire fields or areas as small as 6 to 8 cm in diameter. Plants of all ages are susceptible, but damage is most severe in seedlings growing under cool, moist conditions. In addition, damage may be extensive during late winter and early spring in fall seedings of alfalfa.

Several methods of inoculation have been tested in the greenhouse (6, 44, 48, 70, 236, 305). Use of infected grain inoculum is probably most common. The disease requires high humidity and cool temperatures. Field-grown plants have been inoculated with root wounding (44, 48). Some cultivars in the USA sustain less damage than others (6, 47, 48, 70, 305), but resistance of economic importance is not available (85). It has been reported that some European cultivars have good levels of resistance (235). Certain species of *Medicago*, including *M. rugosa* and *M. scutellata*, are immune to the disease (235).

27-2.2.13 Stagnospora Crown and Root Rot and Leaf Spot [*Stagonospora meliloti* (Lasch) Petr.]

Resistant cultivars are not available. However, differences in susceptibility have been observed among cultivars and germplasm sources (87, 163, 164).

According to Erwin (87), the most successful method for inoculation has been to dip roots of 8-week-old plants in a spore or mycelial suspension for 30 min, keep them moist 2 days, and transplant into soil. Severe symptoms develop in 2 to 3 months. Necrotic lesions with pycnidia can be induced on stems, petioles, and leaves, by spraying with a pycnidiospore suspension. Inoculum can be produced on sterilized oat seed, potato dextrose agar, V-8 agar, or in Houston's medium. The pathogen will grow from 6–24C on V-8 agar, but sporulates best from 18 to 24°C.

27-2.2.14 Stemphylium Leaf Spot (*Stemphylium botryosum* Wallr.)

Severity of Stemphylium leaf spot has been shown to differ among alfalfa cultivars (26, 134, 160, 165, 185). In greenhouse experiments, Hill and Leath (137) reported limited genetic variation for resistance in parents and crosses derived from three cultivars. Significant additive variation for resistance was reported in crosses from two of three cultivars tested. In a fourth cultivar, however, five methods of selection for resistance were unsuccessful in achieving significant improvement (118). In California, separation of resistant plants within cultivars was not successful, in spite of broad-sense heritability estimates for resistance that appeared large enough to justify a selection program (26).

The fungus can be grown on various types of media and is applied in suspensions sprayed on the plant foliage held in moist chambers at 20 to 25°C (110, 137). More recently, different forms of Stemphylium leaf spot have been reported from California and eastern North America. The two biotypes of *S. botryosum* require different temperatures for disease development (55). Differences in virulence among isolates could not be ascribed to pathogenic races (54).

Recurrent phenotypic selection was used to develop populations with high resistance to the California biotype (103, 277). A high degree of resistance also was found in collections of *Medicago cancellata*, *M. falcata*, *M. hemicycla*, and *M. coerulea* (25). Inheritance in *M. hemicycla* was determined by a pair of dominant genes with duplicate effects, with a dosage effect observed in the dominant genes. Inheritance was unclear in *M. falcata* and *M. cancellata* and was not studied in *M. coerulea*.

27-2.2.15 Verticillium Wilt (*Verticillium albo-atrum* Reinke & Berth.)

In general, alfalfa cultivars commonly grown in the USA (171) and the United Kingdom (7, 152) are susceptible to verticillium wilt, as are annual types of *Medicago* (213). Resistant plants have been found in *M. sativa* × *M. falcata* hybrids and in several *Medicago* spp. Also, resistant plants have been selected from 'Alfa', 'DuPuits', 'Flamande', Grimm, 'New Mexico 11-1', 'Nevada Syn F', 'Pegauer', Ranger, Saranac, and 'Tuna' (7, 36, 213, 215, 216, 217). A number of Verticillium wilt resistant cultivars have been developed in Europe, and several resistant proprietary and a few public cultivars (219) are adapted and available in the USA and Canada.

Inoculum can be produced on several laboratory media or on sterilized small grains (152, 213, 215, 229, 320). Biotypes were reported within *Verticillium* spp. (110, 214), and isolates of *V. albo-atrum* differed in virulence (213).

Several inoculation methods have been tried: (i) using the rolled paper technique (7, 215); (ii) placing inoculum in stem or root wounds (153); (iii) adding grain culture to the soil before planting (153, 215); (iv) spraying a spore suspension (213, 215); and (v) immersing injured roots of seedlings in a spore suspension before planting in soil (213, 215, 229). The latter has been most successful. A standard greenhouse evaluation technique was described by Peaden (68).

Differences in field resistance observed in test crosses of inbred parents not selected for resistance, and in diallel crosses among noninbred parents selected for resistance, could be explained largely on the basis of combining ability effects (100). Panton (217) found that both general and specific combining ability effects were significant. Even though the heritability of resistance was low, recurrent phenotypic selection for resistance was effective in the greenhouse in Sweden (216, 217) and in the field in Italy (213). Recurrent selection practiced either in the greenhouse, field, or both, has been used in developing resistant cultivars adapted for use in USA and Canada.

27-2.2.16 Yellow Leaf Blotch *Leptotrochila medicaginis* (Fckl.) Schuepp.)

The pathogen can be cultured, but it does not sporulate in culture. Although infected leaves have been used as a source of inoculum in greenhouse evaluations, little work has been done on creating epidemics

(68, 227, 252). Some field resistance has been observed in clones (158) and cultivars, including Teton, Uinta, and Vernal (1, 122, 223). In a South Dakota test, high resistance was found in cultivars possessing *M. falcata* germplasm in their pedigree, and in six perennial and 22 annual *Medicago* spp. (253).

27-2.2.17 Other Fungal Diseases

Other fungal diseases that may damage alfalfa (109) include: crown-wart caused by *Physoderma alfalfae* (Lagh.) Karling (161, 173, 175, 251); Plenodomus or brown root rot caused by *Plenodomus meliloti* Dearn. & Sanford (44, 250, 290); Cylindrocarpon root rot caused by *Cylindrocarpon ehrenbergii* Wr. (45); Mycoleptodiscus crown and root rot caused by *Mycoleptodiscus terrestris* (Gerd.) Ostazeski (syn. *Leptodiscus terrestris* Gerd.) (101, 102, 206); winter crown rot or snow mold caused by an unidentified low-temperature basidiomycete (47, 49, 50, 52, 105, 179, 180); Phymatotrichum root rot caused by *Phymatotrichum omnivorum* (Shear) Dug. (syn. *Ozonium omnivorum* Shear) (110); Sclerotium or southern blight caused by *Sclerotium rolfsii* Sacc. (110, 154); Phythium seed rot, damping-off, or root rot caused by *Pythium* spp. (32, 41, 111, 119, 120); and root rots caused by *Myrothecium roridum* Tode ex Fr. and *M. verrucaria* (Alb. & Schw.) Ditm. ex Fr. (177). Limited work has been done in identifying sources of resistance and developing germplasm with resistance to these diseases.

27-2.3 Virus Diseases

27-2.3.1 Alfalfa Mosaic

No alfalfa cultivars have been developed for alfalfa mosaic virus (AMV) resistance although plants resistant to most, if not all, AMV strains can be found in many cultivars (57). However, since AMV is a complex of strains that vary in pathogenicity (9, 56, 95), resistance must be obtained to all strains in order to develop germplasm pools with significant levels of resistance.

In the greenhouse, resistant alfalfa plants are isolated by dusting young leaves with fine silicon carbide and gently rubbing them with sap ("juice") from infected plants (95). Infective juice is prepared by triturating AMV-infected plant tissue. Adding neutral phosphate buffer increases infectivity of plant juice. Silicon carbide added to infective plant juice and atomized under 0.303 to 0.404 Mpa pressure onto cotyledons of either 7- to 10-d-old seedlings or leaves of older plants, reduces the number of plants that may escape infection (183). Typical virus symptoms may not be present in AMV-infected alfalfa plants so indicator plants or serological methods are used to identify infected plants. Infection percentages following mechanical and pea aphid [*Acyrthosiphon pisum* (Harris)] inoculation of clones with AMV isolates were generally similar (95).

Resistance to an isolate of AMV was controlled by a single recessive gene in the progeny of crosses between a resistant and a susceptible clone (57). One or more dominant or partially dominant genes conferred resistance to each strain in a diallel series of crosses among five clones tested against four AMV strains (12). It is reasonable to assume that a very large number of genes will be required to confer resistance to all AMV strains in any geographic area.

27–2.3.2 Other Virus Diseases

Other virus diseases that affect alfalfa include: alfalfa enation, a serious disease in southern France and other European countries (110); transient streak, recently observed and studied in Australia (28); alfalfa latent virus, recently discovered in Nebraska, USA, and Australia (28, 296); and witches' broom, which occurs in semiarid areas of USA and Canada where it is of minor importance, Australia, and USSR (27, 104, 133, 162, 194). Little success in identifying sources of resistance or breeding for resistance has been reported.

27–2.4 Abiotic Diseases

27–2.4.1 Ozone Injury

The possibility of developing alfalfa populations with resistance to phytotoxicants is supported by research conducted at Beltsville, MD. Three experimental populations and the cultivar Team were subjected to recurrent selection for resistance to foliar injury in the greenhouse during periods of relatively high levels of air pollutants. After several cycles of selection these four sources were more tolerant to 20 mg L^{-1} O_3 than cultivars developed elsewhere (146).

27–2.4.2 Other Abiotic Diseases

Injury from other air pollutants, nutrient deficiencies and toxicities, winter injury, frost and freezing damage, flooding, and herbicide injury are also recognized as abiotic diseases (110). Increased tolerance to some of these factors may be available in the next few years.

27–2.5 Nematodes

27–2.5.1 Stem Nematode (*Ditylenchus dipsaci* [Kühn] Filipjev)

Stem nematode is the most destructive nematode that attacks alfalfa. Infested alfalfa stands may become economically unproductive within 2 or 3 yr (90).

Smith (265) was the first to compare levels of stem nematode resistance in several alfalfa cultivars. Subsequently, several additional studies comparing cultivar and germplasm resistance were reported (23, 77, 117). In the USA, the first known resistance in alfalfa to stem nematode was reported by Thorne (280) in a Turkestan introduction, FC 19304. 'Nemastan', Lahontan, and 'Washoe' (148) were developed by selection from this introduction (Fig. 27-5). The highly resistant 'Apalachee' was developed from Lahontan, DuPuits, and Flamande germplasm (258). 'Vernema' (219) and several highly resistant germplasm lines from other sources have been released (71, 75, 128, 281). Resistance to stem nematode can be found in varying degrees in most sources of *M. sativa*; and a number of other commercially available cultivars are resistant to stem nematode.

Several reports describing procedures useful in screening and evaluation for stem nematode resistance are available (23, 68, 72, 73, 112, 113, 116, 127, 258, 319). Generally, techniques that depend on the degree of cotyledonary node swelling of 7- to 24-d-old seedlings as a resistance indicator have produced variable results. More reliable techniques delay inoculation until seedlings are about 2-weeks-old, with selection of resistant plants made at 12 weeks or longer.

Nematodes for plant inoculations can be obtained from infected field and greenhouse grown plants. However, the most productive source is from monoxenic cultures of alfalfa tissue growing on a nutrient medium and maintained in the laboratory (21, 22, 91). Nematodes are extracted easily from host tissue using a modified Baermann-funnel technique (8, 68). Extracted nematodes settle in water and are concentrated by siphoning off excess water. Nematodes per milliliter are determined with a stereomicroscope. Standard inoculations are made at the rate of 200 nematodes per seedling (68).

Burkart (33) and Ragonese and Marco (233) first studied the inheritance of stem nematodes resistance assuming disomic inheritance. They concluded that resistance was controlled by a multifactorial system. Grundbacher and Stanford (117) found that a single dominant gene with tetrasomic inheritance conditioned resistance in an Iranian introduction and 'Talent'. However, in an Argentine introduction and the cultivars

Fig. 27-5. Stem-nematode-damaged stands of susceptible 'Ranger'; resistant 'Lahontan' suffered no apparent damage. Seeded 26 Aug. 1953; photographed 12 Apr. 1958 at Reno, NV.

Lahontan and DuPuits they suggested the existence of a second type of inheritance, controlled by one or more minor genes. Elgin (67) subsequently reviewed Burkart's data and presented additional information to show that inheritance to stem nematode resistance is conditioned by two complementary, dominant, tetrasomically inherited genes.

Smith (263) tested seven nematode isolates and proposed that at least two biological races of stem nematode occur on alfalfa. Conversely, Elgin et al. (74) found no evidence of significant differences among nine isolates. Although there is no clear evidence of races, Barker and Sasser (11) were able to distinguish different nematode populations by their reproductive characteristics on garden pea (*Pisum sativum* L.).

27-2.5.2 Root-Knot Nematode (*Meloidogyne* spp.)

Three species of root-knot nematodes are considered of economic importance to alfalfa. Northern root-knot nematode (*Meloidogyne hapla* Chitwood) is found in most areas where dormant or hardy alfalfa is grown. The southern root-knot nematode [*M. incognita* (Kofoid & White) Chitwood] and the Javanese root-knot nematode [*M. javanica* (Treub) Chitwood] are found primarily where nondormant to semidormant alfalfas are adapted.

Resistance to the northern root-knot nematode (*M. hapla*) is available in several sources. Stanford et al. (268) found resistance only in Vernal and 'Hilmar' in a test of 275 alfalfa lines. However, Elgin et al. (77) found resistant plants in several of 179 cultivars tested. Germplasm with high levels of resistance to *M. hapla* have been released (Fig. 27-6) (129, 220).

Resistance to *M. incognita* and *M. javanica* is found in varying de-

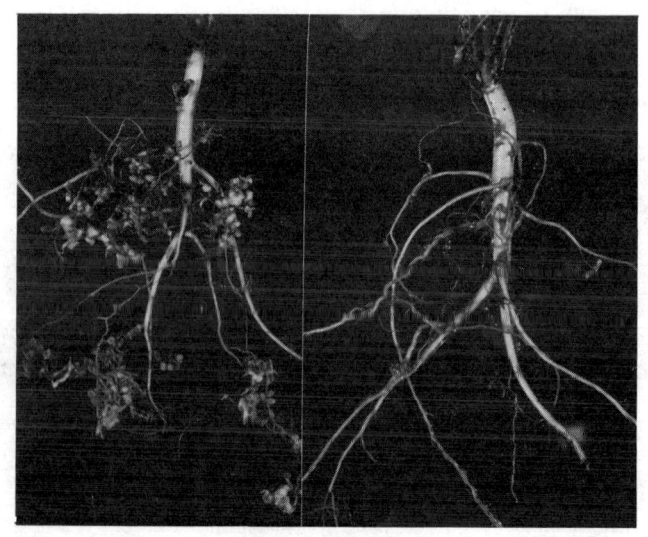

Fig. 27-6. Plant of susceptible 'Lahontan' (left) with northern root-knot nematode galling; plant of resistant Nevada Synthetic XX (right). Sixteen-week-old seedlings.

grees in many alfalfa sources (241, 242). Nondormant cultivars, such as 'African' and 'Indian', are highly resistant to both species and highly resistant germplasm is also available (129). Most dormant alfalfas possess only low frequencies of resistant plants.

Nematodes for screening are cultured on roots of tomato (*Lycopersicon esculentum* Mill.) plants (or some other susceptible host) in the greenhouse. They are extracted from the root gall tissues and concentrated using methods similar to those described earlier for stem nematode (68).

The most satisfactory method for evaluating resistance to root-knot nematodes makes use of nematode larvae applied in an aqueous suspension (68, 115, 242). A less precise method of shredding infected roots of a host plant, like tomato, and mixing them with the soil in which the seedlings are growing, may be satisfactory for screening purposes (149, 268). Optimum temperature for development of *M. hapla* is 25°C (114); whereas optimum temperature for tests with *M. incognita* and *M. javanica* is 30°C. Seedlings screened with nematodes applied in aqueous suspension usually are inoculated about 2 weeks after planting at the rate of 600 nematodes per seedling (76). Roots of seedlings are examined for nematode galls about 12 weeks later (68).

Goplen and Stanford (107) found that inheritance of resistance to *M. hapla* in two Vernal clones, M-7 and M-9, was determined by one tetrasomically inherited dominant gene. These findings were confirmed later by others (149). In the same study, (107) it was found that resistance to *M. javanica* was conditioned by a single dominant gene inherited tetrasomically in M-9 and by duplicate dominant genes in M-7. In further study with S_2 lines of M-9 it was shown that the genes controlling resistance to *M. hapla* and *M. javanica* are separate, with the possibility of linkage between the two genes. Sullivan et al. (274) working with nine different Vernal clones determined that resistance to *M. hapla* was controlled by two loci segregating tetrasomically, requiring the presence of two dominant alleles at both loci for the expression of resistance.

Goplen and Allen (106) identified three physiologic races of *M. incognita*, two of *M. javanica*, and two of *M. hapla* in a test based on five host differentials and 20 collections of root-knot nematodes. Physiologic races of root-knot nematodes were observed to evolve more rapidly on resistant than on susceptible plants (246), and their virulence on otherwise resistant plants was stable (318).

27–2.5.3 Other Nematodes

Other nematode species that are pathogenic on alfalfa include *Pratylenchus penetrans* (Cobb) Filipjev and Schuumans Stekhoven (37, 287) and *P. pratensis* (de Man) Filipjev (269), root lesion nematodes; *Paratylenchus projectus* Jenkins (288), pin nematode; *Tylenchorhynchus clarus* Allen (200), stunt nematode; *Trichodorus christiei* Allen (279), stubbyroot nematode; *Criconemella curvata* (Raski) Luc and Raski (201), ring nematode; and *Xiphinema americanum* Cobb (202, 203), dagger nematode. Research has demonstrated the destructive effect of *P. penetrans* on stand establishment (255). Also variability for *P. penetrans* resistance and screening procedures have been described in alfalfa and resistant germplasm selected (197). Additional research needs to be done with all nematodes to ascertain their economic importance, their role in the alfalfa disease complex, and the potential for developing resistant cultivars.

27–3 SUMMARY

Development of resistant cultivars is the most practical and economical means of controlling most alfalfa diseases and nematodes. Genes for resistance have been isolated and reliable screening procedures developed for many of the major alfalfa diseases and nematodes. Cultivars or germplasm are available with relatively high levels of resistance to bacterial wilt, anthracnose, Verticillium wilt, Fusarium wilt, Phytophthora root rot, rust, downy mildew, stem nematode, and root-knot nematode. Disease resistant cultivars save growers millions of dollars annually. In large areas of the USA, the availability of improved cultivars, with resistance to either a specific pathogen or more likely to a combination of pathogens (multiple pest resistance), represents the difference between economic vs. noneconomic levels of production. Increased emphasis should be placed on developing cultivars with improved levels of multiple pest resistance and in developing reliable screening methods and procedures for other pathogens that have received minimal attention.

REFERENCES

1. Adams, M.W., and G. Semeniuk. 1958. Teton alfalfa, a new multi-purpose variety for South Dakota. S. D. Agric. Exp. Stn. Bull. 469.
2. ----, and ----. 1958. The heritability of reaction in alfalfa to common leafspot. Agron. J. 50:677–679.
3. ----, and ----. 1959. A comparison of predicted with realized gain from selection for leafspot resistance in alfalfa. Agron. J. 51:91–92.
4. ----, and ----. 1959. The use of population subdivision in effecting a stratification of gene frequencies for reaction in alfalfa to *Pseudopeziza medicaginis*. Agron. J. 51:608–610.
5. Agricultural Experiment Station. 1984. Varietal trials. AD-MR-1953. University of Minnesota, St. Paul.
6. Allison, J.L., and C.H. Hanson. 1951. Methods for determining pathogenicity of *Sclerotinia trifoliorum* on alfalfa and *Rhizoctonia solani* on Lotus. Phytopathology 41:1.
7. Aubury, R.G., and H.H. Rogers. 1969. The determination of resistance to verticillium wilt (*V. albo-atrum*) in lucerne. J. Br. Grassl. Soc. 24:235–237.
8. Baermann, G. 1917. Eine einjache method zur auffindung von anklyostomum (nematoden) larven in erdproben. Gennesk. Tijkschr. Nederl. Indic 57:131–137.
9. Bancroft, J.B., E.L. Moorhead, J. Tuite, and H.P. Liu. 1960. The antigenic characteristics and the relationship among strains of alfalfa mosaic virus. Phytopathology 50:34–39.
10. Banttari, E.E., and R.D. Wilcoxson. 1964. Relation of nutrients in inoculum and inoculum concentration to severity of spring black stem of alfalfa. Phytopathology 54:1048–1052.
11. Barker, K.R., and J.N. Sasser. 1959. Biology and control of the stem nematode, *Ditylenchus dipsachi*. Phytopathology 49:664–670.

12. Barnes, D.K., and F.I. Frosheiser. 1971. Unpublished data. USDA-ARS, University of Minnesota, St. Paul.
13. ----, and ----. 1973. Registration of Agate alfalfa. Crop Sci. 13:768-769.
14. ----, C.H. Hanson, F.I. Frosheiser, and L.J. Elling. 1971. Recurrent selection for bacterial wilt resistance in alfalfa. Crop Sci. 11:545-546.
15. ----, S.A. Ostazeski, J.A. Schillinger, and C.H. Hanson. 1969. Effect of anthracnose (*Colletotrichum trifolii*) infection on yield, stand, and vigor of alfalfa. Crop Sci. 9:344-346.
16. ----, D.J. Sarajak, F.I. Frosheiser, and N.A. Anderson. 1980. Registration of alfalfa germplasm with seedling resistance to *Rhizoctaria solani*. Crop Sci. 20:675.
17. Baxter, J.W. 1956. Cercospora black stem of alfalfa. Phytopathology 46:398-400.
18. Bean, G.A., and R.D. Wilcoxson. 1961. Development of spring black stem on alfalfa and red clover. Crop Sci. 1:233-235.
19. Berger, R.D., and E.W. Hanson. 1963. Relation of environmental factors to growth and sporulation of *Cercospora zebrina*. Phytopathology 53:286-294.
20. ----, and ----. 1963. Pathogenicity, host-parasite relationships, and morphology of some forage legume Cercosporae, and factors related to disease development. Phytopathology 53:500-508.
21. Bingefors, S., and S. Bingefors. 1976. Rearing stem nematode inoculum for plant breeding purposes. Swed. J. Agric. Res. 6:13-17.
22. ----, and K.B. Eriksson. 1963. Rearing stem nematode inoculum on tissue culture. Lantbrukshoegsk. Ann. 29:107-118.
23. ----. 1961. Stem nematode in lucerne in Sweden. II. Resistance in lucerne against stem nematode. Kungl. Lantbrukshogskolans Annaler 27:385-398
24. Bordeleau, L.M., and R. Michaud. 1981. Association between resistance to bacterial wilt and symbiotic nitrogen fixation in alfalfa, *Medicago sativa*. Phytoprotection 62:39-43.
25. Borges, O.L., E.H. Stanford, and R.K. Webster. 1976. Sources and inheritance of resistance to Stemphylium leafspot of alfalfa. Crop Sci. 16:458-461.
26. ----, ----, and ----. 1976. Selection for resistance to *Stemphylium botryosum* in alfalfa. Crop Sci. 16:156-160.
27. Bowyer, J.W., J.C. Atherton, D.S. Teakle, and G.A. Alern. 1969. Mycoplasma-like bodies in plants affected by legume little leaf, tomato big bud, and lucerne witches' broom diseases. Aust. J. Biol. Sci. 22.271-274.
28. Blackstock, J. McK. 1978. Lucerne transient streak and lucerne latent, two new viruses of lucerne. Aust. J. Agric. Res. 29:291-304.
29. Brigham, R.D. 1957. Etology of and screening methods for cercospora disease of alfalfa. Iowa State Coll. J. Sci. 32:141-143.
30. ----. 1957. Stem lesions associated with Zanothomonas alfalfae. Phytopathology 47:309-310.
31. Brink, R.A., F.R. Jones, and H.R. Albrecht. 1934. Genetics of resistance to botanical wilt in alfalfa. J. Agric. Res. 49:635-642.
32. Buchholtz, W.F. 1942. Influence of cultural factors on alfalfa seedling infection by *Pythium debaryanum* Hesse. Iowa Agric. Exp. Stn. Res. Bull. 296:569-592.
33. Burkart, A. 1937. Selection of alfalfa resistant to stem nematode *Anguillulina dipsaci* (La seleccion de alfalfa al nematode del tallo). Rev. Argent. Agron. 4:171-196.
34. Campbell, T.A., S.A. Ostazeski, and C.H. Hanson. 1969. Dry inoculum for inoculating alfalfa with *Colletotrichum trifolii*. Crop Sci. 9:845-846.
35. Carnahan, H.L., J.H. Graham, and R.C. Newton. 1962. Quantitative analyses of inheritance of resistance to common leaf spot in alfalfa. Crop Sci. 2:237-240.
36. Carr, A.J.H. 1960. The significance of virus diseases in herbage crops. p. 64-66. Rep. Welsh Plant Breed. Stn., 1959.
37. Chapman, R.A. 1959. Development of *Pratylenchus penetrans* and *Tylenchorhynchus martini* on red clover and alfalfa. Phytopathology 49:357-359.
38. Chi, C.C. 1968. Nutrition stress in relation to fusarium wilt and root rot of alfalfa. Plant Dis. Rep. 52:939-943.
39. ----, and W.R. Childers. 1964. Penetration and infection in alfalfa and red clover by *Pellicularia filimentosa*. Phytopathology 54:750-754.
40. ----, and ----. 1965. Virulence of *Rhizoctonia soloni* on alfalfa and red clover. Plant Dis. Rep. 49:512-515.
41. ----, and E.W. Hanson. 1962. Interrelated effects of environment and age of alfalfa and red clover seedlings on susceptibility to *Pythium debaryanum*. Phytopathology 52:985-989.
42. Childers, W.R., and D.K. Barnes. 1972. Evolution of hybrid alfalfa. Agric. Sci. Rev. 10:11-18.
43. Cope, W.A. 1957. Inheritance of rust resistance in alfalfa. Diss. Abstr. 17:481-482.
44. Cormack, M.W. 1934. On the invasion of roots of Medicago and Melilotus by *Sclerotinia* sp. and *Plendomus meliloti* D. and S. Can. J. Res. 11:474-480.

45. ———. 1937. *Cylindrocarpon ehrenbergi* Wr. and other species as root parasites of alfalfa and sweet clover in Alberta. Can. J. Res. Sect. C:15:403-424.
46. ———. 1937. *Fusarium* spp. as root parasites of alfalfa and sweet clover in Alberta. Can. J. Res. Sect. C:15:493-510.
47. ———. 1942. Varietal resistance of alfalfa and sweet clover to root- and crown-rotting fungi in Alberta. Sci. Agric. 22:775-786.
48. ———. 1946. *Sclerotina sativa*, and related species, as root parasites of alfalfa and sweet clover in Alberta. Sci. Agric. 26:448-449.
49. ———. 1948. Winter crown rot or snow mold of alfalfa, clovers, and grasses in Alberta. I. Occurrence, parasitism, and spread of the pathogen. Can. J. Res. Sect. C:26:71-85.
50. ———. 1952. Winter crown rot or snow mold of alfalfa, clovers, and grasses in Alberta. II. Field studies on host and varietal resistance and other factors related to control. Can. J. Bot. 30:537-548.
51. ———. 1961. Longevity of the bacterial wilt organism in alfalfa hay, pod debris, and seed. Phytopathology 5:260-261.
52. ———, and J.B. LeBeau. 1959. Snow mold infection of alfalfa, grasses, and winter wheat by several fungi under artificial conditions. Can. J. Bot. 37:685-693.
53. ———, R.W. Peake, and R.K. Downey. 1957. Studies on methods and materials for testing alfalfa for resistance to bacterial wilt. Can. J. Plant Sci. 37:1-11.
54. Cowling, W.A., and D.G. Gilchrist. 1980. Influence of the pathogen on disease severity in Stemphylium leafspot of alfalfa in California. Phytopathology 70:1148-1153.
55. ———, and J.H. Graham. 1981. Biotypes of *Stemphylium botryosum* on alfalfa in North America. Phytopathology 71:679-684.
56. Crill, P., D.J. Hagedorn, and E.W. Hanson. 1971. An artificial system for differentiating strains of alfalfa mosaic virus. Plant Dis. Rep. 55:127-130.
57. ———, E.W. Hanson, and D.J. Hagedorn. 1971. Resistance and tolerance to alfalfa mosaic virus in alfalfa. Phytopathology 61:369-371.
58. Davis, R.L. 1951. Study of the inheritance of resistance in alfalfa to common leaf spot. Agron. J. 43:331-337.
59. Davis, W.H., and I.M. Greenblatt. 1967. Cytoplasmic male sterility in alfalfa. J. Hered. 58:301-305.
60. Devine, T.E., C.H. Hanson, S.A. Ostazeski, and T.A. Campbell. 1971. Selection for resistance to anthracnose (*Colletotrichum trifolii*) in four alfalfa populations. Crop Sci. 11:854-855.
61. ———, ———, ———, and O.J. Hunt. 1973. Registration of alfalfa germplasm. Crop Sci. 13:289.
62. ———, R.H. Ratcliffe, C.M. Rincker, D.K. Barnes, S.A. Ostazeski, T.H. Busbice, C.H. Hanson, J.A. Schillinger, G.R. Buss, and R.W. Cleveland. 1975. Registration of Arc alfalfa. Crop Sci. 15:97.
63. Dickason, E.A., C.M. Leach, and A.E. Gross. 1968. Clover root curculio injury and vascular decay of alfalfa roots. J. Econ. Entomol. 61:1163-1168.
64. Donnelly, E.D. 1952. Breeding for wilt resistance in alfalfa. Agron. J. 44:562-568.
65. Dudley, J.W., R.R. Hill, Jr., and C.H. Hanson. 1963. Effects of seven cycles of recurrent phenotypic selection on means and genetic variances of several characters in two pools of alfalfa germ plasm. Crop Sci. 3:543-546.
66. Edmunds, L.K., and E.W. Hanson. 1960. Host range, pathogenicity, and taxonomy of *Ascochyta imperfecta*. Phytopathology 50:105-108.
67. Elgin, J.H., Jr. 1979. Inheritance of stem-nematode resistance in alfalfa. Crop Sci. 19:352-354.
68. ———, et al. 1984. Standard tests to characterize pest resistance in alfalfa cultivars. USDA Misc. Pub. 1434. U.S. Government Printing Office, Washington, DC.
69. ———, D.K. Barnes, T.H. Busbice, G.R. Buss, N.A. Clarke, R.W. Cleveland, R.L. Ditterline, D.W. Evans, S.C. Fransen, R.D. Horrocks, O.J. Hunt, W.R. Kehr, C.C. Lowe, D.A. Miller, M.S. Offutt, R.C. Pickett, E.L. Sorensen, C.M. Taliaferro, M.B. Tesar, and R.W. Van Keuren. 1981. Anthracnose resistance increases alfalfa yields. Crop Sci. 21:457-460.
70. ———, and E.H. Beyer. 1968. Evaluation of selected alfalfa clones for resistance to *Sclerotinia trifoliorum* Eriks. Crop Sci. 8:265-266.
71. ———, and D.W. Evans. 1978. Registration of WDS3P1 and W1S1P1 alfalfa germplasm. Crop Sci. 18:530.
72. ———, ———, and L.R. Faulkner. 1975. Factors affecting the infection of alfalfa seedlings by *Ditylenchus dipsaci*. J. Nematol. 7:380-383.
73. ———, ———, and ———. 1975. Evaluation of alfalfa for stem-nematode resistance. Crop Sci. 15:275-276.
74. ———, ———, and ———. 1977. Response of resistant and susceptible alfalfa cultivars to regional isolates of stem nematode. Crop Sci. 17:957-959.
75. ———, ———, and R.N. Peaden. 1978. Registration of 18 germplasm populations of alfalfa. Crop Sci. 18:529-530.

76. ----, F.A. Gray, R.N. Peaden, L.R. Faulkner, and D.W. Evans. 1973. Optimum inoculum levels for screening alfalfa seedlings for resistance to northern root-knot nematode in a controlled environment. Plant Dis. Rep. 57:657–660.
77. ----, B.J. Hartman, D.W. Evans, B.D. Thyr, L.R. Faulkner, and O.J. Hunt. 1980. Stem nematode and northern root-knot nematode resistance ratings for alfalfa cultivars and experimental lines. USDA-SEA, ARR-NE-7. U.S. Government Printing Office, Washington, DC.
78. ----, R.R. Hill, Jr., and K.E. Zeiders. 1970. Comparison of four methods of multiple trait selection for five traits in alfalfa. Crop Sci. 10:190–193.
79. ----, and S.A. Ostazeski. 1982. Evaluation of selected alfalfa cultivars and related *Medicago* species for resistance to race 1 and race 2 anthracnose. Crop Sci. 22:39–42.
80. ----, and ----. 1983. Inheritance of resistance to race 1 and race 2 of *Colletotrichum trifolii* in Arc and Saranac AR alfalfa. Agron. Abstr. American Society of Agronomy, Madison, WI, p. 62.
81. ----, and ----. 1984. Registration of W10AnWFuPy3, (B2An4 × Arc) AnWFuPy3, and B28 multiple pest resistant alfalfa germplasms. Crop Sci. 24:633.
82. ----, and ----. 1984. Registration of eleven race 1 and race 2 anthracnose resistant alfalfa germplasm lines. Crop Sci. 24:633.
83. ----, J.E. McMurtrey III, B.J. Hartman, B.D. Thyr, E. L. Sorensen, D.K. Barnes, F.I. Frosheiser, R.N. Peaden, R.R. Hill, Jr., and K.T. Leath. 1983. Use of strain crosses in the development of multiple pest resistant alfalfa with improved field performance. Crop Sci. 23:57–64.
84. Elling, L.J., and F.I. Frosheiser. 1960. Reaction of twenty-two alfalfa varieties to bacterial wilt. Agron. J. 52:241–242.
85. Elliott, E.S., R.E. Baldwin, and R.B. Carroll. 1969. Root rots of alfalfa and red clover. W. V. Agric. Exp. Stn. Bull. 585T.
86. Emberger, G., and R.E. Welty. 1983. Evaluation of virulence of *Fusarium oxysporum* f. sp. *medicaginis* and Fusarium wilt resistance in alfalfa. Plant Dis. 67:94–98.
87. Erwin, D.C. 1954. Relation of *Stagonospora, Rhizoctonia* and associated fungi to crown rot of alfalfa. Phytopathology 44:137–144.
88. ----. 1965. Reclassification of the causal agent of root rot of alfalfa from *Phytophthora cryptogea* to *P. megasperma*. Phytopathology 55:1139–1143.
89. ----. 1966. Varietal reaction of alfalfa to *Phytophthora megasperma* and variation in virulence of the causal fungus. Phytopathology 56:653–657.
90. Faulkner, L.R., and W.J. Bolander. 1967. Occurrence of large nematode poulations in irrigation canals of south-central Washington. Nematologica 12:591–600.
91. ----, D.B. Bower, D.W. Evans, and J.H. Elgin, Jr. 1974. Mass culturing of *Ditylenchus dipsaci* to yield large quantities of inoculum. J. Nematol. 6:126–129.
92. Frosheiser, F.I. 1960. A method for screening alfalfa plants for resistance to *Pseudopeziza medicaginis*. Phytopathology 50:568.
93. ----. 1966. Alfalfa cotyledon inoculation with bacterial wilt inoculum prepared from infected alfalfa roots. Phytopathology 56:566–567.
94. ----. 1967. *Phytophthora* root rot of alfalfa in Minnesota. Plant Dis. Rep. 51:679–681.
95. ----. 1969. Variable influence of alfalfa mosaic virus strains on growth and survival of alfalfa and on mechanical and aphid transmission. Phytopathology 59:857–862.
96. ----, and D.K. Barnes. 1980. Registration of alfalfa germplasm with Fusarium wilt resistance. Crop Sci. 20:553.
97. ----, and ----. 1973. Field and greenhouse selection for *Phytophthora* root rot resistance in alfalfa. Crop Sci. 13:735–738.
98. ----, and ----. 1978. Field reaction of artificially inoculated alfalfa populations to the Fusarium and bacterial wilt pathogens alone and in combination. Phytopathology 68:943–946.
99. Fulkerson, J.F. 1960. Pathogenicity and stability of strains of *Corynebacterium insidiosum*. Phytopathology 50:377–380.
100. Fyfe, J.L. 1964. Hereditary variation in resistance to verticillium wilt within cultivated lucerne. J. Agric. Sci. 63:273–276.
101. Gerdemann, J.W. 1953. An undescribed fungus causing a root rot of red clover and other Leguminosae. Mycologia 45:548–554.
102. ----. 1954. Pathogenicity of *Leptodiscus terrestris* on red clover and other Leguminosae. Phytopathology 44:451–455.
103. Gilchrist, D.G., L.R. Teuber, A.M. Martensen, and W.A. Cowling. 1982. Progress in selecting for resistance to *Stemphylium botryosum* (cool-temperature biotype) in alfalfa. Crop Sci. 22:1155–1159.
104. Glover, D.V., and D.R. McAllister. 1960. Transmission studies of alfalfa witches' broom virus in Utah. Agron. J. 52:63–65.
105. Goplen, B.P. 1969. Legume Breeding. p. 225–259. *In* Canadian Forage Crops Symposium. Modern Press, Saskatoon, Canada.

106. ——, and M.W. Allen. 1959. Demonstration of physiological races within three root-knot nematode species attacking alfalfa. Phytopathology 49:653–656.
107. ——, and E.H. Stanford. 1960. Autotetraploidy and linkage in alfalfa—a study of resistance to two species of root-knot nematodes. Agron. J. 52:337–342.
108. Graham, J.H. 1960. Rapid screening of alfalfa for resistance to *Corynebacterium insidiosum* by inoculating petioles. Phytopathology 50:637.
109. ——, R.R. Hill, Jr., D.K. Barnes, and C.H. Hanson. 1965. Effects of three cycles of selection for resistance to common leafspot in alfalfa. Crop Sci. 5:171–173.
110. ——, D.L. Stuteville, F.I. Frosheiser, and D.C. Erwin. 1979. A compendium of alfalfa diseases. American Phytopathological Society, St. Paul.
111. Grandfield, C.O., C.L. Lefebvre, and W.H. Metzger. 1935. Relation between fallowing and the damping-off of alfalfa seedlings. J. Am. Soc. Agron. 27:800–806.
112. Griffin, G.D. 1967. Evaluation of several techniques for screening alfalfa for resistance to *Ditylenchus dipsaci*. Plant Dis. Rep. 51:651–654.
113. ——. 1968. The pathogenicity of *Ditylenchus dipsaci* to alfalfa and the relationship of temperature to plant infection and susceptibility. Phytopathology 58:929–932.
114. ——. 1969. Effects of temperature on *Meloidogyne hapla* in alfalfa. Phytopathology 59:599–602.
115. ——, and J.H. Elgin, Jr. 1977. Penetration and development of *Meloidogyne hapla* in resistant and susceptible alfalfa under differing temperatures. J. Nematol. 9:51–56.
116. Grundbacher, F.J. 1962. Testing alfalfa seedlings for resistance to the stem nematode *Ditylenchus dipsaci* (Kuhn) Filipjev. Proc. Helminthol. Soc. Wash. 29:152–158.
117. ——, and E.H. Stanford. 1962. Genetic factors conditioning resistance in alfalfa to the stem nematode. Crop Sci. 2:211–217.
118. Haag, W.L., and R.R. Hill, Jr. 1974. Comparison of selection methods for autotetraploids. II. Selection for disease resistance in alfalfa. Crop Sci. 14:591–593.
119. Halpin, J.E., and E.W. Hanson. 1958. Effect of age of seedlings of alfalfa, red clover, Ladino white clover, and sweetclover on susceptibility to Pythium. Phytopathology 48:481–485.
120. ——, ——, and J.G. Dickson. 1954. Studies on the pathogenicity of seven species of Pythium on alfalfa, sweetclover, and Ladino clover seedlings. Phytopathology 44:572–574.
121. Hanson, C.H. 1969. Registration of alfalfa germplasm. Crop Sci. 9:526–527.
122. ——, T.H. Busbice, R.R. Hill, Jr., O.J. Hunt, and A.J. Oakes. 1972. Directed mass selection for developing multiple pest resistance and conserving germplasm in alfalfa. J. Environ. Qual. 1:106–111.
123. ——, C.S. Garrison, and H.O. Graumann. 1960. Alfalfa varieties in the United States. USDA-ARS Handb. 177. U.S. Government Printing Office, Washington, DC.
124. ——, C.H. Hanson, F.I. Frosheiser, E.L. Sorensen, R.T. Sherwood, J.H. Graham, L.J. Elling, D. Smith, and R.L. Davis. 1964. Reactions of varieties, crosses, and mixtures of alfalfa to six pathogens and the potato leafhopper. Crop Sci. 4:273–276.
125. ——, G.M. Loper, G.O. Kohler, E.M. Bickoff, K.W. Taylor, W.R. Kehr, E.H. Stanford, J.W. Dudley, M.W. Pedersen, E.L. Sorensen, H.L. Carnahan, and C.P. Wilsie. 1965. Variation in coumestrol content of alfalfa as related to location, variety, cutting, year, stage of growth, and disease. USDA-ARS Tech. Bull. 1333. U.S. Government Printing Office, Washington, DC.
126. ——, and W.K. Smith. 1964. Reactions of some alfalfa varieties and synthetics to downy mildew. Crop Sci. 4:229.
127. Hanna, M.R., and E.J. Hawn. 1964. Seedling inoculation studies with alfalfa stem nematode. Can. J. Plant Sci. 45:357–363.
128. Hartman, B.J., O.J. Hunt, J.H. Elgin, Jr., R.N. Peaden, and B.D. Thyr. 1979. Registration of Washington SNI alfalfa germplasm. Crop Sci. 19:416, 20:4120
129. ——, ——, R.N. Peaden, H.J. Jensen, B.D. Thyr, L. R. Faulkner, and G.D. Griffin. 1979. Registration of Nevada Synthetic YY nondormant alfalfa germplasm. Crop Sci. 19:416.
130. Hawn, E.J. 1963. Transmission of bacterial wilt of alfalfa by *Ditylenchus dipsaci* (Kuhn). Nematologica 9:65–68.
131. ——, and M.R. Hanna. 1967. Influence of stem nematode infestation on bacterial wilt reaction and forage yield of alfalfa varieties. Can. J. Plant Sci. 47:203–208.
132. ——, and J.B. Lebeau. 1962. Antibiosis in bacterial wilt of alfalfa. Phytopathology 52:266–268.
133. Helms, K. 1957. Witches' broom disease of lucerne. I. The occurrence of two strains of the disease and their relationship to big bud of tomato. Aust. J. Agric. Res. 8:135–147.
134. Henderson, R.G., and T.J. Smith. 1948. Relative susceptibility of alfalfa varieties to certain foliage diseases. Phytopathology 38:570.
135. Hijano, E.H., D.K. Barnes, and F.I. Frosheiser. 1983. Inheritance of resistance to Fusarium wilt in alfalfa. Crop Sci. 23:31–34.

136. Hill, R.R., Jr., C.H. Hanson, and T.H. Busbice. 1969. Effect of four recurrent selection programs on two alfalfa populations. Crop Sci. 9:363-365.
137. ----, and K.T. Leath. 1972. Genetic variance for reaction to five foliar pathogens in alfalfa. Crop Sci. 12:813-816.
138. ----, and ----. 1979. Comparison of four methods of selection for resistance to *Leptosphaerulina briosiana* in alfalfa. Can. J. Genet. Cytol. 21:179-186.
139. ----, J.J. Murray, and K.E. Zeiders. 1971. Relationships between clover root curculio injury and severity of bacterial wilt in alfalfa. Crop Sci. 11:306-307.
140. ----, R.C. Newton, K.E. Zeiders, and J.H. Elgin, Jr. 1969. Relationships of the clover root curculio, *Fusarium* wilt, and bacterial wilt in alfalfa. Crop Sci. 9:327-329.
141. ----, R.T. Sherwood, and J.W. Dudley. 1963. Effect of recurrent phenotypic selection on resistance of alfalfa to two physiological races of *Uromyces striatus medicaginis*. Phytopathology 53:432-435.
142. Hohrein, B.A., G.A. Bean, and J.H. Graham. 1983. Greenhouse technique to evaluate alfalfa resistance to *Phytophthora megasperma* f. sp. *medicaginis*. Plant Dis. 67:1332-1333.
143. Houston, B.R. 1949. Dwarf resistant alfalfa. Calif. Agric. 3:3, 10.
144. ----, and E.H. Stanford. 1954. Resistance to the dwarf disease in alfalfa. Phytopathology 44:493.
145. Howe, W.L., W.R. Kehr, and C.O. Calkins. 1965. Appraisal for combined pea aphid and spotted alfalfa aphid resistance. Nebr. Agric. Exp. Stn. Res. Bull. 221.
146. Howell, R.K., T.E. Devine, and C.H. Hanson. 1971. Resistance of selected alfalfa strains to ozone. Crop Sci. 11:114-115.
147. Hunt, O.J., G.D. Griffin, J.J. Murray, M.W. Pedersen, and R.N. Peaden. 1971. The effects of root knot nematodes on bacterial wilt in alfalfa. Phytopathology 61:256-259.
148. ----, R.N. Peaden, H.L. Carnahan, and F.V. Lieberman. 1966. Registration of Washoe alfalfa. Crop Sci. 6:610.
149. ----, R.N. Peaden, L.R. Faulkner, G.D. Griffin, and H.J. Jensen. 1969. Development of resistance to root-knot nematode (*Meloidogyne hapla* Chitwood) in alfalfa (*Medicago sativa* L.). Crop Sci:9:624-627.
150. Irwin, J.A.G., D.P. Maxwell, and E.T. Bingham. 1981. Inheritance of resistance to *Phytophthora megasperma* in diploid alfalfa. Crop Sci. 21:271-276.
151. ----, ----, and ----. 1981. Inheritance of resistance to *Phytophthora megasperma* in tetraploid alfalfa. Crop Sci. 21:277-283.
152. Issac, I. 1957. Wilt of lucerne caused by species of Verticillium. Ann. Appl. Biol. 45:550-558.
153. ----, and A.T.E. Lloyd. 1959. Wilt of lucerne caused by species of Verticillium. II. Seasonal cycle of disease; range of pathogenicity; host-parasite relationship; effect of seed dressings. Ann. Appl. Biol. 47:673-684.
154. Johnson, H.W. 1968. Alfalfa diseases. Miss. Agric. Exp. Stn. Bull. 767.
155. Jones, F.R. 1930. Bacterial wilt of alfalfa. J. Am. Soc. Agron. 22:568-572.
156. ----. 1934. Testing alfalfa for resistance to bacterial wilt. J. Agric. Res. 48:1085-1098.
157. ----. 1945. Winter injury and longevity in unselected clones from four wilt-resistant varieties of alfalfa. J. Am. Soc. Agron. 37:828-838.
158. ----. 1949. Resistance in alfalfa to yellow leaf blotch. Phytopathology 39:1064-1065.
159. Jones, F.R. 1953. Measurement of resistance in alfalfa to common leaf spot. Phytopathology 43:651-654.
160. ----, J.L. Allison, and W.K. Smith. 1941. Evidence of resistance in alfalfa, red clover, and sweet clover to certain fungus parasites. Phytopathology 31:765-766.
161. ----, and C. Dreschler. 1920. Crown wart of alfalfa caused by *Urophlyctis alfalfae*. J. Agric. Res. 20:293-324.
162. ----, and O.F. Smith. 1953. Sources of healthier alfalfa. p. 228-237. USDA Agric. Yearbook.
163. ----, and J.H. Torrie. 1946. Systemic infection of downy mildew in soybean and alfalfa. Phytopathology 36:1057-1059.
164. ----, and J.L. Weimer. 1938. Stagonospora leaf spot and root rot of forage legumes. J. Agric. Res. 57:791-812.
165. Kehr, W.R. 1970. Registration of N.S. 16 alfalfa germplasm. Crop Sci. 10:731.
166. ----, E.C. Conard, M.A. Alexander, and F.G. Owen. 1963. Performance of alfalfas under five management systems. Nebr. Agric. Exp. Stn. Res. Bull. 211.
167. Kernkamp, M.F., J.W. Gibler, and L.J. Elling. 1949. Damping-off of alfalfa cuttings caused by *Rhizoctonia solani*. Phytopathology 39:928-935.
168. ----, and G. Hemerick. 1952. A "deep-freeze" method of maintaining virulent inoculum of the alfalfa wilt bacterium, *Corynebacterium insidiosum*. Phytopathology 42:13.
169. Kochler, B., and F.R. Jones. 1932. Alfalfa wilt as influenced by soil temperature and soil moisture. Ill. Agric. Exp. Stn. Bull. 378.

170. Koepper, J.M. 1942. Relative resistance of alfalfa species and varieties to rust caused by *Uromyces straitus*. Phytopathology 32:1048–1057.
171. Kreitlow, K.W. 1962. Verticillium wilt of alfalfa—a destructive disease in Britain and Europe not yet observed in the U.S. USDA ARS Bull. 34–20. U.S. Government Printing Office, Washington, DC.
172. ———. 1963. Infecting seven-day-old alfalfa seedlings with wilt bacteria through wounded cotyledons. Phytopathology 53:800–803.
173. Leach, C.M., and J.R. Hardison. 1959. Susceptibility of alfalfa varieties to *Physoderma alfalfae*. Plant Dis. Rep. 43:619–621.
174. Leath, K.T. 1971. Quality of light required for sporulation by Leptosphaerulina. Phytopathology 61:70–72.
175. ———. 1978. Crown wart of alfalfa in Pennsylvania. Plant Dis. Rep. 62:621–623.
176. ———, and W.A. Kendal. 1978. Fusarium root rot of forage species: Pathogenicity and host range. Phytopathology 68:826–831.
177. ———, and ———. 1983. *Myrothecium roridum* and *M. verrucaria* pathogenic to roots of red clover and alfalfa. Plant Dis. 67:1154–1155.
178. ———, and R.C. Newton. 1969. Interaction of a fungus gnat, *Bradysia* sp. (Sciaridae) with *Fusarium* spp. on alfalfa and red clover. Phytopathology 59:257–258.
179. Lebeau, J.B., and T.G. Atkinson. 1967. Borax inhibition of cyanogenesis in snow mold of alfalfa. Phytopathology 57:863–865.
180. ———, M.W. Cormack, and J.E. Moffatt. 1959. Measuring pathogenesis by the amount of toxic substance produced in alfalfa by a snow mold fungus. Phytopathology 49:303–305.
181. Lehman, W.F., D.C. Erwin, and E.H. Stanford. 1969. Registration of *Phytophthora*-tolerant alfalfa germplasm, UC 38 and UC 47. Crop Sci. 9:527.
182. ———, and E.H. Stanford. 1970. Notice of release of C937 to alfalfa breeders. Calif. Agric. Exp. Stn. Mimeo.
183. Lindner, R.E., and H.C. Kirkpatrick. 1959. The airbrush as a tool in virus inoculations. Phytopathology 49:507–509.
184. Lu, N.S.-J., D.K. Barnes, and F.I. Frosheiser. 1973. Inheritance of Phytophthora root rot resistance in alfalfa. Crop Sci. 13:714–717.
185. Lucas, L.T., T.H. Busbice, and D.S. Chamblee. 1973. Resistance to Stemphylium leafspot in a new alfalfa variety. Plant Dis. Rep. 57:946–948.
186. Martinez, E.X., and E.W. Hanson. 1963. Factors affecting growth, sporulation, pathogenicity, and dissemination of *Leptosphaerulina briosiana*. Phytopathology 53:938–945.
187. McDonald, W.C. 1961. Note on a method of producing and storing inoculum of the alfalfa black stem fungus, *Ascochyta imperfecta* Pk. Can. J. Plant Sci. 41:447–448.
188. McGuire, J.H., H.J. Walters, and D.A. Slack. 1958. The relationship of root-knot nematodes to the development of Fusarium wilt in alfalfa. Phytopathology 48:344.
189. McKenzie, J.S., and J.G.N. Davidson. 1975. Prevalence of alfalfa crown and root diseases in the Peace River Region of Alberta and British Columbia. Can. Plant Dis. Surv. 55:121–125.
190. McMurtrey, J.E., III, M.K. Aycock, Jr., and J.H. Elgin, Jr. 1979. Examination of the inheritance of rust (*Uromyces striatus* (Schroet.)) resistance in five alfalfa populations. Agron. Abstr., American Society of Agronomy, Madison, WI, p. 69.
191. Mead, H.W. 1963. Comparison of temperatures favoring growth of *Ascochyta imperfecta* Peck and development of spring black stem on alfalfa in Saskatchewan. Can. J. Bot. 41:312–314.
192. ———, and M.W. Cormack. 1961. Studies on *Ascochyta imperfecta* Peck. Parasitic strains among fifty isolates from Canadian alfalfa seed. Can. J. Bot. 39:793–797.
193. Melton, B.A. 1968. Messillo alfalfa. N. M. Agric. Exp. Stn. Bull. 530.
194. Menzies, J.D. 1946. Witches' broom in alfalfa in North America. Phytopathology 36:762–774.
195. Minion, G.D. 1964. Heritability of resistance in alfalfa selections to *Cercospora medicaginis* Ellis and Everhart. Diss. Abstr. 25:2697.
196. Murphy, R.P., and C.C. Lowe. 1966. Registration of Saranac alfalfa. Crop Sci. 6:611.
197. Nelson, D.L., D.K. Barnes, D.H. MacDonald. 1984. Field and growth chamber evaluations for root lesion nematode resistance in alfalfa. Crop Sci. 25:35–39.
198. Nelson, G.A. 1959. Antagonism between strains of *Corynebacterium insidiosum*. Phytopathology 49:547.
199. Nittler, L.W., G.W. McKee, and J.L. Newcomer. 1964. Principles and methods of testing alfalfa seed for varietal purity. N.Y. State Agric. Exp. Stn. Bull. 807.
200. Noel, G.R., and B.F. Lownsbery. 1978. Effects of temperature on the pathogenicity of *Tylenchorhynchus clarus* to alfalfa and observations on feeding. J. Nematol. 10:195–198.
201. ———, and ———. 1984. Pathoginicity of *Crieonemella curvata* to alfalfa. J. Nematol. 10:140–145.

202. Norton, D.C. 1965. *Xiphenema americanum* populations and alfalfa yields as affected by soil treatment, spraying and cutting. Phytopathology 55:615–619.
203. ----. 1963. Population fluctuation of *Xiphinema americanum* in Iowa. Phytopathology 53:66–68.
204. ----. 1969. Meloidogyne hapla as a factor in alfalfa decline in Iowa. Phytopathology 59:1824–1828.
205. O'Rourke, C.J., and R.L. Millar. 1966. Root rot and root microflora of alfalfa as affected by potassium nutrition, frequency of cutting, and leaf infection. Phytopathology 56:1040–1046.
206. Ostazeski, S.A. 1967. An undescribed fungus associated with a root and crown rot of birdsfoot trefoil (*Lotus corniculatus*). Mycologia 59:970–975.
207. ----, D.K. Barnes, and C.H. Hanson. 1969. Laboratory selection of alfalfa for resistance to anthracnose, *Colletotrichum trifolii*. Crop Sci. 9:351–354.
208. ----, and J.H. Elgin, Jr. 1980. Physiologic races of *Colletotrichum trifolii*. Phytopathology 70:691.
209. ----, and ----. 1981. Hypodermic inoculations of *Colletotrichum trifolii* in alfalfa: Rapid race identification and host reaction determination for anthracnose isolates. Phytopathology 71:247.
210. ----, ----, and J.E. McMurtrey III. 1979. Occurrence of anthracnose on formerly anthracnose-resistant 'Arc' alfalfa. Plant Dis. Rep. 63:734–736.
211. Pandey, M.C., and R.D. Wilcoxson. 1967. Effect of carbon and nitrogen nutrition on reproduction in *Leptosphaerulina briosiana*. Am. J. Bot. 54:1170–1175.
212. Pandey, M.C., and R.D. Wilcoxson. 1970. The effect of light and physiologic races on Leptosphaerulina leaf spot of alfalfa and selection of resistance. Phytopathology 60:1456–1462.
213. Panella, A., M. Ribaldi, and F. Lorenzetti. 1969. Screening of alfalfa lines for resistance to verticillium wilt. Phytopathol. Mediterr. 8:116–123.
214. Panton, C.A. 1964. A review of some aspects of the wilt pathogen *Verticillium albo-atrum* Rke. et Berth. Acta Agric. Scand. 14:97–112.
215. ----. 1965. The breeding of lucerne, *Medicago sativa* L., for resistance to *Verticillium albo-atrum* Rke. et Berth. Acta Agric. Scand. 15:85–100.
216. ----. 1966. Genetic control of resistance of lucerne, *Medicago sativa* L., to *Verticillium albo-atrum* Rke. et Berth. p. 741–745. *In* A.G.G. Hill (ed.) Int. Grassland Congr. Proc., 10th, Finland. Finnish Grassland Association, Helsinki.
217. ----. 1967. The breeding of lucerne, *Medicago sativa* L. for resistance to *Verticillium albo-atrum* Rke. et Berth. II. The quantitative nature of the genetic mechanism controlling resistance in inbred and hybrid generations. Acta Agric. Scand. 17:43–52.
218. Patel, M.K. 1926. Study of *Peronospora trifoliorum* DeBy. on species of Leguminosae. Phytopathology 16:72.
219. Peaden, R.N., D.W. Evans, J.H. Elgin, Jr., and C.M. Rincker. 1983. Registration of Vernema alfalfa. Crop Sci. 23:1009.
220. ----, O.J. Hunt, L.R. Faulkner, G.D. Griffin, H.J. Jensen, and E.H. Stanford. 1976. Registration of multiple-pest resistant alfalfa germplasm. Crop Sci. 16:125.
221. Pearson, L.C., and L.J. Elling. 1960. Predicting disease resistance in synthetic varieties of alfalfa from clonal cross data. Agron. J. 52:291–294.
222. Pedersen, M.W., and D.K. Barnes. 1965. Inheritance of downy mildew resistance in alfalfa. Crop Sci. 5:4–5.
223. ----, and D.R. McAllister. 1961. Unita alfalfa. Utah Farm Home Sci. 22:97, 109.
224. Peltier, G.L., and F.R. Schroeder. 1932. The nature of resistance in alfalfa to wilt. Nebr. Agric. Exp. Stn. Res. Bull. 63.
225. ----, and H.M. Tysdal. 1930. The relative susceptibility of alfalfa to wilt and cold. Nebr. Agric. Exp. Stn. Res. Bull. 52.
226. ----, and ----. 1934. Wilt and cold resistance of self fertilized lines of alfalfas. Nebr. Agric. Exp. Stn. Res. Bull. 76.
227. Perisic, M., and D. Stojanovic. 1966. A study of the biology of *Pseudopeziza jonesii* Nannf (*Pyrenopeziza medicaginis* Tuck) the causes of yellow blotches on leaves of alfalfa. Zast. Bilja 17:183–190.
228. Peterson, M.L., and L.E. Melchers. 1942. Studies on black stem of alfalfa caused by *Ascochyta imperfecta*. Phytopathology 32:590–597.
229. Peterson, V.S. 1965. Methods of inoculation and diagnosis of *Verticillium albo-atrum* in lucerne. Den. Kgl. Veterinacr-og Land. Arss. 1965:108–120.
230. Pitcher, R.S. 1963. Role of plant-parasitic nematodes in bacterial diseases. Phytopathology 53:35.
231. ----. 1965. Interrelationships of nematodes and other pathogens of plants. Helmenthol. Abstr. 34:1–17.
232. Powell, N.T. 1963. The role of plant-parasitic nematodes in fungus diseases. Phytopathology 53:28–35.

233. Ragonese, A.E., and Y.P.R. Marco. 1943. Resistencia al nematode del tallo de diversas lineas y procedencias de alfalfas. Rev. Argent. Agron. 10:378-384.
234. Raynal, G. 1978. Artificial inoculation of lucerne with pepper spot agent, *Leptosphaerulina briosiana*: Development, sensibility of 30 cultivars, variation of pathogenicity. Rev. Zool. Agric. Pathol. Veg. 77:1-13.
235. ----. 1981. Red clover and alfalfa crown rot caused by *Sclerotinia trifoliorum* Erik. II. Pathogen variability and plant resistance in controlled conditions. Agronomie 1:573-578
236. ----. 1981. Red clover and alfalfa crown rot caused by *Sclerotinia trifoliorum* Erik. I. Choice of an inoculation method. Agronomie 1:565-572.
237. Reitz, L.P., C.O. Grandfield, M.L. Peterson, G.V. Goodding, M.A. Arneson, and E.D. Hansing. 1948. Reactions of alfalfa varieties, selections, and hybrids to *Ascochyta imperfecta*. J. Agric. Res. 76:307-323.
238. Renfro, B.L., and E.W. Sprague. 1959. Reaction of *Medicago* species to eight alfalfa pathogens. Agron. J. 51:481-483.
239. ----, and R.D. Wilcoxson. 1963. Production and storage of inoculum of *Phoma herbarium* var. *medicaginis*. Plant Dis. Rep. 47:168-169.
240. ----, and ----. 1963. Spring black stem of alfalfa in relation to temperature, moisture, wounding, and nutrients and some observations on pathogen dissemination. Phytopathology 53:1340-1345.
241. Reynolds, H.W. 1955. Varietal susceptibility of alfalfa to two species of root-knot nematodes. Phytopathology 45:70-72.
242. ----, and J.H. O'Bannon. 1960. Reaction of sixteen varieties of alfalfa to two species of root-knot nematodes. Plant Dis. Rep. 44:441-443.
243. Richard, C. 1977. Le *Leptosphaerulina briosiana* sur la luzerne au Quebec. Phytoprotection 58:37-38.
244. ----, R. Michaud, A. Freve, and C. Gagnon. 1980. Selection for root and crown rot resistance in alfalfa. Crop Sci. 20:691-695.
245. Richards, B.L. 1936-1937. Reaction of alfalfa varieties to stem blight. Proc. Utah Acad. Sci., Arts, Lett. 14:33-38.
246. Riggs, R.D., and N.N. Winstead. 1959. Studies on resistance in tomato to root-knot nematodes and on the occurrence of pathogenic biotypes. Phytopathology 49:716-724.
247. Rotar, P.P., and W.R. Kehr. 1963. Nebr. Agric. Exp. Stn. Res. Bull. 209.
248. Rumbaugh, M.D., G. Semeniuk, and H.A. Geise. 1962. Inheritance of reaction of diploid alfalfa clones to two isolates of *Phoma herbarum* var. *medicaginis*. Crop Sci. 2:13-15.
249. Sackett, W.G. 1910. A bacterial disease of alfalfa. Colo. Agric. Exp. Stn. Bull. 158.
250. Sanford, G.D. 1933. A root rot of sweet clover and related crops caused by *Plenodomus meliloti* Dearness and Sanford. Can. J. Res. 8:337-348.
251. Schoth, H.A., L.G. Gentner, and H.H. White. 1952. Talent alfalfa. Oreg. Agric. Exp. Stn. Bull. 511.
252. Semeniuk, G. 1971. Diseased leaves as a source of ascospores of *Leptotrochila medicaginis* for alfalfa inoculations. Phytopathology 61:910.
253. ----, and M.D. Rumbaugh. 1976. Reaction of some perennial and annual *Medicago* species and cultivars to the yellow leafblotch disease caused by *Leptotrochila medicaginis*. Plant Dis. Rep. 60:596-599.
254. Seth, J., and S.T. Dexter. 1958. Root anatomy and growth habit of some alfalfa varieties in relation to wilt resistance and winterhardiness. Agron. J. 50:141-144.
255. Sheaffer, C.C., D.L. Rabas, F.I. Frosheiser, and D.L. Nelson. 1982. Nematicides and fungicides improve legume establishment. Agron. J. 74:536-538.
256. Sherwood, R.T. 1971. Personal communication. USDA-ARS, Regional Pasture Research Laboratory, University Park, PA.
257. ----, J.W. Dudley, T.H. Busbice, and C.H. Hanson. 1967. Breeding alfalfa for resistance to the stem nematode *Ditylenchus dipsaci*. Crop Sci. 7:382-384.
258. ----, ----, ----, and ----. 1967. Breeding alfalfa for resistance to the stem nematode *Ditylenchus dipsaci*. Crop Sci. 7:382-384.
259. Smith, D. 1948. The reaction of strains and varieties of alfalfa to seedling infection by downy mildew. J. Am. Soc. Agron. 40:189-190.
260. ----. 1962. Alfalfa cutting practices. I. Influence of cutting schedule, soil fertility, and insect control on yield and persistence of Vernal and Narragansett alfalfa. p. 3-11. Wis. Agric. Exp. Stn. Res. Rep. 11.
261. Smith, O.F. 1945. Parasitism of *Rhizoctonia solani* from alfalfa. Phytopathology 35:832-837.
262. ----. 1946. Effect of soil temperature on the development of Rhizoctonia root canker of alfalfa. Phytopathology 36:638-642.
263. ----. 1951. Biologic races of *Ditylenchus dipsaci* on alfalfa. Phytopathology 41:189-190.

264. ----. 1955. Breeding alfalfa for resistance to bacterial wilt and the stem nematode. Nev. Agric. Exp. Stn. Bull. 188.
265. ----. 1958. Reactions of some alfalfa varieties to the stem nematode. Phytopathology 48:107.
266. Sorensen, E.L., D.L. Stuteville, and E.K. Horber. 1983. Registration of KS145 alfalfa germplasm. Crop Sci. 23:188-189.
267. Stanford, E.H. 1952. Transfer of disease resistance to standard varieties. Proc. Int. Grassl. Congr., 6th 6:1585-1590.
268. ----, B.P. Goplen, and M.W. Allen. 1958. Sources of resistance in alfalfa to the northern root-knot nematode, *Meloidogyne hapla*. Phytopathology 48:347-349.
269. Stessel, G.H. 1960. Effects of nematocides on *Pratylenchus pratensis*. Phytopathology 50:656.
270. Stivers, R.K., W.A. Jackson, A.J. Ohlrogge, and R.L. Davis. 1956. The relationships of varieties and fertilization to observed symptoms of root rots and wilt of alfalfa. Agron. J. 48:71-73.
271. Stuteville, D.L. 1970. Personal communication. Kansas State University, Manhattan.
272. ----. 1973. Pathogenic specialization in *Peronospora trifoliorum*. Abstract 0715. Int. Congr. Plant Path., 2nd. Sept. 5-12, Univ. of Minnesota, St. Paul.
273. ----, and E.L. Sorensen. 1966. Distribution of leaf spot and damping-off (*Xanthomonas alfalfae*) of alfalfa in Kansas, and new hosts. Plant Dis. Rep. 50:731-734.
274. Sullivan, J.A., B.R. Christie, and J.W. Potter. 1980. Inheritance of northern root-knot nematode resistance in alfalfa. Can. J. Plant Sci. 60:533-537.
275. Sundheim, L., and R.D. Wilcoxson. 1965. *Leptosphaerulina briosiana* on alfalfa: Infection and disease development, host-parasite relationships, ascospore germination and dissemination. Phytopathology 55:546-553.
276. Tamimi, S.A., and M.D. Rumbaugh. 1963. Resistance of diploid alfalfa to *Phoma herbarum* var. *medicaginis* and *Cercospora zebrina*. Crop Sci. 3:227-230.
277. Teuber, L.R., D.G. Gilchrist, A.M. Martensen, W.A. Cowling, S. Buhling, and W.L. Green. 1983. UC1249 and NC1250. Stemphylium leafspot resistant alfalfa germplasm. Crop Sci. 23:805.
278. Theurer, J.C., and J.J. Filing. 1963. Comparative performance of diallel crosses and related second generation synthetics of alfalfa. I. Bacterial wilt resistance. Crop Sci. 3:50-53.
279. Thomason, I.J., and S.A. Sher. 1957. Influence of the stubby-root nematode on growth of alfalfa. Phytopathology 47:159-161.
280. Thorne, G. 1938. Alfalfa resistant to the stem nematode. Proc. Am. Soc. Sugar Beet Technol. 1:71-72.
281. Thyr, B.D., D.K. Barnes, F.I. Frosheiser, O.J. Hunt, R.N. Peaden, and B.J. Hartman. 1979. Registration of $MSE_6SN_3W_3$ and $MSF_6SN_3W_3$ alfalfa germplasms. Crop Sci. 19:417.
282. ----, O.J. Hunt, B.J. Hartman, T.J. McCoy, and T.R. Knous. 1984. Registration of alfalfa germplasms NMP-11, NMP-12, and NMP-13 resistant to the common leafspot fungus *Pseudopeziza medicaginis* (Lib.) Sacc. Crop Sci. 24:387.
283. ----, ----, ----, J.G. Dean, T.J. McCoy, and T.R. Knous. 1984. Registration of alfalfa germplasm NMP-14, with resistance to the common leafspot pathogen *Pseudopeziza medicaginis* (Lib.) Sacc. Crop Sci. 24:388-389.
284. ----, K.T. Leath, R.R. Hill, Jr., O.J. Hunt, B.J. Hartman, T.J. McCoy, and T.R. Knous. 1984. Registration of common leafspot-resistant winter hardy alfalfa germplasms BIC-6 CLS_5, MSF6 CLS_6 and Washington SNI CLS_4. Crop Sci. 24:388.
285. ----, W.F. Lehman, B.J. Hartman, O.J. Hunt, T.J. McCoy, D.E. Rowe, and T.R. Knous. 1984. Registration of CUSN-242 CLS_4 germplasm. Crop Sci. 24:387.
286. Toovey, F.W., J. M. Waterson, and F.T. Brooks. 1936. Observations on the black-stem disease of lucerne in Britain. Ann. Appl. Biol. 23:705-717.
287. Townshend, J.L., and H. Baenziger. 1976. Evidence of resistance to root-knot and root-lesion nematodes in alfalfa clones. Can. J. Plant Sci. 56:977-979.
288. ----, and J.W. Potter. 1982. Forage legume yields and nematode population behavior in microplots infested with *Paratylenchus projectus*. Can. J. Plant Sci. 62:95-100.
289. Tsao, P., and J.M. Menyonga. 1966. Response of Phytophthora spp. and soil microflora to antibiotics in the pimaricin-vancomycin medium. Phytopathology 56:152.
290. Tsukamoto, J.Y. 1965. Phenotypic characteristics of alfalfa tolerant to brown root rot. Can. J. Plant Sci. 45:197-198.
291. Twamley, B.E. 1960. Variety, fertilizer, management interactions in alfalfa. Can. J. Plant Sci. 40:130-138.
292. Tysdal, H.M., and B.H. Crandall. 1948. The polycross progeny performance as an index of the combining ability of alfalfa clones. J. Am. Soc. Agron. 40:293-306.
293. U.S. Department of Agriculture. 1968. Varieties of alfalfa. USDA Farmers Bull. no. 2231.

294. Viands, D.R., and D.K. Barnes. 1980. Inheritance of resistance to bacterial wilt in two alfalfa gene pools: Qualitative analysis. Crop Sci. 20:48–54.
295. ----, ----, and F.I. Frosheiser. 1980. An association between resistance to bacterial wilt and nitrogen fixation in alfalfa. Crop Sci. 20:699–703.
296. Veerisetty, V., and M.K. Brakke. 1977. Alfalfa latent virus, a naturally occurring carlavirus in alfalfa. Phytopathology 67:1202–1206.
297. Waddington, J., W.C. McDonald, and A.K. Storgaard. 1964. A comparative study of methods of inoculating alfalfa with *Leptosphaerulina briosiana*. Can. J. Plant Sci. 44:249–252.
298. Wallace, H.R. 1963. Resistance at the plant surface. p. 194. *In* The biology of plant parasitic nematodes. Edward Arnold Lt., London.
299. Walters, H.J., and D.A. Slack. 1956. Fusarium wilt plus nematodes... hard on alfalfa. Arkansas Farm Res. 5:7.
300. Ward, C.H. 1959. The detached-leaf technique for testing alfalfa clones for resistance to black stem. Phytopathology 49:690–696.
301. Waterhouse, W.L. 1953. Australian rust studies. XII. Specialization within *Chomyces striatus* Schroet. on *Trigonella suavissima* Lindl. and *Medicago sativa* L. Proc. Linn. Soc. N. S. Wales. 78:147–150.
302. Weimer, J.L. 1930. Temperature and soil-moisture relations of *Fusarium oxyporum* var. *medicaginis*. J. Agric. Res. 40:97–103.
303. ----. 1937. The possibility of insect transmission of alfalfa dwarf. Phytopathology 27:697–702.
304. Welty, R.E., K.R. Barker, and D.L. Lindsey. 1980. Effects of *Meloidogyne hapla* and *M. incognita* on Phytophthora root rot of alfalfa. Plant Dis. 64:1097–1099.
305. ----, and T.H. Busbice. 1978. Field tolerance in alfalfa to Sclerotinia crown and stem rot. Crop Sci. 17:508–509.
306. ----, and H.B. Collins. 1979. Loss of viability of spores of *Colletotrichum trifolii* produced on lima bean agar. Plant Dis. Rep. 63:486–489.
307. ----, R.Y. Gurgis, and D.E. Rowe. 1982. Occurrence of race 2 of *Colletotrichum trifolii* in North Carolina and resistance of alfalfa cultivars and breeding lines to race 1 and 2. Plant Dis. 66:48–51.
308. ----, and J.P. Mueller. 1979. Occurrence of a highly virulent isolate of *Colletotrichum trifolii* on alfalfa in North Carolina. Phytopathology 69:537.
309. ----, and ----. 1979. Occurrence of a highly virulent isolate of *Colletotrichum trifolii* on alfalfa in North Carolina. Plant Dis. Rep. 63:666–670.
310. ----, and J.O. Rawlings. 1980. Effects of temperature and light on development of anthracnose on alfalfa. Plant Dis. 64:476–478.
311. Whitehead, M.D., and E.L. Pinnell. 1957. Severe damage from bacterial stem blight of alfalfa in Missouri. Plant Dis. Rep. 41:876–877.
312. Wiant, J.S., and G.H. Starr. 1936. Field studies on the bacterial wilt of alfalfa. Wyo. Agric. Exp. Stn. Bull. 214.
313. Wilcoxson, R.D., D.K. Barnes, F.I. Frosheiser, and D.M. Smith. 1977. Evaluating and selecting alfalfa for reaction to crown rot. Crop Sci. 17:93–96.
314. Willis, W.G., D.L. Stuteville, and E.L. Sorensen. 1969. Effects of leaf and stem diseases on yield and quality of alfalfa forage. Crop Sci. 9:637–640.
315. Wilson, M.C., Jr. 1947. Inheritance of resistance in alfalfa to bacterial wilt. J. Am. Soc. Agron. 39:570–583.
316. Wilson, M.L., and B.A. Melton. 1962. Zia alfalfa. N.M. Agric. Exp. Stn. Bull. 468.
317. ----, ----, and C.E. Watson. 1959. Zia alfalfa. N.M. Agric. Exp. Stn. Bull. 435.
318. Winstead, N.N., and R.D. Riggs. 1963. Stability of pathogenicity of B biotypes of the root-knot nematode *Meloidogyne incognita* on tomato. Plant Dis. Rep. 47:870.
319. Wynne, J.C., and T.H. Busbice. 1968. Effects of temperature and incubation period on the expression of resistance to stem nematode in alfalfa. Crop Sci. 8:179–183.
320. Zaleski, A. 1957. Reactions of lucerne strains to verticillium wilt. Plant Pathol. 6:137–142.

28 Breeding for Insect Resistance

E. L. SORENSEN
USDA-ARS
Manhattan, Kansas

R. A. BYERS
USDA-ARS
University Park, Pennsylvania

E. K. HORBER
Kansas State University
Manhattan, Kansas

Harmful insects stunt, defoliate, and kill alfalfa (*Medicago sativa* L.) plants. Forage and seed losses from insects approximate $250 million annually in the USA. More than 100 species have been recorded as injurious. Recent damage and distribution maps for those important in the USA (66) and their descriptions (see Chapter 22 in this book) are available.

Resistant alfalfa cultivars, when available, provide protection from damage where temperatures keep parasites and predators inactive or prevent the application of insecticides. They contribute to the establishment and maintenance of stands and to forage yields and quality. When large acreages are involved, even the slightest resistance expressed as reduced damage or increased tolerance leads to immense savings. However, resistant cultivars are not a panacea for all pest problems (250). They often may be most effective in integrated control systems that combine cultural, chemical, and biological methods. The entire topic of breeding plants for insect control was recently reviewed (202).

No deleterious effects on forage quality have been found in resistant alfalfa cultivars. Chemical constituents important in animal nutrition, such as protein, carotene, and digestible dry matter, were similar for uninfested cultivars that were characterized as either susceptible or resistant to the pea aphid (*Acyrthosiphon pisum* Harris) and the spotted alfalfa aphid (*Therioaphis maculata* Buckton) (156, 164). Neither digestibility coefficients of the forage nor performance of yearling Holsteins (*Bos taurus* L.) varied significantly when cultivars differing in reaction to the alfalfa weevil (*Hypera postica* Gyllenhal) were compared under weevil infested conditions (9). Also, no toxic compounds were identified in the exudate of glandular-haired *Medicago* spp. that appear to be good sources of resistance to alfalfa insects (335). Total digestibilities and crude

Copyright 1988 © ASA-CSSA-SSSA, 677 South Segoe Road, Madison, WI 53711, USA.
Alfalfa and Alfalfa Improvement—Agronomy Monograph no. 29.

protein content of glandular-haired species were superior or equal to those of alfalfa control populations (185). However, breeders should be aware of the possibility that breeding a new pest-resistant cultivar might inadvertently alter chemical or physical properties of the forage.

28-1 CATEGORIES AND CAUSES OF RESISTANCE

Resistance is conditioned most often by a complex of plant and environmental characteristics. Morphological, anatomical, biochemical, and physiological features interact to determine insect-host relationships. Orientation and host selection by insects result from a chain of interdependent conditioned reflexes whereby visual, chemical, tactile, and other stimuli influence oviposition and feeding and account for the suitability of the host. Interruption in that behavior contributes to resistance.

A comprehensive explanation of types and classification of resistance to insects was provided by Horber (120).

28-1.1 Terms Expressing Degrees of Resistance

Because there may be a complete gradation from extreme resistance to extreme susceptibility, the term *plant resistance* to insects must be qualified in the following manner:
1. *High resistance*: little damage under a given set of conditions.
2. *Intermediate resistance*: moderate infestation and injury under given environmental conditions.
3. *Low resistance*: less damage or infestation than average for the crop considered.
4. *Susceptibility*: average or more damage.
5. *High susceptibility*: severe damage.

28-1.2 Functional Resistance Categories

1. *Pseudoresistance*: transitory resistance in potentially susceptible host plants.
   a. *Escape*: a susceptible plant that is undamaged because of unequal distribution of insects in a plant population.
   b. *Host evasion*: a susceptible cultivar whose susceptible stage does not coincide with infestation by the insect population.
   c. *Induced resistance*: temporary or increased resistance attributable to the specific status of a plant or to environmental conditions, such as a change in the availability of water or soil fertility.
2. *Hypersensitive resistance*: an intense, rapid response characterized by premature death (necrosis) of the infested tissue together with inactivation and localization of the attacking agent (218).
3. *Adult plant resistance*: resistance manifested in maturing plants.
4. *Juvenile resistance*: resistance apparent in the seedling stage.

28-1.3 Mechanisms of Resistance

These terms describe insect or plant response rather than the nature of resistance (248, 249, 250).

1. *Nonpreference*: negative reaction towards or avoidance of a plant by insects during their search for food, oviposition sites, or shelter. Kogan and Ortman (175) proposed substituting *antixenosis* for nonpreference because nonpreference is an insect characteristic. Antixenosis (against guests) is a plant characteristic and is more appropriate when describing plant resistance.
2. *Antibiosis*: all adverse effects exerted by the plant on the insect's biology (survival, development, and reproduction).
3. *Tolerance*: plant responses resulting in the ability to either withstand infestation or to support insect populations that would severely damage otherwise susceptible plants.

28-1.4 Genetic Resistance Categories

Mode of Inheritance
1. *Monogenic*: governed by a single gene.
2. *Oligogenic*: governed by a few genes.
3. *Polygenic*: governed by many genes.
4. *Vertical or Specific*: resistance expressed against only certain biotypes of a pest species (340).
5. *Horizontal or General*: resistance expressed against all biotypes of a pest species (340).

28-1.5 Biochemical and Morphological Bases of Resistance

Chemically based resistance is a major component of the plant's defense arsenal against herbivores. Since antiquity, phytochemicals have been used in human medicine and against predators and parasites on animals and plants. In recent years, knowledge of the chemistry of plants has mushroomed. The effect of phytochemicals on behavior and biology of insects and mites resulting in antibiosis or nonpreference has been explained in several cases. Inorganic, as well as organic compounds (e.g. silica, Se, oxalic acid, DIMBOA (2, 4-dihydroxy-7-methoxy-1, 4-(2H)-benzoxazin-3-one), gossypol ($C_{30}H_{30}O_8$), saponins, condensed tannins), affect insect/plant- relationships. Primary metabolites (e.g., nonprotein amino acids) as well as secondary metabolites (e.g., isoprenoids, acetogenins, alkaloids) play an important role as allelochemicals to determine the suitability of plants as hosts for insects and mites by affecting their behavior and metabolic processes (242).

Morphological (physical) characteristics of plants interfere with locomotion, host selection, feeding, ingestion, digestion, mating, and oviposition. Coloration, trichomes, surface waxes, silication, sclerotization, or lignification of tissues constitute physical barriers or deterrents. Much of the existing resistance in crops is based on morphological characters.

Both chemical and morphological traits may act at a distance, close range, or on contact. Biochemicals and morphological traits may be perceived and responded to as attractants, arrestants, or repellents, as incitants or suppressants, or as stimulants or suppressants (14, 55). Because chemical and morphological resistance factors are intertwined (e.g., in glandular trichomes) it sometimes will be difficult to distinguish their relative contribution to resistance. Examples of resistance attributable to biochemical or morphological differences are discussed under specific insects.

28-2 DURABILITY OF RESISTANCE

28-2.1 Environmental Influences on Resistance

Environment affects the level and expression of plant resistance to insects. Insect responses, including those to olfactory (smell) and gustatory (taste) stimuli, also are subject to modification by the environment. Interactions among the alfalfa plant, insect, and environmental factors are complex. Thus, a knowledge of climatic, soil, and cultural factors affecting resistance is highly desirable in developing and assessing resistant germplasm.

Increased or reduced constant temperatures may lead to the loss of resistance, whereas fluctuating temperatures may stabilize it. Quality, intensity, and duration of light and relative humidity influence the physiological processes of both alfalfa and insects. Low light intensity and high humidity generally have the greatest adverse influence on the expression of resistance. Soil fertility and moisture, herbicides, and plant growth stimulants and retardants are capable of altering the level and expression of genetic resistance.

A detailed discussion of environmental factors influencing resistance of plants to insects was presented by Tingey and Singh (334).

28-2.2 Biotypes

Biotype refers to an individual or a population of an insect species that is distinguished by a difference in its ability to parasitize a host (78).

The useful life of resistant cultivars is directly influenced by the development of biotypes. Resistance may be permanent in their absence. Resistance determined by one to two major genes (vertical resistance) and expressed as antibiosis may lead to the rapid buildup of biotypes. Conversely, moderate to high levels of resistance conditioned by minor genes in a quantitative manner seldom lead to the emergence of biotypes (horizontal resistance). Both genetic systems have been exploited in developing crop cultivars. Vertical resistance is easier for plant breeders to work with than horizontal, but the latter is more stable with regard to biotypes (78).

Tolerance mechanisms usually do not lead to biotypes mainly because little selection pressure is exerted by the plant on the insect population. The gene frequency in the insect remains stable, and biotypes are not likely to develop. For instance, no biotypes have been detected on weevil-tolerant 'Team' or 'Arc' alfalfas in the decade since their release. However, spotted alfalfa aphid biotypes are numerous because of vertical resistance. Gallun and Khush (78) outlined three ways to use vertical resistance to avoid biotypes: (i) sequential release of cultivars with single genes for resistance, (ii) pyramiding the vertical genes into the same cultivar, and (iii) development of multiline cultivars.

28-3 ROLE OF RESISTANCE IN INTEGRATED CONTROL

Host plant resistance can be a part of an integrated pest management system in four ways. It can serve as the principal control method, or can be combined with chemical, biological, and cultural control. Examples of plant resistance as the principal control method in alfalfa are the spotted alfalfa aphid, pea aphid, and blue alfalfa aphid [*Acyrthosiphon kondoi* Shinji] (52, 235).

Plant resistance can be expressed at moderate or low levels and still be useful because it will help to keep population increases at lower levels, and thereby reduce the number of spray applications required for pest control. Team and Arc are moderately tolerant of the alfalfa weevil, and may require fewer sprays than do susceptible cultivars to prevent the population from reaching economic levels (9).

Plant resistance usually has been considered compatible with biological control methods. However, the mechanisms of resistance employed may have an important affect on the impact of parasites and predators. Horber (119) recommended the use of highly tolerant cultivars of alfalfa, which allow subeconomic levels of a pest species while supporting the pest's natural enemies. On the other hand, antibiosis mechanisms, such as allomones present in plants or concentrated in the bodies of their herbivores, glandular hairs, hooked trichomes, and the lack of floral nectaries, may adversely affect attraction of the plant to parasites and predators during the host-seeking phase of behavior (15). The transfer of glandular hairs from other *Medicago* spp. to alfalfa for control of the alfalfa weevil (301, 306) should be studied for its impact on alfalfa weevil parasites such as *Bathyplectes curculionis* and *B. anurus*.

However, plant resistance can be combined with biological control successfully. The resistance of KS 10 (297) to the pea aphid did not adversely affect parasite and predator populations (262). Tolerance mechanisms usually do not involve allomones or physical structures that interfere with parasite behavior.

If there is an undesirable interaction between plant resistance and biological control, this interaction has the potential of contributing to pest outbreaks in companion crops, because alfalfa is an important reservoir for natural enemies in alfalfa cotton (*Gossypium* spp.) or alfalfa-small grain ecosystems (15).

28-4 DEVELOPMENT OF PLANT RESISTANCE

28-4.1 General Concepts

28-4.1.1 Breeding

Alfalfa genetic behavior and breeding procedures appropriate for developing insect resistant cultivars are described in Chapter 25. Breeding for insect resistance, like breeding for disease resistance, involves two biological entities. It is helpful to have information on genetic variation associated with the insect, as well as host reaction, insect-plant interactions, and the extent to which environmental conditions affect plant resistance and insect behavior. However, such information is seldom available at the beginning of a breeding program and is not required for development of resistant cultivars.

Success of a breeding program depends on the effectiveness of isolating resistance and combining it with other desirable germplasm. Resistant germplasm usually has been identified on the basis of intensive testing, because natural populations of alfalfa contain vast amounts of genetic variability. All possible sources of genetically diverse materials should be exploited. Harris (103) indicated that in the search for resistance one should proceed from plants grown locally, to abandoned cultivars, to foreign cultivars, to landraces, wild types, and finally to related species. Alfalfas from the native habitat of the insect may have special merit because through long association of plant populations with insects, plant types may have evolved that resist or tolerate insect damage. Diverse sources and types of resistance are important as protection against the development of biotypes.

Recurrent phenotypic selection in random mating populations has been effective in development of various degrees of resistance to the alfalfa weevil (9), potato leafhopper (*Empoasca fabae* Harris) (64, 92, 299), blue alfalfa aphid (300, 311), pea aphid, and spotted alfalfa aphid (134, 252, 297, 308, 310). Tandem and independent culling procedures were used in a recurrent phenotypic selection program to combine resistance to five diseases and two insects in one germplasm pool (309). Also, strain crosses may be useful in developing multiple-pest-resistant cultivars and populations (67).

28-4.1.2 Screening and Evaluation Technique

Success of a breeding program depends on the effectiveness of screening and evaluating the available sources of resistance and eliminating escapes. These sources are planted in the field or in greenhouses when insect populations are available.

The entomological techniques for development of insect resistant alfalfa cultivars include mass screening of plant populations, followed by testing of individual selections for antibiosis, nonpreference, or tolerance. The final selection of individual plants is based on performance under varied infestation pressures of the test insect and on the level and type of resistance desired. The progeny are tested to determine if the level of resistance is adequate. In an attempt to provide uniform procedures to characterize insect resistant alfalfa germplasm, standard resistant and susceptible control cultivars were designated and effective selection and evaluation techniques have been published for several insects (66).

28-4.2 Specific Insects

28-4.2.1 Spotted Alfalfa Aphid

28-4.2.1.1 Distribution and Importance. The spotted alfalfa aphid (SAA) (*Therioaphis maculata* Buckton), apparently native to India or nearby areas, has spread to many regions of alfalfa production (79, 99, 198, 255, 275, 347, 354). The first report of SAA in the USA was from New Mexico in 1954. In the same year it threatened alfalfa production in the entire Southwest; 1 yr later damage was estimated at $42 million (5). It now has spread over the entire continental USA and into Canada. However, it is a problem in production primarily from Nebraska, south to Texas, and west to the Pacific Coast. Damage from aphid injury may range from stunted and defoliated plants to loss of entire stands.

The contributions of SAA-resistant cultivars are striking. In 1969, Luginbill (191) estimated that $35 million was a conservative figure for the annual savings to growers from using resistant cultivars. Spotted alfalfa aphid-resistant cultivars can be established successfully under heavy aphid infestations that would kill susceptible cultivars (Fig. 28-1) (126, 294, 305). Highly resistant cultivars remain sufficiently free of aphids to provide economic control under conditions that would require repeated insecticide applications to susceptible cultivars. In some situations, such as proximity to nearby fields of susceptible cultivars, resistant cultivars may require insecticide application for successful seed production.

Resistance ensures higher yields and quality of forage. Under epidemic SAA infestations in field plots, forage yields of resistant cultivars exceeded those of susceptible cultivars by 300 to 400% (127). In infested field cages, resistance provided significant protection against losses of protein, carotene, and dry matter (164).

28-4.2.1.2 Development of Resistant Cultivars and Germplasm. Breeding for resistance to SAA has been highly successful. 'Moapa', developed from nondormant 'African', was released to growers in the southwestern USA in 1957, only 3 yr after the aphid was discovered

Fig. 28-1. Resistance to the spotted alfalfa aphid made the difference in the above photograph taken near Bakersfield, CA. Plots 2 and 6 were two resistant cv., Lahontan and Moapa, respectively; plots 4 and 8 were resistant experimental lines. A susceptible variety had been planted in plots 1, 3, 5, and 7.

(294). In the same year, 'Zia', derived principally from Turkistan germplasm, was released in New Mexico (90, 353). 'Cody', developed from Chilean germplasm, was released in 1959 for use in the South Central States (305). Between 1957 and 1983, more than 100 resistant cultivars were developed and released for commercial production and over 50 germplasm releases were made available to plant breeders (211).

Some resistant plants occur in most alfalfas (86, 97, 126, 156, 172, 189, 271, 277, 292, 336, 337) but the highest levels of resistance occur in those of Turkistan origin. 'Lahontan', developed from Turkistan germplasm for resistance to the stem nematode, *Ditylenchus dipsaci* (Kühn) Filipjev, and bacterial wilt [*Corynebacterium insidiosum* (McCull.) H.L. Jens.] had resistance to SAA (292). Some annual (75, 253) and perennial (302) glandular-haired *Medicago* spp. had natural resistance and lines of *M. truncatula* Gaertn. varied in their reaction to aphids (271). Rapid response to selection for resistance indicates that the reaction is highly heritable, but a factorial basis for inheritance has not been proposed (110, 134). Biotypes of SAA developed (228, 229, 230, 236, 241, 261, 314), but combined resistance to them has been reported (232, 237).

Temperature is the most important of several environmental factors that modify plant resistance. Although temperatures of 10 to 15°C generally lower resistance (85, 135, 168, 206, 208, 278), there is evidence of plant genotype and temperature interactions (136). A deficiency of Ca or K or an excess of Mg or N may be associated with loss of resistance, whereas resistant plants deficient in P may become more resistant (167, 208). Photoperiod (208), relative humidity (135), and soil moisture (168) appear to have little influence on resistance.

28-4.2.1.3 Screening and Evaluation Techniques. The SAA can be cultured in the laboratory simply by maintaining an adequate supply of susceptible alfalfa free of other insects and pests. Manglitz and Schalk (197) described a method for excluding parasites. Relative humidity of 25 to 30% (82), fluctuating temperatures with a mean of 23°C and a diurnal range of 17°C (209) contributed to a rapid increase in aphid populations. When culturing holocyclic strains found in northern areas (194), it was necessary to maintain aphids under at least a 14-h photoperiod to prevent formation of sexual forms (279).

Aphid populations have been maintained at high levels for evaluating of resistance in field plots by two general methods: (i) applying selective insecticides to control predators (126), and (ii) excluding predators with cages (110, 125, 127, 156).

A combination of greenhouse and field screening may be used because plant resistance is similar in both environments (110, 187). Lodge and Greenup (190), however, indicated that resistance of established plants in the field was more clearly defined than that of seedlings in glasshouse studies. Field ratings of resistance are based on plant damage and insect numbers. Plant damage may be expressed as percent plant mortality (86, 89, 123, 125, 126, 127, 189, 190, 271, 336), percent plants with different levels of resistance (104, 123), or on a numerical scale (123). Aphid populations have been estimated by number of aphids per plant (31, 110, 125, 189), number of aphids per 10 cm (4 in.) of row (110), number of aphids per leaf (224), number of aphids per stem (31, 126, 190), and with a fork sampling device (47).

Selection for resistance in the greenhouse generally is accomplished by seeding entries in rows in flats and infesting them at the seedling stage (86, 104, 123). In the initial cycle of selection, involving alfalfa germplasm with unknown or low levels of resistance, it is important to control the rate of infestation so resistant plants can be detected.

Surviving plants may be screened further by observing survival and reproduction of aphids on the entire plant or on a plant part confined in small cages (Fig. 28-2) (110, 123). Excised plant parts usually give good results, although Thomas et al. (328) and Thomas and Sorensen (327) have shown somewhat less resistance in excised plant parts vs. intact parts.

Standard resistant and susceptible cultivars have been designated (66) and an evaluation procedure outlined (226) to provide uniform procedures for describing levels of pest resistance in cultivars.

28-4.2.1.4 Mechanisms of Resistance. Antibiosis, nonpreference, and tolerance may be expressed singly or together in the same plant. A combination of mechanisms enhances the value of germplasm or cultivars. Several early studies showed that plant resistance increased the time needed for aphid development and decreased survival of the aphids (61, 104, 127, 227). Subsequently, Kishaba and Manglitz (170) verified high mortality of aphids confined to resistant plants but noted that, when given a choice, aphids left resistant plants to congregate on susceptible ones,

Fig. 28-2. Caged-leaf test to determine spotted alfalfa aphid fecundity and survival on alfalfa.

with no visible ill effects on the aphids. Their observations confirm nonpreference. In this and other studies, aphids confined to resistant plants were restless and died at nearly the same rate as those confined without food (166, 170, 207). Jones et al. (147) demonstrated a wide range of tolerance among alfalfa clones. Simple trichomes on seedling stems were implicated in temporary resistance (195).

Investigators have attempted to determine the nature of resistance to this aphid, but an adequate explanation of the phenomenon is not available (60, 105, 169, 171, 199, 203, 204, 205, 207, 223, 246). Nielson and Don (229) related resistance to the phytoalexin concept of interaction between pathogens and plants.

28-4.2.2 Pea Aphid

28-4.2.2.1 Distribution and Importance. The pea aphid (PA) (*Acyrthosiphon pisum* Harris), is widely distributed throughout the world. In 1963, losses were estimated to cost alfalfa growers in the USA about $60 million annually (42). Relative distribution and severity maps for the USA are available (66). Populations generally peak in spring and fall, and occasionally are high in midsummer. Damage varies from stunting and yellowing of plants to death of entire top growth. In laboratory tests, infestations reduced cold hardiness, crown weight, and the height and weight of alfalfa top growth (100).

Pea aphid resistant cultivars sharply reduce losses (164). In trials where PA damaged susceptible cultivars, but in which top growth was not killed, resistant entries produced two to three times as much forage as susceptible ones (Fig. 28-3) (257, 298). Yields of succeeding crops of susceptible cultivars were reduced because PA weakened the first crop.

Fig. 28–3. Comparative growth of susceptible Cody (*left*) and resistant Kanza (*right*) during a pea aphid infestation in Kansas.

In Kansas, average increases in forage yield of resistant over susceptible cultivars were 211, 188, 107, and 114% for the first, second, third, and fourth cuttings, respectively. Resistant entries produced 78% more forage than did susceptible ones at first harvest in the following year (108). Protein yields of resistant 'Kanza' were almost double and carotene yields triple those of the susceptible cultivars, Buffalo, Ranger, and Vernal during a natural infestation. Aphids per gram of dry matter from Buffalo outnumbered those from Kanza by 9:1 (298).

28–4.2.2.2 Development of Resistant Cultivars and Germplasms. The possibility of breeding alfalfa resistant to the PA was considered in 1934 when resistant plants were first detected by Blanchard and Dudley (17). However, significant breeding progress did not occur until > 20 yr later, following the successful development of resistance to the SAA. The breeding of cultivars with resistance to the PA required the development of suitable greenhouse procedures for mass screening and clonal testing (247). During 1966, 'Washoe' was released for use in the Pacific Coast and Intermountain states (133, 256) and 'Apex' for areas where Flemish cultivars were adapted (211). In 1967, 'Dawson' (153, 156) and 'Mesilla' (208) were released for the North Central States and New Mexico, respectively. Kanza was released in 1969 for the South Central States (304). Between 1969 and 1983, >75 resistant cultivars were released for commercial production (211). The number increases to >100 with the inclusion of moderately resistant cultivars. Also, >40 alfalfa breeding lines with resistance to PA have been released to individual scientists and organizations interested in alfalfa improvement (211).

Resistance to PA is highly heritable (134) and can be incorporated readily into germplasm concurrently with other superior characteristics. Jones et al. (148) indicated that two genes, a dominant and a recessive, conditioned resistance, whereas Glover and Stanford (80) concluded that a tetrasomically inherited dominant gene with modifying factors explained inheritance of resistance. Flemish alfalfa seems to be the best source of resistance, followed by Turkistan. However, resistant plants may be found in many sources (42, 87, 239, 247, 251, 270). Biotypes were recognized first on pea (*Pisum sativum* L.) by Harrington (101, 102). The insect varies with crop species (8, 20, 77, 201, 217) and with alfalfa (21, 43, 130, 161, 179, 180). It has been shown from alfalfa trials exposed to an array of biotypes that cultivars can be developed with resistance over broad geographic areas and environmental conditions. The expression of resistance may be modified by environmental factors (3, 69). Low temperatures tend to reduce levels of resisitance, but genotype × temperature interactions are similar to those noted for the SAA (135, 136). Resistance reduces fecundity and survival (42, 46, 87, 196, 247, 262, 283) and probably feeding of PA (246).

28-4.2.2.3 Screening and Evaluation Techniques. Adequate numbers of PA, needed for selection of resistant plants, can be cultured in the greenhouse throughout the year on succulent susceptible alfalfa or broad bean (*Vicia faba* L.). Broad bean is desirable since many biotypes reproduce well on them (216). Broad bean must be used with caution, however, because aphids may become conditioned to this specific host plant (77, 201). Pea aphids may be reared on a chemically defined diet (7).

Many resistance criteria and methods of selecting resistant plants have been used (22, 42, 87, 109, 122, 247, 293). Ortman et al. (247) identified resistant plants by survival of seedlings (Fig. 28-4), lack of stunting of young plants, antibiosis reaction, low PA numbers on field plants, and healthy appearance in heavily infested fields. Reliable resistance evaluations have been made in field cages (293) and by counting parasitized aphids that remain fastened to the upper surface of alfalfa leaves in the field (106). In isolating resistance under laboratory conditions, low temperatures are recommended because plants that are resistant at low temperatures are likely to be resistant at high temperatures. A seedling resistance evaluation test used in Kansas was described by Sorensen (296).

28-4.2.2.4 Mechanisms of Resistance. Antibiosis, nonpreference,

Fig. 28-4. Survival of seedling alfalfas after infestation with pea aphids. Resistant KS 10 was developed from susceptible Ladak.

and tolerance have been reported (42, 247). Little is known about the physiological basis of resistance. Resistance in Dawson and Washoe was attributed to a factor inhibiting aphid development (177). Saponin content of the foliage did not appear to be related to resistance (259) even though alfalfa lines selected for high foliage saponin were generally more resistant than those selected for low saponin (258). Krzymanska et al. (178) concluded that PA resistance in *Medicago* spp. depended upon differences in the toxicity of the saponins in their tissues.

28-4.2.3 Blue Alfalfa Aphid

28-4.2.3.1 Distribution and Importance. The blue alfalfa aphid (BAA) (*Acyrthosiphon kondoi* Shinji), was first collected near Bakersfield, CA in spring 1974 (315). It closely resembles the PA, but is smaller, more blue green, and has a waxy appearance, whereas the PA is light green and shiny (285). In just 4 yr this aphid had spread to many western states and as far east as Kansas (174). It is found now in Argentina, New Zealand, Australia (315), Chile, and South Africa. Damage to alfalfa by the BAA occurs at lower population levels than for PA. Infested alfalfa shows stunting, leaf curling, yellowing, and slow recovery after cutting. Injection of an insect toxin may contribute to the damage (68).

28-4.2.3.2 Development of Resistant Cultivars and Germplasms. 'CUF 101', the first cultivar developed for resistance to the BAA, was released in 1977 (182). It is a nonhardy alfalfa developed directly from resistant plants selected from heavily infested and severely damaged alfalfa fields. 'Rere' was resistant to BAA in New Zealand (71). Between 1977 and 1986 over 20 certified cultivars were developed with varying degrees of resistance. Recurrent phenotypic selection was effective in developing moderately hardy (300) and nonhardy (311) alfalfa germplasm. KS108GH5, a glandular-haired germplasm, is resistant to the BAA (302).

In studies with a small number of entries, Lehman et al. (181) reported a strong relationship between resistance to BAA and PA. Similar results were reported by Salisbury et al. (276). Additional studies by Nielson et al. (235) and Nielson and Lehman (233) failed to verify the relationship. They did find a few BAA resistant plants in each PA resistant cultivar except Washoe.

28-4.2.3.3 Screening and Evaluation Techniques. Methods used by Nielson et al. (235) to evaluate resistance of seedlings in the greenhouse were similar to those descibed for the SAA (241). Seed of the test entries was planted in rows in flats. A susceptible control entry was planted every third row. The seedlings were initially infested at the unifoliolate-leaf stage at a rate of 6 to 8 mL of aphids per flat. Subsequent infestations were made as necessary to maximize killing of plants. The surviving plants of the test entries were counted after the susceptible control plants had died. Salisbury et al. (276) used a similar method but rated the seedlings on a one to four scale, a rating of zero indicating active growth and no apparent aphid damage and a rating of four indicating death. The susceptibility index was the mean rating over all seedlings in a row. Wellings (345) used first instar nymphs and determined the relative growth rates. A field cage technique to evaluate seedling alfalfa plants was described by Nielson (226).

28-4.2.3.4 Mechanisms of Resistance. Wellings (345) emphasized the need to discriminate between those aspects of aphid performance statistics that are attributable to plant growth patterns and those that are related to host plant resistance. He indicated that tolerance is the mechanism in CUF 101 and 'Trifecta'. Likewise, Lloyd et al. (188) assumed tolerance in the entries they tested. Bishop et al. (16) showed that CUF 101 could tolerate a three- to fourfold higher aphid pressure than susceptible cultivars.

No biotypes have been detected on any of the resistant cultivars in the USA or elsewhere.

28-4.2.4 Alfalfa Weevil and Egyptian Alfalfa Weevil

28-4.2.4.1 Distribution and Importance. The alfalfa weevil (*Hypera postica* Gyllenhal) has spread over most alfalfa growing areas of the USA and southern Canada. It has become one of the most destructive insect pests of alfalfa, and may occasionally cause damage to other legumes. Epidemic outbreaks have been common, especially as the eastern strain moved westward. The Egyptian alfalfa weevil (*Hypera brunneipennis* Boheman) occurs in California (245, 321). In both species, damage is caused by adults and larvae. However, the most significant damage is done by the larvae chewing on the growing plant terminals. This may result in complete defoliation of the first growth of the season and delay of regrowth in subsequent harvests. Shredded foliage gives infested fields a characteristic grayish cast.

Recent evidence suggests that there is only one alfalfa weevil species in the USA (Chapter 22).

28-4.2.4.2 Development of Resistant Cultivars and Germplasm

28-4.2.4.2.1 Alfalfa Weevil

Differences in damage have been observed among experimental strains and commercial cultivars (34, 56, 62, 84, 263, 269, 286, 288, 320, 326, 355).

The first weevil-resistant cultivar (Team) was released in 1969 (9). Its parentage traces to Kansas and Nebraska synthetics, 'Atlantic', 'DuPuits', 'Narrangansett', and 'Rhizoma'. Arc (56, 57) and 'Liberty' released in 1974 and 1977, respectively, were developed from the same germplasm as Team by recurrent phenotypic selection. 'Weevlchek,' released in 1971, was derived from *M. falcata* L. and Indian breeding lines (211). By 1983, over a dozen resistant cultivars were available to growers (211). Resistance of these cultivars is attributed primarily to tolerance (expressed in heavy terminals and axillary branching); however, when Dhaliwal and Grewal (58) compared eight cultivars in the greenhouse and five in the field for preference during feeding and oviposition, Arc was judged highly resistant. Even though weevil feeding occurs, this low to moderate level of resistance may make an appreciable contribution to an integrated control program that includes plant resistance, biological control, good crop management, and judicious choice and use of insecticides (355).

Germplasm possessing gene combinations and special significance for use in breeding programs for alfalfa weevil resistance were listed in Hunt et al. (132) and in Miller and Melton (211).

28-4.2.4.2.2 Egyptian Alfalfa Weevil

UC73, a cultivar originating from Northrup King Blend 9-19, had larval populations and damage scores significantly below those of Moapa (184).

28-4.2.4.3 Screening and Evaluation Techniques

28-4.2.4.3.1 Alfalfa Weevil

Most progress in breeding for resistance to the alfalfa weevil has resulted from field selection. Based on trials from 1966 to 1976 with the Starnes strains, derived from plants selected from a field in North Carolina for less defoliation under natural infestation, Busbice et al. (33) concluded that selection in the field in areas with natural infestation was the only effective method of selection. Tolerance ratings were based on foliage damage relative to larval infestation (62). Preference ratings have been based on larval infestation relative to that of a standard cultivar (62, 268), eggs per stem or per unit of stem (40, 243), or dry matter consumed by adults. The evaluation of annual *Medicago* spp. as sources of alfalfa weevil resistance was described by Barnes and Ratcliffe (12). A suggested standard test to evaluate resistant cultivars was presented by Ratcliffe (268).

Several laboratory and greenhouse screening techniques have been suggested for identifying and combining resistance to different stages of the insect (10, 11, 12, 13, 38, 243, 244, 339). Four successive tests (cotyledon, leaf disk, larval development, and oviposition stimulus) in each cycle of selection were described by Barnes et al. (9).

A nonpreference test, with 2-d-old larvae in a no-choice situation in the laboratory after which larval weights were recorded and expressed as index of resistance, was described by Byrne and Ritterhausen (39).

Techniques for rearing and maintaining alfalfa weevil colonies in the laboratory have been described. Adults and eggs can be stored for long periods at low temperatures to provide a continuous source of insects for testing and screening in the laboratory and greenhouse (37, 44, 45). Diapause is prevented by rearing the larvae under an 8-h photoperiod (131). Although rearing large numbers of weevils requires extensive labor, entomologists at Beltsville, MD, have produced 1000 adults per week by placing 150 to 200 adults in 3.8-L jars with cut stems of alfalfa and adding alfalfa pollen to stimulate oviposition. Eggs are collected every other day. The eggs are placed on paper disks; prior to hatching, the disks are placed in flats of 2- to 3-week-old alfalfa seedlings. Larvae mature on the flats, pupate, and the adults are collected. To prevent using weevil populations that have changed genetically in the laboratory, testing for resistance is done with first generation weevils reared from adults collected in the field. An artificial oviposition technique for the alfalfa weevil using parafilm tubes simulating alfalfa stems as an oviposition medium was described by Hower and Ferrer (128). A recipe for an artificial diet based on wheat germ and alfalfa leaf powder for laboratory rearing of the alfalfa weevil was presented by Hsiao and Hsiao (129). The most decisive advantage of the artificial diet was the feasibility of rearing disease-free laboratory cultures.

28-4.2.4.3.2 Egyptian Alfalfa Weevil

Certain modifications in screening techniques have to be made because of differences in behavior of this weevil and environment factors in California. The germplasm developed for resistance in the eastern USA could not be used in California because of its winter dormancy and susceptibility to the SAA.

Adults were collected through spring and summer at their aestivation sites, stored at room temperature until December, and then held at 4.5°C for testing from January to May.

The selection method used by Lehman and Stanford (183, 184) included the field, leaf disk, and seedling techniques described above. Unlike the seedling or cotyledon test described by Barnes et al. (10, 13), these tests were conducted under greenhouse conditions with mature plants. For the leaf disk test, all disks were weighed before and after the tests. Lehman and Stanford (184) concluded that selections made in the field showed the most promise.

28-4.2.4.4 Mechanisms of Resistance

28-4.2.4.4.1 Alfalfa Weevil

"Tolerance" operates as the main resistance factor in the presently commercially available alfalfa cultivars. Tolerance is expressed in heavy terminals and axillary branching in Team (9, 32), Arc (56), and Liberty. Well-developed axillary buds can continue growth after the stem terminals are destroyed by larval feeding. In addition, the improved vigor and stand persistence of Arc resulting from disease resistance contribute indirectly to this cultivar's tolerance to larval feeding. Although several workers have screened many alfalfa cultivars and other species of *Medicago*, only a few have been reported to demonstrate antibiosis to the weevil (12). Shuster and Tereshchenko (288) reported pronounced antibiosis in alfalfa cultivars, Raduga and Kometa. Clones of individual plants of PI 247790 from Peru exhibited considerable variation in causing convulsions and death of larvae (330). The resistance in *Medicago* spp. has not been increased to a high level. The causes for the low level of larval antibiosis reported have not been identified. Likewise, the causes for nonpreference during oviposition are unknown (58).

While screening annual *Medicago* spp. for alfalfa weevil resistance, Barnes and Ratcliffe (12) detected sources of moderate resistance of which three [*M. rugosa* Desr., *M. scutellata* (L.) Mill., and *M. minima* Bart.] possess glandular trichomes. The exudate from secretory glandular trichomes on *M. disciformis* DC caused antibiosis on alfalfa weevil larvae. They do not survive topical application of the exudate on the first instar or exposure on the plant. At low concentrations of the exudate, larvae showed reduced feeding and rate of development (284). The chemical responsible for the toxic effect has not yet been identified (335). The potential of glandular hairs as a resistance mechanism to young alfalfa weevil larvae and other small insects and mites offers promise (48, 144, 284). Plants selected from the perennial species, *M. prostrata* Jacq. and *M. glandulosa* David. were highly resistant to larvae (48). Adult feeding resistance has been found in glandular-haired annual (142) and perennial (50) *Medicago* spp. Dry matter consumption, leaf damage ratings, and numbers of weevils on the entries were lower for the glandular-haired species than for susceptible *M. sativa* L.

Nonpreference-Resistance to egg laying was observed in greenhouse-grown *M. sativa* var. *gaetula* (PI 239953). These plants had only 4 to 18% as many eggs as those laid in the Atlantic check. Reduction in number of eggs and egg-masses was associated with a decrease in the amount of stem pith, i.e., more solid stems. The effectiveness of this type of nonpreference was not determined under field conditions (40, 132). Similarly, the high resistance of *M. prostrata* and *M. glandulosa* to oviposition may be associated with stem morphology (49). Weevils reared on *M. sativa* laid fewer eggs in stems of *M. blancheana* Boiss., *M. rugosa*, and *M. scutellata* than on *M. sativa* (143). In free-choice arenas, fewer contacts with glandular-haired *M. rugosa* were observed, and the mean duration of visits was shorter than on *M. sativa*. Both precontact cues (olfactory and visual) and contact cues (chemotactic or mechanical) appeared to be responsible in rejection of *M. rugosa* in favor of *M. sativa* (145). Simple hair pubescence, as found in 'Hairy Peruvian' selections, provides some protection from damage by adults (121).

The effects of flavonoids in alfalfa leaves on feeding and physiology were studied by Shuster and Shuster (287). They concluded that the variable concentration of inhibitory and stimulatory flavonoids was responsible for host selection and preference, with concomitant effects on both lipid metabolism and reproduction.

28-4.2.4.4.2 Egyptian Alfalfa Weevil

Summers and Lehman (320) found that resistance levels in alfalfa selected for either resistance to *Hypera postica* or *H. brunneipennis* were similar. Stem size was not considered a factor in resistance, and there was no correlation between number of eggs or egg clusters per stem and stem density. Adults failed to show nonpreference for oviposition among the cultivars evaluated. Some antibiosis may exist in UC73 (320). In laboratory studies, mortality among first and second instars was greater when they were fed on parental clones of UC73 than when fed on Moapa. In addition, developmental time was prolonged in larvae feeding on these clones (Summers, unpublished).

28-4.2.5 Potato Leafhopper and Three-Cornered Alfalfa Hopper

28-4.2.5.1 Distribution and Importance.
The potato leafhopper (*Empoasca fabae* Harris) is one of the most important insect pests of alfalfa in the eastern half of the USA (6). Both adults and nymphs pierce leaves and stems and suck plant juices. Feeding injury to alfalfa is recognized by a yellowing, reddening, or purpling of leaves, wilting of leaves and terminals, and stunting of stems termed "leafhopper burn," which result in reduced stands, yield, and quality (83, 93, 95, 157, 165, 176, 212, 264, 290, 291, 352). Most damage to alfalfa occurs after first harvest. Rate of regrowth after cutting was more rapid where leafhoppers were controlled (352). Numerous reports indicate the extent and scope of interest in the potato leafhopper as a pest of alfalfa (176, 264, 352).

The three-cornered alfalfa hopper, *Spissistilus festinus* (Say) has many host plants, but alfalfa is one of the most important. This insect occurs over the western and southern USA, being more abundant in southern areas. The most serious losses to alfalfa occur from Arkansas to California. Adults and nymphs puncture stems either randomly or in regular lines that completely girdle them. Stems yellow, wilt, break off, or die and the fields may have an unthrifty appearance.

28-4.2.5.2. Development of Resistant Cultivars and Germplasm. Differences in potato leafhopper damage have been reported among cultivars and strains (53, 70, 81, 94, 140, 162, 281, 299, 307, 343, 344, 352). Differential potato leafhopper damage also has been reported among individual plants and clones (53, 70, 88, 138, 157, 165, 221, 299).

Most *M. sativa* germplasms in the nonhardy and Turkistan groups are highly susceptible to the potato leafhopper, but *M. falcata* has contributed a modest degree of resistance to 'Culver', 'Rambler', Rhizoma, and 'Teton'. Plants with varying degrees of resistance can be isolated from most cultivars, as was demonstrated in the development of 'Cherokee' (63, 64) by seven cycles of recurrent phenotypic selection for resistance to yellowing in a broad-based population. Miller and Melton (211) list over 70 cultivars and 19 germplasms with some resistance to the potato leafhopper, and six cultivars with resistance to the threecornered alfalfa hopper.

Kehr et al. (157) found that general combining ability (GCA) was more important than specific combining ability (SCA) for resistance to foliar yellowing and nymphs per gram of dry matter. Similarly, Soper et al. (295) found that GCA was important for feeding damage and the number of nymphs. Also, SCA was significant, an indication of nonadditive variance, only for the number of nymphs. They concluded that resistance to feeding damage and leafhopper oviposition and/or nymphal survival should be considered separately in breeding programs, and concurrent selection for both traits would be desirable.

28-4.2.5.3 Screening and Evaluation Techniques. Selection of plants resistant to the potato leafhopper has been practiced in the field and greenhouse (64, 154, 162, 299, 343, 344). Plants resistant to potato leafhopper yellowing have been used in the development of cultivars. Some were developed from selected clones, whereas others were developed through recurrent selection programs (64, 91, 154, 299, 307).

Resistance to yellowing in the field has been the criterion used in most breeding programs (64, 92, 165). The degree of yellowing caused by leafhopper feeding was adopted as a standard in rating plants from one to nine (137). Schillinger et al. (280) screened plants by measuring their effect on the reproductive biology of the leafhopper, whereas Webster et al. (344) used seedling survival and leafhopper damage as the basis for screening. Newton and Barnes (221) exposed potted plants to field populations of leafhopper to determine antibiosis. A standard evaluation test was developed by Kindler and Kehr (163) and Kehr and Manglitz (155). Seedlings, 6- to 8-weeks-old, were transplanted to the field in the spring. Single-row plots with plants spaced 30 to 90 cm apart were recommended, with four replications of 25 plants. Resistant and susceptible controls were included at least once per replication. Time of planting was crucial so that maximum effect of high populations might be utilized. Plantings were kept weed free and no insecticides were used. The initial growth from transplanting was scored for yellowing. After the first cutting in the year of establishment and in subsequent years, harvest intervals were extended 10 to 20 d beyond the normal cutting interval to maximize infestation. The plants were scored from one to nine based on the degree of yellowing. Jarvis and Kehr (138) found a high correlation between population count and nymphs per gram of plant material when they evaluated 75 alfalfa clones over a 2-yr period. This method appears to have considerable value since long-term effects were measured. Sorensen and Horber (299) subjected seedlings of three alfalfa synthetics to two cycles of recurrent phenotypic selection for resistance to the potato leafhopper in growth chambers. Resistance based on seedling survival after infestation was increased by cyclic selection. Resistance to leafhopper yellowing in the field increased directly with cycles of seedling selection in each synthetic. Spring growth and plant vigor were increased in each synthetic. However, actual forage yield declined in one synthetic, did not change in a second, and increased in a third, indicating the importance of field testing in conjunction with laboratory screening.

Roof et al. (273) evaluated 22 alfalfa clones for potato leafhopper resistance by infesting excised plant parts with adult leafhoppers in a controlled environment. Field and laboratory ratings for yellowing and wilting were significantly correlated, whereas leafhopper yellowing in the field and nymphs per gram of dry plant material were not. There was no association between clonal preferences displayed in the field and laboratory.

Excised plant parts from the field-grown material provided a more reliable assessment of resistance than did similar parts from plants grown in a greenhouse. Clonal yellowing and wilting of excised terminals from field grown plants were correlated highly with yellowing in the field. Excised stem terminals from plants growing in a greenhouse were less resistant to both adult and nymphal leafhopper feeding than were terminals from plants grown in the field.

Production of seed on nonyellowed plants under isolation to obtain progeny for the next cycle of recurrent selection has been an accepted method for isolating potato leafhopper resistance. Moore (213) suggested that only plants with antibiosis should be recombined if one hopes to obtain high degrees of resistance. The incorporation of resistance, which would suppress the leafhopper population, would be an important contribution. Newton and Barnes (221) tested for antibiosis by caging eight newly hatched nymphs on individual plants and recording time of development and survival of the leafhoppers. Schillinger et al. (280) conducted similar tests.

Randolph and Meisch (267) evaluated alfalfa cultivars for resistance to the three-cornered alfalfa hopper. Screening and evaluation for the same insect also was reported by Harville and Green (111) and Kadir et al. (149).

Potato leafhopper colonies have been successfully maintained in the greenhouse on broad beans (221) or on baby lima beans (*Phaseolus* spp.). Schillinger et al. (280) reared them on alfalfa. First-instar nymphs can be obtained in 10 d by exposing broad beans to ovipositing leafhoppers for 24 h. Oviposition is enhanced at 24°C and a 16-h photoperiod (160).

28–4.2.5.4 Mechanisms of Resistance. Although the host range of the potato leafhopper extends beyond legumes, feeding and oviposition are highly selective and apparently related to physical and chemical characteristics of the host (30). This has encouraged considerable research on insect-host interaction. Carlson and Hibbs (41) investigated ovipositional behavior. Decker et al. (54) reported on the sex ratio and fecundity. Newton et al. (222) studied differential injury produced by male and female leafhopper feeding.

Nonpreference was observed by Taylor (325) in 'Hairy Arabian' alfalfa plants that were more resistant to infestation than were glabrous plants. The presence of simple-hair pubescence appeared to interfere with feeding and oviposition. Similarly, resistance to yellowing in Hairy Peruvian (San Pedro) selections increased directly with density of simple hairs (121). Shade et al. (282) observed nonpreference on five glandular-haired *Medicago* spp. Adults spent significantly less time on *M. blancheana* and *M disciformis* than on the glabrous *M. sativa* (Vernal).

Saponins extracted from DuPuits alfalfa (*M. sativa*) reduced the survival of first-instar nymphs. A bioassay of 10 DuPuits saponin fractions showed that two fractions significantly reduced nymphal survival, whereas 10 Lahontan fractions exhibited no significant effect on the longevity of nymphs. Mortality of leafhoppers on diets with DuPuits saponin fractions was attributed to nonpreference of leafhoppers for saponins. When adults were offered a choice between 1% sucrose diets with and without DuPuits saponins, the choice of nonsaponin diet was highly significant at concentrations of 0.2 and 0.01% (272). Fiori and Dolan (76) field tested a collection of 70 *Medicago* introductions for resistance to leafhopper damage. During four seasons of testing they found all *M. sativa* L. ssp. *falcata*, ssp. *coerulea* Schmalh., *M. sativa* × *varia*, and *M. pironae* Vis. to be resistant and contributed most observed resistance to avoidance during oviposition due to tough stems.

Factors contributing to antibiosis to leafhoppers have been sought by several workers (212, 221, 280). Antibiosis, in commercial cultivars and currently registered germplasm, remained elusive until Shade et al. (282) observed that adult mortality occurred at a higher rate on five glandular-haired annual *Medicago* spp. than on the glabrous control. The highest rate of mortality (100%) occurred within three days on *M. noëana* Boiss., while mortality in the control was 53% after 9 d. Total nymphal production, which reflects the overall influence of the plant on the potato leafhopper, was significantly less on all five glandular-haired species than on the control. Glandular-haired species were nearly immune to the first instars, since all nymphs died before the first molt. Antibiosis and antixenosis also were reported by Brewer et al. (24) in glandular-haired perennial species (Fig. 28-5). Clones from *M. glandulosa* (303), *M. glutinosa* M.B. (302), and *M. prostrata* exhibited antixenosis in free-choice tests under two temperature regimes (24). *Medicago glutinosa* and *M. prostrata* clones showed no damage symptoms. In a no-choice situation, all adult leafhoppers died on these clones. In oviposition tests, eggs were rare in stems of clones with high antibiosis (24).

Medicago clones with different levels of resistance were examined histologically and by in vitro digestion in rumen fluid to determine if anatomical features of the stem contribute to resistance to the potato leafhopper (29). A cylinder of hard tissues, composed of lignified xylem elements, phloem fibers, and interfasicular areas surrounded the pith and was associated with reduced oviposition. The hard cylinder was present in the first internode (from apex) of two resistant clones, but only in more distal internodes of other clones. Early lignification of tissues forming the cylinder appeared to contribute to potato leafhopper resistance either by mechanically or chemically deterring or preventing feeding and oviposition.

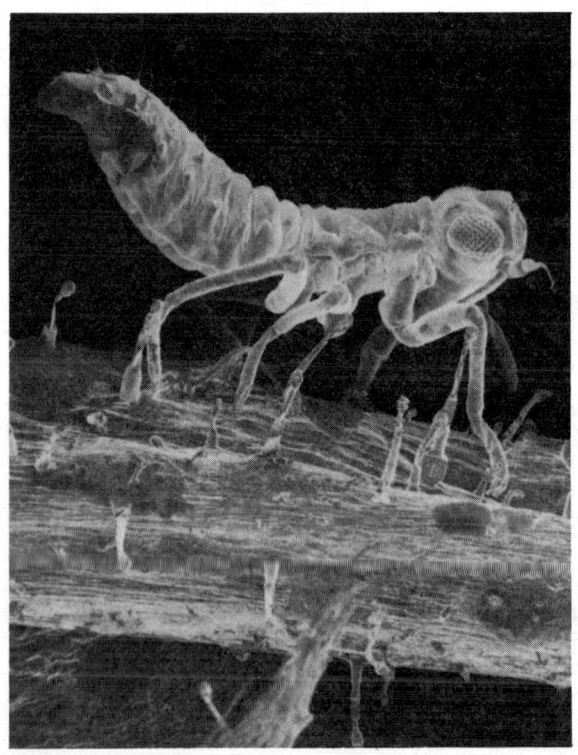

Fig. 28-5. Potato leafhopper nymph trapped by glandular hairs on stem of *Medicago glandulosa*. × 60. Bar = 100μ.

28–4.2.6 Meadow Spittlebug

28–4.2.6.1 Distribution and Importance. The meadow spittlebug (*Philaenus spumarius* L.) occurs on many host plants (342) in humid areas of the world, including eastern and north central USA (66), Europe, Hawaii, and South America. Nymphal feeding increases protein content but decreases stem length, stem weight, and yield at first harvest (254).

28–4.2.6.2 Development of Resistant Cultivars and Germplasms. Culver, Released in 1959, is the only cultivar with reported resistance to meadow spittlebug (350).

28–4.2.6.3 Screening and Evaluation Technique. Alfalfa cultivars have been evaluated in field nurseries with plants in five-row or broadcast plots. Spittlebugs were sampled by sweeping adults in the fall and by counting nymphs in square foot samples in the spring, with rosetting or damage estimated on a scale of one to nine, one indicating little damage and nine indicating severe injury (348). Adequate populations of spittlebugs can be obtained if alfalfa is planted with a small grain companion crop, because spittlebug infestation depends upon stubble and straw for oviposition sites. Tests for antibiosis can be conducted on potted plants in an insectary. Nonpreference has been determined by distributing cut stems infested with nymphs among 7.6-cm tall seedlings in flats and examining them daily for preferences among entries (349).

28–4.2.6.4 Mechanisms of Resistance. Culver alfalfa has tolerance, antibiosis, and nonpreference mechanisms of resistance to meadow spittlebug (349, 351).

28–4.2.7 Alfalfa Seed Chalcid

28–4.2.7.1 Distribution and Importance. The alfalfa seed chalcid (*Bruchophagus roddi* Gussakovsky) occurs in all areas of the world where alfalfa seed is produced. Oviposition takes place directly into young seed, where the insect develops. After development of the adult and eclosion, the chalcid chews an exit hole in the seed, and if necessary the pod, and emerges (312). Seed infestations as high as 85% have been reported (317, 318, 338). The problem is magnified by lack of suitable chemical control that is not hazardous to insect pollinators and seed chalcid parasites in the field when the chalcid is active (98). Cultural methods to destroy larvae overwintering in infested seed and volunteer plants may be beneficial under some conditions. Most control recommendations are similar to those in Johansen and Retan's (141) extension guide.

28–4.2.7.2 Development of Resistant Cultivars and Germplasm. Low levels of seed chalcid resistance are present in *Medicago* spp. (238, 240, 274, 317, 318) but resistance of current alfalfa cultivars is inadequate. Among a broad range of *Medicago* germplasm screened for resistance, light infestations were reported on Hairy Peruvian, Lahontan, 'Mesa-Sirsa', Ranger, 'Sirsa No. 9', Zia, and *M. tianschanica* Vass. var. *agropyretorum* Vass. (124, 240, 317, 318). Nondormant alfalfas, particularly 'Afghanistan', have been suggested as possible sources of resistance (274, 317, 318). Two germplasm sources (C-89 and 68 CC) with resistance to the seed chalcid are listed by Miller and Melton (211).

Medicago spp. with glandular trichomes on seed pods show promise as sources of resistance to the alfalfa seed chalcid (Fig. 28–6) (25, 26, 27). Five annual *Medicago* spp. [*M. disciformis, M. rigidula* (L.) All., *M. ciliaris* All., *M. rugosa*, and *M. rotata* Boiss.] were highly resistant to the chalcid (28).

28–4.2.7.3 Screening and Evaluation Techniques. Evaluation of seed chalcid resistance can be conducted in the field, greenhouse, or laboratory. Field populations of the chalcid fluctuate greatly from year to year, and methods for maintaining uniform populations have not been developed. Chalcid resistance in the field can be evaluated by gathering racemes of comparable maturity from each entry and placing them in

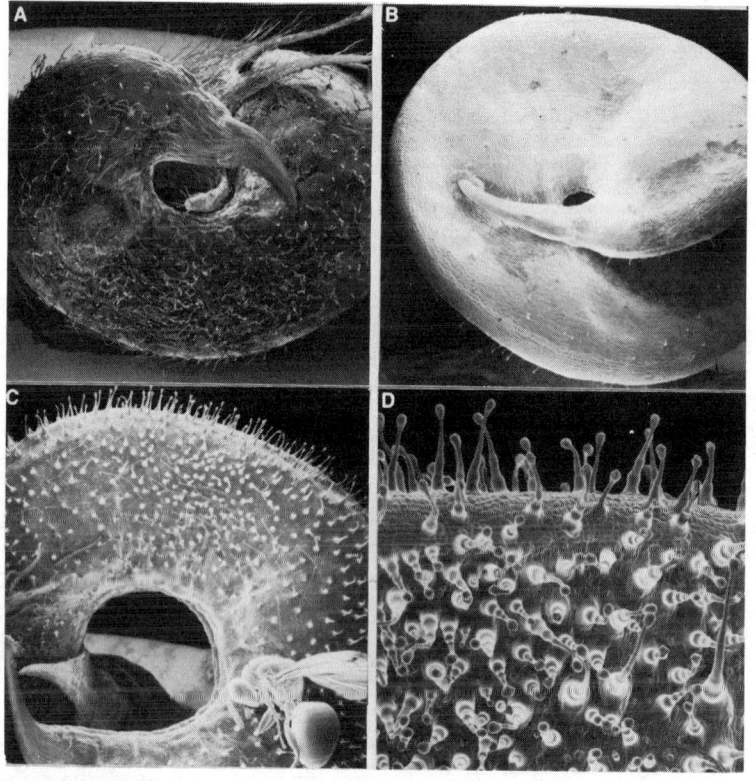

Fig. 28–6. Scanning electron micrograph of seed pods: (a) Simple hairs on *M. sativa* × 10. (b) Nearly glabrous *M. sativa* × 10. (c) Glandular hairs and female alfalfa seed chalcid on *M. sativa praefalcata*. × 10. (d) Glandular hairs on *M. glandulosa*. × 50.

rearing containers until all adults emerge (124, 225). Samples should be taken during a period between July and September, depending on location and climate. In Arizona (226), seed of germplasm to be tested is planted in rows spaced 1 m apart, with 50 to 60 cm between plants within rows. Six replications are usually used. Samples of 25 racemes, with fully developed pods, are taken from each plant and placed in an approximately 0.5 L (pint-sized) ice cream carton. Adult chalcids are allowed to emerge. Statistical comparisons of cultivars are made on the basis of mean number of adults per 25 racemes per plant. Greenhouse evaluations are made by caging 10 adult chalcids on a raceme. From six to eight racemes are caged per plant. The adult chalcids are reared from infested seed screenings held in storage at 10°C. After 2 weeks, the caged racemes are removed and placed in rearing cartons. The number of emerged adult chalcids are counted after 2 weeks.

An alternate method involves determining the percentage of infested seeds. Examining or dissecting individual seed makes the process time consuming, even with Strong's (316) technique. Two chemical techniques, based on measuring insect-induced uric acid content of seed were suggested by Booth (19) and Watts et al. (341).

Selection for resistance in the laboratory begins with collection of infested seed, which may be held in cold storage (4°C) until needed. Adult chalcids emerge 2 to 3 weeks after the infested seed are placed at room temperature (223, 317, 318). Selected plants are grown in pots at the greenhouse and the flowers are hand pollinated. Adults are confined to the developing alfalfa pods by small plastic cages to encourage oviposition. Nielson and Lehman (234) confined 10 adult chalcids on 5 to 10 racemes of each plant for 2 weeks. After removal of the original 10 adults, racemes were clipped and placed in insect rearing containers until adults emerged. Plants that averaged less than one chalcid per raceme were saved for further testing.

28-4.2.7.4 Mechanisms of Resistance. Mechanisms of resistance to the alfalfa seed chalcid were studied by Tingey and Nielsen (331, 332). Nonpreference for oviposition was observed for a period of 8 d in seed pods 10 to 18 d after pollination of resistant plants. No effect on reproductive biology or antibiosis was found in any of the resistant clones. Brewer and Horber (23) found that larval development rates did not significantly differ between *M. sativa* clones. They concluded that variation in field infestation probably depends on factors affecting oviposition. Temperature, daylength, and resource competition affected levels of field infestation in association with diapause, which appeared to be induced in the female parent. Kamm and Fronk (152) found 38 attractants and nine repellents in alfalfa but the relationship of these substances to resistance has not been studied. Kamm and Buttery (150) tested (in an olfactometer) volatiles from alfalfa on seed chalcids flying toward the source and after landing during close-range orientation. These workers also (151) studied the behavior of the chalcids during site selection and oviposition. Volatile components that had an effect on oviposition varied substantially in concentration among parts of alfalfa plants.

Of the plant parts tested in an olfactometer (26), only pods were attractive to the seed chalcid. Attractiveness varied with pod age. Pod aroma stimulated seed chalcid movement to the pods but was not significantly related to resistance or to hair type. Both simple-haired and glandular-haired clones (susceptible and resistant) were attractive for variable periods. Chemical and mechanical tactile cues apparently are used to select oviposition sites. Therefore, the use of an olfactometer to screen *Medicago* lines for seed chalcid resistance is not recommended.

Density and length of erect glandular trichomes on the seed pods of *M. sativa* L. subsp. *praefalcata* Sinsk. and *M. glandulosa* were negatively correlated with seed chalcid infestation (Fig. 28-6), whereas simple hair pubescence on seed pods from Hairy Peruvian (San Pedro) did not affect seed chalcid infestation (25).

Small and Brooks (289) observed resistance to chalcid infestation in tightly coiled pods. Because seed develops on the ventral suture close to the central axis of the coiled pod, the coiling may protect developing seed from the chalcid's short ovipositor. Resistance may be increased by selecting for tightly coiled pods with long, dense glandular hairs that prevent the ovipositor from reaching the seed (27). Although glandular-hair density declines as number of seed per pod increases, pod coils increase directly with number of seed. Thus, it may be possible to develop *Medicago* lines, the seed of which are inaccessible to the alfalfa seed chalcid.

28-4.2.8 Lygus and Other Plant Bugs

28-4.2.8.1 Distribution and Importance. There are 43 known species of *Lygus* in the world, 34 in North America, 7 in Europe, and 2 in China (158). *Lygus hesperus* Knight is an important pest of alfalfa in the West. The tarnished plant bug, *L. lineolaris* (Palisot de Beauvois), the most widespread species, was first discovered in eastern USA, but now occurs in all alfalfa growing areas worldwide.

The alfalfa plant bug, *Adelphocoris lineolatus* Goeze, was discovered in 1922 on Cape Breton Island, Nova Scotia, and in 1926 at Ames, IA. This bug was introduced to Ames in a collection of 700 seed accessions received from Europe. It has spread throughout northern USA and southern Canada (173).

Tingey and Pillemer (333) reviewed the nature of injury caused by *Lygus* bugs and described five types of plant damage: (i) localized wilting and tisssue necrosis, (ii) abcission of fruiting forms, (iii) morphological deformation of fruit, (iv) altered vegetative growth, and (v) tissue malformations. *Lygus* feed by lacerating the plant tissue, and secreting a saliva containing enzymes that digest plant cells and interfere with plant growth hormones. Strong (319) suggests that the pruning of host floral buds by *Lygus* results in the initiation of new buds and additional supplies of their preferred food.

28-4.2.8.2 Development of Resistant Cultivars and Germplasm.
Tingey and Pillemer (333) state that genetic resistance to plant bugs has not been exploited. Resistant cultivars have not been developed although differential resistance has existed for some time. Aamodt and Carlson (1) showed that 'Grimm' flowered despite feeding injury by *Lygus*.

Malcolm (193) noted differences of resistance to several species of *Lygus* among 16 cultivars in the field. 'Hardistan', 'Nemastan', and "Ladak' remained green and had low injury ratings, whereas Ranger turned yellow and rated 7.8 on a scale of 1 to 9 (9 = most injury).

28-4.2.8.3 Screening and Evaluation Techniques.
Radcliffe and Barnes (266) scored 22 alfalfa clones in the field for yellowing by the alfalfa plant bug and detected differences among entries. They concluded that it should be possible to develop resistant cultivars.

Nielson and Schonhorst (239) observed *Lygus* populations for 3 yr, in field nurseries. Zia, Rhizoma, and Stoneville polycross no. 10 had the fewest *L. hesperus* and were considered resistant. They recommended evaluating the growth habit of each entry when grown in the field to avoid pseudoresistance associated with either the physiological condition or growth habits of the plants. Later, Nielson et al. (231) screened 98 alfalfas in the seedling stage for resistance to *L. hesperus*, and separated them, based on cold hardiness, into three classes: hardy, semihardy, and nonhardy. This avoided the problem of plant growth variability in the field. Selection of resistant plants was based on seedling survival. Culver, 'Travois', and 'Nomad' had highest survival percentages (61–70%) for the hardy group; Cody, Kanza, and Lahontan (34–39%) for the semihardy group; and T-3-12, M-5-44 Syn B, and 'California Common' (67-75%) for the nonhardy group.

Lindquist et al. (186) screened alfalfa for resistance to the tarnished plant bug. Percent survival of seedlings infested in the unifoliolate stage varied from 80% for Ranger to 16% for 'Alfa'. Plant progenies from intravariety crosses of selected plants were 23 to 28% superior to their parents in resistance to tarnished plant bug.

28-4.2.8.4 Mechanisms of Resistance.
The mechanism of resistance to *Lygus* has been classed as tolerance in greenhouse tests and nonpreference in field tests. No resistant cultivars have been developed, and no biotypes of *Lygus* reported.

28-4.2.9 Clover Root Curculio

28-4.2.9.1 Distribution and Importance.
The clover root curculio, *Sitona hispidulus* (F.) originally described by Fabricius in 1776 in Europe, was found about the roots of grass in sand dunes at Long Beach, NJ, in 1876 (346). Morrison et al. (214), in a review article, reported *S. hispidulus* as one of the 17 species of *Sitona* in North America. It is also found in Canada, and represents one of the four *Sitona* spp. in Japan. Aeschlimann

(2) reported *S. hispidulus* as one of 20 species in the Mediterranean region. Markkula and Koppa (200) listed *S. hispidulus* among 11 species in Finland.

Principal losses occur from larval injury to roots that predispose the plant to root diseases (59, 115). Neonate (newly hatched) larvae attack root nodules and ingest *Rhizobium* bacteria (51, 146). Nodules seem to be essential to survival of early instars, and late instars feed on lateral roots and tap roots (35, 220).

28-4.2.9.2 Development of Resistant Cultivars and Germplasms. No resistant cultivars have been identified.

28-4.2.9.3 Screening and Evaluation Techniques. Thompson and Willis (329) assessed *Sitona* feeding damage to foliage and roots of 14 *Medicago* spp. by rating the damage 0 to 4 (none to severe). In addition, they tested adult and larval preferences among five forage legumes. Adults preferred white clover (*Trifolium repens* L.) most and birdsfoot trefoil (*Lotus corniculatus* var. *vulgaris* Hoch) least. Larvae preferred three *Trifolium* spp. (red [*T. pratense* L.], alsike [*T. hybridum* L.], and white clovers) over alfalfa and birdsfoot trefoil. In general, *Trifolium* spp. were more susceptible than *Medicago* spp. in the field. Feeding indices on alfalfa ranged from 0 to 1.0 for adults and 1 to 1.6 for larvae.

Neal and Ratcliffe (220) assigned root damage scores to six alfalfa cultivars on a scale of 1 to 10 (slight to severe). Scores ranged from 3.9 for 'Saranac' to 5.6 for Cherokee in one experiment and from 5.8 for Cherokee to 7.0 for 'Williamsburg' in a second experiment. The response of this comparatively small sample of cultivars is not definitive with respect to the presence or absence of resistance to the clover root circulio.

Pesho (260) evaluated 725 plant introductions, representing eight *Medicago* spp. and interspecific hybrids (*M. sativa* × *M. falcata*). Tap roots were divided into eight sections each 2.5 cm long. A longitudinal incision was made in a root section and the epidermis peeled off to obtain a rectangle. Root circumference, surface area, number of lesions, and area damaged were recorded. He found a 1:1 ratio between number of lesions and percentage of tap root area damaged and concluded that evaluation of root surface damaged is unnecessary. By taking data in September 1970 and 1971, April 1972, and October 1972 he showed a steady buildup of *Sitona* populations. The number of escapes were minimized and differential injury among genotypes detected by delaying evaluation of tap root injury until the middle or end of the third growing season.

Byers and Kendall (35) evaluated 12 alfalfa genotypes along with several other legumes in the laboratory for resistance to larval feeding and development. Plants were grown on slant boards, as described by Kendall and Leath (159), and watered with a solution formulated by Jensen (139) to facilitate nodulation of the roots by *Rhizobium* bacteria. Paired *t* tests with a susceptible control of 'Kenstar' red clover showed no differences in susceptibility among alfalfa genotypes. However, three sources of birdsfoot trefoil and crown vetch (*Coronilla varia* L.) were resistant to larval development.

28-4.2.9.4 Mechanisms of Resistance. In the absence of reports on resistance, no information has been generated on mechanisms of resistance in alfalfa. However, Powell et al. (265) located antibiosis to larvae of the clover root curculio in white clover.

28-4.2.10 Alfalfa Blotch Leafminer

28-4.2.10.1 Distribution and Importance. The alfalfa blotch leafminer (ABL), (*Agromyza frontella* Rondani) is found throughout Europe (18). It was first discovered in North America in 1968 in Hampshire County, MA (210, 313). Harcourt (96) reported the leafminer near the Quebec-Vermont border in 1972. Hendrickson and Plummer (113) showed a distribution in 1981 from Maine to North Carolina and west through Ohio. By 1982 it had spread in Canada to western Ontario (65).

The major impact on alfalfa from ABL is a lowering of forage quality. Losses in protein, digestible dry matter, water soluble carbohydrates, and chlorophyll from mining larvae have been documented (4, 36, 324).

28-4.2.10.2 Development of Resistant Cultivars and Germplasms. No resistance to ABL has been found in alfalfa. In 1976, Murphy (219) reported negative results from screening a wide array of germplasm.

28-4.2.10.3 Screening and Evaluation Techniques. Alfalfa plants have been evaluated in field nurseries by counting the number of blotch mines on 10 trifoliolates per plant. Selection has been practiced within entries receiving low ratings with little success (114). Although greenhouse screening has not been attempted, leafminers can be reared on flats or pots of uncaged alfalfa in the greenhouse (112).

28-4.2.10.4 Mechanisms of Resistance. Glandular hairs on related *Medicago* sp. were a deterrent to oviposition by leafminer adults (192). Incorporation of the glandular-haired character into alfalfa should help to control this leafminer. Parasites imported from Europe have brought the leafminer under control over part of its range (113). It is not known whether glandular-haired cultivars will be compatible with these parasites.

28-4.2.11 Other Alfalfa Insects

The polyphagous larvae of several Scarabaeidae (Coleoptera), commonly known as white grubs, feed on the roots of forage and pasture plants. Screening for resistance in alfalfa to cockchafer (*Melolontha vulgaris* F.) was described by Horber (116, 117). Individual alfalfa plants apparently damaged little by heavy, natural infestations were collected from pastures, replanted singly in concrete pipes, and exposed to white grubs to simulate a heavy infestation (40–60/m²). Seed collected from surviving plants was planted in pots or trays and infested with white grubs. Commercial cultivars served as susceptible controls. The protection of roots and the poor growth and development of insects on alfalfa root sections, tested in isolation, were ascribed to the saponin content of the roots (118, 119). Studies on the resistance of pasture plants to feeding by the grass grub, (*Costelytra zealandica* White) revealed resistance in alfalfa (72, 73, 74). Sutherland et al. (323) demonstrated the failure of grass grubs to feed actively on media containing crude root extracts and showed that the sterol-precipitable saponin isolated from such extracts acts as a strong feeding deterrent to third-instar larvae. Considerable variation was found in the saponin levels in roots of 13 alfalfa cultivars. Pure saponin isolated from alfalfa roots reduced feeding by larvae of *C. zealandica* by 50%. A second deterrent was isolated from alfalfa but not identified (322).

Grasshoppers usually are considered polyphagous and voracious insects that can devastate cultivated crops when present in great numbers. Mulkern (215) reviewed grassshopper preferences for different plant species, including alfalfa.

The differential grasshopper (*Melanoplus differentialis* Thomas) is one of the most important species attacking alfalfa. Harvey and Hackerott (107) obtained plants resistant to the differential grasshopper by selecting uninjured plants from a field of Buffalo alfalfa severely damaged by grasshoppers. Resistance was verified in greenhouse tests with either field-collected adults or reared nymphs. Resistance appeared partially effective against other grasshopper species. Plants selected for resistance differed agronomically from susceptible plants and, when grasshoppers were absent, yielded more forage than did susceptible plants. Also, similar differences in grasshopper injury to alfalfa were observed in Switzerland during severe infestations by mixed populations of *Stauroderus* and *Stenobothrus* spp.

The alfalfa flower midge (*Contarina medicaginis* Kieffer) causes severe losses to seed production in eastern Europe by producing galls on flowers. According to Shuster and Tereshshenko (288), the two cultivars Raduga and Kometa showed pronounced antibiosis when compared to other commercial cultivars.

28–5 SUMMARY

Host plant resistance is a significant component of insect control. Resistant cultivars provide insurance against damage when weather keep parasites and predators inactive or prevent applications of insecticides. They contribute to the establishment and maintenance of stands and to improved forage yields and quality. Even moderate levels of resistance are useful in combination with cultural, chemical, and biological control.

Mechanisms for insect resistance in alfalfa include antibiosis, nonpreference, and tolerance. Effective techniques for the selection of alfalfa germplasm with resistance to several insects are available.

Recurrent phenotypic selection has been an effective technique in developing resistance to several insects. More than 100 cultivars resistant to the pea aphid, spotted alfalfa aphid, or both, have been developed by plant breeders and entomologists working as teams. A few cultivars now contain resistance to all three alfalfa aphids. Some success has been achieved in breeding for resistance to the meadow spittlebug, potato leafhopper, and the alfalfa weevil, although much less spectacular than for the aphids.

Resistance of alfalfa to the alfalfa seed chalcid, alfalfa blotch leafminer, clover root curculio, grasshoppers, *Lygus* spp., white grub, and alfalfa flower midge has been studied but resistant cultivars have not been developed.

REFERENCES

1. Aamodt, O.S., and J. Carlson. 1938. Grimm alfalfa flowers in spite of *Lygus* injury. Wis. Agric. Exp. Stn. Bull. 440, Part II:67.
2. Aeschlimann, J.P. 1980. The *Sitona* [Col.:Curculionidae] species occurring on *Medicago* and their natural enemies in the Mediterranean region. Entomophaga 25:139–153.
3. Albrecht, H.R., and T.R. Chamberlin. 1941. Instability of resistance to aphids in some strains of alfalfa. J. Econ. Entomol. 34:551–554.
4. Andaloro, J.T., T.M. Peters, and A.J. Alicandro. 1983. Population dynamics of the alfalfa blotch leafminer, *Agromyza frontella*, and its influence on alfalfa yield in Massachusetts. Environ. Entomol. 12:510–514.
5. Anonymous. 1957. The spotted alfalfa aphid. p. 1–8. USDA–ARS-22-39. U.S. Government Printing Office, Washington, DC.
6. App, B.A., and G.R. Manglitz. 1972. Insects and related pests. *In* C.H. Hanson (ed.) Alfalfa science and technology. Agronomy 15:527–554.
7. Auclair, J.L. 1963. Pea aphid: Rearing on a chemically defined diet. Science (Washington, DC) 142(3595):1068–1069.
8. ----, and P.N. Srivastava. 1977. Distinction des biotypes dans des populations du puceron du pois, *Acyrthosiphon pisum* (Harris) de Amérique du Nord. Can. J. Zool. 55:983–989.
9. Barnes, D.K., C.H. Hanson, R.H. Ratcliffe, T.H. Busbice, J.A. Schillinger, G.R. Buss, W.V. Campbell, R.W. Hemken, and C.C. Blickenstaff. 1970. The development and performance of Team alfalfa: A multiple pest resistant alfalfa with moderate resistance to the alfalfa weevil. USDA–ARS 34–115. U.S. Government Printing Office, Washington, DC.
10. ----, B.L. Norwood, C.H. Hanson, R.H. Ratcliffe, and C.C. Blickenstaff. 1969. A mass screening procedure for isolating alfalfa seedlings with resistance to the alfalfa weevil. J. Econ. Entomol. 62:66–69.
11. ----, and R.H. Ratcliffe. 1967. Leaf disk method of testing alfalfa plants for resistance to feeding by adult alfalfa weevils. J. Econ. Entomol. 60:1561–1565.
12. ----, and ----. 1969. Evaluation of annual species of *Medicago* as sources of alfalfa weevil resistance. Crop Sci. 9:640–642.
13. ----, ----, and C.H. Hanson. 1969. Interrelationship of three laboratory screening procedures for breeding alfalfa resistant to the alfalfa weevil. Crop Sci. 9:77–79.
14. Beck, S.D. 1965. Resistance of plants to insects. Annu. Rev. Entomol. 10:207–232.
15. Bergman, J.M., and W.M. Tingey. 1979. Aspects of interaction between plant genotypes and biological control. Bull. Entomol. Soc. Am. 25:275–279.
16. Bishop, A.L., P.J. Walters, R.H. Holtkamp, and B.C. Dominiak. 1982. Relationship between *Acyrthosiphon kondoi* and damage in three varieties of alfalfa. J. Econ. Entomol. 75:118–122.

17. Blanchard, R.A., and J.E. Dudley, Jr. 1934. Alfalfa plants resistant to the pea aphid. J. Econ. Entomol. 27:262-264.
18. Bollow, H. 1955. Ein Massenauftreten der Miniefliege *Agromyza frontella* Rond. an Luzerne, Pflanzenschutzberichte 7:141-143.
19. Booth, G.M. 1969. Use of uric acid analysis to evaluate seed chalcid infestation in alfalfa seed. Ann. Entomol. Soc. Am. 62:1379-1382.
20. Bournoville, R. 1977. Etude de quelques relations entre le végétal (espèce, variété, stade phenologique) et le puceron du pois, *Acyrthosiphon pisum* Harris (Homoptera-Aphididae). Ann. Zool. Ecol. Anim. 9:87-98.
21. ----. 1980. Varietal characteristics under French conditions of some lucerne cultivars selected for resistance to two pest insects. Bull. OEPP 10:317-322.
22. ----. 1981. Variability of the net reproductive rate of clones of the pea aphid on lucerne. p. 129-132. *In* Bull. SROP, Rep. 2nd, Canterbury, UK.
23. Brewer, G.J., and E. Horber. 1984. Field infestation and alfalfa seed chalcid (Hymenoptera: Eurytomidae) development in different *Medicago* clones. Environ. Entomol. 13:1157-1159.
24. ----, E.K. Horber, and E.L. Sorensen. 1986. Potato leafhopper (Homoptera: Cicadellidae) antixenosis and antibiosis in *Medicago* species. J. Econ. Entomol. 79:421-425.
25. ----, E.L. Sorensen, and E.K. Horber. 1983. Trichomes and field resistance of *Medicago* species to the alfalfa seed chalcid (Hymenoptera: Eurytomidae). Environ. Entomol. 12:247-251.
26. ----, ----, and ----. 1983. Attractiveness of glandular and simple-haired *Medicago* clones with different degrees of resistance to the alfalfa seed chalcid (Hymenoptera: Eurytomidae) tested in an olfactometer. Environ. Entomol. 12:1504-1508.
27. ----, ----, and ----. 1983. Laboratory techniques to evaluate resistance of alfalfa clones to the alfalfa seed chalcid (Hymenoptera: Eurytomidae). Environ. Entomol. 12:1601-1605.
28. ----, ----, and ----. 1985. Alfalfa seed chalcid (Hymenoptera: Eurytomidae) infestation trials in annual *Medicago*. J. Kansas Entomol. Soc. 58:369-371.
29. ----, ----, ----, and G.L. Kreitner. 1986. Alfalfa stem anatomy and potato leafhopper (Homoptera: Cicadellidae) resistance. J. Econ. Entomol. 79:1249-1253.
30. Broersma, D.B., R.L. Bernard, and W.H. Luckmann. 1972. Some effects of soybean pubescence on populations of the potato leafhopper. J. Econ. Entomol. 65:78-82.
31. Brownlee, H., P.D. Cregan, G.M. Lodge, and R.D. Murison. 1984. The evaluation of aphid-resistant lucerne varieties in New South Wales, 1977-81. Tech. Bull. 29, Department of Agriculture, New South Wales, Australia.
32. Busbice, T.H., D.K. Barnes, C.H. Hanson, R.R. Hill, Jr., W.V. Campbell, C.C. Blickenstaff, and R.C. Newton. 1967. Field evaluation of alfalfa introductions for resistance to the alfalfa weevil, *Hypera postica* (Gyllenhal). USDA-ARS U.S. Government Printing Office, Washington, DC.
33. ----, W.V. Campbell, L.V. Bunch, and R.Y. Gurgis. 1978. Breeding alfalfa cultivars resistant to the alfalfa weevil. Euphytica 27:343-352.
34. ----, ----, J.O. Rawlings, D.K. Barnes, R.H. Ratcliffe, and C.H. Hanson. 1968. Developing alfalfa resistant to alfalfa weevil oviposition. Crop Sci. 8:762-767.
35. Byers, R.A., and W.A. Kendall. 1982. Effects of plant genotypes and root nodulation on growth and survival of *Sitona* spp. larvae. Environ. Entomol. 11:440-443.
36. ----, and K. Valley. 1981. Losses in digestible dry matter and crude protein in alfalfa caused by the alfalfa blotch leafminer. Melsheimer Entomol. Ser. 31:8-13.
37. Byrne, H.D. 1965. An improved method for storage of the alfalfa weevil in the laboratory. J. Econ. Entomol. 58:1161.
38. ----, C.C. Blickenstaff, J.L. Huggans, A.L. Steinhauer, and R.E. VanDenburgh. 1967. Laboratory studies of factors determining host plant selection by the alfalfa weevil, *Hypera postica* (Gyllenhal). Md. Agric. Exp. Stn. Bull. A-147
39. ----, and E.L. Ritterhausen. 1970. Screening of clones of alfalfa for resistance to alfalfa weevil larvae. J. Econ. Entomol. 63:682-683.
40. Campbell, W.V., and J.W. Dudley. 1965. Differences among *Medicago* species in resistance to oviposition by the alfalfa weevil. J. Econ. Entomol. 58:245-248.
41. Carlson, O.V., and E.T. Hibbs. 1970. Oviposition of *Empoasca fabae* (Homoptera: Cicadellidae). Ann. Entomol. Soc. Am. 63:516-519.
42. Carnahan, H.L., R.N. Peaden, F.V. Lieberman, and R.K. Petersen. 1963. Differential reactions of alfalfa varieties and selections to the pea aphid. Crop Sci. 3:219-222.
43. Cartier, J.J., A. Isaak, R.H. Painter, and E.L. Sorensen. 1965. Biotypes of pea aphid, *Acyrthosiphon pisum* (Harris), in relation to alfalfa clones. Can. Entomol. 97:754-760.
44. Cothran, W.R., and G.G. Gyrisco. 1966. Influence of cold storage on the viability of alfalfa weevil eggs and feeding ability of hatching larvae. J. Econ. Entomol. 59:1019-1020.

45. ----, and ----. 1966. A container for rearing phytophagous insects with potential application to controlled humidity experiments. J. Econ. Entomol. 59:481-482.
46. Dahms, R.G., and R.H. Painter. 1940. Rate of reproduction of the pea aphid on different alfalfa plants. J. Econ. Entomol. 33:482-485.
47. Daniels, N.E. 1960. Field studies of spotted alfalfa aphid resistance. Tex. Agric. Exp. Stn. Prog. Rep. 2153.
48. Danielson, S.D., G.R. Manglitz, and E.L. Sorensen. 1986. Development of alfalfa weevil larvae when reared on perennial glandular-haired *Medicago* species in the greenhouse. Environ. Entomol. 15:396-398.
49. ----, ----, and ----. 1987. Resistance of perennial glandular-haired *Medicago* species to oviposition by alfalfa weevils (Coleoptera: Curculionidae). Environ. Entomol. 16:195-197.
50. ----, ----, and ----. 1987. Resistance of perennial glandular-haired *Medicago* species to feeding by adult alfalfa weevils (Coleoptera: Curculionidae). Environ. Entomol. 16:708-711.
51. Danthanarayana, W. 1967. Host specificity of *Sitona* beetles. Nature (London) 213:1153-1154.
52. Davis, D.W., N.P. Nichols, and E.J. Armbrust. 1974. The literature of arthropods associated with alfalfa. 1. A bibliography of the spotted alfalfa aphid. Biological Notes no. 87. Illinois Natural History Survey, Urbana.
53. Davis, R.L., and M.C. Wilson. 1953. Varietal tolerance of alfalfa to the potato leafhopper. J. Econ. Entomol. 46:242-245.
54. Decker, G.C., C.A. Kouskolekas, and R.J. Dysart. 1971. Some observations on fecundity and sex ratios of the potato leafhopper. J. Econ. Entomol. 64:1127-1129.
55. Dethier, V.G., L. Barton-Browne, and C.N. Smith. 1960. The designation of chemicals in terms of the responses they elicit from insects. J. Econ. Entomol. 53:134-136.
56. Devine, T.E., R.H. Ratcliffe, T.H. Busbice, J.A. Schillinger, L. Hoffman, G.R. Buss, R.W. Cleveland, F.L. Lukezic, J.E. McMurtrey, and C.M. Rincker. 1977. Arc, a multiple pest resistant alfalfa [*Colletotrichum trifolii, Acyrthosiphon pisum, Hypera postica, Corynebacterium insidiosum*]. USDA Tech. Bull. 1559. U.S. Government Printing Office, Washington, DC.
57. ----, ----, C.M. Rincker, D.K. Barnes, S.A. Ostazeski, T.H. Busbice, C.H. Hanson, J.A. Schillinger, G.R. Buss, and R.W. Cleveland. 1975. Registration of Arc alfalfa. Crop Sci. 15:97.
58. Dhaliwal, J.S., and G.S. Grewal. 1983. Preference of lucerne varieties by *Hypera postica* (Gyllenhal) for feeding and oviposition. Indian J. Agric. Sci. 53:361-364.
59. Dickason, E.A., C.M. Leach, and A.E. Gross. 1968. Clover root curculio injury and vascular decay of alfalfa roots. J. Econ. Entomol. 61:1163-1168.
60. Diehl, S.G., and R.M. Chatters. 1956. Studies on the mechanics of feeding of the spotted alfalfa aphid on alfalfa. J. Econ. Entomol. 49:589-591.
61. Dobson, R.C., and J.G. Watts. 1957. Spotted alfalfa aphid occurrence on seedling alfalfa as influenced by systemic insecticides and varieties. J. Econ. Entomol. 50:132-135.
62. Dogger, J.R., and C.H. Hanson. 1963. Reaction of alfalfa varieties and strains to alfalfa weevil. J. Econ. Entomol. 56:192-197.
63. Dudley, J.W. 1963. Registration of Cherokee alfalfa. Crop Sci. 3:458-459.
64. ----, R.R. Hill, and C.H. Hanson. 1963. Effects of seven cycles of recurrent phenotypic selection on means and genetic variances of several characters in two pools of alfalfa germplasm. Crop Sci. 3:543-546.
65. Elia, F.C. 1982. Distribution of the alfalfa blotch leafminer and two of its introduced parasites. p. 9. *In* Proc. Northeast. Alfalfa Insects Conf., 19th, Wooster, OH. 19-20 October.
66. Elgin, J.H., Jr., D.K. Barnes, R.H. Ratcliffe, F.I. Frosheiser, M.W. Nielson, K.T. Leath, E.L. Sorensen, W.H. Lehman, S.A. Ostazeski, D.L. Stuteville, W.R. Kehr, R.N. Peaden, M.D. Rumbaugh, G.R. Manglitz, J.E. McMurtrey, III, R.R. Hill, Jr., B.D. Thyr, and B.J. Hartman. 1984. Standard tests to characterize pest resistance in alfalfa cultivars. USDA-ARS Misc. Pub. 1434. U.S. Government Printing Office, Washington, DC.
67. ----, J.E. McMurtrey, III, B.J. Hartman, B.D. Thyr, E.L. Sorensen, D.K. Barnes, F.I. Frosheiser, R.N. Peaden, R.R. Hill, Jr., and K.T. Leath. 1983. Use of strain crosses in the development of multiple pest resistant alfalfa with improved field performance. Crop Sci. 23:57-64.
68. Ellsbury, M.M., and M.W. Nielson. 1981. Comparative host plant range studies of the blue alfalfa aphid, *Acyrthosiphon kondoi* Shinji, and the pea aphid, *Acyrthosiphon pisum* (Harris) (Homoptera: Aphididae). USDA Tech. Bull. 1639. U.S. Government Printing Office, Washington, DC.
69. Emery, W.T. 1946. Temporary immunity in alfalfa ordinarily susceptible to attack by the pea aphid. J. Agric. Res. 73:33-43.

70. Farrar, N.D., and C.M. Woodworth. 1939. New strains of alfalfa studied for leafhopper resistance. Ill. Agric. Exp. Stn. Ann. Rep. 50:167–168.
71. Farrell, J.A., and M.W. Stufkens. 1981. Field evaluation of lucerne cultivars for resistance to blue-green lucerne aphid and pea aphid (*Acyrthosiphon* spp.) in New Zealand. N. Z. J. Agric. Res. 24:217–220.
72. ——, and W.J. Sweney. 1972. Plant resistance to the grass grub *Costelytra zealandica* (Col., Scarabaeidae). I. Resistance in pasture legumes. N. Z. J. Agric. Res. 15:904–908.
73. ——, and ——. 1974. Plant resistance to the grass grub *Costelytra zealandica* (Coleoptera: Scarabaeidae). II. Screening for resistance in grasses. N. Z. J. Agric. Res. 17:63–67.
74. ——, and ——. 1974. Plant resistance to the grass grub *Costelytra zealandica* (Coleoptera: Scarabaeidae). III. Resistance in *Lotus* and *Lupinus*. N. Z. J. Agric. Res. 17:69–72.
75. Ferguson, S., E.L. Sorensen, and E.K. Horber. 1982. Resistance to the spotted alfalfa aphid (Homoptera: Aphididae) in glandular-haired *Medicago* species. Environ. Entomol. 11:1229–1232.
76. Fiori, B.J., and D.D. Dolan. 1982. Field tests for *Medicago* resistance against the potato leafhopper *Empoasca fabae* Homoptera Cicadellidae. Can. J. Entomol. 113:1049–1054.
77. Frazer, B.D. 1972. Population dynamics and recognition of biotypes in the pea aphid (Homoptera: Aphididae). Can. Entomol. 104:1729–1733.
78. Gallun, R.L., and G.S. Khush. 1980. Genetic factors affecting expression and stability of resistance. p. 63–86. *In* F.G. Maxwell and P.R. Jennings (ed.) Breeding plants resistant to insects. John Wiley & Sons, New York.
79. Gentry, O.W. 1965. Crop insects in northeast Africa-Southwest Asia. USDA Handb. 273. U.S. Government Printing Office, Washington, DC.
80. Glover, D.V., and E.H. Stanford. 1966. Tetrasomic inheritance of resistance in alfalfa to the pea aphid. Crop Sci. 6:161–165.
81. Graber, L.F. 1941. Recovery after cutting and differentials in the injury of alfalfa by leafhoppers (*Empoasca fabae*). J. Am. Soc. Agron. 33:181–183.
82. Graham, H.M. 1959. Effects of temperature and humidity on the biology of *Therioaphis maculata* (Buckton). Univ. Calif., Publ. Entomol. 16:47–80.
83. Granovsky, A.A. 1928. Alfalfa "yellow top" and leafhoppers. J. Econ. Entomol. 21:261–266.
84. Grewal, G.S., and J.S. Dhaliwal. 1983. Antibiosis and tolerance in lucerne *Medicago sativa* to lucerne weevil *Hypera postica*. Indian J. Agric. Sci. 53:73–77.
85. Hackerott, H.L., and T.L. Harvey. 1959. Effect of temperature on spotted alfalfa aphid reaction to resistance in alfalfa. J. Econ. Entomol. 52:949–953.
86. ——, ——, E.L. Sorensen, and R.H. Painter. 1958. Varietal differences in survival of alfalfa seedlings infested with spotted alfalfa aphids. Agron. J. 50:139–141.
87. ——, E.L. Sorensen, T.L. Harvey, E.E. Ortman, and R.H. Painter. 1963. Reactions of alfalfa varieties to pea aphids in the field and greenhouse. Crop Sci. 3:298–301.
88. Ham, W.E., and H.M. Tysdal. 1946. The carotene content of alfalfa strains and hybrids with different degrees of resistance to leafhopper injury. J. Am. Soc. Agron. 38:68–74.
89. Hamilton, B.A., L.R. Greenup, and G.M. Lodge. 1978. Seedling mortality of lucerne varieties in field plots subjected to spotted alfalfa aphid. J. Aust. Inst. Agric. Sci. 44:54–56.
90. Hanson, C.H. 1960. Registration of varieties and strains of alfalfa, V. Agron. J. 52:405–406.
91. ——. 1969. Registration of alfalfa germplasm. Crop Sci. 9:526–527.
92. ——, T.H. Busbice, R.R. Hill, Jr., O.J. Hunt, and A.J. Oakes. 1972. Directed mass selection for developing multiple pest resistance and conserving germplasm in alfalfa. J. Environ. Quality 1:106–111.
93. ——, G.M. Loper, G.O. Kohler, E.M. Bickoff, K.W. Taylor, W.R. Kehr, F.H. Stanford, J.W. Dudley, M.W. Pederson, E.L. Sorensen, H.L. Carnahan, and C.P. Wilsie. 1965. Variation in coumestrol content of alfalfa as related to location, variety, cutting, stage of growth, and disease. USDA Tech. Bull. no. 1333. U.S. Government Printing Office, Washington, DC.
94. ——, B.I., Norwood, C.C. Blickenstaff, and R.S. VanDenburgh. 1963. Recurrent phenotypic selection for resistance to potato leafhopper yellowing in alfalfa. Agron. Abstr. American Society of Agronomy, Madison, WI, p. 80.
95. Hanson, E.W., C.H. Hanson, F.I. Frosheiser, E.L. Sorensen, R.T. Sherwood, J.H. Graham, L.J. Elling, Dale Smith, and R.L. Davis. 1964. Reactions of varieties, crosses and mixtures of alfalfa to six pathogens and the potato leafhopper. Crop Sci. 4:273–276.
96. Harcourt, D.G. 1973. *Agromyza frontella* (Rond.) (Diptera: Agromyzidae) a pest of alfalfa new to Canada. Ann. Entomol. Soc. Queb. 18:49–51.

97. Harpaz, I. 1955. Bionomics of *Therioaphis maculata* (Buckton) in Israel. J. Econ. Entomol. 48:668–671.
98. ----, 1978. Cultural control of the alfalfa seed chalcid, *Bruchophagus roddi* in Israel. FAO Plant Prot. Bull. 26:158–162.
99. Harper, A.M. 1983. Spotted alfalfa aphid. Agric. No. Agdex 622-17, Agrifax, Alberta, Canada.
100. ----, and S. Freyman. 1979. Effect of pea aphid *Acyrthosiphon pisum* (Homoptera: Aphididae) on cold hardiness of alfalfa. Can. Entomol. 111:635–636.
101. Harrington, C.D. 1941. Influence of aphid resistance in peas upon aphid development, reproduction and longevity. J. Agric. Res. 62:461–466.
102. ----. 1945. Biological races of the pea aphid. J. Econ. Entomol. 38:12–22.
103. Harris, M.K. 1979. Arthropod-plant interactions related to agriculture, emphasizing host plant resistance. p. 21–51. *In* Biology and breeding for resistance to arthropods and pathogens in agricultural plants. Tex. Agric. Exp. Stn. MD-1451.
104. Harvey, T.L., and H.L. Hackerott. 1956. Apparent resistance to the spotted alfalfa aphid selected from seedlings of susceptible alfalfa varieties. J. Econ. Entomol. 49:289–291.
105. ----, and ----. 1958. Spotted alfalfa aphid reaction and injury to resistant and susceptible alfalfa clones reciprocally grafted. J. Econ. Entomol. 51:760–762.
106. ----, and ----. 1967. Use of parasitized pea aphids to evaluate alfalfa for resistance. J. Econ. Entomol. 60:573–575.
107. ----, ----, and ----. 1976,. Grasshopper resistant alfalfa selected in the field. Environ. Entomol. 5:572–574.
108. ----, ----, and E.L. Sorensen. 1971. Pea aphid injury to resistant and susceptible alfalfa in the field. J. Econ. Entomol. 64:513–517.
109. ----, ----, and ----. 1972. Pea aphid resistant alfalfa selected in the field. J. Econ. Entomol. 65:1661–1663.
110. ----, ----, ----, R.H. Painter, E.E. Ortman, and D.C. Peters. 1960. The development and performance of Cody alfalfa, a spotted alfalfa aphid resistant variety. Kans. Agric. Exp. Stn. Tech. Bull. 114.
111. Harville, B.G., and A. Green. 1982. Evaluation of three-cornered alfalfa hopper damage [*Spissistilus festinus*; varieties, breeding lines, resistance, Louisiana]. La. Agric. Exp. Stn. Rep. p. 270–272.
112. Hendrickson, R.M. Jr., and S.E. Barth. 1977. Techniques for rearing the alfalfa blotch leafminer. J. N. Y. Entomol. Soc. 85:153–157.
113. ----, and J.A. Plummer. 1983. Biological control of alfalfa blotch leafminer (Diptera: Agromyzidae) in Delaware. J. Econ. Entomol. 76:757–761.
114. Hill, R.R. Jr., and R.A. Byers. 1979. Allocation of resources in selection for resistance to alfalfa blotch leafminer in alfalfa. Crop Sci. 19:253–257.
115. ----, R.C. Newton, K.E. Zeiders, and J.H. Elgin, Jr. 1969. Relationships of the clover root curculio, *Fusarium* wilt, and bacterial wilt in alfalfa. Crop Sci. 9:327–329.
116. Horber, E. 1959. Verbesserte Methode zur Aufzucht und Haltung von Engerlingen des Feldmaikäfers (*Melolontha vulgaris* F.) im Laboratorium. Landwirtsch Jahrb. Schweiz N.F. 8:361–370.
117. ----. 1961. Versuche zur Verhinderung der vom Maikäferengerling (*Melolontha vulgaris* F.), von der Fritfliege (*Oscinella frit* L.) und vom Maiszünsler (*Pyrausta nubilalis* Hbn.) verursachten Schäden mittels resistenter Sorten. Landwirtsch Jahrb. Schweiz. N.F. 9:635–669.
118. ----. 1965. Isolation of components from the roots of alfalfa (*Medicago sativa* L.) toxic to white grubs (*Melolontha vulgaris* F.). p. 540–541. *In* Proc. 12th Int. Congr. Entomol., London. July, 1964.
119. ----. 1972. Alfalfa saponins significant in resistance to insects. p. 611–628. *In* J.G. Rodriguez (ed.) Insect and mite nutrition. North Holland Publishing Co., Amsterdam.
120. ----. 1980. Types and classification of resistance. p. 15–21. *In* F.G. Maxwell and P.R. Jennings (ed.) Breeding plants resistant to insects. John Wiley and Sons, New York.
121. ----, E.L. Sorensen, and K.J.R. Johnson. 1981. Resistance to alfalfa weevil and potato leafhopper increased by glandular and simple hairs. p. 51. *In* Rep. 27th Alfalfa Improve. Conf. Madison, WI. 8–10 July 1980. ARM-NC-19. USDA-SEA, Peoria, IL.
122. Howe, W.L., W.R. Kehr, and C.O. Calkins. 1965. Appraisal for combined pea aphid and spotted alfalfa aphid resistance in alfalfa. Nebr. Agric. Exp. Stn. Res. Bull. 221.
123. ----, ----, M.E. McKnight, and G.R. Manglitz. 1963. Studies of the mechanisms and sources of spotted alfalfa aphid resistance in Ranger alfalfa. Nebr. Agric. Exp. Stn. Res. Bull. 210.
124. ----, and G.R. Manglitz. 1961. Observations on the clover seed chalcid as a pest of alfalfa in eastern Nebraska. Proc. North Cent. Branch Entomol. Soc. Am. 16:49–51.
125. ----, and G.R. Pesho. 1960. Influence of plant age on the survival of alfalfa varieties differing in resistance to the spotted alfalfa aphid. J. Econ. Entomol. 53:142–144.

126. ----, and ----. 1960. Spotted alfalfa aphid resistance in mature growth of alfalfa varieties. J. Econ. Entomol. 53:234-238.
127. ----, and O.F. Smith. 1957. Resistance to the spotted alfalfa aphid in Lahontan alfalfa. J. Econ. Entomol. 50:320-324.
128. Hower, A.A., and F.R. Ferrer. 1970. An artificial oviposition technique for the alfalfa weevil. J. Econ. Entomol. 63:761-764.
129. Hsiao, T.H., and C. Hsiao. 1974. A practical artifical diet for laboratory rearing of the alfalfa weevil, *Hypera postica* (Gyllenhal). Ann. Entomol. Soc. Am. 67:149-150.
130. Hubert-Dahl, M.L. 1975. Changes of the host selection of three biotypes of *Acyrthosiphon pisum* after rearing on different host plants (Homoptera: Aphididae) Beitr. Entomol. 25:77-83.
131. Huggans, J.L., and C.C. Blickenstaff. 1964. Effects of photoperiod on sexual development in the alfalfa weevil. J. Econ. Entomol. 57:167-168.
132. Hunt, O.J., H. Baenziger, B.J. Hartman, D.H. Heinrichs, E.S. Horner, I.I. Kawaguchi, B.A. Melton, M.H. Schonhorst, and B.D. Thyr. 1978. Improved breeding lines of alfalfa. USDA-SEA, ARM-W-5. USDA-SEA, Berkeley, CA.
133. ----, R.N. Peaden, H.L. Carnahan, and F.V. Lieberman. 1966. Registration of Washoe alfalfa. Crop Sci. 6:610.
134. ----, ----, M.W. Nielson, and C.H. Hanson. 1971. Development of two alfalfa populations with resistance to insect pests, nematodes and diseases. I. Aphid resistance. Crop Sci. 11:73-75.
135. Isaak, A., E.L. Sorensen, and E.E. Ortman. 1963. Influence of temperature and humidity on resistance in alfalfa to the spotted alfalfa aphid and pea aphid. J. Econ. Entomol. 56:53-57.
136. ----, ----, and R.H. Painter. 1965. Stability of resistance to pea aphid and spotted alfalfa aphid in several alfalfa clones under various temperature regimes. J. Econ. Entomol. 58:140-143.
137. Jarvis, J.L. 1964. Potato leafhopper. p. 4. *In* Rep. 19th Alfalfa Improve. Conf. CR-54-64:4, Lafayette, IN. 28 June-1 July.
138. ----, and W.R. Kehr. 1966. Population counts vs. nymphs per gram of plant material in determining degree of alfalfa resistance to the potato leafhopper. J. Econ. Entomol. 59:427-430.
139. Jensen, H.L. 1942. Nitrogen fixation in leguminous plants. I. General characteristics of root-nodule bacteria isolated from species of *Medicago* and *Trifolium* in Australia. Proc. Linn. Soc. N. S. W. 66:98-108.
140. Jewett, H.H. 1929. Leafhopper injury to clover and alfalfa. Ky. Agric. Exp. Stn. Bull. 293:157-172.
141. Johansen, C.A., and A.H. Retan. 1974. The alfalfa seed chalcid. Wash. State Univ. Coop. Ext. Ser. EM2923. Washington State University, Pullman.
142. Johnson, K.J.R., E.L. Sorensen, and E.K. Horber. 1980. Resistance in glandular-haired annual *Medicago* species to feeding by adult alfalfa weevils (*Hypera postica*). Environ. Entomol. 9:133-136.
143. ----, ----, ----. 1980. Resistance of glandular-haired *Medicago* species to oviposition by adult alfalfa weevils [*Hypera postica* (Gyl.)]. Environ. Entomol. 9:241-244.
144. ----, ----, and ----. 1980. Effect of temperature and glandular-haired *Medicago* species on development of alfalfa weevil larvae. Crop Sci. 20:631-633.
145. ----, ----, and ----. 1981. Behavior of adult alfalfa weevils [*Hypera postica* (Gyl.)] on resistant and susceptible *Medicago* species in free-choice preference tests. Environ. Entomol. 10:580-585.
146. Johnson, M.P., and L.E. O'Keeffe. 1981. Presence and possible assimilation of *Rhizobium leguminosarum* in the gut of pea leaf weevil, *Sitona lineatus* larvae. Entomol. Exp. Appl. 29:103-108.
147. Jones, B.F., E.L. Sorensen, and R.H. Painter. 1968. Tolerance of alfalfa clones to the spotted alfalfa aphid. J. Econ. Entomol. 61:1046-1050.
148. Jones, L.G., F.N. Briggs, and R.A. Blanchard. 1950. Inheritance of resistance to the pea aphid in alfalfa hybrids. Hilgardia 20:9-17.
149. Kadir, M.A., B.G. Harville, and C.M. Smith. 1982. Screening for three-cornered alfalfa hopper resistance [*Spississtilus festinus* cultivar resistance]. La. Agric. Exp. Stn. Rep. p. 268-269.
150. Kamm, J.A. and R.G. Buttery. 1983. Response of the alfalfa seed chalcid *Bruchophagus roddi* to alfalfa volatiles. Entomol. Exp. Appl. 33:129-134.
151. ----, and ----. 1986. Ovipositional behavior of the alfalfa seed chalcid (Hymenoptera: Eurytomidae) in response to volatile components of alfalfa. Environ. Entomol. 15:333-391.
152. ----, and W.D. Fronk. 1964. Olfactory response to the alfalfa-seed chalcid, *Bruchophagus roddi* Guss., to chemicals found in alfalfa. Wyo. Agric. Exp. Stn. Bull. 413.
153. Kehr, W.R. 1967. Registration of Dawson alfalfa. Crop Sci. 7:680-681.

154. ————. 1970. Registration of N.S. 16 alfalfa germplasm. Crop Sci. 10:731.
155. ————, and G.R. Manglitz. 1984. Potato leafhopper yellowing resistance. p. 29. In J.H. Elgin, Jr. (ed.) Standard tests to characterize pest resistance in alfalfa cultivars. USDA–ARS Misc. Pub. 1434. U.S. Government Printing Office, Washington, DC.
156. ————, ————, and R.L. Ogden. 1968. Dawson alfalfa, a new variety resistant to aphids and bacterial wilt. Nebr. Agric. Exp. Stn. Bull. 497.
157. ————, R.L. Ogden, and S.D. Kindler. 1970. Diallel analyses of potato leafhopper injury to alfalfa. Crop Sci. 10:584–586.
158. Kelton, L.A. 1975. The lygus bugs (Genus *Lygus* Hahn) of North America (Heteroptera: Miridae). Mem. Entomol. Soc. Can. 95:1–101.
159. Kendall, W.A., and K.T. Leath. 1974. Slant-board culture methods for root observations of red clover. Crop Sci. 14:317–320.
160. Kieckhefer, R.W., and J.T. Medler. 1964. Some environmental factors influencing oviposition by the potato leafhopper, *Empoasca fabae*. J. Econ. Entomol. 57:482–484.
161. Kilian, L., and M.W. Nielsen. 1971. Differential effects of temperature on the biological activity of four biotypes of the pea aphid. J. Econ. Entomol. 64:153–155.
162. Kindler, S.D., and W.R. Kehr. 1970. Field tests of alfalfa selected for resistance to potato leafhopper in the greenhouse. J. Econ. Entomol. 63:1463–1467.
163. ————, and ————. 1974. Evaluating potato leafhopper yellowing resistance. p. 19. In J.H. Elgin, Jr. (ed.) Standard tests to characterize pest resistance in alfalfa cultivars. USDA–ARS NC-19. U.S. Government Printing Office, Washington, DC.
164. ————, ————, and R.L. Ogden. 1971. Influence of pea aphids and spotted alfalfa aphids on the stand, yield of dry matter, and chemical composition of resistant and susceptible varieties of alfalfa. J. Econ. Entomol. 64:653–657.
165. ————, ————, ————, and J.M. Schalk. 1971. Effect of potato leafhopper injury on yield and quality of resistant and susceptible alfalfa clones. J. Econ. Entomol. 66:1298–1302.
166. ————, and R. Staples. 1969. Behavior of the spotted alfalfa aphid on resistant and susceptible alfalfas. J. Econ. Entomol. 62:474–478.
167. ————, and ————. 1970. Nutrients and the reaction of two alfalfa clones to the spotted alfalfa aphid. J. Econ. Entomol. 63:938–940.
168. ————, and ————. 1970. The influence of fluctuating and constant temperatures, photoperiod, and soil moisture on the resistance of alfalfa to the spotted alfalfa aphid. J. Econ. Entomol. 63:1198–1201.
169. Kircher, H.W., R.L. Misiorowski, and F.V. Lieberman. 1970. Resistance of alfalfa to the spotted alfalfa aphid. J. Econ. Entomol. 63:964–969.
170. Kishaba, A.N., and G.R. Manglitz. 1965. Non-preference as a mechanism of sweetclover and alfalfa resistance to the sweetclover aphid and the spotted alfalfa aphid. J. Econ. Entomol. 58:566–569.
171. ————, and ————. 1968. Substances from alfalfa biologically active against the spotted alfalfa aphid. USDA–ARS 33–126. U.S. Government Printing Office, Washington, DC.
172. Klement, W.J., and N.M. Randolph. 1960. The evaluation of resistance of seedling alfalfa varieties and strains to the spotted alfalfa aphid, *Therioaphis maculata*. J. Econ. Entomol. 53:667–669.
173. Knight, H.H. 1941. The plant bugs or Miridae of Illinois. Ill. Nat. Hist. Surv. Bull. 22:1–234.
174. Kodet, R.T., M.W. Nielson, and R.O. Kuehl. 1982. Effect of temperature and photoperiod on the biology of blue alfalfa aphid, *Acyrthosiphon kondoi* Shinji. USDA Tech. Bull. 1660. U.S. Government Printing Office, Washington, DC.
175. Kogan, M., and E.E. Ortman. 1978. Antixenosis—A new term proposed to replace Painter's "nonpreference" modality of resistance. ESA Bull. 24:175–176.
176. Kouskolekas, C., and G.C. Decker. 1968. A quantitative evaluation of factors affecting alfalfa yield reduction caused by the potato leafhopper attack. J. Econ. Entomol. 61:921–927.
177. Kryzmanska, J. 1976. Biochemical aspects of plant resistance to insects. p. 115–132. In Mater. Ses. Nauk. Inst. Ochr. Rosl. 16th Poznan, Poland. 12–14 Luty. Instytutu Ochrony Roslin, Poznan, Poland.
178. ————, J. Rosada, D. Waligora, and W. Jakubiak. 1983. Investigation on the role of saponins in the resistance of alfalfa (*Medicago* sp.) to the pea aphid (*Acyrthosiphon pisum* H.) with the use of radioisotope method. Pr. Nauk. Inst. Ochr. Rosl. 25:207–214.
179. Kugler, J.L. 1981. A pink pea aphid *Acyrthosiphon pisum* biotype on alfalfa. p. 68. In Rep. 27th Alfalfa Improve. Conf., Madison, WI. 8–10 July 1980. ARM–NC-19. USDA–SEA, Peoria, IL.
180. ————, and R.H. Ratcliffe. 1983. Resistance in alfalfa to a red form of the pea aphid (Homoptera: Aphididae). J. Econ. Entomol. 76:74–76.
181. Lehman, W.F., O.J. Hunt, and E.H. Stanford. 1975. Possibilities for varietal resistance

to the blue alfalfa aphid (*Acyrthosipon kondoi*). p. 31-34. *In* Proc. Calif. Alfalfa Symp., 5th Fresno, CA, 10-11 December. Cooperative Extension, University of California, Davis.
182. ——, M.W. Nielson, V.L. Marble, and E.H. Stanford. 1983. Registration of CUF 101 alfalfa. Crop Sci. 23:398.
183. ——, and E.H. Stanford. 1972. Progress in the development of alfalfa varieties with resistance to the Egyptian alfalfa weevil *Hypera brunneipennis* (Boh.). p. 23-27. *In* Proc. Calif. Alfalfa Symp., Fresno, CA. 5-6 December.
184. ——, and ——. 1975. Possibilities for developing alfalfa varieties for resistance to the Egyptian alfalfa weevil *Hypera brunneipennis*. p. 53-59. *In* Proc. Calif. Alfalfa Symp., 5th, Fresno CA. 10-11 December. Cooperative Extension, University of California, Davis.
185. Lenssen, A.W., E.L. Sorensen, and G.L. Posler. 1985. Forage quality of perennial glandular-haired *Medicago* populations. Agron. Abstr., American Society of Agronomy, Madison, WI, p. 61.
186. Lindquist, R.K., R.H. Painter, and E.L. Sorensen. 1967. Screening alfalfa seedlings for resistance to the tarnished plant bug. J. Econ. Entomol. 60:1442-1445.
187. Lloyd, D.L., B.A. Franzmann, and T.B. Hilder. 1983. Resistance of lucerne lines at different stages of plant growth to the spotted alfalfa aphid and blue green aphid. Aust. J. Exp. Agric. Anim. Husb. 23:288-293.
188. ——, J.W. Turner, and T.B. Hilder. 1980. Effects of aphids on seedling growth of lucerne lines in the field. 1. Blue-green aphid (*Acyrthosiphon kondoi* Shinji) in field conditions. Aust. J. Exp. Agric. Anim. Husb. 20:72-76.
189. ——, ——, and ——. 1980. Effects of aphids on seedling growth of lucerne lines in the field. 2. Spotted alfalfa aphid (*Therioaphis trifolii* F. *maculata* (Monell) in field conditions. Aust. J. Exp. Agric. Anim. Husb. 20:452-456.
190. Lodge, G.M., and L.R. Greenup. 1983. Field tests for evaluating spotted alfalfa aphid resistance in lucerne cultivars: A comparison with glasshouse studies. Aust. J. Exp. Agric. Anim. Husb. 23:393-398.
191. Luginbill, P. 1969. Developing resistant plants—the ideal method of controlling insects. USDA Prod. Res. Rep. 111. U.S. Government Printing Office, Washington, DC.
192. Maclean, P.S., and R.A. Byers. 1983. Ovipositional preferences of the alfalfa blotch-leafminer (Diptera: Agromyzidae) among some simple and glandular-haired *Medicago* species. Environ. Entomol. 12:1083-1086.
193. Malcolm, D.R. 1953. Host relationship studies of *Lygus* in south-central Washington. J. Econ. Entomol. 46:485-488.
194. Manglitz, G.R., C.O. Calkins, R.J. Walstrom, S.D. Hintz, S.D. Kindler, and L.L. Peters. 1966. Holocyclic strain of the spotted alfalfa aphid in Nebraska and adjacent states. J. Econ. Entomol. 59:636-639.
195. ——, and W.R. Kehr. 1984. Resistance to spotted alfalfa aphid (Homoptera: Aphididae) in alfalfa seedlings of two plant introductions [*Therioaphis maculata*]. J. Econ. Entomol. 77:357-359.
196. ——, ——, and C.O. Calkins. 1962. Pea-aphid resistant alfalfa now in sight. Nebr. Exp. Stn. Qt. 24:5-6.
197. ——, and J.M. Schalk. 1970. Occurrence and hosts of *Aphelinus semiflavus* Howard (Hymenoptera: Eulophidae) in Nebraska. J. Kansas Entomol. Soc. 43:309-314.
198. Marble, V.L. 1978. Lucerne aphids: A world wide threat. World Farming 20:10-11, 24-26.
199. ——, J.C. Meldeen, H.C. Murray, and F.P. Zscheile. 1959. Studies of free amino acids in the spotted alfalfa aphid, its honeydew, and several alfalfa selections, in relation to aphid resistance. Agron. J. 51:740-743.
200. Markkula, M., and P. Koppa. 1960. The composition of the *Sitona* (Col., Curculionidae) population on grassland legumes and some other leguminous plants. Ann. Entomol. Fenn. 26:246-263.
201. ——, and M. Roukka. 1970. Resistance of plants to the pea aphid, *Acyrthosiphon pisum* Harris (Hom:Aphididae). I. Fecundity of the biotypes on different host plants. Ann. Agric. Fenn. 9:127-132.
202. Maxwell, F.G., and P.R. Jennings. 1980. Breeding plants resistant to insects. John Wiley and Sons, New York.
203. ——, and R.H. Painter. 1959. Factors affecting rate of honeydew deposition by *Therioaphis maculata* (Buck.) and *Toxoptera graminum* (Rond.). J. Econ. Entomol. 52:368-373.
204. ——, and ——. 1962. Auxins in honeydew of *Toxoptera graminum*, *Therioaphis maculata*, and *Macrosiphum pisi*, and their relation to degree of tolerance in host plants. Ann. Entomol. Soc. Am. 55:229-233.
205. ——, and ——. 1962. Auxin content of extracts of certain tolerant and susceptible host plants of *Toxoptera graminum*, *Macrosiphum pisi*, and *Therioaphis maculata* and relation to host plant resistance. J. Econ. Entomol. 55:46-56.

206. McMurtry, J.A. 1962. Resistance of alfalfa to spotted alfalfa aphid in relation to environmental factors. Hilgardia 32:501–539.
207. ———, and E.H. Stanford. 1960. Observations of feeding habits of the spotted alfalfa aphid on resistant and susceptible alfalfa plants. J. Econ. Entomol. 53:714–717.
208. Melton, B.A. 1968. Mesilla alfalfa. N. M. Agric. Exp. Stn. Bull. 530.
209. Messenger, P.S. 1964. The influence of rhythmically fluctuating temperatures on the development and reproduction of the spotted alfalfa aphid, *Therioaphis maculata*. J. Econ. Entomol. 57:71–76.
210. Miller, D.E., and G.L. Jensen. 1970. Agromyzid alfalfa leaf miners and their parasites in Massachusetts. J. Econ. Entomol. 63:1337–1338.
211. Miller, D., and B. Melton. 1983. Description of alfalfa cultivars and germplasm sources. N. M. Agric. Exp. Stn. Spec. Rep. 53.
212. Moore, G.D. 1968. Evaluation of feeding injury to alfalfa by potato leafhopper. *Empoasca fabae* (Harris). Ph.D. diss. Univ. of Minnesota, St. Paul. [Diss. Abst. Int. B29(8):2931].
213. ———. 1971. Selecting of resistance to the potato leafhopper in alfalfa. Proc. North Cent. Branch Entomol. Soc. Am. 26:83–84.
214. Morrison, W.P., B.C. Pass, N.P. Nichols, and E.J. Armbrust. 1974. The literature of arthropods associated with alfalfa. II. A bibliography of the *Sitona* species (Coleoptera: Curculionidae). Biol. Notes no. 88. Ill. Nat. Hist. Surv.
215. Mulkern, G.B. 1967. Food selection by grasshoppers. Annu. Rev. Entomol. 12:59–78.
216. Müller, F.P. 1962. Biotypen und Unterarten der Erbenslaus *Acyrthosiphon pisum* Harris. Z. Pflanzenkrankh Pflanzenschutz 69:129–136.
217. ———. 1971. Isolations-mechanismen zwischen sympathischen bionomischen Rassen am Beispiel der Erbsenblattlaus *Acyrthosiphon pisum* Harris (Homoptera, Aphididae). Zool. Jahrb. Abt. Syst. Ökol. Geogr. Tiere 98:131–152.
218. Muller, K.O. 1959. Hypersensitivity. p. 469–519. *In* J.G. Horsfall and A.E. Diamond (ed.), Plant pathology, Vol. 1. Academic Press, New York.
219. Murphy, R.P. 1976. Preliminary evaluation of alfalfa for resistance to the alfalfa blotch leafminer. p. 23. *In* Rep. 25th Alfalfa Improve. Conf., Ithaca, NY. 13–15 July.
220. Neal, J.W. Jr., and R.H. Ratcliffe. 1975. Clover root curculio. Control with granular carbofuran as measured by alfalfa regrowth, yield, and root damage. J. Econ. Entomol. 68:829–831.
221. Newton, R.C., and D.K. Barnes. 1965. Factors affecting resistance of selected alfalfa clones to the potato leafhopper. J. Econ. Entomol. 58:435–439.
222. ———, R.R. Hill, and J.H. Elgin, Jr. 1970. Differential injury to alfalfa by male and female potato leafhoppers. J. Econ. Entomol. 63:1077–1079.
223. Nickel, J.L., and E.S. Sylvester. 1959. Influence of feeding time, stylet penetration, and developmental instar on the toxic effect of the spotted alfalfa aphid. J. Econ. Entomol. 52:249–254.
224. Nielson, M.W. 1957. Sampling technique studies on the spotted alfalfa aphid. J. Econ. Entomol. 50:385–389.
225. ———. 1967. Procedures for screening and testing alfalfa for resistance to the alfalfa seed chalcid. USDA-ARS 33-120. U.S. Government Printing Office, Washington, DC.
226. ———. 1984. Spotted alfalfa aphid resistance and alfalfa seed chalcid resistance. *In* J.H. Elgin, Jr. (ed.) Standard tests to characterize pest resistance in alfalfa cultivars. USDA-ARS Misc. Pub. 1434. U.S. Government Printing Office, Washington, DC.
227. ———, and W.E. Currie. 1959. Effect of alfalfa variety on the biology of the spotted alfalfa aphid in Arizona. J. Econ. Entomol. 52:1023–1024.
228. ———, and H. Don. 1974. Interaction between biotypes of the spotted alfalfa aphid and resistance in alfalfa. J. Econ. Entomol. 67:368–370.
229. ———, and ———. 1974. Probing behavior of biotypes of the spotted alfalfa aphid on resistant and susceptible alfalfa clones. Entomol. Exp. Appl. 17:477–486.
230. ———, ———, M.H. Schonhorst, W.F. Lehman, and V.L. Marble. 1970. Biotypes of the spotted alfalfa aphid in western United States. J. Econ. Entomol. 63:1822–1825.
231. ———, ———, and J. Zaugg. 1974. Sources of resistance in alfalfa to *Lygus hesperus* Knight. USDA-ARS W-21. USDA-ARS, Berkeley, CA.
232. ———, and R.O. Keuhl. 1982. Screening efficacy of spotted alfalfa aphid biotypes and genic systems for resistance in alfalfa. Environ. Entomol. 11:989–996.
233. ———, and W.F. Lehman. 1977. Multiple aphid resistance in CUF 101 Alfalfa. J. Econ. Entomol. 70:13–14.
234. ———, and ———. 1980. Breeding approaches in alfalfa. p. 277–311. *In* F.G. Maxwell and P.R. Jennings (ed.) Breeding plants resistant to insects. John Wiley and Sons, New York.
235. ———, ———, and R.T. Kodet. 1976. Resistance in alfalfa to *Acyrthosiphon kondoi*. J. Econ. Entomol. 69:471–472.
236. ———, ———, and V.L. Marble. 1970. A new severe strain of the spotted alfalfa aphid in California. J. Econ. Entomol. 63:1489–1491.

237. ----, and D.L. Olson. 1982. Horizontal resistance in 'Lahontan' alfalfa to biotypes of the spotted alfalfa aphid (Homoptera: Aphididae). Environ. Entomol. 11:989-930.
238. ----, and M.H. Schonhorst. 1965. Alfalfa seed chalcid resistance in alfalfa. Prog. Agric. Ariz. 17:20-21.
239. ----, and ----. 1965. Screening alfalfas for resistance to some common insect pests in Arizona. J. Econ. Entomol. 58:147-150.
240. ----, and ----. 1967. Sources of alfalfa seed chalcid resistance in alfalfa. J. Econ. Entomol. 60:1506-1511.
241. ----, ----, H. Don, W.F. Lehman, and V.L. Marble. 1971. Resistance in alfalfa to four biotypes of the spotted alfalfa aphid. J. Econ. Entomol. 64:506-510.
242. Norris, D.M., and M. Kogan. 1980. Biochemical and morphological bases of resistance. p. 23-62. *In* F.G. Maxwell and P.R. Jennings (ed.) Breeding plants resistant to insects. John Wiley and Sons, New York.
243. Norwood, B.L., D.K. Barnes, R.S. VanDenburgh, C.H. Hanson, and C.C. Blickenstaff. 1967. Influence of stem diameter on oviposition preference of the alfalfa weevil and its importance in breeding for resistance. Crop Sci. 7:428-430.
244. ----, R.S. VanDenburgh, C.H. Hanson, and C.C. Blickenstaff. 1967. Factors affecting resistance of field-planted alfalfa clones to the alfalfa weevil. Crop Sci. 7:96-99.
245. Okiwelu, S.N. 1977. Consumption of alfalfa by larvae and adults of the Egyptian alfalfa weevil, *Hypera brunneipennis*. Ann. Entomol. Soc. Am. 70:622-624.
246. Ortman, E.E. 1963. A study of the free amino acids in the spotted alfalfa aphid and pea aphid and their honeydew in relation to fecundity and honeydew excretion of aphids feeding on a range of resistant and susceptible alfalfa selections. Ph.D. diss. Kansas State Univ., Manhattan (Diss. Abstr. 25:4315-4316).
247. ----, E.L. Sorensen, R.H. Painter, T.L. Harvey, and H.L. Hackerott. 1960. Selection and evaluation of pea aphid resistant alfalfa plants. J. Econ. Entomol. 53:881-887.
248. Painter, R.H. 1936. The food of insects and its relation to resistance of plants to insect attack. Am. Nat. 70:547-566.
249. ----. 1941. The economic value and biological significance of plant resistance to insect attack. J. Econ. Entomol. 34:358-367.
250. ----. 1951. Insect resistance in crop plants. Macmillan Publishing Co., New York.
251. ----, and C.O. Grandfield. 1935. Preliminary report on resistance of alfalfa varieties to the pea aphids, *Illinoia pisi* (Kalt). Agron. J. 27:671-674.
252. ----, E.L. Sorensen, T.L. Harvey, and H.L. Hackerott. 1965. Selection for combined resistance in alfalfa, *Medicago sativa* L., to pea aphid, *Acyrthosiphon pisum* (Harris), and spotted alfalfa aphid, *Therioaphis maculata* (Buckton). I:531. *In* Proc. 12th Int. Congr. Entomol., London. 21 July 1964.
253. Pandey, K.C., Singh Amar, and S.A. Faruqui. 1984. Sources of resistance to spotted alfalfa aphid (*Therioaphis maculata* Buckton) in Medics. Indian J. Genet. Plant Breed. 44:1-6.
254. Parman, V.R., and M.C. Wilson. 1982. Alfalfa crop responses to feeding by the meadow spittlebug (Homoptera: Cercopidae). J. Econ. Entomol. 75:481-486.
255. Passlow, T. 1977. The spotted alfalfa aphid, *Therioaphis trifolii* F. *maculata*, a new pest of lucerne in Australia. Queensl. Agric. J. 103:329-330.
256. Peaden, R.N., H.L. Carnahan, O.J. Hunt, and F.V. Lieberman. 1966. Washoe alfalfa. Nev. Agric. Exp. Stn. Circ. 64.
257. ----, O.J. Hunt, and F.V. Lieberman. 1967. Washoe-a new alfalfa variety. Nevada Ranch Home Rev. 3(8):12-13.
258. Pedersen, M.W., D.K. Barnes, E.L. Sorensen, G.D. Griffen, M.W. Nielson, R.R. Hill, Jr., F.I. Frosheiser, R.M. Sonoda, C.H. Hanson, O.J. Hunt, R.N. Peaden, J.H. Elgin, Jr., T.E. Devine, M.J. Anderson, B.P. Goplen, L.J. Elling, and R.E. Howarth. 1976. Effects of low and high saponin selection in alfalfa on agronomic and pest resistance traits and the interrelationship of these traits. Crop Sci. 16:193-199.
259. ----, E.L. Sorensen, and M.J. Anderson. 1975. A comparison of pea aphid-resistant and susceptible alfalfas for field performance, saponin concentration, digestibility, and insect resistance. Crop Sci. 15:254-256.
260. Pesho, G.R. 1975. Clover root curculio: Estimates of larval injury to alfalfa tap roots. J. Econ. Entomol. 68:61-65.
261. ----, F.V. Lieberman, and W.F. Lehman. 1960. A biotype of the spotted alfalfa aphid on alfalfa. J. Econ. Entomol. 53:146-150.
262. Pimentel, D., and A.G. Wheeler. 1973. Influence of alfalfa resistance on a pea aphid population and its associated parasites, predators, and competitors. Environ. Entomol. 2:1-11.
263. Pitre, H.N., V.H. Watson, and C.Y. Ward. 1970. Field evaluation of alfalfa cultivars for resistance to the alfalfa weevil in Mississippi-a preliminary study. Agron. J. 62:678-679.
264. Poos, F.W., and H.W. Johnson. 1936. Injury to alfalfa and red clover by the potato leafhopper. J. Econ. Entomol. 29:325-331.

265. Powell, G.S., W.V. Campbell, W.A. Cope, and D.S. Chamblee. 1983. Ladino clover resistance to the clover root curculio (Coleoptera: Curculionidae). J. Econ. Entomol. 76:264–268.
266. Radcliffe, E.B., and D.K. Barnes. 1970. Alfalfa plant bug injury and evidence of plant resistance in alfalfa. J. Econ. Entomol. 63:1995–1996.
267. Randolph, N.M., and M.V. Meisch. 1970. Evaluation of chemicals and resistant alfalfa varieties for control of the three-cornered alfalfa hopper. J. Econ. Entomol. 63:979–981.
268. Ratcliffe, R.H. 1984. Alfalfa weevil feeding resistance. *In* J.H. Elgin, Jr. (ed.) Standard tests to characterize pest resistance in alfalfa cultivars. USDA-ARS Misc. Pub. 1434. U.S. Government Printing Office, Washington, DC.
269. Reynolds, J.H., C.R. Graves, C.R. Leslie, and S.E. Bennett. 1974. Alfalfa weevil resistance in four alfalfa varieties when sprayed and unsprayed. Tenn. Farm Home Sci. Prog. Rep. 90:24–26.
270. Ridland, P.M., and G.N. Berg. 1981. Seedling resistance to pea aphid of lucerne. Aust. J. Exp. Agric. Anim. Husb. 21:506–511.
271. ----, and ----. 1981. Seedling resistance to spotted alfalfa aphid of lucerne and annual medic species in Victoria. Aust. J. Exp. Agric. Anim. Husb. 21:59–62.
272. Roof, M.E. 1974. Mechanisms of potato leafhopper resistance in alfalfa. Ph.D. diss., Kansas State Univ.
273. ----, E. Horber, and E.L. Sorensen. 1976. Evaluating alfalfa cuttings for resistance to the potato leafhopper. Environ. Entomol. 5:295–301.
274. Rowley, W.A., and B.A. Haws. 1964. Studies on alfalfa resistance to the seed chalcid *Bruchophagus roddi* Gussakovsky (Abstract). Proc. Utah Acad. Sci. Arts Lett. 41:150.
275. Russell, L.M. 1957. Distribution of legume infesting *Therioaphidine* aphids. Plant Prot. Bull. (Rome) 5:78.
276. Salisbury, P.A., R.W. Downes, and W.J. Müller. 1985. Relationship between resistance to blue-green aphid and to pea aphid in lucerne. Aust. J. Exp. Agric. 25:133–137.
277. Sandhu, G.S., and A.S. Nijjar. 1980. Biology of spotted alfalfa aphid, *Therioaphis trifolii* (Monell) on resistant and susceptible varieties/clones of lucerne. Indian J. Entomol. 42:398–402.
278. Schalk, J.M., S.D. Kindler, and G.R. Manglitz. 1969. Temperature and the preference of the spotted alfalfa aphid for resistant and susceptible alfalfa plants. J. Econ. Entomol. 62:1000–1003.
279. ----, and G.R. Manglitz. 1969. Migration of an holocyclic strain of the spotted alfalfa aphid into Nebraska. J. Econ. Entomol. 62:946–947.
280. Schillinger, J.A., F.C. Elliott, and R.F. Ruppel. 1964. A method for screening alfalfa plants for potato leafhopper resistance. Mich. Agric. Exp. Stn. Bull. 46:512–517.
281. Searls, E.M. 1932. A preliminary report on the resistance of certain legumes to certain homopterous insects. J. Econ. Entomol. 25:46–49.
282. Shade, R.E., M.J. Doskocil, and N.P. Maxon. 1979. Potato leafhopper resistance in glandular-haired alfalfa species. Crop Sci. 19:287–289.
283. ----, and L.W. Kitch. 1983. Pea aphid *Acyrthosiphon pisum* Homoptera Aphididae biology on glandular-haired *Medicago* species. Environ. Entomol. 12:237–240.
284. ----, T.E. Thompson, and W.R. Campbell. 1975. An alfalfa weevil larval resistance mechanism detected in *Medicago*. J. Econ. Entomol. 68:399–404.
285. Sharma, R.K., V.M. Stern, and R.W. Hagemann. 1976. Blue alfalfa aphid: A new pest in the Imperial Valley. Calif. Agric. 30:14–15.
286. Shuster, M.M., and E.M. Shuster. 1980. Resistance of lucerne varieties to *Hypera postica*. Skh. Biol. 15:135–138.
287. ----, and ----. 1982. The effect of some phenolic compounds of lucerne on the physiological condition of the lucerne weevil (*Phytonomus variabilis* Hbst.). p. 95–100. *In* N.A. Vilkovoi (ed.) Proc. All-Union Res. Inst. for Plant Protection, Leningrad, USSR. 1979.
288. ----, and N.M. Tereshchenko. 1982. Resistance of lucerne varieties to *Hypera postica* and *Contarinia medicaginis*. Nauchno-Tekh. Byull. Vses. Selektsionno-Genet. Inst. 3:56–59.
289. Small, E., and B.S. Brooks. 1982. Coiling of alfalfa pods in relation to resistance against seed chalcids. Can. J. Plant Sci. 62:131–135.
290. Smith, Dale. 1962. Alfalfa cutting practices. Part 1. Influence of cutting schedule, soil fertility, and insect control on yield and persistence of Vernal and Narragansett alfalfa. Wis. Agric. Exp. Stn. Res. Rep. 11.
291. ----, and J.T. Medler. 1959. Influence of leafhoppers on the yield and chemical composition of alfalfa hay. Agron. J. 51:118–119.
292. Smith, O.F. 1958. Lahontan alfalfa. Nev. Agric. Exp. Stn. Circ. 14.
293. ----, and R.N. Peaden. 1960. A method of testing alfalfa plants for resistance to the pea aphid. Agron. J. 52:609–610.

294. ——, ——, and R.K. Petersen. 1958. Moapa alfalfa. Nev. Agric. Exp. Stn. Circ. 15.
295. Soper, J.F., M.S. McIntosh, and T.C. Elden. 1984. Diallel analysis of potato leafhopper resistance among selected alfalfa clones. Crop Sci. 24:667 670.
296. Sorensen, E.L. 1984. Evaluating pea aphid resistance. p. 28–29. *In* J.H. Elgin, Jr. (ed.) Standard tests to characterize pest resistance in alfalfa cultivars. USDA-ARS Misc. Pub. 1434. U.S. Government Printing Office, Washington, DC.
297. ——, H.L. Hackerott, and T.L. Harvey. 1975. Registration of KS10 pest-resistant alfalfa germplasm. Crop Sci. 15:105.
298. ——, T.L. Harvey, and H.L. Hackerott. 1969. New alfalfa unappetizing to pea aphid. Crops Soils 22(1):22.
299. ——, and E.K. Horber. 1974. Selecting alfalfa seedlings to resist the potato leafhopper. Crop Sci. 14:85–86.
300. ——, ——, and D.L. Stuteville. 1983. Registration of KS80 alfalfa germplasm resistant to the blue alfalfa aphid, pea aphid, spotted alfalfa aphid, Anthracnose, and downy mildew. Crop Sci. 23:599.
301. ——, ——, and ——. 1983. Development of grandular-haired alfalfas with multiple pest resistance. p. 28. *In* Proc. of the 18th Central Alfalfa Improve. Conf., Manhattan, KS. 8–10 June.
302. ——, ——, and ——. 1985. Registration of KS108GH5 glandular-haired alfalfa germplasm with multiple pest resistance. Crop Sci. 25:1132.
303. ——, ——, and ——. 1986. Registration of KS94GH6 glandular-haired alfalfa. Crop Sci. 26:1088.
304. ——, R.H. Painter, H.L. Hackerott, and T.L. Harvey. 1969. Registration of Kanza alfalfa. Crop Sci. 9:847.
305. ——, ——, E.E. Ortman, H.L. Hackerott, and T.L. Harvey. 1961. Cody alfalfa, it resists spotted alfalfa aphids. Kans. Agric. Exp. Stn. Circ. 381.
306. ——, N. Sangduen, and G.H. Liang. 1981. Transfer of glandular hairs from diploid *M. prostrata* to tetraploid *M. sativa*. p. 53. *In* Rep. of the 27th Alfalfa Improve. Conf. Madison, WI. 8–10 July 1980. ARM-NC-19. USDA-SEA, Peoria IL.
307. ——, D.L. Stuteville, and E. Horber. 1978. Registration of Riley alfalfa. Crop Sci. 18:911.
308. ——, ——, and ——. 1981. Registration of K78-10 alfalfa germplasm. Crop Sci. 21:476.
309. ——, ——, and ——. 1983. Registration of KS167 alfalfa germplasm. Crop Sci. 23:1224.
310. ——, ——, and ——. 1985. Registration of KS187 alfalfa germplasm resistant to five diseases and two insects. Crop Sci. 25:889.
311. ——, ——, and ——. 1986. Registration of KS189 alfalfa germplasm with resistance to five diseases and three insects. Crop Sci. 26:204–205.
312. Sorenson, C.J. 1930. The alfalfa seed chalcis-fly in Utah 1926–1929, inclusive. Utah Agric. Exp. Stn. Bull. 218.
313. Spencer, K.A. 1973. Agromyzidae (Diptera) of economic importance. p. 85–87. *In* E. Schimitschek (ed.) Series Entomologica, Vol. 9. W. Junk Publishing, The Hague.
314. Stanford, E.H., and J.A. McMurtry. 1959. Indications of biotypes of the spotted alfalfa aphid. Agron. J. 51:430–431.
315. Stern, V.M., R. Sharma, and C. Summers. 1980. Alfalfa damage from *Acyrthosiphon kondoi* and economic threshold studies in southern California. J. Econ. Entomol. 73:145–148.
316. Strong, F.E. 1960. Sampling alfalfa seed for clover seed chalcid damage. J. Econ. Entomol. 53:611–615.
317. ——. 1962. The reaction of some alfalfas to seed chalcid infestations. J. Econ. Entomol. 55:1004–1005.
318. ——. 1962. Laboratory studies on the biology of the alfalfa seed chalcid *Bruchophagus roddi* Guss. (Hymenoptera: Eurytomidae). Hilgardia 32:229–249.
319. ——. 1968. The selective advantage accruing to lygus bugs that cause blasting of floral parts. J. Econ. Entomol. 61:315–316.
320. Summers, C.G., and W.F. Lehman. 1976. Evaluation of nondormant alfalfa cultivars for resistance to the Egyptian alfalfa weevil. J. Econ. Entomol. 69:29–34.
321. ——, and W.D. McClellan. 1975. Interaction between Egyptian alfalfa weevil feeding and foliar disease: Impact on yield and quality in alfalfa. J. Econ. Entomol. 68:487–490.
322. Sutherland, O.R.W. 1976. A chemical basis for plant resistance to grass grub and a black beetle larvae (Coleoptera: Scarabaeidae). Proc. N. Z. Grassl. Assoc. 37:126–131.
323. ——, N.D. Hood, and J.R. Hillier. 1975. Lucerne root saponins a feeding deterrent for the grass grub *Costelytra zealandica* (Coleoptera: Scarabaeidae). N. Z. J. Zool. 2:93–100.
324. Suzuki, M., and L.S. Thompson. 1981. Effects of alfalfa blotch leafminer on chemical components of alfalfa. Can. J. Plant Sci. 61:595–600.

325. Taylor, N.L. 1956. Pubescence inheritance and leafhopper resistance relationships in alfalfa. Agron. J. 48:78–81.
326. ----, M.K. Anderson, and C. Tutt. 1972. Forage yield and weevil resistance of alfalfa varieties. Ky. Univ. Ext. Misc. 402:5–6.
327. Thomas, J.G., and E.L. Sorensen. 1971. Effect of excision duration on spotted alfalfa aphid resistance in alfalfa cuttings. J. Econ. Entomol. 64:700–704.
328. ----, ----, and R.H. Painter. 1966. Attached vs. excised trifoliolates for evaluation of resistance in alfalfa to the spotted alfalfa aphid. J. Econ. Entomol. 59:444–448.
329. Thompson, L.S., and C.B. Willis. 1971. Forage legumes preferred by the clover root curculio and preferences of the curculio and root lesion nematodes for species of *Trifolium* and *Medicago*. J. Econ. Entomol. 64:1518–1520.
330. Thompson, T.E., R.E. Shade, and J.D. Axtell. 1978. Alfalfa weevil resistance mechanism characterized by larval convulsions. Crop Sci. 18:208–209.
331. Tingey, W.M., and M.W. Nielson. 1974. Alfalfa seed chalcid nonpreference resistance in alfalfa. J. Econ. Entomol. 67:219–221.
332. ----, and ----. 1975. Developmental biology of the alfalfa seed chalcid on resistant and susceptible alfalfa clones. J. Econ. Entomol. 68:167–168.
333. ----, and E.A. Pillemer. 1977. *Lygus* bugs: Crop resistance and physiological nature of feeding injury. Entomol. Soc. Am. Bull. 23:277–287.
334. ----, and S.R. Singh. 1980. Environmental factors influencing the magnitude and expression of resistance. p. 87–113. *In* F.G. Maxwell and P.R. Jennings (ed.) Breeding plants resistant to insects. John Wiley and Sons, New York.
335. Triebe, D.C., C.E. Meloan, and E.L. Sorensen. 1981. The chemical identification of the glandular hair exudate from *Medicago scutellata*. p. 52. *In* Rep. 27th Alfalfa Improve. Conf., USDA–ARS ARM–NC–19. USDA–SEA, Peoria, IL.
336. Turner, J.W., D.L. Lloyd and T.B. Hilder. 1981. Effects of aphids on seedling growth of lucerne lines. 3. Blue-green aphid and spotted alfalfa aphid: A glasshouse study. Aust. J. Exp. Agric. Anim. Husb. 21:227–230.
337. ----, and A.P. Robins. 1982. Aphid resistant lucernes. Queensl. Agric. J. 108:153.
338. Urbahns, T.D. 1920. The clover and alfalfa seed chalcis-fly. USDA Bull. 812. U.S. Government Printing Office, Washington, DC.
339. VanDenburgh, R.S., B.L. Norwood, C.C. Blickenstaff, and C.H. Hanson. 1966. Factors affecting resistance of alfalfa clones to adult feeding and oviposition of the alfalfa weevil in the laboratory. J. Econ. Entomol. 59:1193–1198.
340. VanderPlank, J.E. 1968. Disease resistance in plants. Academic Press, New York.
341. Watts, J.G., C.B. Coleman, and C.R. Glover. 1967. Colorimetric detection of chalcid-infested alfalfa seed. J. Econ. Entomol. 60:59–60.
342. Weaver, C.R. and D.R. King. 1954. Meadow spittlebug. Ohio Agric. Exp. Stn. Res. Bull. 741.
343. Webster, J.A., E.L. Sorensen, and R.H. Painter. 1968. Temperature, plant growth stage, and insect-population effects on seedling survival of resistant and susceptible alfalfa infested with potato leafhoppers. J. Econ. Entomol. 61:142–145.
344. ----, ----, and ----. 1968. Resistance of alfalfa varieties to the potato leafhopper: Seedling survival and field damage after infestation. Crop Sci. 8:15–17.
345. Wellings, P.W. 1983. Growth, development and survival of *Acyrthosiphon kondoi* (Homoptera: Aphididae) on five cultivars of lucerne. J. Aust. Entomol. Soc. 24:155–160.
346. Wildermuth, W.L. 1910. The clover root curculio. p. 29–38. *In* USDA Bur. Entomol. Bull. 85, Part 3. U.S. Government Printing Office, Washington, DC.
347. Wilson, C.G., D.E. Swincer, and K.J. Walden. 1981. The origins, distribution, and host range of the spotted alfalfa aphid, *Therioaphis trifolii* (Monell) F. *maculata*, with a description of its spread in South Australia. J. Entomol. Soc. South. Afr. 44:331–341.
348. Wilson, M.C., and R.L. Davis. 1953. Varietal tolerance of alfalfa to the meadow spittlebug. J. Econ. Entomol. 46:238–241.
349. ----, and ----. 1958. Development of an alfalfa having resistance to the meadow spittlebug. J. Econ. Entomol. 51:219–222.
350. ----, and ----. 1960. Culver alfalfa, a new Indiana variety developed with insect resistance. Proc. North Cent. Br. Entomol. Soc. Am. 15:30–31.
351. ----, and ----. 1966. Host plant resistance research on *Philaenus spumarius* (L.) in alfalfa. p. 1283–1286. *In* Proc. 9th Int. Grassl. Congr., Vol. 2. Sao Paulo, Brazil 8 June 1965.
352. ----, ----, and G.G. Williams. 1955. Multiple effects of leafhopper infestation on irrigated and nonirrigated alfalfa. J. Econ. Entomol. 48:323–326.
353. ----, B.A. Melton, and C.E. Watson. 1959. Zia alfalfa. N. M. Agric. Exp. Stn. Bull. 435.

354. Yano, K., T. Miyake, and S. Hamasaki. 1982. Discovery of a spotted alfalfa-like aphid, *Therioaphis trifolii* (Monell) S. Lat. in Japan. J. J. Appl. Entomol. Zool. 26:35–40.
355. Zavaleta, L.R., and W.G. Ruesink, 1980. Expected benefits from nonchemical methods of alfalfa weevil control. Am. J Agric. Econ. 62:801–805.

29 Alfalfa Tissue Culture

E. T. BINGHAM

University of Wisconsin
Madison, Wisconsin

T. J. McCOY

University of Arizona
Tucson, Arizona

K. A. WALKER

Plant Genetics, Inc.
Davis, California

Alfalfa (*Medicago sativa* L.) tissues were cultured as callus (12, 20) and cell suspensions (20) several years before plant regeneration from cultured cells was reported in 1972 (63). Alfalfa was one of the first major crop plants to be regenerated from tissue culture following the pioneering research a decade earlier with tobacco (*Nicotiana tabacum* L.) (67) and carrot (*Daucus carota* L.) (74). The ability to produce plants from cultured cells (Fig. 29–1 through 3) is necessary for most basic and applied goals involving tissue culture.

Alfalfa plants were initially regenerated from callus derived from anthers, ovaries, internodes, and seedling hypocotyls (63). Most plants had the somatic chromosome number of the donor plant even when they were regenerated from anther-derived callus. Histology of anthers during callus proliferation suggested that most callus was derived from somatic cells in the interlocular connective tissue of the anthers (63). The list of tissues from which plants have been regenerated now includes leaf, petiole, sepal, petal, cotyledon, root, and immature embryos (2, 24, 41, 53, 55, 86, 91). Thus, there is flexibility in the type of tissue that can be cultured and from which plants can be regenerated. These and other general features of alfalfa tissue culture have been reviewed (89, 105, 109).

In the years following the first regeneration, the efficiency of alfalfa regeneration was improved by breeding (7) and by development of optimal culture methods (82, 83, 84). The genetic ability to regenerate plants was scaled down to the diploid level by haploidy (6, 63), and regenerability was found to be a dominant genetic trait (59). Regeneration was reported from suspension cultured cells (2, 46, 49), leaf protoplasts (1, 2, 18, 30, 32, 55), cotyledon protoplasts (40), root protoplasts (55, 86), cell

---

Copyright 1988 © ASA-CSSA-SSSA, 677 South Segoe Road, Madison, WI 53711, USA.
Alfalfa and Alfalfa Improvement—Agronomy Monograph no. 29.

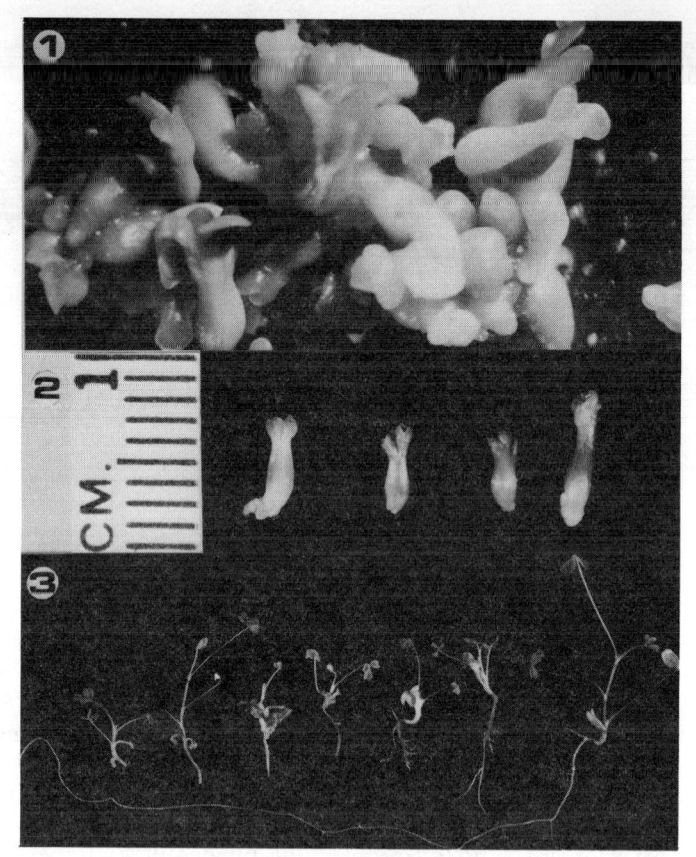

Fig. 29-1 through 29-3. Fig. 29-1. Somatic embryos of alfalfa at 21 d of age (×4.5). The embryos were induced for 3 d on Schenk and Hildebrandt (SH) medium supplemented with 50 µM 2,4-D and 5µM kinetin and subsequently transferred to a growth regulator-free regeneration medium composed of SH basal salts and supplemented with various sources of reduced N (75). Fig. 29-2. Isolated somatic embryos of alfalfa derived from conditions similar to those of Fig. 29-1. Notice the embryos on the left side are fused pairs of embryos. About 70% of these embryos will produce a plantlet if treated carefully. Fig. 29-3. Alfalfa plants that developed from somatic embryos at about 90 d of age. Procedures as in Fig. 29-1. Plants at this stage are transferred from regeneration medium to potting mixture and kept under relatively humid conditions for a few days before being taken to the greenhouse. Notice the plant on the right that appears to be developing a tap root in culture; such roots are ordinarily broken or cut when transferred to potting mixture.

suspension protoplasts (1, 49, 55), and after fusion of protoplasts (80). Moreover, several somatic cell selection experiments and much spontaneous and induced variation have been reported. These events will be reviewed and discussed in detail in the following sections. Advances in alfalfa tissue culture coupled with the importance of the crop make it a model system for tissue culture research.

29-1 ROLE OF THE GENOTYPE IN REGENERATION FROM CALLUS, SUSPENSION, AND PROTOPLASTS

29-1.1 Regeneration from Callus

An effect of genotype on regeneration was observed in the first report of regeneration when only 4 of 34 'Saranac' and 1 of 26 'Vernal' genotypes produced whole plants (63). Ability to regenerate was transferred to the offspring of one Saranac plant, which indicated that regeneration was heritable. Recurrent selection, an effective method of accumulating alleles controlling desired genetic traits, was used to breed 'Regen-S' alfalfa, which has a high frequency of regenerable genotypes (7).

In the development of Regen-S, one 'DuPuits' and four Saranac clones were selected from among 12% of the genotypes that regenerated from cultures derived from seedling hypocotyls in the first generation of selection. The five plants were intercrossed and their progeny (seedling hypocotyls) tested for regeneration from cultures initiated from seedling hypocotyls. The first generation of selection was effective in increasing the frequency of genotypes that regenerated in the second generation to 50%. In the second generation, 25 regenerated plants were selected and intercrossed to form the population for the third and last generation of selection. In this generation, 67% of the genotypes regenerated, and 75 regenerated plants were intercrossed to produce Regen-S.

In just three generations, recurrent selection for regeneration had increased the frequency of regenerable genotypes from 12 to 67%, indicating a highly heritable trait. Another regenerable alfalfa strain, 'Regen-Y' was synthesized by taking advantage of the high heritability of regeneration. Regen-Y was produced by intercrossing yellow-flowered plants regenerated from the Canadian cultivars 'Rhizoma', 'Rambler', and 'Drylander' (7). Several studies have determined that cultivars, lines, and genotypes differ in regeneration ability (7, 10, 51, 54, 108, 110).

The high heritability of regeneration also was exploited to breed a diploid alfalfa clone HG2 for use in suspension culture and somatic cell selection (46). This was accomplished in several steps. Germplasm from Regen-S was reduced first to the diploid level via maternal haploids and triploid bridge-crosses (6). Then, regenerable diploid plants were identified among the derived diploids. Finally, a cross of two regenerable diploids produced an array of F_1 progeny from which HG2 was selected for its exceptional ability to regenerate after suspension culture (46).

Genetic analysis of regeneration in diploid HG2 alfalfa confirmed that regeneration from callus cultures was relatively simply inherited (59). Segregation for regeneration in F_1, F_2, and backcross generations involving crosses between the high regenerator HG2 and plants with very low regeneration indicated that regeneration was dominant and fitted a two-locus model. A dominant allele at each locus appeared necessary to explain high regeneration in which >75% of callus colonies produced buds. Thus alleles at two loci influenced the formation and frequency of budding in this genetic background and specified media. A different regeneration response and genetic interpretation might have been obtained with different media and/or culture conditions. Also, it is not known at this time how many different modes of genetic control may exist for regeneration in alfalfa.

In the past few years, alfalfa genotypes that are capable of regeneration from callus tissue have been identified in numerous cultivars and breeding lines (Table 29–1). Several other cultivars and breeding lines in which regenerator genotypes have not yet been reported may be found in the references cited in Table 29–1. Some general conclusions that apply to regeneration from callus may be drawn from several independent studies, covering essentially the complete range of alfalfa germplasm. These are:

1. Regeneration is genotype specific and highly heritable.
2. Most cultivars, lines, and germplasm sources contain genotypes capable of regeneration.
3. Frequency of regenerating genotypes in most stocks is about 10%.
4. Cultivars with exceptional regeneration ability can be identified, e.g., Ladak and Rangelander.
5. Cultivars with exceptional regeneration ability can be developed by conventional breeding methods, e.g., Regen-S and Regen-Y.
6. Cultivars that form adventitious shoots from roots (creeping rooted alfalfa) may be excellent regenerators.
7. Some genotypes of *Medicago falcata* and *M. coerulea* are capable of regeneration.
8. Essentially any tissue explant from a regenerator genotype that will form callus will regenerate plants.
9. Regeneration efficiency (number of embryos per replicate plate) may be increased by media manipulations.
10. Regeneration is principally via somatic embryogenesis (discussed later).

Thus, there is a wealth of alfalfa germplasm with a wide range in diversity of adaptation, growth habit, and pest resistance that is capable of regeneration from solid callus. Clones may be selected for specialized experiments or for use as parents to breed stocks with a high frequency of regeneration, such as Regen-S. As discussed later, plants that regenerate from solid callus often are an excellent source of material for regeneration after culture as suspensions or from protoplasts.

Creeping rooted cultivars Rambler, Roamer, and Drylander (Table 29–1) are better regenerators as a group than noncreepers. The cultivar Rangelander from which essentially all genotypes regenerate (2) is the best regenerator currently known in alfalfa. Ladak, perhaps the next best with 82% of the genotypes regenerating in one study (51), also was noted to form adventitious shoots from cut root segments in a 1950 study by Smith (68). Root segments of alfalfa plants of Ladak, Grimm, Montana

Table 29-1. Alfalfa cultivars, germplasm sources, and genetic stocks reported to contain 10% or more genotypes capable of regeneration. Cultivars ranked in order of highest reported regeneration following callus culture.

Cultivar	Percentage regeneration and reference		Cultivar	Percentage regeneration and reference		
Rangelander	100%(2)	80%(10)	Regen-Y	35%(90)	25%(10)	
Ladak	82%(51)	9%(10)	Hardistan	23%(10)		
Regen-S	67%(2,7)	48%(10)	Saranac	20%(10)	14%(7)	
Norseman	46%(51)	9%(10)	Anchor	18%(10)		
Kane	45%(10)		Sonora	17%(7)	5%(51)	
SCMF3713	43%(10)		DuPuits	15%(10)	10%(7)	6%(51)
Vernal	41%(7)	18%(7)	Drylander	14%(10)	12%(7)	
Nomad	39%(51)		Saranac	14%(7)		
Rambler	38%(10)	36%(7)	Trek	11%(10)		
Turkestan	37%(51)		Beaver	10%(10)		
Roamer	35%(7)	24%(10)	Algonquin	10%(10)		
Spreader	34%(10)		Cossack	10%(51)		
Average of 2500 proprietary clones (Plant Genetics, Inc., Davis, CA)						12%(92)
Average of 320 southwestern USA breeding lines						11%(92)
Average of proprietary cultivars (Pioneer Hybrid, Int., Johnston, IA)						10%(92)
Special clones and genetic stocks that regenerate Diploid clone HG2 (46) Diploid clones derived from tetraploid alfalfa (88) Tetraploid clones heterozygous for genetic markers (21)						

Common, Ranger, and *M. falcata* cut at 2.54, 7.62, 12.70 cm below the cotyledonary node were potted and kept moist for several weeks. By the third week, adventitious shoots began to emerge only from Ladak and *M. falcata*. Age of plant, stage of growth when roots were cut, and size and length of segment all affected adventitious shooting, which occurred at a maximum of 36% in Ladak (68). This frequency is less than the 82% regeneration from callus of Ladak reported by Mitten et al. (51). It would be interesting to repeat these experiments and challenge specific Ladak genotypes to produce both types of adventitious development. Since adventitious shoots from root segments presumably develop without the intervening callus associated with tissue culture, it also would be interesting to measure somaclonal variation from both types of adventitious development.

The apparent association between regeneration in vitro and adventitious shoot formation in roots of some cultivars merits further study. If genotypes that form adventitious shoots in planta also regenerate in vitro, then the in planta response could provide an efficient method of screening for regeneration.

29-1.2 Suspension Culture

The ability to regenerate plants after alfalfa cells have been grown in liquid suspension cultures is a prerequisite for some somatic-cell selection

Table 29-2. Regeneration of alfalfa clones and cultivars after suspension culture of cells from callus of various tissue explants.

Clone or cultivar	Tissue explant	Reference
Diploid clone HG2	Ovaries	46
Diploid clone HG2	Ovaries, petioles	61
Hybrids and derivatives of clone HG2 × clone N11	Ovaries, hypocotyls	59
Regen-S	Ovaries, petioles, hypocotyls, cotyledons, leaves	2
Regen-S	Hypocotyls, cotyledons	97
Rambler and Rangelander	Hypocotyl, cotyledons, leaves	2
Canadian no. 1 and no. 2	Leaves	33
Czeckoslovakian A-15	Hypocotyls, cotyledons, stems petioles, leaves	53

experiments. In general, suspension cultures are susceptible to significantly lower concentrations of the selective agent. Regeneration after suspension culture has been reported in a number of studies (Table 29-2). Although genotypes that regenerate from solid callus provide good sources of material for suspension culture, they must form soft callus that disperses well when shaken in liquid medium, and retain their ability to regenerate after suspension culture.

About nine of 10 Regen-S plants classified as superior regenerators from solid callus also grew well in suspension culture and retained their ability to regenerate (97). In the same study, a larger group of plants represented a range in efficiency of regeneration from solid callus. From this group, 14 of 18 or 78% were suitable for suspension culture and subsequent regeneration. In a study of the genetics of regeneration ability using the diploid clone HG2, the proportion of diploid hybrids and derivatives that formed suspensions and could be regenerated was only about half that of the group with good regeneration when grown only as solid callus (59). It was suggested that additional genetic factors may be necessary for regeneration from suspension cultured cells. Regeneration from suspension cultured cells has been consistently higher for cells grown for only a short time in culture, although the decline in regeneration over time in culture has been greater in some studies (46) than in others (2). Also, the frequency of chromosome doubling (6, 63), partial doubling (23, 61), and doubling and redoubling (2) has increased over time in suspension culture.

29-1.3 Protoplasts

Plants have been regenerated from alfalfa protoplasts derived from leaf mesophyll (1, 2, 30, 32), cotyledons (2, 40), roots (55, 86), and cell suspensions (2, 49, 55) (Table 29-3). Furthermore, donor plants represent a wide range of cultivars and genotypes. In most studies the protoplasts gave rise to a callus colony that was subsequently regenerated (2, 18, 30, 32, 55), but in two examples embryogenesis occurred directly with little

Table 29-3. Regeneration from alfalfa protoplasts produced from various cells of several genotypes presented in chronological order.

Genotype	Source of protoplasts	References
'Canadian' no. 1 and no. 2	Leaf mesophyll	32
'Europe'	Leaf mesophyll	18
Regen-S clones	Leaf mesophyll	30
'Debinovsky' clone R-54	Cell suspension	49
Medicago coerulea	Leaf mesophyll	1
Europe	Cotyledon	39, 40
Europe	Root	39, 86
M. falcata '318' and Somatic hybrid of 318 and M. sativa 'MSR 12'	Leaf mesophyll	80
Regen-S	Leaf meosphyll	2
Answer	Cotyledon-derived cell suspension	2
'Adriana'	Leaf mesophyll root cell suspension	55

or no intervening callus (32, 55). To the extent that somaclonal variation arises during callus growth it may be desirable in certain experiments to employ culture methods that promote direct embryogenesis. The nature and extent of somaclonal variation arising from protoplasts is discussed later.

29-2 REGENERATION PHYSIOLOGY

29-2.1 Nature of the System

Since the earliest reports of alfalfa regeneration in vitro, shoot and root formation have been interpreted as via organogenesis (32, 82) as well as via somatic embryogenesis (18, 30, 32, 63, 75, 76, 84). It is customary in tissue culture research to assign the patterns of regeneration observed in culture to one of these two categories based largely upon morphological or operational criteria. Some investigators have been highly critical of these criteria, correctly pointing out that in vitro morphology is often highly abnormal and detailed cytological analysis rarely conducted. The problem of delineating mode of regeneration is not unique to alfalfa, but typical of most regeneration systems. Nevertheless, with improvements in regeneration conditions, the morphologies of certain structures within the cultures have been identified more clearly as somatic embryos (Fig. 29-1 through 3) (77). Recent biochemical evidence supports this interpretation. Stuart and his colleagues (77) reported that somatic embryos of alfalfa contain 11S seed storage protein by using antibodies prepared against purified 11S protein isolated form mature seed. In *Brasica napus* L. 11S protein was found in somatic embryos but not in adventitious shoots (13, 14). A recent study (77) showed that the principle pattern of regeneration in vitro for alfalfa is via somatic embryogenesis. This is in contrast to reports of root formation in cultured cells and tissues (82, 83), indicating that root formation and embryo formation represent alternative regeneration pathways. However, at this time there is no definitive proof to support either the view that the target cells for roots and embryos are one and the same or different subpopulations of cells in the culture. It is of interest, however, that among the early reports of somatic embryogenesis in carrots, root formation and somatic embryo formation were seen as alternative events (23).

29-2.2 Measuring Embryogenesis in Vitro

As emphasized earlier, regneration is affected by genotype. the impact of genotype must be considered carefully in the interpretation of results from experiments aimed at understanding regeneration physiology. Mitten et al. (51) brought into focus the problem of quantification of somatic embryogenesis, in reporting on the regenerative potential of alfalfa germplasm sources. In Ladak, for example, where 82% of the genotypes were regenerable, the average number of somatic embryos formed per replicate plate for the regenerable genotypes varied from less than one to over 30. The comparable response for RA3, the highly regenerable genotype isolated from Regen-S, was a mean of 86 embryos per replicate. Other regenerable genotypes from Regen-S varied in their regeneration efficiencies just as did genotypes from Ladak. The practical significance to psysiological studies of this response is realized in two ways. First, the more responsive a genotype, the more likely that subtle treatment effects will be resolved by appropriate experimental design and analysis. As a consequence, there is a critical need for at least one highly responsive clone in any study designed to unravel the physiological complexities of somatic embryogenesis. Investigators must endeavor, as well, to describe regeneration in quantitative terms. Unfortunately, these two requirements seldom have been met, either in alfalfa studies, or in the general area of tissue culture. Second, even though regeneration is heritable, apparently dominant and fitting a two gene model (59), a relationship has not been established between what little genetic information is available and corresponding physiological data. The role of medium composition, especially mineral salts and hormones, in inducing embryogenesis in cell suspensions derived from different plants, has been clearly shown (33). Differences were sometimes absolute. Furthermore, it remains to be established that only one "gene system" is responsible for in vitro regeneration across the entire range of alfalfa germplasm.

29-2.3 Factors Controlling Somatic Embryogenesis

Three cultural factors appear to control alfalfa regeneration via somatic embryogenesis in vitro. These are: (i) the structural quality and quantity of the exogenous auxin source (64, 82); (ii) the concentration of exogenous cytokinin (64, 82); and (iii) the level and nature of reduced N in the regeneration medium (82, 83, 117).

29-2.3.1 Growth Regulator Effects

In one of the early reports on regeneration in vitro, Saunders and Bingham (64) made an observation that was to provide a key point of leverage for subsequent studies of regeneration physiology by Walker and his co-workers. Both groups of investigators observed that in vitro regeneration seemed to be highly dependent upon the exposure of cells to 2,4-dichlorophenoxyacetic acid (2,4-D) prior to regeneration on growth regulator-free medium, whereas naphthaleneacetic acid (NAA) supported cell proliferation but did not share the "morphogenetic" effect of 2,4-D.

Building upon this observation, Walker et al. (83) reported the development of a regeneration system for a highly responsive clone of Regen-S, RA3, which allowed the formation of cultured cells that were competent for regeneration but functionally noninduced to regenerate. This system permitted the temporal separation of induction processes in these cells from regeneration per se, which occurred only upon transfer of cells from induction medium to a regeneration medium lacking growth regulators. The development of this system depended greatly on a critical response to NAA in these cultured cells. Namely, when used at concentrations between 10 and 25 μM, and in combination with kinetin, NAA stimulated cell proliferation only and did not "induce" the cells to regenerate. At higher concentrations, however, NAA induced either embryos or roots.

When grown in this fashion, the tissue would regenerate in response to more effective auxin-type growth regulators after a short period of exposure, c. 3 to 4 d. A wide range of growth regulators could be evaluated in this system for their ability to induce somatic embryogenesis using a variety of concentrations and combinations of test growth regulators (119). Through such hormone-matrix analyses, it was discovered that different synthetic auxins (in combination with different concentrations of cytokinins) not only differ in the induction of somatic embryogenesis, but also differ in inducing the formation of roots.

Five general guidelines for the control of regeneration by growth regulators emerged from this work. First, synthetic auxins and cytokinins appeared to interact to control the quantity and pattern of morphogenesis in vitro. Second, few structural quality differences existed among cytokinins in their effects on in vitro morphogenesis. Third, the structure of the auxin greatly affected the pattern and quantity of morphogenesis. Fourth, although response optima could be found for concentrations of growth regulators, these optima were generally very broad. Hence, a large number of specific concentration combinations of growth regulators would lead to either root or embryo formation, confirming the historical observation in carrot that root formation and embryo formation may be alternative events in vitro. Fifth, although not underscored in original reports, roots and embryos and/or highly abnormal structures were often found together in cultures treated with "borderline" concentrations of growth regulators (119).

29-2.3.2 Reduced Nitrogen Effects

Very early in their studies, Walker et al. (82, 83) observed that the composition of regeneration medium had profound effects on alfalfa morphogenesis in vitro. Walker and Sato (84) subsequently discovered that the NH_4^+ ion was critical to somatic embryogenesis. Somatic embryo formation required at least 12.5 mM NH_4^+ in regeneration medium for optimal expression. Root formation was inhibited at concentrations of NH_4^+ of 50 mM and higher. However, the most interesting observation was that at high NH_4^+ levels, somatic embryos were formed from cells exposed to a root-inducing combination of hormones. Walker and Sato (84) suggested that the growth regulators and exogenously applied NH_4^+ ion comprise an interactive system controlling the pattern of alfalfa morphogenesis.

The manner whereby NH_4^+ affects morphogenesis remains unclear. However, Stuart and Strickland (75, 76) have extended the observations on reduced-nitrogen enhanced somatic embryogenesis to several amino acids. Supplementing regeneration medium with very high levels of proline analogs and other amino acids were found effective, most notably, alanine (Table 29-4). These further stimulations of embryogenesis required some level of NH_4^+ in all tests except those with glutamine (75, 76). Little additional information is available regarding the biochemical events associated with these observations. We speculate that the amino acids simply provide a ready source of NH_4^+ to cells, which could explain their high concentration optima.

29-2.3.3 Conversion of Embryos to Plants

In general, little research has been directed toward improving the efficiency of plantlet formation from somatic embryos. As a rule, because somatic embryos are obtained in satisfactory numbers, and because only one regenerated plant may suffice, high conversion efficiencies are unnecessary. Recently, the issue of conversion efficiency has been addressed

Table 29-4. Active amino acids for somatic embryogenesis in alfalfa (75).

Protein amino acids	Nonprotein amino acids
L-proline	L-ornithine
L-alanine	Amino acid mixtures
L-glutamine	Yeast extract
L-asparagine	Protein hydrolysates
L-serine	L-proline amide
L-lysine	L-proline methyl ester
	L-protyl-L-alanine

(77, 58). Stuart et al., (77) reported that certain amino acids not only increase embryo quantity but also embryo quality. When grown on amino acids such as proline, alanine, arginine and glutamine, embryo lengths and widths nearly doubled in comparison to comparably treated NH_4^+ controls. These "larger" embryos converted to plants at nearly twice the frequency of the smaller NH_4^+ produced embryos. In the course of this work, Stuart et al., (77) reported that lower 2,4-D (10 μM) or 50 μM parachlorophenoxyacetic acid in induction medium improved embryo morphology and conversion, albeit at lower overall embryo yield. In addition, the accumulation of 11S storage protein improved in low 2.4-D-treated embryos to a level of 2.1 μg of protein per embryo on average.

Even though regeneration physiology is still poorly understood, it has entered a new phase of investigation. Much is now known about the nutritional biochemistry of the process. The principle pattern of regeneration has been established as one of somatic embryogenesis. Very recent efforts have turned to the study of embryo conversion to plants, which has focused attention on phases of somatic embryogeny, or development, as opposed to embryogenesis, or initiation. These efforts have led to the use of alfalfa somatic embryos in capsules that may eventually lead to "synthetic seed" (58). It may be possible in the future to use somatic embryos to study aspects of seed metabolism and development. Alfalfa appears to be an ideal species for use in this research. Finally, a greater appreciation of the factors controlling regeneration may require a more complete understanding of the genetic factors involved in the process.

29-3 VARIABILITY AMONG REGENERATED PLANTS

29-3.1 Historical Perspective

Regenerated plants that appear to be variants of the donor plant have been noted in alfalfa beginning with the first report of regeneration in 1972 (63). In this study, most of the 226 regenerated plants were indistinguishable from the Saranac donor clone, but nine plants appeared to be chromosomally doubled (near $8x$), five plants were chlorophyll deficient, one was dwarf, and one was albino (63). Similarly, in 1974 when reporting on the use of spontaneous doubling in culture of tetraploid alfalfa, the authors found a male-sterile and some morphological off-types (6). In the 1975 description of breeding Regen-S for regeneration (7), it was noted that selection for vigor and fertility likely minimized the number of aneuploid variants in each breeding cycle and that no effort was made to salvage variant plants.

In more recent studies, spontaneous variant plants (also designated somaclonal variants) have been observed among plants regenerated from cells grown as callus (6, 7, 63), from suspension culture (47, 59, 61), and from protoplasts (31, 37, 114). It will be shown that variants due to chromosome doubling and partial doubling are the most common types of variants regenerated from either cells (6, 7, 21, 23, 46, 61, 63) or protoplasts (31, 37). Other causes of somaclonal variation that have been observed to date include aneuploidy (31, 60), structural changes in chromosomes (31), qualitative genetic changes (21, 22), quantitative or complex genetic changes (56, 60), and the unstable expression of a single allele (22, 98). Essentially all main classes of somaclonal variation that have been observed or suggested in other plant species (36, 38, 47) have been observed in alfalfa (89), with the possible exception of epigenetic changes (48). Epigenetic changes that have been found in other species likely occur in alfalfa, but none have been documented. In addition to the many variants that have occurred spontaneously in several experiments, variants have been induced by chemical mutagenesis in the diploid HG2 (60, 61, 88). Once it was established that somaclonal variants were produced routinely by a cycle of tissue culture and regeneration, several researchers investigated the nature and frequency of the variants as well as their value as a source of variation for basic research and breeding.

29-3.2 Mechanisms of Variation

29-3.2.1 Chromosome Doubling

In a study to evaluate the use of spontaneous chromosome doubling in alfalfa tissue culture as an alternative to colchicine doubling, it was found that octoploids and near octoploids could be produced efficiently from tetraploid donor tissue (6). However, it was observed that some of the near octoploids were variant for more than chromosome number. Among 50 octoploids produced from the tissue culture of a Saranac clone, at least four regenerated plants differed from plants typical of the group. Two had altered flower color, one had modified leaves, and one had altered flower color and increased susceptibility to a leaf disease. Hence, several types of variation could occur simultaneously and complicate the use of tissue culture for chromosome doubling. Subsequent studies using colchicine to double the chromosome number of tetraploid alfalfa indicated that putative doubled plants from a single clone usually were at least as variable as those produced by spontaneous doubling in tissue culture (88). Most of the variation produced by colchicine or spontaneous doubling probably results from aneuploidy in near doubled plants. The only known method of producing euploid octoploids routinely is by restitution gametes (see Chapter 24 in this book).

Chromosomally doubled plants are especially common after prolonged culture periods. A clear relationship between time in suspension culture and increased polyploidization was established for the diploid clone HG2 (46). In a study where a salt tolerant cell line of tetraploid alfalfa was in culture for 128 weeks the plants that regenerated (72) appeared chromosomally doubled (our interpretation of pictures). Similarly, when tetraploid alfalfa callus was subcultured for >1 yr on media containing increasing concentrations of a fungal culture filtrate, only hexaploid and octoploid plants were regenerated (24). By exposing the same tetraploid donor callus to only one concentration of filtrate, which shortened the time in culture, the same workers were able to recover tetraploid plants.

In studies in which alfalfa plants were regenerated from protoplasts and carefully examined, many plants were doubled or partially doubled. Plants regenerated from mesophyll protoplasts of Europe alfalfa included plants with the normal tetraploid chromosome number of the donor and variant plants that were doubled or partially doubled (37). Of five variants studied, one was hexaploid, one was near heptaploid, and three were near octoploid. Johnson et al. (31) studied a large number of plants regenerated from mesophyll protoplasts of two tetraploid alfalfa clones of Regen-S. Octoploids were common and constituted about one-half of the regenerated plants of one clone and about one-third of the other; however, several other types of variants were produced. These included aneuploids with $2n = 31$ and 33 chromosomes, plants with translocations in each category of chromosome numbers, and a few variants that were not associated with changes in chromosome number changes. One of the tetraploid variants yielded 60% more herbage than the donor at the fourth cutting and showed no outward symptoms of crown rot, whereas the donor was injured by crown rot.

The occurrence of hexaploids among plants regenerated from tetraploid donor cells (24, 37) is difficult to explain. One possibility is that the spindle apparatus fragments during a restitution division, and produces in a partially doubled cell. Another possibility is that hexaploids arise from octoploid cells by gradual loss of chromosomes and stabilize near the hexaploid level because of selective advantage for hexaploidy. Also, hexaploid and several near octoploids were also produced after colchicine treatment of a tetraploid donor clone, and their origin may be similar to those arising in culture (88). It is not known whether hexaploids produced from somatic tetraploid cells are euploid or aneuploid.

Hexaploids also occurred in the culture of diploid alfalfa (60). Two such hexaploids were comparatively fertile and may have been euploid. It is much easier to explain euploid hexaploids arising somatically from diploids than from tetraploids. If a diploid nucleus were to mitotically divide twice, followed by one cytokinesis, which set three nuclei apart from the forth, the result would be an euploid hexaploid cell and a diploid cell (60).

29-3.2.2 Aneuploidy

Aneuploid plants with $2n = 31$ and $2n = 33$ have been regenerated from diploid donor cells (60) as well as from tetraploid cells (21) and protoplasts (31). It is interesting that no aneuploids near the diploid chromosome number were found using diploid donor tissue in a study that identified 66 plants with $2n = 16$, 20 plants with $2n = 32$, 2 plants with $2n = 31$, 2 plants with $2n = 33$, and 1 plant with $2n = 48$ (60).

In another study where a comparatively large number of plants were regenerated from protoplasts of two donor clones, both donors produced about the same frequency of aneuploids with $2n = 31$ (12–15%), but not with $2n = 33$ (31). One donor produced 6% aneuploids with $2n = 33$, whereas the other donor produced none (31).

For a study designed to monitor the loss of chromosomes from tetraploid donors carrying genetic marker genes (21), a stock was developed that was heterozygous for four traits: (i) anthocyanin pigmentation conditioned by the basic color factor gene C_2, (ii) multifoliolate leaves, (iii) ability to regenerate from tissue culture, and (iv) cytoplasmic male-sterility as conditioned by nuclear fertility-restorer genes. A group of 116 regenerated plants were examined for shifts in expression of these traits as well as for changes in vigor, fertility, morphology, or ploidy. At least 11% of regenerates lost one or more chromosomes in one callus culture cycle with observed chromosome numbers of $2n = 28$ to 31. Aneuploids with $2n = 31$ were most common. No aneuploids with $2n = 33$ were identified, although two near octoploids were found. A few aneuploids were found in a group of typical plants not expressing variant characteristics, and about 40% of plants expressing some variant characteristics did not have a change in chromosome number. Genetic changes rather than chromosomal shifts were suspected in the latter case. A white flowered variant arose without loss of the chromosome carrying the dominant C_2 allele. This mutant subsequently was shown to represent to a mutation of the C_2 allele to an unstable recessive allele. When the white-flowered variant or some of its sexual progeny were cultured and regenerated, both white-flowered donor and purple-flowered revertant plants were obtained. Both the original mutation and its reversion are heritable (22). These results are consistent with involvement of a transposable element that frequently transposes in culture.

29–3.2.3 Genetic Variation

Many variants were found among plants regenerated from 91 callus lines of diploid HG2 alfalfa (Fig. 29–4A). Suspension cultured cells were mutagenized with ethyl methane sulfonate (EMS) and selected for resistance to the toxic effects of ethionine (an analog of methionine) (61). Two classes of variants were obtained: variants resistant to the toxic effects of ethionine (Fig. 29–4B–F, see next section), and variants in morphology, growth habit, herbage yield, and fertility (Fig. 29–4F–H). The morphological variants were unexpected and most were not resistant to ethionine. It was hypothesized that the susceptible cells that gave rise to the morphological variants survived the selection regime in chimeral callus composed of resistant and susceptible cells.

A group of 55 variant and donor HG2 plants were cloned by vegetative cuttings and evaluated for changes in morphology and herbage yield in replicated field trials (60). Variants with significantly higher clonal herbage yield, multifoliolate leaves, compact growth habit, and several

Fig. 29–4. Diploid alfalfa HG2, $2n = 16$, (A) and a sample of its variants (B-H) regenerated after cell selection for ethionine resistance (60, 61). Variants R32, $2n = 40$, (B), R39, $2n = 16$, (C), R59, $2n = 16$, (D), and R64, $2n = 16$, (E) were regenerated from callus lines that were resistant to ethionine. Variants R105, $2n = 16$, (F), R85, $2n = 16$, (G), and R121, $2n = 32$, (H) were regenerated from callus lines that were not resistant to ethionine. See text for discussion.

other leaf and floral abnormalities were identified. One variant that arose as a spontaneously doubled variant (NS1) in a control culture yielded threefold more herbage dry matter than diploid and tetraploid HG2 controls. The improved dry matter herbage yield of variant NS1 was confirmed in a subsequent experiment, which also detected significantly improved fertility in NS1 (57). Moreover, the improvement in herbage yield and fertility was transmitted to progeny in testcrosses (57). Herbage yield and fertility are considered to be correlated polygenic traits.

Because NS1 was a variant for polygenic traits and exhibited the euploid chromosome number, $2n = 32$, homologous chromosome substitution was suggested as a possible mechanism for somaclonal improvement in NS1. According to a chromosome substitution model, a chromosome carrying unfavorable genes could be replaced by an extra copy of a homologue carrying more favorable alleles. Although the improvement in NS1 was significant it was not as great as that achieved by selection during inbreeding of tetraploid HG2 (57). The improvement during inbreeding and selection of HG2 can be explained by the accumulation of favorable alleles released by recombination involving all the chromosomes in the genome over several sexual generations (56).

A genetic analysis of several of the most interesting morphological variants of HG2 is in progress. Donor HG2 and its tissue culture derivatives are self-sterile, hence the genetic analysis was based on crosses to cultivated alfalfa at the diploid level (CADL) (see Chapter 24 in this book) 2x026, or its tetraploid counterpart 4x026, which was produced by colchicine treatment. Thus, both diploid and tetraploid variants were crossed as females to the same tester parent at their respective ploidy level.

Variant R85 was particularly interesting because it had more uniform side branching (Fig. 29-4G) and produced significantly more herbage than HG2 in a clonal test (60). The symmetrical side branching of R85 was expressed in about half of the F_1 progeny of R85 × 026; the other half resembled HG2 and 026, whose shoots are similar except that 026 tends to be semiprostrate. When the F_1 plants with R85-type shoots were crossed and backcrossed to new diploid sources, a segregation ratio near 1:1 was observed. Thus, the variant condition in R85 behaved as a dominant genetic trait, which was induced presumably by mutagenesis in the donor line.

The morphology of the dominant R85-trait represents an improvement in HG2; however, in all generations examined, plants possessing the dominant mutant allele were delayed in flowering, and produced only a few flowering racemes on the main shoot and uppermost side branches. There was ample flowering to produce seed for genetic analysis, but not sufficient for commercial seed production. Moreover, about one-fourth of the plants in a typical F_2 population produced by intercrossing F_1 progeny either never flowered or produced only a few flowering racemes near the longest day of the year. The analysis of this trait is continuing to determine whether these nonflowering plants are homozygous for the dominant mutant allele.

Variant R121 (Fig. 29-4H) regenerated as a tetraploid semidwarf plant that flowers profusely and is fully fertile. In genetic analysis, R121 behaved as a dominant trait. The F_1 progeny of R121 × 4x026 as well as R121 crossed with several other cultivated testers typically segregated five semidwarfs to one normal. Evidently, a dominant mutation was induced at the diploid level prior to spontaneous doubling in tissue culture.

Four other variants with altered leaf shape, abnormal raceme and flower structure, and reduced fertility, respectively, were not expressed in the F_1 generation and all were difficult to classify in the F_2 generation. The F_2 segregation of the control HG2 × 026 was quite variable, with a range in leaf shapes, fertility, and an occasional (1/25–1/100) segregate with abnormal racemes or flowers. Thus it was difficult to conclude that certain variant F_2 generations were different from the control. Whereas it had been easy to distinguish a variant of donor HG2 against the uniform clonal background of HG2, it was difficult to distinguish them in the variable F_2 population.

A variant with abnormal flowers (C49), which arose in a control culture of HG2, is of particular interest because it was expressed directly in diploid HG2 but behaved as a complete recessive in crosses to several normal diploid (CADL) testers. Further genetic analyses of donor HG2 and mutant C49 are in progress to determine their allelic structure at the mutant locus.

29-4 SOMATIC CELL SELECTION

Tissue culture media containing compounds that are toxic or inhibitory to normal cells allow selective growth of variant cells possessing resistance to the compounds (42, 43). Such variant cells are known to arise spontaneously in culture and they can be induced by mutagenesis. Selection of somatic plant cells in culture is similar to microbial selection schemes, and alfalfa cells have been selected for resistance to several toxic substances. This has led to expectations for the use of somatic cell selection that are being realized in some experiments but not in others. Useful levels of plant resistance may be recovered by selection for resistant cells in culture when the desired plant resistance reaction is at the cell level, and the reaction is similar in vitro and in planta. Conversely, where the desired plant resistance involves growth habit, anatomy, or complex tissue-membrane interactions, it may be difficult or impossible to select for the desired plant resistance in cell culture.

29-4.1 Selection for Resistance to Ethionine

Plant cell lines with resistance to growth inhibition by amino acid analogs often produce increased amounts of the corresponding amino acid (11, 42, 43). Ethionine (ETH), an analog of methionine, was used to select resistant alfalfa cell lines that were overproducers of methionine (61). Approximately 10^7 suspension cultured cells of diploid alfalfa line HG2 were mutagenized with EMS, centrifuged, washed, resuspended momentarily, and then plated on solid medium containing ethionine. About 20 000 cells and small cell aggregates per plate were cultured. Typically, one to two resistant callus colonies were formed per plate containing ETH. Of 124 cell lines recovered in this fashion, 91 regenerated plants.

In a retest of ETH resistance, callus lines were initiated from 25 regenerated plants, including those whose initial cell lines were ETH resistant and/or enhanced in unbound amino acid concentration. Only seven of the 25 were resistant to ethionine inhibition. Four of the seven, R59, R64, R90, R115 were diploids ($2n = 16$); one R39-3 4X was tetraploid ($2n = 32$), and two, R131 and R144, were aneuploid ($2n = 31$ and $2n = 33$, respectively). As many as three different plant types were regenerated from some cell lines indicating that the lines contained a mixture of cells.

High concentrations of free methionine, cysteine, cystathionine, and glutathione relative to HG2 control callus, were found in some but not all ETH-resistant cell lines. Cell line R32 had a 10-fold increase in soluble methionine, which was the highest level of overproduction found among the variants analyzed. Cell line R32 had the highest methionine concentration, and was the lowest in regeneration ability. When a plant was eventually regenerated (Fig. 29–4B), it was very slow growing, unthrifty, and never flowered over a 2-yr period. The chromosome number of plant R32 was approximately $2n = 40$, but chromosome number alone would not account for its manifold abnormalities because alfalfa plants with $2n = 40$ usually resemble normal alfalfa ($2n = 32$). The most important factor was that the leaves of plant R32 and variant plants R59, R64, and R90 did not contain enhanced levels of unbound or protein-bound methionine. Although somatic cell selection for enhanced methionine was accomplished, the plants regenerated from selected cell lines did not express the desired trait in the herbage (88).

An attempt was made to determine the genetic control of ethionine resistance in variants R59 and R64. The analysis was complicated by the fact that both R59 and R64 (Fig. 29–4) were variant in vegetative and floral morphology (60) and nearly sterile. Nonetheless, a few viable seeds and F_1 progeny were produced by pollinating >1000 flowers per variant with pollen from a normal tester. The ETH resistance of R64 tended to be incompletely dominant and expressed in most F_1 progeny. The resistance of R59 was not expressed in the F_1 progeny and was either recessive or epigenetic. The F_2 analysis of both R59 and R64 was inconclusive, because of a large experimental error for callus growth rate in the presence or absence of ETH (88).

29–4.2 Selection for Disease Resistance

Tissue culture selection of disease resistant plants has been proposed frequently as one of the potential applications of cell culture to plant improvement. Disease resistant plants may be recovered from cell culture either by selection for resistance to toxins(s) produced by the pathogen (either purified toxin, toxic culture filtrates, or toxin analogs), or by regenerating large numbers of plants and screening for resistant somaclonal variants. Either procedure may rely on spontaneous or induced mutations.

Interest in the first method was generated when tobacco (*Nicotiana tabacum* L.) plants resistant to wildfire disease were regenerated from cultures (11) selected for resistance to methionine sulfoximine. Subsequently, maize (*Zea mays* L.) plants resistant to *Helminthosporium maydis* (Nishikado and Miyake) race T were regenerated from cultures selected for resistance to *H. maydis* toxin (9, 19).

Selection of cell lines resistant to culture filtrates of the pathogen (or to purified toxins) followed by plant regeneration from the selected cell lines has resulted in: *Brassica napus* L. plants resistant to *Phoma lingam* (Tode ex. Fr.) Desm (62); potato (*Solanum tuberosum* L.) plants resistant to *Fusarium oxysporum* Schl. f. sp. *tuberosi* (Weimer) Sny. & Hans., (4), and *Phytophthora infestans* (Mont.) de Bary (3, 5); tobacco plants resistant to *Pseudomonas syringae* (Wolf and Foster), and *Alternaria alternata* (81); and alfalfa plants resistant to *F. oxysporum* Schl. f. sp. *medicaginis* (Weimer) Syn. & Hans. (24, 25).

Screening of somaclonal variants has resulted in the identification of regenerated potato plants resistant to *Phytophthora infestans* (65, 66), and *Alternaria solani* (Ellis and Martin) Jones and Groat (45); sugarcane (*Saccharum officinarum* L.) plants resistant to Fiji virus, *Sclerospora sacchari* T. Miyake, and *Drechslera sacchari* (Butl) Subram. and Jain (26, 38); maize plants resistant to *H. maydis* race T (8, 9); and, alfalfa plants resistant to *Verticillium albo-atrum* Reinke and Berth (37).

Of interest is the discovery that hexaploid plants were observed in the population of regenerated alfalfa plants from selection experiments for *F. oxysporum* resistance (24) as well as in the *V. albo-atrum* resistant somaclonal variants (37). Latunde-Dada and Lucas (37) found the variant protoclones (all with elevated ploidy) had the greatest level of resistance to *V. albo-atrum*. They suggested that the increased tolerance in the variants was due to gene dosage effects. Mechanisms for the origin of hexaploids in cultures of tetraploids were discussed earlier.

Cellular selection resulting in disease resistant plants has been realized for alfalfa plants resistant to *Fusarium oxysporum* f. sp. *medicaginis* (24). Following initial screening of clones at the whole plant level, callus cultures were initiated from plants known to be susceptible. These cultures died on a medium containing *Fusarium* culture filtrate. Both long-term selection (60 weeks) and short-term selection (28 weeks) were conducted. The long-term selection used a step-up concentration wherein cultures resistant at low concentration were selected, followed by selection for resistance at high concentration. The short-term experiment utilized the higher concentration throughout. Both procedures resulted in cell lines resistant to the high concentration of filtrate. Most plants regenerated from resistant cell lines were resistant at the whole plant level (24). The few susceptible plants could have originated from susceptible cells maintained in the resistant cell line. Regenerated plants from the long-term selection experiment were at or near the hexaploid level or the octoploid level; whereas short-term selection produced predominantly tetraploid plants. Although the genetic control of the selected resistance has not been determined, a preliminary inheritance study has demonstrated that resistance is genetically transmitted (24).

The isolation of disease resistant plants by cellular selection for resistance to toxins(s) produced by the pathogen is dependent on how well cellular resistance to pathogen-produced toxins corresponds to at least one whole plant resistance mechanism. Whole plant resistance mechanisms include morphological, anatomical, and physiological traits that prevent colonization by the pathogen; as well as complex biochemical interactions between the host and the pathogen that prevent pathogen development following initial penetration. Tissue culture selection will be directed primarily to resistance mechanism(s) expressed at the cellular level.

A separation of resistance mechanisms may have been identified in selection for *Fusarium* resistance. First, resistant clones were identified in Moapa 69 (the resistant check in *Fusarium* evaluation nurseries) and in a Hungarian experimental line bred for *Fusarium* resistance. Then, callus cultures were established from resistant and susceptible clones. Some clones that were resistant as whole plants were resistant at the cellular level, whereas some resistant clones were susceptible at the cellular level (104). This difference in cellular response by resistant plants may correspond to different resistance mechanisms. It is important to note that all clones that exhibited cellular resistance were resistant at the whole plant level, but the reverse is not true. Further genetic studies utilizing whole plant and cellular screening may permit elucidation of various resistance mechanisms. An additional advantage of a tissue culture system is that cellular screening may eliminate escapes that can occur frequently at the whole plant level. This is obviously dependent on finding a perfect correlation between cellular resistance and whole plant resistance.

Cell cultures have been utilized for studies of host-pathogen interactions (17, 27, 28, 29, 50). Several studies using species as diverse as tobacco, white pine (*Pinus alba* L.), and soybean (*Glycine max* L.) have demonstrated that host-pathogen responses at the whole plant level are the same in tissue culture (16, 17, 28, 44). Miller et al. (50) used an alfalfa tissue culture system to study host resistance to the alfalfa pathogen *Phytophthora megasperma* f. sp. *medicaginis* and to the nonhost pathogen *P. megasperma* f. sp. *glycinea*. Callus cultures mirrored the whole plant response as judged by the colonization of callus by the fungus. Callus cultures from plants susceptible to *P. megasperma* f. sp. *medicaginis* were quickly colonized by the fungus, whereas callus cultures from resistant plants were minimally colonized. *Phytophthora megasperma* f. sp. *glycinea* had limited colonization on either alfalfa callus. Although Miller et al. (50) found that fungal growth was not inhibited totally in any interaction, they indicated that callus resistance was similar to that of the seedling resistance. The potential of tissue culture in defining factors that influence expression of disease resistance remains to be determined in alfalfa.

29–4.3 Selection for Salt Tolerance

The importance of salt tolerance in plants and early successes in selecting salt tolerant cell lines in *Nicotiana* (52) encouraged research on selection at the cell level (73). It was essential, however, to establish that salt tolerance in culture is expressed at the whole plant level (69). Plants from four alfalfa cultivars, 'Hunter River,' 'Cuf 101,' 'Hasawi,' and Regen-S (W75RS) were tested for tolerance to three concentrations of NaCl

(70). Of these only Regen-S showed some salt tolerance, with callus growing equally well in the control (no salt) and at 62.5 mM NaCl. Moreover, the tolerance of Regen-S callus was observed in the regenerated whole plant. Regen-S plants were the most tolerant of the cultivars tested at the 62.5 mM level of NaCl (70). The same authors (70) discussed the possibility that Regen-S, which had been selected to perform well in tissue cultures growing on high levels of nutrient salts, may have undergone indirect selection for salt tolerance.

Salt tolerant cell and callus tissue lines of Regen-S can be readily selected (15, 71, 72). However, regeneration has been severely depressed in NaCl tolerant cell lines (71, 72, 73). In one test, regeneration was obtained eventually from a salt tolerant line of Regen-S, which had been in culture for several years (72). This was accomplished by developing a special regeneration medium high in sucrose and with an altered ratio of auxin to cytokinin. Although hundreds of plants were regenerated from the salt tolerant callus after the optimum regeneration media were developed, the plants were stunted, slow growing, and the degree of salt tolerance expressed in the plants was not determined (73).

Many salt tolerant cell lines were selected in a study in which suspension cultured cells of clones of Regen-S, HG2, and CufR3 were plated on media containing several levels of salt (71). Resistant cell lines that grew at up to 250 mM NaCl were recovered from each clone of alfalfa, but plants could be regenerated from only one callus line of Regen-S that grew in 62.5 mM NaCl. Plants regenerated from this callus and new calli from the regenerated plants had the same levels of salt tolerance as plants and calli of unselected Regen-S, but there was no evidence of improvement in level of tolerance. Hence, although a correspondence between cellular and whole plant tolerance was found in an early study using Regen-S, a useful level of improved salt tolerance has not been recovered to date using cultured tissue or cell selection.

29-4.4 Potential in Vitro Selection Strategies

Somatic cell selection is a relatively new and developing area of research. Although several cell selection strategies appear appropriate for alfalfa, they have not been either evaluated and/or reported. For example, selection for herbicide resistance using alfalfa tissue and cell cultures has received considerable effort in industry (90) but the results have not been published. In birdsfoot trefoil (*Lotus corniculatus* L.) a positive correlation exists between 2,4-D tolerance of cells in culture and 2.4-D tolerance of whole plants (78). Moreover, a rapidly growing, 2,4-D tolerant callus line of birdsfoot trefoil was selected in vitro and whole plants regenerated (79). Such selection should be equally effective in alfalfa in isolating genetic resistance to herbicides and in studying modes of herbicide action.

Metal tolerance in crop plants is often desired or required under some soil conditions. A clone of the grass *Agrostis stolonifera* L. that was tolerant to both Zn and Cu as a whole plant was found to express the tolerance in tissue culture (85). The pattern of metal uptake in tissue culture resembled uptake by whole plants in that tolerant tissue took up more metal than nontolerant tissue. Where metal tolerance is judged important, somatic cell selection may be an appropriate selection strategy to isolate genetic resistance.

29-5 SOMATIC CELL FUSION

Somatic cell fusion and regeneration of hybrid plants has been achieved in alfalfa (80). Protoplasts of mesophyll cells of tetraploid *M. sativa* and tetraploid *M. falcata* were subjected to fusion treatment and 28 plants regenerated. Nine plants were analyzed and one was confirmed as a hybrid. The hybrid was obtained without employing a culture regime that selected for growth of hybrid cell lines. Hybridization of *M. sativa* and *M. falcata* often results in heterosis. In addition, it has been shown that heterosis at the cell level results in greater callus growth in alfalfa (34) and tobacco (35). Hence, there may be at least a slight selective advantage for fusion hybrids.

In the next few years somatic cell fusions are expected between more distantly related *Medicago* spp. that have not been hybridized sexually. Great progress in the wide hybridization of *Medicago* spp. has already been made using embryo rescue that involves some aspects of tissue culture (106, 107) (also see Chapter 24 in this book).

29-6 RECENT DEVELOPMENTS

Several studies relating to areas of alfalfa tissue culture reviewed in this chapter have been published recently. In addition, transformation of alfalfa using recombinant DNA has been reported. A brief summary of these studies follows.

Certain diploid and polyploid wild species relatives of alfalfa have been regenerated (89, 102, 108). Additional factors affecting embryogenesis (115) and induction of protoplast division (99) have been reported. A new in vitro assay for disease resistance has been developed (93), and the genetic mechanism for a herbicide resistance was found to be gene amplification (95). Induction of freezing tolerance in alfalfa cell suspension cultures (111), growth and regeneration of alfalfa callus after freezing in liquid N_2 (96), and novel methods of propagation (100, 101) suggest new ways of storing and propagating alfalfa germplasm. Somaclonal variation associated with changes in the genomes of the mitochondria (114) and the chloroplast (118) has been reported. Finally, there has been recent progress in the infection (103) and transformation (94, 116) of *Medicago* by *Agrobacteriaum tumefaciens* and by microinjection of alfalfa protoplasts (112, 113).

29-7 CONCLUSION

Alfalfa is one of the few major crop species where all phases of tissue culture, plant regeneration, and molecular genetic engineering can be performed. These areas of alfalfa science have advanced over the past 15 yr through coordinated research on the genetic basis and the physiology of regeneration. An understanding of the role of the genotype in regeneration has allowed selection of clones with superior regeneration ability for specialized experiments and has been used to breed clones and strains that regenerate. An understanding of regeneration physiology has aided in the development of defined culture media, which enhance the efficiency of regeneration and distinguish embryogenesis and organogenesis. Thus, selection of optimal genotypes and tissue culture procedures has ensured regeneration and genetic analysis of plants following in vitro experiments in alfalfa. In turn, this has helped define mechanisms of disease and herbicide resistance that are suitable for in vitro selection methods. Essentially all known mechanisms of somaclonal variation have been documented among regenerated alfalfa plants. Desirable variation has been separated from undesirable by genetic analysis and breeding. The concept of managing somaclonal variation through genetic analysis and breeding should aid in finding and exploiting variation and transformation events in future experiments.

REFERENCES

1. Arcioni, S., M.R. Davey, A.V.P. dos Santos, and E.C. Cocking. 1982. Somatic embryogenesis in tissues from mesophyll and cell suspension protoplasts of *Medicago coerulea* and *M. glutinosa*. Z. Pflanzenphysiol. 106:105–110.
2. Atanassov, A., and D.C.W. Brown. 1984. Plant regeneration from suspension culture and mesophyll protoplasts of *Medicago sativa* L. Plant Cell Tissue Organ Cult. 3:149–162.
3. Behnke, M. 1979. Selection of potato callus for resistance to culture filtrates of *Phytophthora infestans* and regeneration of resistant plants. Theor. Appl. Genet. 55:69–71.
4. ----. 1980. Selection of dihaploid potato callus for resistance to the culture filtrate of *Fusarium oxysporum*. Z. Pflanzenzuecht. 85:254–258.
5. ----. 1980. General resistance to late blight of *Solanum tuberosum* plants regenerated from callus resistant to culture filtrates of *Phytophthora infestans*. Theor. Appl. Genet. 56:151–152.
6. Bingham, E.T., and J.W. Saunders. 1974. Chromosome manipulations in alfalfa: Scaling the cultivated tetraploid to seven ploidy levels. Crop Sci. 14:474–477.
7. ----, L.V. Hurley, D.M. Kaatz, and J.W. Saunders. 1975. Breeding alfalfa which regenerates from callus tissue in culture. Crop Sci. 15:719–721.
8. Brettell, R.I.S., B.V.D. Goddard, and D.S. Ingram. 1979. Selection of Tms-cytoplasm maize tissue cultures resistant to *Drechslera maydis* T-toxin. Maydica 24:203–213.
9. ----, R. Thomas, and D.S. Ingram. 1980. Reversion of Texas male-sterile cytoplasm maize in culture to give fertile, T-toxin resistant plants. Theor. Appl. Genet. 58:55–58.
10. Brown, D.C.W., and A. Atanassov, 1985. Role of genetic background in somatic embryogenesis in *Medicago*. Plant Cell Tissue Organ Cult. 4:111–122.
11. Carlson, P.S. 1973. Methionine sulfoximine-resistant mutants of tobacco. Science (Washington, DC) 180:1366–1368.
12. Clement, W.M. 1964. Stability of chromosome numbers in tissue cultures of alfalfa, *Medicago sativa* L. Am. J. Bot. (Suppl.) 51:670.

13. Crouch, M.L. 1982. Non-zygotic embryos of *Brassica napus* L. contain embryo specific storage proteins. Planta 156:520-524.
14. ----, and I.M. Sussex. 1981. Development of storage protein synthesis in *Brassica napus* L. embryos *in vitro* and *in vivo*. Planta 153:64-74.
15. Croughan, T.P., S.J. Stavarek, and D.W. Rains. 1978. Selection of a NaCl tolerant line of cultured alfalfa cells. Crop Sci. 18:959-963.
16. Deaton, W.R., G.J. Keyes, and G.B. Collins. 1982. Expressed resistance to black shank among tobacco callus cultures. Theor. Appl. Genet. 63:65-70.
17. Diner, A.M., R.L. Mott, and H.V. Amerson. 1984. Cultured cells of white pine show genetic resistance to axenic blister rust hyphae. Science (Washington, DC) 224:407-408.
18. dos Santos, A.V.P., D.E. Outka, E.C. Cocking, and M.R. Davey. 1980. Organogenesis and somatic embryogenesis in tissues derived from leaf protoplasts and leaf explants of *Medicago sativa*. Z. Pflanzenphysiol. 99:261-270.
19. Gengenbach, B.G., C.E. Green, and C.M. Donovan. 1977. Inheritance of selected pathotoxin resistance in maize plants regenerated from cell culture. Proc. Natl. Acad. Sci. USA 74:5113-5117.
20. Graham, P.H. 1968. Growth of *Medicago sativa* L. and *Trifolium subterraneum* L. in callus and suspension culture. Phyton (Berlin) 25:159-162.
21. Groose, R.W., and E.T. Bingham. 1984. Variation in plants regenerated from tissue culture of tetraploid alfalfa heterozygous for several traits. Crop Sci. 24:655-658.
22. ----, and ----. 1986. An unstable anthocyanin mutation recovered from tissue culture of alfalfa (*Medicago sativa*) 1. High frequency reversion upon reculture. Plant Cell Rep. 5:104-107.
23. Halperin. W. 1966. Alternative morphogenetic events in cell suspensions. Am. J. Bot. 53:443-453.
24. Hartman, C.L., T.J. McCoy, and T.R. Knous. 1984. Selection of alfalfa (*Medicago sativa*) cell lines and regeneration of plants resistant to the toxin(s) produced by *Fusarium oxysporum* f. sp. *medicaginis*. Plant Sci. Lett. 34:183-184.
25. ----, T.R. Knous, and T.J. McCoy. 1984. Field testing and preliminary progeny evaluation of alfalfa regenerated from cell lines resistant to the toxins produced by *Fusarium oxysporum* f. sp. *medicaginis*. Phytopathology 74:818.
26. Heinz, D.J., M. Krishnamurthi, L.G. Nickell, and A. Maretzski. 1977. Cell, tissue and organ culture in sugarcane improvement. p. 1-7. *In* J. Reinert and Y.P.S. Bajaj (ed) Applied and fundamental aspects of plant cell, tissue and organ culture, Springer-Verlag, Berlin.
27. Helgeson, J.P., J.D. Kemp, G.T. Haberlach, and D.P. Maxwell. 1972. A tissue culture system for studying disease resistance: The black shank disease in tobacco callus cultures. Phytopathology 62:1439-1443.
28. ----, G.T. Haberlach, and C.D. Upper. 1976. A dominant gene conferring disease resistance to tobacco plants is expressed in tissue cultures. Phytopathology 65:91-96.
29. Huang, J., and C.G. vanDyke. 1978. Interaction of tobacco callus tissue with *pseudomonas tabaci*, *P. fluorescens*. Physiol. Plant Pathol. 13:65-72.
30. Johnson, L.B., D.L. Stuteville, R.K. Higgins, and D.Z. Skinner. 1981. Regeneration of alfalfa plants from protoplasts of selected Regen-S clones. Plant Sci. Lett. 20:297-304.
31. ----, ----, S.E. Schlarbaum, and D.Z. Skinner. 1984. Variation in phenotype and chromosome number in alfalfa protoclones regenerated from nonmutagenized calli. Crop Sci. 24:948-952.
32. Kao, K.N., and M.R. Michayluk. 1980. Plant regeneration from mesophyll protoplasts of alfalfa. Z. Pflanzenphysiol. 96:135-141.
33. ----, ----. 1981. Embryoid formation in alfalfa cell suspension cultures from different plants. In Vitro 17:645-648.
34. Keyes, G.J., and E.T. Bingham. 1979. Heterosis and ploidy effects on the growth of alfalfa callus. Crop Sci. 19:473-476.
35. ----, W.R. Deaton, G.B. Collins, and P.D. Legg. 1981. Hybrid vigor in callus tissue cultures and seedlings of *Nicotiana tabacum* L. J. Hered. 72:172-174.
36. Lapitan, N.L.V., R.G. Sears, and B.S. Gill. 1984. Translocations and other karyotypic structural changes in wheat rye hybrids regenerated from tissue culture. Theor. Appl. Genet. 68:547-554.
37. Latunde-Dada, A.O., and J.A. Lucas. 1983. Somaclonal variation and reaction to *Verticillium* wilt in *Medicago sativa* L. plants regenerated from protoplasts. Plant Sci. Lett. 32:205-211.
38. Larkin, P.J., and W.R. Scowcroft. 1981. Somaclonal variation—A novel source of variability from cell cultures for plant improvement. Theor. Appl. Genet. 60:197-214.
39. Lu, D.Y., M.R. Davey, D. Pental, and E.C. Cocking. 1982. Forage legume protoplasts: Somatic embryogenesis from protoplasts of seedling cotyledons and roots of *Medicago sativa*. p. 597-598. *In* A. Fujiwara (ed.) Plant tissue culture, Marizen Co., Ltd., Tokyo.

40. ----, D. Pental, and E.C. Cocking. 1982. Plant regeneration from seedling cotyledon protoplasts. Z. Pflanzenphysiol. 107:59-63.
41. Maheswaran, G., and E.G. Williams. 1984. Direct somatic embryoid formation in immature embryos of *Trifolium repens, T. pratense* and *Medicago sativa*, and rapid clonal propagation of *T. repens*. Ann. Bot. (London) 54:201-211.
42. Maliga, P. 1978. Resistant mutants and their use in genetic manipulation. p. 381-392. *In* T.A. Thorpe (ed.) Frontiers of plant tissue culture, Calgary University Press, Calgary.
43. ----. 1984. Isolation and characterization of mutants in plant cell culture. Ann. Rev. Plant Physiol. 35:519-542.
44. Maronek, D.M., and J.W. Hendrix. 1978. Resistance to race 0 of *Phytophthora parasitica* var. *nicotianae* in tissue cultures of a tobacco breeding line with black shank resistance derived from *Nicotiana longiflora*. Phytopathology 68:233-234.
45. Matern, U., G. Strobel, and J. Shepard. 1978. Reaction to phytotoxins in a potato population derived from mesophyll protoplasts. Proc. Natl. Acad. Sci. USA 75:4935-4939.
46. McCoy, T.J., and E.T. Bingham. 1977. Regeneration of diploid alfalfa plants from cells grown in suspension culture. Plant Sci. Lett. 10:59-66.
47. ----, R.L. Phillips, and H.W. Rines. 1982. Cytogenetic analysis of plants regenerated from oat (*Avena sativa*) tissue cultures: High frequency of partial chromosome loss. Can. J. Genet. Cytol. 24:37-50.
48. Meins, F., Jr. 1983. Heritable variation in plant cell culture. Annu. Rev. Plant Physiol. 34:327-346.
49. Mezentsev, A.V. 1981. Mass regeneration of plants from the cells and protoplasts of lucerne. (In Russian.) Dokl. Vses. Akad. Sh. Nauk im. V. I. Lenin 4:22-23.
50. Miller, S.A., L.C. Davidse, and D.P. Maxwell. 1984. Expression of genetic susceptibility, host resistance, and nonhost resistance in alfalfa callus tissue inoculated with *Phytophthora megasperma*. Phytopathology 74:345-348.
51. Mitten, D.H., S.J. Sato, and T.A. Skokut. 1984. In vitro regenerative potential of alfalfa germplasm sources. Crop Sci. 24:943-945.
52. Nabors, M.W., S.E. Gibbs, C.S. Bernstein, and M.E. Meis. 1980. NaCl-tolerant tobacco plants from cultured cells. Z. Pflanzenphysiol. 97:13-17.
53. Novak, F.J., and D. Konecna. 1982. Somatic embryogenesis in callus and cell suspension cultures of alfalfa (*Medicago sativa* L.). Z. Pflanzenphysiol. 105:279-284.
54. Oelck, M.M., and O. Schieder. 1983. Genotypic differences in some legume species affecting the redifferentiation ability from callus to plants. Z. Pflanzenzuecht. 91:312-321.
55. Pezzotti, M., S. Arcioni, and D. Mariotti. 1984. Plant regeneration from mesophyll root and cell suspension protoplasts of *Medicago sativa* cv. Adriana. Genet. Agric. 38:195-208.
56. Pfeiffer, T.W., and E.T. Bingham. 1983. Improvement of fertility and herbage yield by selection within two-allele population of tetraploid alfalfa. Crop Sci. 23:633-636.
57. ----, and ----. 1984. Comparisons of alfalfa somaclonal and sexual derivatives from the same genetic source. Theor. Appl. Genet. 67:263-266.
58. Redenbaugh, M.K., Paasch, B.D., Nichol, J.W. Kossler, M.E. Viss, P.R., and K.A. Walker. 1986. Somatic seeds. Encapsulation of asexual plant embryos. Biotechnology 4:797-801.
59. Reisch, B., and E.T. Bingham. 1980. The genetic control of bud formation from callus cultures of diploid alfalfa. Plant Sci. Lett. 20:71-77.
60. ----, and ----. 1981. Plants from ethionine-resistant alfalfa tissue cultures: Variation in growth and morphological characteristics. Crop Sci. 21:783-788.
61. ----, S.H. Duke, and E.T. Bingham. 1981. Selection and characterization of ethionine-resistant alfalfa (*Medicago sativa*) cell lines. Theor. Appl. Genet. 59:89-94.
62. Sacristan, M.D. 1982. Resistance responses to *Phoma lingam* of plants regenerated from selected cell and embryogenic cultures of haploid *Brassica napus*. Theor. Appl. Genet. 61:193-200.
63. Saunders, J.W., and E.T. Bingham. 1972. Production of alfalfa plants from callus tissue. Crop Sci. 12:804-808.
64. ----, and ----. 1975. Growth regulator effects on bud initiation in callus cultures of *Medicago sativa*. Am. J. Bot. 62:850-855.
65. Secor, G.A., and J.F. Shepherd. 1981. Variability of protoplast-derived potato clones. Crop Sci. 21:102-105.
66. Shepard, J.F., D. Bidney, and E. Shahin. 1980. Potato protoplasts in crop improvement. Science (Washington, DC) 208:17-24.
67. Skoog, F., and C.O. Miller. 1957. Chemical regulation of growth and organ formation in plant tissue cultured *in vitro*. Symp. Soc. Exp. Biol. 11:118-131.
68. Smith, D. 1950. The occurrence of adventitious shoots on severed alfalfa roots. Agron. J. 42:398-401.

69. Smith, M.K., and J.A. McComb. 1981. Effect of NaCl on the growth of whole plants and their corresponding callus cultures. Aust. J. Plant Physiol. 8;267–275
70. ——, and ——. 1981. Use of callus cultures to detect NaCl tolerance in cultivars of three species of pasture legumes. Aust. J. Plant Physiol. 8:437–442.
71. ——, and ——. 1983. Selection for NaCl tolerance in cell cultures of *Medicago sativa* and recovery of plants from a NaCl-tolerant cell line. Plant Cell Rep. 2:126–128.
72. Stavarek, S.J., T.P. Croughan, and D.W. Rains. 1980. Regeneration of plants from long-term cultures of alfalfa cells. Plant Sci. Lett. 19:253–261.
73. ——, and D.W. Rains. 1984. Cell culture techniques: Selection and physiological studies of salt tolerance. p 321–334. *In* R.C. Staples and G.H. Toenniessen (ed.) Salinity tolerance in plants: Strategies for crop improvement, John Wiley and Sons, New York.
74. Steward, F.C., M.O. Mapes, and K. Mears. 1958. Growth and organized development of cultured cells. II. Organization in cultures grown from freely suspended cells. Am. J. Bot. 45:705–708.
75. Stuart, D.A., and S.G. Strickland. 1984. Somatic embryogenesis from cell cultures of *Medicago sativa* L. I. The role of amino acid additions to the regeneration medium. Plant Sci. Lett. 34:165–174.
76. ——, and ——. 1984. Somatic embryogenesis from cell cultures of *Medicago sativa* L. II. The interaction of amino acids with ammonium. Plant Sci. Lett. 34:175–181.
77. ——, J. Nelson, C.M. McCall, S.G. Strickland, and K.A. Walker, 1985. Physiology of the development of somatic embryos in cell cultures of alfalfa and celery. p. 35–47. *In* M. Zaitlin et al. (ed.) Biotechnology in plant science. Academic Press, New York.
78. Swanson, E.B., and D.T. Tomes. 1980. Plant regeneration from cell cultures of *Lotus corniculatus* and the selection and characterization of 2,4-D tolerant cell lines. Can. J. Bot. 58:1205–1209.
79. ——, and ——. 1980. *In vitro* responses of tolerant and susceptible lines of *Lotus corniculatus* L. to 2,4-D. Crop Sci. 20:792–795.
80. Teoule, E. 1983. Somatic hybridization between *Medicago sativa* L. and *Medicago falcata* L. (In French) C. R. Acad. Sci. Paris Ser. III, 297:13–16.
81. Thanutong, P., I. Furusawa, and M. Yamamoto. 1983. Resistant tobacco plants from protoplast-derived calluses selected for their resistance to *Pseudomonas* and *Alternaria* toxins. Theor. Appl. Genet. 66:209–215.
82. Walker, K.A., P.C. Yu, S.J. Sato, and E.G. Jaworski. 1978. The hormonal control of organ formation in callus of *Medicago sativa* L. cultured *in vitro*. Am. J. Bot. 65:654–659.
83. ——, M.L. Wendeln, and E.G. Jaworski. 1979. Organogenesis in callus tissue of *Medicago sativa*. The temporal separation of induction processes from differentiation processes. Plant Sci. Lett. 16:23–30.
84. ——, and S.J. Sato. 1981. Morphogenesis in callus tissue of *Medicago sativa*: The role of ammonium ion in somatic embryogenesis. Plant Cell Tissue Organ Cult. 1:109–121.
85. Wu, L., and J. Antonovics. 1978. Zinc and copper tolerance of *Agrostis stolonifera* L. in tissue culture. Am. J. Bot. 65:268–271.
86. Xu, Z-H, M.R. Davey, and E.C. Cocking. 1982. Organogenesis from root protoplasts of the forage legumes *Medicago sativa* and *Trigonella foenum-graecum*. Z. Pflanzenphysiol. 107:231–235.
87. Atanassov, A., and M. Vlachova. 1985. Somatic embryogenesis in callus and cell suspension cultures of three species of *Medicago*. p. 301–302. *In* R. Henke et al., (ed.) Tissue culture in forestry and agriculture, Plenum Press, New York.
88. Bingham, E.T. Unpublished data, Agronomy Dep., Univ. of Wisconsin, Madison.
89. ——, and T.J. McCoy, 1986. Somaclonal variation in alfalfa. Plant Breed. Rev. 4:123–152.
90. ——, T.J. McCoy, and K.A. Walker. 1984. Personal communication, Univ. of Wisconsin, Madison, Univ. of Arizona, Tucson, and Plant Genetics, Inc., Davis, CA, respectively.
91. ——, ——, and ——. 1984. Unpublished data.
92. ——, et al. 1984. Committee on use of tissue culture research in alfalfa improvement. p. 112–114. *In* Rep. 29th Alfalfa Improve. Conf., Lethbridge, AB, Canada. 27–31 July.
93. Cucazza, J.D., and J. Kao. 1986. *In vitro* assay of excised cotyledons of alfalfa (*Medicago sativa*) to screen for resistance to *Colletotrichum trifolii*. Plant Dis. 70:111–115.
94. Deak, M., B.G. Kiss, C. Koncz, and D. Dudits. 1986. Transformation of *Medicago* by *Agrobacterium* mediated gene transfer. Plant Cell Rep. 5:97–100.
95. Donn, G., E. Tischer, J.A. Smith, and H.M. Goodman. 1984. Herbicide-resistant alfalfa cells: An example of gene amplification in plants. J. Mol. Appl. Genet. 2:621–635.
96. Finkel, B.J., J.M. Ulrich, D.W. Rains, and S.J. Stavarek. 1985. Growth and regeneration of alfalfa callus lines after freezing in liquid nitrogen. Plant Sci. 42:133–140.

97. Groose, R.W., and E.T. Bingham. Unpublished, Agronomy Dep. Univ. of Wisconsin, Madison.
98. Groose, R.W., and E.T. Bingham. 1986. An unstable anthocyanin mutation from tissue culture of alfalfa (*Medicago sativa*). 2. Stable nonrevertants derived from reculture. Plant Cell Rep. 5:108-110.
99. Holbrook, L.A., T.J. Reich, V.N. Iyer, M. Haffner, and B.L. Miki. 1985. Induction of efficient division in alfalfa protoplasts. Plant Cell Rep. 4:229-232.
100. Lupotto, E. 1986. The use of single somatic embryo culture in propagating and regenerating lucerne (*Medicago sativa* L.). Ann. Bot. (London) 57:19-24.
101. Maheswaran, G., and E.G. Williams. 1984. Direct somatic embryoid formation on immature embryos of *Trifolium repens*, *T. pratense* and *Medicago sativa* and rapid clonal propagation of *T. repens*. Ann. Bot (London) 54:201-211.
102. Mariotti, D., S. Arcioni, and M. Pezzotti. 1984. Regeneration of *Medicago arborea* L. plants from tissue and protoplast cultures of different organ origin. Plant Sci. Lett. 37:149-156.
103. Mariotti, D., M.R. Davey, J. Draper, J.P. Freeman, and E.C. Cocking. 1984. Crown gall tumorigenesis in the forage legume *Medicago sativa* L. Plant Cell Physiol. 25:473-482.
104. McCoy, T.J., and T.R. Knous. Unpublished, Plant Science Dep. Univ. of Nevada, Reno.
105. ----, and K. Walker. 1984. Alfalfa. p. 171-192. *In* P.V. Ammirato et al. (ed.) Handbook of plant cell culture, Vol. 3. Macmillan Publishing Co., New York.
106. ----. 1985. Interspecific hybridization of *Medicago sativa* L. and *M. rupestris* M. B. using ovule-embryo culture. Can. J. Genet. Cytol. 27:238-245.
107. ----, and L.Y. Smith. 1986. Interspecific hybridization of perennial *Medicago* species using ovule-embryo culture. Theor. Appl. Genet. 71:772-783.
108. Meijer, E.G.M., and D.C.W. Brown. 1985. Screening of diploid *Medicago sativa* germplasm for somatic embryogenesis. Plant Cell Rep. 4:285-288.
109. Mroginski, L.A., and K.K. Kartha. 1984. Tissue culture of legumes for crop improvement. Plant Breed. Rev. 2:215-264.
110. Nagarajan, P., J.S. McKenzie, and P.D. Walton. 1986. Embryogenesis and plant regeneration of *Medicago* spp. in tissue culture. Plant Cell Rep. 5:77-80.
111. Orr, W., J. Singh, and D.C.W. Brown. 1985. Induction of freezing tolerance in alfalfa cell suspension cultures. Plant Cell Rep. 4:15-18.
112. Reich, T.J., V.N. Iyer, M. Haffner, L.A. Holbrook, and B.B. Miki. 1986. The use of fluorescent dyes in the microinjection of alfalfa protoplasts. Can. J. Bot. 64:1259-1267.
113. ----, ----, B. Scobie, and B.L. Miki. 1986. A detailed procedure for the intranuclear microinjection of plant protoplasts. Can. J. Bot. 64:1255-1258.
114. Rose, R.J., L.B. Johnson, and R.L. Kemble. 1986. Restriction endonuclease studies on the chloroplast and mitochondrial DNAs of alfalfa (*Medicago sativa* L.) protoclones. Plant Mol. Biol. 6:331-338.
115. Schaefer, J. 1985. Regeneration in alfalfa tissue culture: Characterization of intracellular pH during somatic embryo production by solid-state P-31 NMR. Plant Physiol. 79:584-589.
116. Shahin, E.A., A. Spielman, K. Sukhapinda, R.B. Simpson, and M. Yashar. 1986. Transformation of cultivated alfalfa using disarmed *Agrobacterium tumefaciens*. Crop Sci. 26:1235-1240.
117. Skokut, T.A., J. Manchester, and J. Schaefer. 1985. Regeneration in alfalfa tissue culture: Stimulation of somatic embryo production by amino acids and N-15 NMR determination of nitrogen utilization. Plant Physiol. 79:579-583.
118. Smith, S.E., E.T. Bingham, and R.W. Fulton. 1986. Transmission of chlorophyll deficiencies in *Medicago sativa*. Evidence for biparental inheritance of plastids. J. Hered. 77:35-38.
119. Walker, K.A. Unpublished.

30 Pollination Control: Mechanical and Sterility

D. R. VIANDS

Cornell University
Ithaca, New York

P. SUN

Dairyland Research Company, Inc.
Clinton, Wisconsin

D. K. BARNES

USDA-ARS
St. Paul, Minnesota

The alfalfa flower has a specialized structure conducive to natural pollination, primarily by bees. The floral morphology and the tripping mechanism are responsible mainly for the dependence of perennial, cultivated alfalfa on cross-pollination. The morphology and tripping mechanism also require the application of specialized techniques in crossing flowers by hand. Critical genetic experiments may require the emasculation of flowers prior to pollination. Alternatives for controlled pollination include the use of self-incompatibility and various sterility mechanisms to prevent selfing in the production of hybrid seed. This chapter describes controlled pollination techniques.

30-1 FLORAL MORPHOLOGY, TRIPPING, POLLINATION, AND FERTILIZATION

Development of the alfalfa flower begins at the shoot apex with the transition from vegetative to reproductive growth. This transition takes place at about the 10th to 14th node from the crown during spring growth, and at about the 6th to 10th node during summer growth (36, 68). The transition is first recognized by the protuberance of meristematic tissue in the axil of the leaf primordium nearest the shoot apex (Fig. 30–1) (36). Each primordium gives rise to a simple raceme. The shoot is normally indeterminate, and the shoot apex continues to differentiate both vegetative and floral organs until the stem senesces or is removed (see Chapter 4 in this book).

Copyright 1988 © ASA-CSSA-SSSA, 677 South Segoe Road, Madison, WI 53711, USA.
Alfalfa and Alfalfa Improvement—Agronomy Monograph no. 29.

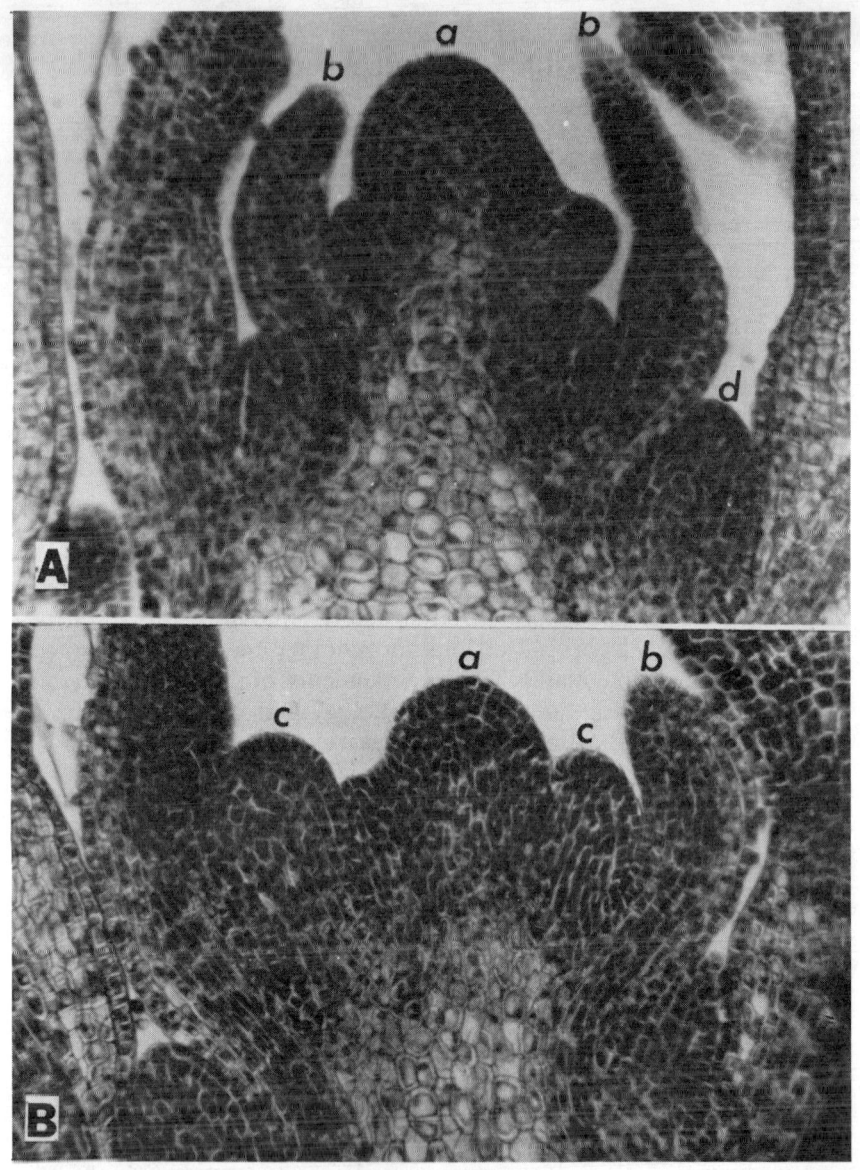

Fig. 30–1. Transition from vegetative shoot (A) to floral shoot; (B) (a) apical meristem, (b) leaf primordia, (c) raceme primordia, and (d) axillary bud meristem [from Dobrenz et al. (36)].

30–1.1 Pollen

Anther dehiscence takes place during the bud stage. Coffman (28) described the stages as straight bud, pointed bud, hooded bud, and erect standard. Armstrong and White (1) found that about 80% of the plants studied, including high and low seed setters, shed their pollen in the pointed bud stage. Dehiscence of pollen occurs only on the inner surfaces of the anthers. As dehiscence proceeds, pollen is forced inward and upward until the stigma is usually completely covered in the erect standard stage (1, 19).

Alfalfa pollen is binucleate and usually has three germination pores (84). Pollen is sticky and readily adheres to most pollinating insects. Pollen is stored naturally in the flower from anthesis until tripping, a period that may last 2 weeks. Hanson (46) reported that no appreciable loss in pollen viability occurred during this period. Alfalfa pollen can be stored artificially. Hanson noted that pollen was viable at 183 d in cork-stoppered vials held at −18°C and in vacuum-dried [20% relative humidity (RH)], sealed vials stored at room temperature. Lehman and Puri (63) stored pollen successfully for 101 d at −10°C. Differential viability of pollen genotypes under various storage conditions has not been studied.

30-1.2 Stigma Morphology

The stigma may not be covered with pollen prior to tripping because of deficiencies in pollen production, high proportions of defective pollen (1), or extension of the stigma above the pollen mass (19, 76). These situations prevent pollination in the untripped flower and reduce selfing after tripping. The major deterrent to pollination prior to tripping is a cuticular membrane that forms a continuous film over the stigma, and ordinarily prevents pollen from making contact with the stigmatic secretion. Kreitner and Sorensen (55, 56) have studied the stigma and stigmatic cuticle in detail. (Alfalfa scientists have used the term *stigmatic membrane*, but Kreitner and Sorensen [56, 57] have proposed *stigmatic cuticle* as a more appropriate term.) The stigma of a young flower is a knob of compact cells. As the developing stigma enlarges, vertical files of cells separate along their lateral walls and leave intercellular spaces apparently filled with lipid secreted from the cells. The stigmatic cuticle is attached to the tangential wall of each cell file and stretches across the intercellular spaces to form a barrier against pollen contact with the stigmatic secretion. The cuticle has a dense, globular inner layer and a homogeneous outer layer. Unicellular hairs develop along the periphery of the stigma (Fig. 30-2).

Kreitner and Sorensen (55, 56, 57) compared the stigmatic cuticles of a perennial, predominately cross-pollinating *Medicago sativa* L. clone with those of annual, self-pollinating *M. scutellata* Mill. They suggested that prevention of self-pollination before tripping in *M. sativa* is probably the result of a stronger and thicker stigmatic cuticle compared to that of *M. scutellata*. In young flowers of *M. sativa*, the stigmatic cuticle was about 120-nm thick compared with about 75 nm in *M. scutellata*. Also, the stigmatic cuticle of *M. scutellata* did not have the definite homogeneous outer layer. Self-pollination before tripping in *M. scutellata* may

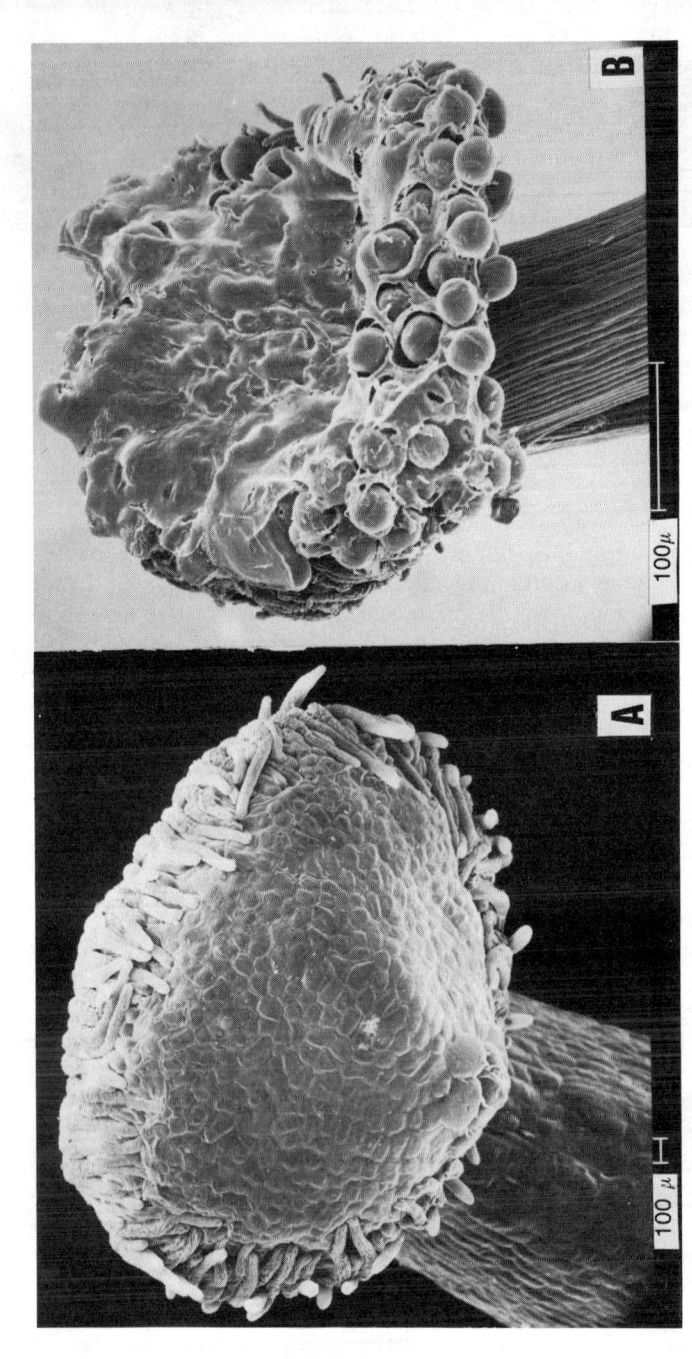

Fig. 30-2. Scanning electron micrograph of alfalfa stigmas in the mature flower stage; (A) stigma with unbroken stigmatic cuticle and peripheral hairs ($\times$ 300), and (B) stigma after smashing against the standard upon tripping the flower; self pollen is embedded in the stigmatic secretion ($\times$ 20) [from Kreitner and Sorensen (56)].

be possible because the stigmatic cuticle is fragile and highly erodable. In prepared specimens, the cuticle was perforated and loosely attached to the cell filaments. The cuticle of *M. sativa* remained intact, but some pollen grains were pressed into the cuticle in contact with secretion. About 20% of the stigmas from untripped *M. sativa* flowers had broken cuticles (56). The cuticle does not appear to be an absolute barrier, possibly explaining why some plants produce autogamous seed in untripped flowers (19, 53, 92).

30-1.3 Floret Tripping

In *M. sativa*, tripping is required, as a rule, to rupture the stigmatic cuticle and effect pollination. The tripping mechanism in the alfalfa flower has been the subject of study for >100 yr. Larkin and Graumann (61) summarized previous studies on tripping and described their own anatomical observations. They described two major forces involved in the tripping mechanism: (i) pressure exerted by the sexual column from cells under tension at the juncture of the staminal tube and the keel; and (ii) the restraining nature of keel petals that cohere because of interlocking finger-like projections of cutinized tissue on the appressed surfaces of the keel petal. Recently, Kreitner and Sorensen (58) examined the keel petals by scanning electron microscopy. The two keels appear to be interlocked by intermeshing of ridges and grooves along the inner surface of the keels (Fig. 3-3). The ridges and grooves appear more exaggerated in perennial alfalfas than in annuals. The fine intermeshing apparently aids annuals in self-pollination by increasing the ease of tripping.

Tripping takes place the instant that the restraint of the appressed keel petals is reduced or becomes less than the pressure exerted by the sexual column. In nature, tripping is effected by mechanical pressure applied by pollinating insects and influenced by environmental factors such as wind, rain, heat, and cold. Differences in ease of tripping are common among alfalfa plants. Larkin and Graumann (61) measured ease of tripping among plants by placing drops of different alcohol concentrations in the throat of the flower. The lesser the alcohol concentration required to trip the flowers of a plant, the easier the flowers are to trip mechanically. Kreitner and Sorensen (58) suggest that both cell turgidity in the keel petals and morphology of the intermeshing ridges on the keel petals control tripping.

A very small percentage of plants in many germplasm sources have flowers that trip without any external stimulus. This self-tripping, or autotripping, mechanism has been described as a heritable mutant. After five generations of selfing, plants have been identified where 100% of the flowers self-tripped. This self-tripping capability was expressed in both the greenhouse and field. Self-tripping appeared to be independent of genetic potential for self-sterility and self-fertility (59, 92).

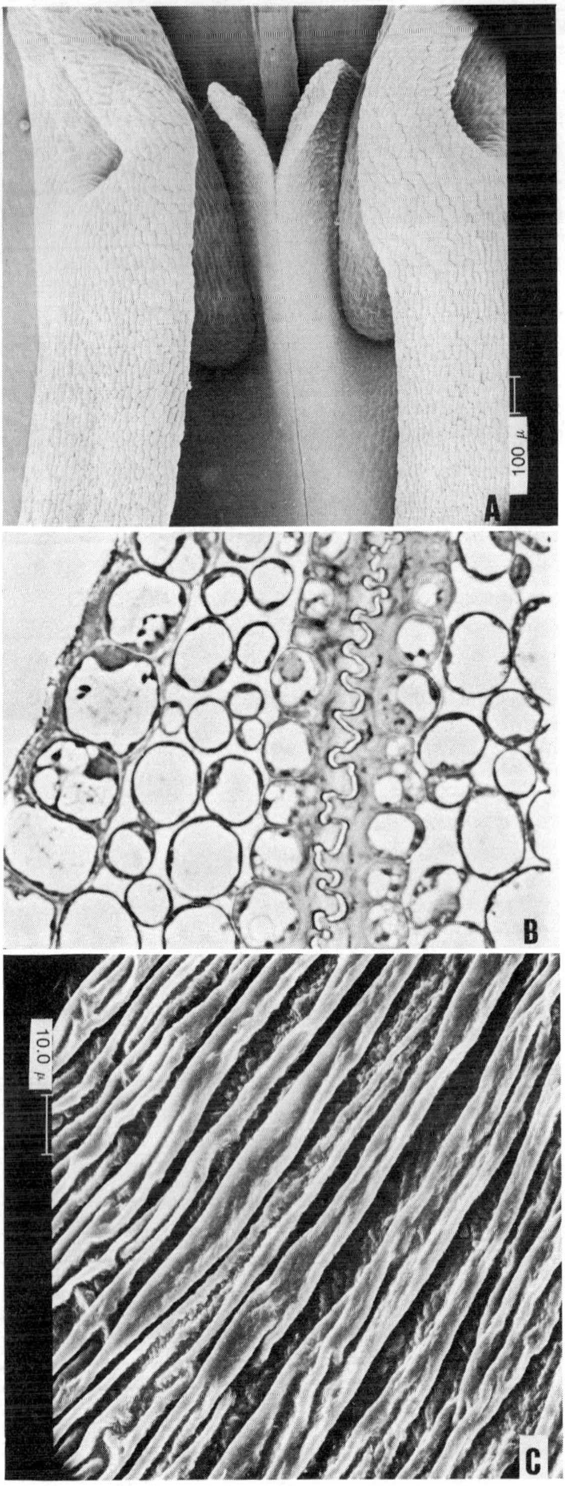
100 μ
A
B
10.0 μ
C

30-1.4 Pollination and Fertilization

When the flower is tripped, the slapping of the stigma onto the standard petal ruptures the stigmatic cuticle and transforms the stigma into a concave shape. Pollen grains embed in the peripheral ring of secretion and are supported by the hairs at the fringe of the stigma (Fig. 30-2) (56). The pollen grains germinate in the stigmatic secretion, which appears to be primarily a source of moisture (19). Nutritional requirements for pollen germination do not appear to be very specific because pollen will germinate in water (12, 19). Sugars and other nutrients in the stigmatic fluid enhance pollen tube development (17, 18). On the tripped stigma, pollen tubes penetrate the core of transmitting tissue at the base of the stigma (55), grow down the hollow style, and enter the ovarian cavity in about 10 to 12 h (88). By fluorescent microscopy, Sangduen et al. (88) examined in vivo pollen germination and tube growth down the style of *M. sativa*. The thread-like tubes in the pistils were lined by callose, and callose plugs developed at irregular intervals along the tubes. The tubes were usually in close proximity to the vascular system of the pistil. Fertilization occurs within 24 to 32 h, depending on pollen tube growth, the position of the ovule (30, 89), and temperature (90).

The mature fruit of cultivated *M. sativa* is a brown indehiscent pod that is curved or coiled with as many as three to four spirals. In contrast, pods of *M. falcata* are usually straight or crescent-shaped and dehiscent. Gradations in pod shape and dehiscence occur in hybrid populations. The pod is pubescent to varying degrees and consists of three distinct regions: epicarp, exocarp, and endocarp (48).

Modifications in floral morphology sometimes occur in alfalfa. Barnes and Hanson (8) summarized inheritance studies conducted on mutants affecting floral morphology. These mutants included two types of branched panicle-like inflorescences; two types of exposed-stigma abnormalities, in which the stigma was not enclosed within the keel petals; two types of modifications that affect the number and size of petals; and a modification characterized by open rather than closed keel petals. Another modification was described as a hornless wing petal mutant. In this mutant, wing petals, lacking the horn-like projections, were wrapped around the keel rather than held apart. The occurrence of this abnormality substantiates the suggestion by Larkin and Graumann (61) that the horn-like projections are not involved with the tripping mechanism, but merely serve to

---

• Fig. 30-3. Micrographs showing the mechanism holding the staminal column inside the keels before tripping; (A) scanning electron micrograph (SEM) of dorsal view of closed keel. Note the horns of the wing petals pressed against the keel petals adjacent to the fused edges of the keels ($\times$ 50) (the horns apparently serve to keep the wing petals spread apart), (B) light micrograph of a vertical section through the union of the keels; note wavy lines where grooves and ridges interlock ($\times$ 600), and (C) SEM of the inner surface of the keel petal at the union between the keels ($\times$ 800). Note the grooves and ridges [from Kreitner and Sorensen (58)].

separate the wing petals. To date, these abnormal floral modifications have proved useful only as gene markers.

30-2 MECHANICAL POLLINATION CONTROL

Barnes (2) described in detail the techniques required for self- and cross-pollinations. Self-pollination requires tripping the flower without introducing foreign pollen. This can be done by hand manipulation in one of four ways: (i) inserting a toothpick in the throat of the flower and tripping the flower; (ii) tripping the flower with a toothpick, the tip of which is covered with emery paper to retain pollen and rupture the stigmatic cuticle; (iii) tripping the flower with folded cardboard or blotter to retain pollen for contact with the stigma, or (iv) gently rolling or squeezing the raceme between fingers. Barnes and Stephenson (10) found that these procedures were equally effective (average seeds per flower tripped), but that rolling racemes between fingers was about three times more efficient (seeds per person-hour) than any other method.

Cross-pollination by hand may be accomplished with or without emasculation. If parent plants have a high degree of self-sterility, emasculation may not be necessary. Crossing may be achieved simply by alternately tripping flowers of the two parents into a folded cardboard or blotter. When this method is used in crossing, 10 to 100% (average about 75%) of the flowers in *M. sativa* develop pods, depending partly on the plant genotypes and the experimenter. In most cultivated types of alfalfa, an average of three to five seeds develop per pod (107). Pedersen and Barnes (79) averaged 1.1 seed per flower from self-pollination and 4.4 from cross-pollination. Three to five grams of seed can be produced per person-hour of cross-pollination if racemes do not require tagging for identification (107).

Without emasculation, cross-pollination usually results in some self-pollination unless the seed parent is male sterile or self-sterile. Plants grown from hybrid seed can be distinguished from selfs by using gene markers. Barnes and Hanson (8) described about 75 simply inherited markers. Since that publication was written, about that many more markers have been described by various researchers (see Chapter 24). Flower color has been used most frequently, but the most useful markers are those that are not detrimental to plant growth and are observable at the seedling stage.

In studies where self-pollination must be avoided, flowers are emasculated before pollination unless the seed parent has some sterility mechanism. Either a suction (52) or alcohol (103) procedure can be used to remove anthers and pollen. In both methods the flower is tripped to avoid breaking the stigmatic cuticle, by first clipping the standard petal and then tripping the flower by gently squeezing it near the base with forceps.

When suction is used for emasculation, it is applied with glass tubing drawn to a 1-mm tip and inserted in a rubber hose attached to a vacuum source. Thoroughness of pollen removal can be determined with a low-powered binocular magnifier attached to the forehead of the operator. Prior to tripping, some stigmas may have a few self pollen grains imbedded in them and in contact with stigmatic secretion (56). Under these circumstances, suction emasculation may not be totally effective in preventing self-pollination.

In alcohol emasculation, the whole raceme is immersed in a small beaker of 57% ethyl alcohol for 10 s, and then washed in water for a few seconds. Tysdal and Garl (103) reported that although emasculation with alcohol was more effective than with suction, the alcohol caused more injury to the stigma. Only 26% of alcohol-emasculated flowers produced pods after foreign pollen was applied to the stigma, compared to 60% for suction-emasculated flowers, and 76% for nonemasculated controls.

When either suction or alcohol emasculation is used, flowers should be pollinated with large amounts of foreign pollen soon after emasculation to reduce competition from any remaining self pollen. In critical studies one should maintain emasculated, unpollinated controls to determine the effectiveness of both procedures and operators. In experiments where hybrid seed production is essential, one should use a male-sterile plant for the seed parent or gene markers for identifying hybrids.

Large quantities of seed can be produced by enclosing parent plants in a cage (in the field or greenhouse) with bees. This method is not reliable for producing hybrid seed, unless there is some type of pollen control, such as self-sterility or male sterility. Hanson et al. (47) reported that excessive amounts of selfing could result from individual bees working a single clone instead of randomly visiting all clones in a cage. In cage plantings with bees, Pedersen (78) observed that 53% of the seed was hybrid, whereas Kehr (50) reported an average of 50%, and a range among cages from 32 to 96%. Male sterility, self-sterility, and variability in amount and time of flowering also influence the extent of cross-pollination.

Barnes (2) summarized the environmental conditions that affect flower production in the greenhouse and growth chamber. In general, flower production is maximized by long photoperiods (at least 14–16 h daylengths) and high light intensity. Supplemental light from high pressure sodium or metal halide lamps promotes profuse flowering and seed production during seasons of short daylength and low sunlight intensity (38).

30-3 SELF-INCOMPATIBILITY AND SELF-STERILITY

In 1914 Piper et al. (82) observed that cross-pollination of alfalfa generally results in a higher percentage of flowers forming pods and a larger number of seeds per pod than does self-pollination. These early observations have been reconfirmed by many investigators (75, 80). Kvasova and Shumnyi (60) attributed the reduction in seed set following self-pollination to at least four mechanisms involving rate of pollen growth on the stigma, down the style, and into the ovary; abortion of ovules; and incomplete formation of the seed. Many alfalfa plants are completely self-sterile, and most appear to be at least partially self-incompatible or self-sterile.

30–3.1 Measuring Self-fertility

Percentage of flowers producing pods, number of seeds per pod, and number of seeds per flower tripped have served as indices of self- and cross-fertility in alfalfa. Gartner and Davis (42) compared all three self-fertility indices on 19 *M. sativa* clones. Pod set ranged from 1.5 to 94%, number of seeds per pod ranged from 0.44 to 3.53, and number of seeds per 100 flowers tripped ranged from 4 to 311. Differences for all three characters were highly significant. Percent pod set was significantly correlated with both number of seeds per pod ($r = 0.88$, 17 df) and number of seeds per 100 flowers tripped ($r = 0.96$, 17 df).

Methods of determining self-fertility of eight alfalfa clones grown in Wyoming, Montana, and Nevada were compared by Melton et al. (71). During the first year, self-fertility values calculated from pods per flower tripped, number of seeds per pod, and seeds per flower tripped were consistent. However, data obtained during the subsequent year in Nevada were inconsistent with previous data. Melton et al. (71) and Melton (70) concluded that although methods of determining self-fertility were equally effective, the self-fertility of clones could be affected differentially by environmental conditions.

Gray et al. (43) compared self-fertility indices on 'Atlantic', 'Buffalo', 'Delta', 'DuPuits', and 'Narragansett'. The indices were equally effective for distinguishing differences in self-fertility, which indicated that selection of an index should be based on considerations such as time, cost, and accuracy. Percentage of flowers producing pods requires the least amount of labor. In uncontrolled pollination studies, number of seeds per pod could be a suitable index. Seeds per flower tripped requires the most labor, but it is the most useful measurement when making critical comparisons over a diverse range of genotypes. Seeds per flower tripped reflects differences in both seeds per pod and percentage of flowers producing pods. Because experimenters may differ in their use of accepted pollination techniques, caution must be taken to avoid confounding different experimenters with other comparisons of self- and cross-fertility.

30–3.2 Self-incompatibility and Self-sterility Terminology

The term *incompatibility* was defined by Mather (66) as "the failure, following mating or pollination, of a male gamete and a female gamete to achieve fertilization, where each of them is capable of uniting with other gametes of the breeding group after similar mating or pollination." Self-incompatibility in higher plants is characterized by failure of self-fertilization as a result of: (i) pollen-stigma interactions, typical of many sporophytic systems of control; (ii) pollen-style reactions, typical of most gametophytically controlled systems; (iii) failure of syngamy within the embryo sac; or (iv) pollen tube-ovule interactions within the ovarian cavity. This last type may be characteristic of *Medicago* spp. Self-incompatibility in most species is measured usually in terms of fruit and seed production. This measurement is not completely satisfactory for alfalfa because of the frequent occurrence of postfertilization abortion of fertilized ovules that may not be associated with self-incompatibility. Histological studies of the pistil following self-pollinations must be done to differentiate between self-incompatibility and postfertilization abortion. Self-incompatibility should be used only to refer to the failure of events before or at the time of fertilization. In most studies with alfalfa, no histological distinction has been made between failure of events preceding fertilization and postfertilization failure. In these studies, the term *self-sterility* should be used to refer to the failure of fruit and seed production from unknown or unspecified causes following self-matings.

30-3.3 Evidence of Incompatibility Systems in Alfalfa

The failure of many pollen tubes to penetrate the style after self-pollination was observed by Bolton and Fryer (15). Sayers and Murphy (89) observed that in some plants, pollen tubes failed to penetrate the ovary after both self- and cross-pollinations. These observations emphasize the importance of stigmatic, stylar, and ovarian factors in determining incompatibility.

Few pollen tubes in *M. sativa* were observed to advance beyond the midregion of the ovary after selfing, whereas after crossing, they frequently reached the base of the ovarian cavity (20). In general, at 30 h after pollination, the longest tubes of the selfed plants had reached the fourth ovule, while in the crossed plants, they had reached the eighth and ninth ovules. At 48 h the self-pollen tubes were between the fifth and sixth ovules, while the cross-pollen tubes were at the 10th ovule. Some ovules reached by pollen tubes were not fertilized (22, 29). The first four ovules were reached with regularity by both self- and cross-pollen tubes. After selfing, however, only 28% of these first four ovules were fertilized, compared with 80% in crossed plants. In crossed plants there was marked evidence of pollen-tube activity near the micropyle, whereas in selfed plants the pollen tubes often were observed to pass directly by the micropyle of an ovule containing an unfertilized egg. Evidence of a similar type of pollen-tube inhibition in self-pollination has been reported in diploid *M. falcata* (73). Pedersen (78) reported that the position of the seed in the pod was not related to whether or not the seed was hybrid or selfed under conditions that provided for open-pollination.

Sayers and Murphy (89) observed that pollen-tube growth in the style after self-pollination, closely paralleled pollen-tube growth after cross-pollination in clones selected for high and low seed set from the cultivar Narragansett. However, selfing had a more pronounced effect on the frequency with which ovules were entered by pollen tubes than it did on pollen-tube growth, especially among plants that were low in self-fertility.

Shipe et al. (91) placed pollen on agar that was diffused with pistil substances. They observed an interaction in pollen tube growth between pollen and pistil substances from different genotypes. In general, pollen tubes grew less with pistil substances compared to the controls, indicating that some inhibitory factor(s) was present in the diffused pistil substances. However, pollen grown with pistil substance from the same plant generally did not grow differently than did foreign pollen. Apparently, other inhibitory factors not diffused from the pistil are responsible for self-incompatibility.

In summary, the self-incompatibility reaction in alfalfa is only partially effective in preventing self-fertilization. Some plants can be found that show virtually no reduction in fertilization following self-pollination. The principal manifestation of self-incompatibility in alfalfa apparently resides in the pollen tube-ovule interactions that occur within the ovarian cavity. More critical evidence is needed to determine the mechanisms of self-incompatibility.

With increased interest in incorporating traits from wild species into cultivated species, there is a need to examine incompatibility systems in making interspecific crosses. Sangduen et al. (88) observed much less pollen germination on stigmas and about half the pollen tube growth rate down styles when crossing annual onto perennial *Medicago* spp. as compared to self-pollinations or intraspecific crosses. Under ultraviolet (UV) fluorescent microscopy, many pollen tube abnormalities were observed in styles in interspecific crosses. In spite of these abnormalities, about 43% of the ovaries had at least one fertilized ovule. However, only one of many thousand interspecific crosses produced seed. Sangduen et al. (87) reported that this postfertilization sterility was caused by the failure of the maternal and embryonic tissues to carry out the timely sequence of metabolism involving lipid, starch, and nucellar crystals (see Chapter 24 in this book).

30–3.4 Evidence for Fertile Ovule Abortion in Self-sterility

In intraspecific pollinations, the frequency of ovule abortion is greater in zygotes and embryos resulting from self-pollination than from cross-pollination. Cooper and Brink (29) conducted histological studies of pistils from seven clones of *M. sativa* following self- and cross-pollinations. The average percentages of collapsed fertilized ovules 48 h after selfing and crossing were 34.4 and 7.1%, respectively. Abnormal growth of the somatic tissue adjacent to the embryo sac causing local hyperplasia of the maternal structures appeared to result in collapse of the endosperm and thereafter the embryo. Previously, Brink and Cooper (21) had proposed the term *somatoplastic sterility* for this type of seed failure. Cooper et al. (30) and Cooper and Brink (29) found no correlation between the percentage of ovules becoming fertile following selfing and the percentage of fertilized ovules collapsing.

A high incidence of postfertilization ovule abortion was observed after selfing clones selected as high, medium, and low seed producers (89). However, the frequency of abortion was higher in the low- and medium-fertility classes than in the high class. During the early stages of zygote development, >80% of the self-fertilized ovules aborted compared with <12% of the cross-fertilized ovules in four clones of diploid *M. falcata* (73). Dattee (33) reported that selfing in *M. sativa* ssp. *coerulea* Schmalh. resulted in many embryo abortions during pod maturation.

30-3.5 Genetic Influences on Self-incompatibility and Self-sterility

Cooper and Brink (29) suggested that abortion of fertile ovules could result from inbreeding. Sayers and Murphy (89) suggested that some fertile ovule abortion may not result from inbreeding since a high degree of abortion also occurs in some clones after crossing. However, most research supports the hypothesis that inbreeding has a major influence on both self- and cross-fertility.

According to Fyfe (41), the fertility of a pollination in both diploid *M. sativa* ssp. *coerulea* and tetraploid *M. sativa* depends upon the degree of inbreeding that is represented in the pollination. He proposed the term *relational incompatibility* for this phenomenon. Furthermore, he suggested that it could result from either interactions between gametophytes before fertilization or from interactions within and between gametic complements after fertilization.

Busbice (24) determined the relationship between seed yield and the coefficient of inbreeding in a study of seven noninbred *M. sativa* clones and their S_1 progeny selfed and intercrossed to obtain seven levels of inbreeding in the developing zygotes. He proposed that reduced seed yield with inbreeding in alfalfa resulted primarily from a loss of heterozygosity in the zygotes causing lethality, not from a failure of gametes to unite in fertilization. As an alternative possibility, Busbice suggested an incompatibility system that increased in potency with inbreeding.

Obajimi and Bingham (77) reported that cross-fertility (seeds per flower) of nearly all of their retetraploidized plants was below 25% of the cross-fertility of the original tetraploids. Meiosis in most of the retetraploidized plants was normal, indicating that fertility problems must have occurred postmeiotically. Because tetraploids derived from colchicine-doubled haploids theoretically are equivalent to 3.8 generations of selfing, inbreeding in the zygote as well as in the maternal gamete could be responsible for low fertility. Additional studies with plants with different levels of heterozygosity provided further evidence for the effect of inbreeding on fertility (37). Haploid-derived tetraploids were used to produce single and double cross populations. Although the gene frequencies were the same, the more inbred populations had substantially lower cross- and self-fertility (seeds per flower) and seed weight than the most heterozygous population (double crosses). These two studies support Busbice's (24) proposal that reduced fertility with inbreeding is the result of a loss of higher order interactions among alleles within loci.

Smith et al. (94) and Smith (93) reported on the associations among the level of inbreeding, cytological abnormalities, and fertility. In partially inbred $3x \times 6x$-derived hexaploids, cross-fertility was negatively associated with pollen mother cells with laggards (94). In tetraploids, increased inbreeding through three generations of selfing decreased self- and cross-fertility and increased the percentage of abnormal quartets originating from univalent formation in either division of meiosis in the pollen mother cells. However, the association between fertility and cytological abnormalities existed only at high levels of the abnormalities (93). Apparently, other factors were more important in reducing fertility upon inbreeding.

Further evidence of the genetic influence on self-sterility was provided by Busbice et al. (25). They reported significant progress from two cycles of selection both for self-sterility and for self-fertility.

To date, attempts to determine the genetic basis of incompatibility in alfalfa have been unsuccessful. Tysdal and Kiesselbach (104) suggested that the absence of well-defined incompatibility relationships and the polyploid nature of alfalfa were complicating factors in determining the genetic basis for self-incompatibility. Diploid *Medicago* spp. and controlled environments have been used to circumvent problems encountered in studies of incompatibility. Nevertheless, Sahni (86) was unable to explain the self-incompatibility system of two diploid clones of *M. falcata* on a single S locus gametophytic self-incompatibility system. Miller (73) studied S_1, F_1, and backcross families derived from three partially self-sterile diploid *M. falcata* clones. The results showed a consistent pattern of intermediate fertility among related matings, a phenomenon Miller termed *semi-cross-fertility*. Semi-cross-fertility was defined as a reduction in seed set, when cross-pollinated seed set with an unrelated pollen parent is used as a reference point. This phenomenon is similar to the "relational self-incompatibility" described by Fyfe (41).

Summarizing the literature in the past decade, Smith (93) computed average declines in self-fertility compared to S_0 plants of 60% in the S_1, 71% in the S_2, and 79% in the S_3 generations. Smith cautioned researchers against studying the effects of inbreeding on self-fertility past the S_3 generation because of inadvertent selection for higher fertility.

30–3.6 Environmental Influences on Self-sterility

The percentage of flowers setting pods even with a highly self-sterile plant may vary considerably, depending partly upon the environment (104). In 1936, Brink and Cooper (19) observed frequent seed set on alfalfa plants without tripping of the flowers during unusually hot, dry weather at Madison, WI. Both Williams (109) and Wilsie (110) found a higher self-fertility index in the greenhouse than in field studies. Wilsie and Skory (111) found that in 1946 considerably more selfed seed was produced in the greenhouse during March than during February, but in 1947 similar differences were not noted. Melton (70) evaluated five clones of alfalfa for self-fertility in the field over a period of 9 yr and found significant differences among clones and clones × years interaction.

Eight clones of *M. sativa* were evaluated for self-fertility and pollen-growth characteristics under four constant-temperature regimes (16, 21, 27, and 32°C) (98). Highly significant differences in self-fertility were found among clones, among temperatures, and for clone × temperature interactions. Three clones increased significantly in self-fertility at 27 compared with 16°C, four clones remained relatively stable across all temperatures, whereas one clone was significantly lower in self-fertility at 27°C compared with 16°C. It may be significant that the three clones that increased in self-fertility at higher temperatures were also the three most self-sterile clones in the study.

Dane and Melton (31) examined temperature effects on both cross- and self-fertility in five *M. sativa* clones. Both self- and cross-fertility (seeds per flower) increased slightly from 21 to 27°C, but decreased at 32°C. Although the clone × temperature interaction was significant, the ranking was generally similar across temperatures. The low fertile clones responded very little to temperature differences. Seed abortion was highest at 32°C (32).

Meeks (69) studied the effect of temperature on the self-sterility of five diploid *M. falcata* clones. Propagules of each clone were grown to flowering under greenhouse conditions and self-pollinated after 2 to 3 d in growth chambers at temperatures of 15, 22.5, and 30°C. Self-fertility was significantly higher in all clones at 30 than at 22.5°C; however, clones varied quantitatively in degree of response to temperature. Extreme temperature sensitivity appeared to be partially dominant over insensitivity. Modification of self-sterility appeared to occur during a 2- to 4-d period after placing the plant in the high-temperature environment. Transfer of clones from the 30°C environment to a 22.5°C chamber resulted in restoration of fertility levels within 24 to 48 h.

30-4 IN VITRO POLLEN GERMINATION AND POLLEN-TUBE GROWTH

Normal pollen germination and pollen-tube growth can be obtained in vitro on a basic medium that includes 14 to 20 g of cane or beet sugar, plus 1.5 to 2.0 g of agar added to 100 mL of water (4, 15, 62, 83).

A pH between 6.0 and 8.0 is optimum for both pollen germination and tube growth (62). Pollen germination and pollen-tube growth are not affected by light vs. dark or by aggregations of pollen grains vs. single pollen grains (74, 83). There is some evidence, however, that a genotype exhibiting good pollen tube growth promotes growth of neighboring pollen of a different genotype (35).

Both temperature and sugar concentrations interact to affect alfalfa pollen germination and pollen-tube growth (6). At low temperatures (18°C) pollen germination and tube growth are generally better on media containing low sugar concentrations (7-14 g). At high temperatures (36°C) pollen-tube growth is reduced on all sugar concentrations, but germination is not affected at sugar concentrations of 14 to 35 g. Generally, requirements for pollen germination are less specific than those necessary for maximum tube growth.

The environment under which an alfalfa plant grows affects pollen quality (15, 31, 64, 74). Pollen germination, pollen-tube growth, pollen-grain size, and pollen staining all varied significantly with sampling dates over a 2-y period (64). Dane and Melton (31) reported that in vitro pollen germination was affected by the temperature in which the plant was grown and not by the temperature in which the pollen was incubated. Differences in pollen germination and pollen-tube growth of greenhouse and field-collected pollen have been reported (15, 74).

Pollen-tube lengths have been measured by stopping pollen-tube growth after 2 h, adding acetocarmine, and measuring the tubes with an ocular micrometer at $\times$ 100 (83). Pollen tubes grown 16 h or more at 20 to 30°C attain their full length, but are usually crooked and intertwined. These can be straightened with a curved needle and measured with an ocular micrometer (4).

Average pollen-tube length in vitro can vary more than twofold among alfalfa clones. Hourly growth rates of all pollen tubes appear similar for the first few hours, but the long pollen tubes have a longer period of sustained growth. Some clones show a bimodal distribution of pollen-tube length, whereas other clones show unimodal or multimodal distributions. Pollen-tube length appears to be under both genetic (4) and environmental (31) control. Inbreeding apparently affects both pollen germination and pollen-tube growth. Bocsa et al. (14) reported 18% average decline in percentage of fertile pollen from the S_0 to the S_4 generation in several cultivars. Froehlich and Barnes (40) observed 10% reduction in in vitro pollen-tube length in S_1 and F_2 progenies. Conversely, heterosis was evident in double crosses.

Comparisons of pollen-tube growth in vitro and in situ indicated that a pollen parent that produced long pollen tubes in vitro fertilized more ovules than a pollen parent that produced short pollen tubes in vitro (5). In general, pollen-tube length in vitro has been better correlated with cross-compatibility reactions in alfalfa than with self-compatibility reactions (74, 85, 98); however, exceptions have been reported (44). Correlations between pollen germination and either self- or cross-compatibility are either very low or nonsignificant (19, 39, 54, 74, 85, 108).

Pollen has been maintained in a sucrose solution for up to 2 h prior to use in liquid pollinations (12). This technique can be used to treat alfalfa pollen with colchicine, chemical mutagens, and labeled nutritional compounds.

30-5 MALE STERILITY

Male sterility, characterized by the failure of the plant to produce viable or functional pollen, is a widely distributed phenomenon in the plant kingdom. It is of particular importance in producing hybrids of cultivated plants, e.g., corn (*Zea mays* L.), cotton (*Gossypium hirsutum* L.), onion (*Allium cepa* L.), rice (*Oryza sativa* L.), sorghum [*Sorghum bicolor. (L.)* Moench], and sugar beet (*Beta vulgaris* L.). Both genetic and cytoplasmic male-sterile systems occur in alfalfa.

30-5.1 Genetic Male Sterility

Childers and McLennan (27) described a form of male sterility characterized by an atypical behavior immediately after the microspores were released from the tetrads. The first indication of abnormal growth was a separation of the tapetum from the inner locule wall. The tapetum, microspore walls, nuclei, and cell organization degenerated rapidly, leaving only a thick-walled endothecium and empty locules. Genetic studies indicated that this type of complete male sterility (20 DRC type) was not conditioned by cytoplasmic factors, but appeared to be controlled by three recessive nuclear genes inherited in a disomic manner. Subsequently, McLennan and Childers (67) transferred the male sterility from the tetraploid to the diploid level and showed that the 20 DRC type of sterility was controlled by one gene (ms_3) rather than three disomic genes.

Another type of male sterility, described by Childers (26), was characterized by swelling and over-growth of the tapetal tissue, and vacuolation of the cytoplasm in adjacent sporogenous cells. A few heavy-walled pollen grains were sometimes present in mature anthers, but the shrunken nature of the anthers precluded anthesis. Inheritance of this type of male sterility was attributed to duplicate genes with disomic inheritance. No subsequent attempt was made to determine whether inheritance was tetrasomic and/or cytoplasmic.

In contrast with 20 DRC genetic male sterility, Suginobu (99) described another type of male sterility that appears to be under complex genetic control, with both recessive and additive genes. Some of these genes are sensitive to and interact with environmental factors, especially temperature.

The recessive jumbo pollen gene (*jp*) conditioning male sterility is discussed in Chapter 24.

30-5.2 Cytoplasmic Male Sterility

Two cytoplasmic male sterility systems have been reported. The first evidence of cytoplasmic male sterility in alfalfa was reported by Davis and Greenblatt (34), Bradner and Childers (16), and Pedersen and Stucker (81). Davis and Greenblatt examined >50 000 alfalfa plants and found one completely sterile and nine partially sterile plants. Four of these plants had a sterility system that could be transferred to F_1 progeny in sterile × fertile crosses. Proof that this was a cytoplasmic-genic type of sterility rather than a genetic type was obtained from reciprocal crosses between partial male-sterile plants and fertile plants. The F_1 progenies of partial male-sterile × fertile crosses were mostly sterile, whereas F_1 progenies from fertile × partial male-sterile crosses were fertile (34, 81). The differential sterility among reciprocal crosses demonstrated that sterility was transmitted through the maternal cytoplasm.

The male sterility described by Davis and Greenblatt (34) and designated as U-1292 by Pedersen and Stucker (81) are controlled by the same cytoplasmic system described by Barnes et al. (3). Also, the LU-18 male-sterility source, reported by Guy (45), belongs to the same system, which is controlled by one homozygous recessive nuclear gene interacting with a sterile cytoplasm. Staszewski (97) designated two sterility systems, NS and ES. The NS system is the same as that reported by David and Greenblatt (34) and Pedersen and Stucker (81). The genotype associated with the NS fertility restoration locus was designated $rf_1\ rf_1\ rf_1\ rf_1$. According to Davis and Greenblatt (34), this cytoplasmic male-sterility system is characterized by incomplete pollen development or abortion of pollen grains after completion of meiosis.

Male sterility in the ES system is inherited like the first, but with different cytoplasmic and nuclear genes (97). The ES-fertility restoration locus does not interact with the NS-fertility restoration locus (97). The male-sterile sources, 20/6 from *M. polychroa* Grossh. (65), as well as Bms 12, Vms 27, Mms 123/28 and Mms 16 (95, 96) belong to the second system. Staszewski (97) designated the genotype of these sources as $rf_2\ rf_2\ rf_2\ rf_2$.

In the genetic male sterility described by Suginobu (99), the degree of sterility in the F_1 plants is highly dependent on the degree of pollen sterility of the paternal parent. In contrast, the degree of male sterility in the cytoplasmic system is independent of the pollen sterility of the paternal parent. At the same degree of sterility, cytoplasmic and genetic male-sterility systems are indistinguishable by observations of pollen and microsporogenesis (99).

Various degrees of partial sterility can occur within the cytoplasmic system. Pollen production can range from a trace to near normal amounts of pollen in the anther. Barnes et al. (3) suggested that this continuous variation in cytoplasmic male sterility must represent dosage effects of restoring genes. Complete cytoplasmic male sterility appears to be controlled by one homozygous recessive nuclear gene interacting with sterile cytoplasm. Incomplete or partial sterility appears to be controlled by the nuclear gene in the simplex, duplex, and possibly triplex conditions. In this genetic model, duplex genotypes have more viable pollen than simplex genotypes when in combination with the sterile cytoplasm, and so forth. In addition, one or more modifier genes may contribute to this variation (100).

Thompson and Axtell (101) demonstrated that factors conditioning cytoplasmic male sterility in alfalfa can be transferred asexually by grafting maintainer scions on male-sterile stocks. Autonomy of scion flowers was not altered, but cytoplasmic male sterility was confirmed in F_1 scion progenies of three graft combinations. One of these combinations produced 25 F_1 scion individuals, 10 of which were cytoplasmic male steriles. In addition, they grafted cytoplasmic male-sterile scions onto fertile maintainer stocks to determine if this combination would increase male fertility in the progeny. A higher proportion of F_1 scion progeny were rated as partially male sterile than were control progeny. However, no progeny plants consistently produced flowers that were male fertile.

A system for classifying the amount of pollen shedding in alfalfa was described by Barnes et al. (9). Flowers are tripped so that pollen is thrown onto a colored pot label or knife blade, and then visually classified as: 1 = no pollen, 2 = trace of pollen, 3 = moderate amount of pollen, and 4 = normal pollen. In their research, populations segregating for degrees of pollen production were classified for amount of dehisced pollen. Differences in mean scores among locations were small but significant. There were no obvious associations, however, between climatic conditions and pollen scores. Differences among observers within locations and populations × locations interactions were nonsignificant (9). Barnes et al. (9) presented a pictorial classification description to help reduce variability among observers in future studies. They suggested that a 200-plant sample is sufficient to accurately determine the level of dehisced pollen in an alfalfa population.

Microscopic observations have been used to measure the amount of normal pollen production (99). Traynor (102) determined pollen production by placing stamens from flowers in the pointed bud stage into stoppered vials. These vials were incubated at 30°C for 36 h, then pollen counted through a haemocytometer.

Pedersen and Barnes (80) studied the relationship between levels of pollen production and percentage crossing after both hand pollinations in the greenhouse and bee pollinations in the field. Greenhouse studies indicated that amounts of sibbed and selfed seed were affected by the level of pollen production in the seed parents. Three-way random crosses among white-flowered complete male steriles, white-flowered sibling partial steriles, and unrelated purple-flowered fertile plants were used to simulate pollination opportunities within a three-way alfalfa hybrid. The type of seed produced on plants classified 1 for pollen production [using the pollen classification developed by Barnes et al. (9)] was 0% self, 2% sib, and 95% hybrid. The amount of self, sib, and hybrid seed produced on Class 2, 3, and 4 plants were 3, 5, and 92%; 12, 27, and 61%; and 18, 40, and 42%; respectively. This classification system has been used in developing certification standards for hybrid alfalfa seed production.

Intraplant variation in pollen production occurred in partially male-sterile clones, but it was not associated with either flower position on the plant or sampling date (7). Anthers within the same flower varied for both number of pollen grains and percentage of normal pollen grains. The average number of pollen grains per anther varied among clones from about 440 to 850 (7). Sun and Yen (100) observed male-sterile, partial male-sterile, and partial fertile florets occurring within a raceme. Intraclonal and interclonal variation of pollen fertility was noted by Staszewski (97) in cytoplasmic male-sterile plants.

Barnes et al. (7) postulated that cytoplasmic male sterility is caused by a chemical system that has a threshold concentration for each level of sterility. Some anthers show less sterility than others in the same flower because they developed slightly earlier. The earlier developing anthers may not be exposed to the same chemical concentration as the later developing anthers. Cytological studies showed that all anthers within a flower are not synchronized for stage of development. Also, there may be inherent interplant differences in time of pollen degeneration.

30–5.3 Genotype × Environment Effects on Male Sterility

Pollen sterility of 20 DRC is quite stable. However, the genetic male-sterility system described by Suginobu (99) was influenced by environmental factors. In his studies, the mean pollen sterility under field conditions was about 10% higher than in the greenhouse. Sensitivity to different levels of light intensity and temperature was variable among the plants.

Staszewski (97) analyzed the expression of cytoplasmic male sterility in 13 alfalfa genotypes, that had different degrees of pollen viability, under three temperature regimes; 15 to 20, 22 to 27, and 35 to 38°C; and under two levels of soil moisture. All first-order interactions and genotype × temperature × water-supply effects on the pollen viability were highly significant. The fertility of pollen grains was higher with higher temperature and lower water supply. Completely male-sterile plants were more stable than partially sterile plants for pollen inviability.

Sun and Yen (100) observed a genotype that had complete cytoplasmic male sterility in a greenhouse during winter in Wisconsin, but that shed some pollen grains in California when grown in the field in summer.

30–6 POLLINATION CONTROL IN HYBRID PRODUCTION IN THE FIELD

Barnes et al. (3) summarized research showing that uncontrolled pollination by natural pollinators results in varying percentages of hybrid seed. Several methods have been suggested for controlling pollination in the production of hybrid seed in the field.

30–6.1 Self-sterility and Self-incompatibility

Tysdal et al. (105) presented a plan for production of double-cross alfalfa cultivars that depended on self-incompatibility for pollination control. Their method has not been used to produce seed of a hybrid cultivar.

Busbice et al. (25) found that selection for self-fertility increased the abundance of pollen, whereas selection for self-sterility reduced the quantity of pollen. Factors controlling pollen abundance may have conditioned ovules in some way because there was a positive correlation between seed-set and abundance of pollen. In their study, mean ovule growth 96 h after selfing was slightly less at the base than elsewhere in the ovary, suggesting a very weak self-incompatibility system.

From the viewpoint of practical plant breeding, intensive selection for self-sterility, without regard to the nature of that sterility, could lead to cultivars with low seed-production potential. On the other hand, intensive selection for self-fertility could increase selfing in seed production, to the extent that the vigor of the cultivar may be reduced. Male-sterility systems seem to provide better alternatives in producing hybrid cultivars.

30–6.2 Genetic Male Sterility

Genetic steriles have not been used to produce hybrid alfalfa cultivars, because genetic male sterility is expressed only in the homozygous recessive genotype. Therefore, most if not all methods of utilizing genetic male sterility in hybrid production requires some vegetative propagation of the male-sterile plants. Unless a dominant type of genetic sterility is found, it probably will be uneconomical to produce hybrids with genetic male sterility as the only source of pollen control.

30–6.3 Cytoplasmic Male Sterility

Barnes et al. (3) stated that cytoplasmic male sterility is the most efficient pollen control method of hybrid production. They presented a plan for making a three-way hybrid. Sun and Yen (100) conducted a study comparing forage productivity of single, three-way, and double-cross hybrids. A few single cross hybrids outyielded the three-way and double cross hybrids by >12%. Conversely, some three-way hybrids outyielded single cross hybrids by >8%. After examining >1000 hybrids, they concluded that genetic background is more important to forage productivity than the type of hybrid.

The discovery of male sterility confronted alfalfa breeders with two major questions. Will pollinating insects such as leafcutter bee (*Megachile rotundata* F.), honey bee (*Apis mellifera* L.), and alkali bee (*Nomia melanderi* Ck11.) consistently visit both male-fertile plants (pollenizers) and male-sterile plants to produce hybrid seed? Does the cytoplasm influence seed set in alfalfa?

Suginobu (99) investigated bee visitation on male-sterile and fertile alfalfa plants, and seed set of these two plant types both in the greenhouse and field. Average seed yield of male-sterile plants was significantly lower than that of male-fertile plants in both environments. In field observations there was no difference, however, between average frequency of bee visitation on male-sterile and fertile plants by either leafcutter bees or honeybees. Also, Beard and Meserve (11) reported no apparent differences for honeybee visitation between male-sterile and fertile clones. Significant differences were noted among fertile plants, but not among male-sterile plants. Conversely, Sun and Yen (100) found significant differences for bee visitation among both male-fertile and male-sterile plants.

Michaud and Busbice (71) studied differences in cross fertility when four alfalfa clones with different cytoplasms, A through D, were used reciprocally as male or female. Plants with Clone C cytoplasm performed poorly as male parents, but normally as female parents. The mode of action of the Clone C cytoplasm on the male function was not determined, but it appeared to differ from the cytoplasmic male-sterile cytoplasm discovered by Davis and Greenblatt (34). Plants containing Clone C cytoplasm produced an abundance of normal-appearing pollen grains. In general, Clone C cytoplasm had a negative effect on seed set; Clone A and D cytoplasm had positive effects, and Clone B cytoplasms had no effect. These results indicate that selection among cytoplasms may be as important as selecting germplasm sources for controlling or improving self- and cross-fertility.

Sun and Yen (100) evaluated several male-sterile and male-fertile plants in topcross fields. Their results show that the sterile cytoplasm reported by Davis and Greenblatt (34) had little or no effect on seed set (Table 30-1). They concluded that nuclear genes play a major role in determining seed yield.

Table 30-1. Topcross seed yield of alfalfa plants from five pollen classes at Sloughhouse, CA, in 1983. From Sun and Yen (100).‡

		Seed yield/experiment					
		Exp. I			Exp. II		
Cytoplasm	Pollen class‡	No. of plants	Mean	Range	No. of plants	Mean	Range
			—— g/plant ——			—— g/plant ——	
Sterile	ms	527	100.2	31–326	127	70.9	8–236
	pms	77	115.5	42–293	45	61.9	12–198
	PF	11	99.1	56–297	39	69.0	20–178
	F				42	63.2	22–140
Normal	F (normal)	100	105.1	11–249	158	59.9	8–236

† Plants were transplanted in fall 1982 for Exp. I and in spring 1983 for Exp. II, and both harvested in fall 1983. Both honeybees and leafcutter bees were used.
‡ Male fertility increases from Class ms = male sterile, to Class F (normal) = completely fertile.

30-6.4 Field Production of Hybrid Seed

The first commercial alfalfa hybrids utilizing cytoplasmic male sterility were marketed by the L. Teweles Seed Company in 1968. The standard commercial production practice was to plant rows spaced 1 m apart as follows: four rows of male sterile, one blank row, four rows of pollenizers, one blank row, four rows of male sterile, etc. Seed yield on male-sterile plants has been a serious economic problem in the production of alfalfa hybrids. On the average, seed yields of male sterile plants have been <50% that of the male-fertile pollenizer. This difference can be explained, in part, by the spacial isolation of male-sterile and male-fertile rows. Usually, male-sterile rows adjacent to pollenizer rows yield more seed than the two nonadjacent rows. Bee preference definitely plays a role.

Sun and Yen (100) examined alternative methods to overcome low seed production on male-sterile plants. They studied the effects of different planting patterns, male-sterile to pollenizer ratios, and different male-sterile lines × pollenizers on seed yield. In all combinations, male-sterile plants in rows adjacent to pollenizers consistently produced more seed than the male-sterile plants in nonadjacent rows. The average yield was 26.0 g per plant for adjacent male-sterile plants and 18.0 g for nonadjacent plants. In all crosses in mixed rows, male-sterile plants produced more seed in mixtures of one male-sterile to one pollenizer plant than of three male-sterile to one pollenizer. The average yield was 39.2 g per plant for the 1:1 ratio and 25.8 g per plant for the 3:1 ratio. The range in seed yield of male sterile plants in relation to pollenizer was 33.4 to 78.7% for the nonadjacent row and 49.4 to 112.3% for the adjacent row; 61.6 to 114.1% for the three male-sterile to one pollenizer mixture and 78.7 to 141.0% for the 1:1 ratio. Similarly, Staszewski (97) found that seed production of male-sterile plants depends both on percentage and system of planting male-sterile and male-fertile components in the field. In addition to planting system, Sun and Yen (100) have shown that seed set of the pollenizer has a pronounced effect on the seed yield of male-sterile lines (Table 30-2). A pollenizer × male-sterile interaction also affects seed yield of male-sterile lines (100).

In hybrid seed production fields, higher percentage of cross-pollination and higher hybrid seed yield can be achieved by mixing male-

Table 30-2. Seed set of pollenizer and cytoplasmic male-sterile lines. From Sun and Yen (100).†

Type of plant	Seed yield/plant				
	g				
Pollenizer range	51-60	61-70	71-80	81-90	91+
Male-sterile mean	32.2	34.4	35.9	37.1	44.0
Male-sterile range	27.5-37.7	33.6-35.0	33.5-38.2	31.5-42.6	42.0-48.2

† This study involved 21 pollenizers and four male-sterile plants in all cross combinations.

Table 30-3. Forage yield of six ratios of hybrid/inbred seed mixtures. From Sun and Yen (100).†

Hybrid/inbred ratio	Forage yield, kg/plot‡
100:0	14.8a
80:20	15.1ab
60:40	14.7abc
40:60	14.5abcd
20:80	14.1de
0:100	14.0de

† Average yield over three seeding rates, three replicates, and two harvests in 1972 and three harvests in 1973 at Clinton, WI. Plot size was 1.2 by 7.6 m.
‡ Means followed by the same letter are not significantly different at the 5% level by Duncan's new multiple range test.

sterile and pollenizer plants within the same row (23, 50). Postharvest separation of hybrid from nonhybrid seed might be accomplished by using genetic markers. Staszewski (97) conducted both greenhouse and field experiments to evaluate variation in seed size of hybrids and parental stock. His results confirm the findings of Pedersen and Barnes (79) and Hunt et al. (49) that seed size could serve as a marker for fractionation of hybrid seeds from a mixed population. Extensive diversity in seed size can be achieved among inbred lines as a result of selection. Differences in seed size between the hybrid and pollenizer can be particularly large if this trait is transferred to F_1 progeny by the male-sterile line. In addition to seed size, Hunt et al. (49) proposed separation of hybrid seeds based on seed coat pigmentation. The use of pollenizer lines with brown or black seed color was proposed.

Sun and Yen (100) investigated the percentage of hybrid seed that would be required in a seed mixture to maximize forage yield. This study consisted of six ratios of hybrid and inbred seed in mixture at each of three seeding rates. It was observed that competition eliminated weak plants from the population. In addition to competition, compensation in the performance of vigorous individual plants provided for maximum forage production of a given genetic combination in the absence of 100% hybrids. Mixtures from 100% down to the mixture of 60% hybrids + 40% inbreds were not statistically different (Table 30-3). However, yields began to decline substantially with an increasing percentage of inbred seed. Yields of the mixture of 20% hybrids + 80% inbreds were significantly lower than the higher percentage hybrid mixtures over three seeding rates. Elimination of inbred plants in mixtures with hybrid plants in competitive stands also has been reported by others (51, 106). If stands are adequate, these studies (51, 100, 106) show that a mixture containing 75% hybrid plants is sufficient for maximum forage productivity of a given genetic population.

30-7 FEMALE STERILITY

Bingham and Hawkins-Pfeiffer (13) found a completely female-sterile plant in their CADL population (cultivated alfalfa at the diploid level).

They transferred this trait into a basically 'Saranac' background at the tetraploid level for microscopic and gametic analyses. Female sterility was caused by incomplete development of the integuments, which left the nucellus and embryo sac unprotected and subsequently destroyed. The trait was expressed in the recessive condition of a single gene, *fs*. Ovaries from simplex plants exhibiting normal seed set (seeds per pod), however, did have *fs*-type ovules mixed with normal ovules. Bingham and Hawkins-Pfeiffer (13) described the advantage of this trait in hybrid seed production where *fs* plants could be interplanted with male-sterile or self-sterile plants to enhance hybrid seed production. Seed could be harvested from the entire field because *fs* plants would produce no seed. Brown and Bingham (23) found that a mixed planting yielded more hybrid seed than plantings in alternate rows, but less seed than that of conventional synthetics. Because no seed can be produced on female-sterile plants, progenies from this type of plant will need to be produced from outcrossing onto female fertile plants. Therefore, lines of pure female-sterile plants will need to be transplanted to the field as vegetative propagules or as plants selected out of large segregating populations. Obviously, this poses a practical limitation to the use of female sterility in large-scale hybrid seed production.

30-8 SUMMARY

Alfalfa has a highly specialized papilionaceous flower with a unique tripping mechanism that limits the types of insects that can effect pollination. Perennial alfalfas are primarily cross-pollinated, because this tripping mechanism is combined with a stigmatic cuticle that usually prevents contact of self pollen with stigmatic secretion prior to tripping of the flower. Partial self-incompatibility and various sterility mechanisms also reduce the prospect of self-pollination in certain alfalfa genotypes.

Many alfalfa genotypes can be self- and cross-pollinated by hand. Cross-pollination can be done either with or without prior emasculation. Cross-pollination with bees may produce both self and hybrid seed in the absence of sterility mechanisms. A few genetic markers are available to distinguish hybrid from self seed.

Cross-pollination usually results in higher seed set (seeds per flower and pods per flower tripped) than does self- pollination. Modest evidence suggests that a partial self-incompatibility system results from interactions between the pollen tube and various parts of the pistil, especially the ovary and ovules. After fertilization, self-sterility may result in the abortion of zygotes and embryos. Both genetics and environment influence the expression of self-sterility. Simple inheritance patterns have not been established for self-incompatibility and self-sterility, but loss of higher order allelic interactions through inbreeding appear to be of major importance.

Pollen germination and pollen-tube growth can be studied in vitro on a simple medium consisting of sugar and agar. Pollen germination and pollen-tube length are influenced by both environment and genetic factors.

Two types of male-sterility systems, genetic and cytoplasmic, have been identified. Two types of genetic male sterility are controlled by different recessive genes. Genetic male sterility will not be very useful in field production of hybrids in the absence of a system controlled by a dominant gene. Cytoplasmic male sterility offers the best mechanism for the production of hybrid seed. Two systems have been identified, both conditioned by single recessive genes interacting with cytoplasmic factors. Presently, hybrid production is not economical. The random distribution of male-sterile plants and pollenizers in the field, increased ratio of pollenizer plants to male-sterile plants, and use of high seed setting or high pollen producing pollenizers are important considerations in increasing seed yield of male-sterile plants.

An alternative method of producing hybrid seed is to use female-sterile plants in pollinating male-sterile plants. Large scale application of this system is limited because female steriles, controlled by a single recessive gene, must be maintained either by outcrossing to female-fertile plants or by vegetative propagules.

REFERENCES

1. Armstrong, J.M., and W.J. White. 1935. Factors influencing seed-setting in alfalfa. J. Agric. Sci. 25:161–179.
2. Barnes, D.K. 1980. Alfalfa. p. 177–188. *In* W.R. Fehr and H.H. Hadley (ed.) Hybridization of crop plants. Crop Science Society of America and American Society of Agronomy, Madison, WI.
3. ----, E.T. Bingham, J.D. Axtell, and W.H. Davis. 1972. The flower, sterility mechanisms, and pollination control. *In* C.H. Hanson (ed.) Alfalfa science and technology. Agronomy 15:123–141.
4. ----, and R.W. Cleveland. 1963. Pollen tube growth of diploid alfalfa *in vitro*. Crop Sci. 3:291–295.
5. ----, and ----. 1963. Genetic evidence for nonrandom fertilization in alfalfa as influenced by differential pollen tube growth. Crop Sci. 3:295–297.
6. ----, and ----. 1972. Interrelationship of temperature, sugar concentration, and pollen parent on alfalfa pollen germination and tube growth *in vitro*. Crop Sci. 12:796–799.
7. ----, and R.A. Garboucheva. 1973. Intra-plant variation for pollen production in male sterile and fertile alfalfa. Crop Sci. 13:456–459.
8. ----, and C.H. Hanson. 1967. An illustrated summary of genetic traits in tetraploid and diploid alfalfa. USDA Tech. Bull. 1370. U.S. Government Printing Office, Washington, DC.
9. ----, M.W. Pedersen, J.H. Elgin, Jr., J.D. Axtell, R.A. Garboucheva, and D.M. Smith. 1974. Pollen production classification in male sterile alfalfa: Location and observer effects. Crop Sci. 14:308–310.
10. ----, and M.G. Stephenson. 1971. Relative efficiencies of four self-pollination techniques in alfalfa. Crop Sci. 11:131–132.
11. Beard, D.F., and J.C. Meserve. 1964. Honey bee visitation to male sterile clones. p. 18–20. *In* Proc. 19th Alfalfa Improve. Conf. Lafayette, IN. 28 June–1 July.
12. Binek, A., and E.T. Bingham. 1969. Liquid pollination of alfalfa, *Medicago sativa*. Crop Sci. 9:605–607.
13. Bingham, E.T., and J. Hawkins-Pfeiffer. 1984. Female sterility in alfalfa *Medicago sativa* due to a recessive trait retarding integument development. J. Hered. 75:231–233.

14. Bocsa, I., R. Kiskeri, and J. Buglos. 1974. The effect of inbreeding on the fertility of lucerne pollen. (English Summary.) Z. Pflanzenzuchtg. 73:287-291.
15. Bolton, J.L., and J.R. Fryer. 1937. Inter-plant variations in certain seed-setting processes in alfalfa. Sci. Agric. (Ottawa) 18:148-160.
16. Bradner, N.R., and W.R. Childers. 1968. Cytoplasmic male sterility in alfalfa. Can. J. Plant Sci. 48:111-112.
17. Brink, R.A. 1924. The physiology of pollen I. The requirements for growth. Am. J. Bot. 11:218-228.
18. ----. 1924. The physiology of pollen. II. Further considerations regarding the requirements for growth. Am. J. Bot. 11:283-294.
19. ----, and D.C. Cooper. 1936. The mechanism of pollination in alfalfa (*Medicago sativa*). Am. J. Bot. 23:678-683.
20. ----, and ----. 1938. Partial self-incompatibility in *Medicago sativa*. Proc. Natl. Acad. Sci. Washington 24:497-499.
21. ----, and ----. 1939. Somatoplastic sterility in *Medicago sativa*. Science (Washington, DC) 90:545-546.
22. ----, and ----. 1940. Double fertilization and development of the seed in angiosperms. Bot. Gaz. (Chicago) 102:1-25.
23. Brown, D.E., and E.T. Bingham. 1984. Hybrid alfalfa seed production using a female-sterile pollenizer. Crop Sci. 24:1207-1208.
24. Busbice, T.H. 1968. Effects of inbreeding on fertility in *Medicago sativa* L. Crop Sci. 8:231-234.
25. ----, Y.G. Ramzy, and H.B. Collins. 1975. Effect of selection for self-fertility and self-sterility in alfalfa and related characters. Crop Sci. 15:471-475.
26. Childers, W.R. 1952. Male sterility in *Medicago sativa* L. Sci. Agric. (Ottawa) 32:351-364.
27. ----, and H.A. McLennan. 1960. Inheritance studies of a completely male sterile character in *Medicago sativa* L. Genet. Cytol. 2:57-65.
28. Coffman, F.A. 1922. Pollination in alfalfa. Bot. Gaz. (Chicago) 74:197-203.
29. Cooper, D.C., and R.A. Brink. 1940. Partial self-incompatibility and the collapse of fertile ovules as factors affecting seed formation in alfalfa. J. Agric. Res. (Washington, DC) 60:453-472.
30. ----, ----, and H.R. Albrecht. 1937. Embryo mortality in relation to seed formation in alfalfa (*Medicago sativa*). Am. J. Bot. 24:203-213.
31. Dane, F., and B. Melton. 1973. Effect of temperature on self- and cross-compatibility and *in vitro* pollen growth characteristics in alfalfa. Crop Sci. 13:587-591.
32. ----, and ----. 1973. Effects of temperature and method of pollination on the frequency of aborted seed in alfalfa. Crop Sci. 13:753-754.
33. Dattee, Y. 1976. Quantitative analysis of self and cross fertility of diploid alfalfa. Ann. Amelior. Plant. 26:419-441.
34. Davis, W.H., and I.M. Greenblatt. 1967. Cytoplasmic male sterility in alfalfa. J. Hered. 58:301-305.
35. Debrand, M., J.M. Cornuvet, Y. Dattee, and P. Guy. 1979. *In vitro* study of pollen competition in some genotypes of lucerne (*Medicago sativa* L.). Ann. Amelior. Plant. 29:63-77.
36. Dobrenz, A.K., M.A. Massengale, and W.S. Phillips. 1965. Floral initiation in alfalfa (*Medicago sativa* L.). Crop Sci. 5:572-575.
37. Dunbier, M.W., and E.T. Bingham. 1975. Maximum heterozygosity in alfalfa: Results using haploid-derived autotetraploids. Crop Sci. 15:527-531.
38. Elgin, J.H., Jr., and J.E. McMurtrey, III. 1977. Effect of four greenhouse supplemental light sources on alfalfa flowering and seed production. Can. J. Plant Sci. 57:1213-1216.
39. Englebert, V. 1932. A study of various factors influencing seed production in alfalfa (*Medicago sativa*). Sci. Agric. (Ottawa) 12:593-603.
40. Froehlich, D.M., and D.K. Barnes. 1983. Effect of inbreeding and heterosis on *in vitro* pollen-tube growth of alfalfa. p. 8. *In* Proc. 18th Central Alfalfa Improve. Conf., Manhattan, KS. 8-10 June.
41. Fyfe, J.L. 1957. Relational incompatibility in diploid and tetraploid lucerne. Nature (London) 179:591-592.
42. Gartner, A., and R.L. Davis. 1966. Effects of self-compatibility on chance crossing in *Medicago sativa* L. Crop Sci. 6:61-63.
43. Gray, E., J.S. Rice, and C.L. Wang. 1969. Comparisons of three indexes of self- and cross-compatibility in alfalfa (*Medicago sativa* L.). Crop Sci. 9:419-420.
44. Gurgis, R.Y., and D.E. Rowe. 1981. Variability of seed-set in alfalfa and correlations with some pollen and ovule characteristics. Can. J. Plant Sci. 61:319-323.
45. Guy, P. 1973. Hypotheses sur le determinisme d'une sterilite male. Proc. Eucarpia Meeting at Kompolt, Hungary. p. 35-38.

46. Hanson, C.H. 1961. Longevity of pollen and ovaries of alfalfa. Crop Sci. 1:114–116.
47. ——, H.O. Graumann, T.J. Elling, J.W. Dudley, H.L. Carnahan, W.R. Kehr, R.L. Davis, F.I. Frosheiser, and A.W. Hovin. 1964. Performance of two-clone crosses in alfalfa and an unanticipated self-pollination problem. USDA Tech. Bull. 1300. U.S. Government Printing Office, Washington, DC.
48. Hayward, H.E. 1938. The structure of economic plants. Macmillan Publishing Co., New York.
49. Hunt, O.J., D.K. Barnes, R.N. Peaden, and B.J. Hartman. 1976. Inheritance of black seed character and its use in alfalfa improvement. p. 33. *In* Proc. 25th Alfalfa Improve. Conf., Cornell Univ., Ithaca, NY. 13–15 July. USDA-SEA, Peoria, IL.
50. Kehr, W.R. 1973. Cross-fertilization of alfalfa as affected by genetic markers, planting methods, locations, and pollinator species. Crop Sci. 13:296–298.
51. ——. 1976. Cross-fertilization in seed production in relation to forage yield of alfalfa. Crop Sci. 16:81–86.
52. Kirk, L.E. 1930. Abnormal seed development in sweet clover species crosses—a new technique for emasculating sweet clover flowers. Sci. Agric. (Ottawa) 10:321–327.
53. ——, and W.J. White. 1933. Autogamous alfalfa. Sci. Agric. (Ottawa) 13:591–593.
54. Koffman, A.J., and C.P. Wilsie. 1961. Effects of inbreeding on various agronomic traits in DuPuits alfalfa. Crop Sci. 1:239–240.
55. Kreitner, G.L., and E.L. Sorensen. 1983. The stigmatic membrane in annual and perennial Medicago species. p. 5–6. *In* Proc. 18th Central Alfalfa Improve. Conf., Manhattan, KS. 8–10 June.
56. ——, and ——. 1984. Stigma development and the stigmatic cuticle of alfalfa, *Medicago sativa* L. Bot. Gaz. (Chicago) 145:436–443.
57. ——, and ——. 1985. Stigma development and the stigmatic cuticle of *Medicago scutellata*. Can. J. Bot. 63:813–818.
58. ——, and ——. 1985. Structure and keel-locking mechanism in insect-pollinated and self-pollinated alfalfa species. Crop Sci. 25:631–634.
59. Kvasova, E.V., and V.K. Shumnyi. 1975. Autotripping in alfalfa and its modification under the effect of inbreeding. Genetika 11:24–30.
60. ——, and ——. 1977. Mechanisms of self-incompatibility in alfalfa. Izv. Sib. Otd. Akad. Nauk SSSR, Ser. Biol. Nauk. Dec. 1977:62–68.
61. Larkin, R.A., and H.O. Graumann. 1954. Anatomical structure of the alfalfa flower and an explanation of the tripping mechanism. Bot. Gaz. (Chicago) 116:40–52.
62. Lehman, W.F., and Y.P. Puri. 1964. Factors affecting germination and tube growth of hand-collected and bee-collected pollen of alfalfa (*Medicago sativa* L.) on agar media. Crop Sci. 4:213–217.
63. ——, and ——. 1967. Rates of germination and tube growth of stored and fresh alfalfa (*Medicago sativa* L.) pollen on agar medium. Crop Sci. 7:272–273.
64. ——, ——, and M.J. Garber. 1969. Effect of environment on quality characteristics of alfalfa (*Medicago sativa* L.) pollen. Crop Sci. 9:560–563.
65. Lubenec, P.A. 1968. High yielding heterotic hybrids of alfalfa. (In Russian.) Proc. Sov. Union Acad. Sci. 9:11–13.
66. Mather, K. 1943. Specific differences in petunia. I. Incompatibility. J. Genet. 45:215–235.
67. McLennan, H.A., and W.R. Childers. 1964. Transfer of genetic male sterility from tetraploid to diploid alfalfa, and inheritance at the diploid level. p. 79. *In* Can. Soc. Agron. 10th Annu. Meeting. Abstr. Proc. Frederickton, New Brunswick. 22–25 June.
68. Medler, J.T., M.A. Massengale, and M. Barrow. 1955. Flowering habit of alfalfa clones during the first and second growth. Agron. J. 47:216–217.
69. Meeks, R.D. 1970. Self-incompatibility in *Medicago falcata* (L.) Ph.D. diss. Purdue Univ., Lafayette, IN (Diss. Abstr. 71–9441).
70. Melton, B. 1970. Effects of clones, generation of inbreeding and years on self-fertility in alfalfa. Crop Sci. 10:497–500.
71. ——, W.A. Riedl, O.J. Hunt, R.N. Peaden, A.E. Slinkard, M.H. Schonhorst, R.V. Frakes, and A.E. Carleton. 1969. Relationship of inbreeding behavior to heterosis in forage crops. N.M. Agric. Exp. Stn. Bull. 550, p. 31–32.
72. Michaud, R., and T.H. Busbice. 1978. Influence of cytoplasm on seed setting in alfalfa. Can. J. Plant Sci. 58:341–346.
73. Miller, J.L. 1960. Sterility in alfalfa. Ph.D. diss. Cornell Univ., Ithaca, NY.
74. Miller, M.K., and M.H. Schonhorst. 1968. Pollen growth of alfalfa *in vitro* as influenced by grouping of grains on the medium and greenhouse versus field sources. Crop Sci. 8:525–526.
75. Myers, W.M., and W. Rudorf. 1955. Kleeartige fuherpflanzen luzerne-arten. Handbuck der Pflanzenzuchtung, 2. Aufl. Band 4, p. 103–217.
76. Nielsen, H.M. 1962. Floral modification in lucerne. 1st Int. Symp. on pollination, Proc. Copenhagen, Aug. 1960. *In* Sveriges Froodlareforbund Meddel. (Swedish Seed Growers' Assoc. Commun.) 7. p. 60–63.

77. Obajimi, A.O., and E.T. Bingham. 1973. Inbreeding cultivated alfalfa in one tetraploid-haploid-tetraploid cycle: Effects on morphology, fertility, and cytology. Crop Sci. 13:36–39.
78. Pedersen, M.W. 1968. Seed number and position in the pod in relation to crossing in alfalfa, *Medicago sativa* L. Crop Sci. 8:263–264.
79. ----, and D.K. Barnes. 1973. Alfalfa seed size as an indicator of hybridity. Crop Sci. 13:72–75.
80. ----, and ----. 1973. Alfalfa pollen production in relation to percentage of hybrid seed produced. Crop Sci. 13:652–656.
81. ----, and R.E. Stucker. 1969. Evidence of cytoplasmic male sterility in alfalfa. Crop Sci. 9:767–770.
82. Piper, C.V., M.W. Evans, R. McKee, and W.J. Morse. 1914. Alfalfa seed production; pollination studies. USDA Bull. 75. U.S. Government Printing Office, Washington, DC.
83. Puri, Y.P., and W.F. Lehman. 1965. Effect of pollen aggregations and sucrose levels on germination of fresh and stored alfalfa (*Medicago sativa* L.) pollen. Crop Sci. 5:465–468.
84. Reeves, R.G. 1930. Nuclear and cytoplasmic division in the microsporogenesis of alfalfa. Am. J. Bot. 17:29–40.
85. Rice, J.S., C.L. Wang, and E. Gray. 1970. Relationship of pollen and pistil characteristics with self- and cross-compatibility in alfalfa. Crop Sci. 10:59–61.
86. Sahni, V.M. 1957. Genetics of self-incompatibility in alfalfa. Ph.D. diss. Purdue Univ., W. Lafayette, IN (Diss. Abstr. 00-22295).
87. Sangduen, N., G.L. Kreitner, and E.L. Sorensen. 1983. Light and electron microscopy of embryo development in an annual × perennial *Medicago* species cross. Can. J. Bot. 61:1241–1257.
88. ----, E.L. Sorensen, and G.H. Liang. 1983. Pollen germination and pollen tube growth following self-pollination and intra- and interspecific pollination of *Medicago* species. Euphytica 32:527–534.
89. Sayers, E.R., and R.P. Murphy. 1966. Seed set in alfalfa as related to pollen tube growth, fertilization frequency, and post-fertilization ovule abortion. Crop Sci. 6:365–368.
90. Sexsmith, J.J., and J.R. Fryer. 1943. Studies relating to fertility in alfalfa (*Medicago sativa* L.). II. Temperature effect on pollen tube growth. Sci. Agric. (Ottawa) 24:145–151.
91. Shipe, E.R., K.H. Quesenberry, and E. Gray. 1971. Influence of excised pistils on *in vitro* growth of alfalfa pollen. Crop Sci. 11:398–399.
92. Shumnyi, V.K., V.I. Kovalenko, E.V. Kvasova, and L.D. Kolosova. 1978. Several genetic and breeding aspects of plant reproductive systems. Sov. Genet. (Engl. Transl.) 14:15–22.
93. Smith, S.E. 1984. Cytological studies of alfalfa, *Medicago sativa* L. (S.L.). I. Chromosome homology in triploid intersubspecific hybrids. II. Relationships between fertility, level of inbreeding, and regularity of microsporogenesis in tetraploid *M. sativa*. Ph.D. diss. Cornell Univ., Ithaca, NY. (Diss. Abstr. 84-15314).
94. ----, R.P. Murphy, and D.R. Viands. 1984. Reproductive characteristics of hexaploid alfalfa derived from 3x × 6x crosses. Crop Sci. 24:169–172.
95. Staszewski, Z. 1970. Investigations on heterosis and male sterile forms in alfalfa. I. Breeding of male sterile lines. (In Polish.) Hodowla Rosl. Aklim. Nasienn. 14:483–498.
96. ----. 1976. Studies on male sterility inheritance and breeding of nonrestoring lines in *Medicago sativa* L. (In Polish.) Hodowla Rosl. Aklim. Nasienn. 20:105–119.
97. ----. 1979. Final technical report: Genetical studies on cytoplasmic male sterility and heterosis in *Medicago sativa* L. and selection of germplasm sources suitable for hybrid breeding. Plant Breeding and Acclimatization Institute, Radzikov, Poland.
98. Straley, C., and B. Melton. 1970. Effect of temperature on self-fertility and *in vitro* pollen growth characteristics of selected alfalfa clones. Crop Sci. 10:326–329.
99. Suginobu, K. 1979. Studies on the application of male sterility to heterosis breeding of alfalfa (*Medicago sativa* L.). Hokkaido Nogyo Shikenjo Kenkyu Hokuku. 124:1–79.
100. Sun, P.L., and S. Yen. 1984. Unpublished data. Dairyland Research International, Clinton, WI.
101. Thompson, T.E., and J.D. Axtell. 1978. Graft-induced transmission of cytoplasmic male sterility in alfalfa. J. Hered. 69:159–164.
102. Traynor, J. 1981. Use of a fast and accurate method for evaluating pollen production of alfalfa and almond flowers. Am. Bee J. 121:23–25.
103. Tysdal, H.M., and J.R. Garl. 1940. A new method for alfalfa emasculation. J. Am. Soc. Agron. 32:405–407.

104. ----, and T.A. Kiesselbach. 1944. Hybrid alfalfa. J. Am. Soc. Agron. 36:649–667.
105. ----, ----, and H.L. Westover. 1942. Alfalfa breeding. Nebr. Agric. Exp. Stn., Res. Bull. 124.
106. Veronesi, F., and F. Lorenzetti. 1983. Productivity and survival of alfalfa hybrid and inbred plants under competitive conditions. Crop Sci. 23:577–580.
107. Viands, D.R. 1984. Unpublished data. Dep. of Plant Breeding and Biometry, Cornell Univ., Ithaca, NY.
108. Whitehead, W.L., and R.L. Davis. 1954. Self- and cross-compatibility in alfalfa, *Medicago sativa*. Agron. J. 46:452–456.
109. Williams, R.D. 1931. Self-fertility in lucerne. Welsh Plant Breed. Stn. Bull., Ser. H., no. 12:217–220.
110. Wilsie, C.P. 1958. Effect of inbreeding on fertility and vigor of alfalfa. Agron. J. 50:182–185.
111. ----, and J. Skory. 1948. Self-fertility of erect and pasture-type alfalfa clones as related to the vigor and fertility of their inbred and outcrossed progenies. J. Am. Soc. Agron. 40:786–794.

31 Seed Physiology, Seedling Performance, and Seed Sprouting

L. N. BASS

USDA-ARS
Ft. Collins, Colorado

C. R. GUNN

USDA-ARS
Beltsville, Maryland

O. B. HESTERMAN

Michigan State University
East Lansing, Michigan

E. E. ROOS

USDA-ARS
Ft. Collins, Colorado

Alfalfa (*Medicago sativa* L.) ovules develop in a one-celled superior ovary. True embryos (2–16 cells) are present in the fertilized alfalfa ovule about 120 h after pollination. The embryo enlarges and differentiates, with two cotyledons developing, one on either side of the epicotyl. As the cotyledons and radicle/hypocotyl axis elongate, the embryo curves in the region where the cotyledons are attached, so that the radicle/hypocotyl is approximately parallel to the cotyledons (23, 39, 63, 83) (see Chapter 4 in this book).

The legume (fruit or pod) of alfalfa is shaped like an open spiral and has two to four, rarely five, coils that are 5 to 9 mm in diameter. As the legume coils during maturation, adjacent seeds may be forced together. Characteristics of mature alfalfa seeds (Fig. 31–1) were compared to those of other *Medicago* spp. often found as contaminants in commercial seed lots by Isely (59). The average seed count for 418 lots representing 39 cultivars was 464.5 seeds/g [13 168/ounce or 210 689/pound (12)]. Alfalfa seeds are composed of two cotyledons, a radicle/hypocotyl axis, an epicotyl in which the first leaves are present (30), endosperm about one-fourth to one-half the volume of the embryo (64), a colored seedcoat with a centrally located hilum, and a lens about 0.4 mm from the hilum, on the side opposite the radicle. The lens (strophiole), a weak point in the palisade layer, is visible as a minute bump on the seedcoat (see Chapter 4 in this book).

Copyright 1988 © ASA-CSSA-SSSA, 677 South Segoe Road, Madison, WI 53711, USA. *Alfalfa and Alfalfa Improvement*—Agronomy Monograph no. 29.

Fig. 31-1. Close up photograph of mature alfalfa seeds. × 10.

31-1 SEED PHYSIOLOGY

Excellent general reviews of the literature on physiology and biochemistry of seed development, dormancy, germination, and storage are available (21, 22, 56, 62, 63, 114). This discussion will be limited largely to alfalfa seeds.

31-1.1 Chemical Composition of Seeds

Longevity of seeds is controlled to a certain extent by chemical composition. Although much research has been conducted on the chemical composition of seeds of numerous species, there is a dearth of literature on alfalfa seeds. Generally, seeds are classified as oily or starchy, according to their principal storage component or industrial use. Alfalfa seeds are classified as starchy. Although the principal storage materials are carbohydrates, varying amounts of other compounds, such as lipids, proteins, and minerals are present.

31-1.2 Water Relations of Seeds

The relationship between water (moisture) content of seeds and environmental conditions involves complicated physical and chemical reactions. Both longevity in storage and germination are dependent upon water relations (86).

The seedcoats of alfalfa seeds may or may not be permeable to water when harvested (75, 76). The permeability of alfalfa seedcoats can change over time. The rapidity and amount of change depends upon such factors as storage temperature, relative humidity, and mechanical abrasions of the seedcoat during harvesting, conditioning, and packaging. According to Harrington (48), it is impossible to predict in advance what proportion of the hard seeds in a sample will germinate in or at a given time. Loss of hard-seededness varied among cultivars over time when stored at 5°C and 40% relative humidity (RH) in the National Seed Storage Laboratory (NSSL) (Table 31-1). Seeds with impermeable coats are commonly referred to as "hard seeds," because when immersed in water or placed in a moist environment they do not imbibe water but remain unchanged or hard.

Lute (76) demonstrated that the thickened outer wall of the palisade cells, not the cuticle, constitutes the moisture barrier. A minute break in this portion of the palisade layer permits uptake of water. Hand-harvested seed lots may be 100% hard-seeded; however, mechnaically harvested lots generally have <60% hard seeds (75).

A weak point in the palisade layer is located in the lens, which is at the apex of the cotyledon in alfalfa seed (see Chapter 4 in this book). This passageway acts as a conduit to the embryo for water or other fluids. Chromic-nitric acid stain graphically illustrates that when the lens is "open," fluids will penetrate the seedcoat, moving under the seedcoat from the lens to the rest of the seed.

Hard seeds of alfalfa may lie dormant in the soil for several weeks, months, or years before absorbing water and germinating. Lute (76) and Leggatt (69) found that high soil temperatures decreased hard-seededness and that fewer hard seeds remained in soil tests than in blotter tests. Soil

Table 31-1. Change in hard seed content in various alfalfa germplasm samples during storage at 5°C and 40% RH.

	Percentage hard seeds		
	Years stored		
Cultivar	0	12	23
Buffalo	16	9	2
Vernal	12	6	6
Cody	35	17	16
Ladak	12	13	4
Nomad	10	10	4
Grimm	11	10	6
Rhizoma	11	10	6
Rambler	11	9	6
	Years stored		
	0	10	21
Travois	41	41	36
Kansas Common	51	14	12
Newstan	8	10	4

aeration and acidity also may help decrease seed impermeability. Alfalfa seeds in a warehouse lose their impermeability slowly.

The percentage of hard seeds is governed by edaphic and climatic factors during and after seed maturation and by genetic factors. Some alfalfa cultivars regularly develop more hard seeds than other cultivars. The same cultivar grown in different parts of the country or different fields (Table 31-2) may exhibit wide variations in hard-seed content. Hard- seed percentages of 40 to 50 are common in northwestern-grown alfalfa; whereas alfalfa seed lots grown in the southwestern USA seldom have more than a 20 to 30% hard-seed content. Although other factors influence hard-seededness, it is thought that temperature during and immediately following maturation plays a major role. When temperatures are high, such as those in southern California, the hard-seed content is low, <20%. In Washington, where temperatures are lower, the hard-seed content is higher. The hard seed content of seeds harvested from the same field year after year will vary as well as the hard seed content of seeds of a cultivar harvested from different fields in a given year (Table 31-2).

Most seedsmen and growers prefer no more than 10% hard seeds, although 20 to 30% may be acceptable under some conditions (98). Hard seeds have little value when planted because these late-germinating seeds seldom contribute to improving the stand. The three most frequently used commercial methods to lower hard-seededness are storage, blending,

Table 31-2. Seasonal trends in hard seed percentages of Vernal alfalfa seed, from the same field-stand, produced from 1956 to 1974 by five growers near Wapato, WA.†

Year	Grower No. 1	2	3	4	5
1956	65	58	--	--	--
1957	60	46	--	--	--
1958	24	26	--	--	--
1959	69	52	--	--	--
1960	65	57	--	--	--
1961	--	--	--	--	--
1962	36	44	--	--	--
1963	40	50	27	--	--
1964	53	61	59	49	--
1965	50	48	42	43	52
1966		47	25	32	32
1967				--	--
1968				--	--
1969					
1970				15	14
1971				33	44
1972				--	--
1973				--	--
1974				31	32

† Data based on 1 440 946 lb of seed produced on five farms during the period taken from C.M. Rincker (113).

and scarification. Normal storage from harvest to planting time will reduce hard-seed percentage, but the exact results are unpredictable. Blending can be used only when a sufficient quantity of seeds with a low hard-seed count is available to blend with high hard-seed count seed lots. Scarification, the mechanical abrasion of seedcoats, reduces hard-seed percentages and also injures seeds. These injuries are reflected in higher abnormal seedling counts and deterioration of vigor and viability (44, 70, 73, 119, 120). A laboratory method of hand-scarifying hard seeds of alfalfa was describeed by Maguire (78).

Efforts to commercially scarify alfalfa seeds without damaging the embryo have led to refined scarification techniques and to the application of other methods. Seed polishing, used to improve the appearance of alfalfa seed lots, may provide a gentle scarification. Seeds are rotated in a cement mixer-like machine lined with elk (*Alces alces*) hide or a similar material. As the seeds are brushed against the elk hide, they are polished and lightly scarified by the tumbiling action. In a modified machine, where emphasis is placed on light scarification rather than polishing, the elk hide is replaced by screening. This increases scarification with a minimum of embryo damage. In light scarification (or brushing) the lens may be opened.

Other methods used to decrease hard-seededness are various combinations of high and low temperature; moisture and high pressure; high-frequency electrical energy; infrared rays; radio-frequency; and gas-plasma.

Early literature on impermeable seeds of alfalfa was reviewed by Lute (76). In general, those studies dealt with percentage of hard seeds in seed lots and possible methods of improving field performance of seed lots containing high percentages of hard seeds. Methods investigated included application of dry heat, soaking in hot water, clipping, abrasion on a grindstone, and treatment with H_2SO_4. Each treatment had an effect on reducing the percentage of hard seeds but all treatments had some deficiency. Varying percentages of hard seeds present at harvest become permeable during storage with most becoming permeable within a few years (76). Subjecting impermeable alfalfa seeds to dry heat between 50 and 80°C for up to 8 h reduced the percentage of impermeable seeds (77). Further studies by Lute (77) showed that not all hard seeds became permeable during 13 yr of storage and that not all hard seeds were viable after storage. Also, there was no great difference in death rate of permeable and impermeable seeds.

Dry freezing (89, 115) as well as high, dry temperatures, 1 to 4 min at 104.4°C (220°F), will reduce hard-seededness (108, 109). Alfalfa seeds exposed to 100°C or above for 1 h showed a significant drop in germination (85). Sliding alfalfa seeds down a metal incline heated to 399 or 510°C (750 or 950°F) effectively increased germination and decreased hard-seed content of alfalfa seed samples (74). Davies (32) applied 202 MPa (2000 atm) of pressure for 1 min and decreased percent hard seeds without appreciable harm to the germination capacity of the sample.

Other studies have shown that subjecting alfalfa seeds to very low temperatures, down to −190°C, reduced the percentage of impermeable seeds without reducing the total viability of a seed lot (25). Several workers (27, 47, 110, 111, 112, 117, 118) have shown that subfreezing storage from −15 to −196°C had no deleterious effect on germination of alfalfa seeds. However, none of the studies reported the percentage germination and percentage hard seeds separately. Apparently freezing and thawing renders seedcoats permeable so that the percentage of seeds that absorb water and sprout is increased without affecting viability.

A study by Eglitis and Johnson (33) on the effect of high-frequency electrical energy was the forerunner of a series of experiments by Nelson and co-workers (93, 94, 97, 98, 99, 100). Radio frequency dielectric heating, gas-plasma radiation, and infrared irradiation were examined as potential methods of reducing hard seeds in alfalfa.

Hard seededness was reduced by radio frequency electrical treatment (10, 95, 96, 100). However, great care had to be taken to maintain seeds at the proper moisture content and to avoid too high a temperature. Best results were obtained when the temperature did not exceed 80°C (10). Storage for up to 4 yr after radio frequency treatment had no effect on seed quality or hard-seed percentage (100). Ballard et al. (10) demonstrated that with radio frequency treatment, the softening of hard alfalfa seeds occurred at the lens (strophiole) and not at random sites. According to Nelson et al. (96), the effectiveness of radio frequency treatment in reducing the hard-seed percentage of alfalfa seed lots increased as seed moisture content decreased within the range of 9.8 to 2.9% moisture content. The quality of radio frequency treated seeds, stored at 4°C and 50% RH remained as good as that of untreated seeds for up to 15 yr.

Alfalfa seeds treated with infrared radiation showed increased germination and decreased hard seededness (109, 129). A direct current that produced a constant filament temperature gave the best results as the intensity of infrared radiation increased within certain limits, beyond which the seeds were killed (129). Treatment at 104.4°C (220°F) for 1.5 min gave best results (109). Treated seeds retained full initial viability and hard-seed percentage declined after 233 d of storage in a warm room (109).

31–1.3 Factors Affecting Respiration of Seeds

Respiration occurs in all living cells. It is an oxidative-reduction process that produces compounds and releases energy partially utilized in various life processes. Respiration combines substances within seeds with O_2 in the presence of enzymes. Over time, respiration uses all food reserves stored in seeds.

The seedcoat has an indirect effect on respiration. Intact seedcoats often restrict absorption of either water, or O_2, or both. Intact seedcoats also provide a poor substrate for the growth of fungi and bacteria. Conversely, ruptured seedcoats provide for the ready entry of water and O_2, which promote mold growth.

At moisture contents in equilibrium, with 75% RH or higher, respiration of seeds and associated fungi proceeds rapidly. Respiration rate increases as temperature increases until limited by inactivation of enzymes by high temperatures, lack of substrate or O_2, or accumulation of CO_2 to inhibitory levels.

Microorganisms, such as fungi, bacteria, and viruses, present on or in seeds at harvest are not removed by seed conditioning. Storage fungi can destroy stored seeds, because these fungi grow at very low moisture levels. In pockets of moist seeds, a rapid growth of fungi can produce localized heating and death of seeds. Such high moisture spots may result from roof leaks, insect activity, or moisture translocation when temperature gradients occur within the seed mass. Deterioration of stored seeds by fungal action is controlled principally be drying the seeds to a safe moisture content and storing in a dry place. Insects are controlled by fumigation (60), low temperatures, and reduced O_2 levels.

31 2 SEED STORAGE

Several factors determine the longevity of seeds stored in either natural or controlled environments. These factors are moisture, temperature, gaseous exchange, seedcoat characteristics, maturity of seed, microflora, and insect infestation. The effects of these factors on seeds are discussed by Barton (11), Brett (25), Crocker and Barton (31), Evans (38), Justice and Bass (62), and Owen (104). Additional papers on seed storage and deterioration are listed by James (60, 61).

31-2.1 Factors That Affect Storage Life

31-2.1.1 Preharvest Factors

Farmers and seedsmen know the perils of excessive moisture and freezing temperatures during the later stages of seed maturation. Although numerous studies have been reported on freezing injury to seeds, none were found that referred to alfalfa seeds. Extremely high temperatures and/or drought conditions during seed maturation can also have serious effects on seed quality and storability, but available literature does not address the effects of such conditions on alfalfa seeds.

Rincker (111, 112) found that alfalfa seeds produced in California showed a significantly greater decline in germination during 20 yr of cold storage than did seeds produced in Washington. Some, but probably not all, of the difference was related to area of production, because the California seeds had more mechanical damage than the Washington seeds. In other species, research has shown that seeds of the same crop produced in different geographical areas deteriorate at different rates under identical storage conditions (111).

In general, *maturity* is defined as the stage at which maximum dry weight is attained. It is important to know the stage of maturity at which seeds should be harvested for best longevity in storage. Unfortunately, most studies on seed maturity have not included storage life. Few studies relating maturity levels and germination, vigor, field emergence, or yield to storage and lifespan, have included alfalfa. Although Erickson (37) reported that both germination prcentage and seedling vigor were higher for large than for small alfalfa seeds, he did not relate either to seed longevity.

When samples from a seed lot were screened to produce five groups having different 1000-seed weight, the lighter weight seeds had lower germination and shorter longevity than did the heavier seeds (42). It is assumed, as a rule, that small seeds in a seed lot are less mature than larger, heavier seeds. If smaller seeds are more immature than the larger seeds, maturity at harvest could possibly have a significant effect on the longevity of a seed lot.

31-2.1.2 Harvest and Conditioning Methods

Damage to seeds has increased with more mechanization of harvesting and conditioning. Brett (25) reported that scarified seeds lost their viability at a significantly greater rate than did unscarified seeds when stored under the same conditions.

All mechanical harvesting, handling, and cleaning equipment damage seeds to some extent. The actual amount of damage depends upon how well or poorly the equipment was adjusted, and whether or not the correct equipment was employed in all operations. As mentioned previously, hand-harvested seeds usually have a higher percentage of hard seeds than mechanically harvested seeds because hand-harvesting methods do not rupture the seedcoat.

31-2.1.3 Postharvest Factors

Factors that affect the percentage of hard seeds in a seed lot were discussed earlier. How hard seed content affects the percentage germination of a seed lot over time could be of considerable importance in germplasm preservation as well as to seedsmen who have to carry seeds over from one year to the next. The data (Table 31-1) for change in hard-seed percentage over time in germplasm samples of a few alfalfa cultivars stored in the NSSL show no uniformity in the rate of decrease during storage.

Mechanical damage to alfalfa seeds usually results from improper adjustment of harvesting or conditioning equipment, as discussed under harvesting. Factors that contribute to the degree of damage are resistance to removal from the pod, seed size and shape, and seed-moisture content. Mechanical damage can seriously reduce the plant-producing potential of a seed lot (60). Battle (19) found that all alfalfa seeds that had been scarified with sandpaper and stored were dead after 14 yr, while unscarified seeds stored under the same conditions germinated 23%. Graber (44), Stevens (119), and Brett (25) reported similar results.

Drying a seed can be accomplished only by evaporation of moisture from its surface, accompanied by the movement of internal moisture to the surface. Moisture evaporation requires the transfer of heat, which may be accomplished by radiation, contact, or convection. Seed must be dried carefully to avoid damage from either too rapid drying or drying at too high a temperature. Seeds will not dry but rather will absorb water if the vapor pressure of the air is greater than that of the seed surface. Methods of drying are summarized by Justice and Bass (63).

Seed-moisture content is a critical factor in seed deterioration (11, 49, 50, 62). The moisture content of seeds at time of harvest is determined principally by existing weather conditions. After harvest, seed-moisture content is dependent on the humidity of the surrounding atmosphere and the type of container. How seed-moisture content is influenced during storage and the effects of seed-moisture content on storage life are discussed extensively by Justice and Bass (62), and briefly in the next section.

31-2.2 Storage Environment

The longevity of stored seeds depends upon their storage environment, temperature, RH (moisture content), surrounding atmosphere, and storage container. The most important factors are seed-moisture content and storage temperature.

31-2.2.1 Temperature

Within limits, storage life of seeds decreases as storage temperature increases. The detrimental effect of increased temperature on higher moisture-content seeds is more rapid in sealed storage than on similar moisture-content seeds in open storage. Subfreezing temperatures are usually superior to warmer temperatures for seed storage. Under all conditions, however, a suitable seed-moisture content is essential for proper storage (62).

Prior to the work of Headdean (52), most authors reported that alfalfa seeds remained viable for only a few years. Headdean found that sound alfalfa seeds will retain their viability for 23 yr. This was confirmed by Nutile (102), who tested a sample of 24-yr-old alfalfa seeds and found a germination of 78%. These samples were stored under dry conditions.

Although alfalfa seeds may be long-lived under natural conditions, the maintenance of highest viability and vigor for several years requires proper storage. This is true for commercial lots that are stored for 1 to 2 yr, or genetic stocks that must be maintained for longer periods. Maximum longevity of alfalfa is obtained by adhering to the following:

1. Using mature seeds with high initial viability and hard-seed count, handled to minimize mechanical injury (24).
2. Storing in an environment with <10% moisture, preferably about

5%, with a temperature near or below 0°C. Low moisture is more important than low temperature (104).

3. Replacing ambient air in sealed containers with CO_2 or N_2 (38).

Hanson and Moore (47) found that the germination of alfalfa seeds could be maintained at the same level for at least 4 yr by storing seeds between -7 to -11°C. Portions of the same seed lots stored in the laboratory dropped about 10% in germination in the 4-yr period.

31-2.2.2 Relative Humidity

Relative humidity (RH) of the surrounding atmosphere is of no consequence when seeds are dried to an appropriate moisture content and stored in sealed moistureproof containers. However, RH is important when seeds are stored in open or porous containers. Seed-moisture content during storage in open or porous containers is determined largely by the RH of the storage area and varies with the kind of seed. A change in RH will change the moisture content of the seed. Seed-moisture content will increase if RH is increased and decrease if RH is decreased (62). Moisture content of seeds in the center of large bins or containers does not change as rapidly as that of seeds at and near the surface. Therefore, seeds stored in bulk should be either stirred or aerated in order to achieve rapid moisture equalization.

Under all storage conditions, the moisture content of seeds will eventually equilibrate with the surrounding RH. The amount of time required for equilibration depends upon the initial seed moisture content, percentage RH, and seed-moisture content at equilibrium. The greater the percentage change in moisture content, the longer the time required for equilibration of the seed.

31-2.2.3 Interaction of Temperature and Relative Humidity

At any given temperature, air will hold a specific amount of water. Relative humidity is a measure of the percentage of the total water-holding capacity utilized at a given temperature. A change in temperature will bring about a change in percent RH when the total amount of available water remains the same. When air is warmed, percent RH is decreased and when air is cooled, percent RH is increased. Therefore, a change in temperature will have an affect on seed-moisture content. However, the temperature effect may not be as great as the effect of a change in RH when temperature remains the same (62).

31-2.2.4 Sealed Storage

Sealed storage has certain advantages and disadvantages. Sealing seeds in moistureproof containers eliminates the need for humidity control and holds seed-moisture content at a uniform level. However, before seeds can be stored in sealed containers, they must be dried to a moisture content (usually 4–7%) that is safe for the highest temperature to which the seeds might be exposed. Care must be exercised in drying to avoid temperatures in excess of 30 to 35°C and too low (usually <3%) a seed-moisture content. Seeds can be stored in sealed glass, metal, or flexible moistureproof containers. Seeds can be sealed in atmospheres other than air; however, research at the NSSL (13, 15, 16, 17, 18, 62) has shown that seed-moisture content at the time of sealing is more important than composition of the surrounding atmosphere. In general, there appears to be little or no advantage to sealing seeds in atmospheres other than air, except perhaps for long storage.

31-2.2.5 Effects of Pests and Chemicals

For alfalfa seed storage, fungi usually do not create a problem, provided the seeds are kept dry. Insects may damage stored seeds, but they can be controlled by fumigation with appropriate chemicals. If seed-moisture content is too high, respiration of seeds, fungi, and/or insects can result in localized temperature buildup to the point where viability is damaged. Chemical treatments to control fungi or insects may or may not be injurious to seeds depending upon seed-moisture content at time of treatment, and whether or not they are applied correctly. Usually, adequately dried seeds are not damaged (62).

Rodents are an important factor in loss of seeds during storage. They can be controlled with either rodent-proof storage facilities or by using traps, poison baits, or fumigation.

31-2.2.6 Packaging

A wide range of packaging materials and filling and closing equipment is available. The packaging materials, methods, and equipment used are those that are most reliable and suitable for an individual user. Decisions on packages may depend largely on convenience and other advantages in storage and marketing. Packages used for commercial storage and marketing usually are not designed to or intended to protect seed viability during long periods of storage (14).

31-3 DETERIORATION

Seed deterioration results from various factors acting alone or in various combinations upon individual seeds. How seeds respond to these factors is a function of their genetic composition, biochemical characteristics, and environment. Inherited traits determine the chemical composition of seeds, which in turn determine the effects of environment on seed longevity.

Much has been written on how factors, such as increased enzyme activity, increased respiration, depleted food reserves, accumulated toxic wastes, increased fat acidity, increased membrane permeability, and similar factors are associated with seed deterioration (1, 31, 62, 114, 131). Most studies on biochemical and physiological changes in seeds have used percent germination as a measure of deterioration. Seedling vigor, the capacity to produce normal seedlings, can be used as a measure of deterioration in seed lots.

Vigor of a seed lot is difficult to measure. There is no single universally accepted method for measuring the vigor of seed lots. A decline in percentage germination of a seed lot is generally accepted as an indication of deterioration. As a rule, the farther in time seeds are removed from harvest or maturation, the greater the expected amount of deterioration. However, old seeds can be vigorous and new seeds can be deteriorated, depending upon how the seeds were handled during harvesting, conditioning, and storage.

Under ideal storage conditions, low temperature and low RH, the effects of mechanical damage are minimized. Under adverse storage conditions, high temperature and high RH, the effects of mechanical damage are maximized. Regardless of the physical condition of seeds, deterioration progresses most rapidly under high temperature and high RH storage conditions. Such conditions promote an increased rate of respiration, which produces more heat within the seed mass. If heat is not dissipated rapidly, it results in hot spots and increases deterioration through heat damage and fungal activity. Any increase in the rate of growth and respiration of fungi will increase the rate of deterioration of seeds. Even under good storage conditions, deterioration will occur with time in storage, although at a much reduced rate when contrasted with unfavorable conditions.

Alfalfa seeds can be quite long-lived under ambient storage conditions in the less humid parts of the USA. For example, alfalfa seeds grown in Idaho in 1934 and stored in a seed company office in a glass display bottle, sealed with a cork, germinated 78% when tested for viability in 1958, 24 yr after storage (102). Alfalfa seeds stored in unheated and uninsulated buildings at Belle Fourche, SD, and Mandan, ND, for 23 to 70 yr showed varying germination percentages. One lot of yellow-flowered alfalfa [*M. sativa* L. ssp. *falcata* (L.) Arcangeli] germinated 27% after 68 yr, and another lot germinated 48% after 66 yr of storage. One lot of *M. sativa* ssp *sativa* germinated 7% and another 30% after 70 and 62 yr of storage, respectively.

There is some evidence to suggest genetic control of the rate of seed deterioration (127). However, none of the proposed theories provide a satisfactory explanation of the causes of seed deterioration. On the other hand, methods for preventing or retarding deterioration are well established. Additional information is available (1, 56, 62, 114, 129).

31-3.1 Seed Testing

The Federal Seed Act (124) and rules promulgated thereunder, plus rules adopted by the Association of Official Seed Analysts (AOSA) (6), and the International Seed Testing Association (ISTA) (58), outline procedures for testing and labeling alfalfa seeds for purity, noxious-weed seed content, and germination.

31-3.2 Purity Analysis

A purity analysis, made on a minimum of 5 g of seeds, consists of separating inert material, other crop seeds, and weed seeds from the alfalfa. By AOSA definition the following alfalfa-like seeds are considered inert matter: (i) chalcid-damaged alfalfa seeds that are puffy, soft, or dry and crumbly; (ii) alfalfa with the seedcoat entirely missing; and (iii) broken alfalfa seeds, one-half or less their original size.

Seed characteristics are of little value in the identification of alfalfa cultivars (102), but they are quite reliable in separating alfalfa seeds from seeds of other *Medicago* spp. Intrageneric seed keys for six species of *Medicago* commonly found in North American commercial seed samples are available, with or without seeds of the related genus *Melilotus* (92, 124). Matthews (84) gives directions for identifying seeds of *Med. hispida* Gaertner (now *M. polymorpha* L. var. *polymorpha*) and *M. sativa*, based on microscopic examination of the seedcoat palisade layer.

Alfalfa seedlings with the first unifoliate leaf fully expanded are separated easily from other common legume crop seedlings in the other genera in tribe Trifolieae: *Lotus, Medicago, Melilotus,* and *Trifolium*. Eifrig (34) discussed and illustrated seedling differences for selected species in these genera.

Brink (26) observed that roots of alfalfa seedlings fluoresced brilliantly, whereas those of sweetclover (*Melilotus*) did not. He suggested the use of this procedure to facilitate the difficult separation of sweetclover and alfalfa seed. Eifrig (35, 36), working with alfalfa and clover (*Trifolium*) seeds, found that the fluorescing substance in alfalfa seed was exuded during imbibition before the seedcoat ruptured.

The Federal Seed Act (124) requires that a certain percentage of the seeds in each container of alfalfa seeds be stained to denote origin when imported into the USA, viz., 10% of seeds orange-red from South America, 1% violet from Canada, and 10% red for country unknown, a mixture of Canadian and South American seeds, or foreign seeds commingled with domestic seeds. Other factors that may affect the color and seedcoat surface of alfalfa seeds are treatment with inoculum (preinoculation) and fungicides. Iron filings or sawdust, the residue of some seed-cleaning processes, may be found on alfalfa seeds, especially on cracked seeds. All of these materials may mask the true color of the seedcoat.

Weed and crop seed impurities that occur in alfalfa seed samples have been enumerated by Hay (51), Hillman and Henry (57), and Woodbridge (128).

31-3.3 Noxious-Weed Seed Examination

In this examination, noxious-weed seeds are removed and counted from a 50-g sample of alfalfa seed. Usually the counts are reported on an ounce or pound basis. Noxious-weed seeds are defined by state (121) and federal (124) laws.

31–3.4 Germination Testing

The AOSA rules (6) specify that alfalfa seeds be "planted" between blotters, in rolled or folded paper towels, or in sand or soil at 20°C, with the first count in 4 d and the final count in 7 d. The ISTA rules (58) also recognize prechilling for fresh and dormant seeds, planting on top of blotters, and using 18 to 20°C. Any after-ripening period for alfalfa seeds (other than overcoming hard-seededness) is short-lived. According to Aughtry (7), freshly harvested alfalfa seeds that germinate poorly should be pretreated at 4°C for 36 h, then regerminated. Larsen (67), using alfalfa seeds to evaluate his thermogradient plate, found the maximum germination for high, medium, and low viability seeds was obtained at 24, 25, and 26°C, respectively. His results are contrary to data gathered in support of the AOSA and ISTA rules.

A normal seedling is one that is considered capable of becoming an independent plant. The classification of normal and abnormal alfalfa seedlings is described and discussed on pages 159 and 160, and illustrated in Fig. 53 of Musil (123). Andersen (4) concluded that abnormal alfalfa seedlings (often the product of mechanically damaged seeds) are of little economic value.

Seedlings of mechanically damaged alfalfa seeds frequently exhibit breaks at the point where the cotyledons are attached to the hypocotyl, with accompanying injury to the epicotyl. Cobb and Jones (29) developed a gross anatomical morphology (GAM) test for recognizing mechanically damaged seeds. They studied the seedcoats of dry seeds with a 20-power stereoscopic microscope. A 400-seed test on newly harvested, non-rain-damaged alfalfa seeds required <30 min. The GAM test is routinely used on alfalfa seed lots certified in California (9). Dyes may be used to emphasize seedcoat fractures. Neither indoxyl acetate (41) nor malachite green depresses germination when used for this purpose.

Twin seedlings, either separate or cojoined, should be considered normal seedlings. Twins, rarely triplets, occur more frequently in alfalfa than in many other small-seeded legume species because of the occurrence of irregular numbers of tetrads and embryo sacs in alfalfa ovules. Greenshields (45) summarized polyembryony in alfalfa seeds.

31–4 SEED AND SEEDLING PERFORMANCE

31–4.1 High-Quality Seed

High-quality seeds are essential in obtaining successful stands under all field conditions. Factors that contribute to high-quality seeds are an ideal environment during pollination, seed maturation, and harvesting;

conditioning to avoid mechanical damage; and maintenance of good storage conditions. After seed lots are conditioned, they may be processed to lower the hard-seed content, inoculated with *Rhizobium meliloti* Dangeard, and treated with fungicides for control of seedling diseases. Fungicides must be selected to avoid destruction of the inoculum. If seeds are preinoculated, care should be taken to ensure the presence and viability of the bacteria. Sound, untreated alfalfa seeds are plump, and usually bright yellow or olive green. The presence of dull brown seeds usually indicates lower vigor and viability (8). Seedcoat dullness and browning are correlated with either physiological or chronological aging. All seed lots should be tested and correctly labeled for purity, germination, and noxious weed content. Germination percentage is especially important if old or discolored seeds are planted.

31–4.2 Seed Size and Weight

Both seed size and weight have been the subject of several studies on seedling emergence and vigor. Beveridge and Wilsie (20) found that there was no advantage to seedling emergence in selectively sizing samples of commercial alfalfa seed. Small seeds, those passing through a 4.23 by 0.98 mm (6 by 26 mesh) sieve should not be in a well-conditioned lot because they were found to be of little field value (37). Niffenegger et al. (101) pointed out the problems inherent in testing and using lightweight seeds. To avoid these problems, alfalfa seed lots should be properly conditioned, sized, graded, treated, and tested to ensure maximum field performance.

31–4.3 Tetrazolium Evaluation

One of the newer techniques for estimating viability, vigor, and storability of seed lots is a tetrazolium evaluation. Hard seeds do not absorb the tetrazolium solution. Their seedcoats must be broken for the solution to penetrate and react with living tissues. Moore (90) stated that sound alfalfa embryos are characterized by normal turgid tissues and a uniform crimson red color. The weak, but still germinative, embryos are revealed by various kinds and degrees of minor imperfections that are not critically located. Deeper than normal red, mottled, or nonstained areas usually reflect the deteriorated areas. Nongerminative embryos are usually characterized by major imperfections or even by minor imperfections in critically located areas. Tissue may appear flaccid or diseased. Broken hypocotyls, or cotyledons that are broken from the embryonic axis, often cause seeds to be nongerminative. Illustrations and detailed discussions on the application of tetrazolium to alfalfa seeds are available (43, 130).

31–5 SEED SPROUTING

Since the early 1970s, the use of germinated alfalfa seeds (sprouts) for human consumption has increased throughout the USA. Sprouts are considered a significant source of minerals, protein, and vitamins, including thiamine, riboflavin, niacin, and ascorbic acid (28, 46, 106). From a nutritional perspective, alfalfa sprouts compare favorably to other salad vegetables (65). On a fresh-weight basis sprouts have greater quantities of vitamin C and iron than raw cabbage (*Brassica oleracea* L. Group *capitata*) or lettuce (*Lactuca sativa* L.) (46). Alfalfa sprouts provide a greater contribution of several other nutrients to the recommended dietary allowance (2) than head lettuce (55).

31-5.1 Economic Importance

In California, use of alfalfa seeds for sprouting increased from 23 000 to 635 000 kg from 1970 to 1979 (54). During the same period, farm value of sprouts sold increased from U.S. $225,000 to $8.5 million. In 1984, it was estimated that 3200 Mg of seeds were used for commercial sprout production in the USA (82). Estimated farm value of sprouts sold was $63 million. Sprout production utilizes approximately 7% of the alfalfa seeds produced and marketed in the USA.

31-5.2 Methods of Sprout Production

Many methods for growing sprouts have been reported (5, 41, 66, 72, 91, 106). The two principal commercial methods are as follows (54): (i) Seeds are placed in large tubs, which are flooded with water and drained several times each day. After a few days, sprouts are harvested and transferred to plastic bags for marketing; and (ii) seeds are germinated in drained plastic trays. Water is applied by overhead sprinklers. When the sprouts reach the appropriate size they are covered with a plastic lid and marketed in the same tray. For home sprout production alfalfa seeds are placed in a special sprouting jar or other suitable container and kept moist by watering and draining until the sprouts are ready to use. Whether produced at home or commercially, sprouts are grown in a controlled environment without soil or added nutrients.

31-5.3 Factors Affecting Sprout Production

31-5.3.1 Cultivar

Six Alfalfa cultivars, representing a broad range in fall dormancy characteristics, were sprouted under various environmental conditions (55). The semidormant cultivars, 'Ranger', 'Tempo', and 'Caliverde 65', produced lower fresh-weight yields and higher percentage protein, whereas the intermediate-dormant cultivars, 'AS-49', '167', and 'Moapa 69', produced higher yields and lower percentage protein. These results agree with those of Larson and Smith (68) and others (53, 126).

Experimental results (55) and commercial experience indicate that the range in the ratio of grams of fresh-weight of sprouts to grams of seeds is 6.5:1 to 13.5:1, with a 9:1 ratio common in commercial sprout production.

31-5.3.2 Seed-lot Characteristics

Seed lot characteristics that are important in sprouting include germination, hard-seed percentage, and seedling vigor (54, 72). For commercial sprout production, germination percentage should be 85 or higher with a low percentage of hard seeds. Hard seeds do not germinate rapidly enough to produce marketable sprouts.

Information on the effect of seed size or weight on sprout production is limited. In one study, more dormant cultivars tended to have lower 100-seed weights and produce lower fresh-weight yields of sprouts than less dormant cultivars (55). Watter and Jensen (126) also suggested that heavier alfalfa seeds produce more vigorous seedlings and, thus, higher sprout yields.

Beveridge and Wilsie (20) found a strong association between alfalfa seed size and seedling vigor, with larger seeds producing more vigorous seedlings. They found, however, that time from planting to seedling emergence could not be predicted by seed size. They concluded that no emergence time advantage could be gained by selecting one seed size over another. Because commercial sprout production depends upon rapid uniform germination and seedling development, it is evident from the work of Beveridge and Wilsie (20) that selecting seed of a particular size may or may not result in more rapid germination and greater sprout yields. To determine if seed size is important to the production of maximum fresh weight of alfalfa sprouts will require research involving comparisons of seed sizes: (i) from a given seed lot: (ii) among cultivars within a dormancy group; (iii) among all cultivars used for sprout production; (iv) within and among cultivars produced in different geographical areas; and (v) within and among cultivars produced in different years in the same and in different geographical areas.

31-5.3.3 Temperature

Temperatures used for sprout production range from 16 to 27°C. Optimum temperature for germination is influenced by seed source, genetic differences (both species and cultivar), and seed age (86). When alfalfa seeds were sprouted at three temperatures, fresh weight of sprouts was lower at 16°C than at either 21 or 27°C (55). Sprouts with the highest protein concentration were produced at 16°C because of the significant negative correlation between fresh weight and percentage protein. Temperatures of 21 and 16°C have been recommended as favorable for producing sprouts with high fresh weight and high nutritional quality, respectively. Greater seedling dry weights and faster germination rates were obtained at 21 and 27°C than at lower temperatures (88, 118, 125). Constant temperatures produce faster germination rates than do corresponding alternating temperatures.

31–5.3.4 Light

Respiration of alfalfa sprouts is increased as light intensity increases (86). Hesterman et al. (55) reported highest fresh-weight yields without light and highest percentage protein with 12 or 24 h of light per day. The decreased fresh weight was attributed to increased respiration with light. Hamilton and Vanderstoep (46) concluded that light during sprouting stimulated ascorbic acid synthesis and increased crude protein concentration in sprouts.

31–5.3.5 Length of Sprouting Period

Length of the sprouting period influences both fresh-weight yield and nutritional quality of alfalfa sprouts. Recommended sprouting time is 1 to 6 d, or when the sprouts are 1 to 5 cm long (5). Fresh weight of sprouts increased as the length of the sprouting period increased up to 8 d (55). However, many unifoliolate leaves, an undesirable market characteristic, were found on the 8-d sprouts. Maximum fresh-weight yield of marketable sprouts was achieved with a 6 d growing period.

Hamilton and Vanderstoep (46) found that moisture, ascorbic acid, and riboflavin content increased as sprouting time was extended from 3 to 5 d. With an increased sprouting period, sprouts contain a lower concentration of crude protein because of the negative correlation between fresh weight and percentage protein.

31–5.3.6 Water

Under favorable moisture conditions, alfalfa seeds imbibe adequate water for germination within 4 to 8 h (89) (see Chapter 11 in this book). Quantity of water applied or application interval may not be as critical as other environmental factors for successful alfalfa sprout production. Only small differences have been reported in fresh weight or percentage protein of sprouts grown under different water-application regimes.

31–5.4 Nutritional Composition of Sprouts

Hamilton and Vanderstoep (46) described the most important changes in nutrient composition that occur during sprout development as follows: increased moisture content; a three- to fourfold increase in total ascorbic acid; and up to a threefold increase in riboflavin. Kylen and McCready (65) found a fresh-weight protein concentration of 51 g/kg for alfalfa sprouts, which can be compared with a range of 18 to 57 g/kg reported by others (46, 55). Some researchers have concluded that alfalfa sprouts contain more protein than alfalfa seeds as a result of leaching of sugars, loss of seedcoats, and protein synthesis during sprouting (46, 65). On the other hand, Hesterman et al. (55) concluded that although percentage protein on a dry-weight basis was greater in sprouts than in ungerminated seed, total protein content decreased as seeds sprouted. Because dry weight decreases during germination (55, 72, 86), conclusions about relative nutrient content during germination must be based on comparisons between the nutrient content of a given amount of seeds with that of sprouts grown from the same amount of seeds from the same seed lot.

Saponin concentration has been evaluated for both laboratory-germinated alfalfa seeds (107) and commercially grown alfalfa sprouts (71). Saponin content of alfalfa sprouts increased rapidly during the first 4 d of germination, and subsequently at a much slower rate. Livingston et al. (71) found saponin concentrations from 15 to 73 g/kg on a dry-weight basis in three alfalfa cultivars, depending upon maturity, compared with saponin concentrations of 0.1 to 0.7 g/kg in some common vegetables. Alfalfa saponins are nontoxic to monkeys (*Macaca* spp.) (79), and alfalfa meal is effective in lowering plasma cholesterol levels and decreasing atherogenesis in monkeys and rabbits (*Oryctolagus cuniculus*) (80).

Alfalfa sprouts do not possess great nutritional value. Alfalfa seeds and sprouts contain approximately 15 g/kg of their dry weight of canavanine, a highly toxic arginine analog that is incorporated into protein in place of arginine (3). Canavanine appears to cause a lupus erythematosus-like syndrome (anemia) in monkeys fed alfalfa sprouts (81).

Patterson and Woodburn (105) found high concentrations of a pathogenic bacterium [*Klebsiella pneumoniae* (Schroeter) Trevisan] on alfalfa sprouts obtained from several retail outlets. They warn of a potential public health hazard if people with low levels of decreased resistance consume large quantities of alfalfa sprouts.

REFERENCES

1. Abdul-Baki, A.A., and J.D. Anderson. 1972. Physiological and biochemical deterioration of seeds. Seed Biol. 2:283–315.
2. Adams, C.F. 1975. Nutritive value of American foods. USDA-ARS Agric. Handb. 456, U.S. Government Printing Office, Washington DC.
3. Ames, B.N. 1983. Dietary carcinogens and anticarcinogens. Science (Washington DC) 221:1256–1264.
4. Anderson, A.M. 1957. Evaluation of normal and questionable seedlings of species of *Meliolotus, Lotus, Trifolium*, and *Medicago* by greenhouse tests. Proc. Int. Seed Test. Assoc. 22(1):237–258.
5. Anonymous. 1974. Sprouts grown in your kitchen. Sunset Magazine 152(2):64.
6. Association of Official Seed Analysts. 1981. Rules for testing seeds. J. Seed Technol. 6(2)(Revised 1984).
7. Aughtry, J.D., Jr. 1948. Effect of genetic factors in *Medicago* on symbiosis with Rhizobium. Cornell Agric. Exp. Stn. Mem. 280.
8. Ayres, J.C. 1929. Relation of plumpness and viability to color of 'Grimm' alfalfa seed. Seed World 25(2):17.
9. Ball, R.B. 1970. Personal communication, California Crop Improvement Associate, Univ. of California, Davis.
10. Ballard, L.A.T., S.O. Nelson, T. Buchward, and L.E. Stetson. 1976. Effects of radiofrequency electric fields on permeability to water of some legume seeds, with special reference to strophiolar conduction. Seed Sci. Technol. 4:257–274.
11. Barton, L.V. 1961. Seed preservation and longevity. Interscience Publishers, New York.
12. Bass, L.N. 1970. Personal communication to C.R. Gunn, Natl. Seed Lab., Ft. Collins, CO.
13. ----. 1978. Sealed storage of crimson clover seed. Seed Sci. Technol. 6:1017–1024.

14. ——, T.M. Ching, and F.L. Winter. 1961. Packages that protect seeds. U.S. Dept. Agric. Yearb. 1961-330–338.
15. ——, D.C. Clark, and E. James. 1962. Vacuum and inert gas storage of lettuce seed. Proc. Assoc. Off. Seed. Anal. 52:116–122.
16. ——, ——, and ——. 1963. Vacuum and inert-gas storage of crimson clover and sorghum seeds. Crop Sci. 3:425–428.
17. ——, ——, and ——. 1963. Vacuum and inert-gas storage of safflower and sesame seeds. Crop Sci. 3:237–240.
18. ——, and P.C. Stanwood. 1978. Long-term preservation of sorghum seed as affected by seed moisture, temperature, and atmospheric environment. Crop Sci. 18:575–577.
19. Battle, W.R. 1948. Effects of scarification on longevity of alfalfa seed. J. Am. Soc. Agron. 40:758–759.
20. Beveridge, J.L., and C.P. Wilsie. 1959. Influence of depth of planting, seed size, and variety on emergence and seedling vigor in alfalfa. Agron. J. 51:731–734.
21. Bewley, J.D., and M. Black. 1978. Physiology and biochemistry of seeds in relation to germination, Vol. 1. Springer-Verlag, New York.
22. ——, and ——. 1982. Physiology and biochemistry of seeds in relation to germination, Vol. 2. Springer-Verlag, New York.
23. Bocquet, G., and J.D. Bersier. 1960. The systemic value of the ovule: Teratological developments. Arch. Sci. 13:475–496.
24. Bolton, J.L. 1962. Alfalfa: Botany, cultivation, and utilization. Interscience Publishers, New York.
25. Brett, C.C. 1952. Factors affecting the viability of grass and legume seed in storage and during shipment. Proc. Int. Grassl. Congr., 6th. 1:878–884.
26. Brink, V.C. 1958. Note on a fluorescence test to distinguish seeds of alfalfa and sweet clover in mixtures. Can. J. Plant Sci. 38:120–121.
27. Busse, W.F. 1930. Effect of low temperatures on germination of impermeable seeds. Bot. Gaz (Chicago) 89:169–179.
28. Chen, L.H., C.E. Wells, and J.R. Fordham. 1975. Germinated seeds for human consumption. J. Food Sci. 40:1290–1293.
29. Cobb, R.D., and L.G. Jones. 1960. Germination of alfalfa (*Medicago sativa* L.) as related to mechanical damage of seed. Proc. Assoc. Off. Seed. Anal. 50:101–108.
30. Cooper, D.C. 1935. Macrosporogenesis and embryology of *Medicago*. J. Agric. Res. (Washington, DC) 51:471–477.
31. Crocker, W., and L.V. Barton. 1953. Physiology of seeds. Chronica Botanica Co., Waltham, MA.
32. Davies, P.A. 1928. The effect of high pressure on the percentages of soft and hard seeds of *Medicago sativa* and *Melilotus alba*. Am. J. Bot. 15:433–436.
33. Eglitis, M., and F. Johnson. 1957. Control of hard seed of alfalfa with high frequency energy. Phytopathology 47:9.
34. Eifrig, H. 1960. Zur artendiagnose bei *Trifolium, Medicago, Melilotus* and *Lotus*. Proc. Int. Seed Test. Assoc. 25:321–327.
35. ——. 1960. Zur methodik der untersheidung von *Trifolium* und *Medicago*-arten auf grund von fluoreszenzersheinungen vorlaufige mitteilung. Saatgut-Wirtsch. 7:194–195.
36. ——. 1967. Moglichkeiten einer identifikation von planzensamen auf grund von fluoreszenzersheinungen. Umschau 19:635.
37. Erickson, L.C. 1946. The effect of alfalfa seed size and depth of seedling upon the subsequent procurement of stand. Agron. J. 38:964–973.
38. Evans, G. 1952. The preservation and maintenance of basic stocks of herbage plants. Proc. Int. Grassl. Congr., 6th. 1:906–911.
39. Farley, H.M., and A.H. Hutchinson. 1941. Seed development in *Medicago* (alfalfa) hybrids. Can. J. Bot. Sect. C. 19:421–437.
40. Fordham, J.R., C.E. Wells, and L.H. Chen. 1975. Sprouting of seeds and nutrient composition of seeds and sprouts. J. Food Sci. 40:552–556.
41. French, R.C., J.A. Thompson, and C.H. Kingsolver. 1962. Indoxyl acetate as an indicator of cracked seed coats of white beans and other light colored legume seeds. Proc. Am. Soc. Hortic. Sci. 80:377–386.
42. Gaspar, S., A. Bus, and J. Banyai. 1981. Relationship between 1000-seed weight and germination capacity and seed longevity in small seeded fabaceae. Seed Sci. Technol. 9:457–467.
43. Grabe, D.F. (ed.). 1970. Tetrazolium testing handbook for agricultural seeds. Assoc. Off. Seed Anal., Handbk. Seed Test. Contrib. 29. Association of Official Seed Analysts, Boise, ID.
44. Graber, L.F. 1922. Scarification as it affects longevity of alfalfa seed. Agron. J. 14:298–302.
45. Greenshields, J.E.R. 1951. Polyembryony in alfalfa. Sci. Agric. (Ottawa) 31:212–222.
46. Hamilton, M.J., and J. Vanderstoep. 1979. Germination and nutrient compostion of alfalfa seeds. J. Food Sci. 44:443–445.

47. Hanson, C.H., and R.P. Moore. 1959. Viability of seeds of eight forage crop plants stored under subfreezing conditions. Agron. J. 51:627-628.
48. Harrington, G.T. 1916. Agricultural value of impermeable seed. J. Agric. Res. (Washington, DC) 6:761-796.
49. Harrington, J.F. 1960. Thumb rules of drying seed. Crops Soils 13(1):16-17.
50. ----. 1972. Seed storage and longevity. Seed Biol. 3:145-256.
51. Hay, W.D. 1930. Impurities commonly found in Montana grown alfalfa seed. Proc. Assoc. Off. Seed Anal. 22:39-42.
52. Headdean, W.P. 1919. The vitality of alfalfa seed as affected by age. Colo. Sci. Soc. Proc. 11:239-249.
53. Heinrichs, D.H. 1967. Rate of germination of alfalfa at four temperatures and relationship to winterhardiness. Can. J. Plant Sci. 47:301-304.
54. Hesterman, O.B., and L.R. Teuber. 1979. Alfalfa sprouts: Methods of production, current research, and economic importance. p. 24-27. *In* Proc. 9th Calif. Alfalfa Symp., Fresno, CA. 12-13 December. Cooperative Extension, University of California, Davis.
55. ----, ----, and A.L. Livingston. 1981. Effect of environment and genotype on alfalfa sprout production. Crop Sci. 21:720-726.
56. Heydecker, W. 1973. Seed ecology. Pennsylvania State University Press, University Park.
57. Hillman, F.H., and H.H. Henry. 1928. The incidental seeds found in commercial seed of alfalfa and red clover. Proc. Int. Seed Testing Assoc. 6:1-22.
58. International Seed Testing Association. 1985. International rules for seed testing 1985. Seed Sci. Technol. 13(2):307-520.
59. Isely, D. 1947. Seed characters of alfalfa and certain other species of *Medicago*. Iowa State coll. J. Sci. 21(2):153-159.
60. James, E. 1961. An annotated bibliography on seed storage and deterioration. A review of 20th century literature reported in the English language. USDA-ARS 34-15-1.
61. ----. 1963. An annotated bibliography on seed storage and deterioration. A review of 20th century literature reported in a foreign language. USDA-ARS 34-15-2.
62. Justice, O.L., and L.N. Bass. 1978. Principles and practices of seed storage. USDA Agric. Handb. no. 506. U.S. Government Printing Office, Washington, DC.
63. Khan, A.A. (ed.). 1982. The physiology and biochemistry of seed development, dormancy and germination. Elsevier Biomedical Press, Amsterdam.
64. Kopooshian, H.A. 1963. Seed character relationships in the Leguminosae. Unpublished Ph.D. diss. Iowa State Univ., Ames.
65. Kylen, A.M., and R.M. McCready. 1975. Nutrients in seeds and sprouts of alfalfa, lentils, mung beans, and soybeans. J. Food Sci. 40:1008-1009.
66. Larimore, B. 1975. Sprouting for all seasons. Horizon Publ., Salt Lake City, UT.
67. Larsen, A.L. 1965. Use of thermogradient plate for studying temperature effects on seed germination. Proc. Int. Seed Test. Assoc. 30(4):861-868.
68. Larson, K.L., and D. Smith. 1963. Association of various morphological characters and seed germination with the winter-hardiness of alfalfa. Crop Sci. 3:234-236.
69. Leggatt, C.W. 1927. The agricultural value of hard seeds of alfalfa and sweet clover under Alberta conditions. Sci. Agric. (Ottawa) 8:243-266.
70. ----. 1931. Further studies on the hard seed problem: Alfalfa and sweet clover. Sci. Agric. (Ottawa) 11:418-427.
71. Livingston, A.L., B.E. Knuckles, L.R. Teuber, O.B. Hesterman, and L.S. Tsai. 1982. Minimizing the saponin content of alfalfa sprouts and leaf protein concentrates. Adv. Exp. Med. Biol. 177:253-268.
72. Lorenz, K. 1980. Cereal sprouts: Composition, nutritive value, food applications. CRC Crit. Rev. Food Sci. Nutr. 11:353-385.
73. Luis de la Loma, J., G. Arguello, and F. Zertuche. 1955. Influencia de la escarificacion de las semillas de trebol y alfalfa en la velocidad de germinacion. Chapingo 8:102-106.
74. Lunden, A.O., and R.C. Kinch. 1957. The effect of high temperature contact treatment on hard seeds of alfalfa. Agron. J. 49:151-153.
75. Lute, A.M. 1927. Alfalfa seeds made permeable by heat. Science (Washington, DC) 65:166.
76. ----. 1928. Impermeable seed of alfalfa. Colo. Agric. Exp. Stn. Bull. 326.
77. ----. 1942. Laboratory germination of hard alfalfa seed as a result of clipping. J. Am. Soc. Agron. 34:90-99.
78. Maguire, J.D. 1970. Viability testing of hard seeds of alfalfa. News Lett. Assoc. Off. Seed Anal. 44(1):16-17.
79. ----, P. McLaughlin, H.K. Naito, L.A. Lewis, and W.P. McNulty. 1978. Effect of alfalfa meal on shrinkage (regression) of atherosclerotic plaques during cholesterol feeding in monkeys. Atherosclerosis 30:27-43.
80. ----, ----, C. Stafford, A.L. Livingston, and G.O. Kohler. 1980. Alfalfa saponins and alfalfa seeds. Atherosclerosis 37:433-438.

81. Malinow, M.R., E.J. Bardana, Jr., B. Pirofsky, S. Craig, and P. McLaughlin. 1982. Systemic lupus erythematosus-like syndrome in monkeys fed alfalfa sprouts: Role of a nonprotein amino acid. Science (Washington, DC) 216:415–417.
82. Marble, V.L. 1985. The past, present, and future of the alfalfa seed industry. p. 6–28. *In* Proc. 16th Annual Interstate Alfalfa Seed Growers Winter Seed School, San Diego, CA. 28–30 January.
83. Martin, A.C. 1946. The comparative internal morphology of seeds. Am. Midl. Nat. 36:513–660.
84. Matthews, D. 1963. Identification of the seeds of *Medicago sativa* and *Medicago hispida*. Proc. Int. Seed Test. Assoc. 28(2):202–206.
85. Maun, M.A. 1977. Response of seeds to dry heat. Can. J. Plant Sci. 57(1):305–307.
86. Mayer, A.M., and A. Poljakoff-Mayber. 1975. The germination of seed. Pergamon Press, UK.
87. McElgunn, J.D. 1973. Germination response of forage legumes to constant and alternating temperatures. Can. J. Plant Sci. 53:797–800.
88. McWilliams, J.R., R.J. Clements, and P.M. Dowbing. 1970. Some factors influencing the germination and early seedling development of pasture plants. Aust. J. Agric. Res. 21:19–32.
89. Midley, A.R. 1926. Effect of alternate freezing and thawing on the impermeability of alfalfa and dodder seeds. J. Am. Soc. Agron. 18:1087–1098.
90. Moore, R.P. 1970. Personal communication, Crop Science Dep., North Carolina State Univ., Raleigh.
91. Munroe, E. 1977. Sprouts to grow and eat. Stephen Green Press, Brattleboro, VT.
92. Musil, A.F.. 1963. Identification of crop and weed seeds. USDA Agric. Handb. 219. U.S. Government Printing Office, Washington, DC.
93. Nelson, S.O. 1965. Electromagnetic radiation effects on seeds. p. 60–63. *In* Conf. Proc.: Electromagnetic Radiation in Agriculture, Roanoke, VA. Illuminating Engineering Society—American Society of Agricultural Engineering, St. Joseph, MI.
94. ----. 1969. A hot foot for hard seed. Crops Soils 22(1):18–19.
95. ----, W.R. Kehr, L.E. Stetson, R.B. Stone, and J.C. Webb. 1977. Alfalfa seed germination response to electrical treatments. Crop Sci. 17:863–866.
96. ----, ----, ----, and W.W. Wolf. 1977. Laboratory germination and sand emergence responses of alfalfa seed to radiofrequency electrical treatment. Crop Sci. 17:534–538.
97. ----, L.E. Stetson, R.B. Stone, J.C. Webb, C.A. Pettibone, D.W. Works, W.R. Kehr, and G.E. VanRiper. 1964. Comparison of infrared, radiofrequency, and gas-plasma treatments of alfalfa seed for hard-seed reduction. Trans. ASAE 7:276–280.
98. ----, ----, and D.W. Works. 1968. Hard seed reduction in alfalfa by infrared and radiofrequency electrical treatments. Trans. ASAE 11(5):728–730.
99. ----, and E.R. Walker. 1961. Effects of radio frequency electrical seed treatment. Agric. Eng. 42:688–691.
100. ----, and W.W. Wolf. 1964. Reducing hard seed in alfalfa by radio-frequency electrical seed treatment. Trans ASAE 7:116–119, 122.
101. Niffenegger, D., R.F. Eslick, and D.J. Davis. 1959. The effect of screening and blowing on the pure line seed concept of alfalfa seed quality. Proc. Assoc. Off. Seed Anal. 49(1):40–45.
102. Nittler, L.W., G.W. McKee, and J.L. Newcomer. Principles and methods of testing alfalfa seed for varietal purity. N.Y. Agric. Exp. Stn. Bull. 807.
103. Nutile, G.E. 1958. Germination of alfalfa and red clover seeds after 24 years of storage. News Lett. Assoc. Off. Seed Anal. 32(1):21–22.
104. Owen, E.B. 1956. The storage of seeds for maintenance of viability. Commonw. Bur. Pasture Field Crops Bull. 43.
105. Patterson, J.E., and M.J. Woodburn. 1980. Klebsiella and other bacteria on alfalfa and bean sprouts at the retail level. J. Food Sci. 45:492–495.
106. Patwardhan, V.N. 1962. Pulses and beans in human nutrition. Am. J. Clin. Nutr. 11:12–30.
107. Pederson, M.W. 1975. Relative quantity and biological activity of saponins in germinated seeds, roots, and foliage of alfalfa. Crop Sci. 15:541–543.
108. Rincker, C.M. 1954. Effect of heat on impermeable seeds of alfalfa, sweet clover, and red clover. Agron. J. 46:247–250.
109. ----. 1957. Heat treating hard alfalfa seed. Univ. Wyo. Agric. Exp. Stn. Bull. 352.
110. ----. 1980. Effect of long-term subfreezing storage of seed on legume forage production. Crop Sci. 20:574–577.
111. ----. 1981. Long-term subfreezing storage of forage crop seeds. Crop Sci. 21:424–427.
112. ----. 1983. Germination of forage crop seeds after 20 years of subfreezing storage. Crop Sci. 23:229–231.
113. ----, Personal communication, Agronomy and Soils Dep. Washington State Univ., Prosser.

114. Roberts, E.H. 1972. Viability of seeds. Chapman and Hall, London.
115. Rodriquez, G.R. 1924. Study of the influence of heat and cold on germination of hard seeds in alfalfa and sweet clover. Proc. Assoc. Off. Seed Anal. 16:75-76.
116. Smoliak, S., A. Johnston, and M.R. Hanna. 1972. Germination and seedling growth of alfalfa, sainfoin, and cicer milkvetch. Can. J. Plant Sci. 52:757-762.
117. Stanwood, P.C. 1980. Tolerance of crop seeds to cooling and storage in liquid nitrogen (−196°C). J. Seed Tech. 5(1):26-31.
118. ----, and L.N. Bass. 1981. Seed germplasm preservation using liquid nitrogen. Seed Sci. and Technol. 9:423-437.
119. Stevens, O.A. 1935. Germination studies on aged and injured seeds. J. Agric. Res. 51:1093-1106.
120. Stewart, G. 1926. Effect of color of seed, of scarification, and of dry heat on the germination of alfalfa seed and some of its impurities. Agron. J. 18:743-760.
121. Streeter, J.W. 1965. Possible mechanisms in the loss of seed viability. News Lett. Assoc. Offic. Seed Anal. 39:27-35.
122. Ueno, M., and D. Smith. 1970. Influence of temperature on seedling growth and carbohydrate composition of three alfalfa cultivars. Agron. J. 62:764-767.
123. U.S. Department of Agriculture. 1952. Testing agricultural and vegetable seeds. USDA Agric. Handb. 30. U.S. Government Printing Office, Washington, DC.
124. ----. 1968. Rules and regulations of the secretary of agriculture under the Federal Seed Act of August 9, 1939 (53 Stat. 1275). USDA Consumer and Marketing Service, Washington, DC.
125. ----. 1986. State noxious-weed seed requirements recognized in the administration of the Federal Seed Act. USDA Agricultural Marketing Service, Seed Branch, Washington, DC.
126. Watter, L.E., and E.H. Jensen. 1970. Effect of environment during seed production on seedling vigor of two alfalfa varieties. Crop Sci. 10:635-638.
127. Wilton, A.C., C.E. Townsend, R.J. Lorenz, and G.A. Rogler. 1978. Longevity of alfalfa seed. Crop Sci. 18:1091-1093.
128. Woodbridge, M.E. 1939. Long list of weed seeds found in alfalfa and clover seed. Farm Res. 5:12.
129. Works, D.W. 1964. Infrared irradiation for water-impermeable seeds. Trans. ASAE 7(3):235-237.
130. Wyttenbach, E. 1955. Der einfluss vershiedener lagerungsfaktoren auf die haltbarkeit von feldsämereien (luzerne, rotklee und gemeinem schotenklee) bei langer dauernder aufbewahrung. Landwirtsch. Jahrb. Schweiz. 4:161-196.
131. Zeleny, L. 1954. Chemical, physical, and nutritive changes during storage. p. 46-76. *In* J.A. Anderson and A.W. Alcock (ed.). Storage of cereal grains and their products. American Association of Cereal Chemists, Inc., St. Paul.

32 Seed Production Practices

CLARENCE M. RINCKER

USDA-ARS
Prosser, Washington

V. L. MARBLE

University of California
Davis, California

D. E. BROWN

Land O' Lakes, Inc.
Caldwell, Idaho

CARL A. JOHANSEN

Washington State University
Pullman, Washington

Successful alfalfa (*Medicago sativa* L.) seed production is favored in regions that are characterized by clear, sunny, warm summer days in combination with little or no rainfall. These climatic conditions promote good flowering of alfalfa and provide an environment conducive to the pollinating activity of bees (Apoidea), two factors that are essential for seed production. In addition to a favorable climate, there are other variables that will influence the yield and quality of alfalfa seed. Three critical factors are as follows: (i) the control of detrimental insects; (ii) the supply of effective pollinators; and (iii) the skillful application of irrigation water when alfalfa plants are flowering (46, 97, 98, 122). Except for adverse weather, the grower can exercise a high degree of control over most of the variables that affect seed yield.

32-1 AREAS OF SEED PRODUCTION

Alfalfa seed is produced in more than 20 states in the USA; however, about 75% of the crop is produced under irrigation on approximately 70 875 ha (175 000 acres) in the five western states of California (35%), Idaho (15%), Washington (12%), Nevada (8%), and Oregon (5%). The states of Oklahoma, Kansas, South Dakota, Montana, and Utah produce about 20% of the crop from 101 250 ha (250 000 acres), mostly under dryland conditions. The USA alfalfa seed crop averaged about 45 417 t/

Copyright 1988 © ASA-CSSA-SSSA, 677 South Segoe Road, Madison, WI 53711, USA. *Alfalfa and Alfalfa Improvement*—Agronomy Monograph no. 29.

yr from 1971 to 1980 (132). In Canada, alfalfa seed is harvested in five provinces, with about 97% of the crop produced in Alberta (48%), Manitoba (38%), and Saskatchewan (11%) (119). Production in Canada averaged 1258 t/yr from 1971 to 1980.

32-2 STAND ESTABLISHMENT

32-2.1 Seedbed Selection and Preparation

Well-drained soils low in alkali and soluble salts with a rooting depth of at least 1 m (40 in.) are suitable for the production of alfalfa seed under irrigation. Deep clay, clay loams, or sandy clay loams with a high water-holding capacity are preferred over sandy soils (74). Gravelly soils or soils underlain with shallow clay pans or hardpan layers should be avoided. A uniform soil texture will help to reduce maturity differences among plants. Alfalfa seed can be grown on subirrigated soils if there is at least 1 m (40 in.) of well-aerated soil and a nonfluctuating water table. Dryland alfalfa seed can be produced successfully, usually at a lower yield per unit area, if there is at least 2 m (80 in.) of soil rooting depth, the soil has a high water-holding capacity, and the crop receives a minimum of 300 mm of precipitation (12 in.) (21, 53, 74, 88).

In California, alfalfa seed grows best if before planting, fields are subsoiled to a depth of 60 cm (24 in.), when sufficiently dry to achieve complete shattering of the subsoil. A preirrigation of 200 to 300 mm (8–12 in.) of water should be made to settle and recharge the top 60 cm of soil profile. Standard land preparation procedures for planting deep rooted row crops, including discing or plowing, leveling, and final discing and/or bed forming for planting should be followed (74).

32-2.2 Weed Control

Weed problems and weed control costs can be minimized through well-planned rotations and cultural practices (29). Fields infested with dodder (*Cuscuta* spp.) or perennial weeds such as bermudagrass [*Cynodon dactylon* (L.) Pers.] johnsongrass [*Sorghum halepense* (L.) Pers.], field bindweed (*Convolvulus arvensis* L.), Canada thistle [*Cirsium arvense* (L.) Scop], Russian knapweed (*Centaurea repens* L.), etc., should be either avoided or treated prior to planting. Proper land preparation and preirrigation will insure a flat seedbed, pregerminate many weed seeds, and enhance effectiveness of herbicides.

Selective herbicides can provide long lasting, effective weed control, and ensure the establishment of new plantings. Soil texture, soil organic matter content, and weed species will influence the choice of chemicals and rates of application. Specific recommendations for herbicides are presented in chapter 23. Selective herbicides have been classified (28, 29) as: (i) *preplant*—incorporated not more than 5 cm (2 in.) into the soil before planting; (ii) *preemergence*—applied on the soil surface after planting but prior to germination of alfalfa and weed seeds; (iii) *postemergence*—applied after the alfalfa and weeds are growing or after the alfalfa is established but prior to weed growth. Rainfall or irrigation is essential after application to activate the herbicides.

Herbicides applied to newly planted seed fields will provide only temporary or short-lived control of weeds. Summer annuals can reinfest fields after cultivation and irrigation. Hand weeding may be needed when cultivation is not practical.

32-2.3 Time of Planting

Alfalfa seed in areas with mild winters can be planted from September to November, or from late January to early March (74). Early fall planting will increase first year seed yields to levels attained in mature stands (98). Low temperatures and wet weather during very late fall and early winter frequently result in poor germination and excessive competition from winter weeds. Spring plantings are recommended in areas where winter temperatures remain below freezing for long periods. Early spring plantings will benefit from residual winter moisture and spring rains (11, 68, 71, 88, 114).

32-2.4 Row vs. Broadcast

Thin, uncrowded stands are recognized as capable of producing higher seed yields than thick stands in solid planted fields (21, 69, 74, 87, 88, 107). In Utah, row plantings produced 685 kg/ha in comparison with dense hay-type stands that yielded 323 kg/ha (88). Low seed production from dense stands can be explained in part by low nectar production, unattractiveness to pollinators, and increased floral abortion. Other advantages of row planting include the following six factors (74):

1. More open, erect plants that favor greater bee access to flowers, light penetration, and higher early season soil and air temperatures.
2. Less lodging and lower humidity in the plant canopy that reduces the incidence of foliar diseases and the amount of water-damaged seeds, in comparison with losses found with the lodging and heating of matted foliage in dense stands.
3. Less stripping of flowers and pods.
4. Better penetration of chemicals used for weed and insect control, and more effective dessication at harvest.
5. Greater flexibility in managing irrigation water and in controlling weeds.
6. More effective control of volunteer alfalfa plants to ensure genetic integrity.

Selection of row width depends on soil depth and texture, total water availability, length of growing season, saltiness of the soil, age of stand, cultivar, and possibly other factors (1, 17, 18, 26, 35, 61, 68, 69, 74, 79,

87, 88, 89, 102, 103, 129). California workers base row width recommendations on soil texture, total water availability, soil salinity, and the anticipated effect that these factors will have on plant growth (43). The development of large plants on sandy soils dictates row spacings from 120 to 150 cm (48–60 in.). Similarly, the development of average size plants on medium textured soils requires row spacings from 75 to 100 cm (30–40 in.), whereas smaller plant development on shallow, clay soils suggests the use of rows on 60 to 90 cm (24–36 in.) centers. In colder areas, such as central Utah, where plant growth is restricted by clay soils, accumulation of salt, and a frost-free period of only 118 d, a near solid stand in 22.5-cm (9-in.) rows was superior to a 90-cm (36-in.) row width. However, on more productive soils in the same state, rows on 60-cm (24-in.) centers were superior to rows spaced either 20 cm (8 in.) or 120 cm (48 in.) apart. In Idaho and Washington, where the climate is milder than Utah, the optimum row spacing varies from 55 to 90 cm (22–36 in.).

Increased seed yields from alfalfa planted in rows can be attributed to both physiological and morphological changes in the plant (69, 87). Spaced plants are shorter, lodge less, and are more robust than plants in dense stands, and they bloom up to 10 d earlier and produce more seed per plant. Furthermore, spaced plantings are characterized by higher early season soil temperatures. In these studies, however, higher seed yield production could be explained largely by the increased attractiveness of alfalfa plants to pollinators. In general, row planted stands have larger seeds and a lower percentage of damaged brown seeds due to heat, humidity, and/or water. Row spacing appears to have no affect on either hard seed or germination percentage.

32–2.5 Seeding Rates

As early as 1928, it was shown that very low rates of seeding increased seed yields per unit area (18). Utah workers demonstrated the superiority of seeding rates as low as 0.55 kg/ha over rates up to 13.2 kg/ha, at row spacings of 20, 60, and 120 cm (8, 24, and 48 in.) (87). In Mexico, seeding rates of 0.5 to 1.0 kg/ha were superior to more densely planted stands (2). In California, optimum seeding rates vary from 0.3 to 2.0 kg/ha. Precision planters, both belt and drill-type, serve to place alfalfa seed clumps of three to four seeds each, with gaps of 15 to 30 cm (6–12 in.) within the row (74, 104). These planters use from 0.25 to 0.75 kg/ha when planting on beds 75 to 100 cm (36–40 in.) apart. Most growers use 1.0 to 1.5 kg/ha, in order to establish from three to seven plants per 30 cm of row, in rows planted on 90-cm centers. In Washington, seeding rates of 0.25 to 0.50 kg/ha have produced seed yields >1000 kg/ha (98).

32–2.6 Plant Spacing within Rows and Thinning

Stand density within the row, because it has an important affect on yield should be controlled by rate of seeding and by thinning. Where precision planting is not possible for obtaining optimum stands, uniform plant populations in excess of five plants per 30 cm should be thinned when plants are in the two to four true leaf stage (74, 107).

Thinning stands in rows will produce the same physiological and morphological changes in plants that are observed in row plantings vs. solid stands. Again, thinned plants are shorter, lodge less, resist frost injury, flower earlier, and are more attractive to bees because of an increase in nectar secretion and nectar sugar concentration. The more upright growth of thinned plants gives bees greater access to flowers (26, 84, 87, 89). In addition, higher seed yields that are obtained by thinning established stands may be associated with the level of carbohydrate reserves in plants (87). Dobrenz and Massengale (25) found that plants with high root carbohydrate reserves produced more seed, more pods per stem, and more seeds per pod, than plants with reduced concentrations of carbohydrates. Similarly, Granfield (42) indicated that under comparable soil moisture conditions less vigorously growing alfalfa, with high root carbohydrate reserves, produced more racemes, flowers, pods, and seeds than plants with low reserves.

Jones and Pomeroy (61) demonstrated that first-, second-, and third-year stands all benefited from thinning within the row. The amount of thinning required to improve seed yields varied with row width and age of stands. New stands were thinned by cross-blocking out 15 or 30 cm (6 or 12 in.) of plants. Thinning can be done by hand hoeing, mechanical thinning with a sugarbeet (*Beta vulgaris* L.) thinner, or by adaptation of cultivating equipment and cross-blocking fields. Cultivars respond differently to thinning (89). Apparently, poor seed-producing cultivars respond more to thinning than high seed-producing cultivars. Spacing between clumps can vary from 15 to 45 cm (6–18 in.) but the wider spacing increases the possibility of lodging and seed losses from failure of harvesting equipment to pick up lodged and matted stems from the furrow. The effect of thinning persisted and, in fact, increased in subsequent production years (87).

32–2.7 Soil Fertility

Small amounts of N and P (15–20 kg/ha of each) applied at or before planting and slightly below or to one side of the drilled seed were beneficial in establishing stands but usually failed to increased seed yields (61, 68). Utah workers (87) noted that seed yield of alfalfa dropped when the sodium bicarbonate-soluble P content of the soil exceeded 17 mg/kg (17 ppm). This depression in association with soil P content has not been observed by workers in other states. California workers have tried unsuccessfully to correlate soil and/or tissue phosphate in leaves with seed yield (102, 103). The deep roots of alfalfa may account for lack of response to fertilizer when P is low in the upper 15 to 30 cm of the soil profile.

Foliar applications of major and minor elements to seed fields in California have failed to increase seed yields (103). Seed yields have been increased in some areas by the application of B (30, 78, 81). The widespread lack of response to applied nutrients may be explained by the application of fertilizers to other crops in the rotation.

32-2.8 Isolation for Certification Purposes

Fields selected for the production of certified alfalfa seed must not have been planted for a period of years to other cultivars or "common" alfalfa seed. Isolation requirements to prevent genetic contamination by other sources of alfalfa pollen are published by each state certifying alfalfa, and the Association of Official Seed Certifying Agencies (AOSCA). The minimum isolation standards for foundation, registered and certified seed fields >2 ha (5 acres) in size are 183 m (600 feet), 91 m (300 feet), and 15 m (50 feet), respectively, from other blooming alfalfa.

Recent research involving large fields has demonstrated that genetic purity for the certified seed class only can be maintained when the isolation requirements are based on the size of field to be certified and the percentage of the field of another cultivar of alfalfa within 50 m (165 feet) (14, 75, 77). This information gives rise to the concept of an isolation zone. The *isolation zone* is that area calculated by multiplying the length of the common border with other alfalfa cultivars by the average width of the certified alfalfa field falling within the 50-m isolation distance requirement (4). When 10% or less of the certified field is within the 50-m isolation zone, measured from each adjacent alfalfa field, no isolation is required except for a 3-m (10-foot) definite separation between fields. If >10% of the field is within the isolation zone, that part of the field must not be harvested as certified seed.

It has been determined that negligible contamination occurred when fields were planted 3 m (10 feet) apart and the total field size qualified for the isolation zone. Evaluation was based on resistance to spotted alfalfa aphid (*Therioaphis maculata* Buckton), pea aphid (*Acyrthosiphon pisum* Harris), anthracnose (*Colletotrichum trifolii* Bain.), winter dormancy, and bacterial wilt (*Corynebacterium insidiosum*) of seed produced in fields ranging from 10 to 65 ha (25–160 acres) in size, using alfalfa leafcutting bees (*Megachile rotundata* Fabricus) and honey-bees (*Aphis mellifera* L.) as pollinators. These isolation standards should be kept in mind when planting alfalfa seed fields so that potential isolation problems can be avoided. Small fields of <10 ha, especially those <4 ha, can be contaminated by out-crossing of up to 7% (62, 82, 83).

32-3 MANAGING ESTABLISHED STANDS

32-3.1 Water Management

Irrigation requirements for alfalfa seed production are dependent upon soil texture and depth, natural precipitation, evaporation, temperature, length of growing season, and cropping practices (18, 42, 74, 97, 98, 107, 129, 130). Highest seed yields are obtained when irrigation practices prevent severe plant stress and promote slow, continuous growth through the entire production period without excessive stimulation of vegetative growth. In northern Utah, first-crop seed did not require additional irrigation when the root zone contained 38 cm (15 in.) of available water at the start of flowering (87, 124). When the soil contained only 18 cm (7 in.), an irrigation following full bloom was necessary. However, in New Mexico, irrigation of 10 cm (4 in.) applied every 20 d resulted in the highest yields of seed (79).

In the San Joaquin Valley of California, where two or three regrowths are produced successively from the crowns and all seed is harvested at one time, water requirements are higher than in Utah and New Mexico. Time of water application, total seasonal requirements, and the proportion of the total water applied as a preirrigation affect the response of alfalfa to irrigation practices. Judgments regarding the application of water must be based on anticipated plant growth, as influenced by soil texture; soil salinity; and plant density. Established stands utilized 122 to 132 cm (48–52 inches) of water (applied, rainfall, and residual) when water was applied in the winter to wet the soil profile to a depth of 2.4 to 3.6 m (49, 50, 51, 92). Many first-year fields may have as much as 41 cm (16 in.) of residual, deep moisture remaining after harvest. Deep soil moisture applied in the winter can partially offset summer requirements. However, enough moisture must be supplied in the spring and summer to maintain growth and overcome the detrimental effects of moisture stress in the upper root zone during June, July, and early August when flowers are pollinated and seed matures. Deep, medium-textured soils should receive approximately half their water requirement in fall, winter, and early spring irrigations, with the other half applied as needed during the growing season. Water requirements on clay-loam soils can be satisfied, as a rule, by one irrigation in early May prior to the onset of blooming and two additional irrigations, each delayed as long as possible, to promote slow, steady growth.

In general, shallow, fine textured clay soils respond to moderate applications of water as needed in three or four applications during the growing season and one postharvest irrigation (43, 107). Taylor and Marble (122) reported that on a shallow silty clay soil in Australia, seed yields of 1105 kg/ha were obtained when the crop was irrigated frequently, based on an accumulated pan evaporation of 75 mm between irrigations. Yields declined to 528 kg/ha as the interval between irrigations increased and the total amount of water applied during bloom was reduced. Soil-water extraction by the crop was confined to 0 to 1.2 m of soil depth, with highest seed yields occurring when soil-water extraction was confined to the 0- to 0.6-m depth by regular irrigation.

Good drainage is essential to prevent weed invasion, root diseases, and excessive vegetative growth. Sandy soils must be monitored to avoid moisture stress. Normally, winter rains will satisfy early spring requirements. Irrigation to saturate the soil to a depth of 1.2 to 2.4 m just before spring "clipback" should be followed by regular applications as needed through the summer.

Sprinkler irrigation is practiced in many areas of the West with satisfactory results. Sprinklers are the most efficient method for applying water on sandy soils. However, Pedersen et al. (87) indicated that sprinkling delayed flowering early in the season, interfered with pollination when plants were sprinkled at the blooming stage, and reduced seed yields by about 15%. They concluded that high humidity and other factors may have affected flower abscission and seed setting.

Sprinkler irrigation is most satisfactory when used on deeper soils. Moisture is stored in the deep soil in late winter or early spring and is supplemented with early season irrigation as necessary. Sprinkler irrigation should be discontinued before the seed crop begins to mature in order to reduce potential damage to seed quality and loss of seed.

Chiseling the soil to depths of 60 cm (24 in.) after harvest has been reported by California growers to increase water penetration and vegetative growth on sandy loam soils (74). In recent experments, chiseling improved seed yields only on soils that did not develop large, deep, surface cracks, and reduced seed yields on clay soils that crack deeply upon drying (105).

Pan evaporation data, as a measure of water loss, has aided irrigation scheduling on variable soils in the Pacific Northwest. These data support the practice of providing adequate water prior to bloom, and the use of one fall irrigation to replenish soil water (44).

32–3.2 Weed Control

Weed control in alfalfa seed fields is important from stand establishment to seed cleaning. Cultural and chemical methods of control are available for both seedling and mature stands (see Chapter 23 in this book).

Weeds reduce stands and yields, slow seed harvest, increase cleaning costs, and may contaminate other crops in the rotation (24, 128). Weeds are easier to control when they are in the seedling stage, with a subsequent reduction in the cost of weed control required to eliminate surviving weeds. Certain noxious weeds should be controlled in seed fields because they are not permitted in seed lots offered for sale, and they are difficult to remove from harvested seed. Volunteer alfalfa plants in certified seed fields should be treated as weeds (74).

If weeds are allowed to mature and their seeds included with the alfalfa seed at harvest, they must be removed at the cleaning plant. The most difficult weed seeds to remove from alfalfa seed include: johnsongrass; mustard (*Brassica* spp.); dodder; sandbur (*Cenchrus* spp.); pigweed (*Amaranthus* spp.); alkali mallow (*Sida hederacea* Torr.); field bindweed; curly dock (*Rumex crispus* L.); and both sweetclover (*Melilotus* spp.) and sourclover (*Melilotus indica* All.). Alfalfa seed may require a rerun with a magnetic separator to remove certain seeds, such as dodder and field bindweed seed. This procedure will usually increase the loss of seed 1 to 3% but may approach 11% or more.

Dodder is the most troublesome weed in many alfalfa seed fields. The profitable production of alfalfa seed is impossible in the absence of dodder control (23, 126, 127). Once a field is infested, dodder can be expected to be a problem for periods of 10 to 20 yr, and its control becomes a season-long operation. Dodder can be spread by contaminated seed, combines, manure or mud adhering to farm implements, animals' hooves, human's shoes, irrigation water, and in other ways (chapter 23).

32-3.3 Soil Fertility

In the major seed-producing areas of the USA and Canada, applications of fertilizers to established stands have not increased seed yields except where severe deficiencies have been identified (21, 61, 68, 71, 114). Known deficiencies should be corrected prior to seeding.

32-3.4 Growth Regulators

Growth-regulating and flower-promoting compounds are not used in the commercial production of alfalfa seed. Repeated large field tests of several growth regulator compounds have failed to show seed increases in California (108) (see chapter 5 in this book).

32-3.5 Timing the Flowering Period

The flowering period should be timed to coincide with the period of least competition from other pollen sources and the greatest pollinator activity. The best period for flowering is not the same at every location because seed is grown over a wide range of climatic conditions where different pollinators predominate (61, 74). In England, first crop seed is advocated (137), so that seed setting coincides with favorable weather, whereas in the Central Plains of the USA the second crop is used for seed (3, 114).

In areas subject to fall frosts, such as Utah, Idaho, and Canada, removal of first growth often reduces seed yield (68, 88). In Idaho, highest seed yields were obtained without clipping (13). However, both alkali bees (*Nomia melanderi* Cockerell) and alfalfa leafcutting bees do not reach their peak activity until after much of the first crop has flowered (54, 80, 115). Consequently, where seed can be matured before fall frost, it is a common practice to either clip or flail or remove a hay crop between 1 and 10 May.

In Washington, spring growth of established alfalfa is clipped usually about 1 May so that the period of bloom will coincide with emergence of wild bee pollinators. Unclipped alfalfa will begin flowering well in advance of wild bee emergence. Furthermore, clipping reduces damage from the alfalfa seed chalcid (*Bruchophagus roddi* Gussakovsky) and the length of time that other detrimental insects must be controlled (97, 98).

Highest seed yields were obtained in California when spring growth was removed between 1 and 21 April after initiation of flowering. Cutting at this time permitted seed fields to reach full bloom 35 to 40 d later, during late May and early June, when pollinators were active. Premature cutting in the bud stage produced sparse flowers on fewer stems on the first regrowth, with seed yields reduced as much as 35% (61, 70, 114). Seeds were produced principally on second regrowth after the first regrowth had matured.

Methods used to remove spring growth include haying, grazing, and rotary or flail chopping. Where seed is grown in rows, flail-type machines are preferred over sickle mowers or rotary cutters. Cutting should be followed by a light spring-tooth harrowing and row cultivation to control weeds and volunteer alfalfa. Special treatment is required if dodder is a problem (see Chapter 23 in this book). In rough fields, cultipacking following cultivation will conserve moisture and reduce the amount of soil picked up in the harvesting operation. Removing soil from field-run seed may represent a loss of from 2 to 10% of the sound seed harvested (74).

Proper timing of the spring cutting (1 through 21 April in California's San Joaquin Valley) helps reduce alfalfa seed chalcid injury (6, 61). Peak emergence of the chalcid is usually over by the time alfalfa plants in clipped fields have seed embryos at the stage for egg deposition. A later cutting, from 15 May through 16 June in California, delays bloom and pod development to a period when the seed crop is subject to severe damage from a rapid increase in chalcid populations.

32–4 INSECT CONTROL

A major concern of the seed grower is the control of those insect pests that reduce alfalfa seed yields. However, the widespread use of insecticides may lead to the develpment of insect strains resistant to chemicals and can eliminate beneficial predators, which would otherwise aid in control. It is for these reasons that some of the most difficult pest problems are encountered in areas of diversified crop production and older seed production areas. The southwestern states have different pest problems than either the northwestern states or high elevation seed production areas because of their warm climate, use of honey bees as the major pollinator, and a different array of pest species. Alfalfa leafcutting bees, which are much more susceptible to most chemicals than honey bees, are the main pollinator in most northern areas, including Alberta, Saskatchewan, and Manitoba, Canada.

32–4.1 Pest Species

In California, the major arthropod pests are lygus bugs (Fig. 32–1), primarily the legume bug, (*Lygus hesperus* Knight); spider mites, mainly

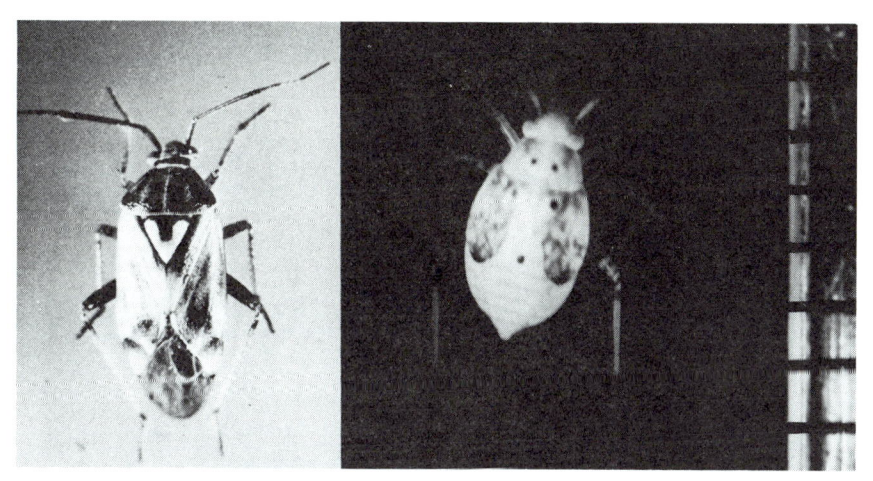

Fig. 32-1. Lygus bugs, (*Lygus hesperus* and *L. elisus*) are among the most damaging insects to an alfalfa seed crop. If not controlled to acceptable levels, very little seed will be produced. Adult lygus (*left*) are 6 to 8 mm long, have wings, and are migratory. Nymphs are wingless, barely visible when they first emerge and do relatively little damage during the first three instar stages of growth. But fourth and fifth instar nymphs (*right*) do more damage per individual than do adult lygus. Scale = 1 mm. Photos by Clarence M. Rincker.

the Pacific mite (*Tetranychus pacificus* McGregor), and the twospotted mite (*T. urticae* Koch); aphids, mainly the spotted alfalfa aphid (*Therioaphis maculata* Buckton), and the pea aphid (*Acyrthosiphon pisum* Harris); and two species of armyworms, the beet armyworm (*Spodoptera exigua* Hubner), and the western yellowstriped armyworm (*S. praefica* Grote) (7, 20, 118).

Minor pests include Say stink bug (*Chlorochroa sayi* Stal), conspers stink bug (*Euschistus conspersus* Uhler), alfalfa seed chalcid, Egyptian alfalfa weevil (*Hypera brunneipennis* Boheman), alfalfa weevil (*H. postica* Gyllenhal), the strawberry mite (*T. turkestani* Ugarov & Nikolski), the blue alfalfa aphid (*Acyrthosiphon kondoi* Shinji), and several species of thrips. In most seasons minor pests are not present in damaging numbers. Alfalfa seed chalcid usually is reduced to low levels by cultural practices. Blue alfalfa aphid and the weevils usually leave fields before the start of the seed crop, whereas the alfalfa weevil is found primarily in higher mountain valleys.

In the Northwest, the major insect pests are the legume bug (Fig. 32-1), the pale legume bug (*Lygus elisus* Van Duzee), and the pea aphid (22, 45, 65, 113) (see also Fig. 32-2, 32-3, and 32-4).

The tarnished plant bug (*L. lineolaris* Palisot de Beauvois), and the alfalfa plant bug (*Adelphocoris lineolatus* Goeze) are pests in eastern Montana, Alberta, and further east. Outbreaks of the spotted alfalfa apid and the blue alfalfa aphid occur in some years, especially in Nevada, Utah, and southern Idaho. Damaging levels of the alfalfa aphid (*Macrosiphum creelii* Davis) have been reported only in Washington. Twospotted mites are mainly a problem in warm, dry years, whereas brown wheat mites

Fig. 32-2. A healthy alfalfa stem (*left*) with normal well-developed floral racemes. Lygus-damaged stem (*right*) shows typical shortening of internodes, blasted flower buds, and stripping of the floral racemes, resulting in little or no seed production. Lygus bugs cause damage by piercing reproductive parts of the plant, injecting salivary secretions that have a localized toxic effect, and removing plant sap. Photo by Clarence M. Rincker.

(*Petrobia latens* Muller) may cause serious damage in Nevada on fields that do not receive adequate fall and winter moisture. In recent years Pacific mites have been a pest in Nevada

In the Northwest, caterpillars in the redbacked cutworm complex, especially *Euxoa septentrionalis* Walker, occasionally cause severe damage to individual fields during the spring. Bertha armyworm (*Mamestra configurata* Walker) causes sporadic damage to fields, whereas western yellowstriped armyworms are seldom, if ever, a significant concern. The alfalfa looper (*Autographa californica* Speyer) has been recorded as damaging in 1 out of 20 yr in Washington. Alfalfa weevils are an occasional pest, especially in areas where pesticide-resistant strains have developed. As in California, the alfalfa seed chalcid is usually controlled by cultural practices, especially where it is feasible to clip back the seed crop during

Fig. 32-3. Healthy alfalfa racemes (*left*) with all florets tripped and/or seed pods forming. Lygus-damaged racemes (*right*) from which most florets have stripped resulting in no seed production. Stripping may also be caused by lack of pollination or excessive moisture stress. Photo by Clarence M. Rincker.

April. Thrips and the pale leafroller (*Clepsis pallorana* Robinson) do not appear to damage seed crops.

There are differences among states and regions in the insecticides recommended for the control of major pests of alfalfa seed. Differences between the Northwest and California can be explained primarily by the use of different pollinators, accelerated pest life cycles in a warm climate, longer residual action in a cool climate, and development in certain areas of resistance to insecticides. Current state and local recommendations are available through Extension Services and other advisory services.

32-4.2 Integration of Chemical and Biological Control in the Northwest

Beneficial insects, namely bigeyed bugs (*Geocoris pallens* Stal and *G. bullatus* Say) and damsel bugs (*Nabis alternatus* Parshley and *N. americoferus* Carayon) were initially tolerant to trichlorfon (Dylox®) [dimethyl (2,2,2-trichloro-1-hydroxyethyl) phosphonate] and became quite resistant to this chemical in eastern Washington by 1970. In addition, several parasites of the pea aphid were also resistant to trichlorfon (47, 117). This paved the way for development of integrated control of lygus bugs, involving the use of trichlorfon in combination with beneficial predators (55). An integrated pest management (IPM) program based on trichlorfon selectivity was used successfully by several Washington seed growers during 1971 to 1972 (58). Since that time, a Northwest Regional Extension program has implemented improved IPM techniques in Washington,

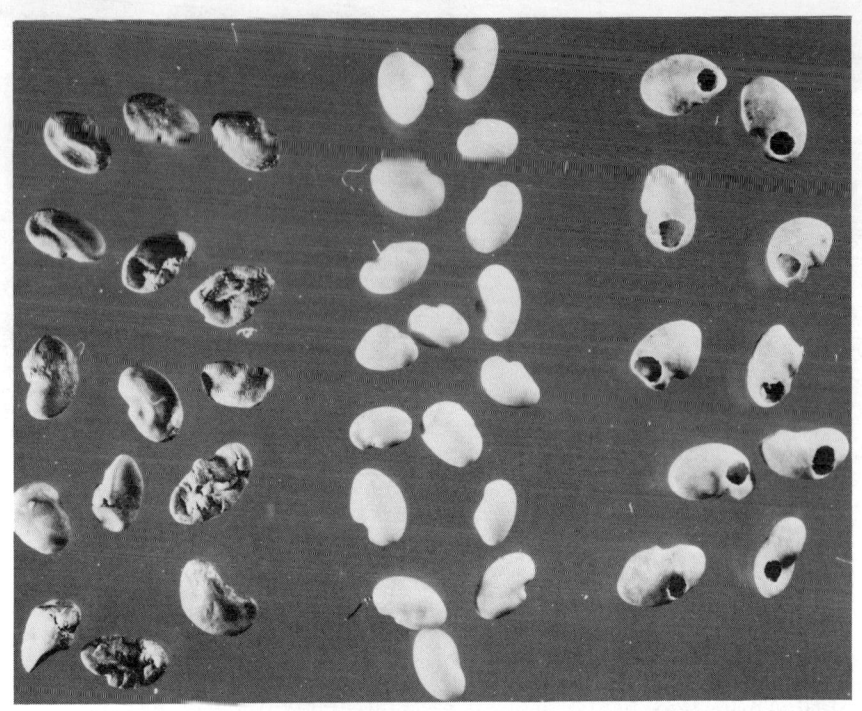

Fig. 32-4. Lygus-damaged seed (*left*) compared with sound, well-matured alfalfa seed (*center*). Lygus-damaged seed are caused by Lygus piercing green seed pods and feeding upon the immature seeds. Chalcid-damaged seed (*right*) showing exit holes made by mature chalcid leaving a hollowed-out empty seed coat. Female chalcid lay one egg in a newly formed seed by inserting her ovipositor through both pod and seed coat. The pods are small and green (10–14 d old) when eggs are laid. Photo by Clarence M. Rincker.

Oregon, Idaho, Nevada, Montana, Utah, and Colorado and in the provinces of Alberta, Saskatchewan, and Manitoba (91, 116, 117, 120, 121, 123).

Typically, a prebloom spray of dimethoate (Cygon® or Defend®) [*0,0*-dimethyl *S*-(methylcarbamoylmethyl) phosphorodithioate], methidathion (Supracide®) [*0,0*,-dimethyl phosphorodithioate *S*-ester with 4-(mercaptomethyl) 2-methoxy-Δ2-1,3,4-thiadiazolin-5-one], carbofuran (Furadan®) [2,3 dihydro-2,2-dimethyl-7-benzofuranyl methylcarbamate] or chlorpyrifos (Lorsban®)[*0,0*-diethyl *0*-(3,5,6-trichloro-2-pyridyl) phosphorothioate] is applied during the first half of May, before either pollinators or most predators are found in seed fields. During bloom, night applications 2100 to 0100 h PDT of trichlorfon plus demeton (Systox®) [*0,0*-diethyl *0* (and *S*)-(2-(ethylthio) ethyl) phosphorothioates] cause minimal losses of beneficial predators, and essential pollinators. Other chemicals may be used for either secondary pests or under certain conditions. Growers in new or isolated seed areas in the Northwest may achieve full-season control of insect pests with the trichlorfon-predator system alone. In current practice, private consultants sample their clients' seed fields on a weekly basis and provide the necessary data for control decisions. Growers on the IPM program usually require one to three insecticide applications per season. Reduced spray requirements result from the following four practices:

1. Better timing of applications based on accurate counts of harmful insects and their stage of development.

2. Minimal disruption of and increased utilization of beneficial predators and parasites in the control system.

3. Elimination of unnecessary treatments such as "fall cleanup applications," that kill pollinators and predators without reducing damage to the seed crop.

4. Improved understanding of factors involved in seed damage, of destructive stages of lygus bugs, and of seasonal variation in maturation of alfalfa seeds (58).

32–4.3 Importance of Prebloom Treatments

If a long-lasting effective treatment is applied during May in the Northwest, lygus bugs may be held below economic levels for the rest of the season. The presence of only small nymphs (first three instars) during the summer is an indication that predators are controlling the pest. Some northwestern growers have obtained good results by making a "split application" of two or more chemicals 7 to 14 d apart during the prebloom period. Because of unfavorable weather conditions, growers may apply maximum dosages of one or more chemicals at a single, favorable time. Although heavy applications are costly, they are cost effective based on long-term pest suppression and protection of pollinating bees.

Growers are advised to assess their prebloom program, especially if lygus bugs have been unusually difficult to control in previous seasons. A decision to reapply this key treatment is warranted if inclement weather occurs soon after application, if lygus bug populations are increasing within 1 or 2 weeks, or if there is any other sign that the spray was only partially effective.

32–4.4 Control of Insect Pests in California

The predatory and parasitic insect species that occur in the Northwest are present in California. They are not sufficiently abundant in alfalfa seed fields, however, to prevent pest species, especially lygus bugs, from exceeding economic levels. Lygus bugs, spotted alfalfa aphid, and spider mites have become resistant to many of the materials commonly used for control in the Northwest. Chemicals required to control these pests are not selective enough to allow a high survival of predators and parasites; thus, California alfalfa seed growers are unable to effectively utilize beneficial species in their insect pest management programs (5).

The timing of insecticide applications for control of lygus bugs is based on population counts and developmental stage of the insect. Three to four insecticide applications are generally required for full season control of lygus bugs. Fields should be monitored weekly throughout the season. The first application is made when lygus bug populations average four to six bugs (adults and/or nymphs) per 180° sweep of an insect sweeping net. This population level is usually reached in the early stages of bloom, about 1 June, before honey bees are placed in the field. During the period of bloom and seed set, lygus bugs are controlled when populations reach levels of eight to 10 bugs per sweep. Although population numbers serve as general guidelines, the developmental stages of lygus bugs are important in timing treatments. Insecticide applications should be timed to coincide with hatching of lygus bug broods. It is important to delay treatment until egg hatch is complete, but applications should be made before nymphs reach the fourth and fifth instars of their development. Late instar (nearly mature) nymphs and adults are less affected by many insecticides than are the younger nymphs.

To minimize the effects of the insecticides on pollinators all applications are made at night or in early morning hours, usually beginning about 2200 h and continuing until about 0400 h PDT.

Little is known about the impact of spider mite populations on seed yields. There is ample evidence that high populations cause yellowing and drying of the leaves, premature leaf drop, and webbing of the flower racemes. Spider mites are most effectively controlled in the early stages of crop development by including an acaricide with the first insecticide application for lygus bug control. Acaricides applied later in the season when high spider mite populations have developed and plant canopies are heavy do not always result in satisfactory control.

Although resistant cultivars of alfalfa are effective against the spotted alfalfa aphid and the pea aphid, many California growers produce seed of susceptible alfalfa cultivars that are grown for forage in other parts of the USA or the world where aphids are not a problem. Biotypes of the spotted alfalfa aphid that are resistant to most insecticides have developed. The most effective chemical controls for the spotted alfalfa aphid on susceptible cultivars in California are a combination of endosulfan (Thiodan®) + methomyl (Lannate®, Nudrin®) or mevinphos (Phosdrin®). Also, experiments have shown that permethrin (Pounce®, Ambush®) is an effective aphicide. With the exception of permethrin, none of the insecticides used alone will control the spotted alfalfa aphid, but combinations apparently produce synergistic effects that result in excellent control. Under California conditions, spotted alfalfa aphid populations can double or triple in 1 week, thus treatments should be applied when the first signs of honeydew are observed on lower leaves.

32-4.5 Use of Cultural Practices and Host Plant Resistance

A number of cultural practices aid in reducing alfalfa seed chalcid damage to low levels. Cutting and removal of hay or clipping the crop during April or May will delay its cycle and reduce populations. Chaff stacks or screenings that contain light seeds should be destroyed or burned before 1 April in the Northwest and in California. Volunteer and waste area alfalfa plants should be removed because they act as a reservoir for the pest. Cultivation and irrigation during the fall reduces overwintering chalcid populations (6). The proportion of chalcid-damaged alfalfa seeds may increase 10% or more each week of the seed setting season. In Oklahoma, chalcid-infested seeds increased from 10.8% for pods set during 7 to 21 June to 53.1% for pods set during 5 to 19 July. Total seed loss in 1979 at El Reno, OK, was 72% (3).

Resistant cultivars of alfalfa have been especially effective in controlling the pea aphid and the spotted alfalfa aphid. For this reason, these aphids are considered to be major pests in California only on susceptible cultivars. There are at least 100 cultivars of alfalfa recorded as carrying resistance to the pea aphid and/or the spotted alfalfa aphid.

32-5 POLLINATION

32-5.1 Attractiveness, Tripping, and Seed Set

Alfalfa must be cross-pollinated to produce commercial amounts of good quality seed. Investigators showed as early as 1935 that "tripping" (Fig. 32-5) of the blossoms by bees constitutes the only important method of cross-pollinating alfalfa (19). The act of tripping is not reversible, since

Fig. 32-5. Untripped and tripped flowers (*left* and *right*). Tripping (release of sexual parts of flower from the keel petal) is necessary for pollination and seed set in alfalfa. Photos courtesy of R.N. Peaden, USDA-ARS, Prosser, WA.

the stigma lodges in the groove of the standard petal. Cross-pollination occurs when the bee that trips the flower is carrying pollen deposited under her "chin" (proboscis fossa) from visiting previous flowers. Later bee visitors to the same flower can no longer cause cross-pollination (9, 134).

Development of managed honey bee pollination of alfalfa enabled California to become a significant producer of seed by 1947 and the leader in 1949 (73, 133). In a region where the honey bee is not an efficient cross-pollinator, research on the alkali bee helped Washington growers become the first to average 560 kg seed/ha in 1950 (80). The third important pollinator of alfalfa, the alfalfa leafcutting bee (Fig. 32-6), was first observed tripping alfalfa flowers in the West in 1957 (10).

32-5.1.1 Pollination by Honey bees

32-5.1.1.1 Biology. Honey bees are unique among insects in a temperate climate because they remain active thoughout the year. During winter they cluster near the center of the hive and feed on honey to provide heat energy. When pollen and nectar become available, workers begin foraging and the queen resumes egg laying (queens may lay eggs through most of the year in the Southwest). As the season progresses and the colony reaches a peak of brood rearing activity, the queen may lay up to 1500 eggs/d. For a variety of reasons, about 15% of the eggs never produce adults during summer. The peak production of new workers is about 1275/d (135, 136).

During late spring, swarms issue from large colonies, especially if there is insufficient hive space for brood rearing and honey storage, and colonies are allowed to heat excessively on sunny days. The old queen and most of the younger worker bees fly out of the hive and form a swarm

Fig. 32-6. The three principal pollinators of alfalfa seed. Honey bees (*right*) are used almost exclusively in California. Growers in the northern alfalfa seed producing areas depend almost entirely on two wild bees. Alkali bees (*center*) nest in the ground and are difficult to move, whereas alfalfa leafcutting bees (*left*) readily nest in wood or other materials with suitable sized holes and are easily moved to seed fields.

hanging from a nearby bush or tree. Previous to issuance of a swarm, workers will have prepared queen cells and started rearing new queens. The first young queens to emerge will either kill sister queens before they emerge, or fight to the death of all but one (32).

On warm, sunny days, virgin queens will begin making 5- to 30-min nuptial flights at 15 to 30 m in the air. Typically, each queen mates with about seven drones per flight, and she will continue to make flights until she has at least 5 million sperm stored in her spermatheca. This usually requires mating with 12 to 14 drones. Queens may live for 3 to 5 yr, but the number that run out of their supply of sperm to produce worker offspring increases with age (31).

Workers live 4 to 6 weeks during the active season, which is divided into 2 to 3 weeks cleaning cells, feeding brood, building comb, and guarding the colony and 2 to 3 weeks foraging in the field for pollen, nectar, water, and propolis (gum gathered from trees and processed to form brownish material to close openings in hive). Strong colonies contain 30 000 to 40 000 workers, a few hundred drones, and one queen at the height of the season (16).

Pollen is a key feature of colony nutrition. It is eaten by newly emerged workers for about 10 d and supplies the protein for royal jelly production. Royal jelly, in turn, supplies the precursor material for production of queen substance (pheromone) by the queen. Queen substance chemicals inhibit the construction of queen cells, attract drones during nuptial flights, and keep swarming workers near the queen. Reduction in queen substance in a colony leads to supersedure: Whenever a queen is failing, workers will drive her out of the hive and attempt to produce new queens (32).

32–5.1.1.2 Pollination in the Northwest. Pollen-collecting honey bees that trip alfalfa flowers only average 0 to 1% of the field force in the northern states and Canada. Nectar collectors contribute to alfalfa pollination by accidental tripping, but such tripping only occurs with 0.2 to 0.3% of the flowers visited in northern states and Canada. Honey bees are more effective in certain isolated valleys of northern Nevada, even at high elevations, where there is less competition from other pollen plants. Growers in these localities have depended on honey bees at 7 to 8 colonies/ha to provide 390 to 670 kg/ha of seed (9, 52, 134).

32–5.1.1.3 Pollination in the Southwest. Honey bees are used as the primary alfalfa pollinator in California, Arizona, and southern Nevada. Pollen collectors may occasionally make up 20 to 100% of the field force in these areas, and accidental tripping by nectar collectors is usually about 2%. Seed yields of >1120 kg/ha have been obtained with honey bees in the Southwest (73, 133).

32–5.1.1.4 Management. A minimum of seven to eight strong honey bee colonies per hectare are recommended in California. Each colony should have at least 6450 cm^2 of brood covered with adult bees and an actively laying queen in a two-story hive. These strong colonies will have eight or more frames covered with bees. Their pollen needs are greater than for colonies with less brood; therefore, more workers become pollen collectors. More than seven colonies per hectare can be helpful, if the seed crop must be set quickly or if there is severe competition from other sources of pollen. Colony strength certification can be obtained by beekeepers or growers at cost from county agricultural commissioners in California (72, 109).

The first colonies normally are moved into fields when the alfalfa is between one-third and one-half bloom. In the San Joaquin Valley this stage is reached about 45 d after a mid-April clip-back of the plants. The full complement of colonies should be moved to the field within 7 d after the crop reaches the one-half bloom stage. If the field begins to look like a flower garden, there are *not* enough bees. This can result from strong competitive bloom in the area, insecticide damage or repellency, or complete field irrigation.

The most effective pollination activity occurs within a 90-m radius of the colony. Decisions concerning colony placement are modified with respect to the insecticides used, accessibility of hives for moving and working, size and shape of the field, and potential insecticide applications on adjacent crops. When bees are placed inside the field, intervals of 160 m are left between bee drives and groups of 12 to 18 colonies are dropped at each location, forming a 160-m grid (Fig. 32–7). Hives may be placed

Fig. 32–7. Hives of honey bees placed inside a California alfalfa seed field for pollination. The most effective pollination activity occurs within a 90-m radius of the colonies. Photo courtesy of Robert Sheesley, Univ. of California, Fresno.

around the edges of small fields but internal colonies are more effective (72, 109, 133).

In California's San Joaquin Valley, alternate block irrigation scheduling and placement of bee watering barrels close to hives encourage honey bees to work close to home. These practices are effective in reducing bee flights to competing sources of bloom and distant sources of water. Since the effective field visitation life span of a healthy honey bee is largely controlled by the condition of her wings, it is assumed that these two management practices result in a more productive pollination life for field worker bees. Seed yields have been consistently higher in fields where these practices have been applied, as contrasted with yields in the same fields prior to the adoption of these practices (106, 111).

Alternate block scheduling of all irrigations following spring clipping can improve honey bee visitation in a field by establishing a staggered bloom-stress condition in the field. Honey bee visitation increases when flowering alfalfa plants are slightly moisture stressed and drops off greatly on freshly irrigated turgid plants or when plants are too dry. By alternate block irrigating, at least half of the plants in a field can be kept attractive to honey bees during the entire blooming period, from 1 June through 10 August for most seed fields in California (110).

Two barrels of water (189 L each) are placed on each 20 ha (45 acres) and refilled each week during the pollination season. This practice is effective in reducing flight distances for water, thus allowing more field bees to gather nectar and pollen. Landing boards placed on the water inside each barrel provides bees easy access to water without drowning. During warm periods these barrels must be serviced twice a week even when alternate water sources are within 0.8 km (0.5 mile) of the colonies (111).

During a heavy nectar flow from the alfalfa or nearby crops, the beekeeper must add honey supers with empty frames. When there are no empty cells for egg laying the colony becomes "honey-bound" and field activity of the bees declines. Also, an extra super provides hive space, which may encourage bees to enter the hive earlier on warm nights and help reduce bee kills from insecticide applications (109).

32–5.1.2 Pollination by Alkali Bees

32–5.1.2.1 Biology. The alkali bee is a soil-nesting solitary bee that occurs naturally in restricted arid areas west of the Rocky Mountains (Fig. 32–6). In nature, nesting is confined to places where the soil is subirrigated over a hardpan layer, which leads to relatively bare alkali spots (10, 58).

Alkali bee adults emerge from the soil in late spring or early summer, depending on temperature and moisture of the soil. In general, males start emerging about 1 week before females. Mating activity is frequent at nesting sites, where the males mate with females as they emerge from the ground (80).

Females begin constructing their nests soon after mating and prefer to dig in existing holes. The nest is a vertical shaft with a lateral tunnel that has oval cells branching from the underside. The first group of six to eight cells is usually constructed at a depth of 10 to 15 cm. If conditions are favorable, the female will construct another series 18 to 30 cm deep in the soil. Average number of healthy larvae per female is seven to nine in managed nesting sites. Females usually live for 4 to 6 weeks and, under good conditions, are active for about 60 d (10, 56).

Each cell is provisioned with an essentially pure ball of pollen (91%) moistened with a small amount of nectar. A single egg, which hatches in 2 to 3 d, is laid on top of each pollen ball. The larva consumes all of the pollen in 7 to 10 d and defecates, which initiates the overwintering prepupa stage. This stage lasts 270 to 300 d until increasing soil temperatures of spring initiate the pupal stage (10, 57).

32–5.1.2.2 Nesting Sites. Four basic conditions must be met and maintained in nesting sites to produce good numbers of alkali bees: (i) the soil is moist throughout the nesting area to a depth of 30 cm; (ii) the soil is firm and compact without either a crusty or fluffy layer at the surface; (iii) the surface is essentially bare with only sparse vegetation; and (iv) the soil is a silt loam with 12 to 24% clay and 10 to 40% sand.

Open-ditched beds have a natural hardpan layer that collects seepage water from a pond, river, or canal to subirrigate the surface layers. In such beds, limited amounts of water can provide excellent conditions for alkali bee nesting (80).

The pipeline bed is the most common type of alkali bee bed in the absence of natural subirrigation on deep soils. Parallel trenches 60 to 75 cm deep and 15 to 20 cm wide are excavated at 2- to 3-m intervals across the bed. Water is supplied through polyethylene drainage tubing placed in a 25-cm layer of gravel in the bottom of each trench.

Plastic-lined beds are an effective way to develop concentrations of alkali bees in selected locations. However, construction costs are high, soil moisture is difficult to manage, and frequent renovation is required.

32–5.1.2.3 Management. Rock salt applied at about 5 kg/m^2 is often needed to seal the surface and maintain moisture in the upper layer of soil. Calcium chloride is used if the surface is too hard and crusty.

Saltgrass (*Distichlis stricta*), bermudagrass, and several alkali tolerant broadleaf weeds are reduced to a sparse level with soil residual herbicides such as bromacil (5-bromo-3-*sec*-butyl-6- methyluracil) and Monoborchlorate (mixture, 66.5% sodium metaborate tetrahydrate + 30% sodium chlorate).

Moisture managment is monitored with a portable tensiometer that provides accurate readings regardless of soil texture. Values in the range of 15 to 25 kPa are optimum for alkali bee nesting. Readings of about 15 kPa must be maintained over the lines in a pipeline bed in order to provide 25-kPa moisture between the lines.

Holes are punched in the surface of the bee bed with a spring-loaded device, which leaves a smooth, unbroken surface. Punching is used to shift concentrations of nests, to spread the nests more uniformly, and to protect females as they start to dig nesting holes.

Transferring alkali bees to prepared beds in new areas is usually done with soil cores 27 L in size, containing the prepupae. Cores are best obtained with a hydraulic cutter mounted on a backhoe. Bee beds can be renovated by removing old soil sections to a depth of 30 cm. Soil removed as cores can be used as decoys or starters in establishing new sites. The excavation must be backfilled with a suitable silt loam soil.

Alkali bee beds with 2.5 million nesting females per hectare can provide excellent pollination for 80 ha of alfalfa (56).

32–5.1.3 Pollination by Alfalfa Leafcutting Bees

32–5.1.3.1 Biology. The exotic alfalfa leafcutting bee is a solitary bee that nests in holes in wood or other materials. It winters as a mature larva (prepupa) in a cell formed from cut leaf pieces. Cells are constructed in series with females developing in the innermost cells of the tunnel and males in the outermost. Adults emerge from cells in late spring or early summer, depending on temperatures. A chilling period is needed to break diapause. Males emerge first, peaking at about 3 d after first emergence. Female emergence peaks at 7 d. The male bee has mandibles with a prominent tooth adapted for chewing through the closing leaf plugs; whereas the female mandible has small teeth adapted for cutting leaf pieces (67).

Females are not receptive to mating as soon as they emerge, but start building cells on the second or third day when sperm are present in the spermatheca. Males cluster in nests or other cavities at night, but male populations dwindle quickly as females begin nesting. Females spend the night in the nest, facing inward. As morning temperatures rise, they turn around and face the entrance, but do not come out and fly until the temperature exceeds 20°C (33).

Females construct thimble-shaped cells from leaf pieces. After a cell is complete, the bee gathers nectar and pollen as food for the larva. The average provision mass is 64% nectar and 36% pollen. The female lays an egg on the surface of the nectar-pollen mass and caps the cell with round leaf pieces, which also form the base of the next cell. She then constructs, gathers provisions, and lays eggs in additional cells until the tunnel is nearly full. An entrance plug is usually formed from round leaf pieces (Fig. 32–8) (66).

Males live for 3 to 4 weeks and females 5 to 6 weeks, but probably <30 d under field conditions. The average number of eggs laid per female under controlled conditions was 28, but 12 is a good average in the field. There usually is a distinct drop in numbers of nesting females at 6 to 7 weeks after starting in the field. In Washington, there is a partial second generation produced by about 15% of the larvae, which are nondiapaus-

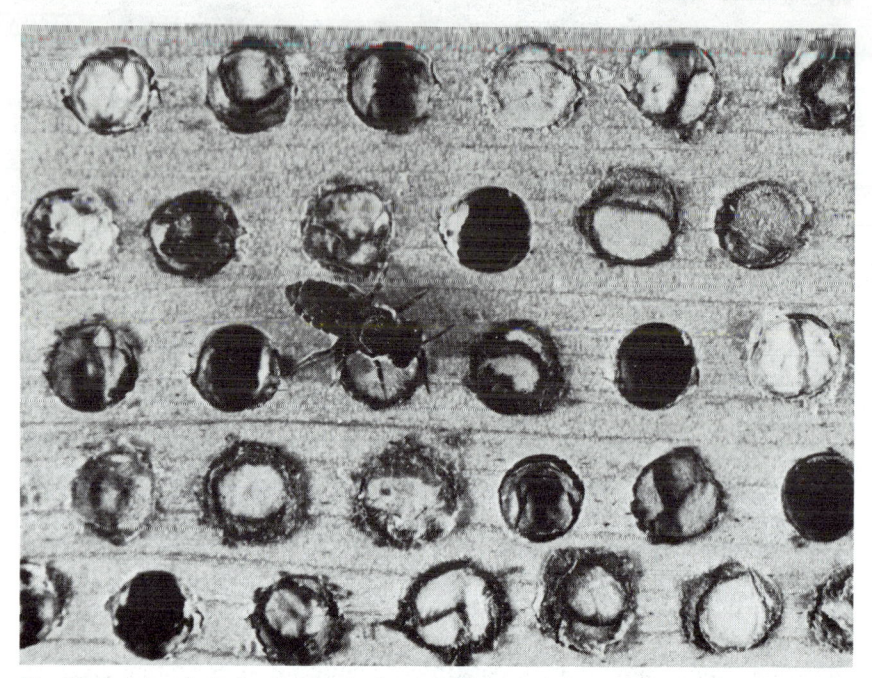

Fig. 32–8. A female leafcutting bee is shown sealing a hole in a drilled wooden board after filling the hole with six to seven cells. The drilled holes are about 5 to 6 mm in diameter and 7 cm deep in a board 8 cm thick. Other filled and capped holes are nearby. Photo by Clarence M. Rincker.

ing. In California, there may be three or more generations per season (67).

32–5.1.3.2 Nesting Materials and Shelters. The first leafcutting bee nesting materials used in the early 1960s were drilled boards (Fig. 32–8), corrugated cardboard, and paper drinking straws. It was soon found that grooved boards or laminates were preferable in the management of bees. The increase in chalkbrood (*Ascosphaera aggregata*) disease in the early 1970s encouraged some growers to switch to nesting materials that could be sterilized. Plastic laminated nests can be disinfected with hypochlorite solutions, and paper nests, which are relatively inexpensive, can be destroyed after the cells are removed (64, 94).

Small field shelters containing 30 000 to 90 000 nesting holes will provide adequate numbers of female bees for 2.5 to 7 ha, but there is a tendency for bees to drift. Larger shelters with 150 000 to 750 000 nesting holes are adequate for 12 to 60 ha, but represent a greater investment and greater exposure to parasites and chalkbrood. Large shelters should be mobile so they can be moved to pollinate other fields on a rotation schedule and to avoid insecticide problems (54, 93).

A good shelter will have a roof with 30- to 45-cm overhang for protection from direct sunlight and sides with an air space of 10 to 20 cm adjacent to the roof to provide ventilation. In most regions, shelters should face slightly northeast so that nests are not in direct sunlight after 1000 h PDT. Some growers attach an awning to the front of the shelter so it can be faced south or southeast to increase interior ventilation by prevailing winds (131).

32-5.1.3.3 Management. The loose cell system is the most efficient method of management. If chalkbrood is present, the cells are dipped in a 0.007 mol/L (1%) sodium hypochlorite solution and allowed to dry. Next, the cells are placed in incubation trays, allowing at least 3.8 cm between trays for air circulation. Temperature is maintained between 28 to 30°C with a relative humidity of 40 to 60% (8, 94). If parasites are present, they are controlled with ultraviolet light traps, a layer of fine sawdust 2.5 cm deep over the cells, vapona (2,2-dichlorovinyl dimethyl phosphate) fumigation strips, and/or a deet (*N,N*-diethyl-m-toluamide) spray on the cells (27, 83).

Field shelters and landing areas should be sprayed with a one percent hypochlorite solution for chalkbrood control before the bees are placed in the field. Used nest materials are dipped in hypochlorite solution or kilned at 120 to 150°C. On the fifth day after bee emergence begins, the incubation trays are placed in the field shelters. Numbers of cells per shelter are calculated to provide at least 17 300 female bees/ha, which under ideal conditions is sufficient to pollinate a field in 10 to 14 d. Bees may have to be moved to another field in bloom to provide sufficient pollen for provisioning later cells (64).

In general, filled nesting materials are removed from the field after mid-August to reduce damage by predators. Cell extraction begins about 2 weeks after removal from the field. Extracted cells are tumbled over screened shaker tables or through screened drums, and pass over an inspection belt to hand sort predators and damaged cells. Cells are stored in metal or hard plastic containers with aeration holes and held at 2 to 3°C (8, 54).

Many growers in the Northwest continue to use drilled wooden boards for bee nesting in combination with improved bee management. Following refrigerated winter storage to keep parasites and nest destroyers inactive, the boards are placed in incubators equipped with phase-out excluder traps. During incubation the parasites and nest destroyers are trapped as the bees exit through the excluder trap and begin nesting in either new or redrilled sanitized boards placed nearby. This management system provides less control of chalkbrood than does the loose cell system.

32-6 HARVESTING

32-6.1 Preparation for Harvest

The two principal methods of harvesting alfalfa seed involve windrow curing followed by threshing with a pickup combine, and spray curing followed by direct combining. The latter is used almost exclusively by farmers in the USA, Canada, Argentina, and Australia (76, 86).

32-6.2 Windrow Curing

Alfalfa seed should be swathed during periods of high humidity or when the leaves are wet with dew after two-thirds to three-fourths of the seed pods have changed to dark brown (59, 112). Swathing is preferable to spray curing where fields are weedy or late maturing, with a high proportion of green but fully formed and plump pods caused by high soil moisture or slow pollination. Physiologically mature seed will ripen in the windrow (74, 99). Spray curing is preferred when fields are relatively dry and where winds are apt to disturb windrows. Cutter bar losses usually do not exceed 5 to 10 kg/ha under optimum conditions. However, windrow losses can exceed 50% in fields subject to strong winds (37, 38). Windrows are ready for threshing when the moisture content of the foliage is from 12 to 18% (wt/wt) (15).

32-6.3 Spray Curing

A chemical desiccant is usually used to prepare seed fields for direct combining. An advantage of direct combining is that harvest can be delayed until nearly all pods are ripe, but once cured, combining must not be delayed or extensive losses may result from shattering (59). Spray curing works best if soil moisture is sufficiently restricted to inhibit new growth from the crowns. Combining should start as soon as the pods and leaves are 15 to 20% (wt/wt) moisture, even though the stems are still relatively green (50% [wt/wt] moisture) (15). It requires 3 to 5 d to cure treated alfalfa in warmer areas of the Southwest and from 5 to 12 d in the Pacific Northwest (12, 59, 74, 99). Seed losses are minimal with good management, ranging from 5 to 20 kg/ha (40).

32-6.4 Chemical Desiccants

Chemicals widely used as desiccants are diquat [6,7-dihydrodipyrido (1,2-a: 2',1'-c)pyrazinediium ion], and endothall [7-oxabicyclo (2,2,1)heptane-2,3-dicarboxylic acid soidum salt]. Dense green stands can be desiccated with aircraft or ground sprayers when daytime temperatures exceed 27°C, and there is little likelihood of exposure to >2 to 6 mm of rainfall. Cool days late in the season will require either heavier application rates or two applications at a more moderate rate, divided by a 2- or 3-d interval (76, 108). It is preferable to apply chemical desiccants when all seed is mature, because green seed will not mature after spraying. Chemical desiccants have not had a significant effect on germination (108).

Diquat is used at 2 L of commercial product (50% a.i.) in 160 to 300 L water/ha when applied with ground equipment, or in 70 to 110 L when applied by air. Addition of 0.5% wetting agent is recommended. Diquat is most effective when applied between sundown and dawn and followed by at least 1 h of darkness. Complete coverage is essential, because incomplete coverage in a dense, green stand will kill only plant parts contacted by the desiccant (76). Recent research in California has shown diquat to be more effective when applied at 1 L/ha followed in 5 d with an application of weed oil at 18 L/ha in 160 to 300 L of water, with harvest about 10 d after original application of diquat.

Endothall is used at 12 L of commercial product in 270 L water/ha when applied in a single application on moderately dense growth by ground equipment. In heavy, dense growth it is better to apply 7 L of product per 270 L water/ha in split applications about 5 to 7 d apart. Aerial applications require the same rates of product applied in smaller amounts of water. Endothall requires 5 to 12 d between application and harvest under cool, fall weather conditions in the Northwest.

32–7 COMBINE HARVESTING

32–7.1 Combine Adjustments

Standard combines can be modified to do a reasonable job of harvesting (Fig. 32–9). In harvesting from windrows, best results are obtained when the header is equipped with either a belt-type pickup device, ground driven at a speed 10 to 15% faster than the ground speed of the machine, or with lifters mounted on the cutter bar. A block mounted on the cutter

Fig. 32–9. Alfalfa seed harvester operating in a California seed field. This harvester is equipped with the air-jet lifter guards. Photo courtesy of Robert Sheesley, Univ. of California, Fresno.

bar edge of the platform will help prevent pods and free seeds from being carried off the front edge of the header (37, 38).

The rasp bar cylinder is in common use. Clearance between the closest cylinder bar and the concave should not be <3.1 mm nor more than 10.0 mm. All cylinder bars should be checked to locate the closest bar and to determine if any are worn. On some machines it is possible to adjust clearance at both the front and rear of the concave. The front clearance should be twice that of the rear clearance (37, 38). Excessive cylinder speed is the greatest cause of seed damage and poor germination, and may overload the harvester shoe by breaking straw into small pieces. Conversely, low cylinder speeds will increase the loss of unthreshed seed. Cylinder speed adjustments are made on the basis of peripheral speed. Recommended speeds for windrow crops are 1280 to 1465 m/min and for spray cured crops, 1220 to 1525 m/min. Lower speeds should be used whenever possible to minimize mechanical damage.

Most combines are equipped with an adjustable lip chaffer. When properly adjusted with openings of 12 to 14 mm (measured at right angles to the axis of the opening), they should operate without plugging the tailings return system. Wind adjustments should be made by starting with excessive air, followed by a gradual reduction to minimize the loss of seed. The loss of free and unthreshed seed should not exceed 0.5 to 2.0% (40). Adjustments should be made for maximum seed recovery rather than for clean seed, with wind adjustment and chaffer extension settings critical to the retention of seed (37, 38).

Forward combine speed depends on both the amount of straw and seed fed into the machine and the humidity of the air. Normally, forward speeds are in the range of 1 to 3 km/h. The forward speed should be varied so the amount of straw fed into the machine remains approximately constant at about 45 to 65 kg of straw and chaff per minute. In harvesting spray cured fields it may be necessary to reduce the auger speed on the platform to 50% of the manufacturer's recommended speed in order to avoid blocking the flow of materials into the machine (37).

32–7.2 Combine Modifications

Heavy crops in spray cured fields require a short, vertical cutter bar (Fig. 32–10) on one end of the platform or header to reduce seed loss. The vertical bar can reduce shattering losses by an average of 50 kg/ha (37, 38). Properly adjusted lifter guards are recommended in row-planted stands. A reel is not necessary except in very light seed crops. The use of a reel can result in the shattering of seed ahead of the cutter bar.

Field studies have documented that large amounts of seed are lost by shattering at the combine header (40). Extended air-jet lifter guards (Fig. 32–11) have been used to reduce these losses by an average of 60 kg/ha (39, 41).

Fig. 32–10. Alfalfa seed harvester with air-jet system installed showing fan (arrow), plenum chamber, and drop tubes. Also shown is a vertical sickle (right edge of header) for opening fields. Reel is raised above crop canopy or removed for harvesting. Photo courtesy of Robert Sheesley, Univ. of California, Fresno.

32–7.3 Seed Damage

Mechanical damage caused by excessive cylinder speed and lack of proper clearance between cylinder bars and concave has been related directly to low seed germination (15, 34, 36, 37, 38, 60, 63). Mechanical damage can be monitored and identified with a hand lens. Damage detection is enhanced if the "India ink" test is used (40). Cracks in the seed coat allow the ink to enter, making the cracks much easier to see. Mechanical damage can be reduced and percentage germination maintained above 90% by the following methods: (i) maintaining proper cylinder speed; (ii) maintaining proper cylinder-concave clearance and clearance between augers and their housings; (iii) installing rubber flights in all elevators; (iv) disengaging the cylinder drive when no plant material is going into the header, i.e. when turning at the end of the field; and (v) maintaining the proper load in the cylinder (37, 38).

Seed damage and loss from rain are heavy when the crop is mature. As little as 5 mm of precipitation can cause serious losses, whereas losses of up to 75% may result from a rain of 10 to 20 mm when seed pods are dry. Rain damage affects both germination and quality. Damaged seeds are discolored, split, and light in weight, resulting in a heavy clean-out in seed conditioning.

Fig. 32–11. Air-jet lifter guards installed on a combine cutter bar have been used to reduce seed shatter losses when harvesting alfalfa seed. The air nozzle (arrow) blows shattered seed up into the header instead of allowing it to fall to the soil surface. Photo courtesy of Robert Sheesley, Univ. of California, Fresno.

32–8 POSTHARVEST CULTURAL OPERATIONS

It is advisable to remove or destroy straw and chaff in seed fields as soon as possible. Debris can be chopped, burned, baled, allowed to decay, or grazed with sheep (*Ovis aries*). The remaining debris should be worked into the soil by cultivation to assist in the control of the alfalfa seed chalcid (43, 74). Fields should be irrigated with 30 to 45 cm of water to maintain plant vigor, begin winter replacement of water, rot chalcid-infested seed and debris, and germinate shattered alfalfa and weed seeds (43, 74).

After alfalfa and weed seeds have germinated, cultivation with a rolling cultivator, springtooth harrow, or a light discing to destroy seedlings should be completed before late fall. The field can then be thinned, furrowed, treated with a preemergence herbicide for control of weeds and volunteer seedlings, or left undisturbed through the winter. Subsequent irrigations, cultivations, spring cut back and clean up, and spring weed control operations should be timed to improve seed yields (74, 76).

32–9 SEED YIELDS AND FACTORS AFFECTING YIELD

32–9.1 Insect Damage

Many factors affect alfalfa seed yields in addition to those mentioned earlier (97, 98). The alfalfa seed chalcid is the one detrimental insect that does not have an effective chemical control. In some years this insect alone can damage over 50% of the seed crop between pollination and harvest (see cultural practices).

32–9.2 Adverse Weather

In general, western states have an ideal climate for alfalfa seed production. However, a cool, late spring may delay emergence of wild bees or keep them inactive during a period when they would be pollinating flowers. Alkali bee populations can suffer devastating reductions from spring rains after the bees emerge. Larvae are destroyed in their cells and nesting females are forced to leave the area. A more serious situation occurs with unseasonal rains in August or September. Seed losses as high as 50% may result from seeds germinating in pods before harvest and from pods splitting and shattering seed with the slightest disturbance. Adverse weather is a production factor over which growers have no control.

32–9.3 Harvest Losses

Seed yields can be adversely affected during harvesting by improper combine adjustments, excessive combine ground speeds, seed leaks in the combine, and harvesting when the crop is not in a proper condition to thresh. These are factors that are of utmost concern to the careful grower.

32–9.4 Components of Seed Yield

Seed yield depends on the number of seeds per unit area and individual seed weight. Seeds are produced in pods and the yield components include: seeds per pod, pods per raceme, racemes per stem, stems per plant, and plants and stems per unit area. Pedersen and Nye (89) compared seed yield components of 'Ranger', 'Lahontan', and 'Uinta' cultivars in Utah. Uinta was the highest seed producer and Lahontan the lowest. Uinta had smaller seeds and pods but more seeds per pod and more pods per raceme than the other cultivars. Stems per unit area, racemes per stem, and racemes per unit area were not affected significantly by cultivars. The reduced seed weight associated with high yield was a function of total flowering and the amount of pollination. Complete pollination resulted in smaller seeds and pods than did one-third pollination (90). Rincker (96) observed that seed weight and seeds per floret (pod) within cultivars were influenced by environmental conditions over a period of 4 yr (Table 32–1). In other studies with spaced plants, seed yields did not differ when plant populations per unit area of 8970, 17 939, and

Table 32-1. Components of alfalfa yield compared with actual seed yields from different years.†

Cultivar	Year	Florets/ raceme	Seeds/ raceme	Seeds/ floret	Wt. of 100 seeds	Seed yield
					mg	kg/ha
Flamande‡	1964	5.6	22.6	4.1	23	202
Flamande	1965	13.2	50.2	3.9	24	577
German Exp.‡	1965	15.1	48.9	3.3	23	557
German Exp	1966	15.6	63.8	4.5	24	1831
Flamande	1966	12.6	46.7	3.7	23	1106
Flamande‡	1966	9.6	39.6	4.2	22	274
German Exp.	1967	14.5	46.8	3.3	21	925
German Exp.‡	1967	12.7	29.4	2.4	24	313
Flamande	1967	10.5	42.8	4.1	23	2110
German Exp.	1968	15.5	58.8	3.8	23	1396
Flamande	1968	12.4	41.8	3.4	21	1094

† Data based on 100-raceme samples obtained from 0.5 to 1.0 ha seed plots grown at Prosser, WA from 1964 to 1968.
‡ Seedling year stands in rows 90 cm apart.

35 879/ha were compared over a 3-yr period (95). In the year after establishment, yields averaged 1158 kg/ha.

A primary goal of many forage breeders is the development of cultivars with the potential to produce high yields of quality forage in humid regions. On occasion this goal is attained without the genetic capacity to achieve economic levels of seed production in the western states. Several studies (48, 100, 101, 125) have been conducted to develop criteria for use by forage breeders in predicting seed yield. It has not been possible to develop reliable criteria, however, to predict seed yield in the western states. Thus, seed yield evaluation of promising genetic materials remains an important step in alfalfa improvement.

32-9.5 Seed Yield Potentials

Alfalfa seed yields range from 0 (crop failures) to a verified yield of 2110 kg/ha of clean seed. Higher yields have been reported by growers but they usually are based on uncleaned seed or on an unverified area of production. In USDA research at Prosser, WA, alfalfa seed yields ranging from 1200 to 2110 kg/ha have been recorded numerous times from both seedling year and mature stands since 1966 (98). In the 10-yr period from 1971 to 1980, California had the highest state average yield of 552 kg/ha, followed by Oregon, 515; Nevada, 489; Idaho, 480; and Washington, 477 (132). Other state averages drop sharply from the foregoing to a low of 48 kg/ha in North Dakota. Since 1980, the California state average yields have increased, through improved technology, from 672 kg/ha in 1982 to 762 kg/ha in 1984.

REFERENCES

1. Abu-Shakra, S., M. Akhtar, and D.W. Bray. 1969. Influence of irrigation interval and plant density on alfalfa seed production. Agron. J. 61:569–571.
2. Aguirre, J. 1974. Personal communication, Plant Genetics, Inc., Davis, CA (formerly Celaya, Mexico).
3. Ahring, R.M., J.O. Moffett, and R.D. Morrison. 1984. Date of pod-set and chalcid fly infestation in alfalfa seed crops. Agron. J. 76:137–140.
4. Association of Official Seed Certifying Agencies. 1971. Certification Handb., Pub. no. 23. Revised Alfalfa Standards, March 1981. Association of Official Seed Certifying Agencies, Raleigh, NC.
5. Bacon, O.G. 1984. Personal communication, University of California, Davis.
6. Bacon, O.G., W.D. Riley, J.R. Russell, and W.C. Batiste. 1964. Experiments on control of the alfalfa seed chalcid, *Bruchophagus roddi*, in seed alfalfa. J. Econ. Entomol. 57:105–110.
7. Bacon, O.G., R.H. James, L.R. Teuber, W.R. Sheesley, and E.T. Natwick. 1983. Insect study results in seed alfalfa. p. 25–29. *In* Proc. Alfalfa Seed Prod. Symp., Fresno, CA. 8 March. University of California Cooperative Extension, Davis.
8. Baird, C.R., and R.M. Bitner. 1982. Managing leafcutting bees with the loose cell method. Idaho Agric. Ext. Serv. CIS 588.
9. Bohart, G.E. 1957. Pollination of alfalfa and red clover. Annu. Rev. Entomol. 2:355–380.
10. ----. 1972. Management of wild bees for the pollination of crops. Annu. Rev. Entomol. 17:287–312.
11. Bolton, J.L. 1956. Alfalfa seed production in the Prairie Provinces. Canada Dep. Agric. Publ. 984.
12. Bovey, R.W., and W.R. Kehr. 1967. New dessicants for alfalfa seed production. Crop Sci. 7:542.
13. Brown, D.E., I.T. Carlson, and M.D. Rumbaugh. 1982. Effects of three management regimes on seed yield of three types of alfalfa cultivars. p. 46. *In* Rep. 28th Alfalfa Improve. Conf., Davis, CA. 13–16 July. Curlie Printing Co., Minneapolis.
14. ----, W.R. Kehr, G.R. Manglitz, J.H. Elgin, Jr., and S. A. Ostazeski. 1980. Crossing contamination along contiguous borders of certified alfalfa seed fields. Crop Sci. 20:405–407.
15. Bunnelle, P.R., L.G. Jones, and J.R. Goss. 1954. Combine performance in small legume seed harvesting. Agric. Eng. 35:554–558.
16. Butler, C.G. 1975. The honey-bee colony—life history. p. 39–74. *In* R.A. Grout (ed.) The hive and the honey bee. Dadant and Sons, Hamilton, IL.
17. Buzi, V.C. 1950. Produzione di sementi foraggere: Alcune aosservazioni ed indirizzi. Nuovo G. Bot. Ital. 57:661–665.
18. Carlson, J.W. 1932. Growing alfalfa seed. Utah Agric. Exp. Stn. Circ. 97.
19. ----. 1935. Alfalfa seed investigations in Utah. Utah Agric. Exp. Stn. Bull. 258.
20. ----. 1946. Pollination, lygus infestation, genotype and size of plants as affecting seed setting and seed production in alfalfa. J. Am. Soc. Agron. 38:502–514.
21. ----, R.J. Evans, M.W. Pedersen, G.L. Stoker, F.V. Lieberman, S.J. Snow, C.J. Sorenson, H.F. Thornley, G.E. Bohart, G.F. Knowlton, W.P. Nye, and F.E. Todd. 1950. Growing alfalfa for seed in Utah. Utah Agric. Exp. Stn. Circ. 125.
22. Cook, W.C. 1963. Ecology of the pea aphid in the Blue Mountain area of eastern Washington and Oregon. USDA Tech. Bull. 1287. U.S. Government Printing Office, Washington, DC.
23. Dawson, J.H., W.O. Lee, and F.L. Timmons. 1969. Controlling dodder in alfalfa. U.S. Dep. Agric. Farmers' Bull. 2211. U.S. Government Printing Office, Washington, DC.
24. ----, and C.M. Rincker. 1982. Weeds in new seedings of alfalfa (*Medicago sativa*) for seed production: Competition and control. Weed Sci. 30:20–25.
25. Dobrenz, A.K., and M.A. Massengale. 1966. Change in carbohydrates in alfalfa (*Medicago sativa* L.) roots during the period of floral initiation and seed development. Crop Sci. 6:604–607.
26. Dovrat, A., D. Levanon, and M. Waldman. 1969. Effect of plant spacing on carbohydrates in roots and on components of seed yield in alfalfa (*Medicago sativa* L.). Crop Sci. 9:33–34.
27. Eves, J., D. Mayer, and C. Johansen. 1980. Parasites, predators, and nest destroyers of the alfalfa leafcutting bee, *Megachile rotundata*. Westrn Reg. Ext. Pub. 32. Washington Agric. Ext. Serv., Pullman.
28. Fischer, B.B. 1980. Control de malezas en la produccion de semilla de alfalfa. IDIA

no. 391-392:23-34. Instituto Nacional de Tecnologia Agropecuaria (INTA), Buenos Aires, Argentina.
29. ———. 1981. Integrated vegetation management in alfalfa seed production. p. 69-73. *In* Proc. Calif. Alfalfa Seed Prod. Symp., Fresno, CA. 8 March. University of California Cooperative Extension, Davis.
30. Fisher, E.H., and K.C. Berger. 1951. Alfalfa seed production as influenced by insecticide and fertilizer application. J. Econ. Entomol. 44:113-114.
31. Gary, N.E. 1963. Observations on mating behaviour in the honeybee. J. Apic. Res. 2:3-13.
32. ———. 1975. Activities and behavior of honey bees. p. 185-264. *In* R.A. Grout (ed.) The hive and the honey bee. Dadant and Sons, Hamilton, IL.
33. Gerber, H.S., and E.C. Klostermeyer. 1972. Factors affecting the sex ratio and nesting behavior of the alfalfa leafcutter bee. Wash. Agric. Exp. Stn. Tech. Bull. 73.
34. Gilden, R.O., C.F. Becker, and C.M. Rincker. 1954. How to cut seed loss when combining legumes. Wyo. Agric. Ext. Serv. Circ. 136.
35. Goplen, B.P. 1970. Alfalfa flower color preference shown by leafcutters. Can. Dep. Agric Forage Notes 16:16-17, 18-19.
36. Goss, J.R. 1971. Performance characteristics and operation of self-propelled combines in California. XII Giornata della Meccanica Agraria, Bari, Italy.
37. ———. 1975. Combine operation and adjustment for harvesting alfalfa seed. p. 18-25. *In* Proc. Calif. Alfalfa Seed Prod. Symp., Fresno, CA. 11 March. University of California Cooperative Extension, Davis.
38. ———. 1980. Cosecha de semilla de alfalfa. IDIA 391-392:64-78. Instituto Nacional de Tecnologia Agropecuaria (INTA), Buenos Aires, Argentina.
39. ———, W.R. Sheesley, and J.J. Mehlschau. 1983. Air-jet lifter attachment effectiveness for alfalfa seed harvesters. p. 37-42. *In* Proc. Calif. Alfalfa Seed Prod. Symp., Fresno, CA. 8 March. University of California Cooperative Extension, Davis.
40. ———, R. Kumar, W.R. Sheesley, and R.G. Curley. 1977. Improvement of harvesting alfalfa seed in California. p. 20-31. *In* Proc. Calif. Alfalfa Seed Prod. Symp., Fresno, CA. 1 March. University of California Cooperative Extension, Davis.
41. ———, W.R. Sheesley, J.J. Mehlschau, and N. Hevrony. 1981. Field tests of the air-jet, lifter guard system. p. 52-60. *In* Proc. Calif. Alfalfa Seed Prod. Symp., Fresno, CA. 10 March. University of California Cooperative Extension, Davis.
42. Grandfield, C.O. 1945. Alfalfa seed production as affected by organic reserves, air temperature, humidity, and soil moisture. J. Agric. Res. 70:123-132.
43. Gregory, E.J., C. Ferris, and V.L. Marble. 1965. Alfalfa seed production in Fresno County. Calif. Agric. Exp. Stn., Fresno County Circ.
44. Griffin, J.H. 1974. Alfalfa seed irrigation. p. 48-51. *In* Proc. 5th Annu. Northwest Alfalfa Seed Growers' Short Course. Walla Walla, WA. Walla Community College, Walla Walla, WA.
45. Gupta, R.K., G. Tamaki, and C. Johansen. 1980. Lygus bug damage, predator-prey interaction and pest management implications on alfalfa grown for seed. Wash. Agric. Res. Tech. Bull. 92.
46. Hageman, R.W., L.S. Willardson, A.W. Marsh, and C.F. Ehlis. 1975. Irrigating alfalfa seed for maximum yield. Calif. Agric. 29(11):14-15.
47. Halfhill, J.E., P.E. Featherston, and A.G. Dickie. 1972. History of the *Praon* and *Aphidius* parasites of the pea aphid in the Pacific Northwest. Environ. Entomol. 1:402-405.
48. Heinrichs, D.H. 1965. Selection for higher seed yield in alfalfa. Can. J. Plant Sci. 45:177-183.
49. Henderson, D.W., L.G. Jones, and H. Yamada. 1966. Alfalfa seed irrigation study. West Side Field Stn. Annu. Rep., University of California, Davis.
50. ———, and H. Yamada. 1967. Alfalfa seed and irrigation study. West Side Field Stn. Annu. Rep., University of California, Davis.
51. ———, and ———. 1968. Irrigation for alfalfa seed production. West Side Field Stn. Annu. Rep., University of California, Davis.
52. Hobbs, G.A., and C.E. Lilly. 1955. Factors affecting efficiency of honey bees (Hymenoptera: Apidae) as pollinators of alfalfa in southern Alberta. Can. J. Agric. Sci. 35:422-432.
53. Hogg, E.S. 1979. Lucerne seed production in South Australia in the 70s. p. 5-7. *In* Proc. Growing Lucerne for Seed in Australia, Melbourne. 19-21 October. Victorian Dep. of Agric., Victoria, Australia.
54. Homan, H.W., L.P. Kish, C.R. Baird, and N.D. Waters. 1982. Alfalfa leafcutting bee management in Idaho. Idaho Agric. Ext. Serv. CIS 588.
55. Johansen, C.A., and J.D. Eves. 1973. Development of a pest management program on alfalfa grown for seed. Environ. Entomol. 2:515-517.
56. ———, D.F. Mayer, and J.D. Eves. 1978. Biology and management of the alkali bee, *Nomia melanderi* Cockerell (Hymenoptra: Halictidae). Melanderia 28:23-46.

57. ----, ----, A. Stanford, and C. Kious. 1982a. Alkali bees: Their biology and management for alfalfa seed production in the Pacific Northwest. Wash. Agric. Ext. Serv. PNW 155.
58. ----, et al. 1982b. Alfalfa seed integrated pest management: The northwest regional program. Wash. Agric. Ext. Serv. Misc. Pub. 50.
59. Jones, L.G. 1952. Preharvest spraying to condition small-seeded legume crops for threshing. Down Earth 8:2-4.
60. ----, and R.D. Cobb. 1960. Water and mechanical damage to seeds. p. 24-28. Proc. Calif. Seed Ind. Conf. Fresno, CA. 17-18 March. University of California Cooperative Extension, Davis.
61. ----, and C.R. Pomeroy. 1962. Effect of fertilizer, row spacing and clipping on alfalfa seed production. Calif. Agric. 16(2):8-10.
62. ----, J.T. Feather, V.L. Marble, and R.B. Ball. 1971. Barriers to outcrossing in alfalfa seed production. Calif. Agric. 25(6):10-12.
63. ----, R.A. Kepner, R. Bainer, and J.P. Fairbank. 1950. Alfalfa seed harvesting. Calif. Agric. 8(4):8-9.
64. Kish, L., N. Waters, and H.W. Homan. 1981. Chalkbrood. Idaho Agric. Ext Serv CIS 477.
65. Klostermeyer, E.C. 1962. The relationship among pea aphids, lygus bugs, and alfalfa seed yields. J. Econ. Entomol. 55:462-465.
66. ----, S.J. Mech, Jr., and W.B. Rasmussen. 1973. Sex and weight of *Megachile rotundata* (Hymenoptera: Megachilidae) progeny associated with provision weights. J. Kans. Entomol. Soc. 46:536-548.
67. ----. 1982. Biology of the alfalfa leafcutting bee. p. 10-19. *In* Proc. Int. Symp. Alfalfa Leafcutting Bee Manage., Saskatoon, SK, Canada. 16-18 August. Univ. of Saskatchewan, Saskatoon.
68. Kolar, J.J., H.R. Roylance, and J.R. Ridley. 1968. Cultural practices in alfalfa seed production. Idaho Current Inform Ser. 65, Agric. Ext. Serv., Idaho Agric. Exp. Stn. University of Idaho, Moscow.
69. ----, and P.J. Torell. 1970. Row planting for alfalfa seed production. Idaho Current Inf. Ser. 122, Agric. Ext. Serv., Idaho Agric. Exp. Stn. University of Idaho, Moscow.
70. Langer, R.H.M. 1967. The lucerne crop. A.H. and A.W. Reed, Wellington, New Zealand.
71. Law, A.G., J.K. Patterson, J. Keene, and H.H. Wolfe. 1957. Producing alfalfa seed in Washington. Wash. Agric. Ext. Serv. Bull. 517.
72. Levin, M.D., and S. Glowska-Konopacka. 1963. Responses of foraging honey-bees in alfalfa to increasing competition from other colonies. J. Apic. Res. 2:33-42.
73. Linsley, E.G., and J.W. MacSwain. 1947. Factors influencing the effectiveness of insect pollinators of alfalfa in California. J. Econ. Entomol. 40:349-357.
74. Marble, V.L. 1970. Producing alfalfa seed in California. University of California-Davis. Div. Agric. Sci. Leafl. 2383. University of California, Davis.
75. ----. 1979. Isolation of alfalfa seed fields to maintain genetic purity. p. 40-48. Proc. Calif. Alfalfa Seed Production Symp., Fresno, CA. 6 March. University of California Cooperative Extension, Davis.
76. ----. 1980. Manejo del cultivo de alfalfa para produccion de semilla. IDIA no. 391-392:6-23. Instituto Nacional de Tecnologia Agropecuaria (INTA), Buenos Aires, Argentina.
77. ----, W.R. Kehr, O.D. McCutcheon, W.R. Sheesley, and G.R. Manglitz. 1979. A study of isolation requirements in alfalfa seed fields pollinated by honeybees, *Aphis mellifera* L.: Spotted alfalfa aphid resistance in seedlots from adjacent susceptible and resistant varieties. Proc. Central Alfalfa Improv. Conf., 16th, University of Nebraska, Lincoln.
78. Medler, J.T., and A.R. Albert. 1953. The relationship between populations of alfalfa insects and soil treatments with boron. J. Econ. Entomol. 46:793-797.
79. Melton, B.V. 1972. Alfalfa seed production studies. N.M. State Agric. Exp. Stn. Bull. 597.
80. Menke, H.F. 1954. Insect pollination in relation to alfalfa seed production in Washington. Wash. Agric. Exp. Stn. Bull. 555.
81. Meyer, R.D., and L. Allen. 1978. Unpublished. Univ. of California, Davis.
82. Pankiw, P. 1975. Effects of isolation distance and border removal on contamination in red clover seed production. Can. J. Plant Sci. 55:391-395.
83. Parker, F.D. 1978. Alfalfa leafcutter bee—reducing parasitism of loose cells during incubation (Hymenoptera: Megachilidae). Pan-Pac. Entomol. 55:90-94.
84. Pedersen, M.W. 1953. Seed production in alfalfa as related to nectar production and honeybee visitation. Bot. Gaz. (Chicago) 115:129-138.
85. ----. 1967. Cross-pollination studies involving three purple-flowered alfalfas, one white-flowered line, and two pollinator species. Crop Sci. 7:59-62.

86. ——, G.E. Bohart, V.L. Marble, and E.C. Klostermeyer. 1972. Seed Production Practices. *In* C.H. Hanson (ed.) Alfalfa science and technology. Agronomy 15:689-720.
87. ——, G.E. Bohart, M.D. Levin, W.P. Nye, S.A. Taylor, and J.L. Haddock. 1959. Cultural practices for alfalfa seed production. Utah Agric. Exp. Stn. Bull. 408.
88. ——, D.R. McAllister, F.V. Lieberman, G.F. Knowlton, G.E. Bohart, W.P. Nye, and M.D. Levin. 1955. Growing alfalfa for seed. Utah Agric. Exp. Stn. Circ. 135.
89. ——, and W.P. Nye. 1962. Alfalfa seed production studies. Utah Agric. Exp. Stn. Bull. 436.
90. ——, H.L. Petersen, G.E. Bohart, and M.D. Levin. 1956. A comparison of the effect of complete and partial cross-pollination of alfalfa on pod sets, seeds per pod, and pod and seed weight. Agron. J. 48:177-180.
91. Perkins, P.V., and T.F. Watson. 1972. *Nabis alternatus* as a predator of *Lygus hesperus*. Ann. Entomol. Soc. Am. 65:625-629.
92. Pomeroy, C.R., and L.G. Jones. 1962. Irrigation trials. West Side Field Stn. Annu. Rep., University of California, Davis.
93. Richards, K.W. 1978. Comparison of nesting materials used for the alfalfa leafcutter bee, *Megachile pacifica*. Can. Entomol. 110:841-846.
94. ——. 1984. Alfalfa leafcutter bee management in western Canada. Agriculture Canada Publ. 1495/E. Research Station, Lethbridge, AB, Canada.
95. Rincker, C.M. 1976. Alfalfa seed yields from seeded rows vs spaced transplants. Crop Sci. 16:268-270.
96. ——. 1964-68. Unpublished USDA Annu. report data. Seed Prod. Investigations. USDA-ARS, Prosser, WA, Forage and Range Res. Branch, CRD.
97. ——. 1979. Alfalfa seed production in the Pacific Northwest. p. 13-19. *In* Proc. Annu. Farm Seed Conf. Am. Seed Trade Assoc., Kansas City, MO. 25 November.
98. ——, C.A. Johansen, and K.J. Morrison. 1987. Alfalfa seed production in Washington. Wash. Agric. Ext. Serv. EB 1406.
99. Roylance, H.B. 1968. Chemical curing of alfalfa seed crops. Current Info. Ser. 69. Agric. Ext. Serv., Idaho Agric. Exp. Stn. University of Idaho, Moscow.
100. Rumbaugh, M.D. 1963. Effects of population density on some components of yield of alfalfa. Crop Sci. 3:423-424.
101. ——, W.R. Kehr, J.D. Axtell, L.J. Elling, E.L. Sorensen, and C.P. Wilsie. 1971. Predicting seed yield of alfalfa clones. S.D. Agric. Exp. Stn. Tech. Bull. 38.
102. Sailsbery, R.L., W.E. Martin, and V.L. Marble. 1962. Personal communication. Univ. of California, Orland.
103. Sarquis, A.V., L.G. Jones, and V.L. Marble. 1960. Personal communication, Univ. of California, Fresno.
104. Sheesley, W.R., R.G. Curley, and C.R. Brooks. 1969. Personal communication. Univ. of California, Fresno.
105. ——. 1970. Personal communication, Univ. of California, Fresno.
106. ——. 1973. Personal communication, Univ. of California, Fresno.
107. ——. 1977. Producing alfalfa seed in a water-short year. p. 32-34. *In* Proc. California Alfalfa Seed Prod. Symp., Fresno, CA. 1 March. University of California Cooperative Extension, Davis.
108. ——. 1981. Alfalfa seed test results and seed production costs in Fresno County. p. 3-9. *In* Proc. Calif. Alfalfa Seed Prod. Symp., Fresno, CA. 10 March. University of California Cooperative Extension, Davis.
109. ——. 1983. Interdependency of pest control, pollination, and irrigation management in alfalfa seed production. p. 3-8. *In* Proc. California Alfalfa Seed Prod. Symp., Fresno, CA. 8 March. University of California Cooperative Extension, Davis.
110. ——. 1985. Alternate block irrigation management for alfalfa seed production. p. 15. *In* Proc. California Alfalfa Seed Prod. Symp., Fresno, CA. 12 March. University of California Cooperative Extension, Davis.
111. ——, and E.L. Atkins. 1985. Value of in-field honeybee watering sites in alfalfa seed pollination. p. 13-14. *In* Proc. Calif. Alfalfa Seed Prod. Symp., Fresno, CA. 12 March. University of California Cooperative Extension, Davis.
112. Smith, G., and B.V. Melton. 1967. Effects of pod maturity on yield and quality of alfalfa seed. N.M. Agric. Exp. Stn. Bull. 516.
113. Sorensen, C.J. 1939. Lygus bugs in relation to alfalfa seed production. Utah Agric. Exp. Stn. Bull. 284.
114. Sorensen, E.L., C.C. Burkhardt, and W. Fowler. 1958. Alfalfa seed production in Kansas. Kans. Agric. Exp. Stn. Circ. 290.
115. Stephen, W.P. 1959. Maintaining alkali bees for alfalfa seed production. Oreg. Agric. Exp. Stn. Bull. 568.
116. Stern, V.M., R.Van den Bosch, and H.T. Reynolds. 1959. Effects of dylox and other insecticides on entomophagous insects attacking field crop pests in California. J. Econ. Entomol. 53:67-72.

117. ----, R.F. Smith, R. Van den Bosch, and K.S. Hagen. 1959. The integrated control concept. Hilgardia 29:81-101.
118. Stitt, L.L. 1940. Three species of the genus lygus and their relationships to alfalfa seed production in southern Arizona and California. USDA Tech. Bull. 741. U.S. Government Printing Office, Washington, DC.
119. Statistical Handbook 81. Canadian Grain Industry, Canada Grain Council, Winnipeg, MB, Canada.
120. Tamaki, G., D.P. Olsen, and R.K. Gupta. 1978. Laboratory evaluation of *Geocoris bullatus* and *Nabis alternatus* as predators of Lygus. J. Entomol. Soc. B. C. 75:35-37.
121. ----, and R.E. Weeks. 1972. Biology and ecology of two predators *Geocoris pallens* Stal and *G. bullatus* (Say). USDA Tech. Bull. 1446. U.S. Government Printing Office, Washington, DC.
122. Taylor, A.J., and V.L. Marble. 1986. Lucerne irrigation and soil water-use during bloom and seed set on a red-brown earth in southeastern Australia. Aust. J. Exp. Agric. 26:577-581.
123. Taylor, E.J. 1949. A life history study of *Nabis alternatus*. J. Econ. Entomol. 42:991.
124. Taylor, S.A., J.L. Haddock, and M.W. Pedersen. 1959. Alfalfa irrigation for maximum seed production. Agron. J. 51:357-360.
125. Taylor, G.R., and R.V. Frakes. 1961. The components and associated variables of seed yield in alfalfa. p. 53-60. *In* Proc. Annu. Meet. Oreg. Seed Growers' League.
126. Torell, P.J. 1967. Dodder control in alfalfa grown for seed. Idaho Current Info. Ser. 39. Agric. Ext. Serv., Idaho Agric. Exp. Stn., University of Idaho, Moscow.
127. ----. 1968. 10 Treatments for controlling dodder in alfalfa seed fields. Idaho Current Info. Ser. 75. Agric. Ext. Serv., Idaho Agric. Exp. Stn. University of Idaho, Moscow.
128. ----, and R.E. Higgins. 1968. Weed control in alfalfa seed crops. Idaho Current Info. Ser. 68. Agric. Ext. Serv., Idaho Agric. Exp. Stn. University of Idaho, Moscow.
129. Tysdal, H.M. 1946. Influence of tripping, soil moisture, plant spacing, and lodging on alfalfa seed production. Agron. J. 38:515-535.
130. ----, and T.A. Kiesselbach. 1941. Alfalfa in Nebraska. Nebr. Agric. Exp. Stn. Bull. 331.
131. Undurraga, J.M., and W.P. Stephen. 1980. Effect of temperature on development and survival in post-diapausing alfalfa leafcutting bee prepupae and pupae [*Megachile rotundata* (Fabricius)]: I. High temperatures. J. Kans. Entomol. Soc. 53:669-676.
132. U.S. Department of Agriculture. 1971-1980. USDA Agricultural Statistics. U.S. Government Printing Office, Washington, DC.
133. Vansell, G.H. 1951. Use of honey bees in alfalfa seed production. USDA Circ. 876. U.S. Government Printing Office, Washington, DC.
134. ----, and F.E. Todd. 1946. Alfalfa tripping by insects. J. Am. Soc. Agron. 38:470-488.
135. Woyke, J. 1964. Causes of repeated mating flights by queen honeybees. J. Apic. Res. 3:17-23.
136. ----. 1977. Cannibalism and brood-rearing efficiency in the honeybee. J. Apic. Res. 16:84-94.
137. Zaleski, A. 1956. Some of the factors affecting lucerne seed production. J. Br. Grassl. Soc. 11:23-33.

33 The Seed Industry

DONALD L. SMITH

Plant Breeder and Consultant
Woodland, California

Seed of alfalfa (*Medicago sativa* L.) may have been a significant commodity of commerce for many centuries because it was the first domesticated forage species. Recorded history makes reference to this special forage as feed for horses (*Equus caballus*) and its importance to the armies and civilizations of the pre-Christian era ranging from the Mediterranean region to China. Although seed was noted infrequently by early authors, it is reasonable to assume that the physical characteristics of alfalfa seed contributed to its ease of transport from region to region. During the pre-Christian period there were no written references to document organized commerce in alfalfa seed; however, some form of movement and interchange must have taken place (2).

The crop did not assume major importance in the USA until the introduction of "Chilean clover" (alfalfa) to California, during the days of the Gold Rush in the period from 1847 to 1850 (7). In retrospect it is an interesting coincidence that the region of the USA where alfalfa first became an important forage crop was to become, approximately one century later, the center of a highly advanced alfalfa seed production and marketing infrastructure.

33-1 DEVELOPMENT OF THE INDUSTRY

The evolution of the alfalfa seed industry can be divided into four periods: 1900 to 1925; 1925 to 1940; 1940 to 1958; and post-1958. The period 1900 to 1925 was a time when many types of seed were made available to farmers in the USA. The seed was of diverse origins, both domestic and foreign, and largely of unknown performance characteristics. In 1920 the domestic seed crop was 8.85 million kilos (13) with another 8.53 million kilos imported from several countries (12). Imports continued to provide a substantial portion of the seed used during the remainder of this era. Much of the imported seed was not adapted and generally gave unfavorable returns to farmers.

Alfalfa first became established in the western USA, and by 1920 this area accounted for > 85% of the total acreage in the country. Since that

---

Copyright 1988 © ASA-CSSA-SSSA, 677 South Segoe Road, Madison, WI 53711, USA.
Alfalfa and Alfalfa Improvement—Agronomy Monograph no. 29.

time, however, the major expansion of forage production has been in the central, eastern, and northeastern regions of the country where the soils and climates were handicaps to alfalfa culture. This major shift in production occurred as a consequence of scientific and educational efforts and in response to the needs of animal agriculture.

The second era in the evolution of the seed industry began with the report in 1925 of the widespread presence of bacterial wilt [*Corynebacterium insidiosum* (McCull.) H. L. Jens.] in the midwestern USA. This discovery was a major concern because none of the available cultivars or land races had significant resistance to the disease (8). New research programs had two primary objectives: (i) to determine the nature of the disease, and (ii) to breed resistant cultivars.

The entire second era, 1925 to 1940, was a period of disappointment in the seed industry because of inadequate supplies of seed of cultivars of known origin and performance, especially in bacterial wilt resistance and winter hardiness. Domestic seed production was usually limited to areas coincident with areas of consumption for forage. The supply was subject to such factors as demand for forage, unfavorable weather for seed production, harmful insects, and lack of a dependable marketing system. Some cultivars that gained popular acceptance and recognition during this era were 'Grimm', 'Ladak', 'Cossack', and a number of variegated types, named after their place of origin.

The third era, 1940 to 1958, was characterized by research and breeding accomplishments resulting from commitments made by the USDA and State Agric. Exp. Stn. in the previous era. The specific milestones that marked the beginning of this era was the release of two bacterial wilt resistant resistant cultivars, 'Ranger' and 'Buffalo', during the period 1940 to 1943. These releases, in conjunction with improved cultural recommendations for forage production, resulted in a major demand for seed of these cultivars. The release of Ranger was a turning point in the culture of alfalfa and the seed industry. This release was a cooperative effort of the USDA and the Nebr. Agric. Exp. Stn. in consultation with the International Crop Improvement Association and under the sponsorship of the Alfalfa Improvement Conference (5).

For the most part the seed industry and especially the seed producers were not prepared for dramatic changes in seed consumer preferences. It was considered necessary to produce seed of the winter hardy types in climates similar to areas of adaptation for forage use. Therefore, seed supplies of these new high-performing cultivars were inadequate and unpredictable for several years after release. Early in this era, it was shown through collaborative research that seed of these and other desirable cultivars could be grown for one generation of seed increase outside their regions of forage adaptation without serious genetic change (10). Once this concept was accepted, the industry evolved rapidly.

The decade 1945 to 1955 was highlighted by the shift in seed production to the western states and the organization of a number of new seed-grower oriented companies to process and market this new source of seed. Major changes in the entire marketing system accompanied the shift in principal areas of production and the development of new production and marketing companies. Public research agencies responded to these changes in the industry by committing more resources to seed production research. For example, research increased the efficiency and reliability of utilizing honey bees (*Apis mellifera* L.) as pollinators of alfalfa. Advances in pollination systems were one of several research contributions that enabled California seed growers to become major sources of seed by the early 1950s and major world suppliers by 1955.

Several organizations, working together toward the common goal of providing a reliable supply of pure seed of cultivars with superior performance, were responsible for dramatic changes in the seed industry during this era. One of the major contributors to this effort was the State Crop Improvement or Seed Certification Agencies in cooperation with the International Crop Improvement Association, presently known as the Association of Official Seed Certifying Agencies (AOSCA). These organizations, in cooperation with research agencies, developed certification standards for seed production outside the region of forage adaptation. In addition, they were charged by their respective state governments with the responsibility of certifying that production requirements and standards were met and that seed was appropriately labeled when offered for sale.

The National Foundation Seed Project, organized by the USDA in 1948, played a critical role in encouraging seed production of public cultivars. This project was charged with ensuring that sufficient quantities of seed stocks were available for the production of certified seed. Shortly after the release of new cultivars in the 1940s, deficiencies in seed stocks had been recognized as a limiting factor in meeting the demand for certified seed. The success of the National Foundation Seed project in meeting this need is recognized as an outstanding example of how state, federal, and private organizations cooperated in the interests of successful alfalfa culture. In 1985, steps were taken to terminate the project because it had served its purpose.

During the period 1952 to 1958, the rapid multiplication of public cultivars supported by improved management practices, especially in California, gave rise to an oversupply of certified seed. Many seed growers had placed a high priority on the production of high quality seed, which resulted not only in improved supplies but also the emergence of professional seed producers as an integral part of the seed industry. In 1955, the supply situation did not deter one seed firm from introducing a proprietary cultivar, 'DuPuits'. This was the first opportunity for the industry to observe the orderly merchandizing of a proprietary cultivar for which the marketer could control the supply of available seed (1). These simultaneous developments provided reinforcement to the concept that private industry could conceivably benefit from its own plant breeding programs. Thus the fourth era, post-1958, began when a number of private seed firms established research programs to develop and produce their own proprietary cultivars (15). These new activities required operational changes, such as field production planning, marketing strategies, and new relationships between seed growers and processors or marketers.

The impact of private breeding programs on the industry since 1958 is evidenced by the fact that proprietary cultivars accounted for $>50\%$ of the total seed produced in the USA and Canada in 1983. An even more dramatic shift occurred in California, where over two-thirds of the production in 1984 was proprietary in origin.

33-2 UNIQUE CHARACTERISTICS OF THE INDUSTRY

The alfalfa seed industry is substantially different from other components of the field seed industry. Much of the seed is produced outside the area of consumption. Furthermore, the climates in these production regions differ from those in areas of seed utilization. A certification procedure for production and conditioning (seed processing) systems is an integral part of the whole process. Crop improvement or seed certifying associations monitor the practices employed from production of stock seed to packaging seed for sale. These associations ensure that seed meets minimum standards for quality, is correctly labeled, and is genetically pure. Seed certification is used by most major seed firms in the USA and Canada for seed sold in domestic and international trade. Unlike most other field seed species, significant quantities of alfalfa seed are exported.

Alfalfa seed organizations are distinguished from other field seed organizations by more diverse marketing systems. The industry produces a wide range of products in terms of quality and cultivars available. Seed quality differs on the basis of germination, primary and secondary noxious weed seed, and percent of hard seed. Product types include seed merchandized as "variety unstated," blends, and both public and private cultivars. Although a large number of superior new cultivars are available to consumers, there is some demand, possibly based on price, for some old (40 yr) cultivars.

33-3 RESEARCH

33-3.1 Development of Industry Research

Most seed company operations involve or are affected by a research component. Dramatic shifts in production areas during the third era, coupled with problems from chronic supply shortages to oversupply, motivated the establishment of research programs. These factors, along with the pioneering success in the production and marketing of a proprietary cultivar, provided the stimulus for private plant breeding research as an integral part of effective seed supply management.

The mission of private breeding programs was to develop and performance-test new cultivars superior in forage and seed yield, disease and insect resistance, as well as winter hardiness. Most companies employ highly trained professional and technical staffs with well-equipped field and laboratory facilities for the development of improved cultivars. Principal research stations were located in Maryland, Indiana, Iowa, Wisconsin, Minnesota, Idaho, Washington, Nevada, and California, with satellite breeding or testing sites in virtually every alfalfa growing region of the USA and Canada. This did represent a new departure for the private seed industry. Many companies had conducted breeding programs for decades, most notable in hybrid corn (*Zea mays* L.). The motivation was the same for alfalfa as it had been in corn; to offer the consumer a superior product and the assurance of an orderly, continuing market. The technical aspects of the alfalfa research programs were more complex than those of the annual crops previously bred by private industry. Furthermore, alfalfa breeding technology was not as highly evolved as was that of most other crops. Thus each company's program could differ in either breeding strategies or germplasm evaluation systems. The particular philosophy of individual breeders also determined the course of specific programs.

In spite of the complexities, projected costs and pay-out timetables, alfalfa seed companies made a commitment to research as the most appropriate response to the introduction of superior cultivars in an orderly marketing system. The record of these research programs is best illustrated by the more than 100 multiple pest resistant proprietary cultivars released and available to consumers during the 1975 to 1985 decade. In addition, the industry responded rapidly to the discovery of the new disease, Verticillium wilt (*Verticillium albo-atrum* Reinke & Berth.), by incorporating resistance in newer cultivars.

As privately supported breeding programs began to release cultivars, it became apparent that the role of public supported alfalfa research should be reassessed. The stated policy of the USDA stresses basic research rather than the development of cultivars. Conversely, a few State Agricultural Experiment Stations have maintained alfalfa breeding research. Although these publicly supported programs compete with private industry, they are less significant than competition among private seed companies. Private programs have established credibility, and a high level of professional respect exists among public and private alfalfa scientists.

33-3.2 Proprietary and Public Cultivars

Maintaining new cultivars as proprietary products was a major concern of seed companies from the outset of their involvement in plant breeding research. It was concluded that for many purposes seed certification would provide a degree of control. In the early phase the American Seed Trade Association, the National Council of Commercial Plant

Breeders, and the American Society of Agronomy provided forums for discussion of "plant breeder's rights." In 1970 the U.S. Congress passed the Plant Variety Protection Act, Public Law 91-577 (11), which granted exclusive rights for a 17-yr period to the breeder or designated owner of new cultivars of seed propagated plant species. As stated in the Act, the owner was allowed to multiply and offer for sale a new cultivar that met the principal criteria, novelty and reproducibility. The Act provided legal protection to the owner for the new development. Title V of the Act provided an additional alternative safeguard that, if selected, required that the new cultivar be merchandised only as a class of certified seed. This gave the owner of a Plant Variety Protection Certificate security and a means for monitoring the proprietary product. The Plant Variety Protection Act, with inclusion of the Title V certification option, was the result of collaborative efforts of the AOSCA, the American Seed Trade Association, the National Council of Commercial Plant Breeders, and the USDA. Title V resulted in the initiation of a new role for certifying agencies that permitted certification for genetic purity only, rather than for a number of other quality factors, which historically had been a part of certification. Between 1975 and 1985, a number of applications for protection under the Act received favorable review and certificates were issued. However, passage of the Plant Variety Protection Act has not had as significant an impact on the level of investment by private industry in alfalfa breeding as it has had on breeding of soybean [*Glycine max* (L.) Merr.] and other crops.

During the decade of the 1960s, there was concern over the role of certification agencies in the seed industry and the failure of some state agencies to comply with the International Crop Improvement Standards (9). The matter of uniform standards was resolved through the development and promulgation of standards and regulations incorporated into the Federal Seed Act. Crop improvement associations led this movement, which contributed substantially to raising the credibility of certification as an integral function in the seed industry. Minimum new standards and regulations for certification were finally enforceable in the USA.

In the late 1950s, alfalfa scientists recognized the need to review the characteristics of publicly and privately developed cultivars and to address related problems of special concern to seed producers and public institutions (6). The Joint Alfalfa Work Conference, organized as an undertaking of the National Alfalfa Improvement Conference, was convened twice during 1960. The Conference prepared a comprehensive report and a recommendation for the formation of an unofficial body, The National Certified Alfalfa Variety Review Board. This Board, now under the leadership of AOSCA, was charged with providing a standardized, uniform review of applications for certification of both public and private alfalfa cultivars. The report of this special conference was adopted by all participating agencies, including regional Alfalfa Improvement Conferences, the American Seed Trade Association, the International Crop Improvement Association (AOSCA), the USDA, and the State Agricultural Experiment Station Committee on Organization and Policy. The first meeting of the review board was in January 1962. As reported by Hanson (6), "this group of alfalfa specialists has had a profound effect on maintaining order and respectability in certification of alfalfa varieties." Another significant result of the Joint Alfalfa Work Conference was the enhancement of relations among public and private plant breeding research groups.

33-3.3 Alfalfa Improvement Conferences

The three regional Alfalfa Improvement Conferences; Northeastern, Central, and Western, in conjunction with the National Alfalfa Improvement Conference [as of 1986 renamed the North American Alfalfa Improvement Conference (NAAIC)] have been instrumental in dealing with concerns of public and private plant breeders. In addition to certification issues, these Conferences have promoted the exchange of seed and other propagating materials through appropriate release policies and procedures. Standards for conducting alfalfa performance trials have been suggested and uniform disease evaluation tests initiated (14). Regular scientific meetings are held throughout the USA and Canada. The first National Alfalfa Improvement Conference was held in 1934, and the 1986 conference of the NAAIC in St. Paul, MN marked 52 yr as a research forum for all interested alfalfa breeders.

33-4 PRODUCTION

33-4.1 Producer-Industry Relationships

The seed industry begins with the producers of raw seed. This portion of the industry is composed of a large number of independent farmers located in at least 20 states and five Canadian provinces. Relationships among these farmers and other components of the industry are dependent upon a number of factors. The principal distinguishing factor is the type of seed conditioning and/or marketing organizations involved. Seed conditioning and marketing organizations may be (i) privately owned by either an individual, small group of individuals, or a closed corporation; (ii) publicly held stock company; (iii) seed producer-owned cooperative; (iv) individual farmer, conditioner, marketer; or (v) a farmer selling on the open market.

The most common arrangement is based on a multiyear production marketing contract that commits the producer to grow a specified number of hectares of a designated cultivar. The producer further agrees to sell the seed harvested from the contracted area to the processor seed organization, which agrees to buy the seed at a guaranteed price. The price is established each year for the term of the contract on the basis of a preset formula. The essential features of the relationships in a farmer-owned cooperative are quite similar except with reference to pricing, which is generally dependent on the actual price received by the cooperative less the cost of operation. The individual farmer/conditioner and/or marketer often has a contract relationship similar to the other producers but the contract includes conditioning the seed for either a wholesale or retail marketer. Most farmers selling on the open market produce seed of public cultivars and sell the crop to the highest bidder, either a seed company or a broker who may resell to a seed company. The most significant feature of this facet of the industry is the contract relationship, which is established before the crop is planted. This enhances management of seed supply and effective merchandising of proprietary cultivars.

33-4.2 Systems of Production

A representative of the seed company or seed conditioner (processor) contacts growers on specific production needs, pricing structure, seed stocks, consulting services, and specific terms of the production contract. The supervision of seed production varies widely depending on the area of production. Most organizations employ field representatives who have grower contact programs and are familiar with the seed producer's operations. They offer suggestions and counsel to help assure maximum seed production. Until recently the field representative operated as a production manager providing specific recommendations on all cultural practices. This pattern of operation has evolved now to one of consultation without recommendations.

33-5 CONDITIONING

33-5.1 Field to Plant

Field representatives maintain close liaison with growers during the entire season. Prior to and at harvest, frequent contacts are made to assist growers in a timely harvest and delivery of the crop. Transport of seed from the field to the receiving location can be in bottom unloading bulk trucks, or in boxes constructed of metal or wood (1.2 m^3) secured to a flatbed truck or trailer.

33-5.2 Receiving Systems

In general, storage of uncleaned seed in warehouses is in either 1.2-m^3 boxes or bulk bins of various capacities. Boxes may be covered with metal, paper, or plastic covers to protect against accidental mixing during handling operations. Handling of bulk shipments at the warehouse usually involves the following equipment: (i) movable belt conveyor; (ii) movable, 4 or 4.8 m in height, electrically powered bucket elevator equipped with plastic buckets and a multiple-spout discharge; (iii) 1.2-m^3 bins, and (iv) a forklift. The conveyor, elevator, and boxes are designed for ease and thoroughness of cleaning. Seed is unloaded on a surface that can be cleaned completely between deliveries.

33–5.3 Cleaning Systems

Although design features of conditioning plants have remained unchanged since 1975, plants and equipment have been modified to improve efficiency. Modifications involved changes in flow patterns and in equipment to improve ease and thoroughness of cleaning. Some advanced designs have been incorporated into weighing and sacking equipment.

Alfalfa seed is conditioned with equipment that utilizes differences in physical characteristics of alfalfa seed and the nonseed fraction, such as particle size, shape, density, and surface texture. Modern seed conditioning plants are equipped with the following separating machines: (i) air-screen cleaner; (ii) specific gravity separator; (iii) velvet rolls separator; and (iv) magnetic separator. Additional specialized equipment may include indent discs, indent cylinders, as well as spirals. Installation and flow patterns within plants are in the order listed.

The air-screen cleaner is the basic machine used with alfalfa. This cleaner makes separations on the basis of particle size, shape, and density. The flow pattern of seed through the machine and the effects upon the seed lot can be divided into three distinct actions. the field-run seed is fed into an air stream that removes lightweight seeds and impurities. The heavier seed and impurities are directed to and flow down or fall through a stacked set of inclined reciprocating screens. The set is usually composed of four screens with different diameter circular and/or rectangular slotted holes. The top screen removes impurities larger than the seed; the second, third, and fourth screens separate the seed with fine particles of soil, broken seed, and other impurities falling through the fourth screen. The cleaned seed is passed through an air stream that removes the remaining light material.

The flow pattern for seed is usually from the air-screen separator to the specific gravity separator. This machine separates on the basis of specific gravity and size of particles. The separating unit of the gravity separator is a quadrangle-shaped, perforated, tilted, and cloth-covered surface. Air is forced up through this surface as the deck mechanism oscillates on both horizontal and vertical planes. Seed is introduced at the lower corner of the deck and moves up the inclined plane to the opposite sides. The path of the seed and foreign material is dependent upon the specific gravity of each particle and the output is divided into a series of distinct fractions.

The velvet roll seed separator is used for more precise cleaning and separation based upon surface texture and shapes. A series of paired velvet covered rolls are mounted with a few degrees tilt from the horizontal with spacing between the rolls so that at the upper end the velvet makes substantial contact and at the lower end the rolls are completely separate. This separator removes dirt, sand, rough-coated weed seeds and other foreign materials that have rough edges that cling to the velvet. In most plants the velvet roll seed separator is used as a finishing machine; however, in a modern installation it is an in-line operation following the gravity separator.

The magnetic separator also makes separations based upon surface texture. Seeds are fed into a screw conveyor with a regulated amount of water and finely ground iron powder. The iron powder clings to rough seed coats, cracks in the seeds, chaff, soil particles, and seeds whose surface becomes sticky when moistened. The mixture is passed over a high-intensity, magnetically charged, rotating drum. Sound seed does not attach to the drum, whereas those particles to which the iron powder adhere are attracted to the drum and removed at a later point in the revolution cycle. The magnetic separator is used extensively to remove rough-coated dodder (*Cuscuta* spp.)

The indent disc and indent cylinder separators are used to remove seeds with the same width or thickness on the basis of length differences. The short seeds are lifted into pockets or indentations in a rotating disc or cylinder. The sizes of the indentations are shallower than alfalfa seed; thus alfalfa seed is not picked up.

The spiral separator is a nonmoving spiral, and its function is based on differences in movement characteristics of particles. Spherical seeds or particles roll down the spiral, whereas the characteristically flat irregular shaped alfalfa seeds tend to slide. Round particles travel faster and discharge at different distances from the center of the spiral.

33-5.4 Quality Control

During the 1980s, quality control has received increased attention, especially in facilities and procedures. This included improvements in monitoring all aspects of field operations, seed storage, and seed conditioning to improve seed quality and to ensure the maintenance of cultivar integrity.

33-6 MARKETING

33-6.1 Systems of Marketing

Marketing systems and strategies in the industry are as varied as the groups involved. The marketing chain includes organizations that range in activity from wholesale only, to direct farmer retail. These organizations can be placed in five major categories: (i) privately owned companies; (ii) publicly owned stock companies; (iii) producer-owned cooperatives; (iv) consumer-owned supply cooperatives; and (v) privately owned direct marketers.

33–6.2 Types of Markets

Marketing systems vary depending on contracting system, product quality, and nature of the product offered for sale in that market. Variables that influence marketing include: (i) retail vs. wholesale; (ii) nature of product; (iii) forward contracted vs. uncontracted; (iv) available supply; (v) seed quality; and (vi) domestic use vs. export. The nature of the product varies with the class of seed offered for sale, i.e., public cultivar, proprietary cultivar, proprietary and nonproprietary blends, common, "variety unstated," and certified or uncertified of either public or proprietary cultivars.

No one marketing strategy or system is unique to a single organization. Some groups employ more than one system because they offer a wide range in products to wholesale and retail markets both domestically and abroad.

33–6.3 Packaging

Advances have been made in seed handling with the development of new marketing systems. Some noteworthy examples include the use of (i) large woven synthetic fiber bags (0.9 × 0.9 × 1.5 m with four loops, one at each corner for lifting) for shipment of cleaned seed, (ii) replacement of wooden pallets with cardboard ship sheets, and (iii) covering palletized sacks with plastic film wrap. Seed for domestic use is packaged in paper bags that are lighter in weight, more durable and nonslip in comparison with older paper bags. Seed for export is packaged in burlap or cotton bags or sealed in bulk containers.

33.6.4 Shipping

Transport of the conditioned seed within the USA and Canada is for the most part by van-type truck. Railroad shipments are now rare except for the movement of seed to selected markets, i.e., Mexico. Since the early 1970s, much of the seed for export has been sealed in ship containers at the point of origin and moved to ports by truck. Loading of vans and containers is highly automated through use of palletized sacks and forklifts.

33–6.5 Regulatory Agencies

Since a large percentage of seed moves either in interstate commerce or is exported, the Federal Seed Act, which is administered by the USDA, is the regulatory basis of labeling for quality, cultivar, and origin or state of production. The label or seed tags must state the following quality factors in percent: purity, inert matter, germination, hard seed, other crop seeds, weeds, noxious weeds, and date of testing. The Seed Act had a strong impact on the industry through the enforcement of appropriate labeling standards as clearly written requirements. Standards established under the Federal Seed Act have contributed, in part, to the continuing demand for alfalfa seed of U.S. origin. Labeling of seed for sale in intrastate commerce is controlled by state seed laws. These state laws contain virtually all the requirements of the Federal Seed Act and are administered, and enforced, by State Departments of Agriculture.

Contracts for export usually require phytosanitary certificates issued by either the USDA or State Departments of Agriculture. The buyer specifies the type of certificate that must be provided. The exporter is required by the U.S. Department of Commerce to provide an export declaration specifying the product exported. Occasionally an export contract will require documentation of the germination and purity analysis by a USDA Federal Seed Laboratory or the State Department of Agriculture Seed Laboratory in the state of origin of the seed. The overall regulation of the industry by state and federal agencies has contributed to furthering the good reputation, integrity, and confidence that nearly all of the industry has developed with their suppliers and customers.

33-7 PRODUCT INFORMATION

33-7.1 Industry-Public Cooperation

Information on product performance is generated through proprietary testing by seed companies, as well as through public testing by State Agricultural Experiment Stations, Extension Services, and the USDA. Data from these tests serve as one of the bases for commercial promotion. The industry employs educational and sales activities similar to those of other segments of the field seed industry, which range from leaflets to advertising in farm magazines and on television. Reliance on public test data to support performance claims varies from state to state depending on the extent and types of tests conducted by public agencies. Faced with the large increase in the availability of new cultivars, few states develop recommended lists but simply characterize cultivars to assist consumers in the selection of adapted cultivars. Many public testing programs accept proprietary entries on a fee basis. Thus the developers of proprietary cultivars can control the test locations where their entries are evaluated. In addition, fee tests have been developed by public agencies for the purpose of characterizing named and experimental cultivars for their reaction to various diseases. These public testing programs have contributed to mutual respect and cooperation among scientists in the public and private sectors.

33-7.2 Certified Alfalfa Seed Council

The Certified Alfalfa Seed Council was organized in 1953 with the objective to promote consumption of certified alfalfa seed by means of the principles set forth in their Bylaws (4):

The Certified Alfalfa Seed Council carries on a single-purpose, non-profit endeavor on behalf of all segments of the certified alfalfa seed industry, including growers, processors, distributors, and the consumer.

The Council endeavors to promote a wider knowledge and better understanding of the production and use of certified improved alfalfa varieties and to assist agricultural leaders in distributing and disseminating information on the value and use of these improved varieties.

The Council does not: 1. attempt to influence legislation or legislative bodies, 2. concern itself with questions of prices or marketing nor does it give advice on such matters, 3. attempt to favor one area or district of production, 4. have any purpose or object except those specifically set out in this declaration of principles.

Funding for the Council's activities is provided by the certified seed growers and usually collected as a portion of the regular certification fee imposed by State Crop Improvement Associations.

Although the effects of the Council's promotional activities are difficult to quantify, the Council has made a significant contribution to strengthening communications and working relationships between the industry and land grant university system. Two projects deserve special mention: special publications on topics of interest to alfalfa growers and The National Alfalfa Symposium. One of the Council's publications, the *Alfalfa Analyst* (3), was produced with colored prints and has descriptions of all major insect pests, diseases, and deficiencies symptoms of alfalfa. It is now in its second printing and >500 000 copies have been distributed. The National Alfalfa Symposium was organized in 1969 and has continued on an annual basis. These symposia are organized by the Council with joint sponsorship provided by a different land grant university in the forage consuming regions of the USA. Farmers, equipment manufacturers, public agricultural scientists, agricultural extension personnel, private agricultural scientists, and seed producers have been regular participants.

33-8 TRADE ORGANIZATIONS

The four organizations that have had a substantial impact upon the evolution of the alfalfa seed industry are the American Seed Trade Association, International Seed Trade Federation, Pacific Seedsmen's Association, and the National Council of Commercial Plant Breeders. It has been through company membership in these organizations that the alfalfa seed industry has participated in the formulation of new state, federal, and international regulations and laws that had an impact on the industry. The National Council of Commercial Plant Breeders has served as the industry's representative on technical issues.

33-9 SUMMARY

Transition and change has characterized the alfalfa seed industry. In the period between 1977 and 1985, total seed production stabilized at between 50 to 55 million kilos, with a more stable and predictable annual market demand. This stability made it possible to develop a balance of research, production, conditioning, and marketing expertise within the various organizations that comprise the industry. As a result, seed of a wide array of high quality, superior alfalfa cultivars are available to consumers world-wide.

REFERENCES

1. Arnold, L.E. 1972. The seed industry. *In* C.H. Hanson (ed.) Alfalfa science and technology. Agronomy 15:721-736.
2. Bolton, J.L., B.P. Goplen, and H. Baeniziger. 1972. World distribution and historical developments. *In* C.H. Hanson (ed.) Alfalfa science and technology. Agronomy 15:1-34.
3. Certified Alfalfa Seed Council, Inc. 1972. Alfalfa analyst. California Crop Improvement Association, University of California, Davis.
4. Crabb, R. 1959. Highlights of the Certified Alfalfa Seed Council, 1953-1959. California Crop Improvement Association, University of California, Davis.
5. Graber, L.F. 1950. A century of alfalfa culture in America. Agron. J. 42:525-535.
6. Hanson, C.H. 1965. The history and operation of the National Certified Alfalfa Variety Review Board. p. 62-70. *In* Annu. Rep. Int. Crop Imp. Assoc. International Crop Improvement Association, Clemson University, Clemson, SC.
7. Hendry, G.W. 1923. Alfalfa in history. J. Am. Soc. Agron. 15:171-176.
8. Jones, F.R. 1925. A new bacterial disease of alfalfas. Phytopathology 15:243-244.
9. Parsons, F.G. 1979. The story of seed certification in California. 1937-1976. Oral History Center, University of California, Davis.
10. Smith, D., and L.F. Graber. 1950. Performance of regional strains of Ranger alfalfa. Wis. Res. Bull. 171.
11. U.S. Congress. 1970. Plant Variety Protection Act, Public Law 91-577. U.S. Government Printing Office, Washington, DC.
12. U.S. Department of Agriculture. 1936. Seed statistics. Hay, feed and seed division. Bureau of Agric. Economics. U.S. Government Printing Office, Washington, DC.
13. ----. 1957. Seed crops by states, 1919-1954. Stat. Bull. 206. U.S. Government Printing Office, Washington, DC.
14. ----. 1984. Standard tests to characterize pest resistance in alfalfa cultivars. USDA Misc. Pub. 1434. U.S. Government Printing Office, Washington, DC.
15. Waterman, W.C. 1983. History of improved alfalfas. p. 4-7. *In* Rep. 28th Alfalfa Improv. Conf., Davis, CA. 13-16 July 1982. National Alfalfa Improvement Conference, University of Minnesota, St. Paul.

34 Future Trends in North America

G. E. CARLSON
USDA-ARS
Peoria, Illinois

A. A. HANSON
W-L Research, Inc.
Highland, Maryland

What is the future of alfalfa in North America? The answer to this question depends on changes in two major areas: agricultural and societal needs, and developments in science and technology that will modify the alfalfa plant, production practices, and demand and thus, cause changes in the various sectors of the "alfalfa industry." It is our opinion that we will be better able to change the alfalfa plant than to accurately predict broad changes in agriculture and society.

34-1 CHANGING DEMANDS FOR ALFALFA

Alfalfa, "Queen of the Forages," has ruled widely and well. Why? Because it filled a need better than its competitors. It is easily established, high in quality, easily modified to resist important diseases and pest insects, and persists and yields well in diverse environments. All of these traits have been discussed in detail in previous chapters. Alfalfa has been used primarily as a high quality feed for dairy and beef cattle (*Bos taurus*). This use will continue. Competitive livestock enterprises will continue to require high quality forage for high levels of animal performance. Indeed, economic pressures will demand increased quality with less input; a need more easily met by changing the plant than by reducing energy and management inputs. If alfalfa is to be changed to meet future needs of the livestock industry, we in the alfalfa industry (research, development, distribution, production, and utilization) must monitor changing trends in the livestock industry, primarily for beef and dairy cattle. Also, processed alfalfa could serve as an important source of protein for direct human consumption (6). For a variety of reasons, however, we do not anticipate any near-term shift away from the primary use of alfalfa as an animal feed.

Copyright 1988 © ASA-CSSA-SSSA, 677 South Segoe Road, Madison, WI 53711, USA. *Alfalfa and Alfalfa Improvement*—Agronomy Monograph no. 29.

34-1.1 Changing Needs for Livestock Products

It is difficult if not impossible to predict the nature of the livestock industry in the last decade of the 20th century. However, it may be instructive to look at recent trends and projections provided by experts in the livestock industry. Participants in an Animal Agricultural Symposium (8) reported that much of the improvement in animal production over the past two to three decades (since the 1950s) has resulted from extensive use of feed grains. However, there is evidence that the importance of grain for feeding ruminant livestock is declining. This trend, the Symposium participants say, is the result of increasing demand for grain exports, leaner meat, and allocation of grain as a fermentation base for liquid fuel; these pressures have been augmented by substantial increases in the cost of energy. Although we have witnessed wide fluctuations in export demand for grain and in the cost of fossil fuels, it is reasonable to assume that in time these predictions will hold true. In addition, forages represent a prime source of energy, and often protein, in feeding ruminant livestock; high quality forage can replace substantial quantities of feed grains (4).

Other experts see the growth in world population and increased consumer buying power creating a demand for 74% more milk, 82% more beef, and 90% more sheep (*Ovis aries*) and goats (*Capra hircus*) in the year 2000, compared with 1970. This increased demand will be met, most likely, by increasing the number of ruminant animals and increasing productivity per animal (4). If the beef cattle industry is to be competitive with the poultry (turkey, *Meleagris gallopavo*; chicken, *Gallus domesticus*) and pork (*Sus scrofa*) industries, it must take advantage of the ability of ruminants to convert forage to meat: "More than 80% of the feed consumed by beef cattle is forage, roughage and byproducts" (2).

A Council of Agricultural Science and Technology (CAST) Task Force projected that by the year 2000, human consumption of meat and milk will increase faster than consumption of cereals (1). A number of factors, including concerns for health, changing societal preferences, and life-style will influence the relative consumption of beef and cereals. Forage-fed beef is acceptable to the consumer if it is young, tender, and tasty. These traits are strongly influenced by the time it takes to finish cattle, which in turn is closely related to forage quality (4). In summary, various projections suggest an increasing demand for high-quality forage, a need that can be realized in many forage/livestock systems through the appropriate use of alfalfa.

34-2 CHANGING SCIENCE AND TECHNOLOGY

The preceding chapters show how science and technology have contributed to elevating alfalfa to its prominent role in agriculture. Thus, we will attempt to relate how present and future developments in science and technology will serve to meet anticipated or known needs to modify the alfalfa plant and alfalfa production systems. Science and technology can change the plant and the environment (management practices) and provide us with information to understand and manipulate the interaction between plants and their environment. There will be a continuing need for alfalfa genotypes adapted to particular environments and to particular uses as a part of a total system, i.e., forage/livestock production.

34-2.1 Changing Crop and Annual Production Practices

With the advent of the computer, the word system has assumed a new dimension in agriculture. Systems, the combination of several interrelated components, always have been a part of agriculture, but today it is possible to define the quantitative relationship among these components, and depend on a computer to manipulate these relationships rapidly and simulate the productivity of an agricultural system. This is a valuable tool, not only for the producer but also for the scientist. Producers will be aided in deciding which crops to plant, the level of production to aim for, and other management strategies based on the output of computer simulated production system models. Scientists will be aided in defining critical problems and in identifying plant parameters for modification based on probability of success and impact on the final product predicted by simulation models. In many livestock production systems, alfalfa has been and will remain a key component.

Models that simulate the growth of alfalfa may serve as a component of a larger agricultural production system model. Model development and validation require input from interdisciplinary teams of scientists; including geneticists, physiologists, agronomists, animal scientists, economists, system analysts, and others. Simulation models will become increasingly important as a tool for developing and understanding all aspects of agriculture, specifically the role of alfalfa in forage-livestock systems.

34-2.2 Exploring New Opportunities: Dinitrogen Fixation

If we are to develop the best management strategies, it will be necessary to understand how basic plant processes respond to the environment. More importantly, we need to know not only which ones *can be changed*, but especially which ones *should be changed.*

Will alfalfa become a major souce of N for crops? If fossil fuels become more expensive and increased emphasis is devoted to reducing N contamination of water supplies, then alfalfa offers both a source of N and improved erosion control in crop rotations. It is expected, therefore, that the symbiotic relationship between the plant and *Rhizobium* will continue to receive considerable attention. Heichel et al. (7) have shown that alfalfa will fix 177 kg of N/ha in the establishment year and as much as 224 kg of N/ha in the best of a 4-yr stand. They also found that the tradeoff between productivity and persistence is important in choosing the appropriate cultivar. Also, most of the N_2 fixed is partitioned to the herbage, and thus management practices will have a significant effect on the amount of N returned to the soil. Partitioning of photosynthetic energy to the root and increasing N_2 fixation could take priority over yield. However, the extent to which N_2 fixation can or should be increased will depend not only on the genetic potential, but also on the relative value of N and forage in the production system.

There are additional benefits from alfalfa and other legumes, particularly as a part of a total cropping system. Residue from these crops, roots and stubble, alter soil organic matter in both kind and amount, soil physical properties, and microbial population and activity. Pressures to reduce costs, questions about the impact of chemical (pesticides, etc.) on the environment, and new research technologies will provide the incentive to reassess the role of alfalfa in cropping systems.

34–2.3 Water and Nutrient Stress Physiology

In many parts of North America where alfalfa is grown, water significantly limits growth during some portion of each year. For some plant species it has been possible to select genotypes that use water more efficiently than others. Bradford et al. (1982) developed a cultivar of Asiatic Bluestem, 'W. W. Spar', that produced more forage with a given amount of water than did others. They combined laboratory measurement of CO_2 and water exchange with field trials in developing new procedures for selecting improved plants. It is likely that differences in water use efficiency exists among alfalfa genotypes. The demand for developing alfalfa cultivars that use water more efficiently will likely come from the more arid parts of the world where a persistent, high quality legume would be a valuable addition to herbage production. Also, one could expect benefits to irrigated agriculture (intensive alfalfa production for hay or seed) where energy and water costs are a major portion of production costs. Sophisticated instruments (such as CO_2 analyzers, nuclear magnetic resonance analyzers, amino acid sequencers, and so forth) coupled with computers have made it possible to examine fundamental physiological, biochemical, and biophysical processes in ways not previously possible, and to relate them to growth and development. Thus, we can expect to see continued advances in our knowledge of research into molecular, subcellular, and cellular processes that will help in understanding and controlling growth and development.

The demand to grow row crops on high value land coupled with increasing awareness of the need to reduce erosion and pollution of streams may be accompanied by attempts to grow alfalfa on more unfavorable sites; for example, on soils that are high in Al. We also may use alfalfa that is grown as part of a "high-residue management systems" in minimum or conservation tillage systems where there is increased competition from weeds, insects, and diseases. A better understanding of why plants differ in their tolerance to either excesses or deficiencies of various elements (such as P, K, Al, etc.) would aid in developing appropriate management strategies as well as procedures for identifying tolerant genotypes. Our understanding of the dynamics of root growth and the environment of the root is considerably less than that of shoot growth and its environment. Future research must focus on the whole plant as an integrated unit; a substantial challenge that is aided today by new technologies.

We can develop cultivars that resist insects and diseases, but we do not understand the nature of plant resistance to pests. How do plants recognize a "foreign" invader and mobilize resources to repel it? Advances in understanding the nature of resistance to insects, diseases, and stress could contribute substantially to improving alfalfa, reducing dependence on pesticides, and reducing public concern over the adverse effects of agricultural practices on the environment.

34-2.4 Expanding Genetic Resources

Biotechnology can mean a variety of things to different people. To some it is molecular biology; to others, genetic engineering; and to others, the application of sophisticated technologies in the development of new biological processes and products. Interest in biotechnology derives from the demonstrated ability to manipulate genes and probe biochemical processes in ways that go substantially beyond conventional genetics, physiology, and biochemistry.

These new technologies (in time they too will become conventional) are an important part of the full array of scientific technologies available for changing germplasm of alfalfa and developing innovative management strategies, but they will be only a part. Although it may be possible to vector genes for a desired trait from one genotype to another, the assessment of progress will depend ultimately upon the application of plant breeding and evaluation techniques.

34-2.5 Genetic Engineering

Biotechnology has captured the imagination of scientists, research administrators, and, thanks to frequent press releases, the general public. This technology offers the prospect of bypassing the constraints of sexual hybridization and facilitating genetic exchange among organisms. Thus, the manipulation of recombinant deoxyribonucleic acid (DNA) through genetic engineering has been suggested as a powerful tool for modifying the alfalfa plant: to isolate resistance to diseases and insect pests; to improve the efficiency of water use and drought tolerance; to develop salt or herbicide tolerance; to improve chemical composition; to eliminate natural inhibitors that restrict plant growth; and to reduce energy requirements, thereby increasing forage yield.

Alfalfa has been selected for use in genetic engineering studies because it is well suited for propagation in cell and tissue cultures, and at least some genotypes can be regenerated from cultures. Somaclonal variation within cultures provides a source of variants for use by breeders (5), and there is the added prospect of practicing large scale selection within cultures to isolate rare genotypes resistant to specific toxins. Also, radical change could be achieved by transformation that involves the injection, via a vector, of genes from one plant to another; or through the creation of new plants by means of protoplast fusion.

The introduction of significant change in a series of genotypes will not obviate the need to retain and/or introduce acceptable levels of multiple pest resistance, attain high levels of forage yield and persistence, and avoid any serious decline in seed set and seed yield. No one can predict the full impact of biotechnology on the improvement and use of alfalfa. It is reasonable to assume, however, that progress could be of limited usefulness in changing quantitatively inherited characteristics (e.g., seed and forage yield, drought resistance, N_2 fixation, etc.). Conversely, the search for methods to aid in the isolation of simply inherited traits (e.g., resistance to certain pests, herbicides, etc.) offers good prospects for success.

34–2.6 Continuing Need for the Conventional Approach

A valuable contribution of biotechnology, in the broad sense, may result from the development of procedures that will increase the efficiency of orthodox breeding methods. Thus, tissue culture could provide a rapid bioassay technique for testing the reaction of selected plants to various toxins and/or chemicals. In this approach, tissue cultures from selected plants could be exposed to a specific toxin, and donor plants either retained or discarded on the basis of tissue reaction. Bioassays of this type would increase the efficiency of breeding programs by reducing the number of escapes encountered in most screening tests as well as the timeframe required to complete a series of tests.

We have sampled and characterized very little of the genetic resources that are available for direct use in improving the alfalfa crop. There are over 2000 alfalfa plant introductions available in the USA for use in breeding programs. However, much of this germplasm requires evaluation, selection for adaptation, and enhancement to improve agronomic characteristics before it can contribute new sources of pest resistance, tolerance to various stress conditions, and improved combining ability for both seed and forage yield. Although most exotic germplasm requires a substantial investment in evaluation and selection, which for the most part has thus far been lacking, it has the potential of making a reasonably predictable contribution to alfalfa improvement. In contrast, we have difficulty in predicting the future course of biotechnology beyond its contribution to an improved understanding of alfalfa biology. This does not suggest that we should deemphasize biotechnology. It simply means that concurrent efforts to bring exotic germplasm into the mainstream of alfalfa improvement should be strengthened. This germplasm base must be expanded substantially, accompanied by a strong commitment to evaluation, in order to ensure stability and future progress in alfalfa improvement.

34-3 CHANGING INSTITUTIONAL RELATIONSHIPS

Improvement of the alfalfa crop in the USA was the sole responsibility of public agencies before the first commercial alfalfa breeding programs were established in the late 1950s. The principal factors that contributed to the initiation of privately supported alfalfa breeding programs are documented elsewhere, but reviewed briefly here. The origin of private research can be traced to the release of the bacterial wilt [*Corynebacterium insidiosum* (McCull.) H. L. Jens.] resistant cultivars, 'Ranger' and 'Buffalo', in 1942; the organization of the National Foundation Seed Project in 1949, to underwrite the cost of producing foundation seed; and the subsequent success of 'Vernal' alfalfa, released in 1953 and included in the National Foundation Seed Project. These developments led to a phenomenal increase in certified seed supplies of hardy, public cultivars, with California producing 28.5 Mt of certified alfalfa seed in 1957.

Production of certified alfalfa seed was so successful that supplies soon surpassed demand. In an effort to encourage use of certified seed, the grower supported Certified Alfalfa Seed Council was organized in 1953. The acreage planted to certified alfalfa seed increased, but seed producers were confronted with a very unstable situation marked by wide fluctuations in seed supply, narrow margins, and price cutting. It is not surprising, therefore, that private alfalfa breeding programs were started to provide distributors with a unique product, namely, proprietary alfalfa cultivars for which the distributors could serve as the sole source of supply.

The first proprietary alfalfa cultivars were composed of parent plants selected with varying degrees of evaluation from available public cultivars. From the very beginning, however, private alfalfa breeders enjoyed the advantages of access to facilities in the western USA, where large numbers of experimental cultivars could be grown for seed under cage isolation.

The resources devoted to private alfalfa breeding have increased substantially since the 1960s with the appearance of programs that employ several plant breeders supported by either full-time or part-time plant pathologists and entomologists. The smaller private alfalfa breeding programs may be limited to one plant breeder plus supporting technicians. All private alfalfa breeding programs have grown in sophistication in response to the increasing complexity encountered in the development of cultivars that are competitive in forage yield, persistence, and multiple pest resistance. The dominant role that has been assumed by private industry is evident in the number of cultivars receiving favorable reviews from the National Certified Alfalfa Variety Review Board. In the 5-yr period from 1979 to 1983, there were 57 cultivars released from private sources and seven from public agencies.

Technology applied within the private sector is based mainly on genetic theory, breeding methods, and techniques for isolating pest resistant germplasm developed by scientists in public institutions. Furthermore, industry has moved rapidly to capitalize on public releases of germplasm and cultivars, especially benchmark cultivars that possess resistance to a new insect pest and/or disease. A notable exception to this pattern is represented by the search for resistance to Verticillium wilt (*Verticillium albo-atrum* Reinke & Berth.), where private and public alfalfa breeders simultaneously released the first successful Verticillium wilt resistant cultivars. In spite of this one example, it is expected that public scientists will remain the primary source for the development of new concepts, procedures, and germplasm that are essential for further advances in alfalfa improvement.

34-3.1 Reassessing Traditional Roles

A number of public institutions have or are in the process of reevaluating their goals, and some have made the decision to stress the fundamental aspects of alfalfa improvement in preference to the release of cultivars. Decision makers hope to benefit consumers by providing information and germplasm to alfalfa breeding programs supported by private industry. There are several options available to public institutions, however, and some will continue to release cultivars while others may decide to shift resources from commodity-oriented research to biotechnology. A few public agencies have elected to obtain plant variety protection for new cultivars, and subsequently arrange for exclusive release to one or more seed companies. The exclusive release of public cultivars involves restrictions similar to those that apply to proprietary cultivars.

Venture capital firms, chemical companies, and other private corporations are exploring the application of genetic engineering in crop improvement, with alfalfa included among the crops selected for study. Projects range from the isolation of cell lines that are resistant to herbicide injury to the development of procedures for encapsulating cultured embryos of superior alfalfa selections. The emergence of privately supported genetic engineering research has added a new dimension to long-established institutional relationships. These changes may involve the initiation of orthodox alfalfa breeding projects within biotechnology firms, as well as active collaboration involving biotechnology companies and either public institutions or private seed companies.

There is no immediate prospect that biotechnology will revolutionize the development of alfalfa cultivars. Nevertheless, the visibility and challenge of biotechnology can be expected to increase in importance within public institutions, with a concomitant reduction in research on breeding methods and germplasm enhancement. Advances in alfalfa breeding demand a major input from public research, and, as public institutions change their emphasis, one might anticipate a decline in the quantity and quality of information that is available for immediate transfer to applied alfalfa breeding programs.

Private plant breeding programs are neither staffed and financed to fill the gap left by the redirection of a substantial portion of publicly supported alfalfa research, nor to conduct the "bridging research" that will be required for the application of many new discoveries. There appears to be a clear need to recognize the complementary nature of public and private research, and thereby act to protect the public resources devoted to the improvement of this remarkable crop.

34-4 SUMMARY

We have not attempted to identify all of the potential changes in the future of alfalfa, nor all of the forces that will bring about change. We believe that (i) alfalfa will increase in importance relative to other feed sources for ruminant livestock (primarily dairy cows), (ii) total alfalfa use could grow in spite of a decrease in the expected rate of increase in use of animal products (health concerns and, in some countries, a decline in disposable income) because of increases in the cost and reduced availability of feed grains; (iii) alfalfa could become increasingly important as a source of N and as a rotation crop for erosion control; (iv) continued improvement of the alfalfa crop will require wider use of germplasm resources, better management, and effective collaboration between public and private organizations; and (v) new technologies will allow us to explore the basic growth and reproduction processes of alfalfa in new, exciting, and useful ways.

REFERENCES

1. Anonymous. 1975. Ruminant as food producers—Now and for the future. Spec. Pub. 4. March 1975. Council of Agricultural Science and Technology, Ames, IA.
2. ——. 1982. The future for beef. A report by the Special Advisory Committee and the National Cattleman's Association. Reprinted from The Beef Business Bull. (March 5), Englewood, CO.
3. Bradford, J.A., P.I. Coyne, and C.L. Dewald. 1982. Leaf-water relations and gas exchange in relation to forage production in four Asiatic bluestems. Crop Sci. 22:1036–1040.
4. Fitzhugh, H.A., H.J. Hodgson, O.J. Scoville, Thanh D. Nguyen, and T.C. Byerly. 1978. The role of ruminants in the support of man. Winrock International Livestock Research and Training Center. April. Morrilton, AR.
5. Groose, R.W., and E.T. Bingham. 1984. Variation in plants regenerated from tissue culture of tetraploid alfalfa heterozygous for several traits. Crop Sci. 24:655–658.
6. Hanson, A.A. 1972. Future trends in the United States. In C.H. Hanson (ed.) Alfalfa science and technology. Agronomy 15:781–791.
7. Heichel, G.H., D.K. Barnes, C.P. Vance, and K.I. Henj. 1984. N_2 fixation, and N and dry matter partitioning during a 4-year alfalfa stand. Crop Sci. 24:811–815.
8. Wedin, W.F., and H.J. Hodgson. 1980. Feed production. p. 153–192. In G. Wilson et al. (ed.) Animal agriculture—Research to meet human needs in the 21st century. Westview Press, Boulder, CO.

APPENDIX

Appendix Table 1. Proportions of alfalfa cultivars approved by National Certified Alfalfa Variety Review Board (NCAVRB) developed by public and private organizations, 1980 to 1985.

Year	Type of organization Public	Private	Total	Percent private
	no.			
1980	2	11	13	85
1981	1	16	17	94
1982	1	7	8	87
1983	1	13	14	93
1984	1	11	12	92
1985	3	33	36	92

Appendix Table 3. Number and dormancy types of alfalfa cultivars approved by the National Certified Alfalfa Variety Review Board (NCAVRB) from 1980 to 1985.

Year	Norseman	Vernal	Ranger	Saranac	Mesilla	Moapa 69	ND	VND	Total no. approved
1980	7	7	15	54	15	--	--	--	13
1981	6	--	23	59	6	--	--	6	17
1982	--	12	--	25	12	12	25	12	8
1983	--	--	14	64	7	7	--	7	14
1984	--	25	42	33	--	--	--	--	12
1985	--	3	33	47	6	6	3	3	36
								Total	100
Avg. %	3	9	25	56	9	5	6	6	

Appendix Table 4. Alfalfa and alfalfa mixtures grown for hay in Canada (1981) (4).†

Province	Acres	Hectares
Newfoundland	543	220
Prince Edward Island	15 545	6 291
Nova Scotia	15 876	6 425
New Brunswick	13 882	5 618
Quebec	421 538	170 594
Ontario	1 458 411	590 211
Manitoba	942 582	381 458
Saskatchewan	1 269 031	513 570
Alberta	1 863 129	753 998
British Columbia	286 445	115 923
Total	6 286 982	2 544 308

† Note: Next census 1991, 1981 figures are reasonably accurate for 1986.

APPENDIX

Appendix Table 2. Alfalfa cultivars released in the USA and Canada since 1975. List includes only those cultivars registered in *Crop Science* and/or favorably reviewed by the National Certified Alfalfa Variety Review Board (NCAVRB).

Cultivar name	Originating or sponsoring agency	Registration No	Registration Reference	Registration Year	Year of review by NCAVRB
A-54	Embro Seed Co.				1979
A-57	Embro Seed Co.	7	CS‡ 15:96	1973	1973
Action	Research Seeds				1985
Admiral	Pick Seed				1985
Adventage	North American Plant Breeders	124	CS 23:799	1983	1981
Agate	Iowa, Mich., Mo. AES† + ARS, USDA	62	CS 13:768-69	1973	1972
Amador	Northrup, King & Co.				1976
Anchor	Rudy-Patrick Seed Division	59	CS 13:286	1973	1971
Anstar	F.F.R. Cooperative				1985
Answer	NAPB	123	CS 23:798-799	1983	1978
Apex	Rudy-Patrick Seed Division	60	CS 13:286	1973	1965
Apollo	NAPB	102	CS 23:178-179	1983	1981
Apollo II	NAPB	118	CS 23:802	1983	1981
Aquarius	Cal/West Seeds				1978
Arc	Md., N.C., Penn., Va. AES + ARS, USDA	76	CS 15:97	1975	1973
Ardente	Ferry-Morse Seed Co.				1975
Armor	NAPB	127	CS 23:800	1983	1981
Arrow	NAPB			1985	
AS-13	Ferry-Morse Seed Co.	74	CS 15:96	1975	1969
AS-13R	Ferry-Morse Seed Co.				1975
AS-49	Ferry-Morse Seed Co.	73	CS 15:96	1975	1975
AS-49R	Ferry-Morse Seed Co.				1976
AS-67	Ferry-Morse Seed Co.				1979
Atlas	NAPB	109	CS 23:179	1983	1976
Baker	Nebr. AES + ARS, USDA	87	CS 18:692	1978	1976
Baron	NAPB	117	CS 23:1010	1983	1982
Bell Ringer	Lovelock Seed Co.				1985
Big 10	Great Lakes Hybrids				1983
Blazer	Land O'Lakes	95	CS 20:283	1980	1978

(continued on next page)

Appendix Table 2. Continued.

Cultivar name	Originating or sponsoring agency	Registration No.	Registration Reference	Registration Year	Year of review by NCAVRB
Caliente	Ferry-Morse Seed Co.	75	CS 15:96–97	1975	1969
Challenger	Northrup King Co.				1983
Citation	NAPB	107	CS 23:178	1983	1974
Classic	F.F.R. Cooperative	117	CS 23:398	1983	1978
Commandor	Northrup King & Co.				1985
Conquest	Pioneer Hi-Bred Int., Inc.				1976
Cuf 101	Calif. AES + ARS, USDA	119	CS 23:398	1983	1976
C/W 61	Cal/West Seeds				1980
C/W 62	Cal/West Seeds				1980
C/W 69	Cal/West Seeds				1980
C/W 327	Cal/West Seeds				1985
C/W 334	Cal/West Seeds				1985
C/W 339	Cargill				1985
C/W 341	Research Seeds				1985
C/W 349	Agway				1985
C/W 940	Cal/West Seeds				1982
C/W 8015	Cal/West Seeds				1981
Dart	NAPB				1985
Decathlon	W-L Research				1982
Defender	Northrup, King and Co.				1980
Deseret	Utah AES + ARS, USDA	78	CS 17:671	1977	1974
Diamond	NAPB				1985
DK-187	DeKalb-Pfizer Genetics				1983
Dona Ana	New Mexico AES	141	CS 25:705	1985	1983
Drummor	Northrup, King and Co.				1983
Drylander	Caracia Dep. of Agric.	84	CS 17:977–978	1977	
DS 309	Dairyland Res. Int.				1985
Duke	NAPB	130	CS 23:801	1983	1981
Eagle	O's Gold Seed Co.				1983
Edge	Research Seeds				1985
Elevation	Land O'Lakes				1984

(continued on next page)

APPENDIX

Appendix Table 2. Continued.

Cultivar name	Originating or sponsoring agency	Registration No.	Registration Reference	Registration Year	Year of review by NCAVRB
El Unico	Arizona AES + ARS, USDA	57	CS 13:129	1973	1967
Endure	NAPB	140	CS 25:705	1985	1983
Epic	Land O'Lakes	154	CS 23:802	1983	1980
Excalibur	Cal/West Seeds				1983
Expo	NAPB	155	CS 23:799	1983	1981
Florida 77	Florida AES + ARS, USDA	79	CS 21:797	1981	1979
Garst 636	Garst Seed Co.				1985
Gladiator	Northrup, King and Co.				1973
Glory	Dairyland Seed Co.				1979
Granada	NAPB	116	CS 23:1010	1983	1982
G 777	Funk Seed Int.				1976
G 2315	W-L Research				1980
G 2318	W-L Research				1982
G-2852	Funk Seed Int.				1985
G 7730	NAPB	139	CS 23:800–801	1983	1980
Heinrichs	Canada Dep. Agric.	159	CS 24:207	1984	1978
HI-PHY	F.F.R. Cooperative	148	CS 23:398	1983	1975
Horgeoye	Cornell Univ. AES	145	CS 23:181	1983	1984
Husky	Lovelock Seed Co.				
Impact	NAPB			1985	1985
Inca	Lovelock Seed Co.				1980
Jubilee	NC + Hybrids				
Kare	Canada Dep. Agric.	58	CS 13:130	1973	
Lew	Ariz. AES	96	CS 21:349	1981	1974
Liberty	N.C. AES + ARS, USDA				1975
LS 9-12, 5	Lovelock Seed Co.				1985
Magnum	Dairyland Seed Co.				1979
Marathon	Northrup, King and Co.				1974

(continued on next page)

Appendix Table 2. Continued.

Cultivar name	Originating or sponsoring agency	Registration No.	Registration Reference	Year of review by NCAVRE
Matador	Northrup, King and Co.			1976
Maverick	NAFB			1981
Maxidor	Northrup, King and Co.	131	CS 23:801	1978
Maxim	Canex			1983
Mercury	NAFB			1981
Milkmaker	Lovelock Seed Co.	126	CS 23:799–800	1984
MNUCXSW	University of Minnesota			1985
Mohawk	Cornell Univ. AES			1985
Multileaf	Cornell Univ. AES			1980
Nugget	NAPB	106	CS 23:178	1974
Olympic	NAPB	110	CS 23:179	1976
Oneida	Cornell Univ. AES			1980
Pacer	Land O'Lakes	82	CS 17:977	1975
Peace	Canada Dep. of Agric.	105	CS 22:1258–1259	
Peak	Land O'Lakes	93	CS 20:283	1978
Perry	Kans., Nebr., S.D., Wis., AES + ARS, USDA	97	CS 21:349	1979
Phytor	Northrup, King and Co.			1977
Pierce	Northrup, King and Co.			1982
Pike	Northrup, King and Co.			1981
Polar I	Northrup, King and Co.			1974
Polar II	Northrup, King and Co.			1980
Preserve	Northrup, King and Co.			1983
Primal	Northrup, King and Co.			1978
Prowler	Northrup, King and Co.			1980
Raidor	Northrup, King and Co.			1980
Ramsey	Minnesota AES + ARS, USDA	63	CS 13:769	1972
Rangelander	Canada Dep. of Agric.	95	CS 20:668	
Riley	Kans AES + ARS, USDA	88	CS 18:911	1977
Rincon	N.M. AES	90	CS 19:741	1978
Roamer	Canada Dep. of Agric.	83	CS 17:977	

(continued on next page)

APPENDIX

Appendix Table 2. Continued

Cultivar name	Originating or sponsoring agency	Registration No.	Registration Reference	Year of review by NCAVRB
Salute	Dairyland Seed Co.			1984
Sapphire	NAPB			1985
Saranac AR	Cornell Univ. AES	116	CS 23:181	1975
Sparta	Land O'Lakes			1984
Spectrum	Cenex Seed Co.			1981
Spredor	Northrup, King and Co.			1974
Spredor II	Northrup, King and Co.			1982
Summit	Stauffer Seeds			1985
Sunrise	Cal/West Seeds			1965
Surpass	Cenex			1985
SX-10	Sexauer Seed Co.	74	CS 15:96	1973
SX-418	Sexauer Seed Co.			1978
Target	Dairyland Seed Co.			1984
Team	Md., N.C. AES + ARS, USDA	64	CS 13:769	1968
Tempo	F.F.R. Cooperative	121	CS 23:798	1969 + 1974
Thorobred	Lovelock Seed Co.			1985
Thunder	NAPB	132	CS 23:801–802	1981
Titan	Rudy-Patrick Seed Division	64	CS 13:286	1968
Tomahawk	Jung Farms			1985
Trek	Canada Dep. of Agric.	77	CS 16:444	1978
Trident	NAPB	128	CS 23:800	1981
Trumpetor	Northrup, King and Co.			
UC Cargo	California AES + ARS, USDA	79	CS 17:671	1975
UC Cibola	California AES	133	CS 23:1216	1982
UC Salton	California ARS	80	CS 17:671–672	1971
Valador	Northrup, King and Co.			1978
Valor	Land O'Lakes	81	CS 17:977	1974
Vancor	Northrup, King and Co.			1980
Vangard	NAPB	111	CS 23:179–180	1976

(continued on next page)

Appendix Table 2. Continued.

Cultivar name	Originating or sponsoring agency	Registration No.	Registration Reference		Year of review by NCAVRB
Vernema	Wash. AES + ARS, USDA	135	CS 23:1009	1983	1981
Verta	NC – Hybrids				1985
Vista	Cal/West Seeds				1975
Voris A-77	NAPB	122	CS 23:798	1983	1978
Weevlchek	F.F.R. Cooperative	120	CS 23:798	1983	1974 + 1979
WL Southern Special	W-L Research				1982
WL 216	W-L Research	54	CS 13:128–139	1973	1971
WL 219	W-L Research				1975
WL 220	W-L Research	91	CS 19:298	1979	1977
WL 221	W-L Research	112	CS 23:180	1983	1979
WL 307	W-L Research	55	CS 13:129	1973	1971
WL 308	W-L Research	56	CS 13:129	1973	1971
WL 309	W-L Research	66	CS 14:337	1974	1972
WL 310	W-L Research	89	CS 19:293	1979	1974
WL 311	W-L Research	85	CS 18:523	1978	1974
WL 312	W-L Research	98	CS 21:349–350	1981	1978
WL 313	W-L Research	113	CS 23:180	1983	1979
WL 314	W-L Research	100	CS 22:1257	1982	1981
WL 315	W-L Research	114	CS 23:180	1983	1980
WL 316	W-L Research	101	CS 22:1257	1982	1981
WL 318	W-L Research	86	CS 18:523	1978	1974
WL 320	W-L Research				1983
WL 450	W-L Research	67	CS 14:805	1974	1972
WL 451	W-L Research	68	CS 14:905	1974	1972
WL 501R	W-L Research	69	CS 14:905	1974	1972
WL 512	W-L Research	102	CS 22:1258	1982	1976
WL 514	W-L Research	103	CS 22:1258	1982	1978
WL 515	W-L Research	104	CS 22:1258	1982	1981
WL 516	W-L Research				1985
WL 600	W-L Research	70	CS 14:905–906	1974	1972
WL 605	W-L Research				1985
Wrangler	Nebraska AES + ARS, USDA	142	CS 26:646	1986	1984

(continued on next page)

APPENDIX

Appendix Table 2. Continued.

Cultivar name	Originating or sponsoring agency	Registration No.	Registration Reference	Year of review by NCAVRB
77-8 Ca B	W-L Research			1985
88	L.L. Olds Seed Co.			1983
120	DeKalb Ag Research	94	CS 20:283	1978
130	DeKalb Ag Research			1980
131	DeKalb Ag Research			1976
141	Hoffman Seeds			1983
167	DeKalb Ag Research			1975
185R	DeKalb Ag Research			1978
521	Pioneer Hi-Bred Int.			1975
524	Pioneer Hi-Bred Int.			1977
526	Pioneer Hi-Bred Int.			1981
530	Pioneer Hi-Bred Int.	65	CS 14:127	1972
531	Pioneer Hi-Bred Int.			1977
532	Pioneer Hi-Bred Int.			1979
545	Pioneer Hi-Bred Int.			1977
555	Pioneer Hi-Bred Int.			1979
572	Pioneer Hi-Bred Int.			1975
581	Pioneer Hi-Bred Int.			1977
624	Dairyland Seed Co.			1984
629	Garst Seed Co.			1984
655	Dairyland Seed Co.			1984
5432	Pioneer Hi-Bred Int.			1985
5444	Pioneer Hi-Bred Int.			1984
5929	Pioneer Hi-Bred Int.			1983

† AES = Agricultural Experiment Station, ARS = Agricultural Research Service, USDA = United States Department of Agriculture.
‡ CS = *Crop Science*.

SUBJECT INDEX

Individual cultivars are listed under the heading Cultivars, rather than separately. Diseases are listed under the headings Diseases, common names and Disease pathogens. Other broad categories include Grasses, Insects, Legumes, other than alfalfa, *Medicago* spp., Nematodes, and Weeds.

Abiotic diseases, 621, 623, 648, 844
Acetylene reduction assay, 245
Acidosis, 544
Adaptation
  adverse conditions, 2
  Canada, 50, 614–617
  cultivar selection, 595–620
  elevation, 601
  fall dormancy, 605–608
  humid regions, 608–614
  irrigation, 603–608
  nonhumid regions, 599–608, 616–617
  rainfall, 597
  United States, 50
Adenosine, 349
Adenosine triphosphate (ATP), 349
ADF. *See* Fiber, acid detergent
Aeration, soil, 478
Afghanistan, 109
Africa, 60–67
Agrochemotaxometry, 97
Air pollution, 499, 622–624, 844
Alfalfa
  crop advisory committee, 4
  improvement conferences (NAAIC), 3, 4, 6, 17–19, 22, 1028–1029
  maceration, 555
  meal, 502, 504, 508
  phenology, 168, 179–182
Alfalfa Hay Quality Committee, 13, 466
Algeria, 61–62
Alkaloids, 508
Alleles, 111–115, 121, 795, 812
Allelochemicals, 861
Allelopathy, 603, 722–723
Alloautoploid, 94, 752, 768, 771
Allogamy, 94, 120
Allozyme. *See* Isozyme
Aluminum, 341–345, 352
American Forage and Grassland Council, 13, 465–466
American Seed Trade Association, 1027–1028, 1035
American Society of Agronomy, 1027
Amines, 545

Amino acids, 240, 274–275, 546, 548–549, 919
Ammonia, 543–545
  dinitrogen fixation, 240
Anaerobic conditions, 167, 170–172, 175, 178
Anatomy, 125–162
  adventitious stems, 146–148
  crown, 138–139
  epicotyl, 126, 143–144, 148
  hypocotyl, 130, 142–146
  leaf, 148–150
  nodules, 234–240
  roots, 132–138
  seed, 125–130
  seedling, 130–132
  vascular transition, 132, 138–140, 142–150, 156
Aneuploidy, 750, 754–755, 914–916
Animal intake, nutrients, 465, 473, 475, 482–483, 515, 523, 525, 528
Animal nutrition, 348, 355, 361, 463–484, 539, 816–821
Animal performance, 463–465, 473, 475, 481
Annual *Medicago*, 93–95, 116–117, 119
Anoxia, 281
Anthers, 932, 950
Anthocyanidins, 98, 110, 499
Antiquality, 463–464, 563–564
Antixenosis, 861, 879–880
AOSCA. *See* Association of Official Seed Certifying Agencies
AOSA. *See* Association of Official Seed Analysts
Apex, shoot, 931
Arabinose, 467–469
Area planted
  Africa, 60–67
  Asia, 68–79
  Canada, 2, 7–8, 20–21, 50, 615
  Central America, 53
  Europe, 32–49
  North America, 49–53
  Oceania, 77–81
  South America, 53–60
  United States, 2–3, 19–21, 50, 595
Argentina, 53–56
Asexual reproduction, 949
Ash, 477, 481
Asia, 68–77, 109
Asparagine, 240
Association of Official Seed Analysts (AOSA), 972–974
Association of Official Seed Certifying

Agencies (AOSCA), 990, 1025, 1028-1029
Attractiveness of flower, 183-184, 989
Australia, 77-80
Autogamy, 94, 935
Autopolyploid, 750, 756-758
Autotetraploid, 20, 113, 812
  confirmation, 756-758
  disequilibrium, 794, 804
  genetic model, 794-798
  heritability, 792, 798-799
  heterosis, 783
  heterozygosity, 756, 759-760
  inbreeding, 804-805
  inheritance, 756-758
  population models, 803
  random mating, 794
  selection, 800-803
Autotoxicity, 309-310, 603
Auxin, 165-166
Avena sativa L. *See* Oat

Backcross, 781
Bacterial diseases. *See* Diseases
Bales, 574-578
  percentage of forage harvested, 567
Band seeding (seeded), 304, 311-312, 317, 322-323, 347, 350
Barley, 318-319, 526
Base temperature, 173, 181
Bees, 777, 779, 783, 1001-1009
  alkali, 951, 1005-1007
  honey, 779, 951, 1002-1005
  leafcutting, 779, 951, 1007-1009
  pollination control, 939
  visitation, 951-952
Belgium, 34
Benchmark cultivars, 1044
Benzoic acids, 498
Biological control, insects, 10, 671, 673, 678, 692-693, 997-999
Biotechnology. *See* Tissue culture
Biotic environment, 163
Biotypes, 862, 866, 870
Bloat, 495-498, 818-819
  foaming agents, 496
  grass mixture, 304
  pasture, 496, 531-532
  prevention, 304, 531-532
  research, 13-14, 497-498
Bolivia, 56
Boron
  availability, 341, 359-360
  concentration, 336-337, 359-360
  deficiency, 306, 359
  rate of application, 335, 361
  role in plant, 359
Brassica napus var. *oleifera*, 261
Brazil, 56-57

Breeding, 783-784, 793
  abiotic diseases, 844
  acidity tolerance, 342
  bloat safety, 14, 497-498
  chemical traits, 816
  constraints and objectives, 777-778
  creeping roots, 148
  cultivar (strain) crosses, 782-783
  digestibility, 817
  dinitrogen fixation, 14-15, 22, 229-230, 240-251, 307, 322, 345
  disease resistance, 9-10, 19-22, 119, 622, 625-626, 630, 633-634, 638, 642, 647, 651, 654, 656, 661, 827-848, 920-922
  fertility, 940, 943
  frequent cutting, 220
  haploids, 747-748, 756-757
  heat tolerance, 179, 285
  herbicide resistance, 221
  heterosis, 783
  hybrid alfalfa, 950, 953-954
  improvement, 5, 16-17, 20-21, 163, 250-251, 809
  insect resistance, 10, 20-21, 119, 671, 673, 677-678, 680, 859-902
  interspecific crosses, 16-17, 116-118
  methods, 777-793, 829
  negative components, 818
  nematode resistance, 638, 654-655, 827, 844-848
  nutrient concentration, 339
  physiological traits, 220-221
  quality, 816-821
  salt tolerance, 119, 164-165, 221, 922-923
  saponins, 505, 819
  stress tolerance, 291
  synthetic cultivars, 780-781, 811-814
  yield, 198, 200, 203, 220-222, 781, 809-816
Buds
  axillary, 130, 132, 146, 170
  basal, 170, 214, 217-218, 326
  flower, 150, 180, 217
  lateral, 144, 146, 153
  stages, 150, 180
Bulgaria, 35

CADL (cultivated alfalfa at the diploid level), 120, 748, 756
Cadmium, 362
Calcium, 306, 507-508
  availability, 343
  Ca/Mg ratio, 355, 819-820
  Ca/P ratio, 508
  concentration, 339, 355, 479
  deficiencies, 341
  effect on nodulation, 232, 244

SUBJECT INDEX

effect on quality, 539–540, 819
Calyx, 151
Canada, 50, 515–518, 614–617, 1046
Canadian Forage Seed Project, 7–8, 20
Canadian Seed Growers Association, 8
Canopy
  development, 386, 413
  resistance, 382, 384
  temperature, 399–400
Cantharidin, 676
Carbamates, 693
Carbohydrates, 213–214, 216–220, 272–273, 387, 477, 481
  cellulose, 177, 468, 503
  cold tolerance, 260
  crowns, 178, 413
  depletion, 218
  diurnal trends, 477
  hemicellulose, 466–470, 503
  mobilization, 216
  nonstructural, 174, 178–179, 216–220, 415, 421–422, 425, 428, 475–476
  pectins, 468
  root reserves, 216–218, 262, 268, 413, 415, 418, 420
  starches, 174, 214, 216–217, 272
  storage, 216–220
  sucrose, 214, 216, 270
  sugars, 214, 216, 270–272
  taproots, 178
  utilization, 213, 216–222
Carbon dioxide
  leaf exchange, 169, 186–199, 202–205, 208, 211–212
  nodule fixation, 241
Carbon metabolism, 195–222
  assimilation, 195–196, 216
  dinitrogen fixation, 240
  exchange, 169, 198, 210–211, 383
  partitioning, 195–196, 213–216
  sinks, 195, 211, 213–216, 218–219
  source-sink concepts, 167, 177, 213–216, 218–219
  utilization, 195–196, 216–219
Carboxylation efficiency, 197
Carotene, 480–481
Carotenoids, 110
CAST. See Council of Agricultural Science & Technology
Cattle feeding, 525
  equipment, 584–592
Caucacus, 93, 109–110
Cell fusion, 120
Cell rupture, 555–556, 564, 819
Cell walls, 465–471, 497, 502, 507
  crude lignin, 177, 469
  digestibility, 470–471
  primary, 466–468, 483
  protein, 468, 494
  secondary, 466–468, 483

Cellulose, 177, 466–467, 469, 503
  digestibility, 469
Center of origin (diversity), 25, 97, 110–115, 120
Central America, 53
Certified Alfalfa Seed Council, 6, 20, 22, 597, 1034–1035
Certified seed, 6, 8, 990
Chalkbrood, 1008–1009
Chemical control
  insects, 321, 329, 693–694, 994–1000
  weeds, 321, 325–328, 344, 986–987, 992–993
Chile, 114–115
Chlorenchyma, 139–141, 148–149
Chlorinated hydrocarbons, 693
Chlorogenic acid, 498
Chlorophyll, 174
  deficiency, 116
Chloroplast membrane, 494
Chloroplastic protein, 494
Chromic-nitric acid, 963
Chromosomes, 116, 737–776
  doubling, 111, 120
  idiograms, 749, 760
  mapping, 754–755
  number, 93–95, 120, 760–761
  rearrangement, 94, 120
Clear seeding, 304, 321, 325
  cutting after, 321–322
  herbicides, 321, 324
  superiority, 324
  use, 321–323
  without companion crop, 321
Climatic effects, 595–620
  cold tolerance, 280–281
  diseases, 624–636, 638–639, 642, 644–645, 648, 650–654
  drying of forage, 571–574
  feeding value, 337–338, 474, 476–480
  pollination, 1015
  resistance to insects, 671, 693, 862, 866, 870
  seed, 964, 1015
Cobalt, 361
Cold tolerance, 260–284
  bound water, 269
  bud development, 268–269
  carbohydrates, 270–273
  critical harvest period. See Fall management
  dehardening, 262, 388
  development, 260
  environmental conditions, 260–265
  fall dormancy, 265–266, 283–284
  growth habit, 265, 269
  growth regulators, 278
  hardening, 260–264
  ice sheets, 281
  leaves, 266–268

light, 260-261
lipids, 277
  measuring, 282-284
  morphological factors, 265-269
  nitrogenous substances, 273
  photoperiod, 260-261
  plant tissues, 266-268
  respiration, 273
  seasonal changes, 261-262
  soil fertility, 264
  survival, 260, 266-268
  temperature, 261-264
  thawing rate, 262-264
  water, 269-270
Collenchyma, 139-142
Combining ability, 786, 814
Common names of alfalfa, 26-27
Compaction, soil, 309, 313-317, 478
Companion crops, 166, 168, 317
  crops used, 318
  reduced seeding rate, 318, 320
  removal, 319-320
  volunteer grain, 328
  weed control, 319
Competition among plants, 18, 319-320
  light, 168, 317-321, 430, 441
  management, 319-320, 450-454
  nitrogen effects, 320
  nutrients, 320-322, 352, 443
  seeding, 320, 328
  temperature, 443
  water, 319-320, 430, 442
  weeds, 309, 325
Competitive ability in mixtures, 166, 320, 446-449
Complementary strain crosses, 782-783
Computer modeling. *See* Modeling
Conductance
  hydraulic, 380
  leaf, 381, 383
  stomata, 381
Coniferyl, 469
Conservation tillage, 1040
Contractile growth, 130-131, 138, 145, 381
Control systems, insects, 694-695
Copper, 361-362, 479
Corn, 310, 333, 345, 515, 547, 780, 783, 785, 788, 812
Corolla, 98, 101, 107, 109, 150
Cotton, 785
Cotyledons (seed leaves), 95, 126-127, 130-132, 138, 142-146, 165-166, 961, 974
Coumarins, 499
Coumestans, 506
Coumestrol, 480-481, 506
Council of Agricultural Science & Technology (CAST), 1038
Creeping-root habit, 146-148, 516, 779, 906

Crop communities, 196, 206, 210
Crop development
  carbohydrate reserves, 414
  dry matter accumulation, 413
  first flower, 413, 417, 419-421
  forage quality, 180, 414
  leaf and stem yield, 413-414
  leaf loss, 413
  morphology, 179-180, 413
  plant maturity, 179-180, 413
  temperature, 337, 419
Crop improvement, certification, 1025-1028
Crop production practices, 1039
Crop rotations, 249-250
Cross fertility, 952
Crown, 138-140
  anatomy, 137-140
  buds, 170, 214, 217-218, 268-269, 421
  depth, 138
  diameter, 169, 177
  diseases, 136-138, 622, 636, 644, 647-649, 651-652
  physical stress, 136, 138
  primary, 132, 138
Cubes, percentage of forage harvested, 567
Cultivar (variety)
  adaptation, 376, 595-620
  approved by NCAVRB, 1980-1985, 1046
  crosses (strain), 782-783
  dormancy, 416, 596, 601
  number of, 597
  release procedures, 597-599
  released in USA and Canada since 1973, 1046-1053
  synthesis, 793
  usage, proprietary, 611-612, 1025-1026
Cultivars. *See also* specific countries
  Afghanistan, 882
  African, 19, 114, 242, 377, 654, 847, 865
  Agate, 21, 248, 378, 839
  Alfa, 842, 885
  Algonquin, 616
  Anchor, 378, 678
  Angus, 616
  Anik, 264, 267
  Apalachee, 782, 837, 845
  Apex, 869
  Apollo, 613
  Apollo II, 22, 608
  Arabian, 879
  Aragon, 376
  Arc, 21, 833-834, 863, 873, 875
  AS-13R+, 607
  AS-49, 976
  Atlantic, 377-378, 873, 940
  Baron, 607
  Beaver, 267, 617, 660, 819
  Beltsville 2-An4, 833

SUBJECT INDEX

Buffalo, 19, 114, 279, 416, 520, 613, 623, 830, 835, 869, 940, 1024, 1043
California Common, 832, 885
California Common 49, 832
Caliverde, 413, 782, 836, 976
Canadian Variegated, 603
Cancreep, 603
Cayuga, 416
Cherna, 603
Cherokee, 805, 840, 877, 886
Chilean, 19, 114-115, 242, 595, 654, 1023
Cimarron, 613
Cody, 171, 383, 613, 832, 835, 866, 885
Common, 603, 617
Cossack, 603, 617, 1024
CUF 101, 21, 605-606, 871-872
Culver, 877, 881, 885
Dawson, 416, 602, 613, 869, 871
Delta, 940
Deseret, 114, 607
DK 135, 22
Dona Ana, 607
Drylander, 603, 617
DuPuits, 114, 412, 416, 419, 520, 612, 616, 842, 845-846, 873, 879, 905, 940, 1025
Europe, 114
Flamande, 842, 845, 1016
Flemish, 114-115, 266
Florida 66, 171, 606, 614
Florida 77, 606, 614
Glacier, 833
Granada, 606
Grimm, 19, 242, 412, 416, 520, 602, 610, 617, 832, 842, 885, 1024
Grimm Sask. 415, 617
Grimm Sask. 666, 617
GT-55, 607
Hairy Peruvian, 113-114, 167, 251, 266, 278, 876, 879, 882, 884
Hardistan, 885
Hayden, 21
Hilmar, 846
Hunter River, 416, 520, 839
Indian, 114, 266, 847
Iroquois, 181, 612, 616, 678, 782
Kane, 603, 617
Kansas Common, 613
Kanza, 20, 613, 832, 869-870, 885
Kometa, 875, 888
Ladak, 19, 114-115, 172, 266, 416, 520, 602-603, 617, 830, 832, 835, 885, 906, 1024
Ladak 65, 165, 603
Ladak 75, 603
Lahontan, 20, 114, 177, 242, 420, 604-605, 607, 839, 845-846, 866, 879, 882, 885, 1015
Lew, 604, 606

Liberty, 873, 875
Maverick, 603
Maxidor, 606
Mesa-Sirsa, 114, 165, 836, 882
Mesilla, 605, 607, 870
Moapa, 20, 114, 168, 172, 177, 181, 242, 419-420, 422, 605, 607, 691, 865, 873, 876, 976
Moapa 69, 416
Monsefu, 114
Narragansett, 242, 416, 612, 831, 836, 873, 940-941
Nemastan, 114, 845, 885
Nevada Syn F, 842
Nevada Syn XX, 829, 846
New Mexico 11-1, 605, 842
Nitro, 15, 22, 322
Nomad, 528, 602, 885
Norseman, 177
Ontario Variegated, 616
Panonia, 501
Pegauer, 842
Peruvian, 114, 266, 416
Pierce, 607
Pike, 607
Pioneer "5" Series, 607
Raduga, 875, 888
Rambler, 167, 183, 263, 271, 274, 520, 602, 617, 877
Ramsey, 416
Rangelander, 617
Ranger, 19, 166, 242, 412, 416, 517, 604, 608, 610, 616-617, 623, 830, 835-836, 842, 845, 869, 882, 885, 976, 1015, 1024, 1043
Rere, 871
Rhizoma, 170, 174, 416, 520, 602, 616, 873, 877, 885
Rincon, 606
Roamer, 271, 276, 603, 617
Roverde, 603
Saranac, 20, 170-171, 263, 271, 276, 321, 386, 394, 416, 610, 612, 616, 678, 782, 829, 833-834, 836, 842, 886, 905
Saranac AR, 416, 612, 833
Sevelra, 416, 602
Sibturk, 603
Sirsa 9, 882
Socheville, 114
Sonora, 114, 271, 273, 276, 383, 607, 836
Sonora 70, 607
Spanish, 595
Spredor, 603
Spredor II, 603
Talent, 845
Team, 20, 833-834, 844, 863, 873, 875
Tempo, 976
Teton, 520, 602-603, 835, 840, 843, 877

Tierra de Campos, 376
Tontana, 170, 174
Travois, 520, 602–603, 835, 885
Trifecta, 872
Trumpetor, 22, 608
Tuna, 114, 842
Turkestan, 266, 829–830, 832, 845
UC 73, 873, 876
UC Cargo, 606
UC Cibola, 605–606
UC Salton, 416, 606
Unita, 836, 843, 1015
Vernal, 20, 168, 170–171, 179, 181, 242, 271, 273, 276, 305, 334, 383, 412, 419, 445, 456, 520, 604, 608, 610, 612, 616–617, 654, 678, 810, 830–833, 836, 839, 843, 846
Vernema, 22, 608, 845, 905
Vertus, 608
Victoria, 603
Washoe, 605, 607, 845, 869, 871–872
Weevlchek, 834, 873
Williamsburg, 426, 613, 834–835, 886
WL 300 Ceries, 607
WL 311, 416
WL 316, 22
WL 318, 378, 416
WL 400 Series, 607
WL 501 R, 607
WL 512, 416, 607
WL 515, 606
Zia, 20, 605, 607, 866, 882, 885
Cultural practices, 303–328, 428
  insect control, 686–687, 690–691
  seed production, 986–994
Cuticle, 125, 127, 130, 385, 472
Cutin, 125, 466, 469
Cutting schedule, 411–426
  calendar date, 417–419, 475–476
  critical harvest period, 266–267
  crown buds, 268–269, 420–421
  cultivar development, 220
  dinitrogen fixation, 249–250
  fall management, 321–322, 411, 423–427
  first flower, 419, 428
  fixed interval and number, 321, 411, 415–421
  forage quality, 321, 335, 414, 474–476
  grasses vs. alfalfa, 452–456
  growth stage, 411, 419–420, 472–476, 483
  height, stubble, 421–422, 428, 456
  historical, 411–412
  insect control, 418, 428–429, 678, 682, 691
  leaf loss, 413, 427
  new crown shoots, 411, 413, 421–422
  number after seeding, 321–322
  optimum, 475–476
  regrowth, 218–219

root reserves, 217–219, 414–415, 420
seed production, spring clipback, 993, 1004
seeding year, 321, 326
spring management, 427–428
stage of growth, 321, 411, 419–420, 475
stand decline, 420, 429
weed control, 429–430
Cytogenetics, 16, 22, 93–94, 109, 116, 737–776
  alloautoploids, 94
  aneuploids, 750, 754–755
  autohexaploids, 94
  autopolyploids, 115, 750, 756–758
  autotetraploids, 750, 756–758
  diploids, 94, 107–113, 120–121, 171, 746–748
  embryo development, 739–748
  gametophyte, 737–739, 741–742, 744, 746
  haploids, 737, 746–748, 756
  heptaploids, 753
  hexaploids, 94, 117, 120, 171, 751–752, 758, 761, 765, 767–768
  hybrids, 109–110, 116–118, 120–121
    interspecific, 762–771
  octoploids, 746, 753, 758–759
  pentaploids, 751
  tetraploids, 94, 107–110, 113–115, 120–121, 171, 750–751
  translocations, 93, 116
  triploids, 748–750
  trisomics, 754–755
Cytokinesis, 744–745
Cytoplasmic sterility, 20, 183
Czechoslovakia, 35–36

Dark period, 167, 177, 182
Daylength, 181, 476. *See also* Photoperiod
DDM. *See* Digestible dry matter
Dehydrated alfalfa for protein, 541, 546
  balanced rations, 548
  beef cattle, 547
  heat damage, 547
  protein, preservation, 550
  protein, solubility, 548
Denmark, 36
Deproteinized juice, 559, 561, 564
Dessicants, 543, 1010–1011
  potassium carbonate, 543, 573
  sodium carbonate, 573
Difference method, 230, 245
Digestibility, 387, 463, 465–466, 481–482, 545–547, 817
  acid detergent fiber, 465, 818
  cell types, 472
  cell wall, 470–472
  cellulose, 469
  decline, 469

SUBJECT INDEX

dry matter, 467
  estimated, 465
  neutral detergent fiber, 465
  rate, 497
  reduction, 474, 481-482, 501
  total nutrients, 416, 476
Digestible dry matter (DDM), 465-466, 473-474, 478, 483
Digestible energy, 464, 475
Dinitrogen fixation, 14-15, 22, 167, 178, 205, 229-257, 273, 311, 322, 343, 345-347, 375-376, 478, 757-759, 1039
  age of stand, 248
  biological cost, 229
  clear seeding, 322
  community performance, 246-247
  crop rotation, 249-250, 345
  cultivar interaction, 242-243
  in establishing stands, 306, 321, 990
  forage harvest, 243
  measurement, 245-246
  soil nitrogen, 244, 349
  temperature, 178
Diploids
  cytology, 146-148
  origin, 111-112
Disease(s)
  abiotic, 621-622, 648, 844
  bacterial, 624-631, 659-662, 830-832
  control
    cultural, 623, 626, 628, 631, 633-634, 638-639, 641-642, 647-649, 657, 661
    fungicides, 481, 622-623, 636, 642, 648, 659, 661-662
    resistance. See Disease resistance
  effect on feeding value, 480-483, 621-623, 625
  effect on phenology, 180-181
  foliar, 480-481
  fungal, 167, 622, 625-626, 640-652, 654, 656-659, 662, 833-843
  leaves, 480, 622-633, 635, 638-640, 656-657, 659-660
  losses, 621-622, 626, 827
  physiogenic, 623-624
  races (biotypes), 634, 636, 830, 833, 835-836, 840-841
  roots, 640
  seedlings, 641
  stems, 622
  vascular system, 656
  virus, 638-640, 843-844
  wilts, 656, 659
  wounding, 631, 633, 644, 647, 659
Disease(s), common names
  alfalfa dwarf, 660, 832
  alfalfa enation, 640, 844
  alfalfa latent, 640, 844

alfalfa mosaic, 638, 843
anthracnose, 9, 21, 119, 268, 412, 603, 622, 633, 647, 815, 833-834
bacterial leafspot, 624, 832
bacterial stemblight, 630, 832
bacterial wilt, 9, 19-20, 119, 137, 168, 305, 412, 452, 596, 603-604, 659, 662, 685, 757, 830-832, 866, 1024, 1043
blackstem and leafspot, 610-611, 834-835
brown root rot, 650, 843
Cercospora leafspot, 481, 631, 835
charcoal rot, 651
common leafspot, 604, 610-611, 625, 835-836
cotton root rot, 604, 649
crown and root rot complex, 645
crown bud rot, 644, 647
crown gall, 649
crownwart, 649, 843
Cylindrocarpon root rot, 650, 843
damping-off, 167, 604, 641-642, 843
downy mildew, 603, 628
dwarf, 660, 832
Fusarium root rot, 836
Fusarium wilt, 9, 221, 412, 604, 607, 610, 654, 656, 757, 815, 837
Leptosphaerulina leafspot, 622, 628, 837-838
low temperature Basidiomycete, 648, 843
Mycoleptodiscus crown and root rot, 651, 843
mycoplasmas, 639
pea early browning virus, 656
Phymatotrichum root rot, 604, 647, 662, 843
Phytophthora root rot, 9, 21, 167, 308, 388, 604, 607-608, 610-611, 613-615, 757, 838-840
Pierce's disease of grapevines, 661
powdery mildew, 630
Pseudopeziza leafspot, 625
Pythium root rot, 843
Rhizoctonia foliar blight, crown, and root rot, 604, 613, 840
rootcanker, 644
rust, 630, 840
scald, 606, 648
Sclerotinia crown and stem blight, 613, 634, 840-841
Sclerotium blight, 843
snow mold, 617, 648, 843
southern blight, 651, 843
spring blackstem and leafspot, 119, 604, 631, 678, 834-835
Stagonospora leafspot and root rot, 604, 628, 647, 841

Stemphylium leafspot, 604, 626, 643, 841–842
summer blackstem and leafspot, 119, 631, 835
Texas root rot, 604, 649
transient streak, 640, 844
Verticillium wilt, 9, 22, 597, 604, 608, 610, 617, 656–659, 662, 842, 1027
violet root rot, 645
virus, 623, 628–640, 843–844
winter crown rot, 617, 648, 843
witches' broom, 639, 844
yellow leaf blotch, 626, 842–843
Disease pathogens
Agrobacterium tumefaciens (Smith & Townsend) Conn., 649, 924
Aphanomyces euteiches Drechs., 641
Ascochyta imperfecta Pk., 505
Basidiomycete, 617, 648
Cercospora medicaginis Ell. & Ev., 481, 610, 631, 835
Colletotrichum destructivum O'Gara, 505
Colletotrichum trifolii Bain., 9, 412, 506, 603, 622, 633, 757, 815, 833–834
Coprinus psychomorbidus Redhead & Traquair, 648
Corynebacterium insidiosum (McCull.) H.L. Jens., 9, 119, 137, 596, 604, 658, 660, 685, 757, 830–832, 1024
Cylindrocarpon ehrenbergi Wr., 650, 843
Cylindrocarpon obtusisporum (Cke. & Harkin.) Wr., 651
Cylindrocarpon olidum Wr., 651
Cylindrocarpon radicicola Wr., 651
Cylindrocladium scoparium Morgan, 506
Cylindrocladium spp., 641
Erysiphe polygoni D.C., 631
Fusarium acuminatum (Ell. & Ev.) Wr., 641
Fusarium culmorum (Smith) Sacc., 641
Fusarium moniliforme Sheldon, 647
Fusarium oxysporum f. sp. *cassia* Armst. & Armst., 656
Fusarium oxysporum f. sp. *vasinfectum* (Atk.) Syn. & Hans., 656
Fusarium oxysporum Schl. f. sp. *medicaginis* (Weimer) Syn. & Hans., 9, 221, 656, 815, 837
Fusarium oxysporum Schlecht., 412, 604, 641, 647, 654, 757, 836
Fusarium roseum Lk. ex Fr. emend. Syn. & Hans., 505, 647, 836
Fusarium solani (Mart.) (Appel. & Wr.) Syn. & Hans., 647, 836
Fusarium tricinctum (Cda.) Syn. & Hans., 647, 836
Helicobasidium purpureum Pat., 645

Leptodiscus terrestris. See Disease pathogens, *Mycoleptodiscus terrestris*
Leptosphaerulina briosiana (Poll.) Graham & Luttrell, 505, 622, 628, 837–838
Leptotrochila medicaginis (Fckl.) Schüepp, 626, 842–843
Macrophomina phaseolina (Tassi) G. Goid., 651
Mycoleptodiscus terrestris (Gerd.) Ostazeski, 651–652, 843
Mycoplasma spp., 505
Myrothecium roridum Tode ex Fr., 843
Myrothecium spp., 647
Myrothecium verrucaria (Alb. & Schw.) Ditm. ex Fr., 843
Pellicularia rolfsii (Cruzi) West, 651
Peronospora trifoliorum dBy., 603, 628, 836
Phoma medicaginis Malbr. & Roum var. *medicaginis* Boerema, 119, 481, 506, 604, 639, 657, 678, 834–835
Phymatotrichum omnivorum (Shear) Dug., 604, 648, 843
Physoderma alfalfae (Lagh.) Karling, 649, 843
Phytophthora cinnamom, 643
Phytophthora cryptogea Pethybr. & Laff., 642
Phytophthora megasperma Drechs., 9, 222, 281, 505, 604, 642, 757
Phytophthora megasperma (Drechs.) f. sp. *medicaginis* Kuan & Erwin, 308, 642, 648, 837
Plendomus meliloti D. & S., 650, 843
Pleospora herbarum (Pers. ex Fr.) Rab., 604
Pseudomonas syringae van Hall, 630, 832
Pseudopeziza medicaginis (Lib.) Sacc., 506, 604, 625, 835–836
Pythium debaryanum Hesse, 641
Pythium hypogynum Middleton, 641
Pythium megasperma f. sp. *medicaginis*, 641
Pythium paroecandrum Drechs., 641
Pythium sylvaticum Camp. & Hend., 641
Pythium torulosum Coker & Patt., 641
Pythium ultimum Trow, 641
Pythium spp., 167, 505, 604, 641, 843
Rhizoctonia crocorum DC ex. Fr., 645
Rhizoctonia solani Kuehn., 447, 453, 505, 604, 613, 641, 840
Rhizoctonia violaceae Tul., 645
Rhizoctonia spp., 604
Sclerotinia sclerotiorum (Lib.) dBy, 650
Sclerotinia trifoliorum Eriks., 613, 634, 840–841
Sclerotium rolfsii Sacc., 505, 651, 843

SUBJECT INDEX

Stagonospora meliloti (Lasch) Petr., 604, 628, 643, 841
Stemphylium botryosum Wallr., 119, 505, 604, 626, 841–842
Trichoderma viride Pers. ex Fr., 505
Uromyces striatus Schroet. var. *medicaginis* (Pass.) Arth., 506, 630, 840
Urophlyctis alfalfae (Lagerh.) Magn., 649
Verticillium albo-atrum Reinke & Berth., 9, 222, 597, 656–657, 842, 1027
Verticillium dahliae Kleb., 657
Xanthomonas alfalfae (Riker, Jones & Davis) Dows., 624, 832
Disease resistance
  alfalfa dwarf, 661, 832
  alfalfa enation, 844
  alfalfa mosaic, 607, 639, 843
  anthracnose, 21, 119, 412, 634, 757, 833–834
  bacterial leafspot, 625, 832
  bacterial stem blight, 631, 832
  bacterial wilt, 19, 119, 168, 305, 412, 604, 660, 757, 830–832, 1024
  black stem, 834–835
  brown root rot, 650
  common leafspot, 626, 835–836
  crown and root rot complex, 647
  cylindrocarpon root rot, 651
  downy mildew, 630, 836
  foliar, 623
  Fusarium root rot, 836. *See* Crown and root rot complex
  Fusarium wilt, 221, 412, 604, 757, 837
  latent virus, 844
  Leptosphaerulina leafspot, 628, 837–838
  ozone tolerance, 624, 844
  Phythium, 843
  Phytophthora root rot, 21, 222, 308, 642, 757, 843
  Rhizoctonia, 644, 840
  rust, 630, 840
  saponins, 505
  Sclerotinia, 636, 840–841
  screening, 622, 829
  sources, 829
  spring blackstem, 633, 834–835
  Stagonospora, 644, 841
  Stemphylium leafspot, 119, 841–842
  summer blackstem, 631, 835
  Texas root rot, 649
  transient streak, 844
  Verticillium wilt, 22, 608, 656–658, 842, 1027
  witches' broom, 639, 844
  yellow leaf blotch, 676, 842–843
Disequilibrium, 794
Distribution, alfalfa, 2, 31–81, 93–95, 112, 115
Dominance, 796, 813

Dormancy, 204, 206, 221, 265–266, 415, 596, 600
  cultivars, 266, 596, 600
  seed, 165
Drainage, 337, 611, 991
Drought tolerance, 119, 602
  cultivar differences, 377
  root growth, 377–379
  stomate density, 381
  transpiration, 380
Dry matter
  digestibility, 465–466, 473–474, 478, 483
  increase (crop), 476
  intake, 466, 523, 525, 528, 540
Dynamic simulation model. *See* Models

East Germany, 36–37
Economics, alfalfa weevil control, 186–187
Ecuador, 59
Egypt, 62
Electron microscopy (ultrastructural analysis), 95–97, 741
Electrophoresis, 98, 111–112
Elevation, 416
Embryo, 125, 739–741, 961, 963, 965, 975
  abortion, 116, 120
  rescue, 764–771
Emergence, 164–166. *See also* Germination
Endosperm, 125–127, 740–741, 747
Energy, 463
  digestible, 463, 465, 817
  metabolic, 195–196
England, 517, 526
Environment
  dinitrogen fixation, 244
  disease screening, 829
  stability of cultivars, 815–816
  winter survival, 280–281
  yield and persistence, 419, 426
Epicotyl, 130, 143–144, 148, 961, 974
Epidermis, 472
Equipment
  forage harvesting, 568–580
    bales, shapes and sizes, 574–578
    baling, 574
    conditioning, mechanical, 572–573
    cubing, 579–580
    harvesters, 578–579
    mower-conditioners, 542, 571–574
    mowing, 568–570
    raking, 570–571
    tedders, 571
    windrowers, 571
  forage storage and handling, 584–592
    accumulators, 580
    bales, 584–587
    chopped haylage, 584
    cubes, 20, 592

SUBJECT INDEX

elevators, 585
greenchop, 590
grinding, 587
haylage, 587–590
loose hay, 580
low-moisture silage, 587–588
plastic covers, 586
silage, 588–589
stacked hay, 590–592
stackers, 580
stackformers, 580
seed harvesting, 1011–1014
seeding, 308–318, 323, 327
cultipacker, 311–318
press wheels, 311–318
soil preparation, 306–310
weed control, 309, 317, 319
Establishment, 986–990
band seeding, 312–318, 450–451
broadcast seeding, 316–317
inoculation, 231–232, 242–243, 311
management, seeding year, 327–329, 986–990
no-till, 325–326, 344
sod seeding, 325–326
weed control, 319–324, 710–725, 986–987
with companion crop, 317–321, 713
without companion crop, 318, 321–325, 716–723
Estrogens, 480, 505–507, 529, 623
Ethionine, 919
Euphorbia cyparissias L., 630
Euploids, origin, 746–753
Europe, 32–49, 333
Evaporation, pan, 391, 395
Evapotranspiration, 171, 178, 377–378, 384, 394–399
climatonomic estimation, 394
dry matter yield, 388–391
Makkink formula, 397
Penman formula, 395
Priestley-Taylor formula, 397
Evolution, 26, 94–95, 110–115, 120

Fall dormancy, 204, 206, 221, 265–266, 268, 603–606
rating system, 172, 596–597
Fall growth, 265–268, 282–284
Fall management
critical period, 266, 423–424
forage quality, 427
improved cultivars, 423–424
plant maturity, 425–426
seeding time effect, 321–322
snow cover, 281, 426
soil fertility, 424, 427
stand persistence, 423
Fat, 481

Federal Seed Act, 1028, 1033–1034
Feeding
beef cattle, 481, 504, 506, 547
dairy cattle, 475, 504, 539–546
feedlot, 547
guinea pig, 479
meadow vole, 481
poultry, 504
rabbits, 477, 479
sheep, 475, 480–481, 483, 504–506
swine, 504
Feeding value, 12–13, 463–484, 539–549, 809, 816–823, 1034
aftermath, 475–476
balanced rations, 549
breeding, 220, 816–823
climatic factors, 476–478
digestibility, 463–465, 817–818
edaphic factors, 478–480
effect of diseases and insects, 480–482, 622, 625, 676, 859–860
efficiency of utilization, 463–465
energy, 540, 542, 817–818
fertilizer, 495
greenchop, 546
growth stage, 180, 472–476, 483, 540–541
hay, 465–466, 542
inhibitors, 818
intake, 463–466, 473, 475, 482, 541
leaf loss, 386–387
meal length, 541
measurement, 12–13
minerals, 348, 355, 361, 479–480, 507–508, 819–821
plant parts, 471–474, 818
postharvest factors, 540, 821
potential forage, 464
pressed forage, 559–560, 564
solubility of proteins, 494, 548
toxicity, 676
weeds, 482–483
Fertility, animal, 506
Fertilization, 937
embryo development, 739–741
in vitro, 120
postfertilization, 941
Fertilizers, 312–313, 333 364, 1040
animal manures, 309, 352, 362–364
application, 307, 346–348, 350–351, 354–355, 358–359, 361–362
banding, 311–318, 347, 350
boron, 306, 361
calcium, 306, 355
copper, 361–362
deproteinized juice, 559, 564
effect on cold survival, 264
factors affecting need, 306–308
gypsum, 306
magnesium, 244, 306, 355

SUBJECT INDEX

micronutrients, 361
molybdenum, 306, 343-344, 361
municipal wastes, 363-364
nitrogen, 307, 346-348, 363, 479
phosphorus, 312-317, 321, 350-351, 479
potassium, 306-307, 321, 350-351, 479-480
soil tests, 306, 340-341
stimulation, 312
sulfur, 306, 358-359, 480
tissue analysis, 340-341
Ferulic acid, 469-470
Fiber, 465-466, 503, 507, 559-560, 818
acid detergent (ADF), 465-466, 478, 481-483, 503, 818
crude, 478, 481
neutral detergent (NDF), 465, 481-482
Fibrils, 467, 469
Finland, 37
Flavones, 499
Flavonoids, 499
Flax, 318-319
Flooding, 388, 642, 644, 648
Floral development, 150-152, 182-183, 931
atrophy, 182
induction, 182, 931
morphology, 150-158, 935-936
structure, 931
Flowering, 182-184, 939
light, 182
photoperiod, 180, 182-183
phytochrome, 181-182
relative humidity, 183
temperature, 181
timing period, 993-994
tripping, 183, 935-938, 1001-1002
Forage grasses. See Grasses, forage
Forage harvesters, 578
Forage quality, 387, 413-414, 427, 463-484, 816-823
France, 37-39
Freezing, 278-281
Frost
injury, 267
leaf loss, 474
Fruit pod, 98-110
Fumaric acid, 498
Fungal diseases. See Diseases
Future trends, 1037-1045

Galactose, 467-468
Gametophyte, 737-739, 741-742, 744, 746
Gametophytic abnormalities, 741-746
Gene
action, 787, 812
forage yield, 812
frequency, 794
nonadditive, 783

Genotype X environment interactions, 815-816
Generations of synthesis, 780, 793
Genes
mapping, 754-755
markers, 149
splicing, 120
Genetic engineering, 120, 1041-1042, 1044. See also Tissue culture
Genetic resources, 115, 118-120, 1045. See also Germplasm
Genetics
autotetraploid, 754-758, 794-798
dinitrogen fixation, 238-239
heterosis, 812
inbreeding, 781, 783
incompatibility, 943-944
markers, 98, 111, 113
quantitative traits, 794
yield, 812-815
Genomic number, 93-94, 120
Geographic adaptation, 595-617
Geographic location, 476, 483
Geographic movement, 27-31
Germany, East, 36-37
Germany, West, 48-49
Germination, 164-166, 974
autotoxicity, effect on, 309
osmotica, 164
salinity, 164-165
seed, 127, 130-132
temperature, 165
toxic effects of ions, 164
water, 385-386
Germplasm, 3-4, 21-22, 118-121, 265-266, 603, 779, 1041-1044
Germplasm Resources Information Project (GRIP), 4
Glandular hairs (trichomes), 10, 21, 107, 109-110, 119, 132-140, 142-148, 859-860, 863, 866, 876, 879-880, 884, 887
Glucose, 216-217, 468
Glutamate synthase, 240, 250-251
Glutamine, 240
Glutamine synthetase, 240
Gossypol, 861
Grass-mixture, 304-305
bloat prevention, 305, 496, 531-532
competition, 352, 446-449
dinitrogen fixation, 230
fertilizer use, 348, 443
management, 454, 456
manure application, 363
persistence, 452-454
production, 452-454
seeded, 304-305, 316
use, first in United States, 305
weed control, 456

Grasses, forage, 465, 467, 469–471, 482–483
Grasses, forage, common names
  bahiagrass, 785
  bermudagrass, 986
  bromegrass, 289, 362
  crested wheatgrass, 363, 451, 515, 517–518
  intermediate wheatgrass, 516, 520, 524, 527
  Kentucky bluegrass, 310, 448–449, 519, 529
  meadow fescue, 452
  orchardgrass, 305, 339, 356, 441, 443–444, 446, 471, 515, 517, 519, 521, 523–527
  perennial ryegrass, 526
  Phalaris (harding grass), 517
  red fescue, 518
  red top, 449
  reed canarygrass, 305, 452, 527
  Russian wildrye, 451, 516, 527–528
  ryegrass, 440–441
  smooth bromegrass, 305, 317, 362, 437, 479, 507, 515, 518, 520–521, 523–528, 725
  sudangrass, 521, 528, 532
  tall fescue, 356, 364, 439, 444, 446, 519, 524–526
  tall wheatgrass, 281
  timothy, 317, 359, 439, 446, 478, 527, 820
Grasses, forage, scientific names
  Agropyron desertorum (Fisch.) Schult., 451, 515, 527–528
  Agropyron elongatum (Host) Beauv., 281
  Agropyron intermedium (Host) Beauv., 516, 520, 524
  Agrostis alba L., 449
  Bromus inermis Leyss., 305, 362, 439, 479, 507, 515, 518, 520–521, 523–528, 725
  Bromus mollis L., 230, 243
  Bromus willenowii Kunth, 452
  Bromus spp., 442–443, 446
  Cynodon dactylon (L.) Pers., 441, 986
  Dactylis glomerata L., 305, 339, 356, 441–442, 444, 446, 515, 517, 519, 521, 523–527
  Elymus junceus Fisch., 451, 516, 527–528
  Festuca arundinacea Schreb., 230, 356, 364, 439, 444, 446, 519, 524–526
  Festuca rubra L., 518
  Lolium perenne L., 230, 526
  Lolium spp., 440–441
  Paspalum notatum Flügge, 785
  Phalaris arundinacea L., 230, 305, 452, 527

Phalaris stenoptera Hack., 517
Pheum pratense L., 317, 359, 439, 527, 820
Poa pratensis L., 230, 310, 448–449, 519, 529
Sorghum sudanense (Piper) Stapf., 521
Grazing
  bloat prevention, 531–532
  carrying capacity, 524
  continuous, 517, 519–521, 525, 528
  cultivars, 516, 520
  fall, 519, 526
  rotational, 517–521, 525, 528
  seedings, 319
  stocking rate, 517–518
  strip, 520–521, 523, 525
Great Britain, 47
Greece, 39
Greenchop, 463, 546
  percentage of forage harvested, 567
  protein bypass, 546
GRIP. *See* Germplasm Resources Information Project
Growth
  buds, 170, 268–269
  cold tolerance, 278
  compensatory, 387
  contractile, 130–131, 138
  correlation with environment, 163–194
  determination of, lateral roots, 136–137
  leaf expansion, 386
  models, 15–16, 184–187, 1039
  moisture stress, 166–167, 171, 175, 178, 386–387
  primary roots, 132–136
  rates, 167
  regulators, 233, 919, 993
  roots, 177–178, 377–379
  salt, effect of, 167, 171, 176
  secondary roots, 133–136
  seedling, 130–132, 166–168
  stage, 179–180, 319, 475, 483
    determination, 180, 475–476
  summer decline, 178–179, 198
  temperature effects, 167–168, 171, 173–174, 176–178, 477–478
  vegetative, 168–179
Guinea pigs, 479

Haploids, isolation, 737, 746–748, 756
Hard seed, grower preference, 183, 963–966, 1033
Hardiness zones, 608
Harvesting. *See also* Cutting schedule
  critical period, 266
  dinitrogen fixation, 243
  effect of disease, 626, 628, 630–631, 633–634, 647, 651
  hay, 542

SUBJECT INDEX

losses, 542, 548
methods, 567
postharvest changes, hay, 541–542
postharvest operations, seed, 1014
seed, 963–964, 968
system, 541
time of, 476–478
Hay, 495
  alfalfa and alfalfa mixtures grown, 1046
  companion crop, 310
  curing, 542
  dessicants, 543, 573
  drying, 541, 571–574
  equipment, 568–578
  feeding value, 472–480, 482–483, 542–544
  grading standards, 13, 465–466, 476
  grinding, 587
  harvesting, 475–476, 541–547, 568–578
  high moisture, 319, 541. *See also* Haylage
  high temperature drying, 502
  maturity, 474–478
  moisture content, 542, 571–574, 588
  moldy, 342, 374
  percentage of forage harvested, 567
  plastic covers, 586
  preservatives, 12, 543
  stack, 590–592
  storage losses, 585–587
  transporting, 580–584
  unique characteristics, 544
  weed infested, 482
Haylage, 495, 543, 578, 584, 587–590
  percentage of forage harvested, 567
Heat
  hardening and recovery, 290
  mechanism of damage, 285–288
  tolerance, 285–291
Heat avoidance, 288
Heat summation, 181
  base temperature, 173, 181
Heaving, 426–428, 640
Hectarage. *See* Area planted
Hemicellulase, 468
Hemicellulose, 466–470, 503
Heptaploids, 753
Herbicides, 705, 708, 712–723, 725–729
  combinations, 712–713, 719–720
  postemergence, 321, 326, 726, 987
  preemergence, 321, 726, 987
  resistance, 221
  soil pH, 344
Herbicides, common names
  amitrole, 713, 722
  benefin, 717
  chlorpropham, 718, 725–726
  2,4-D, 326, 328, 712, 720–723
  dalapon, 712–713, 722

  2,4-DB, 326, 713, 717–720, 722–723, 728
  DCPA, 726
  dicamba, 721–723
  dichlobenil, 726
  dinoseb, 328, 717, 719–720, 722, 726
  diquat, 712
  diuron, 727
  DNBP, 717, 726
  EPTC, 328, 344, 713, 716–717, 719, 728
  fluazifop, 718–722, 724, 728
  glyphosate, 309, 326, 712, 721–722
  haloxyfop, 718–719, 721, 723, 728
  hexazinone, 728
  MCPA, 328, 720
  metribuzin, 728
  paraquat, 325–326, 712, 721–723, 728
  profluralin, 717
  pronamide, 722, 727
  propham, 712, 718, 726
  Roundup (trade name), 309
  sethoxydim, 718, 720–722, 728
  simazine, 727
  terbacil, 727
  triazine, 344
  trifluralin, 728
Heritability, 142, 785, 791–792, 798–799
  disease and nematode resistance, 830–848
Heterosis, 780, 782, 812, 814
Heterozygosity, 746, 759–760, 814
Hexaploids, 94, 120, 171, 751–752, 758, 761, 765, 767–768
High moisture hay, 543–544
Highlights, USA-Canada (1850–1986), 1–24
Hilum, 126–128, 130, 961
Hordeum vulgare L. *See* Barley
Hormones
  cold tolerance, 278
  coumestrol, 480–481, 506
  dinitrogen fixation, 233–234
  growth promoting, 278
  phytohormones, 120
Horse pasture, 531
Horses, 482
Host-pathogen interactions, 828–829
Hour-glass cells, 127, 130
Human food, 564
Hungary, 40–41
Hybridization, 16–17, 94, 109–110, 112–113, 116–118, 120
  barriers, 107, 120
  interspecific, 762–771
Hybrids, 98, 108, 114, 116–118, 120, 830. *See also* Interspecific hybrids
Hydraulic conductance, 380–381, 384
Hydrogenase, 240–241
Hypocotyl, 165, 961, 974–975

SUBJECT INDEX

Ice sheets. *See* Cold tolerance
Imports seed, 1023
Inbred plants, competition, 954
Inbreeding, 758, 781–782, 793, 803–805, 812, 814, 943–944
Incompatibility, relational, 94, 943
India, 68–70
Industry research, 1026–1027
Ineffective nodules, 237–240
Inflorescence, 95, 110, 937–938
   mutants, 937–938
Infrared thermometry, 399–400
Inheritance
   qualitative, 794
   quantitative, 794–805
   resistance
     disease, 757, 830–844
     insect, 861, 870
     nematode, 757, 844–848
Inoculum, for breeding resistance
   disease, 830–844
   nematode, 844–848
Insect(s)
   affecting establishment, 326, 329
   affecting seed production, 686–690, 994–999
   consuming foliage, 671–678
   control, 428–429, 647, 690–695, 994–1000
     insecticides, 326, 693, 873, 997–998
     integrated, 863, 997–998
     resistance. *See* Insect resistance
     seed production, 994–1000
     seeding, 321, 326, 329
     systems, 673–674, 678, 682
     tactics, 690–694
   damage, 180, 483, 671–678, 681–689, 995–1000
   effect on feeding value, 480–482, 676, 678, 681, 688
   feeding on roots, 647, 684–686
   numbers, 671
   pathogens, 673–675, 681, 692
   pollinators, 94
   toxicity, livestock poisoning, 482, 676
Insect(s), beneficial, common names
   alfalfa leafcutting bee, 951, 990, 1002, 1007–1009
   alfalfa leafcutting beetle, 779
   alfalfa weevil parasite, 673, 692
   alkali bee, 951, 1002, 1005–1007
   bigeyed bug, 997
   convergent lady beetle, 692
   damsel bug, 997
   honey bee, 779, 990, 1002–1005, 1025
Insect(s), beneficial, scientific names
   Apis mellifera L., 779, 951, 990, 1002–1005, 1025
   Bathyplectes anurus (Thomson), 863

Bathyplectes curculionis (Thomson), 692, 863
Geocoris bullatus (Say), 997
Geocoris pannens (Stal), 997
Hippodamia convergens Guérin-Méneville, 692
Megachile rotundata (F.), 7, 779, 951, 990, 1002, 1007–1009
Nabis alternatus Parshley, 997
Nabis americoferus Carayon, 997
Nomia melanderi Ckll., 951, 1002, 1005–1007
Insect(s), harmful, common names
   alfalfa aphid, 995, 999–1000
   alfalfa blotch leafminer, 677, 887
   alfalfa bud mite, 684
   alfalfa caterpillar, 674, 691–692
   alfalfa flower midge, 888
   alfalfa looper, 674, 692, 996
   alfalfa plant bug, 688–689, 884, 995
   alfalfa seed chalcid, 10, 686–687, 691, 881–884, 994–996, 1000–1001, 1014
   alfalfa snout beetle, 685–686
   alfalfa webworm, 676
   alfalfa weevil, 19–20, 119, 142, 186–187, 428, 481–482, 597, 605, 611, 671–674, 691, 766, 859, 872–876, 995–996
   aphids, 481, 678–681
   army cutworm, 675
   aster leafhopper, 682
   beet armyworm, 995
   beet webworm, 676
   Bertha armyworm, 996
   black blister beetle, 676
   blister beetles, 482, 676
   blotch leafminer, 10
   blue alfalfa aphid (bluegreen aphid), 10, 21, 506–507, 597, 605, 681, 692, 814, 871–872, 995
   bristly cutworm, 675
   brown wheat mite, 995–996
   carmine spider mite, 684
   caterpillars, 674–676
   chalcid wasp, 994
   clover leaf weevil, 674
   clover leafhopper, 682
   clover root curculio, 612, 660, 684–685, 885–887
   consperse stink bug, 995
   crickets, 689
   cutworms, 675
   differential grasshopper, 677, 888
   dingy cutworm, 675
   Egyptian alfalfa weevil, 418, 605, 672, 872–874, 876, 995
   forage looper, 674
   garden webworm, 676
   granulate cutworm, 675
   grass grub, 888

SUBJECT INDEX

grasshoppers, 603, 610, 677, 691
green cloverworm, 674
leafhopper. *See* Potato leafhopper
leafminers, 612, 677-678, 887
leafroller, 690
lygus bugs, 420, 688, 884-885, 994-997, 999-1000
meadow spittlebug, 481, 610, 683, 692, 881
migratory grasshopper, 677
mirids, 688
mites, 684
Pacific spider mite, 684, 995
painted leafhopper, 682
pale leafroller, 997
pale legume bug, 688, 995
pale western cutworm, 675
pea aphid, 10, 20, 480-481, 505-506, 597, 605, 607, 613, 640, 680-681, 691, 859, 864, 868-871, 995, 997-1001
pea leaf weevil, 685
plant bugs, 688-689
potato leafhopper, 10, 119, 142, 326, 329, 359, 428-429, 480-481, 597, 611-612, 681-682, 691, 864, 876-880
rapid plant bug, 688
red-legged grasshopper, 677
redbacked cutworm, 675, 996
Say sink bug, 689, 995
small grey blister beetle, 676
spotted alfalfa aphid, 10, 20-21, 605-607, 613, 654, 679-680, 692, 859, 863-868, 995, 1000-1001
strawberry spider mite, 684, 995
superb plant bug, 688
tarnished plant bug, 688, 884, 995
three-cornered alfalfa hopper, 613, 683, 876, 878
three-striped blister beetle, 676
thrips, 689
two-spotted spider mite, 684, 995
two-striped grasshopper, 677
variegated cutworm, 675
webworm caterpillars, 676
western potato leafhopper, 682
western yellow-striped armyworm, 676, 995
white grub, 887
white-fringed beetles, 686
yellow clover aphid, 679
yellow-striped armyworm, 675, 995
Insect(s), harmful, scientific names
Aceratagallia curvata Oman, 682
Aceratagallia sanguinolenta (Provancher), 682
Achyra rantalis (Guenée), 676
Acinopterus angulatus Lawson, 682

Acyrthosiphon kondoi Shinji, 10, 507, 597, 681, 863, 871-872, 995
Acyrthosiphon pisum (Harris), 10, 480, 505-506, 597, 680, 995, 997-1000
Adelphocoris lineolatus (Geoze), 688, 859, 868-871, 884-885, 995
Adelphocoris rapidus (Say), 688
Adelphocoris superbus (Uhler), 688
Agalia constricta (Van Duzee), 682
Agromyza frontella (Rondani), 10, 677, 887
Agrotis orthogonia Morrison, 675
Aphelia pallorana (Robinson), 690
Aphis craccivora Kochs, 640
Arachnida (class), 684
Autographa californica (Speyer), 674, 996
Bruchophagus roddi Gussakovsky, 10, 686-687, 881-884, 994-996, 1000-1001, 1014
Caenurgina erechta (Cramer), 674
Camnula spp., 603
Chlorochroa sayi (Stal), 689, 995
Clepsis pallorana Robinson, 997
Colias eurytheme Boisduval, 674
Contarina medicaginis Kieffer, 888
Costelytra zealandica (White), 888
Deltocephalus flavocostatus Van Duzee, 682
Draeculacephla antica Walker, 682
Empoasca abrupta DeLong, 682
Empoasca fabae (Harris), 10, 119, 142, 321, 359, 428, 480, 597, 681-682
Endria inimica (Say), 682
Epicauta immaculata (Say), 682
Epicauta lemniscata (F.), 676
Epicauta maculata (Say), 676
Epicauta murina (Le Conte), 676
Epicauta pennsylvanica (De Geer), 676
Epicauta sabricii (Le Conte), 676
Epicauta sericans (Le Conte), 676
Epicauta spp., 676
Eriphyes medicaginis Keifer, 684
Euschistus conspersus (Uhler), 995
Euxoa auxiliaris (Grote), 675
Euxoa ochrogaster (Guenée), 675
Euxoa septentrionalis Walker, 996
Feltia ducens Walker, 675
Feltia subterranea (F.), 675
Frankiniella spp., 689
Graminella nigrifrons (Forbes), 682
Graphognathus spp., 689
Gryllus spp., 689
Hypera brunneipennis (Boheman), 418, 605, 672, 872-876, 995
Hypera postica (Gyllenhal), 119, 142, 187, 428, 481, 597, 605, 672, 766, 995-996
Hypera punctata (F.), 674
Lacinipolia renigera (Stephens), 675

Liriomyza spp., 677
Lopidea spp., 688
Loxostege commixtalis (Walker), 676
Loxostege sticticalis (L.), 676
Lygus desertus Knight, 688
Lygus elisus Van Duzee, 688, 995–1000
Lygus hesperus Knight, 688, 884–885, 994–1000
Lygus lineolaris Palisot de Beauvois, 688, 884–885, 995
Lygus spp., 10, 19
Macrosiphum creelii Davis, 995
Macrosteles fascifrons (Stal), 682
Mamestra configerata Walker, 996
Melanoplus bivittatus (Say), 677
Melanoplus differentialis (Thomas), 677, 888
Melanoplus femurrubrum (De Geer), 677
Melanoplus sanguinipes (F.), 677
Melanoplus spp., 603
Melolontha vulgaris (F.), 887
Neurocolpus spp., 688
Otiorhynchus ligustici (L.), 685
Peridroma saucia (Hübner), 675
Petrobia latens Muller, 996
Philaenus spumarius (L.), 481, 610, 683, 881
Plagiognathus spp., 688
Plathypena scabra (F.), 674
Poecilocapsus spp., 688
Sitona flavescens Marsham, 685
Sitona hispidulus (F.), 612, 660, 684–685, 885–886
Sitona lineata (L.), 685
Sitona sissifrons Say, 683
Spissistilus festinus (Say), 613, 683, 876, 878
Spodoptera exigua (Hübner), 995
Spodoptera ornithogalli (Guenée), 675
Spodoptera praefica Grote, 676, 995
Stauroderus spp., 888
Stenobothrus spp., 888
Tathorhynehus augustiorata (Grote), 675
Tetranychus cinnabarinus (Boisduval), 684
Tetranychus pacificus McGregor, 684, 995
Tetranychus turkestani Ugarov & Nikolski, 684, 995
Tetranychus urticae Koch, 684, 995
Therioaphis maculata (Buckton), 10, 605, 654, 679–680, 859, 865–868, 995
Therioaphis trifolii (Monell), 679
Tibolium castaneum (Hbst.), 505
Insect resistance, 119, 498, 691, 859–902
  alfalfa blotch leafminer, 887
  alfalfa flower midge, 888
  alfalfa leafcutting bee, 7, 21
  alfalfa plant bug, 884
  alfalfa weevil, 119, 692, 859, 864, 872–876
  alfalfa weevil parasite, 863
  biotypes, 862, 866, 870
  blue alfalfa aphid (bluegreen aphid), 692, 863–864, 871–872
  chalcid, 881–884
  clover root curculio, 885–887
  degrees, 860
  development, 691, 864–866, 869–871, 873, 877, 881–882, 885
  durability, 862
  Egyptian alfalfa weevil, 872–876
  evaluation, 864–867, 870, 872–873, 877, 881–882
  general concepts, 864–865
  glandular hairs, 10, 859–860, 863, 866, 876, 879–880, 884, 887
  grasshoppers, 888
  inheritance, 861, 870, 877
  isolation, 864, 866–867, 870, 872–875, 877–878, 881–883, 885–887
  lygus bugs, 884–885
  meadow spittlebug, 692, 881
  mechanisms, 119, 142, 861, 867–868, 870–872, 875–876, 879–880, 883–885, 887
  nature, 861
  pea aphid, 692, 859, 863–864, 868–871
  potato leafhopper, 119, 692, 864, 876–880
  saponins, 505
  sources, 119
  spotted alfalfa aphid, 692, 859–860, 864–868
  three-cornered alfalfa leafhopper, 876–880
  value, 691, 859–860, 865, 868–869
  white grub, 887–888
Integrated control, 863–864, 997–999
International Crop Improvement Association (now Association of Official Seed Certifying Agencies), 1025
International Seed Testing Association, 973–974
Interspecific hybrids, 762–771, 942
  annual species, 771
  cytogenetics, 16–17, 116–118, 762–771
  embryo rescue, 120, 764–771
  perennial plus annual, 117–118
  perennial species, 764–771
Introductions, 21–22
Iran, 70
Iraq, 70–71
Irrigation
  application systems, 307, 401
  area, 600
  availability, 600–601
  criteria, 392–393

SUBJECT INDEX

crop coefficients, 398
crop temperature, 399
cultivar adaptation, 603
effect on disease, 388, 636, 641-643, 648
effect on feeding value, 338, 478-479
effect on growth, 166
establishment during, 307, 323
flooding, 175-176, 178, 388, 648
frequency and rate, 166, 178, 183
requirement, 392-401, 990-992
scheduling, 392-399
seed production, 182-183, 990-992, 1014
soil measurements, 393
sprinkler, 307-308, 992
wastewater, 363-364
yield per unit area, 171
Isoflavones, 506
Isoflavonoids, 500
Isotope dilution, 230, 246
Isozymes (allozymes), 98, 111-116, 121
Israel, 71-72
Italy, 41

Japan, 72-73

Kenya, 62-63

Latitude, 476
Lead, 362
Leaf area, 169, 207-212, 414, 428
 accumulation, 172-173
 distribution, 207-208
 index (LAI), 206-212, 353, 380, 384, 414, 428
Leaf mass, 169, 386-387, 474
Leaf protein concentrate, 495, 759
Leaves
 anatomy, 148-150
 canopy structure, 206-208
 carbon dioxide exchange, 168, 196-205, 353, 759, 1040
 cell walls, 466, 497
 characteristics, 172-175
 composition, 336, 473
 digestibility, 473-474
 epidermis, 148-149
 frost damage, 474
 mesophyll, 127, 148-149, 175, 353
 morphology, 148-150, 173-174, 202-203
 multifoliolate, 150
 mutual shading, 207
 necrosis, 647
 nutritive value, 472
 orientation, 207-209
 petiole, 130, 145, 147-150

 photosynthesis, 169, 196-199, 759
 physiology, 169, 172-175
 proportion in herbage, 173
 rate of appearance, 174
 salinity, 172
 senescent, 201, 414
 specific leaf weight, 174-175, 203-205
 stem ratio, 172, 175, 466, 473-474, 476-478, 480, 483, 818, 820
 stipules, 145, 147-150
 stomata, 148-159, 353
 unifoliolate, 130-132, 138, 148
 water potential, 172
Lectins, 232
Leghemoglobin, 235, 237, 241, 347
Legumes, other than alfalfa, common names
 alsike, 886
 barrel medic, 506
 bean, 231
 birdsfoot trefoil, 230, 305, 359, 443, 469, 497, 524-525, 529, 721, 886
 broad bean, 870
 cicer milkvetch, 818
 cowpea, 231
 crown vetch, 497
 ladino clover, 305, 446, 523
 lupine, 231
 pea, 231, 319
 red clover, 168, 305, 359, 446, 469, 506, 529, 721, 886
 sainfoin, 497
 sericea lespedeza, 497
 sourclover, 993
 soybean, 231, 273, 287, 310, 344, 381, 385, 547
 subterranean clover, 506, 519, 529
 sweetclover, 992
 white clover, 230, 379, 441, 469, 519, 523, 529-530, 886
Legumes, other than alfalfa, scientific names
 Astragalus cicer L., 818
 Coronilla varia L., 497
 Lespedeza cuneata (Dumont) G. Don, 497
 Lotus spp., 231, 973. See Legumes, birdsfoot trefoil
 Medicago littoralis Rhode. See Legumes, barrel medic; *Medicago* spp.
 Melilotus indica All., 993
 Melilotus officinalis Lam., 992
 Melilotus spp., 231
 Onobrychis viciifolia Scop., 497
 Pisum sativum ssp. *arvense* (L.) Poir., 319
 Trifolium hybridum L., 886
 Trifolium pratense L. See Legumes, red clover

Trifolium repens L. *See* Legumes, ladino clover; Legumes, white clover
Trifolium spp., 886, 973
Trifolium subterraneum L., 243, 506, 517, 529
Trigonella, 231
Trigonella asclersoniana, 97
Vicia faba var. *major* L., 870
Lens, 126-127, 129, 961, 963
Lenticels, 136
Leucine aminopeptidase, 111-115
Libya, 63
Light, 196-197, 206-208, 210
 compensation point, 197, 210
 cytoplasmic male sterility, 20, 183
 daylength, 474, 476. *See also* Light, photoperiod
 feeding value, 476
 illuminance, 476, 483
 intensity. *See* Light, irradiance
 interception, 208-210, 441
 interruption, 177, 182
 irradiance, 166, 168, 182-183, 197, 202
 lodging, 319
 photoperiod, 167-168, 172, 175, 177-178, 180-183
 photosynthesis, 169, 196-197, 209-211
 quality, 166, 168, 181, 261
 seedling establishment, 164, 166, 319
 shading, 170, 174-175, 476-477
Lignification, 467, 469
Lignin, 177, 465-467, 469-472, 476, 478, 481, 483, 502-503, 817-819
Lime, 306, 341-345
 benefits, 315, 342-344, 379
 calcium, 306, 343, 354-355
 coated-seed, 311
 gypsum, 306
 magnesium, 306, 343, 354-355
 method of application, 306, 308, 344-345
 pH effect, 306, 341-342
 rate of application, 306
 rooting, 379
 sources, 306
 surface applied, sod, 306
 time of application, 306, 344-345
Linkage groups, 754-755, 813
Linum usitatissimum L. *See* Flax
Lipids, 277, 466
Livestock industry, 1038
 grain feeding, 1038
 human consumption, 1038
Longevity of stand, 457
Losses
 diseases including nematodes, 621-625, 630, 638
 insects, 671-672, 674-676, 678-679, 681, 683-684, 686
Lucerne (alfalfa), 27

Lycopersicon esculentum Mill. *See* Tomato

Magnesium
 application, 306, 355
 availability, 343
 concentration, 355
 in lime, 306
Maize. *See* Corn
Male sterility
 cytoplasmic, 20, 120, 183, 748, 783, 947-952
 environment, 950
 genetic, 20, 947, 951
Malpighian cells, 127
Manganese, availability, 341-343, 361
Mannose, 468
Manure, 309, 362-363
Mean stage by weight, 180
Medicagenic acid, 503
Medicago genus
 cytogenetics, 762-771
 identification, 95-106
 taxogenetics, 116-118
Medicago spp., 94-96, 98-106, 142, 498, 762-771, 833, 842, 876-879, 881-882, 886, 961, 973
 afghanica Vass., 108
 agropyretorum Vass., 108
 alaschanica Vass., 108
 alatavica Vass., 108
 altissima Grossh., 108-109
 arabica Huds., 104, 117, 761
 arborea L., 94-96, 98, 117, 119, 741, 746, 751-752, 761
 asiatica Sinsk., 108
 asiatioa Sinsk., 108
 beipinensis Vass., 108
 blancheana Boiss., 103, 116-118, 876, 879
 bonarotiana Arcengeli (*M. blancheana* Boiss.), 118, 761
 borealis Grossh., 108-109
 cancellata M.B., 94, 100, 118-119, 746, 751-752, 761-768, 842
 carstiensis Wulf., 95, 99, 117, 761
 caucasica Vass., 108
 ciliaris All., 102, 116-118, 761, 882
 coerulea Less., 108, 842, 906, 909
 constricta Dur., 93-94, 106, 117, 761
 coronata Bart., 104, 117, 761
 cretacea M.B., 99, 117, 761
 daghestanica Rupr., 101, 117-118, 761-765, 768-769
 difalcata Sinsk., 108
 disciformis DC, 104, 117, 119, 761, 875, 879, 882
 doliata Carmign., 106, 119, 761
 dzhawakhetica Bordz., 100, 117-119, 751, 761-765, 769-770

SUBJECT INDEX

erecta Kotov, 108-109
falcata L., 107-109, 118, 152, 164, 168, 260-261, 269, 280, 291, 412, 603, 626, 647, 651, 742, 746, 748, 750, 754, 761-768, 842-843, 873, 877, 906-907, 909, 937, 941, 943-945
gaetula (*M. sativa* var. *gaetula*), 108, 875
glandulosa David., 108, 875-876, 880, 884
glomerata Balb., 107-112, 117-118, 121, 761-766
glutinosa M.B., 107-109, 761, 880
granadensis Willd., 102, 117, 761
grandiflora Vass., 108
grossheimii Vass., 108
gunibica Vass., 108
hemicoerulea, 108, 113
hemicycla Grossh., 108-109, 842
heyniana Greuter, 102, 117, 761
hispida Gaertn., 973
hybrida Trautv., 95-96, 99, 117-118, 761-765, 770-771
hypogaea Small (*Trigonella aschersoniana* Urb.), 97
ignatzii L. & L., 761, 769
intertexta Mill., 102, 116-118, 761
jemenemsis Sinsk., 108
komarovii Vass., 108
kopetdaghi Vass., 108
kultiassovii Vass., 108
laciniata Mill., 105, 116-118, 761
ladak Vass., 108
lanigera Winkl. and Fedtsch., 102, 117, 761
lavrenkoi Vass., 108
leiocarpa Benth., 95-96
lesinsii Small. See *M. murex* Willd.
lethyrus, 146
littoralis Rhode, 106, 116-119, 761
lupulina L., 98, 101, 117, 741, 761, 763
marina L., 99, 117-119, 741, 761-765, 770
media Pers. (*M. sativa* X *M. falcata*), 108, 267, 362, 603
mesopotamica Vass., 108
minima Bart., 105, 117, 119, 761, 875
murex Willd., 93-94, 98, 105, 117-119, 760-761, 763
muricoleptis Tineo., 102, 116-118, 761
noëana Boiss., 103, 117, 142, 761, 879, 879
ochroleuca M. Kult., 108
orbicularis Bart., 102, 117, 761
orientalis Vass., 108
papillosa Boiss., 100, 117-118, 746, 761-765, 769-770
pironae Vis., 100, 117-118, 761, 879
platycarpa Trautv., 99, 117-118, 741, 761, 763

polia Vass., 108
polychroa Grossh., 93, 101, 108-109
polymorpha L., 105, 117, 119, 760-761, 763, 973
praecox DC, 93, 104, 117, 760-761, 763
prostrata Jacq., 100, 107, 110, 117-118, 121, 143, 385, 746, 761-766, 875, 880
quasifalcata Sinsk., 108
radiata L., 102, 117, 761
rhodopea Velen., 95-96, 100, 117-118, 752, 761-768
rigidula (L.) All., 93, 106, 117, 119, 741, 760-761, 763, 882
rivularis Vass., 108
roborovskii Vass., 108
romanica Prod., 108-109
rotata Boiss., 103, 116-117, 761, 882
rugosa Desr., 94, 103, 117, 119, 761, 763, 841, 875-876, 882
rupestris M.B., 100, 117-118, 752, 761-768
ruthenica Ledeb., 99, 117-118, 741, 761, 763
sativa L., 95, 108-109, 117-118, 125, 143, 152, 163-164, 260, 269, 291, 385, 859, 875-877, 879, 883, 933, 937, 945
sativa L. complex, 93, 97, 107-115, 120
sativa ssp. *ambigua* Tutin, 107
sativa ssp. *coerulea* Schmalh., 98, 101, 108-112, 120, 879, 943
sativa ssp. *falcata* Arcengeli, 95, 98-99, 107-115, 119-120, 972
sativa ssp. *faurei*, 100
sativa ssp. *faurel*, 108
sativa ssp. *glomerata* (Balbis) Tutin, 107-108
sativa ssp. *glutinosa* M.B., 101, 107-111, 113, 121
sativa ssp. *praefalcata* Sinsk., 107, 884
sativa ssp. *sativa* (L.) L. & L., 96-98, 101, 107-115, 117, 120, 972
sativa ssp. X *hemicycla* Grossh., 101, 108, 110-112, 114-115, 121
sativa ssp. X *polychroa* Grossh., 108, 110-111, 119, 121
sativa ssp. X *tunetana* Murbeck, 101, 108-109, 111, 121
sativa ssp. X *varia* Arcengeli, 101, 107-108, 110-115, 119, 121
sauvagei Negre, 105, 116-118, 761
saxatilis M.B., 94, 100, 117-118, 746, 751-752, 761-768
schischkinii Sumn., 108
scutellata (L.) Mill., 94, 103, 117-119, 741, 760-765, 771, 841, 933
secundiflora Durieu, 101, 117, 761
shepardii Post, 103, 117, 761
sogdiana Vass., 108

1074 SUBJECT INDEX

soleirolii Duby, 95–98, 105, 116–118, 761
striata Bast., 116
subdicycla Vass., 108
suffruticosa Raymond, 100, 117–119, 761–763, 765, 770
tadzhicorum Vass., 108
tenderensis Opperm., 108–109
tenoreana Ser., 104, 117, 119, 761
tianschanica var. *agropyretorum* Vass., 108, 113, 882
tibetana Vass., 108
tornata Mill., 95, 97–98, 106, 117–118, 761
transoxana Vass., 108
trautvetteri Summ., 107–108
truncatula Gaertn., 98, 106, 116–119, 376, 761, 886
tunetana Vass., 108, 108–109
turbinata (L.) All., 98, 105, 116–118, 761
vardanis Vass., 108
varia Martin, 108–109, 266
virescens Grossh., 108
Megasporogenesis, 737, 739–742
Meiotic abnormalities, 741–746
Membranes, 277, 288
Mesophyll, 466, 472
Metabolic dysfunction, 286
Methionine, 548–549
Methods of breeding. *See* Breeding methods
Mexico, 50–53, 115
Microfibrils, 467
Microorganisms, in digestion, 467, 470–471
Microsporogenesis, 737–746
Milk production, 475, 540, 549
Mineral content, 198, 335, 414, 507–508, 540
Minerals, 463, 466, 479–480, 483
Mitochondria
 efficiency, 201–202, 212
 respiration, 212, 286
Mixtures. *See* Grass-mixtures
Modeling, 179, 184–187, 1039
Models, 15–16
 ALFALFA, 185
 ALFMAN, 185
 ALSIM, 185–186
 GROWIT, 184–185
 LEVEL ZERO, 185, 187
 REGROW, 185
 SIMED, 185–186
 SIMFOY, 184–185
 yield, 185
Mold, 482
Molybdenum
 application, 361
 deficiency, 342–343

 toxicity, 361
Morocco, 63–64
Morphology, 125–162
 adventitious stems, 146–148
 axillary buds, 421, 474
 basal buds, 328
 canopy structure, 206–209
 crown, 138–139
 crown shoots, 413, 415, 421, 474
 developmental, 179–180
 epicotyl, 126, 143–144, 148
 floral, 150–158
 leaves, 148–150, 173–174, 385–387, 476
 measure of cutting date, 477
 nodules, 235
 pollen, 95–97, 101, 103, 120
 pubescence, 385
 roots, 132–138, 377–378, 381
 seed, 125–130
 seedling, 130–132
 stages, 180
 stems, 139–148, 386, 476
 stigma, 153–155
 stomata, 380–383, 387
 vascular development, 132–140, 142–150, 156
Mower conditioners, 571–574
Mowers, 568
Multiple pest resistance, 596–597, 607–608, 610, 612–613, 778, 782, 827, 848
Municipal wastes, 363–364
Mycorrhizae, 350, 376

National Alfalfa Hay Testing Association, 466
National Alfalfa Symposium, 1035
National Certified Alfalfa Variety Review Board (NCAVRB), 6, 20, 597, 1028, 1044
 cultivars approved, 1980–1985, 1046
National Council of Commercial Plant Breeders, 1027, 1035
National Foundation Seed Project, 7, 19, 22, 1025, 1043
National Hay Association, 466
National Seed Storage Laboratory, 399
NCAVRB. *See* National Certified Alfalfa Variety Review Board
NDF. *See* Fiber, neutral detergent
Near infrared reflectance spectroscopy, 13, 21, 466, 816–817
Nectary, 156–158
 nectar reservoir (holder), 157
 stomata, 157
Nematode(s)
 breeding for resistance, 604–605, 607, 614, 757, 829–830, 844–848
 effect on seeding, 326

SUBJECT INDEX   1075

Nematode(s), common names
 Columbia root-knot, 652
 dagger, 605, 847
 Javanese root-knot, 654
 northern root-knot, 652-653, 846-847
 pin, 847
 ring, 605, 847
 root lesion, 605, 655, 847
 root-knot, 606, 846-847
 southern root-knot, 654, 846-847
 stem, 604, 606, 636-637, 757, 844-846, 866
 stubby, 605, 847
 stunt, 605, 847
Nematode(s), scientific names
 Criconemella spp., 655
 Criconemella survata (Raski) Luc & Raski, 847
 Criconemoides spp., 605
 Ditylenchus dipsaci (Kuhn) Filipjev, 604, 636-638, 652, 844-846, 866
 Heliocotylenchus spp., 655
 Hoplolaimus spp., 655
 Meliodogyne chitwoodi Golden et al, 652
 Meloidogyne arenaria Chitwood, 605-606, 654
 Meloidogyne hapla Chitwood, 652, 654, 846-847
 Meloidogyne incognita (Kofoid & White) Chitwood, 604, 606, 652, 654, 846-847
 Meloidogyne javanica (Treub.) Chitwood, 604, 606, 654, 846-847
 Paratrichodorus spp., 655-656
 Paratylenchus penetrans (Cobb) Filipjev & Schu. Stek., 655, 847
 Paratylenchus pratensis (de Man) Filipjev, 847
 Paratylenchus projectus Jenkins, 847
 Paratylenchus spp., 605, 655
 Trichodorus christiei Allen, 847
 Trichodorus spp., 605
 Tylenchorhynchus clarus Allen, 655, 847
 Tylenchorhynchus spp., 605, 655
 Xiphinema americanum Cobb, 656, 847
 Xiphinema spp.605, , 655
Netherlands, 42
Neutral detergent fiber, 465, 481-482, 507
New Zealand, 80-81
Nickel, 362
Nitrate, 307, 346-349, 362, 364, 507-508
Nitrite, 347
Nitrogen
 accumulation in roots, 336
 assimilating enzymes, 240, 250-251
 concentration, 336, 345, 387
 crude protein, 251, 414
 deficiencies, 288
 effect on forage quality, 495
 in established stands, 348
 in establishing stands, 347-348
 fixation. *See* Dinitrogen fixation
 nonprotein, 347-348
 rate of application, 346-348
 removal, 335, 349
 sources, 347
Nitrogenase, 229-251, 347, 375
No-till seeding, 325-327
Nodulation. See *Rhizobium meliloti*
Nodules, 205, 215-216, 231-239
 ineffective, 237-240
 initiation, 234
 morphology, 235-239, 375
 respiration, 205
 structure, 235-239
Nomenclature, scientific and common, 26-27
Nonruminants, 553, 563
North America, 49-53, 115
Norway, 42
Nurse crop. *See* Companion crops
Nutrients, plant
 concentration in forage, 336
 fertilizers, 306, 312-319, 333-362
 needs, 333
 removal, 335
 tissue tests, 336
Nutritive value, 464, 472-473

Oat, 318-319, 523
Oceania, 77-81
Octoploid, 746, 753, 758-759
Organic acids, 477, 480
Organic matter, 322
Organoleptic factors, 466, 483
Organophosphates, 693
Osmotic adjustment, 374
Osmotic potential, 374
Osmotica, 164
Osteosclerids (hour-glass cells), 127
Ovary, 155-157
Overdominance, 797
Ovules, 120, 155, 937, 941-943
 abortion, 942-943
Ozone, 499, 624, 844

Pachytene chromosomes, 738, 756, 760
Pakistan, 73
Palatability, 464
Palisade cells, 126-128, 130, 148-149, 175
Paracoumaric acid, 469-470
Parenchyma, 133, 135, 140, 142, 144-146, 466, 472
Parenchymatous cells. *See* Parenchyma
Pasture production
 adaptation and distribution, 515-516
 beef cattle, 523-526

beef, cow-calf, 326
bloat, 304, 531-532
cultivars, 516, 520-521, 602-603
dairy cattle, 521-523
feeding value, 482
grain feeding, 523, 525-526, 528-529
horses, 531
lambs, 526-527
management, 516-521
poultry, 530-531
sheep, 526-529
swine, 529-530
Pathogens. *See* Disease pathogens
Pectic substances, 468
Pectins, 232, 466-468
Pelleting, 547
Pentaploids, 751
Pentose phosphate cycle, 196
People's Republic of China, 73-75, 93, 115, 119
Pericyte, 132-136
Peroxidase (PRX), 111-115
Peroxyacetyl nitrate (PAN), 624
Persistence, 423, 602, 606
Peru, 59-60
Pest management, 622, 994-1001
  standard tests for resistance, 829
Petals, 151-154
  color, 152-153
  keel, 151-154
  standard (banner), 151, 153, 155, 180
  wing, 151, 153
pH effects
  nodulation, 232, 244, 341-342
  nutrient availability, 341-345
  seeding, 325
  sewage sludge, 364
  toxicity, 341-345
Phenology, 168, 179-182
Phenols, 469-470, 481, 498-500
Phenylpropane, 498
Phloem, 132-133, 140-146, 472
  fibers, 135, 142
  primary, 133, 140-142
  secondary, 135-137
Phosphoenolpyruvate carboxylase, 205, 241, 250
Phosphorus, 306, 312-314, 334-339, 341, 343, 349-352, 363, 507-508
  absorption in seeding, 307, 311-313
  application, 307, 334-335, 350-352, 479
  availability, 337, 343, 351
  cold tolerance, 264
  concentration, 336, 477, 479
  deficiencies, 479
  effect on forage quality, 479, 819-820
  fertilizer, 479
  movement, 350
  radioactive, 311
  role in plant, 349

sources, 351-352
stand establishment, 306-307, 311-313, 989
stand maintenance, 334-335, 351
Photoperiod, 175, 177
  flowering, 180-183
  plant yield, 177
  root growth, 167, 178
Photorespiration, 196, 199-200, 287
Photosensitization, 508
Photosynthate partitioning, 167-168, 172, 174, 213-216, 353, 811
  interorgan competition, 213-215
Photosynthesis, 196-199, 209-211, 287, 289, 353
  canopies, 170, 209-211
  dinitrogen fixation, 241, 243, 311, 322
  heat effects, 289
  inhibition, 199
  leaves, 169, 174-175, 196-199, 623
  metabolic pathways, 196
  moisture deficits, 374-375
  source-sink concepts, 213
Phototropic response, 166
Physiology of growth
  carbon utilization, 213-220, 353
  germination and emergence, 164-166
  internode length, 177
  leaves, 172-175
  light interception, 209
  photosynthesis
    canopies, 209-211
    leaves, 196-200
  roots, 177-178
  seedlings, 166-168
  stem elongation, 172
  stems, 175-177
  stress, 1040
  translocation, 203, 213-216, 353
  vegetative growth, 168-179
Physiology of reproduction, 181-184
Phytoalexins, 500
Phytochrome, 181-182
Pistil, 151, 153
Pith, 140, 144
Plant analysis, 336
Plant food elements. *See* Nutrients, plant
Plant introductions. *See* Germplasm
Plant Variety Protection Act, 599, 1028
Ploidy, 93-94, 111, 120, 171
Pocket gopher, 603
Pods, 98-110
Poland, 42-43
Pollen
  amount, 948-951
  characteristics, 933
  colpi, 155-156
  exine, 155
  germination, 945-946
  grains, 155-156, 937, 950

SUBJECT INDEX

morphology, 95-97, 101, 103
mother cells, 944
size, 946
stain, 946
tubes, 739, 741, 937, 941-942
viability, 933, 950
Pollination, 182-184
  bees, 931, 951-952, 1001-1009
  control, 931-956
  cross, 780, 938
  emasculation, 938-939
  hybrids, 953
  mechanical, 938-939
  moisture stress, 388
  seed production, 1001-1009
  self, 933-935, 938
  techniques, 938-939
  tripping, 183, 935-936, 1001-1002
Poloxalene, 496
Polyembryonic seed, 755, 974
Polyploidy, 94, 111
  dinitrogen fixation, 234
  uses of, 758-760
Polysaccharides, 467, 469-471, 477
Postemergence herbicides, 321, 324, 726, 987
Potash. *See* Potassium
Potassium
  application, 334-335, 354
  cold tolerance, 264
  competition, legumes and grasses, 352
  concentration, 335-338, 341
  dinitrogen fixation, 244
  effect on forage quality, 479-480, 495, 819
  effect on photosynthesis, 198, 353
  establishing stands, 306-307
  maintaining stands, 334-335, 354, 425
  movement, 353
  removal, 335, 352
  role in plant, 340, 353
  sources, 354
Potassium carbonates, 573
Poultry pasture, 530-531
Preemergence herbicides, 726, 987
Preservatives, 12
  chemical, 821
Private research, 5, 610, 1026-1027, 1043-1045
Processing dehydrated products, 558-559
Progeny testing, 785-787, 815
Proline, 274-275
Propionic acid, 12
Proprietary cultivars, 5, 611-612, 1025-1028, 1043-1044
Protease inhibitors, 508-509
Protein, 463, 466, 553, 558, 561-564, 820-822
  bypass, 546, 548
  cell wall, 466-469
  concentrate, 554-555, 562-564
  concentration, 473, 476, 478, 481-482, 494
  digestion, 470
  digestion rate, 497
  dinitrogen fixation, 251
  expression, 556
  extension, 468
  Fraction 1, 494, 496
  Fraction 2, 494-496
  heat damage, 821
  heat shock, 289
  intake (animal), 463
  leaf, 498
  precipitation, 501
  18-S, 494
  separation, 558, 564
  silage, 539
  solubility, 494, 496-497, 543
  synthesis, 275-277
  value, 545
  yield per hectare, 416-417, 473, 546
Protozoa, 468
Public research, 5, 1025-1026
Purines, 508
Pyrimidines, 508

Quadrivalent associations (multivalents), 750-751
Quality. *See* Feeding value

Recurrent selection, 829
Regional adaptation, 595
Regional trends, 595-596
Relative growth rate, 167, 171
Relative humidity, effect on seed, 970
Renovation, 325
Reproduction, sexual, 94
Research
  Africa, 60-67
  Asia, 68-77
  contributions, 1024-1027
  Europe, 32-49
  North America, 49-53, 1024-1027
  Oceania, 77-81
  South America, 53-60
  worldwide, 31-81
Resistance. *See also* Disease resistance; Insect resistance; Nematodes, breeding for resistance
  breeding, 829, 1024, 1027
  evaluation, 480
  inheritance, 830
  severity index, 828
  standard tests, 6, 622
  sources, 119, 829
Respiration
  canopy, 211

cold tolerance, 273
dark, 196, 200–202, 211, 287, 374
dry matter accumulation, 169
losses, 475
maintenance, 167
photo, 196, 199–200, 287, 374
seeds, 966–967
Restriction fragment length polymorphisms (RFLP), 98
Retetraploidized plants, 943
Rhamnose, 468
Rhizobia, 311
Rhizobium, 133, 229–251, 1039
Rhizobium meliloti, 14
acid tolerance, 231, 341
biochemistry, 241
classification, 231
cold tolerance, 273
commercial preparation, 231–232
cross inoculation, 232
cultivar interaction, 242–243
dinitrogen fixation, 229–231, 347
dinitrogen fixed, 205, 1039
genetic control, 231
gum arabic, 311
heat effects, 288
host specificity, 231
ineffective types, 15, 237–238, 242
infection site, 232
initiation, 232–235
inoculation, 231–232, 242–243, 311
molybdenum, 342–344, 361
nodulation, 15, 205, 232–240, 347, 375
regulation, 232–235
strains, 231–232, 242
sulfur, 356
Romania, 43–44
Roots, 132–138
anaerobic conditions, 167, 178
cambium, 133–136, 146–147
casperian strip, 133
cortex, 132–134, 136–138
creeping, 146–148
depth, 167, 177, 339, 377, 394
dinitrogen fixation, 230–238
diseases, 640–656
endodermis, 132–133, 137
epidermis, 133
flooding, 167, 178
growth, 132–138, 167, 377–379
hairs, 132–133, 232–234
lateral, 136–137, 146, 178
length, 377–379
mass, 177–178, 377–379
meristem, 132–133, 136–137, 146–147
nitrogen accumulation, 15
partitioning, 167–168
periderm, 135
phellem, 136
phelloderm, 136, 146

phellogen, 136, 146
photoperiod, 178
primary, 130, 132–137, 146
radicle, 126–127
respiration, 205–206
secondary, 133–136, 146
soil fertility, effects, 340, 343
stele, 132–133
systems, 177–178, 377
vascular elements, 132–136
vascular transition to stem, 132, 138–140, 142–150, 156
Rotational grazing. *See* Grazing
Rye, 275, 319

Salinity, 379
sources of resistance, 119
Salt effects, 164, 167, 171–172, 176
Saponification, 470
Saponins, 503–505, 563, 861, 871, 879
animal nutrition, 563, 819
bloat, 496
breeding, 819
fungal species, 505
insects, 505, 861, 871, 879
Saudi Arabia, 75–76
Schlerenchyma, 472
Scientific names, 25–26
Sclerotia, 635–636, 649, 651
Secale cereale L. *See* Rye
Seed
allogamous, 94
autogamous, 94, 935
coat/testa, 125–127, 129
color, 102, 125, 784, 973
composition, 962
count, 125, 961
damaged, 965, 1013
description, 125–130, 961
deterioration, 967, 971–972
development, 120
dormancy, 165, 707, 963, 974
germination, 127, 130–132, 164–165, 385–386, 974
hard, 127, 183, 311, 963–966
identification, 973
imports, 1023
industry, 5, 20, 22, 597, 1023–1036
lime-coated, 311
maturity, 968
mechanical damage, 968, 1013
micropyle, 126–128, 155
morphology, 102, 125
pests, 641–642, 971
physiology, 962–967
pigmentation, 954
placement, in seeding, 307
preharvest, 967–968
purity, 973

SUBJECT INDEX

quality, 310–311, 974–975, 1026, 1032–1034
radicle, 126–127
reserves, 962
scarification, 965
sealed storage, 970–971
set, 182–184
size, 183, 954, 975
sprouting, 975–979, 1015
storage, 967–971
taxonomy, 95–106
testing, 972–974
usage, 6–7
vigor, 972, 975
water relations, 182, 962–966
weed, 707, 973, 992–993
weight, 183, 975
Seed conditioning (processing), 963, 965, 968, 1030–1032
  blends, 1033
  cleaning equipment, 1031–1032
  flow pattern, 1031
  packaging, 1033
  products, 1033
  quality control, 1032–1034
  receiving, 1030–1031
Seed marketing
  blends, 1026
  Certified Alfalfa Seed Council, 6, 1034–1035
  cooperatives, 1029–1030, 1032
  packaging, 971, 1033
  phytosanitary certificates, 1034
  product information, 1034
  SeCan, 7–8
  shipping, 1033
  systems, 1032
  trade associations, 1028, 1035
  variety unstated, 1026, 1033
Seed production, 6–8, 19–20, 182–184, 387, 985–1016, 1029–1030, 1040
  agronomic practices, 986–994
  areas of production, 964, 985–986
  certification, 6–8, 990, 1025–1026
  contract, 1029–1032
  cultivar differences, 1001
  defoliation, 1010–1011
  equipment, 1011–1014
  eras, 1023–1026
  factors affecting yield, 1014–1016
  fieldmen, 1030
  harvesting, 1009–1014
  hybrids, 783, 950–955
  industry, 1029–1030
  insect control, 994–1000
  integrated control, 997–999
  irrigation, 182–183, 388, 990–992, 1005
  managed, 1029–1030
  National Foundation Seed Project, 1025
  outside area of adaptation, 1024, 1026
  planting, 986–990
  pollination, 388, 1001–1009, 1025
  postharvest practices, 1014
  preharvest factors, 967–968
  quantity produced, 1023, 1036
  rows vs. broadcast, 987
  soil fertility, 989–990, 993
  temperature, 183
  thinning the stand, 988–989
  timing the flowering period, 993–994
  water requirement, 388
  water stress, 388, 991
  weed control, 986–987, 992–993
  yield potential, 183, 1016
Seedbed preparation, 305–311, 325, 986
Seeding, 11, 310–325
  autotoxicity, 309–310, 603
  banding, 307, 312–318, 347, 350
  basal buds, 326
  broadcasting, 316
  clear, 304, 321–324, 716–718
  compaction, effect of, 309
  critical last date, 323
  cuttings, first year, 321–322
  dates, 320, 324, 711, 997
  depth, 309–317
  droughy soils, 322
  equipment, 308–309, 312–318
  fall, 323
  fluid or suspension, 11
  insect control, 326, 329
  management in seeding year, 327–329
  manure, effect of, 309
  methods, 309, 312–318
  mixtures, 304–305
  no-till, 11, 325, 344
  preparation of seedbed, 306–309
  rate, 988
    alone, 305, 320, 324–325
    with grass, 305
  rate and use, 305, 320, 324
  seed placement, 309–318
  seed quality, 310
  sod, 325, 329
  soil moisture, 319
  spring, 320, 322, 326
  straw removal, 328–329
  summer, 309, 320–323, 326
  surviving seedlings, 324
  time and use, 320–326
  tracheid bar, 127–128
  volunteer grain, 328
  weed control, 309, 317, 710–723
  with companion crop, 317–321, 328–329
  without companion crop, 321
Seedling
  development, 130–132
    phenological, 168
    unifoliolate leaf, 130–132
  irrigation, 386

growth, 166-168
  excess water, 167
  phosphorus placement, 307, 312-313
  salt, 167
  temperature, 167
  vigor, 446, 972
Selection methods. See also Breeding
  backcrossing, 781-782
  clonal progeny, 781, 785
  combined, 788
  diallel crosses, 786-787
  ecotype, 778-779
  family, 788-789, 801-802
  grid, 785
  hybrid alfalfa, 783-784
  independent culling, 790-791
  index, 791-792
  mass, 780, 784-785, 801-802, 804
  maternal, 785, 789
  natural, 783, 785
  open-pollinated progeny, 786
  polycross progeny, 781, 786-787, 815
  recurrent phenotypic, 250, 780, 789, 801-802, 829, 864, 871, 873, 877-878
  self-progeny, 781, 786, 815
  strain building, 779-780
  strain crosses, 782-783, 829, 864
  tandem, 790
  threshold traits, 792
  topcross progeny, 786
Selenium, 362, 861
Self-fertilization
  fertility, 940
  fertility index, 944
  incompatibility, 784, 939-945, 951
  sterility, 939-945, 951
  tripping, 94, 935
Shading effects
  feeding value, 466-467
  leaves, 174
  photosynthesis, 172
  stem numbers, 172
Sheep
  feeding, 475, 480-481, 483
  grazing, 526-529
Shoot. See also Stem
  moisture stress, 170, 386
  number, 169-170
  root ratio, 166
  stage by weight, 413. See also Mean stage by weight
  yield, 170-171
Silage, 463, 495, 544-546
  additives, 545
  direct cut, 544
  equipment, 578, 587-590
  feeding value, 544-546
  heat damage, 546
  inoculation, 586

loosses, 544
low moisture, 587-590
odors, 544
optimizing feeding value, 545-546
pressed forage, 560
seedings, 320
transporting, 580
unloaders, 589-591
wilted, 544, 561
Silica, 507, 861
Silos
  horizontal, 589-590
  upright, 588-589
Simulation models. See Models
Sinaphyl, 469
"Sink," 195, 213-216
Snow cover, 261, 281, 426, 635
Sod seeding, 325-327
Sodium, 479
Sodium carbonates, 573
Soil
  acidity
    adaptation, 342, 595, 608
    liming, 306, 341-345, 597, 610-611, 614
    nutrient availability, 342-344
    rhizobia, 231, 244, 342
    root penetration, 344
    tolerance, 342
    winter survival, 264
  compaction, 478
  fertility, 264, 333-364, 379, 424-425, 479-480, 483, 614, 713, 989-990, 993
  moisture, 166, 179, 307-309, 319, 337-338, 478-479, 483
  nutrient supplying capabilities, 339, 599-601
  temperature, 178, 181, 483
  tests, 306, 340-341
  tillage and seeding, 306-309, 325
  type, 312, 478, 483
  water budget models, 391
  water depletion, 377, 379-380, 384, 393-394
  water measurement, 394
  water potential, 182, 380
Solar radiation, 207-210, 396
Somaclonal variation, 913-919
Somatic embryogenesis, 909-911
Sorghum, 783
Sorghum bicolor L. Moench. See Sorghum
South Africa, 64-65
South America, 516
Soviet Union, 47-48
Spain, 44-45
Specific leaf weight (SLW), 174-175, 203-205
Spoilage, 523, 525, 528
Sprouts (human food), 975-979

SUBJECT INDEX

Stachydrine, 508
Stacker, hay, percentage of forage harvested, 567
Stamen, 151, 155
Staminal column (fused filaments), 155–156
Stand
  age, 169, 425, 430
  companion crops, 317–319
  decline, 420, 429, 644
  density, 303–304
  establishment, 986–990
  fertilizers, 306, 333–364
  high yield, 171, 333–335
  intensive management, 321–322
  loss, 424, 606
  maintenance, 454–457
  mixtures, 305–311, 452–454
  persistence, 334–335, 351, 354, 411, 413, 606
  plants per unit area, 169, 304
  seeding, 310–325
  survival, 219, 475
  weed control, 321–326
Standard tests (pest resistance), 6, 21, 829
Starch, 174, 272–273, 466–467, 477
  nodules, 239
  root, 216–217
Stele, 132–133, 145
Stem, 139–148
  adventitious, 146–148
  anatomy, 139–142
  cambium, 142
  casparian strip, 140
  characteristics, 175–177
  composition, 177, 336, 472–473
  cortex, 139–140, 466
  decumbency, 269
  elongation, 172, 386
  epidermis, 139–141
  flooding, 175
  leaf ratios, 172, 175, 414, 466, 473–474, 476–478, 480, 483
  length, 176
  light quality, 168
  mass per stem, 169–171, 474
  meristem, 139, 146–148, 151
  middle lamella, 467–468, 470
  moisture stress, 175, 386
  number, 169–170, 3868
  phyllotaxy, 139, 145–146
  pith, 140, 144, 466
  primary, 130–132, 151
  quality, 414
  salinity, 176
  secondary, 130–132, 150–151
  trichomes, 142–143
  vascular tissue, 472
Sterility
  abortion, 942–943
  cytoplasmic, 947–952
  environment, 944–945, 950
  female, 954–955
  genetic, 947, 951
  partial, 948
  systems, 948
Stigma, 151–155, 933–935
Stigmatic membrane, 153
  cuticle, 153, 157, 933–935
  fluid, 937
  secretion, 933
Stomata, 202–204, 374, 381, 384
  conductance, 353, 381–382, 396
  density, 381
Strain crosses, 829, 864. *See also* Breeding; Selection methods
Style, 153–155
Sudangrass, 521
Sugars, 216–220, 270–272, 466, 468, 477
Sulfur, 198, 306–307
  application, 358–359
  concentration, 336–337, 356, 480
  role in plant, 356
  sources, 307, 358–359, 480
Sulfur dioxide, 624
Summer decline, 178–179
Sweden, 46
Switzerland, 46–47
Symbiosis, 229
Symbiotic bacteria, 229–251, 288
Synthesis of cultivars (varieties), 780–781, 793, 805
Synthetic generations, 780, 793
Systems, forage-livestock, 1039
Systems analysis, 593

Tannins, 469, 498, 500–502, 861
Tanzania, 65–66
Tapetum, 155
Taxogenetics, 108, 116–118
Taxonomy, 95–116
  numerical, 96–97
Temperature effects
  adaptation, 596, 598, 601, 605
  bloom, 477
  carbohydrates, 174
  crop, 399–400
  crude protein, 478
  disease resistance, 626, 635, 642, 652, 659, 828
  feeding value, 477–478, 483
  fluctuating, 281
  freezing, 261, 278
  germination, 164–165
  gradient, 605
  growth, 178–179, 186, 386, 391
  heat tolerance, 285–291
  leaf development, 173–174
  leaf/stem ratio, 474, 477–478

maturation, 477
morphology, 180-181
night, 167, 174, 176
nutrients, 337
optimal, 167, 169, 171-172, 174, 176-177
photosynthesis, 197-198
plant height, 176-177
respiration, 198-201
rhizobia, 229-251
seed set, 183
seed storage, 967-971
seedlings, 167, 171-172
shoots, 169
stem diameter, 477
stems, 176-177
stress tolerance, 291
winter hardening, 206
Tetraploid, origin, 113-115
Tetrazolium test, 975
Tillering, 170
Tissue culture, 120, 221, 752-753, 903-925, 1042
  genotype effects, 905-909
    protoplasts, 908-909
    Regen-S, 905-907
    regeneration from callus, 905-907
    suspension cultures, 907-908
  physiology, 909-913
    conversion of embryos, 912-913
    embryogenesis, 910-911
    growth regulators, 911
    reduced nitrogen, 912
  recent developments, 924
  regeneration from callus, 21
  somatic cell fusion, 919-924
  somatic cell selection, 924
    disease resistance, 920-922
    ethionine resistance, 919-920
    potential, 923
    salt tolerance, 922-923
  variability, 913-919
    aneuploidy, 915-916
    chromosome doubling, 914-915
    genetic, 916-919
Tomato, after alfalfa, 322
Translocation, 203, 213
Transpiration, 197, 384-385, 389. *See also* Evapotranspiration
  hydraulic conductance, 380-381, 384
  leaf conductance, 381, 383
  stomatal conductance, 381, 384
Triarch, 132
Trichome, 132-140, 142-148. *See also* Glandular hairs
Tripping, 183, 1001-1002
Trisomics, 754-755
Triticum aestivum L. *See* Wheat
True feeding value, 464
Tunisia, 66-67

Turkey, 76-77

United Kingdom, 47
United States, historical highlights, 19-22, 29-30
Unreduced gametes, 111, 113, 121
Uruguay, 60
USSR, 47-48

Vascular elements, 132-140, 142-150, 156
Virus disease, 623, 638-640, 843-844. *See also* Diseases
Vitamins, 463
in vitro digestible dry matter, 465, 472, 474, 476-479, 481-483, 817-818
Volatiles, 183-184
Volunteer grain, 328, 721

Water
  application, 391, 401
  bound, 269
  daily use, 390-391, 397
  deficits, 172, 179, 373, 382, 478, 1040
  depletion, 379-380, 391, 394
  dinitrogen fixation, 375-376
  disease and insect control, 167, 828
  excess, 167, 171, 175, 337, 388
  extraction, 379
  field capacity, 393-394
  germination and emergence, 164-165
  movement, 377-380
  photosynthesis, 374
  potential, 165, 172, 373-375, 383, 386, 478
  requirements, 385-388
  seed production, 990-992
  seeds, 962-963
  sprouting, 978
  stress, 166, 170, 175, 182, 280, 386-387, 478-479
  uptake, 379-380
  use efficiency, 197, 338, 391, 603, 1040
Wax, 166
Weeds
  dormancy, 707
  feeding value, 430, 482-483
  losses from, 482-483, 705-707
  production of inhibitors, 706
Weed control, 11, 321-324, 326, 328, 429-430, 705-729
  band seeding, 317
  before seeding, 309, 317
  clean seed, 710
  clear (pure) seeding, 321-322, 716-718
  clipping below buds, 328
  companion crops, 317-319, 713-714
  cultural, 710-714

SUBJECT INDEX

established stands, 708-710, 723-729
fertilizers, 713
flaming, 725
grass in, 328, 718-719
herbicides, 11, 321-322, 324-328, 715-729
  combinations, 344, 719-720
history, 705
mowing, 328, 714, 724-725
no-till, 325, 710
seed production, 986-987, 992-993
seedling stands, 321, 329, 710-723
sod seeding, 325
spring seeding, 309, 317, 321-322, 707-708
summer and fall seeding, 325, 708
volunteer grain, 323, 328, 721
Weeds, common names
  alkali mallow, 992
  annual bromes, 706-707, 727
  barnyard grass, 482
  black mustard, 716
  Canada thistle, 483, 706, 724, 726, 986
  chickweed, 708, 716, 718, 726
  cinquefoils, 710
  curley dock, 483, 706, 710, 717, 993
  dandelion, 305, 482-483, 710, 723, 726-728
  dodder, 707, 709, 726, 986, 992
  downy brome, 482, 727
  field bindweed, 710, 712, 986, 992
  field dodder, 709
  flixweed, 482
  foxtail barley, 707
  giant foxtail, 482
  goldenrod, 710
  hawkweed, 710
  henbit, 717
  hoary alyssum, 483, 727
  horseweed, 706
  Jerusalem artichoke, 482-483
  johnsongrass, 706, 710, 724, 986, 992
  knawel, 717
  kochia, 717
  lambsquarters, 482, 717
  largeseed dodder, 709
  leafy spurge, 712, 725
  milkweed, 710
  mustard, 992
  nutsedge, 716
  Pennsylvania smartweed, 482
  pennycress, 717
  pigweed, 706, 717, 992
  prickly lettuce, 717
  quackgrass, 309, 325, 482-483, 706, 710, 712, 724-727
  ragweed, 482, 707, 717
  redroot pigweeds, 482, 706, 717
  Russian knapweed, 725, 986
  Russian thistle, 717
  sandbur, 992
  shepherd's purse, 482, 708, 716-717
  smallseed dodder, 709
  smartweed, 483, 717
  sorrel, 710
  sourclover, 993
  sow thistle, 727
  thistles, 710
  white cockle, 482-483, 727
  wild barley, 482
  wild garlic, 707
  wild radish, 718
  yellow foxtail, 482
  yellow nutsedge, 708
  yellow rocket, 482, 727
Weeds, scientific names
  Agropyron repens L. Beauv., 309, 482, 706
  Allium veneale L., 707
  Amaranthus retroflexus L., 482, 706, 992
  Amaranthus spp., 992
  Ambrosia artemisiifolia L., 482, 707
  Asclepias spp., 710
  Barbarea vulgaris R. Br., 482, 727
  Berteroa incana (L.) DC., 483, 727
  Brassica nigra L., 716
  Brassica spp., 717, 720, 992
  Bromus spp., 706-708
  Bromus tectorum L., 482
  Capsella bursa-pastoris (L.) Medic., 482, 708
  Cardus spp., 710
  Cenchrus spp., 992
  Centaurea repens L., 725, 986
  Chenopodium album L., 482, 717
  Cirsium arvense (L.) Scop., 483, 706, 710, 724, 986
  Convolvulus arvensis L., 710, 986, 992
  Conyza canadensis (L.) Cronq., 706
  Cuscuta campestris Yunck., 709
  Cuscuta indecora Choisy, 709
  Cuscuta planiflora Tenore, 709
  Cuscuta spp., 706-707, 986, 992
  Cyperus esculentus L., 708
  Descurainia sophia (L.) Webb., 482
  Echinochloa crus-galli (L.) P. Beauv., 482
  Euphorbia esula L., 712
  Helianthus tuberosus L., 482-483
  Hieraicum spp., 710
  Hordeum jubatum L., 707
  Hordeum leporinum Link., 482
  Kochia scoparia L., 717
  Lactuca serriola L., 717
  Lamium amplexicaule L., 717
  Lychnis alba Mill., 482, 727
  Melilotus indica (L.) All., 993
  Plantago spp., 710
  Polygonum coccineum Muhl., 483
  Polygonum pensylvanicum L., 482

Polygonum spp., 717
Potentilla spp., 710
Raphanus raphanistrum L., 718
Rumex crispus L., 483, 706, 992
Rumex spp., 710
Salsola kali L., 717
Scleranthus annuus L., 708
Setaria faberi Herrm., 482
Setaria glauca (L.) P. Beauv., 482
Sida hederacea Torr., 992
Solidago spp., 710
Sonchus arvensis L., 708
Sorghum halepense (L.) Pers., 706, 986, 992
Stellaria media (L.) Cyrillo, 708
Taraxacum officinale Weber, 305, 482, 710
Thlaspi arvense L., 717
Wet fractionation, 463
Wheat, 269, 275, 318–319, 322, 531, 791
Winter injury, 38, 280–284, 412–426, 623
 heaving, 426
 ice sheets, 281
 partitioning of assimilates, 221
 seeding effect on, 317, 323
 water stress, 282
Winterhardiness, 2–3, 119, 206, 260, 265, 278–284, 424, 595–596, 605, 610–611

Xylan, 469

Xylem, 136–137, 140–146, 149
 libriform fibers, 133, 135
 metaxylem, 132, 143–144
 primary, 133, 136–137, 140–142, 145
 protoxylem, 132–133, 136, 143, 145
 secondary, 136
Xylose, 467–468

Yield
 barriers to, 11–12, 22
 carbon dioxide exchange, 211–212
 components, 169–171
 dinitrogen fixation, 229–230, 248
 forage, 212, 219–221, 411–412, 809–812
 genetic increases, 809–812
 high (maximum) forage, 22, 475–476
 level (nutrient need), 334–335
 losses to disease and nematodes, 827
 per unit area, 171
 photoperiod, 172
 seed, 183, 1016
 seeding year, 322
 stems and leaves, 172–173
 year after seeding, 310, 322
Yugoslavia, 49

Zambia, 67
Zea mays L. *See* Corn
Zimbabwe, 67
Zinc, 336–337, 361–362, 479